PREFACE

This is a self-contained exposition of the theory of comparison of statistical experiments.* The idea of comparing experiments by comparing the risks they produce for varying losses, developed gradually after the appearance of Wald's works on statistical decision functions. Contributions by Blackwell, Bohnenblust, Shapley, Sherman and Stein in the years 1949–1953 provided criteria for one experiment being more informative than another. In 1955 Boll showed how the comparison problem may be reduced when invariance conditions are satisfied. In 1965 Strassen generalized the dilation criterion and also provided a variety of other related and interesting results.

In the same period a considerable effort was made to clarify basic statistical concepts such as sufficiency and invariance within the framework introduced by Kolmogorov in 1933. Building on earlier works of Fisher, Neyman, Pearson and Wald these concepts were explored by Bahadur, Blackwell, Halmos and Savage, for example, and, although it was not well known at the time, by Dynkin.

Substantially broadening the scope, LeCam (1959) (see LeCam, 1964) addressed himself to the following question.

Given two experiments $\mathscr{E}$ and $\mathscr{F}$, when are we justified in claiming that only so much information is lost by basing ourselves on the experiment $\mathscr{E}$ rather than on the experiment $\mathscr{F}$?

It turned out that a variety of different approaches to this problem led to the same numerical quantity, the *deficiency* of $\mathscr{E}$ with respect to $\mathscr{F}$. Thus the deficiency has natural interpretations in terms of pointwise comparison of

* With one exception the references to the preface are found at the end of the appropriate chapter. The exception is: Kolmogorov, A.N. 1933. Grundbegriffe der Wahrscheinlichkeitsrechnung. *Erg. Math.* 2, **No. 3**. Springer–Verlag, Berlin.

risks, in terms of maximum risks, in terms of performance functions of decision rules, in terms of Bayes risk and in terms of randomizations.

A substantial part of contemporary asymptotical statistical theory is based on these ideas. Mathematically they are related to, and extend, the fruitful theory of majorization and Schur convexity.

Another achievement of LeCam's contribution was to clarify the distinction between essential statistical problems on the one side and technical problems on the other side. In particular it was shown that the standard measure theoretical framework forces problems upon us which may be avoided by working within the more naive framework of vector lattices. The σ parts of 'σ-additivity' and 'σ-algebra' are technical in nature, while additivity and algebra refers to concepts having statistical significance.

The material in this book is organized as follows.

The notion of a statistical experiment is introduced in chapter 1. After having considered a crude classification of experiments we turn to some problems arising in experiments which do not obey the weak compactness lemma. In such experiments sequences of power functions of tests may converge pointwise to functions which are not power functions of any test. Extending the notion of a random variable we arrive at a notion of generalized variables which enables us to remedy most of the observed pathologies. One might argue that this is much ado about nothing since most of the standard experiments obey the weak compactness lemma. However it is also obvious that the slightest perturbation of any nice experiment may render this lemma invalid.

The space of bounded generalized variables constitutes a linear space called the M-space of the experiment. This space may also be expressed as the dual of another linear space called the L-space of the experiment. The L-space is the linear space of finite measures which are 'dominated' by the experiment. We shall meet these spaces on several occasions later.

New experiments may be derived from old ones by various methods of combination. Thus we may combine experiments independently and thereby obtain their product experiments. In particular we may obtain replicated experiments by combining independently copies of the same experiment. Another possibility is to select experiments according to known probabilities and then perform the selected experiment. Such combinations are called *mixtures* and we shall see in chapter 6 that any experiment may, loosely speaking, be considered as a mixture of experiments permitting boundedly complete and sufficient statistics.

Mixtures and products together define an interesting algebraic structure for experiments.

We may also modify a given experiment by imposing restrictions on the observations or on the unknown parameters. Conversely we may supplement the observations or expand the parameter set, e.g. by considering prior distributions.

In chapter 1 we also review some famous functionals of experiments. Among them are: statistical distance, affinity, Hellinger transform, Kullback–Leibler information, Fisher information and expressions for minimum Bayes risk. These functionals share the property of being affine under mixtures. In fact they possess the stronger property of being definable as integrals of homogeneous functionals on the space of likelihood functions.

Sufficiency in statistics usually refers to the case of a sub-experiment (statistic) being just as informative as the given experiment. With an eye towards later generalizations in chapter 7, we have included an exposition of some basic results on sufficiency and completeness.

Convex analysis is an indispensible tool for statistical theory in general and for this work in particular. In chapter 2 we have provided an exposition of results which all have more or less immediate applications to decision theory. The basic notions are here introduced within the framework of a general linear space. However nontrivial results are usually only proved in the finite dimensional case.

Another important tool is the theory of two-person zero-sum games. Chapter 3 provides an introduction to this topic. The central results are the various minimax (maximin) theorems for mixed games and results on dominating classes of strategies. Some of these results are phrased both generally and within the framework of statistical games.

The basic notions of statistical decision theory are introduced in chapter 4. Thus we consider here loss functions, decision rules, performance functions, risk functions and the Bayesian machinery. Here we encounter again the need for a more flexible set-up. Abstracting the essential properties of a decision rule we arrive at the generalized decision rules (transitions) of LeCam. The discussion here necessitates a few excursions into the realm of finite additivity. We provide also a few conditions to ensure that the generalized decision rules may be represented by ordinary decision rules. As some of the proofs here are quite technical it may be advisable at a first reading just to look over these results.

L-spaces of experiments, M-spaces of experiments as well as a multitude of other linear spaces related to experiments are *vector lattices* (Riesz spaces), i.e. they share the property of possessing a natural lattice ordering which is compatible with the linear structure. As acquaintance with these spaces simplifies many arguments we have provided an exposition of some central features in chapter 5.

After having introduced the general notion of a normed vector lattice we turn to the particular cases of (abstract) L- or M-spaces. Following LeCam any family of non-negative normalized vectors in an L-space is admitted as an (abstract) experiment. The last section provides an exposition on Kakutani's representation theory for these spaces. In particular we shall see that any L-space is isomorphic to the L-space of some totally informative experiment. These results clarify to a large extent what kind of an animal a statistical experiment actually is. The proofs of the representation results are based on the theory of Borel measures in topological spaces. Readers who are not familiar with this theory may choose to proceed, taking the representation results for granted.

The core of this work is chapters 6 and 7. After having introduced deficiencies in terms of risk functions we proceed to derive a variety of results which follow more or less directly from the definition. One important technical aspect of LeCam's definition is that almost everything is simply expressed in terms of restrictions to finite sub-parameter sets. In particular sufficiency in the deficiency sense is ordinary sufficiency for finite subparameter sets and thus amounts to pairwise sufficiency in the ordinary sense.

Also working directly with the definition we provide some answers to naive questions concerning comparison with experiments which are either totally informative or are totally non-informative.

As mentioned above, the deficiencies might just as well have been introduced from other points of view. After having established the equivalence of a variety of risk dependent criteria we apply them to obtain more explicit expressions for deficiencies in general and for testing problems in particular. Thus the deficiency distance between two experiments equals a natural sup-distance metric for the functionals they define on sets of sublinear functions on the likelihood space. There is also a discussion of two kinds of weighted deficiencies and their related distances.

A very important criterion for deficiency is LeCam's randomization criterion. This criterion does not involve risk and provides a very natural generalization of the operational definition of sufficiency. The task of evaluating (or of bounding) deficiencies may be considerably reduced if the experiments we are comparing are invariant for actions originating from the same group. In that case it may be permissible to restrict attention to randomizations (kernels) which are also invariant for these actions. Thus comparison of dominated shift experiments on solvable groups may be expressed in terms of convolution kernels.

Although reduction by invariance is not generally permissible for all groups this works at least for groups possessing invariant means.

As a by-product of our considerations we obtain results on the feasibility of reducing invariant decision problems as well as on the possibility of exchanging the principles of invariance and of sufficiency. The results described may be regarded as generalizations of the Hunt–Stein theorem as well as of results by Hall, Wijsman & Ghosh.

Related to the notion of a deficiency are notions of a *distance*, an *ordering* and an *equivalence*. Distances are obtained by symmetrizing the deficiencies. The ordering is the ordering of zero deficiency. This ordering is also called the information ordering or, even more suggestively, the ordering 'being at least as informative as'. The equivalence is the equivalence of zero deficiency distance and this is also the equivalence of the information ordering. Saying that equivalent experiments are representatives of each other we shall see that for any given set we may construct a set of experiments containing representatives of any experiment having the given set as a parameter set.

Indeed any experiment having a countable parameter set is equivalent to the experiment induced by the posterior distribution for any prior distribution assigning positive mass to each point in the parameter set. If the parameter set is finite and the prior distribution is uniform then this is the standard experiment introduced by Blackwell. Regarding the posterior distribution as a normalized likelihood function we shall see that this generalizes to any parameter set.

A standard experiment is determined by its standard measure which is just the sum of the individual measures comprising the experiment. Considering the standard measures as the functionals they define for homogeneous functions on the likelihood space, for a general experiment we arrive at LeCam's notion of a conical measure.

In this way statements on experiments become statements on standard measures or, more generally, on conical measures. Thus if Θ is finite then the standard experiments of products and mixtures of experiments are, respectively, convolutions for 'normalized' pointwise multiplication and convex combinations of the individual standard measures. The information ordering becomes ordering by dilations of standard measures and equivalence reduces simply to equality of standard measures. The deficiency distance becomes one of the equivalent metrics describing the usual weak convergence of measures.

We shall spend some time trying to characterize functionals of experiments possessing particular properties as being monotone for the information ordering or as being affine under mixtures.

Another topic discussed in chapter 7 is convergence. Convergence for the deficiency distance is not compact unless the parameter set is finite. However the topology of main interest is the topology of deficiency convergence for restrictions to finite sub-parameter sets and this topology is always compact. This may be deduced from LeCam's consistency criterion which does for experiments what Kolmogorov's consistency criterion does for random processes.

Another main topic of chapter 7 is sufficiency and completeness of subexperiments. The discussion is to a large extent carried out in terms of linear spaces associated with the experiment. One highlight is the characterizations of the (always existing) minimal sufficient algebra of LeCam. If the experiment obeys the weak compactness lemma, as it does when it is discrete or dominated, then this establishes the existence of minimal sufficient σ-algebras.

We shall also see that the experiments which, loosely speaking, admit boundedly complete and sufficient statistics, are precisely the experiments which are not obtainable as non-trivial mixtures. One interesting consequence of this and of the compactness results just mentioned is that the totality of finite mixtures of boundedly complete experiments is dense within the totality of all experiments.

The exact evaluation of the deficiency is usually a most demanding task. As remarked above, a reduction to convolution kernels is often possible in the case of translation experiments. Furthermore, optimal convolution kernels are sometimes provided by least favourable distributions for associated testing problems.

This works for linear normal experiments whenever the covariances are known, as well as in most cases where the covariances are known except for a common unknown multiplicative scalar. Indeed these experiments are translation experiments on solvable groups of linear transformations.

Chapter 8 is devoted to comparison of linear models and in particular to linear normal experiments. We provide here a variety of criteria for information orderings as well as for the deficiencies. A most useful tool for this study is the Hellinger transform. When the covariances are known then the information ordering coincides with the ordering provided by this transform, and this in turn coincides with the usual ordering of Fisher information matrices. If the covariances are not known then the information ordering is replication dependent and thus not directly expressible in terms of the Hellinger transform. However the transform is useful even in this case. One feature of most of the results obtained is that they are easily expressed in terms of linear maps. Thus, for the linear normal models just mentioned, the information ordering is just the ordering obtained by restricting attention to kernels given by linear maps. Furthermore, the information ordering coincides in these cases with the generally much stronger ordering of 'being divisible by' for experiment multiplication.

The arguments underlying most results in chapters 6 and 7 work equally well if we allow ourselves to work with general finite (possibly signed) distributions. Noting this, in chapter 9 we deduce several interesting consequences. In particular we obtain a fairly complete theory of local (non asymptotic) comparison of differentiable experiments. The basis for this theory is the observation that if we shrink small neighbourhoods of a given point in the parameter set then the deficiencies of the restricted experiments decrease at a fixed rate. This rate defines a local deficiency and thereby also a local information ordering as well as a local notion of equivalence.

In the one-dimensional case this implies that any differentiable experiment is locally equivalent to a strongly unimodal translation experiment. We shall also see that the local information ordering for those linear normal experiments we considered in chapter 8 is just the usual ordering of the Fisher information matrices.

Another consequence is a theory for comparison of measures providing, e.g. criteria ensuring that one measure is an ε-dilation of another one. Using the general principles we shall also rediscover some famous results on probability distributions with given marginals.

Restricting ourselves to comparing pairs of measures we obtain several results concerning comparison of experiments having a fixed two-point parameter set. Comparison may in this case be expressed in terms of functions describing relations between level of significance and power of most powerful tests. These functions, here called β-functions or Neyman–Pearson functions, are closely related to the Lorenz transformations (curves) of econometrics as well as to the total time on test (TTT) transform of reliability theory.

The view illustrated here shows that the theory of comparison experiments may be considered as a vast generalization of well known theories on majorization, Schur convexity and on the Lorenz ordering of income distributions.

Results and concepts are illustrated with examples. In addition a few complements to the discussion are inserted as problems. Most complements however are collected in chapter 10. They appear there with references to chapters of particular relevance. In addition the introductions to the various chapters provide references to relevant complements in chapter 10. Complements 37–49 provide a study of monotone likelihood ratio.

References are given at the end of each chapter. We do not claim completeness for any of these lists. At the end of the book there is a combined index of authors along with a list of symbols and a subject index. Some works which are not cited may here be found in a list of additional references.

Theorem (proposition ...) $x \cdot y \cdot z$ refers to the unique theorem (proposition ...) in chapter x, section y, bearing the label z.

This work originated from lecture notes which provided material for a one semester course for graduate students, on decision theory with emphasis on comparison of statistical experiments. The students were required to have absorbed a one semester beginners course in measure theory.

A course covering roughly the same ground may be based on chapters 1 and 3, sections 1–4 of chapter 4, sections 1–3 of chapter 5, sections 1–4 of chapter 6 (the material on weighted deficiences and on extensions of support functions may be omitted), sections 1 and 2 of chapter 7 and relevant material from chapters 2 and 10. The assumptions that the experiments obey the weak compactness lemma and that sample spaces are Euclidean may then be imposed whenever they appear to simplify the discussion.

These assumptions may be lifted in a subsequent course covering the remaining sections of chapters 4–7 and selections from chapters 8–10.

Erik Torgersen
Oslo

ACKNOWLEDGMENTS

I have had the good fortune of being a student of inspiring teachers who are also main contributors to the field of statistical inference.

Along with most of a generation of Norwegian statisticians I learned about laws and chance variations from Erling Sverdrup. From him we received an education based on a careful interplay between applications and theory, each being necessary for the other.

I have also had the privilege of being one of a truly international group of statisticians who as students or as visitors have enjoyed the hospitality of the Department of Statistics at the University of California, Berkeley. I am grateful to David Blackwell and Lucien LeCam for writing interesting commentaries on whatever notes I sent them.

As a student and later as a visitor I almost always found the door to LeCam's office open. In spite of the mountain of papers on his desk he gave me generously of his time. A really comprehensive list of references should have included numerous references to conversations with him.

As a teacher here at the University of Oslo I have had the equally good fortune of having inspiring students. Those among them who worked within this speciality have all contributed to its development. My indebtedness to Geir Dahl, Tor Haldorsen, Ole Håvard Hansen, Jon Helgeland, Bo Lindqvist and Anders Rygh Swensen goes beyond what may be inferred from any reference list.

I am in particular grateful to Bo Lindqvist for his excellent job of recording and supplementing my lectures on the comparison of experiments. It is most appropriate to mention this here since the lecture notes which emerged (Statistical Memoirs, No. 1, 1975, University of Oslo) are the starting point for this work.

Thumbing through the pages it should be apparent that it has not been easy to type this book. What is not so apparent are the almost never ending stages of revisions and corrections. I am grateful to Dina Haraldsson and Mai Bente Olsen for making all this into a readable manuscript.

I wish to express my gratitude to David Tranah of Cambridge University Press and to a reviewer for thorough and valuable suggestions.

And I am grateful to my wife, Liv Torgersen, who in addition to shouldering even more of our common responsibilities provided encouragement throughout the writing of this book.

1

Statistical Experiments within the Measure Theoretical Framework

1.1 Introduction

This chapter is intended to serve several purposes. Firstly it contains tools which will be used throughout this work. Thus several of the basic concepts are introduced. Some of the definitions are repeated in later chapters, but others are only given here. It is not necessary to have acquired an understanding of everything here in order to proceed. Indeed some of the material in this chapter anticipates later developments and may not be fully comprehended before these developments have been studied. Thus the reader may just look over the chapter and later return to this material when the need arises.

Another purpose is to provide a self-contained introduction to some ideas which later will play a central role. Thus it is intended to be possible to go through this chapter before (or whilst) reading the later chapters.

A third purpose is to present some 'classical' material which is important as background for the theory of statistical experiments as it is presented here. In particular some of this material indicates the need of extending, as LeCam did, the measure theoretical framework of decision theory.

The notion of a statistical experiment (statistical model) is introduced in section 2. It is argued there that the usual notion of a statistic as a random variable is too narrow and that it should be extended to the more general concept of a consistent family. The optimistic goal is to make everything as simple (or difficult) as it is in the 'finite' case. Difficulties beyond the finite case may then be considered as being of technical nature.

Extending the notion of a bounded variable we arrive, following LeCam (1964), at the notion of the M-space of an experiment. The M-space is introduced as the dual of another space called the L-space. This L-space consists of those finite measures which are absolutely continuous with respect to countably infinite convex combinations of the distributions which constitute the experiment.

L-spaces of experiments are, as will be explained in chapter 4, particular examples of abstract L-spaces as defined in linear analysis. Indeed any abstract L-space is isomorphic to the L-space of a (totally informative) experiment as shown by Kakutani (1941a). The abstract M-spaces which can be represented as M-spaces of experiments are thus the abstract M-spaces which are duals of abstract L-spaces.

Several methods for deriving new experiments from old ones are considered in section 3. This may e.g. be done by reducing (or expanding) the parameter set or the sample space. In particular we may consider experiments obtained by observing certain functions of the basic observations. These functions may involve random mechanisms and may thus not be strictly determined by the observations.

We may also consider independent combinations, i.e. products, of experiments. Another possibility is to observe a random experiment and then carry out the observed experiment. This leads to the notion of a mixture of experiments. Together products and mixtures provide an interesting algebraic structure for statistical experiments.

Using LeCam's weak (or strong) experiment topology we may also derive experiments as limits.

In section 4 we discuss functionals of experiments. In particular we consider functionals which are determined by non-negatively homogeneous functions on the likelihood space. Among them are: statistical distance, minimum Bayes risk, affinity, Hellinger transform, the Kullback–Leibler information number and the Fisher information matrix.

Section 5 is an exposition of some of the 'classical' theory of sufficiency. As shown by Burkholder (1961) this theory exhibits peculiarities which indicate the need for a revision of the distinction between essential statistical problems on the one hand and technical problems on the other. These matters were to a large extent clarified by LeCam's 1964 paper.

It is useful to know that λ-systems of events generated by π-systems are themselves π-systems. As we have not found any other natural place for this we have included an exposition on these systems in an appendix.

Complements in chapter 10 which are of interest in connection with the topics treated in this chapter are numbers 1, 2, 4–8, 10–12, 18, 27, 28, 36, 38, 40, 46 and 47.

We shall often employ the notion of a *Euclidean measurable space* for a measurable space which is Borel isomorphic to a Borel subset of the real line. Thus a measurable space $(\mathscr{X}, \mathscr{A})$ is Euclidean if and only if there is a 1–1 bimeasurable map from $\mathscr{X}$ onto a Borel subset of a complete separable metric space. The isomorphism theorem for Borel subsets of complete separable metric spaces (see e.g. Parthasarathy, 1967) implies that if $\mathscr{X}$ is uncountable then any uncountable Borel subset of any complete separable metric space may be chosen as the range of this bimeasurable map. Thus if $\mathscr{X}$ is uncountable then $(\mathscr{X}, \mathscr{A})$ is Euclidean if and only if $(\mathscr{X}, \mathscr{A})$ is Borel isomorphic to the real line.

Here are some other notions, notations and conventions which will be used throughout.

An *ordering* of a set D is a reflexive and transitive relation $\geqslant$ on D. Thus we do not require that an ordering is antisymmetric, i.e. that $d_1 = d_2$ when $d_1 \geqslant d_2$ and $d_2 \geqslant d_1$. The general terminology concerning ordered sets is described in section 2.4. We shall find it convenient to use the notations $\wedge$ and $\vee$ for inf and sup respectively.

A *directed set* D is an ordered set D such that for any two elements d_1, d_2 in D, there is a third element d_3 in D such that $d_3 \geqslant d_1$ and $d_3 \geqslant d_2$. Thus D is directed if and only if finite subsets of D possess upper bounds.

A *net* (generalized sequence) is a family $(x_d : d \in D)$ where D is a directed set. We often use the simplified notation (x_d) for a net $(x_d : d \in D)$. A net $(y_e : e \in E)$ is called a *subnet* of the net $(x_d : d \in D)$ if there is a function ϕ such that $y_e = x_{\phi(e)}$ for all elements e in E and also to each d_0 in D there corresponds an element e_0 in E such that $\phi(e) \geqslant_D d_0$ whenever $e \geqslant_E e_0$. A subnet of a net (x_d) is often denoted by $(x_{d'})$.

A net $(x_d : d \in D)$ *is in* (*or on*) a set $\mathscr{X}$ if $x_d \in \mathscr{X}$ for all d in D. If $\mathscr{X}$ is a topological space and $x \in \mathscr{X}$ then we say that a net $(x_d : d \in D)$ converges to x, or that x *is the limit of* (x_d), if any neighbourhood V of x contain some *tail* $\{x_d : d \geqslant d_0\}$ of (x_d). If (x_d) converges to x then this may be expressed by writing $\lim_d x_d = x$. Concerning a net $(x_d : d \in D)$ on the extended real line $[-\infty, \infty]$ we define the quantities $\operatorname{liminf}_d x_d$ and $\operatorname{limsup}_d x_d$ by:

$$\operatorname*{liminf}_d x_d = \sup_{d_0} \inf\{x_d : d \geqslant d_0\} \text{ and } \operatorname*{limsup}_d x_d = \inf_{d_0} \sup\{x_d : d \geqslant d_0\}.$$

We refer the reader to, e.g. Kelley (1955) or Dunford & Schwartz (1958) (where nets are called generalized sequences) for the use of nets in point set topology.

We have found it convenient to use the notation $\mathscr{L}(X)$ for the probability distribution of the random variable X. Some additional notation is listed here:

$\# A$ = the number of elements in A.

$o(h)$ = function of the quantity h having the property that $o(h)/h \to 0$ as $h \to 0$.

$\mathrm{d}\mu/\mathrm{d}\nu$ = the Radon–Nikodym density of the ν-absolutely continuous part of μ with respect to ν.

$\mu \gg \nu$ is short for 'the measure μ dominates the measure ν', i.e. ν is absolutely continuous with respect to μ.

$\mu \gg (\nu_\theta : \theta \in \Theta)$ is short for '$\mu \gg \nu_\theta$ when $\theta \in \Theta$'.
$\langle a, b, \dots \rangle$ = the convex hull of $\{a, b, \dots\}$.

Notation for intervals is:

$]a, b[= \{x : a < x < b\}$
R = the real line $=]-\infty, \infty[$
$]a, b] = \{x : a < x \leqslant b\}$
$[a, b[= \{x : a \leqslant x < b\}$
$[a, b] = \{x : a \leqslant x \leqslant b\}$.

1.2 The measure theoretical framework

Leaving mathematical niceties aside for the moment, a statistical experiment may be described as consisting of three parts:

(i) the list $\mathscr{X}$ of possible outcomes. This list is usually called the *sample space*.
(ii) the list Θ of explaining theories. This list is usually called the *parameter set*.
(iii) the rule which to each explaining theory θ in Θ associates the chance mechanism P_θ *explained* by the theory θ. The chance mechanism P_θ is the probability law which governs the random outcome in $\mathscr{X}$ when the explaining theory θ prevails.

Mathematically *a statistical experiment* $\mathscr{E}$ may conveniently be described as a triple $\mathscr{E} = (\mathscr{X}, \mathscr{A}; P_\theta : \theta \in \Theta)$ where $(\mathscr{X}, \mathscr{A})$ is a measurable space, Θ is a set and P_θ, for each θ in Θ, is a probability measure on $\mathscr{A}$. The measurable space $(\mathscr{X}, \mathscr{A})$ is called the *sample space of the experiment* $\mathscr{E}$, while the set Θ is called the *parameter set of* $\mathscr{E}$. You may think of θ as a possible law of nature (see (ii) above). If nature obeys θ then the outcome of the experiment is assumed to be distributed according to P_θ.

The parameter set will, unless otherwise stated, be assumed to be fixed, although arbitrary, throughout. An experiment $\mathscr{E} = (\mathscr{X}, \mathscr{A}; P_\theta : \theta \in \Theta)$ with parameter set Θ is simply a family of probability measures on a common measurable space. The sample space will often be suppressed in the notation of an experiment i.e. we may write $\mathscr{E} = (P_\theta : \theta \in \Theta)$ instead of $\mathscr{E} = (\mathscr{X}, \mathscr{A}; P_\theta : \theta \in \Theta)$.

We shall find it convenient to say that a ($\mathscr{X}$-valued) random variable X *realizes the experiment* $\mathscr{E} = (\mathscr{X}, \mathscr{A}; P_\theta : \theta \in \Theta)$ if the distribution of X is P_θ when θ is the true value of the unknown parameter, i.e. when θ prevails. Making this rigorous we may think of an experiment $(Z, \mathscr{C}; R_\theta : \theta \in \Theta)$ and a measurable map X from $(Z, \mathscr{C})$ to $(\mathscr{X}, \mathscr{A})$ so that $R_\theta X^{-1} \equiv_\theta P_\theta$.

The reader should be aware that an experiment $\mathscr{E} = (P_\theta : \theta \in \Theta)$ is a *family* of probability measures and is therefore quite different from the *set* $\{P_\theta : \theta \in \Theta\}$ of probability measures. Thus if a_1 and a_2 are different numbers then the pairs $(N(a_1,1),N(a_2,1))$ and $(N(a_2,1),N(a_1,1))$ of normal distributions define different experiments. Now $a_1 + a_2 - X$ is distributed as, respectively, $N(a_2,1)$ and $N(a_1,1)$ as X is distributed as $N(a_1,1)$ or as $N(a_2,1)$. It follows that these two pairs, although different, ought to be statistically equivalent for any reasonable concept of statistical equivalence. This, however, is due to a particular property of symmetry which is satisfied by these pairs. We shall see in chapter 6 that ordered pairs (P,Q) and (Q,P) of probability measures on the same measurable space define 'statistically' equivalent experiments if and only if $\mathscr{L}(\mathrm{d}P/\mathrm{d}(P+Q)|\frac{1}{2}(P+Q))$ is symmetric about $\frac{1}{2}$.

Before proceeding let us make some general comments on the structure of an experiment $\mathscr{E} = (\mathscr{X},\mathscr{A};P_\theta:\theta\in\Theta)$.

Firstly the class $\mathscr{A}$ of subsets of the set $\mathscr{X}$ is an algebra and each probability P_θ is an additive non-negative measure on $\mathscr{A}$. This implies essential assumptions on permissible operations of events as well as on behaviour of probabilities. The stronger assumptions, which are inherent in the definition, of the closure of $\mathscr{A}$ under countably infinite unions (intersections) and of the σ-additivity of the probabilities P_θ, are of a more technical nature.

It is important to realize the flexibility of this notion of a sample space. The algebra $\mathscr{A}$ decides what can be observed and what cannot be observed. If it should turn out, for example, that some observations are not performed or not utilized, then this may be described mathematically by decreasing the algebra $\mathscr{A}$, keeping the set $\mathscr{X}$ fixed.

The statistical reasons for using algebras $\mathscr{A}$ which in general are smaller than the class of all subsets of $\mathscr{X}$ should not be confused with the fact that well known measures may not be extended to all subsets. Indeed as the Hahn–Banach theorem implies that any additive probability set function may be extended to all subsets, it is doubtful if this has any statistical relevance.

If we try to classify experiments according to the cardinality of the parameter set Θ then the simplest case is that of experiments having one-point parameter sets. This is the situation where the underlying theory is known. Within the context of comparison of statistical experiments this is essentially a trivial case. However, you lose more than you gain by excluding this case.

Experiments with two-point parameter sets are called *dichotomies*. Although the study of dichotomies is far from trivial, the general structure of dichotomies may be described in relatively simple terms. The richness of the general theory is already apparent in this case. Thus, as we shall see in chapter 9, the basic ideas of majorization and Schur convexity may be phrased within the framework of the theory of dichotomies. The results obtained for dichotomies may, of course, be applied to general experiments by, for example, restricting attention to two-point subsets of the parameter set.

As the next level of generality we may consider the case of experiments possessing a general finite parameter set. Now the theory of comparison of experiments, as it has been developed by LeCam, adheres to the principle that general structures should be approximable by finite ones. General statements on experiments may often be obtained by first establishing statements for experiments with finite parameter sets, and then using some approximation. The difficulties encountered in the last step may appear technical, although far from trivial. For the scope of the theory as well as for many applications, it is important to know how such problems may be handled. We shall therefore elaborate somewhat on this. The nicest approach, as shown by LeCam, is to utilize the theory of vector lattices. We shall not assume any knowledge of this theory here. Later on, in chapter 5, we shall present some results from the theory of vector lattices which will simplify life greatly.

Let us proceed by describing some further collections of experiments which all contain any experiment having a countable parameter set.

In order to describe the first collection we shall need the *statistical distance* between probability distributions P and Q on a common measurable space $(\mathscr{X}, \mathscr{A})$. This distance is the total variation $\|P-Q\| = 2\sup\{P(A)-Q(A): A \in \mathscr{A}\}$ of the difference measure $P - Q$. This important quantity will enter into several of our evaluations. For the time being just note that $\|P - Q\|/4 + \frac{1}{2}$ is the maximum probability of guessing the true distribution in the Bayes estimation problem where the prior distribution assigns probability $\frac{1}{2}$ to P and probability $\frac{1}{2}$ to Q.

Consider now an experiment $\mathscr{E} = (\mathscr{X}, \mathscr{A}; P_\theta : \theta \in \Theta)$. Then the map $(\theta_1, \theta_2) \sim \rightarrow \|P_{\theta_1} - P_{\theta_2}\|$ defines a pseudo-metric on Θ ('pseudo' because the map $\theta \sim \rightarrow P_\theta$ may not be $1 - 1$). The parameter set Θ together with this pseudo-metric becomes a pseudo-metric space. We may then classify experiments with respect to the properties of these spaces. Thus we may call an experiment *totally bounded*, *compact* or *separable* according to whether this pseudo-metric space is, respectively, totally bounded, compact or separable.

Suppose $\mathscr{E} = (\mathscr{X}, \mathscr{A}; P_\theta : \theta \in \Theta)$ is separable and that $\{\theta_1, \theta_2, \ldots\}$ is a dense countable subset of Θ. Put $\pi = \sum_{i=1}^{\infty} 2^{-i} P_{\theta_i}$. Then π is a probability distribution on $\mathscr{A}$ such that each P_θ is absolutely continuous with respect to π. It follows then, by the Radon–Nikodym theorem, that each distribution P_θ has a density with respect to π. Experiments $\mathscr{E} = (\mathscr{X}, \mathscr{A}; P_\theta : \theta \in \Theta)$ such that each probability measure P_θ is absolutely continuous with respect to some common σ-finite measure μ are called *dominated*. If $\mathscr{E}$ is dominated then, as we shall see later, the dominating measure may always be chosen as a probability measure $\pi = \sum_{i=1}^{\infty} 2^{-i} P_{\theta_i}$ where $\theta_1, \theta_2, \cdots \in \Theta$. More generally a family $(\mu_\theta : \theta \in \Theta)$ of finite measures on a common measurable space is called *dominated* if there is a non-negative and σ-finite measure μ such that each μ_θ is absolutely continuous with respect to μ.

A much stronger condition than domination is the requirement that $\mathscr{E} = (P_\theta : \theta \in \Theta)$ is *homogeneous* i.e. that the distributions $P_\theta : \theta \in \Theta$ are mutually absolutely continuous.

Suppose now that $\mathscr{E}$ is dominated by, say, a probability measure π. If the σ-algebra $\mathscr{A}$ has a countable basis, then the space of π-integrable functions is separable for the π-mean of absolute differences as the distance. Identifying each probability measure P_θ with its density with respect to π we find, since $\|P_{\theta_1} - P_{\theta_2}\| = \int |dP_{\theta_1}/d\pi - dP_{\theta_2}/d\pi| d\pi$, that $\mathscr{E}$ is separable for the total variation distance. It follows that an experiment $\mathscr{E} = (\mathscr{X}, \mathscr{A}; P_\theta : \theta \in \Theta)$ where $\mathscr{A}$ has a countable basis is dominated if and only if it is separable.

An interesting characterization of separability within the collection of dominated experiments due to Pfanzagl (1969) is as follows. Suppose $\mathscr{E}$ is dominated by the σ-finite measure μ. Let Θ have the measurability structure given by the σ-algebra consisting of all subsets. Then $\mathscr{E}$ is separable if and only if $[dP_\theta|d\mu]_x$ may be specified to be jointly measurable in (θ, x).

Piecing 'disjoint' dominated experiments together we obtain a still more general collection of experiments. The experiments $\mathscr{E} = (\mathscr{X}, \mathscr{A}; P_\theta : \theta \in \Theta)$ obtained in this way should have the following two properties:

(i) there is a measurable partition $\{\mathscr{X}_t : t \in T\}$ of $\mathscr{X}$ such that a subset A of $\mathscr{X}$ is measurable whenever $A \cap \mathscr{X}_t$ is measurable for each t.
(ii) for each $\theta \in \Theta$ and each $t \in T$, let $P_{\theta t}$ be the restriction of P_θ to measurable subsets of $\mathscr{X}_t$. Then, for each $t \in T$, the family $(P_{\theta t} : \theta \in \Theta)$ of measures on the σ-algebra of measurable subsets of $\mathscr{X}_t$ is dominated.

Experiments satisfying (i) and (ii) are called Σ-*finite* by LeCam. Since it is not usual to call dominated experiments finite we have found it more natural to use the term Σ-*dominated* for experiments satisfying (i) and (ii).

Any dominated experiment is clearly Σ-dominated.

If in the description above we assume that each $\mathscr{X}_t$ is a one-point set and if it is assumed that each P_θ has a countable support, then we obtain another subcollection of the collection of Σ-dominated experiments. These experiments are called discrete. Thus a *discrete experiment* is an experiment $\mathscr{E} = (\mathscr{X}, \mathscr{A}; P_\theta : \theta \in \Theta)$ where $\mathscr{A}$ is the class of all subsets of $\mathscr{X}$ and where each P_θ has a countable support.

Example 1.2.1 (Experiments associated with sampling plans). Consider a population Π which is an enumerable set and which may be any such set. Suppose also that there is a characteristic which, with varying amount (value, degree, ...) is possessed by all individuals in Π. Let $\theta(i)$ be the amount of this characteristic for individual $i \in \Pi$. The function θ on Π defined this way is our parameter of interest. We shall assume that it is *a priori* known that θ belongs to, and may be any element of, a set Θ of functions on Π.

In order to find out about θ we may take a sample from Π and measure the characteristic for each of the individuals in the sample. An essential assumption is now that the sampling is carried out according to a known *sequence sampling plan* α, i.e. a probability distribution on the space Π_s of finite sequences of elements from Π. Before proceeding let us agree that a probability measure on an enumerable set is defined for all subsets. In order to retain the possibility of making no observations at all, we may include the 'empty' sequence $\emptyset$ in Π_s. If the sequence sampling plan α is used and if the characteristics of the sampled individuals are measured without errors, then the outcome $(i_1, \theta(i_1)), \ldots, (i_n, \theta(i_n))$ is obtained with probability $\alpha(i_1, \ldots, i_n)$. Thus we may let our sample space consist of all sequences $(i_1, f_1), (i_2, f_2), \ldots, (i_n, f_n)$ where $(i_1, \ldots, i_n) \in \Pi_s$, $f_1, \ldots, f_n \in \bigcup_\theta \theta[\Pi]$ and where $f_\mu = f_\nu$ whenever $i_\mu = i_\nu$.

Let $P_{\theta,\alpha}$ denote the probability distribution of the outcome when θ prevails and α is used. Thus the sequence sampling plan α determines a discrete statistical experiment $\mathscr{E}_\alpha = (P_{\theta,\alpha} : \theta \in \Theta)$. We may call $\mathscr{E}_\alpha$ *the sequence sampling plan experiment* determined by α.

Let $(I_1, F_1), \ldots, (I_n, F_n)$ be the random outcome and consider the statistics U and G where $U = \{I_1, \ldots, I_n\}$ and G is the function on the set U determined by F. Now $P_{\theta,\alpha}((i_1, f_1), \ldots, (i_n, f_n)) = \alpha(i_1, \ldots, i_n)$ or $= 0$ as $(f_1, \ldots, f_n) = (\theta(i_1), \ldots, \theta(i_n))$ or not. It is easily checked that conditional probabilities given (U, G) may be specified independently of θ. Thus (U, G) is sufficient. The important thing is that the reduction by sufficiency leads to another, and equivalent, discrete experiment $\bar{\mathscr{E}}_{\bar{\alpha}} = (\bar{P}_{\theta,\bar{\alpha}} : \theta \in \Theta)$ which may be described as follows.

Let $\mathscr{U}$ be the class of all finite subsets of Π. If $u \in \mathscr{U}$ and α is a sequence sampling plan on Π then $\bar{\alpha}$ is the probability distribution on $\mathscr{U}$ induced from α by the set-valued map $(i_1, \ldots, i_n) \to \{i_1, \ldots, i_n\}$. Thus $\bar{\alpha}$ is the probability distribution of the sampled subset of Π.

We may then let the sample space, $\bar{\mathscr{X}}$, of $\bar{\mathscr{E}}_{\bar{\alpha}}$ consist of all pairs (u, g) where $u \in \mathscr{U}$ and $g = \theta | u$ for some $\theta \in \Theta$. If α is used then the probability $\bar{P}_{\theta,\bar{\alpha}}((u, g))$ of the outcome (u, g) is $\bar{\alpha}(u)$ or $= 0$ as $g = \theta | u$ or not.

It follows that the structure of experiments $\mathscr{E}_\alpha$ may be identified with a structure of probability measures on the set of finite subsets of the population Π.

In general a probability distribution $\bar{\alpha}$ on the enumerable set of finite subsets of Π may be called a *set sampling plan*. (All classes of finite subsets are then considered measurable.)

Any set sampling plan $\bar{\alpha}$ determines a *set sampling plan experiment* $\bar{\mathscr{E}}_{\bar{\alpha}}$ according to the recipe just given.

The problem of comparing experiments determined by sampling plans is treated in Torgersen (1982) and Milbrodt (1985).

Note that if $\mathscr{E}$ is discrete then each P_θ has a density with respect to the counting measure on $\mathscr{A}$. In general, if $\mathscr{E} = (P_\theta : \theta \in \Theta)$ is Σ-dominated then, as noted above, there are always non-negative measures μ on $\mathscr{A}$ such that each P_θ has a density with respect to μ. This may be seen as follows. Suppose $\mathscr{E}$ satisfies

(i) and (ii) in the definition of Σ-dominatedness. We may assume that for each t there is a point $\theta \in \Theta$ such that $P_\theta(\mathscr{X}_t) > 0$. Then for each t there is a probability measure μ_t on $\mathscr{A}$ such that $P_\theta(A \cap \mathscr{X}_t) \equiv_\theta 0$ when $\mu_t(A) = 0$. We may assume that $\mu_t(\mathscr{X} - \mathscr{X}_t) \equiv_t 0$. Then each distribution P_θ has a density with respect to the measure $\mu = \sum_t \mu_t$. Note that μ is σ-finite if and only if T is countable.

A non-negative measure μ is *majorizing* an experiment $\mathscr{E} = (P_\theta : \theta \in \Theta)$ if each distribution P_θ is given by a density with respect to μ. If so then we shall say that $\mathscr{E}$ is *majorized*. Thus, as we have just seen, Σ-dominated experiments are majorized.

A very useful tool for proving the existence of decision procedures with specified properties is the so-called weak compactness lemma. Let us consider an example where this tool is not available.

Example 1.2.2 (The set of power functions is not necessarily closed). Let $\mathscr{E} = ([0, 1], \text{Borelclass}; P_\theta : \theta \in [0, 1])$ where P_0 is the uniform distribution on $[0, 1]$ and where P_θ is the one-point distribution in θ when $\theta > 0$. Consider the problem of testing the null hypothesis '$\theta > 0$' against the alternative '$\theta = 0$'. If we had the additional information that θ belonged to some specified countable subset C of Θ containing both 0 and positive points θ then the non-randomized test δ_C with rejection region $[0, 1] - C$ has power 1 and level of significance 0. Consider now the class $\mathscr{C}$ of countable subsets of Θ which contain 0. Then $\mathscr{C}$ is directed by inclusion so that $\delta_C = 1 - I_C = C \in \mathscr{C}$ is a net (= generalized sequence) of test functions. Clearly

$$\lim_C E_\theta \delta_C = I_{\{0\}}(\theta); \qquad \theta \in \Theta.$$

Thus the pointwise limit of the power functions looks like the 'ideal' power function. However it is not the power function of any test. On the contrary $\sup_\theta |\beta(\theta) - \lim_C E_\theta \delta_C| \geqslant \frac{1}{2}$ for any power function β.

An experiment $\mathscr{E} = (\mathscr{X}, \mathscr{A}; P_\theta : \theta \in \Theta)$ such that P_{θ_1} and P_{θ_2} are mutually singular (i.e. $\|P_{\theta_1} - P_{\theta_2}\| = 2$) when $\theta_1 \neq \theta_2$ is called *totally informative*. The experiments considered in the last example were clearly totally informative. In a sense, which will be made precise later, any totally informative experiment is as informative as observing the unknown parameter θ itself. 'Pathologies' as those in the last example will then be 'defined away' by the principle that whatever can be achieved by restricting attention to an arbitrarily large finite subset of Θ should also be achievable on all of Θ. If, however, we want to discuss fine points concerning the measurability structure of an experiment, then the totally informative experiments exhibit by and large the same variability as there is within the totality of all experiments.

The example shows that if we want the class of power functions to be closed for pointwise convergence, then we may either generalize our notion of statistical decision procedure or we may limit our attention to experiments satisfying a compactness condition.

An experiment $\mathscr{E} = (\mathscr{X}, \mathscr{A}; P_\theta : \theta \in \Theta)$ will be called *coherent* if to each uniformly bounded net (δ_s) of real valued variables there corresponds a subnet $(\delta_{s'})$ and a real variable δ such that $\int \delta_{s'} h \, \mathrm{d}P_\theta \to \int \delta h \, \mathrm{d}P_\theta$ when $\int |h| \, \mathrm{d}P_\theta < \infty$.

Thus the experiment $\mathscr{E}$ considered in example 1.2.1 is not coherent.

Before proceeding to a study of coherent experiments let us consider some basic results on weak compactness and uniform integrability.

Let $(\mathscr{X}, \mathscr{A}, \pi)$ be a probability space and let $X_t : t \in T$ be a family of real valued variables. Then we shall say that the family $\{X_t : t \in T\}$ is *uniformly integrable* if $\int_{|X_t| \geqslant c} |X_t| \, \mathrm{d}\pi \to 0$ as the constant $c \to \infty$ uniformly in t. If $\{X_t : t \in T\}$ is uniformly integrable then the L_1-norms, $\int |X_t| \, \mathrm{d}\pi$, are uniformly bounded by any of the finite constants $c + \sup_t \int_{|X_t| \geqslant c} |X_t| \, \mathrm{d}\pi$. A family $\{X_t : t \in T\}$ of integrable real variables is called *uniformly absolutely continuous* if $\int_A X_t \, \mathrm{d}\pi \to 0$ as $\pi(A) \to 0$ uniformly in t.

It follows from the inequality $|\int_A X_t \, \mathrm{d}\pi| \leqslant c\pi(A) + \sup_t \int_{|X_t| \geqslant c} |X_t| \, \mathrm{d}\pi$ that any uniformly integrable family of real variables is also uniformly absolutely continuous. Suppose conversely that the family $\{X_t : t \in T\}$ is uniformly absolutely continuous and that $\sup \int |X_t| \, \mathrm{d}\pi < \infty$. Then, since $\pi(|X_t| \geqslant c) \leqslant c^{-1} \sup_t \int |X_t| \, \mathrm{d}\pi$; the integrals $\int_{|X_t| \geqslant c} |X_t| \, \mathrm{d}\pi \to 0$ as $c \to \infty$ uniformly in t. Altogether we have proved the following proposition.

Proposition 1.2.3. *A family $(X_t : t \in T)$ of real valued random variables is uniformly integrable if and only if it is uniformly absolutely continuous and $\sup_t E|X_t| < \infty$.*

Some basic results on weak compactness are collected in the following.

Theorem 1.2.4 (Weak compactness theorem). *Let $(\mathscr{X}, \mathscr{A}, \pi)$ be a probability space and, as usual, let $L_1 = L_1(\pi)$ and $L_\infty = L_\infty(\pi)$ denote the equivalence classes of, respectively, integrable functions and bounded measurable functions. Equip L_1 with its L_∞-topology, i.e. the topology where convergence of a net $(X_t : t \in T)$ of integrable variables to an integrable variable X takes place if and only if $\int X_t Y \, \mathrm{d}\pi \to \int XY \, \mathrm{d}\pi$ whenever $y \in L_\infty$.*

Then the following three conditions on a set V of integrable variables are equivalent:

(i) *V is uniformly integrable.*
(ii) *the closure of V is compact.*
(iii) *each sequence $X_1, X_2, \ldots$ in V has a convergent sub-sequence.*

We shall not prove this theorem here. We refer to, e.g. section IV.2 in Neveu (1965), for a complete proof. See also section V.6 in Dunford & Schwarz (1958).

The statement which is usually called the weak compactness lemma is that (i) ⇒ (iii). For the sake of completeness we provide a direct proof of this statement, as well as of the same statement for nets.

Corollary 1.2.5 (Weak compactness lemma). *Let $(X_t : t \in T)$ be a uniformly integrable net (sequence) of real valued variables on a probability space $(\mathscr{X}, \mathscr{A}, \pi)$. Then there is an integrable variable X and a subnet (sub-sequence) $(X_{t'})$ such that*

$$\lim_{t'} EYX_{t'} = EYX$$

whenever the random variable Y is bounded.

Remark. By the Eberlein Šmulian theorem, theorem V.6.1 in Dunford & Schwartz (1958), a weakly closed subset of a Banach space is weakly compact if and only if it is weakly sequentially compact. This implies in particular the equivalence of the statements for nets and for sequences.

Proof. Put $M_c = \sup_t \int_{|X_t| \geqslant c} |X| \, d\pi$ and define, for each t, a linear functional F_t on $L_\infty = L_\infty(\pi)$ by $F_t(Y) = \int YX_t \, d\pi$; $Y \in L_\infty$. For simplicity of notation let us omit the $d\pi$ in the integral notation and thus write $\int Z$ instead of $\int Z \, d\pi$. Clearly $|F_t(Y)| \leqslant c \int |Y| + \|Y\|_\infty M_c$. Here $\|Y\|_\infty$ is the smallest constant a such that $\pi(a \geqslant |Y|) = 1$. It follows from Tychonoff's theorem that there is a subnet $\{X_{t'}\}$ such that $F(Y) = \lim_{t'} F_{t'}(Y)$ exists for all $Y \in L_\infty$. Put $\lambda(A) = F(I_A)$. (When there is no need we do not distinguish between a random variable X and the equivalence class consisting of those random variables which are a.s. equal to X.) Then λ is a finite, additive set function on $\mathscr{A}$. Furthermore, $\lambda(A) \leqslant c\pi(A) + M_c$. Hence $\limsup_{\pi(A) \to 0} \lambda(A) \leqslant M_c \to 0$ as $c \to \infty$. Thus $\lambda(A) \to 0$ when $\pi(A) \to 0$. It follows that λ is countably additive and absolutely continuous with respect to π. Hence by the Radon–Nikodym theorem there is a variable X such that $F(I_A) = \lambda(A) = \int I_A X$; $A \in \mathscr{A}$. Thus, since F is continuous for uniform convergence and since the measurable indicators are fundamental in L_∞, we have $F(Y) = \int Y \, d\lambda = \int YX$; $Y \in L_\infty$. This proves the 'net part' of the theorem.

Suppose now that $(X_t) = (X_1, X_2, \ldots)$ is a sequence. Let $\mathscr{B}$ be the σ-algebra generated by $X_1, X_2, \ldots$ Then $\mathscr{B}$ has a countable basis $\mathscr{B}_0 = \{B_1, B_2, \ldots\}$. We may assume that $\mathscr{B}_0$ is an algebra. By Cantor's diagonal procedure there is a sub-sequence $(X_{n_1}, X_{n_2}, \ldots)$ such that $F(I_B) = \lim_i F_{n_i}(I_B)$ exists whenever $B \in \mathscr{B}_0$. If $B \in \mathscr{B}$ then for each $\varepsilon > 0$ there is a set $B_\varepsilon \in \mathscr{B}_0$ such that $\pi(B \,\Delta\, B_\varepsilon) < \varepsilon$. Then

$$\begin{aligned} |F_{n_i}(I_B) - F_{n_j}(I_B)| &\leqslant |F_{n_i}(I_{B_\varepsilon}) - F_{n_j}(I_{B_\varepsilon})| + \int |(I_B - I_{B_\varepsilon})(X_{n_i} - X_{n_j})| \\ &\leqslant |F_{n_i}(I_{B_\varepsilon}) - F_{n_j}(I_{B_\varepsilon})| + \sup_{i,j} \int_{B \Delta B_\varepsilon} [|X_{n_i}| + |X_{n_j}|] \to 0 \end{aligned}$$

as $(i,j) \to (\infty, \infty)$ and then $\varepsilon \to 0$. Thus $\lim_i F_{n_i}(I_B)$ exists whenever $B \in \mathscr{B}$. Hence, since any $\mathscr{B}$-measurable bounded function is the uniform limit of $\mathscr{B}$-measurable step functions, $F(Y) = \lim_i F_{n_i}(Y)$ exists for all bounded $\mathscr{B}$-measurable functions Y.

Finally consider a general bounded measurable function Y. By the Radon–Nikodym theorem there is a $\mathscr{B}$-measurable bounded function $E(Y|\mathscr{B})$ such that $\int_B Y = \int_B E(Y|\mathscr{B})$ when $B \in \mathscr{B}$. Then $\int XY = \int XE(Y|\mathscr{B})$ when X is integrable and measurable. In particular $F_t(Y) = F_t(E(Y|\mathscr{B}))$; $t = 1, 2, \ldots$ Hence $F(Y) = \lim_i F_{n_i}(Y)$ exists for all $Y \in L_\infty$.

We have thus shown that the subnet constructed in the first part of the proof actually may be chosen as a sub-sequence when the given net is a sequence. This together with the last part of the 'net part' of the proof completes the proof of the theorem. □

When applying the weak compactness lemma the following simple corollary is useful.

Corollary 1.2.6. *Let $(X_t : t \in T)$ be a uniformly bounded net (sequence) of real random variables on a probability space $(\mathscr{X}, \mathscr{A}, \pi)$. Then there is a subnet (sub-sequence) $(X_{t'})$ and a bounded random variable X such that*

$$\lim_{t'} EYX_{t'} = EYX$$

whenever $E|Y| < \infty$.

Proof. The family $(X_t : t \in T)$ is uniformly integrable since it is uniformly bounded. Hence, by the weak compactness lemma, there is an integrable variable X and a subnet (sub-sequence) $(X_{t'})$ such that $\lim_{t'} \int YX_{t'} = \int YX$, $Y \in L_\infty$. Let c be a constant such that $-c \leqslant X_t \leqslant c$ when $t \in T$. Then $-c\int Y \leqslant \int YX_t \leqslant c\int Y$ when $0 \leqslant Y \in L_\infty$. Hence, by taking the limit as t' runs through the given subset, we obtain the same inequality with X_t replaced by X. It follows that $|X| \leqslant c$ a.s. π. Suppose $Y \in L$, and put $Y_n = Y$ or $= 0$ as $|Y| \leqslant n$ or $|Y| > n$. Then $|\int YX_{t'} - \int YX| \leqslant |\int Y_n(X_{t'} - X)| + 2c\int_{|Y| \geqslant n} |Y|$. Thus $\limsup_{t'} |\int YX_{t'} - \int YX| \leqslant 2c\int_{|Y| \geqslant n} |Y| \to 0$ as $n \to \infty$. □

Weakly convergent sequences do not always behave as one might expect. Consider e.g. a probability space $(\mathscr{X}, \mathscr{A}, \pi)$ and a sequence $A_1, A_2, \ldots$ of events in $\mathscr{A}$. Let X_n, for each n, denote the indicator function I_{A_n} of A_n. If, as $n \to \infty$, X_n converges in measure or converges almost surely or converges in $L_1(\pi)$, then in all cases the limit X is, give or take a null set, also an indicator function of an event. If, however, X_n converges weakly to X, then X need not be equivalent to an indicator function.

Example 1.2.7 (Indicator variables converging to $\frac{1}{2}$). Let π be the Lebesgue measure on $[0, 1]$ and for $n = 1, 2, \ldots$ put $A_n = [0, 1/2n] \cup [2/2n, 3/2n] \cup [4/2n, 5/2n] \cup \ldots$ Then $I_{A_n} \to \frac{1}{2}$ for the $L_\infty(P)$-topology on $L_1(P)$.

It is now fairly easy to show that Σ-dominated experiments are coherent. We shall use the following direct consequence of Tychonoff's theorem. Let $\sigma_t : t \in T$ be a family of compact topologies for a set $\mathscr{X}$ and let $\{x_s\}$ be a net in $\mathscr{X}$. Then there is a subnet $\{x_{s'}\}$ and a family $x^t : t \in T$ in $\mathscr{X}$ such that $x_{s'}$ converges for σ_t to x^t whenever $t \in T$.

Suppose now that the experiment $\mathscr{E} = (\mathscr{X}, \mathscr{A}; P_\theta : \theta \in \Theta)$ allows a measurable partition $(\mathscr{X}_t : t \in T)$ of $\mathscr{X}$ such that

(i) $A \in \mathscr{A}$ whenever $A \cap \mathscr{X}_t \in \mathscr{A}$ for all t, and
(ii) for each t there is a non-negative σ-finite measure π_t on $\mathscr{A}$ such that $\pi_t(A) = 0 \Rightarrow P_\theta(A \cap \mathscr{X}_t) \equiv_\theta 0$.

We may assume that the π_t are probability measures (see the first part of the proof of theorem 1.2.10). Let (δ_s) be a uniformly bounded net of real variables. Then, by the weak compactness theorem for probability spaces and the above remark, there is a subnet $(\delta_{s'})$ and a uniformly bounded family $(\delta^t; t \in T)$ of real variables such that $\int h\delta_{s'}\,\mathrm{d}\pi_t \to \int h\delta^t\,\mathrm{d}\pi_t$ when $h \in L_1(\pi_t)$. Put $\delta = \sum I_{\mathscr{X}_t}\delta^t$. Then δ is bounded measurable and $\int h\delta_{s'}\,\mathrm{d}\pi_t \to \int h\delta\,\mathrm{d}\pi_t$ when $h \in L_1(\pi_t)$. Consider any $h \in L_1(P_\theta)$. Then $\lim_{s'} |\int_{\mathscr{X}_t} [h(\delta_{s'} - \delta)\,\mathrm{d}P_\theta/\mathrm{d}\pi_t]\,\mathrm{d}\pi_t| = 0$, $t \in T$, $|\int_{\mathscr{X}_t} [h(\delta_{s'} - \delta)\,\mathrm{d}P_\theta/\mathrm{d}\pi_t]\,\mathrm{d}\pi_t| \leqslant \sup_{x,s'} |\delta_{s'}(x) - \delta(x)| \int_{\mathscr{X}_t} |h|\,\mathrm{d}P_\theta$ and $\sum_t \int_{\mathscr{X}_t} |h|\,\mathrm{d}P_\theta = \int |h|\,\mathrm{d}P_\theta < \infty$. Hence by the dominated convergence theorem $|\int h\delta_{s'}\,\mathrm{d}P_\theta - \int h\delta\,\mathrm{d}P_\theta| \leqslant \sum_t |\int_{\mathscr{X}_t} h(\delta_{s'} - \delta)\,\mathrm{d}P_\theta| = \sum_t |\int_{\mathscr{X}_t} [h(\delta_{s'} - \delta)\,\mathrm{d}P_\theta/\mathrm{d}\pi_t]\,\mathrm{d}\pi_t| \to 0$. Altogether this shows that $\mathscr{E}$ is coherent.

Theorem 1.2.8. *Σ-dominated experiments are coherent.*

The converse of this theorem is not true. If, however, the cardinality of Θ is not larger than that of the continuum and if $\mathscr{A}$ contains any set A which is completion measurable for each P_θ, then $\mathscr{E} = (\mathscr{X}, \mathscr{A}; P_\theta : \theta \in \Theta)$ is Σ-dominated. This follows from the proof in Fell (1956). Nevertheless the class of Σ-dominated experiments does, from the point of view of comparison of experiments, contain representatives of any experiment. To be more precise we shall show that any experiment, however complicated, is essentially equivalent to a Σ-dominated experiment. This applies also to the apparently more general experiments introduced by LeCam in his 1964 paper.

The reason for using the term 'coherent' may become more apparent after we have linked it with the fairly natural concepts of consistency and of coherence for families of random variables. It will then be seen that the coherent experiments are precisely those experiments such that each consistent family is coherent. However we shall not get satisfactory insight before we have said a few words about vector lattices later, in chapter 5.

Let $\mathscr{E} = (\mathscr{X}, \mathscr{A}; P_\theta : \theta \in \Theta)$ be an experiment and let $(\phi_\theta : \theta \in \Theta)$ be a family of real valued variables on $(\mathscr{X}, \mathscr{A})$. Then the family $(\phi_\theta : \theta \in \Theta)$ is called *consistent* if to each two-point subset F of Θ there corresponds a variable ϕ_F such that $P_\theta(\phi_\theta \neq \phi_F) = 0$ when $\theta \in F$. The family $(\phi_\theta : \theta \in \Theta)$ is called *coherent* if there is a variable ϕ such that $P_\theta(\phi_\theta \neq \phi) = 0$ for all $\theta \in \Theta$. It will follow from the next result that if $(\phi_\theta : \theta \in \Theta)$ is consistent then to each countable subset C of Θ there corresponds a variable ϕ_C such that $\phi_\theta = \phi_C$ a.s. P_θ when $\theta \in C$.

Theorem 1.2.9 (Consistency criterion for coherence). *An experiment $\mathscr{E}$ is coherent if and only if consistent families of real variables in $\mathscr{E}$ are coherent.*

Remark. The notion of a coherent family given here extends the same notion in Hasegawa & Perlman (1974). Their notion of a countably (finitely) coherent family is here replaced (and extended) by the notion of a consistent family.

The theorem states that the notions of a coherent experiment as given here and in Hasegawa & Perlman's paper coincide.

Proof.

1. Suppose $\mathscr{E}$ is coherent. Let $(\phi_\theta : \theta \in \Theta)$ be a consistent family of real random variables. We may without loss of generality assume that $0 \leqslant \phi_\theta \leqslant 1, \theta \in \Theta$. (If not, then replace ϕ_θ with $G(\phi_\theta)$ where G is the cumulative $N(0,1)$ distribution function.) By assumption, for each two-point set F there is a variable ϕ_F such that $\phi_\theta = \phi_F$ a.s. P_θ; $\theta \in F$. Let $F = \{\theta_1, \theta_2, \ldots, \theta_n\}$ be an n-point set where $n \geqslant 3$. Put $s_i = \mathrm{d}P_{\theta_i}/\mathrm{d}\sum_{j=1}^n P_{\theta_j}$. We may assume that $s_1, s_2, \ldots, s_n \geqslant 0$ and $\sum_1^n s_i = 1$. Put $\phi_F = \phi_{\theta_i}$ when $s_1 = \cdots = s_{i-1} = 0$ and $s_i > 0$. Then $P_{\theta_i}(\phi_{\theta_i} \neq \phi_F) = \sum_{j=1}^{i-1} P_{\theta_i}((\phi_{\theta_i} \neq \phi_F); s_1 = \cdots = s_{j-1} = 0, (s_j > 0)) \leqslant \sum_{j=1}^{i-1} P_{\theta_i}((\phi_{\{\theta_i,\theta_j\}} \neq \phi_{\theta_j}) \cap (s_j > 0)) = 0$ since $P_{\theta_j}(\phi_{\{\theta_i,\theta_j\}} \neq \phi_{\theta_j}) = 0$ and P_{θ_j} dominates P_{θ_i} on the set $(s_j > 0)$.

   It follows that to each finite subset F of Θ there corresponds a variable ϕ_F such that $\phi_\theta = \phi_F$ a.s. when $\theta \in F$. Order the finite subsets of Θ by inclusion and consider the corresponding net (ϕ_F). By assumption there is a bounded measurable function ϕ and a subnet $(\phi_{F'})$ such that $\int_A \phi_{F'}\,\mathrm{d}P_\theta \to \int_A \phi\,\mathrm{d}P_\theta$; $A \in \mathscr{A}$. Now $\int_A \phi_{F'}\,\mathrm{d}P_\theta = \int_A \phi_\theta\,\mathrm{d}P_\theta$ when $F' \supseteqq \{\theta\}$. Hence $\int_A \phi_\theta\,\mathrm{d}P_\theta = \int_A \phi\,\mathrm{d}P_\theta$; $A \in \mathscr{A}$ i.e. $\phi_\theta = \phi$ a.s. P_θ.
2. Suppose the condition is satisfied. Let (δ_s) be a net of measurable functions such that $\sup_{x,s} |\delta_s(x)| \leqslant M < \infty$. Denote by H the set of measurable functions δ such that $\sup_x |\delta(x)| \leqslant M$. For each finite non-empty subset F of Θ put $\mu_F = \sum_F P_\theta$. Then, as we saw in the proof of theorem 1.2.8, there is a subnet $(\delta_{s'})$ in H and a family (δ^F) such that $\lim_{s'} \int h\delta_{s'}\,\mathrm{d}\mu_F = \int h\delta^F\,\mathrm{d}\mu_F$ when $h \in L_1(\mu_F)$. Let $A \in \mathscr{A}$ and write δ^θ instead of $\delta^{\{\theta\}}$. Then $\int_A \delta^\theta\,\mathrm{d}P_\theta = \lim_{s'} \int_A \delta_{s'}\,\mathrm{d}P_\theta = \lim_{s'} \int_A \delta_{s'}(\mathrm{d}P_\theta/\mathrm{d}\mu_F)\,\mathrm{d}\mu_F = \int_A \delta^F(\mathrm{d}P_\theta/\mathrm{d}\mu_F)\,\mathrm{d}\mu_F = \int_A \delta^F\,\mathrm{d}P_\theta$ when $\theta \in F$. Hence $\delta^\theta = \delta^F$ a.s. P_θ when $\theta \in F$. Thus $(\delta^\theta : \theta \in \Theta)$ is consistent and consequently, by assumption, coherent. It follows that there is a δ in H such that $\delta^\theta = \delta$ a.s. P_θ for all θ and then $\int h\delta_{s'}\,\mathrm{d}P_\theta = \int h\delta_{s'}\,\mathrm{d}\mu_{\{\theta\}} \to \int h\delta^\theta\,\mathrm{d}\mu_{\{\theta\}} = \int h\delta\,\mathrm{d}P_\theta$ when $h \in L_1(P_\theta)$.

□

The theorem shows that consistent families of variables are essentially ordinary variables when the experiment is coherent. In general consistent families are limits of ordinary variables. Before making this precise let us introduce two important spaces associated with an experiment $\mathscr{E} = (\mathscr{X}, \mathscr{A}; P_\theta : \theta \in \Theta)$.

The L-space of the experiment $\mathscr{E}$ is the space consisting of all finite measures μ on $\mathscr{A}$ such that μ is absolutely continuous with respect to some measure of the form $\sum_{i=1}^{\infty} 2^{-i} P_{\theta_i}$ where $\theta_1, \theta_2, \ldots$ is a sequence in Θ. The L-space of $\mathscr{E}$ will be denoted by $L(\mathscr{E})$. This space is the union of the L-spaces of restrictions of $\mathscr{E}$ to countable subsets of the parameter set Θ. It is easily seen that $L(\mathscr{E})$ is a Banach space for linear combinations defined set wise and for the total variation norm $\| \; \|$.

The space of bounded linear functionals on $L(\mathscr{E})$ is called the *M-space of the experiment* $\mathscr{E}$ and will be denoted by $M(\mathscr{E})$. It is well known that $M(\mathscr{E})$ is also a Banach space for the dual norm defined by $\|u\| = \sup\{u(\lambda) : \|\lambda\| \leqslant 1\}$.

What do these spaces look like in the dominated case? In order to answer this we shall need the fact that the ordering '$\geqslant$ a.e.' of the set of extended real valued measurable functions is complete.

Theorem 1.2.10 (Completeness of the a.e. ordering of measurable functions). *Let $(\mathscr{X}, \mathscr{A}, \mu)$ be a σ-finite measure space, i.e. $(\mathscr{X}, \mathscr{A})$ is a measurable space and μ is a non-negative σ-finite measure on $\mathscr{A}$.*

Then the ordering '$\geqslant$ a.e. μ' for extended real valued measurable functions is complete.

In fact a family $(X_t : t \in T)$ of extended real valued measurable functions on $(\mathscr{X}, \mathscr{A})$ possesses a least upper bound for this ordering of the form $\sup\{X_t : t \in C\}$ where C is a countable subset of T.

Remark. As argued in example 5.5.10, this result is an immediate consequence of the order completeness of L-spaces.

Proof. If $\mu(\mathscr{X}) = \infty$ then we may, since μ is σ-finite, decompose $\mathscr{X}$ as $\mathscr{X} = \bigcup_{i=1}^{\infty} A_i$ where $A_1, A_2, \ldots$ are disjoint sets in $\mathscr{X}$ having finite positive μ-measures. Then μ defines the same null sets as the probability measure P given by:

$$P(A) = \sum_{i=1}^{\infty} 2^{-i} \mu(A \cap A_i)/\mu(A_i); \qquad A \in \mathscr{A}.$$

It follows that to any non-null non-negative σ-finite measure μ there corresponds a probability measure P defining the same null sets as μ.

We may therefore assume without loss of generality that μ is a probability measure.

Choose a continuous and strictly increasing cumulative probability distribution function F on the real line. Extend F to the extended real line by putting $F(-\infty) = 0$ and $F(\infty) = 1$. Let $(X_t : t \in T)$ be a family of extended real valued measurable functions on $(\mathscr{X}, \mathscr{A})$. For each countable subset C of T put $X_C = \sup\{X_t : t \in C\}$. Put also $a = \sup_C E_\mu F(X_C)$. Then there is a sequence $C_1, C_2, \ldots$ of countable subsets of T such that $E_\mu F(X_{C_n}) \uparrow a$. Thus the sup is obtained for

$C_0 = \bigcup_n C_n$ i.e. $a = E_\mu F(X_{C_0})$. If C is countable then, by the maximality of C_0, $E_\mu F(X_{C_0}) = E_\mu F(X_{C_0 \cup C})$. Hence, since $F(X_{C_0 \cup C}) = F(\max(X_{C_0}, X_C)) = \max(F(X_{C_0}), F(X_C)) \geqslant F(X_{C_0})$, $F(X_{C_0 \cup C}) = F(X_{C_0})$ a.e. μ. It follows that $X_{C_0 \cup C} = \max(X_{C_0}, X_C) = X_{C_0}$ a.e. μ. Thus $X_{C_0} \geqslant X_C$ a.e. μ whenever C is a countable subset of T. In particular $X_{C_0} \geqslant X_C$ a.e. μ when $t \in T$. □

We will find the following corollary useful on many occasions.

Corollary 1.2.11 (Dominated experiments are dominated by countable convex combinations). *If $\mathscr{E} = (\mathscr{X}, \mathscr{A}; P_\theta : \theta \in \Theta)$ is a dominated experiment then there is a countable subset Θ_0 of Θ such that if N denotes an event in $\mathscr{A}$ then $P_\theta(N) \equiv_\theta 0$ if and only if $P_\theta(N) = 0$ when $\theta \in \Theta_0$.*

It follows that a dominated experiment $\mathscr{E} = (P_\theta : \theta \in \Theta)$ is dominated by probability measures of the form $P_\kappa = \sum \kappa(\theta) P_\theta$ where κ is a probability distribution on Θ having countable support.

Proof. Let μ be a non-negative σ-finite measure μ dominating the experiment $\mathscr{E} = (\mathscr{X}, \mathscr{A}; P_\theta : \theta \in \Theta)$. Choose for each $\theta \in \Theta$ a non-negative version f_θ of $\mathrm{d}P_\theta/\mathrm{d}\mu$. Then, by the theorem, there is a countable subset Θ_0 of Θ such that $f = \sup\{f_\theta : \theta \in \Theta_0\}$ is a least upper bound of the family $\{f_\theta : \theta \in \Theta\}$ for the ordering '$\geqslant$ a.e. μ'. Let N be an event such that $P_\theta(N) = 0$ when $\theta \in \Theta_0$ and put $g = f I_{N^c}$. Now $f_\theta I_{N^c}$, for each $\theta \in \Theta_0$, is a version of $\mathrm{d}P_\theta/\mathrm{d}\mu$. Thus $f_\theta I_{N^c} = f_\theta$ a.e. μ when $\theta \in \Theta_0$. Hence $g = f I_{N^c} = \sup\{f_\theta I_{N^c} : \theta \in \Theta_0\} = \sup\{f_\theta : \theta \in \Theta_0\} = f$ a.e. μ. Thus g is also a least upper bound of $\{f_\theta : \theta \in \Theta\}$. If θ is any point in Θ then, since $g \geqslant f_\theta$ a.e. μ, $P_\theta(N) = \int_N f_\theta \,\mathrm{d}\mu \leqslant \int_N g \,\mathrm{d}\mu = \int_N 0 \,\mathrm{d}\mu = 0$. Hence $P_\theta(N) \equiv_\theta 0$ whenever $P_\theta(N) = 0$ when $\theta \in \Theta_0$.

The proof may now be completed by showing that if the prior probability distribution κ on Θ has Θ_0 as its minimal support then $P_\kappa = \sum \kappa(\theta) P_\theta$ dominates $\mathscr{E}$. □

We are now ready to characterize the L-spaces and the M-spaces of dominated experiments.

Theorem 1.2.12 (L-spaces and M-spaces of dominated experiments). *Let $\mathscr{E} = (\mathscr{X}, \mathscr{A}; P_\theta : \theta \in \Theta)$ be a dominated experiment.*

Let μ be a non-negative σ-finite measure dominating $\mathscr{E}$ and having the further property that $\mu(N) = 0$ whenever N is an event in $\mathscr{A}$ such that $P_\theta(N) \equiv 0$. (In fact by corollary 1.2.11 *we may put $\mu = P_\kappa = \sum \kappa(\theta) P_\theta$ for some prior probability measure κ on Θ having countable support.)*

Then the L-space $L(\mathscr{E})$ of $\mathscr{E}$ consists precisely of the μ-absolutely continuous finite measures on $\mathscr{A}$. Thus $L(\mathscr{E})$, as an ordered Banach space, is isomorphic to $L_1(\mu)$ by the correspondence which to each measure in $L(\mathscr{E})$ associates the element in $L_1(\mu)$ defined by its Radon–Nikodym derivative with respect to μ.

It follows that the M-space $M(\mathscr{E})$ of $\mathscr{E}$, as an ordered Banach space, is *isomorphic to $L_\infty(\mu)$ by the map which to each $\delta \in L_\infty(\mu)$ associates the linear functional $\mathring{\delta} \in M(\mathscr{E})$ defined by $\mathring{\delta}(\tau) = \int \delta \, d\tau$; $\tau \in L(\mathscr{E})$.*

Proof. Let μ be as described. We have just seen that there is a countable subset Θ_0 of Θ such that $P_\theta(N) \equiv_\theta 0$ whenever $P_\theta(N) = 0$ when $\theta \in \Theta_0$ Let $\theta_1, \theta_2, \ldots$ be a sequence in Θ_0 which runs through all of Θ_0. Then μ and $\sum_i 2^{-i} P_{\theta_i}$ possess the same null sets. Thus any finite measure dominated by μ belongs to $L(\mathscr{E})$. Conversely the definition implies directly that measures in $L(\mathscr{E})$ are dominated by μ.

This, together with the Radon–Nikodym theorem, proves the first two statements of the theorem. The proof may now be completed by using the fact that $L_\infty(\mu)$ and $L_1(\mu)^*$ are, as ordered Banach spaces, isomorphic by the map which to each $\delta \in L_\infty(\mu)$ associates the functional $f \to \int \delta f \, d\mu$ on $L_1(\mu)$. □

Problem 1.2.13 (L-spaces and M-spaces of discrete experiments). The L-space of a discrete experiment $\mathscr{E} = (\mathscr{X}, \mathscr{A}; P_\theta : \theta \in \Theta)$ consists precisely of all finite measures having countable support in N^c where $N = \{x : P_\theta(x) \equiv_\theta 0\}$.

It follows that $M(\mathscr{E})$, as an ordered Banach lattice, is isomorphic to the Banach space of bounded functions on N^c equipped with the sup norm.

Problem 1.2.14 (L-spaces and M-spaces of Σ-dominated experiments). What do the L-spaces and M-spaces of Σ-dominated experiments look like?

Example 1.2.15 (The Rao–Blackwell–Bahadur theorem). The extension of the notion of a random variable which we have considered here throws light on the problem of reversing the Rao–Blackwell theorem in estimation theory. The reader is referred to almost any textbook in statistics for this famous theorem. However he or she might also go to the sources and consult Rao (1945) and Blackwell (1947).

Here we shall permit ourselves to be a bit frivolous in our association with σ-algebras, thus escaping some technicalities.

In this spirit let us agree to say that a statistic t is *quadratically complete* if the only unbiased estimator of zero which is based on t and which has finite variance for all θ, is the constant estimator 0. Say also that a function of the unknown parameter θ is *estimable* if it possesses an unbiased estimator with everywhere finite variance.

If a quadratically complete and sufficient statistic t exists then the Rao–Blackwell theorem tells us not only what the uniformly minimum variance unbiased (UMVU) estimators look like (they are functions of t) but also how they may be obtained from unbiased estimators by conditioning them on t.

Say that an experiment $\mathscr{E}$ has property (Ra–Bl) = (Rao–Blackwell) if any estimable real valued function of θ has a UMVU estimator. Say that $\mathscr{E}$ has property (Ba) = (Bahadur) if it permits a quadratically complete and sufficient statistic.

The Rao–Blackwell theorem may then be expressed as the implication

$$(\text{Ba}) \Rightarrow (\text{Ra–Bl}).$$

The question then naturally arises whether there are experiments which do not allow quadratically complete statistics and still have the property that any parameter possessing an unbiased estimator with everywhere finite variance also has a UMVU estimator.

Bahadur (1957) (see also his 1976 paper and Torgersen (1988)) settled this problem for dominated models by showing that (Ra–Bl) $\Leftrightarrow$ (Ba) in this case. In its full generality and within the usual framework of mathematical statistics, the problem is still open.

If, however, we admit consistent families of random variables as estimators and modify (Ra–Bl) and (Ba) accordingly, then (Torgersen, 1981a) these conditions are equivalent without any regularity conditions whatsoever. Returning to the traditional framework we find in this way that the equivalence (Ra–Bl) $\Leftrightarrow$ (Ba) extends to the case of coherent models.

Consider now an experiment $\mathscr{E}$ and let $v = (\phi_\theta : \theta \in \Theta)$ be a uniformly bounded consistent family of real variables on $\mathscr{E}$. Then, as we have seen, to each countable subset C of Θ there corresponds a variable ϕ_C such that $\phi_\theta = \phi_C$ a.s. P_θ when $\theta \in C$. Let $\lambda \in L(\mathscr{E})$ and choose a sequence $\theta_1, \theta_2, \ldots$ such that $\lambda \ll \sum_{i=1}^{\infty} 2^{-i} P_{\theta_i}$. Put $C = \{\theta_1, \theta_2, \ldots\}$ and $v(\lambda) = \int \phi_C \, d\lambda$. It is then easily checked that the number $v(\lambda)$ does not depend on the choice of the sequence $(\theta_1, \theta_2, \ldots)$. Furthermore $v(\lambda)$ is linear in λ and $|v(\lambda)| \leqslant \sup_{x,\theta} |\phi_\theta(x)| \, \|\lambda\|$. Thus v defines an element of $M(\mathscr{E})$. Say that two families $(\phi_\theta : \theta \in \Theta)$ and $(\psi_\theta : \theta \in \Theta)$ of real variables are *equivalent* if $P_\theta(\phi_\theta \neq \psi_\theta) \equiv_\theta 0$. Then two consistent and uniformly bounded families of real variables define the same functional in $M(\mathscr{E})$ if and only if they are equivalent. The correspondence between equivalence classes of such families and linear functionals is 1–1.

Theorem 1.2.16 (Representation of $M(\mathscr{E})$). *Let $\mathscr{E} = (\mathscr{X}, \mathscr{A}; P_\theta : \theta \in \Theta)$ be an experiment with L-space L and M-space M. Then to each uniformly bounded and consistent family $(\phi_\theta : \theta \in \Theta)$ of real variables there corresponds a unique linear functional v in M such that $v(\lambda) = \int \phi_C \, d\lambda$ whenever $C = \{\theta_1, \theta_2, \ldots\} \subseteq \Theta$, $\lambda \ll \sum_i 2^{-i} P_{\theta_i}$ and $\phi_\theta = \phi_C$ a.s. P_θ when $\theta \in C$. Any linear functional v in M may be obtained from a consistent and uniformly bounded family of real variables this way. The consistent family is determined by v up to equivalence.*

Proof. It remains to show that any v in M may be obtained this way. For each countable subset C of Θ, let $\theta(C)_1, \theta(C)_2, \ldots$ be a sequence in C which runs through all of C. Put $P_C = \sum_i 2^{-i} P_{\theta(C)_i}$. For each $h \in L_1(P_C)$, let λ_h^C be the unique finite measure whose Radon–Nikodym derivative with respect to P_C is h. Finally put $\tilde{v}_C(h) = v(\lambda_h^C)$ when $h \in L_1(P_C)$. Then $\tilde{v}_C$ is a linear functional on $L_1(P_C)$. It is bounded since $|\tilde{v}_C(h)| \leqslant \|v\| \, \|\lambda_h^C\| = \|v\| \int |h| \, \mathrm{d}P_C$. It follows that there is a measurable function ϕ_C such that $\tilde{v}_C(h) = \int h\phi_C \, \mathrm{d}P_C$. ϕ_C may be chosen so that $\sup_x |\phi_C(x)| \leqslant \|\tilde{v}_C\| \leqslant \|v\|$. Let λ be any finite measure dominated by P_C. Then $v(\lambda) = \tilde{v}_C(\mathrm{d}\lambda|\mathrm{d}P_C) = \int [\mathrm{d}\lambda/\mathrm{d}P_C]\phi_C \, \mathrm{d}P_C = \int \phi_C \, \mathrm{d}\lambda$. In particular $\int \phi_{\{\theta\}} \, \mathrm{d}\lambda = v(\lambda) = \int \phi_C \, \mathrm{d}\lambda$ when $\lambda \ll P_\theta$ and $\theta \in C$. Thus $\phi_{\{\theta\}} = \phi_C$ a.s. P_θ when $\theta \in C$. It follows that the family $(\phi_{\{\theta\}} : \theta \in \Theta)$ is consistent and uniformly bounded and that v is derived from this family. □

Each bounded real variable δ defines a consistent family $\{\delta : \theta \in \Theta\}$. The corresponding linear functional $\dot{\delta}$ in $M(\mathscr{E})$ assigns the number $\int \delta \, \mathrm{d}\lambda$ to $\lambda \in L(\mathscr{E})$. If $\mathscr{E}$ is coherent then all functionals in $M(\mathscr{E})$ arise this way. This property characterizes coherence.

Corollary 1.2.17 (M-space characterization of coherence). *$\mathscr{E} = (\mathscr{X}, \mathscr{A}; P_\theta : \theta \in \Theta)$ is coherent if and only if each functional in $M(\mathscr{E})$ is of the form $\dot{\delta}$ for some bounded measurable function δ.*

Proof. If $\mathscr{E}$ is coherent then the representation theorem for $M(\mathscr{E})$ together with the consistency criterion for coherence, imply that each functional in $M(\mathscr{E})$ is of the form $\dot{\delta}$. Suppose conversely that $\mathscr{E}$ has this property and that $\{\phi_\theta : \theta \in \Theta\}$ is a consistent and uniformly bounded family of real variables. Then this family defines a linear functional which, by assumption, is of the form $\dot{\delta}$. Hence, by the last statement of the representation theorem, $\phi_\theta = \delta$ a.s. P_θ; $\theta \in \Theta$. Thus $(\phi_\theta : \theta \in \Theta)$ is coherent so that $\mathscr{E}$ is coherent by the consistency criterion for coherence. □

Problem 1.2.18 (Supports and majorization). Consider an experiment $\mathscr{E} = (\mathscr{X}, \mathscr{A}; P_\theta : \theta \in \Theta)$ having L-space $L = L(\mathscr{E})$ and M-space $M = M(\mathscr{E})$.

According to the terminology established in chapter 5 a functional u in M is called *idempotent* if $u^2 = u$; i.e. if $u \wedge (\dot{1} - u) = 0$. Verify the following statements.

(i) To any measure λ in L there corresponds a smallest idempotent u_λ in M such that $|\lambda|(u_\lambda) = \|\lambda\|$. The functional u_λ is representable as $\dot{I}_A$ for a set A in $\mathscr{A}$ if and only if $|\lambda|(A) = \|\lambda\|$ and $P_\theta(A - B) \equiv_\theta 0$ for any other set B in $\mathscr{A}$ such that $|\lambda|(B) = \|\lambda\|$.

A set A representing u_λ will be called a *smallest support of λ in $\mathscr{A}$*.

(ii) All measures in L possess smallest supports in $\mathscr{A}$ when $\mathscr{E}$ is majorized.

(iii) If each measure P_θ has a smallest support in $\mathscr{A}$ then also each measure in $L(\mathscr{E})$ has a smallest support in $\mathscr{A}$. If so then $\mathscr{E}$ is majorized by $\sum_t \pi_t$ for any maximal family $(\pi_t : t \in T)$ of disjoint probability measures in $L(\mathscr{E})$.

(iv) $\mathscr{E}$ is majorized if and only if each measure P_θ possesses a smallest support in $\mathscr{A}$.

Problem 1.2.19 (Majorized experiments need not be coherent). Show, by e.g. considering infinite sequences of independent and identically distributed random variables, that majorized experiments need not be coherent.

1.3 Derived experiments

There is a large variety of ways of obtaining 'new' experiments from 'old' ones. Thus experiments may be mixed by first drawing an experiment according to a known chance mechanism and then performing the drawn experiment. Conversely experiments may be decomposed as mixtures i.e. they may be shown to be equivalent to experiments obtained by mixing with known mixing distributions. Thus in chapter 7 we shall see that experiments, however complicated, may be regarded as mixtures of experiments admitting complete and sufficient statistics.

Another way of combining experiments is by joining them 'independently' into a large experiment called the product experiment. The converse procedure of 'factorizing' experiments is also of interest as we shall see in chapters 8–10.

Experiments may also be changed by restricting or by expanding the parameter set or the set of available observations.

In this section we shall briefly discuss these methods, as well as a few other methods, for deriving experiments.

Consider an experiment $\mathscr{E} = (\mathscr{X}, \mathscr{A}; P_\theta : \theta \in \Theta)$. If $\mathscr{B}$ is a sub σ-algebra of $\mathscr{A}$ then we may consider the *restricted experiment* (*sub-experiment*) $\mathscr{E}|\mathscr{B} = (\mathscr{X}, \mathscr{A}; P_\theta|\mathscr{B} : \theta \in \Theta) = (P_\theta|\mathscr{B} : \theta \in \Theta)$. Any reasonable concept of statistical information should make $\mathscr{E}|\mathscr{B}$ at most as informative as $\mathscr{E}$. If no information is lost by basing ourselves on $\mathscr{E}|\mathscr{B}$ rather than $\mathscr{E}$ then we may say that $\mathscr{B}$ is sufficient in $\mathscr{E}$ for the chosen notion of information. On the other hand if $\mathscr{B}$ is sufficient in the technical sense that conditional probabilities given $\mathscr{B}$ may be specified independently of the unknown parameter θ, then we would also like $\mathscr{B}$ to be sufficient for any chosen notion of statistical information.

A more general procedure is to consider the *experiment induced by a measurable map* Y from $(\mathscr{X}, \mathscr{A})$ to some other measurable space $(\mathscr{Y}, \mathscr{B})$. This is the experiment $\mathscr{E}Y^{-1} = (\mathscr{Y}, \mathscr{B}; P_\theta Y^{-1} : \theta \in \Theta) = (P_\theta Y^{-1} : \theta \in \Theta)$. If X realizes $\mathscr{E}$ then $Y(X)$ realizes $\mathscr{E}Y^{-1}$.

Even more generally we may consider a *randomization* M (*Markov kernel*) from $(\mathscr{X}, \mathscr{A})$ to $(\mathscr{Y}, \mathscr{B})$, i.e. a rule which to each point x in $\mathscr{X}$ assigns a probability measure $M(\cdot|x)$ on $\mathscr{B}$ such that $M(B|x)$ is measurable in x for each set $B \in \mathscr{B}$.

Let M be a randomization from $(\mathscr{X}, \mathscr{A})$ to $(\mathscr{Y}, \mathscr{B})$ and consider a finite measure λ on $\mathscr{A}$ as well as a bounded measurable function g on $\mathscr{Y}$.

For each set $B \in \mathscr{B}$, putting $\lambda M(B) = \int M(B|\cdot)\,\mathrm{d}\lambda$ we obtain a finite measure λM on $\mathscr{B}$. Putting $(Mg)(x) = \int g(y) M(\mathrm{d}y|x)$ for all points x in $\mathscr{X}$ we obtain a bounded measurable function Mg on $\mathscr{X}$. By a well known generalization of

Fubini's theorem (see e.g. Neveu, 1965) there is also a unique measure $\lambda \times M$ on $\mathscr{A} \times \mathscr{B}$ such that $\int h(x,y)(\lambda \times M)(\mathrm{d}(x,y)) = \int [\int h(x,y) M(\mathrm{d}y|x)]\lambda(\mathrm{d}x)$ for each bounded measurable function h on $\mathscr{X} \times \mathscr{Y}$.

Combining these notions we obtain the expressions

$$(\lambda M)(g) = \int g \,\mathrm{d}\lambda M,$$

$$\lambda(Mg) = \int \mathrm{d}\lambda(Mg),$$

$$(\lambda \times M)(g) = \int g(y)(\lambda \times M)(\mathrm{d}(x,y)).$$

It is useful to be aware of the fact that these expressions all represent the same number. Deleting parentheses in the first two expressions we may without ambiguity denote this number as $\lambda M g$.

Assume now that $\mathscr{E} = (\mathscr{X}, \mathscr{A}; P_\theta : \theta \in \Theta)$ is an experiment with sample space $(\mathscr{X}, \mathscr{A})$ and that M is a randomization from $(\mathscr{X}, \mathscr{A})$ to $(\mathscr{Y}, \mathscr{B})$. Then M induces experiments $\mathscr{E}M$ and $\mathscr{E} \times M$ from $\mathscr{E}$ by

$$\mathscr{E}M = (P_\theta M : \theta \in \Theta)$$

and

$$\mathscr{E} \times M = (P_\theta \times M : \theta \in \Theta).$$

Note that if M is the (non-random) randomization $B \to I_B(Y)$ associated with the measurable map Y from $(\mathscr{X}, \mathscr{A})$ to $(\mathscr{Y}, \mathscr{B})$ then $\mathscr{E}M = \mathscr{E}Y^{-1}$.

If X realizes $\mathscr{E} = (\mathscr{X}, \mathscr{A}; P_\theta : \theta \in \Theta)$ and if the conditional distribution of $Y|X$ is $M(\cdot|X)$ for a randomization M then the statistic Y and the pair (X, Y) of statistics realize the experiments $\mathscr{E}M$ and $\mathscr{E} \times M$ respectively.

As $\mathscr{E} \times M$ is obtained from $\mathscr{E}$ by a known chance mechanism and since $\mathscr{E}$ is obtained from $\mathscr{E} \times M$ by just observing the first coordinate, we would expect that any sensible notion of statistical information would make $\mathscr{E}$ and $\mathscr{E} \times M$ equally informative. Indeed the first coordinate in $\mathscr{E} \times M$ is sufficient.

The experiment $\mathscr{E}M$ is obtained from the experiment $\mathscr{E}$ by using the known chance mechanism M. Thus we would expect that $\mathscr{E}M$ is at most as informative as $\mathscr{E}$ for any reasonable notion of 'being at most as informative as'. Indeed it is trivial to show that the decision theoretical ordering of experiments which we shall consider here makes $\mathscr{E}M$ at most as informative as $\mathscr{E}$. The question then naturally arises whether any experiment $\mathscr{F}$ which is at most as informative as $\mathscr{E}$ is of the form $\mathscr{F} = \mathscr{E}M$ for a randomization M. It was one of the major accomplishments of a line of research which began in the 1940s to clarify that this is almost so. Indeed if the sample space of $\mathscr{F}$ is Euclidean and if $\mathscr{E}$ is coherent then this is so without any 'almost'.

In the general case randomizations are inadequate in the same way as ordinary random variables were shown to be inadequate in the previous section. Extracting the essentials of the notion of a randomization (Markov kernel) LeCam (1964) introduced the notion of a transition. Replacing the concept of a randomization with the more general notion of a transition, he showed in full generality that $\mathscr{F}$ is at most as informative as $\mathscr{E}$ if and only if $\mathscr{F}$ is obtainable from $\mathscr{E}$ by a transition. The notion of a transition has turned out to be an important tool in mathematical statistics. In particular the fundamental notion of a deficiency of one experiment with respect to another experiment has, as we shall see, a direct and revealing interpretation in terms of transitions.

As transitions may be regarded as limits of randomizations we may avoid them, just as we may avoid the notion of the M-space of an experiment. However, if we want completely general as well as lucid statements, then they appear to be indispensible.

Let us look at alterations of the parameter set. If we are informed that θ in fact belongs to a subset Θ_0 then we arrive at the *restricted experiment*

$$\mathscr{E}|\Theta_0 = (P_\theta : \theta \in \Theta_0).$$

As we shall not attempt here to compare experiments with different parameter sets we shall not consider a comparison of $\mathscr{E}|\Theta_0$ and $\mathscr{E}$ when $\Theta_0 \neq \Theta$.

More generally if H is a given set and if ϕ is a function from H to Θ then we may consider the experiment $(P_{\phi(\eta)} : \eta \in H)$. Changes in location and scale parameters are of this form.

It is frequently advantageous to extend the parameter set to sets of prior distributions on Θ. Thus if Λ is a family of probability measures on σ-algebras on Θ such that $P_\theta(A)$ is λ-measurable in θ for each event A, then we may consider the experiment $(\int P_\theta \lambda(\mathrm{d}\theta) : \lambda \in \Lambda)$. In particular we may obtain the *convex extension* of $\mathscr{E}$ by letting Λ be the collection of probability distributions on Θ having countable (finite) supports.

The *mixture* of a countable family $\mathscr{E}_i = (\mathscr{X}_i, \mathscr{A}_i; P_{\theta,i} : \theta \in \Theta)$; $i \in I$ of experiments with respect to a given mixing distribution π on I is defined as the experiment $\mathscr{E} = (\mathscr{X}, \mathscr{A}; P_\theta : \theta \in \Theta)$ where

$$\mathscr{X} = \bigcup_i \{i\} \times \mathscr{X}_i,$$

$$\mathscr{A} = \left\{ A : A = \bigcup_i \{i\} \times A_i \text{ where, for each } i,\ A_i \in \mathscr{A}_i \right\},$$

and

$$P_\theta\left(\bigcup_i \{i\} \times A_i\right) = \sum \pi(i) P_{\theta,i}(A_i) \text{ where, for each } i,\ A_i \in \mathscr{A}_i.$$

The experiment $\mathscr{E}$ is called the *π-mixture of* $\mathscr{E}_i : i \in I$. Self-explanatory notations for the π-mixture of $\mathscr{E}_i : i \in I$ are $\sum_i \pi(i)\mathscr{E}_i$ and $\int \mathscr{E}_i \pi(\mathrm{d}i)$. Assume now that X_i, for each i, realizes $\mathscr{E}_i$ and that S is an I-valued random variable which is independent of $(X_i : i \in I)$ and which is distributed according to π. Then (S, X_S) realizes the mixture $\int \mathscr{E}_i \pi(\mathrm{d}i)$. The π-mixture of $\mathscr{E}_i : i \in I$ is in general strictly more informative than the experiment realized by X_S alone.

Example 1.3.1 (Sampling plans as mixing distributions). Mixtures occur naturally in several contexts in sampling theory. In order to explain this let us use the notation used in example 1.2.1.

For each sequence $(i_1, \ldots, i_n)$ of individuals from Π, let $\mathscr{E}(i_1, \ldots, i_n)$ denote the experiment obtained by observing $((i_1, \theta(i_1)), \ldots, ((i_n, \theta(i_n)))$.

Also, for each finite subset u of Π, let $\bar{\mathscr{E}}_u$ be the experiment obtained by observing the restriction of the function θ to the set u. Then, for each sequence sampling plan α, the equivalent experiments $\mathscr{E}_\alpha$ and $\bar{\mathscr{E}}_{\bar{\alpha}}$ may be decomposed as

$$\mathscr{E}_\alpha = \int \mathscr{E}_{(i_1, \ldots, i_n)} \alpha(\mathrm{d}(i_1, \ldots, i_n))$$

and

$$\bar{\mathscr{E}}_{\bar{\alpha}} = \int \bar{\mathscr{E}}_u \bar{\alpha}(\mathrm{d}u).$$

As $\bar{\mathscr{E}}_u$ contains at least as much information as $\bar{\mathscr{E}}_v$ when $u \supseteqq v$ it is also clear that $\mathscr{E}_\alpha$ is at least as informative as $\mathscr{E}_\beta$ if the sequence sampling plan β chooses subsets of the samples chosen by the sequence sampling plan α.

In other words $\mathscr{E}_\alpha$ is at least as informative as $\mathscr{E}_\beta$ if there are set valued random variables $U \supseteqq V$ such that $Pr(U = u) = \bar{\alpha}(u)$ and $Pr(V = v) = \bar{\beta}(v)$ when $u, v \subseteqq \Pi$. Another way of phrasing this condition is that $E_{\bar{\alpha}} h \geqslant E_{\bar{\beta}} h$ for each monotonically increasing set function h.

If Θ is not too small then, by Torgersen (1982), this may be turned around, i.e. then $\mathscr{E}_\alpha \geqslant \mathscr{E}_\beta$ if and only if there are set valued variables $U \supseteqq V$ having distributions $\bar{\alpha}$ and $\bar{\beta}$ respectively.

If Π is finite then we shall say that a sequence sampling plan α is *population symmetric* if $\alpha(i_1, \ldots, i_n) = \alpha(\tau(i_1), \ldots, \tau(i_n))$ for each finite sequence $(i_1, \ldots, i_n)$ in Π and for each permutation τ of Π. If α is population symmetric then $\bar{\alpha}(u)$ depends on $u \subseteqq \Pi$ only via the number $\#u$ of elements in u. Assuming a finite population let ρ_n denote simple random drawing without replacement of n elements out of Π. More precisely if $\#\Pi = N$ then $\rho_n(i_1, \ldots, i_n) = [N(N-1), \ldots, (N-n+1)]^{-1}$ or $= 0$ as $i_1, \ldots, i_n$ are distinct or not.

Any sampling experiment derived from a population symmetric sampling plan α is equivalent to a mixture of experiments $\mathscr{E}_{\rho_n}$, $n = 1, 2, \ldots$ In fact $\mathscr{E}_\alpha$ is equivalent to the mixture $\sum \bar{\bar{\alpha}}(n) \mathscr{E}_{\rho_n}$ where $\bar{\bar{\alpha}}(n) = \bar{\alpha}(\{u : \#u = n\})$.

Finally let us generalize the whole set up by letting Θ be any set and by associating with each sequence $(i_1,\ldots,i_n)$ of individuals some experiment $\mathscr{G}(i_1,\ldots,i_n)$ with parameter set Θ. Then the sequence sampling plan α determines the experiment

$$\mathscr{G}_\alpha = \sum \alpha(i_1,\ldots,i_n)\mathscr{G}(i_1,\ldots,i_n).$$

Similarly with each finite subset u of Π we may associate an experiment $\mathscr{H}_u$ and then put

$$\mathscr{H}_{\bar{\alpha}} = \sum \bar{\alpha}(u)\mathscr{H}_u$$

for a set sampling plan $\bar{\alpha}$.

Assume e.g. that for each individual $i \in \Pi$ there is available a variable X_i realizing an experiment $\mathscr{F}_i$. Assume also that X_i; $i \in \Pi$ are independent variables. Let $\mathscr{H}_u$, for each finite subset u of Π, be the experiment realized by the sub-family $(X_i : i \in u)$. According to the terminology established later in this section the experiment $\mathscr{H}_u$ is the product of the experiments $\mathscr{F}_i : i \in u$.

The $\bar{\alpha}$ mixture $\mathscr{H}_{\bar{\alpha}}$ of $\mathscr{H}_u : u \subseteqq \Pi$ is then obtained by first observing the finite subset U of Π according to the set sampling plan $\bar{\alpha}$ and then observing $(X_i : i \in U)$. Thus $\mathscr{H}_{\bar{\alpha}}$ is realized by the random family $(X_i : i \in U)$ of random variables.

A model $\mathscr{H}_{\bar{\alpha}}$ of this form is called a *super population model.*

The asymptotic theory of superpopulation models is treated in Milbrodt (1988).

Mixing distributions need not of course be discrete. Thus if, e.g. $(\mathscr{X}, \mathscr{A}; P_{\theta t} : \theta \in \Theta)$; $t \in T$ is a family of experiments and if π is a probability measure on a σ-algebra $\mathscr{S}$ of subsets of T and if $P_{\theta t}(A)$ is measurable in t for all $\theta \in \Theta$ and all $A \in \mathscr{A}$, then we may consider the *π-mixture of* $\mathscr{E}_t : t \in T$. This is the experiment

$$\int \mathscr{E}_t \pi(\mathrm{d}t) = (T \times \mathscr{X}, \mathscr{A} \times \mathscr{S}; \pi \times P_{\theta,\cdot} : \theta \in \Theta) \quad \text{where}$$

$$(\pi \times P_{\theta,\cdot})(S \times A) = \int_S P_{\theta,t}(A)\pi(\mathrm{d}t); \qquad S \in \mathscr{S},\, A \in \mathscr{A},\, \theta \in \Theta.$$

Consider a family $X = (X_i : i \in I)$ of independent random variables $X_i : i \in I$. If X_i, for each $i \in I$, realizes the experiment $\mathscr{E}_i = (\mathscr{X}_i, \mathscr{A}_i; P_{\theta,i} : \theta \in \Theta)$ then the family X realizes the experiment $(\prod_i \mathscr{X}_i, \prod_i \mathscr{A}_i; \prod_i P_{\theta,i} : \theta \in \Theta)$. This experiment is called the *product of the experiments* $(\mathscr{E}_i : i \in I)$ and is denoted by $\prod_i \mathscr{E}_i$ or, if $I = \{1,\ldots,n\}$, as $\mathscr{E}_1 \times \cdots \times \mathscr{E}_n$. The experiment obtained by combining n independent *replicates* of an experiment $\mathscr{E}$ is denoted by $\mathscr{E}^n$. Thus $\mathscr{E}^n = \prod_{i=1}^n \mathscr{E}_i$ where $\mathscr{E}_1 = \cdots = \mathscr{E}_n = \mathscr{E}$. It is sometimes convenient to consider an experiment $\mathscr{E}^0$ which may be any totally non-informative experiment i.e. any experiment of the form $(Q_\theta : \theta \in \Theta)$ where Q_θ does not depend on θ.

Mixtures and products are both commutative and associative up to the information preserving operations of rearranging and permuting observations. In the same sense products are distributive with respect to mixtures. We shall later see how we may operate with equivalence classes of experiments in such a way that the statements above get precisely the algebraic content they should have.

Example 1.3.2 (Poissonization of experiments). Combining the notations developed so far, for any experiment $\mathscr{E}$ and constant $\lambda \geqslant 0$ we may consider the mixture $\mathscr{G}_\lambda = \sum_{n=0}^{\infty} (\lambda^n/n!)e^{-\lambda}\mathscr{E}^n$. Then $\mathscr{G}_\lambda : \lambda \geqslant 0$ constitutes a *semigroup* of experiments in the sense that if λ, $\mu \geqslant 0$ then the experiments $\mathscr{G}_\lambda \times \mathscr{G}_\mu$ and $\mathscr{G}_{\lambda+\mu}$ are equivalent. In particular $\mathscr{G}_\lambda$ is equivalent to $[\mathscr{G}_{\lambda/n}]^n$ so that each experiment $\mathscr{G}_\lambda$ is infinitely divisible.

The experiment $\mathscr{G}_1$ is called *the Poissonization of* $\mathscr{E}$. Poissonized experiments play a role in the central limit theory for experiments which is similar to the role of Poissonized variables in the central limit theory for random variables.

An important contribution of LeCam's 1964 paper, is the idea of deficiency of one experiment with respect to another experiment. Related to the deficiency is a distance, in fact a pseudo-metric, for experiments. If this distance is small for two experiments then they may be considered as carrying approximately the same amount of statistical information. This distance is the least upper bound of all corresponding distances for restrictions to finite subparameter sets. The distance determines a topology which we may call the strong experiment topology. Another interesting topology is the topology of the uniformity generated by distances for restrictions to finite sub-parameter sets. This topology may be called the weak experiment topology. Both topologies are interesting, but the latter is often easier to handle. In particular the weak experiment topology is always compact while the strong experiment topology is not compact unless the parameter set is finite, i.e. when these topologies coincide.

We mention this here since we may also derive experiments as limits. In particular there is now a fairly well worked out central limit theory for experiments, which closely parallels the central limit theory for random variables. The reader may, in addition to LeCam's works, consult e.g. the works by Janssen, Milbrodt & Strasser (1984), Millar (1983) and Strasser (1985).

1.4 Some useful functionals of experiments

In this section we shall briefly discuss some commonly used functionals of experiments. They include: the statistical distance, power of Neyman–Pearson tests, minimum Bayes risk, Hellinger distance, affinities, Hellinger transforms, Kullback–Leibler information numbers, Chernoff information numbers and Fisher information matrices.

Most of these functionals, although not all, are constructed from homogeneous functions of the likelihood function. We shall therefore begin by considering how this construction may be carried out.

If Θ denotes the parameter set under consideration then the set $Z = [0, \infty[^\Theta$ of real valued non-negative functions on Θ constitutes the *set of possible likelihood functions.* Actually the null function 0 might have been deleted from Z. However it is convenient to be able to work with a set Z which is a cartesian product, and then Z must contain the null function.

If the experiment $\mathscr{E} = (\mathscr{X}, \mathscr{A}; P_\theta : \theta \in \Theta)$ is dominated by a non-negative σ-finite measure μ then the *likelihood function based on* μ, for a given choice of densities $\mathrm{d}P_\theta/\mathrm{d}\mu$; $\theta \in \Theta$, is the map which assigns the function $x \to [\mathrm{d}P_\theta/\mathrm{d}\mu]_x$ in Z to each $x \in \mathscr{X}$. Replacing μ with another dominating σ-finite measure ν we see that, for each θ, $(\mathrm{d}P_\theta/\mathrm{d}\mu)(\mathrm{d}\mu/\mathrm{d}\nu)$ is a version of $\mathrm{d}P_\theta/\mathrm{d}\nu$. Thus the effect of replacing μ with ν is to replace the likelihood function with a proportional likelihood function. If we now argue that inference should be based on the likelihood function and if we do not want our decisions to depend on which dominating measure we have chosen, then we are led to decision functions which are constant on rays $\{tz : t > 0\}$ in the likelihood space.

Now there is a natural 1–1 correspondence between vector valued functions on the likelihood space which are constant on these rays, and non-negatively homogeneous functions. As the latter functions appear to present themselves in a more natural way than the first ones, in the sequel we will usually work with homogeneous functions rather than functions which are constant on rays. Let us however keep in mind that the ratio of two such functions, if well defined, is constant on rays, and conversely that real valued functions which are constant on rays are ratios of non-negatively homogeneous functions.

For now we will assume that the measurability structure of Z is given by the product σ-algebra, i.e. the smallest σ-algebra on Z which makes all projections into the coordinate spaces measurable. Later on, in chapter 6, we will consider a more appropriate σ-algebra which does not distinguish between points belonging to the same ray through the origin.

Consider now an extended real valued function h on the likelihood space $Z = [0, \infty[^\Theta$. We shall assume that h is measurable and that it is *non-negatively homogeneous* i.e. that $h(tz) = th(z)$ when $z \in Z$ and $t \in [0, \infty[$.

As we adhere to the convention that $0 \cdot \infty = 0 \cdot (-\infty) = 0$, the requirement that h is non-negatively homogeneous is equivalent to the condition that h is positively homogeneous on $Z - \{0\}$ and that $h(0) = 0$. If some expression defines h as a positive homogeneous function on $Z - \{0\}$ then unless otherwise stated, we will assume that h is the unique non-negatively homogeneous extension to Z obtained by putting $h(0) = 0$.

The measurability condition implies that there are countable subsets Θ_0 of Θ such that $h(z)$ depends solely on the projection $(z_\theta : \theta \in \Theta_0)$ of z onto $[0, \infty]^{\Theta_0}$. In fact if h is non-constant, i.e. if h is not the null function, then there is a smallest such set Θ_0.

If $\mathscr{E} = (\mathscr{X}, \mathscr{A}; P_\theta : \theta \in \Theta)$ is an experiment then we may choose a measure $\mu \geqslant 0$ on $\mathscr{A}$ such that P_θ has a density f_θ with respect to μ when $\theta \in \Theta_0$, where the countable set Θ_0 is adapted to the function h as above. Thus $P_\theta(A) = \int_A f_\theta \, d\mu$ when $A \in \mathscr{A}$ and $\theta \in \Theta_0$. Note that we do not require that μ is σ-finite. If the integral $\int h(f_\theta : \theta \in \Theta_0) \, d\mu$ exists then we shall call it the *integral of h with respect to $\mathscr{E}$* and we shall denote it by $\int h(d\mathscr{E})$ or by $\int h(dP_\theta : \theta \in \Theta)$ or by related expressions. If this integral does not exist then no integral of h with respect to $\mathscr{E}$ shall be defined.

The important point to note is that neither the existence nor the value of the integral $\int h(d\mathscr{E})$ depend on how the set Θ_0 and the measure μ is chosen, provided $h(z)$ depends on z via $(z_\theta : \theta \in \Theta_0)$ and provided μ majorizes $\mathscr{E}|\Theta_0$. If $h = 0$ then this is trivial. If h is not constant then there is a smallest countable set Θ_1 such that $h(z)$ depends on z via $(z_\theta : \theta \in \Theta_1)$. As Θ_1 is countable there is a prior probability distribution κ on Θ_1 such that $\kappa(\theta) > 0$ for all points θ in Θ_1. Put $P_\kappa = \sum_\theta \kappa(\theta) P_\theta$ and $g_\theta = dP_\theta / dP_\kappa$. Consider a pair (Θ_0, μ) such that μ majorizes $\mathscr{E}|\Theta_0$ and such that h depends on z via $(z_\theta : \theta \in \Theta_0)$. Then $\Theta_0 \supseteqq \Theta_1$. For each $\theta \in \Theta_0$, let f_θ be a density of P_θ with respect to μ and put $f_\kappa = \sum_\theta f_\theta \kappa(\theta)$. Then f_κ is a density of P_κ with respect to μ. It follows that $g_\theta f_\kappa$ is a density of P_θ with respect to μ when $\theta \in \Theta_1$. Thus $g_\theta f_\kappa = f_\theta$ a.e. μ when $\theta \in \Theta_1$. Hence

$$\int |h(g_\theta : \theta \in \Theta_1)| \, dP_\kappa = \int |h(g_\theta : \theta \in \Theta_1)| f_\kappa \, d\mu = \int |h(g_\theta f_\kappa : \theta \in \Theta_1)| \, d\mu$$

$$= \int |h(f_\theta : \theta \in \Theta_1)| \, d\mu = \int |h(f_\theta : \theta \in \Theta)| \, d\mu$$

where the absolute value signs may be deleted throughout provided one (and hence all) of the integrals obtained by this deletion exists. This proves that the notion of an integral of h with respect to $\mathscr{E}$ which we have just introduced is independent of our choice of the frame (Θ_0, μ).

Completing this notion of an integral with respect to $\mathscr{E}$ we shall say that h is *$\mathscr{E}$-integrable* if $\int |h|(d\mathscr{E}) < \infty$.

In order to clarify this notion of an integral let us consider probability measures P and Q on the same measurable space $(\mathscr{X}, \mathscr{A})$. Using the above recipe we may construct the integrals:

$$I_1 = \int (\mathrm{d}P/\mathrm{d}Q)\,\mathrm{d}Q$$

$$I_2 = \int (\mathrm{d}P + \mathrm{d}Q) \wedge (2\mathrm{d}P + 3\mathrm{d}Q)$$

and

$$I_3 = \int [(\mathrm{d}P)^2 + (\mathrm{d}Q)^2]^{1/2}.$$

If f and g are the densities of, respectively, P and Q with respect to some majorizing measure μ, e.g. $\mu = P + Q$, then:

$I_1 = \int_{[g>0]} f\,\mathrm{d}\mu = P(g > 0)$
$= $ the total mass of the Q-absolutely continuous part of P.
$I_2 = \int [(f + g) \wedge (2f + 3g)]\,\mathrm{d}\mu$
$=$ minimum Bayes risk for the problem of testing 'P' against 'Q' for the uniform prior distribution on $\{P, Q\}$ and for losses given by the following table:

		True distribution P	Q
Decision	P	2	2
made	Q	4	6

and
$I_3 = \int (f^2 + g^2)^{1/2}\,\mathrm{d}\mu.$

We have now seen that any extended real valued measurable non-negatively homogeneous (take a deep breath) function h on the likelihood space defines a functional $\mathscr{E} \to \int h(\mathrm{d}\mathscr{E})$ on the collection of experiments $\mathscr{E}$ such that $\int |h|(\mathrm{d}\mathscr{E}) < \infty$. Such a functional of experiments is called *representable* and we shall say that the functional is *represented by h.*

As representable functionals are affine under mixtures it is clear that the representable functionals constitute a very particular class of functionals.

Another important property of these functionals is that they are determined by the laws of likelihood ratios and are therefore not affected by sufficiency reductions. Indeed we shall see that they respect the equivalence 'being equally informative' and thus are functionals of equivalence 'classes'.

If we restrict h to the vector lattice generated by the projections into the coordinate spaces of Z, then $\int h(\mathrm{d}\mathscr{E})$ as a function of h is the *Choquet conical measure* of $\mathscr{E}$. This concept, which was introduced by LeCam, generalized Blackwell's notion of a standard measure to the case of a possibly infinite parameter set. The Choquet conical measure (as the standard measure when Θ is finite) is a very convenient way of expressing the laws of likelihood ratios.

The notation $\int h(\mathrm{d}\mathscr{E})$ suggests that there is a 'natural' measure $h(\mathscr{E})$ associated with h and $\mathscr{E}$ whose total mass is $\int h(\mathrm{d}\mathscr{E})$. Indeed this is the case whenever $\int |h|(\mathrm{d}\mathscr{E}) < \infty$ and we can then write $\int h(\mathrm{d}\mathscr{E}) = \int \mathrm{d}h(\mathscr{E})$. The measure $h(\mathscr{E})$ is 'natural' in the sense that the map $h \to h(\mathscr{E})$ is the unique vector lattice homomorphism which maps the projections $z \to z_\theta$; $\theta \in \Theta$ into the measures $P_\theta : \theta \in \Theta$ defining the experiment $\mathscr{E} = (P_\theta : \theta \in \Theta)$.

Thus the statistical distance between probability measures P and Q on the same probability space may be expressed both as $\int |\mathrm{d}P - \mathrm{d}Q|$ or as $\int \mathrm{d}|P - Q|$ where $|P - Q|$ should be interpreted in the usual way, i.e. as the smallest measure majorizing both $P - Q$ and $Q - P$.

The reader may try to work out what the measure $h(\mathscr{E})$ is, and what properties this measure has as a function of the pair $(h, \mathscr{E})$, for some of the homogeneous functions h discussed later in this section.

We shall return to the general theory of representations of experiments in chapter 7. In particular we shall provide there some characterization theory for functionals of experiments.

Let us now turn to some important examples of functionals of experiments which are either representable or are closely related to representable functionals.

Example 1.4.1 (Statistical distance). The statistical distance is an indispensible tool for obtaining and for describing results related to the theory of comparison of experiments. We defined it in section 1.2 as the total variation of differences of probability measures. In other words, the statistical distance is the restriction of the total variation norm metric to the set of probability measures.

Now the linear space of finite measures equipped with the total variation norm is a Banach space having the set of probability measures as a closed subset. It follows that the set of probability measures on a given measurable space is complete for statistical distance.

If the probability measures P and Q have densities f and g with respect to a non-negative majorizing measure μ, then $\|P - Q\| = \int |f - g| \,\mathrm{d}\mu$ which is also the $L_1(\mu)$ distance $\int |f - g| \,\mathrm{d}\mu$ between f and g. Thus, as mentioned above, we may represent this distance as $\int |\mathrm{d}P - \mathrm{d}Q|$. This implies in particular that the statistical distance between the chance mechanisms P_{θ_1} and P_{θ_2} associated with the parameter values $\theta = \theta_1$ and $\theta = \theta_2$ for an experiment $\mathscr{E} = (P_\theta : \theta \in \Theta)$, is a representable functional of $\mathscr{E}$.

The Bayes interpretation of statistical distance mentioned in section 1.2 amounted to describing *the statistical distance between P and Q as $2\pi - 1$*

where $\pi =$ *the maximum probability of guessing correctly the true distribution in the Bayesian guessing problem which prescribes probability* $\frac{1}{2}$ *to* P *and probability* $\frac{1}{2}$ *to* Q.

A related description is in terms of probabilities of errors in the problem of testing, say, the null hypothesis 'P' against the alternative 'Q'. According to this description $\|P - Q\| = 2(1 - b)$ *where* b *is the minimum of possible sums of probabilities of errors.*

Another useful characterization of the statistical distance is in terms of random variables U and V having distributions P and Q respectively. If H is the Hahn set for the difference measure $P - Q$, then we find:

$$Pr^*(U = V) \leqslant Pr(U \notin H) + Pr(V \in H) = 1 - (P(H) - Q(H))$$

so that

$$Pr^*(U = V) \leqslant 1 - \tfrac{1}{2}\|P - Q\|$$

where * indicates outer measure.

We shall now see that the upper bound for this 'outer' probability is obtained for the joint distribution R on $\mathscr{A} \times \mathscr{A}$ which is defined as follows.

Let κ be the map from $(\mathscr{X}, \mathscr{A})$ to $(\mathscr{X} \times \mathscr{X}, \mathscr{A} \times \mathscr{A})$ which maps x, for each $x \in \mathscr{X}$, into the point (x, x) on the diagonal $\{(x, x) : x \in \mathscr{X}\}$ of $\mathscr{X} \times \mathscr{X}$. Put $R = (P \wedge Q)\kappa^{-1} + 2\|P - Q\|^{-1}(P - Q)^+ \times (P - Q)^-$ or $= P\kappa^{-1}$ as $P \neq Q$ or $P = Q$.

A better way of describing this joint distribution is to describe it as $R = P \times M$ where the randomization (Markov kernel) M is given by

$$M(A|\cdot) = \left(\frac{\mathrm{d}Q}{\mathrm{d}P} \wedge 1\right) I_A + 2\|P - Q\|^{-1}\left(1 - \frac{\mathrm{d}Q}{\mathrm{d}P}\right)^+ (P - Q)^-(A); \qquad A \in \mathscr{A}$$

and where

$$R(A \times B) = \int_A M(B|x)P(\mathrm{d}x) \quad \text{when} \quad A, B \in \mathscr{A}.$$

It is readily checked that R has marginals P and Q and that $\|R - P\kappa^{-1}\| = \|R - Q\kappa^{-1}\| = \|P - Q\|$. Let U and V be the projections on the first and the second coordinate spaces of $\mathscr{X} \times \mathscr{X}$ respectively. Consider U and V as random variables on the probability space $(\mathscr{X} \times \mathscr{X}, \mathscr{A} \times \mathscr{A}, R)$ and let G be any measurable set containing the diagonal.

Then $\kappa^{-1}(G^c) = \varnothing$ so that $Pr((U, V) \notin G) = 2\|P - Q\|^{-1}[(P - Q)^+ \times (P - Q)^-](G^c) \leqslant 2\|P - Q\|^{-1}\|(P - Q)^+ \times (P - Q)^-\| = \frac{1}{2}\|P - Q\|$ so that $Pr((U, V) \in G) \geqslant 1 - \frac{1}{2}\|P - Q\|$. Hence $Pr(U = V)^* \geqslant 1 - \frac{1}{2}\|P - Q\|$.

It follows that $1 - \frac{1}{2}\|P - Q\|$ *may be characterized as the largest number of the form* $Pr(U = V)^*$ *for random variables* U *and* V *having distributions* P *and* Q *respectively.*

If $\mathscr{A}$ contains a separable sub σ-algebra which contains all one-point sets then the diagonal, and therefore the set $[U = V]$, is measurable. Thus the star may be deleted in this case.

The statistical distance belongs to an important family of metrics for probability measures. These metrics are related to metrics on the underlying space in a particular way. Disregarding measurability problems, consider any metric d on the set $\mathscr{X}$. If P and Q are probability measures on $\mathscr{X}$, then the greatest lower bound of numbers $\varepsilon \geqslant 0$ such that the inequality

$$P(A) \leqslant Q(\{x : \inf\{\mathrm{d}(x, a) : a \in A\} \leqslant \varepsilon\}) + \varepsilon$$

holds for all events A, is called the *Prohorov distance between P and Q.*

It follows that the *statistical distance between P and Q is twice the Prohorov distance between P and Q for the discrete metric* which assigns distance 1 to any pair of distinct points in $\mathscr{X}$.

Now there are theorems which assert with varying degrees of generality that the Prohorov distance between P and Q associated with the metric d is $\leqslant \varepsilon$ if and only if there are random variables U and V having distributions P and Q respectively such that $Pr(\mathrm{d}(U, V) \geqslant \varepsilon) \leqslant \varepsilon$. We refer the reader to Strassen's 1965 paper for a proof of this theorem for complete separable metric spaces. Another proof is given in chapter 9 where we relate this to the general theory of comparison of experiments.

Returning to the case of a discrete metric we see that in this case we have proved the equivalence of the two characterizations of the Prohorov distance by an explicit construction of an optimal joint distribution. This is of interest since the general theorem of Strassen is an existence theorem which does not provide much information on how optimal joint distributions actually may be found.

The precise behaviour of the statistical distance under replications is usually difficult to figure out. It is fairly simple to show, by telescoping, that if $(P_1, Q_1), \ldots, (P_n, Q_n)$ are dichotomies then

(i) $\|(P_1 \times \cdots \times P_n) - (Q_1 \times \cdots \times Q_n)\| \leqslant \|P_1 - Q_1\| + \cdots + \|P_n - Q_n\|$.

In particular

(ii) $\|P^n - Q^n\| \leqslant n\|P - Q\|; n = 1, 2, \ldots$ for a dichotomy (P, Q).

The last inequality is of course trivial whenever $\|P - Q\| \geqslant 2/n$. Actually, as will be explained in the next example, $\|P^n - Q^n\| \to 2$ with at least exponential speed as $n \to \infty$ provided $P \neq Q$.

If $P_1 = P$, $Q_1 = Q$ and $P_2 = Q_2 = R$ then (i) reduces to:

$$\|P \times R - Q \times R\| \leqslant \|P - Q\|.$$

As any difference $P(A) - Q(A)$ of probabilities may be expressed as

$(P \times R)(A \times B) - (Q \times R)(A \times B)$ where $R(B) = 1$ we conclude that equality holds i.e. that:

$$\|P \times R - Q \times R\| = \|P - Q\|$$

for dichotomies (P, Q) and (R, R). This equality expresses, and indeed is a consequence of, the fact that information is neither gained nor lost by adding to our observations an independent observation whose distribution is known.

Working freely with lattice operations on finite measures we obtain

$$\begin{aligned}\tfrac{1}{2}\|P - Q\| &= \tfrac{1}{2}\||P - Q|\| = \|(P - Q)^+\| = \|(P - Q)^-\| = 1 - \|P \wedge Q\| \\ &= \|P \vee Q\| - 1.\end{aligned}$$

These equalities may be derived from the lattice operations on the real line using the fact that if P and Q have densities f and g with respect to, say, $P + Q$, then $P - Q$, $|P - Q|$, $(P - Q)^+$, $(P - Q)^-$, $P \wedge Q$ and $P \vee Q$ have densities $f - g$, $|f - g|$, $(f - g)^+$, $(f - g)^-$, $f \wedge g$ and $f \vee g$ with respect to the same measure $P + Q$.

The number $\|P \wedge Q\| = \int \mathrm{d}P \wedge \mathrm{d}Q = 1 - \frac{1}{2}\|P - Q\|$ may be regarded as a measure of closeness of the distributions P and Q. It obtains its maximal value 1 if and only if $P = Q$, while it obtains its minimal value 0 if and only if P and Q have disjoint supports. Actually we shall soon see that $\frac{1}{2}\|P \wedge Q\|$ is the minimum Bayes risk in the testing problem of testing 'P' against 'Q' when each of these distributions occur with probability $\frac{1}{2}$ and when the loss $= 0$ or $= 1$ according to whether the decision is right or wrong.

In the next example we shall encounter another measure of association, related to another metric, which is easier to handle for replicated experiments. However this quantity is not so simply interpreted in terms of decision problems.

Example 1.4.2 (Affinities and Hellinger distances). The likelihood may be regarded as a random process with time replaced by the unknown parameter. If this process is defined by densities in the usual way then it may easily happen that the variances for some values of θ are infinite. The standard procedure for remedying this, and thereby obtaining a second order process, is to replace the likelihood function with its (non-negative) square root. As demonstrated by LeCam and others, this simple idea has been extremely useful in asymptotic statistical theory.

In general if P and Q are probability measures on the same measurable space then the number $\int (\mathrm{d}P)^{1/2}(\mathrm{d}Q)^{1/2}$ is called the *affinity between P and Q*. We shall often use the notation $\gamma(P, Q)$ to denote the affinity between P and Q. The affinity is clearly non-negative and the Cauchy–Schwartz inequality implies that it is never greater than 1. Actually $\gamma(P, Q)$ obtains its smallest value 0 if and only if $P \wedge Q = 0$, while it obtains its maximal value 1 if and only if $P = Q$. This follows from the computations

$$2(1 - \gamma(P, Q)) = \int dP + dQ - 2(dP)^{1/2}(dQ)^{1/2} - \int ((dP)^{1/2} - (dQ)^{1/2})^2$$
$$= d^2(P, Q)$$

where $d(P, Q) = [\int ((dP)^{1/2} - (dQ)^{1/2})^2]^{1/2}$.

If we limit ourselves to probability measures which are majorized by a given non-negative measure μ then we see that $d(P, Q)$ is the $L_2(\mu)$ distance between the square roots of the densities of P and of Q. Since any $L_2(\mu)$ is complete, it follows that d is a complete metric on the set of all probability measures P and Q. This metric is called the *Hellinger distance* and the number $d(P, Q)$ is called the *Hellinger distance between P and Q*.

The computations above show that the affinity may be expressed in terms of the Hellinger distance by

$$\gamma(P, Q) = 1 - \tfrac{1}{2}d^2(P, Q).$$

In a moment we will see that the Hellinger distance, as a metric, is equivalent to the statistical distance. Combining this with the expression above we obtain confirmation of our suggestion that the affinity actually measures affinity.

The equivalence of the Hellinger distance and the statistical distance may be argued, using the Cauchy–Schwartz inequality, as follows:

$$\|P \wedge Q\| = \int dP \wedge dQ = \int (dP \wedge dQ)^{1/2}(dP \wedge dQ)^{1/2} \leqslant \int (dP)^{1/2}(dQ)^{1/2}$$
$$= \int (dP dQ)^{1/2} = \int (dP \wedge dQ)^{1/2}(dP \vee dQ)^{1/2}$$
$$\leqslant \|P \wedge Q\|^{1/2}\|P \vee Q\|^{1/2} = \|P \wedge Q\|^{1/2}(2 - \|P \wedge Q\|)^{1/2}.$$

Thus

$$\|P \wedge Q\| \leqslant \gamma(P, Q) \leqslant \|P \wedge Q\|^{1/2}(2 - \|P \wedge Q\|)^{1/2}$$

or equivalently:

$$\tfrac{1}{2}\gamma(P, Q)^2 \leqslant 1 - (1 - \gamma(P, Q)^2)^{1/2} \leqslant \|P \wedge Q\| \leqslant \gamma(P, Q). \qquad (1.4.1)$$

Here the first inequality follows from the inequality $t/2 \leqslant 1 - (1 - t)^{1/2}$ which is valid when $t \leqslant 1$.

These inequalities show that $\|P \wedge Q\| \to 1$ or $\to 0$ as $\gamma(P, Q) \to 1$ or $\to 0$.

It follows from the identities $\|P - Q\| = 2(1 - \|P \wedge Q\|)$ and $d^2(P,Q) = 2(1 - \gamma(P,Q))$ that $\|P - Q\| \to 0$ if and only if $d(P,Q) \to 0$ while $\|P - Q\| \to 2$ if and only if $d(P,Q) \to 2^{1/2}$.

This implies in particular that the *statistical distance and Hellinger distance are equivalent metrics.*

By using the above identities and the inequalities we have just derived, we may express this more precisely by the inequalities

$$\tfrac{1}{2}\|P - Q\| \leqslant 2^{1/2}[1 - (1 - \tfrac{1}{4}\|P - Q\|^2)^{1/2}]^{1/2} \leqslant d(P,Q) \leqslant \|P - Q\|^{1/2}. \tag{1.4.2}$$

Here the first inequality is obtained from the same inequality as we used for establishing the first inequality of (1.4.1).

The affinity as a function of two variables is non-negative definite, since for any experiment $(P_1, \ldots, P_n)$ and for any numbers $c_1, \ldots, c_n$ we have:

$$\sum c_i c_j \gamma(P_i, P_j) = \sum c_i c_j \int (\mathrm{d}P_i)^{1/2}(\mathrm{d}P_j)^{1/2} = \int \left(\sum c_i (\mathrm{d}P_i)^{1/2}\right)^2 \geqslant 0.$$

It follows that for each measurable space $(\mathcal{X}, \mathcal{A})$ there is a second order process $P \to Z_P$ such that $\mathrm{cov}(Z_P, Z_Q) = \gamma(P,Q)$ whenever P and Q are probability measures on $\mathcal{A}$. In particular there is a zero mean Gaussian process with this property.

If $\mathcal{E} = (\mathcal{X}, \mathcal{A}; P_\theta : \theta \in \Theta)$ is an experiment and if $P \to Z_P$ is as above then $\theta \to Z_{P_\theta}$ is a process with covariance operator $(\theta_1, \theta_2) \to \gamma(P_{\theta_1}, P_{\theta_2})$. This has the interesting consequence that properties of the experiment $\mathcal{E}$ which may solely be described in terms of the pseudo-distance $(\theta_1, \theta_2) \to d(P_{\theta_1}, P_{\theta_2})$ on Θ, are precisely the second order properties of mean zero random processes $\theta \to Y_\theta$ such that $EY_{\theta_1} Y_{\theta_2} = \gamma(P_{\theta_1}, P_{\theta_2})$ when $\theta_1, \theta_2 \in \Theta$.

In particular we may inquire about the implications for $\mathcal{E}$ of the continuity and differentiability properties of this process.

We shall return to this aspect of the affinity in example 1.4.7.

Besides being an inner product, the affinity has the very useful property of being multiplicative for experiment multiplication. Thus if $(\mathcal{X}_i, \mathcal{A}_i; P_i; Q_i)$; $i = 1, \ldots, n$ are dichotomies then $\gamma(\prod_i P_i, \prod_i Q_i) = \prod_i \gamma(P_i, Q_i)$.

It follows that $\gamma(P^n, Q^n) = \gamma(P,Q)^n$ for any dichotomy (P,Q). Hence $\gamma(P^n, Q^n) = 1 - \tfrac{1}{2}d^2(P^n, Q^n) \to 0$ as $n \to \infty$ with exponential speed whenever $P \neq Q$. This is one way of expressing how fast the feasibility of distinguishing two specified hypotheses improves as the number of replicates increases.

Combining this with some of the inequalities we have obtained, we find successively that:

$$\gamma(P,Q)^{2n} = \gamma(P^n,Q^n)^2 \leqslant 2\|P^n \wedge Q^n\|$$
$$= 2 - \|P^n - Q^n\| \leqslant 2\gamma(P^n,Q^n) = 2\gamma(P,Q)^n.$$

Thus, assuming $P \neq Q$, we see that $\|P^n - Q^n\| \to 2$ at least as fast as $\gamma^n \to 0$ and at most as fast as $\gamma^{2n} \to 0$ where $\gamma = \gamma(P,Q)$.

Taking this into account it is not likely to be a surprise if the speed of convergence turns out to be exactly exponential. Indeed Chernoff proved that in his 1952 paper where he derived a simple expression for the exponential rate. This result will be discussed further in example 1.4.6.

The corresponding problem for the Hellinger distance is completely trivial, since

$$2^{1/2} - d(P^n,Q^n) = 2\gamma(P,Q)^n/[2^{1/2} + d(P^n,Q^n)].$$

Example 1.4.3 (Power of the Neyman–Pearson test and Kullback–Leibler information). An appealing way of quantifying a distance between probability measures P and Q on the same measurable space $(\mathscr{X},\mathscr{A})$ is to evaluate how well they may be separated by testing. If, say, we test the null hypothesis 'P' against the alternative 'Q' then the Neyman–Pearson lemma tells how we may find a test obtaining maximal power among all tests having a prescribed level α of significance. Let the power of this test at 'Q' be denoted as $\beta(\alpha|P,Q)$. Although most applications require that α is small we shall find it convenient to consider $\beta(\alpha|P,Q)$ for all numbers α in $[0,1]$.

We shall call $\beta(\cdot|P,Q)$ the *β-function*, or the *Neyman–Pearson* (*N–P*) *function, of the dichotomy* (P,Q). A function β of the form $\beta = \beta(\cdot|P,Q)$ for some dichotomy (P,Q) is called a *β-function* or a *Neyman–Pearson* (*N–P*) *function.* The concept of a β-function as used here has no relation to the well established concepts of β-integrals and β-distributions.

β-functions play important roles on the most diverse occasions, not all of them in statistics. They may be considered as particular upper envelope functions of planar compact sets having a point of symmetry. As such they are discussed in section 9.3. The relation to the Lorenz functions of econometrics is the subject of problem 9.7.5. Maximin power and β-functions are discussed in complements 5, 7 and 8 in chapter 10.

It is readily checked that β-functions are concave continuous maps from $[0,1]$ to $[0,1]$ having 1 as a fixed point. Actually any concave function from $[0,1]$ to $[0,1]$ which maps 1 into 1 is continuous on $]0,1]$. Thus the point is that β is continuous from the right at $\alpha = 0$. Conversely any function β from $[0,1]$ to $[0,1]$ having these properties is of the form $\beta = \beta(\cdot|P,Q)$ for some dichotomy (P,Q). Indeed we may just let P be the Lebesgue measure on $[0,1]$ and let Q be the measure on $[0,1]$ which has β as its cumulative distribution function.

Here are three graphs of functions $\beta(\cdot|P,Q)$:

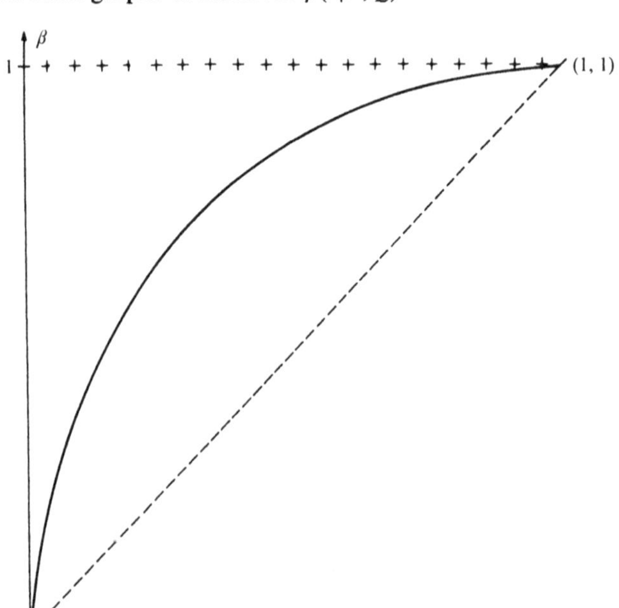

Fig. 1. $++++++$ Graph of $\beta(\cdot|P,Q)$ when (P,Q) is totally informative i.e. $P \wedge Q = 0$. ——— Graph of $\beta(\cdot\, P,Q)$ for some dichotomy (P,Q). ------ Graph of $\beta(\cdot|P,P)$ for a totally non informative dichotomy (P,P).

Note that the graph of $\beta(\cdot|P,Q)$ is piecewise linear when P and Q are measures on a finite algebra. In particular the graph is triangular when (P,Q) is a double dichotomy.

Clearly $\beta(\alpha|P,Q) \geqslant \alpha$ for all numbers $\alpha \in [0,1]$. Furthermore, if $0 < \alpha < 1$ and $P \neq Q$ then $\beta(\alpha|P,Q) > \alpha$.

As $\beta(\alpha|P,P) \equiv \alpha$ we could interpret $\beta(\alpha|P,Q) - \alpha$ as a measure of the separation between P and Q. In fact the statistical distance is determined by these quantities since

$$\sup_{\alpha}\,[\beta(\alpha|P,Q) - \alpha] = \sup_{0<\alpha<1}\,[\beta(\alpha|P,Q) - \alpha] = \|P - Q\|/2.$$

We shall see in chapter 8 that $\beta(\cdot|P,Q)$ determines $\mathscr{L}_P(\mathrm{d}Q|\mathrm{d}P)$ and thus all statistical properties of the dichotomy (P,Q).

It follows that the dichotomy (P,Q) is equivalent to the dichotomy $(\beta(\cdot|P,P), \beta(\cdot|P,Q))$ where we allow ourselves to use the same notation for a cumulative distribution function as for the measure it determines.

Since the set of functions of this form is clearly convex, it is tempting to infer that $\beta(\alpha|\cdot,\cdot)$ is representable. However, convex combinations of these functions do not in general represent the corresponding convex combinations of dichotomies for mixtures. In fact $\beta(\alpha|P,Q)$ is not affine for mixtures and is thus not directly representable as the integral of a homogeneous function on the likelihood space $[0,\infty[\times[0,\infty[$. Such integrals may nevertheless be expressed in terms of $\beta(\cdot|P,Q)$. Thus the affinity between P and Q may be expressed as

$$\gamma(P,Q) = \int (\mathrm{d}P)^{1/2}(\mathrm{d}Q)^{1/2} = \int_0^1 [\beta'(\alpha)]^{1/2}\,\mathrm{d}\alpha = \int_0^1 [K^{-1}(1-\alpha)]^{1/2}\,\mathrm{d}\alpha,$$

where K denotes the cumulative distribution function of $\mathscr{L}_P(\mathrm{d}Q/\mathrm{d}P)$. This may be seen by showing that if λ is the uniform distribution on $[0,1]$ then

$$\mathscr{L}_P(\mathrm{d}Q/\mathrm{d}P) = \mathscr{L}_\lambda(\beta')$$

and

$$1-\beta(\alpha|P,Q) \underset{\alpha}{\equiv} \int_\alpha^1 K^{-1}(1-p)\,\mathrm{d}p = P(Z \geqslant \alpha)$$

where Z is the observed level of significance defined as $Z = 1 - K(\mathrm{d}Q/\mathrm{d}P) + UK(\{\mathrm{d}Q/\mathrm{d}P\})$, with U being independent of $\mathrm{d}Q/\mathrm{d}P$ and uniformly distributed on $[0,1]$. Using the last expression for $1-\beta$ the reader may check that the representation (λ,β) of a dichotomy (P,Q) is induced from (P,Q) by Z, i.e. $PZ^{-1} = \lambda$ and $QZ^{-1} = \beta$.

One would expect that a reasonable concept of statistical information would make $\beta(\cdot|P,Q)$ a monotonically increasing function of information. Actually we shall see in section 3 of chapter 9 that the pointwise ordering of these functions is precisely the ordering of dichotomies (P,Q) with respect to the information they carry, as shown by Blackwell. As these functions constitute a lattice for the pointwise ordering we conclude that the collection of dichotomies is also a lattice for the ordering of 'being at least as informative as'. This property is peculiar to dichotomies and does not extend to the case of a parameter set containing three or more points.

Deficiencies also have very simple interpretations in terms of these functions. Thus we shall see in chapter 9 that the deficiency distance between dichotomies equals twice the Paul Levy diagonal distance between these functions considered as distribution functions.

Let us take a closer look at these quantities in order to see how they behave for replicated experiments. Observe first that $1-\beta(0|P,Q)$ is the total mass of the P-absolutely continuous part of Q. Thus $1-\beta(0|P^n,Q^n) = (1-\beta(0|P,Q))^n$. On the other hand $\beta(1|P,Q) = 1$ for any dichotomy (P,Q). We may therefore restrict our attention to the case $0<\alpha<1$.

A first approximation may be obtained from the inequalities

$$\frac{1}{d}\left[1-\alpha-P\left((\mathrm{d}Q/\mathrm{d}P)\leqslant\frac{1}{d}\right)\right]\leqslant 1-\beta(\alpha|P,Q)\leqslant c(1-\alpha) \quad (1.4.3)$$

which are valid whenever $0<\alpha\leqslant\beta(\alpha|P,Q)<1$, $cd>1$ and $P((\mathrm{d}Q/\mathrm{d}P)\geqslant c)\geqslant\alpha\geqslant P((\mathrm{d}Q/\mathrm{d}P)>c)$.

These inequalities may be argued as follows. Note first that $P=Q$ and $c=1$ when $\alpha=\beta(\alpha|P,Q)$ and then the inequalities are trivial. We may therefore assume that $\alpha<\beta(\alpha|P,Q)$, i.e. that $P\neq Q$.

Let $\mathrm{d}Q/\mathrm{d}P$ be a $[0,\infty]$-valued version of this Radon–Nikodym derivative which is maximal under Q. Then $\mathrm{d}Q/\mathrm{d}P>0$ a.e. Q while $\mathrm{d}Q/\mathrm{d}P<\infty$ a.e. P. Thus $0<c<\infty$. Choose $\gamma\in[0,1]$ so that $1-\alpha=P((\mathrm{d}Q/\mathrm{d}P)<c)+(1-\gamma)P((\mathrm{d}Q/\mathrm{d}P)=c)$. By the Neyman–Pearson lemma

$$1-\beta(\alpha|P,Q)=Q\left(\frac{\mathrm{d}Q}{\mathrm{d}P}<c\right)+(1-\gamma)Q\left(\frac{\mathrm{d}Q}{\mathrm{d}P}=c\right).$$

Hence

$$\begin{aligned}1-\beta(\alpha|P,Q)&=\int_{(\mathrm{d}Q/\mathrm{d}P)<c}\mathrm{d}Q+(1-\gamma)\int_{(\mathrm{d}Q/\mathrm{d}P)=c}\mathrm{d}Q\\&\leqslant\int_{(\mathrm{d}Q/\mathrm{d}P)<c}c\,\mathrm{d}P+(1-\gamma)\int_{(\mathrm{d}Q/\mathrm{d}P)=c}c\,\mathrm{d}P\\&=c(1-\alpha)\end{aligned}$$

and

$$\begin{aligned}1-\beta(\alpha|P,Q)&\geqslant Q\left(\frac{1}{d}<\frac{\mathrm{d}Q}{\mathrm{d}P}<c\right)+(1-\gamma)Q\left(\frac{\mathrm{d}Q}{\mathrm{d}P}=c\right)\\&\geqslant\frac{1}{d}P\left(\frac{1}{d}<\frac{\mathrm{d}Q}{\mathrm{d}P}<c\right)+(1-\gamma)\frac{1}{d}P\left(\frac{\mathrm{d}Q}{\mathrm{d}P}=c\right)\\&=\frac{1}{d}\left[1-\alpha-P\left(\frac{\mathrm{d}Q}{\mathrm{d}P}\leqslant\frac{1}{d}\right)\right].\end{aligned}$$

For each $n=1,2,\ldots$ there is a $c_n>0$ such that

$$P^n(\mathrm{d}Q^n/\mathrm{d}P^n\geqslant c_n)\geqslant\alpha\geqslant P^n(\mathrm{d}Q^n/\mathrm{d}P^n>c_n).$$

Let Z denote a version of $\log(\mathrm{d}Q/\mathrm{d}P)$ and Z_j a version of $\log(\mathrm{d}Q/\mathrm{d}P)$ from the jth trial. Then these inequalities may be written

$$P^n\left(\bar{Z}_n\geqslant\frac{1}{n}\log c_n\right)\geqslant\alpha\geqslant P^n\left(\bar{Z}_n>\frac{1}{n}\log c_n\right)$$

where $\bar{Z}_n=(1/n)(Z_1+\cdots+Z_n)$. Hence, since $\bar{Z}_n\to E_pZ$ in P-probability, we conclude that $(1/n)\log c_n \to E_pZ$ whether Z is P-integrable or not. Note that by Jensen's inequality

$$E_pZ = E_p\log(\mathrm{d}Q/\mathrm{d}P) \leqslant \log E_p(\mathrm{d}Q/\mathrm{d}P) \leqslant \log 1 \leqslant 0.$$

The right hand inequality of (1.4.3) implies that $1 - \beta(\alpha|P^n, Q^n) \leqslant c_n(1-\alpha)$ so that $\limsup_n[1 - \beta(\alpha|P^n, Q^n)]^{1/n} \leqslant \exp(E_pZ)$. If $E_pZ > -\infty$ and $\varepsilon > 0$ then from the left hand inequality of (1.4.3) we obtain $1 - \beta(\alpha|P^n, Q^n) \geqslant [\exp n(E_pZ-\varepsilon)][1-\alpha-P^n(Z_n \leqslant E_pZ-\varepsilon)]$ when n is sufficiently large. Hence $\liminf_n[1 - \beta(\alpha|P^n, Q^n)]^{1/n} \geqslant \exp(E_pZ - \varepsilon) \to \exp(E_pZ)$ as $\varepsilon \to 0$. It follows that $\liminf_n[1 - \beta(\alpha|P^n, Q^n)]^{1/n} \geqslant \exp(E_pZ)$ whether Z is P-integrable or not.

Altogether this shows that

$$[1 - \beta(\alpha|P^n, Q^n)]^{1/n} \to \exp(E_p\log(\mathrm{d}Q/\mathrm{d}P))$$

as $n \to \infty$ whenever $0 < \alpha < 1$.

Chernoff (1956) and Kullback (1959) refer to an unpublished paper by C. Stein for this result.

It follows that $1 - \beta_n(\alpha|P^n, Q^n)$ as $n \to \infty$, behaves roughly as the geometric progression e^{-nK} where $K = -E_p\log(\mathrm{d}Q/\mathrm{d}P)$. Using large deviation theory and assuming that $\mathscr{L}_p(\log(\mathrm{d}Q/\mathrm{d}P))$ is not a lattice distribution, Efron (1967) gives a considerably more accurate asymptotic expression for $1 - \beta(\alpha|P^n, Q^n)$. The case $K = \infty$ is discussed in Janssen (1986).

The point we want to make is the role of the non-negative number $K = -E_p\log(\mathrm{d}Q/\mathrm{d}P)$. Kullback (1959) calls K the *mean information per observation for discriminating in favour of P against Q*.

Like the Fisher information matrix this quantity also behaves additively for experiment multiplication. We refer to Kullback (1959) for other interesting properties of this information number.

Example 1.4.4 (The Hellinger transform). The role of the Hellinger transform in the theory of comparison of experiments has many similarities with the role of characteristic functions in probability theory.

Consider an experiment $\mathscr{E} = (P_\theta : \theta \in \Theta)$ and let Λ denote the set of prior probability distributions on Θ which have finite supports. The *Hellinger transform of $\mathscr{E}$* is the map which assigns the number $\int \prod_\theta \mathrm{d}P_\theta^{t_\theta}$ to each distribution $t \in \Lambda$. The Hellinger transform of $\mathscr{E}$ may be denoted as $H(\cdot|\mathscr{E})$ or just as H, so that

$$H(t) = H(t|\mathscr{E}) = \int \prod_\theta \mathrm{d}P_\theta^{t_\theta}$$

when $t \in \Lambda$.

The Hellinger transform generalizes the affinity. In fact, with the obvious notation

$$\gamma(P, Q) = H(\tfrac{1}{2}, \tfrac{1}{2}|(P, Q))$$

for a dichotomy (P, Q) and

$$\gamma(P_{\theta^1}, P_{\theta^2}) = H(\tfrac{1}{2}\delta_{\theta^1} + \tfrac{1}{2}\delta_{\theta^2}|\mathscr{E})$$

for an experiment $\mathscr{E} = (P_\theta : \theta \in \Theta)$. Here δ_θ denotes the one-point distribution in θ.

The Hellinger transform is closely related to the Laplace transform. Thus if $\theta^0, \theta^1, \ldots, \theta^r \in \Theta$ and if the probability distribution t assigns masses $t_0, t_1, \ldots, t_r$ to $\theta^0, \theta^1, \ldots, \theta^r$ respectively, where $t_0 + t_1 + \cdots + t_r = 1$ and $t_0 > 0$, then $H(t|\mathscr{E}) = E_{\theta^0} \exp(\sum_{i=1}^r t_i Z_i)$ where $Z_i = \log(\mathrm{d}P_{\theta^i}/\mathrm{d}P_{\theta^0})$; $i = 1, \ldots, r$.

It follows that the Hellinger transform determines the law of the likelihood process $\theta \to \mathrm{d}P_\theta/\mathrm{d}P_{\theta^0}$ under P_{θ^0}. Although there is a slight difficulty for non-homogeneous experiments (see chapter 6), it may be fairly clear from this that the statistical properties of an experiment are determined by its Hellinger transform.

The Hellinger transform is in fact a monotonically decreasing function of the statistical information carried by $\mathscr{E}$. It is also affine for mixtures. The reader may check that:

(i) $\underline{H} \leqslant H(\cdot|\mathscr{E}) \leqslant 1$ where $\underline{H}(t) = 1$ or $= 0$ as t is a one-point distribution or not.
(ii) $H(\cdot|\mathscr{E}) = \underline{H}$ if and only if $\mathscr{E} = (P_\theta : \theta \in \Theta)$ is totally informative i.e. if and only if $P_{\theta^1} \wedge P_{\theta^2} = 0$ when $\theta^1 \neq \theta^2$.
(iii) $H(\cdot|\mathscr{E}) = 1$ if and only if $\mathscr{E} = (P_\theta : \theta \in \Theta)$ is totally non informative i.e. if P_θ does not depend on θ.
(iv) $H(t|\mathscr{E})$ is convex in t. (Use Hölder's inequality.)

In spite of its relation to the Laplace transform and in spite of (iv) the Hellinger transform is not continuous throughout its domain Λ unless $\mathscr{E}$ is homogeneous.

The basic property of the Hellinger transform is that it is multiplicative for experiment multiplication, i.e.

$$H(\cdot|\mathscr{E} \times \mathscr{F}) = H(\cdot|\mathscr{E})H(\cdot|\mathscr{F})$$

for experiments $\mathscr{E}$ and $\mathscr{F}$ having the same parameter set. This may be deduced directly from the definition by writing out the integrals involved and then using Fubini's theorem.

Using multiplicativity and affinity (under mixtures) we find in particular that the Hellinger transform of the Poissonization $\hat{\mathscr{E}}$ of an experiment $\mathscr{E}$ has

Hellinger transform

$$H(\cdot|\hat{\mathscr{E}}) = \sum_{n=0}^{\infty} \frac{1}{n!} e^{-1} H(\cdot|\mathscr{E})^n = \exp[H(\cdot|\mathscr{E}) - 1],$$

(see example 1.3.2).

Although not previously stated, it should be noted that we have adhered to the standard convention of interpreting 0^0 as 1. Slightly different multiplicative transforms appear for non-homogeneous experiments if we also allow some of the expressions 0^0 to take the value 0. In chapter 7 we shall see that if Θ is finite then except for this variation, all non-constant real valued representable multiplicative functionals are of the form $\mathscr{E} \to H(t|\mathscr{E})$ for some prior distribution $t \in \Lambda$.

The multiplicativity of the affinity is of course a consequence of the multiplicativity of the Hellinger transform.

The *Kullback–Leibler information number* is linked to the Hellinger transform since

$$-\frac{\mathrm{d}}{\mathrm{d}t}\left[\int \mathrm{d}P^{1-t}\,\mathrm{d}Q^t\right]_{t=0} = -E_p \log \frac{\mathrm{d}Q}{\mathrm{d}P}.$$

Thus the additivity of the Kullback–Leibler information number for experiment multiplication also follows from the multiplicativity of the Hellinger transform.

The Hellinger transform may be employed for convergence problems for experiments, just as characteristic functions are employed for convergence problems for distributions. Following LeCam, we shall see in chapter 7 that there is a natural decision theoretical pseudo-metric Δ for experiments having the same parameter set Θ. If Θ is finite then Δ is compact and convergence may then be described as the usual weak convergence of laws of likelihood ratios. From the relationship between Laplace transforms for laws of likelihoods and Hellinger transforms, it is then not surprising that Δ-convergence of experiments with a common finite parameter set Θ corresponds to pointwise convergence of Hellinger transforms.

If Θ is infinite then the restrictions of the pseudo-metric Δ to finite subsets of Θ provide a compact uniformity for experiments, and the topology of this uniformity coincides with the topology of pointwise convergence of Hellinger transforms.

A particular and almost immediate application of these ideas to infinitely divisible experiments is as follows. Consider an experiment $\mathscr{E}$ which for each positive integer n has an nth root, i.e. is equivalent to a product $\mathscr{E}_n^n$ for some experiment $\mathscr{E}_n$. Then $H(\cdot|\mathscr{E}_n) = H(\cdot|\mathscr{E})^{1/n}$ so that if $\hat{\mathscr{E}}_n$ is the Poissonization of $\mathscr{E}_n$ then $H(\cdot|\hat{\mathscr{E}}_n^n) = \exp[n[H(\cdot|\mathscr{E}_n) - 1]] \to H(\cdot|\mathscr{E})$ as $n \to \infty$.

Referring to the limit concept described above (the one which is equivalent to pointwise convergence of Hellinger transforms) we find that any infinitely divisible experiment is the limit of a sequence of products of Poissonized experiments. On the other hand, Poissonized experiments are always infinitely divisible, as we saw in example 1.3.2.

It follows that if we accept the general facts on convergence just mentioned then we have arrived almost effortlessly at the complete characterization of infinitely divisible experiments as limits of products of Poissonized experiments.

Example 1.4.5 (Minimum Bayes risk. b-functions = dual β-functions of dichotomies). Consider the problem of choosing a decision t belonging to a set T of possible decisions, on the basis of an observation X realizing the experiment $\mathscr{E} = (\mathscr{X}, \mathscr{A}; P_\theta : \theta \in \Theta)$. We shall assume that the loss incurred by taking decision t when the true value of the unknown parameter is θ may be quantified as a number $L_\theta(t)$.

In order to avoid technicalities, we will also assume that the parameter set Θ and the set T of possible decisions are finite sets.

If the decision rule $x \to \tau(x)$ is used and if θ prevails, then the loss we suffer is the random variable $L_\theta(\tau(X))$. The expected value under P_θ of this loss is called *the risk of using the decision rule τ when θ prevails.* As a function of θ the risk is called the risk functon of τ and we shall here denote it as $r(\cdot|\tau)$ so that $r(\theta|\tau) = E_\theta L_\theta(\tau(X))$. We usually want a decision rule τ with as small risk as possible. As smallness of the risk $r(\theta|\tau)$ for the various θs in general are conflicting requirements we have to seek a compromise. One possibility is to try to minimize the maximum risk $\max_\theta r(\theta|\tau)$. Another possibility is to minimize weighted averages of the risk function. As weight distributions on the parameter set may be interpreted as prior distributions, the resulting methods are often called Bayes methods. Consider now any system $(\lambda(\theta) : \theta \in \Theta)$ of non-negative weights. *The λ-weighted risk* is then $\sum_\theta r(\theta|\tau)\lambda(\theta)$.

If P_θ, for each θ, has density f_θ with respect to some non-negative measure μ, then the risk of τ at θ is $r(\theta|\tau) = \int L_\theta(\tau(x))f_\theta(x)\mu(\mathrm{d}x)$ so that the λ-weighted risk becomes

$$\sum_\theta r(\theta|\tau)\lambda(\theta) = \int \sum_\theta L_\theta(\tau(x))\lambda(\theta)f_\theta(x)\mu(\mathrm{d}x)$$

$$\geqslant \int \left[\min_t \sum_\theta L_\theta(t)\lambda(\theta)f_\theta(x)\right]\mu(\mathrm{d}x)$$

with equality when $t = \tau(x)$, for each x, minimizes $\sum_\theta L_\theta(t)\lambda(\theta)f_\theta(x)$. Any such procedure therefore minimizes the λ-weighted risk and the *minimum (Bayes) λ-weighted risk* is

$$b(\lambda|\mathscr{E}) = \int \min_t \left(\sum_\theta L_\theta(t)\lambda(\theta)\,\mathrm{d}P_\theta\right).$$

Thus we see that minimum Bayes risk, as a function of $\mathscr{E}$, is representable by a concave non-negative homogeneous function.

Furthermore, this concave function is of a very general form. In fact any finite concave and non-negatively homogeneous function is the infimum of a countable family of linear functionals. The form of the minimum Bayes risk reveals that any minimum of a finite family of linear functionals on the likelihood space represents a minimum Bayes risk, even for a given set $(\lambda(\theta) : \theta \in \Theta)$ of positive weights.

Now any form for minimum risk ought to decrease monotonically as information increases. This leads to the important observation that if h is concave then we ought to have $\int h(\mathrm{d}\mathscr{E}) \leqslant \int h(\mathrm{d}\mathscr{F})$ whenever $\mathscr{E}$ is at least as informative as $\mathscr{F}$. In particular we would like the Hellinger transform to decrease monotonically with increasing information.

We shall see in chapter 6 that we have actually arrived at a complete criterion for 'being at least as informative' since we shall see there that $\mathscr{E}$ is at least as informative as $\mathscr{F}$ if and only if $\int h(\mathrm{d}\mathscr{E}) \leqslant \int h(\mathrm{d}\mathscr{F})$ when h is non-negatively homogeneous and concave.

We shall often find it more convenient to work with utility and in particular with maximum Bayes expected utility, rather than with loss and minimum Bayes risk. Mathematically this just amounts to a multiplication by -1.

As a function of the prior distribution λ the minimum Bayes risk is the 'lower support function' of the set of available risk functions.

Let us consider a few particular cases of minimum Bayes risks and of maximum Bayes expected utilities.

Assume first that $\Theta = \{1, 2\}$ and consider the problem of testing '$\theta = 1$' against '$\theta = 2$' when the loss is 0 or 1 according to whether the decision is right or wrong. If the weights on $\theta = 1$ and $\theta = 2$ are, respectively, λ_1 and λ_2, then the minimum λ-weighted risk ($=$ minimum Bayes risk for the prior λ) is the number $\|\lambda_1 P_1 \wedge \lambda_2 P_2\| \equiv \int \lambda_1\,\mathrm{d}P_1 \wedge \lambda_2\,\mathrm{d}P_2$.

Putting $\lambda_1 = \lambda_2 = 1$ we obtain the measure of association between P_1 and P_2 which we related to the statistical distance. In particular this confirms the Bayes interpretation of the statistical distance mentioned in section 1.2.

Considering the minimum Bayes risk $\|(1 - \lambda)P_1 \wedge \lambda P_2\|$ as a function $b(\cdot|\mathscr{E}) = b(\cdot|P_1, P_2)$ of the prior probability λ of the event '$\theta = 2$', we obtain another interesting function on $[0, 1]$. This function is here called the *b-function*, or the dual Neyman–Pearson (N–P) function, *of the dichotomy* (P_1, P_2).

Clearly $0 \leqslant b(\lambda|P_1, P_2) \leqslant (1 - \lambda) \wedge \lambda$ and it may be shown that any concave non-negative function b on $[0, 1]$ which is majorized by the triangular function $\lambda \to \lambda \wedge (1 - \lambda)$ is the b-function of some dichotomy (P_1, P_2).

Here are three graphs of b-functions $b(\cdot|P, Q)$.

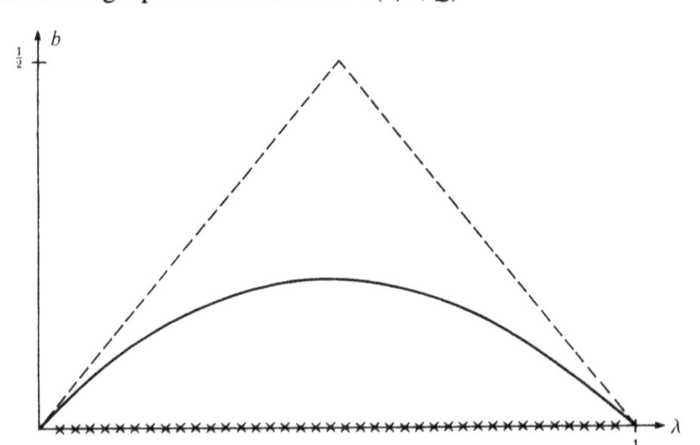

Fig. 2. × × × × Graph of $b(\cdot|P, Q)$ when (P, Q) is totally informative i.e. when $P \wedge Q = 0$. ——— Graph of $b(\cdot|P, Q)$ for a dichotomy (P, Q). - - - - - - Graph of $b(\cdot|P, Q)$ for a totally non informative dichotomy (P, P).

We mentioned in example 1.4.3 that many of the basic features of dichotomies could be directly described in terms of the functions β of the form $\beta = \beta(\cdot|P_1, P_2)$ for dichotomies (P_1, P_2). This is also the case for the functions b which are of the form $b = b(\cdot|P_1, P_2)$ for dichotomies (P_1, P_2).

In particular the ordering 'being at most as informative as' for dichotomies is just the pointwise ordering of the functions b. Thus the lattice operations on dichotomies correspond with the lattice operations of the reversed pointwise ordering of functions b. This implies that the b-function of a supremum of a family of dichotomies is the infimum of the corresponding family of b-functions. On the other hand the β-function of an infimum of a family of dichotomies equals the pointwise infimum of the corresponding family of β-functions.

As we mentioned in example 1.4.3, the half deficiency distance between two dichotomies equals the Paul Levy diagonal distance between the corresponding functions β. In terms of the minimum Bayes risk functions this is just the sup norm distance between the corresponding functions b.

In contrast to the functionals $(P_1, P_2) \to \beta(\alpha|P_1, P_2)$; $0 < \alpha < 1$ the functionals $(P_1, P_2) \to b(\lambda|P_1, P_2)$ are representable and thus affine under mixtures.

The Hellinger transform of a dichotomy (P_1, P_2) may be obtained directly in terms of the function $b(\cdot|P_1, P_2)$ since

$$H(1 - t, t|P_1, P_2)/(1 - t)t = \int_0^1 \{\|(1 - \lambda)P_1 \wedge \lambda P_2\|/[(1 - \lambda)^{2-t}\lambda^{1+t}]\}\, d\lambda$$

$$= \int_0^\infty [\|xP_1 \wedge P_2\|/x^{t+1}]\, dx \quad \text{when} \quad 0 < t < 1.$$

This may be proved by e.g. first checking that it holds when $(\mathscr{X}, \mathscr{A}; P_1, P_2)$ is a *double dichotomy* i.e. when $\mathscr{A}$ is generated by a two set partition of $\mathscr{X}$ (i.e. $\# \mathscr{A} = 4$). Now we shall see in chapter 7 that any dichotomy is a mixture of double dichotomies. The identity then follows by appealing to this and by utilizing the fact that the expressions are affine under mixtures.

Alternatively we may use the fact that if f_1 and f_2 are non-negative numbers and if $0 < t < 1$ then

$$f_1^{1-t} f_2^t/(1-t)t = \int_0^1 \{[(1-\lambda)f_1 \wedge \lambda f_2]/[(1-\lambda)^{2-t}\lambda^{t+1}]\}\, d\lambda.$$

In particular it follows that the affinity $\gamma(P_1, P_2) = \int [dP_1\, dP_2]^{1/2}$ may be expressed as

$$\gamma(P_1, P_2) = \frac{1}{4}\int_0^1 \{\|(1-\lambda)P_1 \wedge \lambda P_2\|/[(1-\lambda)\lambda]^{3/2}\}\, d\lambda.$$

Rewriting $b(\lambda | P_1, P_2)$ as $b(\lambda | P_1, P_2) = 1 - \|(1-\lambda)P_1 \vee \lambda P_2\|$ we arrive at the functional $\lambda \to \|(1-\lambda)P_1 \vee \lambda P_2\|$ which to each $\lambda \in [0, 1]$ assigns the maximal Bayes probability of guessing the right distribution for the prior $(1-\lambda, \lambda)$. We may extend this construction to a general experiment $\mathscr{E} = (\mathscr{X}, \mathscr{A}; P_\theta : \theta \in \Theta)$ since $\|\bigvee_\theta \lambda_\theta P_\theta\| = \int \bigvee_\theta (\lambda_\theta\, dP_\theta)$ is the maximal Bayes probability of correctly guessing the true value of θ for the prior distribution λ. Letting λ run through the set Λ of prior probability distributions with finite supports, we obtain another transform which also characterizes the laws of likelihood and thus the statistical properties of $\mathscr{E}$.

Another interesting (and characterizing) transform of an experiment $\mathscr{E} = (P_\theta : \theta \in \Theta)$ is the transform which associates to each finite measure a with finite support in Θ the total variation $\|\sum_\theta a_\theta P_\theta\| = \int |\sum_\theta a_\theta\, dP_\theta|$ of the linear combination $\sum_\theta a_\theta P_\theta$.

This transform describes how $\mathscr{E}$ behaves for testing problems, as we shall see.

We shall comment on the asymptotic behaviour of minimum Bayes risk under replications in the next example.

Example 1.4.6 (Minimum Bayes risk and Chernoff's information number). In the previous example we considered the minimum Bayes risk for testing '$\theta = 1$' against '$\theta = 2$' in a general dichotomy $\mathscr{D} = (P_1, P_2)$ when the prior probability of the event '$\theta = 2$' was $\lambda \in [0, 1]$ and when the loss was 0 or 1 according to whether the decision made was right or wrong. Let us continue to use the notation $b(\lambda | P_1, P_2)$ or $b(\lambda | \mathscr{D})$ for this risk.

What is the asymptotic behaviour of the minimum Bayes risk $b(\lambda | \mathscr{D}^n)$ for n-times replicated dichotomies $\mathscr{D}^n$ as $n \to \infty$?

As $b(0 | \mathscr{D}) = b(1 | \mathscr{D}) = 0$ for any dichotomy $\mathscr{D}$ we may assume that $0 < \lambda < 1$.

Note first that $b(\lambda|P_1,P_2) = \int (1-\lambda)\,dP_1 \wedge \lambda\,dP_2 \leqslant \int dP_1 \wedge dP_2 \leqslant \gamma(P_1,P_2)$.

Using the right hand inequality we obtain from the multiplicativity of the affinity that $b(\lambda|P_1^n, P_2^n) \leqslant \gamma(P_1,P_2)^n$.

It follows that $b(\lambda|P_1^n, P_2^n) \to 0$ at least as fast as $\gamma(P_1,P_2)^n \to 0$ when $P_1 \neq P_2$.

A completely trivial way of obtaining an exponential upper bound is to write

$$b(\lambda|\mathscr{D}^n) \leqslant C(\lambda|\mathscr{D})^n, \qquad n = 1,2,\ldots$$

where $C(\lambda|\mathscr{D}) = \sup_n b(\lambda|\mathscr{D}^n)^{1/n}$.

Although it may appear a bit optimistic we shall see now that provided $P_1 \neq P_2$ the constant $C(\lambda|P_1,P_2)$ determines the exact exponential rate of convergence of $b(\lambda|P_1^n, P_2^n)$ as $n \to \infty$. Furthermore this constant turns out to be independent of λ as long as $0 < \lambda < 1$.

These facts may be deduced from some easily obtained inequalities. Consider first two dichotomies $\mathscr{D}_1 = (P_1,P_2)$ and $\mathscr{D}_2 = (Q_1,Q_2)$. Putting $\mu = P_1 + P_2$, $\nu = Q_1 + Q_2$, $f_i = dP_i/d\mu$, and $g_i = dQ_i/d\nu$ we find

$$\begin{aligned} b(\lambda|\mathscr{D}_1 \times \mathscr{D}_2) &= \int \{[(1-\lambda)(f_1 \otimes g_1)] \wedge [\lambda(f_2 \otimes g_2)]\}\,d(\mu \times \nu) \\ &\geqslant \int \{[(1-\lambda)f_1 \wedge \lambda f_2] \otimes [g_1 \wedge g_2]\}\,d(\mu \times \nu) \\ &= \int [(1-\lambda)f_1 \wedge \lambda f_2]\,d\mu \int g_1 \wedge g_2\,d\nu \end{aligned}$$

so that:

(i) $b(\lambda|\mathscr{D}_1 \times \mathscr{D}_2) \geqslant b(\lambda|\mathscr{D}_1)2b(\tfrac{1}{2}|\mathscr{D}_2)$.

Furthermore $\int dP_1 \wedge dP_2 \geqslant \int (1-\lambda)\,dP_1 \wedge \lambda\,dP_2 \geqslant [(1-\lambda)\wedge\lambda]\int dP_1 \wedge dP_2$, i.e.

(ii) $2b(\tfrac{1}{2}|\mathscr{D}) \geqslant b(\lambda|\mathscr{D}) \geqslant 2[(1-\lambda)\wedge\lambda]b(\tfrac{1}{2}|\mathscr{D})$.

Hence, by (i):

(iii) $b(\lambda|\mathscr{D}_1 \times \mathscr{D}_2) \geqslant b(\lambda|\mathscr{D}_1)b(\xi|\mathscr{D}_2)$ for dichotomies $\mathscr{D}_1$ and $\mathscr{D}_2$ and for numbers λ, ξ in $[0,1]$.

Using (iii) we find $b(\lambda|\mathscr{D}^{m+n}) \geqslant b(\lambda|\mathscr{D}^m)b(\lambda|\mathscr{D}^n)$ for integers $m, n \geqslant 0$. Putting $a_n = \log b(\lambda|\mathscr{D}^n)$ we see that the sequence $\{a_n\}$ is *super-additive*, i.e. that

$$a_{m+n} \geqslant a_m + a_n$$

when $m, n \geqslant 1$.

In general any superadditive sequence a_n, $n = 1,2,\ldots$ enjoys the property that $a_n/n \to \sup_n(a_n/n)$. This may be seen as follows.

Put $a_0 = 0$ and choose numbers $k < \sup_n(a_n/n)$, $\varepsilon > 0$ and a positive integer

n_0 such that $k < a_{n_0}/n_0$. Let $n_1 > n_0$ be sufficiently large to ensure that $(a_i - ik)/n > -\varepsilon$ when $i < n_0$ and $n \geqslant n_1$. If $n > n_1$ then we write $n = n_0 j + i$ where $j \geqslant 1$ and $0 \leqslant i < n_0$. This yields

$$a_n/n \geqslant (ja_{n_0} + a_i)/n = (a_{n_0}/n_0)(jn_0/n) + (a_i/n)$$
$$\geqslant [k(n-i) + a_i]/n > k - \varepsilon$$

proving the asserted convergence of a_n/n.

It follows that $b(\lambda|\mathscr{D}^n)^{1/n} \to C(\lambda|\mathscr{D})$.

The inequalities in (ii) imply that $2b(\frac{1}{2}|\mathscr{D}^n) \geqslant b(\lambda|\mathscr{D}^n) \geqslant [(1-\lambda)\wedge\lambda]b(\frac{1}{2}|\mathscr{D}^n)$. Hence $C(\frac{1}{2}|\mathscr{D}) = C(\lambda|\mathscr{D})$ for all numbers $\lambda \in]0,1[$.

As the quantity $C(\lambda|\mathscr{D})$ does not depend on $\lambda \in]0,1[$ from now on we shall write $C(\mathscr{D})$ or $C(P_1,P_2)$ for a dichotomy $\mathscr{D} = (P_1,P_2)$ instead of $C(\lambda|\mathscr{D})$ or $C(\lambda|P_1,P_2)$.

Using the Hellinger transform we find the following upper bound

$$b(\lambda|\mathscr{D}) = \int (1-\lambda)\,\mathrm{d}P_1 \wedge \lambda\,\mathrm{d}P_2$$
$$= \int [(1-\lambda)\,\mathrm{d}P_1 \wedge \lambda\,\mathrm{d}P_2]^{1-t}[(1-\lambda)\,\mathrm{d}P_1 \wedge \lambda\,\mathrm{d}P_2]^t$$
$$\leqslant \int [(1-\lambda)\,\mathrm{d}P_1]^{1-t}[\lambda\,\mathrm{d}P_2]^t = (1-\lambda)^{1-t}\lambda^t H(1-t,t|\mathscr{D}).$$

Hence, by the multiplicativity of the Hellinger transform

$$b(\lambda|\mathscr{D}^n)^{1/n} \leqslant [(1-\lambda)^{1-t}\lambda^t]^{1/n} H(1-t,t|\mathscr{D}) \leqslant H(1-t,t|\mathscr{D}).$$

Thus $C(\mathscr{D}) \leqslant \inf_t H(1-t,t|\mathscr{D})$.

Using large deviation theory Chernoff (1952) showed that this upper bound is actually the precise value of $C(\mathscr{D})$, i.e. that

$$C(\mathscr{D}) = \inf_t H(1-t,t|\mathscr{D}).$$

The constant $C(\mathscr{D})$ enters into a large variety of asymptotic expansions related to replicated experiments. It would therefore be useful to have a name for it and we shall define *the Chernoff information number of the dichotomy* $\mathscr{D}$ as the number $C(\mathscr{D})$.

The asymptotic evaluations we have just carried out may then be summarized by

$$C(\mathscr{D}) \geqslant b(\lambda|\mathscr{D}^n)^{1/n} \to C(\mathscr{D}).$$

If the distribution of $\mathrm{d}P_2/\mathrm{d}P_1$ under P_1 is not a lattice distribution and if $\inf_t H(1-t,t|\mathscr{D})$ is obtained for $t = t_0 \in]0,1[$ then we may obtain a better approximation by taking the curvature at $t = t_0$ of the graph of $t \to H(1 - t, t|\mathscr{D})$ into account, as shown in Efron & Truax (1968).

In fact under these circumstances

$$\int \phi(\mathrm{d}P_2^n/\mathrm{d}P_1^n)\,\mathrm{d}P_1^n = \int_0^\infty [\phi(x)/x^{t_0+1}]\,\mathrm{d}x(2\pi n\tau^2)^{-1/2}C(\mathscr{D})^n(1 + o(1))$$

where $\tau^2 = C(\mathscr{D})^{-1}(\mathrm{d}^2/\mathrm{d}t^2)[H(1 - t, t|\mathscr{D})]_{t=t_0}$ provided ϕ is continuous a.e. on $[0, \infty[$ and provided ϕ and the function $x \to \phi(x)/x$ are bounded (Torgersen, 1981b). The prior distribution $(1 - t_0, t_0)$ on $\Theta = \{1, 2\}$ determined by the minimizing value $t = t_0$ turns out to be asymptotically least favourable for the problem of testing '$\theta = 1$' against '$\theta = 2$' for the 0–1 loss function and for the n-times replicated dichotomy $\mathscr{D}^n = (P_1, P_2)^n = (P_1^n, P_2^n)$.

In any case the deficiency distance between $\mathscr{D}^n$ and a totally informative dichotomy behaves, to a first approximation, as $C(\mathscr{D})^n$. In fact simple expressions involving Chernoff information numbers provide first approximations of minimum Bayes risks in replicated experiments with finite parameter sets.

If we proceed to experiments $\mathscr{E}$ with countably infinite parameter sets then there still is a constant $\sigma(\mathscr{E})$ so that LeCam's deficiency distance between $\mathscr{E}^n$ and total information behaves as $\sigma(\mathscr{E})^n$. However no general explicit expression for $\sigma(\mathscr{E})$ is known in this case. If Θ is uncountable this problem loses some of its appeal since then the deficiency distance between $\mathscr{E}$ and a totally informative experiment assumes its maximal possible value, 2, whenever $\mathscr{E}$ is dominated.

Related results and discussions and examples in Torgersen (1976), may be found in Torgersen (1981b) and in the last chapter of Heyer (1982).

Example 1.4.7 (Continuity and differentiability). An experiment $\mathscr{E} = (\mathscr{X}, \mathscr{A}; P_\theta : \theta \in \Theta)$ is a family in the Banach space $\mathrm{ca}(\mathscr{X}, \mathscr{A})$ obtained by equipping the linear space of finite measures on $\mathscr{A}$ with the total variation norm. One possibility for studying continuity and differentiability properties of $\mathscr{E}$ is to do so for the map $\theta \to P_\theta$ from Θ to $\mathrm{ca}(\mathscr{X}, \mathscr{A})$. In particular this leads to a notion of differentiability which we shall discuss in chapter 8.

Another interesting approach is based on the observation that the affinity $\gamma(P_{\theta^1}, P_{\theta^2})$ is non-negative definite in (θ^1, θ^2) and thus there is a second order process $\theta \to Y_\theta$ on some probability space $(\Omega, \mathscr{S}, \pi)$ such that $EY_{\theta^1}Y_{\theta^2} = \gamma(P_{\theta^1}, P_{\theta^2})$ when θ^1, $\theta^2 \in \Theta$. If, in particular, the experiment $\mathscr{E}$ is dominated, then we have seen that it is dominated by a probability measure P on $\mathscr{A}$. In that case we may put $(\Omega, \mathscr{S}, \pi) = (\mathscr{X}, \mathscr{A}, P)$ and $Y_\theta \equiv_\theta (\mathrm{d}P_\theta/\mathrm{d}P)^{1/2}$. In any case we may inquire about the consequences for $\mathscr{E}$ of continuity and differentiability of the map $\theta \to Y_\theta$ from Θ to $L_2(\pi)$.

Consider for example the continuity of this process for a specified topology on Θ and at a particular point θ^0 in Θ. The second order process $\theta \to Y_\theta$ is continuous at $\theta = \theta^0$ if and only if $d(P_\theta, P_{\theta^0}) \to 0$ as $\theta \to \theta^0$. If P_{θ^0} dominates $\mathscr{E}$ then the second order process $\theta \to (\mathrm{d}P_\theta/\mathrm{d}P_{\theta^0})^{1/2}$ from Θ to $L_2(P_{\theta^0})$ have exactly the same product moments as the process Y. In any case the map $\theta \to (\mathrm{d}P_\theta/\mathrm{d}P_{\theta^0})^{1/2}$ is a second order process in $L_2(P_{\theta^0})$ and the requirements of continuity in quadratic mean are the same for these processes. In fact the total mass of the P_{θ^0} singular part of P_θ is $1 - \int (\mathrm{d}P_\theta/\mathrm{d}P_{\theta^0})\,\mathrm{d}P_{\theta^0}$ and this quantity approaches null as $(\mathrm{d}P_\theta/\mathrm{d}P_{\theta^0})^{1/2} \to 1$ in $L_2(P_{\theta^0})$.

The topological equivalence of the Hellinger distance d and statistical distance implies that continuity of the second order process $\theta \to Y_\theta$ at $\theta = \theta^0$ is also equivalent to the condition that $\theta \to P_\theta$ is continuous for statistical distance at $\theta = \theta^0$.

Yet another way to phrase this is to require that the map $\theta \to \mathrm{d}P_\theta/\mathrm{d}P_{\theta^0}$ from Θ to $L_1(P_{\theta^0})$ is continuous at $\theta = \theta^0$.

Although the two distances lead to the same notion of continuity, even for homogeneous experiments they lead to different notions of differentiability.

In order to explain this let us consider differentiability at a given interior point θ^0 of Θ where $\Theta \subseteq \mathbb{R}^k$. Then under these circumstances we shall say that *$\mathscr{E}$ is differentiable (in the first mean) at* $\theta = \theta^0$ if $\|P_\theta - P_{\theta^0} - \sum_i (\theta - \theta^0)_i \dot{P}_{\theta^0,i}\| / \|\theta - \theta^0\| \to 0$ as $\theta \to \theta^0$ for bounded measures $\dot{P}_{\theta^0,1}, \ldots, \dot{P}_{\theta^0,k}$.

A more sophisticated way of expressing this condition is to say that there should be a linear operator L_{θ^0} from $\mathbb{R}^k$ to the Banach space $\mathrm{ca}(\mathscr{X}, \mathscr{A})$ of finite measures on $\mathscr{A}$ such that $\|P_\theta - P_{\theta^0} - L_{\theta^0}(\theta - \theta^0)\| / \|\theta - \theta^0\| \to 0$ as $\theta \to \theta^0$. Replacing $\mathbb{R}^k$ and $\mathrm{ca}(\mathscr{X}, \mathscr{A})$ with arbitrary Banach spaces this is precisely the general notion of Frechet differentiability at an interior point $\theta = \theta^0$ of Θ.

As linear maps from $\mathbb{R}^k$ into Banach spaces are always bounded (i.e. continuous) one might be tempted to insert the requirement of boundedness in the general case. This assumption need not, however, be a part of the notion of a Frechet derivative. We refer to e.g. Brown & Page (1970) and Dieudonné (1960) for expositions on Frechet derivatives.

The requirement on the linear operator L_{θ^0} may be expressed as

$$\lim_{t \to 0} \sup_{\|v\|=1} \left\| \frac{1}{t}(P_{\theta_0 + tv} - P_{\theta^0}) - L_{\theta^0}(v) \right\| = 0.$$

This implies that

$$\sup_{\|v\|=1} \left\| \frac{1}{t_1}(P_{\theta^0 + t_1 v} - P_{\theta^0}) - \frac{1}{t_2}(P_{\theta^0 + t_2 v} - P_{\theta^0}) \right\| \to 0 \text{ as } t_1, t_2 \to 0$$

and by linearity also that

$$\|(P_{\theta^0 + tv^1} - P_{\theta^0}) + (P_{\theta^0 + tv^2} - P_{\theta^0}) - 2(P_{\theta^0 + (1/2)t(v^1 + v^2)} - P_{\theta^0})\| / t \to 0$$

as $t \to 0$ for any pair (v^1, v^2) of vectors in $\mathbb{R}^k$.

Conversely the last two conditions imply differentiability at $\theta = \theta^0$ by completeness, i.e. that

$$\lim_{t\to 0} \sup_{\|v\|=1} \left\| \frac{1}{t}(P_{\theta^0+tv} - P_{\theta^0}) - L_{\theta^0}(v) \right\| = 0.$$

These comments on differentiability do not in any way exploit the particular properties of the Banach spaces involved, $\mathbb{R}^k$ and $\mathrm{ca}(\mathscr{X}, \mathscr{A})$. They are therefore generally valid for maps $\theta \to P_\theta$ from a subset Θ of a Banach space $\tilde{\Theta}$ to another Banach space, provided θ^0 is an interior point of Θ.

If, however, the range space is a Hilbert space, then it may be easier to work with the squares of the norms which we have just considered. Doing that we infer that whether the map $\theta \to F_\theta$ from Θ to a Hilbert space is differentiable or not depends solely on the function which to each pair $(\theta^1, \theta^2) \in \Theta$ associates the inner product of F_{θ^1} and F_{θ^2}.

As we stated above $\mathscr{E} = (P_\theta : \theta \in \Theta)$ is differentiable if and only if the map $\theta \to P_\theta$ from $\mathbb{R}^k$ to $\mathrm{ca}(\mathscr{X}, \mathscr{A})$ is Frechet differentiable at $\theta = \theta^0$. The reader may have observed that we have not completely described $\mathbb{R}^k$ as a Banach space. However this does not matter since any two norms on $\mathbb{R}^k$ are equivalent.

It is a matter of checking that $\mathscr{E}$ is differentiable at $\theta = \theta^0$ if and only if the map $\theta \to \mathrm{d}P_\theta/\mathrm{d}P_{\theta^0}$ from Θ to $L_1(P_{\theta^0})$ is Frechet differentiable. If so then the partial derivatives $\dot{P}_{\theta^0,i}$, $i = 1, \ldots, k$ are absolutely continuous with respect to P_{θ^0} and

$$\int \left| \mathrm{d}P_\theta/\mathrm{d}P_{\theta^0} - 1 - \sum_i (\theta - \theta^0)_i \,\mathrm{d}\dot{P}_{\theta^0,i}/\mathrm{d}P_{\theta^0} \right| \mathrm{d}P_{\theta^0} \Big/ \|\theta - \theta^0\| \to 0$$

as $\theta \to \theta^0$. Conversely this limiting relation for finite measures $\dot{P}_{\theta^0,i}$; $i = 1, \ldots, k$ implies differentiability at $\theta = \theta^0$. If so and if $\dot{P}_{\theta^0,i}$ is P_{θ^0} absolutely continuous then $\dot{P}_{\theta^0,i}$ is the partial (Frechet) derivative of the map $\theta \to P_\theta$ with respect to θ_i at $\theta = \theta^0$.

In chapter 9 we shall use this notion of differentiability as a basis for a local (non-asymptotic) theory for comparison of statistical experiments.

It is the notion of differentiability associated with the Hellinger distance which appears to be appropriate for asymptotic statistical theory. There is a slight difficulty which appears when we discuss differentiability in quadratic mean and which does not enter the discussion of continuity, and this is the fact that it does not quite suffice to consider the second order process $\theta \to (\mathrm{d}P_\theta/\mathrm{d}P_{\theta^0})^{1/2}$ in $L_2(P_{\theta^0})$. We shall also need to ensure that the P_{θ^0} singular part of P_θ is sufficiently negligible when θ is close to θ_0. If P_{θ^0} dominates $\mathscr{E}$ then of course this is no problem.

Consider the general case and for any experiment $(Q_\theta : \theta \in \Theta)$ let the Q_{θ^0} singular (absolutely continuous) part of Q_θ be denoted as $Q_{\theta,s}(Q_{\theta,a})$. Then, since $(P_\theta^n)_a = (P_{\theta,a})^n$, we find $\|(P_\theta^n)_a\| = \|P_{\theta,a}\|^n = (1 - \|P_{\theta,s}\|)^n$. Substituting $\theta = \theta^0 + h/n^{1/2}$ we find that $\|(P_\theta^n)_a\| \to 1$ as $n \to \infty$ if and only if $n\|P_{\theta,s}\| \to 0$

i.e. if and only if $\|P_{\theta,s}\|/\|\theta-\theta^0\|^2 \to 0$. It follows that there is no hope for a homogeneous limit experiment for $(P^n_{\theta^0+h/n^{1/2}} : \|h\| \leqslant b)$, as $n \to \infty$ unless $\|P_{\theta,s}\| = o(\|\theta-\theta^0\|^2)$ as $\theta \to \theta^0$. If this condition is satisfied then the process $\theta \to Y_\theta$ is differentiable in quadratic mean at $\theta=\theta^0$ if and only if this is the case for the process $\theta \to (\mathrm{d}P_\theta/\mathrm{d}P_{\theta^0})^{1/2}$ when $\theta=\theta^0$ prevails. In fact

$$EY_{\theta^1}Y_{\theta^2} - E_{\theta^0}(\mathrm{d}P_{\theta^1}/\mathrm{d}P_{\theta^0})^{1/2}(\mathrm{d}P_{\theta^2}/\mathrm{d}P_{\theta^0})^{1/2} = E_{\theta^0}(\mathrm{d}P_{\theta^1})^{1/2}(\mathrm{d}P_{\theta^2})^{1/2}I_{\{0\}}(\mathrm{d}P_{\theta^0}) \leqslant \|P_{\theta^1,s}\|\,\|P_{\theta^2,s}\|.$$

It follows that we may write

$$0 \leqslant EY_{\theta^1}Y_{\theta^2} - E_{\theta^0}(\mathrm{d}P_{\theta^1}/\mathrm{d}P_{\theta^0})^{1/2}(\mathrm{d}P_{\theta^2}/\mathrm{d}P_{\theta^0})^{1/2} \leqslant g(\theta^1,\theta^2)\,\|\theta^1-\theta^0\|\,\|\theta^2-\theta^0\|$$

where g is bounded and $g(\theta^1,\theta^2) \to 0$ as θ^1, $\theta^2 \to 0$.

Writing out the conditions for differentiability in quadratic mean we find that the process $\theta \to Y_\theta$ is differentiable in quadratic mean at $\theta=\theta^0$ if and only if the process $\theta \to (\mathrm{d}P_\theta/\mathrm{d}P_{\theta^0})^{1/2}$ is differentiable in quadratic mean at $\theta=\theta^0$ when P_{θ^0} is the underlying distribution. Here we shall say that an experiment $\mathscr{E}=(P_\theta : \theta\in\Theta)$ is *differentiable in quadratic mean at* $\theta=\theta^0$ if the two processes mentioned are so and if, in addition, the total mass of the P_{θ^0} singular part of P_θ is $o(\|\theta-\theta^0\|^2)$ as $\theta\to\theta^0$.

Assuming that $\theta \to (\mathrm{d}P_\theta/\mathrm{d}P_{\theta^0})^{1/2}$ is differentiable in quadratic mean at $\theta=\theta^0$ when θ^0 prevails we may expand $g_\theta = (\mathrm{d}P_\theta/\mathrm{d}P_{\theta^0})^{1/2}$ as

$$g_\theta = 1 + \sum(\theta-\theta^0)_i\dot{g}_{\theta^0,i} + \|\theta-\theta^0\|^2 r_\theta$$

where

$$\int(\dot{g}_{\theta^0,i})^2\,\mathrm{d}P_{\theta^0} < \infty; \qquad i=1,\ldots,k$$

and where

$$\int r_\theta^2\,\mathrm{d}P_{\theta^0} \to 0 \quad \text{as} \quad \theta\to\theta^0.$$

Hence

$$g_\theta^2 = 1 + \sum(\theta-\theta^0)_i 2\dot{g}_{\theta^0,i} + \|\theta-\theta^0\|\,\tilde{r}_\theta$$

where

$$\int|\tilde{r}_\theta|\,\mathrm{d}P_{\theta^0} \to 0 \quad \text{as} \quad \theta\to\theta^0.$$

Thus differentiability in quadratic mean implies differentiability in the first mean.

Say that an experiment $\mathscr{E} = (P_\theta : \theta \in \Theta)$ *is k-wise at least as (at most as) informative as the experiment* $\mathscr{F} = (Q_\theta : \theta \in \Theta)$ if $\mathscr{E}|\Theta_0$ is at least as (as most as) informative as $\mathscr{F}|\Theta_0$ whenever the subset Θ_0 of Θ contains at most k points. If $k = 2$ or $k = 3$ then we may say pairwise or triplewise instead of, respectively, 2-wise or 3-wise.

Differentiability is spread among experiments according to the following rules.

(D.1) If $\mathscr{E}$ is differentiable in the first mean (quadratic mean) at $\theta = \theta^0$ and if $\mathscr{F}$ is triplewise at most as informative as $\mathscr{E}$, then $\mathscr{F}$ is also differentiable in the first mean (quadratic mean) at $\theta = \theta^0$.

(D.2) $\mathscr{E}_1 \times \cdots \times \mathscr{E}_r$ is differentiable in the first mean (quadratic mean) at $\theta = \theta^0$ if and only if all experiments $\mathscr{E}_i$; $i = 1, \ldots, r$ are differentiable in the first mean (quadratic mean) at $\theta = \theta^0$.

(D.3) If $p_1, \ldots, p_r > 0$ and if $p_1 + \cdots + p_r = 1$ then $p_1\mathscr{E}_1 + \cdots + p_2\mathscr{E}_2$ is differentiable in the first mean (quadratic mean) at $\theta = \theta^0$ if and only if the experiments $\mathscr{E}_i$; $i = 1, \ldots, r$ are all differentiable in the first mean (quadratic mean) at $\theta = \theta^0$.

(D.1) may be derived by observing that:

(i) the criteria for differentiability may, as we have seen, be expressed in terms of functionals which involve at most three values of θ at a time.

(ii) the functionals mentioned in (i) are representable by convex homogeneous functions on the likelihood space.

As explained in example 1.4.5 (and proved in chapter 6) convex functions on the likelihood space define functionals of experiments which are monotonically increasing for the ordering 'being at least as informative as'.

This also explains the 'only if' in (D.2). The 'if' in (D.2) follows by observing that if $q = 1$ or $q = 2$ and if $(Y_\theta : \theta \in \Theta)$ and $(Z_\theta : \theta \in \Theta)$ are independent families in $L_q(\pi)$ for some probability space $(\Omega, \mathscr{S}, \pi)$ and if the maps $\theta \to Y_\theta$ and $\theta \to Z_\theta$ are both Frechet differentiable at $\theta = \theta^0$ then $\theta \to Y_\theta Z_\theta$ is also a map from Θ to $L_q(\pi)$ which is Frechet differentiable at $\theta = \theta^0$.

Assume now that $\mathscr{E}$ is differentiable in quadratic mean at $\theta = \theta^0$ and that the Fisher information matrix is nonsingular at $\theta = \theta^0$. Then LeCam's central limit theory for experiments implies that the replicated experiment $\mathscr{E}^n$ behaves within neighbourhoods of θ^0 with diameter constant $n^{-1/2}$ asymptotically, as $n \to \infty$, as a normal translation experiment on $\mathbb{R}^k$.

Fisher information and asymptotic normality will be discussed further in the next example.

Example 1.4.8 (Local comparison, Fisher information and asymptotic normality). There is a variety of interesting, and generally non-equivalent, ways of defining Fisher information. A fairly complicated one proceeds by considering partial derivatives of densities with respect to the unknown parameter. Here we shall take an easier way out which avoids regularity conditions on densities and which is directly related to well known characteristics of experiments.

Assume that the parameter set Θ is a subset of $\mathbb{R}^k$. Let θ^0 be an interior point of Θ and consider an experiment $\mathscr{E} = (P_\theta : \theta \in \Theta)$.

Disregarding regularity conditions the Fisher information matrix $\Gamma(\theta^0, \mathscr{E})$ of $\mathscr{E}$ at θ^0 is usually defined as the covariance matrix under P_{θ^0} of the k variables

$$\frac{\partial}{\partial\theta_1}[\log \mathrm{d}P_\theta/\mathrm{d}P_{\theta^0}]_{\theta=\theta^0}, \ldots, \frac{\partial}{\partial\theta_k}[\log \mathrm{d}P_\theta/\mathrm{d}P_{\theta^0}]_{\theta=\theta^0}.$$

If $\mathrm{d}P_\theta/\mathrm{d}P_{\theta^0} = s_\theta$ then

$$\frac{\partial}{\partial\theta_i}[\log \mathrm{d}P_\theta/\mathrm{d}P_{\theta^0}]_{\theta=\theta^0} = \frac{\partial}{\partial\theta_i}[s_\theta]_{\theta=\theta^0} = \lim_{t\to 0}(s_{\theta^0+te_i} - s_{\theta^0})/t$$

where $e_i = (\overset{(1)}{0}, \ldots, \overset{(i)}{1}, \ldots, \overset{(k)}{0})$. Thus

$$\Gamma(\theta^0, \mathscr{E})_{ij} = \int \lim_{t\to 0}[(s_{\theta^0+te_i} - s_{\theta^0})(s_{\theta^0+te_j} - s_{\theta^0})/t^2]\,\mathrm{d}P_{\theta^0} = \int h_{ij}(\mathrm{d}P_\theta : \theta \in \Theta)$$

where

$$h_{ij}(z) = \lim_{t\to 0}(z_{\theta^0+te_i} - z_{\theta^0})(z_{\theta^0+te_j} - z_{\theta^0})/t^2.$$

Here we shall introduce Fisher information only for differentiable experiments. Assuming that $\mathscr{E}$ is differentiable at $\theta = \theta^0$ we may expand P_θ as

$$P_\theta = P_{\theta^0} + \sum(\theta - \theta^0)_i \dot{P}_{\theta^0,i} + H_{\theta^0,\theta}$$

where $\|H_{\theta^0,\theta}\|/\|\theta - \theta^0\| \to 0$ as $\theta \to \theta^0$. If $P_{\theta^0}(A) = 0$ or $= 1$ then necessarily $\dot{P}_{\theta^0,i}(A) = 0$ so that the measures $\dot{P}_{\theta^0,1}, \ldots, \dot{P}_{\theta^0,k}$ are all absolutely continuous with respect to P_{θ^0}. Putting $f_\theta = \mathrm{d}P_\theta/\mathrm{d}P_{\theta^0}$ we find

$$f_\theta = 1 + \sum(\theta - \theta^0)_i s_{\theta^0,i} + h_{\theta^0,\theta}$$

where $s_{\theta^0,i} = \mathrm{d}\dot{P}_{\theta^0,i}/\mathrm{d}P_{\theta^0}$ and $\int |h_{\theta^0,\theta}|\,\mathrm{d}P_{\theta^0}/\|\theta - \theta^0\| \to 0$ as $\theta \to \theta^0$. As mentioned in example 1.4.2, the local non-asymptotic properties of $\mathscr{E}$ within 'small' neighbourhoods of θ^0 are determined by the distribution $F(\cdot|\theta^0, \mathscr{E})$ of $s_{\theta^0} = (s_{\theta^0,1}, \ldots, s_{\theta^0,k})$ under P_{θ^0}. The distributions $F(\cdot|\theta, \mathscr{E})$ of differentiated log likelihoods enjoy the following properties.

(L.1) $\int x F(\mathrm{d}x|\theta^0, \mathscr{E}) = 0$.

(L.2) If $\mathscr{E}$ is at least as informative as $\mathscr{F}$ then $F(\cdot|\theta^0, \mathscr{E})$ is at least as spread out as $F(\cdot|\theta^0, \mathscr{E})$ around θ^0. Three equivalent ways of making 'at least as spread out as' precise are:

(i) $\int \phi \, dF(\cdot|\theta^0, \mathscr{E}) \geqslant \int \phi \, dF(\cdot|\theta^0, \mathscr{F})$ for all convex functions ϕ on R^k;

(ii) $F(\cdot|\theta^0, \mathscr{E}) = DF(\cdot|\theta^0, \mathscr{F})$ for an expectation preserving Markov kernel (dilation) D;

and

(iii) $P_{\theta^0}T = Q_{\theta^0}$ and $\dot{P}_{\theta^0,i}T = \dot{Q}_{\theta^0,i}; i = 1, \ldots, k$ for a transition T from $L(\mathscr{E})$ to $L(\mathscr{F})$.

(L.3) If $\mathscr{E}_1, \ldots, \mathscr{E}_n$ are differentiable at $\theta = \theta^0$ and if $\mathscr{E} = \mathscr{E}_1 \times \cdots \times \mathscr{E}_n$ then $F(\cdot|\theta^0, \mathscr{E}) = F(\cdot|\theta^0, \mathscr{E}_1) * \cdots * F(\cdot|\theta^0, \mathscr{E}_n)$ where $*$ denotes convolution.

(L.4) If $\mathscr{E}_1, \ldots, \mathscr{E}_r$ are differentiable at $\theta = \theta^0$ and if $\mathscr{E} = \sum_{i=1}^r p_i \mathscr{E}_i$ where $p_1, \ldots, p_r \geqslant 0$ and $p_1 + \cdots + p_r = 1$ then

$$F(\cdot|\theta^0, \mathscr{E}) = \sum_{i=1}^{r} p_i F(\cdot|\theta^0, \mathscr{E}_i).$$

Property (L.1) provides a full characterization of the set of such distributions since, as we shall see in chapter 9, any probability distribution on $\mathbb{R}^k$ having zero expectation is of the form $F(\cdot|\theta^0, \mathscr{E})$ for some differentiable experiment $\mathscr{E}$.

If either of the equivalent conditions (i), (ii) and (iii) of (L.2) are satisfied then we shall say that $\mathscr{E}$ is locally at least as informative as $\mathscr{F}$ at $\theta = \theta^0$. This local ordering is related to a notion of local deficiency in the same way as the global ordering 'being at least as informative as' is related to LeCam's notion of a (global) deficiency.

Condition (L.3) strengthens condition (L.2) when $\mathscr{F}$ is a factor experiment of $\mathscr{E}$. If so then (L.3) tells us that $F(\cdot|\theta^0, \mathscr{E})$ may be obtained from $F(\cdot|\theta^0, \mathscr{F})$ by a convolution kernel.

Conditions (L.2), (L.3) and (L.4) indicate how we may construct functionals which are additive for multiplication or are monotone or are affine under mixture. In fact the covariance matrix of $F(\cdot|\theta^0, \mathscr{E})$ has all three properties provided we restrict attention to those experiments $\mathscr{E}$ which are differentiable at $\theta = \theta^0$ and which also satisfy the condition

$$\int \|x\|^2 F(dx|\theta^0, \mathscr{E}) < \infty.$$

If these conditions are satisfied then we shall define the *Fisher information matrix* $\Gamma(\theta^0, \mathscr{E})$ *of* $\mathscr{E}$ *at* $\theta = \theta^0$ as the covariance matrix of $F(\cdot|\theta^0, \mathscr{E})$. If $\mathscr{E}$ is not differentiable at $\theta = \theta^0$ or if $\int \|x\|^2 F(dx|\theta^0, \mathscr{E}) = \infty$ then the Fisher information matrix will not be defined. Thus in the one-dimensional case we refrain from expressing ourselves in terms of infinite Fisher information.

If $\mathscr{E}$ is differentiable at $\theta = \theta^0$ and if $\mathscr{F}$ is at most as informative as $\mathscr{E}$ then $\mathscr{F}$ is also differentiable at $\theta = \theta^0$ and hence, by property (L.2)

$$\int \langle a, x \rangle^2 F(dx|\theta^0, \mathscr{E}) \leqslant \int \langle a, x \rangle^2 F(dx|\theta^0, \mathscr{F})$$

for all vectors $a \in R^k$. Thus:

(F.1) if $\Gamma(\theta^0, \mathscr{E})$ exists then $\Gamma(\theta^0, \mathscr{F})$ exists whenever $\mathscr{F}$ is at most as informative as $\mathscr{E}$ and then $\Gamma(\theta^0, \mathscr{E}) \geqslant \Gamma(\theta^0, \mathscr{F})$ in the sense that $\Gamma(\theta^0, \mathscr{E}) - \Gamma(\theta^0, \mathscr{F})$ is non-negative definite.

Furthermore, the Fisher information matrix is additive for experiment multiplication, i.e.

(F.2) $\Gamma(\theta^0, \mathscr{E}_1 \times \cdots \times \mathscr{E}_r)$ exists if and only if $\Gamma(\theta^0, \mathscr{E}_1), \ldots, \Gamma(\theta^0, \mathscr{E}_r)$ exist and then $\Gamma(\theta^0, \mathscr{E}_1 \times \cdots \times \mathscr{E}_r) = \Gamma(\theta^0, \mathscr{E}_1) + \cdots + \Gamma(\theta^0, \mathscr{E}_r)$.

Affinity under mixtures may be expressed similarly by

(F.3) if $p_1, \ldots, p_r > 0$ and if $p_1 + \cdots + p_r = 1$ then $\Gamma(\theta^0, p_1\mathscr{E}_1 + \cdots + p_r\mathscr{E}_r)$ exists if and only if $\Gamma(\theta^0, \mathscr{E}_1)$, $\ldots$, $\Gamma(\theta^0, \mathscr{E}_r)$ exist, and then $\Gamma(\theta^0, p_1\mathscr{E}_1 + \cdots + p_r\mathscr{E}_r) = p_1\Gamma(\theta^0, \mathscr{E}_1) + \cdots + p_r\Gamma(\theta^0, \mathscr{E}_r)$.

If $\mathscr{E} = (P_\theta : \theta \in \Theta)$ is differentiable in quadratic mean at $\theta = \theta^0$ then the Fisher information matrix at $\theta = \theta^0$ always exists. The Fisher information matrix may be derived in this case by expanding the affinity for values of the parameter which are close to θ^0. In fact then

$$\gamma(P_\theta, P_\eta) = 1 - \tfrac{1}{8}(\theta - \eta)'\Gamma(\theta^0, \mathscr{E})(\theta - \eta) + o(\|\theta - \theta^0\|^2 + \|\eta - \theta^0\|^2).$$

In particular

$$\gamma(P_\theta, P_{\theta^0}) = 1 - \tfrac{1}{8}(\theta - \theta^0)'\Gamma(\theta^0, \mathscr{E})(\theta - \theta^0) + o(\|\theta - \theta^0\|^2).$$

In order to see this let us consider a second order process $\theta \to Y_\theta$ having product moments given by $EY_{\theta^1}Y_{\theta^2} = \gamma(P_{\theta^1}, P_{\theta^2})$ when $\theta^1, \theta^2 \in \Theta$. By quadratic differentiability Y_θ may be expanded as

$$Y_\theta = Y_{\theta^0} + \Sigma\,(\theta - \theta^0)_i \dot{Y}_{\theta^0, i} + \hat{Y}_\theta$$

where $E(\hat{Y}_\theta)^2/\|\theta - \theta^0\|^2 \to 0$ as $\theta \to \theta^0$. Hence

$$\begin{aligned} 2[1 - \gamma(P_\theta, P_\eta)] &= E(Y_\theta^2 + Y_\eta^2 - 2Y_\theta Y_\eta) = E(Y_\theta - Y_\eta)^2 \\ &= E[\textstyle\sum (\theta - \eta)_i \dot{Y}_{\theta^0, i} + \hat{Y}_\theta - \hat{Y}_\eta]^2 \\ &= \tfrac{1}{4}(\theta - \eta)'\Gamma(\theta^0, \mathscr{E})(\theta - \eta) + o(\|\theta - \theta^0\|^2 + \|\eta - \theta^0\|^2). \end{aligned}$$

Applying this to $\mathscr{E}^n$, for $\theta = \theta^0 + (h_n/n^{1/2})$ and $\eta = \theta^0 + (k_n/n^{1/2})$, where the sequences $\{h_n\}$ and $\{k_n\}$ are bounded, we find that

$$\begin{aligned} \gamma(P_\theta^n, P_\eta^n) = \gamma(P_\theta, P_\eta)^n &= \left[1 - \frac{1}{8n}(h_n - k_n)'\Gamma(\theta^0, \mathscr{E})(h_n - k_n) + \frac{1}{n}o(1)\right]^n \\ &= [\exp(-\tfrac{1}{8}(h_n - k_n)'\Gamma(\theta^0, \mathscr{E})(h_n - k_n))] + o(1). \end{aligned}$$

Now it may be checked that if $\Gamma(\theta^0, \mathscr{E})$ is non-singular then the last bracket is the affinity between the normal distributions $N(h_n, \Gamma(\theta^0, \mathscr{E})^{-1})$ and $N(k_n, \Gamma(\theta^0, \mathscr{E})^{-1})$ (see chapter 8). Thus the above expansion confirms the central limit theorem of LeCam which was mentioned at the end of the previous example.

As shown by LeCam, the assumption of quadratic differentiability at $\theta = \theta^0$ leads to an asymptotic expansion of $\log(\mathrm{d}P_\theta^n/\mathrm{d}P_{\theta^0}^n)$ which may be described as follows.

Let X_1, X_2, ... be independent observations which all realize $\mathscr{E} = (P_\theta : \theta \in \Theta)$. Thus the n-tuple $(X_1, \ldots, X_n)$ realizes $\mathscr{E}^n$. Put $s_{\theta^0, i} = \mathrm{d}\dot{P}_{\theta^0, i}/\mathrm{d}P_{\theta^0}$; $i = 1, \ldots, k$ and $s_{\theta^0} = (s_{\theta^0,1}, \ldots, s_{\theta^0,k})$. Also put $T_n = T_n(X_1, \ldots, X_n) = \sum_{j=1}^n s_{\theta^0}(X_j)/n^{1/2}$. Let $h, h_{(1)}, h_{(2)}, \ldots$ be vectors in $\mathbb{R}^k$ such that $h_{(n)} \to h$ as $n \to \infty$ and put $\theta = \theta^0 + h_{(n)}/n^{1/2}$. Writing $P_{\theta,n}$ instead of P_θ^n and Γ_{θ^0} instead of $\Gamma(\theta^0, \mathscr{E})$ we may expand the log likelihood ratio as

$$\log(\mathrm{d}P_{\theta,n}/\mathrm{d}P_{\theta^0,n}) = h'T_n - \tfrac{1}{2}h'\Gamma_{\theta^0}h + Z_n \tag{1.4.4}$$

where $Z_n \to 0$ in $P_{\theta^0,n}$ probability as $n \to \infty$.

The point is now that important conclusions on the asymptotic behaviour of a sequence $\mathscr{E}_n = (P_{\theta,n} : \theta \in \Theta)$ of experiments may be drawn solely on the basis of (1.4.4). As (1.4.4) does not in any way require that the basic observations are independent and identically distributed, these conclusions do not depend on such assumptions. Here we shall mention two such conclusions.

However let us first restate that we are now dealing with a sequence $\mathscr{E}_n = (P_{\theta,n}; \theta \in \Theta)$; $n = 1, 2, \ldots$ of experiments such that there is a non-negative definite matrix Γ_{θ^0} and sequences $\{T_n\}$ and $\{Z_n\}$ of statistics so that (1.4.4) holds.

The first conclusion states that for each constant $b > 0$ we may truncate T_n and thereby obtain an asymptotically equivalent statistic T_n^* so that $\sup_{\|h\| \leqslant b} \|P_{\theta^0 + h/n^{1/2}, n} - c_n[\exp(h'T_n^*)]P_{\theta^0,n}\| \to 0$, where $c_n = c_n(h)$ is chosen so that $c_n[\exp(h'T_n^*)]P_{\theta^0,n}$ is a probability distribution. Here, as elsewhere, we adhere to the convention that if λ is a measure and g is a function then $g\lambda$ denotes the measure (if it exists) whose Radon–Nikodym derivative with respect to λ is g.

In particular this tells us that $\mathscr{E}_n$ is locally approximable by an exponential family at $\theta = \theta^0$. It follows also that the statistic T_n^*, and thus T_n, is locally asymptotically sufficient in the strong sense that the insufficiencies of these statistics (they are defined at the end of section 1.5) are asymptotically negligible.

The other conclusion we may draw is that for each $b > 0$ the experiment $(P_{\theta^0 + h/n^{1/2}, n} : \|h\| \leqslant b)$ converges according to LeCam's deficiency distance to the linear normal experiment $(N(\Gamma_{\theta^0}h, \Gamma_{\theta^0}) : \|h\| \leqslant b)$. If in addition Γ_{θ^0} is non-singular, then the last experiment may be replaced by the restricted translation experiment $(N(h, \Gamma_{\theta^0}^{-1}) : \|h\| \leqslant b)$. Thus (1.4.4) implies that $\mathscr{E}_n$ is asymptotically locally Gaussian as $n \to \infty$.

In addition to LeCam's own works the reader may also consult Millar (1983), Roussas (1972) and Strasser (1985) for information on asymptotic behaviour of experiments.

Problem 1.4.9. Let Ω be a point not belonging to Θ and for each θ let P_θ assign probabilities $1 - \xi_\theta$ and ξ_θ to θ and Ω respectively. In this case write out explicit expressions for:

(i) the statistical distance and the related measure of association for P_{θ^1} and P_{θ^2};
(ii) the Hellinger distance and the affinity between P_{θ^1} and P_{θ^2};
(iii) the Hellinger transform;
(iv) the Chernoff information number and the Kullback–Leibler information number for $(P_{\theta^1}, P_{\theta^2})$;
(v) $\beta(\alpha|P_1, P_2)$ and $b(\lambda|P_1, P_2)$ when $\Theta = \{1, 2\}$ and $\alpha, \lambda \in [0, 1]$.

Denoting this experiment by $\mathscr{E}_\xi$ show that the experiments $\mathscr{E}_\xi \times \mathscr{E}_\eta$ and $\mathscr{E}_{\xi\eta}$ are 'statistically' equivalent. Use this to exemplify some of the statements in this chapter on replicated experiments.

1.5 Sufficient algebras of events

There are several interesting ways of introducing the concept of sufficiency. The most usual method is the original due to Fisher (1920) which says that a statistic S is sufficient if conditional probabilities given S are themselves statistics, i.e. do not depend on the unknown parameters.

That sufficient statistics really are 'sufficient' for statistical decision problems may then be argued by saying that the original experiment may be recovered from S by using the known random mechanism determined by the conditional distribution, given S.

Relying on decision theory we may introduce sufficiency in terms of operational characteristics. Thus we may say that S is sufficient provided that to each decision rule there corresponds another decision rule depending on S only and having the same operational characteristic.

Closely related is the 'risk definition' of sufficiency which says that S is sufficient provided every obtainable risk function is dominated by a risk function obtainable from S.

According to LeCam's theory on approximate sufficiency, this amounts to saying that the experiment defined by S is 0-deficient with respect to the given experiment.

The appropriate notion of sufficiency within this set up is the notion of pairwise sufficiency, as we shall see in chapter 7.

Finally we would also like to mention that sufficiency may be introduced within a Bayesian context by saying that S is sufficient provided the unknown parameter and the set of observations are independent given S.

Other and equivalent ways of defining conditional independence (i.e. the Markov property) lead to equivalent definitions of Bayesian sufficiency. Thus we may say that S is sufficient if the posterior distribution depends on S only.

The purpose of this section is to provide an exposition of some basic results which may be deduced fairly directly from the definition of sufficiency in terms of conditional probabilities.

The reader should at some point consult some of the classical papers by e.g. Fisher (1922), Halmos & Savage (1949), Lehmann & Scheffé (1950, 1955, 1956), Dynkin (1951), Bahadur (1954, 1955) and Burkholder (1961). An exposition of some of the central ideas in these works may be found in Sverdrup (1966).

Working within the framework of Wald's (1950) treatise, Bahadur established the equivalence of the decision theoretical concept of sufficiency and the 'operational' concept in terms of conditional probabilities. We mention this since the persistence of this equivalence within the theory of comparison of experiments is at the core of the main theme of this work. Indeed one version of it is LeCam's fundamental randomization criterion for deficiency.

We shall avoid the problem of minimal sufficiency of statistics. The theory presented here is adequate provided a minimal sufficient statistic is defined as a statistic whose induced σ-algebra is minimal sufficient. Thus we do not consider all measurable inverse images of sets, only inverse images of measurable sets. The reader may get an impression of the difficulties which may occur from the paper by Landers (1974) and the references there.

The problem of existence and construction of minimal pairwise sufficient σ-algebras is treated in Siebert (1979) and in Ghosh, Morimoto & Yamada (1981).

Another interesting problem, which is not discussed here, is to what extent replicated experiments are reducible by sufficiency. In this connection the exponential (Darmois–Koopman) experiments appear to play an important role. Replicates of these experiments admit 'natural' minimal sufficient statistics having a finite dimension not depending on the number of replicates. A variety of results is available which, under various regularity conditions, imply that homogeneous replicated experiments permitting 'smooth' sufficient statistics having a fixed dimension are exponential. The reader may, e.g., consult the works by Fisher (1934), Darmois (1935), Koopman (1936), Pitman (1936), Dynkin (1951), Barankin & Maitra (1963), Brown (1964), Barndorff-Nielsen & Pedersen (1968), Denny (1969, 1970), Andersen (1970), Pfanzagl (1972) and Hipp (1974).

As we shall mostly be concerned with algebras of events rather than statistics, let us begin by deriving some important facts about classes (algebras) of events described by statistics. Here *statistic* is considered as synonymous with measurable function. If the domain of definition of a measurable function might be interpreted as a sample space then we might prefer to 'describe' this function as a statistic.

Consider a statistic Y from the measurable space $(\mathscr{X}, \mathscr{A})$ to the measurable space $(\mathscr{Y}, \mathscr{B})$. Then the sub σ-algebra $Y^{-1}\mathscr{B} = \{Y^{-1}[B] : B \in \mathscr{B}\}$ of $\mathscr{A}$ is called the *σ-algebra induced by Y.* Another notation for this σ-algebra is $\sigma(Y)$.

If a function Φ on a space containing the range of Y is applied to Y then we obtain the composite function ΦY (or $\Phi(Y)$).

In general a function U is a composed function $U = \Phi(Y)$ with Y as kernel if and only if U induces a partitioning of the common domain $\mathscr{X}$ of U and Y which is coarser than the partitioning induced by Y. If, in addition, Φ is measurable, then U is also measurable and the σ-algebra induced by U is contained in the σ-algebra induced by Y. In fact if Φ is a measurable function from $(\mathscr{Y}, \mathscr{B})$ to some measurable space $(T, \mathscr{S})$ and $U = \Phi Y$ then $U^{-1}\mathscr{S} = Y^{-1}\Phi^{-1}\mathscr{S} \subseteqq Y^{-1}\mathscr{B}$.

The important thing to note is that the converse holds whenever the range space $(T, \mathscr{S})$ of U is Euclidean. This is established in the following theorem.

Theorem 1.5.1 (Reduced statistics). *Let $(\mathscr{X}, \mathscr{A})$, $(\mathscr{Y}, \mathscr{B})$ and $(T, \mathscr{S})$ be measurable spaces and let $Y : (\mathscr{X}, \mathscr{A}) \to (\mathscr{Y}, \mathscr{B})$ and $U : (\mathscr{X}, \mathscr{A}) \to (T, \mathscr{S})$ be measurable functions. Then:*

(i) *if $U = \Phi(Y)$ for a measurable function Φ from $\mathscr{Y}$ to T then $\sigma(U) \subseteqq \sigma(Y)$.*
(ii) *if $(T, \mathscr{S})$ is Euclidean then $\sigma(U) \subseteqq \sigma(Y)$ if and only if $U = \Phi(Y)$ for some measurable function Φ from $\mathscr{Y}$ to T.*
(iii) (ii) *remains true if the condition that $(T, \mathscr{S})$ is Euclidean is replaced by the condition that the range $Y[\mathscr{X}]$ of Y is measurable* (i.e. *belongs to $\mathscr{B}$*) *and that the contours $U^{-1}(t)$; $t \in T$ of U are measurable* (i.e. belong to $\mathscr{A}$).

Proof. We proved (i) above. It therefore remains to prove the 'only if' parts of statements (ii) and (iii). Assume then that $(T, \mathscr{S})$ is Euclidean and that $\sigma(U) \subseteqq \sigma(Y)$. By the Borel isomorphism theorem for Euclidean spaces we may assume that T is the real line and that $\mathscr{S}$ is the class of Borel subset of T. Suppose U is an indicator function of the set $A \in \mathscr{A}$. Then $A = [U = 1] = Y^{-1}[B]$ for some set $B \in \mathscr{B}$ and then $U = I_B(Y)$. It follows that $U = \Phi(Y)$ for some measurable function Φ whenever U is a measurable step function. In the general case $U = \lim_n U_n$ for a sequence $\{U_n\}$ of $\sigma(Y)$ measurable functions. If $U_n = \Phi_n(Y)$ then $U = \Phi(Y)$ where $\Phi = \limsup_n \Phi_n$.

The 'if' part of (iii) is proved by showing that the unique function Φ on the range of Y satisfying $U = \Phi(Y)$ is measurable. □

In particular it follows that real valued (or 'Euclidean' valued) statistics induce the same σ-algebras of events if and only if they are obtainable from each other by measurable transformations.

Let us turn to the definition of sufficiency in terms of conditional probabilities.

Consider an experiment $\mathscr{E} = (\mathscr{X}, \mathscr{A}; P_\theta : \theta \in \Theta)$ and a sub σ-algebra $\mathscr{B}$ of $\mathscr{A}$. We shall then say that *$\mathscr{B}$ is sufficient (for $\mathscr{A}$ or in $\mathscr{E}$)* if for each event A in $\mathscr{A}$ there is a common version of all the conditional probabilities $P_\theta(A|\mathscr{B})$; $\theta \in \Theta$. Thus $\mathscr{B}$ is sufficient if and only if conditional probabilities given $\mathscr{B}$ are coherent in $\mathscr{E}|\mathscr{B}$.

In other words $\mathscr{B}$ is sufficient if and only if to each event A in $\mathscr{A}$ there corresponds a $\mathscr{B}$-measurable variable $\Gamma(A|\cdot)$ such that

$$\int_B \Gamma(A|\cdot)\,\mathrm{d}P_\theta = P_\theta(A \cap B); \qquad B \in \mathscr{B}.$$

In general the class of events A in $\mathscr{A}$ such that $P_\theta(A|\mathscr{B})$ may be specified independently of θ, is a Λ-system.

It follows that it suffices to check that the conditional probability of A given the σ-algebra $\mathscr{B}$ may be specified independently of θ for all events A belonging to a given π-system of events generating $\mathscr{A}$.

The reader is referred to the appendix to this chapter (section 1.6) for the basic interplay theorem for λ-systems and π-systems.

Furthermore, if $\mathscr{B}$ is sufficient then conditional expectations of bounded (or, say, non-negative) random variables given $\mathscr{B}$ may be specified independently of θ.

Note by the way that sufficiency is a 'set property' of experiments in the sense that whether $\mathscr{B}$ is sufficient depends solely on the set $\{P_\theta : \theta \in \Theta\}$. Thus we might, as several authors do, discuss sufficiency with respect to *sets* of probability measures rather than with respect to *families* of probability measures.

A statistic (i.e. a measurable function) from the sample space $(\mathscr{X}, \mathscr{A})$ of $\mathscr{E}$ to the measurable space $(\mathscr{Y}, \mathscr{B})$ is called *sufficient* if the σ-algebra $\sigma(Y)$ induced by Y is sufficient.

It is readily checked that Y is sufficient if and only if to each event A in $\mathscr{A}$ there corresponds a measurable function $Q(A|\cdot)$ on $\mathscr{Y}$ which is a mixed conditional probability of A given Y for all θ, i.e.

$$\int_B Q(A|\cdot)\,\mathrm{d}P_\theta Y^{-1} = P_\theta(A \cap (Y \in B))$$

for all sets B in $\mathscr{B}$.

Problem 1.5.2 (Sufficiency in sampling models). Verify that the statistic (U, G) in example 1.2.1 actually is sufficient.

This problem is discussed further in example 7.3.5 where it is shown that the σ-algebra induced by (U, G) is minimal sufficient provided the set Θ of functions on Π is not too small.

Problem 1.5.3 (Sufficient equivalence relations in discrete experiments). Let $\mathscr{E} = (\mathscr{X}, \mathscr{A}; P_\theta : \theta \in \Theta)$ be a discrete experiment.

Let '$\sim$' be an equivalence relation on $\mathscr{X}$ such that whenever $x \sim y$ then there is a positive constant k such that $P_\theta(x) \equiv_\theta kP_\theta(y)$. Let $\bar{x}$ denote the equivalence class containing x.

Let $\mathscr{B}$ be the σ-algebra of events A in $\mathscr{A}$ which are saturated with respect to this equivalence relation.

(i) Show that $\mathscr{B}$ is sufficient by showing that $P_\theta(A|\bar{x})$ does not depend on θ provided x belongs to the minimal support of P_θ.
(ii) Apply this to the previous problem.

If sample points x and y are equivalent if and only if $P_.(x)$ and $P_.(y)$ are positively proportional, then the σ-algebra $\mathscr{B}$ is minimal sufficient (see example 7.3.4).

A treatment of sufficiency in discrete experiments may be found in Morimoto (1972).

Assume now that $\mathscr{B}$ is sufficient and that $\Gamma(A|\cdot)$, for each event A in $\mathscr{A}$, is a common version of $P_\theta(A|\mathscr{B})$; $\theta \in \Theta$. Then Γ is an '*almost randomization*' in the sense that:

(i) $0 \leqslant \Gamma(A|\cdot) \leqslant 1$ a.s. P_θ when $A \in \mathscr{A}$ and $\theta \in \Theta$;
(ii) $\Gamma(\varnothing|\cdot) = 0$ a.s. $P_\theta : \theta \in \Theta$;
(iii) $\Gamma(A_1 \cup A_2 \cup \cdots|\cdot) = \Gamma(A_1|\cdot) + \Gamma(A_2|\cdot) + \cdots$ a.s. $P_\theta : \theta \in \Theta$ for each sequence $A_1, A_2, \ldots$ of disjoint events in $\mathscr{A}$.

If $\mathscr{A}$ is generated by a countable partition of $\mathscr{X}$ then Γ may be regularized so that the qualification a.s. P_θ may be deleted in all these statements. If $\mathscr{X}$ is the real line equipped with its class $\mathscr{A}$ of Borel subsets then, following the standard procedure found in most probability textbooks, we may first obtain a conditional cumulative distribution function given $\mathscr{B}$ and then define conditional probabilities by this conditional distribution function. Thus we may delete the qualification 'a.e. P_θ' in this case also.

This extends to Euclidean sample spaces $(\mathscr{X}, \mathscr{A})$ and, more generally, to the situation where $\mathscr{A}$ as a σ-algebra is 'σ-isomorphic' to the σ-algebra of Borel subsets of a Borel subset of the real line.

It follows that under these circumstances the conditional probabilities given $\mathscr{B}$ may be given by a randomization (Markov kernel) Γ from the sample space $(\mathscr{X}, \mathscr{B})$ of $\mathscr{E}|\mathscr{B}$ to the sample space $(\mathscr{X}, \mathscr{A})$ of $\mathscr{E}$. The experiment $\mathscr{E}$ may thus be recovered from the sub-experiment $\mathscr{E}|\mathscr{B}$ by using the known chance mechanism

Γ i.e.

$$\mathscr{E} = (\mathscr{E}|\mathscr{B})\Gamma.$$

In fact $[(P_\theta|\mathscr{B})\Gamma](A) = \int \Gamma(A|\cdot)d(P_\theta|\mathscr{B}) = \int P_\theta(A|\mathscr{B})\,\mathrm{d}P_\theta = P_\theta(A)$ when $\theta \in \Theta$ and $A \in \mathscr{A}$.

Let us consider the implications of this for statistical decision problems.

In general a statistical decision problem involves, in addition to an experiment $\mathscr{E}$, both a decision space and a family $(L_\theta : \theta \in \Theta)$ of loss functions on the decision space. If the *decision space* is the measurable space $(T, \mathscr{S})$ then the *loss functions* are extended real valued $\mathscr{S}$ measurable functions on T. The value $L_\theta(t)$ may be interpreted as the loss incurred by taking decision $t \in T$ when θ prevails.

Assuming that $\mathscr{E}$ is realized by the observation(s) X the statistical decision problem may be described as follows.

Find a rule which for each possible value x of X tells us which decision t in T should be made if $X = x$ is observed. Here we shall greatly simplify the mathematics by allowing for the possibility that x only determines the chance mechanism producing the decision $t \in T$. Thus we arrive at the notion of a *decision rule* as a randomization from the sample space $(\mathscr{X}, \mathscr{A})$ of $\mathscr{E}$ to the decision space $(T, \mathscr{S})$.

If decisions are chosen according to the decision rule ρ, then the probability of choosing a decision in a measurable set S of decisions is $\int \rho(S|\cdot)\,\mathrm{d}P_\theta$. Thus the distribution of the chosen distribution is $P_\theta\rho$ when θ prevails.

The map which to each theory θ in Θ assigns the probability distribution $P_\theta\rho$ of the chosen decision when θ prevails and ρ is used, is called the *performance function* (*operational characteristic*) *of the decision rule* ρ.

Furthermore, if loss functions $L_\theta : \theta \in \Theta$ on $(T, \mathscr{S})$ are involved, then the expected losses, i.e. risks, are obtained by integrating with respect to the distributions on the decision space given by the performance function. Thus the *risk under* θ *incurred by using the decision rule* ρ is $E_{P_{\theta\rho}} L_\theta = \int L_\theta(t)(P_\theta\rho)(\mathrm{d}t)$ provided this integral exists. If L_θ is bounded from below or if L_θ is bounded then we may also write this risk as $\int [\int L_\theta(t)\rho(\mathrm{d}t|x)] P_\theta(\mathrm{d}x)$ or just as $P_\theta\rho L_\theta$.

It follows that *decision rules possessing identical performance functions also possess identical risk functions regardless of how the losses are specified*, provided of course that the risks are well defined. This implies that the performance function is a more fundamental quantity than the risk function.

If an experiment $\mathscr{F}$ is obtained from the experiment $\mathscr{E}$ by using a known chance mechanism, then of course $\mathscr{F}$ should be considered as being at most as informative as $\mathscr{E}$. In particular one would expect that every decision rule in $\mathscr{F}$ may be parried by a decision rule in $\mathscr{E}$ not yielding greater risk. In fact everything achievable in $\mathscr{F}$ ought also to be achievable in $\mathscr{E}$. Thus we should be able to conclude that any performance function obtainable in $\mathscr{F}$ is also obtainable in $\mathscr{E}$.

That this is indeed so is mathematically quite simple to see. Assuming that $\mathscr{E} = (\mathscr{X}, \mathscr{A}; P_\theta : \theta \in \Theta)$ and that $\mathscr{F} = (\mathscr{Y}, \mathscr{B}; Q_\theta : \theta \in \Theta)$ where $Q_\theta \equiv_\theta P_\theta M$ for a randomization M we find that the decision rule σ in $\mathscr{F}$ possesses the same performance function as the decision rule $\rho = M\sigma$ in $\mathscr{E}$ i.e. $P_\theta \rho = P_\theta M \sigma = Q_\theta$; $\theta \in \Theta$.

It follows that risk functions in $\mathscr{F}$ are also risk functions in $\mathscr{E}$, regardless of how losses are specified.

Note that performance functions are also experiments. Choosing the identity function as the decision rule we find that among the performance functions which are obtainable in $\mathscr{E}$ is the experiment $\mathscr{E}$ itself. It follows that *an experiment $\mathscr{F}$ is obtainable from $\mathscr{E}$ by a randomization if and only if $\mathscr{F}$ is a performance function in $\mathscr{E}$.*

In chapter 6 we shall see that if we admit transitions as legitimate chance mechanisms, then $\mathscr{F}$ is at most as informative as $\mathscr{E}$ if and only if $\mathscr{F}$ is a performance function in $\mathscr{E}$. This is a particular case of LeCam's randomization criterion for comparison of experiments.

As we have seen, these considerations apply in particular to the situation where $\mathscr{B}$ is a sufficient sub σ-algebra of $\mathscr{A}$ for the experiment $\mathscr{E} = (\mathscr{X}, \mathscr{A}; P_\theta : \theta \in \Theta)$ provided the common conditional probabilities given $\mathscr{B}$ may be regularized. However the last assumption is not needed at all when the decision space $(T, \mathscr{S})$ is Euclidean.

In fact we may then proceed as follows.

Let ρ be a decision rule from $(\mathscr{X}, \mathscr{A})$ to $(T, \mathscr{S})$. For each set $S \in \mathscr{S}$ choose a version $\tilde{\rho}(S|\cdot)$ of $E_\theta(\rho(S|\cdot)|\mathscr{B})$ which does not depend on θ. Then regularize $\tilde{\rho}$ according to the usual procedure. (We may assume that $\mathscr{S}$ is the Borel class on the real line.) If $S \in \mathscr{S}$ then $\tilde{\rho}(S|\cdot)$ is still a version of $E_\theta(\rho(S|\cdot)|\mathscr{B})$ so that $E_\theta \tilde{\rho}(S|\cdot) = E_\theta \rho(S|\cdot)$. Thus $\tilde{\rho}$ and ρ have the same performance function.

This completes the proof of the following theorem.

Theorem 1.5.4 (Sufficiency of sufficiency). *Let $\mathscr{B}$ be a sufficient sub σ-algebra of $\mathscr{A}$ for the experiment $\mathscr{E} = (\mathscr{X}, \mathscr{A}; P_\theta : \theta \in \Theta)$.*

Consider also a decision space $(T, \mathscr{S})$.

Suppose that the sample space $(\mathscr{X}, \mathscr{A})$ or the decision space $(T, \mathscr{S})$ is Euclidean.

Then to each decision rule σ in $\mathscr{E}$ there corresponds a decision rule ρ in $\mathscr{E}|\mathscr{B}$ possessing the same performance function as σ. In fact ρ may be chosen so that $\rho(S|\cdot)$ is a common version of $E_\theta(\sigma(S|\cdot)|\mathscr{B})$; $\theta \in \Theta$.

Having argued the 'sufficiency' of sufficient σ-algebras of events (following Bahadur (1954)), we shall now derive a few important features of sufficiency which follow fairly directly from the definition.

From here on we shall consider a fixed experiment $\mathscr{E} = (\mathscr{X}, \mathscr{A}; P_\theta : \theta \in \Theta)$ and sub σ-algebras $\mathscr{B}, \mathscr{C}, \ldots$ of $\mathscr{A}$.

Let us begin by making the trivial observation that if $\mathscr{B}$ is sufficient then $\mathscr{B}$ is also sufficient with respect to the experiment obtained from $\mathscr{E}$ by restricting θ to a non-empty subset Θ_0 of Θ. In particular if $\mathscr{B}$ is sufficient then $\mathscr{B}$ is sufficient for every dichotomy of the form $(P_{\theta_1}, P_{\theta_2})$ where $\theta_1, \theta_2 \in \Theta$. What is interesting is that under regularity conditions this may be turned around. This is all the more interesting since it is this kind of sufficiency which, as we shall see, is the appropriate kind within the theory of comparison of experiments.

To describe this situation we say that a sub σ-algebra $\mathscr{B}$ of $\mathscr{A}$ is *pairwise sufficient* if $\mathscr{B}$ is sufficient for each pair $(P_{\theta_1}, P_{\theta_2})$ where $\theta_1, \theta_2 \in \Theta$. Thus $\mathscr{B}$ is pairwise sufficient if and only if conditional probabilities given $\mathscr{B}$ are consistent according to the definition of consistency given in section 1.2. If, in addition, $\mathscr{E}|\mathscr{B}$ is coherent, then theorem 1.2.9 implies that $\mathscr{B}$ actually is sufficient.

Theorem 1.5.5 (Pairwise sufficiency and sufficiency). *Any sufficient sub σ-algebra is pairwise sufficient.*

If $\mathscr{E}|\mathscr{B}$ is coherent then $\mathscr{B}$ is sufficient if and only if $\mathscr{B}$ is pairwise sufficient.

Some of the essential features of sufficiency in the dominated case are collected in the following theorem.

Theorem 1.5.6 (Sufficiency in the dominated case). *Assume that the experiment $\mathscr{E} = (\mathscr{X}, \mathscr{A}; P_\theta : \theta \in \Theta)$ is dominated by the non-negative σ-finite measure μ.*

Let κ be a prior probability distribution on Θ with countable support such that the probability measure $P_\kappa = \sum_\theta \kappa(\theta) P_\theta$ dominates $\mathscr{E}$ i.e. $P_\theta(A) \equiv_\theta 0$ if and only if $P_\theta(A) = 0$ when $\kappa(\theta) > 0$.

Then the following conditions are equivalent for a sub σ-algebra $\mathscr{B}$ of $\mathscr{A}$.

(i) *$\mathscr{B}$ is sufficient.*

(ii) *$\mathscr{B}$ is pairwise sufficient.*

(iii) $\mathrm{d}P_\theta/\mathrm{d}\mu$ *may be specified as a product hg_θ for each $\theta \in \Theta$, where g_θ is $\mathscr{B}$-measurable and where h does not depend on θ.*

(iv) $\mathrm{d}P_\theta/\mathrm{d}P_\kappa$ *may be specified $\mathscr{B}$-measurable for each θ.*

(v) *$P_\kappa(A|\mathscr{B})$ is a version of $P_\theta(A|\mathscr{B})$ for each θ.*

(vi) *$\mathscr{L}(\mathrm{d}P_\theta/\mathrm{d}P_\kappa : \theta \in \Theta | P_\kappa) = \mathscr{L}(\mathrm{d}Q_\theta/\mathrm{d}Q_\kappa : \theta \in \Theta | Q_\kappa)$ where $Q_\theta \equiv_\theta P_\theta|\mathscr{B}$ and $Q_\kappa = \sum_\theta \kappa(\theta) Q_\theta = P_\kappa|\mathscr{B}$.*

(vii) *If h is a non-negatively homogeneous measurable function on the likelihood space then $\int h(\mathrm{d}P_\theta : \theta \in \Theta) = \int h(\mathrm{d}(P_\theta|\mathscr{B}) : \theta \in \Theta)$ in the sense that the left hand side exists if and only if the right hand side exists, and if so then equality holds.*

(viii) $\gamma(P_{\theta_1}, P_{\theta_2}) = \gamma(P_{\theta_1}|\mathscr{B}, P_{\theta_2}|\mathscr{B})$; $\theta_1, \theta_2 \in \Theta$.

(ix) *To any (non-randomized) test function there corresponds a $\mathscr{B}$-measurable test function having the same power function.*

Remark 1. Condition (iii) is the *factorization criterion for sufficiency*. It was stated by Fisher (1922) and later by Neyman (1935). The validity of the criterion for general dominated experiments was established in Bahadur (1954). This improved the factorization criterion in Halmos & Savage (1949) by removing unnecessary integrability conditions. Extensions to experiments which are majorized by given measures which are not necessarily dominated may be found in Ghosh, Morimoto & Yamada (1981).

Criterion (iv) may also be found in Bahadur (1954), and this also was an improvement of a similar result in the paper by Halmos & Savage.

The equivalence of sufficiency and pairwise sufficiency in dominated experiments was however fully established in Halmos & Savage's paper.

Remark 2. Criterion (viii) may be found in LeCam, 1986. This condition states that the Hellinger distances $d(P_{\theta_1}, P_{\theta_2})$ and $d(Q_{\theta_1}, Q_{\theta_2})$ are equal whenever $\theta_1, \theta_2 \in \Theta$. If so then, by (vii), the statistical distances $\|P_{\theta_1} - P_{\theta_2}\|$ and $\|Q_{\theta_1} - Q_{\theta_2}\|$ are also equal whenever $\theta_1, \theta_2 \in \Theta$. However this is not sufficient for sufficiency. The reader is invited to provide a counterexample.

If we try to argue as in the proof of criterion (viii) for sufficiency with $\gamma(P_{\theta_1}, P_{\theta_2})$ and $\gamma(P_{\theta_1}|\mathscr{B}, P_{\theta_2}|\mathscr{B})$ replaced by $\|P_{\theta_1} \wedge P_{\theta_2}\|$ and $\|(P_{\theta_1}|\mathscr{B}) \wedge (P_{\theta_2}|\mathscr{B})\|$ respectively, then we see that the reason why the proof works for Hellinger distance but not for statistical distance is that $[x(1-x)]^{1/2}$ is strictly concave in $x \in]0,1[$, while $x \wedge (1-x)$ is merely concave in $x \in]0,1[$.

Remark 3. Criterion (ix) implies in particular that if the experiment $\mathscr{E}$ is dominated and has Euclidean sample space, then $\mathscr{E}$ is obtainable from the sub-experiment $\mathscr{E}|\mathscr{B}$ by a randomization if and only if $\mathscr{B}$ is sufficient.

Proof. We know by theorem 1.5.5 that conditions (i) and (ii) are equivalent. If condition (iii) is satisfied with g_θ and h as described and if $g_\kappa = \sum \kappa(\theta) g_\theta$, then $\mathrm{d}P_\theta/\mathrm{d}P_\kappa = g_\theta/g_\kappa$ so that (iv) holds. On the other hand (iv) $\Rightarrow$ (iii) since $\mathrm{d}P_\theta/\mathrm{d}P_\mu = (\mathrm{d}P_\kappa/\mathrm{d}\mu)(\mathrm{d}P_\theta/\mathrm{d}P_\kappa)$. Thus conditions (iii) and (iv) are also equivalent.

Assume that condition (i) is satisfied. Put $\phi_\theta = \mathrm{d}P_\theta/\mathrm{d}P_\kappa$ and $\bar{\phi}_\theta = E(\phi|\mathscr{B})$. If $\Gamma(A|\cdot)$ is a version of $P_\theta(A|\mathscr{B})$ for all θ then

$$\int_A \bar{\phi}_\theta \,\mathrm{d}P_\kappa = \sum_{\theta'} \kappa(\theta') \int_A \bar{\phi}_\theta \,\mathrm{d}P_{\theta'} = \sum_{\theta'} \kappa(\theta') \int \Gamma(A|\cdot)\bar{\phi}_\theta \,\mathrm{d}P_{\theta'} = \int \Gamma(A|\cdot)\bar{\phi}_\theta \,\mathrm{d}P_\kappa$$

$$= \int \Gamma(A|\cdot)\phi_\theta \,\mathrm{d}P_\kappa = \int \Gamma(A|\cdot)\,\mathrm{d}P_\theta = P_\theta(A).$$

It follows that $\bar{\phi}_\theta = \mathrm{d}P_\theta/\mathrm{d}P_\kappa$. Thus (i) $\Rightarrow$ (iv).

If (iv) holds and if A and B are events such that $A \in \mathscr{A}$ and $B \in \mathscr{B}$ then

$$\int_B P_\kappa(A|\mathscr{B})\,dP_\theta = \int_B P_\kappa(A|\mathscr{B})[dP_\theta/dP_\kappa]\,dP_\kappa = \int_B E_\kappa(I_A\,dP_\theta/dP_\kappa|\mathscr{B})\,dP_\kappa$$

$$= \int_B [I_A\,dP_\theta/dP_\kappa]\,dP_\kappa = P_\theta(A \cap B).$$

Hence $P_\kappa(A|\mathscr{B})$ is a version of $P_\theta(A|\mathscr{B})$ for each $A \in \mathscr{A}$ and each $\theta \in \Theta$. Thus (iv) $\Rightarrow$ (v).

Noting that the implication (v) $\Rightarrow$ (i) is trivial, we see that the implications (v) $\Rightarrow$ (i) $\Leftrightarrow$ (ii) $\Rightarrow$ (iii) $\Leftrightarrow$ (iv) $\Rightarrow$ (v) are all verified.

It follows that the first five conditions are equivalent.

In the remainder of the proof let us use the notation Q_θ and Q_κ for the restricted measures $P_\theta|\mathscr{B}$ and $P_\kappa|\mathscr{B}$ respectively.

If $\mathscr{B}$ is sufficient then by (iv) dP_θ/dP_κ may be specified $\mathscr{B}$-measurable and then $dQ_\theta/dQ_\kappa = E_\kappa(dP_\theta/dP_\kappa|\mathscr{B}) = dP_\theta/dP_\kappa$ a.s. P_κ. It follows that condition (vi) is satisfied when $\mathscr{B}$ is sufficient. Furthermore the implication (vi) $\Rightarrow$ (vii) follows directly from the definition of the integrals involved, while the implication (vii) $\Rightarrow$ (viii) follows by specialization. If δ is a test function and if $\mathscr{B}$ is sufficient then there is a $\mathscr{B}$-measurable function $\tilde{\delta}$ which is a common version of $E_\theta(\delta|\mathscr{B})$; $\theta \in \Theta$. Then $E_\theta\delta \equiv_\theta E_\theta\tilde{\delta}$ so that δ and $\tilde{\delta}$ have the same power functions.

We have so far argued the implications (ix) $\Leftarrow$ (i–v) $\Rightarrow$ (iv) $\Rightarrow$ (vi) $\Rightarrow$ (vii) $\Rightarrow$ (viii). It remains to show that (ix) $\Rightarrow$ (viii) $\Rightarrow$ (i–v).

If (ix) is satisfied then

$$\begin{aligned}\|(1-\lambda)P_{\theta_1} \wedge \lambda P_{\theta_2}\| &= \inf\{(1-\lambda)E_{\theta_1}\delta + \lambda E_{\theta_2}(1-\delta) : 0 \leqslant \delta \leqslant 1\}\\ &= \inf\{(1-\lambda)P_{\theta_1}(A) + \lambda P_{\theta_2}(A^c) : A \in \mathscr{A}\}\\ &= \|(1-\lambda)Q_{\theta_1} \wedge \lambda Q_{\theta_2}\|\end{aligned}$$

when $0 \leqslant \lambda \leqslant 1$. Hence, by the relationship between the Hellinger transform and minimum Bayes risks which we found in example 1.4.5, $\gamma(P_{\theta_1}, P_{\theta_2}) = \gamma(Q_{\theta_1}, Q_{\theta_2})$. Hence (ix) $\Rightarrow$ (viii).

Finally, assume that $\gamma(Q_{\theta_1}, Q_{\theta_2}) = \gamma(P_{\theta_1}, P_{\theta_2})$. Put $P = \frac{1}{2}(P_{\theta_1} + P_{\theta_2})$, $Q = \frac{1}{2}(Q_{\theta_1} + Q_{\theta_2})$, $2g = dP_{\theta_2}/dP$ and $2h = dQ_{\theta_2}/Q$ where we may assume that $0 \leqslant g \leqslant 1$ and $0 \leqslant h \leqslant 1$. Let $M(\cdot|\cdot)$ be a regular version of the conditional distribution of g given $\mathscr{B}$ under P. Then $dP_{\theta_1}/dP = 2(1-g)$ and $dQ_{\theta_1}/dP = 2(1-h)$. If $B \in \mathscr{B}$ then $\int_B h\,dP = \frac{1}{2}Q_{\theta_2}(B) = \frac{1}{2}P_{\theta_2}(B) = \int_B g\,dP$ so that $h = E_P(g|\mathscr{B})$. Thus we can assume that $h(x) \equiv_x \int zM(dz|x)$. Putting $\phi(z) = z^{1/2}(1-z)^{1/2}$ we find successively that

$$0 = \frac{1}{2}\gamma(P_{\theta_1}, P_{\theta_2}) - \frac{1}{2}\gamma(Q_{\theta_1}, Q_{\theta_2}) = \int \phi(g)\,\mathrm{d}P - \int \phi(h)\,\mathrm{d}Q$$
$$= E_P[\phi(g) - \phi(h)] = E_P[E_P(\phi(g)|\mathscr{B}) - \phi(E_P(g|\mathscr{B}))]$$
$$= \int \left[\int \phi(z) M(\mathrm{d}z|x) - \phi\left(\int z M(\mathrm{d}z|x) \right) \right] P(\mathrm{d}x)$$

where the last bracket is non-negative by Jensen's inequality. Thus $\int \phi(z) M(\mathrm{d}z|x) = \phi(\int z M(\mathrm{d}z|x))$ when $x \notin N$ where $P(N) = 0$. Since ϕ is strictly concave on $]0, 1[$, it follows that $M(\cdot|x)$ is the one-point distribution in $\int z M(\mathrm{d}z|x) = h(x)$ when $x \notin N$.

Thus $E_p|g - h| = E_P E_P(|g - h||\mathscr{B}) = \int [\int |z - h(x)| M(\mathrm{d}z|x)] P(\mathrm{d}x) = 0$ so that $g = h$ a.e. P. It follows that the Radon–Nikodym derivative $2g$ may be specified $\mathscr{B}$-measurable. Thus, by the equivalence of conditions (i) and (iv), the σ-algebra $\mathscr{B}$ is pairwise sufficient. □

An important consequence of this theorem is that the functionals of the form $h \to \int h(\mathrm{d}\mathscr{E})$ which we defined in section 1.2 is not altered by reduction by pairwise sufficiency, and therefore is not altered by sufficiency reductions. This indicates that it is the notion of pairwise sufficiency which is the appropriate notion of sufficiency within the theory of comparison of experiments. Here is another indication.

Example 1.5.7 (Pairwise sufficiency does not imply sufficiency). Consider again the totally informative and non-coherent experiment $\mathscr{E}$ in example 1.2.2.

Let $\mathscr{B}$ be the σ-algebra consisting of the Borel subsets of $[0, 1]$ which are either countable or have countable complements in $[0, 1]$. Then any pair $(P_{\theta_1}, P_{\theta_2})$ of distinct probability measures may be separated by $\mathscr{B}$. Thus the sub-experiment $\mathscr{E}|\mathscr{B}$ is also totally informative. In spite of this the σ-algebra $\mathscr{B}$ is not sufficient. In fact there is no common version of the conditional probabilities $P_\theta([0, \frac{1}{2}[|\mathscr{B}) : \theta \in [0, 1])$.

On the other hand it is easily checked that if $\mathscr{E} = (\mathscr{X}, \mathscr{A}; P_\theta : \theta \in \Theta)$ is totally informative, then a sub σ-algebra $\mathscr{B}$ of $\mathscr{A}$ is pairwise sufficient if and only if $\mathscr{E}|\mathscr{B}$ is also totally informative.

Some immediate consequences of theorem 1.5.6 concerning pairwise sufficiency are stated in the following corollary.

Corollary 1.5.8 (Pairwise sufficiency). *Given an experiment $\mathscr{E} = (\mathscr{X}, \mathscr{A}; P_\theta : \theta \in \Theta)$ the following conditions are equivalent for a sub σ-algebra $\mathscr{B}$ of $\mathscr{A}$:*

(i) *$\mathscr{B}$ is pairwise sufficient.*
(ii) *$\mathscr{B}$ is sufficient in the restricted experiment $\mathscr{E}|\Theta_0$ whenever Θ_0 is a subset of Θ such that $\mathscr{E}|\Theta_0$ is dominated.*

(iii) *if h is a non-negatively homogeneous measurable function on the likelihood space then* $\int h(\mathrm{d}P_\theta : \theta \in \Theta) = \int h(\mathrm{d}(P_\theta|\mathscr{B}) : \theta \in \Theta)$ *in the sense that the left hand side exists if and only if the right hand side exists and if so then equality holds.*

(iv) $\gamma(P_{\theta_1}, P_{\theta_2}) = \gamma(P_{\theta_1}|\mathscr{B}, P_{\theta_2}|\mathscr{B})$; $\theta_1, \theta_2 \in \Theta$.

The inclusion ordering, even in regular cases, is usually too fine to permit the existence of a smallest sufficient σ-algebra. However there is another closely related and interesting ordering which permits the existence of smallest sufficient σ-algebras in coherent experiments. We shall now describe this ordering.

Let $\mathscr{E} = (\mathscr{X}, \mathscr{A}; P_\theta : \theta \in \Theta)$ be an experiment and consider sub σ-algebras $\mathscr{B}$ and $\mathscr{C}$ of $\mathscr{A}$. Then we shall say that $\mathscr{B}$ *is $\mathscr{E}$-essentially contained in* $\mathscr{C}$ if to each event B in $\mathscr{B}$ there corresponds an event $C \in \mathscr{C}$ such that $P_\theta(B \,\Delta\, C) \equiv_\theta 0$.

Here $S \,\Delta\, T$ for sets S and T denotes their *symmetric difference* $(S - T) \cup (T - S)$. Thus $I_{S\Delta T} = |I_S - I_T|$.

If $\mathscr{B}$ is $\mathscr{E}$-essentially contained in $\mathscr{C}$ then we may denote this by writing $\mathscr{B} \leqslant \mathscr{C}$. It is easily checked that the relation '$\leqslant$' for sub σ-algebras of $\mathscr{A}$ is a partial ordering.

Furthermore $\mathscr{B} \leqslant \mathscr{C}$ if and only if to each $\mathscr{B}$-measurable extended real valued function g there corresponds a $\mathscr{C}$-measurable real valued function h such that $g = h$ a.s. P_θ for all $\theta \in \Theta$.

It is convenient to express this ordering in terms of closures for the pseudo-metric $A_1, A_2 \to \sup_\theta P_\theta(A_1 \,\Delta\, A_2)$ on $\mathscr{A}$. If A belongs to the closure of $\mathscr{B}$ for this pseudo-metric then $\sup_\theta P_\theta(A \,\Delta\, B_n) \to 0$ for some sequence $B_1, B_2, \ldots$ in $\mathscr{B}$. We may always assume that convergence is fast enough to ensure that $\sum_n \sup_\theta P_\theta(A \,\Delta\, B_n) \to 0$ and then $I_{B_n} \to I_A$ a.e. P_θ for all θ. Thus $P_\theta(A \,\Delta\, (\limsup_n B_n)) \equiv_\theta 0$ so that $A \,\Delta\, B$ is an $\mathscr{E}$ null set for some set B in $\mathscr{B}$.

Thus the closure of the sub σ-algebra $\mathscr{B}$ of $\mathscr{A}$ for this pseudo-metric consists precisely of the events A in $\mathscr{A}$ such that $P_\theta(A \,\Delta\, B) \equiv_\theta 0$ for some event B in $\mathscr{B}$. We shall call this class the (strong) closure of $\mathscr{B}$ and denote it as $\bar{\mathscr{B}}$. It is then easily checked that $\bar{\mathscr{B}}$ is also a sub σ-algebra of $\mathscr{A}$. Furthermore the ordering $\leqslant$ for σ-algebras reduces to the inclusion ordering $\subseteqq$ for closed σ-algebras i.e. $\mathscr{B} \leqslant \mathscr{C}$ if and only if $\bar{\mathscr{B}} \subseteqq \bar{\mathscr{C}}$.

We shall say that σ-algebras $\mathscr{B}$ *and* $\mathscr{C}$ *are $\mathscr{E}$-equivalent* if they are equivalent for the ordering '$\leqslant$' i.e. if $\mathscr{B} \leqslant \mathscr{C}$ and $\mathscr{C} \leqslant \mathscr{B}$. If $\mathscr{B}$ and $\mathscr{C}$ are $\mathscr{E}$-equivalent then we may write this '$\mathscr{B} \sim \mathscr{C}$'.

It follows that $\mathscr{B}$ and $\mathscr{C}$ are $\mathscr{E}$-equivalent if and only if the closures $\bar{\mathscr{B}}$ and $\bar{\mathscr{C}}$ are equal.

Note next that sufficiency respects this equivalence relation for σ-algebras i.e. if $\mathscr{B} \sim \mathscr{C}$ and if $\mathscr{C}$ is sufficient then $\mathscr{B}$ is also sufficient. In other words $\mathscr{B}$ is sufficient if and only if $\bar{\mathscr{B}}$ is sufficient.

A sub σ-algebra $\mathscr{B}$ of $\mathscr{C}$ is called *minimal sufficient* if it is sufficient and if it is $\mathscr{E}$-essentially contained in any other sufficient σ-algebra $\mathscr{C}$. Also the notion of minimal sufficiency respects equivalence i.e. $\mathscr{B}$ is minimal sufficient if and only if $\bar{\mathscr{B}}$ is the smallest closed sufficient σ-algebra.

If the σ-algebra $\mathscr{A}$ itself is minimal sufficient then we shall say that *$\mathscr{E}$ is minimal sufficient.*

The ordering '$\leqslant$' for sufficient σ-algebras has the interesting property that local minima are also global minima, i.e. $\mathscr{B}$ is minimal sufficient if and only if $\mathscr{B}$ is minimal for the ordering $\leqslant$ for sufficient σ-algebras. In other words a sufficient σ-algebra $\mathscr{B}$ is minimal sufficient if and only if it is *irreducible by sufficiency* i.e. $\mathscr{C} \sim \mathscr{B}$ whenever $\mathscr{C} \leqslant \mathscr{B}$ and $\mathscr{C}$ is sufficient.

This was shown by Burkholder (1961). In the same paper it was also shown that minimal sufficient σ-algebras need not exist.

In spite of this we shall see in chapter 7 that minimal sufficient σ-algebras exist whenever $\mathscr{E}$ is coherent. Here we shall content ourselves with establishing this result due to Bahadur (1954) for dominated experiments.

Theorem 1.5.9 (Minimal sufficiency in the dominated case). *Let $\mathscr{E} = (\mathscr{X}, \mathscr{A}; P_\theta : \theta \in \Theta)$ be a dominated experiment. Choose a prior distribution κ on Θ having countable support such that the probability measure $P_\kappa = \sum \kappa(\theta) P_\theta$ dominates $\mathscr{E}$.*

Then any σ-algebra $\mathscr{B}$ generated by versions $\mathrm{d}P_\theta/\mathrm{d}P_\kappa$; $\theta \in \Theta$ is minimal sufficient.

Proof. Choose versions $g_\theta = \mathrm{d}P_\theta/\mathrm{d}P_\kappa$ and let $\mathscr{B}$ be the σ-algebra generated by $(g_\theta : \theta \in \Theta)$. Then $\mathscr{B}$ is sufficient by theorem 1.5.6. Assume also that the σ-algebra $\mathscr{C}$ is sufficient. It follows from theorem 1.5.6 that for each θ there is a $\mathscr{C}$-measurable version h_θ of $\mathrm{d}P_\theta/\mathrm{d}P_\kappa$. Then, since $g_\theta = h_\theta$ a.e. P_κ for all θ in Θ, the σ-algebra $\mathscr{B}$ is contained in the P_κ closure (i.e. the closure) of the σ-algebra $\mathscr{C}$. □

Example 1.5.10 (Minimal sufficiency of the order statistic in replicated translation experiments). Let the translation experiment $\mathscr{E} = (P_\theta : -\infty < \theta < \infty)$ on the real line be realized by a random variable having density $f(x - \theta)$; $-\infty < x < \infty$ with respect to Lebesgue measure. If f is rational then the theorem may be applied to show that the order statistic is minimal sufficient in $\mathscr{E}^n$ for $n = 1, 2, \ldots$. Actually, by Torgersen (1965), the arguments work for any meromorphic density f having at least one pole (zero) such that the real parts of all other poles (zeros) having the same imaginary part are either bounded from below or are bounded from above.

It would be interesting to know if there is a convenient way to describe the set of all densities f making the order statistic minimal sufficient.

Another interesting criterion for minimal sufficiency is in terms of a variant of the important notion of completeness.

Consider an experiment $\mathscr{E} = (\mathscr{X}, \mathscr{A}; P_\theta : \theta \in \Theta)$ along with a sub σ-algebra

$\mathscr{B}$ of $\mathscr{A}$. Then, following Lehmann & Scheffé (1950), we shall say that $\mathscr{B}$ is *boundedly complete* if bounded $\mathscr{B}$-measurable functions are determined by their expectations.

In other words $\mathscr{B}$ is boundedly complete if and only if $P_\theta(\delta = 0) \equiv_\theta 1$ whenever δ is a bounded $\mathscr{B}$-measurable function such that $E_\theta \delta \equiv_\theta 0$.

If the σ-algebra $\mathscr{A}$ is boundedly complete then we shall say that the experiment $\mathscr{E}$ is boundedly complete. We shall see in chapter 6 that (loosely speaking in the non-coherent case) $\mathscr{E}$ permits a boundedly complete reduction if and only if $\mathscr{E}$ is extreme for mixtures.

Theorem 1.5.11 (Minimal sufficiency from completeness). *Boundedly complete and sufficient sub σ-algebras are minimal sufficient.*

Remark. As boundedly complete and sufficient σ-algebras are clearly irreducible by sufficiency this is an immediate consequence of Burkholder's irreducibility criterion for minimal sufficiency, mentioned just before theorem 1.5.9.

Proof. Let $\mathscr{B}$ be boundedly complete and sufficient and let $\mathscr{C}$ be another sufficient σ-algebra. Consider an event B in $\mathscr{B}$ and let $X = I_B$ denote the indicator variable of B. By assumption there is a common version Y of $E_\theta(X|\mathscr{C}) : \theta \in \Theta$. The version Y may be chosen so that $0 \leqslant Y \leqslant 1$. It follows, since $\mathscr{B}$ is sufficient, that there is a $\mathscr{B}$-measurable function Z such that $0 \leqslant Z \leqslant 1$ and such that $Z = E_\theta(Y|\mathscr{B})$ a.s. P_θ for all θ. Then $E_\theta Z \equiv_\theta E_\theta Y \equiv_\theta E_\theta X$. Hence, since $\mathscr{B}$ is boundedly complete, $X = Z$ a.s. P_θ for all θ. This yields

$$E_\theta X^2 = E_\theta E_\theta(X^2|\mathscr{C}) \geqslant E_\theta Y^2 = E_\theta E_\theta(Y^2|\mathscr{B}) \geqslant E_\theta Z^2.$$

Hence, since $X = Z$ a.e. P_θ, equality holds throughout here. It follows, since $E_\theta(X^2|\mathscr{C}) \geqslant Y^2$ a.s. P_θ, that $E_\theta(X^2|\mathscr{C}) = Y^2$ a.s. P_θ. Thus $E_\theta(X - Y)^2 = E_\theta E_\theta[(X - Y)^2|\mathscr{C}] = E_\theta[E_\theta(X^2|\mathscr{C}) - Y^2] = 0$ so that $P_\theta(X = Y) \equiv_\theta 1$. Hence $P_\theta(B \,\Delta\, C) \equiv_\theta 0$ where $C = [Y = 1] \in \mathscr{C}$. Thus $B \in \bar{\mathscr{C}}$. It follows that $\mathscr{B} \leqslant \mathscr{C}$. □

Completeness may also be used to establish independence of random variables. Consider an experiment $\mathscr{E} = (\mathscr{X}, \mathscr{A}; P_\theta : \theta \in \Theta)$ and a boundedly complete and sufficient sub σ-algebra $\mathscr{B}$ of $\mathscr{A}$. Let C be any *ancillary event*, i.e. an event whose probability $P_\theta(C) = a$ does not depend on θ. Let $\Gamma(C|\cdot)$ be any common version of $P_\theta(C|\mathscr{B})$; $\theta \in \Theta$. If $B \in \mathscr{B}$ then we find $P_\theta(B \cap C) \equiv_\theta E_\theta I_B \Gamma(C|\cdot)$. In particular $a \equiv_\theta P_\theta(C) \equiv_\theta E_\theta \Gamma(C|\cdot)$ so that $\Gamma(C|\cdot) = a$ a.s. P_θ for all θ. Thus $P_\theta(B \cap C) \equiv_\theta E_\theta I_B a \equiv_\theta (E_\theta I_B) a \equiv_\theta P_\theta(B) \cdot P_\theta(C)$. It follows that the event C is independent of $\mathscr{B}$.

Conversely assume that C is independent of a sufficient σ-algebra $\mathscr{B}$. Again let $\Gamma(C|\cdot)$ be a common version of $P_\theta(C|\mathscr{B})$. Independence implies that $P_\theta(C)$, for each θ, is a version of $P_\theta(C|\mathscr{B})$. Thus $P_\theta(C) = \Gamma(C|\cdot)$ a.s. P_θ; $\theta \in \Theta$. If $P_{\theta_1}(C) \neq P_{\theta_2}(C)$ then this implies that the measures P_{θ_1} and P_{θ_2} are disjoint. It follows that if $P_{\theta_1} \wedge P_{\theta_2} \neq 0$ for all pairs θ_1, θ_2 then C is ancillary. Altogether this proves the following useful results due to Basu (1955, 1958).

Theorem 1.5.12 (Statistics which are independent of a given sufficient statistic). *Let $\mathscr{B}$ be a sufficient sub σ-algebra of $\mathscr{A}$ in the experiment $\mathscr{E} = (\mathscr{X}, \mathscr{A}; P_\theta : \theta \in \Theta)$. Let $\mathscr{C}$ be another sub σ-algebra of $\mathscr{A}$. Then:*

(i) *if $\mathscr{C}$ is ancillary then $\mathscr{C}$ is independent of $\mathscr{B}$ provided $\mathscr{B}$ is boundedly complete.*
(ii) *if $\mathscr{C}$ is independent of $\mathscr{B}$ and if P_{θ_1} and P_{θ_2} are not disjoint for distinct points θ_1 and θ_2 in Θ then $\mathscr{C}$ is ancillary.*

Remark 1. $\mathscr{C}$ is called *ancillary* if $\mathscr{E}|\mathscr{C}$ is *totally non-informative*, i.e. if $P_\theta(C)$ does not depend on θ when $C \in \mathscr{C}$.

Remark 2. Sub σ-algebras $\mathscr{B}$ and $\mathscr{C}$ of $\mathscr{A}$ are here called *independent* if they are independent for P_θ for each $\theta \in \Theta$.

Example 1.5.13. Let $X_1, \dots, X_n$ be independent $N(0, 1)$ distributed and consider the statistics $\bar{X} = (1/n)\sum_i X_i$ and $S^2 = (1/n)\sum_i (X_i - \bar{X})^2$.

It is well known that under these circumstances the statistics $\bar{X}$ and S^2 are independent. This may be concluded directly from Basu's theorem since $\bar{X}$ is complete and sufficient while S^2 is ancillary when $X_1, \dots, X_n$ are independent $N(\theta, 1)$-distributed with $\theta \in \Theta =]-\infty, \infty[$.

Intuitive notions on sufficiency may not be disturbed by the results described so far. However the measure theoretical approach used here leads also to results which may contradict what common sense appears to dictate. For example, we might expect that σ-algebras of events containing sufficient σ-algebras should themselves be sufficient.

Indeed if $\mathscr{E}$ is dominated then this is a consequence of theorem 1.5.6. If $\mathscr{E}$ is not dominated however then, as shown by Burkholder, this need no longer be so. Another intriguing result in Burkholder's 1961 paper is that minimal sufficient σ-algebras need not exist in non-coherent experiments.

How relevant are these results to the theory of comparison of statistical experiments as it is described here? We mentioned before that this theory adheres to the principle that general structures should be approximable by finite (dimensional) ones. As we shall see in chapter 7, this implies that an experiment $\mathscr{E} = (\mathscr{X}, \mathscr{A}; P_\theta : \theta \in \Theta)$ 'carries' the same amount of information as the restricted experiment $\mathscr{E}|\mathscr{B}$ for a sub σ-algebra $\mathscr{B}$ of $\mathscr{A}$ if and only if $\mathscr{B}$ is pairwise sufficient for $\mathscr{A}$. On the other hand if $\mathscr{E}$ is dominated then this yields the notion of sufficiency introduced in this section. In general, however, these concepts differ and then it is the notion of pairwise sufficiency which is of primary interest here.

How do Burkholder's results appear from the point of view of pairwise sufficiency?

Considering the first pathology we see that it simply disappears since, by theorem 1.5.6, any σ-algebra containing a pairwise sufficient σ-algebra is itself pairwise sufficient.

The possibility of non-existence of a minimal sufficient reduction persists, however. Thus example 7.3.6, adapted from Siebert (1979), exhibits an experiment $\mathscr{E}$ not permitting a minimal pairwise sufficient σ-algebra. If $\mathscr{E}$ is majorized however then, as shown in Siebert's paper, minimal pairwise sufficient σ-algebras exist. We should at this point be aware that in general the relevant ordering of σ-algebras differs from the ordering considered here. In fact the relevant ordering is the ordering corresponding to the inclusion ordering of the associated M-spaces.

However this is not the whole story, since LeCam in his 1964 paper established, without any regularity conditions whatsoever, that there is always an algebra which deserves to be called minimal sufficient or rather, the smallest sufficient algebra. This does not contradict the example mentioned since LeCam's algebra involves events which need not belong to the σ-algebra $\mathscr{A}$. What are these events? Firstly, as we may now expect, they are consistent families of ordinary events. Secondly, as we have seen here in the dominated case, they may be expressed in terms of Radon–Nikodym derivatives $\mathrm{d}P_{\theta_2}/(\mathrm{d}P_{\theta_1} + \mathrm{d}P_{\theta_2})$; θ_1, $\theta_2 \in \Theta$. These Radon–Nikodym derivatives are then minimal non-negative within the M-space $M(\mathscr{E})$ of the experiment $\mathscr{E}$.

If the experiment $\mathscr{E}$ is coherent then LeCam's minimal sufficient algebra is representable as a minimal sufficient σ-algebra of events. Independently of LeCam's work, Hasegawa & Perlman (1974) established the existence of a minimal sufficient σ-algebra in this case.

We mentioned before that the decision theoretical way of expressing the sufficiency of a sub σ-algebra $\mathscr{B}$ is to say that the sub-experiment $\mathscr{E}|\mathscr{B}$ is at least as informative as, and thus equally informative as, the experiment $\mathscr{E}$. One might be misled by this to conclude that sufficiency is just a very particular case of ordering of experiments. That this is not so follows almost directly from LeCam's randomization criterion for the ordering 'being at least as informative as'. Restricting ourselves to dominated experiments having Euclidean sample spaces, this criterion says that the experiment $\mathscr{E} = (\mathscr{X}, \mathscr{A}; P_\theta : \theta \in \Theta)$ is at least as informative as the experiment $\mathscr{F} = (\mathscr{Y}, \mathscr{B}; Q_\theta : \theta \in \Theta)$ if and only if $\mathscr{F} = \mathscr{E}M$ for a randomization M from $\mathscr{E}$ to $\mathscr{F}$. Along with the experiments $\mathscr{E}$ and $\mathscr{F}$ we then also have the 'big' experiment $\mathscr{G} = \mathscr{E} \times \mathscr{F} = (\mathscr{X} \times \mathscr{Y}, \mathscr{A} \times \mathscr{B}; P_\theta \times M : \theta \in \Theta)$. The experiments $\mathscr{E}$ and $\mathscr{F}$ may then be considered as imbedded in $\mathscr{G}$. In fact they are obtained by observing the first coordinate X and the second coordinate Y of $(X, Y) \in \mathscr{X} \times \mathscr{Y}$ where (X, Y) realizes $\mathscr{G}$. Since the conditional distribution $M(\cdot|X)$ of Y given X does not depend on θ, it also follows that the first coordinate X is sufficient. Thus $\mathscr{E}$ is at least as informative as $\mathscr{F}$ if and only if $\mathscr{E}$ and $\mathscr{F}$ may be imbedded as sub-experiments of an experiment $\mathscr{G}$ such that the imbedding of $\mathscr{E}$ is sufficient in $\mathscr{G}$.

Using LeCam's notion of deficiency we may measure the degree of sufficiency of a sub σ-algebra $\mathscr{B}$ by evaluating the deficiency of the experiment $\mathscr{E}|\mathscr{B}$ with respect to the experiment $\mathscr{E}$. However, in a 1974 paper (1974a) LeCam showed that it is often simpler, and in itself useful, to measure the lack of sufficiency by a quantity named insufficiency. According to this concept *the insufficiency of the sub σ-algebra $\mathscr{B}$ of $\mathscr{A}$* is the infimum of all numbers $\varepsilon \geqslant 0$ such that there is another experiment $\mathscr{F} = (\mathscr{X}, \mathscr{A}; Q_\theta : \theta \in \Theta)$ with the same sample space as $\mathscr{E} = (\mathscr{X}, \mathscr{A}; P_\theta : \theta \in \Theta)$ such that $\mathscr{B}$ is pairwise sufficient in $\mathscr{F}$ while $\|P_\theta - Q_\theta\| \leqslant \varepsilon$ for all points θ in Θ.

Thus any pairwise sufficient σ-algebra $\mathscr{B}$ has insufficiency zero. Conversely any sub σ-algebra whose insufficiency is zero is necessarily pairwise sufficient.

If η denotes the insufficiency of $\mathscr{B}$ then

$$\tfrac{1}{2}\gamma \leqslant \eta \leqslant \gamma$$

where γ is obtained from the general randomization criterion for deficiency by restricting attention to those randomizations which behave as conditional probabilities given $\mathscr{B}$, i.e. as projections. We refer the reader to LeCam (1986) for general properties and bounds for insufficiencies.

1.6 Appendix: λ-systems and π-systems of events

Suppose we want to show that sets satisfying a certain requirement belong to a given σ-algebra $\mathscr{A}$ of subsets of a given set χ. If we know that the class of sets satisfying this requirement constitutes a σ-algebra $\mathscr{C}$ of subsets of χ, then it suffices to show that $\mathscr{A}$ contains a basis for $\mathscr{C}$. Thus if we want to check that the map Y from the measurable space $(\mathscr{X}, \mathscr{A})$ to the measurable space $(\mathscr{Y}, \mathscr{B})$ is measurable, then it suffices to show that $\mathscr{A}$ contains all inverse images $Y^{-1}[B]$ of sets B belonging to a given basis for $\mathscr{B}$.

However, it is not always sufficient to consider sets belonging to a basis. Distinct probability measures P and Q on a measurable space $(\mathscr{X}, \mathscr{A})$ may assign the same probabilities to sets A belonging to a given basis of $\mathscr{A}$. If so then the class of events A such that $P(A) = Q(A)$ is not a σ-algebra. In fact, as we shall see shortly, this class can't even be closed under the formation of finite intersections.

In general if $\mathscr{E} = (\mathscr{X}, \mathscr{A}; P_\theta : \theta \in \Theta)$ is an experiment then the class $\mathscr{B}$ of ancillary events (i.e. events whose probabilities do not depend on θ) satisfies the following conditions:

(λ_1) $\varnothing \in \mathscr{B}$.
(λ_2) $B^c \in \mathscr{B}$ when $B \in \mathscr{B}$.
(λ_3) $B_1 \cup B_2 \in \mathscr{B}$ when $B_1, B_2 \in \mathscr{B}$ and $B_1 \cap B_2 = \varnothing$.
(λ_4) $\bigcup_{i=1}^{\infty} B_i \in \mathscr{B}$ when $B_1 \subseteqq B_2 \subseteqq \cdots$ and $B_1, B_2, \cdots \in \mathscr{B}$.

A class $\mathscr{B}$ of subsets of a set $\mathscr{X}$ which satisfies (λ_1), (λ_2), (λ_3) and (λ_4) is called a *λ-system of subsets of $\mathscr{X}$*, or just a *λ-system.*

Thus the class of ancillary events in an experiment $\mathscr{E}$ is a λ-system.

λ-systems are closed under the formation of proper differences. In fact if $\mathscr{B}$ is a λ-system of subsets of $\mathscr{X}$ and if B_1 and B_2 are sets in $\mathscr{B}$ such that $B_1 \supseteqq B_2$ then $(B_1 - B_2)^c = B_1^c \cup B_2 \in \mathscr{B}$ by (λ_2) and (λ_3) and thus $B_1 - B_2 \in \mathscr{B}$ by (λ_3). We state this as a proposition.

Proposition 1.6.1. *λ-systems are closed under the formation of proper differences of sets.*

Remark. A difference $S - T$ of sets is called *proper* if $S \supseteqq T$.

λ-systems need not be closed under the formation of intersections. On the other hand, bases of sets in measure theory, as elsewhere, are often closed under intersections. The fact that it is often solely this property we need is due to an interesting interplay between the property of being closed under intersections on the one hand and the property of being a λ-system on the other.

In order to express this it is convenient to introduce the notion of a *π-system of sets* or just a *π-system* as a class $\mathscr{B}$ of subsets of a given set $\mathscr{X}$ such that:

(π) $B_1 \cap B_2 \in \mathscr{B}$ when $B_1, B_2 \in \mathscr{B}$.

Before proceeding let us recall that a σ-algebra of subsets of $\mathscr{X}$ is a class $\mathscr{A}$ such that:

(σ_1) $\varnothing \in \mathscr{A}$.
(σ_2) $A^c \in \mathscr{A}$ when $A \in \mathscr{A}$.
(σ_3) $\bigcup_{n=1}^{\infty} A_n \in \mathscr{A}$ when $A_1, A_2, \cdots \in \mathscr{A}$.

Here (σ_3) may be replaced by:

(σ_4) $\bigcap_{n=1}^{\infty} A_n \in \mathscr{A}$ when $A_1, A_2, \cdots \in \mathscr{A}$.

It is easily seen that σ-algebras are λ-systems as well as π-systems. Assume now that a class $\mathscr{A}$ of subsets of $\mathscr{X}$ is both a π-system and a λ-system. Then, by (λ_1) and (λ_2), $\mathscr{A}$ satisfies (σ_1) and (σ_2). Furthermore (π) implies that $A_1 \cap \cdots \cap A_n \in \mathscr{A}$ when $A_1, \ldots, A_n \in \mathscr{A}$. Assume that $A_1, A_2, \cdots \in \mathscr{A}$ and put $B_n = [A_1 \cap \cdots \cap A_n]^c$, $n = 1, 2, \ldots$ Then (π) and (λ_2) imply that $B_n \in \mathscr{A}$, $n = 1, 2, \ldots$ Hence, since $B_1 \subseteq B_2 \subseteq \ldots$, ($\lambda_3$) implies that $\bigcup_n B_n \in \mathscr{A}$. Thus, by ($\lambda_2$), $\bigcap_n A_n = [\bigcup_n B_n]^c \in \mathscr{A}$. This proves the following proposition.

Proposition 1.6.2. *A class of subsets of a set $\mathscr{X}$ is a σ-algebra if and only if it is both a π-system and a λ-system.*

For each class $\mathscr{H}$ of subsets of the set $\mathscr{X}$ let us denote the smallest σ-algebra containing $\mathscr{H}$ by $\sigma(\mathscr{H})$. Similarly the smallest π-system containing $\mathscr{H}$ and the smallest λ-system containing $\mathscr{H}$ will be denoted by $\pi(\mathscr{H})$ and $\lambda(\mathscr{H})$ respectively.

Assume we know that any set in a π-system $\mathscr{H}$ has a certain property of interest. If we then know that the class of all sets having this property is a λ-system then we are entitled to conclude that any set in the σ-algebra $\sigma(\mathscr{H})$ also has the property. This important result may be neatly phrased as in the following theorem.

Theorem 1.6.3 (Interplay of λ-systems and π-systems). *If λ, π and σ operate on the classes of subsets of χ as defined above then*

$$\sigma = \lambda\pi,$$

i.e. $\sigma(\mathscr{H}) = \lambda(\pi(\mathscr{H}))$ for all classes $\mathscr{H}$ of subsets of χ. In particular $\sigma(\mathscr{H}) = \lambda(\mathscr{H})$ for any π-system $\mathscr{H}$ of subsets of χ.

Proof. Let $\mathscr{H}$ be a π-system of sets. We shall then argue that $\lambda(\mathscr{H})$ is also a π-system so that $\lambda(\mathscr{H})$ is in fact a σ-algebra. If $H \in \mathscr{H}$ then it is readily checked that the class of sets B such that $H \cap B \in \lambda(\mathscr{H})$ is a λ-system containing $\mathscr{H}$ and thus contains $\lambda(\mathscr{H})$. Hence $H_1 \cap H_2 \in \lambda(\mathscr{H})$ when $H_1 \in \mathscr{H}$ and $H_2 \in \lambda(\mathscr{H})$. It follows that if H belongs to $\lambda(\mathscr{H})$ then the class of sets B such that $H \cap B \in \lambda(\mathscr{H})$ is also a λ-system containing $\mathscr{H}$ and thus also contains $\lambda(\mathscr{H})$. Hence $\lambda(\mathscr{H})$ is a π-system whenever $\mathscr{H}$ is a π-system.

Returning to the case of a general class $\mathscr{H}$ of subsets of $\mathscr{X}$ we conclude that $\lambda(\pi(\mathscr{H}))$ is both a π-system and a λ-system and thus it is also a σ-algebra. Hence, $\lambda(\pi(\mathscr{H})) \supseteqq \sigma(\mathscr{H})$. On the other hand, by proposition 1.6.2 again, $\pi(\mathscr{H}) \subseteqq \sigma(\mathscr{H})$ so that $\lambda(\pi(\mathscr{H})) \subseteqq \lambda(\sigma(\mathscr{H})) = \sigma(\mathscr{H})$. □

Using the above theorem and observing that the class of ancillary events is a λ-system we obtain the following.

Theorem 1.6.4 (π-system bases are determining). *Probability measures P and Q on the same σ-algebra $\mathscr{A}$ of sets are equal if they assign the same probabilities to all events belonging to a π-system generating $\mathscr{A}$.*

Remark. Say that a measure μ on $\mathscr{A}$ *is σ-finite on a subclass $\mathscr{B}$ of $\mathscr{A}$* if $\mathscr{X} = \bigcup_{i=1}^{\infty} B_i$ for a sequence $(B_1, B_2, \ldots)$ of sets in $\mathscr{B}$ such that $|\mu|(B_i) < \infty$; $i = 1, 2, \ldots$ Thus μ is σ-finite if and only if μ is σ-finite on $\mathscr{A}$.

By modifying the above arguments we see that measures μ and ν on $\mathscr{A}$ are equal provided they agree on a π-system generating $\mathscr{A}$ such that both μ and ν are σ-finite on this π-system.

Another important application is the following.

Theorem 1.6.5 (Independence in terms of bases). *Consider a probability space* $(\mathscr{X}, \mathscr{A}, P)$.

Then classes $\mathscr{B}_t : t \in T$ *of events (in* $\mathscr{A}$*) are independent if and only if the* λ*-systems* $\lambda(\mathscr{B}_t) : t \in T$ *are independent.*

In particular, σ*-algebras* $\sigma(\mathscr{B}_t) : t \in T$ *of events are independent whenever* $\mathscr{B}_t : t \in T$ *are independent* π*-systems.*

Remark. Events $B_t : t \in T$ in the probability space $(\mathscr{X}, \mathscr{A}, P)$ are called *independent* if $P(\bigcap_F B_t) = \prod_F P(B_t)$ for each non-empty finite subset F of T.

Classes of events $\mathscr{B}_t : t \in T$ are called *independent* if the events $(B_t : t \in T)$ are independent whenever $B_t \in \mathscr{B}_t$ for all $t \in T$.

Strictly speaking, these notions of independence refer to families. Thus we use the expression 'The events (classes of events) ... are independent' as synonymous with 'The family ... of events (classes of events) is independent'.

References

Andersen, E. B. 1970. Sufficiency and exponential families for discrete sample spaces. *J. Am. Stat. Soc.* **65**, pp. 1248–55.

Bahadur, R. R. 1954. Sufficiency and statistical decision functions. *Ann. Math. Statist.* **25**, pp. 423–62.

— 1955. A characterization of sufficiency. *Ann. Math. Statist.* **26**, pp. 286–93.

— 1957. On unbiased estimates of uniformly minimum variance. *Sankhyā*, **18**, pp. 211–44.

— 1976. A note on UMV estimates and ancillary statistics. In memorial volume dedicated to J. Hajék, Charles University, Prague.

Barankin, E. W. & Maitra, A. P. 1963. Generalization of the Fisher–Darmois–Koopman–Pitman theorem on sufficient statistics. *Sankhyā*, **25**, pp. 217–44.

Barndorff-Nielsen, O. & Pedersen, K. 1968. Sufficient data reduction and exponential families. *Math. Scand.* **22**, pp. 197–202.

Basu, D. 1955. On statistics independent of a complete sufficient statistic. *Sankhyā*, **15**, pp. 377–80.

— 1958. On statistics independent of a sufficient statistic. *Sankhyā*, **20**, pp. 223–6.

Blackwell, D. 1947. Conditional expectation and unbiased sequential estimation. *Ann. Math. Statist.* **18**, pp. 105–10.

Brown, A. L. & Page, A. 1970. *Elements of functional analysis*. Van Nostrand, London.

Brown, L. D. 1964. Sufficient statistics in the case of independent random variables. *Ann. Math. Statist.* **35**, pp. 1456–74.

Burkholder, B. L. 1961. Sufficiency in the undominated case. *Ann. Math. Statist.* **32**, pp. 1191–1200.

Chernoff, H. 1952. A measure of asymptotic efficiency for tests of a hypothesis based on the sum of observations. *Ann. Math. Statist.* **23**, pp. 493–507.

— 1956. Large sample theory: "Parametric case". *Ann. Math. Statist.* **27**, pp. 1–22.

Darmois, G. 1935. Sur les lois de probabilité à estimation exhaustive. *C. R. Acad. Sci. Paris*, **260**, pp. 1265–6.

Denny, J. L. 1969. Note on a theorem of Dynkin on the dimension of sufficient statistics. *Ann. Math. Statist.* **40**, pp. 1474–6.

— 1970. Cauchy's equation and sufficient statistics on arcwise connected spaces. *Ann. Math. Statist.* **41**, pp. 401–11.

Dieudonné, J. 1960. *Foundation of modern analysis*. Academic Press, New York.

Dunford, N. & Schwartz, J. T. 1958. *Linear operators. Part I*. Interscience Publishers, New York.

Dynkin, E. B. 1951. Necessary and sufficient statistics for a family of probability distributions (in Russian). *Uspetri Mat. Nauk.* (*N.S.*) 6 *no.* 1, **41**, pp. 68–90. English translation in *Select. Transl. Math. Statist. Prob.* **1** (1961) pp. 23–41.

Efron, B. 1967. The power of the likelihood ratio test. *Ann. Math. Statist.* **38**, pp. 802–6.

Efron, B. & Truax, D. 1968. Deviation theory in exponential families. *Ann. Math. Statist.* **39**, pp. 1402–24.

Fell, J. M. G. 1956. A note on abstract measure. *Pacific Journal of Mathematics*, **VI**, pp. 43–5.

Fisher, R. A. 1920. A mathematical examination of the methods of determining the accuracy of an observation by the mean error, and by the mean square error. Monthly Notices, *R. Astron. Soc.* **80**, pp. 758–70.

— 1922. On the mathematical foundation of theoretical statistics. *Philos. Trans. Roy. Soc. London Sec. A*, **222**, pp. 309–68.

— 1934. Two new properties of mathematical likelihood. *Proc. R. Soc. A*, **144**, pp. 285–307.

Ghosh, J. K., Morimoto, H. & Yamada, S. 1981. Neyman factorization and minimality of pairwise sufficient subfields. *Ann. Statist.* **9**, pp. 514–30.

Halmos, P. & Savage, L. J. 1949. Application of the Radon–Nikodym theorem to the theory of sufficient statistics. *Ann. Math. Statist.* **20**, pp. 225–41.

Hasegawa, M. & Perlman, M. D. 1974. On the existence of a minimal sufficient subfield. *Ann. Statist.* **2**, pp. 1049–55.

Heyer, H. 1982. *Theory of statistical experiments.* Springer–Verlag, New York.

Hipp, C. 1974. Sufficient statistics and exponential families. *Ann. Statist.* **2**, pp. 1283–92.

Janssen, A. 1986. Asymptotic properties of Neyman–Pearson tests for infinite Kullback–Leibler information. *Ann. Statist.* **14**, pp. 1068–79.

Janssen, A., Milbrodt, H. & Strasser, H. 1984. Infinitely divisible experiments. *Lecture notes in statistics.* Springer–Verlag, Berlin.

Kakutani, S. 1941a. Concrete representation of abstract L-spaces and the mean ergodic theorem. *Ann. of Math.* (2) **42**, pp. 523–37.

— 1941b. Concrete representation of abstract M-spaces. (A characterization of the space of continuous functions.) *Ann. of Math.* (2) **42**, pp. 994–1024.

Kelley, J. L. 1955. *General topology.* Van Nostrand, London.

Koopman, B. O. 1936. On distributions admitting a sufficient statistic. *Trans. Amer. Math. Soc.* **39**, pp. 399–409.

Kullback, S. 1959. *Information theory and statistics.* (1968 edition). J. Wiley and Sons, Dover.

Landers, D. 1974. Minimal sufficient statistics for families of product measures. *Ann. Statist.* **2**, pp. 1335–9.

LeCam, L. 1964. Sufficiency and approximate sufficiency. *Ann. Math. Statist.* **35**, pp. 1419–55.

— 1974a. On the information contained in additional observations. *Ann. Statist.* **2**, pp. 630–49.

— 1974b. *Notes on asymptotic methods in statistical decision theory.* Centre de Recherches Mathematique, Université de Montréal.

— 1986. *Asymptotic methods in statistical decision theory.* Springer–Verlag, New York.

Lehmann, E. L. & Scheffé, H. 1950, 1955, 1956. Completeness, similar regions, and unbiased estimation. *Sankhyā*, **10**, pp. 305–40, **15**, pp. 219–36, **17** (correction), p. 250.

Milbrodt, M. 1985. Local asymptotic normality of sampling experiments. *Z. Wahrscheinlichkeitstheorie verw. Geb.* **70**, pp. 131–56.

— 1988. Limits of experiments associated with sampling plans. *Journal of statistical planning and inference.* **19**, pp. 1–29.

Millar, P. W. 1983. The minimax principle in asymptotic statistical theory. *Lecture notes in statistics.* **976**, pp. 74–265. Springer–Verlag, Berlin.

Morimoto, H. 1972. Statistical structure of the problem of sampling from finite populations. *Ann. Statist.* **43**, pp. 490–7.

Neveu, J. C. 1965. *Mathematical foundations of the calculus of probability.* Holden Day, San Francisco.

Neyman, J. 1935. Sur un teorema concernente le cosidette statistiche sufficienti. *Giorn. Ist. Ital. Att.* **6**, pp. 320–34.

Parthasarathy, K. R. 1967. *Probability measures on metric spaces.* Academic Press, New York.

Pfanzagl, J. 1969. On the existence of product measurable densities. *Sankhyā Ser. A.* **31**, pp. 13–18.

— 1972. Transformation groups and sufficient statistics. *Ann. Math. Statist.* **43**, pp. 553–68.

Pitman, E. J. G. 1936. Sufficient statistics and intrinsic accuracy. *Proc. Camb. Phil. Soc.* **32**, pp. 567–79.

Rao, C. R. 1945. Information and accuracy attainable in estimation of statistical parameters. *Bull. Calcutta Math. Soc.* **37**, pp. 81–91.

Roussas, G. G. 1972. *Contiguity of probability measures: Some applications in statistics.* Cambridge University Press.

Siebert, E. 1979. Statistical experiments and their conical measures. *Z. Wahrscheinlichkeitstheorie verw. Geb.* **45**, pp. 247–58.

Strassen, V. 1965. The existence of probability measures with given marginals. *Ann. Math. Statist.* **36**, pp. 423–39.

Strasser, H. 1985. *Mathematical theory of statistics.* Walter de Gruyter, Berlin.

Sverdrup, E. 1966. The present state of the decision theory and the Neyman–Pearson theory. *Review of the Int. Statist. Institute.* **34**, pp. 309–33.

Torgersen, E. 1965. Minimal sufficiency of the order statistic in the case of translation and scale parameters. *Scand. Aktuarietidsskrift*, pp. 16–21.

— 1976. Deviations from total information and from total ignorance as measures of information. *Stat. Res. Rep.* University of Oslo.

— 1981a. On complete sufficient statistics and uniformly minimum variance unbiased estimators. *Istituto Nazionale Di Alta Matematica. Symposia Mathematica* **XXV**, pp. 137–53.

— 1981b. Measures of information based on comparison with total information and with total ignorance. *Ann. Statist.* **9**, pp. 638–57.

— 1982. Comparison of some statistical experiments associated with sampling plans. *Probability and Mathematical Statistics.* **3**, pp. 1–17.

— 1988. On Bahadur's converse of the Rao–Blackwell theorem. Extension to majorized experiments. *Scand. Jour. of Statistics.* **15**, pp. 273–80.

Wald, A. 1950. *Statistical decision functions.* Wiley, New York.

2

Convexity

2.1 Introduction

Convex analysis is an indispensible tool of mathematical statistics. The purpose of this chapter is to provide a self-contained body of central results which are useful for decision theory in general and which are particularly useful for the topic covered in this book.

Several excellent textbooks are available in this area. Some well known books in decision theory, e.g. Blackwell & Girshick (1954), Ferguson (1967) and LeCam (1986), also contain quite an amount of convex analysis. Another highly relevant work is the book by Marshall & Olkin (1979).

References to original sources, with a few exceptions, are not given. They may be found either in the cited textbooks or in some of the other textbooks which are included in the references. The references are mainly books which have been particularly useful to the author.

The basic concepts and notions are introduced within the framework of a general real linear space, in section 2.1. However results which do not follow fairly directly from the definitions are usually only given in the finite dimensional case. An exception is the Hahn–Banach theorem. The spadework contained in the 'standard' proof of that theorem is finite dimensional. As we shall see, an inspection of this proof provides criteria for measurability of the guaranteed extension and this yields a substantial part of an important decomposition theorem in Strassen (1965). The complete proof of this theorem is provided in complement 22 in chapter 10. Some notation and conventions which will be used are:

$\mathbb{R}$	the real line.
functional	real valued function.
V'	the algebraic dual of $V =$ the linear space of linear functionals on the linear space V.

$\mathring{A}$	the interior of the set A.
$\bar{A}$	the closure of the set A.
A^c	the complement of A, with respect to a given set.
$A + B$	the algebraic sum $\{a + b : a \in A, b \in B\}$ of A and B.
λA	$\{\lambda a : a \in A\}$ when $\lambda \in \mathbb{R}$.

An upper semicontinuous function f is an extended real valued function f such that all the sets $[f < \alpha]$; $\alpha \in \mathbb{R}$ are open.

A lower semicontinuous function f is an extended real valued function f such that all the sets $[f > \alpha]$; $\alpha \in \mathbb{R}$ are open.

f *is continuous from above (below)* = f is upper (lower) semicontinuous.

Some useful facts on semicontinuity are also given in theorem 3.2.11.

Finally, vectors in a given linear space are often referred to as points.

2.2 Convexity

Convex sets are sets which are closed for convex combinations, just as affine sets and linear sets are sets which are closed for, respectively, affine combinations and linear combinations. It will therefore be convenient to begin this chapter on convexity by considering some analogous notions related to combinations of vectors which, by decreasing generality, are linear, affine and convex.

It will be assumed throughout that the vectors are taken from a given real linear space V.

Let W be any subset of V. A vector v is called a *linear combination of vectors from* W if $v = \sum_{i=1}^{r} \alpha_i w_i$ for vectors $w_1, \ldots, w_r$ and scalars $\alpha_1, \ldots, \alpha_r$. If v is a linear combination $v = \sum_{i=1}^{r} \alpha_i w_i$ of vectors $w_1, \ldots, w_r$ and scalars $\alpha_1, \ldots, \alpha_r$ such that $\alpha_1 + \cdots + \alpha_r = 1$ then we shall say that v is an *affine combination of vectors from* W. A vector v is called a *convex combination of vectors from* W if $v = \sum_{i=1}^{r} \alpha_i w_i$ for vectors $w_1, \ldots, w_r$ from W and for non-negative scalars $\alpha_1, \ldots, \alpha_r$ such that $\alpha_1 + \cdots + \alpha_r = 1$.

The last notion is clearly probabilistic and we may describe it by saying that the convex combinations of vectors from W are the expectations of vector valued random variables taking their values in given finite subsets of V.

Corresponding to the three varieties of combinations are two (non-stochastic) notions of independence (dependence).

Firstly a subset W of V is called *linearly independent* if linear combinations $\sum_{i=1}^{r} \alpha_i w_i$ and $\sum_{i=1}^{r} \beta_i w_i$ of distinct vectors $w_1, \ldots, w_r$ from W are equal if and only if $\alpha_i = \beta_i$; $i = 1, \ldots, r$. Otherwise W is called *linearly dependent*.

In other words W is linearly independent (dependent) if it is impossible (possible) to express the null vector as a linear combination $0 = \sum_{i=1}^{r} \alpha_i w_i$ of distinct vectors $w_1, \ldots, w_r$ from W and with coefficients $\alpha_1, \ldots, \alpha_r$ not all being zero.

A subset W of V is called *affinely independent* if affine combinations $\sum_{i=1}^r \alpha_i w_i$ and $\sum_{i=1}^r \beta_i w_i$ of distinct vectors $w_1, \ldots, w_r$ from W are equal if and only if $\alpha_i = \beta_i$; $i = 1, \ldots, r$. Otherwise W is called *affinely dependent.*

In other words W is affinely independent (dependent) if it is impossible (possible) to express the null vector as a linear combination $0 = \sum_{i=1}^r \alpha_i w_i$ of distinct vectors $w_1, \ldots, w_r$ from W and with coefficients $\alpha_1, \ldots, \alpha_r$ not all being zero and satisfying $\sum_{i=1}^r \alpha_i = 0$.

It may be checked that W is affinely independent if and only if $\{1\} \times W$ is linearly independent in $\mathbb{R} \times V$.

The analogous notion of convex independence coincides with affine independence.

Theorem 2.2.1. *A subset W of V is affinely independent if and only if convex combinations $\sum_{i=1}^r \lambda_i w_i$ and $\sum_{i=1}^r \mu_i w_i$ of vectors $w_1, \ldots, w_r$ in W are distinct whenever $\lambda_i \neq \mu_i$ for at least one i.*

Proof. If W is affinely independent and if $\sum \lambda_i w_i = \sum \mu_i w_i$ and $\sum_i \lambda_i = \sum_i \mu_i = 1$ then $\sum_i (\lambda_i - \mu_i) w_i = 0$ and $\sum_i (\lambda_i - \mu_i) = 1 - 1 = 0$ so that, by affine independence, $\lambda_i = \mu_i$; $i = 1, \ldots, r$. Therefore assume that the condition is satisfied and let $\alpha_1, \ldots, \alpha_r$ be numbers so that $\sum_i \alpha_i = 0$ and $\sum_i \alpha_i w_i = 0$. If $\beta = \sum \alpha_i^+ = \sum \alpha_i^-$ is positive then $\sum_i (\alpha_i^+/\beta) w_i = \sum_i (\alpha_i^-/\beta) w_i = 0$ and thus, by the assumptions, $\alpha_i = \beta[(\alpha_i^+/\beta) - (\alpha_i^-/\beta)] = 0$; $i = 1, \ldots, r$ which is impossible when $\beta > 0$. Hence $\sum \alpha_i^+ = \sum \alpha_i^- = 0$ so that $\alpha_i = 0$; $i = 1, \ldots, r$. It follows that W is affinely independent. □

Using the arguments in the proof we readily obtain the following criterion for affine dependence.

Corollary 2.2.2. *A subset W of a linear space V is affinely dependent if and only if it may be decomposed as $W = W_1 \cup W_2$ where $W_1 \cap W_2 = \emptyset$ such that at least one vector is both a convex combination of vectors from W_1 and a convex combination of vectors from W_2.*

Linear combinations may be reduced to linear combinations of linearly independent subsets. The analogous facts for affine and convex combinations are stated in the following proposition.

Proposition 2.2.3. *Any affine combination of vectors from a set W is also an affine combination of vectors from any given maximal affinely independent subset of W.*

Any convex combination of vectors from W is also a convex combination of an affinely (convexly) independent subset of W.

Remark. An essential difference between the two statements is that the appropriate affinely independent subset of W in the first statement does not depend on the particular combination under consideration, while in the second statement such independence is not guaranteed.

Proof. If W_0 is a maximal affinely independent subset of W and if $w \in W$ then $W_0 \cup \{w\}$ is affinely dependent. It follows that w is an affine combination of vectors from W_0. Hence also any affine combination of vectors from W is an affine combination of vectors from W_0. Consider a vector v which can be expressed as a convex combination of vectors from W. Let $v = \sum_{i=1}^{r} c_i w_i$ be a representation of v where $c_1, \ldots, c_r \geqslant 0$, $\sum_{i=1}^{r} c_i = 1$, $w_1, \ldots, w_r \in W$ and where r is minimal in the sense that there is no such representation with fewer than r terms. Then $c_1, \ldots, c_r > 0$. The proof will be completed by showing that $w_1, \ldots, w_r$ are affinely independent. In fact if $\sum_{i=1}^{r} \alpha_i w_i = 0$ where $\sum_{i=1}^{r} \alpha_i = 0$ and $\sum_i |\alpha_i| > 0$ then the closed convex set T of real numbers t such that $c_i + t\alpha_i \geqslant 0$; $i = 1, \ldots, r$ is a proper subset of $\mathbb{R}$ containing 0. It follows that T is a closed interval having at least one finite end point τ. Then $c_i + \tau\alpha_i = 0$ for at least one index i and, since $v = \sum_i c_i w_i = \sum_i c_i w_i + \tau \sum_i \alpha_i w_i = \sum_{i=1}^{r} (c_i + \tau\alpha_i) w_i$, this contradicts the minimality of r. □

A subset W of V is called *linear* (*linear subspace, subspace*) if linear combinations of vectors from W also belong to W, i.e. if W is a linear space in its own right.

A subset W of V is called *affine*, or a *flat*, if affine combinations of vectors from W also belong to W.

A subset W of V is called *convex* if convex combinations of vectors from W also belong to W.

Clearly subspaces are affine and affine sets are convex.

When verifying that a subset W of V has one of these properties we may restrict our attention to pairs of vectors from W.

Theorem 2.2.4. *Let W be a subset of a linear space V. Then*

(i) *W is a subspace of V if and only if $\alpha_1 w_1 + \alpha_2 w_2 \in W$ whenever $w_1, w_2 \in W$ and $\alpha_1, \alpha_2 \in \mathbb{R}$.*
(ii) *W is affine if and only if $(1 - \alpha)w_1 + \alpha w_2 \in \mathbb{R}$ when $w_1, w_2 \in W$ and $\alpha \in \mathbb{R}$.*
(iii) *W is convex if and only if $(1 - \alpha)w_1 + \alpha w_2 \in \mathbb{R}$ when $w_1, w_2 \in W$ and $\alpha \in \,]0, 1[$.*

Proof. The 'ifs' follow by induction and the 'only ifs' follow directly from the definitions. □

A map T from a linear space V to another linear space $\tilde{V}$ is called *linear* if it maps linear combinations $\sum_{i=1}^{r} \alpha_i v_i$ of vectors $v_1, \ldots, v_r$ in V into the same linear combinations $\sum_{i=1}^r \alpha_i T(v_i)$ of the image vectors $T(v_1), \ldots, T(v_r)$. A linear map to $\mathbb{R}$ is called a *linear functional.*

As in the previous theorem, in order to establish linearity it suffices to verify that $T(\alpha_1 v_1 + \alpha_2 v_2) = \alpha_1 T(v_1) + \alpha_2 T(v_2)$ whenever $v_1, v_2 \in V$ and $\alpha_1, \alpha_2 \in \mathbb{R}$.

A map A from a linear space V to another linear space $\tilde{V}$ is called *affine* if it maps affine combinations $\sum_{i=1}^r \alpha_i v_i$ of vectors $v_1, \ldots, v_r$ in V into the same affine combinations $\sum_{i=1}^r \alpha_i A(v_i)$ of the image vectors $A(v_1), \ldots, A(v_r)$. An affine map to $\mathbb{R}$ is called an *affine functional.*

As in the previous theorem, in order to verify the affinity of a map A it suffices to show that $A((1 - \alpha)v_1 + \alpha v_2) = (1 - \alpha)A(v_1) + \alpha A(v_2)$ when $v_1, v_2 \in V$ and $\alpha \in [0, 1]$.

It is readily verified that a map A from a linear space V to another linear space $\tilde{V}$ is affine if and only if there is a vector $\tilde{v}_0$ in $\tilde{V}$ and a linear map T such that $A(v) = \tilde{v}_0 + T(v)$ when $v \in V$. Indeed necessarily we have $\tilde{v}_0 = A(0)$ and $T(v) = A(v) - A(0)$ and it is a matter of checking that T defined this way is in fact linear.

If v_0 is a vector in V then the affine map $v \to v_0 + v$ is called a *translation.* It follows that a map is affine if and only if it is obtained by letting a translation act on a linear map.

A *translate* of a subset W of a linear space V is a subset of the form $v + W$ where $v \in V$. Affine sets which are translates of each other are called *parallel.* Any affine set is parallel to a unique linear space.

The *dimension of a linear space* is the unique cardinality of maximal independent bases. We shall use the notation $\dim V$ for the dimension of a linear space V.

The dimension of an affine space A is defined as the dimension of the unique linear space parallel to A. We shall denote the dimension of an affine space A by $\dim \operatorname{aff} A$.

It is readily verified that just as direct or inverse images of linear sets by linear maps are linear, also direct or inverse images of affine (convex) sets by affine maps are affine (convex).

Prominent among the affine sets are the *lines,* i.e. the one-dimensional affine sets. In fact a set W is affine if and only if it contains any line joining any two points in W.

A subset L of V is a line if and only if it is the image of $\mathbb{R}$ for an affine 1–1 map from $\mathbb{R}$ to V.

Prominent among the convex sets are the *line segments* i.e. the images of $[0, 1]$ by affine maps from $\mathbb{R}$ to V. If $v, w \in V$ then the line segment $\{(1 - t)v + tw; t \in [0, 1]\}$ will be called the line segment joining v and w. In fact a set W is convex if and only if it contains any line segment joining any two of its points.

The line segment joining v and w will be denoted as $[v, w]$. The vectors v

and w are called the *end points* of $[v, w]$. It is readily checked that the two-point set $\{v, w\}$ of end points is uniquely determined by the segment $[v, w]$.

According to this notation a one-point set $\{v\}$ may also be considered as a (degenerate) line segment $[v, v]$.

The collections of linear sets, affine sets and of convex sets are all closed under the formation of non-empty intersections. As V itself is convex it follows that each non-empty subset G of V is contained in a smallest subspace as well as in a smallest affine set and as well as in a smallest convex set. The smallest subspace containing G is called the *linear span of* G and is denoted as $[G]$. The smallest affine set containing G is called the *affine hull of* G and is denoted as aff G. The smallest convex set containing G is called the *convex hull of* G and is denoted as $\langle G \rangle$.

Direct descriptions of these smallest sets are simple.

Theorem 2.2.5. *Let G be a non-empty subset of the linear set V. Then the linear span $[G]$, the affine hull* aff G *and the convex hull $\langle G \rangle$ are, respectively, the set of linear combinations of vectors from G, the set of affine combinations of vectors from G and the set of convex combinations of vectors from G.*

Proof. The three sets last mentioned are, respectively, linear, affine and convex. □

It follows that the line segment $[v, w]$ joining v and w is the convex hull $\langle \{v, w\} \rangle$. If $v \neq w$ then the line through v and w is the affine hull $\operatorname{aff}\{v, w\}$.

Convex hulls of finite sets of vectors are called *polytopes*.

Convex hulls of finite and affinely independent sets of vectors are called *simplices*. Thus a simplex is a set of the form $\langle \{g_1, \ldots, g_r\} \rangle$ where $g_1, \ldots, g_r$ are affinely independent. It follows from Proposition 2.2.3 that the convex hull $\langle G \rangle$ of a set G may be expressed as a union of simplices.

Two natural dimensions which may be associated with a subset U of V are

(i) the dimension $\dim[U]$ of the linear span $[U]$ of U, and
(ii) the dimension $\dim \operatorname{aff} U$ of the affine hull $\operatorname{aff} U$ of U.

(This notation is consistent since $\dim \operatorname{aff} U = \dim[U]$ when $\operatorname{aff} U = [U]$ i.e. when $0 \in \operatorname{aff} U$.)

The dimension $\dim \operatorname{aff} U$ is sometimes called the *geometric dimension of* U while the dimension $\dim[U]$ is usually just called the *dimension of* U.

If U is finite dimensional then $\dim[U] = (\dim \operatorname{aff} U) + 1$ or $= \dim \operatorname{aff} U$ as $0 \notin \operatorname{aff} U$ or $0 \in \operatorname{aff} U$.

A *face* of a convex set C is a convex subset D of C such that $y, z \in D$ whenever $y, z \in C$ and $\frac{1}{2}(y + z) \in D$. Thus D is a face of C if any line segment in C having its centre in D is entirely contained in D. If so then any line segment in C having an interior point in D is entirely contained in D.

Proposition 2.2.6. *A convex subset D of a convex set C is a face of C if and only if $x_1,\ldots,x_r \in D$ whenever $x_1,\ldots,x_r \in C$ and $\sum_{i=1}^r \lambda_i x_i \in D$ for positive numbers $\lambda_1,\ldots,\lambda_r$ such that $\sum_{i=1}^r \lambda_i = 1$.*

Proof. The last condition is clearly sufficient so suppose D is a face of C. Consider the set M of integers $r \geqslant 1$ such that $x_1,\ldots,x_r \in D$ whenever $\sum_{i=1}^r \lambda_i x_i \in D$ for some positive constants $\lambda_1,\ldots,\lambda_r$ such that $\lambda_1 + \cdots + \lambda_r = 1$. Trivially $1 \in M$. If $x_1,\ldots,x_r \in C$, $\sum \lambda_i x_i \in D$ where $\lambda_1,\ldots,\lambda_r > 0$ and $\lambda_1 + \cdots + \lambda_r = 1$ then $\lambda_i x_i + (1-\lambda_i)\sum_{j\neq i}(\lambda_j/(1-\lambda_i))x_j \in D$ so that $x_i \in M$; $i = 1,\ldots,r$ provided $2 \in M$. It remains therefore to show that $2 \in M$.

Consider vectors y,z in C such that $(1-\lambda)y + \lambda z \in D$ for some $\lambda \in]0,1[$. Let T be the interval $\{t : 0 \leqslant t \leqslant 1, (1-t)y + tz \in D\}$. Then T has the property that $t_1,t_2 \in T$ whenever $t_1,t_2 \in [0,1]$ and $\frac{1}{2}(t_1+t_2) \in T$. (In other words T is a face of $[0,1]$.) If $t \in T$ and if $t \leqslant \frac{1}{2}$ then, since $t = \frac{1}{2}(0+2t)$ we find that $0, 2t \in T$. If $t \in T$ and if $t \geqslant \frac{1}{2}$ then, since $t = \frac{1}{2}(2t-1+1)$ we find that $2t-1, 1 \in T$.

If $0 \notin T$ then this shows that necessarily $\lambda > \frac{1}{2}$ and thus that $2\lambda - 1 \in T$. Hence $2\lambda - 1 > \frac{1}{2}$ and $4\lambda - 3 \in T$. Proceeding this way we arrive at the contradiction that all the numbers $2\lambda-1, 4\lambda-3, 8\lambda-7,\ldots$ are $>\frac{1}{2}$. Thus $0 \in T$ so that $y \in D$ and, by the same argument, $z \in D$. □

The faces of a given convex set constitute a lattice.

Proposition 2.2.7. *If $D_i : i \in I$ is a family of faces of the convex set C then their intersection $D = \bigcap_i D_i$ is also a face of C.*

Furthermore, faces of faces are faces, i.e. if $D_1 \subseteqq D_2 \subseteqq \cdots \subseteqq D_r$ are convex sets such that D_i is a face of D_{i+1} when $i = 1,2,\ldots,r-1$ then D_1 is a face of D_r.

Proof. The first statement follows directly from the definition and the second statement follows by induction on r. □

A point x is called an *extreme point* of the convex set C if the one-point set $\{x\}$ is a face of C. Thus if $x \in C$ then x is an extreme point of C if and only if $x = y = z$ whenever $x = \frac{1}{2}(y+z)$ where $y,z \in C$. If so then $x_1 = \cdots = x_r = x$ whenever $x = \sum_{i=1}^r \lambda_i x_i$ where $x_1,\ldots,x_r \in C$, $\lambda_1,\ldots,\lambda_r > 0$ and $\lambda_1 + \cdots + \lambda_r = 1$.

The set of extreme points of a convex set C will be denoted as $\operatorname{ext} C$.

If we want to express a given convex set C as the convex hull $\langle G \rangle$ of a subset G then $\operatorname{ext} C$ is a smallest candidate.

Proposition 2.2.8. *If $C = \langle G \rangle$ then $G \supseteqq \operatorname{ext} C$.*

Proof. This follows readily from the characterization of extreme points given above. □

Convex sets need not have extreme points and if they have then they may not generate C.

If C is a finite dimensional compact convex set then, as we shall see in the next section, C is the convex hull of its extreme points, i.e. $C = \langle \text{ext}\, C \rangle$. The Krein–Milman theorem asserts that if C is a compact convex set in a locally convex linear space V then C is the closure of $\langle \text{ext}\, C \rangle$.

Here we shall just consider the simple case of finitely generated convex sets i.e. polytopes.

Proposition 2.2.9. *$C = \langle \text{ext}\, C \rangle$ when C is a polytope.*

Proof. Express C as $C = \langle G \rangle$ where $G = \{g_1, \ldots, g_r\}$. We may assume that G is minimal in the sense that $C \supset \langle H \rangle$ whenever $H \subset G$. Assume $x = \frac{1}{2}(y + z) \in G$ where $y, z \in C$. Put $y = \sum_i \lambda_i g_i$ and $z = \sum \mu_i g_i$ where $\lambda_1, \ldots, \lambda_r, \mu_1, \ldots, \mu_r \geqslant 0$ and $1 = \sum \lambda_i = \sum \mu_i$. Then $x = \sum \frac{1}{2}(\lambda_i + \mu_i) g_i$. As $x \in G$ we may write $x = g_s$ where $1 \leqslant s \leqslant r$. Then $\frac{1}{2}(\lambda_s + \mu_s) = 1$ since otherwise g_s would be expressible as a convex combination of vectors from $G - \{g_s\}$ and this contradicts the minimality of G. Hence $\lambda_s = \mu_s = 1$ so that $y = z = x$. It follows that any point in G is an extreme point of C. □

Let us move on to establish some standard terminology on separation of convex sets by hyperplanes and by linear functionals.

A *hyperplane* A in V is a set of the form $A = [f = \alpha]$ for some non-zero linear functional f on V and some constant $\alpha \in \mathbb{R}$. Hyperplanes are clearly affine and an affine set A is a hyperplane if and only if the linear space W which is parallel to A has co-dimension 1, i.e. $W \cup \{u\}$ spans V for some vector $u \notin W$.

If f is a non-zero linear functional and if $\alpha \in \mathbb{R}$ then the linear space parallel to the hyperplane $[f = \alpha]$ is the hyperplane $[f = 0]$.

The representation of a hyperplane as a contour $[f = \alpha]$ of a non-zero linear functional f is unique up to a multiplicative scalar. In other words if f and g are both non-zero linear functionals and if α and β are constants then the hyperplanes $[f = \alpha]$ and $[g = \beta]$ are equal if and only if there is a constant κ such that $g = \kappa f$ and $\beta = \kappa\alpha$.

If the hyperplane A is represented as $A = [f = \alpha]$ for a non-zero linear functional f and a constant α then the set of points which does not belong to A may be separated into two parts (sides). One part is the 'open halfspace' $[f < \alpha]$ and the other part is the 'open halfspace' $[f > \alpha]$. This decomposition of A^c clearly does not depend on which representation of A we are using.

Now consider two non-empty subsets S and T of V, along with a hyperplane A.

We shall say that *S and T are weakly separated by A* if they are not both contained in A and if they are contained in different parts of the decomposition described above.

We shall say that *A separates S and T strongly* if A may be represented as the contour $A = [f = \alpha]$ of a linear functional f such that either $\sup\{f(v) : v \in S\} < \alpha < \inf\{f(v) : v \in T\}$ or $\inf\{f(v) : v \in S\} > \alpha > \sup\{f(v) : v \in T\}$. If this condition is satisfied for one representation of A then it is satisfied for all other representations of A.

It is often more convenient to express separation directly in terms of linear functionals. Thus if S and T are non-empty subsets of V and if f is a linear functional on V then we shall say that f *separates S and T weakly* if f is not constant on $S \cup T$ and if either $\sup\{f(v) : v \in S\} \leqslant \inf\{f(v) : v \in T\}$ or if $\inf\{f(v) : v \in S\} \geqslant \sup\{f(v) : v \in T\}$. We shall say that a linear functional f *separates S and T strongly* if either $\sup\{f(v) : v \in S\} < \inf\{f(v) : v \in T\}$ or if $\inf\{f(v) : v \in S\} > \sup\{f(v) : v \in T\}$.

It follows that a linear functional f separates S and T weakly (strongly) if and only if some hyperplane $[f = \alpha]$ separates S and T weakly (strongly).

The problem of separating two given sets by hyperplanes, or by linear functionals, may be reduced to the problem of separating their convex hulls. In fact a hyperplane A (a linear functional f) separates S and T weakly (strongly) if and only if it separates the convex hulls $\langle S \rangle$ and $\langle T \rangle$ weakly (strongly). Actually we may even reduce the problem of separating S and T to the problem of separating the origin from the convex set $\langle S \rangle - \langle T \rangle = \langle S - T \rangle$. We state these readily verified facts as a proposition.

Proposition 2.2.10. *A linear functional f separates weakly (strongly) the non-empty sets S and T if and only if it separates weakly (strongly) $\{0\}$ and $\langle S - T \rangle$ where $S - T = \{v - w : v \in S, w \in T\}$.*

We shall return to concrete criteria for separation of finite dimensional convex sets in the next section.

Let us proceed from convexity of sets to convexity of functions. Following Rockafellar (1970) we shall say that a function f from V to the extended real line $[-\infty, \infty]$ is *convex* if its *epigraph* $\{(v, t) : v \in V, t \in \mathbb{R}, f(v) \leqslant t\}$ is a convex subset of $V \times \mathbb{R}$. If $-f$ is convex then we shall say that f is *concave*.

The *effective domain* of a convex function f on V is the convex set $\{v : v \in V, f(v) < \infty\}$.

The standard definition of a convex function is consistent with the definition.

Proposition 2.2.11 (Jensen's inequality* I*). *A function f from V to $]-\infty, \infty]$ is convex if and only if*

$$f((1 - \theta)v + \theta w) \leqslant (1 - \theta)f(v) + \theta f(w)$$

whenever $\theta \in]0, 1[$ and $v, w \in V$.

If f is convex then $f(\lambda_1 v_1 + \cdots + \lambda_r v_r) \leqslant \lambda_1 f(v_1) + \cdots + \lambda_r f(v_r)$ whenever $v_1, \ldots, v_r \in V$, $\lambda_1, \ldots, \lambda_r \geqslant 0$ and $\lambda_1 + \cdots + \lambda_r = 1$.

Remark. The inequalities in this proposition are versions of Jensen's inequality. In the next section we shall establish this inequality for general probability distributions on finite dimensional vector spaces.

If f is defined merely on a convex set C and if the epigraph $\{(v,t): v \in C, t \in \mathbb{R}, f(v) \leqslant t\}$ of f on C is convex then the extension of f to all of V is obtained by putting $f(v) = +\infty$ when $v \notin C$ is convex. In particular we see that a subset K of V is convex if and only if $\delta_K = \infty \cdot I_{K^c}$ is convex. The function δ_K, rather than the usual indicator function I_K, appears to be the 'correct' choice of an 'indicator' function of K in convex analysis.

If f is a convex function on V then the level sets $\{v: f(v) \leqslant \alpha\}$; $\alpha \in \mathbb{R}$ are convex. This property is shared by many other functions e.g. all functions of the form $\phi(f)$ where f is a linear functional and where ϕ is monotone. According to the definition in section 3.2, an extended real valued function f on V is quasiconvex if all the level sets $\{v: f(v) \leqslant \alpha\}$ are convex. If $-f$ is quasiconvex then f is quasiconcave.

Thus convex (concave) functions are quasiconvex (quasiconcave). The class of quasiconvex functions may be considered as being substantially larger than the class of convex functions (see section 3.2 and Roberts & Varberg (1973)).

Pointwise maxima of linear functionals play a particularly important role in the theory of comparison of experiments. A real valued function ψ on V which is a pointwise maxima of linear functionals is both convex and non-negatively homogeneous or, in other words, it is subadditive and non-negatively homogeneous. Functions having these properties are called *sublinear*.

Thus a real valued function ψ on V is sublinear if and only if:

(i) $\psi(v + w) \leqslant \psi(v) + \psi(w)$; $v, w \in V$

and

(ii) $\psi(tv) = t\psi(v)$; $v \in V$; $t \geqslant 0$.

If ψ is sublinear then:

(iii) $|\psi(v) - \psi(w)| \leqslant \psi(v - w) \vee \psi(w - v)$; $v, w \in V$.

It follows from (iii) that a sublinear function, in particular a linear functional, is continuous for a vector topology on V if and only if ψ is continuous at the origin, and if so then ψ is automatically uniformly continuous.

We shall see in a moment that sublinear functionals are maxima of linear functionals.

For now, note that the class of all sublinear functionals on V constitutes a convex cone having the particular property that it is closed under the formation of suprema of families which are pointwise bounded from above.

A sublinear functional ψ which is reflection symmetric with respect to the origin, i.e. $\psi(v) \equiv_v \psi(-v)$, is called a *pseudonorm*.

It follows from (iii) above that pseudonorms are always non-negative. Pseudonorms which are positive outside the origin are called *norms*.

Consider a sublinear function ψ along with a given vector $v_0 \in V$. Then $tv_0 \to t\psi(v_0)$ defines a linear functional $\tilde{l}$ on the space $[\{v_0\}]$ spanned by v_0. If $t \geqslant 0$ then $\tilde{l}(tv_0) = \psi(tv_0)$ while $\tilde{l}(tv_0) = -|t|\psi(v_0) \leqslant |t|\psi(-v_0) = \psi(tv_0)$ when $t \leqslant 0$. Thus $\tilde{l} \leqslant \psi$ on $[\{v_0\}]$. The Hahn–Banach theorem, stated below, ensures that there is a linear function l which extends $\tilde{l}$, i.e., $l(v_0) = \tilde{l}(v_0)$, and which is majorized by ψ. Thus any real valued sublinear function ψ is a pointwise maximum of linear functionals.

If we replace the 0–1 dimensional space $\{tv_0 : t \in \mathbb{R}\}$ by an arbitrary linear subspace W and consider an arbitrary linear function $\tilde{l}$ on W such that $\tilde{l} \leqslant \psi | W$ then the Hahn–Banach theorem ensures that we may extend $\tilde{l}$ to a linear functional l on V such that $l \leqslant \psi$.

Sometimes we would like to know more than just that such an extension exists. In particular we might want to inquire about the measureability of the extension with respect to $(\tilde{l}, \psi)$. For this reason, let us take a look at the 'standard' proof of this theorem. Another interesting proof is given in Valentine (1964).

Now let ψ, W and $\tilde{l}$ be given as above and consider any vector $v_0 \in V$. An extension of $\tilde{l}$ to a linear functional on $W + [v_0]$ is necessarily of the form $w + tv_0 \to \tilde{l}(w) + tc_0$; $w \in W, t \in \mathbb{R}$ for some constant c_0. (If $v_0 \in W$ then $c_0 = \tilde{l}(v_0)$. Otherwise there is no restriction on the number c_0.)

If we want to preserve majorization by ψ then we require that $\tilde{l}(w) + tc_0 \leqslant \psi(w + tc_0)$ whenever $w \in W$ and $t \in \mathbb{R}$. Dividing through by $|t|$ when $t \neq 0$ we see that c_0 must satisfy the inequalities

$$\sup\{\tilde{l}(w) - \psi(w - v_0) : w \in W\} \leqslant c_0 \leqslant \inf\{\psi(w + v_0) - \tilde{l}(w) : w \in W\}.$$

The existence of at least one such number c_0 follows by sublinearity (actually subadditivity) since

$$\tilde{l}(w_1) - \psi(w_1 - v_0) = \psi(w_2 + v_0) - \tilde{l}(w_2) + [\ ]$$

where

$$\begin{aligned} [\ ] &= \tilde{l}(w_1 + w_2) - [\psi(w_1 - v_0) + \psi(w_2 + v_0)] \\ &\leqslant \tilde{l}(w_1 + w_2) - \psi((w_1 - v_0) + (w_2 + v_0)) \\ &= \tilde{l}(w_1 + w_2) - \psi(w_1 + w_2) \leqslant 0 \end{aligned}$$

when $w_1, w_2 \in W$.

If the vector space V is finite dimensional then after a finite number of steps this procedure provides a linear extension l to all of V such that l is majorized by ψ. If V has a countable basis then the same procedure yields an induction proof for the existence of such an extension l on $\tilde{l}$.

In the last case we may express W as $W = \bigcup_n W_n$ where $W_1 \subseteqq W_2 \subseteqq \cdots$ are finite dimensional subspaces of W. If $\tilde{W}_n$, for each n, is a countable dense subset of W_n, then the infimum and the supremum defining the end points of the intervals of possible constants c_0 may be taken over the countable set $\bigcup_n \tilde{W}_n$.

In particular we see that if $\tilde{l} = \tilde{l}_t$ and $\psi = \psi_t$ and both depend measurably on a parameter t then we may in this case ensure that the extension $l = l_t$ also depends measurably on t.

In the general case the existence of a linear extension of $\tilde{l}$ to all of V, which is majorized by ψ, follows from Zorn's lemma. In fact, by the above argument, any maximal majorized extension must involve all of V.

Theorem 2.2.12 (The Hahn–Banach theorem). *Let ψ be a sublinear function on a (real) vector space V and let $\tilde{l}$ be a linear functional on a linear subspace W of V such that*

$$\tilde{l}(w) \leqslant \psi(w) : w \in W.$$

Then $\tilde{l}$ may be extended to a linear functional l on all of V such that

$$l(v) \leqslant \psi(v) : v \in V.$$

Sublinear functionals are pointwise maxima of linear functionals.

Corollary 2.2.13 ('Tangents' of sublinear functions). *If ψ is a sublinear function on V and if $v_0 \in V$ then there is a linear functional l on V such that $l \leqslant \psi$ and such that $l(v_0) = \psi(v_0)$.*

The possibility of a measurable extension in the case of a separable Banach space is implied by the following corollary.

Corollary 2.2.14. *Consider a family $\{\psi_t : t \in T\}$ of continuous sublinear functionals on a separable Banach space V.*

Let W be a linear subspace of V and let $\tilde{l}_t$, for each t, be a linear functional on W such that $\tilde{l}_t \leqslant \psi_t$ on W.

Then there is a family $l_t : t \in T$ of linear functionals on V such that:

(i) $l_t | W = \tilde{l}_t;\ t \in T$
(ii) $l_t \leqslant \psi_t;\ t \in T$

and

(iii) *$l_t(v)$ for each $v \in V$, is measurable in t for any σ-algebra of subsets of T making $\psi_t(v)$ and $\tilde{l}_t(w)$ measurable in t when $v \in V$ and $w \in W$.*

Proof. Let $\mathscr{S}$ be a σ-algebra of subsets of T making $\psi_t(v)$ and $\tilde{l}_t(w)$ measurable in t when $v \in V$ and $w \in W$. By separability V contains a countable dense subset $D = \{v_1, v_2, \ldots\}$ such that $D \cap W$ is dense in W. Put $V_0 = [D]$ and $W_0 = [D \cap W]$. By the remarks made before the statement of the Hahn–Banach theorem, there is for each $t \in T$ a linear functional $\tilde{l}_t$ on V_0 such that

$\hat{l}_t \leqslant \psi_t | V_0$, $\hat{l}_t | W_0 = \tilde{l}_t | W_0$ and such that $\hat{l}_t(v)$ is measurable in t whenever $v \in V_0$. As $\hat{l}_t$ is continuous on V_0 it has a unique continuous extension l_t to all of V and it is readily checked that the functionals $l_t : t \in T$ have the desired properties. □

Sublinear functionals are often called support functions. In fact if H is any set of linear functionals on V then the *support function* ψ_H of H is defined as the extended real valued function

$$v \to \sup\{l(v) : l \in H\}.$$

Support functions are sublinear but not necessarily real valued. Furthermore support functions cannot distinguish between a set and its convex hull since clearly $\psi_H = \psi_{\langle H \rangle}$ for any set H of linear functionals.

Support functions may be used to convert set operations on convex sets into pointwise operations on their support functions.

Proposition 2.2.15 (Computations with support functions). *If H, H_1 and H_2 are subsets of the algebraic dual V' of the vector space V and if λ_1 and λ_2 are non-negative constants then:*

(i) $\psi_{H_1 + H_2} = \psi_{H_1} + \psi_{H_2}$

(ii) $\psi_{\lambda H} = \lambda \psi_H$

and

(iii) $\psi_{H_1 \cup H_2} = \psi_{H_1} \vee \psi_{H_2}$.

Furthermore $\psi_{H_1} \leqslant \psi_{H_2}$ whenever $H_1 \subseteqq H_2 \subseteqq V'$.

A closer examination of support functions on finite dimensional vector spaces will be given at the end of the next section.

2.3 Convexity in finite dimensional spaces

A finite dimensional vector space V has the peculiar property that there is just one Hausdorff topology on V which converts V into a linear topological space (see e.g. Valentine (1964) for a proof). In terms of a linearly independent basis $(v_1, \ldots, v_n)$ this is the topology obtained by identifying vectors in V with the coordinate vectors they defined in $\mathbb{R}^n$. It follows in particular that if V is a finite dimensional linear subspace of a Hausdorff linear topological vector space U then the relativized topology from U on V is unique.

Moreover, any finite dimensional linear subset of a Hausdorff linear topological space is closed so that closedness and closures of finite dimensional sets do not depend on their possibly infinite dimensional surroundings. Keeping this in mind the results below for finite dimensional vector spaces may be translated into results for finite dimensional subsets of general linear spaces.

On the other hand, if the underlying space is finite dimensional then, since finite dimensional Hausdorff linear topological spaces having the same dimensions are isomorphic, its Hausdorff topology is the topology of any given norm, e.g. the norm defined by a given inner product. The main advantage of using inner products is the familiar representation of linear functionals as vectors from the same space V.

Let us begin by making the trivial observation that interiors of interesting convex sets may be empty for the reason that they are imbedded in spaces of higher dimension than the flat (affine space) they support. Thus for example a triangle in 3-dimensional Euclidean space has no interior points.

The appropriate notion of an interior in such situations is then the interior relative to the affine hull spanned by the set. According to the usual terminology this relative interior is called *the relative interior*. Thus we shall here use the term '*the relative interior of* C' as synonymous with 'the interior of C relative to aff C'. Similarly the boundary of C relative to aff C will be called *the relative boundary of* C. We shall say that C is *relatively open* if it is open relative to its affine hull.

The interior and closure of a set C will here be denoted as $\bar{C}$ and $\mathring{C}$ respectively. Thus the boundary of C is the set $\operatorname{bd} C = \bar{C} - \mathring{C}$. The relative interior and the relative boundary of a set C will be denoted as ri C and rbd C respectively. Thus $\operatorname{rbd} C = \bar{C} - \operatorname{ri} C$. If $\mathring{C} \neq \varnothing$ then the two interiors are the same, i.e. $\operatorname{ri} C = \mathring{C}$.

A basic fact on relative interiors of convex sets is the following theorem.

Theorem 2.3.1. *The relative interior of a non-empty convex set is non-empty.*

Proof. Let C be a convex set containing a point v. Choose a maximal set of linearly independent vectors $w_1, \ldots, w_r$ in $C - v$. Put $T(\xi) = x + \sum_i \xi_i(w_i - v) = (1 - \sum_i \xi_i)v + \sum_i \xi_i v_i$ when $\xi = (\xi_1, \ldots, \xi_r) \in \mathbb{R}^r$. Then T maps $\mathbb{R}^r$ homeomorphically onto aff C.

If D denotes the open set $\{\xi : \xi_1 > 0, \ldots, \xi_r > 0, \sum_i \xi_i < 1\}$ then $T[D]$ is an open subset of aff C which is entirely contained in C. □

In addition to the notation $[v, w]$ for the line segment between v and w we shall use the notation $[v, w[$ for the 'half open' segment $[v, w] - \{w\}$ and $]v, w]$ for the 'half open' segment $[v, w] - \{v\}$. Removing both end points we obtain the 'open' segment $[v, w] - \{v, w\}$ which will be denoted as $]v, w[$.

Theorem 2.3.2 (Characterizations of the relative interior). *The following conditions are equivalent for a vector v belonging to a convex set C.*

(i) $v \in \operatorname{ri} C$.
(ii) *If $w \in \operatorname{aff} C$ and $w \neq v$ then $]v, w[\cap C \neq \varnothing$.*
(iii) *If $w \in C$ and $w \neq v$ then $v \in]u, w[$ for some $u \in C$.*

Proof. If $v \in \operatorname{ri} C$ and $w \in C$ then, since $\lim_{t\to 0}(1-t)v + tw = v$, the vector $(1-t)v + tw$ is in C provided t is sufficiently small. Thus (i) $\Rightarrow$ (ii). If (ii) is satisfied and if $w \in C$ then, since $v - (w - v) = 2v - w \in \operatorname{aff} C$, there is a number α in $]0,1[$ so that $u = (1-\alpha)v + \alpha(2v - w) = (1+\alpha)v - \alpha w \in C$. Then $v = (1-t)u + tw$ where $t = \alpha/(1+\alpha)$. Hencc (ii) $\Rightarrow$ (iii). Assume finally that (iii) holds. Choose a maximal set of linearly independent vectors $w_1 - v, \ldots, w_r - v$ in $C - v$. Then $w_1, \ldots, w_r \in C$. Let T be the homeomorphism between $\mathbb{R}^r$ and $\operatorname{aff} C$ defined by putting $T(\xi) = v + \sum_j \xi_j(w_j - v) = (1 - \sum \xi_j)v + \sum_j \xi_j w_j$ where $\xi = (\xi_1, \ldots, \xi_r) \in \mathbb{R}^r$. By (iii) for each $j = 1, \ldots, r$ there is a number α_j in $]0,1[$ and a vector u_j in C such that $v = (1-\alpha_j)u_j + \alpha_j w_j$. Then $\alpha_j(w_j - v) = -(1-\alpha_j)(u_j - v)$; $j = 1, \ldots, r$. Letting the summation symbols $\sum^+$ and $\sum^-$ indicate, respectively, summation on $\{j : \xi_j \geqslant 0\}$ and $\{j : \xi_j < 0\}$ we may now rewrite $T(\xi)$ as

$$
\begin{aligned}
T(\xi) &= v + \textstyle\sum^+ \xi_j(w_j - v) + \sum^- \xi_j(w_j - v) \\
&= v + \textstyle\sum^+ \xi_j(w_j - v) + \sum^- - \xi_j \dfrac{1-\alpha_j}{\alpha_j}(u_j - v) \\
&= \left(1 - \textstyle\sum^+ |\xi_j| - \sum^- |\xi_j| \dfrac{1-\alpha_j}{\alpha_j}\right) v + \textstyle\sum^+ |\xi_j| w_j + \sum^- |\xi_j| \dfrac{1-\alpha_j}{\alpha_j} u_j
\end{aligned}
$$

and this is a convex combination of vectors from C provided $\sum^+ |\xi_j| + \sum^- |\xi_j|(1-\alpha_j)/\alpha_j \leqslant 1$.

Since the left hand side is majorized by $\kappa \sum |\xi_j|$ where $\kappa = \max\{1, (1-\alpha_1)/\alpha_1, \ldots, (1-\alpha_r)/\alpha_r\}$ it follows that this provision is satisfied when $\sum_j |\xi_j| < 1/\kappa$. Hence T maps the open set $\{\xi : \sum_j |\xi_j| < 1/\kappa\}$ onto an open subset of $\operatorname{aff} C$ which is entirely contained in C and which contains the point $v = T(0)$. Hence $v \in \operatorname{ri} C$ and (iii) $\Rightarrow$ (i). □

Corollary 2.3.3. *If v and w are vectors in the convex set C and if, say, $v \in \operatorname{ri} C$, then $[v, w[\subseteqq \operatorname{ri} C$.*

Remark. We shall soon see that C and $\bar{C}$ have the same relative interior points. Thus $[v, w[\subseteqq \operatorname{ri} C$ whenever $v \in C$ and $w \in \bar{C}$.

Proof. Let $v_0 \in \operatorname{ri} C$, $v_1 \in C$ and put $v_\theta = (1-\theta)v_0 + \theta v_1$ where $\theta \in]0,1[$. If $w \in \operatorname{aff} C$ and if $w \neq 0$ then there is a $t \in]0,1[$ such that $v_t = (1-t)v_0 + tv_1 \in C$. Putting $u = (1-\theta t)^{-1}[(1-t)v_\theta + (1-\theta)w] = (1-\theta)(1-\theta t)^{-1}[(1-t)v_0 + tw] + \theta(1-t)(1-\theta t)^{-1}v_1$ we find that $u \in]v_\theta, w[\cap C$. It follows that $v_\theta \in \operatorname{ri} C$. □

Corollary 2.3.4. *If v and w are points in a convex set C and if $v \in \mathring{C}$, then $[v, w[\subseteqq \mathring{C}$.*

Proof. This follows from the previous corollary and the fact that $\operatorname{ri} C = \mathring{C}$ when $\mathring{C} \neq \emptyset$. □

It follows that interiors, as well as relative interiors, of convex sets are convex. Furthermore, by the continuity of the vector operations, closures of convex sets are also convex.

Theorem 2.3.5. *If C is convex then the sets* $\operatorname{ri} C$, $\mathring{C}$ *and* $\bar{C}$ *are all convex.*

In general, seven different sets (but not more) may be obtained from a given set A by successive applications of the closure operation and the interior operation. However, if the set A is convex, then at most three sets may be so obtained. This is a consequence of the following theorem.

Theorem 2.3.6. *If C is convex then:*

(i) $\operatorname{ri} \bar{C} = \operatorname{ri} C$
(ii) $\bar{C} = \overline{\operatorname{ri} C}$
(iii) $\mathring{\bar{C}} = \mathring{C}$
and
(iv) $\bar{C} = \bar{\mathring{C}}$ *when* $\mathring{C} \neq \emptyset$.

Proof. As $\operatorname{ri} C = \mathring{C}$ when $\mathring{C} \neq \emptyset$ the last two statements follow from the two first ones. Since $\operatorname{aff} C$ is closed it follows that $\operatorname{aff} C = \operatorname{aff} \bar{C}$. Thus $\operatorname{ri} C \subseteqq \operatorname{ri} \bar{C}$. On the other hand, consider a point v in $\operatorname{ri} \bar{C}$. Choose a maximal set of linearly independent vectors $w_1 - v, \dots, w_k - v$ in $\bar{C} - v$. Then $w_1, \dots, w_k \in \bar{C}$. If δ is sufficiently small then the vectors $v + \delta(w_i - v)$; $i = 1, \dots, k$ and the vector $v - \delta \sum_{i=1}^k (w_i - v)$ all belong to $\bar{C}$. Replacing w_i with $\tilde{w}_i = v + \delta(w_i - v)$; $i = 1, \dots, k$ we see that we may as well assume that $\delta = 1$ i.e. that the vectors $w_1, \dots, w_k$ and $v - \sum_i (w_i - v)$ all belong to $\bar{C}$. Then there are sequences $\{u_{n,i}\}$; $i = 1, \dots, k$ and $\{u_n\}$ in C such that $u_{n,i} \to w_i$; $i = 1, \dots, k$ and $u_n \to v - \sum_i (w_i - v)$.

Since $w_1 - v, \dots, w_k - v$ are linearly independent, we may assume that the vectors $u_{n,1} - v, \dots, u_{n,k} - v$ are linearly independent for each n. We may then, for each n, express $u_n - v$ as a linear combination $u_n - v = \sum_i \alpha_{n,i}(u_{n,i} - v)$. It follows that $\alpha_{n,i} \to -1$; $i = 1, \dots, k$. Thus we may also assume that $\alpha_{n,i} < 0$ for each n and for each i. Now the vectors $u_{n,1} - v, \dots, u_{n,k} - v$ and $u_n - v$ all belong to the convex set $C - v$. Hence the convex combination $0 = (1 - \sum_i \alpha_{n,i})^{-1}[(u_n - v) - \sum_i \alpha_{n,i}(u_{n,i} - v)]$ also belongs to $C - v$ i.e. $v \in C$.

It follows that $\operatorname{ri} \bar{C} \subseteqq C$. Hence, since $\operatorname{ri} C$ is the largest subset of C which is open relative to $\operatorname{aff} C$, we find that $\operatorname{ri} \bar{C} \subseteqq \operatorname{ri} C$ so that $\operatorname{ri} \bar{C} = \operatorname{ri} C$.

The monotonicity of the closure operation immediately yields that $\overline{\operatorname{ri} C} \subseteqq \bar{C}$. On the other hand, consider a point w in $\bar{C}$ and also a point v in

$\operatorname{ri}\bar{C}$. Then, by corollary 2.3.3, $(1-t)v + tw \in \operatorname{ri}\bar{C} = \operatorname{ri}C$ when $t < 1$. Hence, letting $t \to 1$, we see that $w \in \overline{\operatorname{ri}C}$. It follows that $\bar{C} \subseteqq \overline{\operatorname{ri}C}$ so that $\bar{C} = \overline{\operatorname{ri}C}$. □

Clearly images of convex sets by affine maps are convex. Furthermore affine images of relatively open sets are relatively open.

Corollary 2.3.7 (Affine images of convex sets). *If T is an affine map from one finite dimensional linear space into another finite dimensional linear space then* $\operatorname{ri}T[C] = T[\operatorname{ri}C]$ *for any convex subset C of the domain of T.*

In particular $\operatorname{ri}(C_1 - C_2) = \operatorname{ri}C_1 - \operatorname{ri}C_2$ *for convex subsets C_1 and C_2 of the same finite dimensional vector space V.*

Proof. Let T be an affine map from a linear space V onto a linear space W and let C be a convex subset of V. Trivially $\operatorname{ri}C \subseteqq C \subseteqq \bar{C}$. Hence, since $\overline{\operatorname{ri}C} = \bar{C}$, $T[\operatorname{ri}C] \subseteqq T[C] \subseteqq T[\bar{C}] = T[\overline{\operatorname{ri}C}] \subseteqq \overline{T[\operatorname{ri}C]}$. Thus the term on the extreme right is the closure of the term on the extreme left. Hence $\overline{T[C]} = \overline{T[\operatorname{ri}C]}$ so that by theorem 2.3.6, $\operatorname{ri}T[C] = \operatorname{ri}\overline{T[C]} = \operatorname{ri}\overline{T[\operatorname{ri}C]} = \operatorname{ri}T[\operatorname{ri}C] \subseteqq T[\operatorname{ri}C]$. Thus $\operatorname{ri}T[C] \subseteqq T[\operatorname{ri}C]$.

Consider a point v_2 in $T[\operatorname{ri}C]$. We may then express v_2 as $v_2 = T(v_1)$ where $v_1 \in \operatorname{ri}C$. Consider also a point $w_2 \in \operatorname{aff}T[C]$. Express w_2 as $w_2 = T(w_1)$ where $w_1 \in \operatorname{aff}C$. By theorem 2.3.2 the open segment $]v_1, w_1[$ intersects C so that $]v_2, w_2[\cap T[C] = T[\,]v_1, w_1[\,] \cap T[C] \supseteqq T[\,]v_1, w_1[\cap C] \neq \varnothing$. Thus, by theorem 2.3.2 again, $v_2 \in \operatorname{ri}T[C]$. It follows that $T[\operatorname{ri}C] \subseteqq \operatorname{ri}T[C]$ so that $T[\operatorname{ri}C] = \operatorname{ri}T[C]$.

The last statement of the theorem follows by applying this to the convex set $C_1 \times C_2$ and the difference map $T : (v_1, v_2) \to v_1 - v_2$ on $V \times V$. □

The convex hull of a one-dimensional set is the union of closed intervals having their end points in the set. Similarly the convex hull of a two-dimensional set is the union of all triangles having their vertices in the set. Points on the boundary of this two-dimensional convex hull belong to line segments having their end points in the set. Except for the last statement, we proved this in the previous section. Repeating and using the information on relative boundaries of convex sets which we have obtained we arrive at Caratheodory's theorem (Caratheodory, 1911).

Theorem 2.3.8 (Convex hulls of finite dimensional sets). *Let S be a subset of an m-dimensional linear space V. Then any vector belonging to the convex hull $\langle S \rangle$ of S belongs to a simplex with vertices in S and thus may be represented as a convex combination of $m + 1$ or fewer vectors in S.*

Furthermore any vector in $\langle S \rangle$ which does not belong to the relative interior of $\langle S \rangle$ belongs to a simplex having at most m vertices which are all in S.

Remark. If $C = \langle \{v_0, v_1, \ldots, v_m\} \rangle$ is a polytope then any vector $v \in C$ may be represented as $v = \sum_{i=0}^{m} \lambda_i(v) v_i$ where $\lambda_i(v) \geqslant 0$, $i = 0, \ldots, m$ and

$\lambda_0(v) + \cdots + \lambda_m(v) = 1$. It may be shown (see Grünbaum 1967, p. 37) that the functions $\lambda_0, \lambda_1, \ldots, \lambda_m$ may be chosen so that all of them are continuous on C and that one pre-assigned among them is convex.

Proof. The first statement is a consequence of proposition 2.2.3. Consider a vector v in $\langle S \rangle$ but not in $\mathrm{ri}\langle S \rangle$. Then $v \in \langle\{v_0, v_2, \ldots, v_h\}\rangle$ for affinely independent vectors $v_0, v_1, \ldots, v_h$ in S. If $h < m + 1$ then there remains nothing to prove. If $h = m + 1$ then we may express v as $v = \sum_{i=0}^{m+1} \lambda_i v_i$ where we may assume that $\lambda_0, \ldots, \lambda_{m+1} > 0$ and that $\lambda_0 + \cdots + \lambda_{m+1} = 1$. Then $v \in \mathrm{ri}\langle\{v_0, \ldots, v_{m+1}\}\rangle$. On the other hand, since now $\dim \mathrm{aff}\{v_0, \ldots, v_{m+1}\} = m = \dim \mathrm{aff}\, S$, we find that $\mathrm{aff}\{v_0, \ldots, v_{m+1}\} = \mathrm{aff}\, S$. Since $\mathrm{ri}\langle\{v_0, \ldots, v_{m+1}\}\rangle \subseteq \mathrm{ri}\langle S \rangle$, it follows that $v \in \mathrm{ri}\langle S \rangle$, contradicting our assumptions. □

Simplices may be 'triangulated' as shown in the following proposition.

Proposition 2.3.9 (Triangulation of simplices). *Let v be a point in the m-dimensional simplex $C = \langle\{v_0, v_1, \ldots, v_m\}\rangle$. Represent v as $v = \lambda_0 v_0 + \cdots + \lambda_m v_m$ where $\lambda_0, \ldots, \lambda_m \geqslant 0$ and $\lambda_0 + \cdots + \lambda_m = 1$. For $i = 0, \ldots, m$, let $C_{i,v}$ be the convex hull of $\{v\} \cup \{v_j : j \neq i\}$. Then C may be decomposed as*

$$C = \bigcup \{C_{i,v} : \lambda_i > 0\}$$

where the sets $C_{i,v} : \lambda_i > 0$ are all simplices.

Proof. It is readily checked that the set $\{v\} \cup \{v_j : j \neq i\}$ is affinely independent when $\lambda_i > 0$. If $w \neq v$ and $w \in C$ then there is a largest ε such that $u = w + \varepsilon(w - v) \in C$. Indeed $w + \varepsilon(v - w) \in C$ if and only if $\mu_i + \varepsilon(\mu_i - \lambda_i) \geqslant 0$; $i = 0, \ldots, m$. Here equality holds for at least one i for the largest ε satisfying these inequalities. Furthermore, since $\mu_0, \ldots, \mu_m \geqslant 0$, this largest ε is non-negative. Thus $u = \sum \kappa_i v_i$ where $\kappa_i = 0$ for at least one i and then $u \in C_{i,v}$ provided $\lambda_i > 0$. Expressing w as the convex combination $w = (1/(1 + \varepsilon))u + (\varepsilon/(1 + \varepsilon))v$ we find in this case that $w \in C_{i,v}$. This proves the statement of the proposition when $\lambda_i > 0$ for all $i = 0, \ldots, m$.

We proceed now by induction on m. If $m \leqslant 1$ then the proposition is trivial. Assume the proposition is proved for $m < n$ and consider the case $m = n$. By the above arguments we may assume that $\lambda_i = 0$ for at least one i. Put $I = \{i : \lambda_i > 0\}$ and decompose $w \in C$ as $w = (1 - \mu)w_1 + \mu w_2$ where $w_1 \in \langle\{v_i : i \in I\}\rangle$ and $w_2 \in \langle\{v_i : i \notin I\}\rangle$. The induction hypothesis implies that $w_1 \in C_{i,v}$ for some $i = i_0 \in I$ and clearly $w_2 \in C_{i,v}$ whenever $\lambda_i > 0$. Thus $w \in C_{i,v}$ when $i = i_0$. □

The notion of convex hulls of sets is not adequate for representation of points in unbounded convex sets. A comprehensive theory which works just as well for unbounded sets as for bounded sets may be obtained by considering convex hulls of unions of sets of points and sets of directions, as in Rockafellar (1970).

Convex hulls of closed sets need not be closed. However, convex hulls of compact sets are always closed (Caratheodory, 1907).

Theorem 2.3.10 (Convex hulls of compact sets and of relatively open sets). *Convex hulls of compact sets are compact and convex hulls of relatively open sets are relatively open.*

In particular the closure operation and the convex hull operation commute for bounded sets, i.e. $\langle \bar{S} \rangle = \overline{\langle S \rangle}$ *when S is bounded.*

Remark. The reader is invited to find two-dimensional closed sets having convex hulls which are not closed.

Proof. Consider the convex hull $\langle S \rangle$ of a subset S of an m-dimensional space. If $\{v_n\}$ is a sequence in $\langle S \rangle$ then v_n may, for each n, be represented as a convex combination $v_n = \sum_{i=0}^{m} \lambda_{n,i} v_{n,i}$ where $v_{n,i} \in S$, $\lambda_{n,i} \geqslant 0$ and $\sum_i \lambda_{n,i} = 1$. Then $z_n = (\lambda_{n,0}, \ldots, \lambda_{n,m}, v_{n,0}, \ldots, v_{n,m})$; $n = 1, 2, \ldots$ is a sequence in the set $K = \{(\lambda_0, \ldots, \lambda_m) : \lambda_0, \ldots, \lambda_m \geqslant 0, \lambda_0 + \cdots + \lambda_m = 1\} \times S^{m+1}$. If S is compact then K is also compact and then $\{z_n\}$ has a sub-sequence $\{z_{n'}\}$ converging to a point $z = (\lambda_0, \ldots, \lambda_m, v_0, \ldots, v_m)$ in K. Then $v_{n'} = \sum_i \lambda_{n',i} v_{n',i} \to \sum_i \lambda_i v_i \in \langle S \rangle$. It follows that $\langle S \rangle$ is compact when S is compact.

For any set S we have trivially that $\overline{\langle S \rangle} \supseteqq S$ and thus, since $\overline{\langle S \rangle}$ is convex and closed, $\overline{\langle S \rangle} \supseteqq \langle \bar{S} \rangle$. On the other hand if S is bounded then $\bar{S}$ is compact and hence, as we have just seen, $\langle \bar{S} \rangle$ is also both convex and closed. It follows that $\langle \bar{S} \rangle \supseteqq \overline{\langle S \rangle}$.

Assume that S is open relatively $\operatorname{aff} S = \operatorname{aff}\langle S \rangle$. If $v \in \langle S \rangle$ then we may represent v as $v = \sum_{i=0}^{m} \lambda_i v_i$ where $v_0, \ldots, v_m \in S$, $\lambda_0, \ldots, \lambda_m \geqslant 0$ and $\sum_i \lambda_i = 1$. Let w be any point in aff S and let $t \in]0, 1[$. Then $(1 - t)v + tw = \sum_i \lambda_i((1 - t)v_i + tw)$ and the points $(1 - t)v_i + tw$ are all in S when t is sufficiently small and then $(1 - t)v + tw \in \langle S \rangle$. Thus, by theorem 2.3.2, $v \in \operatorname{ri}\langle S \rangle$. □

The apparently innocent looking corollary 2.2.2 provides the basic step for proving Helly's intersection theorem.

Theorem 2.3.11 (Helly's intersection theorem). *Let* $\mathscr{C}$ *be a finite class of convex subsets of an m-dimensional linear space V. Then* $\bigcap \{C : C \in \mathscr{C}\} \neq \emptyset$ *provided* $C_1 \cap C_2 \cap \cdots \cap C_s \neq \emptyset$ *whenever* $C_1, \ldots, C_s \in \mathscr{C}$ *and* $s \leqslant m + 1$.

The assumption that $\mathscr{C}$ *is finite may be deleted provided the sets in* $\mathscr{C}$ *are closed and at least one set in* $\mathscr{C}$ *is bounded.*

Remark. The example $\mathscr{C} = \{]0, 1/n]; n = 1, 2, \ldots\}$ shows that the assumption that $\mathscr{C}$ is finite cannot be deleted unless other conditions are imposed. Other conditions than those used here can be used.

Proof. The last statement follows from the first since classes of compact sets having the finite intersection property (f.i.p) have non-empty intersections.

We shall prove the first statement by induction on $r = \#\mathscr{C}$. If $r \leqslant m+1$ then the statement is trivial. Assume that we have proved the first statement when $r \leqslant n$ and consider the case $r = n+1$. We may then write $\mathscr{C} = \{C_0, C_1, \ldots, C_n\}$. For each $i \in \{0, 1, \ldots, n\}$ choose a point v_i in the non-empty set $\bigcap \{C_j : j \neq i\}$ and put $S = \langle\{v_0, v_1, \ldots, v_n\}\rangle$. If $v_0, \ldots, v_n$ are affinely independent then $n \leqslant m$ and then it follows directly from the basic assumption that $\bigcap \mathscr{C} \neq \varnothing$. On the other hand if $v_0, v_1, \ldots, v_m$ are affinely dependent then by corollary 2.2.2 we may decompose the set $M = \{v_0, \ldots, v_n\}$ as $M = M_1 \cup M_2$ such that $M_1 \cap M_2 = \varnothing$ and such that $\langle M_1\rangle \cap \langle M_2\rangle \neq \varnothing$. Let v be a point in $\langle M_1\rangle \cap \langle M_2\rangle$ and let $j \in \{0, \ldots, n\}$. Considering v as a point in $\langle M_1\rangle$ or in $\langle M_2\rangle$ according to whether $v_j \notin M_1$ or $v_j \notin M_2$, we see that we may represent v as a convex combination $v = \sum_{i=0}^{n} \lambda_i v_i$ such that $\lambda_j = 0$. If $i \neq j$ then $v_i \in C_j$. Hence $v \in C_j$ for any $j \in \{0, \ldots, n\}$. □

Here is the crucial result on separation of convex subsets of a finite dimensional linear space.

Proposition 2.3.12. *A non-empty convex subset C of V may be separated strongly from the origin if and only if $0 \notin \bar{C}$.*

Proof. If f is a linear functional on V and if $0 \in \bar{C}$ then $\inf\{f(v) : v \in C\} \leqslant f(0) = 0$ proving the 'only if'.

Assume that $0 \notin \bar{C}$. We may without loss of generality suppose that V is an inner product space with norm $\|\cdot\|$ and inner product $(\cdot,\cdot)$. Put $\delta = \inf\{\|v\| : v \in C\} = \inf\{\|v\| : v \in \bar{C}\} = \text{distance}(C, 0)$. Then $\delta = \|v_0\|$ for some v_0 in $\bar{C}$. If $v \in C$ then the map $t \to \|(1-t)v_0 + tv\|^2 = \|v_0\|^2 + 2(v_0, v - v_0)t + \|v - v_0\|^2 t^2$ obtains its minimum on $[0, 1]$ for $t = 0$, and this is only possible when the coefficient $(v_0, v - v_0)$ of t is non-negative. Putting $f(v) = (v_0, v)$ we find that $f(v) \geqslant f(v_0) = \delta > 0$ when $v \in C$. □

The basic criterion for strong separation is the following theorem.

Theorem 2.3.13 (Strong separation). *Non-empty convex sets S and T may be strongly separated by a linear functional if and only if $0 \notin \overline{S - T}$. This condition is fulfilled whenever the sets S and T are disjoint and both are closed and one of them is bounded (compact).*

Proof. This follows from propositions 2.2.10 and 2.3.12. □

Disjoint finite dimensional non-empty convex sets may always be weakly separated.

Theorem 2.3.14 (Weak separation). *Non-empty convex sets S and T may be weakly separated by a linear functional if and only if their relative interiors* ri S *and* ri T *are disjoint.*

Proof. By corollary 2.3.7, letting '$-$' express algebraic difference, $\text{ri}\, S - \text{ri}\, T = \text{ri}(S - T)$. Therefore it suffices to consider the case where one of the sets, say T, is the one point set $\{0\}$. If $0 \in \text{ri}\, S$ and if f is bounded from below or from above on S, then $f = \text{constant} = 0$ on $\text{aff}\, S = [S]$. Thus $0 \notin \text{ri}\, S$ whenever S may be weakly separated from the origin.

Consider a convex set S such that $0 \notin \text{ri}\, S$. If $0 \notin \bar{S}$ then 0 and S may be strongly separated, by proposition 2.3.12. We may therefore assume that $0 \in \bar{S}$. Then $\text{aff}\, S$ is linear $= [S]$. As $\text{ri}\, \bar{S} = \text{ri}\, S$ we may also assume that S is a closed set having the origin on its relative boundary. Let v_0 be any point in $\text{ri}\, S$. If $n = 1, 2, \ldots$ then $-(1/n)v_0 \notin S$ since otherwise $0 \in]-(1/n)v_0, v_0[\subseteqq \text{ri}\, S$. By strong separation there is, for each n, a linear functional f_n on $[S]$ such that $f_n(-(1/n)v_0) < f_n(v)$ when $v \in S$. We may assume that $\|f_n\| = 1$ for some norm on the dual space of $[S]$. By compactness there is a sub-sequence $\{f_{n'}\}$ such that $f_{n'}(v)$ converges when $v \in [S]$. Putting $f(v) = \lim_{n'} f_{n'}(v)$ we see that f is a linear functional on $[S]$ such that $\|f\| = 1$ (and thus $f \neq 0$) and such that $f(v) = \lim_{n'} f_{n'}(v) \geqslant \lim_{n'} f_{n'}((-1/n')v_0) = f(0) = 0$ when $v \in S$. Extending f in any way to a linear functional on V we obtain a separating linear functional.

□

We saw in the previous section that polytopes are the convex hulls of their extreme points. This extends to any finite dimensional compact convex set by the following result due to Minkowski.

Theorem 2.3.15. *If C is a finite dimensional compact convex set then* $C = \langle \text{ext}\, C \rangle$.

Remark 1. If we extend the notion of a convex hull as described before theorem 2.3.10, then any closed convex set which contains no lines is the convex hull of all its extreme points and all its extreme directions (see Rockafellar, 1970).

Remark 2. It follows from this and proposition 2.2.8 that a finite dimensional convex set is a polytope if and only if it is a compact convex set having a finite number of extreme points.

Remark 3. The Krein–Milmann theorem asserts that C is the closed convex hull of $\text{ext}\, C$. A further argument is needed to ensure that $\langle \text{ext}\, C \rangle$ is closed.

Proof. Put $d = \dim \text{aff}\, C$.

If $d \leqslant 1$ then the theorem is trivial. Assume that the theorem is proved whenever $d \leqslant n$ and consider the case $d = n + 1$.

We shall show that any point $v \in C$ is a convex combination of extreme points of C. As v may be expressed as a convex combination of at most two points on the relative boundary $C \cap [\text{ri}\, C]^c$ of C we may assume that $v \notin \text{ri}\, C$. By weak separation there is a linear functional f which is not constant on C such that $f(v) = \kappa$ where $\kappa = \sup\{f(w) : w \in C\}$.

The set $F = [f = \kappa] \cap C$ is a closed face of C. Furthermore, since f is not constant on C, F is a proper subset of C. Indeed, for the same reason, aff F is a proper subset of aff C. Thus dim aff $F \leqslant n$. Hence, by the induction hypothesis and since $v \in F$, v is a convex combination of extreme points of F. However, extreme points of F are also extreme points of C, by proposition 2.2.7. □

Two-dimensional convex polygons are intersections of halfplanes. In order to generalize this to any finite dimensional linear space we shall need the following extension of the notion of a halfplane.

A *halfspace* in a linear space V is a set which is either of the form $[l \geqslant \alpha]$ or of the form $[l > \alpha]$ for a non-zero linear functional l and a constant α.

Clearly a set is a halfspace if and only if it is either of the form $[l \leqslant \alpha]$ or of the form $[l < \alpha]$ for a non-zero linear functional l and a constant α. The halfspaces defined by the weak inequalities are the closed halfspaces, while the halfspaces defined by the strong inequalities are the open halfspaces. All halfspaces are clearly convex. It follows that intersections of halfspaces are convex and clearly a large variety of convex sets may be constructed by intersecting halfspaces. However, not all convex sets may be obtained in this way. Thus any halfplane in $\mathbb{R}^2$ containing the convex set $C = \{(x_1, x_2): 0 \leqslant x_1 \leqslant 1,\ 0 \leqslant x_2 < 1\} \cup \{(x_1, x_2): 0 \leqslant x_1 \leqslant \frac{1}{2}, x_2 = 1\}$ also contains the line segment $\{(x_1, x_2): \frac{1}{2} < x_1 < 1, x_2 = 1\}$ which is entirely outside C. However, all closed convex sets may be obtained by intersecting halfspaces.

Theorem 2.3.16 (Intersections of closed halfspaces). *A proper subset of a finite dimensional linear space is closed and convex if and only if it is an intersection of closed halfspaces.*

Proof. Let C be a closed convex proper subset of V. If $v \notin C$ then by strong separation there is a closed halfspace containing C but not containing v. □

Sets which are intersections of finite families of closed halfspaces are called *polyhedral.* Clearly, affine sets are polyhedral and we shall see that bounded sets are polyhedral if and only if they are polytopes. Convex sets having an infinite number of extreme points cannot be polyhedral.

Proposition 2.3.17. *The set of extreme points of a polyhedral set is finite.*

Proof. By induction on dim aff A, we shall argue that ext A is finite when A is polyhedral. If dim aff $A \leqslant 1$ then this is trivial so suppose that this holds whenever dim aff $A < d$. If $v_0 \in \operatorname{ri} A$ then $0 \in \operatorname{ri}(A - v_0)$ and $\operatorname{ext}(A - v_0) = (\operatorname{ext} A) - v_0$. Furthermore, translates of polyhedral sets are polyhedral. It follows that we may assume that $0 \in \operatorname{ri} A$. Passing from V to $[A]$ we see that we may also assume that $0 \in \mathring{A}$. Express A as an intersection $A = \bigcap_{i=1}^{r} \{v : f_i(v) \leqslant t_i\}$ for non-zero linear functionals $f_1, \ldots, f_r$ and numbers

$t_1, \ldots, t_r$. Then $D_i = A \cap \{v : f_i(v) = t_i\}$ is a face of A. Since $0 \in \mathring{A}$ it follows that each face D_i is a proper face of A and thus $\dim \operatorname{aff} D_i < d$; $i = 1, \ldots, r$. If $f_i(v) < t_i$; $i = 1, \ldots, r$, then, by continuity, $v \in \mathring{A}$. Hence $\operatorname{ext} A \subseteqq \operatorname{bd} A \subseteqq \bigcup_i D_i$. However, if $v \in (\operatorname{ext} A) \cap D_i$, then $v \in \operatorname{ext} D_i$. Hence $\operatorname{ext} A \subseteqq \bigcup_i \operatorname{ext} D_i$, so that, by the induction hypothesis, $\# \operatorname{ext} A \leqslant \sum_i \# \operatorname{ext} D_i < \infty$. □

Combining this with theorem 2.3.15 we obtain the following corollary.

Corollary 2.3.18. *Bounded polyhedral sets are polytopes.*

Remark. Conversely, as we shall see later, polytopes are bounded polyhedral sets.

We have noted before that a finite dimensional linear space possesses a unique Hausdorff topology making the vector operations continuous. A finite dimensional space also possesses a natural measurability structure which may e.g. be defined by the Borel class, i.e. the σ-algebra generated by the open sets. As V as a linear topological space is isomorphic to some space $\mathbb{R}^n$ it is Borel isomorphic to the real line, as is $\mathbb{R}^n$. Unless otherwise mentioned it is this measurability structure which will be employed when discussing random variables in V.

We shall say that a random variable X taking its values in V is integrable if $l(X)$ is integrable for any linear functional l on V. It suffices of course to ensure that the coordinates of X relative to a fixed basis are integrable. If X is integrable then by reflexivity there is a unique vector EX in V such that $El(X) = l(EX)$ for all linear functionals l on V.

Several useful inequalities in probability theory may be derived from the following theorem.

Theorem 2.3.19. *If X is a V-valued integrable random variable and if C is a convex subset of V such that $X \in C$ a.s. then $EX \in C$.*

If in addition $Pr(X \in \operatorname{ri} C) > 0$ then $EX \in \operatorname{ri} C$.

Remark 1. We do not assume that C is a Borel set. Our assumptions imply however that the event $[X \in C]$ is completion measurable for the underlying probability space.

Remark 2. In view of the fact that closed convex sets are intersections of halfplanes it is not surprising that $EX \in \bar{C}$. However it is not quite so obvious that $EX \in C$. In fact if $C =]0, 1]$ then we may have $\int x P(dx) = \int_0^1 P(]t, 1])\, dt = 0$ for a finitely additive probability set function on the Borel class of $\mathbb{R}$.

On the other hand, if we give up finite dimensionality then we may be in trouble here even for countably additive distributions. Consider e.g. a random variable Z which is uniformly distributed on $[0, 1[$ and put $X_n = 0$ or $= n$ as $Z < 1 - 1/n$ or $Z \geqslant 1 - 1/n$. Then $X = (X_1, X_2, \ldots)$ belongs to the convex set C of sequences converging to 0 although $EX = (EX_1, EX_2, \ldots) = (1, 1, \ldots) \notin C$.

Proof. If $EX \notin \bar{C}$ then by strong separation there is a linear functional l on V such that $l(EX) < l(v)$ when $v \in C$. If so then $l(X - EX) = l(X) - l(EX) > 0$ a.s. although $El(X - EX) = l(0) = 0$. Thus $EX \in \bar{C}$.

Let $d = \dim \operatorname{aff} C$ be the geometric dimension of C. If $d = 0$ then C is a one-point set, $C = \{v\}$ say, and then $EX = v$. If $d = 1$ then there are distinct vectors v_0 and v_1 in C and then $\operatorname{aff} C = \{v_0 + t(v_1 - v_0) : t \in \mathbb{R}\}$. Put $v_t = v_0 + t(v_1 - v_0)$. Since $Pr(X \in C) = 1$ we may define a random variable T by putting $X = v_0 + T(v_1 - v_0)$. Then $EX = v_0 + ET(v_1 - v_0)$. Let I denote the interval of real numbers t such that $v_t \in C$. Then $Pr(T \in I) = 1$. Using the fact that the expectation of a strictly positive random variable is positive we deduce easily that $ET \in I$ and thus that $EX \in C$.

Now assume that the first statement of the theorem is established when $d \leqslant r$ and consider the case $d = r + 1$.

If $EX \notin C$ then, by weak separation, there is a linear functional l which is not constant on $\{EX\} \cup C$ and such that $l(EX) \leqslant l(v)$ when $v \in C$. Then $l(X) - l(EX) \geqslant 0$ a.s. On the other hand, since $E[l(X) - l(EX)] = l(EX) - l(EX) = 0$, $l(X) = l(EX)$ a.s. It follows that $X \in \tilde{C}$ a.s. where $\tilde{C} = \{v : l(v) = l(EX)\} \cap C$. As l is not constant on $\operatorname{aff} C$ we conclude that $\operatorname{aff} \tilde{C}$ is a proper subset of $\operatorname{aff} C$ and thus $\dim \operatorname{aff} \tilde{C} \leqslant r$. Hence, by the induction hypothesis, $EX \in \tilde{C} \subseteqq C$. The first statement of the theorem follows by induction.

Put $\pi = Pr(X \in \operatorname{ri} C)$. If $\pi = 1$ then $EX \in \operatorname{ri} C$ by what we have proved. If $0 < \pi < 1$ then we may decompose the underlying probability distribution Pr as $Pr = \pi Q_1 + (1 - \pi) Q_2$ where Q_1 is obtained from Pr by conditioning on the event $[X \in \operatorname{ri} C]$ while Q_2 is obtained from Pr by conditioning on the event $[X \notin \operatorname{ri} C]$. Taking expectations we find that $EX = \pi E_{Q_1} X + (1 - \pi) E_{Q_2} X$. The first statement of the theorem implies that $E_{Q_1} X \in \operatorname{ri} C$ and that $E_{Q_2} X \in C$. Hence, by corollary 2.3.3, the vector EX belongs to $\operatorname{ri} C$. □

Turning to convex functions, note first that any non-negative function defined on a closed circular disk in the plane and which is zero inside the disk, is convex. It follows that convex functions may exhibit almost any imaginable kind of irregular behaviour on the boundaries of their effective domain. In particular convex functions, just as convex sets, may not be (Borel) measurable.

However, we shall see that a convex function is always continuous on the interior of its effective domain. Furthermore, if the boundary looks like a finite union of polytopes locally then the function is automatically upper semi-continuous, in particular measurable, on its effective domain.

The local property of the effective domain which we shall need may be expressed as follows for any subset S of V.

Each point v in S possesses a neighbourhood U in V such that the S-relative neighbourhood $U \cap S$ of v is contained in a finite union $S_1 \cup \cdots \cup S_m$ of simplices which are all contained in S. A set S having this property is called *locally simplicial.* Concerning the definition, since simplices are closed sets, we may restrict our attention to simplices S_i containing v. By proposition 2.3.9 we may also assume that v is a vertex of each of the simplices S_i. Finally, since polytopes are finite unions of simplices, we may relax the requirement that the sets S_i are simplices, to the requirement that these sets are polytopes.

Clearly polytopes are locally simplicial and we shall see later that in fact all polyhedral sets are locally simplicial. It follows readily that finite unions of polyhedral sets are locally simplicial.

Any point v belonging to the relative interior $\operatorname{ri} C$ of a convex set C belongs to $\operatorname{ri} S$ for a simplex S contained in C. Thus relatively open convex sets are locally simplicial.

Now consider a convex function f on V and a locally simplicial subset S of the effective domain of f. Then $f(v) < \infty$ whenever $v \in S$. Let v be any point in S and consider a sequence $\{v_n\}$ of points in S which converges to v. Let U be a neighbourhood of v such that $S \cap U \subseteq S_1 \cup \cdots \cup S_m \subseteq S$ where $S_1, \ldots, S_m$ are simplices which all have v as one of the vertices. If σ is an accumulation point for the sequence $\{f(v_n)\}$ in $[-\infty, \infty[$ then $\sigma = \lim f(v_{n'})$ for some subsequence $\{v_{n'}\}$ of $\{v_n\}$. We may assume that the points $v_{n'}$ are all in the same simplex S_i. Let the vertices of S_i be $v, w_1, \ldots, w_m$. Then $v_{n'}$ may be represented as $v_{n'} = \lambda_{0,n'}v + \lambda_{1,n'}w_1 + \cdots + \lambda_{m,n'}w_m$ where $\lambda_{0,n'}, \ldots, \lambda_{m,n'} \geqslant 0$ and $\lambda_{0,n'} + \cdots + \lambda_{m,n'} = 1$. Since $v_{n'} \to v$, from the uniqueness and the continuity of representation it follows that $\lambda_{0,n'} \to 1$ while $\lambda_{1,n'} \to 0, \ldots, \lambda_{m,n'} \to 0$. By convexity $f(v_{n'}) \leqslant \lambda_{0,n'}f(v) + \lambda_{1,n'}f(w_1) + \cdots + \lambda_{m,n'}f(w_m)$. Hence, by passing to the limit, we find that $\sigma \leqslant f(v)$. It follows that $\limsup_n f(v_n) \leqslant f(v)$. This proves the following theorem.

Theorem 2.3.20. *If f is a convex function from the convex set C to $[-\infty, \infty[$ then any restriction of f to a locally simplicial subset of C is upper semicontinuous.*

Lower semicontinuity may be obtained from the following important existence theorem for tangent hyperplanes.

Theorem 2.3.21 (Existence of tangents). *Let f be a real valued convex function on a convex set C and let $v_0 \in \operatorname{ri} C$. Then there is a linear functional l on V such that*

$$f(v) \geqslant f(v_0) + l(v - v_0)$$

for all $v \in C$.

Remark. A linear functional l having this property is in Rockafellar's book (1970) called a subgradient of f at v_0.

Proof. Clearly $(v_0, f(v_0))$ belongs to the convex subset $D = \{(v, y) : v \in C, y \in \mathbb{R}, f(v) \leqslant y\}$ of $V \times R$. On the other hand, since $(v_0, f(v_0) - \varepsilon) \notin D$ when $\varepsilon > 0$, the point $(v_0, f(v_0))$ is not in $\operatorname{ri} D$. Hence, by weak separation, there is a linear functional l on V and a constant β such that $l(v_0) - \beta f(v_0) \geqslant l(v) - \beta y$ when $(v, y) \in D$ and such that the map $(v, y) \to l(v) - \beta y$ is not constant on D.

Letting $y \to \infty$ we find that $\beta \geqslant 0$. If $\beta = 0$ then, by the last property above, l is not constant on C although $l(v_0) \geqslant l(v)$ when $v \in C$. As $v_0 \in \operatorname{ri} C$ this is impossible. It follows that $\beta > 0$ and then we may just as well assume that $\beta = 1$. As $(v, f(v)) \in D$ when $v \in C$ we obtain the desired inequality in the form

$$l(v_0) - f(v_0) \geqslant l(v) - f(v). \qquad \square$$

It follows that a real valued convex function defined on a relatively open convex set C is a supremum of a family of affine linear functionals. In particular such a function is lower semicontinuous on C. Combining this with theorem 2.3.20 we obtain the following continuity theorem for convex functions.

Theorem 2.3.22. *A real valued convex function on a convex set C is continuous on* ri *C.*

We refer to Rockafellar (1970) and Roberts & Varberg (1973) for sharpenings and extensions of theorems 2.3.20–2.3.22. In particular it may be shown that a real valued convex function on a convex set C is Lipschitzian on compact subsets of C.

A stronger result holds if f is sublinear.

Proposition 2.3.23. *Any real valued sublinear function on V is Lipschitzian.*

Proof. Let ψ be sublinear on V and let $v_1, v_2, \ldots, v_k$ be a linearly independent basis for V. We may then norm V by putting $\|v\| = \max |\xi_i|$ where $\xi_1, \ldots, \xi_k \in \mathbb{R}$ are the coordinates of v, i.e. $v = \sum \xi_i v_i$. If so then $\psi(v) \leqslant \sum_i \psi(\xi_i v_i) \leqslant \sum_i |\xi_i| [\psi(v_i) \vee \psi(-v_i)] \leqslant \|\psi\| \, \|v\|$ where $\|\psi\| = \sum_i \psi(v_i) \vee \psi(-v_i)$. If $v, w \in V$ then this yields $\psi(v) - \psi(w) = \psi(v - w + w) - \psi(w) \leqslant \psi(v - w) \leqslant \|\psi\| \, \|v - w\|$. Hence, since the right hand term is symmetric in (v, w), $|\psi(v) - \psi(w)| \leqslant \|\psi\| \, \|v - w\|$ so that $\|\psi\|$ is a Lipschitz constant for ψ. $\square$

An important application of the tangent theorem 2.3.21 to probability theory is as follows.

Theorem 2.3.24 (Jensen's inequality II). *Let f be a measurable real valued convex function defined on a convex set C. Then $f(EX) \leqslant Ef(X)$ for any integrable V-valued random vector X such that $Pr(X \in C) = 1$.*

Moreover, if $Ex \in$ ri C then strict inequality holds unless $Pr(f(X) = a(X)) = 1$ for an affine function a on V.

Proof. The vector EX belongs to C, by theorem 2.3.19. Thus $f(EX)$ is defined. Let $v_0 \in \operatorname{ri} C$. By theorem 2.3.21 there is a linear functional l on V such that $f(v) \geqslant f(v_0) + l(v - v_0)$ when $v \in C$. Then $f(X) \geqslant f(v_0) + l(X - v_0)$ a.s. Hence, since $f(x_0) + l(X - v_0)$ is integrable, the expectation $Ef(X)$ exists and is in $]-\infty, \infty]$. If $Ef(X) = \infty$ then the inequality of the theorem is trivial. Thus we may assume that $f(X)$ is integrable. Now the random vector $(X, f(X))$ in $V \times \mathbb{R}$ belongs, with probability 1, to the convex set $D = \{(v, y) : v \in V, y \in \mathbb{R}, f(v) \leqslant y\}$. Hence, by theorem 2.3.19 again, $(EX, Ef(X)) \in D$ so that $f(EX) \leqslant Ef(X)$.

Assume that $EX \in \operatorname{ri} C$ and that $f(EX) = Ef(X)$. Then, by theorem 2.3.21 again, $f(X) \geqslant f(EX) + l(X - EX)$ for some linear functional l. Since both sides of this inequality have the same expectation, it follows that $f(X) = f(EX) + l(X - EX)$ a.s. □

Example 2.3.25 (Hölder's inequality). *Let* $t_1, \ldots, t_r$ be non-negative constants such that $t_1 + \cdots + t_r = 1$. Then the function $(x_1, \ldots, x_r) \to x_1^{t_1}, \ldots, x_r^{t_r}$ is concave on $[0, \infty[^r$. Hence, by Jensen's inequality

$$EX_1^{t_1}X_2^{t_2}, \ldots, X_r^{t_r} \leqslant (EX_1)^{t_1}(EX_2)^{t_2}, \ldots, (EX_r)^{t_r}$$

when $X_1, \ldots, X_r$ are non-negative variables. In particular if f and g are non-negative measurable functions on a measure space $(\mathscr{X}, \mathscr{A}, \mu)$ where $\mu \geqslant 0$ then

$$\int fg \, \mathrm{d}\mu \leqslant \left(\int f^p \, \mathrm{d}\mu\right)^{1/p} \left(\int g^q \, \mathrm{d}\mu\right)^{1/q}$$

when $p, q \geqslant 1$ and $1/p + 1/q = 1$. The last inequality is the familiar form of Hölder's inequality.

A particular implication of interest here is the convexity of the natural parameter space of an exponential family. We shall return to Hölder's inequality in example 2.3.38.

Let us now turn to some important relationships for convex sets and functions. We shall find it convenient to present them within the framework of a fixed finite dimensional inner product space V.

One advantage of this framework is that the inner product provides a natural isomorphism between the vector space V and its algebraic dual V'. If, however, we want to study these and similar relationships for general vector spaces, then the framework of paired linear spaces, as in Kelly & Namioka (1961), appears to be advantageous. Paired vector lattices were introduced in statistics in LeCam's 1964 paper.

From here on we shall assume that V is a finite dimensional linear space equipped with an inner product $(\cdot, \cdot)$ and with the associated norm $\|\cdot\|$. Thus if $v, w \in V$ then (v, w) is the inner product of v and w while $\|v\| = (v, v)^{1/2}$ is the norm (length) of the vector v.

To any non-empty subset A of V we may associate a function ψ_A on V by putting

$$\psi_A(w) = \sup\{(v, w) : v \in A\}.$$

The quantity $\psi_A(w)$ for a unit vector w measures how far *the set A extends in the direction of w.*

We shall call ψ_A *the support function of A*. In general a function ψ on V is called a *support function* if it is of the form $\psi = \psi_A$ for a non-empty subset A of V.

Now we introduced support functions in the previous section without employing any inner product. The terminology introduced here is consistent with the terminology of the previous section provided we identify the algebraic dual V' of V with V as described above.

Set operations may often be studied in terms of pointwise operations on support functions.

Theorem 2.3.26 (Computations with support functions). *Let A with or without subscripts denote non-empty subsets of V. Then:*

(i) $\psi_A = \psi_B$ *where* $B = \overline{\langle A \rangle}$ *is the closed convex hull of* A.
(ii) $\psi_{A_1} \leqslant \psi_{A_2}$ *if and only if* $\overline{\langle A_1 \rangle} \subseteqq \overline{\langle A_2 \rangle}$.
(iii) $\psi_{\alpha A} = \alpha \psi_A$; $0 \leqslant \alpha < \infty$.
(iv) $\psi_{A_1 + A_2} = \psi_{A_1} + \psi_{A_2}$.
(v) *if* $A = \bigcup_i A_i$ *then* $\psi_A = \bigvee_i \psi_{A_i}$.

Furthermore, a closed convex set A may be recovered from its support function by

$$A = \{v : (v, w) \leqslant \psi_A(w); w \in V\}.$$

Proof. (i), (iii), (iv) and (v) and the 'if' in (ii) follow directly from the definition.

Assume that $\psi_{A_1} \leqslant \psi_{A_2}$ and that $v \notin \overline{\langle A_2 \rangle}$. Then, by strong separation, there is a $w \in V$ so that $(v, w) > \psi_{A_2}(w) \geqslant \psi_{A_1}(w) = \psi_{B_1}(w)$ where $B_1 = \overline{\langle A_1 \rangle}$. Hence $v \notin \overline{\langle A_1 \rangle}$. It follows that $\overline{\langle A_1 \rangle} \subseteqq \overline{\langle A_2 \rangle}$.

The last statement follows from (ii) since $\psi_A(w) = (v, w)$ when $A = \{v\}$. □

Generalizing the notion of a sublinear function from the previous section we shall say that a function ψ from V to $]-\infty, \infty]$ is *sublinear* if:

it is *subadditive*, i.e. $\psi(w_1 + w_2) \leqslant \psi(w_1) + \psi(w_2)$; $w_1, w_2 \in V$
and
it is *non-negatively homogeneous*, i.e. $\psi(tw) = t\psi(w)$; $w \in V, 0 \leqslant t < \infty$.

It is readily checked that a function ψ from V to $]-\infty, \infty]$ is sublinear if and only if it is convex and non-negatively homogeneous.

Support functions are clearly sublinear. In addition, since they are suprema of linear functions, they are also lower semicontinuous. In general nothing more may be claimed of a support function.

Proposition 2.3.27. *A lower semicontinuous sublinear function ψ from V to $]-\infty, \infty]$ is the support function of the set*

$$A = \{v : (v, w) \leqslant \psi(w); w \in W\}.$$

It follows in particular that a function from V to $]-\infty, \infty]$ is lower semicontinuous and sublinear if and only if it is the pointwise supremum of a countable set of linear functionals.

Remark. If ψ is real valued (i.e. $\psi(v) < \infty$ for all $v \in V$), then, by corollary 2.2.13, this is a consequence of the Hahn–Banach theorem.

Proof. Let ψ be a lower semicontinuous sublinear function and define A as in the proposition. If $v \in A$ then $(v, w) \leqslant \psi(w)$; $w \in W$ so that $\psi_A \leqslant \psi$. Let B be the effective domain of ψ i.e. $B = \{w : \psi(w) < \infty\}$. B is not empty since $0 \in B$. Let w_0 be any point in $\operatorname{ri} B$. The tangent theorem 2.3.21 implies that $\psi(w) \geqslant \psi(w_0) + (v, w - w_0)$ for all $w \in V$ for some $v \in V$. $w = 0$ yields $(v, w_0) \geqslant \psi(w_0)$. On the other hand, putting $w = tw_0$ where $t > 0$, dividing by t and then letting $t \to \infty$ we find that $(v, w_0) \leqslant \psi(w_0)$ so that $(v, w_0) = \psi(w_0)$ and thus $(v, w) \leqslant \psi(w)$ for all $w \in V$. It follows that $v \in A$ and thus in particular that the set A is not empty. Furthermore, since $(v, w_0) = \psi(w_0)$, $\psi_A(w_0) = \psi(w_0)$. Hence $\psi_A = \psi$ on $\operatorname{ri} B$. Consider any vector $w_1 \in \bar{B}$ and again let $w_0 \in \operatorname{ri} B$. Then $w_t = (1 - t)w_0 + tw_1 \in \operatorname{ri} B$ when $t \in [0, 1[$. The functions $t \to \psi_A(w_t)$ and $t \to \psi(w_t)$ are both lower semicontinuous and convex on $[0, 1]$. Hence

$$\begin{aligned}\psi(w_1) &\leqslant \liminf_{t \to 1} \psi(w_t) \leqslant \liminf_{t \to 1} \psi_A(w_t) \\ &\leqslant \liminf_{t \to 1} [(1 - t)\psi_A(w_0) + t\psi_A(w_1)] \leqslant \psi_A(w_1)\end{aligned}$$

so that $\psi_A(w_1) = \psi(w_1)$. Thus $\psi = \psi_A$ on $\bar{B}$.

Finally consider a vector w_2 outside $\bar{B}$. By strong separation there is a vector v such that $(v, w_2) > \sup\{(v, w) : w \in B\}$. If $w \in B$ then clearly also $tw \in B$ when $0 \leqslant t < \infty$. Thus

$$(v, w_2) > 0 = \sup\{(v, w) : w \in B\}.$$

Let v_* be any vector in A and let $0 \leqslant t < \infty$. If $w \in B$ then we find $(v_* + tv, w) = (v_*, w) + t(v, w) \leqslant (v_*, w) \leqslant \psi(w)$. Thus $(v_* + tv, w) \leqslant \psi(w)$ for all w i.e. $v_* + tv \in A$. Hence

$$\begin{aligned}\psi_A(w_2) &= \sup\{(u, w_2) : u \in A\} \\ &\geqslant \sup_{t > 0} (v_* + tv, w_2) = \sup_{t > 0} [(v_*, w_2) + t(v, w_2)].\end{aligned}$$

Since $(v, w_2) > 0$, it follows that $\psi_A(w_2) = \infty = \psi(w_2)$.

Altogether we have shown that $\psi = \psi_A$.

The last statement follows now by noting that $\psi = \psi_D$ for any countable dense subset of A. □

Combining this with theorem 2.3.26 we obtain the following theorem.

Theorem 2.3.28. *The map which to each non-empty closed convex set C associates its support function ψ_C is a 1–1 order preserving map onto the class of lower semicontinuous sublinear functions from V to $]-\infty, \infty]$.*

The inverse of this map is the map which to each lower semicontinuous sublinear function ψ from V to $]-\infty, \infty]$ assigns the set

$$C = \{v : (v, w) \leqslant \psi(w); w \in V\}.$$

Say that a sublinear function ψ is *positive* if $\psi(w) > 0$ when $w \neq 0$. The positive lower semicontinuous sublinear functions may be characterized as follows.

Corollary 2.3.29. *The support function ψ_C of a non-empty closed convex set C is positive if and only if C is a neighbourhood of the origin.*

Proof. Denote the closed unit ball $\{v : \|v\| \leqslant 1\}$ by B. If $C \supseteq \varepsilon B$ then $\psi_C \geqslant \psi_{\varepsilon B} = \varepsilon\psi_B = \varepsilon\|\cdot\|$ so that ψ_C is positive. Conversely if ψ_C is positive then to each unit vector w there correspond vectors v' and v'' in C so that $(v', w) > 0 > (v'', w)$. This shows firstly that C spans V and secondly, by strong separation, that $0 \in \operatorname{ri} C = \mathring{C}$. □

Real valued sublinear functionals correspond to non-empty compact convex sets and vice versa.

Corollary 2.3.30. *The map which to each non-empty compact convex set K associates its support function ψ_K is a 1–1 order preserving map from the class of non-empty compact convex sets onto the class of real valued sublinear functions on V.*

Convergence of compact convex sets is very conveniently expressed in terms of support functions. In fact the Hausdorff distance for these sets is equivalent to a natural distance for sublinear functionals.

Theorem 2.3.31. *The Hausdorff distance between non-empty compact convex sets K_1 and K_2 for the metric of the inner product is equal to the quantity* $\sup\{|\psi_{K_1}(w) - \psi_{K_2}(w)|/\|w\| : w \neq 0\} = \sup\{|\psi_{K_1}(w) - \psi_{K_2}(w)| : \|w\| = 1\}$.

Remark. The Hausdorff distance between bounded subsets A and B of a metric space (X, d) is the infimum of numbers $r > 0$ such that $A \subseteqq \{x : \operatorname{dist}(x, B) < r\}$ and $B \subseteqq \{x : \operatorname{dist}(x, A) < r\}$.

Proof. Let B be the closed unit ball $\{v : \|v\| \leqslant 1\}$. Then $\sup\{|\psi_1(w) - \psi_2(w)| : \|w\| = 1\} \leqslant \varepsilon$ if and only if $\psi_1 \leqslant \psi_2 + \varepsilon\|\cdot\|$ and $\psi_2 \leqslant \psi_1 + \varepsilon\|\cdot\|$. If $\psi_1 = \psi_{K_1}$

and $\psi_2 = \psi_{K_2}$ where K_1 and K_2 are compact and convex then this may be expressed as

$$K_1 \subseteqq K_2 + \varepsilon B$$

and

$$K_2 \subseteqq K_1 + \varepsilon B. \qquad \square$$

Corollary 2.3.32. *A non-empty compact convex set K is the Hausdorff limit of an increasing sequence of polytopes having their vertices in* ext K.

Proof. By theorem 2.3.15 $K = \langle \text{ext}\, K \rangle$.

Let $\{v_n\}$ be a sequence running through a countable dense subset of ext K and put $K_n = \langle \{v_1, \ldots, v_n\} \rangle$. Then $\psi_K = \psi_{\text{ext}\, K} = \sup_n (v_n, \cdot) = \sup_n \psi_{K_n} = \lim_n \psi_{K_n}$ so that $\psi_{K_n} \to \psi_K$ uniformly on compact sets. $\square$

Compact convex neighbourhoods of the origin may be characterized by another interesting functional. If K is such a neighbourhood of the origin then for each vector $v \neq 0$ there is a smallest $t > 0$ so that $(1/t)v \in K$. We shall denote this smallest t by $\omega_K(v)$ and we shall put $\omega_K(0) = 0$. Thus

$$\omega_K(v) = \inf\left\{t : t > 0, \frac{1}{t} v \in K\right\}.$$

Using the convexity of K it is readily verified that ω_K is a positive real valued sublinear functional on V. This functional is called *the Minkowski functional of* K. The set K is recovered from ω_K as the 'unit ball' $[\omega_K \leqslant 1]$.

If K is symmetric then ω_K is a norm and any norm on V may be obtained that way.

Theorem 2.3.33. *The map which to each compact convex neighbourhood K of the origin associates its Minkowski functional, is a* 1–1 *and order reversing map onto the class of real valued positive sublinear functionals on V.*

The inverse of this map assigns the set $[\omega \leqslant 1]$ *to a given Minkowski functional* ω.

The reason we mention Minkowski functionals here is that they provide an interesting application of a very useful notion of duality for compact convex neighbourhoods of the origin. For two compact convex neighbourhoods B_1 and B_2 of the origin this duality may be expressed by the requirement that the Minkowski functional of B_1 is the support function of B_2. If so then, as we shall see, the Minkowski functional of B_2 is the support function of B_1.

Considering two positive real valued sublinear functionals ψ_1 and ψ_2 on V we may now inquire whether:

(i) ψ_1 and ψ_2 are support functions of dual compact convex neighbourhoods of the origin;

or

(ii) ψ_1 and ψ_2 are Minkowski functionals of dual compact convex neighbourhoods of the origin;

or

(iii) ψ_1 is the support function of a set which has ψ_2 as its Minkowski functional.

We shall see in a moment that conditions (i), (ii) and (iii) are in fact all equivalent to

(iv) $\psi_2(v) \equiv_v \sup\{(v, w)/\psi_1(w) : w \neq 0\}$.

From (iv) we obtain the interesting inequality

$$(v, w) \leqslant \psi_1(v) \cdot \psi_2(w); \; v, w \in V.$$

If ψ_1 is the norm $\|\cdot\|$ associated with the inner product then $\psi_2 = \psi_1$ so that this reduces to the Cauchy–Schwarz inequality.

Besides substantiating these claims we shall also verify some statements made earlier in this section on polyhedral sets and on locally simplicial sets. The reader may consult e.g. Brønsted (1982) and Rockafellar (1970) for other interesting features of this notion of duality.

The fundamental concept is that of a polar of a set. In general the *polar* of a subset A of V is the set

$$\Pi(A) = \{w : (v, w) \leqslant 1; v \in A\}.$$

Thus the polar of a non-empty set A is the set of points w such that $\psi_A(w) \leqslant 1$ i.e. $\Pi(A) = [\psi_A \leqslant 1]$. It follows that polars are intersections of closed half-spaces containing the origin and thus a polar belongs to the class of closed convex sets which contain the origin. Furthermore, as we soon shall see, any such set is the polar of its own polar so that this is precisely the class of polars.

If $0 \in \overline{\langle A \rangle}$, so that $\psi_A \geqslant 0$, then the problem of finding the maximum of a linear form $v \to (v, w)$ on A may be converted to the problem of finding the minimum shrinking (if any) of the vector w which is contained in $\Pi(A)$ since then

$$\psi_A(w) = \inf\{t : t > 0, w/t \in \Pi(A)\}.$$

Indeed if $t > 0$ then $w/t \in \Pi(A)$ if and only if $(v, w/t) \leqslant 1$ when $v \in A$, i.e. if and only if $\psi_A(w) \leqslant t$.

The polar of the empty set is V and conversely the polar of V is empty. It is also readily checked that polars of closed balls (for the norm $\|\cdot\|$) with centre at the origin are also closed balls with centre at the origin. Restricting our attention to balls with centre at the origin we find that the closed unit ball is its own polar and that the polar of the closed ball with radius r is the ball with radius $1/r$.

As mentioned above, the polar $\Pi(A)$ of a set A is a closed convex set containing the origin. Furthermore, as is readily checked, $A \subseteqq \Pi(\Pi(A))$ so that $\overline{\langle A \cup \{0\}\rangle} \subseteqq \Pi(\Pi(A))$. On the other hand if $v_0 \notin \overline{\langle A \cup \{0\}\rangle}$ then, by strong separation, there is a vector w in V such that $(v_0, w) > \sup\{(v, w) : v \in A \cup \{0\}\}$. Since $(0, w) = 0$, the supremum is non-negative and thus (v_0, w) is positive. Hence we may modify w so that $(v_0, w) = 2 > 1 \geqslant (v, w)$; $v \in A$ and thus, since $w \in \Pi(A)$, $v_0 \notin \Pi(\Pi(A))$. It follows that $\Pi(\Pi(A)) = \overline{\langle A \cup \{0\}\rangle}$. This proves the fifth statement of the following theorem.

Theorem 2.3.34 (Behaviour of polars of sets). *Let A with or without subscripts denote subsets of V. Then:*

- (i) *$\Pi(A)$ is a closed convex set containing the origin and any closed convex set containing the origin is the polar of its polar.*
- (ii) $\Pi(A) = \Pi(\overline{\langle A \cup \{0\}\rangle})$.
- (iii) *If $A_1 \subseteqq A_2$ then $\Pi(A_1) \supseteqq \Pi(A_2)$ and the converse holds if A_1 and A_2 are closed convex sets containing the origin.*
- (iv) $\Pi(tA) = (1/t)\Pi(A)$; $t > 0$.
- (v) $\Pi(\Pi(A)) = \overline{\langle A \cup \{0\}\rangle}$. *In particular $\Pi(\Pi(A)) = A$ when A is a closed convex set containing the origin.*
- (vi) $\Pi(\bigcup_i A_i) = \bigcap_i \Pi(A_i)$.
- (vii) *$\Pi(A)$ is a neighbourhood of the origin if A is bounded.*
- (viii) *$\Pi(A)$ is bounded if A is a neighbourhood of the origin.*
- (ix) *$\Pi(A)$ is polyhedral when A is finite.*
- (x) *$\Pi(A)$ is a polytope when A is a polyhedral neighbourhood of the origin.*

Proof. Statements (ii), (iv) and (vi) follow directly from the definition, while we proved (v) above. Clearly $\Pi(A_1) \supseteqq \Pi(A_2)$ when $A_1 \subseteqq A_2$. Conversely, if $\Pi(A_1) \supseteqq \Pi(A_2)$ then this yields $\overline{\langle A_1 \cup \{0\}\rangle} = \Pi(\Pi(A_1)) \subseteqq \Pi(\Pi(A_2)) = \overline{\langle A_2 \cup \{0\}\rangle}$ so that (iii) holds. The first part of (i) follows directly from the definition, while the second part follows from (v).

Let B denote the closed unit ball $\{v : \|v\| \leqslant 1\}$. If A is bounded then $A \subseteqq tB$ for some $t > 0$ and then $\Pi(A) \supseteqq \Pi(tB) = (1/t)\Pi(B) = (1/t)B$ so that $\Pi(A)$ is a neighbourhood of the origin. On the other hand if A is a neighbourhood of the origin then $A \supseteqq tB$ for some $t > 0$ and then $\Pi(A) \subseteqq \Pi(tB) = (1/t)B$ so that $\Pi(A)$ is bounded.

If A is finite then $\Pi(A)$ is the intersection of the halfspaces $\{w : (v, w) \leqslant 1\}$; $v \in A$ and is thus polyhedral by the definition of a polyhedral set.

Finally assume that A is a polyhedral neighbourhood of the origin. Then $A = \bigcap_{i=1}^r \{v : (v, w_i) \leqslant t_i\}$ for non-zero vectors $w_1, \ldots, w_r$ and for numbers $t_1, \ldots, t_r$. Since each hyperplane $\{v : (v, w_i) \leqslant t_i\}$ is a neighbourhood of the origin, it follows that $t_1, \ldots, t_r > 0$. Then, by replacing w_i with $t_i w_i$, we may assume without loss of generality that $t_1 = \cdots = t_r = 1$ so that $A = \Pi(\{w_1, \ldots, w_r\})$ and thus $\Pi(A) = \Pi(\Pi(\{w_1, \ldots, w_r\})) = \overline{\langle\{0, w_1, \ldots, w_r\}\rangle}$.

□

Here we state six of the many noteworthy corollaries to this theorem.

Corollary 2.3.35. *Polars of compact convex neighbourhoods of the origin are also compact convex neighbourhoods of the origin and any such set is the polar of its polar.*

Proof. Let A be a compact convex neighbourhood of the origin. Then $\Pi(A)$ is a compact convex neighbourhood of the origin by (i), (vii) and (viii) and $A = \Pi(\Pi(A))$ by (v). □

It follows that polarity provides a notion of duality for compact convex neighbourhoods of the origin. This is the notion of duality described before the definition of polars.

Corollary 2.3.36. *Compact convex neighbourhoods B_1 and B_2 of the origin are polars of each other if and only if the Minkowski functional of B_1 is the support function of B_2, i.e. $\omega_{B_1} = \psi_{B_2}$.*

Proof. If $B_1 = \Pi(B_2)$ then

$$\begin{aligned}\omega_{B_1}(w) &= \inf\{t : t > 0,\ w/t \in B_1\}\\ &= \inf\{t : t > 0,\ (v, w/t) \leqslant 1; v \in B_2\}\\ &= \inf\{t : t > 0,\ (v, w) \leqslant t; v \in B_2\}\\ &= \inf\{t : t > 0,\ \psi_{B_2}(w) \leqslant t\} = \psi_{B_2}(w).\end{aligned}$$

Conversely if $\omega_{B_1} = \psi_{B_2}$ then

$$w \in B_1 \Leftrightarrow \omega_{B_1}(w) \leqslant 1 \Leftrightarrow \psi_{B_2}(w) \leqslant 1 \Leftrightarrow w \in \Pi(B_2)$$

so that $B_1 = \Pi(B_2)$. □

The four ways of expressing duality for positive sublinear functions are equivalent.

Corollary 2.3.37 (Duality of Minkowski functionals). *The following conditions are equivalent for positive real valued sublinear functionals ψ_1 and ψ_2.*

(i) *ψ_1 and ψ_2 are support functions of compact convex neighbourhoods of the origin which are polars of each other.*
(ii) *ψ_1 and ψ_2 are Minkowski functionals of compact convex neighbourhoods of the origin which are polars of each other.*
(iii) *ψ_1 is the support function of a compact convex neighbourhood of the origin which has ψ_2 as its Minkowski functional.*
(iv) $\psi_2(v) = \sup\{(v, w)/\psi_1(w) : w \neq 0\} = \sup\{(v, w) : \psi_1(w) = 1\}$.

If these conditions are satisfied then $(v, w) \leqslant \psi_1(w)\psi_2(v)$; $v \in V$, $w \in V$.

Proof. We may express ψ_1 and ψ_2 as $\psi_1 = \psi_A = \omega_B$ and $\psi_2 = \psi_C = \omega_D$ for compact convex neighbourhoods A, B, C, D of the origin. By the previous corollary $B = \Pi(A)$ and $D = \Pi(C)$. Hence $C = \Pi(A) \Leftrightarrow D = \Pi(B)$ which establishes the equivalence of (i) and (ii) and also that these conditions are equivalent to (iii).

If conditions (i), (ii) and (iii) are satisfied and if $v, w \neq 0$ then $w/\psi_1(w) \in B$ and $v/\psi_2(v) \in D$ so that $(w/\psi_1(w), v/\psi_2(v)) \leqslant 1$ and thus $(v, w) \leqslant \psi_1(w)\psi_2(v)$ whenever $v, w \in V$. In particular $\psi_2(v) \geqslant \sup\{(v, w)/\psi_1(w) : w \neq 0\}$. On the other hand if $t = \sup\{(v, w)/\psi_1(w) : w \neq 0\}$ and if $v \neq 0$ then $t > 0$ and $1 = \sup\{(v/t, w/\psi_1(w)) : w \neq 0\}$ so that $v/t \in \Pi(B) = A$ and thus $t \geqslant \omega_A(v) = \psi_C(v) = \psi_2(v)$. Hence (i)–(iii) imply (iv) and thus also the inequality stated at the end of the corollary.

Finally, assume that (iv) holds and put $\psi_2' = \psi_B$. Then ψ_1 and ψ_2' are related as ψ_1 and ψ_2 are in (i) and thus, by what we have shown, $\psi_2'(v) = \sup\{(v, w)/\psi_1(w) : w \neq 0\} = \psi_2(v)$; $v \in V$ so that $\psi_2' = \psi_2$ and thus (i)–(iii) hold. □

Example 2.3.38 (Dual norms). The dual of a Banach space V normed by the norm $\| \ \|$ is the linear space V^* of continuous linear functionals equipped with the dual norm $\| \ \|^*$ defined by

$$\|v\|^* = \sup\{v^*(v) : \|v\| = 1\}.$$

Comparing this with (iv) of the previous corollary we see that if V is normed by the norm ψ_1 then the 'dual' functional ψ_2 is also the dual norm on $V'(= V^*)$ provided we identify a linear functional f on V with the unique vector v such that $f(w) = (v, w)$; $w \in V$.

Let us agree to say that positive sublinear functionals ψ_1 and ψ_2 are duals (of each other) if the equivalent conditions (i)–(iv) of the previous corollary are satisfied.

As polars of symmetric sets are symmetric it is clear that if ψ_1 and ψ_2 are dual and if one of them is a norm then also the other one is a norm. Thus we have a notion of duality for norms which is consistent with the notion of the dual of a normed space.

Now consider a measure space $(\mathscr{X}, \mathscr{A}, \mu)$ where the measure μ is non-negative. If f is a measurable function on $\mathscr{X}$ and if $p \geqslant 1$ then the L_p *norm* $\|f\|_p$ is defined as $(\int |f|^p \, \mathrm{d}\mu)^{1/p}$ or as $\mathrm{ess}_\mu \sup|f|$ as $p < \infty$ or $p = \infty$. For each $p \in [1, \infty]$ the functional $\| \ \|_p$ defines a norm for the linear space of μ equivalence classes of functions f such that $\|f\|_p < \infty$. The normed linear space defined this way is the Banach space $L_p = L_p(\mathscr{X}, \mathscr{A}, \mu)$.

If $p, q \in [1, \infty]$ are conjugate in the sense that $1/p + 1/q = 1$ and if $g \in L_q$ then, by Hölder's inequality, $f \to \int fg \, \mathrm{d}\mu$ defines a continuous linear functional on L_p. If $p < \infty$ or if $\mathscr{X}$ is a finite set, then all continuous linear functionals on L_p may be obtained that way. Furthermore, under these conditions, the Banach spaces L_p^* and L_q are isomorphic by this identification. The duality of the norms is expressed by the identity

$$\|g\|_q = \sup\left\{\int fg\,\mathrm{d}\mu : \int |f|^p\,\mathrm{d}\mu = 1\right\}$$

which is valid when $g \in L_q$.

Specializing to $\mathscr{X} = \{1,\ldots,n\}$ we obtain the norms $\|\ \|_p : x \to (\sum_i |x_i|^p)^{1/p}$; $1 \leqslant p < \infty$ and the norm $\|\ \|_\infty : x \to \max_i |x_i|$. Two norms $\|\ \|_p$ and $\|\ \|_q$ are thus dual of each other when $p, q \geqslant 1$ and $1/p + 1/q = 1$.

Returning to our inner product space V with norm $\|\ \|$ we may for each positive definite self adjoint operator A on V define dual norms ψ_1 and ψ_2 by putting

$$\psi_1(v) = (v, Av)^{1/2} = \|A^{1/2}v\|$$

and

$$\psi_2(v) = (v, A^{-1}v)^{1/2} = \|A^{-1/2}v\|$$

when $v \in V$.

Polytopes and bounded polyhedral sets are the same objects.

Corollary 2.3.39. *The following conditions are equivalent for a convex subset A of V.*

(i) *A is a polytope.*
(ii) *A is compact and convex and* ext *A is a finite set.*
(iii) *A is a bounded polyhedral set.*

Proof. The equivalence of (i) and (ii) was argued in remark 2 after theorem 2.3.15. Furthermore the implication (iii) $\Rightarrow$ (i) is another way of stating corollary 2.3.18. It remains to show that (i) $\Rightarrow$ (iii).

Let A be a polytope. By replacing A with $A - v$ where $v \in \operatorname{ri} A$ we see that without loss of generality we may assume that $0 \in \operatorname{ri} A$. Reducing V to the linear span $[A]$ of A we see also that we may assume that the origin is an interior point of A. Under these circumstances theorem 2.3.34 guarantees that $\Pi(A) = \Pi(\operatorname{ext} A)$ is a bounded polyhedral neighbourhood of the origin. If so then, by corollary 2.3.18, $\Pi(A)$ is a polytope and thus $A = \Pi(\Pi(A))$ is polyhedral. □

It may not be obvious from the definition that intersections of finite classes of polytopes are polytopes. On the other hand, the definition of a polyhedral set as a finite intersection of closed halfspaces implies immediately that the class of polyhedral sets is a π-system, i.e. is closed under finite intersections.

However we have just seen that bounded polyhedral sets and polytopes are the same objects.

Corollary 2.3.40. *If A and B are polytopes then their intersection $A \cap B$ is also a polytope.*

This implies in turn that the class of locally simplicial sets constitute a π-system.

Corollary 2.3.41. *If S and T are locally simplicial subsets of V then their intersection $S \cap T$ is also locally simplicial.*

Proof. Let $v \in S \cap T$ where the sets S and T are both locally simplicial. Then there are simplices $S_1, \ldots, S_a$ contained in S and simplices $T_1, \ldots, T_b$ contained in T such that $S \cap U' = (S_1 \cup \cdots \cup S_a) \cap U'$ and $T \cap U'' = (T_1 \cup \cdots \cup T_b) \cap U''$ for neighbourhoods U' and U'' of v. Replacing U' and U'' with $U' \cap U''$ we see that we may assume that $U' = U'' = U$. Then $(\bigcup_{i,j} S_i \cap T_j) \cap U = (\bigcup_i S_i \cap \bigcup_j T_j) \cap U = S \cap U \cap T \cap U = S \cap T \cap U$ and the sets $S_i \cap T_j$ are all polytopes, by the previous corollary. It follows that $S \cap T$ is locally simplicial. □

Now consider a vector v belonging to a closed halfspace H. If S is a simplex having v as an interior point then $v \in S \cap H$ and the set $S \cap H$ is a union of simplices by corollary 2.3.39. It follows that closed halfspaces are locally simplicial. However polyhedral sets are finite intersections of closed halfspaces.

Corollary 2.3.42. *Polyhedral sets are locally simplicial.*

2.4 References

Beckenback, E. & Bellman, R. 1965. *Inequalities.* Springer–Verlag, Berlin.

Blackwell, C. & Girshick, M. A. 1954. *Theory of games and statistical decisions.* Wiley, New York.

Bonnesen, T. and Fenchel, W. 1934. *Theorie der Konvexen Körper.* Springer–Verlag, Berlin.

Brøndsted, A. 1982. *An introduction to convex polytopes.* Springer–Verlag, Berlin.

Caratheodory, C. 1907. Uber den variabilitätsbereich der koeffizienten von Potenzreihen, die gegebene werte annamhen. *Math. Ann.* **64**, pp. 95–115.

— 1911. Ueber den variabilitätsbereich der Fourierschen konstanten von positiven harmonischen funktionen. *Rend. Circ. Mat. Palermo.* **32**, pp. 193–217.

Ferguson, T. S. 1967. *Mathematical Statistics. A decision theoretic approach.* Academic Press, New York and London.

Grünbaum, B. 1967. *Convex Polytopes.* Wiley, New York.

Hardy, G. H., Littlewood, J. E. & Polya, G. 1934. *Inequalities.* Cambridge University Press.

Helly, E. 1923. Uber Mengen convexen körper mit gemeinschaftlichen Punkten. *Jahrb. Deut. Math. Verein.* **32**, pp. 175–6.

Kelley, J. L. & Namioka, I. 1961. *Linear topological spaces.* Van Nostrand, London.

LeCam, L. 1964. Sufficiency and approximate sufficiency. *Ann. Math. Stat.* **35**, pp. 1419–55.

— 1986. Asymptotic methods in mathematical statistics. Springer-Verlag, Berlin.

Lyusternik, L. A. 1962. *Convex figures and Polyhedra.* Dover, New York.

Marshall, A. W. and Olkin, I. 1979. *Inequalities: Theory of majorization and its applications.* Academic Press, New York.

McMullen, P. & Shephard, G. C. 1971. *Convex polytopes and the upper bound conjecture.* Cambridge University Press.
Roberts, A. W. & Varberg, D. E. 1973. *Convex functions.* Academic Press, New York.
Rockafellar, R. T. 1970. *Convex analysis.* Princeton University Press.
Strassen, V. 1965. The existence of probability measures with given marginals. *Ann. of Math. Statist.* **36**, pp. 423–39.
Valentine, F. A. 1964. *Convex sets.* McGraw-Hill, New York.
Yaglom, I. M. & Boltyanskii, V. G. 1961. *Convex figures.* Holt, Rinehart and Winston, New York.

3

Two-Person, Zero-Sum Games

3.1 Introduction

Game theory and statistical decision theory, as treated here, are closely connected. Although the origins of most theories are hard to trace, the original major contributions to these theories are von Neumann & Morgenstern's book, *Theory of games and economic behaviour*, 1944 and Wald's book *Statistical decision functions*, 1950. Here we shall give a short introduction to those parts of game theory which have been particularly useful for decision theory. The author learned about most of the results which are presented here from the lectures and writings of Bickel, Blackwell and LeCam.

Sections 3.2 and 3.3 are intended as an introduction to the general theory of two-person, zero-sum games. Some of the standard tools for establishing minimaxity and admissibility are described in section 3.4. We have found it convenient to phrase these results in terms of a general functional rather than in terms of Bayes risk. The principles are elementary as well as useful. We have therefore organized the material so that this section may be read independently of the previous two sections as well as of chapter 1.

3.2 Values and strategies

A *two-person, zero-sum game* is a triple $\Gamma = (A, B, M)$ where A and B are arbitrary sets and M is a function from $A \times B$ to $[-\infty, \infty]$. The game involves two players, *player* I *and player* II. The elements of A and B are called the (pure) *strategies* of player I and player II, respectively. We assume that the players choose their strategies independently of each other. If player I uses the strategy $a \in A$ and player II uses the strategy $b \in B$, then player II 'pays' player I an amount $M(a, b)$. The function M is called the *pay-off* function of Γ. Obviously, the sum of gain and loss is zero, as is indicated in the term 'two-person, zero-sum'.

Although the sets A and B need not be disjoint (indeed they are often equal) we shall express ourselves as if they were disjoint. Thus if the letter b, with or without affixes, denotes a strategy as well as an element in the set B, then b is considered as a strategy for player II.

This notion of a game may at first sight appear too narrow to be applicable to, say, the usual games of cards or pieces. It may however be argued that a game between two persons such that the gain of one player equals the loss of the other, may be reduced to this form. The notion of a strategy in the reduced form then involves all moves or responses which might be considered. A strategy for, say, white in chess, should tell white exactly how to respond to any situation which might occur on the chessboard.

We refer the interested reader to Blackwell & Girchick, 1954 and to Jones, 1980 for a general background in game theory. However, such knowledge is not needed for an understanding of this exposition.

All games considered here are two-person, zero-sum games. We shall therefore permit ourselves to use the term 'game' as synonymous with the term 'two-person, zero-sum game'.

Many statistical problems may be regarded as two-person, zero-sum games. 'Nature' then takes the role of player I, choosing a parameter $\theta \in \Theta$, where Θ is the space of parameters. Without knowing the strategy chosen by nature the statistician makes a decision d, which may for example be an estimate of θ. As a consequence of these choices, the statistician 'loses' an amount $M(\theta, d)$. We call M the loss function. In the problem of estimating a real parameter θ the pay-off may e.g. be given as $(\theta - d)^2$ (quadratic loss). Usually observations are available for the statistician. The strategies are then decision procedures δ which determine for each possible outcome X of an experiment, which decision $\delta(X)$ to make. The pay-off is then defined to be the expected loss and is called *risk*. We shall return to this later on.

A game $\Gamma = (A, B, M)$ where M is real valued and A and B are finite sets is called a *matrix game*. It is then customary to represent M by a matrix (table) where a strategy a for player I defines the row $(M(a,b) : b \in B)$, while a strategy b for player II defines the column $(M(a,b) : a \in A)$.

In the following we consider a given game $\Gamma = (A, B, M)$. Of course each player wants to maximize his/her gain. Player I is therefore interested in the behaviour of $M(a_0, b)$ as a function of b for each strategy a_0 in A, while player II is interested in $M(a, b_0)$ as a function of a for each strategy b_0 in B.

Let a_1 and a_2 be strategies for player I. We shall then say that a_1 *dominates* a_2 if $M(a_1, b) \geqslant M(a_2, b)$ for all $b \in B$. More generally, if $A_1 \subseteq A$ and $A_2 \subseteq A$ then we say that A_1 *dominates* A_2 if for each $a_2 \in A_2$ there exists $a_1 \in A_1$ such that a_1 dominates a_2. Similarly if b_1 and b_2 are strategies for player II then we say that b_1 *dominates* b_2 if $M(a, b_1) \leqslant M(a, b_2)$ for all $a \in A$. If $B_1, B_2 \subseteq B$ then we say that B_1 *dominates* B_2 if for all $b_2 \in B_2$ there is $b_1 \in B_1$ such that b_1 dominates b_2.

Strategies a_1, $a_2 \in A$ ($b_1, b_2 \in B$) are called *equivalent* if $a_1(b_1)$ dominates $a_2(b_2)$ and $a_2(b_2)$ dominates $a_1(b_1)$ i.e. if $M(a_1, \cdot) = M(a_2, \cdot)(M(\cdot, b_1) = M(\cdot, b_2))$. A strategy $a_0 \in A$ ($b_0 \in B$) is called *admissible* if there is no strategy in $A(B)$ which dominates $a_0(b_0)$ and which is not equivalent to $a_0(b_0)$. The set of all admissible strategies in $A(B)$ will be denoted by $A_0(B_0)$.

Dominance is clearly a partial ordering and the notion of equivalence is the equivalence relation associated with this ordering. Admissibility is maximality with respect to the domination ordering.

A subset $A_1(B_1)$ of $A(B)$ is called *essentially complete* if $A_1(B_1)$ dominates $A(B)$. A subset $A_1(B_1)$ of $A(B)$ is called *complete* if to each $a \in A - A_1$ ($b \in B - B_1$) there corresponds a strategy $a_1 \in A_1$ ($b_1 \in B_1$) which dominates $a(b)$ and which is not equivalent to $a(b)$.

Any complete set of strategies is clearly essentially complete and the two notions of completeness coincide when the equivalence relation for strategies reduces to equality.

The set $A_0(B_0)$ of admissible strategies is clearly complete if and only if it is essentially complete.

In general if a player considers two strategies then usually neither dominates the other one, i.e. the ordering of dominance is in general not total. The player may, however, obtain total orderings of the sets of strategies by considering real valued functionals on the pay-off functions defined by the strategies. Thus the player might average, minimize or maximize. Considering the worst possibilities the players might utilize the functions

$$M_{\mathrm{I}}(a) = \inf_{b \in B} M(a, b); \qquad a \in A$$

and

$$M_{\mathrm{II}}(b) = \sup_{a \in A} M(a, b); \qquad b \in B.$$

Using the strategy $a \in A$, player I is certain to receive an amount of at least $M_{\mathrm{I}}(a)$, and is not guaranteed any larger amount. $M_{\mathrm{I}}(a)$ is thus a measure of how good each strategy $a \in A$ is and M_{I} defines an ordering of strategies of player *I*. $M_{\mathrm{II}}(b)$ is the maximum loss of player II using strategy b, and M_{II} defines an ordering of the strategies of player II.

M_{I} and M_{II} will be called, respectively, *the lower function of* Γ and *the upper function of* Γ.

The quantity $\sup_a M_{\mathrm{I}}(a)$ is called the *lower value of* Γ and will be denoted as $\underline{V}(\Gamma)$ or as $\underline{V}$. Similarly the quantity $\inf_b M_{\mathrm{II}}(b)$ is called the *upper value of* Γ and is denoted by $\bar{V}(\Gamma)$ or by $\bar{V}$.

A strategy $a_0 \in A$ such that $M_{\mathrm{I}}(a_0) = \underline{V}$ is called a *maximin strategy* for player I. A maximin strategy $a_0 \in A$ maximizes player I's minimal (or rather infimal) gain.

Similarly a *minimax strategy* for player II is a strategy b_0 such that $M_{\mathrm{II}}(b_0) =$

$\bar{V}$. Thus a minimax strategy $b_0 \in B$ minimizes player II's maximal (or rather supremal) loss. If $A(B)$ is infinite then maximin (minimax) strategies may not exist. Then we may find strategies a in A (b in B) such that the quantities $M_{\mathrm{I}}(a)(M_{\mathrm{II}}(b))$ are arbitrarily close to $\underline{V}(\bar{V})$.

The justifications for the terms lower and upper are given in the following proposition.

Proposition 3.2.1. *If $a \in A$ and $b \in B$ then*

$$M_{\mathrm{I}}(a) \leqslant M(a,b) \leqslant M_{\mathrm{II}}(b)$$

and

$$M_{\mathrm{I}}(a) \leqslant \underline{V} \leqslant \bar{V} \leqslant M_{\mathrm{II}}(b).$$

Proof. The second set of inequalities follows from the first, which in turn follows directly from the definitions of M_{I} and M_{II}. □

The game Γ is said to have the *value* $V(\Gamma)$ if $\underline{V}(\Gamma) = \bar{V}(\Gamma) = V(\Gamma)$. We may write V instead of $V(\Gamma)$.

By using a maximin strategy (or a strategy that is approximately a maximin strategy) player I is guaranteed a gain of at least $\underline{V}$. On the other hand if $s > \underline{V}$ then to any strategy a for I there corresponds a strategy b for II rendering the gain $M(a,b)$ for I less than s. If II uses a minimax strategy then the gain of I is at most $\bar{V}$. Suppose now that the game has a value V, that a_0 is a maximin strategy for player I and that b_0 is a minimax strategy for player II.

If a and b are strategies for player I and player II respectively, then by proposition 3.2.1

$$M(a,b_0) \leqslant M_{\mathrm{II}}(b_0) = V = M_{\mathrm{I}}(a_0) \leqslant M(a_0,b).$$

In particular $M(a_0,b_0) \leqslant V \leqslant M(a_0,b_0)$ so that

$$M(a,b_0) \leqslant M(a_0,b_0) \leqslant M(a_0,b) \tag{3.2.1}$$

when $a \in A$ and $b \in B$. A pair (a_0,b_0) of strategies satisfying (3.2.1) is called a *saddlepoint* for the game. Clearly (a_0,b_0) is a saddlepoint for $\Gamma = (A,B,M)$ if and only if a_0 is I's best strategy when II uses b_0 and b_0 is II's best strategy when I uses a_0.

Let a_0 and b_0 be strategies for I and II respectively. Then we shall say that *a_0 is optimal for b_0* if $M(a,b_0) \leqslant M(a_0,b_0)$ when $a \in A$. Similarly b_0 is called *optimal for a_0* if $M(a_0,b_0) \leqslant M(a_0,b)$ when $b \in B$.

Thus (a_0,b_0) is a saddlepoint if and only if a_0 and b_0 are optimal with respect to each other.

For an example of a saddlepoint of an actual saddle consider the point $(0,0)$ and the graph of $M(a,b) \equiv b^2 - a^2$.

If Γ is a matrix game such that A defines the rows while B defines the columns, then (a_0, b_0) is a saddlepoint if and only if $M(a_0, b_0)$ is both the maximum of the b_0 column $(M(a, b_0) : a \in A)$ and the minimum of the a_0 row $(M(a_0, b) : b \in B)$.

Now suppose that the game $\Gamma = (A, B, M)$ has a saddlepoint (a_0, b_0). Then, by (3.2.1) $M_{\mathrm{I}}(a_0) = M(a_0, b_0) = M_{\mathrm{II}}(b_0)$. Hence, by proposition 3.2.1 we have the following theorem.

Theorem 3.2.2. *The game $\Gamma = (A, B, M)$ has a value, player* I *has maximin strategy a_0 and player* II *has minimax strategy b_0 if and only if (a_0, b_0) is a saddlepoint of Γ. If this is so then the value of Γ is $M(a_0, b_0)$.*

Example 3.2.3. Let Γ_i; $i = 1, 2$ be given by the matrices $\begin{pmatrix} -1, & -2 \\ 2, & -1 \end{pmatrix}$ and $\begin{pmatrix} 0, & 1 \\ 1, & 0 \end{pmatrix}$ respectively. Then Γ_1 has the saddlepoint $(2, 2)$ while Γ_2 has no saddlepoint. In fact $V(\Gamma_1) = -1$ while $\underline{V}(\Gamma_2) = 0$ and $\bar{V}(\Gamma_2) = 1$.

Consider Γ_2 and suppose the players permit themselves to use randomized strategies. Let $\xi_i(\eta_i)$ be the probability of player I (II) choosing the ith row (column). If $M^*(\xi, \eta)$ denotes the expected pay-off, then $M^*(\xi, \eta) = \xi_1\eta_2 + \xi_2\eta_1$. Thus $M^*(\xi, \eta) = 1/2$ if and only if $\xi_2 = 1/2$ or $\eta_2 = 1/2$ and any such pair (ξ, η) constitutes a saddlepoint for the game $\Gamma^* = (\{\xi\}, \{\eta\}, M^*)$.

The last example suggests that even if the game Γ does not have a value, it may be converted into a game possessing a value by permitting the players to use randomized strategies. Actually, von Neumann (1928) proved that this holds for any matrix game.

Thus under general conditions we may expect that games Γ where the original (pure) strategies dominate the randomized (mixed) strategies, have values. However, the simple example where $A = B =]-\infty, \infty[$ and $M(a, b) \equiv a + b$ shows the need for some limitations.

We shall continue our investigation by giving precise descriptions of those games where randomization is superfluous.

In the following we shall need to consider convex combinations of pay-offs, so we must ensure that these combinations are well defined. We shall assume *that the range of the pay-off function contains at most one of the infinities $-\infty$ and $+\infty$*, although this condition is somewhat too strong.

Let ξ be a probability distribution on A with finite support. Suppose A chooses a strategy according to ξ. Then the expected pay-off to A when B chooses strategy b is $\sum_a M(a, b)\xi(a)$. Thus the strategy $a_0 \in A$ is at least as good as this mixture if

$$M(a_0, b) \geqslant \sum_a M(a, b)\xi(a); \qquad b \in B. \tag{3.2.2}$$

We shall say that A is *mixed* or *concave* (relatively Γ) if to each probability distribution ξ on A with finite support there corresponds a strategy $a_0 \in A$ such that (3.2.2) holds. If, in addition, a_0 may always be chosen so that (3.2.2) holds with '=' instead of '$\geqslant$' then A is called *affine* (relatively Γ). Similarly B is called *mixed* or *convex* if to each probability distribution η on B with finite support there corresponds a $b_0 \in B$ such that

$$M(a, b_0) \leqslant \sum_b M(a, b)\eta(b); \qquad a \in A. \tag{3.2.3}$$

If in addition b_0 may always be chosen so that (3.2.3) holds with '=' instead of '$\leqslant$' then B is called *affine* (relatively Γ).

If A and B are both mixed (affine) then the game Γ itself is called *mixed or concave–convex* (*affine*).

When establishing that A or B is mixed or affine we need only consider two-point distributions.

Proposition 3.2.4. *Let $\Gamma = (A, B, M)$ be a game. Then:*

(i) *A is mixed if and only if to each two-point distribution ξ on A there corresponds an $a_0 \in A$ such that*

$$M(a_0, b) \geqslant \sum M(a, b)\xi(a); \qquad b \in B.$$

(ii) *A is affine if and only if to each two-point distribution ξ on A there corresponds an $a_0 \in A$ such that*

$$M(a_0, b) = \sum_a M(a, b)\xi(a); \qquad b \in B.$$

(iii) *B is mixed if and only if to each two-point distribution η on B there corresponds a $b_0 \in B$ such that*

$$M(a, b_0) \leqslant \sum M(a, b)\eta(b); \qquad a \in A.$$

(iv) *B is affine if and only if to each two-point distribution η on B there corresponds a $b_0 \in B$ such that*

$$M(a, b_0) = \sum M(a, b)\eta(b); \qquad a \in A.$$

Proof. The 'only if' is clear in each case and the 'if' follows by induction on the number of strategies which occur with positive probabilities. □

Example 3.2.5 (Concave–convex pay-off functions). Let A and B be convex subsets of linear spaces. Then $\Gamma = (A, B, M)$ is mixed provided $M(a, b)$ is concave in a for fixed b and convex in b for fixed a.

For any set C let the set of probability distributions on C with finite support be denoted by C^*. The representation of each $c \in C$ by the one-point distribution in c is the natural imbedding of C in C^*. Now consider a game $\Gamma = (A, B, M)$. Suppose players I and II, independently of each other, choose strategies according to probability distributions $a^* \in A^*$ and $b^* \in B^*$ respectively. Then the expected pay-off to player I is $\int M \mathrm{d}(a^* \times b^*) = \sum_{a,b} M(a,b)a^*(a)b^*(b)$. Thus we have a new game $\Gamma^* = (A^*, B^*, M^*)$ where $M^*(a^*, b^*) = \sum_{a,b} M(a,b)a^*(a)b^*(b)$. The game Γ^* is clearly affine and, consequently, mixed. We shall call Γ^* the *finitely mixed extension* of Γ. By von Neumann's minimax theorem Γ^* has a value whenever the sets A and B are both finite and if M is real valued. The lower and upper values (functions) of a game and its finitely mixed extension are related as follows.

Theorem 3.2.6. *Let $\Gamma^* = (A^*, B^*, M^*)$ be the finitely mixed extension of the game $\Gamma = (A, B, M)$. Then*:

(i) $M_{\mathrm{I}}^*(a^*) = \inf_b M^*(a^*, b);\ a^* \in A^*$
$M_{\mathrm{II}}^*(b^*) = \sup_a M^*(a, b^*);\ b^* \in B.$

(ii) $M_{\mathrm{I}}^* | A = M_{\mathrm{I}}$

$M_{\mathrm{II}}^* | B = M_{\mathrm{II}}.$

(iii) $\underline{V}(\Gamma) \leqslant \underline{V}(\Gamma^*) \leqslant \bar{V}(\Gamma^*) \leqslant \bar{V}(\Gamma)$.

(iv) *If Γ has value V then V is also the value of Γ^*.*

Proof. By the definition,

$$M_{\mathrm{I}}^*(a^*) = \inf_{b^*} M^*(a^*, b^*) = \inf_{b^*} \sum_b M^*(a^*, b)b^*(b) = \inf_b M^*(a^*, b).$$

Similarly $M_{\mathrm{II}}^*(b^*) = \sup_a M^*(a, b^*)$.

(ii)–(iv) are direct consequences of (i) and proposition 3.2.1. □

Before proceeding let us note that general statements on one of the two players may be converted into general statements on the other player.

Proposition 3.2.7. *Associated with any game $\Gamma = (A, B, M)$ the game $\Gamma' = (B, A, M')$ where $M'(b, a) = -M(a, b)$ when $a \in A$ and $b \in B$. Then $(\Gamma')' = \Gamma$ and the following statements hold*:

(i) $M_{\mathrm{I}}'(b) = -M_{\mathrm{II}}(b),\ M_{\mathrm{II}}'(a) = -M_{\mathrm{I}}(a)$.

(ii) *b_0 is a maximin strategy for player* I *in Γ' if and only if b_0 is a minimax strategy for player* II *in Γ.*

(iii) *a_0 is a minimax strategy for player* II *in Γ' if and only if a_0 is a maximin strategy for player* I *in Γ.*

(iv) $\underline{V}(\Gamma') = -\bar{V}(\Gamma)$ *and* $\bar{V}(\Gamma') = -\underline{V}(\Gamma)$.

(v) *Γ' has a value if and only Γ has a value and then $V(\Gamma') = -V(\Gamma)$.*

Remark. Games such that $\Gamma = \Gamma'$ are called *symmetric*.

Proof. This follows readily from the formula $\sup(-f) = -\inf f$ which is valid for any extended real valued function f. □

The next theorem provides a useful stepping stone for proving, at least theoretically, that certain games have values.

Theorem 3.2.8. *$\Gamma = (A, B, M)$ has a value if and only if the sets $\{a : M(a,b) \geqslant \tau\}$; $b \in B$ have a non-empty intersection whenever $\tau < \bar{V}$.*

Remark. If $\tau < \underline{V}$ or if $\tau = -\infty$ then, trivially, $\bigcap_b \{a : M(a,b) \geqslant \tau\} \neq \varnothing$.

Proof.

(i) Suppose Γ has the value V and let $\tau < \bar{V} = V = \underline{V}$. Then, since $\tau < \underline{V}$, $M(a,b) \geqslant \tau$ for all $b \in B$ for some $a \in A$.

(ii) Suppose the condition is satisfied and let $\tau < \bar{V}$.
Then there is an $a \in A$ such that $M_{\text{I}}(a) \geqslant \tau$. Hence $\underline{V} \geqslant \tau$ when $\tau < \bar{V}$.
Thus $\underline{V} \geqslant \bar{V}$ so that $\underline{V} = \bar{V}$. □

Note that the sets $\{a : M(a,b) \geqslant \tau\}$ are necessarily non-empty whenever $\tau < \bar{V}$. (If not, then $\tau \geqslant \bar{V}$.) Thus we have to show, under given conditions, that the intersections of these sets, for $\tau < \bar{V}$, are non-empty. There is a tool from topology which offers itself here. Following the terminology from general topology we shall say that a class (family) of sets has the *finite intersection property*, abbreviated *f.i.p*, if each finite subclass (sub-family) has a non-empty intersection. The complementary version of the covering criterion for compactness implies that if a class of closed compact subsets of a topological space has the f.i.p. property then the intersection of this class is non-empty. Turning to finite intersections we shall need the following basic result, which is compiled from information from lectures given by Blackwell. The reader might notice that most of the complications in the proof below are due to our desire to permit player II to receive $+\infty$ as pay-off. If M is assumed finite then the proof below reduces to a straightforward application of the weak separation theorem in section 2.3.

Theorem 3.2.9 (Finite intersection theorem for mixed games). *Let $\Gamma = (A, B, M)$ be a mixed game where $-\infty \leqslant M < \infty$ and assume that $M(a,b) > -\infty$ for all $a \in A$ when $M_{\text{II}}(b) = \infty$.*

Then for any given number $\tau < \bar{V}$ the class of sets of the form $\{a : M(a,b) > \tau\}$ with b in B has the finite intersection property.

Remark 1. Heyer (1979) pointed out that, contrary to what was claimed in the original statement of this theorem, the last assumption need not hold for mixed games having finite values but admitting possibilities of infinite pay-offs.

Remark 2. The condition $\tau < \bar{V}$, whether Γ is mixed or not, is equivalent to the condition that all sets $\{a : M(a,b) > \tau\}$ where $b \in B$ are non-empty.

Proof. Since $\tau < \bar{V}$ it follows that there is a number $\tau_1 \in]\tau, \bar{V}[$. Thus we may assume that $\tau > -\infty$. Let $b_1, b_2, \ldots, b_m$ be strategies for player II and denote by S the set of all m-tuples $(M(a,b_1), M(a,b_2), \ldots, M(a,b_m))$ where $a \in A$.

Then $S \subseteq [-\infty, \infty[^m$. Put $H = [\tau, \infty[^m$. Suppose $\bigcap_{i=1}^m \{a : M(a,b_i) \geqslant \tau\} = \varnothing$, i.e. $S \cap H = \varnothing$. Let $T = \{y : y \in [-\infty, \infty[^m$ and $y \leqslant x$ for some $x \in S\}$ ($y \leqslant x$ is defined coordinatewise). Clearly $S \subseteq T \subseteq [-\infty, \infty[^m$.

Let $y^1, \ldots, y^n \in T$, $\xi_1, \ldots, \xi_n \geqslant 0$, $\sum \xi_i = 1$. Then $y^i \leqslant x^i$ for $x^i \in S$, $i = 1, \ldots, n$.

Hence $\sum \xi_i y^i \leqslant \sum \xi_i x^i$. We may write $x_j^i = M(a_i, b_j)$ so that $\sum \xi_i x_j^i = \sum \xi_i M(a_i, b_j)$. Since A is mixed, it follows that there is an $a \in A$ such that $\sum \xi_i M(a_i, b_j) \leqslant M(a, b_j) = z_j$ for all j. Then $\sum \xi_i y^i \leqslant \sum \xi_i x^i \leqslant z \in S$. Hence T is closed for convex combinations. In particular it follows that $T' = T \cap]-\infty, \infty[^m$ is a convex subset of $]-\infty, \infty[^m$.

Next let us show that the set T' is non-empty and that $T' \cap H = \varnothing$.

If T' were empty then each vector in S would have at least one coordinate which would be $-\infty$. This would imply that $\sum_{i=1}^m M(a, b_i) = -\infty$ for all $a \in A$.

Since B is mixed this would in turn imply the existence of a $\tilde{b} \in B$ such that $M(a, \tilde{b}) \leqslant (1/m) \sum_{i=1}^m M(a, b_i) = -\infty$ for all $a \in A$. Thus we have the contradiction $-\infty = M_{\mathrm{II}}(\tilde{b}) \geqslant \bar{V} > \tau$. It follows that $T' \neq \varnothing$. If $y \in T' \cap H$ then $y \leqslant x \in S \cap H$, contradicting the assumption that $S \cap H = \varnothing$. Hence $T' \cap H = \varnothing$.

We now proceed with the proof of the theorem. Since H and T' are disjoint convex subsets of $]-\infty, \infty[^m$, it follows from the weak separation theorem in section 2.3 that there are numbers $l_1, \ldots, l_m$, not all equal to 0, such that $\sum l_i x_i \geqslant \sum l_i y_i$ whenever $x \in H$, $y \in T'$. If we fix $y \in T'$ and let some $x_i \to \infty$ then we observe that necessarily $l_i \geqslant 0$ for $i = 1, \ldots, m$. Then $\sum_i l_i > 0$ and we may assume that $\sum l_i = 1$.

Since $(\tau, \ldots, \tau) \in H$ we have $\sum l_i y_i \leqslant \tau$ for any $y \in T'$. If $\sum l_i y_i \leqslant \tau$ for all $y \in S$ then since B is mixed, there exists a strategy b such that

$$M(a,b) \leqslant \sum l_i M(a, b_i) \leqslant \tau$$

for all $a \in A$. As $\sup_a M(a,b) = M_{\mathrm{II}}(b) \geqslant \bar{V} > \tau$ this is impossible. It follows that there are vectors $\tilde{y}$ in $S - T'$ such that $\sum l_i \tilde{y} > \tau$.

As $S \subseteqq T'$ when M is finite this argument completes the proof under the additional assumption that M is finite.

Put $I = \{i : \tilde{y}_i > -\infty$ for all $\tilde{y}$ in $S - T'$ such that $\sum l_i \tilde{y}_i > \tau\}$. Then $I \subset \{1, \ldots, m\}$ and, since $i \notin I$ implies $l_i = 0$, I is not empty. Let $0 < p < 1$ and put $p_i = l_i p$ for all $i \in I$. Obviously $\sum_{i \in I} p_i = p$. Put $p_i = (1-p)/(m - \#(I))$ when $i \notin I$. Then $\sum_{i \notin I} p_i = 1 - p$ so that $p_1, \ldots, p_m \geqslant 0$ and $\sum_{i=1}^m p_i = 1$.

Since B is mixed there is a strategy b_p in B such that for all $a \in A$ $M(a, b_p) \leqslant \sum_{i=1}^m p_i M(a, b_i) = p \sum_{i \in I} l_i M(a, b_i) + (1-p)/(m - \#(I)) \sum_{i \notin I} M(a, b_i)$.

If $\sum_i l_i M(a, b_i) \leqslant \tau$, this yields

$$M(a, b_p) \leqslant p\tau + \frac{1-p}{m - \#(I)} \sum_{i \notin I} M(a, b_i) \leqslant p\tau + \frac{1-p}{m - \#(I)} \sum_{i \notin I} M_{\text{II}}(b_i).$$

If $\sum l_i M(a, b_i) > \tau$, then there is some i such that $M(a, b_i) = -\infty$ and $l_i = 0$. Then $i \notin I$ and $M(a, b_p) = -\infty$.

It follows that $M(a, b_p) \leqslant p\tau + (1-p)/(m - \#(I)) \sum_{i \notin I} M_{\text{II}}(b_i)$ for all $a \in A$ so that $\bar{V} \leqslant M_{\text{II}}(b_p) \leqslant p\tau + (1-p)/(m - \#(I)) \sum_{i \notin I} M_{\text{II}}(b_i)$.

Let $i \notin I$. Then there is a $\tilde{y} \in S - T'$ such that $\sum l_i \tilde{y}_i > \tau$ and $\tilde{y}_i = -\infty$. If we write $\tilde{y} = (M(\tilde{a}, b_1), \ldots, M(\tilde{a}, b_m))$ then $M(\tilde{a}, b_i) = -\infty$. Hence, by the assumptions in the theorem, $M_{\text{II}}(b_i) < \infty$ when $i \notin I$. Therefore, by letting $p \to 1$, we get the contradiction $\tau < \bar{V} \leqslant \tau$. □

Example 3.2.10. Put $A_1 =]-\infty, \infty[\times]-\infty, 0[$, $A_2 = \{-\infty\} \times]-\infty, \infty[$, $A = A_1 \cup A_2$, $B = [0, 1]$ and $M(a, b) = (1-b)a^1 + ba^2$ when $a = (a^1, a^2) \in A$ and $b \in B$. It is easily checked that the game $\Gamma = (A, B, M)$ is mixed, even affine, and that $-\infty \leqslant M < \infty$. Furthermore $\underline{V} = 0$ while $\bar{V} = \infty$. Here the conditions of the theorem are violated since $M_{\text{II}}(0) = \infty$ while $M(a, 0) = -\infty$ when $a \in A_2$.

In this case $\{a : M(a, 0) \geqslant \tau\} \cap \{a : M(a, 1) \geqslant \tau\} = \emptyset$ when $0 < \tau < \bar{V}$. Thus the condition that $M(a, b) > -\infty$ for all $a \in A$ when $M_{\text{II}}(b) = \infty$ cannot be deleted from the statement of the theorem.

A handy tool for studying games, and in particular games in statistics, is the concept of semicontinuity. An extended real valued function f on a topological space is called *upper semicontinuous* if, for each constant τ, the set $[f < \tau]$ is open. Similarly an extended real valued function f on a topological space is called *lower semicontinuous* if, for each constant τ, the set $[f > \tau]$ is open.

It is easily seen that an extended real valued function f is continuous if and only if it is both upper semicontinuous and lower semicontinuous.

Some useful facts on semicontinuity are collected in the following theorem.

Theorem 3.2.11. *Let $f, g, \ldots$ be extended real valued functions on a topological space χ. Then:*

(i) *f is upper semicontinuous if and only if $-f$ is lower semicontinuous.*

(ii) *if $f_i : i \in I$ is a family of upper (lower) semicontinuous functions then the function $\inf_i f_i$ (the function $\sup_i f_i$) is upper (lower) semicontinuous.*

(iii) *if $f_i : i \in I$ is a finite family of upper (lower) semicontinuous functions then the functions $\sup_i f_i$ and $\inf_i f_i$ are both upper (lower) semicontinuous.*

(iv) *if $f_i : i \in I$ is a finite family of upper (lower) semicontinuous functions from χ to $[-\infty, \infty[$ (to $]-\infty, \infty]$) and if $c_i : i \in I$ are non-negative real constants then $\sum_i c_i f_i$ is upper (lower) semicontinuous.*

(v) *if χ is compact and if f is upper (lower) semicontinuous on χ then there is an $x_0 \in \chi$ such that $f(x_0) = \sup_x f(x)$ (such that $f(x_0) = \inf_x f(x)$).*

Remark. It may be shown (see e.g. Nachbin, 1965) that if χ is completely regular then f is upper (lower) semicontinuous if and only if f equals the infimum (supremum) of all continuous functions $\geqslant f$ ($\leqslant f$). Theorems 2.3.20 and 2.3.22 provide conditions ensuring upper semicontinuity and continuity of convex functions respectively.

Proof. Let f_1 and f_2 be upper semicontinuous functions from χ to $[-\infty, \infty[$. Then $[f_1 + f_2 < \tau] = \bigcup_{\tau_1+\tau_2=\tau} [f_1 < \tau_1] \cap [f_2 < \tau_2]$ is open. If f is upper semicontinuous and $\tau_0 = \sup_x f(x)$ then the class $\{[f \geqslant \tau]; \tau < \tau_0\}$ is a class of compact sets having f.i.p. Hence, by compactness, there is an x_0 in the intersection of these sets. Then $f(x_0) = \tau_0$. The proofs of the other statements are left to the reader. □

The notion of semicontinuity permits the statement of the following useful minimax theorem.

Theorem 3.2.12. *Let $\Gamma = (A, B, M)$ be a mixed game where $-\infty \leqslant M < \infty$. Assume that there is a topology on A such that A is compact and such that $M(\cdot, b)$ is upper semicontinuous for each $b \in B$. Then the game has a value and player I has a maximin strategy.*

Proof. Let $\tau < \bar{V}$. It follows from theorem 3.2.11 that $M_{\mathrm{II}}(b) < \infty$ for all $b \in B$. Hence, by theorem 3.2.9, the family $\{a : M(a,b) \geqslant \tau\}; b \in B$ has f.i.p. These sets are closed and compact, since A is compact and M is upper semicontinuous on A. It follows that the intersection of these sets is non-empty so that Γ has a value V, by theorem 3.2.8. The function M_{I} is upper semicontinuous by theorem 3.2.11 so that, by the same theorem, player I has a maximin strategy. □

Corollary 3.2.13. *Let $\Gamma = (A, B, M)$ be a game where A and B are convex subsets of linear spaces and where M is real valued and concave on A and convex on B. Suppose $A(B)$ possesses a topology making $A(B)$ compact and M upper (lower) semicontinuous on $A(B)$. Then Γ has a value and player* I (II) *has a maximin (minimax) strategy. If A and B are both compact topological spaces such that M is upper semicontinuous on A and lower semicontinuous on B then Γ has a saddlepoint.*

Proof. The concave–convexity of M implies that Γ is mixed. Thus the conclusion follows directly from the theorem. □

Corollary 3.2.14. *Let $\Gamma = (A, B, M)$ be a game where M is real valued and A is finite. Then the finitely mixed extension Γ^* of Γ has a value and player* I *has a maximin strategy in Γ^*.*

Proof. If A^* is equipped with the topology of setwise convergence then A^* becomes a compact set such that $M^*(\cdot, b^*)$ is continuous for each $b^* \in B^*$. □

The first part of the following corollary is von Neumann's minimax theorem for matrix games.

Corollary 3.2.15. *If $\Gamma = (A, B, M)$ is a matrix game where M is real valued, then its extension $\Gamma^* = (A^*, B^*, M^*)$ has a saddlepoint $(\xi_0, \eta_0) \in A^* \times B^*$. If A has m points and B has n points then ξ_0 and η_0 may be chosen so that their supports contain at most $m \wedge n$ strategies.*

Proof. The existence of saddlepoints follows directly from the previous corollary since A^* and B^* are compact sets and M^* is continuous on $A^* \times B^*$. A strategy $\xi \in A^*$ is maximin in Γ^* if and only if $M^*(\xi, b) = \sum_a M(a, b)\xi(a) \geqslant V(\Gamma^*)$ when $b \in B$. Thus it is only the vector $\sum_a M(a, \cdot)\xi(a)$ in $]-\infty, \infty[^B \equiv]-\infty, \infty[^n$ which matters. Let S denote the set of all vectors in $]-\infty, \infty[^B$ of the form $(M^*(\xi, b); b \in B)$ for some strategy $\xi \in A^*$. Then S is the convex hull spanned by the m vectors $(M(a, b); b \in B)$ where $a \in A$.

Clearly $\min_b s(b) \leqslant V(\Gamma^*)$ when $s \in S$. Furthermore $s \geqslant (V(\Gamma^*), \ldots, V(\Gamma^*))$ when s is produced by a maximin solution ξ. Thus s belongs to the boundary of S in the latter case.

Hence, since $S \subseteqq]-\infty, \infty[^B$, s may be produced by a $\xi \in A^*$ whose support contains at most n strategies. □

Consider a game $\Gamma = (A, B, M)$ having a saddlepoint (a_0, b_0) and let h be a monotonically increasing function on $[-\infty, \infty]$. Then (a_0, b_0) is clearly also a saddlepoint of the game $(A, B, h(M))$. Thus the condition of concave–convexity in corollary 3.2.13, or the condition that Γ is mixed in the general case, is far from being necessary.

At this point it is interesting to note that there is a natural extension of the concept of a convex function which is invariant for monotonically increasing transformations. The idea is that many general properties of convex functions depend solely on the fact that the level sets $[f \leqslant c]$ of a convex function f are convex. In general an extended real valued function f on a linear space is called *quasiconvex* (*quasiconcave*) if the level set $[f \leqslant c]$ (the level set $[f \geqslant c]$) is convex for each constant c. If h is real valued and monotonically increasing and if f is quasiconvex (quasiconcave) then $h(f)$ is quasiconvex (quasiconcave). We refer to Roberts & Varberg (1973) for further information on this interesting class of functions.

In view of these remarks, it is most interesting that Sion (1958) showed that the game $\Gamma = (A, B, M)$ has a saddlepoint provided A and B are compact convex subsets of $]-\infty, \infty[^m$ and $]-\infty, \infty[^n$ respectively, M is upper semicontinuous and quasiconcave on A and lower semicontinuous and quasiconvex on B. Actually von Neumann (1928) in his proof of the first part of corollary 3.2.15 begins by noting that he will only utilize the fact that the sets $\{a : M(a,b) \geqslant \tau\}$ and $\{b : M(a,b) \leqslant \tau\}$ are compact convex sets.

Another interesting approach to minimax theory, and more generally to N-person games, is based on fixed point theorems. An exposition of this method, which is due to Nash (1950) may be found in Jones (1980). An application of this idea, using general additive probability set functions as mixed strategies, is given by Fenstad (1967).

Several interesting weakenings of the basic assumptions are discussed in a survey of minimax theory by Irle (1981).

What conditions on A and M ensure the existence of a topology possessing the properties listed in theorem 3.2.12? In order to provide such conditions we shall need the concept of a *net*, which generalizes the concept of a sequence. The reader is referred to the definition in the introduction to chapter 1 and to the references given there. It suffices to consider the coarsest topology on A for which the functions $M(\cdot, b)$, $b \in B$ are upper semicontinuous on A. This is the topology τ on A generated by the sets $A_{b,t} = \{a \in A : M(a,b) < t\}, b \in B$, $t \in \mathbb{R}$. Thus the sets $A_{b,t}$ constitute a sub-basis for τ, i.e. the intersections of finite families of sets $A_{b,t}$ constitute a basis for τ. Convergence with respect to this topology may be described as follows.

Proposition 3.2.16. *A net (a_α) in A converges to a point $a \in A$ in the topology τ if and only if*

$$\limsup_{\alpha} M(a_\alpha, b) \leqslant M(a,b)$$

for all b in B.

Proof. Assume $a_\alpha \to a$. As the set $A_{b,t} = \{a : M(a,b) < t\}$ is open we find that $M(a_\alpha, b) < t$ when α is sufficiently large provided $a \in A_{b,t}$. It follows that $\limsup_\alpha M(a_\alpha, b) \leqslant M(a,b)$ for all b. On the other hand if the condition of the proposition is fulfilled and if $a \in A_{b_1,t_1} \cap \cdots \cap A_{b_m,t_m}$ then $M(a_\alpha, b_i) < t$ for all $i = 1, \ldots, m$ provided α is sufficiently large. □

Corollary 3.2.17. *A is compact in the topology τ if and only if for each net (a_α) of strategies in A there is a strategy a in A such that*

$$\liminf_{\alpha} M(a_\alpha, b) \leqslant M(a,b)$$

for all b in B.

Proof. Assume A is compact and let (a_α) be a net in A. Then by compactness there is a subnet $(a_{\alpha'})$ of (a_α) which converges to a point $a \in A$. Then $\liminf_\alpha M(a_\alpha, b) \leqslant \liminf_{\alpha'} M(a_{\alpha'}, b) \leqslant \limsup_{\alpha'} M(a_{\alpha'}, b) \leqslant M(a,b)$.

Now assume that the condition of the corollary is fulfilled and let (a_α) be a net in A. By Tychonoff's theorem there is a subnet $(a_{\alpha'})$ such that $\lim_{\alpha'} M(a_{\alpha'}, b)$

exists in $[-\infty, \infty]$ for all b in B. Then there is a strategy a in A such that:

$$\limsup_{\alpha'} M(a_{\alpha'}, b) = \liminf_{\alpha'} M(a_{\alpha'}, b) \leqslant M(a, b)$$

for all strategies b in B. Thus any net has a convergent subnet. □

The first part of the following theorem is as an immediate consequence of this corollary and theorem 3.2.12.

Theorem 3.2.18. *Let $\Gamma = (A, B, M)$ be a mixed game with $-\infty \leqslant M < \infty$. Then each of the following two conditions ensure that Γ has a value and that player* I *has a maximin strategy.*

(i) *To each net $\{a_\alpha\}$ in A there corresponds an $a \in A$ such that $M(a, b) \geqslant \liminf_\alpha M(a_\alpha, b)$ for all $b \in B$.*
(ii) *There is a countable subset B_c of B such that $M_1(a) = \inf\{M(a, b) : b \in B_c\}$ for all $a \in A$. Furthermore, to each sequence $\{a_n\}$ in A there corresponds a strategy a in A such that $M(a, b) \geqslant \liminf_n M(a_n, b)$ for all $b \in B_c$.*

Proof of the last part of the theorem. Let $b_1, b_2 \ldots$ be an enumeration of B_c and let $\bar{V} > \tau_n \uparrow \bar{V}$ as $n \to \infty$. By theorem 3.2.9, for each n there is a strategy $a_n \in A$ such that $M(a_n, b_i) \geqslant \tau_n$ when $n \geqslant i$. By assumption there is an $a \in A$ such that $M(a, b_i) \geqslant \liminf_n M(a_n, b_i) \geqslant \bar{V}$; $i = 1, 2 \ldots$. Hence $\underline{V} \geqslant M_1(a) = \inf_i M(a, b_i) \geqslant \bar{V}$. □

Now let us turn to the problem of completeness of sets of strategies. We shall begin by considering the sets A_0 and B_0 of admissible strategies. As we naturally want to achieve large reductions of the game we would like to restrict our attention to small complete sets of strategies. A smallest complete set of strategies is called *minimal complete.* Thus a set of strategies is minimal complete if and only if it is complete and is contained in any other complete set of strategies.

The only candidate for a minimal complete set of strategies is the set of admissible strategies.

Theorem 3.2.19. *The following conditions are equivalent for a subset A_1 of A.*

(i) *$A_1 = A_0$ and A_0 is complete;*
(ii) *A_1 is complete and contains no proper complete subset;*
(iii) *A_1 is minimal complete.*

Furthermore any complete set A_1 of strategies contains A_0.

Remark. The statement of the theorem remains true if the letter A is replaced by the letter B throughout.

Proof of the theorem. Suppose A_1 is complete and $a_0 \in A_0 - A_1$. Then there is an $a_1 \in A_1$ dominating a_0 and not equivalent to a_0. However, this contradicts the admissibility of a_0. Hence $A_0 \subseteq A_1$, proving the last statement. This also proves the implication (i) ⇒ (iii), while the implication (iii) ⇒ (ii) is trivial. Suppose finally that (ii) holds. Then, as we have seen, $A_1 \supseteq A_0$. Let $a_1 \in A_1 - A_0$. Then since $a_1 \notin A_0$, there is an $a_2 \in A$ dominating a_1 and not equivalent to a_1. Since A_1 is complete, we may assume that $a_2 \in A_1$. This implies that $A_1 - \{a_1\}$ is also complete, contradicting (ii). Thus $A_1 = A_0$ so that (i) holds. □

It is easy to construct examples of affine games where the sets of admissible strategies are empty and, henceforth, not complete. However, the same topological assumptions which were used in theorem 3.2.12 ensure that A_0 is complete.

Theorem 3.2.20 (Completeness of the set of admissible strategies). *Let $\Gamma = (A, B, M)$ be a game where A is a compact topological space such that $M(\cdot, b)$ is upper semicontinuous on A for each $b \in B$. Then the set A_0 of admissible strategies for player I is complete and hence minimal complete.*

Proof. Let $a_1 \in A$ and consider the set A_1 of all strategies which dominate a_1. Equipped with the domination ordering, this set becomes a partially ordered set. Let A_2 be a totally ordered subset of A_1. The class of closed compact sets $\{a : a \text{ dominates } a_2\}$; $a_2 \in A_2$ has f.i.p. and hence has a non-empty intersection. A strategy a_3 in this intersection is clearly an upper bound for A_2. It follows then from Zorn's lemma (see e.g. Kelley (1955) or Halmos (1960)) that there are maximal strategies in A_1. Any maximal strategy in A_1 is maximal in A, i.e. is admissible. □

Let a_0 and b_0 be strategies for player I and player II respectively and let $\varepsilon \geqslant 0$. Then we shall say that *a_0 is ε-optimal with respect to b_0* if $M(a_0, b_0) \geqslant M(a, b_0) - \varepsilon$ for all $a \in A$. Similarly b_0 is called *ε-optimal with respect to a_0* if $M(a_0, b_0) \leqslant M(a_0, b) + \varepsilon$ for all b in B. If $\varepsilon = 0$ then this is just the notion of optimality we defined in connection with the description of a saddlepoint. A strategy $a_0 \in A$ is called *extended optimal* if a_0 is ε-optimal for some $b_\varepsilon \in B$ for each $\varepsilon > 0$. The set of all extended optimal strategies for player I will be denoted by $\tilde{A}$. Similarly a strategy $b_0 \in B$ is called *extended optimal* if it is ε-optimal for some $a_\varepsilon \in A$ for each $\varepsilon > 0$. The set of all extended optimal strategies for player II will be denoted by $\tilde{B}$.

Maximin and minimax strategies are extended optimal if the game has a finite value.

Proposition 3.2.21. *Let $\Gamma = (A, B, M)$ be a game with a finite value. Then each maximin strategy for player* I *and each minimax strategy for player* II *is extended optimal.*

Proof. Let a_0 be a maximin strategy and, for each $\varepsilon > 0$, let b_ε be such that $M_{\mathrm{II}}(b_\varepsilon) \leqslant V + \varepsilon$. Then $M(a_0, b_\varepsilon) \geqslant M_{\mathrm{I}}(a_0) = V \geqslant M_{\mathrm{II}}(b_\varepsilon) - \varepsilon$. □

It is an important fact that the players may often restrict their attention to extended optimal strategies.

Theorem 3.2.22. *Let $\Gamma = (A, B, M)$ be a mixed game where $-\infty \leqslant M < \infty$. Assume that B is affine relatively Γ and that A possesses a topology making A compact and each function $M(\cdot, b)$; $b \in B$ upper semicontinuous.*

Then the set $\tilde{A}$ of extended optimal strategies for player I *is complete and consequently, by theorems* 3.2.19 *and* 3.2.20, *contains the minimal complete set A_0 of admissible strategies.*

Proof. Let a_0 be an admissible strategy for player I and put $B_f = \{b : M(a_0, b) > -\infty\}$. If $B_f = \varnothing$ then $M(a_0, b) \equiv_b -\infty$ so that, since a_0 is admissible, $M(a, b) \equiv -\infty$ and then trivially $a_0 \in \tilde{A}$. Suppose next that $B_f \neq \varnothing$ and put $\Gamma_f = (A, B_f, M_f)$ where $M_f(a, b) = M(a, b) - M(a_0, b)$ when $a \in A$ and $b \in B_f$. Then Γ_f satisfies the conditions of the theorem and, furthermore, $M_f(a_0, b) \equiv_b 0$. By theorem 3.2.12 the game Γ_f has a value $V(\Gamma_f)$ and player I has a maximin strategy a_1. The supremum in $M_{\mathrm{II}}(b) = \sup_a M(a, b)$ is obtained so that $V(\Gamma_f) < \infty$. Hence $0 \leqslant V(\Gamma_f) < \infty$. It follows that $M_f(a_1, b) \geqslant 0$ when $b \in B_f$. Hence $M(a_1, b) \geqslant M(a_0, b)$ for all $b \in B$. Furthermore a_1 is extended optimal in Γ_f and hence in Γ as well. Thus in any case a_0 is dominated by an extended optimal strategy. Since A_0 is complete it follows that $\tilde{A}$ is also complete. □

3.3 Coordinate games

For a strategy b for player II in the game (A, B, M), the only feature which matters is the function $M(\cdot, b)$ on A. If $M(\cdot, b_1) = M(\cdot, b_2)$ then b_1 is just as good (or bad) as b_2. Thus we might replace the set B of strategies for player II with the subset $G = \{M(\cdot, b) : b \in B\}$ of the function space $[-\infty, \infty]^A$. If player I plays $a \in A$ and player II plays $g \in G$ then the pay-off to I from II is the value $g(a)$ of g at a. With the case of a finite A in mind, we may also consider $g(a)$ as the 'ath' coordinate of $g \in G$. Of course the same considerations also apply to player I.

A game (A, B, M) is called a *coordinate game* if either A is a set of functions on B and $M(a, b) \equiv_{a,b} a(b)$, or if B is a set of functions on A and $M(a, b) \equiv_{a,b} b(a)$.

Coordinate games enter naturally in statistics, by the following assumptions and considerations. Imagine that the chance mechanism determining the random outcome of the experiments we are considering is determined by an unknown theory θ belonging to a known set Θ of possible theories on nature's behaviour. Using the terminology introduced in section 1.2, the set Θ may be called the *parameter set*.

The statistician's task is to recommend (take) a decision (action). In order to achieve this the statistician needs a method of decision making which is applicable to the situation in hand. At this point we need not discuss the general properties of statistical methods. The essential thing here is that the short-comings of any particular method ρ may, when θ prevails, be judged in terms of an extended real valued quantity $r_\rho(\theta)$. This quantity $r_\rho(\theta)$ is called *the risk of ρ when θ is used* and the function r_ρ is called *the risk function of ρ*. Again, at this point we need not worry about how risk functions are determined. The essential thing is that they are there. Thus they need not be expressed as expected loss, although this is what we shall do later on.

We shall assume that a method ρ_1 is judged to be at least as good as another method ρ_2 if $r_{\rho_1} \leqslant r_{\rho_2}$. If ρ_1 and ρ_2 determine the same risk functions then they are considered to be equivalent. Thus the set of strategies for the statistician may be reduced to a subset $\mathscr{R}$ of the function space $[-\infty, \infty]^\Theta$.

In order to avoid tedious struggles with $\infty - \infty$, we shall assume throughout this section that $r(\theta) > -\infty$ when $r \in \mathscr{R}$ and $\theta \in \Theta$. It will also be assumed that the statistician has at least one strategy. Hence

$$\emptyset \neq \mathscr{R} \subseteq]-\infty, \infty]^\Theta.$$

If not otherwise stated, topological properties of $\mathscr{R}$ are with respect to the topology of pointwise (= coordinatewise) convergence.

The opponent of the statistician will be called *nature* and nature's set of strategies is the parameter set Θ. If nature chooses the strategy $\theta \in \Theta$ and the statistician chooses the risk function r then we imagine that the statistician must pay nature an amount $r(\theta)$. Thus we have arrived at a coordinate game $(\Theta, \mathscr{R}, M)$ where $M(\theta, r) \equiv r(\theta)$ and where nature is player I and the statistician is player II.

Let us translate some of the material we have developed into this framework.

Note first that the set $\mathscr{R}$ of strategies for the statistician is convex (mixed) if and only if to each pair $(r_1, r_2) \in \mathscr{R}^2$ and each $\alpha \in]0, 1[$ there corresponds an $r_3 \in \mathscr{R}$ such that $(1 - \alpha)r_1 + \alpha r_2 \geqslant r_3$. A subset $\mathscr{R}$ of $]-\infty, \infty]^\Theta$ with this property is called *subconvex*. The set of strategies for the statistician is affine if and only if $\mathscr{R}$ is convex in the sense that $(1 - \alpha)r_1 + \alpha r_2 \in \mathscr{R}$ whenever $(r_1, r_2) \in \mathscr{R}^2$ and $\alpha \in]0, 1[$. This condition is usually verified by observing that the risk is affine with respect to a convexity structure on the set of methods. Unfortunately the well established notion of an affine subset of a linear space as a translate of a linear set does not fit well with the notion of an affine set of strategies.

Example 3.3.1. It is desired to estimate a real valued function $g(\theta)$ of θ on the basis of a realization X of an experiment $\mathscr{E} = (\mathscr{X}, \mathscr{A}; P_\theta : \theta \in \Theta)$. The set $\mathscr{R}$ of possible risk functions of the form $E_\theta(\delta(X) - g(\theta))^2$, where δ belongs to a given convex set of real valued measurable functions, is subconvex.

We shall not assume that the set of strategies for nature is concave. However, it is very convenient to allow mixed strategies for nature. We shall employ the following notations.

$\Lambda = \Theta^* =$ the set of probability distributions on Θ with finite supports. A probability distribution on a σ-algebra of subsets of Θ is called an *a priori distribution* or a *prior distribution.* This is part of the Bayesian terminology which we shall explain later. The main point here is that the set of mixed strategies for nature is just the set of prior distributions with finite supports. If λ is a prior distribution and $\int f\,d\lambda$ exists then we may write $\int f\,d\lambda = f(\lambda)$ or $\int f\,d\lambda = \lambda(f)$ according to whether we want to express that f acts on a set of prior distributions or that λ acts on a set of functions. If f is a risk function and $\int f\,d\lambda$ exists, then this quantity is called the *Bayes risk of f with respect to λ.*

The extension of the game $(\Theta, \mathscr{R}, M)$ obtained by permitting nature to use any prior distribution in Λ as a strategy is then

$$\Gamma(\mathscr{R}) = (\Lambda, \mathscr{R}, B)$$

where $B(\lambda, r)$, for each $\lambda \in \Lambda$ and $r \in \mathscr{R}$, is the Bayes risk of r with respect to λ. The 'upper' and 'lower' functions of this game are $B_{\text{I}}(\lambda) = \inf_r r(\lambda)$; $\lambda \in \Lambda$ and $B_{\text{II}}(r) = \sup_\theta r(\theta)$; $r \in \mathscr{R}$. The function B_{I} is called the *lower envelope* of $\mathscr{R}$. By proposition 3.2.1

$$\sup_\lambda B_{\text{I}}(\lambda) = \underline{V}(\Gamma(\mathscr{R})) \leqslant \bar{V}(\Gamma(\mathscr{R})) = \inf_r \sup_\theta r(\theta).$$

A minimax strategy r_0 for the statistician is also called a *minimax solution.* Thus r_0 is minimax if and only if $\sup_\theta r_0(\theta) \leqslant \sup_\theta r(\theta)$ when $r \in \mathscr{R}$. A maximin strategy λ_0 for nature is also called *a least favourable prior distribution.*

A strategy $r_0 \in \mathscr{R}$ is admissible if and only if it is minimal for the pointwise ordering, i.e. $r = r_0$ when $r \leqslant r_0$ and $r \in \mathscr{R}$. The set of all admissible risk functions in $\mathscr{R}$ will be denoted by $\mathscr{R}_0$.

If $\varepsilon \geqslant 0$ then a strategy $r_\varepsilon \in \mathscr{R}$ for the statistician is called *ε-Bayes* if it is *ε-optimal* for some $\lambda_\varepsilon \in \Lambda$ i.e. if $r_\varepsilon(\lambda_\varepsilon) \leqslant B_{\text{I}}(\lambda_\varepsilon) + \varepsilon$. A strategy $r_0 \in \mathscr{R}$ is called a *Bayes solution for λ_0* if $r_0(\lambda_0) = B_{\text{I}}(\lambda_0)$. Thus $r_0 \in \mathscr{R}$ is a *Bayes solution* if and only if it is 0-Bayes. A strategy $r \in \mathscr{R}$ for the statistician is called *extended Bayes* if it is ε-Bayes for each $\varepsilon > 0$.

A subset $\mathscr{R}_1$ of $\mathscr{R}$ is complete if it dominates $\mathscr{R}$, i.e. if to each $r \in \mathscr{R}$ there corresponds an $r_1 \in \mathscr{R}_1$ such that $r \geqslant r_1$. The notions of completeness and essential completeness for sets of strategies for the statistician coincide, since equivalent risk functions are equal. If follows from theorem 3.2.19, or directly, that any complete subset $\mathscr{R}_1$ of $\mathscr{R}$ contains $\mathscr{R}_0$ and that $\mathscr{R}_0$ is the only possible candidate for a minimal complete set of strategies for the statistician.

We shall occasionally find it convenient to extend the set of strategies for the statistician by considering the set $\hat{\mathscr{R}}$ of all functions $\hat{r}$ on Θ such that $\hat{r} \geqslant r$ for some $r \in \mathscr{R}$.

The reader is now advised to consider the geometrical interpretations of the notions we have introduced so far when $\Theta = \{1,2\}$, and then to do so for the remaining results in this section. Thus the set of points (r_1, r_2) in $]-\infty, \infty]^2$ with finite maximin risk c is the boundary of the corner $\{(r_1, r_2): r_1 \leqslant c, r_2 \leqslant c\}$. The set of points (r_1, r_2) in $]-\infty, \infty]^2$ yielding the same finite Bayes risk b for the prior distribution (λ_1, λ_2) is the line $\{(r_1, r_2): \lambda_1 r_1 + \lambda_2 r_2 = b\}$ in $]-\infty, \infty[^2$ when $\lambda_1 > 0$ and $\lambda_2 > 0$. The set of points (r_1, r_2) in $]-\infty, \infty]^2$ yielding the same finite Bayes risk b for the prior distribution $(1,0)$ consists of the vertical line $\{(r_1, r_2): r_1 = b\}$ in $]-\infty, \infty]^2$, including the point (b, ∞) at infinity.

The following two propositions and the subsequent theorem are translations of theorems 3.2.2 and 3.2.8 and of the first part of theorem 3.2.18.

Proposition 3.3.2. *$\Gamma(\mathscr{R})$ has a value, $\lambda_0 \in \Lambda$ is least favourable and $r_0 \in \mathscr{R}$ is minimax if and only if $r(\lambda_0) \geqslant r_0(\lambda_0) \geqslant r_0(\theta)$ when $r \in \mathscr{R}$ and $\theta \in \Theta$. If this condition is satisfied then $V(\Gamma(\mathscr{R})) = r_0(\lambda_0)$.*

Proposition 3.3.3. *$\Gamma(\mathscr{R})$ has a value if and only if $\bigcap_\theta \{r: r(\theta) \leqslant \sigma\} \neq \emptyset$ when $\sigma > \underline{V}(\Gamma(\mathscr{R}))$.*

Theorem 3.3.4. *Suppose $\mathscr{R}$ is subconvex and that $\inf_r r(\theta) > -\infty$ for all $\theta \in \Theta$. Assume that there is a sequence $\theta_1, \theta_2, \ldots,$ in Θ such that $\sup_\theta r(\theta) = \sup_n r(\theta_n)$ when $r \in \mathscr{R}$. Assume also that to each sequence $r_1, r_2, \ldots$ in $\mathscr{R}$ there corresponds an $r \in \mathscr{R}$ such that $\limsup_n r_n(\lambda) \geqslant r(\lambda)$ when $\lambda \in \Lambda$. Then $\Gamma(\mathscr{R})$ has a value and the statistician has a minimax strategy.*

Remark. This condition is satisfied whenever Θ is finite and $\mathscr{R}$ is a closed subconvex subset of $]-\infty, \infty]^\Theta$.

The basic compactness condition may be described as follows.

Proposition 3.3.5. *The following three conditions on $\mathscr{R}$ are equivalent:*

(i) *$\hat{\mathscr{R}}$ is a compact subset of $]-\infty, \infty]^\Theta$.*
(ii) *$\mathscr{R}$ is compact for the coarsest topology on $\mathscr{R}$ making $r(\theta)$ lower semicontinuous in r for each $\theta \in \Theta$.*
(iii) *to each net $\{r_\alpha\}$ in $\mathscr{R}$ there corresponds an $r \in \mathscr{R}$ such that $\limsup_\alpha r_\alpha(\theta) \geqslant r(\theta)$ when $\theta \in \Theta$.*

Proof. Suppose (i) holds. Let $\{r_\alpha\}$ be a net in $\mathscr{R}$. Then there is a subnet $\{r_{\alpha'}\}$ and an $\hat{r} \in \hat{\mathscr{R}}$ such that $\lim r_{\alpha'} = \hat{r}$. Hence $\limsup_\alpha r_\alpha(\theta) \geqslant \hat{r}(\theta) \geqslant r(\theta)$ for some $r \in \mathscr{R}$. Thus (i) $\Rightarrow$ (iii) and the converse follows easily from the compactness of $[-\infty, \infty]^\Theta$. The equivalence (ii) $\Leftrightarrow$ (iii) follows from corollary 3.2.17. □

Theorems 3.2.12 and 3.2.20 yield the next two theorems.

Theorem 3.3.6. *Suppose $\mathscr{R}$ is subconvex and that $\hat{\mathscr{R}}$ is compact. Then $\Gamma(\mathscr{R})$ has a value and the statistician has a minimax solution.*

Theorem 3.3.7. *$\mathscr{R}_0$ is minimal complete when $\hat{\mathscr{R}}$ is compact.*

If $\mathscr{R}$ is subconvex and $\hat{\mathscr{R}}$ is compact then the set $\hat{\mathscr{R}}$ is characterized by the lower envelope function.

Theorem 3.3.8. *Suppose $\mathscr{R}$ is subconvex and $\hat{\mathscr{R}}$ is compact. Then an extended real valued function $\hat{r}$ on Θ belongs to $\hat{\mathscr{R}}$ if and only if $\hat{r}(\lambda) \geqslant \inf\{r(\lambda) : r \in \mathscr{R}\}$ when $\lambda \in \Lambda$.*

Proof. Suppose $\hat{r}(\lambda) \geqslant \inf_r r(\lambda)$ when $\lambda \in \Lambda$. Put $\Theta_0 = \{\theta : \hat{r}(\theta) < \infty\}$ and for each finite non-empty subset F of Θ_0, let the set of prior distributions in Λ which is supported by F be denoted by Λ_F. Then $0 \geqslant \sup_{\lambda \in \Lambda_F} \inf_r [r(\lambda) - \hat{r}(\lambda)] = \inf_r \sup_{\lambda \in \Lambda_F} [r(\lambda) - \hat{r}(\lambda)] = \sup_{\lambda \in \Lambda_F} [r_F(\lambda) - \hat{r}(\lambda)] = \sup_{\theta \in F} [r_F(\theta) - \hat{r}(\theta)]$ for some $r_F \in \mathscr{R}$. The interchange of inf and sup and the existence of r_F follow from theorem 3.2.12. Consider $\{r_F\}$ as a net in $\hat{\mathscr{R}}$ where the subsets F of Θ_0 are ordered by inclusion. By assumption there is an $r \in \mathscr{R}$ such that $\limsup_F r_F \geqslant r$. Then, since $r_F \leqslant \hat{r}$ on F, $r \leqslant \hat{r}$ on Θ_0 so that $r \leqslant \hat{r}$. Hence $\hat{r} \in \hat{\mathscr{R}}$. If $\Theta_0 = \varnothing$ then $\hat{r}(\theta) \equiv \infty$ and, consequently, $\hat{r} \in \hat{\mathscr{R}}$. □

Under general conditions the statistician does not lose anything by restricting himself to extended Bayes solutions or limits of Bayes solutions.

Theorem 3.3.9. *If $\mathscr{R}$ is subconvex and $\hat{\mathscr{R}}$ is compact then any admissible risk function is extended Bayes as well as a limit of Bayes solutions.*

In particular it follows that the set of limits of Bayes solutions and the set of extended Bayes solutions are both (essentially) complete.

Remark. If Θ is finite then the proof below shows that any everywhere finite admissible risk function is a limit of Bayes solutions for strictly positive prior distributions.

Proof. Let $r_0 \in \mathscr{R}_0$ and put $\Theta_0 = \{\theta : r_0(\theta) < \infty\}$. The fact that r_0 is extended Bayes follows from theorem 3.2.22. If Θ_0 is empty then $\mathscr{R} = \{\infty\} = \{r_0\}$ and r_0 is Bayes. Assume therefore that $\Theta_0 \neq \varnothing$. Again let Λ_F for each finite non-empty subset F of Θ_0 be the set of prior distributions in Λ which are supported by F. Put $\Lambda_{F,\varepsilon} = \{\lambda : \lambda \in \Lambda_F \text{ and } \lambda(\theta) \geqslant \varepsilon \text{ when } \theta \in F\}$ when $0 < \varepsilon \leqslant 1/\#F$.

Consider the game $G_{F,\varepsilon} = (\Lambda_{F,\varepsilon}, \mathscr{R}, M_{F,\varepsilon})$ where $M_{F,\varepsilon}(\lambda, r) = r(\lambda) - r_0(\lambda)$ when $r \in \mathscr{R}$ and $\lambda \in \Lambda_{F,\varepsilon}$. Then, by say theorem 3.2.12, there is a $\lambda_{F,\varepsilon} \in \Lambda_{F,\varepsilon}$ and an $r_{F,\varepsilon} \in \mathscr{R}$ such that $M_{F,\varepsilon}(\lambda_{F,\varepsilon}, r) \geqslant M_{F,\varepsilon}(\lambda_{F,\varepsilon}, r_{F,\varepsilon}) \geqslant M_{F,\varepsilon}(\lambda, r_{F,\varepsilon})$ when $r \in \mathscr{R}$ and $\lambda \in \Lambda_{F,\varepsilon}$. Hence $r_{F,\varepsilon}(\lambda) - r_0(\lambda) = M_{F,\varepsilon}(\lambda, r_{F,\varepsilon}) \leqslant M_{F,\varepsilon}(\lambda_{F,\varepsilon}, r_0) = 0$ when $\lambda \in \Lambda_{F,\varepsilon}$.

Choose a sequence $\varepsilon_1 > \varepsilon_2 > \cdots > 0$ such that $\lim_n \varepsilon_n = 0$, $\lim_n r_{F,\varepsilon_n} = r_F \in \hat{\mathcal{R}}$ and $\lim_n \lambda_{F,\varepsilon_n} = \lambda_F \in \Lambda_F$. Let $\lambda \in \Lambda_F$ assign positive mass to each $\theta \in F$. Then $\lambda \in \Lambda_{F,\varepsilon_n}$ when n is sufficiently large. Hence $r_{F,\varepsilon_n}(\lambda) \leqslant r_0(\lambda)$ when n is sufficiently large. If follows that $r_F(\lambda) \leqslant r_0(\lambda)$ for all $\lambda \in \Lambda_F$ so that $r_F \leqslant r_0$ on F. Furthermore $r(\lambda_{F,\varepsilon}) \geqslant r_{F,\varepsilon}(\lambda_{F,\varepsilon})$ for each $r \in \mathcal{R}$ so that $r_{F,\varepsilon}$ is a Bayes solution for a strictly positive prior. Choose a subnet $\{r_{F'}\}$ of $\{r_F\}$ such that $\lim_{F'} r_{F'} = s$ exists in $\hat{\mathcal{R}}$. Then $s \leqslant r_0$ on Θ_0 and hence on all of Θ. Hence, since r_0 is admissible, $s = r_0$ so that $\lim_F r_F = r_0$. Let $\mathcal{R}_b$ denote the set of Bayes solutions in $\mathcal{R}$ and let $\bar{\mathcal{R}}_b$ be the closure of $\mathcal{R}_b$ in $\hat{\mathcal{R}}$. Then, as we have just seen, $r_F \in \bar{\mathcal{R}}_b$ for each F. Hence $r_0 = \lim_F r_F \in \bar{\mathcal{R}}_b$. □

Corollary 3.3.10. *Suppose Θ is finite and that the conditions of theorem 3.3.9 are satisfied (i.e. $\mathcal{R}$ is subconvex and $\hat{\mathcal{R}}$ is compact).*

Assume also that any admissible risk function is everywhere finite. Then the following conditions are equivalent for a risk function $r \in \mathcal{R}$:

(i) *r is a Bayes solution.*
(ii) *r is a limit of Bayes solutions.*
(iii) *r is an extended Bayes solution.*

Furthermore, any admissible risk function satisfies these conditions. In particular it follows that the set of Bayes solutions is complete.

Proof. Let $\mathcal{R}_b$ and $\tilde{\mathcal{R}}$ denote, respectively, the set of Bayes solutions and the set of extended Bayes solutions. Denoting closure for pointwise convergence by a bar, we see that the complete statement of the corollary amounts to

$$\mathcal{R}_0 \subseteqq \mathcal{R}_b = \bar{\mathcal{R}}_b = \tilde{\mathcal{R}}.$$

In view of the theorem it suffices to show that $\tilde{\mathcal{R}} \cup \bar{\mathcal{R}}_b \subseteqq \mathcal{R}_b$. Let $r_0 \in \tilde{\mathcal{R}}$. Then for each integer $n \geqslant 1$ there is a prior distribution λ_n such that $r_0(\lambda_n) \leqslant r(\lambda_n) + 1/n$; $r \in \mathcal{R}_0$. Let $\{\lambda_{n'}\}$ be a sub-sequence converging to the prior distribution λ_0. Then $r_0(\lambda_0) \leqslant r(\lambda_0)$; $r \in \mathcal{R}_0$ so that r_0 is a Bayes solution for λ_0. Hence $\tilde{\mathcal{R}} \subseteqq \mathcal{R}_b$. Suppose finally that $r_0 \in \bar{\mathcal{R}}_b$. Then $r_0 = \lim_n r_n$ where r_n, for each n, is a Bayes solution for the prior distribution λ_n. We may assume without loss of generality that $\lambda_n \to \lambda_0$. Let $r \in \mathcal{R}_0$. Then $r_n(\lambda_n) \leqslant r(\lambda_n)$; $n = 1, 2, \ldots$ so that $r_0(\lambda_0) \leqslant r(\lambda_0)$. Hence $r_0 \in \mathcal{R}_b$ so that $\mathcal{R}_b$ is closed. □

Example 3.3.11 (An everywhere finite admissible decision rule which is not a Bayes solution). Let $\Theta = \{1, 2\}$ and define the compact convex subset $\mathcal{R} = \hat{\mathcal{R}}$ by putting $\mathcal{R} = \mathcal{R}_1 \cup \mathcal{R}_2$ where $\mathcal{R}_1 = \{r : r \geqslant 0\} \cup \{r : -1 \leqslant r(1) < 0,\ r(2) \geqslant r(1)^2\}$ and $\mathcal{R}_2 = \{r : r(1) = \infty,\ r(2) \geqslant -1\}$. Then $\mathcal{R} = \hat{\mathcal{R}}$ is compact and $(0, 0)$ is admissible in $\mathcal{R}$. However, it is easily checked that $(0, 0)$ is not a Bayes solution for any prior distribution λ. Note that $(\infty, -1)$ is an admissible risk function. Thus the assumption in the corollary that admissible risk functions are everywhere finite is not superfluous.

3.4 Bayesian tools

When considering particular decision problems, or other particular problems related to game theory, minimaxity or admissibility may often be established by using a few simple devices. Although these devices are more or less contained in our exposition of game theory, we shall now state and prove them explicitly, without resorting to the general theory. This section may therefore be read independently of the previous material on game theory. In order to make this section self contained we have repeated a few definitions.

Let $\mathscr{R}$ be any non-empty subset of the set $[-\infty, \infty]^\Theta$ of extended real valued functions on Θ and let us agree to call functions in $\mathscr{R}$ risk functions.

A function r_0 in $\mathscr{R}$ is called *minimax* if $\sup_\theta r_0(\theta) \leqslant \sup_\theta r(\theta)$ when $r \in \mathscr{R}$.

A function r_0 in $\mathscr{R}$ is called *admissible* if it is minimal in $\mathscr{R}$ with respect to the pointwise ordering, i.e. if there is no $r \neq r_0$ in $\mathscr{R}$ such that $r \leqslant r_0$.

A probability distribution on a σ-algebra of subsets of Θ making each $r \in \mathscr{R}$ measurable is called a *prior distribution* or just a *prior*. If λ is a prior distribution and $\varepsilon \geqslant 0$ then $r_\varepsilon \subset \mathscr{R}$ is called *ε-Bayes for λ* if $r_\varepsilon(\lambda) \leqslant r(\lambda) + \varepsilon$ when $r \in \mathscr{R}$. Here $r(\lambda)$ for $r \in \mathscr{R}$ and λ a prior distribution, is short for $\int r \, d\lambda$. The notation implies that this integral should exist.

A risk function $r_0 \in \mathscr{R}$ is called *Bayes for the prior distribution* λ if $r_0(\lambda) \leqslant r(\lambda)$ when $r \in \mathscr{R}$, i.e. if r_0 is 0-Bayes for λ.

A risk function is called *extended Bayes* if it is ε-Bayes for each $\varepsilon > 0$.

This terminology is consistent with the terminology used in the previous sections.

In section 4.4 we shall see how Bayes solutions may actually be found by a straightforward probability analysis.

Here is the first device.

Theorem 3.4.1. *A risk function r_0 is minimax provided there is an extended real valued functional Φ_0 on $\mathscr{R}$ such that:*

(i) $\Phi_0(r_0) \leqslant \Phi_0(r) \leqslant \sup_\theta r(\theta)$; $r \in \mathscr{R}$

and

(ii) $\Phi_0(r_0) = \sup_\theta r_0(\theta)$.

If (r_0, Φ_0) *satisfies* (i) *and* (ii) *then* $\Phi_0(r) \geqslant \Phi_0(r_0) \geqslant \Phi(r_0)$ *whenever* $r \in \mathscr{R}$ *and the functional* Φ *satisfies* (i).

Remark. If Φ_0 is a probability distribution λ_0 on the σ-algebra of subsets of Θ induced by $\mathscr{R}$ then theorem 3.4.1 is usually phrased: if the risk function r_0 is Bayes for the prior λ_0 and if r_0 is λ_0 equivalent to the constant function $\sup_\theta r_0(\theta)$ then r_0 is minimax and λ_0 is least favourable. Clearly $\sup_\theta r_0(\theta) = \max_\theta r_0(\theta)$ in this case, justifying the 'max' part of the word 'minimax'.

Proof. $\sup_\theta r(\theta) \geqslant \Phi_0(r) \geqslant \Phi_0(r_0) = \sup_\theta r_0(\theta) \geqslant \Phi(r_0)$ when $r \in \mathscr{R}$ and Φ satisfies (i). □

Here are two useful applications which utilize the theorem for functionals Φ_0 which are not necessarily probability distributions.

Corollary 3.4.2. *Let $\mathscr{U}$ be a σ-algebra of subsets of Θ such that $r(\theta)$ is measurable in θ for each $r \in \mathscr{R}$. Then each of the following two conditions ensure that $r_0 \in \mathscr{R}$ is minimax.*

(i) *There is a net (r_n) in $\mathscr{R}$ and a net (λ_n) of probability distributions on $\mathscr{U}$ such that r_n, for each n, is Bayes for λ_n and such that $\sup_\theta r_0(\theta) \leqslant \limsup_n r_n(\lambda_n)$.*
(ii) *r_0 is constant and extended Bayes.*

Proof. Apply the theorem with $\Phi_0(r)$ defined as, respectively, $\limsup_n \lambda_n(r)$ and $\limsup_{\varepsilon \to 0} \lambda_\varepsilon(r)$ where, in case (ii), $r_0(\lambda_\varepsilon) \leqslant r(\lambda_\varepsilon) + \varepsilon$ when $\varepsilon \geqslant 0$. □

The theorem and its corollary indicate how we may proceed to find minimax solutions. Thus we may look for a constant risk function c and then try to find a prior distribution such that c is Bayes (ε-Bayes) for this prior distribution. If c is Bayes for λ then we know by theorem 3.4.1 that λ is necessarily least favourable.

Alternatively we may consider various prior distributions and then see if some of them yield a constant risk Bayes solution.

If $r_0 \in \mathscr{R}$ is an everywhere positive and finite Bayes risk function for the prior distribution λ_0 and if $r_0(\lambda_0) < \infty$ then $1 \leqslant \sup_\theta [r(\theta)/r_0(\theta)]$ when $r \in \mathscr{R}$. Thus the constant function 1 is minimax with respect to $\mathscr{R}/r_0 = \{r/r_0 : r \in \mathscr{R}\}$. The replacement of $\mathscr{R}$ with $\mathscr{R}/r_0$ may be achieved in the statistical decision problems we shall discuss later by replacing the loss L_θ, which we suffer when θ prevails, by $L_\theta/r_0(\theta)$.

Minimax solutions may be expected to have constant risk for least favourable priors provided inf and sup may be interchanged.

Theorem 3.4.3. *Let Φ run through a set Π of extended real valued functionals on $\mathscr{R}$ such that $\sup_\Phi \Phi(r) = \sup_\theta r(\theta)$ when $r \in \mathscr{R}$.*

Suppose also that $\sup_\Phi \inf_r \Phi(r) = \inf_r \sup_\Phi \Phi(r)$ and that the supremum on the left hand side is obtained by $\Phi_0 \in \Pi$. Then $\Phi_0(r_0) = \sup_\theta r_0(\theta)$ for any minimax solution r_0.

Remark. A quick look at the definitions in section 3.2 of a game and the value of a game will convince the reader that $\sup_\Phi \inf_r \Phi(r) = \inf_r \sup_\Phi \Phi(r)$ if and only if the game $\Gamma = (\Pi, \mathscr{R}, M)$ where $M(\Phi, r) \equiv r(\Phi)$ has a value. If Φ_0 is a prior distribution and if $-\infty < \Phi_0(r_0) = \sup_\theta r_0(\theta) < \infty$ then, necessarily, $r_0 = \sup_\theta r_0(\theta)$ a.s. Φ_0.

Proof of the theorem. Let r_0 be minimax. Then $\sup_\theta r_0(\theta) = \inf_r \sup_\theta r(\theta) = \inf_r \sup_\Phi \Phi(r) = \sup_\Phi \inf_r \Phi(r) = \inf_r \Phi_0(r) \leqslant \Phi_0(r_0) \leqslant \sup_\theta r_0(\theta)$ so that '=' holds throughout. □

Admissibility may sometimes be decided by the following theorem.

Theorem 3.4.4. *Let Φ be a monotone and extended real valued functional on $\mathscr{R}$. Then a risk function in $\mathscr{R}$ is admissible provided it is minimal within the set of all Φ minimizing risk functions in $\mathscr{R}$.*

Remark. Thus r_0 is admissible if it is a minimal Bayes risk function for some prior distribution λ_0.

Proof. If r_0 satisfies the condition of the theorem and if $r_0 \geqslant r \in \mathscr{R}$ then, by minimality, $r = r_0$. □

The reader is referred to LeCam (1986) and Farrell (1968a, b), for particular criteria for admissibility.

3.5 References

Bickel, P. 1968–9. Lectures.
Blackwell, D. & Girchick, M. A. 1954. *Theory of games and statistical decisions.* Wiley, New York.
Farrell, R. H. 1968a. Towards a theory of generalised Bayes tests. *Ann. Math. Statist.* **39**, pp. 1–22.
— 1968b. On a necessary and sufficient condition for admissibility of estimators when strictly convex loss is used. *Ann. Math. Statist.* **39**, pp. 23–8.
Fenstad, J. E. 1967. Good strategies in general games. *Math. Zeitschr.* **101**, pp. 322–30.
Halmos, P. 1960. *Naive set theory.* Van Nostrand, Princeton, New Jersey.
Heyer, H. 1979. Personal communication.
Irle, A. 1981. Minimax theorems in convex situations. In: *Game theory and mathematical economics.* North–Holland, Amsterdam.
Jones, A. J. 1980. *Game theory: mathematical models of conflict.* Ellis Horwood Ltd, Chichester.
Kelley, J. L. 1955. *General topology.* Van Nostrand, Princeton, New Jersey.
LeCam, L. 1986. *Asymptotic methods in statistical decision theory.* Springer–Verlag, Berlin.
Nachbin, L. 1965. *The Haar integral.* Van Nostrand, Princeton, New Jersey.
Nash, J. F. 1950. Equilibrium points in *n*-persons games. *Proc. Nat. Acad. Sci. Wash.* **36**, pp. 48–9.
Roberts, A. W. & Varberg, D. E. 1973. *Convex functions*, Academic Press, New York.
Sion, M. 1958. On general minimax theorems. *Pacific J. Math.* **8**, pp. 171–6.
von Neumann, J. 1928. Zur Theorie der Gesellschaftspiele. *Math. Annalen.* **100**, pp. 295–320.
von Neumann, J. & Morgenstern, O. 1944 (revised 1947). *Theory of games and economic behaviour.* Princeton University Press.
Wald, A. 1950. *Statistical decision functions.* Wiley, New York.

4

Statistical Decision Problems

4.1 Introduction

The main task of any statistical theory is to investigate how inference can be obtained from observations. However there is no clear cut distinction between 'observing' on the one side and 'deciding' on the other. We may have to decide which observations our final decisions should be based upon. The problem of deciding which observations should constitute the basis for inference, will not be considered in this chapter. We shall assume that we somehow have been able to describe the model or experiment $\mathscr{E} = (\mathscr{X}, \mathscr{A}; P_\theta : \theta \in \Theta)$ linking the random outcome X of our observations to the set Θ of states of nature. We shall find it convenient to say that *the random variable X realizes this experiment $\mathscr{E}$ if the probability distribution of X is P_θ when θ describes the true state of nature.*

It should be noted that this framework does not suffice to describe sequential statistical problems where after having made a random number of intermediate decisions on the direction of investigations, we must make a final (terminal) decision. If, however, the procedures for making these intermediate decisions are prescribed, then we are within the set-up of this chapter. Thus we may apply this framework to sequential statistical decision problems with given stopping rules.

Statistical decision rules, their performance functions and their risk functions are introduced in section 4.2. Section 4.3 provides a short discussion of randomized decision rules. The Bayesian machinery is introduced in section 4.4. In particular it is shown there how optimal (ε-optimal) Bayes decision rules may be obtained from the posterior loss. Section 4.5 is heavy reading. It can be avoided if the reader is willing to restrict his/her attention to 'regular' cases. However, one should then note that most of the relevant regularity conditions are destroyed by the slightest perturbation. If we want to ensure, in the general case, that limits of risk functions are risk functions, then we are forced to extend the notion of a decision rule. Following LeCam we derive the notion of a generalized decision rule ('transition' in LeCam's terminology) by a straightforward observation of what decision rules actually do.

4.2 Decision rules and their risk functions

Let us assume that we are given the list T of all possible decisions (actions) which we might want to consider. As we shall work with probability distributions on T, it will be convenient to equip T with a measurability structure. Thus we define a *decision space* $(T, \mathscr{S})$ as a set T together with a σ-algebra $\mathscr{S}$ of subsets of T. Mathematically a decision space is just a measurable space.

A decision space $(T, \mathscr{S})$ is called finite if $\mathscr{S}$ is finite. If $\mathscr{S}$ contains 2^k sets where $k < \infty$ then $(T, \mathscr{S})$ is called a k-decision space. If the set T of decisions is finite and if $\mathscr{S}$ is not explicitly defined then it shall be assumed that $\mathscr{S}$ contains all subsets of T. The naming of the individual decisions is of course irrelevant. Thus the typical k-decision space is the set $T_k = \{1, 2, \ldots, k\}$ equipped with its 2^k subsets.

The statistician's task is to construct a rule which for each realization x of X tells us which decision should be made. This amounts to finding a function τ from the sample space $(\mathscr{X}, \mathscr{A})$ to the decision space $(T, \mathscr{S})$ such that the decision $\tau(x)$ is taken when x is the realization of X. It will be assumed that τ is measurable, i.e. that $\tau^{-1}[S] \in \mathscr{A}$ when $S \in \mathscr{S}$. The function τ is then called a *non-randomized decision rule (procedure)*.

Although it may appear artificial from a practical point of view, we shall find it convenient to admit randomized decision rules. Using a randomized procedure the eventual decision is obtained as the realization of a random T-valued variable whose distribution depends on the realization x of X only. As the non-randomized decision rules may be regarded as a degenerate kind of randomized decision rules, we shall use the term 'decision rule' instead of 'randomized decision rule'. Thus a *decision rule (procedure)* ρ *from the experiment* $\mathscr{E} = (\mathscr{X}, \mathscr{A}; P_\theta : \theta \in \Theta)$ *to the decision space* $(T, \mathscr{S})$ *is a map* ρ *from* $\mathscr{X}$ *to the set of probability measures on* $(T, \mathscr{S})$ *such that* $\rho(S|\cdot)$ *is measurable for each set* S *in* $\mathscr{S}$.

In other words, a decision rule ρ from $\mathscr{E}$ to $(T, \mathscr{S})$ is a Markov kernel from the measurable space $(\mathscr{X}, \mathscr{A})$ to the measurable space $(T, \mathscr{S})$. The decision rule is called *non-randomized* if there is a function τ from $\mathscr{X}$ to T (necessarily measurable) such that $\rho(S|x) = 1$ or $= 0$ as $\tau(x) \in S$ or $\tau(x) \notin S$.

A decision rule ρ as described carries finite measures on $\mathscr{A}$ onto finite measures on $\mathscr{S}$, and carries bounded measurable functions on T onto bounded measurable functions on $\mathscr{X}$. These mappings may be described as follows.

Let ρ be a decision rule from $\mathscr{E}$ to $(T, \mathscr{S})$. Then to each finite measure λ on $\mathscr{A}$ there corresponds a unique finite measure $\lambda \times \rho$ on $\mathscr{A} \times \mathscr{S}$ such that

$(\lambda \times \rho)(A \times S) = \int_A \rho(S|x)\,\mathrm{d}\lambda$ when $A \in \mathscr{A}$ and $S \in \mathscr{S}$. The first marginal of this joint distribution (measure) is λ and the second marginal will be denoted by $\lambda\rho$. Thus ρ maps λ onto the finite measure $\lambda\rho$. The measure $\lambda\rho$ is called *the measure on S induced from λ by ρ*. Next let g be a measurable function on $\mathscr{X}$, which is bounded from below, say. Then the expectation $\int g(t)\rho(\mathrm{d}t|x)$, as a function of x, is a measurable function on $\mathscr{X}$ which is bounded from below. This function will often be denoted as ρg. If $\lambda(\geqslant 0)$ is a finite measure on $\mathscr{A}$ and if g is a measurable function on T which is bounded (bounded from below) then

$$\lambda\rho(g) = \lambda(\rho g) = (\lambda \times \rho)(g)$$

and this quantity may, without ambiguity, be denoted as $\lambda\rho g$.

Suppose now that the statistician uses the decision rule ρ. Then the probability of taking a decision within the (measurable) set S of decisions is $(P_\theta\rho)(S)$ when θ prevails. The function which to each state θ in the parameter set Θ assigns the probability distribution of the decision when the decision rule ρ is used, is called the *operational characteristic of ρ* or the *performance function of ρ*. The performance function of ρ is thus the experiment $(P_\theta\rho; \theta \in \Theta)$ on $(T, \mathscr{S})$ induced from $\mathscr{E}$ by the kernel ρ.

Which decision rule should we choose? Suppose first that no observations are available. If the true state of nature is θ and if the statistician chooses the decision t then there might be some consequences which should be considered. Here we shall imagine that there is a quantity $L_\theta(t)$ in $[-\infty, \infty]$ which roughly summarizes the consequences of taking the decision t when θ prevails. It will be convenient to consider the loss as a pay-off to nature in the game (Θ, T, L) between nature (player I) and the statistician (player II). The choice of the loss function L is necessarily somewhat arbitrary. See e.g. Gauss (1821) for a discussion of squared error as a measure of loss. A translation into English may be found in Lehmann (1950).

Economic factors and mathematical convenience, as well as other aspects, may be considered. It is known that certain rules for 'rational behaviour' imply that this behaviour should be based on a numerical quantity, usually called utility, and which may be considered as $-$loss. We refer the reader to Blackwell & Girchick (1954), Ferguson (1967) or to De Groot (1970) for some expositions of utility theory. Although this theory is quite interesting it is not needed as a justification for a decision theory based on the concept of loss.

Now consider a general decision space $(T, \mathscr{S})$. Then a *loss function on the decision space $(T, \mathscr{S})$ is a function from Θ to the set of measurable and extended real valued functions on $(T, \mathscr{S})$*. If L is a loss function then it maps each θ in Θ into L_θ where $L_\theta(t)$ should be interpreted as the loss we suffer if we take the decision t and θ is the state of nature. *Throughout we shall assume that the function L_θ, for each θ, is bounded from below, i.e. that* $\inf_t L_\theta(t) > -\infty$. Thus $L_{\cdot}(t) \in\,]-\infty, \infty]^\Theta$ for each $t \in T$.

A statistical decision problem shall here be considered as completely specified by the triple $(\mathscr{E}, (T, \mathscr{S}), L)$ *where* $\mathscr{E}$ *is an experiment and* $(T, \mathscr{S})$ *is a decision space with loss function L.* $-$*loss is called utility. Thus a utility function U on a decision space* $(T, \mathscr{S})$ *is a function which to each theory* θ *in the parameter set* Θ *associates an extended real valued and measurable function* U_θ. The quantity $U_\theta(t)$, for a given pair (θ, t), may be considered as a measure of the utility of t when θ is true. It will be assumed that $\sup_t U_\theta(t) < \infty$ for each $\theta \in \Theta$. Using the decision rule ρ in the statistical decision problem $(\mathscr{E}, (T, \mathscr{S}), L)$ the actual loss becomes an extended real valued random variable whose expectation exists. Expected loss is called *risk*. Thus we define *the risk of the decision rule* ρ *at* $\theta \in \Theta$ as the quantity $P_\theta \rho L_\theta$. Alternative notions for the risk are $r_\mathscr{E}(\theta : \rho, L)$, $r_\mathscr{E}(\theta : \rho)$ or just $r(\theta : \rho)$. However, remember that the analysis of risk functions in chapter 3 does not require that risk is defined as expected loss. As mentioned there, that analysis does not even require that loss is defined.

The function $r(\cdot : \rho) = P_. \rho L_.$ is called the *risk function of* ρ. Here we shall regard *the risk function as the fundamental entity for judging the quality of a decision rule.*

The definition of risk implies that the risk of ρ at θ may be written as

$$r(\theta : \rho) = (P_\theta \times \rho)(L_\theta)$$

or as

$$r(\theta : \rho) = (P_\theta \rho)(L_\theta)$$

or as

$$r(\theta : \rho) = P_\theta(\rho L_\theta).$$

The second formula expresses the risk in terms of the performance function of ρ. We see in particular that decision rules having identical performance functions have the same risk functions, regardless of how the loss function is defined. It follows that the performance function is a more fundamental quantity than the risk function.

The last expression for the risk shows how the risk may be found by first finding ρL_θ, i.e. the conditional risk given the observations, and then taking overall expectation.

It is quite conceivable that the actual loss might also depend on the observations. Then let $L_\theta(t, x)$ be the loss we suffer when decision t is taken, x is observed and θ is the true value of the parameter. In that case the risk, i.e. the expected loss, does not depend on the performance function alone, but on the complete joint distribution $P_\theta \times \rho$ of observations made and the decision taken. Thus we may still write the risk as $(P_\theta \times \rho)(L_\theta) = P_\theta(\rho L_\theta)$ where $(\rho L_\theta)_X = \int L_\theta(t, X)\rho(\mathrm{d}t | X)$ denotes the conditional risk given the observations X.

Sometimes the total loss we suffer by taking the decision t when θ prevails and the observation is X is a sum of two terms, $L_\theta(t)$ and $C_\theta(X)$ where L is a loss function as defined above, while $C_\theta(X)$ is the cost of observing X when θ prevails. The risk of a decision rule ρ then becomes $P_\theta \rho L_\theta + \int C_\theta \,\mathrm{d}P_\theta$. Since the cost term does not depend on ρ, it follows that comparison of decision rules in terms of risk does not involve the cost term. In sequential problems the cost usually depends on the number of 'coordinates' of X and this, in turn, is governed by the stopping rule which is a part of the total decision rule. However, if the stopping rule is fixed then a partitioning of the total risk as above may be feasible.

Here, if not otherwise stated, we shall assume that the loss only depends on the true value of the parameter and of the decision taken.

Although considerations of risk go far back in time, the foundation of a theory of statistical decision rules based on the concept of risk was laid by Wald in his book *Statistical decision functions* (1950). At this point, the reader may benefit from reading the first chapter of this book.

Example 4.2.1 (Estimation problems). Suppose we want to estimate $g(\theta)$ where g is a given known function on Θ. We may then consider a decision space $(T, \mathscr{S})$ where T contains the range $g[\Theta]$ of g. A decision rule ρ from $\mathscr{E}$ to $(T, \mathscr{S})$ is called an *estimator of* $g(\theta)$. The performance function of an estimator is its probability distribution as a function of θ. A 'good' estimator of $g(\theta)$ should with 'large' probabilities be 'close' to $g(\theta)$. Thus if T is equipped with a distance then we might look for an estimator ρ such that the expected distance from the estimator to $g(\theta)$ is small for each θ. We might, of course, get a perfect fit at $\theta_0 \in \Theta$ if δ is the one-point distribution in $g(\theta_0)$. This estimator, however, behaves badly when $g(\theta)$ is far from $g(\theta_0)$. In general the requirements of a good fit at θ for the various possible values of θ in Θ are conflicting and we are faced with the task of finding a compromise solution.

Referring to some chosen distance on the decision space T we may consider the distance between t and $g(\theta)$ as the loss incurred by the decision t when θ prevails. In that case the risk of an estimator equals the expected distance between this estimator and $g(\theta)$. In general the loss $L_\theta(t)$ of the decision t when θ holds may be considered as a measure of how close t is to $g(\theta)$. Here are a few loss functions for the case where g is real valued.

(i) $L_\theta(t) = |t - g(\theta)|$; $\theta \in \Theta$, $t \in \,]-\infty, \infty[$. This loss function is called *absolute error* and the risk function of an estimator ρ for this loss function is called *the expected absolute error of* ρ.

(ii) $L_\theta(t) = (t - g(\theta))^2$; $\theta \in \Theta$, $t \in \,]-\infty, \infty[$. This loss function is called *squared error* and the risk of an estimator for this loss function is called *expected squared error.*

(iii) $L_\theta(t) = 0$ or $= 1$ as $|t - g(\theta)| \leqslant c$ or $|t - g(\theta)| > c$ where $c \geqslant 0$ is a constant. The risk of an estimator for this loss function is the probability that the deviation from $g(\theta)$ exceeds c. If $c = 0$ then this is the probability that the estimator differs from $g(\theta)$.

Example 4.2.2 (Testing problems). A decision problem where the set T of possible decisions contains exactly two decisions t_0 and t_1, and where, of course, all four subsets are measurable, is called a *testing problem*. Typically the parameter set may be partitioned into three parts as $\Theta = \Theta_0 \cup \Theta_I \cup \Theta_1$ where decision t_0 is preferred on Θ_0, decision t_1 is preferred on Θ_1 and where it does not matter which decision we are making when $\theta \in \Theta_I$. The set Θ_I is then called the *set of indifference*. The decisions t_0 and t_1 may be identified with the hypotheses '$\theta \in \Theta_0$' and '$\theta \in \Theta_1$' respectively. If the erroneous rejection of one of these hypotheses is considered to be more serious than the erroneous rejection of the other hypothesis, then it is customary to call the first hypothesis *the null hypothesis* and to call the second hypothesis *the alternative (hypothesis)*. We may then assume that the subscripts are arranged so that '$\theta \in \Theta_0$' is the null hypothesis and '$\theta \in \Theta_1$' is the alternative.

A useful modification of these considerations is to let H_0 denote the hypothesis '$\theta \in \Theta_0$' and then let t_0 and t_1 represent, respectively, the decisions: 'There is insufficient evidence for rejecting H_0' and 'Reject H_0'.

A decision rule ρ in the two-decision problem where $T = \{t_0, t_1\}$ is determined by the probability $\rho(t_1|x)$ which, as a function of x, may be any measurable function from $(\mathscr{X}, \mathscr{A})$ to $[0, 1]$. A measurable function with range contained in $[0, 1]$ will be called a *test function*, or just a *test*. Thus there is a 1–1 correspondence between decision rules ρ and test functions $\delta = \rho(t_1|\cdot)$. The test function δ, for each $x \in \mathscr{X}$, provides us with the probability of concluding with t_1 and thus rejecting t_0.

The *power function* of a test δ is its expectation, $E_\theta\delta$, as a function of θ. The number $E_\theta\delta$ will be called the power of the test δ at θ. The power function determines the performance function of the corresponding decision rule ρ, since $1 - (P_\theta\rho)(t_0) = (P_\theta\rho)(t_1) = E_\theta\delta$ is the probability of rejecting t_0 as a function of θ. If Θ is partitioned as above then, whether the indicated modification is valid or not, we would like to have little or much power, according to whether $\theta \in \Theta_0$ or $\theta \in \Theta_1$. Just as in the case of estimation problems we are again faced with conflicting requirements. There are two kinds of errors which we may commit. Firstly we may take the decision t_1 while actually '$\theta \in \Theta_0$'. Secondly we may take the decision t_0 while actually '$\theta \in \Theta_1$'. These errors are usually called, respectively, *an error of the first kind* and *an error of the second kind*.

If the test δ is used then the probability of an error of the first kind is $E_\theta\delta$ when $\theta \in \Theta_0$, while the probability of an error of the second kind is $1 - E_\theta\delta$ when $\theta \in \Theta_1$.

One way to proceed is to insist on a safety bound for probabilities of errors of the first kind, and then to try to choose δ such that $E_\theta\delta$ is large when $\theta \in \Theta_1$. A number $\alpha \in [0, 1]$ is called *a level of significance* for the test δ if $E_\theta\delta \leqslant \alpha$ when $\theta \in \Theta_0$. A test with level of significance α will also be called a *level α test*. The smallest level of significance of the test δ is clearly $\sup_{\theta\in\Theta_0} E_\theta\delta$ and this number is called the *size* of δ. We might then look for level α tests with sufficiently high power when $\theta \in \Theta_1$. One possibility is to look for a test which maximizes the minimum, or rather infimum, power on Θ_1 among all level α tests. A test with this property is called a *maximin level α test.* Maximin tests exist whenever $\mathscr{E}$ is coherent. Some properties of maximin tests are discussed in complements 4–10 in chapter 10. This and several other principles for choosing tests are discussed in Lehmann's book *Testing statistical hypotheses* (1986). One of the most important and beautiful results of statistical testing theory is the fundamental lemma of Neyman & Pearson (1933). The reader could consult Lehmann's book for an exposition. This lemma characterizes the most powerful level α tests for testing a simple null hypothesis '$\theta \in \theta_0$' against a simple alternative hypothesis '$\theta \in \theta_1$'.

A loss function L in the two-decision case is determined by the two functions $L_{.}(t_0)$ and $L_{.}(t_1)$. The risk of a test δ is then $r(\theta:\delta) = L_\theta(t_0)(1 - E_\theta\delta) + L_\theta(t_1)E_\theta\delta$. If, for $i = 0$ and $i = 1$, $L_\theta(t_i) = 0$ or $= 1$ as $\theta \in \Theta_i$ or $\theta \notin \Theta_i$ then the risk coincides with the probabilities of errors when $\theta \in \Theta_0 \cup \Theta_1$.

Example 4.2.3 (Multiple decision problems). A decision problem with finite decision space is called a *multiple decision problem.* If the decision space is a k-decision space then the decision problem may be called a *k-decision problem.* The case $k = 1$ does not involve any decision making and the case $k = 2$ was described in the previous example.

Consider a general finite decision space $(T, \mathscr{S})$ where T is finite and $\mathscr{S}$ consists of all subsets of T. Call a vector $z = (z_t : t \in T)$ a *probability vector on* T if $z_t \geqslant 0$ when $t \in T$ and $\sum_t z_t = 1$. Then a probability distribution Q on T may be identified with the probability vector $(Q(t); t \in T)$. It follows that a decision rule ρ may be identified with the map which to each x in χ assigns the probability vector $(\rho(t|x); t \in T)$ on T. Similarly the performance function of ρ may be identified with the map which to each θ in Θ assigns the probability vector $(P_\theta\rho(t); t \in T)$ on T. If L is a loss function on T then the risk of the decision rule ρ at $\theta \in \Theta$ is

$$r(\theta : \rho) = \sum_t L_\theta(t)(P_\theta\rho)(t) = E_\theta\left[\sum_t L_\theta(t)\rho(t|\cdot)\right].$$

It follows easily that if $\mathscr{E}$ is coherent then the set of all functions g in $]-\infty, \infty]^\Theta$ such that $g \geqslant r(\cdot : \rho)$ for some decision rule ρ is compact. If, in addition, L is finite, then the set of all risk functions is compact. However, we have seen that there are experiments $\mathscr{E}$, not necessarily coherent, such that the set of power functions is not closed.

What can we say about the structure of the set $\mathscr{R}$ of risk functions of a general statistical decision problem $(\mathscr{E},(T,\mathscr{S}),L)$? If the set of permissible decision rules is convex, as the sets of all decision rules are, then $\mathscr{R}$ is convex. This follows since the risk is an affine function of the decision rule. Assuming that $\mathscr{R}$ is convex, we may try to apply the theory in sections 3.3 and 3.4 of chapter 3. The basic compactness condition there, which is described in proposition 3.3.5, amounts to the assumption that to any function g in the closure $\bar{\mathscr{R}}$ of $\mathscr{R}$ for the topology of pointwise convergence there corresponds an $r \in \mathscr{R}$ such that $r \leqslant g$. Using the notation of proposition 3.3.5, compactness may then be achieved by replacing $\mathscr{R}$ by any convex subset $\mathscr{R}_1$ of $]-\infty, \infty[^\Theta$ such that $\hat{\bar{R}}_1 = \hat{\bar{\mathscr{R}}}$. A convex set $\mathscr{R}_1$ satisfies this condition if and only if it possesses the set of minimal functions in $\bar{\mathscr{R}}$ as a minimal complete class. The largest and the smallest convex sets with this property are, respectively, $\hat{\bar{\mathscr{R}}}$ and the convex set generated by the minimal elements of $\bar{\mathscr{R}}$. An intermediate candidate for $\mathscr{R}_1$ is the closure $\bar{\mathscr{R}}$ of $\mathscr{R}$. We shall later see how the concept of a decision rule may be extended so that the correspondingly increased set of 'risk functions' satisfies the compactness condition described in proposition 3.3.5.

Let us make a few remarks (disregarding measure theoretical niceties) on the connection between decision theory and prediction theory. Assume that the joint distribution of two random variables X and Y is determined by the unknown parameter $\theta \in \Theta$. Suppose that Y is not (yet) observed but that X is observable (observed). We shall consider the problem of predicting Y on the basis of X. A *predictor* p is then a rule (Markov kernel) which to each possible value of X assigns a probability distribution on the sample space of Y. Thus if X is observed and the predictor p is used then the probability distribution of our guess is $p(\cdot|X)$.

Furthermore, assume that the loss we suffer when we predict that $Y = y'$ while actually $Y = y$ is quantified as $L_\theta(y', y, X)$. The risk is then the expectation of $\int L_\theta(y', Y, X)p(dy'|X)$. Thus we are again faced with a set of risk functions and the theory in sections 3.3 and 3.4 of chapter 3 apply.

Example 4.2.4 (Prediction for squared error loss). Suppose X and Y are independent random variables and that Y is real valued and such that $E_\theta Y^2 < \infty$ when $\theta \in \Theta$. Let us restrict our search to non-randomized predictors $g(X)$ such that $E_\theta g(X)^2 < \infty$; $\theta \in \Theta$. The expected squared error is then

$$E_\theta(g(X) - Y)^2 = E_\theta(g(X) - E_\theta Y)^2 + \operatorname{Var}_\theta Y.$$

It follows that this prediction problem is equivalent to the problem of estimating $E_\theta Y$ on the basis of X and for squared error loss. More generally, if X

and Y are not assumed independent then this prediction problem may be phrased as the problem of 'estimating' the parameter dependent random variable $E_\theta(Y|X)$.

Convexity of the loss function implies here, by Jensen's inequality, that the set of non-randomized predictors is complete.

4.3 Randomized decision rules

There is another way of defining randomized decision procedures which we will only mention briefly here. In the previous section we defined a decision rule ρ as a map from the sample space into the set of probability distributions on T. Thus we observe X and then choose a decision t in T according to the distribution $\rho(\cdot|X)$. This method of randomization is called *randomization after experimentation.* In contrast to this we might also consider *randomization before experimentation*, where we first choose a non-randomized decision rule τ according to a specified probability distribution π on a set of non-randomized decision rules, and then plug our observation X into τ, yielding the decision $\tau(X) \in T$.

In the latter case, disregarding measurability problems for the moment, the decision procedure is completely described by the probability distribution π. Risk = expected loss is still determined by the performance function which to each θ assigns the probability distribution of $\tau(X)$ in T. If π is used, then the probability of concluding with a decision in $S \in \mathscr{S}$ is $E_\theta\pi(\{\tau : \tau(X) \in S\})$, provided this expression is well defined. It follows then that the decision rule ρ has the same performance function as π has, provided

$$\rho(S|x) = \pi(\{\tau : \tau(x) \in S\})$$

for P_θ almost all x when $S \in \mathscr{S}$. Of course there is no guarantee that $\tau(x)$ is measurable in τ nor that $\pi(\{\tau : \tau(x) \in S\})$, if defined, is measurable in x. However these assumptions are fulfilled whenever $\tau(x)$ is jointly measurable in (τ, x). They are also satisfied when $\mathscr{E}$ is discrete and $\tau(x)$, for each x, is measurable in τ.

On the other hand suppose that the decision rule ρ is given. Then we might try to construct a probability distribution π on the set of measurable functions from $\mathscr{X}$ to T such that the 'x marginal' of π for each x is $\rho(\cdot|x)$. If $\mathscr{E}$ is discrete then we might just let π be the product measure $\prod_x \rho(\cdot|x)$ on $(T^{\mathscr{X}}, \mathscr{S}^{\mathscr{X}})$. Thus the two methods of randomization are equivalent when $\mathscr{E}$ is discrete.

Let us consider the particular case where both $\mathscr{X}$ and T are finite sets. Put $n = \#\mathscr{X}, r = \#T$ and let A^* denote the set of probability distributions on the finite set A. The set of non-randomized procedures is then $D = T^{\mathscr{X}}$ so that the set of 'before experimentation' randomizations is $(T^{\mathscr{X}})^*$ which is of dimension

$r^n - 1$ as a subset of $\mathbb{R}^D$. The set of decision rules, i.e. 'after experimentation' randomizations, is the set $(T^*)^{\mathscr{X}}$ of functions from $\mathscr{X}$ to T^*. This set has dimension $n(r-1)$. It follows that the set of 'before' procedures has greater dimension than the set of 'after' procedures whenever $r \geqslant 2$ and $n \geqslant 2$. Nevertheless, we have just seen that the corresponding sets of performance functions are the same. The reason is, of course, that it is only the various 'x marginals' which matter for 'before' procedures π in D^*.

If T is a two-point set, i.e. if $r = 2$, and Θ is a finite set containing m points, then the set of performance functions may be identified with the set of power functions of tests. This set, being a compact convex subset of R^Θ, is the convex hull of its extreme points. If the power function of a test function δ is extreme then, since $\delta = \frac{1}{2}(\delta + \delta \wedge (1-\delta)) + \frac{1}{2}(\delta - \delta \wedge (1-\delta))$, δ is equivalent to an indicator function. Thus, by theorem 2.3.8, any power function is the power function of a convex combination of at most $(m+1)$ non-randomized tests. (If the power function is on the boundary of the set of power functions, then '$m+1$' may be replaced by 'm'.) Quite similar considerations show that when Θ and T are finite, any decision rule is equivalent to a finite convex combination of non-randomized decision rules.

The general problem of the equivalence of the two methods of randomization is quite difficult. Important results in this direction have been obtained by Dvoretzky, Wald & Wolfowitz in their joint papers (1951). One of their basic assumptions is that each measure P_θ should be atomless. This may always be achieved by replacing each P_θ with a product measure $P_\theta \times Q$ where Q is atomless and does not depend on θ. We could simply let Q be the uniform distribution on $[0, 1]$. It is easily seen that the set of performance functions for the experiments $\mathscr{E} = (P_\theta : \theta \in \Theta)$ and $\mathscr{F} = (P_\theta \times Q : \theta \in \Theta)$ are the same, for any decision space. Dvoretzky, Wald & Wolfowitz based their investigations on a result of Liapounoff (1940) concerning the range of a vector valued measure. Liapounoff showed that if the vector measure is an n-tuple of finite measures then the range is always closed and it is convex when each of the measures are atomless. This implies that any power function is the power function of a non-randomized test when θ is finite and each P_θ is atomless. Dvoretzky, Wald & Wolfowitz extend this by showing that any performance function is the performance function of a non-randomized decision rule provided again that Θ is finite and each P_θ is atomless.

Later we shall see that decision rules minimizing Bayes risk under general conditions may be chosen non-randomized. Another situation where randomization may be eliminated is estimation problems with convex loss. Then Jensen's inequality, theorem 2.3.24, may ensure that a given decision rule is dominated by a non-randomized decision rule. To be more specific, let ρ be a decision rule and suppose the loss $L_\theta(t, x)$ is convex in $t \in \mathbb{R}^m$ where $m < \infty$. Then $\int L_\theta(t, x)\rho(\mathrm{d}t|x) \geqslant L_\theta(\int t\rho(\mathrm{d}t|x), x)$ provided $\int |t|\rho(\mathrm{d}t|x) < \infty$.

4.4 Bayes risk

The main idea of Bayesian statistics is that we usually have some more or less vague knowledge about the unknown parameter before experimentation, and that this knowledge may be summarized by a probability distribution on the parameter set. This probability distribution is called the *prior distribution* or just the *prior*.

The prior distribution determines a joint distribution of the parameter and the outcome of the experiment. After the result of the experiment has been obtained, the Bayesian statistician adjusts the prior distribution by replacing it with the conditional distribution of the parameter given the outcome of the experiment. This distribution is called the *posterior distribution.* If another experiment is carried out independently of the first one, and if the posterior distribution in the latter is derived by using the posterior distribution determined by the outcome of the first experiment, then we arrive at the posterior distribution given the outcome of the combined experiment for the original prior distribution. Thus the procedure leading to the posterior distribution may be carried out stepwise for independent observations.

The *Bayesian risk* of a decision rule is its risk function averaged with respect to the given prior distribution. Ordering the risk functions according to their Bayesian risk we obtain a total order of risk functions, and it is this ordering which is the basis for comparison of decision procedures within the Bayesian framework.

Bayes risk may be obtained in terms of posterior loss and posterior risk. The *posterior loss* specifies, for each possible decision, the expected loss we suffer if this decision is taken and the parameter is distributed according to the posterior distribution. Thus the posterior loss does not depend on the decision rule. The *posterior risk* of a decision rule specifies, for each possible outcome of the experiment, the expected loss we suffer if this decision rule is used and if this particular outcome is obtained. It follows that posterior risk is the conditional expectation of the posterior loss given the outcome of the experiment. A decision rule which for each outcome minimizes the posterior risk is called a *Bayes rule.* A Bayes rule minimizes the Bayes risk for the given prior.

The Bayesian method is consistent with the likelihood principle, i.e. if outcomes in possibly unrelated experiments yield equivalent likelihood functions then they yield the same posterior loss and consequently the same conclusions, provided posterior losses are minimized in the same way. Therefore general objections to the likelihood principle are also objections to the Bayesian method. Although the likelihood principle does not appear convincing to everyone, the usefulness of Bayesian concepts and methods are generally acknowledged. We do not, however, attach belief to a particular prior but will feel free to use them all.

If we consider the game between nature and the statistician where nature chooses a law θ and the statistician independently chooses a risk function r and then pays nature the amount $r(\theta)$, then the prior distributions are nature's mixed strategies.

We shall now make these definitions and results precise within the measure theoretical framework.

Let the experiment $\mathscr{E} = (\mathscr{X}, \mathscr{A}; P_\theta : \theta \in \Theta)$ be realized by observing the variable X. We have defined a *prior* (prior distribution) λ as a probability distribution on some σ-algebra $\mathscr{U}$ of subsets of Θ. It will be assumed throughout this section that $\mathscr{U}$ is linked to $\mathscr{E}$ so that $P_\theta(A)$ is measurable in θ for each set A in $\mathscr{A}$.

The prior distribution λ determines a joint distribution $\lambda \times P$ on $\mathscr{U} \times \mathscr{A}$ by

$$(\lambda \times P)(U \times A) = \int_U P_\theta(A)\lambda(\mathrm{d}\theta); \; U \in \mathscr{U}, A \in \mathscr{A}.$$

The distribution of X, i.e. the second marginal of this measure, is the probability distribution $\lambda P(A) = (\lambda \times P)(\Theta \times A); A \in \mathscr{A}$.

After the variable X realizing the experiment $\mathscr{E}$ is observed, the Bayesian statistician adjusts the prior distribution by replacing λ with a conditional distribution of θ given X. If, say, $(\theta, \mathscr{U})$ is Euclidean, then such a conditional distribution exists. A conditional distribution π of θ given X with respect to $\lambda \times P$ is called a *posterior* distribution of θ given X. Mathematically a posterior distribution π of θ given X is a Markov kernel: $(x, U) \to \pi(U|x)$ from $(\mathscr{X}, \mathscr{A})$ to $(\Theta, \mathscr{U})$ such that $\pi(U|X)$, for each $U \in \mathscr{U}$, is a version of the $\lambda \times P$ conditional probability of $U \times \mathscr{X}$ given X.

The latter condition may be phrased

$$\int_A \pi(U|x)(\lambda P)(\mathrm{d}x) = (\lambda \times P)(U \times A); \; U \in \mathscr{U}, A \in \mathscr{A}.$$

Assume now that $\mathscr{E}$ is dominated by a σ-finite measure μ so that, for each θ, P_θ has a density f_θ with respect to μ. Let us also assume that $f_\theta(x)$ is jointly measurable in (θ, x). (The last assumption, as shown by Pfanzagl (1969) is satisfied for suitably specified densities if and only if $\mathscr{E}$ is separable and $P_\theta(A)$ is measurable in θ for each set A in $\mathscr{A}$.) If follows from Fubini's theorem that $P_\theta(A)$ is measurable in θ when $A \in \mathscr{A}$. If λ is a prior distribution on $\mathscr{U}$ then $f_\theta(x)$ as a function of (θ, x) is a density of $\lambda \times P$ with respect to $\lambda \times \mu$. It follows that the probability distribution with density $f_\theta(X)/\int f_\theta(X)\lambda(\mathrm{d}\theta)$ with respect to λ is a version of the posterior distribution $\pi(\cdot|X)$ given X.

This adjustment procedure for prior distributions satisfies the following consistency requirement. Suppose that, in addition to the statistic X realizing $\mathscr{E}$, there is also available a statistic Y independent of X and realizing the experiment $\mathscr{F} = (\mathscr{Y}, \mathscr{B}; Q_\theta : \theta \in \Theta)$. Suppose also that Q_θ, for each θ, has a density g_θ with respect to a σ-finite measure ν and that $g_\theta(y)$ is jointly measurable in (θ, y). If we adjust the posterior distribution given X described above after Y has been observed, and proceeding as above, then we arrive at a distribution $\tilde{\pi}(\cdot|X, Y)$ with density $g_\theta(Y)/\int g_\theta(Y)\pi(d\theta|X)$ with respect to $\pi(\cdot|X)$. Writing this out we see that $\tilde{\pi}(\cdot|X, Y)$ has density $f_\theta(X)g_\theta(Y)/\int f_\theta(X)g_\theta(Y)\lambda(d\theta)$ with respect to λ and consequently, by this method of construction, is the posterior distribution of θ given the combined statistic (X, Y).

It follows that the posterior distribution of θ given the n-tuple $(X_1, X_2, \ldots, X_n)$ of independent statistics may be obtained stepwise by first adjusting for X_1, then for $X_2, \ldots$ and finally for X_n. We assume then that the experiment defined by X_i, for each i, is dominated, that the densities may be specified jointly measurable and that the posterior distributions are defined in terms of these densities as described above.

Let us next consider decision problems within the Bayesian framework. The following discussion does not require that $\mathscr{E}$ is dominated except when it is explicitly mentioned. Thus we are given an experiment $\mathscr{E} = (\mathscr{X}, \mathscr{A}; P_\theta : \theta \in \Theta)$, a prior distribution λ on a σ-algebra $\mathscr{U}$ of subsets of Θ making $P_\theta(A)$ measurable in θ for each A in $\mathscr{A}$. Let $(T, \mathscr{S})$ be a decision space equipped with a loss function $L_\theta(t, x)$ which is jointly measurable in (θ, x, t). In order to ensure the existence of integrals, we shall assume that $L_\theta(t, x) \geqslant a_\theta(x)$ where $\int [\int a_\theta(x)P_\theta(dx)]\lambda(d\theta) > -\infty$. If ρ is a decision rule from $\mathscr{E}$ to $(T, \mathscr{S})$ then there exists a unique joint distribution for $(\tilde{\theta}, X, \tilde{t})$ on $(\Theta, \mathscr{U}) \times (\mathscr{X}, \mathscr{A}) \times (T, \mathscr{S})$ such that:

(i) $\tilde{\theta}$ has distribution λ.
(ii) the conditional distribution of X given $\tilde{\theta}$ is $P_{\tilde{\theta}}$.
(iii) the conditional distribution of $\tilde{t}$ given $\tilde{\theta}$ and X is $\rho(\cdot|X)$.

Let Pr and E without subscript denote, respectively, probability and expectation for $\lambda \times P \times \rho$. The most important feature of the distribution $\lambda \times P \times \rho$ is that it is the distribution of a three term Markov process, i.e. $\tilde{\theta}$ and $\tilde{t}$ are conditionally independent given X.

The performance function of ρ now becomes the conditional distribution of $\tilde{t}$ given $\tilde{\theta}$, while the posterior distribution is still the conditional distribution of θ given X. The risk at $\tilde{\theta}$ becomes the conditional expected loss given $\tilde{\theta}$, i.e.

$$r(\tilde{\theta} : \rho) = E(L|\tilde{\theta}) = E[E(L|\tilde{\theta}, X)|\tilde{\theta}]$$

where $L = L_{\tilde{\theta}}(\tilde{t}, X)$. The conditional expectations given $(\tilde{\theta}, X)$ and given $\tilde{\theta}$ are here defined by the conditional distributions $\rho(\cdot|X)$ and $P_{\tilde{\theta}} \times \rho$ respectively. If our prior belief is summarized by the prior distribution λ then the overall expected risk incurred by using the decision procedure ρ is

$$EL = E(E(L|\tilde{\theta})) = \int r(\theta : \rho)\lambda(\mathrm{d}\theta).$$

This quantity, which is simply λ-weighted risk, is denoted by $r(\lambda : \rho, L)$ or $r(\lambda : \rho)$ and is called *the Bayes risk of* ρ *for the prior* λ. A decision procedure ρ_0 is called *a Bayes solution for the prior* λ if ρ_0 minimizes Bayes risk for λ, i.e. if

$$r(\lambda : \rho_0) = \inf_{\rho} r(\lambda : \rho).$$

More generally, if $\varepsilon \geqslant 0$ then a *decision procedure* ρ_ε *is called an* ε*-Bayes solution for the prior* λ if

$$r(\lambda : \rho_\varepsilon) \leqslant \inf_{\rho} r(\lambda : \rho) + \varepsilon.$$

Thus, according to the terminology in section 3.4, a decision rule is a Bayes (an ε-Bayes) solution if and only if its risk function is Bayes (ε-Bayes).

This definition of a (an ε-) Bayes solution does not require the existence of a posterior distribution. Mathematically the Bayes risk is just expected loss for $\lambda \times P \times \rho$.

Let us suppose that a posterior distribution $\pi(\cdot\,|X)$, i.e. a regular conditional distribution of $\tilde{\theta}$ given X, is available. Put $L(t|x) = \int L_\theta(t, x)\pi(\mathrm{d}\theta|x)$ when $t \in T$ and $x \in \mathscr{X}$. Then $L(t|X)$ is the expected loss we suffer by taking the decision t when the observations X are given. The statistic $L(t|X)$ is called the *posterior loss incurred by t*. The posterior loss does not depend on the decision rule ρ. It is completely determined by the decision problem and the specification of the posterior distribution on Θ.

If the decision rule ρ is used and if the observations X are given, then the expected loss is

$$\int L(t|X)\rho(\mathrm{d}t|X) = \int\left[\int L_\theta(t, X)\rho(\mathrm{d}t|X)\right]\pi(\mathrm{d}\theta|X).$$

This statistic is called the *posterior risk of* ρ.

Conditioning with respect to the observations X and then using the Markov property, we arrive at the following important expression for Bayes risk

$$r(\lambda : \rho) = E(EL|X) = \int\left[\int L(t|x)\rho(\mathrm{d}t|x)\right](\lambda P)(\mathrm{d}x).$$

It follows that the problem of finding decision rules with small Bayes risks may be reduced to the problem of finding decisions t making $L(t|x)$ small for each given realization x of X. For each $x \in \mathscr{X}$, let $\underline{L}(x) = \inf_t L(t|x)$ be the infimal posterior loss given $X = x$. The following basic rule for finding Bayes solutions is now immediate.

Theorem 4.4.1. *Let* ρ_0 *be a non-randomized decision rule such that* $L(t|x)$, *for* λP *almost all* x, *is minimized by* $t = \rho_0(x)$. *Then*

(i) ρ_0 *is a Bayes solution.*
(ii) *a decision rule* ρ *is a Bayes solution if and only if*

$$\rho(\{t : L(t|x) = \underline{L}(x)\}|x) = 1 \textit{ for } \lambda P \textit{ almost all } x.$$

If the decision space is finite then a Bayes solution with these properties certainly exists.

Let us say that a decision t is ε-*optimal for the realization* x of X if $\varepsilon \geqslant 0$ and $L(t|x) \leqslant \underline{L}(x) + \varepsilon$. Theorem 4.4.1 may then be generalized as follows.

Theorem 4.4.2. *A non-randomized decision rule* ρ_ε *is an* ε-*Bayes solution if* $\rho_\varepsilon(x)$, *for* λP *almost all* x, *is* ε-*optimal for the realization* x.

Proof. Let ρ be another decision rule. Then $\int L(t|x)\rho_\varepsilon(\mathrm{d}t|x) = L(\rho_\varepsilon(x)|x) \leqslant \underline{L}(x) + \varepsilon \leqslant \int L(t|x)\rho(\mathrm{d}t|x) + \varepsilon$. The conclusion follows by integrating with respect to λP. □

When applying these ideas we should note that the existence of a posterior distribution, i.e. a regular conditional distribution of $\tilde{\theta}$ given X, is not essential. What matters is that we are allowed to write

$$\int \hat{L}(t|X)\rho(\mathrm{d}t|X) = E(L|X) \quad \text{a.s.}$$

where $\hat{L}(t|x)$ does not depend on ρ and is jointly measurable in (t, x). Then for each t, $\hat{L}(t|X)$ is necessarily a version of $E(L_{\tilde{\theta}}(t, X)|X)$. If $\hat{L}$ has these properties then theorems 4.4.1 and 4.4.2 remain valid if $\hat{L}(t|x)$ is permitted as posterior loss, i.e. if $L(t|x)$ is replaced throughout by $\hat{L}(t|x)$. If T is countable then $\hat{L}(t|X) = E(L_{\tilde{\theta}}(t, X)|X)$, for each $t \in T$, may be arbitrarily specified. If the loss L is of the form $L_\theta(t, x) = \hat{L}_{g(\theta)}(t, x)$ where g is a measurable map from Θ to some Euclidean space, then we may put $\hat{L}(t|X) = \int \hat{L}_g(t, X)\sigma(\mathrm{d}g|X)$ where $\sigma(\cdot|X)$ is a regular conditional distribution of $g(\tilde{\theta})$ given X.

If the construction in the last theorem is feasible for each $\varepsilon > 0$, as it is whenever $\mathscr{E}$ is discrete or T is countable, then the minimum Bayes risk is $\int \underline{L}(x)\lambda P(\mathrm{d}x)$.

The general problem of the feasibility of these constructions of Bayes and ε-Bayes solutions involves difficult measurability problems of dubious statistical significance. Fortunately these difficulties can usually be overcome in special situations.

If $\mathscr{E}$ is dominated by a σ-finite measure μ such that $[\mathrm{d}P_\theta/\mathrm{d}\mu]_x$ may be specified jointly measurable in (θ, x), then the posterior loss may be specified as

$$L(t|X) = \int L_\theta(t, X)f_\theta(X)\lambda(\mathrm{d}\theta) \Big/ \int f_\theta(X)\lambda(\mathrm{d}\theta).$$

Hence

$$r(\lambda : \rho) = \int \left\{ \int \left[\int L_\theta(t, x) f_\theta(x) \lambda(\mathrm{d}\theta) \right] \rho(\mathrm{d}t | x) \right\} \mu(\mathrm{d}x).$$

It follows that a non-randomized decision rule ρ_0 is a Bayes solution whenever

$$\int L_\theta(\rho_0(x), x) f_\theta(x) \lambda(\mathrm{d}\theta) = \min_t \int L_\theta(t, x) f_\theta(x) \lambda(\mathrm{d}\theta)$$

for μ almost all x.

Example 4.4.3 (Estimation. Continuation of example 4.2.1). If the loss is given as absolute error or as quadratic error then by convexity we may restrict our attention to non-randomized estimators. Let σ be a regular conditional distribution of $g(\theta)$ given X. Let $\hat{g}(x)$, for each x, be a specification of the median of $\sigma(\cdot | x)$. We may assume that $\hat{g}$ is measurable. Let $\tilde{g}$ be a measurable real valued function on $\mathscr{X}$ such that $\tilde{g}(x) = \int g\sigma(\mathrm{d}g | x)$ whenever $\int g^2 \sigma(\mathrm{d}g | x) < \infty$. Consider now a non-randomized decision rule ρ. Then

$$E|\rho(X) - g(\tilde{\theta})| = E[E|\rho(X) - g(\tilde{\theta})| | X] \geqslant E|\hat{g}(X) - g(\tilde{\theta})|$$

and

$$E(\rho(X) - g(\tilde{\theta}))^2 = E[E(\rho(X) - g(\tilde{\theta}))^2 | X] \geqslant E(\tilde{g}(X) - g(\tilde{\theta}))^2.$$

It follows that $\hat{g}$ and $\tilde{g}$ are Bayes solutions in these two cases.

The structures of these inequalities reveal that:

(i) if the loss is measured by absolute error and if there is an estimator with finite Bayes risk, then a non-randomized estimator is a Bayes solution if and only if it is, modulus λP null sets, a median of $g(\tilde{\theta})$ given X.

(ii) if the loss is measured by quadratic error and if there is an estimator with finite Bayes risk, then $E(g(\tilde{\theta}) | X)$ is, modulus λP null sets, the unique Bayes solution.

Note that the Bayes risk of 0 is $\int |g| \, \mathrm{d}\lambda$ under (i), while it is $\int g^2 \, \mathrm{d}\lambda$ under (ii). Thus the existence conditions in (i) and (ii) are satisfied if, respectively, $\int |g| \, \mathrm{d}\lambda < \infty$ and $\int g^2 \, \mathrm{d}\lambda < \infty$.

Suppose now that loss is measured by quadratic error and that $\int g^2 \, \mathrm{d}\lambda < \infty$. Then minimum Bayes risk is

$$E \operatorname{Var}(g(\tilde{\theta}) | X) = \operatorname{Var} g(\tilde{\theta}) - \operatorname{Var}(E(g(\tilde{\theta}) | X)).$$

Here $\operatorname{Var} g(\tilde{\theta})$ would be the minimum Bayes risk if no observations were available, or if $\mathscr{E}$ was totally non-informative. Thus the quantity $\operatorname{Var}(E(g(\tilde{\theta}) | X))$ may be considered as a measure of the increase of precision, and hence information, we obtain by observing X.

Assume in addition that the Bayes solution $\phi(X) = E(g(\tilde{\theta}) | X)$ may be specified unbiased a.e. λ. In other words we require that $E_\theta\phi(X) = g(\theta)$ for λ almost all θ. Then $E\phi(X)^2 = E[E(\phi(X)g(\tilde{\theta})|X)] = E\phi(X)g(\tilde{\theta}) = E[E(\phi(X)g(\tilde{\theta})|\tilde{\theta})] = Eg(\tilde{\theta})^2$ so that $\int E_\theta(\phi(X) - g(\theta))^2\lambda(d\theta) = E(\phi(X) - g(\tilde{\theta}))^2 = E\phi(X)^2 - 2E\phi(X)g(\tilde{\theta}) + Eg(\tilde{\theta})^2 = 0$. Thus there is a λ-null set N such that $\phi(X) = g(\theta)$ a.s. P_θ when $\theta \notin N$. This implies that P_{θ_1} and P_{θ_2} are mutually singular whenever $\theta_1, \theta_2 \notin N$ and $g(\theta_1) \neq g(\theta_2)$. It follows that the Bayes solution cannot be almost everywhere unbiased if $P_{\theta_1} \wedge P_{\theta_2} \neq 0$ when $\theta_1 \neq \theta_2$ and g is not λ-equivalent to any constant.

An experiment $\mathscr{E} = (P_\theta : \theta \in \Theta)$ is called *pairwise imperfect* if $P_{\theta_1} \wedge P_{\theta_2} \neq 0$ when $\theta_1 \neq \theta_2$. Thus if $\mathscr{E}$ is pairwise imperfect then the Bayes solution $E(g(\tilde{\theta})|X)$ cannot be unbiased unless g is λ-equivalent to a constant.

Example 4.4.4 (Testing theory. Continuation of example 4.2.2). The risk function of a test δ in the two-decision case is $r(\theta : \delta) = L_\theta(t_0)(1 - E_\theta\delta) + L_\theta(t_1)E_\theta\delta$; $\theta \in \Theta$. It follows that the Bayes risk may be written

$$r(\lambda : \rho) = E[(1 - \delta(X))E(L_{\tilde{\theta}}(t_0)|X) + \delta(X)E(L_{\tilde{\theta}}(t_1)|X)].$$

Hence a test δ is a Bayes solution if and only if, up to null sets, $\delta(X) = 1$ or $=0$ according to whether $E(L_{\tilde{\theta}}(t_0)|X)$ is greater than or is less than $E(L_{\tilde{\theta}}(t_1)|X)$.

For $i = 0, 1$, if $L_\theta(t_i) = 0$ or $=1$ as $\theta \in \Theta_i \cup \Theta_I$ or $\theta \notin \Theta_i \cup \Theta_I$ then δ is a Bayes solution if and only if, up to null sets, $\delta(X) = 1$ or $=0$ according to whether $Pr(\tilde{\theta} \in \Theta_1|X)$ is greater than or less than $Pr(\tilde{\theta} \in \Theta_0|X)$.

Finally let us consider the case of a completely specified null hypothesis $\Theta_0 = \{\theta_0\}$ and a completely specified alternative $\Theta_1 = \{\theta_1\}$. A hypothesis is called *completely specified* or *simple* if it is equivalent to a statement which singles out one particular value of θ as the 'true' value of θ. Assume for simplicity that Θ is the two-point set $\{\theta_0, \theta_1\}$ so that the region Θ_I of indifference is empty. Let μ be a σ-finite measure dominating $\mathscr{E}$, say $\mu = P_{\theta_0} + P_{\theta_1}$, and let f_{θ_i}; $i = 0, 1$ be the density of P_{θ_i} with respect to μ. Then $\delta(X)$ is a Bayes solution if and only if, up to null sets, $\delta(X) = 1$ or $=0$ according to whether $\lambda(\theta_1)f_{\theta_1}(X)$ is greater than or less than $\lambda(\theta_0)f_{\theta_0}(X)$. Minimum Bayes risk becomes

$$\int (\lambda(\theta_0)f_{\theta_0} \wedge \lambda(\theta_1)f_{\theta_1})\, d\mu = \|\lambda(\theta_0)P_{\theta_0} \wedge \lambda(\theta_1)P_{\theta_1}\|$$

where $\|\ \ \|$ denotes total variation, and where $\lambda(\theta_0)P_{\theta_0} \wedge \lambda(\theta_1)P_{\theta_1}$ is the largest measure which is $\leqslant \lambda(\theta_0)P_{\theta_0}$ and is $\leqslant \lambda(\theta_1)P_{\theta_1}$. *This quantity may also be considered as the minimum probability of making a wrong guess of the true value for the prior* λ.

It follows directly from the Neyman–Pearson lemma that the Bayes solution $\delta(X)$ is a most powerful level α test for testing 'θ_0' against 'θ_1' where $\alpha = E_{\theta_0}\delta$. In general the value of α is not determined by λ alone, but also by the P_{θ_0} integral of δ on the set $[\lambda(\theta_1)f_{\theta_1} = \lambda(\theta_0)f_{\theta_0}]$. However, this set has μ-measure zero except for at most a countable set of priors λ.

Conversely, by the Neyman–Pearson lemma again, any most powerful level α test is a Bayes solution for a suitably chosen prior distribution λ.

Example 4.4.5 (Multiple decision problems. Continuation of example 4.2.3). We saw that the risk of a decision procedure ρ at θ was $E_\theta \sum_t L_\theta(t)\rho(t|\cdot)$. Hence

$$r(\lambda:\rho) = E\sum_t L_{\tilde{\theta}}(t)\rho(t|X) = E\sum_t (EL_{\tilde{\theta}}(t)|X)\rho(t|X) \geqslant E\bigwedge_{t'} E(L_{\tilde{\theta}}(t)|X)$$

where the last term is the minimum Bayes risk. The decision rule ρ is a Bayes solution if and only if $\rho(\cdot|X)$ almost everywhere assigns probability 1 to the set $\{t: E(L_{\tilde{\theta}}(t)|X) = \bigwedge_{t'} E(L_{\tilde{\theta}}(t')|X)\}$.

Let us consider the particular case of multiple classification. Suppose $\{\Theta_t: t \in T\}$ is a measurable partition of Θ. Put $L_\theta(t) = 0$ or $=1$ according to whether $\theta \in \Theta_t$ or not. Then $E(L_\theta(t)|X) = Pr(\tilde{\theta} \notin \Theta_t|X) = 1 - Pr(\tilde{\theta} \in \Theta_t|X)$. Thus a Bayes solution selects a $t \in T$ such that the set Θ_t has maximal posterior probability. Now assume in addition that $\Theta = T$ and that $\Theta_t = \{t\}$. Then $L_\theta(\theta') = 0$ or $=1$ according to whether $\theta' = \theta$ or $\theta' \neq \theta$. Let f_θ, for each θ, be a density of P_θ with respect to some σ-finite measure μ dominating $\mathscr{E}$. Then a non-randomized decision rule (estimator) $\hat{\theta}(X)$ is a Bayes solution if and only if

$$\lambda(\hat{\theta}(X))f_{\hat{\theta}(X)}(X) = \max_\theta \lambda(\theta)f_\theta(X) \quad \text{a.e. } \lambda P.$$

If λ is the uniform distribution on Θ, i.e. if $\lambda(\theta) = 1/\#\Theta$ when $\theta \in \Theta$, then the non-randomized estimator is a Bayes solution if and only if

$$f_{\hat{\theta}(X)}(X) = \max_\theta f_\theta(X) \quad \text{a.e. } P_\theta;\ \theta \in \Theta.$$

An estimator having this property, whether Θ is finite or not, is called a *maximum likelihood estimator* of θ. Hence in this case a non-randomized estimator is a Bayes solution if and only if it is a maximum likelihood estimator.

We shall need an expression for the minimum Bayes risk in the case of a general prior λ, when Θ is finite and $L_\theta(\theta') = 0$ or $=1$ as $\theta = \theta'$ or not. We find

$$\min_\rho r(\lambda:\rho) = 1 - \int\left(\bigvee_\theta \lambda_\theta f_\theta\right)\mathrm{d}\mu = 1 - \left\|\bigvee_\theta \lambda_\theta P_\theta\right\|$$

where $\|\ \|$ denotes total variation and $\bigvee_\theta \lambda_\theta P_\theta$ is the smallest measure which is $\geqslant \lambda_\theta P_\theta$ for all $\theta \in \Theta$.

4.5 Generalized decision rules and transitions

It is mathematically unsatisfactory that limits of risk functions may not be dominated by any risk function. In this section we shall see how this problem may be overcome, following LeCam, by generalizing the concept of a decision rule. Before doing so we should observe that the scope of this problem is in no way limited to decision problems. Mathematically, decision rules are simply Markov kernels, and the purpose of this section is to study certain linear maps extending the notion of a Markov kernel. As we shall see these maps will be of great help in many situations where it is not natural to consider them as related to particular decision problems.

Throughout this section we shall assume that we are given an experiment $\mathscr{E} = (\mathscr{X}, \mathscr{A}; P_\theta : \theta \in \Theta)$ and a decision space $(T, \mathscr{S})$. Equipping the linear space of bounded measurable functions on T with the supremum norm we obtain a Banach space which will be denoted by $M(T, \mathscr{S})$.

The generalized decision rules may be considered as:

> linear maps from $L(\mathscr{E})$ to the linear space of bounded additive set functions on $\mathscr{S}$;

or as:

> additive $M(\mathscr{E})$-valued probability set functions on $\mathscr{S}$;

or as:

> bilinear maps on the product space $L(\mathscr{E}) \times M(T, \mathscr{S})$.

The decision rules of the Markov kernel variety have the particular property that they map measures in $L(\mathscr{E})$ into σ-additive set functions. Generalized decision rules possessing this property will be called *σ-continuous*. Thus, generalized decision rules which are definable in terms of Markov kernels are σ-continuous. The converse, however, does not hold, not even in the case where Θ is a one-point set. In fact any probability space $(T, \mathscr{S}, P)$ together with a sub σ-algebra $\mathscr{S}_0$ of $\mathscr{S}$ such that there is no regular conditional probability distribution given $\mathscr{S}_0$, provides a counter example. If the decision space is a k-decision space where $k < \infty$ then any generalized decision rule is σ-continuous. However, it is only when the experiment is coherent that we can be sure that all generalized decision rules are definable in terms of Markov kernels when $k \geqslant 2$.

The second interpretation of a generalized decision rule indicates how it may be expressed by kernel-like functions. Thus if $\mathscr{E}$ is coherent and the decision space is Euclidean then any σ-continuous generalized decision rule is definable by a Markov kernel. If $\mathscr{E}$ is Σ-dominated and null set complete then this extends to even more general decision spaces.

Later on we shall study situations where the performance function of a generalized decision rule may be considered as an approximation of another experiment $\mathscr{F}$ with the decision space as sample space. In this situation it can always be assumed that the generalized decision rule maps $L(\mathscr{E})$ into $L(\mathscr{F})$ and, therefore, is σ-continuous.

The proof of this assertion is based on some basic facts on the linear space of bounded additive set functions. These facts are best comprehended by considering them within the framework of abstract vector lattices. Therefore they will be deduced in the next chapter which provides an elementary introduction to vector lattices. The reader who is not acquainted with the notion of an abstract vector lattice can just accept this result here.

It will become apparent from the interpretation of a generalized decision rule as a bilinear functional, that the set of all generalized decision rules is compact for the topology of pointwise convergence on $L(\mathscr{E}) \times M(T, \mathscr{S})$. We shall see that the set of 'uniformly finitely supported' decision rules constitutes a dense subset of the set of all generalized decision rules for this topology. It follows that the set of generalized decision rules may be considered as the completion of the set of decision rules in roughly the same sense as the set of real numbers may be considered as the completion of the set of rational numbers. (In the first case the completion is with respect to the uniformity of pointwise convergence – in the latter with respect to the distance metric.)

The performance function of a generalized decision rule is defined as before. The risk is then obtained by integrating the loss with respect to the performance function. This definition of risk, as well as some of the other considerations in this section, necessitates an easy excursion into the realm of integration with respect to bounded additive set functions. The extension to integrals of functions which are bounded below is obtained by approximation from below by bounded functions. As pointed out by LeCam, the definition ensures that the risk is an affine and lower semicontinuous function of the generalized decision rule.

Let us begin by reminding ourselves that a *Markov kernel ρ from a measurable space $(\mathscr{X}, \mathscr{A})$ to a measurable space $(T, \mathscr{S})$* is a map $x \to \rho(\cdot|x)$ from the set $\mathscr{X}$ to the set of probability measures on $\mathscr{S}$ such that $\rho(S|x)$ is measurable in x for each set S in $\mathscr{S}$. In section 4.2 we defined a decision rule ρ from the experiment $\mathscr{E} = (\mathscr{X}, \mathscr{A}; P_\theta : \theta \in \Theta)$ to the decision space $(T, \mathscr{S})$ as a Markov kernel from $(\mathscr{X}, \mathscr{A})$ to $(T, \mathscr{S})$. The probability measure P_θ did not enter into this definition. If λ is any finite measure on $\mathscr{A}$ then ρ maps λ into the finite measure $\lambda\rho$ on $\mathscr{S}$. We shall find it convenient to use the notation $\text{ca}(T, \mathscr{S})$ for the linear space of finite measures on $\mathscr{S}$.

Setwise limits of uniformly bounded sequences of measures in $\text{ca}(T, \mathscr{S})$ are bounded as well as additive. However, they are not necessarily measures, i.e. countably additive.

For this and other reasons we shall need a few facts on bounded additive set functions and the integrals they determine. It will be shown in the next chapter how these facts follow easily from elementary results on vector lattices. However, here we have attempted to make this exposition independent of vector lattice theory. The material developed below is more or less covered in most introductory texts on measure theory.

Let $\mathrm{ba}(T, \mathscr{S})$ denote the linear space of bounded additive set functions on $\mathscr{S}$. The symbol S, with or without subscripts, in this section will denote a set in $\mathscr{S}$. Consider a particular set function μ in $\mathrm{ba}(T, \mathscr{S})$. Put

$$\mu^+(S) = \sup\{\mu(S') : S' \subseteqq S; S \in \mathscr{S}\}.$$

Let S_1 and S_2 be disjoint. Then $\mu^+(S_1 \cup S_2) = \sup\{\mu(S') : S' \subseteqq S\} = \sup\{\mu(S_1' \cup S_2') : S_1' \subseteqq S_1, S_2' \subseteqq S_2\} = \sup\{\mu(S_1') + \mu(S_2') : S_1' \subseteqq S_1, S_2' \subseteqq S_2\} = \mu^+(S_1) + \mu^+(S_2)$. It follows that $\mu^+ \in \mathrm{ba}(T, \mathscr{S})$. Clearly $\mu^+ \geqslant \mu$ and $\mu^+ \geqslant 0$. If v is any other additive set function on $\mathscr{S}$ such that $v \geqslant \mu$ and $v \geqslant 0$ then $v \geqslant \mu^+$. Hence μ^+ *is the smallest non-negative measure which majorizes* μ *for the setwise ordering.* Also put $\mu^- = (-\mu)^+$ and $|\mu| = \mu^+ + \mu^-$. Then

$$-\mu^-(S) = \inf\{\mu(S') : S' \subseteqq S\} = \inf\{\mu(S - S') : S' \subseteqq S\} = \mu(S) - \mu^+(S).$$

It follows that μ may be decomposed as a difference $\mu = \mu^+ - \mu^-$ of the non-negative set functions μ^+ and μ^- in $\mathrm{ba}(T, \mathscr{S})$. This decomposition is called the *Jordan decomposition of* μ. It is minimal in the sense that if $\mu = \mu_1 - \mu_2$ where μ_1 and μ_2 are non-negative set functions in $\mathrm{ba}(T, \mathscr{S})$ then $\mu_1 \geqslant \mu^+$ and $\mu_2 \geqslant \mu^-$.

It may be checked that $|\mu|$ *is the smallest set function in* $\mathrm{ba}(T, \mathscr{S})$ *which majorizes both* μ *and* $-\mu$.

If $S \in \mathscr{S}$ then $\mu^+(S)$, $\mu^-(S)$ and $|\mu|(S)$ are called, respectively, *the upper variation of* μ *at* S, *the lower variation of* μ *at* S and *the total variation of* μ *at* S. If $S = T$ then the 'at S' may be dropped. Thus the numbers $\mu^+(T)$, $\mu^-(T)$ and $|\mu|(T)$ are called, respectively, *the upper variation of* μ, *the lower variation of* μ and *the total variation of* μ.

The total variation of μ at S is the supremum of all numbers of the form $\sum_{i=1}^{r} |\mu(S_i)|$ where $(S_1, S_2, \ldots, S_r)$ is a partition of S. Actually it suffices to consider the case $r = 2$, i.e. we may write

$$|\mu|(S) = \sup\{|\mu(S')| + |\mu(S - S')| : S' \subseteqq S\}.$$

The total variation of μ is also called *the total variation norm of* μ and may be denoted by $\|\mu\|$. Thus

$$\|\mu\| = |\mu|(T).$$

The terminology is justified since $\|\cdot\|$ actually defines a norm on $\mathrm{ba}(T, \mathscr{S})$, i.e.

(i) $\|\mu\| = 0 \Leftrightarrow \mu = 0$;
(ii) $\|\mu_1 + \mu_2\| \leqslant \|\mu_1\| + \|\mu_2\|$ when $\mu_1, \mu_2 \in \mathrm{ba}(T, \mathscr{S})$;
(iii) $\|c\mu\| = |c| \, \|\mu\|$ when $c \in \mathbb{R}$ and $\mu \in \mathrm{ba}(T, \mathscr{S})$.

The total variation norm is equivalent to the supremum norm since

(iv) $\sup_S |\mu(S)| \leqslant \|\mu\| \leqslant 2\sup_S |\mu(S)|$.

The inequalities in (iv) imply that the linear space $\mathrm{ba}(T, \mathscr{S})$, equipped with the total variation norm, is a complete normed space, i.e. a Banach space. Note that $\|\mu\| = \mu(T)$ whenever $\mu \geqslant 0$.

How do we integrate with respect to a bounded additive set function? If $g = \sum_{i=1}^n c_i I_{S_i}$ where $c_1, c_2, \ldots, c_n \in \mathbb{R}$ then, of course, we put $\int g \, \mathrm{d}\mu = \sum_{i=1}^n c_i \mu(S_i)$. As is shown in the beginning of most textbooks on measure theory, this defines the integral as a linear functional on the linear space $M_0(T, \mathscr{S})$ of $\mathscr{S}$-measurable step functions. Now $M(T, \mathscr{S})$ is a Banach space for the supremum norm and $M_0(T, \mathscr{S})$ is a dense subset of $M(T, \mathscr{S})$. If g is a step function then $|\int g \, \mathrm{d}\mu| \leqslant \|g\| \, \|\mu\|$ where $\|g\| = \sup_t |g(t)|$ is the supremum norm of g. Thus the integral as a function of g is a bounded linear functional on $M_0(T, \mathscr{S})$ and thus has a unique extension as a bounded linear functional on $M(T, \mathscr{S})$. We define *the integral of the bounded measurable function g on $(T, \mathscr{S})$ with respect to the bounded additive set function μ as the value of this functional at g*. As usual, this integral will be denoted as $\int g \, \mathrm{d}\mu$. We shall also feel free to use notation such as $g\mu$, $\mu(g)$ and μg for this integral.

Let $\mu \in \mathrm{ba}(T, \mathscr{S})$. Then

$$\|\mu\| = \sup\{|\mu(g)| : g \in M(T, \mathscr{S}); \|g\| \leqslant 1\}.$$

It follows that the total variation of μ equals the norm of the linear functional μ defined, by its integral, on $M(T, \mathscr{S})$. For the moment denote this functional by Φ_μ. Any bounded linear functional Φ on $M(T, \mathscr{S})$ is of the form Φ_μ where $\mu(S) = \Phi(I_S)$ when $S \in \mathscr{S}$. *It follows that the map: $\mu \to \Phi_\mu$ is an isometry from* $\mathrm{ba}(T, \mathscr{S})$ *onto the dual space $M(T, \mathscr{S})^*$ of $M(T, \mathscr{S})$*. (The dual space V^* of a normed space V is the linear space of bounded linear functionals on V equipped with the dual norm

$$\|v^*\| = \sup\{|v^*(v)| : v \in V; \|v\| \leqslant 1\}; \qquad v^* \in V^*.$$

Here $\|v\|$ denotes the norm of the vector v in V. It is easily checked that the dual space V^* is complete, whether V is complete or not.)

The integral $\int g \, \mathrm{d}\mu$ clearly defines a bilinear functional on the product space $M(T, \mathscr{S}) \times \mathrm{ba}(T, \mathscr{S})$. This bilinear functional is non-negative in the sense that $\int g \, \mathrm{d}\mu \geqslant 0$ when $g \geqslant 0$ and $\mu \geqslant 0$.

Note also that the assumption that $\mathscr{S}$ is a σ-algebra can be relaxed to the assumption that $\mathscr{S}$ is an algebra provided $M(T, \mathscr{S})$ is replaced by the closure of the linear space $M_0(T, \mathscr{S})$ of step functions with steps in $\mathscr{S}$. If $\mathscr{S}$ is a σ-algebra then, as remarked above, this closure is $M(T, \mathscr{S})$.

Next let us consider integrals of functions which are not necessarily bounded. We shall restrict our attention to the case where the function g we are integrating is measurable, extended real valued and bounded from below, and where the set function μ we are integrating with respect to is bounded, additive and non-negative. Then we define the integral of g with respect to μ as the supremum of all numbers $\int h \, d\mu$ where $g \geq h \in M(T, \mathscr{S})$. This integral may also be denoted as $\int g \, d\mu$, $g\mu$, μg or $\mu(g)$. It suffices to consider the bounded measurable functions $g \wedge n$; $n = 1, 2, \ldots$ since

$$\int g \, d\mu = \sup_n \int (g \wedge n) \, d\mu.$$

This implies that this notion of an integral is consistent with the usual one, when μ is a measure.

Let $g, g_1, g_2, \ldots, g_r$ be measurable functions which are bounded from below, let $\mu, \mu_1, \mu_2, \ldots, \mu_r$ be bounded non-negative additive set functions and let $c_1, c_2, \ldots, c_r$ be non-negative constants. Then

$$\int \left(\sum_i c_i g_i \right) d\mu = \sum_i c_i \int g_i \, d\mu$$

and

$$\int g \, d \sum c_i \mu_i = \sum_i c_i \int g \, d\mu_i.$$

The first equation may be proved by using the fact that if $h, h_1, \ldots, h_r$ are real valued non-negative measurable functions on T such that $h \leq \sum_{i=1}^r h_i$, then $h = \sum_{i=1}^r \tilde{h}_i$ where the functions $\tilde{h}_1, \ldots, \tilde{h}_r$ are measurable and $0 \leq \tilde{h}_i \leq h_i$; $i = 1, 2, \ldots, r$.

Considering first step functions and then bounded functions, we obtain the well known formula

$$\int g \, d\mu = \int_0^\infty \mu(g \geq x) \, dx - \int_{-\infty}^0 \mu(g \leq x) \, dx$$

which is valid whenever the additive set function μ and the measurable function g are both bounded. If μ is non-negative then this extends to any measurable function g which is bounded from below.

We shall say that a *net* (μ_n) *in* $\mathrm{ba}(T, \mathscr{S})$ *converges setwise to* μ in $\mathrm{ba}(T, \mathscr{S})$ if $\lim_n \mu_n(S) = \mu(S)$ when $S \in \mathscr{S}$. This is just the notion of convergence for the topology of setwise convergence in $\mathrm{ba}(T, \mathscr{S})$. Since the space of measurable step functions is dense in $M(T, \mathscr{S})$, it follows that the net (μ_n) in $\mathrm{ba}(T, \mathscr{S})$ converges setwise to μ in $\mathrm{ba}(T, \mathscr{S})$ if and only if $\lim_n \mu_n(g) = \mu(g)$ when $g \in M(T, \mathscr{S})$. If g is measurable and bounded from below then this implies that the integral $\int g \, d\mu$ is a lower semicontinuous function of μ on the cone of non-negative set functions in $\mathrm{ba}(T, \mathscr{S})$.

As mentioned before, one reason for wanting to go beyond σ-additivity was that setwise limits of measures may not be countably additive. On the other hand, by Tychonoff's theorem, it is clear that closed and bounded sets of additive set functions are compact. The topology is then the topology of setwise convergence, and boundedness is with respect to the total variation norm.

We shall see that the closure of the space $\mathrm{ca}(T,\mathscr{S})$ of finite measures is actually $\mathrm{ba}(T,\mathscr{S})$. This is a particular case of a fairly trivial approximation result which holds for any, not necessarily real valued, additive set function on $\mathscr{S}$. The minimal requirement we are forced to meet, in order that the notion of an additive set function should be meaningful, is that the range of the set function is a commutative semigroup $(K, +)$ with a null element 0 (i.e. $k + 0 = k$ whenever $k \in K$).

A set function μ from $\mathscr{S}$ to K is called additive if $\mu(\emptyset) = 0$ and if $\mu(S_1 \cup S_2) = \mu(S_1) + \mu(S_2)$ when S_1 and S_2 are disjoint sets in $\mathscr{S}$. The notation for induced distributions carries over to finitely additive set functions as follows. Let μ_1 be an additive set function on the σ-algebra $\mathscr{S}_1$ of subsets of the set T_1. Let τ be a measurable map from the measurable space $(T_1, \mathscr{S}_1)$ into the measurable space $(T_2, \mathscr{S}_2)$. Define a set function μ_2 on $\mathscr{S}_2$ by putting, for each $S_2 \in \mathscr{S}_2$, $\mu_2(S_2) = \mu_1(\tau^{-1}[S_2])$. Then μ_2 is an additive set function on $\mathscr{S}_2$. The usual notation for the set function μ_2 is $\mu_1\tau^{-1}$ and this set function is called *the set function on $\mathscr{S}_2$ induced from the set function μ_1 by the map τ*. The set function $\mu_1\tau^{-1}$ is a bounded (non-negative) additive set function whenever μ_1 is a bounded (non-negative) additive set function. The change of variables rule extends as it should since

$$\mu_1(g \circ \tau) = \mu_1\tau^{-1}(g)$$

whenever μ_1 is a bounded (non-negative) additive set function and g is a measurable function on $(T,\mathscr{S})$ which is bounded (from below).

The map $\mu_1 \to \mu_1\tau^{-1}$ from $\mathrm{ba}(T_1,\mathscr{S}_1)$ into $\mathrm{ba}(T_2,\mathscr{S}_2)$ is linear, non-negative and preserves total mass. Later on we shall extend the notion of a transition so that such maps, without reservation, may be called transitions. Considering the map $\mu_1 \to \mu_1\tau^{-1}$ as an extension of the map τ we shall occasionally find it convenient to write $\mu_1\tau$ instead of $\mu_1\tau^{-1}$.

The simplest additive set functions are those which are supported by finite sets. We shall now see that any additive set function is the setwise limit of a net of additive set functions with finite supports. This implies that problems involving general additive set functions may sometimes be reduced to the same problems for additive set functions which are supported by finite sets.

An additive set function μ on $\mathscr{S}$ is said to be *supported by the finite subset T_0 of T* if $\mu(S) = 0$ whenever $S \in \mathscr{S}$ and $S \cap T_0 = \emptyset$. If the additive set function μ on $\mathscr{S}$ is supported by some finite set then we shall say that μ is *finitely supported*. Note that if τ is a measurable map from the measurable space

$(T_1, \mathscr{S}_1)$ to the measurable space $(T_2, \mathscr{S}_2)$ and τ has finite range, then $\mu_1 \tau^{-1}$ is finitely supported whenever μ_1 is an additive set function on $\mathscr{S}_1$.

It may be checked that an additive and K-valued set function μ is supported by the finite set T_0 if and only if there are quantities $m(t) : t \in T_0$ in K such that

$$\mu(S) = \sum \{m(t) : t \in ST_0\}.$$

Finitely supported additive set functions are clearly σ-additive provided we agree that $k_1 + k_2 + \cdots + k_n + 0 + 0 + \cdots = k_1 + \cdots + k_n$ when $k_1, k_2, \ldots, k_n \in K$.

Let us consider how general additive set functions may be approximated by finitely supported additive set functions. One way to proceed is to partition T by very fine finite partitions. Let N denote the set of all pairs $n = (\Pi, \xi)$ where $\Pi = (S_1, \ldots, S_r)$ is an ordered partition of T into $\mathscr{S}$-measurable sets and $\xi = (t_1, \ldots, t_r)$ is an r-tuple such that $t_i \in S_i$; $i = 1, \ldots, r$. Direct N by defining $n_1 = (\Pi_1, \xi_1) \geqslant n_2 = (\Pi_2, \xi_2)$ to mean that Π_1 is at least as fine as Π_2, i.e. each set in the partition defining Π_1 is contained in some set in the partition defining Π_2. For each $t \in T$ and each $n \in N$, put $\sigma_n(t) = t_i$ when $n = ((S_1, \ldots, S_r), (t_1, \ldots, t_r))$ and $t \in S_i$. Let $\mathscr{S}_0$ be a finite sub-algebra of $\mathscr{S}$ and assume $n_0 = (\pi_0, \xi_0)$ where Π_0 is at least as fine as the partition defined by $\mathscr{S}_0$. Then $g \circ \sigma_n = g$ whenever g is $\mathscr{S}_0$-measurable and $n \geqslant n_0$. In fact $\|g \circ \sigma_n - g\| \to 0$ if and only if g is bounded and measurable. This implies of course that $\mu\sigma_n^{-1} g \to \mu g$ for any bounded additive set function μ and any bounded measurable function g.

If $\mu \geqslant 0$ and g is bounded from below and if $\mu(g > k) > 0$ for all constants k then $\sum_i g(t_i)\mu(S_i)$ may be made arbitrarily large for any given partition $(S_1, \ldots, S_r)$. Thus convergence does not take place in this case unless $\int g \, d\mu = \infty$. In order to remedy this, we may modify the construction as follows.

Let D consist of all pairs $d = (G, \varepsilon)$ where $G = \{g : g \in G\}$ is a finite set of measurable functions which are all bounded from below, and $\varepsilon > 0$. Order D by defining $d_1 = (G_1, \varepsilon_1) \geqslant d_2 = (G_2, \varepsilon_2)$ when $G_1 \supseteqq G_2$ and $\varepsilon_1 \leqslant \varepsilon_2$. Then '$\geqslant$' converts D into a directed set. Let $d = (G, \varepsilon) \in D$. Then the boundedness condition implies that there is a finite subset T_d of T such that to each $t \in T$ there corresponds an element $\tau_d(t) \in T_d$ such that $g(t) \geqslant g(\tau_d(t)) - \varepsilon$ when $g \in G$. We can always arrange things so that τ_d becomes measurable as a map from $(T, \mathscr{S})$ to the set T_d equipped with the (σ-)algebra of all subsets. Then τ_d is also measurable as a map from $(T, \mathscr{S})$ to $(T, \mathscr{S})$.

The construction can be carried out as follows. For each subset U of G put $T_U = \{t : g(t) < \infty \text{ or } = \infty \text{ as } g \in U \text{ or } g \notin U\}$. For each $t \in T_U$ put $w_U^t = (g(t) : g \in U)$. Also put $W_U = \{w_U^t : t \in T_U\}$. Then $W_U \subseteqq \,]-\infty, \infty]^U$. By the boundedness condition, the closure of W_U in $]-\infty, \infty]^U$ is compact. Finally, for each $t \in T_U$ put $Z_t = \prod \{]g(t) - \varepsilon, \infty] : g \in U\}$. Then $\{Z_t : t \in T_U\}$ is an open covering of the closure of W_U. It follows that $W_U \subseteqq \bigcup \{Z_t : t \in T_U^0\}$ where

T_U^0 is a finite subset of T_U. Let $t \in T_U$. Then there is a $t^0 \in T_U^0$ such that $w^t \in Z_{t^0}$. Hence $g(t) > g(t^0) - \varepsilon$ when $g \in U$, so that $g(t) \geqslant g(t^0) - \varepsilon$ for all $g \in G$. The construction is now completed by putting $T_d = \bigcup \{T_U^0 : U \subseteqq G\}$.

Suppose g is bounded and that $d_0 = (G_0, \varepsilon_0)$ where $-g, g \in G_0$. Then $\|g - g \circ \tau_d\| \leqslant \varepsilon_0$ when $d \geqslant d_0$. Again it follows that to each finite subalgebra $\mathscr{S}_0$ of $\mathscr{S}$ there corresponds a $d_0 \in D$ such that $g \circ \tau_d = g$ whenever $d \geqslant d_0$ and g is constant on $\mathscr{S}_0$ atoms. If b is a finite constant and $d \geqslant d_0 = (G_0, \varepsilon_0)$ where $G_0 = \{g, -(g \wedge b)\}$, then $g \wedge b - \varepsilon_0 \leqslant g \circ \tau_d \leqslant g + \varepsilon_0$. Altogether this proves the following proposition.

Proposition 4.5.1. *The set of finitely supported set functions in* $\text{ba}(T, \mathscr{S})$ *is dense for the topology of setwise convergence.*

The set of non-negative and finitely supported set functions in $\text{ba}(T, \mathscr{S})$ *constitutes a dense subset of the set of non-negative set functions in* $\text{ba}(T, \mathscr{S})$ *for the topology of pointwise convergence on the set of measurable functions which are bounded from below.*

Remark. Let L_0 and M_0 be separable subsets of $\text{ba}(T, \mathscr{S})$ and $M(T, \mathscr{S})$ respectively.

Then the proposition implies that there are sequences of the form $\sigma_{n_1}, \sigma_{n_2}, \ldots$ and of the form $\tau_{d_1}, \tau_{d_2}, \ldots$ such that $\lim_{i \to \infty} \lambda \sigma_{n_i}^{-1} g = \lambda g$ and $\lim_{i \to \infty} \lambda \tau_{d_i}^{-1} g = \lambda g$ when $\lambda \in L_0$ and $g \in M_0$.

This is by and large all we need to know about finitely additive set functions.

Returning to the main theme we observe that the map from $L(\mathscr{E})$ to $\text{ba}(T, \mathscr{S})$ defined by a decision rule has the following easily established properties.

Theorem 4.5.2. *Let ρ in the statements below designate the linear map: $\lambda \to \lambda\rho$ from $L(\mathscr{E})$ to* $\text{ba}(T, \mathscr{S})$ *defined by the decision rule ρ from the experiment $\mathscr{E} = (P_\theta : \theta \in \Theta)$ to the decision space $(T, \mathscr{S})$. Then:*

(i) *ρ is non-negative, i.e. $\lambda\rho \geqslant 0$ when $0 \leqslant \lambda \in L(\mathscr{E})$.*
(ii) *ρ preserves total mass, i.e. $(\lambda\rho)(T) = \lambda(\mathscr{X})$ when $\lambda \in L(\mathscr{E})$.*
(iii) *ρ maps $L(\mathscr{E})$ into* $\text{ca}(T, \mathscr{S})$.

Furthermore, two decision rules ρ_1 and ρ_2 define the same map from $L(\mathscr{E})$ to $\text{ba}(T, \mathscr{S})$ *if and only if $\rho_1(S|\cdot) = \rho_2(S|\cdot)$ a.e. P_θ when $\theta \in \Theta$.*

Using the terminology introduced in LeCam's 1964 paper we shall call any linear map ρ from $L(\mathscr{E})$ to $\text{ba}(T, \mathscr{S})$ which satisfies conditions (i) and (ii) of theorem 4.5.2 *a transition from $L(\mathscr{E})$ to* $\text{ba}(T, \mathscr{S})$ or just *a transition from $\mathscr{E}$ to* $(T, \mathscr{S})$. When we want to emphasize that we are dealing with a decision problem we can use the term *generalized decision rule* instead of the term 'transition'.

If a transition is defined by a decision rule ρ as described in the first sentence of the theorem then we shall say that *the transition is representable by (defined by)* ρ.

Transitions possessing property (iii), i.e. transitions which map measures in $L(\mathscr{E})$ into measures on $\mathscr{S}$, are here called *σ-continuous.*

Thus, by the theorem, any transition which is representable (defined) by a decision rule (Markov kernel) is σ-continuous. We shall see later that the converse holds provided the decision problem satisfies regularity conditions of a general nature.

We saw in section 4.2 that a decision rule ρ defined an adjoint map from $M(T, \mathscr{S})$ into the set $M(\mathscr{X}, \mathscr{A})$ of bounded measurable functions on $\mathscr{A}$. It also defined a bilinear functional $(\lambda, g) \to \lambda\rho g$ on $L(\mathscr{E}) \times M(T, \mathscr{S})$. The generalizations of these definitions to transitions are now straightforward. If ρ is a transition from $L(\mathscr{E})$ to $\text{ba}(T, \mathscr{S})$ then its adjoint maps $M(T, \mathscr{S})$ into $M(\mathscr{E})$. The image ρ^*g of a bounded measurable function g by the adjoint map ρ^* of ρ, is the linear functional: $\lambda \to (\lambda\rho)(g)$ on $L(\mathscr{E})$. If ρ is defined by a decision rule ρ then ρ^*g may be represented by the function $\rho g : x \to \int g(t)\rho(dt|x)$ in $M(\mathscr{X}, \mathscr{A})$. The definition of the adjoint map ρ^* of the transition ρ implies that $(\lambda\rho)(g) = \lambda(\rho^*g)$ and this number will usually be written $\lambda\rho g$. We shall usually delete the asterisk on ρ and write ρg instead of ρ^*g. Thus the notation ρg for a decision rule ρ and a bounded measurable function g should sometimes be interpreted as a bounded measurable function and sometimes as the functional in $M(\mathscr{E})$ defined by this function.

Before proceeding let us remind ourselves that a functional v in $M(\mathscr{E})$ is called *non-negative* if $\lambda(v) \geqslant 0$ when $0 \leqslant \lambda \in L(\mathscr{E})$. More generally if v_1 and v_2 are functionals in M then we shall write $v_1 \geqslant v_2$ (or $v_2 \leqslant v_1$) and say that v_1 *is at least as large as v_2 (or v_2 is at most as large as v_1)* if $v_1 - v_2$ is non-negative. Thus v is non-negative if and only if $v \geqslant 0$, and $v_1 \geqslant v_2$ if and only if $v_1 - v_2 \geqslant 0$. The relation '$\geqslant$' defined this way is the *ordering* of $M(\mathscr{E})$. This ordering is *compatible with the linear structure* in the sense that $c_1v_1 + c_2v_2 \geqslant 0$ when $c_1, c_2 \geqslant 0$ are real numbers and $v_1, v_2 \geqslant 0$ belong to $M(\mathscr{E})$.

If a family $v_t : t \in T$ in $M(\mathscr{E})$ possesses an upper bound in $M(\mathscr{E})$, then this family possesses a least upper bound in $M(\mathscr{E})$. The least upper bound is denoted as $\sup_t v_t$ or as $\bigvee_t v_t$. It is determined by

$$\lambda\left(\sup_t v_t\right) = \sup_t \lambda^+(v_t) - \sup_t \lambda^-(v_t); \qquad \lambda \in L(\mathscr{E}).$$

It follows from this, or directly, that if the family $v_t : t \in T$ in $M(\mathscr{E})$ possesses a lower bound then it possesses a greatest lower bound which we shall denote as $\inf_t v_t$ or as $\bigwedge_t v_t$. It is determined by

$$\lambda\left(\inf_t v_t\right) = \inf_t \lambda^+(v_t) - \inf_t \lambda^-(v_t); \qquad \lambda \in L(\mathscr{E}).$$

A bounded measurable function f on $(\mathscr{X}, \mathscr{A})$ determines a linear functional $\dot{f}: \lambda \to \lambda(f)$ on $L(\mathscr{E})$. We shall usually write f instead of $\dot{f}$. In particular the real constant c determines the constant function c on $(\mathscr{X}, \mathscr{A})$ as well as the functional $\dot{c}: \lambda \to c\lambda(\mathscr{X})$ on $L(\mathscr{E})$.

$M(\mathscr{E})$ becomes a Banach space for the dual norm $\|\cdot\|^*$ of the total variation norm $\|\cdot\|$. If f is a bounded measurable function on $(\mathscr{X}, \mathscr{A})$ then $\|\dot{f}\|^*$ is the smallest number a such that $P_\theta(|f| > a) = 0$ for all θ in Θ. If $f = (f_\theta : \theta \in \Theta)$ is a uniformly bounded and consistent family of measurable functions, then the norm of the functional in $M(\mathscr{E})$ determined by f is the smallest number a such that $P_\theta(|f_\theta| > a) = 0$ for all θ in Θ. Thus this norm is also equal to $\sup_\theta \|f_\theta : L_\infty(P_\theta)\|$ where $\|f_\theta : L_\infty(P_\theta)\|$ denotes the $L_\infty(P_\theta)$ norm of f_θ. It follows easily from these considerations that the dual norm of a functional v in $M(\mathscr{E})$ is the smallest constant a such that $\pm v \leqslant \dot{a}$. We shall usually omit the asterisk on the dual norm.

The defining properties of a transition may be expressed simply in terms of its adjoint.

Theorem 4.5.3. *Let ρ be a transition from $L(\mathscr{E})$ to $M(T, \mathscr{S})$ and let ρ^* be the restriction of the adjoint of ρ to $M(T, \mathscr{S})$. The linear map ρ^* from $M(T, \mathscr{S})$ to $M(\mathscr{E})$ has the following properties:*

(i) *ρ^* is non-negative, i.e. $\rho^* g \geqslant 0$ when $g \geqslant 0$.*
(ii) *$\rho^* 1 = \dot{1}$.*

Conversely any linear map ρ^* from $M(T, \mathscr{S})$ to $M(\mathscr{E})$ satisfying (i) *and* (ii) *is the adjoint of a unique transition ρ from $L(\mathscr{E})$ to* $\text{ba}(T, \mathscr{S})$.

Proof. Let $0 \leqslant g \in M(T, \mathscr{S})$ and $0 \leqslant \lambda \in L(\mathscr{E})$. Then, since ρ is non-negative, $\lambda(\rho^* g) = (\lambda\rho)g \geqslant 0$. Furthermore, $\lambda(\rho^* 1) = (\lambda\rho)(1) = (\lambda\rho)(T) = \lambda(\mathscr{X})$ since ρ preserves total mass. If the linear map ρ^* satisfies (i) and (ii), then it is the adjoint of the transition ρ which for each $\lambda \in L(\mathscr{E})$ maps λ into the set function $\lambda\rho : S \to \lambda(\rho^* I_S)$ on $\mathscr{S}$. □

This last theorem suggests that the image $\rho^* g$ in $M(\mathscr{E})$ of g in $M(T, \mathscr{S})$ by the adjoint ρ^* may be considered as being obtained by integrating g with respect to the $M(\mathscr{E})$-valued set function $S \to \rho^* I_S$. In order to make this precise, put $\kappa_\rho(S) = \rho^* I_S$ when ρ is a transition from $L(\mathscr{E})$ to $\text{ba}(T, \mathscr{S})$ and $S \in \mathscr{S}$. Call an $M(\mathscr{E})$-valued set function κ on $\mathscr{S}$ an *$M(\mathscr{E})$-valued probability set function* if it is non-negative, additive and maps the set T into $\dot{1}$. Thus κ is an $M(\mathscr{E})$-valued probability set function on $\mathscr{S}$ if and only if it is a map from $\mathscr{S}$ to $M(\mathscr{E})$ such that:

(i) $\kappa(S_1 \cup S_2) = \kappa(S_1) + \kappa(S_2)$ when $S_1, S_2 \in \mathscr{S}$ and $S_1 \cap S_2 = \emptyset$.
(ii) $\kappa(S) \geqslant 0$ when $S \in \mathscr{S}$.
(iii) $\kappa(T) = \dot{1}$.

It is now easily checked that for each transition ρ, κ_ρ is an $M(\mathscr{E})$-valued probability set function on $\mathscr{S}$.

Furthermore:

(iv) $\rho^*(\sum_i c_i I_{S_i}) = \sum_i c_i \kappa_\rho(S_i)$ when $c_1, c_2, \ldots, c_n \in \mathbb{R}$ and $S_1, S_2, \ldots, S_n \subset \mathscr{S}$.
(v) $\|\rho^* g\| \leqslant \|g\|$ when $g \in M(T, \mathscr{S})$.

Hence $\|\rho^* g_n - \rho^* g\| \to 0$ when (g_n) is a sequence in $M(T, \mathscr{S})$ converging uniformly to g. Thus $\rho^* g$ may be obtained as the limit, $\lim_n \rho^* g_n$, for any sequence (g_n) of $\mathscr{S}$-measurable step functions which converges uniformly to g.

It is perhaps abundantly clear at this point that we have created much ado about a simple fact which may be described as follows. Let $\mathscr{S}$ be an algebra of subsets of a set T and let V be a Banach space. Let $\overline{M}_0(T, \mathscr{S})$ be the closure, for the supremum norm, of the space of real valued step functions with steps in $\mathscr{S}$. Then the bounded additive and V-valued set functions on $\mathscr{S}$ may be identified with the bounded linear V-valued maps on $\overline{M}_0(T, \mathscr{S})$. If V is equipped with an order structure which is compatible with the linear structure on V, then we can make some additional remarks. Restricting our attention to the case $V = M(\mathscr{E})$ we have by and large proved the following theorem.

Theorem 4.5.4. *The map $\rho \to \kappa_\rho$ from the set of transitions to the set of $M(\mathscr{E})$-valued probability set functions is affine, $1-1$ and onto. If κ is an $M(\mathscr{E})$-valued probability set function on $\mathscr{S}$ then $\kappa = \kappa_\rho$ where $(\lambda\rho)(S) = \lambda(\kappa(S))$ when $S \in \mathscr{S}$ and $\lambda \in L(\mathscr{E})$.*

σ-continuity of transitions amounts to σ-continuity from above for the associated $M(\mathscr{E})$-valued probability set functions.

Theorem 4.5.5. *Let ρ be a transition from $L(\mathscr{E})$ to* $\text{ba}(T, \mathscr{S})$. *Then the following conditions are equivalent:*

(i) *ρ is σ-continuous.*
(ii) *if $g_1 \geqslant g_2 \geqslant \cdots$ is a monotonically decreasing sequence in $M(T, \mathscr{S})$ such that $\inf_n g_n = 0$ then $\inf_n \rho^*(g_n) = 0$.*
(iii) *if $S_1 \supseteqq S_2 \supseteqq \cdots$ is a monotonically decreasing sequence of sets in $\mathscr{S}$ such that $\bigcap_n S_n = \emptyset$ then $\inf_n \kappa_\rho(S_n) = 0$.*
(iv) *the set functions $P_\theta \rho : \theta \in \Theta$ are all σ-additive.*

Proof. Suppose ρ is σ-continuous and let the sequence $\{g_n\}$ be as in (ii). Let $0 \leqslant \lambda \in L(\mathscr{E})$. Then, since $\lambda^\pm \in \text{ca}(T, \mathscr{S})$

$$\lambda\left(\inf_n \rho^* g_n\right) = \inf_n \lambda^+ \rho g_n - \inf_n \lambda^- \rho g_n = 0 - 0 = 0.$$

Hence (i) $\Rightarrow$ (ii). The implication (ii) $\Rightarrow$ (iii) follows by specialization. Let $0 \leqslant \lambda \in L(\mathscr{E})$ and assume that (iii) holds. If the sequence $\{S_n\}$ in $\mathscr{S}$ is as in (iii), then

$$\inf_n \lambda\rho(S_n) = \inf_n \lambda(\kappa_\rho(S_n)) = \lambda\left(\inf_n \kappa_\rho(S_n)\right) = \lambda(\emptyset) = 0.$$

Thus (iii)$\Rightarrow$(i) and hence (iv). Finally suppose that (iv) holds and let $0 \leqslant \lambda_0 \in L(\mathscr{E})$. Then $\lambda_0 \ll \sum_n 2^{-n} P_{\theta_n} = \pi$ for some sequence $\{\theta_n\}$ in Θ. Clearly $\pi\rho = \sum 2^{-n} P_{\theta_n}\rho$ so that $\pi\rho$ is σ-additive. Let $\{S_\nu\}$ be a sequence in $\mathscr{S}$ such that $S_1 \supseteqq S_2 \supseteqq \cdots$ and $\bigcap_\nu S_\nu = \emptyset$. Then, since $(P_{\theta_1}, P_{\theta_2}, \ldots)$ is coherent, for each ν there is a bounded non-negative measurable function g_ν on χ such that $\lambda\rho(S_\nu) = \lambda g_\nu$ when $\lambda \ll \pi$. Then $g_1 \geqslant g_2 \geqslant \cdots$ a.e. π so that $0 = \lim_\nu(\pi\rho(S_\nu)) = \lim_\nu \pi(g_\nu) = \pi(\inf_\nu g_\nu)$. Hence $\inf_\nu g_\nu = 0$ a.s. π. Then $0 = \lim_\nu \lambda_0(g_\nu) = \lim_\nu \lambda_0\rho(S_\nu)$ so that $\lambda_0\rho$ is σ-additive. □

In the following we shall often find it convenient to write $\rho(S)$ instead of $\kappa_\rho(S) = \rho^* I_S$ when ρ is a transition from $L(\mathscr{E})$ to $\mathrm{ba}(T, \mathscr{S})$.

The $M(\mathscr{E})$-valued 'probabilities', since they are bounded linear functionals on $L(\mathscr{E})$, are representable by consistent families of measurable functions. As $M(\mathscr{E})$-valued probabilities are bounded by $\dot{0}$ and $\dot{1}$ these functions may all be chosen as test functions, i.e. measurable functions bounded by 0 and 1. The interpretation of transitions as $M(\mathscr{E})$-valued probability set functions leads in this way to representations in terms of 'kernel like' functions.

Theorem 4.5.6 ('Kernel like' representations of transitions). *Let ρ be a transition from $L(\mathscr{E})$ to $\mathrm{ba}(T, \mathscr{S})$. For each set S in $\mathscr{S}$ consider a representation of $\rho(S)$ as a consistent family $\{\rho_\theta(S|\cdot) : \theta \in \Theta\}$ of measurable functions. Then these functions satisfy:*

(i) *if S_1 and S_2 are disjoint sets in $\mathscr{S}$ then*

$$\rho_\theta(S_1 \cup S_2|\cdot) = \rho_\theta(S_1|\cdot) + \rho_\theta(S_2|\cdot) \quad \text{a.s. } P_\theta \quad \text{when } \theta \in \Theta.$$

(ii) *if S is a set in $\mathscr{S}$ then*

$$\rho_\theta(S|\cdot) \geqslant 0 \quad \text{a.s.} \quad P_\theta \text{ when } \theta \in \Theta.$$

(iii) *$\rho_\theta(T|\cdot) = 1$ a.s. P_θ when $\theta \in \Theta$.*

Conversely, any rule which to each set S in $\mathscr{S}$ assigns a consistent family $\{\rho_\theta(S|\cdot) : \theta \in \Theta\}$ of measurable functions satisfying conditions (i), (ii) *and* (iii), *determines a transition ρ such that the functional $\rho(S)$ in $M(\mathscr{E})$, for each set S in $\mathscr{S}$, is represented by the family $\{\rho_\theta(S|\cdot) : \theta \in \Theta\}$. The transition ρ is then σ-continuous if and only if these functions also satisfy:*

(iv) *if $S_1, S_2, \ldots$ are disjoint sets in $\mathscr{S}$ then*

$$\rho_\theta(S_1 \cup S_2 \ldots|\cdot) = \rho_\theta(S_1|\cdot) + \rho_\theta(S_2|\cdot) + \cdots \quad \text{a.s. } P_\theta \quad \text{when } \theta \in \Theta.$$

Proof. Condition (i) expresses the additivity of the $M(\mathscr{E})$-valued set function κ_ρ which maps $S \in \mathscr{S}$ into $\rho(S)$. (ii) states that κ_ρ is non-negative while (iii) states that $\kappa_\rho(T) = \dot{1}$. Finally condition (iv) expresses that ρ is σ-continuous, by condition (iii) of theorem 4.5.5. □

If $\mathscr{E}$ is coherent then conditions (i)–(iv) simplify as follows.

Corollary 4.5.7. *If $\mathscr{E}$ is coherent then $\rho_\theta(S|\cdot)$, for each set S in $\mathscr{S}$, may be specified independently of θ. Thus the subscript θ may then be dropped in statements* (i)–(iv).

Proof. In this case consistent families of real valued measurable functions are coherent. □

If, in addition, the decision space $(T, \mathscr{S})$ is Euclidean then we get representations in terms of Markov kernels by the following corollary.

Corollary 4.5.8. *If $\mathscr{E}$ is coherent and the decision space $(T, \mathscr{S})$ is Euclidean, then any σ-continuous transition is representable by a decision rule (Markov kernel).*

Proof. We may assume, without loss of generality, that T is the real line and that $\mathscr{S}$ is the class of Borel subsets of T. (Any Euclidean measurable space is Borel isomorphic to some Borel subset of the real line.) Let the map $S \to \rho(S|\cdot)$ represent the transition ρ as described in the previous corollary. The bothersome null sets appearing in statements (i)–(iv) may then be disposed of by the usual regularization procedure for conditional probabilities. □

We do not need to make any assumptions on the structure of the decision space when $\mathscr{E}$ is discrete.

Corollary 4.5.9. *If $\mathscr{E}$ is discrete, then any σ-continuous transition is representable by a decision rule (Markov kernel).*

Proof. Put $N = \{x : P_\theta(x) \equiv_\theta 0\}$. Then $P_\theta(N) \equiv_\theta 0$. Let ρ be a σ-continuous transition from $\mathscr{E}$ to $(T, \mathscr{S})$ and let Q be a probability measure on $\mathscr{S}$. If $\rho(S|\cdot) \in M(\mathscr{X}, \mathscr{A})$ represents $\rho^* I_S$, then $\tilde{\rho}(S|\cdot) = \rho(S|\cdot) I_{N^c} + Q(S) I_N$ also represents $\rho^* I_S$. Therefore we may as well assume that $\rho(S|x) = Q(S)$ when $x \in N$. Then $x \to \rho(\cdot|x)$ is a Markov kernel representing ρ. □

If $\mathscr{E}$ is Σ-dominated and null set complete in the sense that $A \in \mathscr{A}$ whenever $A \subseteqq N$ where $P_\theta(N) \equiv_\theta 0$, then regularization may be obtained from the lifting theorem of Ionescu Tulcea (1961). The part of this theorem which we will need may be described as follows.

Let $(\mathscr{X}, \mathscr{A}, P)$ be a probability space. We have defined $M(\mathscr{X}, \mathscr{A})$ above as the Banach space consisting of the linear space of bounded measurable functions on $\mathscr{X}$ normed by the supremum norm. The Banach space consisting of the linear space of the P-equivalence classes of functions in $M(\mathscr{X}, \mathscr{A})$ together with the L_∞ norm, is as usual denoted by $L_\infty(P)$. Let $\tilde{g}$ denote the equivalence class in $L_\infty(P)$ determined by $g \in M(\mathscr{X}, \mathscr{A})$.

A linear isometry U from $L_\infty(P)$ to $M(\mathscr{X}, \mathscr{A})$ is called a *linear lifting* on $(\mathscr{X}, \mathscr{A}, P)$ if:

(i) $U(g) \geqslant 0$ when $0 \leqslant g \in L_\infty(P)$.
(ii) $U(\tilde{c}) = c$ when $-\infty < c < \infty$.
(iii) $g \in L_\infty(P)$ implies $U(g) \in g$.

The lifting theorem ensures that the probability space $(\mathscr{X}, \mathscr{A}, P)$ has a linear lifting provided it is complete. A particularly nice proof, based on the martingale convergence theorem, may be found in Paul A. Meyer's book, *Probability and potentials* (1966).

Combining this with the representation theorem for bounded linear functionals on the Banach space of continuous functions on a compact space, we obtain conditions ensuring representability of transitions.

Theorem 4.5.10. *Let $\mathscr{E} = (\mathscr{X}, \mathscr{A}; P_\theta : \theta \in \Theta)$ be a Σ-dominated experiment such that if $N \in \mathscr{A}$ and $P_\theta(N) \equiv_\theta 0$ then $\mathscr{A}$ contains all subsets of N. Suppose also that T is a Baire subset of a compact Hausdorff space H and that $\mathscr{S}$ is the σ-algebra of Baire subsets of T.*

Then any σ-continuous transition ρ from $L(\mathscr{E})$ to $\mathrm{ba}(T, \mathscr{S})$ *is representable by a Markov kernel.*

Remark. We refer to e.g. Halmos (1950), for an exposition of integration on locally compact spaces. We use the usual notation $C(Z)$ for the Banach space of bounded continuous functions on a topological space Z.

Proof. The lifting result we described implies that there is a linear isometry U from $M(\mathscr{E})$ into $M(\mathscr{X}, \mathscr{A})$ such that $U(\dot{1}) = 1$, $U(\dot{g}) \geqslant 0$ when $0 \leqslant g \in M(\mathscr{X}, \mathscr{A})$ and $U(\dot{g}) \in \dot{g}$ for all g in $M(\mathscr{X}, \mathscr{A})$. Let ρ be a σ-continuous transition from $L(\mathscr{E})$ to $\mathrm{ba}(T, \mathscr{S})$. Represent ρ by a family $\rho(S|\cdot)$; $S \in \mathscr{S}$ as in corollary 4.5.7. If we fix $x \in \mathscr{X}$ then the map $f \to [U(\rho^* f)]_x$ on $C(H)$ is linear, non-negative and maps the constant function 1 into the number 1. Hence, by the representation theorem mentioned just before the statement of the theorem, there is a unique Baire measure $\kappa(\cdot|x)$ on the Baire class on H such that $[U(\rho^* f)]_x = \int f(t)\kappa(\mathrm{d}t|x)$ when $f \in C(H)$. The definition of κ implies directly that $\int f(t)\kappa(\mathrm{d}t|x)$ is measurable in x when $f \in C(H)$. If $f \in C(H)$ then $\int (1 \wedge nf^+)\mathrm{d}\kappa(\cdot|x) \uparrow \kappa([f > 0]|x)$. Thus $\kappa(S|x)$ is measurable in x for each Baire subset S of H. Furthermore, if $f \in C(H)$ then $\int [\int f(t)\kappa(\mathrm{d}t|x)]\lambda(\mathrm{d}x) = \lambda(U(\rho^* f)) = \lambda(\rho^* f) = \lambda\rho(f)$ when $\lambda \in L(\mathscr{E})$. By σ-additivity this extends to all bounded functions which are measurable with respect to the Baire class on H. In particular $\int \kappa(T|x)\lambda(\mathrm{d}x) = (\lambda\rho)(T) = \lambda(\mathscr{X})$ when $\lambda \in L(\mathscr{E})$. Hence $\kappa(T|\cdot) = 1$ a.s P_θ when $\theta \in \Theta$. Redefining κ on $\{x : \kappa(T|x) \neq 1\}$ we obtain a Markov kernel from $(\mathscr{X}, \mathscr{A})$ to $(T, \mathscr{S})$ which represents ρ. □

Now, in addition to the experiment $\mathscr{E} = (\mathscr{X}, \mathscr{A}; P_\theta : \theta \in \Theta)$ consider an experiment $\mathscr{F} = (T, \mathscr{S}; Q_\theta : \theta \in \Theta)$ with $(T, \mathscr{S})$ as sample space. The last experiment may for example be the performance function of a σ-continuous transition from $\mathscr{E}$ to $(T, \mathscr{S})$. Suppose we want to enquire how far $\mathscr{F}$ is from being the performance function of a transition from $\mathscr{E}$ to $(T, \mathscr{S})$. If ρ is a transition from $\mathscr{E}$ to $(T, \mathscr{S})$ then the goodness of the approximation achieved by ρ may be judged by the distances $\|P_\theta \rho - Q_\theta\|; \theta \in \Theta$. The purpose of the following result is to show that in this situation we can always modify ρ such that it maps $L(\mathscr{E})$ into $L(\mathscr{F})$ without increasing any distance $\|\lambda\rho - \mu\|$ between a measure $\mu \in L(\mathscr{F})$ and a measure $\lambda\rho$ where $\lambda \in L(\mathscr{E})$. In the proof we shall use some ideas from the theory of vector lattices. These ideas are covered in the next chapter. The reader who does not want to pursue vector lattices at this point can simply accept the following result.

Theorem 4.5.11. *Let $\mathscr{E} = (\mathscr{X}, \mathscr{A}; P_\theta : \theta \in \Theta)$ and $\mathscr{F} = (T, \mathscr{S}; Q_\theta : \theta \in \Theta)$ be two experiments and let ρ be a transition from $L(\mathscr{E})$ to* $\text{ba}(T, \mathscr{S})$. *Then there is a transition $\tilde{\rho}$ from $L(\mathscr{E})$ to* $\text{ba}(T, \mathscr{S})$ *which maps $L(\mathscr{E})$ into $L(\mathscr{F})$ such that $\|\lambda\tilde{\rho} - \mu\| \leqslant \|\lambda\rho - \mu\|$ whenever $\lambda \in L(\mathscr{E})$ and $\mu \in L(\mathscr{F})$. Since $L(\mathscr{F}) \subseteqq$* $\text{ca}(T, \mathscr{S})$, *the transition $\tilde{\rho}$ is necessarily σ-continuous.*

Proof. Let Q be any probability measure in $L(\mathscr{F})$. Any μ in $\text{ba}(T, \mathscr{S})$ has a unique decomposition as $\mu = \hat{\mu} + \bar{\mu}$ where $\hat{\mu} \in L(\mathscr{F})$ and $|\bar{\mu}| \wedge \tau = 0$ whenever $0 \leqslant \tau \in L(\mathscr{F})$. The map $\mu \to \hat{\mu}$ is linear and maps non-negative set functions into non-negative set functions. For each $\lambda \in L(\mathscr{E})$ define a set function $\lambda\tilde{\rho}$ in $L(\mathscr{F})$ by

$$\lambda\tilde{\rho} = \widehat{\lambda\rho} + [\lambda(\mathscr{X}) - \widehat{\lambda\rho}(T)]Q.$$

Then $\tilde{\rho}$ defines a transition from $L(\mathscr{E})$ to $L(\mathscr{F})$. If $\lambda \in L(\mathscr{E})$ and $\mu \in L(\mathscr{F})$, then putting $\tau = \lambda\rho - \mu$ we find $\|\lambda\tilde{\rho} - \mu\| = \|\widehat{\lambda\rho} + [\lambda(\mathscr{X}) - \widehat{\lambda\rho}(T)]Q - \mu\| = \|\hat{\tau} + \bar{\tau}(T)Q\| \leqslant \|\hat{\tau}\| + \|\bar{\tau}(T)Q\| \leqslant \|\hat{\tau}\| + |\bar{\tau}(T)| \leqslant \|\hat{\tau}\| + \|\bar{\tau}\| = \|\hat{\tau} + \bar{\tau}\| = \|\tau\| = \|\lambda\rho - \mu\|$ since $\|\tau\| = \|\hat{\tau}\| + \|\bar{\tau}\|$ for any τ in $\text{ba}(T, \mathscr{S})$. □

So far we have investigated transitions from the experiment $\mathscr{E}$ to the decision space $(T, \mathscr{S})$ by considering them either as linear maps from $L(\mathscr{E})$ to $\text{ba}(T, \mathscr{S})$ or as $M(\mathscr{E})$-valued probability set functions on $\mathscr{S}$. The third aspect we mentioned at the beginning of this section was that of transitions as bilinear functionals. We shall now see how this last interpretation may be utilized to discuss convergence of transitions.

Let us denote the bilinear functional on $L(\mathscr{E}) \times M(T, \mathscr{S})$ associated with the transition ρ from $L(\mathscr{E})$ to $\text{ba}(T, \mathscr{S})$ by B_ρ. Thus

$$B_\rho(\lambda, g) = \lambda\rho g \qquad \text{when } \lambda \in L(\mathscr{E}) \text{ and } g \in M(T, \mathscr{S}).$$

Bilinear functionals arising this way may be characterized as follows.

Theorem 4.5.12. *The map $\rho \to B_\rho$ from the set of transitions to the set of real valued bilinear functionals on $L(\mathscr{E}) \times M(T, \mathscr{S})$ is affine and $1 - 1$. A bilinear functional B on $L(\mathscr{E}) \times M(T, \mathscr{S})$ is of the form B_ρ for a necessarily unique transition ρ if and only if:*

(i) $B(\lambda, g) \geqslant 0$ *when* $0 \leqslant \lambda \in L(\mathscr{E})$ *and* $0 \leqslant g \in M(T, \mathscr{S})$.

and

(ii) $B(\lambda, 1) = \lambda(\mathscr{X})$ *when* $\lambda \in L(\mathscr{E})$.

The transition ρ is σ-continuous if and only if $B_\rho(\lambda, g_n) \to B_\rho(\lambda, g)$ whenever $\lambda \in L(\mathscr{E})$ and (g_n) is a uniformly bounded sequence in $M(T, \mathscr{S})$ converging pointwise to g.

Proof. The map $\rho \to B_\rho$ is clearly affine and $1 - 1$. Furthermore, any bilinear functional B_ρ satisfies (i) and (ii). The last condition is equivalent to the condition that $\lambda\rho$ is σ-additive whenever $\lambda \in L(\mathscr{E})$. Let B be any bilinear functional satisfying (i) and (ii). Then $B = B_\rho$ where $(\lambda\rho)(S) = B(\lambda, I_S)$ when $S \in \mathscr{S}$. □

It follows that the set of transitions may be identified with a subset of the product space $\prod \{[-\|\lambda\| \, \|g\|, \|\lambda\| \, \|g\|] : \lambda \in L(\mathscr{E}), g \in M(T, \mathscr{S})\}$. As this set is compact for pointwise convergence, and since the set of bilinear functionals satisfying (i) and (ii) is closed for this topology, we get the following useful result.

Theorem 4.5.13. *The set of transitions from the experiment $\mathscr{E}$ to the decision space $(T, \mathscr{S})$ is compact for the topology of pointwise convergence on $L(\mathscr{E}) \times M(T, \mathscr{S})$.*

Furthermore, if the set of transitions is topologized that way, if $0 \leqslant \lambda \in L(\mathscr{E})$ and if the measurable function g on $(T, \mathscr{S})$ is bounded from below, then $\lambda\rho g$ is a lower semicontinuous function of the transition ρ.

We shall need the following application.

Corollary 4.5.14. *Let $\mathscr{E} = (\mathscr{X}, \mathscr{A}; P_\theta : \theta \in \Theta)$ and $\mathscr{F} = (T, \mathscr{S}; Q_\theta : \theta \in \Theta)$ be two experiments and let ε_θ; $\theta \in \Theta$ be non-negative constants. Suppose that to each finite subset F of Θ there corresponds a transition ρ_F from $L(\mathscr{E})$ to* ba$(T, \mathscr{S})$ *such that, for each F:*

$$\|P_\theta \rho_F - Q_\theta\| \leqslant \varepsilon_\theta \qquad \text{when } \theta \in F.$$

Then there is a transition ρ from $L(\mathscr{E})$ to ba$(T, \mathscr{S})$ *which maps $L(\mathscr{E})$ into $L(\mathscr{F})$ and which satisfies*

$$\|P_\theta \rho - Q_\theta\| \leqslant \varepsilon_\theta; \qquad \theta \in \Theta.$$

Remark. The conclusion of the theorem may also be phrased: ρ_F may be chosen independently of F and such that it maps $L(\mathscr{E})$ into $L(\mathscr{F})$.

Proof. Consider (ρ_F) as a net where the finite sets F are ordered by inclusion. By theorem 4.5.13 this net has a subnet $(\rho_{F'})$ converging to a transition ρ_0. Then since $\|P_\theta\rho - Q_\theta\|$ is lower semicontinuous in ρ,

$$\|P_\theta\rho_0 - Q_\theta\| \leq \liminf_{F'} \|P_\theta\rho_{F'} - Q_\theta\| \leq \epsilon_\theta \quad \text{when } \theta \in \Theta.$$

The conclusion follows now from theorem 4.5.11. □

As we have already noted, in general the set of decision rules is not compact for the topology of pointwise convergence on $L(\mathscr{E}) \times M(T, \mathscr{S})$. Actually any generalized decision rule may be obtained as a limit of a net of decision rules which are 'supported' by finite sets. The terminology is then as follows. Say that a *transition (generalized decision rule)* ρ *is supported by the finite subset* T_0 *of* T if $\lambda\rho$ is supported by T_0 whenever $\lambda \in L(\mathscr{E})$. We shall say that a *Markov kernel (decision rule)*: $x \to \rho(\cdot|x)$ *from* $(\mathscr{X}, \mathscr{A})$ *to* $(T, \mathscr{S})$ *is supported by the finite subset* T_0 *of* T if $\rho(\cdot|x)$ is supported by T_0 whenever $x \in \mathscr{X}$. If ρ is a Markov kernel (decision rule) or a transition (generalized decision rule) then we shall say that ρ *is uniformly finitely supported* if ρ is supported by some finite subset T_0 of T. Clearly a transition defined by a uniformly finitely supported Markov kernel is itself uniformly finitely supported.

Proposition 4.5.15. *Let* $\mathscr{E} = (\mathscr{X}, \mathscr{A}; P_\theta : \theta \in \Theta)$ *be an experiment and let* $(T, \mathscr{S})$ *be a decision space.*

A transition ρ *from* $L(\mathscr{E})$ *to* ba$(T, \mathscr{S})$ *is uniformly finitely supported (supported by the finite set* T_0*) if and only if the* $M(\mathscr{E})$*-valued probability set function* $(\rho(S) : S \in \mathscr{S})$ *is finitely supported (supported by* T_0*).*

If the transition ρ *is supported by the finite set* T_0 *and if* ρ *is definable by a decision rule (Markov kernel) then* ρ *is definable by a decision rule (Markov kernel) which is supported by* T_0.

Proof. The statements concerning a transition ρ follow directly from the definitions. Suppose that the transition ρ is supported by T_0 and is representable by the Markov kernel σ. We may assume that $\mathscr{S}$ distinguishes points of T_0. (If not then replace T_0 by $T_1 \subseteqq T_0$ where T_1 has exactly one representative from each $\mathscr{S} \cap T_0$ atom.) The $M(\mathscr{E})$-valued probability set function κ_ρ associated with ρ is also supported by T_0. Thus we may write

$$\kappa_\rho(S) = \sum_{t \in T_0} I_S(t) v_t; \qquad S \in \mathscr{S}$$

where $v_t = \kappa_\rho(S_t)$ for disjoint measurable sets $(S_t : t \in T_0)$ such that $t \in S_t$ when $t \in T_0$. The functional v_t is represented by the function $g_t = \sigma(S_t|\cdot)$. Hence $\sum\{g_t : t \in T_0\}$ represents $\sum\{v_t : t \in T_0\} = \kappa_\rho(T) = \dot{1}$. Put $N = [\Sigma\{g_t : t \in T_0\} \neq 1]$. Then $P_\theta(N) \equiv_\theta 0$. Choose a point $t_0 \in T_0$ and define a Markov kernel τ by putting:

$$\tau(S|x) = \begin{cases} \sum\{I_S(t)g_t(x) : t \in T_0\} & \text{when } S \in \mathscr{S} \text{ and } x \notin N. \\ I_S(t_0) & \text{when } S \in \mathscr{S} \text{ and } x \in N. \end{cases}$$

Then $\tau(\cdot|x)$ is supported by T_0 for each $x \in \chi$ and for each $S \in \mathscr{S}$ the function $\tau(S|\cdot)$ represents the functional $\kappa_\rho(S)$. □

Corollary 4.5.16. *If $\mathscr{E}$ is coherent then any transition which is supported by a finite set T_0 is definable by a Markov kernel supported by T_0.*

Proof. This follows from the first part of the theorem by modifying the arguments used in the proof of the last part. □

Earlier in this section we saw how bounded additive set functions can be approximated by finitely supported set functions. The same reasoning implies that transitions may be approximated by uniformly finitely supported transitions. We may even ensure that the approximations are Markov kernels, whether $\mathscr{E}$ is coherent or not. This follows readily from the properties of the maps $\sigma_n : n \in N$ and $\tau_d : d \in D$ which we used to prove proposition 4.5.1. Indeed if ρ is a transition from $L(\mathscr{E})$ to $\mathrm{ba}(T, \mathscr{S})$ then for each $n \in N$ and $d \in D$ $\rho\sigma_n$ and $\rho\tau_d$ define uniformly finitely supported transitions. However, they need not be definable by Markov kernels when $\mathscr{E}$ is not coherent. We remedy this as follows. Put $\tilde{N} = N \times \mathscr{C}$ and $\tilde{D} = D \times \mathscr{C}$ where $\mathscr{C}$ is the class of countable subsets of Θ. Order $\mathscr{C}$ by inclusion and then order $\tilde{N}$ and $\tilde{D}$ by the product orderings. Thus $(n_1, C_1) \geqslant (n_2, C_2)$ if $(n_1, C_1), (n_2, C_2) \in \tilde{N}$, $n_1 \geqslant n_2$ and $C_1 \supseteqq C_2$. Similarly $(d_1, C_1) \geqslant (d_2, C_2)$ if $(d_1, C_1), (d_2, C_2) \in \tilde{D}$, $d_1 \geqslant d_2$ and $C_1 \supseteqq C_2$. Then $\tilde{N}$ and $\tilde{D}$ with these orderings become directed sets. Let $C \in \mathscr{C}$. Then, since $\mathscr{E}|C$ is coherent, the transitions $\rho\sigma_n$ and $\rho\tau_d$ restricted to $L(\mathscr{E}|C)$ are definable in terms of uniformly finitely supported Markov kernels $\rho_{n,C}$ and $\rho_{d,C}$. Let the measurable functions g and h on $(T, \mathscr{S})$ be, respectively, bounded and bounded from below. Then, by the discussion preceding proposition 4.5.1, $\lambda\rho_{n,C}g \to \lambda\rho g$ when $\lambda \in L(\mathscr{E})$, while $\lambda\rho_{d,C}h \to \lambda\rho h$ when $0 \leqslant \lambda \in L(\mathscr{E})$.

This proves the following theorem.

Theorem 4.5.17. *Let $\mathscr{E}$ be an experiment and let $(T, \mathscr{S})$ be a decision space. Then:*

(i) *the set of uniformly finitely supported decision rules (Markov kernels) constitutes a dense subset of the set of all transitions for the topology of pointwise convergence on $L(\mathscr{E}) \times M(T, \mathscr{S})$.*

(ii) *the set of uniformly finitely supported decision rules (Markov kernels) constitutes a dense subset of the set of all transitions for the topology of pointwise convergence on $L_+(\mathscr{E}) \times \hat{M}(T, \mathscr{S})$ where $L_+(\mathscr{E}) = \{\lambda : 0 \leqslant \lambda \in L(\mathscr{E})\}$ and $\hat{M}(T, \mathscr{S})$ is the set of extended real valued measurable functions on $(T, \mathscr{S})$ which are bounded from below.*

Remark. As $\|g \circ \sigma_n - g\| \to 0$ when $g \in M(T, \mathscr{S})$ it is clear that $\lambda\rho\sigma_n g \to \lambda\rho g$ uniformly on bounded subsets of $L(\mathscr{E})$, as $\mathscr{E}$ itself, provided g is restricted to a strongly compact subset of $M(T, \mathscr{S})$. A far-reaching approximation theorem in LeCam, 1986 (or 1974) implies that approximations by uniformly finitely supported transitions may be arranged uniformly for pairs (λ, g) in $L(\mathscr{E}) \times M(T, \mathscr{S})$ provided λ and g are restricted to weakly compact sets.

So far we have not defined risk for generalized decision rules. The obvious generalizations of the concepts of risk and performance function are as follows. Let ρ be a generalized decision rule (transition) from the experiment $\mathscr{E}$ to the decision space $(T, \mathscr{S})$. Then the *performance function of the generalized decision rule* ρ is the family $(P_\theta\rho : \theta \in \Theta)$ of finitely additive 'probability' set functions on $\mathscr{S}$. If ρ is σ-continuous then for each θ, $P_\theta\rho$ is σ-additive. It follows that the performance function of a σ-continuous generalized decision rule is an experiment. If ρ is a generalized decision rule and if we are given a loss function L then *the risk of* ρ *at* θ is the quantity $(P_\theta\rho)L_\theta$ in $]-\infty, \infty]$. This risk of ρ at θ is denoted by $r(\theta : \rho)$ or $P_\theta\rho L_\theta$ so that

$$r(\theta : \rho) = P_\theta\rho L_\theta = \sup_n P_\theta(L_\theta \wedge n).$$

Considering the risk at θ as a function of the generalized decision rule ρ we see that it is affine, as well as lower semicontinuous for the topology of pointwise convergence on $L(\mathscr{E}) \times M(T, \mathscr{S})$. It is clear that if the generalized decision rule is definable in terms of a decision rule of the Markov kernel variety, then its performance function and its risk function are, respectively, the performance function and the risk function of this decision rule. The *Bayes risk of the generalized decision rule with respect to the prior distribution* μ is the integral $\int r(\theta : \rho)\mu(\mathrm{d}\theta)$, provided this integral exists. As before, we shall denote the Bayes risk of ρ with respect to the prior μ by $r(\mu : \rho)$. Thus $r(\mu : \rho) = \sum_\theta r(\theta : \rho)\mu(\theta)$ when μ has finite support. If ρ is definable by a decision rule then the Bayes risk of ρ with respect to μ is equal to the Bayes risk of this decision rule with respect to μ.

The main reason for introducing generalized decision rules (transitions) was that in general the set $\mathscr{R}$ of available risk functions is not compact for the coarsest topology which, for each $\theta \in \Theta$, makes $r(\theta)$ lower semicontinuous in r. By introducing the generalized decision rules we have achieved compactness by enlarging the set of risk functions. Let us make this more precise by introducing the following three affine sets of risk functions associated with the decision problem $(\mathscr{E}, (T, \mathscr{S}), L)$:

$\mathscr{R}_f$ = the set of risk functions of uniformly finitely supported decision rules.
$\mathscr{R}$ = the set of risk functions of decision rules.
$\mathscr{R}_g$ = the set of risk functions of generalized decision rules.

Trivially

$$\mathscr{R}_f \subseteq \mathscr{R} \subseteq \mathscr{R}_g$$

and by theorem 4.5.17

$$\bar{\mathscr{R}}_f = \bar{\mathscr{R}} = \bar{\mathscr{R}}_g$$

where the bar indicates closure with respect to the topology of pointwise convergence on Θ. It follows from theorem 4.5.13 that $\mathscr{R}_g$ is compact for the coarsest topology which, for each $\theta \in \Theta$, makes $r(\theta)$ lower semicontinuous in $r \in \mathscr{R}_g$. Hence, using the notation of proposition 3.3.5, $\hat{\mathscr{R}}_g = \hat{\bar{\mathscr{R}}}_g = \hat{\mathscr{R}}_g = \hat{\bar{\mathscr{R}}}$. As noted just after example 4.2.3 this tells us that the set of minimal functions in $\bar{\mathscr{R}}$ is a minimal complete class in $\mathscr{R}_g$, and hence *the* minimal complete class in $\mathscr{R}_g$.

Thus the basic compactness assumption used in section 3.3 of chapter 3 is satisfied for the set of risk functions of generalized decision rules. Hence theorems 3.3.6–3.3.9 apply.

Theorem 4.5.18. *Consider a decision problem $(\mathscr{E}, (T, \mathscr{S}), L)$ and define the sets $\mathscr{R}_f$ and $\mathscr{R}_g$ of risk functions as above. Then:*

(i) *the coordinate game $\Gamma(\mathscr{R}_g)$ has a value and the statistician has a minimax strategy in $\mathscr{R}_g$.*

(ii) *let g be an extended real valued function on Θ. Then there is a risk function r in $\mathscr{R}_g$ dominating g, i.e. $g \geqslant r$, if and only if $\sum_\theta g(\theta)\lambda(\theta) \geqslant \min\{\sum \tilde{r}(\theta)\lambda(\theta) : \tilde{r} \in \mathscr{R}_g\} = \inf\{\sum \tilde{r}(\theta)\lambda(\theta) : \tilde{r} \in \mathscr{R}_f\}$ whenever λ is a prior distribution on Θ with finite support.*

(iii) *the set of admissible risk functions in $\mathscr{R}_g$ is minimal complete for the game $\Gamma(\mathscr{R}_g)$.*

(iv) *any admissible risk function in $\mathscr{R}_g$ is extended Bayes as well as a limit of Bayes solutions.*

Example 4.5.19 (Bayes risk in totally informative experiments). Let the experiment $\mathscr{E} = (\mathscr{X}, \mathscr{A}; P_\theta : \theta \in \Theta)$ be totally informative, i.e. P_{θ_1} and P_{θ_2} are mutually singular when $\theta_1 \neq \theta_2$. Suppose that to each θ there corresponds a 'best' decision t_θ for this θ, i.e. $L_\theta(t_\theta) = \min_t L_\theta(t)$. Let us, for simplicity, assume that $\inf_t L_\theta(t) \equiv_\theta 0$. Thus $L_\theta(t_\theta) \equiv 0$. Let ρ be the unique generalized decision rule which maps $\lambda \in L(\mathscr{E})$ into the distribution on $(T, \mathscr{S})$ which assigns its total mass $\lambda(\mathscr{X})$ to t_θ when $\lambda \ll P_\theta$. Then $P_\theta \rho L_\theta \equiv_\theta 0$ so that the minimum Bayes risk of a generalized decision rule is 0 whatever the prior distribution might be. If, however, the prior distribution does not have countable support, then it may happen that the Bayes risk of any decision rule of the Markov kernel variety may differ considerably from 0.

In order to make this precise, consider the two-decision problem $(\mathscr{E}, (T, \mathscr{S}), L)$ where:

$\mathscr{E} \quad = ([0,1], \text{Borel class } P_\theta : \theta \in [0,1]),$
$T \quad = \text{the two-point set } \{0,1\},$
$L_\theta(1) = 1 \text{ or } = 0 \text{ as } \theta > 0 \text{ or } \theta = 0,$
$L_\theta(0) = 0 \text{ or } = 1 \text{ as } \theta > 0 \text{ or } \theta = 0,$
$P_0 \quad = \text{the uniform distribution on } [0,1],$
$P_\theta \quad = \text{the one-point distribution in } \theta \text{ when } \theta > 0,$
$\mu \quad = \frac{1}{2}P_0 + \frac{1}{2}Q \text{ where } Q \text{ is the one-point distribution in } 0.$

Here $t_\theta = 0$ or $= 1$ as $\theta > 0$ or $\theta = 0$. The generalized decision rule ρ defined above maps $\lambda \in L(\mathscr{E})$ into $\lambda\rho$ where $(\lambda\rho)(0)$ and $(\lambda\rho)(1)$ are, respectively, the total mass of the (Lebesgue) singular part of λ and the total mass of the (Lebesgue) absolutely continuous part of λ. This generalized decision rule therefore has Bayes risk 0 for any prior distribution. However, simple calculations show that the Bayes risk of any decision rule with respect to the prior distribution μ is $\frac{1}{2}$.

It follows from corollary 4.5.8 that phenomena like this cannot occur when the decision space is Euclidean and the prior distribution is concentrated on a measurable subset U of Θ such that $\mathscr{E}|U$ is coherent. Although this is reassuring, it must be maintained that changing the model by much less than a microscopic amount may destroy coherence. We can therefore ask whether the limitation to decision rules of the Markov kernel variety is mathematically sensible. In this exposition we permit ourselves to pass to the 'limit' and use the generalized decision rules (transitions) of LeCam (1964).

4.6 References

Blackwell, D. & Girchick, M. A. 1954. *Theory of games and statistical decision.* Wiley, New York.

De Groot, M. A. 1970. *Optimal statistical decisions.* McGraw–Hill, New York.

Dvoretzky, A., Wald, A. & Wolfowitz, J. 1951. Elimination of randomization in certain statistical decision procedures and zero-sum two person games. *Ann. Math. Statist.* **22**, pp. 1–21.

— 1951. Relations among certain ranges of vector measures. *Pacific J. Math.* **1**, pp. 59–74.

Ferguson, T. S. 1967. *Mathematical statistics.* Academic Press, New York and London.

Gauss, C. F. 1821. *Teorie der den kleinsten Fehlern unterworfenen Combination der Beobachtungen. Abhandling zur methode der kleinsten Quadrate.* Berlin, 1887.

Halmos, P. 1950. *Measure Theory.* Van Nostrand, Princeton, New Jersey.

Ionescu Tulcea, C. A. 1961. On the lifting property 1. *J. Math. Anal. Appl.* **3**, pp. 537–46.

LeCam, L. 1964. Sufficiency and approximate sufficiency. *Ann. Math. Statist.* **35**, pp. 1419–55.

— 1974. *Notes on asymptotic methods in statistical decision theory.* Centre de Recherches Math. Univ. de Montréal.

— 1986. *Asymptotic methods in statistical decision theory.* Springer–Verlag, New York.

Lehmann, E. L. 1950. *Notes on the theory of estimation.* Chapters 1–5, University of California, Berkeley.

— 1986. *Testing statistical hypotheses.* Wiley, New York.

Liapounoff, A. 1940. Sur les fonctions vecteurs complètement additives. *Bull. Acad. Sci. URSS Ser. Math.* **4**, pp. 465–78.

Meyer, P. A. 1966. *Probability and potentials.* Blaisdell Publishing Company, Waltham, Mass.

Neyman, J. & Pearson, E. S. 1933. On the problem of the most efficient tests of statistical hypotheses. *Phil. Trans. Roy. Soc. Ser. A.* **231**, pp. 289–337.
Pfanzagl, J. 1969. On the existence of product measurable densities. *Sankhyā. Ser. A.* **31**, pp. 13–18.
Wald, A. 1950. *Statistical decision functions.* Wiley, New York.

5

Vector Lattices

5.1 Introduction

A vector lattice (Riesz space) is a linear space equipped with a lattice ordering which is 'compatible' with the linear structure. An ordering is here called a lattice ordering if finite sets possess greatest lower bounds and smallest upper bounds. The theory of statistical experiments abounds with linear spaces which are vector lattices for their 'natural orderings'. Without specifying the structures we mention a few examples of such spaces.

The space of (equivalence classes) of bounded variables.
The space of (equivalence classes) of real valued variables.
The space of (bounded) continuous real valued functions on a topological space.
The space of differences of sublinear functionals on a linear space.
The M-space $M(\mathscr{E})$ of an experiment $\mathscr{E}$.
The space of bounded additive set functions on a given algebra of sets.
The space of finite measures on a given σ-algebra of sets.
The L-space $L(\mathscr{E})$ of an experiment $\mathscr{E}$.

The industrious reader may define the 'natural' structures in these examples and check that they really deserve to be called vector lattices.

In this chapter we will give a short introduction to the theory of vector lattices. The theory is illustrated throughout by various spaces which are important to the theory of statistical experiments.

The nuts and bolts are collected in sections 5.2–5.6. Actually it might suffice at a first reading to study sections 5.2, 5.3, theorem 5.4.1 and example 5.4.2, and then look over the contents of sections 5.5 and 5.6. The reader might then proceed and return to this chapter when the need arises.

Section 5.2 contains several useful, mostly elementary, identities and inequalities. The important Riesz theorem on band decompositions of vector lattices is proved in section 5.3. The 'natural' ordering of the dual space, the dual ordering, is the main theme of section 5.4. Here there are also some results on order convergence and on the evaluation map.

Several of the vector lattices we shall encounter are normed. Norms which behave as the L_1-norm are called L-norms. Norms which behave as the supremum norm are called M-norms. Vector lattices which are complete for L-norms (M-norms) are called L-spaces (M-spaces) and it is these spaces which constitute the main topic of sections 5.5–5.7. This terminology would have been very badly organized if the L-space (M-space) of an experiment had not been the prominent example of an L-space (M-space).

The definitions imply directly that an experiment is a family in its own L-space. Following LeCam, from section 5.6 onwards we admit any family of non-negative and normalized 'vectors' in an L-space, as an abstract experiment.

By Kakutani's representation theory, which is established in section 5.7, any L-space is isomorphic to an L-space of finite measures. The conclusion is then that abstract experiments are no more general than traditional experiments. However, this does not in any way contradict the fact that the idea of an abstract experiment is a very useful one.

The basic result (Kakutani's) in section 5.7 is that any M-space is isomorphic to the M-space $C(Z)$ consisting of the continuous functions on an essentially unique compact topological space Z. If in addition the M-space is the dual of an L-space, as $M(\mathscr{E})$ is the dual of $L(\mathscr{E})$, then L may be identified with the band of regular Borel measures on Z which assigns mass zero to any Borel set of the first category. The exposition of these results is strongly influenced by the appendix in Kelley & Namioka (1963).

The material in this chapter is well known and is covered in several books. A possibility for further study and for reference is Schaefer (1974). However, my main source of inspiration has been LeCam's 1964 paper and several of his later writings. We refer to his 1974 and 1986 books for further study of the role of vector lattices in mathematical statistics.

5.2 Definitions, notation and basic properties

The linear spaces occurring in connection with statistical experiments often have a natural order structure which is compatible with the linear structure. To be more precise, let V be a linear space over the real field $\mathbb{R}$ and let '$\geqslant$' be an antisymmetric ordering of V. (An ordering '$\geqslant$' is called *antisymmetric* if $v_1 \geqslant v_2$ and $v_2 \geqslant v_1$ implies $v_1 = v_2$.) Let V_+ denote the set of elements v in V which are $\geqslant 0$. Then $(V, \geqslant)$ is called an *ordered linear space* if

(i) $v_1 \geqslant v_2$ if and only if $v_1 - v_2 \in V_+$

and

(ii) $c_1 v_1 + c_2 v_2 \in V_+$ when $c_1, c_2 \in \mathbb{R}_+$ and $v_1, v_2 \in V_+$.

Example 5.2.1 (Vector lattices of functions). Let $\mathscr{F}$ be a linear space of real valued functions on a set Z. Define an ordering $\geqslant$ on $\mathscr{F}$ by agreeing that $f_1 \geqslant f_2$ iff $f_1(z) \geqslant f_2(z)$; $z \in Z$. This ordering is called the *pointwise* ordering of $\mathscr{F}$. Particular cases of great interest are:

(i) $(Z, \mathcal{O})$ is a topological space and $\mathscr{F}$ is the space $C(X)$ of bounded continuous functions on Z.

(ii) $(Z, \mathscr{C})$ is a measurable space and $\mathscr{F}$ is the class of bounded measurable functions on Z.

(iii) $Z = \mathscr{A}$ is an algebra (field) of subsets of a set X and $\mathscr{F}$ is the class of bounded additive set functions on $\mathscr{A}$.

(iv) $Z = \mathscr{A}$ is a σ-algebra (σ-field) of subsets of a set X and $\mathscr{F}$ is the class of finite measures on $\mathscr{A}$.

Variations may be obtained by considering subspaces of $\mathscr{F}$. Thus in (i) we may restrict our attention to those functions which have compact supports, while in (iv) we may restrict our attention to those finite measures which belong to the L-space of an experiment $\mathscr{E}$.

Let us fix some terminology and some notation concerning ordered sets and ordered linear spaces. Consider first an ordering $\geqslant$ on a set V. Where convenient we can write $v_1 \leqslant v_2$ instead of $v_2 \geqslant v_1$. Consider a subset W of V. An element $v \in V$ is called *an upper bound for* W if $w \leqslant v$ when $w \in W$. Similarly an element $v \in V$ is called *a lower bound for* W if $v \leqslant w$ when $w \in W$. A subset W of V is called *order bounded* if there are elements v_1 and v_2 in V such that $v_1 \leqslant w \leqslant v_2$ when $w \in W$.

If $(v_n : n \in N)$ is a net in V then $v_n\uparrow (v_n\downarrow)$ indicates that the net is monotonically increasing (decreasing) i.e. that $v_{n_1} \leqslant v_{n_2} (v_{n_1} \geqslant v_{n_2})$ when $n_1 \leqslant n_2$.

A set of the form $\{v : x \leqslant v \leqslant y\}$ where $x, y \in V$ is called an *order interval*. Thus a set is order bounded if and only if it is contained in some order interval. We may use the notation $[x, y]$ for the order interval $\{v : x \leqslant v \leqslant y\}$ in V determined by $x, y \in V$.

An *infimum* (*inf*) of a subset W of V is a lower bound of W which is an upper bound for all lower bounds for W. The notation for this element, if it exists, is $\inf W$ or $\wedge W$. If $(w_t : t \in T)$ is a family of elements in V then the infimum of the *set* $\{w_t : t \in T\}$ may be denoted by $\inf_t w_t$ or $\bigwedge_t w_t$. The infimum of a finite family $(w_1, w_2, \ldots, w_r)$ of elements in V may be denoted by $w_1 \wedge w_2 \wedge \ldots \wedge w_r$.

Similarly a *supremum* (*sup*) of a subset W of V is an upper bound for W which is a lower bound for all upper bounds for W. The notation for this element, if it exists, is $\sup W$ or $\vee W$. If $(w_t : t \in T)$ is a family of elements in V, then the supremum of the *set* $\{w_t : t \in T\}$ may be denoted by $\sup_t w_t$ or $\bigvee_t w_t$. The supremum of a finite family $(w_1, w_2, \ldots, w_r)$ may be denoted by $w_1 \vee w_2 \vee \ldots \vee w_r$.

An ordered set W is called a *lattice* if each two-point subset, and hence each finite subset, has an infimum as well as a supremum. The ordered set V is called *order complete* if any non-empty subset W of V having an upper bound has a supremum. If so, then any subset W of V having a lower bound has an infimum.

Now consider a subset A of an ordered linear space $(V, \geqslant)$. Then

$$\sup(A + b) = (\sup A) + b$$

in the sense that the left hand side is defined if and only if the right hand side is defined, and then they are equal. In the same sense

$$\inf(A + b) = (\inf A) + b.$$

If the subsets A and B of V both have suprema in V then

$$\sup(A + B) = (\sup A) + (\sup B).$$

Similarly

$$\inf(A + B) = (\inf A) + (\inf B)$$

if the right hand side is defined. Infima may be converted to suprema and vice versa since

$$\inf A = -[\sup(-A)]$$

whenever one side of this formula is well defined.

An ordered linear space V which, as an ordered space, is a lattice, is called a *vector lattice* or a *Riesz space.* Actually, by the identities $v \vee w = v + (w - v) \vee 0$ and $v \wedge w = -((-v) \vee (-w))$, an ordered linear space $(V, \geqslant)$ is a vector lattice iff $(v \vee 0)$ exists for each $v \in V$. The following notation is standard for vector lattices $(V, \geqslant)$

$$v^+ = v \vee 0$$

$$v^- = (-v) \vee 0$$

$$|v| = v \vee (-v)$$

The vectors v^+, v^- and $|v|$ are called, respectively, *the non-negative part of* v, *the non-positive part of* v and *the absolute value of* v.

Example 5.2.2 (Lattice operations). The space of bounded continuous functions on a topological space, as well as the space of bounded measurable functions on a measurable space, are both vector lattices for the pointwise ordering. In both cases the lattice operations are carried out pointwise. On the other hand, the ordered linear space of polynomials on [0, 1] is not a lattice.

The space of bounded additive set functions on an algebra of subsets of a given set, as well as the space of finite measures on a σ-algebra of subsets of a given set, are vector lattices. In these two cases the orderings may be called *setwise orderings*. However, then the lattice operations are generally not setwise operations.

Example 5.2.3 $\mathbb{R}^n$ as a vector lattice. The Euclidean space $\mathbb{R}^n$ is a vector lattice for the coordinate-wise ordering. Another ordering converting $\mathbb{R}^n$ into a vector lattice is the *lexicographic ordering*, where $x = (x_1, x_2, \ldots, x_n) \geqslant y = (y_1, y_2, \ldots, y_n)$ iff either $x = y$ or $x \neq y$ and $x_i > y_i$ where i is the smallest i such that $x_i \neq y_i$. Then $(0, 1, 0, \ldots, 0) \leqslant (1/n)\,(1, 0, \ldots, 0)$; $n = 1, 2, \ldots$ although $(0, 1, 0, \ldots, 0) \nleqslant (0, 0, \ldots, 0) = \lim_n 1/n(1, 0, \ldots, 0)$.

$\mathbb{R}^n$ with the lexicographic ordering is clearly totally ordered and it may be shown (see Schaefer, 1974) that any n-dimensional totally ordered vector lattice is isomorphic to $\mathbb{R}^n$ with the lexicographic ordering.

A vector lattice V is called *Archimedian* if $v \leqslant 0$ whenever v, $w \in V$ and $v \leqslant (1/n)w$, $n = 1, 2\ldots$ All the vector lattices mentioned above, except for $\mathbb{R}^n$ with the lexicographic ordering, are Archimedian.

A vector lattice $(V, \geqslant)$ is called *order complete* if the ordered set $(V, \geqslant)$ is ordered complete. Any order complete vector lattice is Archimedian. To see this, consider vectors v and w in an order complete vector lattice such that $v \leqslant (1/n)w$, $n = 1, 2, \ldots$ Then $nv^+ \leqslant w^+$, $n = 1, 2, \ldots$ so that $\sup_n nv^+$ exists. Then $2 \sup_n nv^+ = \sup_m 2mv^+ \leqslant \sup_n nv^+$ so that $v^+ \leqslant \sup_n nv^+ \leqslant 0$. Thus $v \leqslant 0$.

Some useful formulas which are valid for vector lattices are:

$$
\begin{aligned}
&v_1 + v_2 = v_1 \vee v_2 + v_1 \wedge v_2 \\
&\lambda(v_1 \vee v_2) = (\lambda v_1) \vee (\lambda v_2) \text{ if } \lambda \geqslant 0 \\
&\lambda(v_1 \vee v_2) = (\lambda v_1) \wedge (\lambda v_2) \text{ if } \lambda \leqslant 0 \\
&(\lambda v)^+ = \lambda(v^+), (\lambda v)^- = \lambda(v^-) \text{ and } |\lambda v| = \lambda |v| \text{ if } \lambda \geqslant 0 \\
&v^+ = \tfrac{1}{2}(v + |v|), v^- = \tfrac{1}{2}(v - |v|) \\
&|v| = v^+ + v^-, v = v^+ - v^-
\end{aligned}
$$

The decomposition $v = v^+ - v^-$ is minimal in the sense that $v_1 \geqslant v^+$ and $v_2 \geqslant v^-$ whenever $v = v_1 - v_2$ and $v_1 \geqslant 0$ and $v_2 \geqslant 0$.

Furthermore, the triangular inequalities

$$\big||v| - |w|\big| \leqslant |v + w| \leqslant |v| + |w|$$

are valid for any pair (v, w) of vectors in the vector lattice V.

Also we have

$$\left(\bigvee_t w_t\right) \wedge w = \bigvee_t (w_t \wedge w) \quad \text{when } \bigvee_t w_t \text{ exists}$$

and

$$\left(\bigwedge_t w_t\right) \vee w = \bigwedge_t (w_t \vee w) \quad \text{when } \bigwedge_t w_t \text{ exists.}$$

Other useful formulas are

$$v \vee w = \tfrac{1}{2}(v + w + |v - w|)$$

and

$$v \wedge w = \tfrac{1}{2}(v + w - |v - w|)$$

and (consequently)

$$v \vee w - v \wedge w = |v - w|.$$

Vectors v and w are called *disjoint* if $|v| \wedge |w| = 0$. By the formulas above, this is equivalent to $|v| + |w| = |v| \vee |w|$ or equivalent to $||v| - |w|| = |v| \vee |w|$.

The set of vectors v which are disjoint with any w in a given set W will be denoted by $W^{\S}$. The set $W^{\S}$, for any subset W of V, is a sub-vector lattice of V which has the additional property that $v_1 \in W^{\S}$ whenever $|v_1| \leqslant |v_2|$ and $v_2 \in W^{\S}$.

Clearly v^+ and v^- are disjoint for any vector v in the vector lattice V. Suppose conversely that $v = v_1 - v_2$ where v_1 and v_2 are disjoint and non-negative. Then $v^+ = (v_1 - v_2) \vee 0 = v_1 \vee v_2 - v_2 = v_1 + v_2 - v_2 = v_1$ and $v^- = v^+ - v = v_1 - v = v_2$. Thus the decomposition $v = v^+ - v^-$ is the unique decomposition of v as a difference between two disjoint non-negative vectors. It follows that $(v + w)^{\pm} = v^{\pm} + w^{\pm}$ and $|v + w| = |v| + |w|$ when v and w are disjoint.

It is worth noting that disjoint summation in V_+ coincides with disjoint maximization. In particular it follows that $\wedge$ acts distributively with respect to disjoint summation in V_+.

The following fact on decomposition of non-negative vectors is very useful.

Theorem 5.2.4 (Decomposition theorem). *Let $v_1, \ldots, v_r$ be non-negative vectors in a vector lattice V. Then any non-negative vector $w \leqslant v_1 + \cdots + v_r$ may be decomposed as $w = w_1 + \cdots + w_r$ where $0 \leqslant w_i \leqslant v_i$; $i = 1, \ldots, r$.*

Proof. Put $w_1 = w \wedge v_1$ and $w_2 = w - w_1$ when $r = 2$. The general case follows by induction on r. □

Corollary 5.2.5. *Let $v_1, v_2, \ldots, v_r$, $w_1, \ldots, w_s$ be non-negative vectors in a vector lattice V. Suppose $\sum v_i = \sum w_j$. Then there are non-negative vectors u_{ij};*

$i = 1, \ldots, r$, $j = 1, \ldots, s$ such that $v_i = \sum_j u_{ij}$ and $w_j = \sum_i u_{ij}$; $i = 1, \ldots, r$, $j = 1, \ldots, s$.

Proof. The case '$s = 1$' is trivial. The case '$s = 2$' follows from the theorem and the general case follows by induction on s. □

It follows directly from the definition that a linear subspace W of an ordered linear space V is itself an ordered linear space with respect to the restricted ordering. As we have seen, infima and suprema of subsets of W may differ according to whether they are considered as subsets of the ordered set W or as subsets of V. If V is a vector lattice then a linear subspace W is called a *sub-vector lattice* if infima and suprema for finite subsets of W remain the same whether they are considered as subsets of W or as subsets of V. Thus a linear subspace W of V is a sub-vector lattice of V if and only if $w^+ \in W$ whenever $w \in W$.

5.3 Band decomposition of order complete vector lattices

We shall find it useful to decompose ordered linear spaces as direct sums of ordered linear spaces. Thus we shall say that the ordered linear space $(V, \geqslant)$ is the direct sum of the ordered linear subspaces $(W_i, \geqslant)$ $i = 1, \ldots, r$ if the linear space V is the direct sum of the linear spaces W_j; $j = 1, \ldots, r$ and if any vector $v \geqslant 0$ in V has a (necessarily unique) decomposition as $v = w_1 + \cdots + w_r$ where $0 \leqslant w_i \in W_i$; $i = 1, \ldots, r$. Equivalently we may require that the map $(w_1, \ldots, w_r) \to w_1 + \cdots + w_r$ on $W_1 \times \cdots \times W_r$ (with coordinatewise ordering) is an isomorphism onto V. If so, and if V is a vector lattice, then each W_i is also a vector lattice and $(w_1 + \cdots + w_r) \vee (\tilde{w}_1 + \cdots + \tilde{w}_r) = (w_1 \vee \tilde{w}_1) + \cdots + (w_r \vee \tilde{w}_r)$ when $w_i, \tilde{w}_i \in W_i$, $i = 1, \ldots, r$.

We shall soon see that there is a theory available for decompositions of order complete vector lattices as direct sums of sub-vector lattices. This theory has similarities with the decomposition theory for Hilbert spaces based on the notion of orthogonality. Here are some of the concepts we shall need.

A subset W of a vector lattice V is called *solid* if $w \in W$ whenever $|w| \leqslant |v|$ and $v \in W$. A solid sub-vector lattice is called an *ideal*. Ideals are the kernels of vector lattice homomorphisms. Thus if f is a vector lattice homomorphism from a vector lattice V to a vector lattice W, then the kernel of f, i.e. $f^{-1}\{0\}$, is an ideal I of V. Conversely, if I is an ideal of V then the quotient space V/I may be organized into a vector lattice by demanding that $v + I \geqslant 0$ whenever $v - v' \in I$ for some $v' \geqslant 0$. The canonical map $v \to v + I$ from V to V/I then becomes a vector lattice homomorphism with kernel I. A vector lattice homomorphism f from a vector lattice V to a vector lattice W may then be decomposed as the product of the canonical map from V to $V/f^{-1}\{0\}$ and the induced vector lattice isomorphism from $V/f^{-1}\{0\}$ to W.

Note that the set $W^\S$ of vectors in a vector lattice V which are disjoint from the set W of vectors in V is always an ideal. If the vector lattice V is order complete then a subset W is called a *band* if it is an ideal containing all suprema of its subsets. Thus W is a band if and only if:

(i) W is a sub-vector lattice of V.
(ii) $w \in W$ whenever $|w| \leqslant |v|$ and $v \in W$.
(iii) if $A \subseteq W$ and $\sup A$ exists then $\sup A \in W$.

Although the definition makes sense for general vector lattices, we shall only consider bands in order complete vector lattices.

The class of bands in V is clearly closed under intersections and contains V. It follows that to any subset A of V there corresponds a smallest band containing A. This band will be called *the band generated by* A. It will be denoted as $\mathrm{Band}(A)$ and A will be called a generator of the band $B = \mathrm{Band}(A)$. We shall give two characterizations of this band.

Theorem 5.3.1 (First characterization of Band(A)). *Let A be a subset of an order complete vector lattice V. Then a vector v in V belongs to* $\mathrm{Band}(A)$ *iff* $|v| \leqslant \sup W$ *where W is a set such that to each w in W there corresponds a finite sequence $(a_1, a_2, \ldots, a_m)$ in A such that $w \leqslant \sum_{i=1}^{m} |a_i|$.*

Proof. It is a matter of checking that the set of vectors satisfying the above requirements is a band containing A. On the other hand, any vector of the form $\sup W$, where W satisfies the requirements of the theorem, is in the band generated by A. □

For any set A, we defined $A^\S$ as the set of vectors which are disjoint from A. It is easily checked that $A^\S$ is a band.

If B is a subspace of a Hilbert space H, then the projection theorem states that H may be decomposed as the algebraic sum of B and the space of vectors which are perpendicular on B. Replacing perpendicularity with disjointness, the analogous result holds in vector lattices (Riesz, 1940).

Theorem 5.3.2 (Riesz's decomposition of a vector lattice). *Let B be a band in an order complete vector lattice V. Then V is the direct sum of the ordered linear spaces B and $B^\S$.*

Proof. Let $v \in V_+$ and put $b_0 = \sup\{w : w \in B, 0 \leqslant w \leqslant v\}$. Then $b_0 \in B$ and $0 \leqslant b_0 \leqslant v$. Let $b \in B$ and put $c = |b| \wedge (v - b_0)$. Then, since $0 \leqslant c \leqslant |b|$, $c \in B$ and clearly $v \geqslant b_0 + c$. It follows that $b_0 + c \leqslant b_0$ so that $c \leqslant 0$. Hence $c = 0$. Thus $v - b_0 \in B^\S$. Hence any $v \in V$ is a sum of two vectors, one from B and one from $B^\S$. These vectors are necessarily unique, and they are non-negative if v is non-negative. □

Corollary 5.3.3 (Second characterization of Band(A)). *Let A be a subset of an order complete vector lattice. Then*

$$\mathrm{Band}(A) = (A^{\S})^{\S}.$$

Proof. Let B be the band generated by A. Then, by theorem 5.3.1, $B^{\S} = A^{\S}$ so that $(B^{\S})^{\S} = (A^{\S})^{\S}$. Furthermore, by the Riesz decomposition theorem, $V = B + B^{\S} = B^{\S} + (B^{\S})^{\S}$. Hence, since $(B^{\S})^{\S} \supseteqq B$, $(B^{\S})^{\S} = B$. □

Consider a band B in an order complete vector lattice V. Then for each $v \in V$ there is a unique $b \in B$ such that $v - b$ is disjoint from B. This vector b is called *the projection of v on B*. The map which to each $v \in V$ assigns the projection of v onto B is called *the projection of V on B*. The projection of V on B is clearly a vector lattice homomorphism.

Suppose now that the band B is generated by a finite set $a_1, a_2, \ldots, a_m$ of vectors. Then, by theorem 5.3.1, B is also generated by the single vector $\sum_i |a_i|$. A band which is generated by a single vector is called a *principal band*. Thus a band is a principal band if and only if it is finitely generated. The principal band generated by the vector a will be denoted by B_a.

Let v and a be non-negative vectors in an order complete vector lattice V. How do we find the projection of v on the principal band B_a generated by a? According to theorem 5.3.2 this projection is the supremum of all vectors $b \in B_a$ such that $0 \leqslant b \leqslant v$. Now $b = \sup W$ where each $w \in W$ is $\leqslant n_w a$ for some positive integer n_w. Thus $b = \sup\{w : w \in W\} = \sup\{w \wedge v : w \in W\} \leqslant \sup_w (n_w a \wedge v) \leqslant \sup_n (na \wedge v)$. Furthermore, $na \wedge v \in B_a$ and $0 \leqslant na \wedge v \leqslant v$. Hence $b \geqslant \sup_n (na \wedge v)$.

Proposition 5.3.4 (Projection onto a principal band). *Let a and v be non-negative elements in an order complete vector lattice V. Then the projection of v onto the band generated by a is $\sup_n (na \wedge v)$.*

Decomposition of an element a in an order complete vector lattice as a finite sum of disjoint elements yields an order directed decomposition of the corresponding principal bands. Thus if $a_1, a_2, \ldots, a_r$ are disjoint, then $B_{a_1 + \cdots + a_r}$ is the order direct sum of $B_{a_1}, B_{a_2}, \ldots, B_{a_r}$. Furthermore, if the band B is contained in the principal band B_a, then B is also a principal band, since $B = B_{\tilde{a}}$ where $\tilde{a}$ is the projection of a onto B.

Example 5.3.5. Let $\mathscr{X}$ be any set and let F be the vector lattice consisting of all real valued functions on $\mathscr{X}$. Then F is clearly order complete. For each $A \subseteqq \mathscr{X}$, let F_A be the subset of F consisting of those functions f which vanish on A^c. Then F_A is a band in F, and the projection on F_A is just multiplication with I_A. Consider now an arbitrary band G in F and put $A = \bigcap\{[g \neq 0] : g \in G\}$. Then $G \subseteqq F_A$. Suppose $f \in F_A$. Then $f = g + h$ where $g \in G$ and h is disjoint with any vector in G. If $x \in A$ then $g(x) \neq 0$ for some $g \in G$. Hence, since $|g| \wedge |h| = 0$, $h(x) = 0$. Thus $h \in F_{A^c} \cap F_A = 0$ so that $f = g \in G$.

It follows that the map $A \to F_A$ from the class of subsets of $\mathscr{X}$ to the class of bands in F is 1–1 and onto. In particular, there are 2^k bands in $\mathbb{R}^k$ when $\mathbb{R}^k$ is equipped with the coordinatewise ordering.

5.4 The order dual. Order convergence

Linear maps from one ordered linear space to another, are called *non-negative* if they map non-negative elements onto non-negative elements. Consider in particular *the algebraic dual* V' consisting of the linear functionals on an ordered linear space V. The linear space V' becomes an ordered linear space if the functionals in V' are ordered according to the vectorwise ordering on $V_+ = \{v : 0 \leqslant v \in V\}$. This ordering of V' is called *the dual ordering of the ordering of* V. Note that a linear functional on V is non-negative with respect to the dual ordering if and only if it is non-negative according to the definition first given. *The order dual of the ordered linear space* V is the ordered linear subspace of V' generated by the non-negative functionals on V. The order dual of V is denoted by $V^{\#}$.

According to the definition in section 5.2, a subset A of an ordered vector space V is order bounded if $v \leqslant A \leqslant w$ for vectors v and w in V. A linear map from one ordered linear space into another is called *order bounded* if images of order bounded sets are order bounded. It is easily seen that linear functionals belonging to the order dual of V are order bounded and, as we now shall see, the converse holds if V is a vector lattice.

Theorem 5.4.1. *Let V be a vector lattice. Then a linear functional on V belongs to the order dual of V if and only if it is order bounded. The order dual equipped with the vectorwise ordering is an order complete vector lattice and the following formulas hold for linear functionals $f, g, \ldots$ in $V^{\#}$ and non-negative vectors $v \in V$:*

(i) $f^+(v) = \sup\{f(w) : 0 \leqslant w \leqslant v\}$.
(ii) $f^-(v) = \sup\{f(w) : -v \leqslant w \leqslant 0\}$.
(iii) $|f|(v) = \sup\{f(w) : |w| \leqslant v\}$.
(iv) $(f_1 \vee \ldots \vee f_r)(v) = \sup\{f_1(w_1) + \cdots + f_r(w_r) : w_1, \ldots, w_r \geqslant 0;$ $w_1 + \cdots + w_r = v\}$.
(v) $(f_1 \wedge \ldots \wedge f_r)(v) = \inf\{f_1(w_1) + \cdots + f_r(w_r) : w_1, \ldots, w_r \geqslant 0;$ $w_1 + \cdots + w_r = v\}$.

Proof. Let f be an order bounded linear functional and put $g(v) = \sup\{f(w) : 0 \leqslant w \leqslant v\}$ when $v \geqslant 0$. Then g is finite and additive on V_+ by the decomposition theorem, i.e. $g(v_1 + v_2) = \sup\{f(w) : 0 \leqslant w \leqslant v_1 + v_2\} = \sup\{f(w_1 + w_2) : 0 \leqslant w_1 \leqslant v_1,\ 0 \leqslant w_2 \leqslant v_2\} = g(v_1) + g(v_2)$. Furthermore, $g(\lambda v) = \lambda g(v)$ when $\lambda \geqslant 0$ and $g(v) \geqslant f(0) = 0$ for all $v \in V_+$.

Define $g(v)$ for any v by putting $g(v) = g(v^+) - g(v^-)$. It is then easily checked that $g \in V^{\#}$. The definition of g implies directly that $g \geqslant f$ and that $g \geqslant 0$. Let h be another linear functional such that $h \geqslant f$ and $h \geqslant 0$. Let $0 \leqslant w \leqslant v$. Then $h(v) \geqslant h(w) \geqslant f(w)$ so that $h(v) \geqslant g(v)$. Hence $h \geqslant g$ so that $g = f \vee 0$. It follows that the linear space of order bounded linear functionals is a vector lattice. Decomposing an order bounded linear functional f as $f = f^+ - f^-$ we see that any order bounded linear functional belongs to the order dual $V^{\#}$ of V. Furthermore, we have proved (i). When $r = 2$ (iv) follows by writing $f_1 \vee f_2 = f_1 + (f_2 - f_1)^+$. The case of a general r follows then by induction. (v) follows by writing $\bigwedge_i f_i = -\bigvee_i -f_i$ while (ii) and (iii) follow from the formulas $f^- = (-f)^+$ and $|f| = f \vee (-f)$, respectively.

It remains to show that $V^{\#}$ is order complete. Consider a family $f_t : t \in T$ and an f in $V^{\#}$ such that $f_t \leqslant f$ for each t. Let $\mathscr{S}$ be the class of finite non-empty subsets of T and for each $S \in \mathscr{S}$ put $f_S = \sup\{f_t : t \in S\}$. Then $f_S \leqslant f$ for each $S \in \mathscr{S}$. For each $v \in V_+$, put $\tilde{g}(v) = \sup_s f_s(v)$. Then $\tilde{g}$ is additive and positive homogeneous on V_+. Thus $\tilde{g}$ extends to a linear functional g on V which, for any $S \in \mathscr{S}$, satisfies $f_S \leqslant g \leqslant f$. It follows that g is order bounded. Hence, by what we have proved, g belongs to $V^{\#}$. It is easily checked that $g = \sup_t f_t$. Hence $V^{\#}$ is order complete. □

Example 5.4.2 (Additive set functions). Let $M(\mathscr{X}, \mathscr{A})$ denote the Banach space of bounded measurable functions on the measurable space $(\mathscr{X}, \mathscr{A})$. Unless otherwise stated, the norm on this space is the supremum norm. As a Banach space $M(\mathscr{X}, \mathscr{A})$ has a dual $M(\mathscr{X}, \mathscr{A})^*$ consisting of the bounded linear functionals on $M(\mathscr{X}, \mathscr{A})$. We saw in section 4.5 how the dual space $M(\mathscr{X}, \mathscr{A})^*$ might be identified with the Banach space of bounded additive set functions on $\mathscr{A}$, equipped with the total variation norm.

As noted in example 5.2.2 the linear space $M(\mathscr{X}, \mathscr{A})$ is a vector lattice for the pointwise ordering. Therefore we may also consider the order dual $M(\mathscr{X}, \mathscr{A})^{\#}$ of $M(\mathscr{X}, \mathscr{A})$. Let us compare the duals $M(\mathscr{X}, \mathscr{A})^*$ and $M(\mathscr{X}, \mathscr{A})^{\#}$.

Consider first a functional Φ in $M(\mathscr{X}, \mathscr{A})^*$ and let $h_0 \in M(\mathscr{X}, \mathscr{A})_+$. Then

$$|\Phi(h)| \leqslant \|\Phi\| \, \|h\| \leqslant \|\Phi\| \, \|h_0\| \quad \text{when} \quad |h| \leqslant |h_0|.$$

Thus $\Phi \in M(\mathscr{X}, \mathscr{A})^{\#}$. Consider next a functional Ψ in $M(\mathscr{X}, \mathscr{A})^{\#}_+$. Then

$$|\Psi(h)| \leqslant \Psi(|h|) \leqslant \|h\| \Psi(1) \quad \text{when} \quad h \in M(\mathscr{X}, \mathscr{A}).$$

It follows that $\|\Psi\| = \Psi(1)$ so that $\Psi \in M(\mathscr{X}, \mathscr{A})^*$. Thus

$$M(\mathscr{X}, \mathscr{A})^* = M(\mathscr{X}, \mathscr{A})^{\#}$$

in the sense that the underlying linear spaces of the two sides are the same.

It follows then from theorem 5.4.1 that $M(\mathscr{X}, \mathscr{A})^*$ is an order complete vector lattice for the setwise ordering. On the other hand, we know from section 4.5 that the linear space of bounded additive set functions constitutes a vector lattice for the setwise ordering. For each bounded additive set function μ on $\mathscr{A}$, let $\Phi_\mu : f \to \int f \, d\mu$ denote the linear functional defined by μ. It is then easily checked that the map $\mu \to \Phi_\mu$ is a vector lattice isomorphism from the vector lattice of bounded additive set functions onto $M(\mathscr{X}, \mathscr{A})^\#$. In particular it follows that the notation introduced in section 4.5 is consistent with the notation we are using in this section, i.e. that $[\Phi_\mu]^\pm = \Phi_{\mu^\pm}$ and $|\Phi_\mu| = \Phi_{|\mu|}$.

Convergence of sequences of real numbers may be expressed in terms of the vector lattice $\mathbb{R}$. Carrying this over to the general case we shall say that *the set $(v_n : n \in D)$ in the vector lattice V order converges to the vector $v \in V$* if there is a monotonically decreasing net $(w_n : n \in D)$ such that $|v_n - v| \leqslant w_n$; $n \in D$ and $\bigwedge_n w_n = 0$. If $(v_n : n \in D)$ order converges to v, then we shall write this as $\operatorname{olim}_n v_n = v$.

It is easily checked that a net $(v_n : n \in D)$ in an order complete vector lattice V order converges to $v \in V$ if and only if $\bigwedge_{N \in D} \bigvee_{n \geqslant N} |v_n - v| = 0$.

If V is the order complete vector lattice of equivalence classes of real variables on a probability space, then a sequence $(v_n; n = 1, 2, \ldots)$ order converges if and only if it converges almost surely. Thus the notion of order convergence is not, in general, a topological notion of convergence.

Some basic properties of order convergence are collected in the following proposition.

Proposition 5.4.3. *Let $(v_n : n \in D)$ and $(w_n : n \in D)$ be nets in a vector lattice V such that $\operatorname{olim}_n v_n = v$ and $\operatorname{olim}_n w_n = w$. Then $\operatorname{olim}_n tv_n = tv$; $t \in \mathbb{R}$, $\operatorname{olim}_n (v_n + w_n) = v + w$, $\operatorname{olim}_n (v_n \vee w_n) = v \vee w$, $\operatorname{olim}_n (v_n \wedge w_n) = v \wedge w$.*

If V is Archimedean then $\operatorname{olim}_n (t_n v) = (\lim_n t_n) v$ when the net $(t_n : n \in D)$ in $\mathbb{R}$ converges in $\mathbb{R}$. If V is not Archimedean then there is a vector $v \in V$ such that $(v/n; n = 1, 2, \ldots)$ does not order converge to 0.

If V is sequentially order complete in the sense that any countable subset of V possessing an upper bound possesses a least upper bound, then a sequence $(v_n : n = 1, 2, \ldots)$ in V order converges to $v \in V$ if and only if $\bigwedge_n \bigvee_{m \geqslant n} |v_m - v| = 0$.

Proof. Suppose $|v_n - v| \leqslant x_n$, $|w_n - w| \leqslant y_n$ where $x_n \downarrow$, $y_n \downarrow$, $\wedge x_n = 0$ and $\bigwedge_n y_n = 0$. Then $|(v_n + w_n) - (v + w)| \leqslant x_n + y_n$ and $\bigwedge_n (x_n + y_n) = 0$. Thus $\operatorname{olim}_n (v_n + w_n) = v + w$. The other limits are established similarly. If $t_n \to t$ then $|t_n v - tv| \leqslant |t_n - t||v| \leqslant s_n |v|$ where $s_n \downarrow 0$. Hence, provided V is Archimedean, $\bigwedge_n s_n |v| = 0$ so that $\operatorname{olim}_n (t_n v) = (tv)$. If V is not Archimedean then there are vectors $v > 0$ and $w > 0$ in V such that $nw \leqslant v, n = 1, 2, \ldots$ Then v/n; $n = 1, 2, \ldots$, does not order converge to 0.

The last two claims follow almost directly from the definitions. □

We shall need a few facts on order continuity of linear functionals. Let us begin by considering the sequential case.

Proposition 5.4.4. *Let f be a linear functional on a vector lattice V. Then the following two conditions are equivalent:*

(i) *if $(v_n : n = 1, 2, \ldots)$ is an order convergent sequence in V then* $\lim_n f(v_n) = f(\operatorname{olim}_n v_n)$.
(i′) *if $(v_n : n = 1, 2, \ldots)$ order converges to 0 then* $\lim_n f(v_n) = 0$.

If V is Archimedean then these conditions imply that f is order bounded.

Proof. The equivalence (i)⇔(i′) follows directly from the definitions. Suppose V is Archimedean and that f is not order bounded. Then there is a $w \geqslant 0$ such that $\sup\{|f(v)| : |v| \leqslant w\} = \infty$. It follows that there is a sequence v_n; $n = 1, 2, \ldots$ such that $|v_n| \leqslant w$ and $|f(v_n)| \geqslant n$; $n = 1, 2, \ldots$ Then $\operatorname{olim}_n (v_n/n) = 0$ although $1 \leqslant |f(v_n/n)|$; $n = 1, 2, \ldots$ □

An order bounded linear functional f on a vector lattice V satisfying the equivalent conditions (i) and (i′) of proposition 5.4.4 is called *sequentially order continuous.*

Often sequences are not adequate and we need general nets.

Proposition 5.4.5. *Let f be a linear functional on a vector lattice V. Then each of the following four conditions are equivalent:*

(i) *if $v \in V$ and $\mathscr{F}$ is a filter in V (see remark below) containing a family of order intervals with intersection $\{v\}$ then to each $\varepsilon > 0$ there corresponds a set $B \in \mathscr{F}$ such that $f[B] \subseteqq [f(v) - \varepsilon, f(v) + \varepsilon]$.*
(i′) *as* (i), *with $v = 0$.*
(ii) *if the net $(v_n : n \in D)$ order converges to v then* $\lim_n f(v_n) = f(v)$.
(ii′) *as* (ii), *with $v = 0$.*

Remark. A class $\mathscr{F}$ of non-empty subsets of a set S is called a *filter* in S if $B_1 \cap B_2 \in \mathscr{F}$ when $B_1 \in \mathscr{F}$ and $B_2 \in \mathscr{F}$ and if $B \in \mathscr{F}$ whenever $A \subseteqq B$ and $A \in \mathscr{F}$.

Proof. The equivalences (i)⇔(i′) and (ii)⇔(ii′) are trivial. Suppose (i′) is satisfied. Let $(v_n : n \in D)$ and $(w_n : n \in D)$ be nets in V such that $|v_n| \leqslant w_n$; $n \in D$, $w_n \downarrow$ and $\wedge\, w_n = 0$. If $\varepsilon > 0$ then $f(v_n) \in f[-w_n, w_n] \subseteqq [-\varepsilon, \varepsilon]$ when n is sufficiently large. Thus (ii′) holds.

Finally, assume that (ii′) is satisfied. Let $\mathscr{F}$ be a filter on V containing a family $[-a_t, b_t]$, $t \in T$ of order intervals such that $\bigwedge_t [-a_t, b_t] = \{0\}$. Then $\bigwedge_t a_t = \bigwedge_t b_t = 0$. Putting $a_S = \bigwedge_S a_t$ and $b_S = \bigwedge_S b_t$ for each finite subset S of T, we see that we may assume without loss of generality, that $a_t : t \in T$ and $b_t : t \in T$ are monotonically decreasing nets. Then $\bigwedge_t c_t = 0$ where $c_t \equiv a_t \vee b_t$.

Thus we may also assume that $a_t \equiv b_t$. Let D be the set of all pairs (x, t) such that $x \in [-a_t, a_t]$. For each pair $d = (x, t)$ in D put $y_d = x$ and $z_d = a_t$. Then $|y_d| \leqslant z_d$ and $\bigwedge_d z_d = 0$. If we direct D by defining $d' = (x', t') \geqslant d = (x, t)$ to mean that $t' \geqslant t$, then $z_d \downarrow$. Let $\varepsilon > 0$. By assumption there is a $d^0 = (x^0, t^0) \in D$ such that $|f(y_d)| \leqslant \varepsilon$ when $d \geqslant d^0$. Then $|f(x)| \leqslant \varepsilon$ when $|x| \leqslant a_t$ and $t \geqslant t^0$. Thus $f[-a_t, a_t] \subseteqq [-\varepsilon, \varepsilon]$ when $t \geqslant t^0$. □

An order bounded linear functional f on a vector lattice V satisfying one of the four equivalent conditions of proposition 5.4.5 is called *order continuous.*

Order continuity of non-negative linear functionals f may be described in terms of monotone nets.

Proposition 5.4.6. *A non-negative linear functional f on a vector lattice V is order continuous (sequentially order continuous) if and only if $f(v_n) \downarrow 0$ whenever $(v_n : n \in D)$ is a monotonically decreasing net (sequence) in V such that $\bigwedge_n v_n = 0$.*

Proof. The necessity of the condition is clear from the definition. Suppose $f \geqslant 0$ satisfies the given condition. Let $(v_n : n \in D)$ and $(w_n : n \in D)$ be nets (sequences) in V such that $|v_n| \leqslant w_n$; $n \in D$, $w_n \downarrow$ and $\bigwedge_n w_n = 0$. Then $|f(v_n)| \leqslant f(|v_n|) \leqslant f(w_n) \downarrow 0$. □

In general, order continuity may be decided by considering monotone nets only, since f^+ and f^- are order continuous if f is order continuous. This and other basic facts on the set of order continuous (sequentially order continuous) linear functionals follows from the following theorem.

Theorem 5.4.7. *The set of order continuous (sequentially order continuous) linear functionals on a vector lattice V is a band in $V^{\#}$.*

Proof. Let $V_g^{\#}$ $(V_s^{\#})$ be the set of order continuous (sequentially order continuous) linear functionals on V. Trivially $V_g^{\#} \subseteqq V_s^{\#}$ and both sets are linear. Let $f \in V_g^{\#}$ and consider f^+. Let $|v_n| \leqslant w_n$ where $w_n \downarrow$ and $\bigwedge_n w_n = 0$. We shall show that $f^+(v_n) \to 0$. Now $|f^+(v_n)| \leqslant f^+(|v_n|) \leqslant f^+(w_n) = \sup\{f(x): 0 \leqslant x \leqslant w_n\}$. Let D be the set of all pairs (x, n) such that $0 \leqslant x \leqslant w_n$. Direct D by defining $(x', n') \geqslant (x, n)$ to mean $n' \geqslant n$. For each $d = (x, n) \in D$, put $y_d = x$ and $z_d = w_n$. Then $z_d \downarrow$, $\bigwedge_d z_d = 0$ and $|y_d| \leqslant z_d$; $d \in D$. Hence $f(y_d) \to 0$. Let $\varepsilon > 0$. Then there is a $d_0 = (x^0, n_0)$ such that $d \geqslant d_0 \Rightarrow |f(y_d)| \leqslant \varepsilon$. Then $|f(x)| \leqslant \varepsilon$ when $|x| \leqslant w_n$ and $n \geqslant n_0$. In particular $f^+(w_n) \leqslant \varepsilon$ when $n \geqslant n_0$. Hence $f^+(w_n) \downarrow 0$. Suppose next that $f \in V_s^{\#}$. If the nets (v_n) and (w_n) above are sequences then for each n, we choose an $a_n \in [0, w_n]$ such that $f(a_n) \geqslant f^+(w_n) - 1/n$. Then $f(a_n) \to 0$ so that $f^+(w_n) \downarrow 0$. In both cases it follows that $f^+(v_n) \to 0$. Thus $V_g^{\#}$ and $V_s^{\#}$ are both vector lattices. Then, as is easily seen, the definitions imply that $V_g^{\#}$ and $V_s^{\#}$ are both solid and hence ideals in $V^{\#}$.

It remains to show that these ideals are order complete. Let $(f_\alpha : \alpha \in A)$ be a monotonically increasing and bounded net (sequence) of non-negative functionals in $V_g^{\#}(V_s^{\#})$. Put $f = \sup_\alpha f_\alpha$. Let $(v_n : n \in D)$ and $(w_n : n \in D)$ be nets (sequences) in v such that $|v_n| \leqslant w_n; n \in D$, $w_n \downarrow$ and $\bigwedge_n w_n = 0$. Then $|f(v_n)| \leqslant f(|v_n|) \leqslant f(w_n)$. Now, since $f_\alpha \in V_g^{\#}(V_s^{\#})$, $f_\alpha(w_n) \downarrow 0$ for each $\alpha \in A$. Let $\varepsilon > 0$ and fix an element $n_0 \in D$. Then there is an $\alpha_0 \in A$ such that $f_{\alpha_0}(w_{n_0}) \geqslant f(w_{n_0}) - \varepsilon$. Choose next an $n_1 \geqslant n_0$ such that $f_{\alpha_0}(w_{n_1}) \leqslant \varepsilon$. Let $n \geqslant n_1$. Then $f(w_n) \leqslant f(w_{n_1}) = (f - f_{\alpha_0})(w_{n_1}) + f_{\alpha_0}(w_{n_1}) \leqslant (f - f_{\alpha_0})(w_{n_0}) + \varepsilon \leqslant 2\varepsilon$. Thus $f(w_n) \downarrow 0$. Hence $f \in V_g^{\#}(V_s^{\#})$. □

If V is a vector lattice then we may form its order dual $V^{\#}$ and then the order dual $V^{\#\#}$ of $V^{\#}$. For each $v \in V$, let $e(v)$ be the linear functional on $V^{\#}$ which to each f in $V^{\#}$ assigns its value $f(v)$ at v. Then $e(v) = e(v^+) - e(v^-)$ so that $e(v) \in V^{\#\#}$. The map e is called the *evaluation map*. The evaluation map is clearly linear and non-negative. In some situations the evaluation provides an imbedding, i.e. it is 1–1, of V in $V^{\#\#}$. In other situations $V^{\#\#}$ may be too small to make imbedding feasible.

Example 5.4.8 (Lexicographic ordering). Let $V = \mathbb{R}^n$ with the lexicographic ordering as described in example 5.2.3. Identifying the algebraic dual V' of V with $\mathbb{R}^n$ in the usual way, we see that the dual ordering is the ordering on $\mathbb{R}^n$ where $x = (x_1, \ldots, x_n) \geqslant y = (y_1, \ldots, y_n)$ if and only if $x_1 \geqslant y_1$ and $x_i = y_i$ when $i \geqslant 2$. The order dual $V^{\#}$ of V is the one-dimensional subspace $\{(a, 0, \ldots, 0); a \in \mathbb{R}\}$. In the same way the second order dual becomes the real line $\mathbb{R}$ with the usual ordering. Thus the vector lattices $V^{\#}$ and $V^{\#\#}$ are isomorphic. If $v \in \mathbb{R}^n$ then the evaluation at v assigns the number av_1 to the vector $(a, 0, \ldots, 0)$ in $V^{\#}$. Thus $e(v) = e(w)$ if and only if $v_1 = w_1$. It follows that the evaluation is not 1–1 when $n \geqslant 2$.

Example 5.4.9. Let $(\mathscr{X}, \mathscr{A}, P)$ be a probability space and consider the vector lattice W of P-equivalence classes of real valued measurable functions. This vector lattice may be more precisely described as follows. Consider the linear space V of real valued measurable functions equipped with the pointwise ordering. Then V is a vector lattice having the class I of real valued null functions as an ideal. The vector lattice W may be identified with the quotient vector lattice V/I. In example 5.5.10 we shall see that W is order complete and, hence, Archimedean. The fact that W is Archimedean may also be readily demonstrated directly. However, the order dual $W^{\#}$ of W possesses non-zero elements if and only if P possesses atoms.

It is interesting that the evaluation map, whether it is an imbedding or not, is always a vector lattice homomorphism of V into $V^{\#\#}$.

Proposition 5.4.10 (The basic property of the evaluation map). *Let V be a vector lattice. Then the evaluation map is a vector lattice homomorphism from V into $V^{\#\#}$.*

Proof. Since e is linear it suffices to show that $e(v^+) = [e(v)]^+$ when $v \in V$. Let $f \in V_+^{\#}$ and let $0 \leqslant g \leqslant f$. Then $g(v) \leqslant g(v^+) \leqslant f(v^+)$ so that $\sup\{g(v); 0 \leqslant g \leqslant f; g \in V^{\#}\} \leqslant f(v^+)$. For each $w \in V$, put $p(w) = f(w^+)$ and let V_0 be the linear subspace generated by the vector v. Define the linear functional g on V_0 by putting $g(cv) = cf(v^+)$ when $c \in \mathbb{R}$. Then $g \leqslant p$ on V_0 and p is sublinear on V. Hence, by the Hahn–Banach theorem, theorem 2.2.12, g may be extended to a linear functional on V such that $g(w) \leqslant p(w) = f(w^+)$ for all $w \in V$. Then $g(w) \leqslant 0$ when $w \leqslant 0$. Hence $g \geqslant 0$ on V. If $w \geqslant 0$ then $g(w) \leqslant f(w)$. Thus $0 \leqslant g \leqslant f$ and $g(v) = f(v^+)$. Hence $f(v^+) = \sup\{g(v) : 0 \leqslant g \leqslant f; g \in V^{\#}\}$ when $f \in V_+^{\#}$ i.e. $e(v^+) = [e(v)]^+$. □

The dual formulas of those in theorem 5.4.1 are collected in the following corollary.

Corollary 5.4.11. Let V be a vector lattice with order dual $V^{\#}$. Let $v, v_1, \ldots, v_r \in V$ and let $f \in V_+^{\#}$. Then

(i) $f(v^+) = \sup\{g(v) : 0 \leqslant g \leqslant f\}$.
(ii) $f(v^-) = \inf\{g(v) : -f \leqslant g \leqslant 0\}$.
(iii) $f(|v|) = \sup\{g(v) : |g| \leqslant f\}$.
(iv) $f(v_1 \vee \ldots \vee v_r) = \sup\{g_1(v_1) + \cdots + g_r(v_r) : g_1, \ldots, g_r \geqslant 0, g_1 + \cdots + g_r = f\}$.
(v) $f(v_1 \wedge \ldots \wedge v_r) = \inf\{g_1(v_1) + \cdots + g_r(v_r) : g_1, \ldots, g_r \geqslant 0, g_1 + \cdots + g_r = f\}$.

Corollary 5.4.12. *The space of order continuous linear functionals on the order dual $V^{\#}$ of the vector lattice V is a band in $V^{\#\#}$ which contains V, by imbedding.*

Proof. Let $v \in V$ and suppose $(f_n : n \in D)$ and $(g_n : n \in D)$ are nets in $V^{\#}$ such that $|f_n| \leqslant g_n; n \in D$, $g_n \downarrow$ and $\bigwedge_n g_n = 0$. Then $|f_n(v)| \leqslant |f_n|(|v|) \leqslant g_n(|v|) \downarrow 0$. □

By definition a linear functional f is non-negative if and only if $f(v) \geqslant 0$ when $v \geqslant 0$. The dual statement that a vector v in V is non-negative if and only if $f(v) \geqslant 0$ when $f \in V_+^{\#}$ holds if and only if the evaluation map is 1–1.

Corollary 5.4.13. *Let V be a vector lattice and let e be the evaluation map from V to $V^{\#\#}$. Then the following conditions are equivalent:*

(i) *e is a vector lattice isomorphism.*
(ii) *if $v \in V$ and if $f(v) \geqslant 0$ whenever $f \in V_+^{\#}$ then $v \geqslant 0$.*
(iii) *e is* 1–1.

Proof. Suppose (iii) holds and that $f(v) \geqslant 0$ when $f \in V_+^{\#}$. Let $f \in V_+^{\#}$. Then $g(v) \leqslant 0$ when $-f \leqslant g \leqslant 0$. Hence, by corollary 5.4.11, $f(v^-) \leqslant 0$. On the other hand, since $v^- \geqslant 0$, $f(v^-) \geqslant 0$. Hence $f(v^-) = 0$ when $f \in V_+^{\#}$. It follows that $f(v^-) = 0$ for all $f \in V^{\#}$ so that $e(v^-) = 0$. Hence, since e is assumed to be 1–1,

$v^- = 0$ so that $v \geqslant 0$. Thus (iii) $\Rightarrow$ (ii). The equivalence (i) $\Leftrightarrow$ (iii) follows from the proposition. Suppose finally that (ii) holds and suppose $e(v) = 0$. Then $f(v) = 0$ for all $f \in V^{\#}$. Hence, by (ii), $v \geqslant 0$. Similarly $-v \geqslant 0$ so that $v = 0$. Thus (ii) $\Rightarrow$ (iii). □

5.5 Normed vector lattices. L-spaces and M-spaces

Most of the vector lattices which we shall encounter are equipped with a norm $\| \ \|$. Let us remind ourselves that a *norm*, $\| \ \|$, on a linear space V is a function from V to $[0, \infty[$ which is:

(i) sublinear, i.e. $\|v_1 + v_2\| \leqslant \|v_1\| + \|v_2\|$ and $\|tv\| = |t|\,\|v\|$ when $v_1, v_2, v \in V$ and $t \in \mathbb{R}$.
(ii) positive definite, i.e. $\|v\| > 0$ when $0 \neq v \in V$.

The norm $\| \ \|$ on V determines a metric $(v_1, v_2) \to \|v_1 - v_2\|$ on V. The norm is called *complete* if this metric is complete, i.e. if Cauchy sequences for this metric converge. A linear space equipped with a norm is called a normed linear space. If the norm is complete then the normed linear space is called a *Banach space*.

A norm $\| \ \|$ on a vector lattice is called a *lattice norm* if $\|v_1\| \leqslant \|v_2\|$ whenever $|v_1| \leqslant |v_2|$. A vector lattice equipped with a lattice norm is called a *normed vector lattice*. If, in addition, this norm is complete, then the normed vector lattice is called a *Banach lattice*.

Let v_1, v_2, w_1 and w_2 be vectors in a normed vector lattice V. Then the vectors $|v_1 \wedge w_1 - v_2 \wedge w_2|$ and $|v_1 \vee w_1 - v_2 \vee w_2|$ are both majorized by $|v_1 - v_2| + |w_1 - w_2|$. Hence $\|v_1 \wedge w_1 - v_2 \wedge w_2\|$ and $\|v_1 \vee w_1 - v_2 \vee w_2\|$ are both at most equal to $\|v_1 - v_2\| + \|w_1 - w_2\|$. It follows that the lattice operations are continuous in normed vector lattices.

By the following theorem, bands in order complete normed vector lattices contain their accumulation points.

Theorem 5.5.1. *A band in an order complete normed vector lattice is closed.*

Proof. Let B be a band in the order complete normed vector lattice V. Suppose $\lim_n v_n = v$ where $v_1, v_2 \ldots \in B$. Decompose v as $v = \hat{v} + \tilde{v}$ where $\hat{v} \in B$ and $\tilde{v} \in B^{\S}$. Then $|v - v_n| = |\hat{v} - v_n + \tilde{v}| = |\hat{v} - v_n| + |\tilde{v}| \geqslant |\tilde{v}|$. Hence $\|\tilde{v}\| \leqslant \|v - v_n\| \to 0$ so that $\tilde{v} = 0$ and thus $v = \hat{v} \in B$. □

We have seen that bands possessing a finite set of generators are always principal bands. If V is an order complete Banach lattice then this may be extended.

Theorem 5.5.2. *Let V be an order complete normed vector lattice. Then any band possessing a countable number of generators is a principal band.*

Proof. Let $\{\mu_1, \mu_2, \ldots\}$ generate the band B in V. We may assume, without loss of generality, that $\mu_i \geqslant 0$ and $\|\mu_i\| = 1, i = 1, 2, \ldots$ Put $\pi = \sum_i 2^{-i}\mu_i$. Then π is well defined and belongs to B since B is complete. Hence $B_\pi \subseteqq B$. On the other hand, $\mu_i \leqslant 2^i\pi$, $i = 1, \ldots$ so that $\mu_1, \mu_2, \ldots \in B_\pi$. Then $B \subseteqq B_\pi$ so that $B = B_\pi$. □

If $(V, \| \ \|)$ is a normed linear space (in particular if it is a Banach space) then the *dual normed linear space* of V is the linear space V^* of bounded linear functionals on V equipped with the dual norm $\| \ \|^*$ defined by $\|f\|^* = \sup\{|f(v)| : \|v\| \leqslant 1\}$. The dual norm $\| \ \|^*$ is always complete so that $(V, \| \ \|^*)$ is a Banach space, even if the original norm $\| \ \|$ is not complete. Suppose now that $(V, \| \ \|)$ is a Banach lattice. Then any continuous linear functional is bounded on norm bounded sets, and hence on order bounded sets. Hence $V^* \subseteqq V^\#$. On the other hand, let f be a non-negative linear functional on V and put $c = \sup\{f(v) : 0 \leqslant v, \|v\| \leqslant 1\}$. If $c = \infty$ then there are vectors $v_1, v_2 \ldots$ in V_+ such that, for each n, $\|v_n\| \leqslant 1$ and $f(v_n) \geqslant n^3$. By completeness the element $v = \sum_{n=1}^\infty v_n/n^2$ is well defined in V_+. Hence $f(v) \geqslant f(v_n/n^2) = f(v_n)/n^2 \geqslant n$, $n = 1, 2 \ldots$ Since $f(v) < \infty$ this is a contradiction. It follows that $c < \infty$. Let $\|v\| \leqslant 1$. Then $\||v|\| \leqslant 1$ so that $|f(v)| \leqslant f(|v|) \leqslant c$. Hence $\|f\|^* \leqslant c < \infty$ so that $f \in V^*$. This proves most of the following theorem.

Theorem 5.5.3. *If $(V, \| \ \|)$ is a normed vector lattice then the space V^* of bounded linear functionals on V is a sub-vector lattice of the order dual $V^\#$. Equipped with the dual norm $\| \ \|^*$, the space V^* becomes a Banach lattice.*

If $(V, \| \ \|)$ is a Banach lattice then the order dual $V^\#$ and the Banach space dual V^* consist of precisely the same linear functionals.

Remark. The asterisk appearing in the notation for the dual norm is usually dropped.

Proof. We argued the inclusion $V^* \subseteqq V^\#$ without using completeness. Let $f \in V^*$. Then $|f|(v) = \sup\{f(w) : |w| \leqslant v\}$ when $v \geqslant 0$. If $|w| \leqslant v$ then $\|w\| \leqslant \|v\|$. Hence $|f|(v) \leqslant \|f\| \|v\|$ when $v \geqslant 0$. It follows that $\||f|\| \leqslant \|f\| < \infty$ so that $|f| \in V^*$. This shows that V^* is a vector lattice and that $\||f|\| \leqslant \|f\|$; $f \in V^*$. On the other hand, $|f(v)| \leqslant |f|(|v|) \leqslant \||f|\| \|v\|$; $v \in V$ so that $\|f\| \leqslant \||f|\|$. Hence $\|f\| = \||f|\|$ when $f \in V^*$. Suppose $0 \leqslant f_1 \leqslant f_2 \in V^*$ and that $v \geqslant 0$. Then $f_1(v) \leqslant f_2(v) \leqslant \|f_2\| \|v\|$ so that $\|f_1\| \leqslant \|f_2\|$. Thus, since $\| \ \|^*$ is complete, $(V^*, \| \ \|^*)$ is a Banach lattice. □

Example 5.5.4 (Additive set functions. Continuation of example 5.4.2). The vector lattice $M(\mathscr{X}, \mathscr{A})$ of bounded measurable functions on $(\mathscr{X}, \mathscr{A})$ equipped with the supremum norm, is a Banach lattice. Hence, by the theorem, we find again that $M(\mathscr{X}, \mathscr{A})^\# = M(\mathscr{X}, \mathscr{A})^*$.

By the Hahn–Banach theorem the evaluation map of any Banach space into its second dual is an isometry. Thus, by corollary 5.4.13, the evaluation provides an imbedding of the Banach lattice V in the Banach lattice $V^{\#\#}$. If $(V, \| \ \|)$ is a normed vector lattice then a vector $v_0 \in V$ is called a *unit* of V if $\|v_0\| \leqslant 1$ and if $v \leqslant v_0$ whenever $\|v\| \leqslant 1$. Thus a unit is the largest vector in the ball $\{v : \|v\| \leqslant 1\}$ around 0 in V. A unit, if it exists, is clearly unique and non-negative.

The L_1-norm on a measure space and the supremum norm on a function space exemplify two important types of lattice norms.

A norm $\| \ \|$ on a vector lattice V is called an *L-norm* (*M-norm*) if $\|v_1 + v_2\| = \|v_1\| + \|v_2\|$ ($\|v_1 \vee v_2\| = \|v_1\| \vee \|v_2\|$) when $v_1, v_2 \in V_+$ and if, furthermore, $\|v\| = \| |v| \|$ when $v \in V$. It is easily seen that L-norms and M-norms are lattice norms. Normed lattices equipped with L-norms are called *L-normed spaces*, while normed lattices equipped with M-norms are called *M-normed spaces*. A complete L-normed space is called an *L-space* while a complete M-normed space is called an *M-space*.

Example 5.5.5 (Bounded additive set functions. Continuation of example 5.5.4). Let $(\mathscr{X}, \mathscr{A})$ be a measurable space and, using the notation of section 4.5, let us denote the linear space of bounded additive set functions on $\mathscr{A}$ by $ba(\mathscr{X}, \mathscr{A})$. Then $ba(\mathscr{X}, \mathscr{A})$ equipped with the set-wise ordering and the total variation norm is an L-space. This L-space is isomorphic to the Banach lattice $M(\mathscr{X}, \mathscr{A})^{\#}$ equipped with the norm of $M(\mathscr{X}, \mathscr{A})^*$. Thus $ba(\mathscr{X}, \mathscr{A})$ and $M(\mathscr{X}, \mathscr{A})^{\#}$ (or $M(\mathscr{X}, \mathscr{A})^*$) are isomorphic L-spaces. $M(\mathscr{X}, \mathscr{A})^{\#}$ is order complete by theorem 5.4.1. Hence $ba(\mathscr{X}, \mathscr{A})$ is also order complete.

The linear space $ca(\mathscr{X}, \mathscr{A})$ of finite measures on $\mathscr{A}$ constitutes a closed linear subspace of $ba(\mathscr{X}, \mathscr{A})$. If $\mu \in ca(\mathscr{X}, \mathscr{A})$ then μ possesses a Hahn set $H \in \mathscr{A}$ so that $\mu(A) \geqslant 0$ or $\leqslant 0$ according to whether $A \subseteqq H$ or $A \subseteqq H^c$. It follows that $\mu^+(A) = \mu(AH)$ when $A \in \mathscr{A}$. In particular μ^+ (and hence μ^- and $|\mu|$) belong to $ca(\mathscr{X}, \mathscr{A})$. It follows that $ca(\mathscr{X}, \mathscr{A})$ equipped with the total variation norm constitutes an L-space in its own right.

The vector lattice $ca(\mathscr{X}, \mathscr{A})$ is an ideal in $ba(\mathscr{X}, \mathscr{A})$ since $\mu \in ca(\mathscr{X}, \mathscr{A})$ whenever $\mu \in ba(\mathscr{X}, \mathscr{A})$ is such that $|\mu| \leqslant \nu$, where $\nu \in ca(\mathscr{X}, \mathscr{A})$. Let $\mu_t : t \in T$ and μ be non-negative finite measures such that $\mu_t \leqslant \mu$ when $t \in T$. Then $\sup_t \mu_t$ exists in $ba(\mathscr{X}, \mathscr{A})$, since $ba(\mathscr{X}, \mathscr{A})$ is order complete. Furthermore, $\sup_t \mu_t \in ca(\mathscr{X}, \mathscr{A})$ since $ca(\mathscr{X}, \mathscr{A})$ is solid and $0 \leqslant \sup_t \mu_t \leqslant \mu \in ca(\mathscr{X}, \mathscr{A})$. Hence $ca(\mathscr{X}, \mathscr{A})$ is a band in $ba(\mathscr{X}, \mathscr{A})$.

It follows from theorem 5.5.1 that any band in an L-space is closed and consequently is an L-space in its own right. In particular any band in $ca(\mathscr{X}, \mathscr{A})$ is an L-space. In a sense this covers the most general L-space since (by Kakutani, 1941) any L-space is isomorphic to a band in $ca(\mathscr{X}, \mathscr{A})$ for some measurable space $(\mathscr{X}, \mathscr{A})$.

Example 5.5.6 (M-normed function spaces). Let F be a vector lattice of real valued functions on a set $\mathscr{X}$. The ordering is then assumed to be the pointwise ordering. Then F equipped with the supremum norm becomes an M-normed space. If F contains the constants then the constant function 1 is a unit in F. As we shall see in section 5.7, Kakutani's representation theory implies that any M-space with a unit is isomorphic to the M-space of continuous functions on a compact Hausdorff space $\mathscr{X}$.

Example 5.5.7. Let V be an Archimedean vector lattice and let $v \in V_+$. Then the ideal A generated by v consists precisely of all vectors a such that $|a| \leqslant nv$ for some integer n. To each vector a in A assign the number $\|a\|_v = \inf\{t : t \geqslant 0, |a| \leqslant tv\}$. Then $\| \ \|_v$ is an M-norm on A (see Schaefer 1974). The M-normed space $(A, \| \ \|_v)$ is an M-space and the imbedding is continuous whenever V is a Banach lattice.

We shall see that band decomposition of L-spaces is a useful tool. The fact that this tool is available in any L-space follows from the following proposition.

Proposition 5.5.8. *Any L-space is order complete. If $(\mu_\theta : \theta \in \Theta)$ is a family of vectors in L which is bounded from above then there is a countable subset C of Θ such that $\sup_\Theta \mu_\theta = \sup_C \mu_\theta$. If $\mu_1, \mu_2, \ldots$ is an infinite sequence of vectors then $\mu_1 \vee \mu_2 \vee \ldots \vee \mu_n \to \sup_n \mu_n$ provided $\sup_n \mu_n$ exists.*

Remark. M-spaces are not necessarily order complete. If, however, we know that the M-space M is the dual of an L-normed space then we also know, by theorems 5.4.1 and 5.5.3, that M is order complete.

Proof. Let L be an L-space and let $(\mu_\theta : \theta \in \Theta)$ be a family in L. Suppose $\mu_\theta \leqslant \lambda \in L; \theta \in \Theta$. Suppose also for the moment that $\mu_{\theta_0} = 0$ for some $\theta_0 \in \Theta$. Let $\mathscr{F}$ be the class of finite subsets of Θ which contains θ_0. Put $\mu_F = \sup\{\mu_\theta : \theta \in F\}$ when $F \in \mathscr{F}$. Also put $\alpha = \sup_F \|\mu_F\|$. Then, since $0 \leqslant \mu_F \leqslant \lambda$ when $F \in \mathscr{F}$, $\alpha \leqslant \|\lambda\|$ so that $\alpha < \infty$. Let $F_1 \subseteqq F_2 \subseteqq \cdots$ be a sequence in $\mathscr{F}$ such that $\|\mu_{F_n}\| \uparrow \alpha$. Put $C = \bigcup_n F_n$. By the L-norm property, $\|\mu_{F_{n+m}} - \mu_{F_m}\| = \|\mu_{F_{m+n}}\| - \|\mu_{F_m}\| \leqslant \alpha - \|\mu_{F_m}\| \to 0$ as $m \to \infty$. Hence, since L is complete, there is a $\mu \in L$ such that $\mu_{F_n} \to \mu$ as $n \to \infty$. Then $\mu_{F_n} - \mu_{F_{n+m}} \to \mu_{F_n} \to \mu$ as $m \to \infty$. Hence $0 = (\mu_{F_n} - \mu_{F_{n+m}})^+ \to (\mu_{F_n} - \mu)^+$ so that $\mu_{F_n} \leqslant \mu; n = 1, 2, \ldots$ If $\theta \in \Theta$ then $\|\mu\| = \alpha \geqslant \|\mu_{F_n \cup \{\theta\}}\| = \|\mu_{F_n} \vee \mu_\theta\| \to \|\mu \vee \mu_\theta\|$ so that $\|\mu\| \geqslant \|\mu \vee \mu_\theta\|$. Hence $\|\mu \vee \mu_\theta - \mu\| = \|\mu \vee \mu_\theta\| - \|\mu\| \leqslant 0$. It follows that $\mu = \mu \vee \mu_\theta \geqslant \mu_\theta; \theta \in \Theta$. Thus μ is an upper bound for the family $(\mu_\theta : \theta \in \Theta)$. If σ is another upper bound then $\mu_{F_n} \leqslant \sigma$, $n = 1, 2 \ldots$ so that $\mu \leqslant \sigma$. Hence $\mu = \sup_\Theta \mu_\theta = \sup_C \mu_\theta$.

Suppose $\Theta = \{1, 2, \ldots\}$ and that $\sup_\theta \mu_\theta$ exists. Then by the arguments used above $\mu_1 \vee \ldots \vee \mu_n \to \sup_\theta \mu_\theta$.

We assumed that $\mu_{\theta_0} = 0$ for some $\theta_0 \in \Theta$. The general case follows by choosing a $\theta_0 \in \Theta$ and then replacing each μ_θ with $\mu_\theta - \mu_{\theta_0}$. □

Here are some useful facts on bands in L-spaces.

Proposition 5.5.9 (Bands in L-spaces). *Let L be an L-space. Then:*

(i) *an ideal V in L is a band if and only if it is closed,*

(ii) *let the family $(V_i : i \in I)$ of ideals in L be directed for inclusion. Then $\bigcup \overline{V_i}$ is a band.*

(iii) *let the family $(V_i : i \in I)$ of bands in L be countably directed for inclusion. Thus we assume that to each sequence $i_1, i_2, \ldots$ in I there corresponds an $i \in I$ such that $V_{i_1} \cup V_{i_2} \cup \ldots \subseteqq V_i$. Then $\bigcup_i V_i$ is a band.*

(iv) *let the band generated by a family $v_\theta : \theta \in \Theta$ be denoted as* $\text{Band}(v_\theta : \theta \in \Theta)$. *Then*

$$\text{Band}(\mu_\theta : \theta \in \Theta) = \bigcup\{\text{Band}(\mu_\theta : \theta \in C)\text{: } C \textit{ is a countable subset of } \Theta\}$$

$$= \textit{the closure of } \bigcup\{\text{Band}(\mu_\theta : \theta \in F)\text{: } F \textit{ is a finite subset of } \Theta\}$$

$$= \bigcup\{B_\pi : \pi = \textstyle\sum 2^{-n}|\mu_{\theta_n}|/(\|\mu_{\theta_n}\| + 1) \textit{ for some sequence } \theta_1, \theta_2, \ldots \text{ in } \Theta\}.$$

(v) *the band generated by a family $(\mu_\theta : \theta \in \Theta)$ of vectors in L is a principal band if and only if it equals the band generated by $(\mu_\theta : \theta \in C)$ for some countable subset C of Θ.*

Proof. (i) follows from theorem 5.5.1 and proposition 5.5.8. (ii) and (iii) follow from (i) by noting that $\bigcup_i V_i$ constitutes an ideal in both cases and that this set is closed under the assumption of (iii). (iv) is a direct consequence of (i)–(iii) and theorem 5.5.2. Finally the 'if' part of (v) follows from theorem 5.5.2 while the 'only if' part of (v) follows from (iv). □

Example 5.5.10 ($L_1(\mu)$ and essential suprema). Let μ be a non-negative measure on the measurable space $(\mathscr{X}, \mathscr{A})$. The L_1-norm of an integrable function f, or of a μ-equivalence class f of integrable functions, is the number $\int |f| \, d\mu$. It is well known that the linear space of μ-equivalence classes of integrable functions equipped with the L_1-norm, is a Banach space (see any introductory book in measure theory). This Banach space is usually denoted as $L_1(\mu)$. The L_1-norm is clearly additive on $[L_1(\mu)]_+$. Hence $L_1(\mu)$ is an L-space.

We shall see later that any L-space is isomorphic to an L-space of the form $L_1(\mu)$.

Suppose now that μ is σ-finite and let π be a probability measure on $\mathscr{A}$ possessing the same null sets as μ. Choose a continuous and strictly increasing function G from $[-\infty, \infty]$ to $[0, 1]$.

Let $\{f_\theta : \theta \in \Theta\}$ be a family of extended real valued measurable functions on $(\mathscr{X}, \mathscr{A})$. Then $\{G(f_\theta) : \theta \in \Theta\}$ is a family of π-integrable functions. Hence, by proposition 5.5.8, there is a countable subset C of Θ such that, for each θ,

$G(f_\theta) \leqslant \sup\{G(f_{\theta'}) : \theta' \in C\}$ a.s. π. It follows that

$$f_\theta \leqslant \sup\{f_{\theta'} : \theta' \in C\} \quad \text{a.e. } \mu$$

when $\theta \in \Theta$. The function $g = \sup\{f_{\theta'} : \theta' \in C\}$ is then a supremum for the family $\{f_\theta : \theta \in \Theta\}$ for the 'a.e. μ' ordering of extended real valued measurable functions. Such a supremum is necessarily unique up to sets of μ-measure zero. A supremum for this ordering is called a *μ-essential supremum* and we write $g = \text{ess}_\Theta\sup f_\theta$. Similarly an infimum for this ordering is called a *μ-essential infimum* and we write $\text{ess}_\Theta\inf f_\theta$ for the μ-essential infimum of a family $(f_\theta : \theta \in \Theta)$.

In particular it follows that the linear space of μ-equivalence classes of real valued measurable functions equipped with the ordering derived from the '$\geqslant$ a.e. μ' ordering, is an order complete vector lattice. Using a different, although related argument, we established this in theorem 1.2.10.

Example 5.5.11 ($L_1(\mu)$ as an L-space of measures). Again let $(\mathscr{X}, \mathscr{A})$ be a measurable space and let μ be a *σ-finite non-negative measure on $\mathscr{A}$*. We defined the L-space $L_1(\mu)$ of μ-equivalence classes of integrable functions in the previous example. Now with each 'function' f in $L_1(\mu)$ associate the measure $f\mu$ on $\mathscr{A}$ defined by

$$(f\mu)(A) = \int_A f \, \mathrm{d}\mu; A \in \mathscr{A}.$$

Then by the Radon–Nikodym theorem, the map $f \to f\mu$ is a map from the L-space $L_1(\mu)$ onto the L-space of μ-absolutely continuous finite measures equipped with the total variation norm. It is a matter of checking to verify that this map is actually an isomorphism.

Example 5.5.12 (Bands in $ca(\mathscr{X}, \mathscr{A})$. Continuation of example 5.5.5). $ca(\mathscr{X}, \mathscr{A})$ is clearly a closed ideal in $ba(\mathscr{X}, \mathscr{A})$. Hence, since $ba(\mathscr{X}, \mathscr{A})$ is an L-space, by proposition 5.5.9 $ca(\mathscr{X}, \mathscr{A})$ is a band in $ba(\mathscr{X}, \mathscr{A})$. We argued this directly in example 5.5.5.

Now consider a non-negative and finite measure π in $ca(\mathscr{X}, \mathscr{A})$. What is the (principal) band B_π generated by π? Let $0 \leqslant \mu \in B_\pi$. Then $\mu \wedge n\pi \to \mu$ as $n \uparrow \infty$. Hence, since $\mu \wedge n\pi \leqslant n\pi$, $\mu(A) = 0$ whenever $\pi(A) = 0$. Thus μ is absolutely continuous with respect to π.

Conversely, suppose that the finite measure μ is absolutely continuous with respect to π. Then, by the Radon–Nikodym theorem, $\mu(\text{A}) \equiv_A \int_A f \, \mathrm{d}\mu$ for a unique $f \in L_1(\mu)$. By the previous example $\| |\mu| - |\mu| \wedge n\pi \| = \int [|f| - |f| \wedge n] \mathrm{d}\pi \to 0$ as $n \to \infty$. Hence $|\mu|$, and therefore μ, belongs to B_π.

Altogether we have shown that B_π consists precisely of those finite measures on $\mathscr{A}$ which are absolutely continuous with respect to π.

If μ is a σ-finite measure then the space of finite measures which are absolutely continuous with respect to μ equals the principal band B_π whenever the finite non-negative measure π possesses the same null sets as μ.

The Riesz decomposition theorem implies that we may decompose $ba(\mathscr{X}, \mathscr{A})$ as $ba(\mathscr{X}, \mathscr{A}) = ca(\mathscr{X}, \mathscr{A}) + ca(\mathscr{X}, \mathscr{A})^\S$ and $ca(\mathscr{X}, \mathscr{A})$ as $ca(\mathscr{X}, \mathscr{A}) = B_\pi + B_\pi^\S$. Now two finite measures λ_1 and λ_2 in $ca(\mathscr{X}, \mathscr{A})$ are disjoint in the vector lattice $ca(\mathscr{X}, \mathscr{A})$ if and only if there is a set N in $\mathscr{A}$ such that $|\lambda_1|(N^c) = 0$ while $|\lambda_2|(N) = 0$. Measures λ_1 and λ_2, such that a set N with these properties exists, are called *mutually singular*.

We have shown that if the non-negative measure μ on $\mathscr{A}$ is σ-finite then any finite measure may be decomposed in a unique way as a sum of two measures, such that one is absolutely continuous with respect to μ and the other is μ-singular. This result is usually called the Lebesgue decomposition theorem.

If V is an L-normed space then the non-negative linear functional I on V^* defined by $I(v) = \|v^+\| - \|v^-\|$; $v \in V$ is the unit of V^*. Using the defining property of an L-norm and provided V is an L-normed space, for non-negative linear functionals f and g we find that

$$\|f \vee g\| = \sup\{f(v_1) + g(v_2) : v_1 \geqslant 0, v_2 \geqslant 0,$$

$$\|v_1 + v_2\| \leqslant 1\} = \sup_{\substack{\alpha,\beta \geqslant 0 \\ \alpha+\beta=1}} \sup_{\substack{\|v_1\| \leqslant \alpha \\ v_1 \geqslant 0}} \sup_{\substack{\|v_2\| \leqslant \beta \\ v_2 \geqslant 0}} (f(v_1) + g(v_2))$$

$$= \sup_{\substack{\alpha,\beta \geqslant 0 \\ \alpha+\beta=1}} [\alpha\|f\| + \beta\|g\|] = \|f\| \vee \|g\|.$$

Suppose next that V is an M-normed space and that $\lambda_1, \lambda_2 \in V_+^*$. Choose an $\varepsilon > 0$ and elements $v_1, v_2 \in V_+$ such that $\|v_i\| \leqslant 1$ and $\lambda_i(v_i) \geqslant \|\lambda_i\| - \varepsilon_i$, $i = 1, 2$. Then $\|v_1 \vee v_2\| = \|v_1\| \vee \|v_2\| \leqslant 1$ and $\lambda_i(v_1 \vee v_2) \geqslant \|\lambda_i\| - \varepsilon$, $i = 1, 2$ so that $(\lambda_1 + \lambda_2)(v_1 \vee v_2) \geqslant \|\lambda_1\| + \|\lambda_2\| - 2\varepsilon$. By letting $\varepsilon \to 0$ it follows that $\|\lambda_1 + \lambda_2\| = \|\lambda_1\| + \|\lambda_2\|$. Combining this with the remark after proposition 5.5.8 we find that duals of L-spaces are M-spaces and vice versa.

Proposition 5.5.13. *The dual of an L-normed space is an order complete M-space with a unit. The dual of an M-normed space is an L-space.*

Example 5.5.14 (The dual of $L_1(\mu)$. Continuation of example 5.5.11). Let $(\mathscr{X}, \mathscr{A})$ be a measurable space and let μ be a σ-finite measure on $\mathscr{A}$. In example 5.5.11 we saw how $L_1(\mu)$ might be considered as an L-space. Let $L_\infty(\mu)$ be the Banach space of bounded μ-equivalence classes of measurable functions equipped with the L_∞-norm. The L_∞-norm of a real valued measurable function f is the smallest constant $a \geqslant 0$ such that $\mu(|f| > a) = 0$. As μ-equivalent functions have the same L_∞-norm this norm is well defined on $L_\infty(\mu)$. It is easily checked that $L_\infty(\mu)$, with the lattice operations defined as in $L_1(\mu)$ and with this norm, is an M-space.

Now any ϕ in $L_\infty(\mu)$ defines a bounded linear functional S_ϕ on $L_1(\mu)$ by $S_\phi(f) = \int \phi f \, d\mu$; $f \in L_1(\mu)$. The Radon–Nikodym theorem implies that the map $\phi \to S_\phi$ maps $L_\infty(\mu)$ onto $L_1(\mu)^*$. Furthermore, this map is a linear isometry. These facts are usually summarized by saying that $L_\infty(\mu)$ is the dual of $L_1(\mu)$. Now, as we have seen, $L_\infty(\mu)$ ought to be an M-space for the dual ordering and for the dual norm of $L_1(\mu)$. However, this ordering and this norm are precisely those we just claimed made $L_\infty(\mu)$ an M-space.

Example 5.5.15 ($M(\mathscr{X}, \mathscr{A})$ and its dual $ba(\mathscr{X}, \mathscr{A})$). Continuation of examples 5.5.4 and 5.5.5). We noted in example 5.5.6 that $M(\mathscr{X}, \mathscr{A})$ is an M-space with the constant function 1 as a unit. It was stated in example 5.5.5 that $ba(\mathscr{X}, \mathscr{A})$ is an L-space. This fact is also a consequence of proposition 5.13 since $ba(\mathscr{X}, \mathscr{A})$ may be identified with $M(\mathscr{X}, \mathscr{A})^*$, by example 5.5.5.

Example 5.5.16 ($\mathbb{R}^n$ with the pointwise ordering). Define the norm $\|x\|_c$ of an n-tuple $(x_1, x_2, \ldots, x_n)$ in $\mathbb{R}^n$ by $\|x\|_c = \sum_{i=1}^n c_i |x_i|$ where $c = (c_1, c_2, \ldots, c_n)$ is an n-tuple of positive numbers. Considering the measure space $(\mathscr{X}, \mathscr{A}, c)$ where $\mathscr{X} = \{1, 2, \ldots, k\}$, $\mathscr{A} =$ the class of all subsets of $\mathscr{X}$ and $c(i) = c_i$; $i = 1, \ldots, k$, we see that this is a particular case of example 5.5.11.

The norm in the dual space is the sup norm $x \to \bigvee_i |x_i|$. However, it is not difficult to show that the dual norm of the norm given here is $x \to \bigvee_i |x_i|/c_i$. How do we reconcile this? The difficulty disappears as soon as we realize that we are working with two different representations of the dual space of $L_1(c)$. In the first case the n-tuple $(a_1, a_2, \ldots, a_n)$ represents the functional $(x_1, \ldots, x_n) \to \sum c_i a_i x_i$ while in the second it represents the functional $(x_1, \ldots, x_n) \to \sum a_i x_i$.

Note that the vector lattice $\mathbb{R}^n$ equipped with the coordinatewise ordering is an M-space for the norm $x \to \bigvee_i |x_i|$ while it is an L-space for the norm $x \to \sum_i |x_i|$.

The real line $\mathbb{R}$ equipped with the norm $x \to |x|$ is both an L-space and an M-space. It may be shown that any non-null Banach lattice which is both an M-space and an L-space is isomorphic to $\mathbb{R}$ (see Jameson, 1970).

5.6 Generalized experiments. Transitions

Let us identify two important vector lattices associated with a statistical experiment $\mathscr{E} = (\mathscr{X}, \mathscr{A}; P_\theta : \theta \in \Theta)$.

First consider the L-space, $L(\mathscr{E})$, of $\mathscr{E}$. This space was defined as the linear space consisting of all finite measures μ on $\mathscr{A}$ such that $\mu \ll \sum_{n=1}^\infty 2^{-n} P_{\theta_n}$ for some sequence $\theta_1, \theta_2, \ldots$ in Θ. Hence, by example 5.5.12 and proposition 5.5.9, the linear space $L(\mathscr{E})$ consists of exactly the same measures as the band generated by $(P_\theta : \theta \in \Theta)$. Thus we may consider the L-space of $\mathscr{E}$ as the band in $ca(\mathscr{X}, \mathscr{A})$ generated by $\mathscr{E}$. In particular it follows by proposition 5.5.9 that $\mathscr{E}$ is dominated if and only if $L(\mathscr{E})$ is a principal band. This, in turn, is equivalent to the condition that there is a sequence $\theta_1, \theta_2 \ldots$ in Θ such that $L(\mathscr{E}) = B_\pi =$ the set of finite measures which are dominated by the probability measure $\pi = \sum_n 2^{-n} P_{\theta_n}$.

Note that $L(\mathscr{E})$ is not in general the smallest sub L-space of $ca(\mathscr{X}, \mathscr{A})$ containing $\mathscr{E}$. This smallest sub-space coincides with the closure of the vector lattice generated by $\mathscr{E}$. We shall clarify the significance of this vector lattice later on.

The M-space, $M(\mathscr{E})$, of an experiment $\mathscr{E} = (\mathscr{X}, \mathscr{A}; P_\theta : \theta \in \Theta)$ was defined as the space of bounded linear functionals on $L(\mathscr{E})$. It is then clear from proposition 5.5.13 that $M(\mathscr{E})$ is an order complete M-space with a unit. Each bounded measurable function h on $\mathscr{X}$, i.e. each function in $M(\mathscr{X}, \mathscr{A})$, determines a linear functional $\dot{h} : \lambda \to \int h \, d\lambda$ on $L(\mathscr{E})$. These functionals are clearly bounded and thus belong to $M(\mathscr{E})$.

The set $\dot{M}(\mathscr{X}, \mathscr{A})$ of functionals $\dot{h}$ where $h \in M(\mathscr{X}, \mathscr{A})$ constitutes a sub-vector lattice of $M(\mathscr{E})$ which is also an M-space for the inherited norm. Clearly $\mathscr{E}$ is coherent if and only if $\dot{M}(\mathscr{X}, \mathscr{A}) = M(\mathscr{E})$.

A natural and useful generalization of the concept of an experiment is to consider as *an abstract* (*generalized*) experiment with parameter set Θ any map $\theta \to P_\theta$ from Θ to an L-space L such that, for each θ, $P_\theta \geqslant 0$ and $\|P_\theta\| = 1$. More generally we might, as we shall in chapter 9, study maps $\theta \to \mu_\theta$ from Θ to an L-space L. In this abstract set-up the role of the bounded random variables are played by the bounded linear functionals on the band generated by the μ_θs.

As we shall see, from Kakutani's representation theorem it follows that abstract experiments are essentially no more general than traditional experiments. It is nevertheless useful to have a general notion of an experiment as a family in an L-space. This is more or less the same thing as the fact that it is useful to have a general concept of an n-dimensional linear space, although any such space is isomorphic to $\mathbb{R}^n$.

What happens to the notion of a decision rule (transition) within this abstract experiments are essentially no more general than traditional experiments. It is nevertheless useful to have a general notion of an experiment from one L-space to another L-space a *transition* if it preserves norms of non-negative elements. Thus a map ρ from an L-space L_1 to another L-space L_2 is a transition from L_1 to L_2 if and only if:

(i) ρ is linear;
(ii) $\lambda\rho \geqslant 0$ when $0 \leqslant \lambda \in L_1$;
(iii) $\|\lambda\rho\| = \|\lambda\|$ when $0 \leqslant \lambda \in L_1$.

If $L_1 = L(\mathscr{E})$ is the L-space of the experiment $\mathscr{E}$ and if $L_2 = M(T, \mathscr{S})^* \cong \text{ba}(T, \mathscr{S})$ where $(T, \mathscr{S})$ is a decision space, then the definition just given reduces to the definition in section 4.5.

It follows from (i)–(iii) that $\|\lambda\rho\| = \|\lambda^+\rho - \lambda^-\rho\| \leqslant \|\lambda^+\rho\| + \|\lambda^-\rho\| = \|\lambda^+\| + \|\lambda^-\| = \|\lambda^+ + \lambda^-\| = \|\lambda\|$ if ρ is a transition from L_1 to L_2 and $\lambda \in L_1$. Hence any transition is a bounded linear map with norm $= 1$, provided of course that its domain contains non-zero vectors.

Identifying generalized decision rules with the bilinear functionals they define, we have seen (theorem 4.5.13) that the set of all generalized decision rules is compact for pointwise convergence. This important result extends as follows.

Theorem 5.6.1. *Let Γ be an M-normed space. Then the set of transitions from an L-space L_1 to the L-space $L_2 = \Gamma^*$ is compact for the topology of pointwise convergence on $L_1 \times \Gamma$.*

Remark. The particular case $\Gamma = M(T, \mathscr{S})$ yields theorem 4.5.13.

Proof. Let $\{\rho_n\}$ be a net of transitions from L_1 to L_2. By Tychonoff's theorem there is a subnet $\{\rho_{n'}\}$ such that $\lim_{n'} \lambda\rho_{n'}\gamma = \Phi(\lambda, \gamma)$ exists when $\lambda \in L_1$ and $\gamma \in \Gamma$. Then Φ is bilinear on $L_1 \times \Gamma$ and the map $\rho : \lambda \to \Phi(\lambda, \cdot)$ is the desired transition. □

The restriction on L_2 is not as severe as it may appear. Actually, as we shall see in the next section, any L-space L constitutes by evaluation a band in Γ^* where $\Gamma = L^*$. Thus we might skip the assumption on L_2 provided we weaken the assumption that ρ is a transition from L_1 into L_2 to the assumption that ρ is a transition from L_1 to L_2^{**}. It follows then from the generalization of theorem 4.5.11 given below that ρ may be converted into a transition from L_1 into L_2 without increasing any approximation bound $\|\mu\rho - \nu\|$ where $\mu \in L_1$ and $\nu \in L_2$.

Theorem 5.6.2 (Transitions from projections). *Let $\hat{L}$ be a band in an L-space L and let τ_0 be a non-negative vector in $\hat{L}$ such that $\|\tau_0\| = 1$. Denote the projection of a vector ν in L on $\hat{L}$ by $\hat{\nu}$ and let I be the unit of the M-space L^*. Then the map*

$$\sigma : \nu \to \hat{\nu} + (\nu - \hat{\nu})(I)\tau_0$$

is a transition from L to $\hat{L}$ leaving the vectors in $\hat{L}$ fixed. In particular

$$\|\nu\sigma - \tau\| \leqslant \|\nu - \tau\|$$

when $\nu \in L$ and $\tau \in \hat{L}$.

Proof. σ is linear since the projection is linear. If $\nu \geqslant 0$ then $\hat{\nu} \geqslant 0$ and $\nu - \hat{\nu} \geqslant 0$ so that $\nu\sigma \geqslant 0$ and then $\|\nu\sigma\| = \hat{\nu}(I) + (\nu - \hat{\nu})(I) = \nu(I) = \|\nu\|$. Clearly $\nu = \hat{\nu} = \nu\sigma$ when $\nu \in \hat{L}$. If $\nu \in L$ and $\tau \in \hat{L}$ then this implies that $\|\nu\sigma - \tau\| = \|(\nu - \tau)\sigma\| \leqslant \|\nu - \tau\|$. □

Corollary 5.6.3 (Generalization of corollary 4.5.14). *Let $(\mu_\theta : \theta \in \Theta)$ and $(\nu_\theta : \theta \in \Theta)$ be families in the L-space L_1 and the L-space L_2 respectively. For each $\theta \in \Theta$, let ε_θ be a quantity in $[0, \infty]$. Suppose that to each finite non-empty subset F of Θ there corresponds a transition ρ_F from L_1 to L_2 such that*

$$\|\mu_\theta \rho_F - \nu_\theta\| \leqslant \varepsilon_\theta$$

when $\theta \in F$.

Then $\rho = \rho_F$ may be chosen independently of F and such that it maps L_1 into the band in L_2 generated by the family $(\nu_\theta : \theta \in \Theta)$.

Proof. It follows from theorem 5.6.1 that $\|\mu_\theta \tilde{\rho} - \nu_\theta\| \leqslant \varepsilon_\theta$ for all $\theta \in \Theta$ for some transition $\tilde{\rho}$ from L_1 to L_2^{**}. Using the fact that L_2 is a band in L_2^{**} (which will be established in the next section), from theorem 5.6.2 we obtain the existence of a transition σ from L_2^{**} to L_2 leaving L_2 fixed. Then the transition $\rho = \sigma \circ \tilde{\rho}$ has the desired property. □

5.7 The spectrum of an M-space. Representation of M-spaces and L-spaces.

In this section we will give a short introduction to the representation theory for M-spaces and L-spaces. The exposition is strongly influenced by the last chapter in Kelley & Namioka, 1963. We refer to chapter 2 for concepts and notation related to convexity.

Let us begin by characterizing the set of extreme points of the positive part of the unit ball of an L-normed space.

Proposition 5.7.1. *Let L be an L-normed space and put $U_+ = \{\lambda : \lambda \in L_+, \|\lambda\| \leqslant 1\}$. Then a non-zero vector λ in U_+ is an extreme point of the convex set U_+ if and only if $\|\lambda\| = 1$ and for each vector μ satisfying $0 \leqslant \mu \leqslant \lambda$ there is a scalar t such that $\mu = t\lambda$.*

Remark. The condition that $0 \leqslant \mu \leqslant \lambda$ implies $\mu = t\lambda$ for some real t whenever $0 \leqslant \mu \leqslant \lambda$ is clearly equivalent to the condition that the band B_λ generated by λ is one-dimensional.

Proof.

(i) Suppose $\lambda \in \text{ext}\, U_+$. If $\|\lambda\| \leqslant 1$ and $\varepsilon \in]0, 1[$ is sufficiently small then $(1 \pm \varepsilon)\lambda \in U_+$. Hence, since $\lambda = \frac{1}{2}(1 - \varepsilon)\lambda + \frac{1}{2}(1 + \varepsilon)\lambda$, $(1 \pm \varepsilon)\lambda = \lambda$, i.e. $\lambda = 0$. It follows that $\|\lambda\| = 1$ when $\lambda \neq 0$. Let $0 < \mu < \lambda$. Then $\lambda = \|\mu\|[\mu/\|\mu\|] + \|\lambda - \mu\|[(\lambda - \mu)/\|\lambda - \mu\|]$ and $1 = \|\mu + (\lambda - \mu)\| = \|\mu\| + \|\lambda - \mu\|$. Hence, since $\mu/\|\mu\|$, $(\lambda - \mu)/\|\lambda - \mu\| \in U_+$, $\mu/\|\mu\| = (\lambda - \mu)/\|\lambda - \mu\| = \lambda/\|\lambda\| = \lambda$ so that $\mu = \|\mu\|\lambda$.

(ii) Suppose λ satisfies the given conditions and that $\lambda = \frac{1}{2}\lambda_1 + \frac{1}{2}\lambda_2$ where $\lambda_1, \lambda_2 \in U_+$. Then $1 = \|\lambda\| = \frac{1}{2}\|\lambda_1\| + \frac{1}{2}\|\lambda_2\| \leqslant \frac{1}{2}\cdot 1 + \frac{1}{2}\cdot 1 = 1$ so that $\|\lambda_1\| = \|\lambda_2\| = 1$. Furthermore, since $\lambda \geqslant \frac{1}{2}\lambda_1 \geqslant 0$, $\frac{1}{2}\lambda_1 = t\lambda$ for some $t \geqslant 0$. Hence $\frac{1}{2} = \frac{1}{2}\|\lambda_1\| = t\|\lambda\| = t$ so that $\lambda_1 = 2t\lambda = \lambda = \lambda_2$. It follows that $\lambda \in \operatorname{ext} U$. □

Now consider an M-normed space M with unit I. Then $\|\lambda\| = \lambda^+(I) + \lambda^-(I)$ when $\lambda \in M^*$. Equip the dual space $L = M^*$ with the weak* topology, i.e. the pointwise topology on M.

Let Z be the subset of L consisting of all vector lattice homomorphisms from M to the real line $\mathbb{R}$ having norm 1. Then the *spectrum of* M is this set Z equipped with the relativized weak* topology. As Z is strongly bounded as well as weak* closed it is clear that the spectrum Z is a compact Hausdorff space. Referring to this particular L-space L we encountered the spectrum in proposition 5.7.1.

Theorem 5.7.2 (Characterizations of the spectrum of an M-normed space with unit). *Let M be an M-normed space with unit. Then the spectrum of M is a compact Hausdorff space. Furthermore, each of the following two conditions on a non-negative functional $\lambda \in M^*$ are equivalent to the condition that λ belongs to the spectrum of M:*

(i) *$\lambda \neq 0$ and λ is an extreme point of $\{\mu : 0 \leqslant \mu, \|\mu\| \leqslant 1\}$*
(ii) *$\|\lambda\| = 1$ and $0 \leqslant \mu \leqslant \lambda \Rightarrow \mu = t\lambda$ for some $t \in \mathbb{R}$.*

Remark. Condition (ii) may be phrased:

(ii′) $\|\lambda\| = 1 = \dim B_\lambda$. Here B_λ, as usual, denotes the principal band determined by λ.

Proof. The equivalence of (i) and (ii) was established in the previous proposition. Suppose $\lambda \in Z$ and that $0 \leqslant \mu \leqslant \lambda$. Then, since λ is a lattice homomorphism, $|\mu(f)| \leqslant \mu(|f|) \leqslant \lambda(|f|) = |\lambda(f)|$ when $f \in M$. Hence $\mu(f) = 0$ whenever $\lambda(f) = 0$. Thus $\mu = t\lambda$ for some scalar t. Suppose next that conditions (i) and (ii) hold. Let $u, v \in M_+$ be disjoint and non-zero. Then u and v are clearly linearly independent and $(su + tv)^+ = (su)^+$ when $s, t \in \mathbb{R}$. Define μ on the linear space A spanned by $\{u, v\}$ by putting $\mu(su + tv) = \lambda(su)$ for scalars s and t. Then $\mu(w) \leqslant \lambda(w^+)$ when $w \in A$. Hence by the Hahn–Banach theorem μ can be extended to a linear functional on all of M such that $\mu(w) \leqslant \lambda(w^+)$ when $w \in M$. Then $0 \leqslant \mu \leqslant \lambda$. It follows that $\mu = c\lambda$ for some $c \geqslant 0$. Hence $\lambda(u) = \mu(u) = c\lambda(u)$ and $0 = \mu(v) = c\lambda(v)$. It follows that $\lambda(u) \wedge \lambda(v) = 0$. If $u = 0$ or $v = 0$ then trivially $\lambda(u) \wedge \lambda(v) = 0$. Consider now two elements u_1 and u_2 in M. Put $u = u_1 - u_1 \wedge u_2$ and $v = u_2 - u_1 \wedge u_2$. Then $u, v \geqslant 0$ and $u \wedge v = 0$. Hence $0 = \lambda(u_1 - u_1 \wedge u_2) \wedge \lambda(u_2 - u_1 \wedge u_2) = \lambda(u_1) \wedge \lambda(u_2) - \lambda(u_1 \wedge u_2)$ so that $\lambda(u_1 \wedge u_2) = \lambda(u_1) \wedge \lambda(u_2)$. It follows that λ is a vector lattice homomorphism. □

A vector in an M-space with unit is determined by the function it defines on the spectrum of M.

Theorem 5.7.3. *Let M be an M-normed space with unit I and spectrum Z. Equip M^* with the topology of pointwise convergence on M. Then the convex hull of $Z \cup \{0\}$ is dense in the non-negative part $\{\lambda : \lambda \in M^*, \lambda \geqslant 0, \|\lambda\| \leqslant 1\}$ of the unit ball in M^*.*

Proof. Put $U_+ = \{\lambda : \lambda \in M^*, \lambda \geqslant 0, \|\lambda\| \leqslant 1\}$. Then U_+ is compact Hausdorff and convex. As we have just seen, the set of extreme points of U_+ is $Z \cup \{0\}$. Therefore the statement follows directly from the Krein–Milman theorem. □

For an exposition of the Krein–Milman theorem, the reader may consult one of several standard textbooks in functional analysis, e.g. Dunford & Schwartz (1958) and Kelley & Namioka (1963).

We are now ready for Kakutani's representation theorem for M-spaces.

Theorem 5.7.4 (Kakutani's representation of M-spaces with units). *Let M be an M-normed space with unit I and spectrum Z. Let ϕ be the evaluation map from M into the M-space $C(Z)$ of continuous real valued functions on Z. Thus if $h \in M$ then $\phi(h)$ is the function $z \to z(h)$ on Z.*

Then $\phi(I) = 1$ and ϕ is a vector lattice isomorphism and an isometry from M into a dense sub-vector lattice of $C(Z)$. If M is an M-space, i.e. if M is complete, then ϕ is an isomorphism from the M-space M onto the M-space $C(Z)$.

Remark. The symbol '1' in the equation '$\phi(I) = 1$' signifies the constant function 1 on Z.

Proof of the theorem. If $z \in Z$ then $e(I)_z = z(I) = \|z\| = 1$. Thus $\phi(I) = 1$. e is clearly linear. If $h \in M$ then $\phi(h^+)_z = z(h^+) = (z(h))^+ = [\phi(h)_z]^+$. Hence $\phi(h^+) = (\phi(h))^+$ so that ϕ is a vector lattice homomorphism. Furthermore, if $h \in M$, then $\|\phi(h)\| = \sup_z |\phi(h)_z| = \sup_z \phi(|h|)_z = \sup_z z(|h|) = \sup\{\lambda(|h|) : \lambda \in \langle Z \cup \{0\}\rangle\} = \sup\{\lambda(|h|) : \lambda \geqslant 0;\ \|\lambda\| \leqslant 1\}$ (by the previous theorem) $= \sup\{|\lambda(|h|)| : \|\lambda\| \leqslant 1\} = \||h|\| = \|h\|$. Thus ϕ is also an isometry. The range of ϕ is a sub-vector lattice of $C(Z)$ which contains the constant functions and distinguishes points of Z. Then it follows from the vector lattice form of the Stone–Weierstrass approximation theorem that the range is dense in $C(Z)$ (see Kelley & Namioka, 1963). The last statement follows directly from completeness. □

The spectrum may be any compact Hausdorff space.

Theorem 5.7.5. *Let $\mathscr{X}$ be a compact Hausdorff space and let $C(\mathscr{X})$ be the M-space of continuous functions on $\mathscr{X}$. For each $x \in \mathscr{X}$, let δ_x be the one-point distribution in x. Then the map $x \to \delta_x$ is a homomorphism from $\mathscr{X}$ onto the spectrum of $C(\mathscr{X})$.*

Proof. It is easily seen that the map $x \to \delta_x$ is a homomorphism from $\mathscr{X}$ into the spectrum of $C(\mathscr{X})$. Let λ be any element in the spectrum of $C(\mathscr{X})$. Considering λ as a regular Borel measure on $\mathscr{X}$, and using the characterization theorem for the spectrum, we see that $\lambda(A) = 0$ or $= 1$ for any Borel set A. By compactness there is a smallest closed set A_0 in $\mathscr{X}$ such that $\lambda(A_0) = 1$. Then, by regularity, $\lambda(B) = 0$ for each Borel set $B \subset A_0$. If $x_1, x_2 \in A_0$ and $x_1 \neq x_2$ then $1 = \lambda(A_0) = \lambda(\{x_1\}) + \lambda(A_0 - \{x_1\}) = 0 + 0 = 0$. Thus A_0 is a one-point set x_0 and then $\lambda = \delta_{x_0}$. □

The representation of an M-space M in the form $M \cong C(\mathscr{X})$ for a compact Hausdorff space $\mathscr{X}$ determines $\mathscr{X}$ up to a homomorphism.

Theorem 5.7.6. *M-spaces are isomorphic if and only if their spectra are homomorphic. In particular, compact Hausdorff spaces $\mathscr{X}$ and $\mathscr{Y}$ are homomorphic if and only if the M-spaces $C(\mathscr{X})$ and $C(\mathscr{Y})$ are isomorphic.*

Proof. Isomorphisms between M-spaces induce isomorphisms between their dual L-spaces and thus also between their spectra. On the other hand it is clear that M-spaces $C(\mathscr{X})$ and $C(\mathscr{Y})$ are isomorphic whenever the topological spaces $\mathscr{X}$ and $\mathscr{Y}$ are homomorphic. □

Example 5.7.7 (The M-space of bounded measurable functions). Let $(\mathscr{X}, \mathscr{A})$ be a measurable space and let $M = M(\mathscr{X}, \mathscr{A})$ be the M-space of bounded measurable functions. Then the spectrum Z of $M(\mathscr{X}, \mathscr{A})$ consists of all $\{0, 1\}$-valued additive probability set functions on $\mathscr{A}$.

For each $x \in \mathscr{X}$, let δ_x be the one-point distribution in x. Then the set $\{\delta_x : x \in \mathscr{X}\}$ is a dense subset of Z. Say that $\mathscr{A}$ *distinguishes* points in $\mathscr{X}$ if to each pair (x, y) of distinct points in $\mathscr{X}$ there corresponds a set $A \in \mathscr{A}$ such that $x \in A$ while $y \notin A$. This condition is trivially satisfied when the one-point sets are measurable. It is then clear that the map $x \to \delta_x$ provides an imbedding of $\mathscr{X}$ into a dense subset of Z if and only if $\mathscr{A}$ distinguishes points in $\mathscr{X}$. In general we can identify points x and y such that $\delta_x = \delta_y$ and proceed with this identification.

If $h \in M(\mathscr{X}, \mathscr{A})$ then the evaluation map maps h into the function $\bar{h}$ on Z defined by $\bar{h}(z) = \int h \, dz$. In particular $\bar{h}(\delta_x) = h(x)$; $x \in \mathscr{X}$. Thus any bounded measurable function on $(\mathscr{X}, \mathscr{A})$ may be considered as the restriction to $\mathscr{X}$ of a bounded continuous function on Z. Conversely, by the representation theorem, any bounded continuous function on Z is the extension of a bounded measurable function on $\mathscr{X}$. In particular the measurable sets are the intersections with $\mathscr{X}$ of those subsets of Z which are both open and closed.

What becomes of the bounded additive set functions under this representation? The answer follows readily from the representation theorem for bounded linear functionals on the Banach space of continuous functions on a compact space. (See Halmos (1950) or Dunford & Schwartz (1958).) If μ is a bounded additive set function on $\mathscr{A}$ then the functional $h \to \int h \, d\mu$ on $M(\mathscr{X}, \mathscr{A})$ defines a finite regular Borel measure on the Borel class of Z. Of course, this *does not* imply that additive set functions are measures.

Example 5.7.8 (The Stone–Čech compactification). Let $\mathscr{X}$ be a Tychonoff space, i.e. a completely regular topological space such that the one-point sets are closed. (A topological space $\mathscr{X}$ is called completely regular if to each point x in $\mathscr{X}$ and to each neighbourhood V of x there corresponds a continuous function f from $\mathscr{X}$ to $[0, 1]$ such that $f(x) = 0$ while $f(y) = 1$ whenever $y \notin V$.) If F denotes the class of continuous functions from $\mathscr{X}$ to $[0, 1]$ then together these requirements are equivalent to the single requirement that the evaluation map $e : x \to (f(x) : f \in F)$ from $\mathscr{X}$ to $[0, 1]^F$ is a homomorphism. Then the closure of the range $e[\mathscr{X}]$ of e in $[0, 1]^F$, together with the imbedding e, provides a compactification of $\mathscr{X}$ which is called the Stone–Čech compactification of $\mathscr{X}$. It is essentially characterized by the property that any continuous bounded function on $\mathscr{X}$ has a continuous extension to this compactification (see e.g. Kelley, 1955). Actually any continuous function from $\mathscr{X}$ to a compact Hausdorff space has a continuous extension to the Stone–Čech compactification.

The Stone–Čech compactification may be described in terms of the representation theorem as follows.

Let $C(\mathscr{X})$ be the M-space of bounded continuous functions on $\mathscr{X}$ and let Z be the spectrum of this M-space. For each $x \in \mathscr{X}$, let δ_x be the one-point distribution in x. Then the map $x \to \delta_x$ provides an imbedding of $\mathscr{X}$ into a dense subset of Z. (If $z_0 \in V \subseteqq Z$ and V is open in Z then there is a $h \in C(Z)$ such that $0 \leqslant h \leqslant 1$, $h(z_0) = 0$ and $h[V^c] = \{1\}$. If h is the image of $f \in C(\mathscr{X})$ then $h(z) \equiv_z z(f)$. In particular $h(\delta_x) \equiv_x f(x)$. It follows, since $h \neq 1$, that $f \neq 1$ so that, for some x, $0 \leqslant f(x) < 1$. Then, necessarily $\delta_x \in V$.) If $f \in C(\mathscr{X})$ then its image by the representation map is a continuous extension of f to all of Z. Identifying the functionals in Z with their restrictions to F we see that Z, together with this imbedding, is the Stone–Čech compactification of $\mathscr{X}$.

It follows from the representation theorem that any M-space may be equipped with the same algebraic structure as a space $C(\mathscr{X})$ where $\mathscr{X}$ is a compact Hausdorff space.

Corollary 5.7.9 (Multiplication on M-spaces). *Let M be an M-space with unit I. Then there is a unique bilinear multiplication $(u, v) \to u \cdot v$ from $M \times M$ to M which is non-negative in the sense that $u \cdot v \geqslant 0$ when $u \geqslant 0$ and $v \geqslant 0$ and* *furthermore satisfies* $u \cdot I = I \cdot u = u$ *for all* $u \in M$. *This multiplication is commutative and may be derived from the vector lattice operations by the identities*:

(i) $u \cdot u = \sup\{2au - a^2 I; a \in \mathbb{R}\}$

and

(ii) $4(u \cdot v) = (u + v)\cdot(u + v) - (u - v)\cdot(u - v)$

which are valid for all elements u, v *in* M.

Proof. We may assume, without loss of generality, that $M = C(Z)$ for a compact Hausdorff space Z. Then pointwise multiplication is a multiplication on M with the desired properties. Suppose the multiplication $(u, v) \to u * v$ also had the desired properties. Then $|u * v| \leqslant |u| * |v| \leqslant \|v\|\,|u|$, $u, v \in M$. Choose a $z \in Z$ and a $v \in M$ and define the linear functional $F_{v,z}$ on M by $F_{v,z}(u) = (u * v)(z)$. Then $|F_{v,z}(u)| \leqslant \|v\|\,|u(z)|$ so that $F_{v,z}(u) = 0$ whenever $u(z) = 0$. Thus $F_{v,z}(u) = C(v, z)u(z)$ for some constant $C(v, z)$. Substituting $u = I$ we find $v(z) = C(v, z)$ so that $(u * v)(z) = F_{v,z}(u) = v(z)\cdot u(z)$. □

If M is an M-space with unit then the multiplication whose existence was established in the last corollary will be called *the multiplication of* M. Note that if H is a closed sub-vector lattice of M containing the unit I then the restriction to $H \times H$ of the multiplication on M is a multiplication on H. Hence, by uniqueness, the multiplication on H is the restriction to $H \times H$ of the multiplication on M. Thus M becomes a Banach algebra with unit I and a closed subset H of M which contains I is a subalgebra if and only if it is a vector lattice.

If the M-space M is the dual of an L-space L then it is necessarily order complete. Thus $C(Z)$, where Z is the spectrum of M, is also order complete. The following theorem implies that this entails peculiar properties of Z.

Theorem 5.7.10. *Let* $\mathscr{X}$ *be a compact Hausdorff space such that* $C(\mathscr{X})$ *is order complete. Then the interiors of closed subsets of* $\mathscr{X}$ *are closed and closures of open subsets of* $\mathscr{X}$ *are open. In particular the class of subsets of* $\mathscr{X}$ *which are both open and closed constitutes a basis for the topology. Furthermore to any Borel subset* A *of* $\mathscr{X}$ *there corresponds an open and closed set* B *such that the symmetric difference* $(A - B) \cup (B - A)$ *is of the first category in* $\mathscr{X}$.

Remark 1. A subset of a compact Hausdorff space $\mathscr{X}$ is called a Borel set if it belongs to the smallest σ-algebra containing the open (compact) subsets of $\mathscr{X}$.

Remark 2. A subset of a topological space $\mathscr{X}$ is called *nowhere dense* if its closure has empty interior. A subset of a topological space is called a *set of the first category* in $\mathscr{X}$ if it is a countable union of nowhere dense sets. All other subsets of $\mathscr{X}$ are called sets of *the second category*. A most interesting discussion of category is Oxtoby (1971).

Let $\mathscr{E}$ and $\mathscr{F}$ be two experiments having the same parameter set Θ. Then we shall say that $\mathscr{E}$ is at least as informative as $\mathscr{F}$, if to any decision problem and any decision rule in $\mathscr{F}$ there is a decision rule in $\mathscr{E}$ which is at most as risky for any θ in Θ.

In this way we arrive at the partial ordering 'being at least as informative as' for experiments having the same parameter set Θ. The concept of sufficiency corresponds to the case where $\mathscr{E}$ is a subexperiment of $\mathscr{F}$.

With this kind of a definition it is to be expected that the ordering is not total. In fact in general we may expect that two experiments having the same parameter set Θ are not comparable with respect to this ordering. Thus we are led to the following generalization of the fundamental question: How much do we lose, under the worst possible circumstances, by using $\mathscr{E}$ instead of $\mathscr{F}$?

As we shall see, an answer to this problem may be given by a non-negative number; the deficiency of $\mathscr{E}$ with respect to $\mathscr{F}$.

Closely associated with the notion of deficiency is a distance for experiments or, equivalently, for the (undefined) amounts of information carried by the experiments.

Finally we may restrict our attention to certain types of decision problems. This leads to deficiencies and distances relative to the relevant type of decision problems.

The definitions and some of their immediate consequences are explored in the next section. One important technical aspect is that general comparison problems may be reduced to the case of experiments having finite parameter sets. Among the particular topics discussed in this section are: comparison for k-decision problems, comparison with experiments which are either totally informative or are totally non-informative, relationship to maximin testing, orderings, equivalence relations and weighted deficiencies.

The concept of deficiency may be regarded as an extension of the notion of sufficiency. It is therefore quite satisfying that a statistic defines an equivalent experiment if and only if it is pairwise sufficient.

Although deficiencies satisfy the triangular inequality they are not symmetric and therefore do not qualify as pseudometrics. Symmetrizing the deficiencies we obtain LeCam's deficiency distances and they are fully-fledged pseudometrics. In particular these distances yield very interesting concepts of convergence.

As mentioned above, the deficiency expresses how much information may be lost by using this experiment rather than that experiment. This loss is quantified in section 6.2 by considering differences of risks. However, there are many other, and natural, ways of comparing decision rules and the question arises to what extent we are forced to work with different concepts of deficiency.

It is a remarkable fact that we arrive at the same concept of deficiency for a large variety of possibilities for comparing decision rules. Thus we shall see in section 6.3 that the definition might just as well have been phrased in terms of maximum risks or in terms of Bayes risks (for a fixed prior distribution) or directly in terms of performance functions.

In particular we will find it convenient to work with maximal Bayes utilities. The essential aspects of utilities are expressed by sublinear functionals on the likelihood space. These functionals turn out to be very useful tools and several results are provided concerning them and the measures of information they define.

Section 6.4 is devoted to the important randomization criterion of LeCam. In order to motivate this criterion let us remind ourselves that a statistic Y is usually called sufficient if conditional probabilities given Y may be defined independently of the unknown parameter. One justification for this terminology is that on the basis of a sufficient statistic Y we may re-create the experiment itself by using the known chance mechanism (randomization) provided by the conditional distribution given Y. More generally, we may say that Y is approximately sufficient if it is possible, on the basis of Y and a known chance mechanism, to obtain an experiment which is close to the given experiment, in the sense of statistical distances. Replacing Y with any experiment we arrive at the randomization criterion. According to this criterion the deficiency measures how well an experiment may be approximated on the basis of another given experiment by using a known chance mechanism.

Now any experiment may trivially be regarded as a performance function obtainable from itself and, furthermore, decision rules are randomizations. Therefore it follows that the randomization criterion may be regarded as an extension of the performance function criterion, and this is the idea of the proof.

If the experiments we are comparing are invariant under actions of a group, then one might hope that it would suffice to consider invariant randomizations or at least invariant transitions. Although this is not true in general we shall see in section 6.4 that if we allow transitions as decision rules then this is true for any solvable group and, more generally, for any group permitting finitely additive and invariant set functions. The roots of the problems concerning the involved groups are closely related to the Banach–Tarski paradox, and in addition to the references mentioned in section 6.5 the reader may also benefit from Wagon (1985).

The passing from invariant transitions to invariant randomizations requires some doing. Here, inspired by Boll (1955), we have chosen a variation of the traditional approach. Another approach was chosen by LeCam (1986) who used lifting theory.

The general results are in examples applied to comparison of translation experiments (obtaining the convolution criterion), to maximin testing (the

Hunt–Stein theorem) and, generalizing a result of Hall, Wijsman and Ghosh (1965) to the interplay between the two reduction principles: sufficiency and invariance.

Complements in chapter 10 of interest in connection with the topics covered in this chapter are numbers 4–13, 15–18, 22, 23, 26, 29–34, 37, 41, 44–46 and 48.

6.2 Deficiencies

The deficiency concept provides a definite meaning for statements such as: '$\mathscr{E}$ is ε-more informative than $\mathscr{F}$' or '$\mathscr{E}$ is, except for at most an ε-amount of risk, better than $\mathscr{F}$'.

Consider two experiments $\mathscr{E} = (\mathscr{X}, \mathscr{A}; P_\theta : \theta \in \Theta)$ and $\mathscr{F} = (\mathscr{Y}, \mathscr{B}; Q_\theta : \theta \in \Theta)$ having the *same* parameter set Θ, a non-negative function $\varepsilon = (\varepsilon_\theta : \theta \in \Theta)$ on Θ and a k-decision space $(T, \mathscr{S})$ where k is a positive integer. The norm $\|g\|$ applied to a function g on T, unless otherwise stated, is the supremum norm $\|g\| = \sup_t |g(t)|$.

We shall say that *$\mathscr{E}$ is ε-deficient with respect to $\mathscr{F}$ for k-decision problems* if to each finite subset Θ_0 of Θ, to each decision rule σ from $\mathscr{F}$ to $(T, \mathscr{S})$ and to each loss function L on $(T, \mathscr{S})$ there corresponds a decision rule ρ from $\mathscr{E}$ to $(T, \mathscr{S})$ such that

$$P_\theta \rho L_\theta \leqslant Q_\theta \sigma L_\theta + \varepsilon_\theta \|L_\theta\| \tag{6.2.1}$$

when $\theta \in \Theta_0$.

If $\mathscr{E}$ is ε-deficient with respect to $\mathscr{F}$ for k-decision problems for any positive integer k then we shall simply say that *$\mathscr{E}$ is ε-deficient with respect to $\mathscr{F}$.*

In order to avoid unnecessary duplication of statements we shall use the sentence '*$\mathscr{E}$ is ε-deficient with respect to $\mathscr{F}$ for ∞-decision problems*' as synonymous with '$\mathscr{E}$ is ε-deficient with respect to $\mathscr{F}$'. Thus ε-deficiency for k-decision problems is defined whenever k is a positive integer or $k = \infty$.

The reader may object to this definition because it is too broad and deficiency therefore may depend on decision problems of little interest. Indeed one might be interested only in a single decision space and a single loss function. Comparison reduces then to comparison of pairs of sets of risk functions and this may be expressed in terms of minimum Bayes risk. For replicated experiments we may then apply the adjustment procedure for posterior distributions given in section 4.4. We refer to complement 37 in chapter 10 for some general results on comparison for given loss functions.

The terms 'totally non-informative' and 'totally informative' as applied to experiments may now be justified as in the following example.

Example 6.2.1 (Totally non-informative and totally informative experiments). Let $\mathscr{M}_i$ and $\mathscr{M}_a$ denote, respectively, a totally non-informative experiment and a totally informative experiment. Thus $\mathscr{M}_i = (P_\theta : \theta \in \Theta)$ where $P_\theta = P$ does not depend on θ while $\mathscr{M}_a = (Q_\theta : \theta \in \Theta)$ where Q_{θ_1} and Q_{θ_2} are mutually singular when $\theta_1 \neq \theta_2$.

Let $\mathscr{E} = (P_\theta : \theta \in \Theta)$ be an arbitrary experiment and let σ be a decision procedure from $\mathscr{M}_i$ to the k-decision space $(T, \mathscr{S})$. The performance function of σ in $\mathscr{M}_i$ is then the (constant) family $(P\sigma : \theta \in \Theta)$. Choose the decision procedure ρ from $\mathscr{E}$ to $(T, \mathscr{S})$ so that the distribution of $t \in T$ is $P\sigma$ regardless of the outcome of $\mathscr{E}$. Then ρ in $\mathscr{E}$ and σ in $\mathscr{M}_i$ have the same performance functions. Next consider an arbitrary decision rule ρ in $\mathscr{E}$. The disjointness of the probability measures $(Q_\theta : \theta \in \Theta)$ implies that provided Θ is countable we may define a decision rule τ in $\mathscr{M}_a$ such that $Q_\theta \tau \equiv_\theta \pi_\theta \rho$. For a general parameter set these considerations imply that:

(i) any experiment is 0-deficient with respect to $\mathscr{M}_i$.
(ii) $\mathscr{M}_a$ is 0-deficient with respect to any experiment.

We shall continue to use the notation $\mathscr{M}_i$ and $\mathscr{M}_a$ for, respectively, a totally non-informative experiment and a totally informative experiment. Note that if Θ_0 is a non-empty subset of Θ then $(\mathscr{M}_i)_{\Theta_0}$ and $(\mathscr{M}_a)_{\Theta_0}$ are experiments with parameter set Θ_0 which are, respectively, totally non-informative and totally informative. For this reason we shall usually omit reference to the parameter set in the notation for totally non-informative and totally informative experiments. Thus when there is no chance of confusion, we can write $\mathscr{M}_i$ and $\mathscr{M}_a$ instead of $(\mathscr{M}_i)_{\Theta_0}$ and $(\mathscr{M}_a)_{\Theta_0}$.

The single experiment with parameter set Θ having a given one-point set $\mathscr{X}$ as sample space is totally non-informative and may thus be denoted as $\mathscr{M}_I$.

For each θ, let δ_θ be the one-point distribution in θ. Equip Θ with a σ-algebra $\mathscr{U}$ of subsets making the one-point sets measurable. Then $\mathscr{M}_a = (\Theta, \mathscr{U}; \delta_\theta : \theta \in \Theta)$ may be a convenient choice of a totally informative experiment.

It is only in exceptionally simple cases that we are able to establish deficiencies directly from the definition. We shall later develop several methods yielding lower bounds and upper bounds for deficiences. Even with these tools to hand, the problem of evaluating deficiencies between the standard models occurring in statistics is usually difficult. Therefore the following example is more to the point than the previous one.

Example 6.2.2 (The Poisson approximation to the binomial distribution). Let $\mathscr{E}$ be the experiment of observing a binomially distributed variable according to n trials and success parameter θ. Let $\mathscr{F}$ be the experiment of observing a Poisson distributed variable with expectation $n\theta$. Assume n is known and that $\Theta = [0, \varepsilon]$ where $0 < \varepsilon \leqslant 1$. The variances of the UMVU estimators of θ in these two models are, respectively $\theta(1 - \theta)/n$ and θ/n. As the latter quantity is larger than the first when $\theta > 0$ it is tempting to conclude that $\mathscr{F}$ is not 0-deficient with respect to $\mathscr{E}$. On the other hand, the parametric function

$e^{-\theta}$ possesses an unbiased estimator in $\mathscr{F}$ but not in $\mathscr{E}$. This supports the conjecture that $\mathscr{E}$ is not 0-deficient with respect to $\mathscr{F}$. We shall later see that these arguments are valid so that neither one of the two experiments is 0-deficient with respect to the other experiment.

How far are these experiments from each other in terms of deficiencies? As the Poisson approximation is known to be good when θ is small we should expect that they are closed in the sense of deficiencies also. We shall make this precise a little later.

The requirement on ε in order that $\mathscr{E}$ should be ε-deficient with respect to $\mathscr{F}$ for k-decision problems is a heavy burden. If the models are non-parametric then this requirement may be too restrictive in order to yield non-trivial results. However we will see that interesting statements are usually available for parametric models. We could of course relax the condition that all loss functions L and all decision rules σ should be considered. Thus we could, e.g. compare linear models in terms of variances of UMVU estimators. In general, we could restrict our attention to particular types of decision problems.

As this might, at a first glance, appear too complicated to consider, it might be comforting to know, as we shall soon see, that the decision rule ρ may *as a transition* be chosen independently of Θ_0 and L. As shown by LeCam, in the case of ∞-decision problems the decision rule ρ may be obtained as the composition $M\sigma$ of σ with a transition M from $\mathscr{E}$ to $\mathscr{F}$ which may be chosen independently of Θ_0, σ, L and k.

Later we will see that there are several apparently different and natural criteria for deficiency. If we had chosen one of these as our definition, then the definition just given would, of course, be converted to a criterion. We have chosen to start out from the given definition since this is the most natural approach from the point of view of statistical decision theory.

Let us begin our investigation by commenting on how the original statement may be simplified or modified.

Firstly we should check that the concept of ε-deficiency for k-decision problems is well defined. This amounts to verifying, for $k < \infty$, that if the statement of the definition holds for one k-decision space then it holds for all others, and in particular for $T_k = \{1, 2, \ldots, k\}$.

What would be the effect on the definition if we admitted generalized decision rules in $\mathscr{E} = (P_\theta : \theta \in \Theta)$, i.e. transitions from $L(\mathscr{E})$ to $\text{ba}(T, \mathscr{S}) = \text{ca}(T, \mathscr{S})$? Actually the basic inequality (6.2.1) is meaningful whenever ρ is a generalized decision rule in the restricted experiment $\mathscr{E}_{\Theta_0} = (P_\theta : \theta \in \Theta_0)$. However, if ρ is a generalized decision rule in $\mathscr{E}_{\Theta_0}$ then it may be extended to a generalized decision rule $\tilde{\rho}$ in $\mathscr{E}$ by putting $\lambda\tilde{\rho} = \hat{\lambda}\rho + (\lambda - \hat{\lambda})(\mathscr{X})\tau$ where τ is any probability measure on $\mathscr{S}$ and $\hat{\lambda}$ is the projection of $\lambda \in L(\mathscr{E})$ on $L(\mathscr{E}_{\Theta_0})$. Thus it does not matter whether we consider ρ as a generalized decision rule in $\mathscr{E}$ or in $\mathscr{E}_{\Theta_0}$.

Now the subset Θ_0 of Θ and the class $\mathscr{S}$ of subsets of T are both finite. It follows that any generalized decision rule in $\mathscr{E}_{\Theta_0}$ can be represented by a Markov kernel, i.e. a decision rule.

Similar considerations of course apply to $\mathscr{F}$. Thus it does not matter whether we choose to interpret the term 'decision rule' in the narrow sense, as we do throughout this work, or if we also admit transitions as decision rules.

We may also impose limitations on the range of the loss function L. Suppose for example that (6.2.1) may be realized when L is bounded. Then to each positive integer N, each finite subset Θ_0 of Θ and each loss function L there corresponds a decision rule ρ_N in $\mathscr{E}_{\Theta_0}$ such that (6.2.1) holds for ρ_N and L replaced by $L \wedge N$. If ρ is a limit of a convergent sub-sequence of $\{\rho_N\}$ then (6.2.1) holds for ρ and L. In particular it follows that we may assume without loss of generality that L is finite, i.e. that L_θ is bounded for each θ.

Furthermore, note that if the defining inequality (6.2.1) holds for the loss function L then it is also satisfied for any loss function $\hat{L}$ such that $\hat{L}_\theta \equiv_\theta a_\theta L_\theta$ where the a_θs are positive constants. We may also add constants b_θ to L_θ provided we modify the deficiency term $\varepsilon_\theta \|L_\theta\|$ accordingly.

The definition of ε-deficiency involved all loss functions on a given finite decision space $(T, \mathscr{S})$. Of course we may also study the more general situation where our interest is limited to a specified set of loss functions. If $\mathscr{L}$ is any collection of loss functions on not necessarily the same decision space, then we may say that the experiment $\mathscr{E} = (P_\theta : \theta \in \Theta)$ is ε-deficient with respect to the experiment $\mathscr{F} = (Q_\theta : \theta \in \Theta)$ for the loss functions in $\mathscr{L}$ if to each loss function $L \in \mathscr{L}$ and to each generalized decision rule σ from $\mathscr{F}$ to the decision space underlying L there corresponds a generalized decision rule ρ from $\mathscr{E}$ to the same decision space such that

$$P_\theta \rho L_\theta \leqslant Q_\theta \sigma L_\theta + \varepsilon_\theta \|L_\theta\|; \qquad \theta \in \Theta.$$

If all the loss functions in $\mathscr{L}$ are loss functions on finite decision spaces then $\mathscr{E}$ is ε-deficient with respect to $\mathscr{F}$ for the loss functions in $\mathscr{L}$ if and only if to each loss function L in $\mathscr{L}$ and each *decision rule* σ from $\mathscr{F}$ to the decision space underlying L and to each finite subset Θ_0 of Θ there corresponds a decision rule ρ from $\mathscr{E}$ to the same decision space such that

$$P_\theta \rho L_\theta \leqslant Q_\theta \sigma L_\theta + \varepsilon_\theta \|L_\theta\|; \qquad \theta \in \Theta.$$

The class $\mathscr{L}$ may of course consist of a single loss function L on some decision space $(T, \mathscr{S})$. Usually, however, along with a given loss function L, one also wants to consider any loss function of the form $aL + b$ where $a > 0$ and b are constants. Thus one might not want the comparisons to be influenced by possible changes in scale and origin for measurements of losses. Let us introduce the notation Lin (L) for the class of all loss functions $\tilde{L}$ of the form $\tilde{L}_\theta \equiv_\theta a_\theta L_\theta + b_\theta$ where the a_θs and the b_θs are constants and all the a_θs are positive. Lin (L) is then the equivalence class determined by L for the equivalence relation which states that loss functions $\tilde{L}$ and L are equivalent if they are related as above. For simplicity assume that L is bounded. If we insist that $\mathscr{E}$ should be ε-deficient with respect to $\mathscr{F}$ for all loss functions $\tilde{L}$ of the form $\tilde{L}_\theta \equiv_\theta L_\theta - b_\theta$ where the b_θs are constants then, by putting $b_\theta = \frac{1}{2}[\bigvee_t L_\theta(t) + \bigwedge_t L_\theta(t)]$, we see that to any generalized decision rule σ in $\mathscr{F}$ there corresponds a generalized decision rule ρ in $\mathscr{E}$ such that

$$P_\theta \rho L_\theta \leqslant Q_\theta \sigma L_\theta + \tfrac{1}{2}\varepsilon_\theta \left[\bigvee_t L_\theta(t) - \bigwedge_t L_\theta(t) \right]; \qquad \theta \in \Theta.$$

Conversely this inequality implies that

$$P_\theta \rho \tilde{L}_\theta \leqslant Q_\theta \sigma \tilde{L}_\theta + \varepsilon_\theta \|\tilde{L}_\theta\|; \qquad \theta \in \Theta$$

for any loss function $\tilde{L}$ in Lin (L).

In particular it follows that the norm $\|L_\theta\|$ in (6.2.1) may be replaced by the pseudonorm $\frac{1}{2}[\bigvee_t L_\theta(t) - \bigwedge_t L_\theta(t)]$. The latter norm has the advantage, not shared by the supremum norm, of being additive for direct summation. If we restrict our attention to non-negative loss functions however, then the supremum norm also becomes additive for direct summation.

For a given loss function L, define loss functions $\hat{L}$ and $\bar{L}$ by $\hat{L}_\theta(t) \equiv \frac{1}{2}(L_\theta(t) + \|L_\theta\|)$ and $\bar{L}_\theta(t) = 2L_\theta(t) - \|L_\theta\|$. Then $\hat{L}_\theta \geqslant 0$ and $\|\hat{L}_\theta\| \leqslant \|L_\theta\|$. If L is non-negative then also $\|\bar{L}_\theta\| \leqslant \|L_\theta\|$. This implies that in (6.2.1) we may assume that L is non-negative provided ε_θ is replaced by $\frac{1}{2}\varepsilon_\theta$. Transforming the loss $L_\theta(t)$ to $[L_\theta(t) - \bigwedge_t L_\theta(t)]/[\bigvee_t L_\theta(t) - \bigwedge_t L_\theta(t)]$ whenever the denominator is positive, we see that we may also assume that $L_\theta(t) \in \{0, 1\} \cup S$ for a $(k-2)$-point subset S of $[0, 1]$ depending on L.

Several equivalent versions of the original definition may be obtained by such considerations.

Proposition 6.2.3. *Equivalent versions of the definition of ε-deficiency for k-decision problems are obtained if $\|L_\theta\|$ is replaced by $\frac{1}{2}[\bigvee_t L_\theta(t) - \bigwedge_t L_\theta(t)]$ or if one of the following holds:*

(i) *it is assumed that $\bigvee_t L_\theta(t) + \bigwedge_t L_\theta(t) = 0$ when $\theta \in \Theta$.*
(ii) *it is assumed that $1 = \bigvee_t L_\theta(t) = -\bigwedge_t L_\theta(t)$ when $\theta \in \Theta$.*
(iii) *the inequality* (6.2.1) *is replaced by the inequality*

$$P_\theta \rho L_\theta \leqslant Q_\theta \sigma L_\theta + \varepsilon_\theta \|L\|$$

where $\|L\| = \sup_\theta \|L_\theta\|$.
(iv) *it is assumed that $L \geqslant 0$ and* (6.2.1) *is replaced by*

$$P_\theta \rho L_\theta \leqslant Q_\theta \sigma L_\theta + \tfrac{1}{2}\varepsilon_\theta \|L_\theta\|.$$

(v) *it is assumed, for each θ, that the range of L_θ is contained in a k-point* *subset of* $[0,1]$ *containing* 0 *and* 1 *and* (6.2.1) *is replaced by*

$$P_\theta \rho L_\theta \leqslant Q_\theta \sigma L_\theta + \tfrac{1}{2}\varepsilon_\theta.$$

(vi) *it is assumed that* $L \geqslant 0$ *and* (6.2.1) *is replaced by*

$$P_\theta \rho L_\theta \leqslant Q_\theta \sigma L_\theta + \tfrac{1}{2}\varepsilon_\theta \|L\|$$

where $\|L\| = \sup_\theta \|L_\theta\|$.

Equivalent versions of other statements may be obtained by similar considerations. However, we shall not usually state them explicitly.

In order to be able to construct examples of ε-deficiency we establish now the easy part of LeCam's fundamental transition (randomization) criterion for ε-deficiency.

Theorem 6.2.4 (ε-deficiency from Markov kernels). *Let* $\mathscr{E} = (\mathscr{X}, \mathscr{A}; P_\theta : \theta \in \Theta)$ *and* $\mathscr{F} = (\mathscr{Y}, \mathscr{B}; Q_\theta : \theta \in \Theta)$ *be experiments and let* $\varepsilon = (\varepsilon_\theta : \theta \in \Theta)$ *be a nonnegative function on* Θ. *Let* M *be a Markov kernel from* $(\mathscr{X}, \mathscr{A})$ *to* $(\mathscr{Y}, \mathscr{B})$, *or more generally, a transition from* $L(\mathscr{E})$ *to* $\operatorname{ba}(\mathscr{Y}, \mathscr{B})$.

Then $\mathscr{E}$ *is* ε*-deficient with respect to* $\mathscr{F}$ *provided* $\varepsilon_\theta \geqslant \|P_\theta M - Q_\theta\|$ *when* $\theta \in \Theta$.

Remark. By theorem 4.5.11, we may always assume that M maps $L(\mathscr{E})$ into $L(\mathscr{F})$.

Proof. Let T_k denote the k-decision space $\{1, \ldots, k\}$ and let σ be a decision rule from $\mathscr{F}$ to T_k. Define a transition ρ from $L(\mathscr{E})$ to T_k by putting $(\lambda\rho)(i) = \int \sigma(i|\cdot)\,\mathrm{d}\lambda M$ when $\lambda \in L(\mathscr{E})$. Let Θ_0 be a finite subset of Θ. Then the restriction of ρ to $L(\mathscr{E}_{\Theta_0})$ is definable by a Markov kernel, which we also shall denote by ρ. If $\theta \in \Theta_0$ then $\|P_\theta\rho - Q_\theta\sigma\| = \|(P_\theta M)\sigma - Q_\theta\sigma\| \leqslant \|P_\theta M - Q_\theta\|$ so that $P_\theta \rho L_\theta - Q_\theta \sigma L_\theta \leqslant \|P_\theta M - Q_\theta\|\,\|L_\theta\|$ for any loss function L. □

Corollary 6.2.5. *Suppose the experiments* $\mathscr{E} = (P_\theta : \theta \in \Theta)$ *and* $\mathscr{F} = (Q_\theta : \theta \in \Theta)$ *have the same sample space. Then* $\mathscr{E}$ *is* ε*-deficient with respect to* $\mathscr{F}$ *provided* $\varepsilon_\theta \geqslant \|P_\theta - Q_\theta\|$ *when* $\theta \in \Theta$.

Proof. Apply the theorem with M being the identity map. □

Example 6.2.6 (Example 6.2.2 continued). It was shown by Khintchin (1933) and Prohorov (1953) (see Hodges and LeCam (1960) or Sheu (1984)) that the total variation of the difference between the (n, θ) binomial distribution and the Poisson distribution with expectation $n\theta$ is at most $\min(2n\theta^2, 3\theta)$. Hence $\mathscr{E}$ is 3ε-deficient with respect to $\mathscr{F}$ and $\mathscr{F}$ is 3ε-deficient with respect to $\mathscr{E}$ in this case.

If we admit transitions as decision rules then the 'to each finite subset Θ_0 of Θ' occurring in the definition of ε-deficiency for k-decision problems when $k < \infty$ may be deleted.

Proposition 6.2.7. *Let $\mathscr{E} = (P_\theta : \theta \in \Theta)$ and $\mathscr{F} = (Q_\theta : \theta \in \Theta)$ be experiments and let $\varepsilon = (\varepsilon_\theta : \theta \in \Theta)$ be a non-negative function on Θ. Let $(T, \mathscr{S})$ be a k-decision space where $k < \infty$.*

Then $\mathscr{E}$ is ε-deficient with respect to $\mathscr{F}$ for k-decision problems if and only if to each (generalized) decision rule σ from $\mathscr{F}$ to $(T, \mathscr{S})$ and to each loss function L there corresponds a generalized decision rule ρ from $\mathscr{E}$ to $(T, \mathscr{S})$ such that

$$P_\theta \rho L_\theta \leqslant Q_\theta \sigma L_\theta + \varepsilon_\theta \|L_\theta\|$$

when $\theta \in \Theta$.

Proof. The 'if' follows immediately from the remark after the definition of ε-deficiency. Suppose $\mathscr{E}$ is ε-deficient with respect to $\mathscr{F}$ for k-decision problems. Let σ be a generalized decision rule in $\mathscr{F}$ and let L be a loss function. Then to each finite subset Θ_0 of Θ there corresponds a transition ρ_{Θ_0} from $L(\mathscr{E})$ to $(T, \mathscr{S})$ such that $P_\theta \rho_{\Theta_0} L_\theta \leqslant Q_\theta \sigma L_\theta + \varepsilon_\theta \|L_\theta\|$ when $\theta \in \Theta_0$. Applying the weak compactness lemma to the net $\{\rho_{\Theta_0}\}$, or using theorem 4.5.13, we obtain a transition ρ such that $P_\theta \rho L_\theta \leqslant Q_\theta \sigma L_\theta + \varepsilon_\theta \|L_\theta\|$. □

If $\mathscr{E}$ is coherent then any generalized decision rule from $\mathscr{E}$ to $(T, \mathscr{S})$ where $\mathscr{S}$ is finite is equivalent to a decision rule. Hence the words 'to each finite subset Θ_0 of Θ' occurring in the definition of ε-deficiency for k-decisions are superfluous in this case.

Corollary 6.2.8 (ε-deficiency when $\mathscr{E}$ is coherent). *Let $\mathscr{E} = (P_\theta : \theta \in \Theta)$ and $\mathscr{F} = (Q_\theta : \theta \in \Theta)$ be experiments and suppose $\mathscr{E}$ is coherent. Let ε be a non-negative function on Θ and let $(T, \mathscr{S})$ be a k-decision space where $k < \infty$. Then $\mathscr{E}$ is ε-deficient with respect to $\mathscr{F}$ for k-decision problems if and only if to each decision rule σ from $\mathscr{F}$ to $(T, \mathscr{S})$ there corresponds a decision rule ρ from $\mathscr{E}$ to $(T, \mathscr{S})$ such that*

$$P_\theta \rho L_\theta \leqslant Q_\theta \sigma L_\theta + \varepsilon_\theta \|L_\theta\| \quad \textit{when} \quad \theta \in \Theta.$$

The statement of the corollary is not necessarily true when $\mathscr{E}$ is not coherent. Thus a totally informative experiment $\mathscr{E}$ is 0-deficient with respect to any other experiment $\mathscr{F}$. However, the conclusion of the proposition does not hold when $\mathscr{E}$ and T are specified as in example 4.5.19 and $Q_\theta \delta L_\theta \equiv_\theta 0$.

Corollary 6.2.9 (Comparison by power functions of tests). *Let $\mathscr{E} = (P_\theta : \theta \in \Theta)$ be a coherent experiment. Then $\mathscr{E}$ is ε-deficient with respect to $\mathscr{F}$ for testing problems if and only if to each testing problem '$\theta \in \Theta_0$' against '$\theta \notin \Theta_0$' and to each power function Π_2 in $\mathscr{F}$ there corresponds a power function Π_1 in $\mathscr{E}$ such that*

$$\Pi_1(\theta) \leqslant \Pi_2(\theta) + \tfrac{1}{2}\varepsilon_\theta \quad \textit{when} \quad \theta \in \Theta_0$$

while

$$\Pi_1(\theta) \geqslant \Pi_2(\theta) - \tfrac{1}{2}\varepsilon_\theta \quad when \quad \theta \notin \Theta_0.$$

Proof. This follows from the previous corollary and part (v) of proposition 6.2.3 which ensures that we may restrict our attention to (0–1)-valued loss functions. □

The concept of a maximin test may be expressed in terms of deficiencies.

Corollary 6.2.10 (Maximin tests in terms of deficiencies). *Let $\mathscr{E} = (\mathscr{X}, \mathscr{A}; P_\theta : \theta \in \Theta)$ be a coherent experiment and consider a partitioning of Θ into three subsets:*

$\Theta_0 =$ the set of theories θ such that the decision $t = 0$ is preferable.
$\Theta_1 =$ the set of theories θ such that the decision $t = 1$ is preferable.
$\Theta_i =$ the set of theories θ such that neither decision is preferable to the other. (Thus Θ_i is the region of indifference.)

Let α and β be constants in $[0, 1]$ and put $\varepsilon_\theta = 2\alpha$ or $\varepsilon_\theta = 2$ or $\varepsilon_\theta = 2(1 - \beta)$ according to whether $\theta \in \Theta_0$, $\theta \in \Theta_i$ or $\theta \in \Theta_1$.

Consider the problem of testing the null hypothesis '$\theta \in \Theta_0$' on the basis of an observation of $\mathscr{E}$.

Take $T = \{0, 1\}$ as the decision space and let $\mathscr{F} = (Q_\theta : \theta \in \Theta)$ be an ideal experiment for this decision problem, i.e.

$$Q_\theta(1) = 1 - Q_\theta(0) = 0 \quad when \quad \theta \in \Theta_0$$

$$Q_\theta(1) = 1 - Q_\theta(0) = 1 \quad when \quad \theta \in \Theta_1.$$

Then there is a test in $\mathscr{E}$ with level of significance α and with power at least β in any alternative $\theta \in \Theta_1$ if and only if $\mathscr{E}$ is ε-deficient with respect to $\mathscr{F}$.

Proof. Let ρ be any decision rule and let δ denote the test function $\delta(x) \equiv_x \rho(1|x)$. Then δ has the desired properties if and only if $\|P_\theta\rho - Q_\theta\| \leqslant \varepsilon_\theta$; $\theta \in \Theta$.

The 'only if' now follows from theorem 6.2.4. Define the loss function L by putting $L_\theta(0) = 0$ or $= 1$ according to whether $\theta \in \Theta_0$ or $\theta \notin \Theta_0$ and by putting $L_\theta(1) = 0$ or $= 1$ according to whether $\theta \in \Theta_0$ or $\theta \notin \Theta_0$. Note that $\int L_\theta(t)Q_\theta(dt) = 0$ when $\theta \in \Theta_0 \cup \Theta_1$. The 'if' now follows by applying the corollary with σ being the identity function on T. □

We shall find this result helpful in section 6.5 when we apply the theory of invariant experiments to the Hunt–Stein theorem. The previous corollary may be generalized by considering a general partitioning of Θ together with an ideal experiment of observing precisely (thus not more than) the set in the partitioning which contains the unknown θ. However, this is just a complicated way of phrasing the problem of evaluating estimators of a given function on Θ by measuring how successful they are in producing approximations to the ideal experiment of observing precisely this function. Here is a result in this direction.

Corollary 6.2.11. *Let $\mathscr{E} = (\mathscr{X}, \mathscr{A}; P_\theta : \theta \in \Theta)$ be a coherent experiment and consider a partitioning $\{\Theta_1, \Theta_2, \ldots, \Theta_r\}$ of the set Θ such that the decision $t = i$ is preferable to all other decisions when $\theta \in \Theta_i$. Let Q_θ assign mass 1 to i when $\theta \in \Theta_i$. Then $\mathscr{E}$ is ε-deficient with respect to $(Q_\theta; \theta \in \Theta)$ if and only if there is a decision rule ρ from $\mathscr{E}$ to $T = \{1, \ldots, r\}$ such that $P_\theta \rho(i) \geqslant 1 - \varepsilon_\theta/2$ when $\theta \in \Theta$.*

The definition of ε-deficiency is formulated in order to permit reduction to finite sub-parameter sets. Although this is trivial it deserves to be stated as a theorem.

Theorem 6.2.12 (Reduction to finite parameter sets). *The experiment $\mathscr{E}$ is ε-deficient with respect to the experiment $\mathscr{F}$ for k-decision problems if and only if $\mathscr{E}_{\Theta_0}$ is $\varepsilon|\Theta_0$ deficient with respect to $\mathscr{F}_{\Theta_0}$ for k-decision problems for each non-empty finite subset Θ_0 of Θ.*

Furthermore, if $\mathscr{E}$ is ε-deficient with respect to $\mathscr{F}$ for k-decision problems then $\mathscr{E}_{\Theta_0}$ is $\varepsilon|\Theta_0$ deficient with respect to $\mathscr{F}_{\Theta_0}$ for any non-empty subset Θ_0 of Θ.

This theorem will be used frequently and usually without reference.

Now consider a totally informative experiment $\mathscr{E}$ and another experiment $\mathscr{F}$. Then, as we observed in example 6.2.1, $\mathscr{E}_{\Theta_0}$ is 0-deficient with respect to $\mathscr{F}_{\Theta_0}$ for any finite subset Θ_0 of Θ. It follows that $\mathscr{E}$ is 0-deficient with respect to $\mathscr{F}$. However, the following example shows that this does not necessarily imply that any pair (σ, L) where σ is a decision rule from $\mathscr{F}$ to a two-decision space $(T, \mathscr{S})$ equipped with a loss function L may be matched by a decision rule ρ in $\mathscr{E}$.

Example 6.2.13 (Continuation of examples 4.5.19 and 6.2.1). Let the experiments $\mathscr{E} = ([0,1]$, Borel class, $P_\theta : \theta \in [0,1])$ and $\mathscr{F} = ([0,1]$, Borel class, $Q_\theta : \theta \in [0,1])$ be given by:

Q_θ = the one-point distribution in $\theta; \theta \in \Theta = [0,1]$;
P_θ = Q_θ when $\theta > 0$;
P_0 = the uniform distribution on $[0,1]$.

Then both $\mathscr{E}$ and $\mathscr{F}$ are totally informative experiments with parameter set $\Theta = [0,1]$. Thus, by example 6.2.1, they are mutually 0-deficient with respect to each other. Fitting nicely in with this is the fact that if ρ is a decision rule from any experiment with parameter set Θ to any decision space, then its performance function defines the unique decision rule σ in $\mathscr{F}$ possessing the same performance function as ρ.

The experiment $\mathscr{E}$, however, is far from being that generous. Consider, for example, the testing problem in example 4.5.19, where the decision space was $T = \{0, 1\}$ and where the loss function was given by:

$$L_\theta(1) = 1 \quad \text{or} \quad = 0 \quad \text{as} \quad \theta > 0 \quad \text{or} \quad \theta = 0$$

$$L_\theta(0) = 0 \quad \text{or} \quad = 1 \quad \text{as} \quad \theta > 0 \quad \text{or} \quad \theta = 0.$$

Let σ be the perfect decision rule in $\mathscr{F}$ defined by putting $\sigma(1|y) = 1$ or $= 0$ as $y = 0$ or $y > 0$. Then $Q_\theta \sigma L_\theta \equiv_\theta 0$. Suppose now that $\mathscr{E}$ admitted a decision rule ρ such that $P_\theta \rho L_\theta \leqslant Q_\theta \sigma L_\theta + \varepsilon_\theta/2$ for all θ. Then $P_\theta \rho L_\theta = \rho(1/\theta) \leqslant \varepsilon_\theta/2$ when $\theta > 0$ while $\varepsilon_0/2 \geqslant P_0 \rho L_0 = \int_0^1 [1 - \rho(1|\theta)]\,d\theta \geqslant \int_0^{1*} (1 - \varepsilon_\theta/2)\,d\theta$ where $*$ signifies upper integral. In particular this implies that $\varepsilon \geqslant 1$ when $\varepsilon_\theta = \varepsilon$ does not depend on θ.

Thus we see that the words 'for all finite subsets Θ_0 of Θ' can *not* in general be deleted in the definition of ε-deficiency for k-decision problems.

If the sample space $(\mathscr{X}, \mathscr{A})$ of $\mathscr{E}$ and the integer k are both finite then we may restrict our attention to subsets Θ_0 of Θ containing at most $N(k-1)+1$ points where $2^N = \#\mathscr{A}$. This may be deduced from Helly's intersection theorem, theorem 2.3.11, as follows. Suppose $\mathscr{E}_{\Theta_0}$ is $(\varepsilon|\Theta_0)$-deficient with respect to $\mathscr{F}_{\Theta_0}$ for k-decision problems whenever $\#\Theta_0 \leqslant N(k-1)+1$.

Choose a k-decision space T equipped with a finite loss function L. Let σ be a decision rule in $\mathscr{E}$. Organize the set of decision rules ρ in $\mathscr{E}$ as a subset C of $\mathbb{R}^{N(k-1)}$ in the natural way. Let C_θ denote the set of decision rules ρ in C such that $P_\theta \rho L_\theta \leqslant Q_\theta \sigma L_\theta + \varepsilon_\theta \|L_\theta\|$. Then C_θ is a non-empty compact set. The assumption implies that $\bigcap \{C_\theta : \theta \in \Theta_0\} \neq \emptyset$ when $\#\Theta_0 \leqslant \dim(C) + 1$. Hence, by Helly's theorem, $\bigcap_\theta C_\theta \neq \emptyset$, so that $\mathscr{E}$ is ε-deficient with respect to $\mathscr{F}$ for k-decision problems.

Theorem 6.2.14. *The experiment $\mathscr{E} = (\mathscr{X}, \mathscr{A}; P_\theta : \theta \in \Theta)$ is ε-deficient with respect to $\mathscr{F}$ for k-decision problems if and only if $\mathscr{E}_{\Theta_0}$ is $(\varepsilon|\Theta_0)$-deficient with respect to $\mathscr{F}_{\Theta_0}$ for k-decision problems whenever $\#\Theta_0 \leqslant N(k-1)+1$ where $N = \log_2 \#\mathscr{A}$.*

Let us take a look at the set of functions ε such that $\mathscr{E}$ is ε-deficient with respect to $\mathscr{F}$ for k-decision problems.

Proposition 6.2.15. *Let $D_k = D_k(\mathscr{E}, \mathscr{F})$ denote the set of functions $\varepsilon \geqslant 0$ such that the experiment $\mathscr{E}$ is ε-deficient with respect to the experiment $\mathscr{F}$ for k-decision problems. Then, for $k = 1, \ldots, \infty$:*

(i) *D_k is convex as a subset of $[0,\infty]^\Theta$, i.e. $\sum_{i=1}^r p_i \varepsilon^i \in D_k$ whenever $\varepsilon^1, \ldots, \varepsilon^r \in D_k$, $p_1, \ldots, p_r \geqslant 0$ and $p_1 + \cdots + p_r = 1$.*
(ii) *D_k is increasing in the sense that $\eta \in D_k$ whenever $\eta \geqslant \varepsilon \in D_k$.*
(iii) *$\varepsilon \in D_k$ if and only if $\varepsilon \wedge 2 \in D_k$.*
(iv) *D_k is compact for the topology of pointwise convergence on Θ.*

Furthermore,

$$D_1 = [0,\infty]^\Theta \supseteqq D_2 \supseteqq D_3 \supseteqq \cdots \supseteqq D_\infty \supseteqq [2,\infty]^\Theta \quad \textit{and} \quad \bigcap \{D_k : k < \infty\} = D_\infty.$$

Remark. In view of (iii) it is clear that we could have limited our considerations to functions ε such that $0 \leqslant \varepsilon \leqslant 2$. The main reason for considering general non-negative functions is that the number 2 need no longer be an upper bound if we pass to families of measures which are not necessarily probability measures.

Proof. It follows directly from the definition and theorem 6.2.12 that we may assume that $\#\Theta < \infty$ and $k < \infty$. If σ is a decision rule from $\mathscr{F}$ to $T_k = \{1,\ldots,k\}$, L is a loss function on T_k and $P_\theta \rho_i L_\theta \leqslant Q_\theta \sigma L_\theta + \varepsilon_\theta^i \|L_\theta\|$; $i = 1,\ldots,k$ then $P_\theta \rho L_\theta \leqslant Q_\theta \sigma L_\theta + (\sum_i p_i \varepsilon_\theta^i)\|L_\theta\|$ where $\rho = \sum p_i \rho_i$. This proves (i). (ii) follows directly from the definition while (iii) follows from (ii) and the trivial inequality $\|P_\theta \rho L_\theta - Q_\theta \sigma L_\theta\| \leqslant 2\|L_\theta\|$. The compactness assertion (iv) follows readily from the weak compactness lemma (corollary 1.2.5).

If $k = 1$ then $P_\theta \rho L_\theta \equiv_\theta Q_\theta \sigma L_\theta$ for all choices of ρ, σ and L. Thus $D_1 = [0, \infty]^\Theta$. The trivial inequality mentioned above shows that $D_k \supseteqq [2, \infty]^\Theta$ while the definition implies directly that $D_1 \cap D_2 \cap \cdots = D_\infty$. It remains to show that $D_k \supseteqq D_{k+1}$.

Suppose $\varepsilon \in D_{k+1}$. Let L be a loss function on T_k and let σ be a decision rule from $\mathscr{F}$ to $T_k \subseteqq T_{k+1}$. Extend L to T_{k+1} by putting $L_\theta(k+1) = L_\theta(k)$. Then, since $\varepsilon \in D_{k+1}$, there is a decision rule $\hat{\rho}$ from $\mathscr{E}$ to T_{k+1} such that $P_\theta \hat{\rho} L_\theta \leqslant Q_\theta \sigma L_\theta + \varepsilon_\theta \|L_\theta\|$. Define a decision rule ρ from $\mathscr{E}$ to T_k by putting $\rho(i|\cdot) = \hat{\rho}(i|\cdot)$ or $= \hat{\rho}(k|\cdot) + \hat{\rho}(k+1|\cdot)$ as $i < k$ or $i = k$. Then $P_\theta \rho L_\theta = \sum_{i<k} (P_\theta \rho)(i) L_\theta(i) + (P_\theta \rho)(k) L_\theta(k) = \sum_{i=1}^{k+1} (P_\theta \hat{\rho})(i) L_\theta(i) = P_\theta \hat{\rho} L_\theta \leqslant Q_\theta \sigma L_\theta + \varepsilon_\theta \|L_\theta\|$. Hence $\varepsilon \in D_k$. □

Of course, the inclusion $D_k \supseteqq D_{k+1}$ amounts to the statement of the following corollary.

Corollary 6.2.16. *The experiment $\mathscr{E}$ is ε-deficient with respect to $\mathscr{F}$ for k-decision problems whenever it is ε-deficient with respect to $\mathscr{F}$ for $(k+1)$-decision problems.*

ε-deficiency enjoys a kind of subadditivity.

Proposition 6.2.17. *Let $\mathscr{E}$, $\mathscr{F}$ and $\mathscr{G}$ be experiments with the same parameter set Θ. Suppose $\mathscr{E}$ is ε-deficient with respect to $\mathscr{F}$ for k-decision problems and that $\mathscr{F}$ is η-deficient with respect to $\mathscr{G}$ for k-decision problems. Then $\mathscr{E}$ is $(\varepsilon + \eta)$-deficient with respect to $\mathscr{G}$ for k-decision problems.*

Proof. We may assume without loss of generality that $\#\Theta < \infty$ and that $k < \infty$. Let L be a loss function on T_k and let $r^{(2)}$ be an available risk function in $\mathscr{G}$. Then, by assumption, there is a risk function $r^{(1)}$ available in $\mathscr{F}$ such that $r^{(1)}(\theta) \leqslant r^{(2)}(\theta) + \eta_\theta \|L_\theta\|$; $\theta \in \Theta$. Again, by assumption, there is a risk function $r^{(0)}$ available in $\mathscr{E}$ such that $r^{(0)}(\theta) \leqslant r^{(1)}(\theta) + \varepsilon_\theta \|L_\theta\| \leqslant r^{(2)}(\theta) + (\varepsilon_\theta + \eta_\theta)\|L_\theta\|$; $\theta \in \Theta$. □

Using the notation of proposition 6.2.15, proposition 6.2.17 implies that $D_k(\mathscr{E},\mathscr{F}) \supseteqq D_k(\mathscr{E},\mathscr{G}) + D_k(\mathscr{G},\mathscr{F})$ for any three experiments $\mathscr{E}$, $\mathscr{F}$ and $\mathscr{G}$. Hence, if $\mathscr{M}_i$ is totally non-informative and $\mathscr{M}_a$ is totally informative then

$$[0,\infty]^\Theta = D_k(\mathscr{E},\mathscr{E}) \supseteqq D_k(\mathscr{E},\mathscr{F}) \supseteqq D_k(\mathscr{E},\mathscr{M}_i) + D_k(\mathscr{M}_i,\mathscr{M}_a) + D_k(\mathscr{M}_a,\mathscr{F})$$

$$= D_k(\mathscr{M}_i,\mathscr{M}_a) \supseteqq D_\infty(\mathscr{M}_i,\mathscr{M}_a), \quad \text{(by example 6.2.1).}$$

It follows that the largest deficiency set $D_k(\mathscr{E},\mathscr{F})$ is $[0,\infty]^\Theta$ while the smallest deficiency set, for fixed k, is $D_k(\mathscr{M}_i,\mathscr{M}_a)$. Thus $D_k(\mathscr{M}_i,\mathscr{M}_a)$ consists precisely of the set of all functions ε such that any experiment $\mathscr{E}$ is ε-deficient with respect to any other experiment $\mathscr{F}$ for k-decision problems, i.e. $D_k(\mathscr{M}_i,\mathscr{M}_a)$ is the set of trivial deficiency functions for k-decision problems. The set $D_\infty(\mathscr{M}_i,\mathscr{M}_a)$ is the set of all 'totally' trivial deficiency functions, i.e. the set of functions ε such that any experiment is ε-deficient with respect to any other experiment for k-decision problems for any k. It is perhaps a little surprising that this set is not $[2,\infty]^\Theta$. A moment's reflection reveals that the lower bound of ε_{θ_0}, for fixed $\theta_0 \in \Theta$, as ε runs through any set $D_k(\mathscr{E},\mathscr{F})$, is zero. Indeed there is always a Markov kernel from $\mathscr{E}$ to $\mathscr{F}$ which yields a perfect fit for $\theta = \theta_0$. Alternatively we might argue this by saying that if there is only one value θ_0 of θ which matters, then we may always ensure that our decision rules yield minimum risk for $\theta = \theta_0$.

The price we have to pay for being so biased towards a particular θ_0 is that usually we are then forced to make ε_θ large for other values of θ. Thus in the extreme case where $\mathscr{E} = \mathscr{M}_i$ and $\mathscr{E} = \mathscr{M}_a$ then perfect fit at $\theta = \theta_0$ implies that $\varepsilon_\theta \geqslant 2$ for all other values of θ. As it is of some interest to know the set of trivial deficiency functions we shall determine the sets $D_k(\mathscr{M}_i,\mathscr{M}_a)$.

Proposition 6.2.18 (Deficiency of $\mathscr{M}_i$ with respect to $\mathscr{M}_a$). *$\mathscr{M}_i$ is ε-deficient with respect to $\mathscr{M}_a$ for k-decision problems if and only if $\sum_{\theta \in F}(1 - \varepsilon_\theta/2)^+ \leqslant 1$ whenever F is a subset of Θ containing at most k elements.*

Remark 1. It suffices to consider subsets of Θ containing exactly $r = k \wedge \#\Theta$ elements. If $\varepsilon_{\theta_1} \leqslant \varepsilon_{\theta_2} \leqslant \cdots \leqslant \varepsilon_{\theta_r} \leqslant \varepsilon_\theta : \theta \in \Theta$ for some r-point set $\{\theta_1,\ldots,\theta_r\}$ then it suffices to check that $\sum_{i=1}^r (1 - \varepsilon_{\theta_i}/2)^+ \leqslant 1$.

Remark 2. Suppose only non-randomized decision rules were permitted. Then ε would belong to the corresponding modification of $D_k(\mathscr{M}_i,\mathscr{M}_a)$, when $k \geqslant 2$, if and only if there were at most one θ such that $\varepsilon_\theta < 2$.

Proof. Suppose $\varepsilon \in D_k(\mathscr{M}_i,\mathscr{M}_a)$ and let F be an r-point subset of Θ. Then $\varepsilon \in D_r(\mathscr{M}_i,\mathscr{M}_a)$. Consider $T = F$ as an r-decision space equipped with the loss function L given by $L_\theta(t) = 0$ or $= 1$ as $t = \theta$ or $t \neq \theta$. Clearly $\mathscr{M}_a$ admits a decision rule with risk 0 when $\theta \in F$. Thus there is a probability distribution ρ on T, i.e. a performance function of a decision rule in $\mathscr{M}_i$, such that

$\rho L_\theta \leqslant \varepsilon_\theta/2$ when $\theta \in F$. Then $\rho(\theta) \geqslant 1 - \varepsilon_\theta/2$; $\theta \in F$ so that $\sum_F (1 - \varepsilon_\theta/2)^+ \leqslant \sum_F \rho(\theta) = 1$.

Conversely, suppose that ε satisfies the given condition. Let T be a k-decision space equipped with a loss function $L \geqslant 0$. We may assume that Θ is finite. Let Θ_t, for each $t \in T$, denote the set of parameter points θ in Θ such that t is the optimal decision for θ. Then $\bigcup \{\Theta_t : t \in T\} = \Theta$. Let $\tilde{\Theta}_t : t \in T$ be disjoint sets such that $\bigcup \{\tilde{\Theta}_t : t \in T\} = \Theta$ and $\tilde{\Theta}_t \subseteqq \Theta_t$; $t \in T$. In each set $\tilde{\Theta}_t$ choose a point θ_t such that $\varepsilon_{\theta_t} = \bigwedge \{\varepsilon_\theta : \theta \in \tilde{\Theta}_t\}$. Put $T_0 = \{t : \tilde{\Theta}_t \neq \emptyset\}$. Then $1 \geqslant \sum \{1 - (\varepsilon_{\theta_t}/2)^+ : t \in T_0\}$. It follows that there is a probability distribution ρ on T_0 such that $\rho(t) \geqslant 1 - \varepsilon_{\theta_t}/2$ when $t \in T_0$. Let $\theta \in \tilde{\Theta}_{t_0}$. Then $L_\theta(t_0) = \min_t L_\theta(t)$ so that $\rho L_\theta = \rho(t_0)L_\theta(t_0) + \sum \{\rho(t)L_\theta(t) : t \neq t_0\} \leqslant \bigwedge_t L_\theta(t) + (1 - \rho(t_0))\|L_\theta\| \leqslant \bigwedge_t L_\theta(t) + (\varepsilon_{\theta_t}/2)\|L_\theta\| \leqslant \bigwedge_t L_\theta(t) + (\varepsilon_\theta/2)\|L_\theta\| \leqslant r(\theta) + (\varepsilon_\theta/2)\|L\|$ for any risk function r in any experiment. □

Two problems which now naturally pose themselves are:

(i) which deficiency functions ε are universal (trivial) for k-decision problems for a given experiment $\mathscr{F}$ in the sense that $\mathscr{E}$ is ε-deficient with respect to $\mathscr{F}$ for k-decision problems for any experiment $\mathscr{F}$?

and

(ii) which deficiency functions ε are universal (trivial) for k-decision problems for a given experiment $\mathscr{F}$ in the sense that any experiment is ε-deficient with respect to $\mathscr{F}$ for k-decision problems?

It follows directly from proposition 6.2.17 that the somewhat unsatisfactory answers to these problems are, respectively, $D_k(\mathscr{E}, \mathscr{M}_a)$ and $D_k(\mathscr{M}_i, \mathscr{F})$. There remains the problem of describing these sets. $D(\mathscr{E}, \mathscr{M}_a)$ may be described as follows.

Proposition 6.2.19. *The experiment $\mathscr{E} = (P_\theta : \theta \in \Theta)$ is ε-deficient with respect to $\mathscr{M}_a$ if and only if $\mathscr{E}$ permits a generalized estimator $\hat{\theta}$ such that*

$$P_\theta(\hat{\theta} \neq \theta) \leqslant \varepsilon_\theta/2$$

for all $\theta \in \Theta$.

Remark. The term 'generalized estimator' may here be interpreted as a transition from $L(\mathscr{E})$ to Θ equipped with some σ-algebra making all one-point sets measurable. If Θ is countable then any such estimator is equivalent to a traditional one.

Proof. We may assume without loss of generality that Θ is finite. Take $T = \Theta$ as a decision space and put $L_\theta(t) = 0$ or $= 1$ as $\theta = t$ or $\theta \neq t$. Suppose $\varepsilon \in D(\mathscr{E}, \mathscr{M}_a)$. Then, since 0 is an attainable risk function in $\mathscr{M}_a$, there is an estimator (possibly randomized) such that $r_{\hat{\theta}}(\theta) = P_\theta(\hat{\theta} \neq \theta) \leqslant \varepsilon_\theta/2$; $\theta \in \Theta$.

Conversely, assume that $P_\theta(\hat{\theta} \neq \theta) \leqslant \varepsilon_\theta/2 : \theta \in \Theta$ for some estimator $\hat{\theta}$. Then $\hat{\theta}$ defines a Markov kernel M from $\mathscr{E}$ to $\mathscr{M}_a = (\delta_\theta : \theta \in \Theta)$ such that $\|P_\theta M - \delta_\theta\| = 2(P_\theta M)(\{\theta\}) = 2P_\theta(\hat{\theta} \neq \theta) \leqslant \varepsilon_\theta; \theta \in \Theta$. Thus $\mathscr{E}$ is ε-deficient with respect to $\mathscr{M}_a$ by the randomization criterion. □

Corollary 6.2.20. *$\mathscr{E} = (P_\theta : \theta \in \Theta)$ is ε-deficient with respect to $\mathscr{M}_a$ where $\varepsilon_\theta/2 \equiv_\theta \sum_{\theta' \neq \theta} \inf_{0<t<1} \int \mathrm{d}P_{\theta'}^{1-t}\, \mathrm{d}P_\theta^t$.*

Proof. We may assume that Θ is finite. Put $\mu = \sum_\theta P_\theta$ and $f_\theta = \mathrm{d}P_\theta/\mathrm{d}\mu$; $\theta \in \Theta$. Choose a maximum likelihood estimator $\hat{\theta}$ of θ. Then

$$P_\theta(\hat{\theta} \neq \theta) \leqslant \sum_{\theta' \neq \theta} P_\theta(\hat{\theta} = \theta') \leqslant \sum_{\theta' \neq \theta} P_\theta(f_{\theta'} \geqslant f_\theta)$$

$$\leqslant \sum_{\theta' \neq \theta} \int f_{\theta'} \wedge f_\theta \,\mathrm{d}\mu = \sum_{\theta' \neq \theta} \int (f_{\theta'} \wedge f_\theta)^{1-t} (f_{\theta'} \wedge f_\theta)^t \,\mathrm{d}\mu$$

$$\leqslant \sum_{\theta' \neq \theta} \int f_{\theta'}^{1-t(\theta')} f_\theta^{t(\theta')} \,\mathrm{d}\mu = \sum_{\theta' \neq \theta} \int \mathrm{d}P_{\theta'}^{1-t(\theta')}\, \mathrm{d}P_\theta^{t(\theta')}$$

where for each $\theta' \neq \theta$, $t(\theta')$ is an arbitrary number in $[0, 1]$. □

Proposition 6.2.19 may be extended to the case of k-decision problems.

Proposition 6.2.21. *$\mathscr{E} = (\mathscr{X}, \mathscr{A}; P_\theta : \theta \in \Theta)$ is ε-deficient with respect to $\mathscr{M}_a$ for k-decision problems if and only if to each function g on Θ taking at most k values there corresponds an estimator $\hat{\theta}$ of θ such that*

$$P_\theta(g(\hat{\theta}) \neq g(\theta)) \leqslant \varepsilon_\theta/2; \qquad \theta \in \Theta.$$

Remark. Put $m = \#\Theta$. As any function g on Θ takes at most m values it is clear that a function on Θ takes at most k values if and only if it takes at most r values where $r = k \wedge m$. Now g takes at most r values if and only if $g = h \circ \tilde{g}$ where $\tilde{g}$ takes exactly r values. Thus we may restrict our attention to functions g taking exactly r values.

Finally, note that an estimator $\hat{\theta}$ satisfies the inequalities $P_\theta(g(\hat{\theta}) \neq g(\theta)) \leqslant \varepsilon_\theta/2$; $\theta \in \Theta$ if and only if $P_\theta(\hat{\theta} \in F) \leqslant \varepsilon_\theta/2$ whenever $\theta \in F$ and F belongs to the partition (algebra) generated by g. Thus it is only the partition induced by g which matters. If $k \geqslant m$ then $r = m$ and the only partition of Θ containing exactly m sets is the decomposition of Θ into its individual points. Thus we obtain the previous proposition as a particular result.

Proof of the proposition. We may assume without loss of generality that Θ is finite. Suppose $\mathscr{E}$ is ε-deficient with respect to $\mathscr{M}_a$ for k-decision problems and let g be a function on Θ taking r values $t_1, \ldots, t_r$ where $r = (\#\Theta) \wedge k$. Take $T = \{t_1, \ldots, t_r\}$ as a decision space and define the loss function L by putting $L_\theta(t) = 0$ or $= 1$ as $t = g(\theta)$ or $t \neq g(\theta)$. Then 0 is a possible risk function in $\mathscr{M}_a = (\delta_\theta : \theta \in \Theta)$. Here δ_θ, as usual, denotes the one-point distribution in θ.

By assumption there is a decision rule ρ from $\mathscr{E}$ to T such that $P_\theta \rho L_\theta \leqslant \varepsilon_\theta/2$; $\theta \in \Theta$. To each $t \in T$ assign a particular $h(t) \in \Theta$ such that $g(h(t)) = t$. Let $\hat{\theta}$ assign probability $\rho(t|x)$ to $h(t)$ when x is the observed outcome of $\mathscr{E}$. Let $\theta \in \Theta$ and put $t = g(\theta)$. Then

$$\begin{aligned} P_\theta(g(\hat{\theta}) \neq g(\theta)) &= E_\theta \sum \{\rho(t'|\cdot) : g(h(t')) \neq t\} \\ &= E_\theta \sum \{\rho(t'|\cdot) : t' \neq t\} = P_\theta \rho L_\theta \leqslant \varepsilon_\theta/2. \end{aligned}$$

Conversely, suppose that the condition described in the proposition is fulfilled for a given function ε on Θ. Let $L \geqslant 0$ be a loss function on a k-decision space T. Choose a function g from Θ to T such that $g(\theta)$, for each θ, is the optimal decision when θ prevails. By assumption there is an estimator $\hat{\theta}$ of θ such that $P_\theta(g(\hat{\theta}) \neq g(\theta)) \leqslant \varepsilon_\theta/2$; $\theta \in \Theta$. Let M be a Markov kernel defining $\hat{\theta}$ and let ρ be the composition of $\hat{\theta}$ with g. Let $\theta_0 \in \Theta$. Then

$$\begin{aligned} P_{\theta_0} \rho L_{\theta_0} &= E_{\theta_0} \sum_t L_{\theta_0}(t) \rho(t|\cdot) \\ &= E_{\theta_0} \sum_t L_{\theta_0}(t) \sum \{M(\theta|\cdot) : g(\theta) = t\} \\ &= E_{\theta_0} \sum_\theta L_{\theta_0}(g(\theta)) M(\theta|\cdot) \\ &= L_{\theta_0}(g(\theta_0)) P_{\theta_0}(g(\hat{\theta}) = g(\theta_0)) \\ &\quad + \sum \{L_{\theta_0}(g(t)) P_{\theta_0}(\hat{\theta} = \theta) : g(\theta) \neq g(\theta_0)\} \\ &\leqslant \bigwedge_t L_{\theta_0}(t) + \|L_{\theta_0}\| P_{\theta_0}(g(\hat{\theta}) \neq g(\theta_0)) \\ &\leqslant \bigwedge_t L_{\theta_0}(t) + (\varepsilon_{\theta_0}/2) \|L_{\theta_0}\| \\ &\leqslant r(\theta_0) + (\varepsilon_{\theta_0}/2) \|L_{\theta_0}\| \end{aligned}$$

for any $\theta_0 \in \Theta$ and for any risk function r in any experiment. □

A satisfactory description of deficiencies of $\mathscr{M}_i$ with respect to a general experiment $\mathscr{F}$ requires a particular case of the complete randomization criterion. As this particular case has all the essential features of the general case we could have postponed this discussion until after the proof of LeCam's randomization theorem. However, we have found it desirable to discuss problem (ii) here, i.e. $D_k(\mathscr{M}_i, \mathscr{F})$, in order to make it easier to compare this discussion with the discussion of problem (i). The proof of the complete randomization theorem may therefore seem familiar to the reader, when it appears later. Problem (ii) may be answered as follows.

Proposition 6.2.22. *$\mathscr{M}_i$ is ε-deficient with respect to $\mathscr{F} = (\mathscr{Y}, \mathscr{B}; Q_\theta : \theta \in \Theta)$ for k-decision problems, where $k < \infty$, if and only if to each subalgebra $\mathscr{B}_0$ of $\mathscr{B}$ containing at most 2^k sets there corresponds a probability measure π on $\mathscr{B}_0$ such* *that*

$$\|Q_\theta|\mathscr{B}_0 - \pi\| \leqslant \varepsilon_\theta; \qquad \theta \in \Theta.$$

$\mathscr{M}_i$ is ε-deficient with respect to $\mathscr{F} = (Q_\theta; \theta \in \Theta)$ if and only if there is an additive probability set function π on $\mathscr{B}$ such that $\|Q_\theta - \pi\| \leqslant \varepsilon_\theta$; $\theta \in \Theta$.

Remark. If there exists an additive probability set function π satisfying the inequalities $\|Q_\theta - \pi\| \leqslant \varepsilon_\theta$; $\theta \in \Theta$, then it may always be chosen within $L(\mathscr{F})$ (see theorem 4.5.11).

Proof. We may again assume without loss of generality that Θ is finite. Consider first the case of k-decision problems where $k < \infty$. Suppose $\mathscr{M}_i$ is ε-deficient with respect to $\mathscr{F}$ for k-decision problems. Let $\mathscr{B}_0$ be a subalgebra of $\mathscr{B}$ containing 2^r sets where $r \leqslant k$. Let $\{B_1, \ldots, B_r\}$ be the r-partition of $\mathscr{Y}$ which generates $\mathscr{B}_0$. Take $T = \{1, \ldots, r\}$ as an r-decision space and let L be a finite loss function on T. Let σ be the decision rule in $\mathscr{F}$ which maps y into i when $y \in B_i$. Then $Q_\theta \sigma L_\theta = \sum_t L_\theta(t) Q_\theta(B_t)$. By assumption there is a probability distribution π_L on T such that $\sum_t L_\theta(t)\pi_L(t) \leqslant \sum_t L_\theta(t) Q_\theta(B_t) + \varepsilon_\theta \|L_\theta\|$. Thus $\bigvee_L \bigwedge_\pi \sum_\theta [\sum_t L_\theta(t)\pi(t) - \sum_t L_\theta(t) Q_\theta(B_t) - \varepsilon_\theta \|L_\theta\|] \leqslant 0$ where L runs through the set of all finite loss functions on T while π runs through the set of all probability distributions on T. The minimax theory in chapter 3 (say theorem 3.2.12) implies that $\bigvee_L$ and $\bigwedge_\pi$ may be interchanged so that $\bigvee_L \sum_\theta [\;] \leqslant 0$ for some probability measure π on τ. Thus $\pi_L = \pi$ may be chosen independently of L. Putting $L_\theta(t) = -1$ or $+1$ as $\pi(t) - Q_\theta(B_t) \leqslant 0$ or > 0 we see that $\sum_t |\pi(t) - Q_\theta(B_t)| \leqslant \varepsilon_\theta$ when $\theta \in \Theta$.

Conversely, suppose that to each subalgebra $\mathscr{B}_0$ of $\mathscr{B}$ containing at most 2^k sets there corresponds a probability measure π on $\mathscr{B}_0$ such that $\|\pi - Q_\theta|\mathscr{B}_0\| \leqslant \varepsilon_\theta$; $\theta \in \Theta$. Take $T = \{1, \ldots, k\}$ as a k-decision space and let σ be a decision rule in $\mathscr{F}$. If σ is non-randomized then it may be described by the partitioning $\sigma^{-1}[\{i\}]$; $i = 1, \ldots, k$ of $\mathscr{Y}$. The assumption implies then that there is a probability distribution π_σ on T such that $\|\pi_\sigma - Q_\theta \sigma\| \leqslant \varepsilon_\theta$; $\theta \in \Theta$. If σ is any decision rule then its performance function is a convex combination of performance functions of non-randomized decision rules. This, together with the convexity of the total variation norm, implies that to any decision rule σ in $\mathscr{F}$ there corresponds a decision rule ρ in $\mathscr{M}_i$ such that $\|\rho - Q_\theta \sigma\| \leqslant \varepsilon_\theta$; $\theta \in \Theta$. If L is a loss function on T then this implies that $|\rho L_\theta - Q_\theta \sigma L_\theta| \leqslant \varepsilon_\theta \|L_\theta\|$. Thus we have completed the proof of the first statement.

The 'if' part of the last statement follows directly from the part of the randomization criterion which we have proved. Finally, suppose that $\mathscr{M}_i$ is ε-deficient with respect to $\mathscr{F}$. Proceeding as in section 4.5 we let N denote the set of pairs (Δ, ξ) where Δ is a finite ordered $\mathscr{B}$-measurable partition $(B_1, \ldots, B_k)$ of $\mathscr{Y}$ and ξ is a k-tuple $(y_1, \ldots, y_k)$ such that $y_i \in B_i$; $i = 1, \ldots, k$.

Direct N by defining $(\Delta_1, \xi_1) \geqslant (\Delta_2, \xi_2)$ when Δ_1 is finer than Δ_2 (see section 1.1). If $n = ((B_1, \ldots, B_k), (y_1, \ldots, y_k)) \in N$ and Q is a probability measure on $\mathscr{B}$ then we put $Q_n(B) = \sum_i I_B(y_i) Q(B_i)$; $B \in \mathscr{B}$. By assumption, for each n, there is a probability measure $\pi^{(n)}$ on $\{1, \ldots, k\}$ such that $\|\pi_n - (Q_\theta)_n\| \leqslant \varepsilon_\theta : \theta \in \Theta$ where $\pi_n(B) \equiv_B \sum I_B(y_i) \pi^{(n)}(B_i)$. Let $\{\pi_{n'}\}$ be a subset of $\{\pi_n\}$ such that $\pi(B) = \lim_{n'} \pi_{n'}(B)$ exists when $B \in \mathscr{B}$. Then, since $Q_{\theta,n}(B) = Q_\theta(B)$ whenever B is a union of sets from the partition part of n, $\|\pi - Q_\theta\| \leqslant \varepsilon_\theta$; $\theta \in \Theta$. □

Proposition 6.2.22 implies that $\mathscr{M}_i$ is ε-deficient with respect to $\mathscr{F} = (\mathscr{Y}, \mathscr{B}; Q_\theta : \theta \in \Theta)$ for k-decision problems, where $k < \infty$, if and only if $\mathscr{M}_i$ is ε-deficient with respect to each experiment $\mathscr{F} | \mathscr{B}_0$ where $\mathscr{B}_0$ is a subalgebra of $\mathscr{B}$ containing at most 2^k sets. Similarly proposition 6.2.21 implies that $\mathscr{E}$ is ε-deficient with respect to $\mathscr{M}_a = (\delta_\theta; \theta \in \Theta)$ for k-decision problems, where $k < \infty$, if and only if $\mathscr{E}$ is ε-deficient with respect to $\mathscr{M}_a | \sigma\{\Theta_1, \ldots, \Theta_k\}$ for any partitioning of Θ into k-sets $\Theta_1, \ldots, \Theta_k$. In the next section we shall see that this extends to the comparison of any pair $(\mathscr{E}, \mathscr{F})$ of experiments.

Corollary 6.2.23. *$\mathscr{M}_i$ is ε-deficient with respect to $\mathscr{F} = (Q_\theta; \theta \in \Theta)$ for k-decision problems if and only if $\mathscr{M}_i$ is $\varepsilon | \Theta_0$ deficient with respect to $\mathscr{F}_{\Theta_0}$ for k-decision problems whenever $\# \Theta_0 \leqslant k$.*

In particular $\mathscr{M}_i$ is ε-deficient with respect to $\mathscr{F}$ for 2-decision problems if and only if $\varepsilon_{\theta_1} + \varepsilon_{\theta_2} \geqslant \|Q_{\theta_1} - Q_{\theta_2}\|$ when $\theta_1, \theta_2 \in \Theta$.

Proof. The first statement follows from theorem 6.2.14 with $N = 1$. Thus it remains to prove the last statement when $\Theta = \{\theta_1, \theta_2\}$. Suppose $\mathscr{M}_i$ is ε-deficient with respect to $\mathscr{F}$ for 2-decision problems and let $B \in \mathscr{B}$. Then by proposition 6.2.22, there is a number π_B in $[0, 1]$ such that $|Q_\theta(B) - \pi_B| \leqslant \varepsilon_\theta/2$; $\theta \in \Theta$. Hence $|Q_{\theta_1}(B) - Q_{\theta_2}(B)| \leqslant (\varepsilon_{\theta_1} + \varepsilon_{\theta_2})/2$. Thus $\|Q_{\theta_1} - Q_{\theta_2}\| \leqslant \varepsilon_{\theta_1} + \varepsilon_{\theta_2}$. Conversely, suppose that $\|Q_{\theta_1} - Q_{\theta_2}\| \leqslant \varepsilon_{\theta_1} + \varepsilon_{\theta_2}$. Put $\pi = (\varepsilon_{\theta_2} Q_{\theta_1} + \varepsilon_{\theta_1} Q_{\theta_2})/(\varepsilon_{\theta_1} + \varepsilon_{\theta_2})$ or $= Q_{\theta_1}$ as $\varepsilon_{\theta_1} + \varepsilon_{\theta_2} > 0$ or $\varepsilon_{\theta_1} + \varepsilon_{\theta_2} = 0$. Then $\|\pi - Q_{\theta_i}\| \leqslant \varepsilon_{\theta_i}$; $i = 1, 2$. Hence $\mathscr{M}_i$ is ε-deficient with respect to $\mathscr{F}$ (for 2-decision problems). □

In particular we will be interested in constant deficiency functions ε. It follows from proposition 6.2.15 that there is a smallest constant $\varepsilon \geqslant 0$ such that the experiment $\mathscr{E}$ is ε-deficient with respect to the experiment $\mathscr{F}$ for k-decision problems. This constant will be denoted by $\delta_k(\mathscr{E}, \mathscr{F})$ and is called *the deficiency of $\mathscr{E}$ with respect to $\mathscr{F}$ for k-decision problems.* We can write $\delta(\mathscr{E}, \mathscr{F})$ instead of $\delta_\infty(\mathscr{E}, \mathscr{F})$ and call this number simply *the deficiency of $\mathscr{E}$ with respect to $\mathscr{F}$.*

As the deficiency will play an important role in the following we should dwell a little on its interpretations. Consider, for simplicity, k-decision problems where k is finite and where the loss function is bounded by, say, 1. Suppose we have been told that $\delta_k(\mathscr{E}, \mathscr{F})$ is a small number. Then by (iii) of proposition 6.2.3 we know that whatever we may achieve in $\mathscr{F}$ may be achieved almost as well in $\mathscr{E}$. What about a large deficiency number $\delta_k(\mathscr{E}, \mathscr{F})$? In this case our conclusion must be qualified. We know then that there are situations where it is impossible for $\mathscr{E}$ to beat a particular decision rule in $\mathscr{F}$. On the other hand a large deficiency $\delta_k(\mathscr{E}, \mathscr{F})$ does not by any means exclude the possibility that $\mathscr{E}$ is immensely better than $\mathscr{F}$ for a particular loss function L.

Another way of expressing these considerations is to say that the deficiency $\delta_k(\mathscr{E}, \mathscr{F})$ measures, relative to the sup norm measure of the loss function, what we lose by basing ourselves on $\mathscr{E}$ rather than on $\mathscr{F}$ under the *worst* possible conditions for this comparison.

If we do not restrict ourselves to coherent experiments $\mathscr{E}$ then these considerations are based on the assumption that generalized decision rules are permitted. It will be apparent later that these comments are valid for general, not necessarily finite, decision spaces.

If $\delta_k(\mathscr{E}, \mathscr{F}) = 0$ then we will write this $\mathscr{E} \geqslant_k \mathscr{F}$ and say that *$\mathscr{E}$ is at least as informative as $\mathscr{F}$ for k-decision problems.* We may write $\mathscr{E} \geqslant \mathscr{F}$ instead of $\mathscr{E} \geqslant_\infty \mathscr{F}$ and express this relation in words by saying that *$\mathscr{E}$ is at least as informative as $\mathscr{F}$.*

The interpretation of the relation $\mathscr{E} \geqslant_k \mathscr{F}$ is that whatever we can achieve in $\mathscr{F}$ in a k-decision problem we can achieve just as well in $\mathscr{E}$. We need not require that the loss function is bounded although the remark on coherence and generalized decision rules made above is in force. Example 6.2.1 is a case in point.

If $\mathscr{E} \geqslant_k \mathscr{F}$ and also $\mathscr{F} \geqslant_k \mathscr{E}$ then we will write this $\mathscr{E} \sim_k \mathscr{F}$ and say that *$\mathscr{E}$ and $\mathscr{F}$ are equally informative for k-decision problems* or that *$\mathscr{E}$ and $\mathscr{F}$ are equivalent for k-decision problems.* We may write $\mathscr{E} \sim \mathscr{F}$ instead of $\mathscr{E} \sim_\infty \mathscr{F}$ and express this relation by saying that *$\mathscr{E}$ and $\mathscr{F}$ are equally informative* or just that *$\mathscr{E}$ and $\mathscr{F}$ are equivalent.* If $\mathscr{E} \sim \mathscr{F}$ then $\mathscr{E}$ may be called a *representative* of $\mathscr{F}$. Clearly $\mathscr{E}$ is a representative of $\mathscr{F}$ if and only if $\mathscr{F}$ is a representative of $\mathscr{E}$.

In view of the above remarks, the interpretation of the relation $\mathscr{E} \sim_k \mathscr{F}$ is that whatever can be achieved in one of the two experiments $\mathscr{E}$ and $\mathscr{F}$ can be achieved in the other. Clearly $\mathscr{E} \sim_1 \mathscr{F}$ for any pair $\mathscr{E}, \mathscr{F}$ of experiments and we shall see that the equivalences '$\sim_k$'; $k = 2, \ldots, \infty$ are all the same.

Before proceeding we should show due respect to the logicians and point out that there is no such mathematical object as the set of all experiments– even in the case of one-point parameter sets. The reader might here consult the 'nothing contains everything' theorem in Halmos' book *Naive set theory*, (1958). However, there are sets which nearly qualify as sets 'containing' all experiments with parameter set Θ in the sense that they contain representatives of any experiment. This may be deduced fairly directly from the definition, as follows. For each experiment $\mathscr{E}$, each finite non-empty subset U of Θ, each positive integer k and each real valued loss function L on $T_k \times U$ let us consider the set $\hat{\mathbb{R}}(\mathscr{E} | U, k, L)$ consisting of those vectors v in $\mathbb{R}^U$ such that $v \geqslant r$ for some risk function in the decision problem $(\mathscr{E}, U, T_k, L)$. For each experiment with parameter set Θ let $\xi(\mathscr{E})$ denote the map which to each triple (U, k, L), where U is a finite subset of Θ, k is a positive integer and L is a loss function on $T_k \times U$, assigns the subset $\hat{\mathbb{R}}(\mathscr{E}|U, k, L)$ of $\mathbb{R}^U$. Then, by the definition of 0-deficiency, the experiments $\mathscr{E}$ and $\mathscr{F}$ are equivalent if and only if $\xi(\mathscr{E}) = \xi(\mathscr{F})$. The collection of all possible functions ξ is a subset Ξ of the set of all functions from the set of permissible triples into the class of sets which are subsets of $\mathbb{R}^U$ for some finite subset U of Θ. Following LeCam the collection of all experiments which are equivalent to a given experiment $\mathscr{E}$ is called *the type of* $\mathscr{E}$. From the above considerations it follows that the collection of types of experiments with parameter set Θ may be identified by ξ with the set Ξ. Later we shall see that there are much more convenient ways of identifying types of experiments. The point to be made here is only that it can be done and, consequently, that we may consider the collection of types of experiments with parameter set Θ as a set.

Keeping this in mind we shall allow ourselves to proceed as if there were a set containing all experiments. Thus we shall work with e.g. distances, orderings and functionals of various kinds for experiments.

The deficiency $\delta_k(\mathscr{E}, \mathscr{F})$ is different in general from the deficiency $\delta_k(\mathscr{F}, \mathscr{E})$. We remedy this asymmetry by defining the *deficiency distance for k-decision problems between the experiments* $\mathscr{E}$ *and* $\mathscr{F}$ as the largest of the numbers $\delta_k(\mathscr{E}, \mathscr{F})$ and $\delta_k(\mathscr{F}, \mathscr{E})$. This number will be denoted by $\Delta_k(\mathscr{E}, \mathscr{F})$. Thus $\Delta_k(\mathscr{E}, \mathscr{F}) = \delta_k(\mathscr{E}, \mathscr{F}) \vee \delta_k(\mathscr{F}, \mathscr{E})$. We may write $\Delta(\mathscr{E}, \mathscr{F})$ instead of $\Delta_\infty(\mathscr{E}, \mathscr{F})$ and call this number simply *the deficiency distance between* $\mathscr{E}$ *and* $\mathscr{F}$.

We list some of the elementary and basic properties of these quantities and relations below. Unless otherwise stated, the symbol k denotes either a positive integer or ∞.

Theorem 6.2.24 (Elementary properties of deficiencies). *Let* $\mathscr{E}$, $\mathscr{F}$ *and* $\mathscr{G}$ *be experiments with the same parameter set* Θ. *Then:*

(i) $0 \leqslant \delta_k(\mathscr{E}, \mathscr{F}) \leqslant \delta_k(\mathscr{M}_i, \mathscr{M}_a) = 2 - 2[k \wedge \#\Theta]^{-1}$.

(ii) $\delta_k(\mathscr{E}, \mathscr{E}) = 0$.

(iii) $\delta_k(\mathscr{E}, \mathscr{G}) \leqslant \delta_k(\mathscr{E}, \mathscr{F}) + \delta_k(\mathscr{F}, \mathscr{G})$.

(iv) $\delta_1(\mathscr{E}, \mathscr{F}) = 0$ *and* $\delta_k(\mathscr{E}, \mathscr{F}) \uparrow \delta(\mathscr{E}, \mathscr{F})$ *as* $k \uparrow \infty$.

(v) $\delta_k(\mathscr{E}, \mathscr{F}) = \sup \delta_k(\mathscr{E}_{\Theta_0}, \mathscr{F}_{\Theta_0})$ *where the* sup *is over all non-empty finite subsets* Θ_0 *of* Θ.

(vi) $\delta_k(\mathscr{E}, \mathscr{F}) \geqslant \delta_k(\mathscr{E}_{\Theta_0}, \mathscr{F}_{\Theta_0})$ *when* $\emptyset \subset \Theta_0 \subseteqq \Theta$.

(vii) *if* $\mathscr{E} = (P_\theta; \theta \in \Theta)$ *and* $\mathscr{F} = (\mathscr{Y}, \mathscr{B}; Q_\theta : \theta \in \Theta)$ *then* $\delta_k(\mathscr{E}, \mathscr{F}) \leqslant \sup_\theta \|P_\theta M - Q_\theta\|$ *for any transition* M *from* $L(\mathscr{E})$ *to* $(\mathscr{Y}, \mathscr{B})$, *i.e. to* $\mathrm{ba}(\mathscr{Y}, \mathscr{B})$.

Remark 1. (vii) may be phrased:

(vii′) $\delta_k(\mathscr{E}, \mathscr{F}) \leqslant \inf_M \sup_\theta \|P_\theta M - Q_\theta\|$ where M runs through all transitions from $L(\mathscr{E})$ to $\mathrm{ba}(\mathscr{Y}, \mathscr{B})$.

The inequality (vii), and hence (vii′), is an immediate consequence of Theorem 6.2.4.

Remark 2. As mentioned above, the deficiency $\delta_k(\mathscr{E},\mathscr{F})$ measures what we may lose by basing ourselves on $\mathscr{E}$ rather than on $\mathscr{F}$ under the most unfavourable circumstances for $\mathscr{E}$ for this comparison. Therefore it would usually be of great interest if the evaluation of a deficiency was accompanied by an explicit description of these circumstances. Thus the deficiency $\delta_k(\mathscr{M}_i, \mathscr{M}_a)$ given in (i) is obtained from the k-decision problem (T, L) where $\#[T \cap \Theta] = k \wedge \#(\Theta)$ and $L_\theta(t) = 0$ or $= 1$ according to whether $t = \theta$ or not.

Proof of the theorem. (i) is a consequence of proposition 6.2.18. (ii), the first part of (iv), and (vi) follow directly from the definition. (iii) is a consequence of corollary 6.2.16. The last part of (iv) follows from proposition 6.2.15. (v) follows from theorem 6.2.12, while (vii) follows from theorem 6.2.4. □

Although δ_k satisfies the triangular inequality (iii) it is not quite a pseudometric since it is not symmetric unless $\#\Theta = 1$. However, the deficiency distance Δ_k is symmetric and shares with δ all the properties of the theorem, except (vii). Thus $\Delta_1, \Delta_2, \ldots, \Delta_\infty$ are all pseudometrics. However, even the Δ_is are not metrics, since we may easily have $\Delta_i(\mathscr{E},\mathscr{F}) = 0$ although the experiments $\mathscr{E}$ and $\mathscr{F}$ are not equal.

Corollary 6.2.25 (Elementary properties of deficiency distances). *All the statements of theorem 6.2.24, except (vii), remain valid if δ is replaced by Δ. In addition Δ_k is symmetric, i.e.*

$$\Delta_k(\mathscr{E},\mathscr{F}) = \Delta_k(\mathscr{F},\mathscr{E})$$

for any experiments $\mathscr{E}$ and $\mathscr{F}$ with the same parameter set Θ.

The following facts on the relation '$\geqslant_k$' follow either from the theorem or directly from the definition.

Corollary 6.2.26 (Elementary properties of the ordering '$\geqslant_k$' for experiments). *Let $\mathscr{E}$, $\mathscr{F}$ and $\mathscr{G}$ be experiments with the same parameter set Θ. Then:*

(i) $\mathscr{E} \geqslant_k \mathscr{E}$.
(ii) $\mathscr{E} \geqslant_k \mathscr{F}$ *and* $\mathscr{F} \geqslant_k \mathscr{G} \Rightarrow \mathscr{E} \geqslant_k \mathscr{G}$.
(iii) $\mathscr{E} \geqslant \mathscr{F}$ *if and only if* $\mathscr{E} \geqslant_k \mathscr{F}$; $k = 1, 2, \ldots$
(iv) $\mathscr{E} \geqslant_k \mathscr{F}$ *if and only if* $\mathscr{E}_{\Theta_0} \geqslant_k \mathscr{F}_{\Theta_0}$ *for each finite subset* Θ_0 *of* Θ.
(v) $\mathscr{E} \geqslant_k \mathscr{F} \Rightarrow \mathscr{E}_{\Theta_0} \geqslant \mathscr{F}_{\Theta_0}$ *for each subset* Θ_0 *of* Θ.
(vi) *if* $\mathscr{E}_1 \geqslant_k \mathscr{E}_2$ *and* $\mathscr{F}_1 \leqslant_k \mathscr{F}_2$ *then* $\delta_k(\mathscr{E}_1, \mathscr{F}_1) \leqslant \delta_k(\mathscr{E}_2, \mathscr{F}_2)$.
(vii) *if* $\mathscr{E} = (\mathscr{X}, \mathscr{A}; P_\theta : \theta \in \Theta)$ *and* $\mathscr{A}_0$ *is a sub* σ*-algebra of* $\mathscr{A}$ *then* $\mathscr{E} \geqslant (\mathscr{X}, \mathscr{A}_0; P_\theta | \mathscr{A}_0 : \theta \in \Theta)$.
(viii) *if* $\mathscr{E} = (P_\theta : \theta \in \Theta)$ *and* M *is a* σ*-continuous transition from* $L(\mathscr{E})$ *to* $(\mathscr{Y}.\mathscr{B})$ *then* $\mathscr{E} \geqslant (\mathscr{Y}, \mathscr{B}; P_\theta M\,; \theta \in \Theta)$.

Remark. (viii) is the easy part of LeCam's randomization criterion for being more informative. The assumption that M is σ-continuous is only needed to ensure that the set functions $P_\theta M$ are σ-additive so that $(P_\theta M : \theta \in \Theta)$ is an experiment as defined here. If we permit abstract experiments then this assumption is not needed.

It follows that '$\geqslant_k$' is actually an ordering. Hence '$\sim_k$' is an equivalence, i.e. (i), (ii) and (iii) below are satisfied. As mentioned above we shall see later that the non-trivial equivalences '$\sim_k$', $k = 2, 3, \ldots, \infty$ are all the same. The facts collected below follow from theorem 6.2.24 and its corollaries.

Corollary 6.2.27 (Basic properties of the equivalence '$\sim_k$'). *Let $\mathscr{E}$, $\mathscr{F}$ and $\mathscr{G}$ be experiments with the same parameter set Θ. Then:*

(i) $\mathscr{E} \sim_k \mathscr{E}$.
(ii) $\mathscr{E} \sim_k \mathscr{F} \Leftrightarrow \mathscr{F} \sim_k \mathscr{E} \Leftrightarrow \Delta_k(\mathscr{E}, \mathscr{F}) = 0$.
(iii) $\mathscr{E} \sim_k \mathscr{F}$ *and* $\mathscr{F} \sim_k \mathscr{G} \Rightarrow \mathscr{E} \sim_k \mathscr{G}$.
(iv) $\mathscr{E} \sim \mathscr{F}$ *if and only if* $\mathscr{E} \sim_k \mathscr{F}$; $k = 2, 3, \ldots$
(v) $\mathscr{E} \sim_1 \mathscr{F}$.
(vi) $\mathscr{E} \sim_k \mathscr{F} \Leftrightarrow \mathscr{E}_{\Theta_0} \sim_k \mathscr{F}_{\Theta_0}$ *for all non-empty finite subsets* Θ_0 *of* Θ.
(vii) $\mathscr{E} \sim_k \mathscr{F} \Rightarrow \mathscr{E}_{\Theta_0} \sim_k \mathscr{F}_{\Theta_0}$ *for each non-empty subset* Θ_0 *of* Θ.
(viii) *suppose* $\mathscr{E}_1 \sim_k \mathscr{E}_2$ *and that* $\mathscr{F}_1 \sim_k \mathscr{F}_2$. *Then*

$$\delta_k(\mathscr{E}_1, \mathscr{F}_1) = \delta_k(\mathscr{E}_2, \mathscr{F}_2),$$

$$\Delta_k(\mathscr{E}_1, \mathscr{F}_1) = \Delta_k(\mathscr{E}_2, \mathscr{F}_2),$$

$$\mathscr{E}_1 \geqslant_k \mathscr{F}_1 \Leftrightarrow \mathscr{E}_2 \geqslant_k \mathscr{F}_2,$$

and

$$\mathscr{E}_1 \sim_k \mathscr{F}_1 \Leftrightarrow \mathscr{E}_2 \sim_k \mathscr{F}_2.$$

Experiments $\mathscr{E}$ and $\mathscr{F}$ need not be equivalent although $\mathscr{E}_{\Theta_0} \sim \mathscr{F}_{\Theta_0}$ for all proper non-empty subsets Θ_0 of Θ. By (iv) this can only happen when Θ is finite. However, if it is known that $\mathscr{E} \geqslant \mathscr{F}$ then, as we shall see later, $\mathscr{E} \sim \mathscr{F}$ provided $\mathscr{E}_{\Theta_0} \sim \mathscr{F}_{\Theta_0}$ for all two-point subsets Θ_0 of Θ. This has the interesting consequence that the sufficiency concept based on deficiencies amounts to pairwise sufficiency for the usual definition of sufficiency. We shall here proceed the other way round by giving a direct proof of this consequence. Later on, this and the complete randomization criterion will be used to establish the more general statement.

Proposition 6.2.28 (Sufficiency in terms of conditional expectations and in terms of deficiencies). *Let $\mathscr{E} = (\mathscr{X}, \mathscr{A}; P_\theta : \theta \in \Theta)$ be an experiment and let $\mathscr{B}$ be a sub σ-algebra of $\mathscr{A}$. If $k \geqslant 2$ then the experiments $\mathscr{E}$ and $\mathscr{E}|\mathscr{B}$ are equivalent for k-decision problems if and only if $\mathscr{B}$ is pairwise sufficient.*

In particular, if $\mathscr{E}|\mathscr{B}$ is coherent then $\mathscr{E} \sim \mathscr{E}|\mathscr{B}$ if and only if $\mathscr{B}$ is sufficient.

Remark. We shall see later that equivalence for testing problems, and hence for k-decision problems for some $k \geqslant 2$, always implies equivalence.

Proof. We may assume without loss of generality that Θ is finite. Then $\mathscr{E}|\mathscr{B}$ is coherent. Assume $\mathscr{B}$ is pairwise sufficient. Then any family $(P_\theta(A|\mathscr{B}) : \theta \in \Theta)$ where $A \in \mathscr{A}$ is consistent and, consequently, coherent. Hence $\mathscr{B}$ is sufficient. If σ is a decision rule from $\mathscr{E}$ to $T_k = \{1, \ldots, k\}$ then a decision rule ρ from $\mathscr{E}|\mathscr{B}$ to T_k having the same performance function as ρ is obtained by specifying $\rho(i|\cdot) = E_\theta[\sigma(i|\cdot)|\mathscr{B}]$; $i = 1, \ldots, k$ independently of Θ and such that $\rho(i|\cdot) \geqslant 0$; $i = 1, \ldots, k$ and $\sum_i \rho(i|\cdot) = 1$.

Assume that $\mathscr{E}|\mathscr{B} \sim \mathscr{E}$. We may then assume without loss of generality that Θ is the two-point set $\{0, 1\}$. Put $\pi = \frac{1}{2}P_0 + \frac{1}{2}P_1$ and $g = \mathrm{d}(\frac{1}{2}P_1)/\mathrm{d}\pi$. Then $0 \leqslant g \leqslant 1$ a.e. π, $2g = \mathrm{d}P_1/\mathrm{d}\pi$ and $2(1 - g) = \mathrm{d}P_0/\mathrm{d}\pi$.

Take $T = \{0, 1\}$ as the decision space and let L be the (0–1)-loss function given by $L_0(0) = L_1(1) = 0$ and $L_0(1) = L_1(0) = 1$. Thus we are considering the testing problem 'P_0' against 'P_1'. Identify each decision rule with the test function δ which describes the conditional probabilities of taking decision 1 given the outcome of the experiment. The risk function of δ is then the pair $(E_0\delta, 1 - E_1\sigma)$. Since $\mathscr{E}|\mathscr{B} \sim \mathscr{E}$ it follows that to any test δ there corresponds a $\mathscr{B}$-measurable test ϕ such that $(E_0\phi, 1 - E_1\phi) \leqslant (E_0\delta, 1 - E_1\delta)$. Fix a number $\lambda \in]0, 1[$ such that $\pi(g = 1 - \lambda) = 0$. The Bayes risk for the prior distribution $(1 - \lambda, \lambda)$ is the number $(1 - \lambda)E_0\delta + \lambda(1 - E_1\delta) = \lambda - 2\int \delta[\lambda g - (1 - \lambda)(1 - g)]\,\mathrm{d}\pi = \lambda - 2\int \delta(g - (1 - \lambda))\,\mathrm{d}\pi \geqslant \lambda - 2\int [g - (1 - \lambda)]^+\,\mathrm{d}\pi$ where '=' holds if and only if δ is π-equivalent to the indicator function of the set $[g > 1 - \lambda]$. Now the assumption that $\mathscr{E} \sim \mathscr{E}|\mathscr{B}$ implies that the Bayes solution δ may be chosen to be $\mathscr{B}$-measurable. Let $\bar{\mathscr{B}}$ denote the closure of $\mathscr{B}$ within $\mathscr{A}$ with respect to $\mathscr{E}$, i.e. $\bar{\mathscr{B}} = \{A : A \in \mathscr{A}$ and $\pi(A \,\Delta\, B) = 0$ for some set $B \in \mathscr{B}\}$. Then the sets $[g > 1 - \lambda]$ where $\pi(g = 1 - \lambda) = 0$ all belong to $\bar{\mathscr{B}}$. Since the set of exceptional numbers λ is countable, it follows that g is $\bar{\mathscr{B}}$-measurable. Let A be any event in $\mathscr{A}$ and let $B \in \mathscr{B}$. Then

$$
\begin{aligned}
\int_B \pi(A|\mathscr{B})\,\mathrm{d}P_\theta &= \int_B E_\pi(I_A|\mathscr{B})[(1 - \theta)2g + \theta 2(1 - g)]\,\mathrm{d}\pi \\
&= \int_B E_\pi(I_A[(1 - \theta)2g + \theta 2(1 - g)]|\mathscr{B})\,\mathrm{d}\pi \\
&= \int_B I_A[(1 - \theta)2g + \theta 2(1 - g)]\,\mathrm{d}\pi \\
&= \int_B I_A\,\mathrm{d}P_\theta = P_\theta(AB).
\end{aligned}
$$

Hence $\pi(A|\mathscr{B})$ is a version of $P_\theta(A|\mathscr{B})$ for each θ so that $\mathscr{B}$ is sufficient. $\square$

We shall say that a property of experiments *respects informational equivalence* if an experiment $\mathscr{E}$ has this property whenever $\mathscr{E} \sim \mathscr{F}$ and $\mathscr{F}$ has the property. Similarly we shall say that a functional on experiments *respects informational equivalence* if it assigns the same values to equivalent experiments.

Even more generally we may say that an expression involving experiments *respects informational equivalence* if the quantity it defines remains unchanged if some of the involved experiments are replaced by new experiments which are equivalent to the old ones.

The prominent examples of such expressions are the deficiencies $\delta_k(\mathscr{E}, \mathscr{F})$ and the deficiency distances $\Delta_k(\mathscr{E}, \mathscr{F})$. The property of an experiment $\mathscr{E}$ of being at least as informative as a given experiment $\mathscr{F}$ also respects informational equivalence.

Properties of experiments which do not respect informational equivalence are:

coherence,
Σ-dominatedness
and
'possessing a finite sample space'.

Example 6.2.29 (Totally informative experiments. Continuation of example 6.2.1). Let $\mathscr{F} = (Q_\theta : \theta \in \Theta)$ be totally informative and suppose $\mathscr{E} \geqslant_2 \mathscr{F}$. Take $T_2 = \{1, 2\}$ as a decision space and let $\theta_1 \neq \theta_2$ be points in the parameter set Θ. Then since Q_{θ_1} and Q_{θ_2} are disjoint, there is a decision rule σ in $\mathscr{F}$ such that $(Q_{\theta_1}\sigma)(1) = 1 = (Q_{\theta_2}\sigma)(2)$. Define the loss function L such that $L_{\theta_i}(i) = 0$; $i = 1, 2$ while $L_{\theta_2}(1) = L_{\theta_1}(2) = 1$. Then $Q_\theta \sigma L_\theta = 0$; $\theta = \theta_1, \theta_2$. Hence, by assumption, there is a decision rule ρ in $\mathscr{E}$ such that $P_\theta \rho L_\theta = 0$; $\theta = \theta_1, \theta_2$. Then $(P_{\theta_2}\rho)(1) = (P_{\theta_1}\rho)(2) = 0$. The set $A = \{x : \rho(1|x) > 0\}$ then has P_{θ_1}-measure 1 and P_{θ_2}-measure 0. Hence P_{θ_1} and P_{θ_2} are disjoint when $\theta_1 \neq \theta_2$ so that $\mathscr{E}$ is also totally informative.

It follows that the property of being totally informative respects informational equivalence.

Furthermore, as we saw in example 6.2.1, any two totally informative experiments are equivalent. Thus $\mathscr{M}_a$ may be considered as a notation for the totality of totally informative experiments.

Example 6.2.30 (Totally non-informative experiments. Continuation of example 6.2.1). Let $\mathscr{E} = (P : \theta \in \Theta)$ be totally non-informative and suppose that $\mathscr{E} \geqslant_{(2)} \mathscr{F} = (Q_\theta : \theta \in \Theta)$. If $Q_{\theta_1} \neq Q_{\theta_2}$ for $\theta_1, \theta_2 \in \Theta$ then there is a decision rule σ from $\mathscr{F}$ to $T_2 = \{1, 2\}$ such that $(Q_{\theta_1}\sigma)(1) > (Q_{\theta_2}\sigma)(1)$. Put $L_{\theta_i}(i) = 0$; $i = 1, 2$ and $L_{\theta_1}(2) = L_{\theta_2}(1) = 0$. Then there is a decision rule ρ in $\mathscr{E}$ such that

$$P\rho L_{\theta_1} = (P\rho)(2) \leqslant Q_{\theta_1}\sigma L_{\theta_1} = (Q_{\theta_1}\sigma)(2)$$

and

$$P\rho L_{\theta_2} = (P\rho)(1) \leqslant Q_{\theta_2}\sigma L_{\theta_2} = (Q_{\theta_2}\sigma)(1) < (Q_{\theta_1}\sigma)(1).$$

However, this yields the contradiction

$$1 = (P\rho)(1) + (P\rho)(2) < (Q_{\theta_1}\sigma)(1) + (Q_{\theta_1}\sigma)(2) = 1.$$

Thus $Q_{\theta_1} = Q_{\theta_2}$ when $\theta_1, \theta_2 \in \Theta$, i.e. $\mathscr{F}$ is also totally non-informative.

It follows that the property of being totally non-informative respects informational equivalence.

Furthermore, as we saw in example 6.2.1, any two totally non-informative experiments are equivalent. Thus $\mathscr{M}_i$ may be considered as a notation for the totality of totally non-informative experiments.

The facts established in example 6.2.1 may be expressed as informative inequalities.

Theorem 6.2.31. *For any experiment $\mathscr{E}$*

$$\mathscr{M}_i \leqslant \mathscr{E} \leqslant \mathscr{M}_a.$$

By the following theorem, the property of dominatedness respects informational equivalence.

Theorem 6.2.32. *The experiment $\mathscr{F}$ is dominated provided $\mathscr{E} \geqslant_2 \mathscr{F}$ and $\mathscr{E}$ is dominated.*

Proof. Write $\mathscr{E} = (\mathscr{X}, \mathscr{A}; P_\theta : \theta \in \Theta)$ and $\mathscr{F} = (\mathscr{Y}, \mathscr{B}; Q_\theta : \theta \in \Theta)$. Suppose $\mathscr{E}$ is dominated and $\mathscr{E} \geqslant_2 \mathscr{F}$. Then there is a countable subset Θ_0 of Θ such that $P_\theta(A) \equiv_\theta 0$ when $P_\theta(A) = 0$; $\theta \in \Theta_0$. Let $B \in \mathscr{B}$ and suppose $Q_\theta(B) = 0$; $\theta \in \Theta_0$. Let $\theta_1 \in \Theta - \Theta_0$. Define a loss function L on $T_2 = \{1,2\}$ by putting $L_\theta(1) = 1 - L_\theta(2) = 0$ or $= 1$ as $\theta \in \Theta_0$ or $\theta \notin \Theta_0$. Let σ be the decision rule in $\mathscr{F}$ which assigns decision 2 or 1 to $y \in \mathscr{Y}$ according to whether $y \in B$ or $y \notin B$. Then the risk of σ at θ is $Q_\theta(B)$ or $Q_\theta(B^c)$ according to whether $\theta \in \Theta_0$ or $\theta \notin \Theta_0$. Hence, by corollary 6.2.9, there is a decision rule ρ in $\mathscr{E}$ such that $(P_\theta\rho)(2) = P_\theta\rho L_\theta \leqslant Q_\theta(B) = 0$ when $\theta \in \Theta_0$, while $(P_{\theta_1}\rho)(1) = P_{\theta_1}\rho L_{\theta_1} \leqslant Q_{\theta_1}(B^c)$. Then $(P_\theta\rho)(2) \equiv_\theta 0$ so that $(P_{\theta_1}\rho)(1) = 1$ yielding $1 \leqslant Q_{\theta_1}(B^c)$ so that $Q_{\theta_1}(B) = 0$. Hence $\mathscr{F}$ is dominated by $\sum_\theta \mu_\theta Q_\theta$ for each prior distribution μ on Θ with countable support such that $\mu_\theta > 0$ when $\theta \in \Theta_0$. □

The randomization criterion can be used to obtain upper bounds for deficiencies. It is a consequence of the complete randomization criterion that the best possible approximation can always be obtained in that way. Lower bounds for deficiencies may be found by considering particular decision problems. If one of the two experiments under comparison is either totally informative or totally non-informative then the following results may be relevant.

Proposition 6.2.33. *If* $\mathscr{E} = (P_\theta : \theta \in \Theta)$ *is an experiment then:*

(i) $\frac{1}{2}$ *diameter* $(\mathscr{E}) = \delta_2(\mathscr{M}_i, \mathscr{E}) \leqslant \delta(\mathscr{M}_i, \mathscr{E}) \leqslant (1 - 1/m)$ *diameter* $(\mathscr{E})$ *where* $m = \#\Theta$ *and diameter* $(\mathscr{E}) = \bigvee_{\theta_1, \theta_2} \|P_{\theta_1} - P_{\theta_2}\|$.

(ii) $\frac{1}{2}\delta(\mathscr{E}, \mathscr{M}_a)$ *is the minimax probability of wrongly guessing the true value of* θ, *i.e.* $\frac{1}{2}\delta(\mathscr{E}, \mathscr{M}_a) = \bigwedge_{\hat\theta} \bigvee_\theta P_\theta(\hat\theta \neq \theta)$ *where* $\hat\theta$ *runs through the set of all estimators of* θ.

Remark. Any transition from $L(\mathscr{E})$ to Θ equipped with some σ-algebra of subsets containing the one-point sets is admitted as an estimator of θ.

Proof. The equality in (i) follows from corollary 6.2.23, while the left hand inequality follows from part (iv) of corollary 6.2.23. If $m = \infty$ then the right hand inequality follows from proposition 6.2.22. If $m < \infty$ then the right hand inequality follows from proposition 6.2.22, together with the inequalities

$$\left\| P_{\theta_0} - \frac{1}{m}\sum_\theta P_\theta \right\| = \left\| \frac{1}{m}\sum_\theta (P_{\theta_0} - P_\theta) \right\| \leqslant \frac{1}{m}\sum_{\theta \neq \theta_0} \|P_\theta - P_{\theta_0}\|$$

$$\leqslant \left(1 - \frac{1}{m}\right) \bigvee_\theta \|P_\theta - P_{\theta_0}\|; \qquad \theta_0 \in \Theta.$$

Finally (ii) follows directly from proposition 6.2.19. □

Corollary 6.2.34. *If* $\mathscr{E}$ *is dominated and* Θ *is uncountable then* $\delta(\mathscr{E}, \mathscr{M}_a) = 2$.

Proof. Let $\mathscr{E} = (P_\theta : \theta \in \Theta)$ and let $\Theta_0 = \{\theta_1, \theta_2, \ldots\}$ be a countable subset of Θ such that $\pi = \sum_m 2^{-m} P_{\theta_m}$ dominates $\mathscr{E}$. Suppose $\delta(\mathscr{E}, \mathscr{M}_a) < 2$. Then, by proposition 6.2.33, there is an estimator $\hat\theta$ of θ such that $\bigvee_\theta P_\theta(\hat\theta \neq \theta) \leqslant 1 - \tau$ where $\tau > 0$. Then $P_\theta(\hat\theta = \theta) \geqslant \tau$; $\theta \in \Theta$. Let Q_θ be the distribution induced from P_θ by the transition $\hat\theta$. We may assume that the distributions Q_θ are all discrete probability distributions on Θ. In the proof of theorem 6.2.32 we saw that the experiment $\mathscr{F} = (Q_\theta : \theta \in \Theta)$ is dominated by $\tilde\pi = \sum_m 2^{-m} Q_{\theta_m}$. Now, since $Q_\theta(\{\theta\}) > 0$ for all θ, we must also have that $\tilde\pi$ assigns positive mass to each one-point set. If Θ were uncountable then this would contradict the fact that $\tilde\pi$ is discrete. □

Corollary 6.2.35 (Hellinger bound for $\delta(\mathscr{E}, \mathscr{M}_a)$***).***

$$\tfrac{1}{2}\delta(\mathscr{E}, \mathscr{M}_a) \leqslant [\#(\Theta) - 1] \sup_{\theta_1 \neq \theta_2} \inf_{0 \leqslant t \leqslant 1} \int dP_{\theta_1}^{1-t} dP_{\theta_2}^t.$$

Example 6.2.36 (The approach to total information as the numbers of observations increase). Consider an experiment $\mathscr{E} = (\mathscr{X}, \mathscr{A}; P_\theta : \theta \in \Theta)$ realized by an observation X. The product experiment $\mathscr{E}^n = (\mathscr{X}^n, \mathscr{A}^n; P_\theta^n : \theta \in \Theta)$ is then the experiment obtained by combining n independent observations of X. As $\mathscr{E}^n$, up to an identification, is a sub-experiment of $\mathscr{E}^{n+1}$, it is clear that

$\mathscr{E} \leqslant \mathscr{E}^2 \leqslant \mathscr{E}^3 \leqslant \cdots \leqslant \mathscr{M}_a$. One might then ask for conditions ensuring that $\mathscr{E}^n$ converges to $\mathscr{M}_a$, as $n \to \infty$, for the Δ-distance. Then a fundamental restriction is that the unknown parameter is *identifiable in* $\mathscr{E}$, i.e. that $P_{\theta_1} \neq P_{\theta_2}$ when $\theta_1 \neq \theta_2$. (If θ is not identifiable in $\mathscr{E}$ then $\delta(\mathscr{E}^n, \mathscr{M}_a) \geqslant \delta_2(\mathscr{E}^n, \mathscr{M}_a) \equiv_n 1$.) However, this condition is far from being sufficient. Thus, by corollary 6.2.34 $\delta(\mathscr{E}^n, \mathscr{M}_a) \equiv_n 2$ when Θ is uncountable and $\mathscr{E}$ is dominated. If Θ is countably infinite and Θ is identifiable then convergence may or may not take place. However, if Θ is finite then $\delta(\mathscr{E}^n, \mathscr{M}_a) \downarrow 0$ as $n \uparrow \infty$ provided θ is identifiable in $\mathscr{E}$. Actually by corollary 6.2.35,

$$\tfrac{1}{2}\delta(\mathscr{E}^n, \mathscr{M}_a) \leqslant (m-1) \sup_{\theta_1 \neq \theta_2} \inf_{0<t<1} \int \mathrm{d}(P^n_{\theta_1})^{1-t}\mathrm{d}(P^n_{\theta_2})^t = (m-1)C(\mathscr{E})^n$$

where $m = \#\Theta$ and $C(\mathscr{E}) = \sup_{\theta_1 \neq \theta_2} \inf_{0<t<1} \int \mathrm{d}P^{1-t}_{\theta_1}\,\mathrm{d}P^t_{\theta_2}$. The quantity $C(\mathscr{E})$ is certainly < 1 whenever $m < \infty$ and θ is identifiable.

It follows that $\delta(\mathscr{E}^n, \mathscr{M}_a) \downarrow 0$ as $n \uparrow \infty$ with at least exponential speed in this case. Actually, as we shall see later, $(\delta(\mathscr{E}^n, \mathscr{M}_a))^{1/n} \to C(\mathscr{E})$ whenever Θ is finite. Thus the speed is exponential provided $\mathscr{E} \not\sim \mathscr{M}_a$.

Note that this entails statements on the asymptotic behaviour of estimators of θ. If $\hat{\theta}_n$, for each n, is the maximum likelihood estimator of θ then, by the proof of corollary 6.2.20,

$$P^n_\theta(\hat{\theta}_n \neq \theta) \leqslant (m-1)C(\mathscr{E})^n.$$

The role of the maximum likelihood estimator is not unique here since $(\max_\theta P^n_\theta(\hat{\theta}_n \neq \theta))^{1/n} \to C(\mathscr{E})$ for a large variety of estimators. However, $C(\mathscr{E})$ determines the optimal exponential speed of convergence, i.e. $\liminf_n (\max_\theta P^n_\theta(\hat{\theta}_n \neq \theta))^{1/n} \geqslant C(\mathscr{E})$ for any sequence $\{\hat{\theta}_n\}$ of estimators.

Example 6.2.37 (The information in an additional observation). Consider an experiment $\mathscr{E}$ and, for each $n = 1, 2, \ldots$, the experiment $\mathscr{E}^n$ obtained by combining n independent replications of $\mathscr{E}$. Having carried out n observations we may ask if it is worthwhile carrying out one more observation to obtain the experiment $\mathscr{E}^{n+1}$. The deficiency $\delta(\mathscr{E}^n, \mathscr{E}^{n+1})$ may throw light on this problem. If this deficiency is small compared with the cost of a single observation then another observation may not be worthwhile. On the other hand, if this deficiency is large then an additional observation may result in a considerably smaller risk.

The problem of evaluating this deficiency is discussed in Torgersen (1972), LeCam (1974, 1986), Helgeland (1982) and Mammen (1984, 1986). Here, following Helgeland, we shall determine an upper bound for this deficiency in the case of multinomial trials.

Consider then independent trials and s events such that:

(i) one and only one of the events occurs in each trial.
(ii) the probability θ_j of the occurrence of the jth event $j = 1, \ldots, s$ in the ith trial does not depend on i.

Put $X_j^{(i)} = 1$ or $= 0$ according to whether the jth event occurred at the ith trial or not. Also put $X^{(i)} = (X_1^{(i)}, \ldots, X_s^{(i)})$ and $Y^{(n)} = \sum_{i=1}^{n} X^{(i)}$. Then $X^{(1)}, X^{(2)}, \ldots$ may be considered as independent observations of X where $P_\theta(X = (0, \ldots, 1^{(j)}, \ldots, 0)) = \theta_j$; $j = 1, \ldots, s$. We shall assume that the probability vector $\theta = (\theta_1, \ldots, \theta_s)$ is completely unknown. Thus the parameter set Θ consists of all s-tuples $(\theta_1, \ldots, \theta_s)$ of non-negative numbers such that $\theta_1 + \cdots + \theta_s = 1$. Let $\mathscr{E}$ denote the experiment realized by observing X. Then $\mathscr{E}^n$ is realized by observing the n observations $X^{(1)}, \ldots, X^{(n)}$ of X. As $Y^{(n)}$ is sufficient for $(X^{(1)}, \ldots, X^{(n)})$ the experiment $\mathscr{E}^n$ is equivalent to the experiment obtained by realizing $Y^{(n)}$ where

$$P_\theta^n(Y^{(n)} = (y_1, \ldots, y_s)) = \frac{n!}{y_1! \ldots y_s!} \theta_1^{y_1} \ldots \theta_s^{y_s}$$

when $(y_1, \ldots, y_s)$ is an s-tuple of non-negative integers with sum n. Let $\mathscr{Y}^{(n)}$ denote this set of s-tuples.

Now consider the problem of simulating $Y^{(n+1)}$ on the basis of $Y^{(n)}$. As $Y^{(n+1)} = Y^{(n)} + X^{(n+1)}$ this problem is equivalent to the problem of simulating $X^{(n+1)}$ on the basis of $Y^{(n)}$. Now, since $P_\theta(X^{(n+1)} = (0, \ldots, 1^{(v)}, \ldots, 0)) = \theta_v$ and since $\hat{\theta}_v = (1/n) Y_v^{(n)}$ is the usual estimator for θ_v, we might require that our Markov kernel M is given by

$$M(y^{(n+1)} | y^{(n)}) = y_v^{(n)}/n$$

if $y^{(n+1)} = y^{(n)} + (0, \ldots, 1^{(v)}, \ldots, 0)$; $v = 1, \ldots, s$. As $\sum_y M(y | y^{(n)}) = 1$ when $y^{(n)} \in \mathscr{Y}^{(n)}$ this defines a legitimate kernel.

Let $Q_\theta^{(n)}$ be the distribution of $Y^{(n)}$ and assume $\theta_v > 0$ for all v. Then $Q_\theta^{(n)} M(z) = \sum_y M(z | y) Q_\theta^{(n)}(y) = \sum_{v=1}^{s} M(z | z - e^{(v)}) Q_\theta^{(n)}(z - e^{(v)})$ where $e^{(v)} = (0, \ldots, 1^{(v)}, \ldots, 0)$; $v = 1, \ldots, s$. Let $z \in \mathscr{Y}^{(n+1)}$. Then

$$\begin{aligned}(Q_\theta^{(n)} M)(z) &= \sum_{v : z_v > 0} \frac{z_v - 1}{n} \frac{n!}{z_1! \ldots (z_v - 1)! \ldots z_s!} \theta_1^{z_1} \ldots \theta_v^{z_v - 1} \ldots \theta_s^{z_s} \\ &= Q_\theta^{(n+1)}(z) \sum \frac{z_v(z_v - 1)}{(n + 1)n \cdot \theta_v}.\end{aligned}$$

It follows that

$$\begin{aligned}\|Q_\theta^{(n)} M - Q_\theta^{(n+1)}\| &= \sum_z Q_\theta^{(n+1)}(z) \left| 1 - \sum_v \frac{z_v(z_v - 1)}{(n + 1)n \cdot \theta_v} \right| \\ &= E_\theta \left| 1 - \sum_v Y_v[Y_v - 1]/(n + 1)n\theta_v \right|\end{aligned}$$

where we write Y instead of $Y^{(n+1)}$.

Hence

$$\|Q_\theta^{(n)}M - Q_\theta^{(n+1)}\| = E_\theta\left|\sum_\nu \frac{Y_\nu}{n+1}\left(1 - \frac{Y_\nu - 1}{n\theta_\nu}\right)\right|$$

$$\leqslant E_\theta\left|\sum_\nu \frac{Y_\nu}{n+1}\left(1 - \frac{Y_\nu}{(n+1)\theta_\nu}\right)\right|$$

$$+ E_\theta\left|\sum_\nu \frac{Y_\nu}{(n+1)\theta_\nu}\left(\frac{Y_\nu}{n+1} - \frac{Y_\nu - 1}{n}\right)\right|.$$

Now

$$E_\theta\left|\sum_\nu \frac{Y_\nu}{n+1}\left(1 - \frac{Y_\nu}{(n+1)\theta_\nu}\right)\right| = E_\theta \sum_\nu \frac{1}{\theta_\nu}\left(\frac{Y_\nu}{n+1} - \theta_\nu\right)^2$$

$$= \sum_\nu \frac{1}{\theta_\nu}\frac{\theta_\nu(1-\theta_\nu)}{n+1} = \frac{s-1}{n+1}$$

while

$$E_\theta\left|\sum_\nu \frac{Y_\nu}{(n+1)\theta_\nu}\left(\frac{Y_\nu}{n+1} - \frac{Y_\nu - 1}{n}\right)\right| = E_\theta \sum_\nu \frac{Y_\nu(n+1-Y)_\nu}{n(n+1)^2\theta_\nu}$$

$$= \sum_\nu \frac{(n+1)n\theta_\nu(1-\theta_\nu)}{n(n+1)^2\theta_\nu} = \frac{s-1}{n+1}.$$

Thus $\|Q_\theta^{(n)}M - Q_\theta^{(n+1)}\| \leqslant 2(s-1)/(n+1)$ when $\theta_\nu > 0$; $\nu = 1, 2, \ldots, s$. By continuity this extends to all $\theta \in \Theta$. Hence, by theorem 6.2.4,

$$\delta(\mathscr{E}^n, \mathscr{E}^{n+1}) \leqslant 2\frac{s-1}{n+1}.$$

Example 6.2.38 (The memory of a process). Consider a random process $X_0, X_1, X_2, \ldots$ with state space Θ. How fast does this process forget its initial state? A possible way of interpreting memory is as follows. For each n, let T_n denote the tail $(X_n, X_{n+1}, \ldots)$ of the process. *Then the process may be considered as having a good (bad) memory at time n according to whether the conditional distribution of T_n given the initial state $X_0 = \theta$ varies much (little) with θ.* For each $\theta \in \Theta$ and each $n = 1, 2, \ldots$, let $Q_\theta^{(n)}$ denote the conditional distribution of T_n given $X_0 = \theta$. Finally, let $\mathscr{E}_n$ denote the experiment $(Q_\theta^{(n)} : \theta \in \Theta)$. Trivially $\mathscr{E}_1 \geqslant \mathscr{E}_2 \geqslant \cdots$

Later we will see that any monotone sequence (net) of experiments converges in the Δ sense to an experiment when Θ is finite. Thus $\Delta(\mathscr{E}_n, \mathscr{E}_\infty) \downarrow 0$ as $n \uparrow \infty$ for an experiment $\mathscr{E}_\infty$. In the general case there is always an experiment $\mathscr{E}_\infty$ such that $\Delta((\mathscr{E}_n)_{\Theta_0}, (\mathscr{E}_\infty)_{\Theta_0}) \to 0$ when Θ_0 is a finite non-empty subset of Θ. The experiment $\mathscr{E}_\infty$ is unique up to equivalence and actually represents a greatest lower bound for the experiments $\mathscr{E}_1, \mathscr{E}_2, \ldots$

It is interesting that these limiting statements do not involve any of the usual regularity conditions appearing in the theory of random processes. Thus we need not distinguish between periodic and aperiodic chains, say. The possibility of using the sequence $\mathscr{E}_n$, $n = 1, 2, \ldots$ to describe the memory of the process in the case where this process is a Markov process, has been investigated in a series of papers by Bo Lindqvist.

In this case all the information about the initial state which is stored in T_n is actually stored in its first term X_n, since by the Markov property X_n is sufficient for T_n. We limit ourselves here to mentioning the following result due to Lindqvist (1977).

Suppose $X_0, X_1, X_2, \ldots$ is a Markov chain with finite state space Θ and with stationary transition probabilities given by the Markov matrix P. Then there are positive numbers a and A such that $an^{\tau-1}\rho^n \leqslant \Delta(\mathscr{E}_\infty, \mathscr{E}_n) \leqslant An^{\tau-1}\rho^n$; $n = 1, 2, \ldots$ where ρ is the largest number less than 1 which is the modulus of a characteristic root of P, while τ is the largest index of characteristic roots of P with modulus ρ. In particular it follows that $(\Delta(\mathscr{E}_n, \mathscr{E}_\infty))^{1/n} \to \rho$.

The possibility of using weighted deficiencies is discussed in the last part of LeCam's 1964 paper. Let us take a look at these quantities in the case of k-decision problems.

If λ is a prior distribution on Θ with finite support then the *λ-weighted deficiency of $\mathscr{E}$ with respect to $\mathscr{F}$ for k-decision problems* is the greatest lower bound of all constants $\sum \lambda_\theta \varepsilon_\theta$ where $\mathscr{E}$ is ε-deficient with respect to $\mathscr{F}$ for k-decision problems. These numbers will be denoted by $\delta_k(\mathscr{E}, \mathscr{F} | \lambda)$. We may write $\delta(\mathscr{E}, \mathscr{F} | \lambda)$ instead of $\delta_\infty(\mathscr{E}, \mathscr{F} | \lambda)$ and this number is simply called the λ-weighted deficiency of $\mathscr{E}$ with respect to $\mathscr{F}$. It follows from proposition 6.2.15 that $\delta_k(\mathscr{E}, \mathscr{F} | \lambda)$ is actually the smallest number of the form $\sum_\theta \lambda_\theta \varepsilon_\theta$ where $\mathscr{E}$ is ε-deficient with respect to $\mathscr{F}$ for k-decision problems. The relationship between λ-weighted deficiency and minimum Bayes risks for the prior λ will be considered when we have more criteria for deficiency at our disposal.

Weighted distances are derived from weighted deficiencies by symmetrization. Thus the *λ-weighted deficiency distance between $\mathscr{E}$ and $\mathscr{F}$ for k-decision problems* is the number $\delta_k(\mathscr{E}, \mathscr{F} | \lambda) \vee \delta_k(\mathscr{F}, \mathscr{E} | \lambda)$. This number is written $\Delta_k(\mathscr{E}, \mathscr{F} | \lambda)$ and we may write $\Delta(\mathscr{E}, \mathscr{F} | \lambda)$ instead of $\Delta_\infty(\mathscr{E}, \mathscr{F} | \lambda)$. $\Delta(\mathscr{E}, \mathscr{F} | \lambda)$ will be called the *λ-weighted deficiency distance between $\mathscr{E}$ and $\mathscr{F}$*. Some properties of λ-weighted deficiencies and distances which follow from the definitions and the results proved so far are collected in the following proposition.

Proposition 6.2.39 (First properties of weighted deficiencies). *Let $\mathscr{E}, \mathscr{F}, \mathscr{G}, \ldots$ be experiments with the same parameter set Θ and let λ be a prior distribution on Θ with finite support $\Theta_\lambda = \{\theta : \lambda(\theta) > 0\}$. Let k be a positive integer or $= \infty$. Then:*

(i) (a) $0 \leqslant \delta_k(\mathscr{E}, \mathscr{F} | \lambda) = \delta_k(\mathscr{E}_{\Theta_\lambda}, \mathscr{F}_{\Theta_\lambda} | \lambda) \leqslant \delta_k(\mathscr{E}, \mathscr{F})$.
(b) $\delta_k(\mathscr{E}, \mathscr{E} | \lambda) = 0$.
(c) $\delta_k(\mathscr{E}, \mathscr{G} | \lambda) \leqslant \delta_k(\mathscr{E}, \mathscr{F} | \lambda) + \delta_k(\mathscr{F}, \mathscr{G} | \lambda)$.
(d) $0 = \delta_1(\mathscr{E}, \mathscr{F} | \lambda) \leqslant \delta_k(\mathscr{E}, \mathscr{F} | \lambda) \uparrow \delta(\mathscr{E}, \mathscr{F} | \lambda)$ *as* $k \uparrow \infty$.
Statements (a)–(d) remain valid if δ is replaced by Δ throughout.
(ii) *if* $\mathscr{E} = (P_\theta : \theta \in \Theta)$ *and* $\mathscr{F} = (Q_\theta : \theta \in \Theta)$ *then* $\delta(\mathscr{E}, \mathscr{F} | \lambda) \leqslant \sum_\theta \|P_\theta M - Q_\theta\| \lambda_\theta$ *for any transition* M *from* $L(\mathscr{E})$ *to* $L(\mathscr{F})$.
(iii) $\delta_k(\mathscr{E}, \mathscr{F} | \lambda) = 0 \Leftrightarrow \mathscr{E}_{\Theta_\lambda} \geqslant \mathscr{F}_{\Theta_\lambda}$.
(iv) $\Delta_k(\mathscr{E}, \mathscr{F} | \lambda) = \Delta_k(\mathscr{F}, \mathscr{E} | \lambda)$.
(v) $\Delta_k(\mathscr{E}, \mathscr{F} | \lambda) = 0 \Leftrightarrow \mathscr{F}_{\Theta_\lambda} \sim_k \mathscr{F}_{\Theta_\lambda}$.

Proof. (d) follows from proposition 6.2.15 while (a), (b) and (iv) follow directly from the definitions. (c) is a consequence of proposition 6.2.17 while (ii) follows from the part of the randomization criterion which is given in theorem 6.2.4. (v) follows from (iii) which in turn is a consequence of (i). □

In the next section we will see how the weighted deficiencies may be expressed in terms of minimum Bayes risk for a given prior distribution.

Deficiencies may be expressed simply in terms of weighted deficiencies.

Proposition 6.2.40.

$$\delta_k(\mathscr{E}, \mathscr{F}) = \sup_\lambda \delta_k(\mathscr{E}, \mathscr{F} | \lambda)$$

and

$$\Delta_k(\mathscr{E}, \mathscr{F}) = \sup_\lambda \Delta_k(\mathscr{E}, \mathscr{F} | \lambda)$$

where suprema are taken for all prior distributions λ on Θ with finite support.

Furthermore, $\delta_k(\mathscr{E}, \mathscr{F} | \lambda) = \delta_k(\mathscr{E}, \mathscr{F})$ *for a particular* λ *if and only if* $\delta_k(\mathscr{E}, \mathscr{F} | \lambda) = \sum_\theta \varepsilon_\theta \lambda_\theta$ *where* $\mathscr{E}$ *is* ε*-deficient with respect to* $\mathscr{F}$ *for k-decision problems and* $\varepsilon_\theta = \sup\{\varepsilon_{\theta'} : \theta' \in \Theta\}$ *when* $\lambda(\theta) > 0$.

Proof. The minimax theory (say corollary 3.2.13) and proposition 6.2.15 yield $\sup_\lambda \delta_k(\mathscr{E}, \mathscr{F} | \lambda) = \sup_\lambda \inf_{\varepsilon \in D_k} \sum \varepsilon_\theta \lambda_\theta = \inf_{\varepsilon \in D_k} \sup_\lambda \sum \varepsilon_\theta \lambda_\theta = \inf_{\varepsilon \in D_k} \sup_\theta \varepsilon_\theta = \delta_k(\mathscr{E}, \mathscr{F})$ where $D_k = D_k(\mathscr{E}, \mathscr{F})$ is the set of functions ε such that $\mathscr{E}$ is ε-deficient with respect to $\mathscr{F}$ for k-decision problems. Symmetrization then yields $\sup_\lambda \Delta_k(\mathscr{E}, \mathscr{F} | \lambda) = \Delta_k(\mathscr{E}, \mathscr{F})$. If $\delta_k(\mathscr{E}, \mathscr{F} | \lambda) = \sum_\theta \varepsilon_\theta \lambda_\theta$ where $\varepsilon \in D_k$ and $\bigvee_\theta \varepsilon_\theta = \varepsilon_\theta$ when $\lambda_\theta > 0$ then $\delta_k(\mathscr{E}, \mathscr{F} | \lambda) = \bigvee_\theta \varepsilon_\theta \geqslant \delta_k(\mathscr{E}, \mathscr{F})$ so that $\delta_k(\mathscr{E}, \mathscr{F} | \lambda) = \delta_k(\mathscr{E}, \mathscr{F})$. Conversely, if $\delta_k(\mathscr{E}, \mathscr{F} | \lambda) = \delta_k(\mathscr{E}, \mathscr{F})$ then $\delta_k(\mathscr{E}, \mathscr{F} | \lambda) = \sum_\theta \varepsilon_\theta \lambda_\theta$ when $\varepsilon_\theta \equiv_\theta \delta_k(\mathscr{E}, \mathscr{F})$. □

The weighted deficiencies $\delta_k(\mathscr{E}, \mathscr{F} | \lambda)$ determine the set $D_k(\mathscr{E}, \mathscr{F})$.

Proposition 6.2.41. $\mathscr{E}$ *is* ε*-deficient with respect to* $\mathscr{F}$ *for k-decision problems if and only if* $\sum_\theta \varepsilon_\theta \lambda_\theta \geqslant \delta_k(\mathscr{E}, \mathscr{F} | \lambda)$ *for each prior (probability) distribution* λ *with finite support.*

Proof. The 'only if' follows from the definition while the 'if' follows by the reduction theorem 6.2.12 and the strongly separating hyperplane theorem, theorem 2.3.13. □

Example 6.2.42 (Deficiencies with respect to $\mathscr{M}_a$). It follows from proposition 6.2.19 that, up to a factor $\frac{1}{2}$, the 'admissible deficiency functions for $\mathscr{E}$ with respect to $\mathscr{M}_a$, are precisely the admissible risk functions in the estimation problem where the loss is 0 or 1 according to whether the estimator hits or not. In particular, if Θ is finite then by corollary 3.3.10 the Bayes rules constitute an essentially complete class. If λ is a prior distribution on Θ with finite support then $\hat{\theta}$ is a Bayes solution if and only if it assigns $\sum_\theta \lambda_\theta P_\theta$ almost everywhere mass 1 to the set of points θ which maximizes $\lambda_\theta \, \mathrm{d}P_\theta/\mathrm{d}\sum_\theta \lambda_\theta P_\theta$. As we saw in section 3.4, the minimum Bayes risk is then $1 - \|\bigvee_\theta \lambda_\theta P_\theta\|$. It follows that

$$\tfrac{1}{2}\delta(\mathscr{E}, \mathscr{M}_a|\lambda) = 1 - \left\| \bigvee_\theta \lambda_\theta P_\theta \right\|.$$

Thus we have shown that the maximal probability of correctly guessing the true value of θ for the prior λ is $\|\bigvee_\theta \lambda_\theta P_\theta\|$. The functional which to each prior distribution λ with finite support assigns the number $\|\bigvee_\theta \lambda_\theta P_\theta\|$ respects equivalence, by construction. Actually this functional characterizes the experiment $\mathscr{E}$ up to equivalence, as shown by Morse & Sacksteder (1966).

The inequality $\|\bigvee_\theta \lambda_\theta P_\theta\| \geqslant \bigvee_\theta \lambda_\theta$ implies that $\delta(\mathscr{E}, \mathscr{M}_a|\lambda) \leqslant 2(1 - \bigvee_\theta \lambda_\theta)$ and clearly equality holds if $\mathscr{E} \sim \mathscr{M}_i$. Equality holds in other cases also and in fact equality holds if and only if $\lambda_{\theta_0} P_{\theta_0} \geqslant \lambda_\theta P_\theta$ for all θ for some $\theta_0 \in \Theta$. This may be seen by noting that $\|\bigvee_\theta \lambda_\theta P_\theta - \lambda_{\theta_0} P_{\theta_0}\| = 0$ when $\lambda_{\theta_0} = \bigvee_\theta \lambda_\theta$ and $\|\bigvee_\theta \lambda_\theta P_\theta\| = \lambda_{\theta_0}$.

Proposition 6.2.40 now yields

$$\tfrac{1}{2}\delta(\mathscr{E}, \mathscr{M}_a) = 1 - \bigwedge_\lambda \left\| \bigvee_\theta \lambda_\theta Q_\theta \right\|.$$

Here $\bigwedge_\lambda \|\bigvee_\theta \lambda_\theta P_\theta\| = \bigwedge_{\hat{\theta}} \bigvee_\theta P_\theta(\hat{\theta} = \theta)$ is the minimax probability of correctly guessing the true value of θ.

We can utilize these expressions to show that if Θ is finite and if $\delta(\mathscr{E}, \mathscr{M}_a)$ takes its maximal value $\delta(\mathscr{M}_i, \mathscr{M}_a)$ then $\mathscr{E} \sim \mathscr{M}_i$. Put $m = \#\Theta$ and suppose $\delta(\mathscr{E}, \mathscr{M}_a) = \delta(\mathscr{M}_i, \mathscr{M}_a) = 2(1 - 1/m)$. Then there is a prior distribution λ such that $\|\bigvee_\theta \lambda_\theta P_\theta\| = 1/m$. It follows that $1/m \geqslant \bigvee_\theta \lambda_\theta$ so that $\lambda_\theta \equiv 1/m$. Thus $\|\bigvee_\theta P_\theta\| = 1$, yielding $\|\bigvee_\theta P_\theta - P_\theta\| \equiv_\theta 0$. Hence P_θ does not depend on θ, i.e. $\mathscr{E} \sim \mathscr{M}_i$.

If Θ is infinite, countable or not, then we may easily have $\delta(\mathscr{E}, \mathscr{M}_a) = \delta(\mathscr{M}_i, \mathscr{M}_a) = 2$ although $\mathscr{E} \nsim \mathscr{M}_i$. Thus, by corollary 6.2.34, $\delta(\mathscr{E}, \mathscr{M}_a) = 2$ whenever Θ is uncountably infinite and $\mathscr{E}$ is dominated.

Example 6.2.43 (Deficiencies of $\mathcal{M}_i$). Consider the λ-weighted deficiency of $\mathcal{M}_i$ with respect to $\mathcal{F} = (Q_\theta : \theta \in \Theta)$. By proposition 6.2.22 this deficiency is $\inf_Q \sum_\theta \lambda_\theta \|Q_\theta - Q\|$ where Q runs through the set of all probability distributions in $L(\mathcal{F})$.

The inequality $\mathcal{F} \leqslant \mathcal{M}_a$ implies that

$$\delta(\mathcal{M}_i, \mathcal{F}) \leqslant \delta(\mathcal{M}_i, \mathcal{M}_a)$$

where equality holds of course when $\mathcal{F} \sim \mathcal{M}_a$. If Θ is infinite then equality holds for a large variety of experiments $\mathcal{F}$ which are not totally informative. However, if Θ is finite then the equality $\delta(\mathcal{M}_i, \mathcal{F}) = \delta(\mathcal{M}_i, \mathcal{M}_a)$ forces $\mathcal{F}$ to be totally informative. To see this let $m = \#\Theta < \infty$. Then equality implies that $2 - 2/m = \delta(\mathcal{M}_i, \mathcal{F}|\lambda)$ for some λ. Then $\sum_\theta \lambda_\theta \|Q_\theta - Q\| \geqslant 2 - 2/m$ whenever Q is a probability measure in $L(\mathcal{F})$. In particular $\sum_\theta \lambda_\theta [\|Q_\theta - \bar{Q}\| - (2 - 2/m)] \geqslant 0$ where $\bar{Q} = \sum_\theta Q_\theta / m$. Then, since $\|Q_\theta - \bar{Q}\| \leqslant 2 - 2/m$ for all θ, we must conclude that $\|Q_\theta - \bar{Q}\| = 2 - 2/m$ when $\lambda_\theta > 0$. Suppose $\lambda_{\theta_0} = 0$. Then we obtain the contradiction

$$2\left(1 - \frac{1}{m}\right) \leqslant \delta(\mathcal{M}_i, \mathcal{F}|\lambda) \leqslant \delta(\mathcal{M}_i, \mathcal{F}_{\{\theta_0\}^c}) \leqslant 2\left(1 - \frac{1}{m-1}\right).$$

Hence $\lambda_\theta > 0$ for all θ and thus $\|Q_\theta - \bar{Q}\| \equiv_\theta 2 - 2/m$. It follows that $Q_\theta \wedge \sum\{Q_{\theta'} : \theta' \neq \theta\} = 0$ for all θ, i.e. $\mathcal{F} \sim \mathcal{M}_a$.

Example 6.2.44 (Weighted deficiencies of $\mathcal{M}_i$ with respect to $\mathcal{M}_a$). Let $\tilde{D}_k(\mathcal{E}, \mathcal{F})$ denote the set of functions q on Θ such that $q_\theta \equiv 1 - \frac{1}{2}\varepsilon_\theta$ for some $\varepsilon \in D_k(\mathcal{E}, \mathcal{F})$ such that $0 \leqslant \varepsilon_\theta \leqslant 2$; $\theta \in \Theta$. Then $\frac{1}{2}\delta_k(\mathcal{E}, \mathcal{F}|\lambda) = 1 - \sup_{\tilde{D}_k} \sum q_\theta \lambda_\theta$. We shall utilize this to determine $\delta_k(\mathcal{M}_i, \mathcal{M}_a|\lambda)$, i.e. the maximal possible λ-weighted deficiency $\delta_k(\mathcal{E}, \mathcal{F}|\lambda)$ as well as the maximal possible λ-weighted deficiency distance $\Delta_k(\mathcal{E}, \mathcal{F}|\lambda)$.

Assume that $k \leqslant \#(\Theta) = m < \infty$. Write $\Theta = \{1, 2, \ldots, m\}$ and for any m-tuple $x = (x_1, \ldots, x_m)$ of real numbers let $x_{[1]} \geqslant x_{[2]} \geqslant \cdots \geqslant x_{[m]}$ denote the decreasing arrangement of $x_1, \ldots, x_m$. We shall use an inequality due to Hardy, Littlewood & Polya (1973) (see chapter 10 on inequalities) which states that $\sum x_i y_i \leqslant \sum x_{[i]} y_{[i]}$ for any two m-tuples x and y in $\mathbb{R}^m$. It follows that $\sum_i \lambda_i q_i \leqslant \sum_i \lambda_{[i]} q_{[i]}$. Now, by proposition 6.2.17, $q \in \tilde{D}_k$ if and only if $q_{[1]} + \cdots + q_{[k]} \leqslant 1$. Hence $\sup_q \sum \lambda_i q_i = \sup_q \sum \lambda_{[i]} q_{[i]} = \sup_q [\lambda_{[1]} q_1 + \cdots + \lambda_{[k-1]} q_{k-1} + (\lambda_{[k]} + \cdots + \lambda_{[m]}) q_k]$ where the last sup is taken for all probability vectors $(q_1, \ldots, q_k)$ such that $q_1 \geqslant q_2 \geqslant \cdots \geqslant q_k$. The extreme distributions of this set of probability vectors are $(1, 0, \ldots, 0)$, $(\frac{1}{2}, \frac{1}{2}, 0, \ldots, 0)$, $(\frac{1}{3}, \frac{1}{3}, \frac{1}{3}, 0, \ldots, 0), \ldots$ $(1/k, 1/k, \ldots, 1/k)$. Hence $\sup_q \sum \lambda_i q_i = \lambda_{[1]} \vee \frac{1}{2}(\lambda_{[1]} + \lambda_{[2]}) \vee \cdots \vee 1/(k-1)(\lambda_{[1]} + \cdots + \lambda_{[k-1]}) \vee 1/k = \lambda_{[1]} \vee 1/k$.

It follows from this and from example 6.2.42 that

$$\delta_k(\mathcal{M}_i, \mathcal{M}_a|\lambda) = 2\left(1 - \bigvee_\theta \lambda_\theta \vee \frac{1}{k}\right)$$

without any restrictions on Θ, k and λ.

It is only the obvious pair which yields the maximal deficiency when Θ is finite.

Proposition 6.2.45. *If Θ is finite then $\delta(\mathscr{E}, \mathscr{F}) \leqslant \delta(\mathcal{M}_i, \mathcal{M}_a)$ and equality holds if and only if $\mathscr{E} \sim \mathcal{M}_i$ and $\mathscr{F} \sim \mathcal{M}_a$.*

Proof. If $\delta(\mathscr{E}, \mathscr{F}) = \delta(\mathcal{M}_i, \mathcal{M}_a)$ then $\delta(\mathcal{M}_i, \mathscr{F}) \geqslant \delta(\mathscr{E}, \mathscr{F}) = \delta(\mathcal{M}_i, \mathcal{M}_a) \geqslant \delta(\mathscr{E}, \mathcal{M}_a)$ so that $\delta(\mathcal{M}_i, \mathscr{F}) = \delta(\mathscr{E}, \mathcal{M}_a) = \delta(\mathcal{M}_i, \mathcal{M}_a)$ and these equalities imply, as we have seen in examples 6.2.42 and 6.2.43, that $\mathscr{E} \sim \mathcal{M}_i$ and $\mathscr{F} \sim \mathcal{M}_a$. □

6.3 Equivalent conditions for deficiency

The notion of ε-deficiency was introduced in terms of pointwise comparison of risk functions. A Bayesian might prefer a notion of deficiency in terms of Bayes risk while others might prefer to compare maximum risks. One might also consider measures of the difficulty of constructing approximations to $\mathscr{F}$ on the basis of $\mathscr{E}$. Thus we might arrive at a confusing variety of notions. Fortunately this is not so and the notion of deficiency described in section 6.2 could just as well have been introduced from any of these points of view. This demonstrates clearly the importance of the concept of deficiency and the clarification of these facts is one of the main contributions of this theory. We collect the spadework in a theorem.

Theorem 6.3.1 (Equivalence of criteria). *Let $\mathscr{E} = (P_\theta : \theta \in \Theta)$ and $\mathscr{F} = (Q_\theta : \theta \in \Theta)$ be two experiments with the same finite parameter set Θ and let ε be a non-negative function on Θ. Consider also a prior distribution λ on Θ such that $\lambda(\theta) > 0$ for each $\theta \in \Theta$ and a decision rule σ from $\mathscr{F}$ to a k-decision space $(T, \mathscr{S})$ where $k < \infty$. Unless explicitly stated, we do not assume that λ is a probability distribution, i.e. that $\sum_\theta \lambda_\theta = 1$.*

Then the following conditions are all equivalent.

(i) Pointwise comparison of risks.
To each loss function L on $(T, \mathscr{S})$ there corresponds a decision rule ρ from *$\mathscr{E}$ to $(T, \mathscr{S})$ such that*

$$P_\theta \rho L_\theta \leqslant Q_\theta \sigma L_\theta + \varepsilon_\theta \|L_\theta\|; \qquad \theta \in \Theta.$$

(ii) Comparison of maximum risks.
To each non-negative function L on $(T, \mathscr{S})$ there corresponds a decision rule ρ from $\mathscr{E}$ to $(T, \mathscr{S})$ such that

$$\bigvee_\theta P_\theta \rho L_\theta \leqslant \bigvee_\theta \left[Q_\theta \sigma L_\theta + \tfrac{1}{2}\varepsilon_\theta \|L_\theta\|\right].$$

(iii) Comparison of Bayes risks.
To each loss function L on $(T, \mathscr{S})$ there corresponds a decision rule ρ from $\mathscr{E}$ to $(T, \mathscr{S})$ such that

$$\sum_\theta (P_\theta \rho L_\theta)\lambda_\theta \leqslant \sum_\theta (Q_\theta \sigma L_\theta)\lambda_\theta + \sum_\theta \varepsilon_\theta \|L_\theta\| \lambda_\theta.$$

(iv) Comparison of performance functions.
There is a decision rule ρ from $\mathscr{E}$ to $(T, \mathscr{S})$ such that

$$\|P_\theta \rho - Q_\theta \sigma\| \leqslant \varepsilon_\theta; \qquad \theta \in \Theta.$$

Remark 1. Note that the Bayes criterion for deficiency does not depend on which prior distribution we use, provided it assigns positive mass to each point in Θ.

Remark 2. Five other and equivalent versions of (i) may be obtained and actually were obtained, in proposition 6.2.3. The same arguments also show that in (iii) we may assume that $\bigvee_\theta L_\theta + \bigwedge_\theta L_\theta = 0$ or that $L \geqslant 0$ provided we replace ε_θ by $\varepsilon_\theta/2$.

Note also that (iii) may be converted into a Bayes utility criterion if we substitute $L = -U$ where U is a utility function.

Proof. Suppose (iii) holds. Remember that $\mathscr{S}$ is finite. Consider the two-person zero-sum game $(\mathscr{L}, \mathscr{R}, M)$ where $\mathscr{L}$ is the set of all loss functions considered as a subset of $\mathbb{R}^n$ where $n = (\#\Theta)k$, $\mathscr{R}$ is the set of decision rules in $\mathscr{E}$ and $M(L, \rho) = \sum_\theta (P_\theta \rho L_\theta - Q_\theta \sigma L_\theta - \varepsilon_\theta \|L_\theta\|)\lambda_\theta$. (iii) implies that $\inf_\rho M(L, \rho) \leqslant 0$ for each $L \in \mathscr{L}$. This game is clearly concave in L and affine in ρ. If we equip $\mathscr{L}$ with the topology of pointwise convergence on $L(\mathscr{E}) \times M(T, \mathscr{S})$ then by the weak compactness lemma, $\mathscr{L}$ becomes compact. Furthermore $M(L, \cdot)$, for each L, is continuous with respect to this topology. Hence by the minimax theory (say corollary 3.2.13) there is a decision rule ρ_0 such that

$$\sup_L M(L, \rho_0) = \inf_\rho \sup_L M(L, \rho) = \sup_L \inf_\rho M(L, \rho) \leqslant 0.$$

Let $\theta_0 \in \Theta$ and choose a bounded measurable function g on $(T, \mathscr{S})$. Define a loss function L by putting $L_{\theta_0} = g$ and $L_\theta = 0$ when $\theta \neq \theta_0$. Then $(P_{\theta_0}\rho_0 g - Q_{\theta_0}\sigma g - \varepsilon_{\theta_0}\|g\|)\lambda_{\theta_0} = M(L, \rho_0) \leqslant 0$ so that $(P_{\theta_0}\rho_0 - Q_{\theta_0}\sigma)g \leqslant \varepsilon_{\theta_0}\|g\|$. Hence, since this holds for all finite measurable functions g, $\|P_\theta \rho_0 - Q_{\theta_0}\sigma\| \leqslant \varepsilon_{\theta_0}$ so that (iv) holds. Since the implication (iv) $\Rightarrow$ (i) follows from the randomization criterion, theorem 6.2.4, and since the implication (i) $\Rightarrow$ (iii) is trivial, it follows that (i), (iii) and (iv) are equivalent. As the implication (i) $\Rightarrow$ (ii) is trivial in view of remark 2, it remains to show that (ii) $\Rightarrow$ (i).

Suppose then that (ii) holds and let L be a non-negative loss function. Put $C_\theta = Q_\theta \sigma L_\theta + \frac{1}{2}\varepsilon_\theta \|L_\theta\| + \eta$ where $\eta > 0$ and define the loss function $\tilde{L}$ by putting $\tilde{L}_\theta = L_\theta / C_\theta$. By (ii) there is a decision rule ρ_η such that

$$\bigvee_\theta (P_\theta \rho_\eta L_\theta)/C_\theta = \bigvee_\theta P_\theta \rho_\eta \tilde{L}_\theta \leqslant \bigvee_\theta [Q_\theta \sigma \tilde{L}_\theta + \tfrac{1}{2}\varepsilon_\theta \|\tilde{L}_\theta\|]$$

$$= \bigvee_\theta [Q_\theta \sigma L_\theta + \tfrac{1}{2}\varepsilon_\theta \|L_\theta\|]/C_\theta \leqslant 1.$$

Hence $P_\theta \rho_\eta L_\theta \leqslant Q_\theta \sigma L_\theta + \frac{1}{2}\varepsilon_\theta \|L_\theta\| + \eta$ when $\theta \in \Theta$. Letting ρ be an accumulation point for $(\rho_\eta : \eta > 0)$ as $\eta \downarrow 0$ we get $P_\theta \rho L_\theta \leqslant Q_\theta \sigma L_\theta + \frac{1}{2}\varepsilon_\theta \|L_\theta\|$; $\theta \in \Theta$.

□

If we do not assume that σ is fixed, and do not begin each of the four statements with: 'To each decision rule σ from $\mathscr{F}$ to $(T, \mathscr{S})$...' then (i) becomes the definition of ε-deficiency for k-decision problems while (ii)–(iv) become criteria for ε-deficiency for k-decision problems.

Corollary 6.3.2 (Equivalent conditions for ε-deficiency for k-decision problems). *Let $\mathscr{E}$, $\mathscr{F}$, ε, λ, $(T, \mathscr{S})$ and k be as in the theorem. Thus we assume that both Θ and $\mathscr{S}$ are finite sets. Then the following conditions are all equivalent.*

(i) Pointwise comparison of risk functions.
$\mathscr{E}$ is ε-deficient with respect to $\mathscr{F}$ for k-decision problems.

(ii) Comparison of maximum risks.
To each decision rule σ in $\mathscr{F}$ and to each non-negative loss function L on $(T, \mathscr{S})$ there corresponds a decision rule ρ in $\mathscr{E}$ such that

$$\bigvee_\theta P_\theta \rho L_\theta \leqslant \bigvee_\theta [Q_\theta \sigma L_\theta + \tfrac{1}{2}\varepsilon_\theta \|L_\theta\|].$$

(iii) Comparison of Bayes risks.
To each decision rule σ in $\mathscr{F}$ and to each non-negative loss function L on $(T, \mathscr{S})$ there corresponds a decision rule ρ in $\mathscr{E}$ such that

$$\sum_\theta (Q_\theta \rho L_\theta)\lambda_\theta \leqslant \sum_\theta (Q_\theta \sigma L_\theta)\lambda_\theta + \sum_\theta \varepsilon_\theta \|L_\theta\| \lambda_\theta.$$

(iv) Comparison of performance functions.
To each decision rule σ in $\mathscr{F}$ there corresponds a decision rule ρ in $\mathscr{E}$ such that

$$\|P_\theta \rho - Q_\theta \sigma\| \leqslant \varepsilon_\theta; \qquad \theta \in \Theta.$$

Remark. The remarks to the preceding theorem apply to this corollary without alteration.

An experiment $\mathscr{E} = (P_\theta : \theta \in \Theta)$ may be convexisized by taking the set Λ of all prior distributions on Θ with finite support as the parameter space, and by putting $P_\lambda = \sum_\theta P_\theta \lambda(\theta)$ when $\lambda \in \Lambda$. The convexisization of $\mathscr{E}$ is then the experiment $(P_\lambda : \lambda \in \Lambda)$. Using the last statement of the corollary we obtain the following counterpiece to theorem 6.2.12.

Corollary 6.3.3 (Deficiencies for convex extensions). *$\mathscr{E} = (P_\theta : \theta \in \Theta)$ is ε-deficient with respect to $\mathscr{F} = (Q_\theta : \theta \in \Theta)$ for k-decision problems if and only if $(P_\lambda : \lambda \in \Lambda)$ is $(\sum \varepsilon_\theta \lambda_\theta : \lambda \in \Lambda)$-deficient with respect to $(Q_\lambda : \lambda \in \Lambda)$ for k-decision problems.*

The Bayes risk criterion (ii) may be expressed very conveniently in terms of minimum Bayes risks or in terms of maximum Bayes utilities. Before doing so let us consider some handy notation for these quantities.

Consider an experiment $\mathscr{E} = (\mathscr{X}, \mathscr{A}; P_\theta : \theta \in \Theta)$ with a not necessarily finite parameter set Θ, a prior distribution λ on Θ with finite support, a k-decision space $(T, \mathscr{S})$ where $k < \infty$ and a finite loss function L on $(T, \mathscr{S})$. Define a non-negative homogeneous function $\phi_\lambda(\cdot, L)$ on $\mathbb{R}^\Theta$ by putting

$$\phi_\lambda(z, L) = \bigwedge_t \sum_\theta L_\theta(t) z_\theta \lambda_\theta; \qquad z \in \mathbb{R}^\Theta.$$

Denote the minimum Bayes risk in this decision problem by $b_{\mathscr{E}}(\lambda, L)$. Let the non-negative and σ-finite measure μ dominate P_θ when $\lambda_\theta > 0$. Denoting the density of P_θ with respect to μ by f_θ we see that the Bayes risk of a procedure ρ satisfies the inequality

$$\begin{aligned}\sum_\theta \lambda_\theta P_\theta \rho L_\theta &= \int\left\{\int\left[\sum_\theta L_\theta(t) f_\theta(x) \lambda_\theta\right]\rho(\mathrm{d}t|x)\right\}\mu(\mathrm{d}x)\\ &\geqslant \int\left\{\bigwedge_t\left[\sum_\theta L_\theta(t) f_\theta(x)\lambda_\theta\right]\right\}\mu(\mathrm{d}x)\end{aligned}$$

where equality holds if and only if $\rho(\cdot|x)$, for μ-almost all x, assigns mass 1 to the set of decisions minimizing $\sum_\theta L_\theta(t) f_\theta(x)\lambda_\theta$. Thus the minimum Bayes risk is the total mass of the measure $\bigwedge_t \sum_\theta L_\theta(t)\lambda(\theta)P_\theta$. Hence the minimum Bayes risk, as a function of $\mathscr{E}$, is the functional determined by the function $\phi_\lambda(\cdot, L)$, i.e.

$$b_{\mathscr{E}}(\lambda, L) = \int \phi_\lambda(\mathrm{d}\mathscr{E}, L).$$

Now a functional l on $\mathbb{R}^\Theta$ is linear if and only if $l(z) \equiv_z \sum l(e^\theta) z_\theta$ where $e^\theta(\theta') = 1$ or $= 0$ as $\theta' = \theta$ or $\theta' \neq \theta$ while z_θ, for each θ, is the θth coordinate of z. In particular the set of points θ in Θ such that $l(e^\theta) \neq 0$ is necessarily finite. The function $\phi_\lambda(\cdot, L)$ is a minimum of at most k linear functionals on $\mathbb{R}^\Theta$. Conversely any function on $\mathbb{R}^\Theta$ which is a minimum of k linear functionals is of the form $\phi_\lambda(\cdot, L)$ for a suitably chosen loss function L.

Instead of working with minimum Bayes risk it is sometimes more convenient to work with maximum Bayes utility. Thus if $U_\theta(t)$ is the utility of decision t when θ prevails then the expected utility of the decision rule U_θ is $P_\theta \rho U_\theta$. The Bayes utility of ρ, using the same notation as above, is

$$\sum_\theta \lambda_\theta P_\theta U_\theta \leqslant \int \left\{ \bigvee_t \left[\sum_\theta U_\theta(t) f_\theta(x) \lambda_\theta \right] \right\} \mu(dx)$$

where equality holds if and only if $\rho(\cdot|x)$, for μ-almost all x, assigns mass 1 to the set of decisions maximizing $\sum_\theta U_\theta(t) f_\theta(x) \lambda_\theta$. Define the sublinear function $\psi_\lambda(\cdot, U)$ on $\mathbb{R}^\Theta$ by putting

$$\psi_\lambda(z, U) = \bigvee_t \sum_\theta U_\theta(t) z_\theta \lambda_\theta.$$

Let us denote the maximum Bayes utility by $\text{bu}_\mathscr{E}(\lambda, U)$. Then

$$\text{bu}_\mathscr{E}(\lambda, U) = \int \psi_\lambda(d\mathscr{E}, U).$$

Minimum Bayes risk may be converted to maximum Bayes utility, and conversely, by the formulas

$$-\phi_\lambda(z, L) = \psi_\lambda(z, -L)$$

and

$$-b_\mathscr{E}(\lambda, L) = \text{bu}_\mathscr{E}(\lambda, -L).$$

Let e^θ be the function on Θ which assigns the value 1 to θ and 0 to any $\theta' \in \Theta - \{\theta\}$. Thus $e^\theta = I_{\{\theta\}}$. Note that

$$\psi_\lambda(\pm e^\theta, U) = \bigvee_t \pm U_\theta(t)\lambda_\theta; \qquad \theta \in \Theta$$

so that

$$\psi_\lambda(-e^\theta, U) \vee \psi_\lambda(e^\theta, U) = \|U_\theta\| \lambda_\theta; \qquad \theta \in \Theta.$$

The function $\psi_\lambda(\cdot, U)$ is the support function of the convex hull in $\mathbb{R}^\Theta$ determined by the k points $(U_.(t)\lambda_.; t \in T)$. It is a maximum of k-linear functionals. Conversely, any function ψ on $\mathbb{R}^\Theta$ such that $\psi(z)$ depends on z only via $(z_\theta : \lambda(\theta) > 0)$ and which is a maximum of at most k linear functionals is of the form $\psi_\lambda(\cdot, U)$ for a suitably chosen utility function U.

For a general parameter set Θ, we shall use the notation $\Psi_k(\Theta)$ or just Ψ_k for the set of functions ψ on $\mathbb{R}^\Theta$ which are of the form

$$\psi(x) = \bigvee_{i=1}^k \sum_\theta a_{\theta i} x_\theta; \qquad x \in \mathbb{R}^\Theta$$

where the constants $a_{\theta i}$ are all zero when θ does not belong to a certain finite set.

The set of sublinear functions ψ on $\mathbb{R}^\Theta$ such that $\psi(z)$ depends on z via

$(z_\theta : \theta \in F)$ for some finite set F not depending on z will be denoted by $\Psi(\Theta)$ or just by Ψ. We may sometimes write Ψ_∞ or $\Psi_\infty(\Theta)$ instead of Ψ or $\Psi(\Theta)$.

The fact that $\int \psi(\mathrm{d}\mathscr{E}) = \sum_\theta \psi(e^\theta)$ for any experiment $\mathscr{E}$ when ψ is a linear functional on $\mathbb{R}^\Theta$, i.e. $\psi \in \Psi_1$, may often be used to obtain several versions of a given identity or inequality.

Let us write out the Bayes criterion for ε-deficiency in terms of minimum Bayes risk.

Theorem 6.3.4 (Minimum Bayes risk criterion for ε-deficiency). *Let $\mathscr{E} = (P_\theta : \theta \in \Theta)$ and $\mathscr{F} = (Q_\theta : \theta \in \Theta)$ be two experiments with the same finite parameter set Θ, let ε be a non-negative function on Θ and let k be a positive integer. Let λ be a prior distribution on Θ such that $\lambda(\theta) > 0$ for all $\theta \in \Theta$. Then each of the following conditions are equivalent.*

(i) *$\mathscr{E}$ is ε-deficient with respect to $\mathscr{F}$ for k-decision problems.*

(ii) *For all loss functions L of T_k,*

$$b_{\mathscr{E}}(\lambda, L) \leqslant b_{\mathscr{F}}(\lambda, L) + \sum_\theta \varepsilon_\theta \|L_\theta\| \lambda(\theta).$$

(iii) *For all utility functions U on T_k,*

$$\mathrm{bu}_{\mathscr{E}}(\lambda, U) \geqslant \mathrm{bu}_{\mathscr{F}}(\lambda, U) - \sum_\theta \varepsilon_\theta \|U_\theta\| \lambda(\theta).$$

(iv) $\int \psi(\mathrm{d}\mathscr{E}) \geqslant \int \psi(\mathrm{d}\mathscr{F}) - \sum_\theta \varepsilon_\theta[\psi(-e^\theta) + \psi(e^\theta)]/2;\ \psi \in \Psi_k.$

Remark 1. Equivalent versions of (iv) are:

(iv′) $\int \psi(\mathrm{d}\mathscr{E}) \geqslant \int \psi(\mathrm{d}\mathscr{F}) - \sum_\theta \varepsilon_\theta[\psi(e^\theta) \vee \psi(-e^\theta)];\ \psi \in \Psi_k.$

(iv″) $\int \psi(\mathrm{d}\mathscr{E}) \geqslant \int \psi(\mathrm{d}\mathscr{F}) - \sum_\theta \varepsilon_\theta \psi(e^\theta)$ whenever $\psi \in \Psi_k$ satisfies $\psi(-e^\theta) \equiv_\theta \psi(e^\theta)$.

(iv‴) $\int \psi(\mathrm{d}\mathscr{E}) \geqslant \int \psi(\mathrm{d}\mathscr{F}) - \sum_\theta \varepsilon_\theta \psi(-e^\theta)/2$ whenever $\psi \in \Psi_k$ is monotonically decreasing.

This may either be deduced directly from proposition 6.2.3 or as follows. Firstly the inequalities $[\psi(-e^\theta) + \psi(e^\theta)]/2 \leqslant \psi(-e^\theta) \vee \psi(e^\theta);\ \theta \in \Theta$ show that (iv) $\Rightarrow$ (iv′). Restricting ψ as in (iv″) we see that (iv′) $\Rightarrow$ (iv″). If $l \in \Psi_1$ then $\psi - l$ satisfies the condition in (iv″) when $l(e^\theta) \equiv_\theta [\psi(e^\theta) - \psi(-e^\theta)]/2$. If, in addition, ψ is decreasing, then $\psi(e^\theta) \leqslant \psi(0) = 0$. Thus (iv″) implies that $\int \psi(\mathrm{d}\mathscr{E}) \geqslant \int \psi(\mathrm{d}\mathscr{F}) - \sum_\theta \varepsilon_\theta[\psi(e^\theta) - l(e^\theta)] = \int \psi(\mathrm{d}\mathscr{F}) - \sum_\theta \varepsilon_\theta[\psi(-e^\theta) + \psi(e^\theta)]/2 \geqslant \int \psi(\mathrm{d}\mathscr{F}) - \sum_\theta \varepsilon_\theta \psi(-e^\theta)/2$. Hence (iv″) $\Rightarrow$ (iv‴). Assume finally that (iv‴) holds and that $\psi \in \Psi_k$. Then $\psi - l$ is monotonically decreasing if $l \in \Psi_1$ satisfies $l(e^\theta) \equiv_\theta \psi(e^\theta)$. This may be seen by just writing out an explicit expression for ψ as a maximum of k-linear functionals. Hence, by (iv‴), $\int \psi(\mathrm{d}\mathscr{E}) \geqslant \int \psi(\mathrm{d}\mathscr{F}) - \sum_\theta \varepsilon_\theta \tilde{\psi}(-e^\theta)/2 = \int \psi(\mathrm{d}\mathscr{F}) - \sum \varepsilon_\theta[\psi(-e^\theta) + \psi(e^\theta)]/2$.

Remark 2. Note that the set of non-negative homogeneous functions ψ on $\mathbb{R}^\Theta$ which are bounded on bounded sets, and which satisfies the inequality in (iv), is a cone. Furthermore, a non-negative homogeneous function $\tilde{\psi}$ satisfies (iv) if ψ satisfies (iv) and $\tilde{\psi} = \psi$ on $[0, \infty[^\Theta$ while $\psi(-e^\theta) \leqslant \tilde{\psi}(-e^\theta)$ for each $\theta \in \Theta$. We shall make use of this fact later on in our discussion of dichotomies.

Proof of the theorem. (ii) follows from (i) by first minimizing with respect to ρ and then with respect to σ in the inequality of part (iii) of corollary 6.3.2. Conversely, (i) follows from (ii) since minimum Bayes risk is obtained. (iii) follows from (ii) and vice versa by multiplication by (-1). Finally (iv′), for $\psi_\lambda(\cdot, U)$ is just (iii). □

Later we will try to figure out the geometrical meaning of conditions such as (iv). Here let us just re-formulate this condition in order to get rid of the assumption that k is finite as well as the assumption of a finite parameter set.

Corollary 6.3.5 (Sublinear function criterion for ε-deficiency). *Let $\mathscr{E} = (P_\theta : \theta \in \Theta)$ and $\mathscr{F} = (Q_\theta : \theta \in \Theta)$ be experiments and let $(\varepsilon_\theta : \theta \in \Theta)$ be a non-negative function on Θ. Let k be a positive integer or $= \infty$.*

Then $\mathscr{E}$ is ε-deficient with respect to $\mathscr{F}$ for k-decision problems if and only if

$$\int \psi(\mathrm{d}\mathscr{E}) \geqslant \int \psi(\mathrm{d}\mathscr{F}) - \sum_\theta \varepsilon_\theta[\psi(-e^\theta) + \psi(e^\theta)]/2$$

for all functions $\psi \in \Psi_k(\Theta)$.

Remark. The modification described in remark 1 after theorem 6.3.4 may also be made here, without alteration.

Proof. If $k < \infty$ then this follows from the reduction theorem 6.2.12 and the definition of Ψ_k. If the inequality holds for all $\psi \in \Psi$ then, since $\Psi \supseteq \Psi_k$; $k = 1, 2, \ldots$, $\mathscr{E}$ is ε-deficient with respect to $\mathscr{F}$ for k-decision problems for any $k < \infty$ and hence for $k = \infty$. Finally, suppose that $\mathscr{E}$ is ε-deficient with respect to $\mathscr{F}$. Then the inequality holds when $\psi \in \Psi_1 \cup \Psi_2 \cup \cdots$ Let $\psi \in \Psi$. Then, by proposition 2.3.27, there is a monotonically increasing sequence $\{\psi_k\}$ in Ψ such that, for each k, $\psi_k \in \Psi_k$ and $\sup_k \psi_k = \psi$. By the theorem the inequality holds for each ψ_k. Hence, by letting $k \to \infty$, we obtain the inequality for ψ as well. □

Corollary 6.3.6 (Testing criterion for ε-deficiency). *$\mathscr{E} = (P_\theta : \theta \in \Theta)$ is ε-deficient with respect to $\mathscr{F} = (Q_\theta : \theta \in \Theta)$ for testing problems if and only if*

$$\left\| \sum_\theta a_\theta P_\theta \right\| \geqslant \left\| \sum_\theta a_\theta Q_\theta \right\| - \sum_\theta |a_\theta| \varepsilon_\theta$$

for each finite measure a on Θ having finite support.

Let $\mathscr{B}(\mathscr{E})$ and $\mathscr{B}(\mathscr{F})$ denote, respectively, the closed linear subspace of $L(\mathscr{E})$ spanned by $(P_\theta : \theta \in \Theta)$ and the closed linear subspace of $L(\mathscr{F})$ spanned by $(Q_\theta : \theta \in \Theta)$.

Then $\mathscr{E}$ is at least as informative as $\mathscr{F}$ for testing problems (i.e. $\mathscr{E} \geqslant_2 \mathscr{F}$) if and only if there is a (necessarily unique) linear contraction from $\mathscr{B}(\mathscr{E})$ into $\mathscr{B}(\mathscr{F})$ which maps P_θ into Q_θ for each θ.

In particular it follows that $\mathscr{E}$ and $\mathscr{F}$ are equivalent for testing problems if and only if the correspondence $P_\theta \leftrightarrow Q_\theta$ extends to a linear isometry between $\mathscr{B}(\mathscr{E})$ and $\mathscr{B}(\mathscr{F})$.

If $\mathscr{E}$ and $\mathscr{F}$ are coherent then they are equivalent for testing problems if and only if they possess the same power functions.

Remark 1. Later on we will see that if the correspondence $P_\theta \leftrightarrow Q_\theta$ is extendable to a linear isometry between $\mathscr{B}(\mathscr{E})$ and $\mathscr{B}(\mathscr{F})$ then it may always be extended further to an isometric vector lattice isomorphism between the vector lattice spanned by $\mathscr{E}$ and the vector lattice spanned by $\mathscr{F}$ (LeCam, 1964). Using the ψ function criterion for equivalence this implies readily that $\mathscr{E}$ and $\mathscr{F}$ are equivalent. Thus the equivalences '$\sim_2$','$\sim_3$',... and '$\sim$' are all the same.

Remark 2. If Θ is finite then the map $a \to \sup\{\sum_\theta a_\theta P_\theta(\delta) : \delta \text{ is a test function}\} = \|(\sum_\theta a_\theta P_\theta)^+\| = \frac{1}{2}[\|\sum_\theta a_\theta P_\theta\| + \sum_\theta a_\theta]$ is the support function of the (compact convex) set of power functions in $\mathscr{E}$. Furthermore the map $a \to \frac{1}{2}\sum_\theta |a_\theta|\varepsilon_\theta$ is the support function of the cube $\prod_\theta[-\frac{1}{2}\varepsilon_\theta, \frac{1}{2}\varepsilon_\theta]$.

Thus the inequality of the corollary is equivalent to the set inequality

$$\tau(\mathscr{E}) + \sum_\theta [-\tfrac{1}{2}\varepsilon_\theta, \tfrac{1}{2}\varepsilon_\theta] \supseteqq \tau(\mathscr{F})$$

where $\tau(\mathscr{E})$ and $\tau(\mathscr{F})$ denote, respectively, the set of power functions in $\mathscr{E}$ and the set of power functions in $\mathscr{F}$.

It is easily checked that this set inequality is just the performance function criterion for ε-deficiency for testing problems (part (iv) of corollary 6.3.2).

Remark 3. If Θ is a two-point set then general comparison may be reduced to comparison for testing problems, as will be shown in section 3 of chapter 9. Indeed then $\mathscr{E}$ is ε-deficient with respect to $\mathscr{F}$ if and only if $\mathscr{E}$ is ε-deficient with respect to $\mathscr{F}$ for testing problems.

Proof of the corollary. This criterion for ε-deficiency for testing problems follows by noting that any sublinear function in Ψ_2 is of the form $x \to |\sum_\theta a_\theta x_\theta| + \sum_\theta b_\theta x_\theta$ for finite measures a and b on Θ possessing finite supports. The last statement is the performance function criterion for equivalence for testing problems. □

The theorem and its corollary may be used to obtain somewhat more explicit expressions for deficiencies. First consider the situation of a finite decision space as well as a finite parameter set.

Corollary 6.3.7 (Minimum Bayes risk expression for deficiency). *Let $\mathscr{E}$ and $\mathscr{F}$ be experiments with the same finite parameter set Θ and let $(T, \mathscr{S})$ be a k-decision space where $k < \infty$. Consider a prior distribution λ on Θ such that $\lambda(\theta) > 0$ for each θ, and for each loss (utility) function L, put $\|L\|_\lambda = \sum_\theta \|L_\theta\| \lambda_\theta$. Then the deficiency $\delta_k(\mathscr{E}, \mathscr{F})$ may be expressed as*

$$\delta_k(\mathscr{E}, \mathscr{F}) = \sup\{b_{\mathscr{E}}(\lambda, L) - b_{\mathscr{F}}(\lambda, L) : \|L_\lambda\| = 1\}$$
$$= \sup\{\mathrm{bu}_{\mathscr{F}}(\lambda, U) - \mathrm{bu}_{\mathscr{E}}(\lambda, U) : \|U_\lambda\| = 1\}$$

while the deficiency distance $\Delta_k(\mathscr{E}, \mathscr{F})$ may be expressed as

$$\Delta_k(\mathscr{E}, \mathscr{F}) = \sup\{|b_{\mathscr{E}}(\lambda, L) - b_{\mathscr{F}}(\lambda, L)| : \|L\|_\lambda = 1\}$$
$$= \sup\{|\mathrm{bu}_{\mathscr{E}}(\lambda, U) - \mathrm{bu}_{\mathscr{F}}(\lambda, U)| : \|U\|_\lambda = 1\}.$$

Proof. This is an immediate consequence of the definition of deficiency and parts (ii) and (iii) of the theorem. □

Using the functional criterion (iv) we find expressions for the deficiency as well as for the deficiency distance which are valid for all Θ and for all $k = 1, 2, \ldots, \infty$.

Corollary 6.3.8 (Sublinear function criterion for deficiency). *Let $\mathscr{E}$ and $\mathscr{F}$ be experiments with the same parameter set Θ and let k be a positive integer or ∞. For each function ψ in Ψ, put $\|\psi\|_\Sigma = \sum_\theta [\psi(-e^\theta) + \psi(e^\theta)]/2$. Then the deficiency may be expressed as*

$$\delta_k(\mathscr{E}, \mathscr{F}) = \sup\left\{\int \psi(d\mathscr{F}) - \int \psi(d\mathscr{E}) : \|\psi\|_\Sigma = 1\right\}$$

while the deficiency distance may be expressed as

$$\Delta_k(\mathscr{E}, \mathscr{F}) = \sup\left\{\left|\int \psi(d\mathscr{E}) - \int \psi(d\mathscr{F})\right| : \|\psi\|_\Sigma = 1\right\}.$$

Remark. The statements of the corollary remain valid if $\|\psi\|_\Sigma$ is replaced by $\sum_\theta \psi(-e^\theta) \vee \psi(e^\theta)$. In fact we may in any case restrict our attention to functions ψ which are symmetric on $\{\pm e^\theta : \theta \in \Theta\}$, i.e. we may assume that $\psi(-e^\theta) \equiv_\theta \psi(e^\theta)$. This follows readily from remark 1 after theorem 6.3.4.

Proof. This follows directly from corollary 6.3.7. □

Corollary 6.3.9 (Deficiency distances as Hausdorff distances). *Let $(T, \mathscr{S})$ be a k-decision space where $k < \infty$. Define the distance $d(\alpha, \beta)$ between two families (experiments, performance functions), $\alpha = (\alpha_\theta : \theta \in \Theta)$ and $\beta = (\beta_\theta : \theta \in \Theta)$ of* *finite measures on the k-decision space $(T, \mathscr{S})$ by*

$$d(\alpha, \beta) = \sup_\theta \|\alpha_\theta - \beta_\theta\|.$$

Then $\delta_k(\mathscr{E}, \mathscr{F}) = \sup_\beta \inf_\alpha d(\alpha, \beta)$ where α (β) runs through the set of performance functions in $\mathscr{E}$ $(\mathscr{F})$. In particular it follows that $\Delta_k(\mathscr{E}, \mathscr{F})$ is the Hausdorff distance between the set $\{\alpha\}$ and the set $\{\beta\}$ for the metric d.

Remark. If d is any bounded pseudometric on a set Z then the *Hausdorff distance* between subsets Z_1 and Z_2 of Z is the maximum of the numbers

$$\sup_{z_2} \inf_{z_1} d(z_1, z_2) \quad \text{and} \quad \sup_{z_1} \inf_{z_2} d(z_1, z_2)$$

where z_1 (z_2) runs through Z_1 (Z_2).

Alternatively the Hausdorff distance h between Z_1 and Z_2 may be described as the smallest number $h \geqslant 0$ such that the 'h ball' $Z_1^{(h)} = \{z : \inf_{z_1} d(z, z_1) \leqslant h\}$ around Z_1 contains Z_2 and the 'h ball' $Z_2^{(h)}$ around Z_2 contains Z_1.

The inequality in corollary 6.3.5 deserves several comments. In order to simplify these comments let us assume that Θ is finite. Firstly we should note that the numbers $\int \psi(\mathrm{d}\mathscr{E})$ and $\int \psi(\mathrm{d}\mathscr{F})$ depend only on the behaviour of ψ on the likelihood space $Z = [0, \infty[^\Theta$, while the numbers $\psi(-e^\theta) : \theta \in \Theta$ depend on the behaviour of ψ outside Z. Thus in order to increase the accuracy of the inequality we could try to reduce ψ as much as possible outside Z while keeping the values in Z fixed. If $\psi(x) \equiv_x \max_t \sum_\Theta U_\theta(t) x_\theta$ for the utility function U on the finite decision space T, then this amounts to deleting superfluous decisions. In general a decision t in T may be called *superfluous* if to any decision rule from any experiment $\mathscr{E}$ to T there corresponds a decision rule from $\mathscr{E}$ to $T - \{t\}$ which is just as good in terms of risk. Actually it suffices to check this for a totally non-informative experiment and then it boils down to the following criterion.

The decision $t_0 \in T$ is superfluous for the utility function U if and only if $U_\theta(t_0) \leqslant \sum_t \pi_t U_\theta(t)$ for *all* θ and for *some* probability distribution π on T not assigning all its mass to t_0. We can then arrange things so that $\pi_{t_0} = 0$. If ρ is any decision rule from any experiment to T then the decision rule which is obtained from ρ by redistributing the mass in t_0 in proportion to π yields an expected utility which is at least as large as that of ρ. The ψ-function associated with the utility function U is $\psi_e(z, U) \equiv_z \bigvee_{t \in T} \sum_\theta U_\theta(t) z_\theta$ and the minimal sublinear function on $\mathbb{R}^\Theta$ which coincides with $\psi_e(\cdot, U)$ on Z is $\psi_e(z, \hat{U}) = \bigvee_{t \in \hat{T}} \sum_\theta U_\theta(t) z_\theta$ where $\hat{T}$ is the set of non-superfluous decisions.

More generally, if H is any bounded subset of $\mathbb{R}^\Theta$ and if H_m is the set of maximal vectors in $\langle H \rangle$ then $x \to \bigvee \{(x, h) : h \in H_m\}$ is the minimum sublinear function on $\mathbb{R}^\Theta$ which coincides with $x \to \bigvee \{(x, h) : h \in H\}$ on Z. As any finite sublinear function ψ on $\mathbb{R}^\Theta$ has a representation as $\psi(x) \equiv \bigvee\{(h,x): h \in H\}$ where $\langle H\rangle$ is determined by ψ, in principle this solves the problem for such functions. However, it is in general cumbersome to derive the minimal function by first determining the set $\langle H\rangle = \{h:(h,x) \leqslant \psi(x) \text{ for all } x\}$, then finding the set H_m of maximal points in $\langle H\rangle$ and finally evaluating the function $x \to \bigvee\{(h,x): h \in H_m\}$. We shall therefore express the minimal function explicitly in terms of the restriction of ψ to Z. Before doing so we shall require some auxiliary results on support functions.

We shall begin by generalizing the theory of support functions for bounded sets given in chapter 2, to the case of arbitrary non-empty subsets of $\mathbb{R}^\Theta$. The presentation is strongly influenced by Rockafellar (1970). It will be assumed throughout these considerations that the set Θ is finite. The following basic result is modelled after theorem 13.1 in Rockafellar (1970).

Theorem 6.3.10 (The support function theorem). *Let $\mathscr{S}$ denote the class of non-empty convex and closed subsets of $\mathbb{R}^\Theta$ where Θ is finite.*

Let Ψ denote the set of all sublinear convex functions from $\mathbb{R}^\Theta$ to $]-\infty,\infty]$ which are continuous from below.

With each set $S \in \mathscr{S}$ associate the function $\psi_S = \sup\{(s,x): s \in S\}$.

With each function $\psi \in \Psi$ associate the set $S_\psi = \{s:(s,x) \leqslant \psi(x): x \in \mathbb{R}^\Theta\}$.

Then:

(i) *the map $S \to \psi_S$ is a 1–1 map from $\mathscr{S}$ onto Ψ and if $S_1, S_2 \in \mathscr{S}$ then $S_1 \subseteq S_2$ if and only if $\psi_{S_1} \leqslant \psi_{S_2}$.*

(ii) *the map $\psi \to S_\psi$ is a 1–1 map from Ψ onto $\mathscr{S}$ and if $\psi_1, \psi_2 \in \Psi$ then $\psi_1 \leqslant \psi_2$ if and only if $S_{\psi_1} \subseteq S_{\psi_2}$.*

(iii) *the maps in* (i) *and* (ii) *are mutually inverse.*

(iv) *if H is a subset of $\mathbb{R}^\Theta$ then $\sup_{h \in H}(h,x) \equiv_x \psi_S(x)$ where S is the smallest set in $\mathscr{S}$ containing H, i.e. $S = \overline{\langle H\rangle}$.*

Proof. Clearly $\psi_{S_1} \leqslant \psi_{S_2}$ when $S_1 \subseteq S_2$ so suppose $\psi_{S_1} \leqslant \psi_{S_2}$. Let $s_1 \in S_1$ and suppose $s_1 \in S_2$. Then, by strong separation, there is a $\xi \in \mathbb{R}^\Theta - \{0\}$ such that $(\xi, s_1) > \sup\{(\xi,s): s \in S_2\} = \psi_{S_2}(\xi)$. Then $\psi_{S_1}(\xi) > \psi_{S_2}(\xi_2)$, contradicting the assumption that $\psi_{S_1} \leqslant \psi_{S_2}$. Thus $S_1 \subseteq S_2$ if and only if $\psi_{S_1} \leqslant \psi_{S_2}$. In particular the map $S \to \psi_S$ is 1–1. Furthermore, $\psi_S \in \Psi$ for each $S \in \mathscr{S}$. Now let $\psi \in \Psi$ and write $S = S_\psi = \{s:(s,x) \leqslant \psi(x): x \in \mathbb{R}^\Theta\}$. Clearly $\psi_S \leqslant \psi$. Let $x_0 \in \mathbb{R}^\Theta$ and assume first that $x_0 \in \text{ri}[\psi < \infty]$, i.e. that x_0 is a relative interior point of the cone of finiteness of ψ. Then, by weak separation, there is a vector $\xi \in \mathbb{R}^\Theta$ and a number b such that $b\psi(x_0) - (\xi, x_0) \leqslant bz - (\xi, x)$ for all $(x,z) \in \mathbb{R}^\Theta \times \mathbb{R}$ such that $z \geqslant \psi(x)$. Substituting $x = x_0$ we see that $b \geqslant 0$. As we may assume that the map $(x,z) \to bz - (\xi,x)$ is non-constant on $\text{aff}[\{(x,z): z \geqslant \psi(x)\}]$ and as we have assumed that $x_0 \in \text{ri}[\psi < \infty]$ we must have $b > 0$. (ri C denotes the set of relative interior points of a convex set C.) We may then assume without loss of generality that $b = 1$. Then $\psi(x_0) - (\xi, x_0) \leqslant \psi(x) - (\xi, x)$ for all x. The homogeneity of ψ implies that $\psi(x) \geqslant (\xi, x)$ for all x so that $\xi \in S$. Substituting $x = 0$ we find $\psi(x_0) \leqslant (\xi, x_0)$ so that actually

$$\psi(x_0) = \max\{(s, x_0) : s \in S_\psi\}$$

when $x_0 \in \operatorname{ri}[\psi < \infty]$.

Suppose next that x_0 belongs to the closure $\overline{[\psi > \infty]}$ of the cone $[\psi < \infty]$. Choose an $x_1 \in \operatorname{ri}[\psi < \infty]$ and consider the two functions of $\lambda \in [0, 1]$

$$\lambda \to \psi((1 - \lambda)x_0 + \lambda x_1)$$

and

$$\lambda \to \sup\{((1 - \lambda)x_0 + \lambda x_1, s); s \in S\}.$$

These functions coincide when $\lambda > 0$ and they are both convex and lower semicontinuous on $[0, 1]$. Hence they assign the same value to $\lambda = 0$ so that

$$\psi(x_0) = \sup\{(s, x_0) : s \in S\}$$

in this case.

Suppose that $x_0 \in \overline{[\psi < \infty]}$. Then, by strong separation, there is a $\xi \in \mathbb{R}^\Theta - \{0\}$ such that $(\xi, x_0) > \sup\{(\xi, x) : \psi(x) < \infty\}$. Homogeneity then implies that $(\xi, x_0) > 0 \geqslant (\xi, x)$ when $\psi(x) < \infty$. Let $s_0 \in S$. Then $s_0 + t\xi \in S$ whenever $t \geqslant 0$. Thus

$$\sup\{(s, x_0) : s \in S\} \geqslant \sup_{t > 0}\, [(s_0, x_0) + t \cdot (\xi, x_0)] = \infty$$

so that

$$\psi(x_0) = \sup\{(s, x_0) : s \in S\} = \infty$$

in this case. The facts established so far prove (i), (ii) and (iii). (iv) follows now from the identity

$$\sup_{h \in H}\, (h, x) = \sup_{h \in S}\, (h, x)$$

where $S = \overline{\langle H \rangle}$. □

Corollary 6.3.11 (Support functions on the likelihood space). *Let $\mathscr{C}$ denote the class of non-empty closed convex and decreasing subsets of $\mathbb{R}^\Theta$ where Θ is finite. (A subset A of a partially ordered set S is called* decreasing *if $a \in A$ whenever $a \in S$ and $a \leqslant b \in A$.)*

Let $\mathscr{V}$ denote the class of lower semicontinuous and sublinear functions from the likelihood space $Z = [0, \infty[^\Theta$ to $]-\infty, \infty]$.

With each set $C \in \mathscr{C}$ associate the function $v_C : z \to \sup\{(c, z) : c \in C\}$ on Z.

With each function $v \in \mathscr{V}$ associate the set $C_v = \{c : c \in \mathbb{R}^\Theta, (x, z) \leqslant v(z)$ for all $z \in Z\}$. Then:

(i) *the map $C \to v_C$ is a 1–1 map of $\mathscr{C}$ onto $\mathscr{V}$ and if $C_1, C_2 \in \mathscr{C}$ then $C_1 \subseteqq C_2$ if and only if $v_{C_1} \leqslant v_{C_2}$.*

(ii) *the map* $v \to C_v$ *is a* 1–1 *map from* $\mathscr{V}$ *onto* $\mathscr{C}$.
(iii) *the maps in* (i) *and* (ii) *are mutually inverse.*
(iv) *if* H *is a non-empty subset of* $\mathbb{R}^\Theta$ *then the closure of the set of all vectors in* $\mathbb{R}^\Theta$ *which are majorized by vectors in the convex hull of* H *is the smallest set* C *in* $\mathscr{C}$ *which contains* H. *If so then* $v_C = \sup\{(h, x) : h \in H\}$.

Remark. If the set (cone) of points $z \in Z$ such that $v(z) < \infty$ has interior points, then each vector $c \in C$ is majorized by a maximal vector $\tilde{c} \in C$. This may be seen by choosing a vector z^0 in the interior of $[v < \infty]$ and then putting $\gamma(x) \equiv_x \sup\{(c', x) : c' \in C, c' \geqslant c\}$. Then z^0 is a point in the interior of the domain of finiteness of γ so that, by weak separation, $\gamma(z^0) = (\tilde{c}, z^0)$ for some $\tilde{c} \geqslant c$ belonging to C. Then $\tilde{c}$ is maximal in C and $c \leqslant \tilde{c}$.

Proof. This follows readily from the support function theorem by noting that if $v \in \mathscr{V}$ then the function ψ which assigns the values $v(x)$ or ∞ to $x \in \mathbb{R}^\Theta$ according to whether $x \in Z$ or not, is a support function on $\mathbb{R}^\Theta$. □

We shall also need a few facts on directional derivatives. If f is an extended real valued function on a subset D of $\mathbb{R}^\Theta$ and if $\lim_{\lambda \downarrow 0} (f(x + \lambda y) - f(x))/\lambda$ exists then this limit is called *the directional derivative of* f *at* x *in the direction of* y. This directional derivative is denoted by $f'(x, y)$ and its existence implies that $x + \lambda y \in D$ when λ is sufficiently small positive. If f is a convex function on $\mathbb{R}^\Theta$ then the directional derivative exists at each point $x \in \mathbb{R}^\Theta$ such that $f(x)$ is finite and then

$$f'(x, y) = \inf_{\lambda > 0} \frac{f(x - \lambda y) - f(x)}{\lambda}.$$

Consider a convex function f from $\mathbb{R}^\Theta$ to $]-\infty, \infty]$ and let $x \in \mathbb{R}^\Theta$ and $a \in \mathbb{R}^\Theta$. Then a is called a *subgradient* for f at x if $f(z) \geqslant (z - x, a) + f(x)$ for all $z \in \mathbb{R}^\Theta$. Following the notation in Rockafellar (1970) we denote the set of subgradients of f at x by $\partial f(x)$. Suppose now that $x \in \mathrm{ri}[f < \infty]$. Then $\partial f(x)$ is a closed convex set having the map $y \to f'(x, y) = \sup\{(a, y) : a \in \partial f(x)\}$ as its support function. The set $\partial f(x)$ is bounded if and only if x is an interior point of the domain $[f < \infty]$ of finiteness of f. We refer to Rockafellar (1970) for an exposition of these facts.

This yields the following proposition.

Proposition 6.3.12 (Derivatives of support functions). *Let* S *be a non-empty closed convex set in* $\mathbb{R}^\Theta$, *where* Θ *is finite, and let* x *be a point where* $\psi_S(x) < \infty$. *Then*:

(i) $\partial \psi_S(x) = \{s : s \in S$ *and* $(s, x) = \psi_S(x)\}$.
(ii) *if* x *is a relative interior point of* $[\psi_S < \infty]$ *then* $\psi'_S(x, y) = \sup\{(s, y) : s \in \partial \psi_S(x)\}$.
(iii) x *is an interior point of* $[\psi_S < \infty]$ *if and only if* $\partial \psi_S(x)$ *is bounded.*

Corollary 6.3.13 (Derivatives of support functions on the likelihood space). *Suppose Θ is finite. Let C be a non-empty closed convex and decreasing subset of $\mathbb{R}^\Theta$ and put $v_C(z) = \sup\{(c,z) : c \in C\}$. Let x be a vector in Z having strictly positive coordinates. Suppose $v_C(x) < \infty$. Then:*

(i) $\partial v_C(x) = \{c : c \in C \text{ and } (c,x) = v_C(x)\}$.
(ii) *if x is a relative interior point of $[v_C < \infty]$ then* $v_C'(x,y) = \sup\{(c,y) : c \in \partial v_C(x)\}$.
(iii) *x is an interior point of $[v_C < \infty]$ if and only if $\partial v_C(x)$ is bounded.*

Let us agree to use the notation $x >_s y$ for vectors x and y in $\mathbb{R}^\Theta$ to denote that $x_\theta > y_\theta$ for all θ.

Explicit solutions to the problem of minimal extensions may now be described as follows.

Theorem 6.3.14 (Minimal extensions of support functions on the likelihood space). *Assume that Θ is finite. Let v be a sublinear and lower semicontinuous function from $Z = [0,\infty]^\Theta$ to $]-\infty,\infty]$ such that $v(x) < \infty$ when $x >_s 0$.*

Put $C = \{c : (c,x) \leqslant v(x) \text{ when } x \geqslant 0\}$ and let $\tilde{C}$ denote the set of maximal vectors in C.

Then v possesses a smallest sublinear extension ψ on $\mathbb{R}^\Theta$. This function ψ is a lower semicontinuous function from $\mathbb{R}^\Theta$ to $]-\infty,\infty]$ and its value at a given $y \in \mathbb{R}^\Theta$ may be expressed as

$$\begin{aligned}\psi(y) &= \sup\{v(x+y) - v(x) : x \underset{s}{>} 0 \text{ and } x + y \geqslant 0\} \\ &= \sup\{v(x) - v(x-y) : x \geqslant 0 \text{ and } x \underset{s}{>} y\} \\ &= \sup\{v'(x,y) : x \underset{s}{>} 0\} \\ &= \sup\{-v'(x,-y) : x \underset{s}{>} 0\} \\ &= \sup\{(c,y) : c \in \tilde{C}\}.\end{aligned}$$

Proof. Let us begin by arguing that the first four expressions for ψ are equal. Define $\psi(y)$ by

$$\begin{aligned}\psi(y) &= \sup\{v(x+y) - v(x) : x \underset{s}{>} 0 \text{ and } x + y \geqslant 0\} \\ &= \sup\{v(x) - v(x-y) : x \geqslant 0 \text{ and } x \underset{s}{>} y\}.\end{aligned}$$

Clearly $\psi(y) > -\infty$ for all $y \in \mathbb{R}^\Theta$. Furthermore,

$$\begin{aligned}\psi(ty) &= \sup\{v(x+ty) - v(x) : x \underset{s}{>} 0 \text{ and } x + ty \geqslant 0\} \\ &= t \sup\left\{v\left(\frac{1}{t}x + y\right) - v\left(\frac{1}{t}x\right) : \frac{1}{t}x \underset{s}{>} 0 \text{ and } \frac{1}{t}x + y \geqslant 0\right\} \\ &= t\psi(y) \text{ when } t > 0\end{aligned}$$

while $\psi(0)=0$. Let $x >_s 0$ and $x+\lambda y \geqslant 0$ where $\lambda > 0$. Then $(v(x+\lambda y)-v(x))/\lambda \leqslant (\psi(\lambda y))/\lambda = \psi(y)$. Hence $v'(x, y) \leqslant \psi(y)$ so that $\psi(y) \geqslant \sup\{v'(x, y) : x >_s 0\}$. We always have $v'(x, y) + v'(x, -y) \geqslant 0$ when $v(x) < \infty$ so that $\sup\{v'(x, y) : x >_s 0\} \geqslant \sup\{-v'(x, -y) : x >_s 0\}$. Put $\delta(y) = \sup\{-v'(x, -y) : x >_s 0\}$. Then $v'(x, -y) > -\delta(y)$ when $x >_s 0$ so that $(v(x - \lambda y) - v(x))/\lambda \geqslant -\delta(y)$ when $x >_s 0$ and $x - \lambda y \geqslant 0$. In particular, $v(x-y) - v(x) \geqslant -\delta(y)$ when $x >_s 0$ and $x \geqslant y$. Hence $\delta(y) + v(x - y) \geqslant v(x)$ when $x >_s 0$ and $x \geqslant y$. Suppose now that $x_0 \geqslant 0$ and that $x_0 \geqslant_s y$. Let $x_1 >_s 0$ and put $x_\varepsilon = x_0 + \varepsilon x_1$ when $\varepsilon \geqslant 0$. Then $x_\varepsilon >_s 0$ and $x_\varepsilon \geqslant y$ when $\varepsilon > 0$. Hence $\delta(y) + v(x_\varepsilon - y) \geqslant v(x_\varepsilon)$ when $\varepsilon > 0$. The lower semicontinuity of v implies that $v(x_\varepsilon - y)$ and $v(x_\varepsilon)$ are both continuous from the right at $\varepsilon = 0$. Hence $\delta(y) + v(x_0 - y) \geqslant b(x_0)$ so that $\delta(y) \geqslant \sup\{v(x) - v(x - y) : x \geqslant 0$ and $x >_s y\} = \psi(y)$. This proves the equality of the first four expressions for ψ. It also shows that ψ is a supremum of lower semicontinuous sublinear functions on $\mathbb{R}^\Theta$ and, consequently, is lower semicontinuous and sublinear.

Let $y \geqslant 0$. Choose a vector $z >_s 0$ and put $y_\varepsilon = y + \varepsilon z$. Then $y_\varepsilon >_s 0$ and $y_\varepsilon + y_\varepsilon \geqslant 0$ so that $\psi(y_\varepsilon) \geqslant v(2y_\varepsilon) - v(y_\varepsilon) = v(y_\varepsilon)$. Letting $\varepsilon \downarrow 0$ we find $\psi(y) \geqslant v(y)$. On the other hand $v(x + y) - v(x) \leqslant v(x) + v(y) - v(x) = v(y)$ when $x >_s 0$ so that $\psi(y) \leqslant v(y)$. This shows that ψ extends v. Let $\tilde{\psi}$ be any extension of v to a sublinear function on all of $\mathbb{R}^\Theta$. Let $y \in \mathbb{R}^\Theta$, $x >_s 0$ and $x + y \geqslant 0$. Then $v(x + y) - v(x) = \tilde{\psi}(x + y) - \tilde{\psi}(x) \leqslant \tilde{\psi}(x) + \tilde{\psi}(y) - \tilde{\psi}(x) = \tilde{\psi}(y)$ so that $\psi(y) \leqslant \tilde{\psi}(y)$. Thus ψ is the smallest extension. Finally consider the function $\gamma(y) \equiv_y \sup\{(c, y) : c \in \tilde{C}\}$. Trivially $\gamma(y) \leqslant \sup\{(c, y) : c \in C\} = v(y)$ when $y \geqslant 0$. If $y >_s 0$ then, by weak separation, there is a vector $c \in C$ such that $(c, y) = v(y)$. (We actually proved this in the support function theorem.) Then, necessarily, $c \in \tilde{C}$. Thus $\gamma(y) = v(y)$ when $y >_s 0$. Hence, since γ and v are both lower semicontinuous and convex, γ extends v so that $\gamma \geqslant \psi$. Now ψ is the support function $y \to \sup\{(y, h) : h \in H\}$ of some closed convex set H. The fact that ψ and v coincide on Z implies that C is the closure of the set of vectors which are majorized by vectors in H. Let $\tilde{c} \in \tilde{C}$ and suppose $z >_s 0$. Then for each $\varepsilon \in]0, 1]$ there is a vector $h_\varepsilon \in H$ so that $\tilde{c} \leqslant h_\varepsilon + \varepsilon z$. Let $y >_s 0$. Then $(\tilde{c}, y) \leqslant (h_\varepsilon, y) + \varepsilon(z, y) \leqslant v(y) + \varepsilon(z, y) \leqslant v(y) + (z, y)$. Thus, since $v(y) < \infty$, $\{(h_\varepsilon, y); \varepsilon \in]0, 1]\}$ is bounded when $y >_s 0$. Hence the family $\{h_\varepsilon; \varepsilon \in]0, 1]\}$ is bounded. It follows that there is an $h \in H$ and a sequence $\varepsilon_n \downarrow 0$ such that $h_{\varepsilon_n} \to h$. Then $\tilde{c} \leqslant h$. The maximality of $\tilde{c}$ then implies that $\tilde{c} = h$. Hence $\tilde{C} \subseteqq H$ so that $\gamma \leqslant \psi$. □

Let v be an extended real valued function on $\mathbb{R}^\Theta$ and let x be a vector such that $v(x) < \infty$. Put $g(t) = v(x + ty)$ where $y \in \mathbb{R}^\Theta$. Then the left hand derivative of g at $t = 0$ is $-v'(x, -y)$ while the right hand derivative of g at $t = 0$ is $v'(x, y)$. Clearly the minimal extension of the theorem equals $\sup\{\hat{v}(x, y) : x >_s 0\}$ whenever $\hat{v}(x, y)$, for each y, is between these two numbers.

Assume now that $\mathscr{E}$ is ε-deficient with respect to $\mathscr{F}$ and that ψ is a sublinear and lower semicontinuous function from $\mathbb{R}^\Theta$ to $]-\infty,\infty]$. (We are still assuming that Θ is finite.) Let $C = \{c:(c,x) \leqslant \psi(x): x \in \mathbb{R}^\Theta\}$ and let $\{c_1, c_2, \ldots\}$ be a countable dense subset of C and put $\psi_n(x) \equiv_x \sup\{(c_i,x): i \leqslant n\}$. Then $\psi_n \uparrow \psi$ and, by corollary 6.3.5, $\sum_\theta \frac{1}{2}\varepsilon_\theta[\psi_n(-e^\theta) + \psi_n(e^\theta)] + \int \psi_n(\mathrm{d}\mathscr{E}) \geqslant \int \psi_n(\mathrm{d}\mathscr{F})$, $n = 1,2,\ldots$ Hence $\sum_\theta \frac{1}{2}\varepsilon_\theta[\psi(-e^\theta) + \psi(e^\theta)] + \int \psi(\mathrm{d}\mathscr{E}) \geqslant \int \psi(\mathrm{d}\mathscr{F})$. Combining this with the extension theorem we get the following sharpening of the inequality in Corollary 6.3.5.

Corollary 6.3.15 (ε-deficiency in terms of support functions on the likelihood space). *Let Θ be a finite set and let k be a positive integer. Then the experiment $\mathscr{E}$ is ε-deficient with respect to $\mathscr{F}$ (for k-decision problems) if and only if*

$$\sum \frac{1}{2}\varepsilon_\theta[v(e^\theta) + \psi(-e^\theta)] + \int v(\mathrm{d}\mathscr{E}) \geqslant \int v(\mathrm{d}\mathscr{F})$$

for any sublinear and lower semicontinuous function v from Z to $]-\infty,\infty]$ which is finite on $Z^0 =]0,\infty[^\Theta$ (and which is a maximum of k-linear functionals on Z).

Here ψ denotes the minimal sublinear extension of v to $\mathbb{R}^\Theta$ such that $\psi(-e^\theta) \equiv_\theta \sup\{v(x+e^\theta) - v(x): x >_s 0,\ x + y \geqslant 0\} \equiv_\theta -\inf\{(\partial v/\partial x_\theta)(x): x >_s 0\} \equiv_\theta -\inf\{c_\theta : c \in \tilde{C}\}$. The quantity $(\partial v/\partial x_\theta)(x)$, for each θ, may be any number between the left and right hand side partial derivative with respect to x_θ at x while $\tilde{C}$ is the set of maximal vectors in $C = \{c:(c,x) \leqslant v(x)$ when $x \geqslant 0\}$.

Remark 1. The conventions $0\cdot\infty = 0$ and $c\cdot\infty = \infty$ when $c > 0$ are in force concerning the terms $\varepsilon_\theta[v(e^\theta) + \psi(-e^\theta)]$. It follows that the bracket $[v(e^\theta) + \psi(-e^\theta)]$ equals $\sup\{c_\theta : c \in \tilde{C}\} - \inf\{c_\theta : c \in \tilde{C}\} = \sup\{c_\theta^1 - c_\theta^2 : c^1, c^2 \in \tilde{C}\}$ which is the width of $\tilde{C}$ in the direction of e^θ.

Remark 2. As the partial derivative $\partial v/\partial x_\theta(x)$ increases with the θth coordinate we could have replaced this expression by $\lim_{x_\theta \downarrow 0}(\partial v/\partial x_\theta)(x) = \inf_{t>0}(\partial v/\partial x_\theta)(x - x_\theta e^\theta + te^\theta)$. These inequalities are written out in section 9.3, for the case where $\#\Theta = 2$, i.e. in the case of dichotomies.

Sublinear functions on $[0,\infty[^\Theta$ may be 'localized' to convex functions on the probability simplex Λ of $[0,\infty[^\Theta$. Thus if ϕ is any convex function from Λ to $]-\infty,\infty]$ then $x \to v(x) \equiv \sum x\phi(x/\sum_\theta x_\theta)$ is a sublinear function from $Z = [0,\infty[^\Theta$ to $]-\infty,\infty]$.

The map $\phi \to v$ is a 1–1 map from the class of convex functions from Λ to $]-\infty,\infty]$ onto the class of sublinear functions from $[0,\infty[^\Theta$ to $]-\infty,\infty]$. The function ϕ is lower semicontinuous if and only if v is lower semicontinuous.

Directional derivatives may be localized to Λ without difficulty, since $\psi'(x,y) = \psi'(tx,y)$ if $t > 0$, $\psi'(x,y)$ exists and ψ is positive homogeneous.

If $v \in \mathscr{V}$ corresponds to ϕ on Λ, if $0 <_s x \in \Lambda$ and if $v(x) < \infty$ then

$$v'(x, y) = \phi'(x, y - (\sum y)) + \sum y\phi(x)$$

where $\sum y = \sum y_\theta$. Hence $v'(x, e^\theta) = \phi'(x, e^\theta - x) + \phi(x)$. If ϕ is the restriction to Λ of a function which is differentiable in x then

$$v'(x, e^\theta) = \sum_{\theta' \in \Theta} x_{\theta'} \left[\frac{\partial \phi}{\partial x_\theta} - \frac{\partial \phi}{\partial x_{\theta'}} \right](x) + \phi(x).$$

This ends our discussion of the inequality in corollary 6.3.5. From now on, unless otherwise stated, we will *not* assume that Θ is finite.

We shall now see how ε-deficiency for k-decision problems for $k < \infty$ can be expressed in terms of ε-deficiencies (for $k = \infty$). Consider first a comparison with respect to an experiment having finite sample space.

Proposition 6.3.16. *Let $\mathscr{E} = (P_\theta : \theta \in \Theta)$ and $\mathscr{F} = (\mathscr{Y}, \mathscr{B}, Q_\theta : \theta \in \Theta)$ be experiments and suppose that $\mathscr{B}$ is generated by a partitioning of $\mathscr{Y}$ into k or fewer sets, where $k < \infty$. Thus we assume that $\# \mathscr{B} \leqslant 2^k$.*

Then $\mathscr{E}$ is ε-deficient with respect to $\mathscr{F}$ if and only if $\mathscr{E}$ is ε-deficient with respect to $\mathscr{F}$ for k-decision problems.

Proof. We may assume, without loss of generality, that Θ is a finite set. By the definition of deficiency, the 'only if' holds without any restrictions on $\mathscr{F}$. Suppose $\mathscr{E}$ is ε-deficient with respect to $\mathscr{F}$ for k-decision problems and let $(B_1, B_2, \ldots, B_k)$ be a measurable partition generating $\mathscr{B}$. Consider $(\mathscr{Y}, \mathscr{B})$ as a decision space and let σ be the identity map on $\mathscr{Y}$ considered as a decision rule. Thus $Q_\theta \sigma \equiv_\theta Q_\theta$. By part (iv) of corollary 6.3.2, there is a Markov kernel (decision rule) M from $\mathscr{E}$ to $(\mathscr{Y}, \mathscr{B})$ such that $\|P_\theta M - Q_\theta\| \leqslant \varepsilon_\theta; \theta \in \Theta$. Hence, by the $\frac{1}{2}$ Markov kernel criterion (i.e. theorem 6.2.4), $\mathscr{E}$ is ε-deficient with respect to $\mathscr{F}$. □

Now consider any prior distribution λ with finite support and a given loss function L on the k-decision space $T_k = \{1, \ldots, k\}$. Let $\mathscr{F} = (\mathscr{Y}, \mathscr{B}, Q_\theta : \theta \in \Theta)$ be an experiment and let $\mathscr{B}_0$ be a subalgebra of $\mathscr{B}$ generated by a non-randomized Bayes solution σ in $\mathscr{F}$. Then $\# \mathscr{B}_0 \leqslant 2^k$ and $b_{\mathscr{F}}(\lambda, L) = b_{\mathscr{F}|\mathscr{B}_0}(\lambda, L)$, i.e. $\int \phi_\lambda(\mathrm{d}\mathscr{F}) = \int \phi_\lambda(\mathrm{d}(\mathscr{F}|\mathscr{B}_0))$. It follows that to each $\psi \in \Psi_k$ where $k < \infty$ there corresponds a subalgebra $\mathscr{B}_0$ of $\mathscr{B}$ containing at most 2^k sets such that $\int \psi(\mathrm{d}\mathscr{F}) = \int \psi(\mathrm{d}(\mathscr{F}|\mathscr{B}_0))$. Using this observation and proposition 6.3.16 we may reduce comparison for k-decision problems to comparison for ∞-decision problems.

Theorem 6.3.17 (Reduction to infinite decision problems). *Let $\mathscr{E} = (P_\theta : \theta \in \Theta)$ and $\mathscr{F} = (\mathscr{Y}, \mathscr{B}, Q_\theta : \theta \in \Theta)$ be experiments and let ε be a non-negative function on Θ. Let k be a positive integer. Then $\mathscr{E}$ is ε-deficient with respect to $\mathscr{F}$ for k-decision problems if and only if $\mathscr{E}$ is ε-deficient with respect to each experiment $\mathscr{F}|\mathscr{B}_0$ where $\mathscr{B}_0$ is a subalgebra of $\mathscr{B}$ containing at most 2^k sets.*

Proof. Suppose $\mathscr{E}$ is ε-deficient with respect to $\mathscr{F}$ for k-decision problems and let $\mathscr{B}_0$ be a subalgebra of $\mathscr{B}$ containing at most 2^k sets. Then, since $\mathscr{F}$ is 0-deficient with respect to $\mathscr{F}|\mathscr{B}_0$, $\mathscr{E}$ is $\varepsilon + 0 = \varepsilon$-deficient with respect to $\mathscr{F}|\mathscr{B}_0$ by proposition 6.2.17. Conversely, suppose that the condition of the theorem is satisfied. We may assume that Θ is finite. Let $\psi \in \Psi_k$. The considerations immediately before this theorem imply that $\int \psi(\mathrm{d}\mathscr{F}) = \int \psi(\mathrm{d}\mathscr{F}|\mathscr{B}_0)$ for some subalgebra $\mathscr{B}_0$ of $\mathscr{B}$ such that $\#\mathscr{B}_0 \leqslant 2^k$. Hence, by assumption, $\mathscr{E}$ is ε-deficient with respect to $\mathscr{F}|\mathscr{B}_0$ so that $\int \psi(\mathrm{d}\mathscr{E}) \geqslant \int \psi(\mathrm{d}(\mathscr{F}|\mathscr{B}_0)) - \sum_\theta \varepsilon_\theta[\psi(-e^\theta) + \psi(e^\theta)]/2 = \int \psi(\mathrm{d}\mathscr{F}) - \sum_\theta \varepsilon_\theta[\psi(-e^\theta) + \psi(e^\theta)]/2$. □

Corollary 6.3.18 (Deficiencies for k-decision problems from deficiencies for infinite decision problems). *Let $\mathscr{E} = (P_\theta : \theta \in \Theta)$ and $\mathscr{F} = (\mathscr{Y}, \mathscr{B}, Q_\theta : \theta \in \Theta)$ be experiments and let k be a positive integer. Then*

$$\delta_k(\mathscr{E}, \mathscr{F}) = \sup\{\delta(\mathscr{E}, (\mathscr{F}|\mathscr{B}_0)) : \mathscr{B}_0 \text{ is a subalgebra of } \mathscr{B} \text{ containing at most } 2^k \text{ sets}\}.$$

In particular $\delta_k(\mathscr{E}, \mathscr{F}) = \delta(\mathscr{E}, \mathscr{F})$ when $k \geqslant \log_2 \#\mathscr{B}$.

Applying the sublinear function criterion, i.e. corollary 6.3.5, we obtain the following result on deficiencies for products of experiments.

Theorem 6.3.19 (Comparison of products). *Let $\mathscr{E}_1, \ldots, \mathscr{E}_n$ and $\mathscr{F}_1, \ldots, \mathscr{F}_n$ be $2n$ experiments having the same parameter set Θ. Suppose $\mathscr{E}_i$ is ε^i-deficient with respect to $\mathscr{F}_i$ for k-decision problems for all $i = 1, \ldots, n$. Then $\prod_i \mathscr{E}_i$ is $\sum_i \varepsilon^i$-deficient with respect to $\prod_i \mathscr{F}_i$ for k-decision problems.*

Proof. Put $\mathscr{H}_0 = \mathscr{E}_1 \times \cdots \times \mathscr{E}_n$, $\mathscr{H}_i = \mathscr{F}_1 \times \cdots \times \mathscr{F}_i \times \mathscr{E}_{i+1} \times \cdots \times \mathscr{E}_n$; $i = 1, \ldots, n-1$ and $\mathscr{H}_n = \mathscr{F}_1 \times \cdots \times \mathscr{F}_n$. By proposition 6.2.17 we must show that $\mathscr{H}_{i-1}$ is ε^i-deficient with respect to $\mathscr{H}_i$; $i = 0, 1, \ldots, n$. We may assume without loss of generality that k is finite and that Θ is a finite set. Put $\mathscr{E}_i = (P_{\theta,i} : \theta \in \Theta)$, $\mathscr{F}_i = (Q_{\theta,i} : \theta \in \Theta)$, $\mu_i = \sum_\theta P_{\theta,i}$, $\nu_i = \sum_\theta Q_{\theta,i}$, $f_{\theta,i} = \mathrm{d}P_{\theta,i}/\mathrm{d}\mu_i$ and $g_{\theta,i} = \mathrm{d}Q_{\theta,i}/\mathrm{d}\nu_i$. Also put $H_{\theta,i} = \prod_{j \leqq i} Q_{\theta,j} \times \prod_{j>i} P_{\theta,j}$, $\kappa_i = \prod_{j \leqq i} \nu_j \times \prod_{j>i} \mu_j$ and $h_{\theta,i} = \mathrm{d}H_\theta/\mathrm{d}\kappa_i$. Then $\mathscr{H}_i = (H_{\theta,i} : \theta \in \Theta)$. We shall also use the notation $f \otimes g$ for the tensor product of two real valued functions f and g. If f has domain $\mathscr{X}$ and g has domain $\mathscr{Y}$ then $f \otimes g$ is the function on $\mathscr{X} \times \mathscr{Y}$ which, for each pair $(x, y) \in \mathscr{X} \times \mathscr{Y}$, assigns the number $f(x)g(y)$ to (x, y). Then, by Fubini's theorem, we get

$$\int \psi(\mathrm{d}\mathscr{H}_i)$$

$$= \int \psi(h_{\theta,i} : \theta \in \Theta)\,\mathrm{d}\kappa_i$$

$$= \int \left[\int \psi(g_{\theta,1} \otimes \cdots \otimes g_{\theta,i-1} \otimes \mathrm{d}\mathscr{F}_i \otimes f_{\theta,i} \otimes \cdots \otimes f_{\theta,n}) \mathrm{d}\left[\prod_{j<i} \nu_j \times \prod_{j>i} \mu_j \right]. \right.$$

Similarly

$$\int \psi(\mathrm{d}\mathscr{H}_{i-1})$$

$$= \int \psi(h_{\theta,i-1} : \theta \in \Theta)\,\mathrm{d}\kappa_i$$

$$= \int \left[\int \psi(g_{\theta,1} \otimes \cdots \otimes g_{\theta,i-1} \otimes \mathrm{d}\mathscr{E}_i \otimes f_{\theta,i+1} \otimes \cdots \otimes f_{\theta,n}) \mathrm{d}\left[\prod_{v<j} v_j \times \prod_{v>j} \mu_j \right] \right].$$

Corollary 6.3.5 implies:

$$\int \psi(g_{\theta,1} \otimes \cdots \otimes g_{\theta,i-1} \otimes \mathrm{d}\mathscr{E}_i \otimes f_{\theta,i+1} \otimes \cdots \otimes f_{\theta,n})$$

$$\geqq \int \psi(g_{\theta,1} \otimes \cdots \otimes g_{\theta,i-1} \otimes \mathrm{d}\mathscr{F}_i \otimes f_{\theta,i+1} \otimes \cdots \otimes f_{\theta,n})$$

$$- \sum_\theta \varepsilon_\theta^i g_{\theta,1} \otimes \cdots \otimes g_{\theta,i-1} \otimes \frac{1}{2}[\psi(-e^\theta) + \psi(e^\theta)] \otimes f_{\theta,i+1} \otimes \cdots \otimes f_{\theta,n}.$$

Integration with respect to $\prod_{j<i} v_j \times \prod_{j>i} \mu_j$ on both sides of this inequality yields

$$\int \psi(\mathrm{d}\mathscr{H}_{i-1}) \geqq \int \psi(\mathrm{d}\mathscr{H}_i) - \sum_\theta \varepsilon_\theta^i[\psi(-e^\theta) + \psi(e^\theta)]/2.$$

Hence, by corollary 6.3.5 again, $\mathscr{H}_{i-1}$ is ε^i-deficient with respect to $\mathscr{H}_i$. □

Corollary 6.3.20 (Deficiencies for products). *Let $\mathscr{E}_1, \ldots, \mathscr{E}_n$ and $\mathscr{F}_1, \ldots, \mathscr{F}_n$ be $2n$ experiments having the same parameter set Θ. Then*

$$\delta_k\left(\prod_{i=1}^n \mathscr{E}_i, \prod_{i=1}^n \mathscr{F}_i\right) \leqq \sum_{i=1}^n \delta_k(\mathscr{E}_i, \mathscr{F}_i).$$

In particular, $\delta_k(\mathscr{E}^n, \mathscr{F}^n) \leqslant n\delta_k(\mathscr{E}, \mathscr{F})$ if the experiments $\mathscr{E}$ and $\mathscr{F}$ have the same parameter set Θ and n is a positive integer.

These inequalities remain valid if 'δ' is replaced by 'Δ' throughout.

It follows that multiplication of experiments is a Δ_k-continuous operation.

The corresponding result for mixtures follows almost directly from the affinity of the linear functionals which are associated with non-negative homogeneous functions. Thus if $\mathscr{E}_i$; $i = 1, 2, \ldots$, is ε_i-deficient with respect to $\mathscr{F}_i$ for k-decision problems then $\int \psi(\mathrm{d}\mathscr{E}_i) \geqslant \int \psi(\mathrm{d}\mathscr{F}_i) - \sum_\theta \varepsilon_i(\theta)[\psi(-e^\theta) + \psi(e^\theta)]/2$ when $\psi \in \Psi_k$. If π is a probability distribution on the integers then

$$\int \psi\left(\mathrm{d}\left(\sum_i \pi(i)\mathscr{E}_i\right)\right)$$
$$= \sum_i \pi(i) \int \psi(\mathrm{d}\mathscr{E}_i)$$
$$\geqslant \sum_i \pi(i)\left\{\int \psi(\mathrm{d}\mathscr{F}_i) - \sum_\theta \varepsilon_i(\theta)[\psi(-e^\theta) + \psi(e^\theta)]/2\right\}$$
$$= \int \psi\left(\mathrm{d}\left(\sum_i \pi(i)\mathscr{F}_i\right)\right) - \sum_\theta\left(\sum_i \pi(i)\varepsilon_i(\theta)\right)[\psi(-e^\theta) + \psi(e^\theta)]/2.$$

This proves the analogue of theorem 6.3.31 for mixtures of experiments.

Theorem 6.3.21 (Comparison of mixtures). *Let $(\mathscr{E}_i, \mathscr{F}_i)$; $i = 1, 2, \ldots$ be a sequence of pairs of experiments with the same parameter set Θ and let π be a probability distribution on the positive integers. Suppose, for each $i = 1, 2, \ldots$, that $\mathscr{E}_i$ is ε_i-deficient with respect to $\mathscr{F}_i$ for k-decision problems.*

Then the π-mixture $\sum_i \pi(i)\mathscr{E}_i$ of the experiments $\mathscr{E}_1, \mathscr{E}_2, \ldots$ is $\sum_i \pi(i)\varepsilon_i$-deficient with respect to the π-mixture $\sum \pi(i)\mathscr{F}_i$ of the experiments $\mathscr{F}_1, \mathscr{F}_2, \ldots$ for k-decision problems.

It follows that averaging experiments is a Δ_k-continuous operation.

Corollary 6.3.22 (Deficiencies for mixtures). *Let $(\mathscr{E}_i, \mathscr{F}_i)$; $i = 1, 2, \ldots$ be a sequence of pairs of experiments with the same parameter set Θ and let π be a probability distribution on the positive integers. Then*

$$\delta_k\left(\sum_i \pi(i)\mathscr{E}_i, \sum \pi(i)\mathscr{F}_i\right) \leqslant \sum_i \pi(i)\delta_k(\mathscr{E}_i, \mathscr{F}_i).$$

In particular, for any pair $(\mathscr{E}, \mathscr{F})$ of experiments with the same parameter set

$$\delta_k\left(\sum_i \pi(i)\mathscr{E}^i, \sum_i \pi(i)\mathscr{F}^i\right) \leqslant \sum_i \pi(i)\delta_k(\mathscr{E}^i, \mathscr{F}^i) \leqslant \sum_i \pi(i)i\delta_k(\mathscr{E}, \mathscr{F}) = EN\delta(\mathscr{E}, \mathscr{F})$$

where N is distributed according to π.

These inequalities remain valid if 'δ' is replaced by 'Δ' throughout.

If in corollary 6.3.7 we replace the norm $\|L\|_\lambda = \sum \lambda_\theta \|L_\theta\|$ of a loss function L by the sup norm $\sup\{\|L_\theta\| : \lambda_\theta > 0\}$ then we get another type of deficiency. The supremum for all k-decision spaces for all k now yields the λ-weighted deficiency and the λ-weighted deficiency distance. However, if we limit our considerations to k-decision spaces where k is fixed and finite then this might not hold. However, we get lower bounds for the weighted deficiencies.

Consider two experiments $\mathscr{E} = (P_\theta : \theta \in \Theta)$ and $\mathscr{F} = (Q_\theta : \theta \in \Theta)$ and a prior probability distribution λ on Θ with finite support. For each loss function L

on a decision space $(T, \mathscr{S})$ denote the supremum norm of L on the minimal support of λ by $\|L : \lambda\|$. Thus

$$\|L : \lambda\| = \sup\{\|L_\theta\| : \lambda_\theta > 0\}.$$

This norm has to be distinguished from the norm $\|L\|_\lambda = \sum_\theta \|L_\theta\| \lambda_\theta$ used in corollary 6.3.7.

Let k be a positive integer and let L run through the set of all loss functions on a given k-decision space such that $\|L : \lambda\| = 1$. Then *the lower λ-weighted deficiency of $\mathscr{E}$ with respect to $\mathscr{F}$ for k-decision problems* is the number $\sup\{b_{\mathscr{E}}(\lambda, L) - b_{\mathscr{F}}(\lambda, L)\}$. This number will be denoted as $\underline{\delta}_k(\mathscr{E}, \mathscr{F} | \lambda)$. Similarly *the lower λ-weighted deficiency distance of $\mathscr{E}$ with respect to $\mathscr{F}$ for k-decision problems* is the number $\sup|b_{\mathscr{E}}(\lambda, L) - b_{\mathscr{F}}(\lambda, L)|$. This number is denoted by $\underline{\Delta}_{,k}(\mathscr{E}, \mathscr{F} | \lambda)$.

If $k = \infty$ then we may, as usual, omit reference to k. We shall define *the lower λ-weighted deficiency of $\mathscr{E}$ with respect to $\mathscr{F}$ (for ∞-decision problems)* as the number $\sup\{b_{\mathscr{E}}(\lambda, L) - b_{\mathscr{F}}(\lambda, L) : \|L : \lambda\| = 1\}$ where the supremum is taken for all loss functions L on finite decision spaces such that $\|L : \lambda\| = 1$. This number is denoted as $\underline{\delta}(\mathscr{E}, \mathscr{F} | \lambda)$ or as $\underline{\delta}_{,\infty}(\mathscr{E}, \mathscr{F} | \lambda)$.

Similarly the number $\sup\{|b_{\mathscr{E}}(\lambda, L) - b_{\mathscr{F}}(\lambda, L)| : \|L : \lambda\| = 1\}$ is called *the lower λ-weighted deficiency distance of $\mathscr{E}$ with respect to $\mathscr{F}$ (for ∞-decision problems)*. The supremum is then taken again for all loss functions L on finite decision spaces such that $\|L : \lambda\| = 1$. This number will be denoted as $\underline{\Delta}(\mathscr{E}, \mathscr{F} | \lambda)$ or as $\underline{\Delta}_{,\infty}(\mathscr{E}, \mathscr{F} | \lambda)$.

If $k = 1$ or $k = \infty$ then $\underline{\delta}_{,k}(\mathscr{E}, \mathscr{F} | \lambda) = \delta_k(\mathscr{E}, \mathscr{F} | \lambda)$ and $\underline{\Delta}_{,k}(\mathscr{E}, \mathscr{F} | \lambda) = \Delta_k(\mathscr{E}, \mathscr{F} | \lambda)$. This is trivial if $k = 1$. The case $k = \infty$ needs a closer investigation. These results are related to the fact that it is precisely for these values of k that $\Psi_k(\Theta)$ is a cone when $\# \Theta \geqslant 2$.

The justification of the term 'lower λ-weighted deficiency' is partly that the lower λ-weighted deficiencies are always smaller than the previously defined λ-weighted deficiencies and partly that they behave like the weighted deficiencies. Thus proposition 6.2.39, the two formulas of proposition 6.2.40 and proposition 6.2.41 all remain valid if the weighted deficiencies there are replaced by the corresponding lower λ-weighted deficiencies.

The relationship between the lower weighted deficiencies and the weighted deficiencies may be described in terms of the effect of adding up the ψ functions, related to loss functions. Now it is clear that the set of non-negative homogeneous functions ψ from $\mathbb{R}^\Theta$ to $]-\infty, \infty]$ which satisfies the inequality

$$\sum_\theta \frac{1}{2} \varepsilon_\theta [\psi(-e^\theta) + \psi(e^\theta)] + \int \psi(\mathrm{d}\mathscr{E}) \geqslant \int \psi(\mathrm{d}\mathscr{F})$$

for given experiments $\mathscr{E}$ and $\mathscr{F}$, constitutes a cone. Furthermore, we know that this inequality has a definite statistical interpretation in terms of Bayes utilities when ψ is a support function depending on a finite set of coordinates. Thus we should try to clarify the statistical meaning of summation of such functions.

Suppose for simplicity that Θ is finite and that $\psi_i(x) = \sup\{(x, s_i) : s_i \in S_i\}$; $i = 1, \ldots, r$ when $x \in \mathbb{R}^\Theta$. Here the sets S_i are non-empty subsets of $\mathbb{R}^\Theta$. If we put $\psi = \psi_1 + \cdots + \psi_r$ then $\psi(x) \equiv_x \sup\{(x, s) : s \in S\}$ where $S = S_1 + \cdots + S_r = \{s_1 + \cdots + s_r : s_1 \in S_1, \ldots, s_r \in S_r\}$. Furthermore, the upper boundary in S (i.e. the admissible utility functions in S) is the algebraic sum of the upper boundaries in S_i, $i = 1, \ldots, r$. Let us express these constructions directly in terms of utility (loss functions) for general parameter sets Θ.

Now consider utility functions $U^{(i)}$ on decision spaces $(T^{(i)}, \mathscr{S}^{(i)})$; $i = 1, \ldots, r$. In multiple decision theory we are faced with the problem of choosing a decision $t^{(i)}$ from the decision set $T^{(i)}$ for each $i = 1, \ldots, r$. In some cases it might then make sense to add the utilities incurred. This leads to the decision space $(T, \mathscr{S})$ and the utility function U where $T = T^{(1)} \times T^{(2)} \times \cdots \times T^{(r)}$, $\mathscr{S} = \mathscr{S}^{(1)} \times \mathscr{S}^{(2)} \times \cdots \times \mathscr{S}^{(r)}$ and $U_\theta(t^{(1)}, \ldots, t^{(r)}) \equiv_\theta \sum_{i=1}^r U_\theta^{(i)}(t^{(i)})$. We may then say that the utility function U is the *direct sum* of the utility functions $U^{(i)}$, $i = 1, 2, \ldots, r$ and write this

$$U = U^{(1)} \oplus U^{(2)} \oplus \cdots \oplus U^{(r)}.$$

Similarly if $L^{(i)}$, $i = 1, \ldots, r$ is a loss function on $(T^{(i)}, \mathscr{S}^{(i)})$ then we may form its direct sum $L = L^{(1)} \oplus \cdots \oplus L^{(r)}$ by putting $L_\theta(t^{(1)}, \ldots, t^{(r)}) \equiv_\theta \sum_{i=1}^r L_\theta^{(i)}(t^{(i)})$.

Now assume that to each i and each decision rule $\sigma^{(i)}$ from $\mathscr{F}$ to $(T^{(i)}, \mathscr{S}^{(i)})$ there corresponds a decision rule $\rho^{(i)}$ in $\mathscr{E}$ such that

$$P_\theta \rho^{(i)} L_\theta^{(i)} \leqslant Q_\theta \sigma^{(i)} L_\theta^{(i)} + \frac{1}{2} \varepsilon_\theta \left[\bigvee_t L_\theta^{(i)}(t) - \bigwedge_t L_\theta^{(i)}(t) \right]$$

when $\theta \in \Theta$. Let σ be any decision rule from $\mathscr{F}$ to $(T, \mathscr{S}) = \prod (T^{(i)}, \mathscr{S}^{(i)})$ and put $L = L^{(1)} \oplus \cdots \oplus L^{(r)}$. Then $Q_\theta \sigma L_\theta \equiv_\theta \sum_\theta Q_\theta \sigma^{(i)} L_\theta^{(i)}$ where, for each i, $\sigma^{(i)}$ is the 'ith marginal' of σ. Our assumption then assures the existence of decision rules $\rho^{(i)}$; $i = 1, \ldots, r$ satisfying the inequalities above. Then $P_\theta \rho L_\theta \leqslant Q_\theta \sigma L_\theta + \frac{1}{2}\varepsilon_\theta [\bigvee_t L_\theta(t) - \bigwedge_t L_\theta(t)]$ for any decision rule ρ from $\mathscr{E}$ to $(T, \mathscr{S})$ having 'marginals' $\rho^{(i)}$, $i = 1, \ldots, r$.

Here we could have replaced the range pseudonorm with the sup norm provided $L^{(i)} \geqslant 0$; $i = 1, 2, \ldots, r$.

The ϕ and ψ functions, and hence minimum Bayes risk and maximum Bayes utility, behave additively for direct summation. This follows immediately from the identities

$$\phi_\lambda(z; L^{(1)} \oplus \cdots \oplus L^{(r)}) \underset{z}{\equiv} \sum \phi_\lambda(z; L^{(i)})$$

and

$$\psi_\lambda(z; U^{(1)} \oplus \cdots \oplus U^{(r)}) \underset{z}{\equiv} \sum \psi_\lambda(z; U^{(i)})$$

which are valid for prior distributions λ with finite support and for loss functions $L^{(i)}$ (utility functions $U^{(i)}$) on finite decision spaces $T^{(i)}$; $i = 1, \ldots, r$. This and corollary 6.3.5 show how ε-deficiency with respect to given loss functions extends to the direct sums of these loss functions.

Proposition 6.3.23 (Deficiencies for direct sums of decision problems). *If the experiment $\mathscr{E}$ is ε-deficient with respect to the experiment $\mathscr{F}$ for k-decision problems then $\mathscr{E}$ is also ε-deficient with respect to the experiment $\mathscr{F}$ for finite direct sums of k-decision problems.*

The converse problem of decomposing, up to some equivalence, a given loss function as a direct sum of simpler loss functions, is in general a very difficult problem. Chapter 15 in Grünbaum's book on convex polytopes (1967) may give an idea of the difficulties which are involved. However it is known that any bounded convex polygon in the plane is an algebraic sum of segments and triangles. This implies that it suffices to compare dichotomies for three-decision problems. Now our interest in such polygons is mainly restricted to their upper boundaries, i.e. the sets of maximal or admissible utility vectors. As we shall show later, this implies that it actually suffices to compare dichotomies for two-decision problems. If we consider general pseudo-dichotomies (i.e. pairs of finite and not necessarily non-negative measures) then the behaviour of the functionals $\mathscr{E} \to \int \psi(d\mathscr{E})$ depends on the behaviour of ψ on all of $\mathbb{R}^2$. Reduction to two-decision problems may then not be feasible while reduction to three-decision problems is always possible.

If the parameter set Θ contains three or more points then in general it is not possible to reduce general comparison to comparison for k-decision problems for a fixed integer k.

We shall use the following notation.

If $k < \infty$ then $\mathscr{L}_k$ denotes the set consisting of all finite loss (utility) functions on a given k-decision space. We put $\mathscr{L}_\infty$ (or $\mathscr{L}$) $= \bigcup_{k=1}^{\infty} \mathscr{L}_k$. The set of loss functions which are direct sums of loss functions in $\mathscr{L}_k$ will be denoted as $\mathscr{L}_k^\Sigma$ and we may write $\mathscr{L}^\Sigma$ instead of $\mathscr{L}_\infty^\Sigma$.

If the set Θ is finite then the sets $\Psi_k(\Theta)$; $k = 1, 2, \ldots, \infty$ were defined earlier as the set of (sublinear) real valued functions on $\mathbb{R}^\Theta$ which were maxima of at most k linear functionals. The set $\Psi_\infty(\Theta)$ was also written as $\Psi(\Theta)$. The set of functions which are finite sums of functions in $\Psi_k(\Theta)$ will be denoted as $\Psi_k^\Sigma(\Theta)$ and we may write $\Psi^\Sigma(\Theta)$ instead of $\Psi_\infty^\Sigma(\Theta)$. Since $\Psi_\infty(\Theta)$ is a cone the superscript Σ may be deleted when $k = \infty$. If Θ is a finite set and if k is an integer then $\Psi_k^\Sigma(\Theta)$ consists precisely of the set of functions $\psi_e(\cdot, L)$ where $e_\theta \equiv_\theta 1$ and $L \in \mathscr{L}_k^\Sigma$.

The definitions of the lower weighted deficiencies may be summarized as follows

$$\underline{\delta}_{,k}(\mathscr{E},\mathscr{F}\,|\,\lambda) = \sup\{b_{\mathscr{E}}(\lambda, L) - b_{\mathscr{F}}(\lambda, L) : L \in \mathscr{L}_k, \|L : \lambda\| = 1\}$$
$$\underline{\Delta}_{,k}(\mathscr{E},\mathscr{F}\,|\,\lambda) = \underline{\delta}_{,k}(\mathscr{E},\mathscr{F}\,|\,\lambda) \vee \underline{\delta}_{,k}(\mathscr{F},\mathscr{E}\,|\,\lambda)$$
$$\underline{\delta}(\mathscr{E},\mathscr{F}\,|\,\lambda) = \sup\{\underline{\delta}_{,k}(\mathscr{E},\mathscr{F}\,|\,\lambda) : k = 1, 2, \ldots\}$$
$$\underline{\Delta}(\mathscr{E},\mathscr{F}\,|\,\lambda) = \sup\{\underline{\Delta}_{,k}(\mathscr{E},\mathscr{F}\,|\,\lambda) : k = 1, 2, \ldots\}.$$

Here is a collection of results concerning weighted and non-weighted deficiencies.

Theorem 6.3.24 (Upper and lower weighted deficiencies). *Let $\mathscr{E} = (P_\theta : \theta \in \Theta)$ and $\mathscr{F} = (\mathscr{Y}, \mathscr{B}, Q_\theta : \theta \in \Theta)$ be experiments. Let k be a positive integer or $= \infty$, let Λ denote the set of prior probability distributions on Θ with finite support and let λ, with or without affixes, denote a particular distribution in Λ. Then:*

(i) *$\delta_k(\mathscr{E},\mathscr{F}\,|\,\cdot)$ is the smallest concave function on Λ which majorizes $\underline{\delta}_{,k}(\mathscr{E},\mathscr{F}\,|\,\cdot)$. The function $\underline{\delta}(\mathscr{E},\mathscr{F}\,|\,\cdot)$ is concave so that*

$$\delta(\mathscr{E},\mathscr{F}\,|\,\lambda) = \underline{\delta}(\mathscr{E},\mathscr{F}\,|\,\lambda)$$

and

$$\Delta(\mathscr{E},\mathscr{F}\,|\,\lambda) = \underline{\Delta}(\mathscr{E},\mathscr{F}\,|\,\lambda).$$

(ii) *for any k and any λ*

$$\underline{\delta}_{,k}(\mathscr{E},\mathscr{F}\,|\,\lambda) \leqslant \delta_k(\mathscr{E},\mathscr{F}\,|\,\lambda) \leqslant \#\{\theta : \lambda_\theta > 0\}\,\underline{\delta}_{,k}(\mathscr{E},\mathscr{F}\,|\,\lambda).$$

Furthermore, $\underline{\delta}_{,k}(\mathscr{E},\mathscr{F}\,|\,\lambda) \uparrow \underline{\delta}(\mathscr{E},\mathscr{F}\,|\,\lambda) = \delta(\mathscr{E},\mathscr{F}\,|\,\lambda)$ as $k \uparrow \infty$. Here δ may be replaced by Δ throughout.

(iii) *let $k < \infty$. Then*

$$\underline{\delta}_{,k}(\mathscr{E},\mathscr{F}\,|\,\lambda) = \sup_\sigma \inf_\rho \sum_\theta \|P_\theta\rho - Q_\theta\sigma\|\,\lambda_\theta$$

where $\rho(\sigma)$ runs through the set of all decision rules from $\mathscr{E}(\mathscr{F})$ to $T_k = \{1, \ldots, k\}$.

(iv) *let $k < \infty$ and assume $\#\mathscr{B} \leqslant 2^k$. Then*

$$\underline{\delta}_{,k}(\mathscr{E},\mathscr{F}\,|\,\lambda) = \delta_k(\mathscr{E},\mathscr{F}\,|\,\lambda) = \delta(\mathscr{E},\mathscr{F}\,|\,\lambda).$$

(v) *let $k < \infty$ and let $\mathscr{B}_0$ run through the set of sub σ-algebras of $\mathscr{B}$ containing at most 2^k sets. Then*

$$\underline{\delta}_{,k}(\mathscr{E},\mathscr{F}\,|\,\lambda) = \sup_{\mathscr{B}_0} \delta(\mathscr{E}, (\mathscr{F}\,|\,\mathscr{B}_0)\,|\,\lambda).$$

(vi) *$\delta_k(\mathscr{E},\mathscr{F}) = \sup_\lambda \delta_k(\mathscr{E},\mathscr{F}\,|\,\lambda) = \sup_\lambda \underline{\delta}_{,k}(\mathscr{E},\mathscr{F}\,|\,\lambda)$. Here δ may be replaced by Δ throughout.*

(vii) *put $\Theta_\lambda = \{\theta : \lambda_\theta > 0\}$. Then*

$$\underline{\delta}_{,k}(\mathscr{E},\mathscr{F}\,|\,\lambda) = \sup\left\{\left[\int \psi(\mathrm{d}\mathscr{F}) - \int \psi(\mathrm{d}\mathscr{E})\right] \Big/ \bigvee_{\theta : \lambda_\theta > 0} [\psi(-e^\theta) \vee \psi(e^\theta)]\lambda_\theta^{-1} : \psi \in \Psi_k(\Theta_\lambda)\right\}$$

while

$$\delta_k(\mathscr{E},\mathscr{F}\,|\,\lambda) = \sup\left\{\left[\int \psi(\mathrm{d}\mathscr{F}) - \int \psi(\mathrm{d}\mathscr{E})\right] \Big/ \bigvee_{\theta:\lambda_\theta>0} [\psi(-e^\theta) \vee \psi(e^\theta)]\lambda_\theta^{-1} : \psi \in \Psi_k^\Sigma(\Theta_\lambda)\right\}$$

where the terms $\psi(-e^\theta) \vee \psi(e^\theta)$; $\theta \in \Theta$ *in both expressions may be replaced by* $\frac{1}{2}[\psi(-e^\theta) + \psi(e^\theta)]$; $\theta \in \Theta$. *We may also restrict our attention to functions* ψ *such that* $\psi(-e^\theta) \equiv \psi(e^\theta)$.

In terms of minimum Bayes risk the last expression for $\delta_k(\mathscr{E},\mathscr{F}\,|\,\lambda)$ *may be written*

$$\delta_k(\mathscr{E},\mathscr{F}\,|\,\lambda) = \sup\{[b_{\mathscr{E}}(\lambda, L) - b_{\mathscr{F}}(\lambda, L)] : L \in \mathscr{L}_k^\Sigma, \|L:\lambda\| = 1\}.$$

The equalities listed here remain valid if δ *is replaced by* Δ *throughout and the differences* $b_{\mathscr{E}}(\lambda, L) - b_{\mathscr{F}}(\lambda, L)$ *and* $\int \psi(\mathrm{d}\mathscr{F}) - \int \psi(\mathrm{d}\mathscr{E})$ *are replaced by their absolute values.*

Remark 1. In general $\underline{\delta}_{,k}(\mathscr{E},\mathscr{F}\,|\,\cdot)$ is not concave when $2 \leqslant k < \infty$. Then $\underline{\delta}_{,k}(\mathscr{E},\mathscr{F}\,|\,\lambda) < \delta_k(\mathscr{E},\mathscr{F}\,|\,\lambda)$ for some λ. By (v) this implies that

$$\delta_k(\mathscr{E},\mathscr{F}\,|\,\lambda) > \sup\{\delta(\mathscr{E},(\mathscr{F}\,|\,\mathscr{B}_0)|\lambda) : \#\mathscr{B}_0 \leqslant 2^k\}.$$

Remark 2. Let α denote one of the symbols $\underline{\delta}_{,k}$, δ_k, $\underline{\Delta}_{,k}$, Δ_k, δ and Δ. Then $\alpha(\mathscr{E},\mathscr{F}\,|\,\cdot)$ satisfies the Lipschitz condition

$$|\alpha(\mathscr{E},\mathscr{F}\,|\,\lambda^1) - \alpha(\mathscr{E},\mathscr{F}\,|\,\lambda^2)| \leqslant 2\|\lambda^1 - \lambda^2\|$$

when $\lambda^1, \lambda^2 \in \Lambda$.

Proof of the theorem. Note first that we may assume without loss of generality that Θ is finite, and that in all statements except (i) and (vi), we may assume that $\lambda_\theta > 0$ for all $\theta \in \Theta$. When this last assumption is in force then the norms $\|L:\lambda\|$ and $\|L\| = \sup_\theta \|L_\theta\|$ coincide.

Partial proof of (ii). Any loss function L on $T_k = \{1, \ldots, k\}$ may be extended to a loss function $\tilde{L}$ on T_{k+1} by putting $\tilde{L}_\theta(k+1) = \|L_\theta\|$. Then $\|L\| = \|\tilde{L}\|$ and $b_{\mathscr{E}}(\lambda, L) = b_{\mathscr{E}}(\lambda, \tilde{L})$ for any experiment $\mathscr{E}$. It follows that $\underline{\delta}_{,k}(\mathscr{E},\mathscr{F}\,|\,\lambda) \leqslant \underline{\delta}_{,k+1}(\mathscr{E},\mathscr{F}\,|\,\lambda)$ so that, by the definitions $\underline{\delta}_{,k}(\mathscr{E},\mathscr{F}\,|\,\lambda) \uparrow \underline{\delta}(\mathscr{E},\mathscr{F}\,|\,\lambda)$ as $k \uparrow \infty$. The definition of $\underline{\delta}$ and $\underline{\Delta}$ and (d) in proposition 6.2.39 now imply that it suffices to consider the inequalities for finite ks. Assume $k < \infty$. Theorem 6.3.4 implies that $\delta_k(\mathscr{E},\mathscr{F}\,|\,\lambda) = \inf\{\sum_\theta \lambda_\theta \varepsilon_\theta : \sum_\theta \lambda_\theta \varepsilon_\theta \|L_\theta\| \geqslant b_{\mathscr{E}}(\lambda, L) - b_{\mathscr{F}}(\lambda, L) : L \in \mathscr{L}_k\}$. Hence $\delta_k(\mathscr{E},\mathscr{F}\,|\,\lambda) \geqslant \inf\{\sum_\theta \lambda_\theta \varepsilon_\theta : \sum_\theta \lambda_\theta \varepsilon_\theta \|L\| \geqslant b_{\mathscr{E}}(\lambda, L) - b_{\mathscr{F}}(\lambda, L) : L \in \mathscr{L}_k\} = \underline{\delta}_{,k}(\mathscr{E},\mathscr{F}\,|\,\lambda)$. The definition of $\underline{\delta}_{,k} = \underline{\delta}_{,k}(\mathscr{E},\mathscr{F}\,|\,\lambda)$ implies that $b_{\mathscr{E}}(\lambda, L) - b_{\mathscr{F}}(\lambda, L) \leqslant \underline{\delta}_{,k}\|L\| \leqslant \underline{\delta}_{,k} \sum_\theta \|L_\theta\| = \sum_\theta (\underline{\delta}_{,k}/\lambda_\theta)\|L_\theta\|\lambda_\theta$. Thus, by theorem 6.3.4 again, $\mathscr{E}$ is $(\underline{\delta}_{,k}/\lambda_\theta : \theta \in \Theta)$-deficient with respect to $\mathscr{F}$ for k-decision problems. Hence, by the definition of $\delta_k(\mathscr{E},\mathscr{F}\,|\,\lambda)$, $\delta_k(\mathscr{E},\mathscr{F}\,|\,\lambda) \leqslant \sum_\theta (\underline{\delta}_{,k}/\lambda_\theta)\lambda_\theta = \#(\Theta)\underline{\delta}_{,k}$. The corresponding statements with δ replaced by Δ follow by symmetrization. Thus we have proved all the statements of (ii) except the assertions that $\underline{\delta}(\mathscr{E}, \mathscr{F} \mid \lambda) = \delta(\mathscr{E}, \mathscr{F} \mid \lambda)$ and that $\underline{\Delta}(\mathscr{E}, \mathscr{F} \mid \lambda) = \Delta(\mathscr{E}, \mathscr{F} \mid \lambda)$.

Proof of (iii). We get successively

$$\begin{aligned}\underline{\delta}_{,k}(\mathscr{E}, \mathscr{F} \mid \lambda) &= \sup_{\|L\| \leqslant 1} \inf_{\rho} \sup_{\sigma} \left[\sum_{\theta} \lambda_\theta P_\theta \rho L_\theta - \sum_{\theta} \lambda_\theta Q_\theta \sigma L_\theta \right] \\ &= \sup_{\sigma} \sup_{\|L\| \leqslant 1} \inf_{\rho} \sum_{\theta} \lambda_\theta (P_\theta \rho - Q_\theta \sigma) L_\theta \\ &= \sup_{\sigma} \inf_{\rho} \sup_{\|L\| \leqslant 1} \sum_{\theta} \lambda_\theta (P_\theta \rho - Q_\theta \sigma) L_\theta \\ &= \sup_{\sigma} \inf_{\rho} \sum_{\theta} \lambda_\theta \sup_{\|L_\theta\| \leqslant 1} (P_\theta \rho - Q_\theta \sigma) L_\theta \\ &= \sup_{\sigma} \inf_{\rho} \sum_{\theta} \lambda_\theta \|P_\theta \rho - Q_\theta \sigma\|.\end{aligned}$$

Since the sum is affine in L as well as in ρ, the third equality follows from minimax theory, say theorem 3.2.12.

Proof of (vii). Let e denote the distribution on Θ which assigns mass $e_\theta = 1$ to each point θ in Θ. Then $b_{\mathscr{E}}(\lambda, L) = b_{\mathscr{E}}(e, \lambda L)$ for any experiment $\mathscr{E}$ and any loss function L. It follows that

$$\begin{aligned}\underline{\delta}_{,k}(\mathscr{E}, \mathscr{F} \mid \lambda) &= \sup\left\{[b_{\mathscr{E}}(e, L) - b_{\mathscr{F}}(e, L)] : L \in \mathscr{L}_k, \bigvee_{\theta} \|L_\theta\| \lambda_\theta^{-1} = 1\right\} \\ &= \sup\left\{[bu_{\mathscr{F}}(e, U) - bu_{\mathscr{E}}(e, U)] : U \in \mathscr{L}_k, \bigvee_{\theta} \|U_\theta\| \lambda_\theta^{-1} = 1\right\}\end{aligned}$$

and writing $bu_{\mathscr{E}}(e, U) = \int \psi_e(d\mathscr{E}, U)$ and $bu_{\mathscr{F}}(e, U) = \int \psi_e(d\mathscr{F}, U)$, this is the '$\psi$-expression' for $\underline{\delta}_{,k}$.

Put $\tau = \sup\{[b_{\mathscr{E}}(\lambda, L) - b_{\mathscr{F}}(\lambda, L)] : L \in \mathscr{L}_k^{\Sigma}, \|L\| = 1\}$. Then

$$\begin{aligned}\tau &= \sup\left\{[bu_{\mathscr{F}}(e, U) - bu_{\mathscr{E}}(e, U)] : U \in \mathscr{L}_k^{\Sigma}, \bigvee_{\theta} \|U_\theta\| \lambda_\theta^{-1} = 1\right\} \\ &= \sup\left\{\left[\int \psi(d\mathscr{F}) - \int \psi(d\mathscr{E})\right] \Big/ \bigvee_{\theta} [\psi(-e^\theta) \vee \psi(e^\theta)] \lambda_\theta^{-1} : \psi \in \Psi_k^{\Sigma}\right\} \\ &= \sup\left\{\left[\int \psi(d\mathscr{F}) - \int \psi(d\mathscr{E})\right] \Big/ \bigvee_{\theta} [\psi(-e^\theta) + \psi(e^\theta)](2\lambda_\theta)^{-1} : \psi \in \Psi_k^{\Sigma}\right\},\end{aligned}$$

i.e. the three asserted expressions for $\delta_k(\mathscr{E}, \mathscr{F} \mid \lambda)$ are all equal to τ. Hence $\omega \geqslant \tau \Leftrightarrow \bigwedge_\theta [\int \psi(d\mathscr{F}) - \int \psi(d\mathscr{E}) - \sum [\psi(-e^\theta) + \psi(e^\theta)](2\lambda_\theta)^{-1}\omega] \leqslant 0$ when $\psi \in \Psi_k^{\Sigma}$. Letting κ run through Λ and letting ψ run through Ψ_k^{Σ} this may be stated as $\omega \geqslant \tau \Leftrightarrow \bigvee_\psi \bigwedge_\kappa [\ ] \leqslant 0$ where $[\ ] = \int \psi(d\mathscr{F}) - \int \psi(d\mathscr{E}) -$

$\sum_\theta \kappa_\theta \frac{1}{2}[\psi(-e^\theta) + \psi(e^\theta)]\lambda_\theta^{-1}\omega$. Using the fact that Ψ_k^Σ and Λ are both convex and that [ ] is affine in ψ as well as in κ the minimax theory, say theorem 3.2.12, implies that $\bigvee_\psi \bigwedge_\kappa [\ ] = \min_\kappa \bigvee_\psi [\ ]$. By theorem 6.3.4, $\omega \geqslant \tau \Leftrightarrow \min_\kappa \bigvee_\psi [\ ] \leqslant 0 \Leftrightarrow (\kappa_\theta \lambda_\theta^{-1}\omega : \theta \in \Theta)$ belongs to D_k for some $\kappa \in \Lambda$. Here D_k denotes the set of functions ε such that $\mathscr{E}$ is ε-deficient with respect to $\mathscr{F}$ for k-decision problems. It follows that there is a $\kappa \in \Lambda$ such that $(\kappa_\theta \lambda_\theta^{-1}\tau : \theta \in \Theta)$ is in D_k. Hence, by the definition of $\delta_k(\mathscr{E}, \mathscr{F}|\lambda) : \delta_k(\mathscr{E}, \mathscr{F}|\lambda) \leqslant \sum_\theta \kappa_\theta \lambda_\theta^{-1}\tau\lambda_\theta = \tau$. On the other hand $\delta_k(\mathscr{E}, \mathscr{F}|\lambda) = \sum \lambda_\theta \varepsilon_\theta$ for some $\varepsilon \in D_k$. Putting $\kappa_\theta \equiv_\theta \lambda_\theta(\varepsilon_\theta + \gamma)/\sum_\theta \lambda_\theta(\varepsilon_\theta + \gamma)$ where $\gamma > 0$ we find that $\varepsilon_\theta + \gamma \equiv_\theta \kappa_\theta \lambda_\theta^{-1}[\delta_k(\mathscr{E}, \mathscr{F}|\lambda) + \gamma]$. Hence $\delta_k(\mathscr{E}, \mathscr{F}|\lambda) + \gamma \geqslant \tau$. Letting $\gamma \downarrow 0$ we find that $\delta_k(\mathscr{E}, \mathscr{F}|\lambda) \geqslant \tau$ so that $\delta_k(\mathscr{E}, \mathscr{F}|\lambda) = \tau$.

If we had known that $\delta_k(\mathscr{E}, \mathscr{F}|\lambda)$ was positive then we could have put $\gamma = 0$. We introduced the positive quantity γ in order to avoid division by 0 when $\delta_k(\mathscr{E}, \mathscr{F}|\lambda) = 0$.

The proof of (vii) may now be completed by symmetrization.

Proof of (i) and completion of the proof of (ii). The function $\delta_k(\mathscr{E}, \mathscr{F}|\cdot)$ is an infimum of linear functions and hence concave and we know by (ii) that $\underline{\delta}_{,k}(\mathscr{E}, \mathscr{F}|\cdot) \leqslant \delta_k(\mathscr{E}, \mathscr{F}|\cdot)$. Let g be a concave function on Λ which majorizes $\underline{\delta}_{,k}(\mathscr{E}, \mathscr{F}|\cdot)$. Let $\lambda^0 \in \Lambda$ and put $\Theta_0 = \{\theta : \lambda_\theta^0 > 0\}$. We must show that $g(\lambda^0) \geqslant \delta_k(\mathscr{E}, \mathscr{F}|\lambda^0)$. The theorem of weak separation of convex sets implies that there are constants $\tau_\theta : \theta \in \Theta_0$ such that $g(\lambda^0) = \sum\{\lambda_\theta^0 \tau_\theta : \theta \in \Theta\}$ while $g(\lambda) \leqslant \sum\{\lambda_\theta \tau_\theta : \theta \in \Theta_0\}$ when $\lambda \neq \lambda^0$. Thus we might as well have assumed that $\lambda_\theta^0 > 0$ for all θ, so that $\Theta_0 = \Theta$, and that $g(\lambda) \equiv \sum_\theta \lambda_\theta \tau_\theta$ for, necessarily non-negative, constants $\tau_\theta : \theta \in \Theta$. Again letting e denote the prior distribution which assigns mass 1 to all points in Θ we find, by the definition of $\underline{\delta}_{,k}(\mathscr{E}, \mathscr{F}|\lambda)$ that $b_\mathscr{E}(\lambda, L) - b_\mathscr{F}(\lambda, L) \leqslant \bigvee_\theta \|L_\theta\| \sum_\theta \tau_\theta \lambda_\theta$ when $L \in \mathscr{L}_k$ and $\lambda \in \Lambda$, i.e. $b_\mathscr{E}(e, L) - b_\mathscr{F}(e, L) \leqslant \bigvee_\theta \|L_\theta\| \lambda_\theta^{-1} \sum_\theta \tau_\theta \lambda_\theta$; $L \in \mathscr{L}_k$ when $\lambda_\theta > 0$ for all $\theta \in \Theta$. Assuming $\|L_\theta\| > 0$ for all $\theta \in \Theta$ we may put $\lambda_\theta \equiv_\theta \|L_\theta\|/\sum_\theta \|L_\theta\|$ yielding $b_\mathscr{E}(e, L) - b_\mathscr{F}(e, L) \leqslant \sum_\theta \tau_\theta \|L_\theta\|$; $L \in \mathscr{L}_k$. Hence $\mathscr{E}$ is τ-deficient with respect to $\mathscr{F}$ for k-decision problems so that $\delta_k(\mathscr{E}, \mathscr{F}|\lambda^0) \leqslant \sum_\theta \tau_\theta \lambda_\theta^0 = g(\lambda^0)$ for all $\lambda \in \Lambda$. This proves the first statement of (i). Now $\Psi = \Psi(\Theta)$ is a cone so that $\Psi^\Sigma = \Psi$. Hence (vii) implies that $\underline{\delta}(\mathscr{E}, \mathscr{F}|\cdot) = \delta(\mathscr{E}, \mathscr{F}|\cdot)$ so that $\underline{\delta}(\mathscr{E}, \mathscr{F}|\cdot)$ is concave. The proof of (i) and hence the proofs of the remaining statements of (ii) follow from this and the usual symmetrization argument.

Proof of (iv). The number of sets in $\mathscr{B}$ is 2^h where $h \leqslant k$. Choosing σ as the identity map on $(\mathscr{Y}, \mathscr{B})$ and letting $\rho = M$ run through the set of Markov kernels from $\mathscr{E}$ to $(\mathscr{Y}, \mathscr{B})$ we find that $\underline{\delta}_{,k}(\mathscr{E}, \mathscr{F}|\lambda) \geqslant \underline{\delta}_{,h}(\mathscr{E}, \mathscr{F}|\lambda) \geqslant \inf_M \sum_\theta \|P_\theta M - Q_\theta\| \lambda_\theta$. The '$\frac{1}{2}$ Markov kernel criterion', theorem 6.2.4, and the definition of $\delta(\mathscr{E}, \mathscr{F}|\lambda)$ imply that $\inf_M \sum_\theta \|P_\theta M - Q_\theta\| \lambda_\theta \geqslant \delta(\mathscr{E}, \mathscr{F}|\lambda) \geqslant \delta_k(\mathscr{E}, \mathscr{F}|\lambda)$. Hence $\underline{\delta}_{,k}(\mathscr{E}, \mathscr{F}|\lambda) \geqslant \delta_k(\mathscr{E}, \mathscr{F}|\lambda)$ so that $\underline{\delta}_{,k}(\mathscr{E}, \mathscr{F}|\lambda) = \delta_k(\mathscr{E}, \mathscr{F}|\lambda)$.

Proof of (v). This follows from (iv) and (vii) and the observation made after proposition 6.3.16 that to each $\psi \in \Psi_k$ there corresponds a $\mathscr{B}_0$ such that $\int \psi(\mathrm{d}\mathscr{F}) = \int \psi(\mathrm{d}(\mathscr{F}|\mathscr{B}_0))$.

Proof of (vi). The first equality in (vi) is taken from proposition 6.2.40. Then by (ii) it suffices to show that $\delta_k(\mathscr{E}, \mathscr{F}|\cdot) \leqslant \sup_\lambda \underline{\delta}_{,k}(\mathscr{E}, \mathscr{F}|\lambda)$. However, since $\underline{\delta}_{,k}(\mathscr{E}, \mathscr{F}|\cdot)$ is trivially majorized by the constant $\sup_\lambda \underline{\delta}_{,k}(\mathscr{E}, \mathscr{F}|\lambda)$, this is a consequence of (i). □

Corollary 6.3.25 (Similarities between weighted deficiencies and lower weighted deficiencies). *Proposition 6.2.39, the two formulas in proposition 6.2.40 and proposition 6.2.41 remain valid if all weighted deficiencies (deficiency distances) are replaced by the corresponding lower weighted deficiencies (deficiency distances), i.e. if*

$$\delta_k(\mathscr{E}, \mathscr{F}|\lambda) \text{ is replaced by } \underline{\delta}_{,k}(\mathscr{E}, \mathscr{F}|\lambda) \text{ throughout}$$

and

$$\Delta_k(\mathscr{E}, \mathscr{F}|\lambda) \text{ is replaced by } \underline{\Delta}_{,k}(\mathscr{E}, \mathscr{F}|\lambda) \text{ throughout.}$$

There are a few experiments, besides $\mathscr{M}_i$ and $\mathscr{M}_a$, which do not yield better experiments when they are combined with one or more independent replications of themselves. Any such experiment is of the form $\mathscr{M}_a|\pi$ where π is a partitioning of Θ. Thus $\mathscr{M}_a|\pi$ is, and may be any, experiment $(Q_\theta : \theta \in \Theta)$ such that $Q_{\theta_1} = Q_{\theta_2}$ or $Q_{\theta_1} \wedge Q_{\theta_2} = 0$ according to whether θ_1 and θ_2 are equivalent with respect to π or not. If we choose any particular experiment of this form for a given π then another experiment will have this property if and only if it is equivalent to the chosen experiment. This may be argued as we argued the similar statements for the experiments $\mathscr{M}_i$ and $\mathscr{M}_a$ in examples 6.2.29 and 6.2.30. The experiments $\mathscr{M}_i$ and $\mathscr{M}_a$ are particular cases obtained, respectively, by letting π consist of the set Θ only or by letting π consist of all one-point subsets of Θ. In order to simplify the notation we will write $\mathscr{M}_\pi$ instead of $\mathscr{M}_a|\pi$.

The characterization first mentioned implies readily that

$$\mathscr{M}_\pi \times \mathscr{M}_\pi \sim \mathscr{M}_\pi.$$

Conversely, if $\mathscr{F} = \{Q_\theta : \theta \in \Theta\}$ is any experiment such that $\mathscr{F} \times \mathscr{F} \sim \mathscr{F}$ then $\mathscr{F}$ is of this form with π being generated by the equivalence relation on Θ induced by $\mathscr{F}$. This may be seen as follows. Suppose $Q_{\theta_1} \neq Q_{\theta_2}$ and let $\mathscr{E}$ be the dichotomy $(Q_{\theta_1}, Q_{\theta_2})$. Then $\mathscr{E}^2 \sim \mathscr{E}$. Hence by induction, using theorem 6.3.19, we find that $\mathscr{E}^n \sim \mathscr{E}$ for all $n = 1, 2, \ldots$ Thus $\delta(\mathscr{E}^n, \mathscr{M}_a) \equiv_n \delta(\mathscr{E}, \mathscr{M}_a)$. However, we saw in example 6.2.36 that $\delta(\mathscr{E}^n, \mathscr{M}_a) \downarrow 0$ as $n \uparrow \infty$. Hence $\delta(\mathscr{E}, \mathscr{M}_a) = 0$ so that $\mathscr{E} \sim \mathscr{M}_a$. It follows that Q_{θ_1} and Q_{θ_2} are disjoint.

An experiment $\mathscr{F}$ is called *idempotent* if $\mathscr{F}^2 \sim \mathscr{F}$. Thus an experiment is idempotent if and only if it is of the form $\mathscr{F}_\pi$ for some partitioning π of Θ.

Consider any experiment $\mathscr{F} = \{Q_\theta : \theta \in \Theta\}$ and let π be the partitioning of Θ induced by $\mathscr{F}$. Thus two points θ_1, θ_2 in Θ belong to the same set in π if and only if $Q_{\theta_1} = Q_{\theta_2}$. Then $\Delta((\mathscr{F}^n|\Theta_0), \mathscr{M}_\pi|\Theta_0) \to 0$ for any finite subset Θ_0 of Θ. In order to see this, we can reduce Θ_0 to $\tilde{\Theta}_0$ where $\tilde{\Theta}_0$ is a subset of Θ_0 which contains exactly one representative of each point θ in Θ_0. Then, as shown in example 6.2.36, $\Delta((\mathscr{F}^n|\Theta_0), \mathscr{M}_\pi|\Theta_0) = \Delta((\mathscr{F}^n|\tilde{\Theta}_0), \mathscr{M}_a) \downarrow 0$. It follows that to any experiment $\mathscr{F}$ there corresponds an idempotent experiment $\mathscr{M}_\pi$ such that $\Delta((\mathscr{F}^n|\Theta_0), (\mathscr{M}_\pi|\Theta_0)) \to 0$ as $n \to \infty$ for each finite subset Θ_0 of Θ. Thus an experiment $\mathscr{G}$ is idempotent if and only if there is an experiment $\mathscr{F}$ such that $\Delta(\mathscr{F}^n|\Theta_0, \mathscr{G}|\Theta_0) \to 0$ for all finite subsets Θ_0 of Θ.

Let us consider deficiencies with respect to idempotent experiments, i.e. experiments of the form $\mathscr{M}_\pi$.

Proposition 6.3.26 (Comparison with respect to idempotent experiments). *Let π be a partitioning of Θ. Then the experiment $\mathscr{E} = (P_\theta : \theta \in \Theta)$ is ε-deficient with respect to $\mathscr{M}_\pi$ if and only if it admits an estimator $\hat{\theta}$ of θ such that*

$$P_\theta(\hat{\theta} \underset{\pi}{\sim} \theta) \geqslant 1 - \varepsilon_\theta/2; \qquad \theta \in \Theta.$$

If λ is a prior distribution on Θ with finite support then

$$\frac{1}{2}\delta(\mathscr{E}, \mathscr{M}_\pi|\lambda) = 1 - \left\| \bigvee_{U \in \pi} \sum_{\theta \in U} \lambda_\theta P_\theta \right\|$$

and

$$\frac{1}{2}\underline{\delta}_{,k}(\mathscr{E}, \mathscr{M}_a|\lambda) = 1 - \min_\pi \left\| \bigvee_{U \in \pi} \sum_{\theta \in U} \lambda_\theta P_\theta \right\|$$

where π in the last expression runs through the class of partitionings of Θ containing at most k sets.

Proof. It suffices to consider the case where Θ is finite. Take $T = \Theta$ as the decision space and put $L_\theta(t) = 0$ or $= 1$ according to whether $t \sim_\pi \theta$ or not. Then, since $L_\theta(\theta) \equiv_\theta 0$, $\mathscr{M}_\pi$ admits a procedure with risk $= 0$. Thus $\mathscr{E}$ admits a procedure (estimator) with risk $\leqslant \varepsilon/2$. Conversely, such an estimator yields a Markov kernel M from $\mathscr{E}$ to Θ such that $\|P_\theta M - (\delta_\theta \pi)\| \leqslant \varepsilon_\theta$; $\theta \in \Theta$. This completes the proof of the first statement. The expression for $\delta(\mathscr{E}, \mathscr{M}_\pi|\lambda)$ follows by direct evaluation of the minimum Bayes risk in the described decision problem. Finally the expression for $\underline{\delta}_{,k}(\mathscr{E}, \mathscr{M}_a|\lambda)$ follows from this and part (v) of theorem 6.3.24. □

Example 6.3.27 (Lower weighted deficiencies may be strictly less than weighted deficiencies). Consider $\mathscr{M}_i$ and $\mathscr{M}_a$ when $\Theta = \{1, 2, 3\}$. Let λ be a prior distribution on Θ and assume that $\lambda_1 \geqslant \lambda_2 \geqslant \lambda_3$. We saw in example 6.2.44 that $\frac{1}{2}\delta_2(\mathcal{M}_i, \mathcal{M}_a|\lambda) = \frac{1}{2} \wedge (1 - \lambda_1)$. The proposition implies that

$$\begin{aligned}
&\tfrac{1}{2}\underline{\delta}_{,2}(\mathcal{M}_i, \mathcal{M}_a|\lambda) \\
&\quad = 1 - \{[\lambda_1 \vee (\lambda_2 + \lambda_3)] \wedge [\lambda_2 \vee (\lambda_1 + \lambda_3)] \wedge [\lambda_3 \vee (\lambda_1 + \lambda_2)]\} \\
&\quad = 1 - \{[\lambda_1 \vee (\lambda_2 + \lambda_3)] \wedge [\lambda_1 + \lambda_3] \wedge [\lambda_2 + \lambda_3]\} \\
&\quad = 1 - \{[\lambda_1 \vee (\lambda_2 + \lambda_3)] \wedge [\lambda_1 + \lambda_3]\} \\
&\quad = 1 - \{\lambda_1 \vee (\lambda_2 + \lambda_3)\} = (1 - \lambda_1) \wedge \lambda_1.
\end{aligned}$$

Hence $\underline{\delta}_{,2}(\mathcal{M}_i, \mathcal{M}_a|\lambda) < \delta_2(\mathcal{M}_i, \mathcal{M}_a|\lambda)$ when $\lambda_1 < \frac{1}{2}$.

6.4 The randomization criterion

A random variable S on an experiment $\mathscr{E}$ is usually called sufficient if conditional probabilities given S do not depend on the unknown parameter θ. The underlying idea is that the experiment may then be recovered from S by the known chance mechanism defined by the conditional probabilities given S. If we consider observations X and Y which are not necessarily related to the same experiment, and therefore need not have a joint distribution, then this idea still makes sense. Here, as usual, we are assuming that the distributions of X and Y are known up to the unknown parameter θ. We may then say that 'X is at least as informative as Y' or that 'X is sufficient for Y' if it is possible, using a known chance mechanism depending on X only, to derive from X a random variable $\tilde{Y}$ having the same distribution as Y. This chance mechanism may of course be deterministic, i.e. $\tilde{Y}$ may be a function of X.

If it is not possible to find a chance mechanism with this property then we could look for a chance mechanism, depending on X only, yielding a variable $\tilde{Y}$ whose distribution is close to Y in total variation. Thus we arrive at a notion of approximate sufficiency. Now following LeCam, we have already developed a concept of approximations for statistical experiments. How is this concept related to the 'conditional probability' notion of approximate sufficiency?

The conditional probability description of sufficiency given here (the approximative as well as the non-approximative) is sometimes called *the operational approach* to sufficiency. No decision theory is involved in this description. In contrast to the operational approach we have *the approach based on decision theory* and in particular on the concept of risk. This was the main theme of the previous two sections. It is not at all obvious that these approaches should not yield contradictory results. Thus it is most noteworthy and satisfactory that by and large the two approaches yield equivalent notions for comparison and, therefore, for sufficiency, as shown by LeCam. Here is a formulaton of this basic result.

Theorem 6.4.1 (The randomization criterion). *The experiment* $\mathscr{E} = (P_\theta : \theta \in \Theta)$ *is* ε*-deficient with respect to the experiment* $\mathscr{F} = (\mathscr{Y}, \mathscr{B}; Q_\theta : \theta \in \Theta)$ *if and only if there is a transition* M *from* $L(\mathscr{E})$ *to the L-space of bounded additive set functions* $\text{ba}(\mathscr{B})$ *on* $\mathscr{B}$ *such that*

$$\|P_\theta M - Q_\theta\| \leqslant \varepsilon_\theta; \qquad \theta \in \Theta.$$

If a transition with this property exists then it may always be modified so that it maps $L(\mathscr{E})$ *into* $L(\mathscr{F})$ *and therefore is* σ*-continuous.*

Remark. If $\mathscr{E}$ is coherent and if the sample space of $\mathscr{F}$ is Euclidean then by corollary 4.5.8, M can be represented by a Markov kernel from the sample space of $\mathscr{E}$ to the sample space of $\mathscr{F}$.

Proof of the theorem. The 'if' part of the first statement and the last statement was established in theorem 6.2.4 and in the remark made after the formulation of this theorem.

Suppose $\mathscr{E}$ is ε-deficient with respect to $\mathscr{F}$. Then by the definition of deficiency, $\mathscr{E}$ is ε-deficient with respect to $\mathscr{F}$ for k-decision problems for any integer k. By corollary 4.5.14, we may assume that the parameter set Θ is finite. As in the proof of proposition 4.5.1, let N denote the set of all pairs $n = (\Pi, \xi)$ where $\Pi = (B_1, \dots, B_k)$ is an ordered partition of $\mathscr{B}$ into $\mathscr{B}$-measurable sets and $\xi = (y_1, \dots, y_k)$ is a k-tuple such that $y_i \in B_i$; $i = 1, \dots, k$. Direct N by defining $n_1 = (\Pi_1, \xi_1) \geqslant n_2 = (\Pi_2, \xi_2)$ to mean that Π_1 is at least as fine as Π_2. For each $n = ((B_1, \dots, B_k), (y_1, \dots, y_k)) \in N$, let $\mathscr{B}_n$ be the algebra generated by $B_1, \dots, B_k$. Also let σ_n be the map from $\mathscr{Y}$ to $\mathscr{Y}$ which maps y onto y_i when $y \in B_i$; $i = 1, \dots, k$. Then σ_n may be considered as a (non-randomized) decision rule from $\mathscr{F}$ to the k-decision space $(\mathscr{Y}, \mathscr{B}_n)$. Hence, by corollary 6.3.2, there is a decision rule ρ_n from $\mathscr{E}$ to $(\mathscr{Y}, \mathscr{B}_n)$ such that $\|P_\theta \rho_n - Q_\theta \sigma_n\| \leqslant \varepsilon_\theta$ when $\theta \in \Theta$. As usual let $M(\mathscr{Y}, \mathscr{B})$ denote the space of bounded measurable functions on $(\mathscr{Y}, \mathscr{B})$. As noted in the proof of proposition 4.5.1 we have $\|g \circ \sigma_n - g\| \to 0$ when $g \in M(\mathscr{Y}, \mathscr{B})$. Let M_n be the Markov kernel from $\mathscr{E}$s sample space $(\mathscr{X}, \mathscr{A})$ to $(\mathscr{Y}, \mathscr{B})$ defined by

$$M_n(B|x) = \sum_{i=1}^{k} \rho_n(B_i|x) I_B(y_i)$$

when $B \in \mathscr{B}$ and $x \in \mathscr{X}$. If $g \in M(\mathscr{Y}, \mathscr{B})$ then

$$P_\theta M_n(g) = P_\theta \rho_n(g \circ \sigma_n)$$

and

$$Q_\theta(g \circ \sigma_n) = Q_\theta \sigma_n(g \circ \sigma_n).$$

Hence $|(P_\theta M_n)(g) - Q_\theta(g \circ \sigma_n)| \leqslant \|P_\theta \rho_n - Q_\theta \sigma_n\| \, \|g \circ \sigma_n\| \leqslant \|P_\theta \rho_n - Q_\theta \sigma_n\| \, \|g\| \leqslant \varepsilon_\theta \|g\|$ when $\theta \in \Theta$.

By the compactness theorem 4.5.13 there is a transition M from $L(\mathscr{E})$ to $(\mathscr{Y}, \mathscr{B})$ and a subnet $(M_{n'})$ of $(M_n : n \in N)$ such that $\lambda M_{n'} g \to \lambda M g$ when the finite measure λ is absolutely continuous with respect to $\sum_\theta P_\theta$ and $g \in M(\mathscr{Y}, \mathscr{B})$. Let $g \in M(\mathscr{Y}, \mathscr{B})$. Then $|P_\theta M g - Q_\theta g| = \lim_{n'} |P_\theta M_{n'} g - Q_\theta(g \circ \sigma_{n'})| \leqslant \varepsilon_\theta \|g\|$ when $\theta \in \Theta$. Hence $\|P_\theta M - Q_\theta\| \leqslant \varepsilon_\theta$ when $\theta \in \Theta$. □

It is sometimes convenient to work with non-negative measures with total variation (mass) being at most 1. Thus Helly's compactness theorem tells us that the set of such measures on the real line is compact for the topology of convergence for continuous functions with compact supports. In general a non-negative measure μ on a σ-algebra $\mathscr{A}$ of subsets of a set $\mathscr{X}$ is called a *sub-probability* on $\mathscr{A}$ if $\mu(\mathscr{X}) \leqslant 1$. Similarly a *sub-Markov kernel* (*or sub-randomization*) from a measurable space $(\mathscr{X}, \mathscr{A})$ to the measurable space $(\mathscr{Y}, \mathscr{B})$ is a map $x \to M(\cdot|x)$ from $\mathscr{X}$ to the set of sub-probabilities on $(\mathscr{Y}, \mathscr{B})$ such that $M(B|\cdot)$ is measurable for each $B \in \mathscr{B}$.

If M is a sub-Markov kernel from $(\mathscr{X}, \mathscr{A})$ to $(\mathscr{Y}, \mathscr{B})$ then it defines a non-negative and linear contraction from the set of finite measures on $(\mathscr{X}, \mathscr{A})$ into the set of finite measures on $(\mathscr{Y}, \mathscr{B})$. In general a non-negative and linear contraction from an L-space L_1 to another L-space L_2 will be called a *sub-transition*. Thus T is a sub-transition from L_1 to L_2 if and only if T is linear, $T((L_1)_+) \subseteqq (L_2)_+$ and $\|\mu T\| \leqslant \|\mu\|$ when $\mu \in L_1$.

In chapter 3 we saw how under regularity conditions transitions could be representable by Markov kernels. The same arguments show that under the same regularity conditions sub-transitions can be representable by sub-Markov kernels.

Let $\mathscr{Y}$ be a σ-compact locally compact Hausdorff topological space and let $\mathscr{B}$ be the class of Borel subsets of $\mathscr{Y}$. (The class of Borel sets $=$ the σ-algebra generated by the open sets.) Let $C_0(\mathscr{Y})$ denote the M-normed space of continuous functions with compact supports. Then, by the appropriate Riesz representation theorem, the set of sub-transitions from the L-space L to $(\mathscr{Y}, \mathscr{B})$ is compact for the coarsest topology which for each $\mu \in L$ and each $g \in C_0(\mathscr{Y})$ makes the functional $T \to \mu T g$ continuous. However, the set of transitions is not compact for this topology unless $\mathscr{Y}$ is compact. Thus, if compactness is considered an advantage, then it might be an advantage to permit sub-transitions. In the theory of comparison of invariant experiments we will also see that there may be invariant sub-transitions with the desired properties but no invariant transitions with these properties. In such circumstances the following alternative formulation of the Markov kernel criterion may be helpful.

Corollary 6.4.2 (The sub-randomization criterion for deficiencies). *The experiment $\mathscr{E} = (P_\theta : \theta \in \Theta)$ is ε-deficient with respect to the experiment $\mathscr{F} = (\mathscr{Y}, \mathscr{B}, Q_\theta : \theta \in \Theta)$ if and only if there is a sub-transition M from $L(\mathscr{E})$ to* ba$(\mathscr{Y}, B)$ *such that*

$$\|(P_\theta M - Q_\theta)^-\| \leqslant \varepsilon_\theta/2; \qquad \theta \in \Theta.$$

Proof. Suppose $\tilde{M}$ is a sub-transition from $L(\mathscr{E})$ to $\mathrm{ba}(\mathscr{Y},\mathscr{B})$ such that $\|(P_\theta\tilde{M} - Q_\theta)^-\| \leqslant \varepsilon_\theta/2$; $\theta \in \Theta$. Then $\tilde{M}$ is a linear non-negative contraction from $L(\mathscr{E})$ to $(\mathscr{Y},\mathscr{B})$. Let $\tilde{M}^*$ denote the adjoint map of $\tilde{M}$ from $M(\mathscr{Y},\mathscr{B})$ to $M(\mathscr{E})$. Let N be any transition from $L(\mathscr{E})$ to $\mathrm{ba}(\mathscr{Y},\mathscr{B})$ and denote by N^* its adjoint map from $M(\mathscr{Y},\mathscr{B})$ to $M(\mathscr{B})$. Finally, for each $\mu \in L(\mathscr{E})$ and each $v \in M(\mathscr{E})$ let the linear functional $h \to \mu(hv)$ on $M(\mathscr{E})$ be denoted as $v\mu$. Trivially $v\mu \in L(\mathscr{E})^{**} = M(\mathscr{E})^*$. However, since the image of L by the evaluation map is actually a band and since $v\mu$ clearly belongs to the principal band generated by μ, it follows that $v\mu \in L$.

We may now define a transition M by putting

$$\mu M = \mu\tilde{M} + [(\dot{1} - \tilde{M}^*1)\mu]N$$

when $\mu \in L(\mathscr{E})$. Here $\dot{1}$ is the unit in $M(\mathscr{E})$.

Let $\theta \in \Theta$. Then $(P_\theta M - Q_\theta)^- \leqslant (P_\theta\tilde{M} - Q_\theta)^-$ so that $\|(P_\theta M - Q_\theta)^-\| \leqslant \varepsilon_\theta/2$. Hence, since M is a transition $\|P_\theta M - Q_\theta\| = 2\|(P_\theta M - Q_\theta)^-\| \leqslant \varepsilon_\theta$.

The corollary follows now by the trivial observation that a transition is a particular kind of a sub-transition. □

If $\mathscr{E}$ is coherent and if the sample space of $\mathscr{F}$ is Euclidean then the deficiency of $\mathscr{E}$ with respect to $\mathscr{F}$ may be described directly in terms of the existence of random variables realizing these experiments.

Corollary 6.4.3 (Deficiencies in terms of random variables). *Let the experiment $\mathscr{E} = (\mathscr{X},\mathscr{A}; P_\theta : \theta \in \Theta)$ be coherent and assume also that the sample space $(\mathscr{Y},\mathscr{B})$ of the experiment $\mathscr{F} = (\mathscr{Y},\mathscr{B}; Q_\theta : \theta \in \Theta)$ is Euclidean. Then $\mathscr{E}$ is ε-deficient with respect to $\mathscr{F}$ if and only if there is an experiment and three random variables X, Y and $\tilde{Y}$ defined on this experiment such that:*

(i) *X realizes $\mathscr{E}$, i.e $\mathscr{L}(X|\theta) \equiv_\theta P_\theta$*
(ii) *Y realizes $\mathscr{F}$, i.e. $\mathscr{L}(Y|\theta) \equiv_\theta Q_\theta$*
(iii) *$Pr(Y \neq \tilde{Y}|\theta) \leqslant \varepsilon_\theta/2; \theta \in \Theta$*

and

(iv) *X is sufficient for $\tilde{Y}$, i.e. the conditional distribution of $\tilde{Y}$ given X may be specified independently of θ.*

Remark. An instructive way of expressing that an experiment $\mathscr{F}$ is at most as informative as an experiment $\mathscr{E}$ realized by observing a variable X is as follows.

Assume first that a variable U which is independent of X and has a known distribution is available. (U may have been produced by some chance mechanism having a known probability law.) Then any experiment $\mathscr{F}$ which is obtained by observing a variable of the form $h(X, U)$ for a (jointly measurable) function h is clearly at most as informative as $\mathscr{E}$.

Conversely, any experiment $\mathscr{F}$ with Euclidean sample space which is at most as informative as $\mathscr{E}$ is obtainable in this way, provided $\mathscr{E}$ is coherent. In fact we can always arrange matters so that U becomes uniformly distributed on $[0, 1]$. This follows from:

(i) any non-countable Euclidean sample space is Borel isomorphic to $[0, 1]$;

and

(ii) if F is a cumulative distribution function on the real line then $F^{-1}(U) = \inf\{x : F(x) \geqslant U\}$ has distribution F provided U is uniformly distributed on $[0,1]$.

In particular it follows that the variable $\tilde{Y}$ in the corollary may be assumed to be of the form $\tilde{Y} = h(X, U)$ where U is uniformly distributed on $[0, 1]$.

Proof. (i) Assume such variables exist. Let M be a conditional distribution of $\tilde{Y}$ given X, which does not depend on θ. If $B \in \mathscr{B}$ then $(P_\theta M)(B) - Q_\theta(B) = Pr(\tilde{Y} \in B|\theta) - Pr(Y \in B|\theta) \leqslant E_\theta|I_B(\tilde{Y}) - I_B(Y)| \leqslant Pr(Y \neq \tilde{Y}|\theta) \leqslant \varepsilon_\theta/2$. Thus $\|P_\theta M - Q_\theta\| \leqslant \varepsilon_\theta$; $\theta \in \Theta$. (ii) Assume $\mathscr{E}$ is ε-deficient with respect to $\mathscr{F}$. Then the regularity conditions, together with the theorem imply that $\|P_\theta M - Q_\theta\| \leqslant \varepsilon_\theta$; $\theta \in \Theta$ for some Markov kernel M from $(\mathscr{X}, \mathscr{A})$ to $(\mathscr{Y}, \mathscr{B})$. From the joint distribution characterization of statistical distance in chapter 1 it follows that there is a joint distribution H_θ on $\mathscr{B} \times \mathscr{B}$, with marginals $P_\theta M$ and Q_θ which assigns at least mass $1 - \varepsilon_\theta/2$ to the diagonal $\{(y, y) : y \in \mathscr{Y}\}$. Let S_θ be a regular conditional distribution of the second marginal given the first marginal fitted to the joint distribution H_θ. Define random variables X, $\tilde{Y}$ and Y on some experiment such that $\mathscr{L}(X|\theta) \equiv_\theta P_\theta$, $\mathscr{L}(\tilde{Y}|X) = M(\cdot|X)$ and $\mathscr{L}(Y|\tilde{Y}, X) = S_\theta(\cdot|\tilde{Y})$. Then $\mathscr{L}(Y, \tilde{Y}|\theta) = H_\theta$ so that $Pr(Y \neq \tilde{Y}|\theta) \leqslant \varepsilon_\theta/2$ and $\mathscr{L}(Y|\theta) = Q_\theta$ when $\theta \in \Theta$. □

Corollary 6.4.4 (Comparison for general decision spaces). *The experiment $\mathscr{E} = (P_\theta : \theta \in \Theta)$ is ε-deficient with respect to $\mathscr{F} = (Q_\theta : \theta \in \Theta)$ if and only if to each decision space $(T, \mathscr{S})$ and each loss function L and each generalized decision rule σ in $\mathscr{F}$ there corresponds a decision rule ρ in $\mathscr{E}$ such that*

$$P_\theta \rho L_\theta \leqslant Q_\theta \sigma L_\theta + \varepsilon_\theta \|L_\theta\|; \qquad \theta \in \Theta.$$

If $\mathscr{E}$ is ε-deficient with respect to $\mathscr{F}$ then ρ may be chosen independently of L so that

$$\|P_\theta \rho - Q_\theta \sigma\| \leqslant \varepsilon_\theta; \qquad \theta \in \Theta.$$

Proof. The 'if' follows from proposition 6.2.7. Suppose now that $\mathscr{E}$ is ε-deficient with respect to $\mathscr{F}$. Then $\|P_\theta M - Q_\theta\| \leqslant \varepsilon_\theta$; $\theta \in \Theta$ for a transition M from $L(\mathscr{E})$ to $L(\mathscr{F})$. Put $\rho = M\sigma$. Then ρ is a transition from $L(\mathscr{E})$ to $(T, \mathscr{S})$ such that $\|P_\theta \rho - Q_\theta \sigma\| = \|(P_\theta M)\sigma - Q_\theta \sigma\| \leqslant \|P_\theta M - Q_\theta\| \leqslant \varepsilon_\theta$; $\theta \in \Theta$. □

Restricting our attention to constant functions ε on Θ we obtain the following minimax expression for the deficiency.

Corollary 6.4.5 (Deficiencies in terms of transitions). *The deficiency $\delta(\mathscr{E},\mathscr{F})$ of the experiment $\mathscr{E} = (P_\theta : \theta \in \Theta)$ with respect to the experiment $\mathscr{F} = (Q_\theta : \theta \in \Theta)$ is*

$$\delta(\mathscr{E},\mathscr{F}) = \min_M \sup_\theta \|P_\theta M - Q_\theta\|$$

where M runs through the set of all transitions from $L(\mathscr{E})$ to $L(\mathscr{F})$.

Proof. This follows from theorem 6.4.1 and proposition 6.2.15 which ensures that $\inf_M = \min_M$. □

The randomization criterion, together with proposition 6.2.15 and theorem 6.3.24, yields the following characterization of weighted deficiencies.

Corollary 6.4.6 (Weighted deficiencies in terms of transitions). *Let $\mathscr{E} = (P_\theta : \theta \in \Theta)$ and $\mathscr{F} = (\mathscr{Y},\mathscr{B}; Q_\theta : \theta \in \Theta)$ be two experiments and let λ be a prior probability distribution on Θ with finite support. Then the λ-weighted deficiency of $\mathscr{E}$ with respect to $\mathscr{F}$ and the lower λ-weighted deficiency of $\mathscr{E}$ with respect to $\mathscr{F}$ are both equal to the number $\min_M \sum_\theta \lambda_\theta \|P_\theta M - Q_\theta\|$, i.e.*

$$\delta(\mathscr{E},\mathscr{F}\,|\,\lambda) = \underline{\delta}(\mathscr{E},\mathscr{F}\,|\,\lambda) = \min_M \sum_\theta \lambda_\theta \|P_\theta M - Q_\theta\|.$$

Here the transition M varies freely within the set of all transitions from $L(\mathscr{E})$ to $L(\mathscr{F})$.

Using the fact that $\sum \lambda_\theta \|P_\theta M - Q_\theta\|$ is affine–convex in (λ, M) we obtain one more corollary.

Corollary 6.4.7. *Let $\mathscr{E}$, $\mathscr{F}$ and λ be as in the previous corollary. Assume that the prior distribution λ and the transition M are such that*

$$\|P_\theta M - Q_\theta\| = \sup_\theta \|P_\theta M - Q_\theta\|$$

when $\lambda_\theta > 0$. Then

$$\delta(\mathscr{E},\mathscr{F}) = \delta(\mathscr{E},\mathscr{F}\,|\,\lambda) = \sup_\theta \|P_\theta M - Q_\theta\|.$$

If Θ is finite then the deficiency $\delta(\mathscr{E},\mathscr{F})$ may always be found in this way.

6.5 Reduction by invariance

The task of determining, or estimating, deficiencies on the basis of the randomization criterion is usually a formidable one. This is so even for apparently quite ‘simple’ experiments. Sometimes however the problem may be considerably reduced by invariance. The idea is that if we are comparing experiments which exhibit the same kind of symmetries then we might expect that our search for transitions might be limited to transitions exhibiting similar symmetries.

If this reduction is permissible then we shall see that invariant procedures may be matched with invariant procedures. The results described here therefore generalize in some respect the theory in the Hall, Wijsman and Ghosh paper (1965) on the exchangeability of reduction by invariance and reduction by sufficiency. Another natural example of the theory is the Hunt–Stein theorem on minimax tests.

The first general result on reduction by invariance within the framework of comparison of experiments may be found in Boll's thesis (1955). The exposition given here owes much to this paper, but also to LeCam's 1964 paper, a paper by Day (1961), and Greenleaf's book (1969). A useful survey of several aspects of the general theory of invariance in statistics is the paper by Bondar & Milnes (1981).

In order to provide some motivation we shall begin by giving an informal description of *invariant decision problems.*

Assume we are given a statistical decision problem consisting of an experiment $\mathscr{E}$ which is realized by observing an $\mathscr{X}$-valued variable X, a set T of possible decisions and a loss function L which to each theory $\theta \in \Theta$ and each decision $t \in T$ assigns the loss $L_\theta(t)$.

Let $g_{\mathscr{X}}$ and g_T be 1–1 maps of, respectively, $\mathscr{X}$ and T onto themselves and let g_Θ be a map from Θ into itself. In order to simplify the notation we will write $g(X)$, $g(t)$ and $g(\theta)$ instead of $g_{\mathscr{X}}(X)$, $g_T(t)$ and $g_\Theta(\theta)$. Assume that $\mathscr{E}$ satisfies the invariance condition

$$\mathscr{L}(g(X)|\theta) \underset{\theta}{\equiv} \mathscr{L}(X|g(\theta))$$

and that L satisfies the invariance condition

$$L_{g(\theta)}(g(t)) \underset{\theta}{\equiv} L_\theta(t); \qquad t \in T.$$

Then a statistician having observed X, facing a theory θ and a decision t is essentially in the same situation as a statistician having observed $g(X)$, facing a theory $g(\theta)$ and a decision $g(t)$. If the first statistician chooses the decision rule ρ_1 and the other statistician chooses the decision rule ρ_2 then a requirement of *consistency in method* may be formalized by the equation $g_T \circ \rho_1 = \rho_2$. If we also want *consistency in conclusion*, up to known chance mechanisms, then we arrive at the requirement $\rho_1 \circ g_{\mathscr{X}} = \rho_2$. Combining these two principles we obtain the *invariance condition, or invariance principle*, that

$$g_T \circ \rho_1 = \rho_1 \circ g_{\mathscr{X}}$$

i.e. that the 'operation g' and the decision rule ρ_1 commute. Of course the invariance condition may also be written

$$\rho_1 = g_T^{-1} \circ \rho_1 \circ g_{\mathscr{X}}.$$

The full effect of the invariance principle is obtained by applying it to groups of transformations. As we want to keep the group structure separated from the decision problem we shall assume that the maps $g_{\mathscr{X}}$, g_T and g_Θ are images of a group element g belonging to a specified group G. We shall express this by saying that G acts on the sets $\mathscr{X}$, T and Θ by, respectively, the maps $g \to g_{\mathscr{X}}$, $g \to g_T$ and $g \to g_\Theta$. The group operations should then correspond to functional composition on the spaces $\mathscr{X}$ and T, i.e. we require that

$$(gh)_{\mathscr{X}} = g_{\mathscr{X}} \circ h_{\mathscr{X}} \quad \text{and} \quad (gh)_T = g_T \circ h_T$$

when $g, h \in G$.

Before dipping into the mathematics let us remind ourselves that a map ϕ from a measurable space $(Z_1, \mathscr{C}_1)$ onto a measurable space $(Z_2, \mathscr{C}_2)$ is called a *Borel isomorphism from* $(Z_1, \mathscr{C}_1)$ onto $(Z_2, \mathscr{C}_2)$ if ϕ is a 1–1 measurable map from Z_1 onto Z_2 possessing a measurable inverse. A Borel isomorphism of a measurable space $(Z, \mathscr{C})$ onto $(Z, \mathscr{C})$ is also called a *Borel isomorphism of* $(Z, \mathscr{C})$ *onto itself*. It is easily checked that the set of Borel isomorphisms of $(Z, \mathscr{C})$ onto itself constitutes a subgroup of the group of 1–1 maps of Z onto itself. As we shall have to deal with homomorphisms from or to such groups let us stick with the standard notation $\phi \circ \psi$ for the composition of the map ψ from Z_1 to Z_2 with the map ϕ from Z_2 to Z_3. Thus $(\phi \circ \psi)(z_1) = \phi(\psi(z_1))$ when $z_1 \in Z_1$. We may sometimes find it more convenient to write $z_1 \psi \phi$ or $\phi \psi z_1$ instead of $\phi \circ \psi(z_1)$. The point to notice is that a map J from a semigroup G into a semigroup of maps of a given set Z into itself is called a *homomorphism* if $J(g_1 g_2) = J(g_1) \circ J(g_2)$ when $g_1, g_2 \in G$. If J instead satisfies the requirement $J(g_1 g_2) = J(g_2) \circ J(g_1)$ then J is called an *antihomomorphism*. Using the notation $z\psi\phi$ for $\phi(\psi(z))$ a homomorphism will appear as an antihomomorphism and vice versa.

If G is a semigroup and J is a functon from G to the set of functions of a set Z into itself then we shall say that *G acts on Z by J*. If G acts on Z by J then a subset of Z is called an *orbit* if it is of the form $\{J(g)(z) : g \in G\}$ for some $z \in Z$. A function on Z is called *orbit invariant* if it is constant on orbits.

Now consider two experiments $\mathscr{E} = (\mathscr{X}, \mathscr{A}; P_\theta : \theta \in \Theta)$ and $\mathscr{F} = (\mathscr{Y}, \mathscr{B}; Q_\theta : \theta \in \Theta)$ and a group G. Assume that G acts on the sets $\mathscr{X}$, $\mathscr{Y}$ and Θ by maps $g \to g_{\mathscr{X}}$, $g \to g_{\mathscr{Y}}$ and $g \to g_\Theta$ such that:

(i) the map $g \to g_{\mathscr{X}}$ is a homomorphism from G to the group of Borel isomorphisms of $\mathscr{X}$.

(ii) the map $g \to g_{\mathscr{Y}}$ is a homomorphism from G to the group of Borel isomorphisms of $\mathscr{Y}$.

(iii) $P_\theta g_{\mathscr{X}}^{-1} = P_{g_\Theta(\theta)}$, and $Q_\theta g_{\mathscr{Y}}^{-1} = Q_{g_\Theta(\theta)}$ when $\theta \in \Theta$ and $g \in G$.

We shall then say that *the pair* $(\mathscr{E}, \mathscr{F})$ *of experiments is invariant under G acting by the maps* $g \to g_{\mathscr{X}}$, $g \to g_{\mathscr{Y}}$ *and* $g \to g_{\Theta}$.

In order to simplify the notation, let us agree to delete the subscripts $\mathscr{X}$, $\mathscr{Y}$ and Θ on $g \in G$ when misunderstandings are not likely to occur. Thus we may write $g(x)$, $g(y)$ and $g(\theta)$ instead of $g_{\mathscr{X}}(x)$, $g_{\mathscr{Y}}(y)$ and $g_{\Theta}(\theta)$. The basic requirement (iii) then becomes

$$P_\theta g^{-1} = P_{g(\theta)} \quad \text{and} \quad Q_\theta g^{-1} = Q_{g(\theta)}, \quad \text{when } \theta \in \Theta \text{ and } g \in G.$$

Let us now assume that $(\mathscr{E}, \mathscr{F})$ is invariant under G as described above.

If M is a Markov kernel from $(\mathscr{X}, \mathscr{A})$ to $(\mathscr{Y}, \mathscr{B})$ and $g \in G$ then we may define another Markov kernel M^g from $(\mathscr{X}, \mathscr{A})$ to $(\mathscr{Y}, \mathscr{B})$ by putting

$$M^g(B|x) = M(g[B]|g(x))$$

when $B \in \mathscr{B}$ and $x \in \mathscr{X}$.

Thus M^g is obtained as the resultant of the three kernels

$$x \to g(x) \qquad \text{from } \mathscr{X} \text{ to } \mathscr{X}$$

$$x \to M(\mathrm{d}y|x) \qquad \text{from } \mathscr{X} \text{ to } \mathscr{Y}$$

and

$$y \to g^{-1} \qquad \text{from } \mathscr{Y} \text{ to } \mathscr{Y}.$$

In other words,

$$M^g = g^{-1} \circ M \circ g.$$

We might note in passing that the fact that M^g is a Markov kernel rests on the assumption that the map $B \to g[B]$ is an isomorphism of the σ-algebra $\mathscr{B}$ into itself.

Let μ be any finite measure on $\mathscr{A}$ and let ν be any finite measure on $\mathscr{B}$. Then $\mu M^g(B) = ((\mu g^{-1})M)(g[B])$, while $\nu(B) = (\nu g^{-1})(g[B])$. Hence $\|\mu M^g - \nu\| = \|(\mu g^{-1})M - \nu g^{-1}\|$. In particular $\|P_\theta M^g - Q_\theta\| = \|P_{g(\theta)}M - Q_{g(\theta)}\| \leqslant \sup_g \|P_{g(\theta)}M - Q_{g(\theta)}\|$.

Assume now that $\|P_\theta M - Q_\theta\| \leqslant \varepsilon_\theta$; $\theta \in \Theta$ where ε is *invariant*, i.e. $\varepsilon_{g(\theta)} = \varepsilon_\theta$ when $g \in G$ and $\theta \in \Theta$. Then $\|P_\theta M^g - Q_\theta\| \leqslant \varepsilon_\theta$ for all $\theta \in \Theta$. Thus M^g is just as good as M for deciding ε-deficiency in this case.

Before proceeding note that $M^{gh} = [M^g]^h$ and that the map $M \to M^g$ is affine.

For simplicity suppose that the group G is a finite group possessing n elements. Put $\overline{M} = (1/n)\sum\{M^g : g \in G\}$. Then $\overline{M}$ has the invariance property

$$\overline{M}^h = \overline{M} \quad \text{for each} \quad h \in H.$$

It follows from the convexity of the norm that $\|P_\theta \overline{M} - Q_\theta\| \leqslant \sup_g \|P_{g(\theta)}M - Q_{g(\theta)}\|$. Hence in this case we could restrict our attention to

Markov kernels M possessing the invariance property

$$M^h = M \quad \text{when} \quad h \in H.$$

Let us agree to say that a Markov kernel possessing this property is *invariant*.

In general we would like to permit ourselves to use the term $\mathscr{E}$-invariant for Markov kernels which induce invariant transitions. Although so far we have not defined invariant transitions, we might as well give the definition of an $\mathscr{E}$-invariant Markov kernel now, as follows.

A Markov kernel M from $(\mathscr{X}, \mathscr{A})$ to $(\mathscr{Y}, \mathscr{B})$ is called *$\mathscr{E}$-invariant* if for each $g \in G$, M^g induces the same transition as M. Thus M is $\mathscr{E}$-invariant if and only if $M(g[B]|g(x)) = M(B|x)$ for P_θ-almost all x whenever $\theta \in \Theta$, $B \in \mathscr{B}$, and $g \in G$. By the way, it is easily seen that the exceptional set may be chosen independently of B when $\mathscr{B}$ is separable. If there is a set $N \in \mathscr{A}$ such that:

(i) N is invariant, i.e. $g[N] \equiv_g N$
(ii) N is an $\mathscr{E}$-null set, i.e. $P_\theta(N) \equiv_\theta 0$
(iii) $M^g \equiv_g M$ on N^c, i.e. $M^g(B|x) \equiv_g M(B|x)$ when $B \in \mathscr{B}$ and $x \notin N$

then we shall say that M is *uniformly $\mathscr{E}$-invariant*. Thus M is uniformly $\mathscr{E}$-invariant if the exceptional $\mathscr{E}$-null set may be chosen invariant and independent of $B \in \mathscr{B}$ and $g \in G$. The requirement that M is invariant amounts to the requirement that M is uniformly $\mathscr{E}$-invariant with the exceptional set being the empty set.

Observe that if there is at least one invariant Markov kernel (there might be none) then any uniformly $\mathscr{E}$-invariant Markov kernel is equivalent to an invariant Markov kernel. Note also that the notions of $\mathscr{E}$-invariance and of uniform $\mathscr{E}$-invariance are both meaningful whether or not an experiment $\mathscr{F}$ with $(\mathscr{Y}, \mathscr{B})$ as sample space is given. The notion of invariance of a Markov kernel M from $(\mathscr{X}, \mathscr{A})$ to $(\mathscr{Y}, \mathscr{B})$ does not even require that experiments are defined.

What can we do with invariant kernels? First of all they can be used to match invariant procedures with invariant procedures. Thus suppose again that the pair $(\mathscr{E}, \mathscr{F})$ of experiments is invariant under the actions of the group G as described above and that $\|P_\theta M - Q_\theta\| \leqslant \varepsilon_\theta$; $\theta \in \Theta$ for some invariant kernel M and some orbit invariant function ε on Θ. Let $(T, \mathscr{S})$ be any decision space such that G acts on T by the homomorphism $g \to g_T$ from G to the group of Borel isomorphisms of $(T, \mathscr{S})$ to itself. Assume now that $\mathscr{F}$ permits a particular invariant decison rule σ. Then $\rho = \sigma \circ M$ is an invariant decision rule in $\mathscr{E}$ whose performance function is ε close to the performance function of σ, i.e.

$$\|P_\theta \rho - Q_\theta \sigma\| \leqslant \varepsilon_\theta; \qquad \theta \in \Theta.$$

The following two notions will be convenient when we discuss non-randomized Markov kernels.

Let $(Z, \mathscr{C})$ be a measurable space. Then we shall say that $\mathscr{C}$ separates the points $z_1, z_2 \in Z$ if there is a set $C \in \mathscr{C}$ such that $z_1 \in C$ while $z_2 \notin C$. If $\mathscr{C}$ separates z_1, z_2 whenever z_1 and z_2 are distinct points in Z then we shall say that $\mathscr{C}$ *separates (distinguishes) the points of* Z. Points z_1 and z_2 in Z are called *equivalent with respect to* $\mathscr{C}$ if $\mathscr{C}$ does not separate them. We shall write $z_1 \sim z_2$ *mod* $(\mathscr{C})$ if $\mathscr{C}$ does not separate z_1 and z_2. Thus $z_1 \sim z_2 \bmod (\mathscr{C})$ if and only if $g(z_1) = g(z_2)$ whenever g is a real valued measurable function on Z. This notion of equivalence is then a proper equivalence relation. Clearly $\mathscr{C}$ distinguishes the points of Z if and only if all the equivalence classes for this equivalence relation are one-point sets.

Let us write out these notions explicitly for non-randomized Markov kernels, i.e. for measurable functions. Thus if s is a measurable function from $(\mathscr{X}, \mathscr{A})$ to $(\mathscr{Y}, \mathscr{B})$ then we shall say that s is invariant or uniformly $\mathscr{E}$-invariant or uniformly invariant according to whether the Markov kernel determined by s has these properties. Now if the Markov kernel M is defined by the function s, i.e. $M(B|x) = I_B(s(x))$ when $x \in \mathscr{X}$ and $B \in \mathscr{B}$, then M^g is defined by the function $g^{-1} \circ s \circ g$. Thus, assuming that $\mathscr{B}$ distinguishes the points of $\mathscr{Y}$ we arrive at the following notions.

(i) s is *$\mathscr{E}$-invariant* if and only if to each $g \in G$ there corresponds an $\mathscr{E}$-null set $N \in \mathscr{A}$ such that $g(s(x)) \sim s(g(x)) \bmod (\mathscr{B})$ when $x \notin N$.

(ii) s is *uniformly $\mathscr{E}$-invariant* if and only if there is an $\mathscr{E}$-null set $N \in \mathscr{A}$ such that $g(s(x)) \sim s(g(x)) \bmod (\mathscr{B})$ for all $g \in G$ when $x \notin N$. (The set N is then necessarily $\mathscr{E}$-invariant, i.e. $g[N] \equiv_g N$.)

(iii) s is *invariant* if and only if G commutes with s in the sense that $g(s(x)) \sim s(g(x)) \bmod (\mathscr{B})$ when $x \in \mathscr{X}$ and $g \in G$.

If there is at least one measurable and invariant function from $(\mathscr{X}, \mathscr{A})$ to $(\mathscr{Y}, \mathscr{B})$ then any measurable and uniformly $\mathscr{E}$-invariant function is equivalent to a measurable invariant function.

If $\mathscr{B}$ distinguishes the points in $\mathscr{Y}$ then '$\sim$' may here be replaced by '$=$' provided that at the same time all the qualifications 'mod $(\mathscr{B})$' are deleted.

The reader should here note that the terminology used in the literature varies. Several authors prefer the notion of equivariance to our notion of invariance. These authors usually reserve the notion of invariance for the particular case when G acts on the range space by the identity map, i.e. when $g(y) \equiv_y y; g \in G$. The notion of almost invariance is then usually used instead of our notion of $\mathscr{E}$-invariance. In the particular case where G acts on $\mathscr{Y}$ by the identity map, any constant function becomes invariant. Thus any uniformly $\mathscr{E}$-invariant Markov kernel (measurable function) is in this case equivalent to an invariant Markov kernel (measurable function), i.e. a Markov kernel (measurable function) which is constant up to $\mathscr{B}$-equivalence on orbits of G in $\mathscr{X}$.

Here we shall use the convention that *if the action of G on $\mathscr{Y}$ is not specifically mentioned then it should be assumed that G acts on $\mathscr{Y}$ by the identity map, i.e. that $g(y) \equiv_y y$ when $g \in G$.*

Later on, when simultaneously considering sufficiency and invariance, we will need to know how invariance appears in terms of the induced σ-algebras. The proposition below is quite technical, although not difficult, and the reader may do well to skip it until it is needed.

The situation we shall consider is the following.

Let the group G act on the measurable space $(\mathscr{X}, \mathscr{A})$ by the homomorphism $g \to g_{\mathscr{X}}$ from G to the group of Borel isomorphisms of $(\mathscr{X}, \mathscr{A})$. Say that a sub σ-algebra $\mathscr{S}$ of $\mathscr{A}$ *respects* G if $g_{\mathscr{X}}$, for each $g \in G$, is a Borel isomorphism of $(\mathscr{X}, \mathscr{S})$ into itself, i.e. if

$$g_{\mathscr{X}}^{-1}\mathscr{S} = \mathscr{S}; \qquad g \in G.$$

Note that if $\mathscr{S}$ respects G then $g(x_1) \sim g(x_2) \bmod (\mathscr{S})$ whenever $x_1 \sim x_2 \bmod (\mathscr{S})$.

Proposition 6.5.1 (Invariance in terms of induced set algebras). *Let G act on the measurable space $(\mathscr{X}, \mathscr{A})$ as described above and let s be a measurable function from $(\mathscr{X}, \mathscr{A})$ to the measurable space $(\mathscr{Y}, \mathscr{B})$. Let $\mathscr{S} = s^{-1}(\mathscr{B})$ be the sub σ-algebra of $\mathscr{A}$ induced by s.*

Then $\mathscr{S}$ respects G whenever G acts on $(\mathscr{Y}, \mathscr{B})$ by a homomorphism $g \to g_{\mathscr{Y}}$ from G to the group of Borel isomorphisms of $(\mathscr{Y}, \mathscr{B})$ such that s becomes invariant, i.e. that $g_{\mathscr{Y}}(s(x)) \sim s(g_{\mathscr{X}}(x)) \bmod (\mathscr{B})$ when $g \in G$ and $x \in \mathscr{X}$. In fact if the map s is onto and if $\mathscr{B}$ distinguishes the points of $\mathscr{Y}$ then there is such a homomorphism $g \to g_{\mathscr{Y}}$ if and only if $\mathscr{S} = s^{-1}\mathscr{B}$ respects G.

In the general case $\mathscr{S} = s^{-1}\mathscr{B}$ respects G if and only if there is a, necessarily unique, antihomomorphism $g \to g_{\mathscr{B}}$ from G to the group of algebra isomorphisms of the algebra $\mathscr{B} \cap s[\mathscr{X}]$ onto itself, such that s is invariant with respect to these actions of G, i.e. that for all $B \in \mathscr{B}$

$$s(g_{\mathscr{X}}(x)) \in B \Leftrightarrow s(x) \in g_{\mathscr{B}}(B \cap s[\mathscr{X}]).$$

Proof. If $g_{\mathscr{X}}(x_1) \in S \in \mathscr{S}$ and $x_2 \sim x_1 \bmod (\mathscr{S})$ then $x_1 \in g_{\mathscr{X}}^{-1}[S] \in \mathscr{S}$ so that $x_2 \in g_{\mathscr{X}}^{-1}[S]$, i.e. $g_{\mathscr{X}}(x_2) \in S$. This proves the first statement. Assume $\mathscr{S} = s^{-1}(\mathscr{B})$ for the measurable function s from $(\mathscr{X}, \mathscr{A})$ to $(\mathscr{Y}, \mathscr{B})$. Let G act on $(\mathscr{Y}, \mathscr{B})$ as described, let $S \in \mathscr{S}$ and $g \in G$. We may then write $S = s^{-1}[B]$ where $B \in \mathscr{B}$. Then $g^{-1}[S] = g^{-1}[s^{-1}[B]] = s^{-1}[g^{-1}[B]] \in \mathscr{S}$. Thus $\mathscr{S}$ respects G. On the other hand, if $\mathscr{S}$ respects G and if s maps $\mathscr{X}$ onto $\mathscr{Y}$ and if $\mathscr{B}$ distinguishes points of $\mathscr{Y}$, then we may define $g_{\mathscr{Y}}$ by putting $g(s(x)) = s(g_{\mathscr{X}}(x))$. Note that if $B \in \mathscr{B}$ then $g^{-1}[B] = \tilde{B}$ where $s^{-1}[\tilde{B}] = g^{-1}[s^{-1}[B]]$. In general if $\mathscr{S}$ respects G then we may define the maps $g_{\mathscr{B}}$ by putting $g_{\mathscr{B}}(B \cap s[\mathscr{X}]) = \tilde{B} \cap s[\mathscr{X}]$ where $s^{-1}[\tilde{B}] = g^{-1}[s^{-1}[B]]$. It is then easily checked that $g \to g_{\mathscr{B}}$ is an antihomomorphism as required. Conversely, if such an antihomomorphism exists then it is unique since $g_{\mathscr{B}}(B \cap s[\mathscr{X}]) = \{s(x) : s(x) \in g_{\mathscr{B}}(B \cap s[\mathscr{X}])\} = \{s(x) : s(g(x)) \in B\}$. This also implies that $s^{-1}(\mathscr{B})$ respects G since then

$$g^{-1}[s^{-1}[B]] = \{x : s(g(x)) \in B\} = s^{-1}[g_{\mathscr{B}}(B \cap s[\mathscr{X}])]. \qquad \square$$

Disregarding for the moment the fact that we might have to consider somewhat more general objects than Markov kernels, and also disregarding the distinction between invariance, $\mathscr{E}$-invariance and uniform $\mathscr{E}$-invariance, we may phrase the reduction problem as follows.

To what extent may comparison within invariant pairs of experiments be based on invariant kernels?

The purpose of this section is to show that nothing is lost by this reduction provided a suitable generalization of the averaging procedure which we just carried out for finite groups is available. Assume for the moment that everything which at first glance ought to work, actually works. We could then proceed as follows.

Let m be a probability distribution on G which is invariant for right shifts. If M is a Markov kernel from $(\mathscr{X}, \mathscr{A})$ to $(\mathscr{Y}, \mathscr{B})$ then we put $\overline{M}(B|x) = \int M^g(B|x)m(dg)$ when $B \in \mathscr{B}$ and $x \in \mathscr{X}$. Then $\overline{M}^h(B|x) = \int M^g(h[B]|h(x))m(dg) = \int [M^g]^h(B|x)m(dg) = \int M^{gh}(B|x)m(dg) = \int M^g(B|x)m(dg) = \overline{M}(B|x)$ when $B \in \mathscr{B}$ and $x \in \mathscr{X}$. Thus $\overline{M}$ is invariant and since it might be considered as an average $\int M^g m(dg)$ it should again follow that nothing is lost by a reduction to invariant kernels. However, there is quite a bit of mathematics here which should be sorted out. Besides the obvious measurability problems there is the problem of the existence of a shift invariant probability distribution m. Even in the case of simple groups, like the additive group of integers or the additive group of vectors in $\mathbb{R}^n$, there might be no shift invariant probability distribution. However, this difficulty can often be overcome by omitting the requirement of σ-additivity. In fact any solvable (in particular Abelian) group possesses additive probability set functions which are shift invariant.

Perhaps at this point we should note that the non σ-additive probabilities which occur here do not express a need for the luxury of unrestricted generality, but rather the commonsense desire to be able to deal with some of the most commonly occurring groups.

Following a well established terminology we shall often find it convenient to use the shorter term *mean* instead of 'additive probability set function'. By the Hahn–Banach theorem, a mean on an algebra of subsets of a set G may always be extended to a mean on the class of all subsets of G. If the domain of a mean m is the algebra of all subsets of G, then we shall say simply that *m is a mean on G.*

The existence of shift invariant additive probability set functions does not guarantee the existence of invariant Markov kernels. They do however guarantee the existence of the required invariant sub-Markov kernels, provided measurability is taken care of. Thus it might be too much to require both invariance and full mass preservation.

In order that the reader should appreciate the need for some of the mathematics we would like to make a quick excursion to five examples before we establish the main results.

Example 6.5.2 (The need for sub-Markov kernels). Let $\mathscr{X}$ be a one-point set and let $\mathscr{Y}$ be the set of integers equipped with the σ-algebra $\mathscr{B}$ of all its subsets. Also let Θ be the real line and let Q be a probability distribution on $\mathscr{Y}$. For each $\theta \in \Theta$, let Q_θ be the right θ translate of Q. Take the group G as the additive group of integers and define its actions by $g_{\mathscr{X}}(x) = x$, $g_{\mathscr{Y}}(y) = y + g$, and $g_\Theta(\theta) = \theta + g$, when $x \in \mathscr{X}$, $y \in \mathscr{Y}$ and $g \in G$.

Then $(\mathscr{E}, \mathscr{F})$ is invariant under G and trivially $\mathscr{E}$ is 2-deficient with respect to $\mathscr{F}$. However there is no $\mathscr{E}$-invariant Markov kernel M from $(\mathscr{X}, \mathscr{A})$ to $(\mathscr{Y}, \mathscr{B})$. As any transition is representable by a Markov kernel this cannot be remedied by passing to transitions. Note however that $\|(P_\theta M - Q_\theta)^-\| \leqslant 1$ where the invariant sub-Markov kernel M is given by $M(i|x) \equiv_{i,x} 0$.

Example 6.5.3 (Convolution kernels). Let $\mathscr{X}$ be any measurable group, i.e. a group equipped with a σ-algebra $\mathscr{A}$ of subsets making the map $(g, h) \to gh^{-1}$ from $G \times G$ to G measurable.

Consider an $\mathscr{X}$-valued random variable Z distributed according to the probability distribution P on $\mathscr{A}$. Assume that we are only able to observe Z up to an unknown (right) multiplicative factor $\theta \in \mathscr{X}$. Thus the outcome of our experiment is a variable X possessing the same distribution as $Z\theta$. The probability distribution P_θ of X is then given by

$$P_\theta(A) = P(A\theta^{-1}).$$

The experiment $\mathscr{E}_P = (P_\theta : \theta \in \mathscr{X})$ is then called *the right translation experiment determined by P.*

Let the group $G = \mathscr{X}$ act on the sample set $\mathscr{X}$ and on the parameter set $\Theta = \mathscr{X}$ by right multiplication, i.e.

$$g_{\mathscr{X}}(x) = xg; \qquad x \in \mathscr{X}, \quad g \in G$$

$$g_\Theta(\theta) = \theta g; \qquad \theta \in \Theta, \quad g \in G.$$

Then any right translation experiment is invariant under G, i.e.

$$P_\theta g^{-1} = P_{g(\theta)}; \qquad \theta \in \Theta, \quad g \in G.$$

What do the invariant Markov kernels look like in this case? If K is an invariant Markov kernel then $K(Bg|xg) = K(B|x)$ when $B \in \mathscr{A}$, $x \in \mathscr{X}$ and $g \in G$. Thus $K(B|x) = K(Bx^{-1}|e)$ where e is the unit of G. Put $M(B) = K(B|e)$

when $B \in \mathscr{A}$. Then M is a probability measure on $\mathscr{A}$ and $K(B|x) = M(Bx^{-1}) = M_x(B)$; $B \in \mathscr{A}$. Conversely, if M is any probability measure on $\mathscr{A}$ then $(x, B) \to M_x(B)$ is a right shift invariant Markov kernel. A kernel of this form is called a *convolution kernel.* If this kernel is applied to a probability measure P then we obtain the measure $B \to \int M_x(B)P(\mathrm{d}x) = \int I_B(yx)M(\mathrm{d}y)P(\mathrm{d}x) = M * P$. In particular if M is concentrated in $y \in \mathscr{X}$ then we get the left translate ${}_yP$ of P.

Now consider two translation experiments $\mathscr{E}_P$ and $\mathscr{E}_Q$. If K_M is the convolution kernel determined by M then $P_\theta K_M = M * P * \delta_\theta$ where δ_θ is the one-point distribution in θ. Thus $\|P_\theta K_M - Q_\theta\| = \|M * P * \delta_\theta - Q * \delta_\theta\| = \|M * P - Q\|$. It follows that

$$\delta(\mathscr{E}_P, \mathscr{E}_Q) \leqslant \inf_M \|M * P - Q\| \text{ where } M \text{ runs through the set of all probability measures on } \mathscr{A}. \tag{6.5.1}$$

Let us see how the quantity $\|M * P - Q\|$ may be minimized. Consider the problem of testing the hypothesis '$({}_\theta P : \theta \in \Theta)$' against the alternative 'Q'. Here ${}_\theta P(A) = P(\theta^{-1}A)$ when $\theta \in \Theta$ and $A \in \mathscr{A}$.

Suppose that there exists a probability distribution M_0 such that to each $\alpha \in [0, 1]$ there corresponds a test δ_α with level of significance α which is the most powerful level α test at Q among all tests for '$M_0 * P$' against Q. Then

$$\int \delta_\alpha d(M * P) = \int \left(\int \delta_\alpha \, \mathrm{d}_\theta P \right) M(\mathrm{d}\theta) \leqq \int \alpha \, \mathrm{d}M = \alpha$$

so that

$$\begin{aligned} \|Q - M * P\| &\geqq 2 \sup_\alpha \left[\int \delta_\alpha \mathrm{d}Q - \int \delta_\alpha d(M * P) \right] \\ &\geqq 2 \sup_\alpha \left(\int \delta_\alpha \mathrm{d}Q - \alpha \right) \\ &= \|Q - M_0 * P\|. \end{aligned}$$

It follows that $\|M * P - Q\|$ is minimized for $M = M_0$.

Later we will see that the general theory implies that (6.5.1) holds with '=' and with 'inf' replaced by 'min' provided $\mathscr{X}$ is a second countable locally compact topological group, $\mathscr{A}$ = the class of Borel subsets of $\mathscr{X}$, P is absolutely continuous (with respect to Haar measure) and that there exist shift invariant means on $\mathscr{A}$.

Assuming that these conditions are satisfied we obtain in particular the convolution (factorization) criterion for 'being at least as informative as'. The convolution criterion states simply that $\mathscr{E}_P \geqslant \mathscr{E}_Q$ if and only if P is a right convolution factor in Q. (The 'if' holds without any of the stated conditions.) This has the interesting consequence that a substantial part of the convolution theory for probability measures has straightforward informational interpretations. Thus if P and Q are probability distributions on $\mathbb{R}^k$, where $k < \infty$, such that Q is a normal distribution and P is absolutely continuous, then $\mathscr{E}_P \geqslant \mathscr{E}_Q$ if and only if P is a normal distribution with covariance matrix $\leqslant$ the covariance matrix of Q. Here the ordering $\leqslant$ is the ordering where '$A \leqslant B$' is short for '$B - A$ is non-negative definite'.

In a later chapter we will see how the theory of invariance may be used to gain insight into the problem of comparison of linear normal models.

If P is not absolutely continuous then convolution kernels may be inadequate for the evaluation of $\delta(\mathscr{E}_P, \mathscr{E}_Q)$.

Example 6.5.4 (Translation and scale experiments). Suppose we are observing a real valued random variable $X = \xi + \sigma U$ where $\xi \in \mathbb{R}$ and $\sigma > 0$ are unknown parameters and where the unobservable variable U is distributed according to a known absolutely continuous distribution F.

The natural group to consider here is the group G of invertible linear transformations on the real line. This group contains the translation group G_1 as a normal subgroup and the factor group G_2/G_1 is Abelian. It follows that this group is solvable.

Let M be an invariant Markov kernel from $\mathbb{R}$ to $\mathbb{R}$. Then $\int \phi(g(y))M(\mathrm{d}y|x) \equiv_{x,g} \int \phi(y)M(\mathrm{d}y|g(x))$ for each bounded measurable function ϕ. If ϕ is continuous and if $g(z) \equiv_z x + (1/n)(z - x)$ where $n \to \infty$ then we obtain $\phi(x) \equiv_x \int \phi(y)M(\mathrm{d}y|x)$.

Hence invariance forces M to be the identity map.

This has the somewhat surprising consequence that if we compare such experiments in terms of invariant Markov kernels then the experiments obtained by considering two different choices of F are at the maximal distance 2 from each other. Actually the theory which we shall describe shows that this is indeed the case.

This 'phenomenon' is clearly related to the difficulty of estimating the pair (σ, ξ) on the basis of a single observation. However, if more observations are available then, as will become apparent in the section on comparison of linear normal models, other and more refined results are available.

Example 6.5.5 (Invariant maximin tests. The Hunt–Stein theorem). In Corollary 6.2.10 we saw how the existence of level α tests with specified bounds on minimum power could be expressed in terms of deficiencies. If the testing problem is invariant then the associated comparison problem is also invariant. It follows from the general theory, as described here, that if the group G possesses invariant means (mean = real valued additive probability set function) then we might express the comparison problem in terms of invariant kernels. This in turn yields the Hunt–Stein theorem which, somewhat loosely phrased, guarantees that nothing is lost in terms of minimax power if we choose to base our testing on invariant tests. Now Lehmann's book on testing theory (1986), contains a simple example, due to Stein, where invariant tests are hopelessly inadequate. The group involved is the group of all $r \times r$ non-singular matrices where $r \geqslant 2$. As noted by Lehmann, it then follows that this group cannot possess invariant means.

As will be explained thoroughly later, this example shows that there are 'non-pathological' situations where comparison of invariant experiments cannot be expressed in terms of (almost) invariant kernels.

We will return to these matters shortly, after describing a few general results.

Example 6.5.6 (Sufficiency and invariance). The problem of the exchangeability of invariance and sufficiency may be considered as a particular case of the following problem.

Let $\mathscr{E} = (\mathscr{X}, \mathscr{A}; P_\theta : \theta \in \Theta)$ and $\mathscr{F} = (\mathscr{Y}, \mathscr{B}; Q_\theta : \theta \in \Theta)$ be two experiments and let ε be a non-negative function on Θ. Let the group G act on $\mathscr{X}$, $\mathscr{Y}$ and Θ by the maps $g \to g_{\mathscr{X}}$, $g \to g_{\mathscr{Y}}$ and $g \to g_\Theta$. Assume as usual that the maps $g \to g_{\mathscr{X}}$ and $g \to g_{\mathscr{Y}}$ are homomorphisms from G to the groups of Borel isomorphisms of $(\mathscr{X}, \mathscr{A})$ to itself, and $(\mathscr{Y}, \mathscr{B})$ to itself respectively. Define the sub σ-algebra $\mathscr{A}_I$ of $\mathscr{A}$ as the class of invariant sets in $\mathscr{A}$, i.e. $\mathscr{A}_I = \{A : A \in \mathscr{A} \text{ and } g[A] \equiv_g A\}$. Also define the sub σ-algebra $\mathscr{B}_I$ of $\mathscr{B}$ as the class of invariant sets in $\mathscr{B}$, i.e. $\mathscr{B}_I = \{B : B \in \mathscr{B} \text{ and } g[B] \equiv_g B\}$.

Now suppose that the pair $(\mathscr{E}, \mathscr{F})$ is invariant under G, that ε is constant on orbits of G in Θ and that $\mathscr{E}$ is ε-deficient with respect to $\mathscr{F}$. Are we then permitted to conclude that the 'invariant part' $\mathscr{E}|\mathscr{A}_I$ of $\mathscr{E}$ is ε-deficient with respect to the 'invariant part' $\mathscr{F}|\mathscr{B}_I$ of $\mathscr{F}$?

Later we will see that the answer is yes, provided G possesses appropriate invariant means and provided certain other regularity conditions are satisfied. However if G does not possess appropriate invariant means then Stein's example, mentioned in the previous example, shows that the answer to this problem may be negative.

Note that the answer to this question is trivially yes whenever G acts transitively on Θ. However the problem of approximating $\mathscr{F}$ by invariant randomizations on $\mathscr{E}$ could still be interesting and non-trivial.

The particular case of sufficiency is obtained by letting $\mathscr{A}$ be a sufficient sub σ-algebra of $\mathscr{B}$. In that case the assumption of the existence of invariant means is not needed. The reason is that we may then concentrate our efforts on the Markov kernels defined by the conditional probabilities given $\mathscr{A}$ and these are automatically $\mathscr{E}$-invariant. The reader may consult Hall, Wijsman & Ghosh (1965) for a thorough treatment of this problem. One of their results will be deduced later in this section.

Let us return to the general theory and consider first a pair $(\mathscr{E}, \mathscr{F})$ of experiments which is invariant under a group G as described above.

In order to simplify the notation let us write $L_1 = L(\mathscr{E})$, $M_1 = M(\mathscr{E})$, $L_2 = L(\mathscr{F})$ and $M_2 = M(\mathscr{F})$ for the L-spaces and M-spaces of $\mathscr{E}$ and $\mathscr{F}$. Thus L_1 and L_2 are L-spaces with duals M_1 and M_2.

Call a map from one L-space to another (possibly the same) L-space an *L-space isomorphism* if it is a vector lattice isomorphism as well as an isometry. It is easily checked that a map from one L-space L_1 onto another L-space L_2 is an L-space isomorphism if and only if it is 1–1 and a transition which maps the non-negative part $(L_1)_+$ of L_1 onto the non-negative part $(L_2)_+$ of L_2.

For each measure $\mu \in L_i$ and each $g \in G$ *let the measure* μg^{-1} be denoted as $g_{L_i}(\mu)$. We may also write μg_{L_i} instead of $g_{L_i}(\mu)$. Note that $g_{L_i}(\mu)$ as a function of μ is an L-space isomorphism of L_i onto itself. The map $g \to g_{L_i}$ is a homomorphism of G into the group of L-space isomorphisms of L_i onto itself. Thus G acts on the L-spaces L_1 and L_2 by the homomorphisms $g \to g_{L_1}$ and $g \to g_{L_2}$. Actually the mental effort may be considerably eased once we recognize that for the subsequent mathematical development we can concentrate on these actions on L-spaces and forget about the experiments $\mathscr{E}$ and $\mathscr{F}$, as well as the Borel isomorphisms on their sample spaces.

Thus the homomorphisms into groups of Borel isomorphisms could be replaced by antihomomorphisms into the groups of algebra isomorphisms of the involved set algebras. The reduction to homomorphisms from G into the group of L-space isomorphisms of, say, $L(\mathscr{E})$, is then obtained by considering the L-space isomorphisms induced by the algebra isomorphisms.

By the way, the reader is excused if at this point he or she feels lost in a mist of σ-algebras, (Borel) isomorphisms and (anti-) homomorphisms.

We shall now embark on the proof of the main result in this section. As the proof provides some background to the theorem it is given before the formal statement. Thus the reader could jump directly from here to the theorem and then, if there is any energy to spare, go through the discussion below.

The essential items before us now are two L-spaces L_1 and L_2 having the duals M_1 and M_2 respectively, and a group G which acts on L_1 and L_2 by maps $g \to g_{L_1}$ and $g \to g_{L_2}$ such that

> the map $g \to g_{L_i}$, for $i = 1, 2$, is a homomorphism from G to the group of L-space isomorphisms of L_i onto itself.

The map $g \to g^*_{L_i}$ then becomes an antihomomorphism from G to the group of M-space isomorphisms of M_i onto itself. (An M-space isomorphism of an M-space is again just an isometric vector lattice isomorphism.) Now consider a sub-vector lattice $M_{2,0}$ of M_2 which is dense in M_2 for the topology of weak* convergence, i.e. for the pointwise topology on L_2. Letting I_2 denote the unit of M_2 we shall also assume that $w \wedge I_2 \in M_{2,0}$ when $0 \leqslant w \in M_{2,0}$. In a concrete case $M_{2,0}$ might be some vector lattice of point functions, e.g. the vector lattice of bounded measurable functions or the vector lattice of continuous functions with compact support, etc. Of course we could always take $M_{2,0} = M_2$, but that choice might be inconvenient. The requirement that $M_{2,0}$ is dense amounts to the requirement that $M_{2,0}$ is total with respect to L_2, i.e. that $\nu = 0$ if $\nu \in L_2$ and $(\nu, v) = 0$ when $v \in M_{2,0}$.

In addition we shall also require that $M_{2,0}$ is invariant under G, i.e. that $g_{L_2}^*[M_{2,0}] \subseteqq M_{2,0}$ when $g \in G$.

The requirement that $M_{2,0}$ is weak* dense is also equivalent to the condition that the map which to each measure $\nu \in L_2$ assigns its restriction $\tilde{\nu}$ to $M_{2,0}$ is an imbedding (i.e. is 1–1) of L_2 into the L-space $M_{2,0}^*$. Now by theorem 5.7.19, the closure of a convex set in the dual $M = L^*$ of an L-space L is the same for the weak* topology as for the topology of uniform convergence on order bounded subsets of L. Furthermore, the latter topology makes the vector lattice operations continuous. Thus if $\nu \in L_2$ then $\tilde{\nu}^+(v) = \sup\{\nu(w) : 0 \leqslant w \leqslant v, w \in M_{2,0}\} = \sup\{\nu(w) : 0 \leqslant w \leqslant v, w \in M_2\} = \nu^+(v) = \widetilde{\nu^+}(v)$ when $0 \leqslant v \in M_{2,0}$. It follows that the imbedding of L_2 in $M_{2,0}^*$ is a vector lattice isomorphism. Finally let $|v| \leqslant I_2$ where $v \in M_2$ and I_2 is the unit of $M_{2,0}$. By assumption there is a net $\{w_t\}$ in $M_{2,0}$ which converges to v uniformly on order bounded subsets of L_2. Since $w \wedge I_2 \in M_{2,0}$ when $w \in M_{2,0}$, we may assume that $|w_t| \leqslant I_2$ for all t. Let $\nu \in L_2$. Then $\nu(v) = \lim_t \nu(w_t) \leqslant \|\tilde{\nu}\|$ so that $\|\nu\| \leqslant \|\tilde{\nu}\|$. Now the definition of $\tilde{\nu}$ implies directly that $\|\tilde{\nu}\| \leqslant \|\nu\|$. Thus $\|\tilde{\nu}\| = \|\nu\|$. It follows that the imbedding is an isometry as well as a vector lattice isomorphism.

Furthermore, the image $\tilde{L}_2$ of L_2 by this imbedding is, as we shall now see, a band in $M_{2,0}^*$. In order to prove this, it suffices to show that if $\nu \in L_2$ and $\sigma \in M_{2,0}^*$ and $0 \leqslant \sigma \leqslant \tilde{\nu}$ then $\sigma = \tilde{\nu}_1$ for some $\nu_1 \in L_2$. So suppose ν and σ satisfy these conditions. Then $\sigma(v) \leqslant \sigma(v^+) \leqslant \nu(v^+)$ when $v \in M_{2,0}$. The Hahn–Banach theorem implies that σ may be extended to a linear functional $\bar{\sigma}$ on M_2 such that $\bar{\sigma}(v^+) \leqslant \nu(v^+)$ when $v \in M_2$. Then $\bar{\sigma}(v) \leqslant 0$ when $v \leqslant 0$ so that $\bar{\sigma}$ is non-negative. Furthermore, $|\bar{\sigma}(v)| \leqslant \bar{\sigma}(|v|) \leqslant \nu(|v|) \leqslant \|\nu\| \|v\|$ so that $\|\bar{\sigma}\| \leqslant \|\nu\|$. Hence $\bar{\sigma} \in M_2^* = L_2^{**}$. Clearly $0 \leqslant \bar{\sigma} \leqslant \nu$. Hence, since L_2 is a band in L_2^{**}, $\bar{\sigma}$ is the evaluation at some $\nu_1 \in L_2$, i.e. $\bar{\sigma}(v) = (\nu_1, v)$; $v \in M_2$. Then $\tilde{\nu}_1(v) = \bar{\sigma}(v) = \sigma(v)$ when $v \in M_{2,0}$. Hence $\tilde{\nu}_1 = \sigma$.

Finally let us extend each map g_{L_2} on L_2 to a map $\tilde{g}_{L_2}$ on $M_{2,0}^*$ by putting $\tilde{g}_{L_2} = [g_{L_2}^* | M_{2,0}]^*$, i.e. $\tilde{g}_{L_2}$ is the dual of the restriction of $g_{L_2}^*$ to $M_{2,0}$. Each map $\tilde{g}_{L_2}$ then becomes an L-space isomorphism of $M_{2,0}^*$. The invariance of $M_{2,0}$ implies that $\tilde{g}_{L_2}(\tilde{\nu}) = g_{L_2}(\nu)$ when $\nu \in L_2$. It is then easily checked that the map $g \to \tilde{g}_{L_2}$ is a homomorphism from G to the group of L-space isomorphisms of $M_{2,0}^*$.

The main advantage of passing from L_2 to $M_{2,0}^*$ is that the set of transitions from L_2 to $M_{2,0}^*$ is compact for the topology of pointwise convergence on $L_2 \times M_{2,0}$. If $M_{2,0} = M_2$ then the above imbedding is simply the imbedding of L_2 in its second dual L_2^{**}. If $M_{2,0}$ is the set of bounded measurable functions on the measurable space $(\mathscr{Y}, \mathscr{B})$ then $M_{2,0}^*$, up to an identification, is the set of additive bounded set functions on $\mathscr{B}$.

For any pair (L_1, L_2) of L-spaces let the convex set of subtransitions from L_1 to L_2 be denoted as SUBTRANS(L_1, L_2). Thus $T \in$ SUBTRANS(L_1, L_2) if and only if T is a non-negative linear contraction from L_1 to L_2, i.e.:

(i) T is linear;
(ii) $T[(L_1)_+] \subseteq (L_2)_+$;
(iii) $\|T\| \leqslant 1$.

Let us also use the notation TRANS(L_1, L_2) for the convex set of transitions from the L-space L_1 to the L-space L_2. Thus TRANS(L_1, L_2) consists of those maps in SUBTRANS(L_1, L_2) which preserve the norms of non-negative elements.

Topologize SUBTRANS$(L_1, M_{2,0}^*)$ by the topology of pointwise convergence on $L_1 \times M_{2,0}$. Then SUBTRANS$(L_1, M_{2,0}^*)$ becomes a compact Hausdorff space having TRANS$(L_1, M_{2,0}^*)$ as a compact subset.

Consider a subtransition T from L_1 to $M_{2,0}^*$ and an element g of the group G. Following the method of Day's paper (1961) we define another subtransition T^g by putting

$$\mu T^g = \mu g_{L_1} T \tilde{g}_{L_2}^{-1}$$

when $\mu \in L_1$. Thus $T^g = \tilde{g}_{L_2}^{-1} \circ T \circ g_{L_1}$. The map $T \to T^g$ is then clearly affine as well as continuous for the topology of pointwise convergence on $L_1 \times M_{2,0}$. Furthermore, $T^{gh} = [T^g]^h$ when $g, h \in G$.

Let $\mathscr{P}(G)$ denote the convex set of means (i.e. additive real valued probability set functions) on the class of all subsets of G. Equip $\mathscr{P}(G)$ with the topology of setwise convergence. If $m \in \mathscr{P}(G)$ and $T \in$ SUBTRANS$(L_1, M_{2,0}^*)$ then for each pair $(\mu, v) \in L_1 \times M_{2,0}$ we can form the integral $\int (\mu T^g v) m(\mathrm{d}g)$. As a function of (μ, v) this integral is bilinear. It is non-negative if $\mu \geqslant 0$ and $v \geqslant 0$ and then $\int \mu T^g v m(\mathrm{d}g) \leqslant \int \|\mu\| \cdot \|v\| m(\mathrm{d}g) = \|\mu\| \|v\|$. It follows that there is a unique subtransition T^m from L_1 to $M_{2,0}^*$ such that

$$\int (\mu T^g v) m(\mathrm{d}g) = \mu T^m v$$

when $(\mu, v) \in L_2 \times M_{2,0}$.

The map $m \to T^m$ is affine as well as continuous. If m is the one-point distribution in $g \in G$ then $T^m = T^g$. If m has finite support $\{g_1, \ldots, g_r\}$ then $T^m = \sum_i m(g_i) T^{g_i}$. The sets TRANS$(L_1, M_{2,0}^*)$, SUBTRANS$(L_1, L_2)$ and SUBTRANS$(L_1, M_{2,0}^*)$ are mapped into themselves by any map $T \to T^m$ where $m \in \mathscr{P}(G)$. If m has finite support then the set TRANS(L_1, L_2) is also mapped into itself by this map. In the general case 'mass' may disappear by the formation of T^m where $T \in$ TRANS(L_1, L_2) so that $T^m \in$ TRANS$(L_1, M_{2,0}^*) -$ TRANS(L_1, L_2).

For each $m \in \mathscr{P}(G)$ and each $h \in G$ the mean m_h denotes the mean induced from m by a right shift of $h \in G$. Thus $m_h(U) = m(Uh^{-1})$ when $U \subseteqq G$ and $h \in G$. Following Day, consider a $T \in \mathrm{SUBTRANS}(L_1, M_{2,0}^*)$, a mean $m \in \mathscr{P}(G)$ and an element h of the group G.

If m assigns masses $m(g_1), \ldots, m(g_r)$ to $g_1, \ldots, g_r$ where $m(g_1) + \cdots + m(g_r) = 1$ then $T^{m_h} = \sum_i m(g_i) T^{g_i h} = \sum m(g_i)[T^{g_i}]^h = [\sum m(g_i) T^{g_i}]^h = [T^m]^h$. In the general case let $\{m^\alpha\}$ be a net of probability measures on G with finite supports which converges setwise to m. Then m_h^α converges setwise to m_h so that

$$T^{m_h} = \lim_\alpha T^{m_h^\alpha} = \lim_\alpha [T^{m^\alpha}]^h = \left[\lim_\alpha T^{m^\alpha}\right]^h = [T^m]^h.$$

Call a subtransition T from L_1 to $M_{2,0}^*$ *invariant* if $T^h = T$ when $h \in G$. Call a mean m on G *right (shift) invariant* if $m_h = m$ when $h \in G$. It follows from the above computations that T^m is invariant for any $T \in \mathrm{SUBTRANS}(L_1, M_{2,0}^*)$ when m is right invariant.

In chapter 4 we saw how a transition could be modified in order to map the L-space of an experiment $\mathscr{E}$ into the L-space of $\mathscr{F}$. Thus we could consider here how a transition T from L_1 to $M_{2,0}^*$ could be modified so that we could obtain a transition from L_1 to L_2 which is invariant if T is invariant. However, in the case of invariance we may be forced to accept subtransitions. The procedure is as follows.

Let Π be the projection of $M_{2,0}^*$ onto the band $\tilde{L}_2$. If T is a subtransition from L_1 to $M_{2,0}^*$ then $T\Pi(= \Pi \circ T)$ is a subtransition from L_1 to L_2. In addition if $m \in \mathscr{P}(G)$ then we put $T_m = T^m \Pi$. Let $\kappa \in M_{2,0}^*$. Then $\kappa = \kappa\Pi + \kappa - \kappa\Pi$. Hence $\kappa\tilde{g}_{L_2} = \kappa\Pi\tilde{g}_{L_2} + (\kappa - \kappa\Pi)\tilde{g}_{\tilde{L}_2}$. Here $\kappa\Pi\tilde{g}_{L_2} = \kappa\Pi g_{L_2} \in L_2$ (or rather $\tilde{L}_2$) while $|(\kappa - \kappa\Pi)|\tilde{g}_{L_2} \wedge |\tilde{v}|\tilde{g}_{L_2} = [|\kappa - \kappa\Pi| \wedge |\tilde{v}|]\tilde{g}_{L_2} = 0\tilde{g}_{L_2} = 0 \in M_{2,0}^*$ when $v \in L_2$ and $g \in G$. Hence Π and $\tilde{g}_{L_2}$ commute, i.e. $\kappa\tilde{g}_{L_2}\Pi = \kappa\Pi\tilde{g}_{L_2}$ when $\kappa \in M_{2,0}^*$. It follows that $[T_m]^h = h_{L_1}[T_m]\tilde{h}_{L_2}^{-1} = h_{L_1}[T^m]\Pi\tilde{h}_{L_2}^{-1} = h_{L_1}[T^m]\tilde{h}_{L_2}^{-1}\Pi = [T^m]^h\Pi = T^{m_h}\Pi = T_{m_h}$ when T is a subtransition from L_1 to $M_{2,0}^*$. Hence both T^m and T_m are invariant when m is right shift invariant.

Actually right shift invariance of m might be too strong a condition. The formula $\mu T^{m_h} v = \int (\mu T^g v) m_h(dg) = \int (\mu T^{gh} v) m(dg)$ shows that it suffices to require that $\int (\mu T^{gh} v) m(dg)$ does not depend on h when $\mu \in L_1$ and $v \in M_{2,0}$.

Let us compare the effects of the subtransitions T, T^g, T^m and T_m when $T \in \mathrm{SUBTRANS}(L_1, M_{2,0}^*)$, $g \in G$ and $m \in \mathscr{P}(G)$. First consider T^g. Let $\mu \in L_1$ and $\kappa \in M_{2,0}^*$. Then, since $\tilde{g}_{L_2}$ is a vector lattice isomorphism as well as an isometry, we find $\|(\mu T^g - \kappa)^\pm\| = \|(\mu g_{L_1} T \tilde{g}_{L_2}^{-1} - \kappa\tilde{g}_{L_2}\tilde{g}_{L_2}^{-1})^\pm\| = \|(\mu g_{L_1} T - \kappa\tilde{g}_{L_2})^\pm \tilde{g}_{L_2}^{-1}\| = \|(\mu g_{L_1} T - \kappa\tilde{g}_{L_2})^\pm\|$. Furthermore, if $m \in \mathscr{P}(G)$ and $w \in M_{2,0}$ then $\mu T^m w - (\kappa, w) = \int [\mu T^g w - (\kappa, w)] m(dg)$. Thus $(\mu T^m - \kappa)^\pm(v) \leqslant \int (\mu T^g - \kappa)^\pm(v) m(dg)$ when $v \geqslant 0$. Hence $\|(\mu T^m - \kappa)^\pm\| \leqslant \int \|(\mu T^g - \kappa)^\pm\| m(dg)$. In particular if $v \in L_2$ then $\|(\mu T_m - \tilde{v})^\pm\| = \|[(\mu T^m - \tilde{v})\Pi]^\pm\| = \|(\mu T^m - \tilde{v})^\pm \Pi\| \leqslant \|(\mu T^m - \tilde{v})^\pm\| \leqslant \int \|(\mu T^g - \tilde{v})^\pm\| m(dg)$

$= \int \|(\mu g_{L_1} T - \tilde{\nu g}_{L_2})^{\pm}\| m(\mathrm{d}g) \leqslant \sup_g \|(\mu g_{L_1} T - \tilde{\nu g}_{L_2})^{\pm}\|$ where all $\pm$ signs in exponential positions may be dropped simultaneously.

Next assume that there is an invariant transition S from L_1 to L_2. Assume also that the mean m on G is such that $\int z(gh)m(\mathrm{d}g)$ does not depend on $h \in G$ when z is of the form $g \to \mu g_{L_1} T g_{L_2}^{-1} v$ where $\mu \in L_1$ and $v \in M_{2,0}$. Put $\mu \hat{T} = \mu T_m + [(I_1 - T_m^* I_2)\mu]S$ where I_1 and I_2 are the units of M_1 and M_2 respectively. (Here $v\mu$ for an element $\mu \in L_1$ and $v \in L_1^*$ denotes the element of L_1 whose Radon–Nikodym derivative with respect to μ is v, i.e. $(v\mu) = \mu(vw)$ when $w \in L_1^*$.) Then $\hat{T}$ is a transition from L_1 to L_2 and it is invariant since $g_{L_1}^*(T_m^* I_2) \equiv_g T_m^* I_2$ and since the maps $g_{L_1}^*$; $g \in G$, are algebraic isomorphisms of M_1 onto itself. If $0 \leqslant \mu \in L_1$ and $\nu \in L_2$ then $\mu\hat{T} - \nu \geqslant \mu T_m - \nu$ so that

$$\|(\mu\hat{T} - \nu)^-\| \leqslant \|(\mu T_m - \nu)^-\| \leqslant \sup_g (\mu g_{L_1} T - \tilde{\nu g}_{L_2})^-\|.$$

If K is a transition from L_1 to L_2, $\mu \in L_1$ and $\nu \in L_2$ then $\|(\mu K - \nu)^+\| = \|(\mu K - \nu)^-\| + \mu(I_1) - \nu(I_2)$ and $\|(\mu K - \nu)\| = 2\|(\mu K - \nu)^-\| + \mu(I_1) - \nu(I_2)$. It follows that the left hand '$-$' sign as well as the right hand '$-$' sign appearing in exponential positions in our last set of inequalities, provided T is a transition, may either both be deleted or both be replaced by the $+$ sign.

This completes the proof of the main result in this chapter.

Theorem 6.5.7 (Reduction to invariant (sub) transitions). *Let the group G act on the L-spaces L_1 and L_2 by homomorphisms $g \to g_{L_1}$ and $g \to g_{L_2}$ from G to, respectively, the group of L-space isomorphisms of L_1 and to the group of L-space isomorphisms of L_2.*

Let $M_{2,0}$ be a sub-vector lattice of $M_2 = L_2^*$ which is total with respect to L_2, i.e. $\nu = 0$ whenever $\nu \in L_2$ and $(\nu, v) = 0$ when $v \in M_{2,0}$. Assume also that $v \wedge I_2 \in M_{2,0}$ when $0 \leqslant v \in M_{2,0}$ and I_2 is the unit of M_2. Identifying any $\nu \in L_2$ with its restriction to $M_{2,0}$ the L-space L_2 becomes a band in $M_{2,0}^*$.

Let Π denote the projection of $M_{2,0}^*$ onto L_2.

Assume that $M_{2,0}$ is G-invariant, i.e. that $g_{L_2}^*[M_{2,0}] \subseteqq M_{2,0}$ when $g \in G$.

Say that a subtransition T from L_1 to $M_{2,0}^*$ is invariant if $g_{L_2}^{-1} \circ T \circ g_{L_1} \equiv_g T$.

Let T be a (sub) transition from L_1 to $M_{2,0}^*$ such that there is a mean m on G which satisfies the invariance condition

$$\int z(gh)m(\mathrm{d}g) \underset{h}{\equiv} \int z(g)m(\mathrm{d}g)$$

whenever z is a function on G of the form

$$g \to \mu g_{L_1} T g_{L_2}^{-1} v \quad \text{where} \quad \mu \in L_1 \quad \text{and} \quad v \in M_{2,0}.$$

Define the (sub) transition $\bar{T}$ from L_1 to $M_{2,0}^*$ by putting

$$\mu\bar{T}v = \int [\mu g_{L_1} T g_{L_2}^{-1} v] m(\mathrm{d}g)$$

when $\mu \in L_1$ and $v \in M_{2,0}$.

Put $\underline{T} = \Pi \circ \bar{T}$. *Then the (sub) transition* $\bar{T}$ *and the subtransition* $\underline{T}$ *are both invariant and*

$$\|(\mu\underline{T} - \nu)^{\pm}\| \leqslant \|(\mu\bar{T} - \nu)^{\pm}\| \leqslant \sup_g \|(\mu g_{L_1} T - \nu g_{L_2})^{\pm}\|$$

when $\mu \in L_1$ *and* $\nu \in L_2$.

Finally assume that T *is a transition and that there is at least one invariant transition* S *from* L_1 *to* L_2. *For each* $\mu \in L_1$, *put* $\mu\hat{T} = \mu\underline{T} + [(I_1 - T^*I_2)\mu]S$ *where* I_1 *and* I_2 *are the units of* M_1 *and* M_2 *respectively (see remark 2). Then* $\hat{T}$ *is an invariant transition from* L_1 *to* L_2 *such that* $\|(\mu\hat{T} - \nu)^{\pm}\| \leqslant \sup_g \|(\mu g_{L_1} T - \nu g_{L_2})^{\pm}\|$ *when* $0 \leqslant \mu \in L_1$ *and* $\nu \in L_2$.

Remark 1. The '$\pm$' signs which appear in exponential positions in the inequalities of the theorem may either all be replaced by the '$+$' sign, or all be replaced by the '$-$' sign or all be deleted.

Remark 2. If L is an L-space and $M = L^*$ is the dual of L then, by the definition in section 5.5, the element I of M given by $(\mu, I) = \|\mu^+\| - \|\mu^-\|$; $\mu \in L$ is the unit of M. Furthermore for $\mu \in L$ and $v \in M$, $v\mu$ denotes the element α of L defined by $(\alpha, w) = (\mu, v \cdot w)$ when $w \in M$.

Remark 3. Let us comment on the condition of the existence of a right shift invariant mean on a linear space of bounded functions on G. Assume that we are given a linear space Z of bounded functions on G. Then, by the Hahn–Banach theorem, theorem 2.2.12, a linear functional m on Z may be extended to a mean on G if and only if

$$m(z) \leqslant \sup\{z(g) : g \in G\}; \qquad z \in Z. \tag{6.5.2}$$

Now assume in addition that Z is right shift invariant in the sense that any map on G of the form $g \to z(gh)$ where $h \in G$ and $z \in Z$ belongs to Z. Then G permits a mean m on G such that $\int z(gh)m(dg)$ does not depend on h when $z \in Z$ if and only if Z possesses a linear functional satisfying (6.5.2) and which is right shift invariant in the sense that, for all $z \in Z$ and all $h \in G$, it assigns the same numbers to the map z as to the map $g \to z(gh)$ on G.

If $M^*_{2,0} = L_2$ then $\bar{T} = \underline{T}$ and the statement of the theorem can be simplified as follows.

Corollary 6.5.8. *Let the group* G *act on the* L*-space* L_1 *by the homomorphisms* $g \to g_{L_1}$ *from* G *to the group of* L*-space isomorphisms of* L_1. *Assume also that* G *acts on the* M*-normed space* Γ *by an antihomomorphism* $g \to g_\Gamma$ *from* G *to the group of isomorphisms of the* M*-space* Γ.

Say that a subtransition T *from* L_1 *to* Γ^* *is invariant if* $[g^*_\Gamma]^{-1} \circ T \circ g_{L_1} \equiv_g T$.

Let T *be a (sub) transition from* L_1 *to* Γ^* *such that there is a mean* m *on* G *which satisfies the invariance condition* $\int z(gh)m(dg) \equiv_h \int z(g)m(dg)$ *whenever* z

is a function on G of the form $g \to \mu g_{L_1} T g_\Gamma^{-1} \gamma$ *where* $\mu \in L_1$ *and* $\gamma \in \Gamma$. *Define the (sub) transition* $\bar{T}$ *from* L_1 *to* Γ^* *by putting* $\mu\bar{T}\gamma = \int [\mu g L_1 T g_\Gamma^{-1}\gamma] m(dg)$ *when* $\mu \in L_1$ *and* $\gamma \in \Gamma$.

Then $\bar{T}$ *is invariant and* $\|(\mu\bar{T} - \nu)^\pm\| \leqslant \sup_g \|(\mu g_{L_1} T - \nu g_\Gamma^*)^\pm\|$ *when* $\mu \in L_1$ *and* $\nu \in \Gamma^*$.

Proof. With one exception, the assumptions of the theorem are satisfied with $L_2 = \Gamma^*$, $M_{2,0} = \Gamma$ and $g_{L_2} \equiv_g g_\Gamma^*$. The exception is the assumption that $v \wedge I_2 \in M_{2,0}$ when $0 \leqslant v \in M_{2,0}$. However, this assumption was only needed to ensure that L_2 (after imbedding) is a band in $M_{2,0}^*$. As $M_{2,0}^* = L_2$ this condition is superfluous here. □

Let us return to experiments and, using the framework of the theorem, assume that $L_1 = L(\mathscr{E})$ is the L-space of an experiment $\mathscr{E} = (P_\theta : \theta \in \Theta)$ which is invariant under G in the sense that $g_{L_1}(P_\theta) \equiv_\theta P_{g(\theta)}$ when $g \in G$ for some map $(g, \theta) \to g(\theta)$ from $G \times \Theta$ to Θ. Then the inequalities of the theorem imply that

$$\|(P_\theta \underline{T} - \nu)^\pm\| \leqslant \|(P_\theta \bar{T} - \nu)^\pm\| \leqslant \sup_g \|(P_{g(\theta)} T - \nu g_{L_2})^\pm\|$$

where $\theta \in \Theta$ and $\nu \in L_1$. Furthermore, if T is a transition and if an invariant transition from L_1 to L_2 exists, then the invariant transition $\hat{T}$ satisfies $\|(P_\theta \hat{T} - \nu)^\pm\| \leqslant \sup_g \|(P_{g(\theta)} T - \nu g_{L_2})^\pm\|$ when $\theta \in \Theta$ and $\nu \in L_1$. If $L_2 = L_2(\mathscr{F})$ where the experiment $\mathscr{F}$ is invariant in the same sense as $\mathscr{E}$ and if invariant transitions exist, then they are adequate for comparing $\mathscr{E}$ with respect to $\mathscr{F}$.

Corollary 6.5.9 (Comparison by invariant transitions). *Let* $\mathscr{E} = (P_\theta : \theta \in \Theta)$ *and* $\mathscr{F} = (Q_\theta : \theta \in \Theta)$ *be two experiments.*

Let G be a group acting on the L-spaces $L(\mathscr{E}) = L_1$ *and* $L(\mathscr{F}) = L_2$ *by the homomorphisms* $g \to g_{L_1}$ *and* $g \to g_{L_2}$ *from G to the group of L-space isomorphisms of* $L(\mathscr{E})$ *and* $L(\mathscr{F})$ *respectively. Let* $M_0(\mathscr{F})$ *be an invariant sub-vector lattice of* $M(\mathscr{F})$ *which is total with respect to* $L(\mathscr{F})$ *and which contains* $v \wedge 1$ *whenever* $0 \leqslant v \in M_0$.

Assume that G acts on Θ *by the map* $g \to g_\Theta$ *from G to the set of maps of* Θ *into itself.*

Finally assume that the pair $(\mathscr{E}, \mathscr{F})$ *is invariant under G in the sense that for each* $\theta \in \Theta$ *and each* $g \in G$

$$g(P_\theta) = P_{g(\theta)} \quad \text{and} \quad g(Q_\theta) = Q_{g(\theta)}$$

where as usual we drop the various subscripts on g.

Let T be a subtransition from $L(\mathscr{E})$ *to* $L(\mathscr{F})$ *such that the group G possesses a mean m making all integrals* $\int [\mu(gh) T (gh)^{-1} v] m(dg)$ *where* $\mu \in L(\mathscr{E})$ *and* $v \in M_0(\mathscr{F})$ *independent of* $h \in G$.

Say that a subtransition S from $L(\mathscr{E})$ *to* $L(\mathscr{F})$ *is invariant if* $g \circ S \equiv_g S \circ g$, *i.e.* $S \equiv_g g_{L_2}^{-1} \circ S \circ g_{L_1}$.

Then there is an invariant subtransition S from $L(\mathscr{E})$ to $L(\mathscr{F})$ such that

$$\|(P_\theta S - Q_\theta)^\pm\| \leqslant \sup_g \|(P_{g(\theta)}T - Q_{g(\theta)})^\pm\|$$

when $\theta \in \Theta$.

If T is a transition and if there is at least one invariant transition from $L(\mathscr{E})$ to $L(\mathscr{F})$, then S may be chosen as a transition.

This and corollary 6.4.2 yield the following criterion for ε-deficiency.

Corollary 6.5.10 (ε-deficiency in terms of invariant (sub) transitions). *Assume that the pair $(\mathscr{E}, \mathscr{F})$ of experiments is invariant under the group G where G acts as described in the previous corollary. In addition to the assumptions made in the previous corollary assume that ε is a non-negative function on Θ which is constant on orbits in Θ. Suppose also that G possesses a mean m such that $\int z(gh)m(\mathrm{d}g)$ does not depend on h when z is of the form $g \to \mu g T g^{-1} v$ where $\mu \in L(\mathscr{E})$, $v \in M_0(\mathscr{F})$ and T is any transition from $L(\mathscr{E})$ to $L(\mathscr{F})$.*

Then $\mathscr{E}$ is ε-deficient with respect to $\mathscr{F}$ if and only if there is an invariant subtransition S from $L(\mathscr{E})$ to $L(\mathscr{F})$ such that

$$\|(P_\theta S - Q_\theta)^-\| \leqslant \varepsilon_\theta/2; \qquad \theta \in \Theta.$$

If, in addition, there is at least one invariant transition from $L(\mathscr{E})$ to $L(\mathscr{F})$, then S may be chosen as a transition and then $\|P_\theta S - Q_\theta\| \leqslant \varepsilon_\theta$; $\theta \in \Theta$.

Let us apply the same averaging procedure to prove the theorem on invariant decision problems.

Consider a statistical decision problem consisting of an experiment $\mathscr{E} = (\mathscr{X}, \mathscr{A}; P_\theta : \theta \in \Theta)$, a decision space $(T, \mathscr{S})$ and a loss function L.

Let G be a group acting on $\mathscr{X}$, Θ and T by, respectively, maps $g \to g_{\mathscr{X}}$, $g \to g_\Theta$ and $g \to g_T$ such that:

the map $g \to g_{\mathscr{X}}$ is a homomorphism from G to the group of Borel isomorphisms from $(\mathscr{X}, \mathscr{A})$ to itself.
the map $g \to g_\Theta$ is a map from G to the set of maps of Θ into itself.
the map $g \to g_T$ is a homomorphism from G to the group of Borel isomorphisms from $(T, \mathscr{S})$ to itself.

Assume that the decision problem $(\mathscr{E}, (T, \mathscr{S}), L)$ is invariant under G as explained in the beginning of this section. Thus we assume that

$$P_\theta g^{-1} = P_{g(\theta)} \quad \text{and} \quad L_{g(\theta)} \circ g = L_\theta$$

when $\theta \in \Theta$ and $g \in G$. Let ρ be a generalized decision rule (i.e. a transition) from $L(\mathscr{E})$ to $\mathrm{ba}(T, \mathscr{S})$ and let ρ^g, for each $g \in G$, denote the generalized decision rule $g^{-1} \circ \rho \circ g$, i.e.

$$\mu \rho^g v = (\mu g^{-1}) \rho (v \circ g^{-1})$$

for each measure $\mu \in L(\mathscr{E})$ and each bounded measurable function v on $(T, \mathscr{S})$. If $r(\theta : \delta)$ denotes the risk we are exposed to when θ prevails and we are using the generalized decision rule δ, then

$$r(\theta : \rho^{g}) \underset{\theta}{\equiv} r(g(\theta) : \rho).$$

If m is any mean on G, invariant or not, and if we define the generalized decision rule ρ^m by putting $\mu\rho^m v = \int (\mu\rho^g v) m(\mathrm{d}g)$ when $\mu \in L(\mathscr{E})$ and $v \in M(T, \mathscr{S})$ then

$$\begin{aligned} r(\theta : \rho^m) &= \sup_N P_\theta \rho^m (L_\theta \wedge N) \\ &= \sup_N \int [P_{g(\theta)} \rho (L_{g(\theta)} \wedge N)] m(\mathrm{d}g) \\ &\leqslant \int [P_{g(\theta)} \rho L_{g(\theta)}] m(\mathrm{d}g) = \int r(g(\theta) : \rho) m(\mathrm{d}g) \end{aligned}$$

with '=' holding provided

$$\int [P_{g(\theta)} \rho (L_{g(\theta)} \wedge N)] m(\mathrm{d}g) \uparrow \int (P_{g(\theta)} \rho L_{g(\theta)}) m(\mathrm{d}g).$$

The latter condition holds automatically when $\sup_g \|L_{g(\theta)}\| < \infty$ or if m is countably additive on a σ-algebra making $g \to P_{g(\theta)}\rho(v \circ g)$ measurable in g when $v \in M(T, \mathscr{S})$.

In any case we obtain the inequalities

$$r(\theta : \rho^m) \leqslant \sup_g r(g(\theta) : \rho); \qquad \theta \in \Theta.$$

If m is right shift invariant on the right shift invariant space of functions on G spanned by the functions of the form $g \to (\mu g^{-1})\rho(v \circ g)$ where $\mu \in L(\mathscr{E})$ and $v \in M(T, \mathscr{S})$, then ρ^m is invariant so that the risk function of ρ^m is constant on orbits of G in Θ, i.e.

$$r(\theta : \rho^m) = r(g(\theta) : \rho^m)$$

when $\theta \in \Theta$ and $g \in G$.

In particular it follows that if the risk function of ρ is bounded above by an orbit invariant function b on Θ then this also holds for the generalized invariant decision rule ρ^m. Thus, under these conditions, minimax rules may be chosen invariant. If the decision space is a 2-point set then we arrive at the Hunt–Stein theorem for invariant testing problems (see examples 6.5.6 and 6.5.15).

There remains the problem of representing, if possible, the generalized decision rules by decision rules, i.e. Markov kernels. Following LeCam the solution could be to consider the original decision rule ρ as a transition from

$L(\mathscr{E})$ to Γ^* for some M-normed space Γ, and then work within the corresponding (compact) set of bilinear functionals on $L(\mathscr{E}) \times \Gamma$. In the case of invariance we could require that Γ is invariant under the actions of G on Γ. If $\Gamma = M(T, \mathscr{S})$ is the space of bounded measurable functions on $(T, \mathscr{S})$ then $\Gamma^* = \text{ba}(T, \mathscr{S})$ is the set of bounded additive set functions on $\mathscr{S}$. A transition from $L(\mathscr{E})$ to Γ^* is in this case just a generalized decision rule as defined in chapter 4. If T is a second countable locally compact Hausdorff space equipped with its class $\mathscr{S}$ of Borel subsets and if Γ is the set of continuous functions on T which vanish at infinity, then Γ^* may be identified with the space of finite regular Borel measures on $\mathscr{S}$. Having somehow obtained kernel representations of invariant generalized decision rules we could then proceed to try to convert it into an invariant kernel, using theorem 6.5.11.

The crucial condition of theorem 6.5.7 and of its corollaries, is the requirement that G possesses means which are sufficiently invariant. Let us begin our comments on this condition by first considering the simplest case where G possesses a mean which is right shift invariant on the class of all subsets of G. Actually this is equivalent to the requirement that there is a left (left and right) shift invariant mean on the class of all subsets of G.

We refer to Bondar & Milnes (1981) and Greenleaf (1969) for verification of this statement, as well as for the verification of other statements from group theory.

A group such that there is a shift invariant mean on the class of all subsets of G is called *amenable*. Quite a few groups are amenable. First the unique Haar probability measure provides a shift invariant mean whenever G is finite. There is a simple argument based on the Hahn–Banach theorem which shows that any Abelian group is amenable. Furthermore, several of the standard ways of constructing 'new' groups from 'old' groups yield amenable groups when the 'old' groups are amenable. Thus subgroups and homomorphic images of amenable groups are amenable. If a group G possesses an amenable normal subgroup and the factor group with respect to this subgroup is amenable, then G itself is amenable. Thus, since Abelian groups are amenable, any solvable group is amenable. It also follows that direct products of amenable groups are amenable.

The prominent example of a group which is not amenable is the free group with two generators. This example also appears to be of statistical significance since, by Bondar & Milnes (1981), there is a 'counterexample' to the Hunt–Stein theorem for this group.

If the group G is amenable then the invariance conditions in the theorem and its corollaries are automatically fulfilled for all choices of $M_{2,0}$ and for all subtransitions T.

Before proceeding, let us give the definitions of a topological group and of a measurable group. They may be formulated as follows.

A group G equipped with a topology is called a *topological group* if it is a Hausdorff topological space such that the map $(g, h) \to gh^{-1}$ from $G \times G$ to G is continuous. Similarly a group G is called a *measurable group* if the same map $(g, h) \to gh^{-1}$ from $G \times G$ to G is measurable. Thus a group G is a measurable group (topological group) if the group operations are measurable (continuous) and the topology is Hausdorff.

If G acts on $L(\mathscr{F})$ by Borel isomorphisms on $(\mathscr{Y}, \mathscr{B})$ then the space $M(\mathscr{Y}, \mathscr{B})$ of bounded $\mathscr{B}$-measurable functions is invariant (it is trivially total). Thus we might then take $M_{2,0} = M(\mathscr{Y}, \mathscr{B})$. If in addition G is a measurable group such that $g(x)$ and $g(y)$ are both jointly measurable and if M is a Markov kernel from $(\mathscr{X}, \mathscr{A})$ to $(\mathscr{Y}, \mathscr{B})$ then $M^g(B|x)$ is jointly measurable in (x, g) for each $B \in \mathscr{B}$. Hence $(\mu M^g)(B)$ is measurable in g for each B. It follows that it suffices to require that there is an additive probability set function on G which is right (or left) shift invariant on measurable sets. This condition is fulfilled by the Haar probability measure whenever G is a compact topological group equipped with the σ-algebra generated by the compact (open, closed) sets.

Locally compact groups which permit an invariant mean on the bounded measurable functions may be derived from other such groups according to rules similar to (and generalizing) those we mentioned for amenable groups. If we restrict our attention to even smaller classes of functions then we can obtain invariant means even in cases where the topological group is not amenable. Thus there is always an invariant mean on the class of functions, which vanishes at infinity.

Before leaving this topic let us just mention a possibly more familiar way of describing amenability. Let G be a group equipped with a σ-algebra $\mathscr{U}$ of subsets which is invariant under right shifts. Let Z be a linear subspace of the space of bounded measurable functions on G. Then Z permits a right shift invariant mean if and only if there is a net $\{m_t\}$ of probability measures on $\mathscr{U}$ which converges setwise to right shift invariance, i.e. $\lim_t [m_t(Uh^{-1}) - m_t(U)] = 0$ whenever $U \in \mathscr{U}$ and $h \in G$. This follows readily from Tychonoff's theorem and the fact that any mean on $\mathscr{U}$ is the setwise limit of probability measures with finite support (see chapter 4). If G is a second countable locally compact topological group then there is an invariant mean on $C(G)$ if and only if there is a sequence $\{m_n\}$ of probability measures which converges strongly to right invariance in the sense that $\sup_U |m_n(Uh^{-1}) - m_n(U)| \to 0$ as $n \to \infty$ for any $h \in G$. (See the note on theorem 3.6.2 in Greenleaf (1969).)

If G is the additive group of $\mathbb{R}^k$ or of the integers, then we may take $m_n(U) = \mu(U \cap [-n, n])/\mu[n, n]$ where μ is, respectively, Lebesgue measure and counting measure.

Theorem 6.5.7 and its corollaries provide conditions ensuring that comparison may be expressed in terms of invariant transitions. When we are handling concrete experiments (as linear normal models) then it is also important to know whether comparison may be expressed in terms of invariant kernels. We should therefore consider the problem of representing invariant (sub)transitions by invariant kernels.

Therefore let us return to the situation where the L-space isomorphisms are induced by Borel isomorphisms. Assume then, as before, that the group G acts on the experiment $\mathscr{E} = (\mathscr{X}, \mathscr{A}; P_\theta : \theta \in \Theta)$ and on the measurable space $(\mathscr{Y}, \mathscr{B})$ by the maps $g \to g_{\mathscr{X}}, g \to g_{\mathscr{Y}}$ and $g \to g_\Theta$ from G to, respectively, the set of Borel isomorphisms of $(\mathscr{X}, \mathscr{A})$ into itself, the set of Borel isomorphisms of $(\mathscr{Y}, \mathscr{B})$ into itself and the set of maps of Θ into itself. We continue to assume that the maps $g \to g_{\mathscr{X}}$ and $g \to g_{\mathscr{Y}}$ are homomorphisms and that $\mathscr{E}$ is invariant under G, i.e. that $P_\theta g^{-1} \equiv_{\theta,g} P_{g(\theta)}$. In that case the L-space isomorphism g_{L_1} of $L(\mathscr{E}) = L_1$ into itself, for each $g \in G$, is given by

$$g_{L_1}(\mu) = \mu g_{\mathscr{X}}^{-1}; \qquad \mu \in L(\mathscr{E}).$$

Similarly the Borel isomorphism $g_{\mathscr{Y}}$, for each $g \in G$, defines an L-space isomorphism g_{L_2} on the L-space of finite measures on $\mathscr{B}$ by putting $g_{L_2}(\nu) = \nu g_{\mathscr{Y}}^{-1}$ for each finite measure ν on $\mathscr{B}$. As noted before, the maps $g \to g_{L_1}$ and $g \to g_{L_2}$ are then both homomorphisms.

In this case the invariant subtransitions can be described in terms of kernel functions as follows.

Let T be a subtransition from $L(\mathscr{E})$ to $\text{ba}(\mathscr{Y}, \mathscr{B})$. Then the adjoint map T^* is a non-negative and linear map from $\text{ba}(\mathscr{Y}, \mathscr{B})^*$ to $M(\mathscr{E})$ which maps $1 \in \text{ba}(\mathscr{Y}, \mathscr{B})^*$ into the non-negative element $T^*1 = 1$ in $M(\mathscr{E})$. Now for each $B \in \mathscr{B}$, T^*I_B may be represented as a consistent family $M(B|\cdot) = (M_\theta(B|\cdot) : \theta \in \Theta)$ of test functions on $(\mathscr{Y}, \mathscr{B})$. These families are connected by the characterizing properties

$$1 \geqslant M_\theta(B_1 \cup B_2|\cdot) = M_\theta(B_1|\cdot) + M_\theta(B_2|\cdot) \geqslant 0 \quad \text{a.e.} \quad P_\theta$$

for all $\theta \in \Theta$ when B_1 and B_2 are disjoint sets in $\mathscr{B}$.

If we put $M_\theta^g(B|x) = M_{g(\theta)}(g[B]|g(x))$ then M^g also satisfies the characterizing properties and, in fact, represents the transition T^g. Thus T is invariant if and only if $M_{g(\theta)}(g[B]|g) = M_\theta(B|\cdot)$ a.e. P_θ whenever $\theta \in \Theta, B \in \mathscr{B}$ and $g \in G$.

In particular it follows that if T is defined by a Markov kernel M then T^g is represented by the Markov kernel M^g. Hence T is invariant if and only if M is $\mathscr{E}$-invariant.

If the experiment $\mathscr{E} = (\mathscr{X}, \mathscr{A}; P_\theta : \theta \in \Theta)$ is coherent and if the measurable space $(\mathscr{Y}, \mathscr{B})$ is Euclidean then, as explained in chapter 4, any transition from $L(\mathscr{E})$ to the L-space of finite measures on $\mathscr{B}$ is representable by a Markov kernel. The same arguments then show that any subtransition from $L(\mathscr{E})$ to the L-space of finite measures on $\mathscr{B}$ is representable by a sub-Markov kernel. Here a sub-Markov kernel from a measurable space $(\mathscr{X}, \mathscr{A})$ to a measurable space $(\mathscr{Y}, \mathscr{B})$ is a function $x \to M(\cdot|x)$ from $\mathscr{X}$ to the set of non-negative measures on $\mathscr{B}$ with total mass $\leqslant 1$ such that $M(B|x)$ is measurable in x for each $B \in \mathscr{B}$.

For each sub-Markov kernel M from $(\mathscr{X}, \mathscr{A})$ to $(\mathscr{Y}, \mathscr{B})$ and for each element of the group G, define the sub-Markov kernel M^g by putting $M^g(B|x) = M(g[B]|g(x))$ when $x \in \mathscr{X}$ and $B \in \mathscr{B}$. Generalizing the concepts of ε-invariance, uniform ε-invariance and invariance to sub-Markov kernels, we shall say that:

> the sub-Markov kernel M from $(\mathscr{X}, \mathscr{A})$ to $(\mathscr{Y}, \mathscr{B})$ is *$\mathscr{E}$-invariant* if $M^g(B|\cdot) = M(B|\cdot)$ a.e. P_θ for all $\theta \in \Theta$, all $B \in \mathscr{B}$ and all $g \in G$. The sub-Markov kernel M from $(\mathscr{X}, \mathscr{A})$ to $(\mathscr{Y}, \mathscr{B})$ is called *uniformly $\mathscr{E}$-invariant* if there is an orbit invariant $\mathscr{E}$-null set N in $\mathscr{A}$ such that $M^g \equiv_g M$ on N^c. Finally the sub-Markov kernel M is called *invariant* if $M^g \equiv_g M$.

We have just seen that if a subtransition T is induced by a sub-Markov kernel M, then T^g, for each g, is induced by the sub-Markov kernel M^g. In particular it follows that if M is induced by T, then T is invariant if and only if M is $\mathscr{E}$-invariant. We shall now see that provided certain regularity conditions are satisfied, any invariant subtransition can be represented by invariant sub-Markov kernels.

Theorem 6.5.11 (From $\mathscr{E}$-invariance to invariance). *Let there be given an experiment $\mathscr{E} = (\mathscr{X}, \mathscr{A}; P_\theta : \theta \in \Theta)$, a measurable space $(\mathscr{Y}, \mathscr{B})$ and a group G. Assume that G acts on $\mathscr{X}$, $\mathscr{Y}$ and Θ by maps $g \to g_{\mathscr{X}}$, $g \to g_{\mathscr{Y}}$ and $g \to g_\Theta$ such that the maps $g \to g_{\mathscr{X}}$ and $g \to g_{\mathscr{Y}}$ are homomorphisms from G to the group of Borel isomorphisms of, respectively, $(\mathscr{X}, \mathscr{A})$ onto itself and $(\mathscr{Y}, \mathscr{B})$ onto itself. The map $g \to g_\Theta$ is a map from Θ to the set of maps of Θ, into itself.*

Assume that $\mathscr{E}$ is invariant under G, i.e. that

$$P_\theta g^{-1} = P_{g(\theta)}; \qquad \theta \in \Theta, \quad g \in G.$$

If $\mathscr{E}$ is not discrete then we shall also assume that:

(i) *G is a measurable group possessing a non-null σ-finite measure $\tau \geqslant 0$ such that $\tau(Ug) = 0$ whenever $\tau(U) = 0$ and $g \in G$.*
(ii) *$\mathscr{B}$ is separable.*
(iii) *$g_{\mathscr{X}}(x)$ is jointly measurable in $(g, x) \in G \times \mathscr{X}$ and $g_{\mathscr{Y}}(y)$ is jointly measurable in $(g, y) \in G \times \mathscr{Y}$.*

Let M be an $\mathscr{E}$-invariant sub-Markov kernel from $\mathscr{E}$ to $(\mathscr{Y}, \mathscr{B})$. Then M is equivalent to (i.e. induces the same transition as) an invariant sub-Markov kernel.

If M is an $\mathscr{E}$-invariant (non-randomized) Markov kernel M from $\mathscr{E}$ to $(\mathscr{Y}, \mathscr{B})$ then M is equivalent to a uniformly $\mathscr{E}$-invariant (non-randomized) Markov kernel. If, in addition, there is at least one invariant (non-randomized) Markov kernel *from $(\mathscr{X}, \mathscr{A})$ to $(\mathscr{Y}, \mathscr{B})$ then M is equivalent to an invariant (non-randomized) Markov kernel.*

Finally assume that $\mathscr{B}$ distinguishes the points of $\mathscr{Y}$. Then any $\mathscr{E}$-invariant measurable function is equivalent to a uniformly $\mathscr{E}$-invariant measurable function. If, in addition, there is at least one invariant measurable function, then any $\mathscr{E}$-invariant measurable function is equivalent to an invariant measurable function.

Remark 1. At a first glance it might appear surprising that we do not need the usual assumptions on the structures of $(\mathscr{Y}, \mathscr{B})$ and $\mathscr{E}$ which are frequently used to regularize conditional probabilities. The explanation for this is that we are assuming from the start that M is regularized, i.e. that M is a sub-Markov kernel.

Remark 2. The separability assumption (ii) ensures that some unions of null sets may be replaced by countable unions of null sets and thus by single null sets.

Another useful consequence is that $\{0,1\}$-valued probability distributions on $\mathscr{B}$ may be localized as one-point distributions. Thus if Q is $\{0,1\}$-valued, $B_0 = \bigcap\{B : Q(B) = 1\}$ and $\mathscr{B}_0$ is a countable algebra generating $\mathscr{B}$, then $B_0 = \bigcap\{B : B \in \mathscr{B}_0, Q(B) = 1\}$ is a $\mathscr{B}$-atom with Q-measure 1 and $Q(B) \equiv_B I_B(y)$ whenever $y \in B_0$. This may be seen by checking that $\mathscr{D} = \{D : B_0 \subseteqq D$ or $B_0 \subseteqq D^c\}$ is a monotone class containing $\mathscr{B}_0$ and, consequently, containing $\mathscr{B}$. (The general atomic decomposition of separable σ-algebra may be found e.g. in Loéve (1977).)

It follows that any Markov kernel which, as a real valued map on $\mathscr{X} \times \mathscr{B}$, only attains the values 0 and 1, is determined by a measurable function, provided $\mathscr{B}$ is separable.

Remark 3. If G is a σ-compact locally compact topological group then we can choose τ in (i) as a right shift invariant Haar measure. Thus condition (i) is automatically satisfied in this case.

Proof of the theorem. Let M be an $\mathscr{E}$-invariant sub-Markov kernel from $(\mathscr{X}, \mathscr{A})$ to $(\mathscr{Y}, \mathscr{B})$. Then the set $N_{B,g} = \{x : M^g(B|x) \neq M(B|x)\}$ is $\mathscr{E}$-null for each $B \in \mathscr{B}$ and each $g \in G$. If $\mathscr{E}$ is discrete and $\mathscr{X}_0 = \{x : P_\theta(x) > 0$ for at least one $\theta\}$ then $N = \mathscr{X}_0^c$ is an invariant $\mathscr{E}$-null set containing all other $\mathscr{E}$-null sets. It follows that any $\mathscr{E}$-invariant sub-Markov kernel is actually uniformly $\mathscr{E}$-invariant in this case. This proves the 'discrete part' of the theorem. For the remaining part of the proof assume that conditions (i), (ii) and (iii) are all satisfied.

We may assume without loss of generality that τ is a probability measure. Combining the maps $(g, y) \to (g^{-1}, y)$ and $(g, y) \to g(y)$ we see that the map $(g, y) \to g^{-1}(y)$ is also jointly measurable. It follows that $M^g(B|x) = \int I_B(g^{-1}(y)) M(dy|g(x))$, for each $B \in \mathscr{B}$, is jointly measurable in (g, x).

Let V denote the set of all pairs (g, x) such that $M^g(B|x) \neq M(B|x)$ for some $B \in \mathscr{B}$. By (iii) there is a countable algebra $\mathscr{B}_0$ generating $\mathscr{B}$. Then $(g, x) \in V$ if and only if $M^g(B|x) \neq M(B|x)$ for some $B \in \mathscr{B}_0$. It follows that V is a measurable subset of $G \times \mathscr{X}$.

The assumption of $\mathscr{E}$-invariance implies that $P_\theta(\{x : (g, x) \in V\}) \equiv_g 0$. Hence, by Fubini's theorem, $(P_\theta \times \tau)(V) = \int \tau(\{g : (g, x) \in V\}) P_\theta(\mathrm{d}x) = 0$. Let S be the set of points $x \in \mathscr{X}$ such that $\tau(\{g : (g, x) \in V\}) > 0$. Then S is $\mathscr{E}$-null, i.e. $P_\theta(S) \equiv_\theta 0$. Let A be the set of points $x \in \mathscr{X}$ such that $M^g(B|x)$ is constant for τ-almost all g when $B \in \mathscr{B}$. Put $\hat{M}(B|x) = \int M^g(B|x)\tau(\mathrm{d}g)$. Then $\hat{M}$ is a Markov kernel from $(\mathscr{X}, \mathscr{A})$ to $(\mathscr{Y}, \mathscr{B})$. A point $x \in \mathscr{X}$ belongs to the set A if and only if $M^g(B|x) = \hat{M}(B|x)$ for τ-almost all g when $B \in \mathscr{B}_0$. (This may be seen by noting that the class of all sets B in $\mathscr{B}$ such that $M^g(B|x) = \hat{M}(B|x)$ for almost all g is monotone for fixed x.) It follows that A is measurable, i.e. that $A \in \mathscr{A}$.

Consider an element g_0 in the group G, a point x in the set A and a set B in the σ-algebra $\mathscr{B}$. Then $M^g(B|g_0(x)) = (M^g)^{g_0}(g_0^{-1}[B]|x) = M^{gg_0}(g_0^{-1}[B]|x)$. Now $\tau(U) = 0$ where $U = \{g : M^g(g_0^{-1}[B]|x) \neq \hat{M}(g_0^{-1}[B]|x)\}$. Hence, by assumption (i), $\tau(Ug_0^{-1}) = 0$. The set Ug_0^{-1} may be described as follows:

$$\begin{aligned} Ug_0^{-1} &= \{gg_0^{-1} : M^g(g_0^{-1}[B]|x) \neq \hat{M}(g_0^{-1}[B]|x)\} \\ &= \{g : M^{gg_0}(g_0^{-1}[B]|x) \neq \hat{M}(g_0^{-1}[B]|x)\} \\ &= \{g : M^g(B|g_0(x)) \neq \hat{M}(g_0^{-1}[B]|x)\}. \end{aligned}$$

Thus $M^g(B|g_0(x)) = \hat{M}(g_0^{-1}[B]|x)$ for τ-almost all g. Since this holds for all $B \in \mathscr{B}$, we find that $g_0(x) \in A$.

It follows that A is invariant under G, i.e. $g[A] \equiv_g A$. Hence $N = A^c$ is also invariant under G.

Let x be any point in $\mathscr{X} - S$. Then $\tau(\{g : (g, x) \in V\}) = 0$ so that $M^g(B|x) = M(B|x)$ for τ-almost all g for all $B \in \mathscr{B}$. Thus $x \in A$. Hence $S^c \subseteqq A$ so that $N = A^c \subseteqq S$. It follows that N is an $\mathscr{E}$-null set, i.e. that $P_\theta(N) \equiv_\theta 0$. Let μ be a measure on $\mathscr{A}$ such that $0 \leqslant \mu \leqslant P_\theta$. Let $B \in \mathscr{B}$. Then, since M is $\mathscr{E}$-invariant,

$$\begin{aligned} (\mu\hat{M})(B) &= \int \hat{M}(B|x)\mu(\mathrm{d}x) = \int\left(\int M^g(B|x)\tau(\mathrm{d}g)\right)\mu(\mathrm{d}x) \\ &= \int\left(\int M^g(B|x)\mu(\mathrm{d}x)\right)\tau(\mathrm{d}g) = \int (\mu M^g)(B)\tau(\mathrm{d}g) \\ &= \int (\mu M)(B)\tau(\mathrm{d}g) = \mu M(B). \end{aligned}$$

Thus $\hat{M}$ is equivalent to M.

Now consider a triple (g_0, x, B) where $g_0 \in G$, $x \in \mathscr{X}$ and $B \in \mathscr{B}$. We saw above that $M^g(B|g_0(x)) = \hat{M}(g_0^{-1}[B]|x)$ for τ-almost all g. Hence $\hat{M}(B|g_0(x)) = \hat{M}(g_0^{-1}[B]|x)$ in this case. Substituting $g_0[B]$ for B we find that $\hat{M}(g_0[B]|g_0(x))$

$= \hat{M}(B|x)$ when $x \notin A$. Thus $\hat{M}$ is invariant on A. This completes the proof of the first assertion of the theorem.

If $x \in A$ then for each set $B \in \mathscr{B}_0$ there is a τ-null set $U_B \subseteq G$ so that $M^g(B|x) = \hat{M}(B|x)$ when $g \notin U_B$. Put $U_0 = \bigcup \{U_B : B \in \mathscr{B}_0\}$. Then $M^g(B|x) \equiv_B \hat{M}(B|x)$ when $g \notin U_0$. The set $G - U_0$ cannot be empty, since $\tau(U_0) = 0$. Thus $\hat{M}(B|x) \equiv_B M^g(B|x)$ for some g when $x \in A$. Hence $\hat{M}$ is $\{0,1\}$-valued on A whenever M is $\{0,1\}$-valued.

The proof follows now from remark 3 and by noting that $\hat{M}$ is a Markov kernel when M is a Markov kernel. □

Assuming that G acts on $\mathscr{Y}$ by the identity map, we obtain the following corollary.

Corollary 6.5.12 (From almost orbit constancy to orbit constancy). *Let there be given an experiment $\mathscr{E} = (\mathscr{X}, \mathscr{A}; P_\theta : \theta \in \Theta)$ and a measurable space $(\mathscr{Y}, \mathscr{B})$ such that $\mathscr{B}$ is separable.*

Assume that G acts on $\mathscr{E}$ by the maps $g \to g_{\mathscr{X}}$ and $g \to g_\Theta$ such that the map $g \to g_{\mathscr{X}}$ is a homomorphism from G to the group of the Borel isomorphisms of $(\mathscr{X}, \mathscr{A})$. The map $g \to g_\Theta$ is a map from GU to the set of maps of Θ into itself.

Assume also that $g_{\mathscr{X}}(x)$ is jointly measurable in (g, x) and that G possesses a σ-finite measure τ such that $\tau(Ug) = 0$ whenever $\tau(U) = 0$ and $g \in G$.

Then any $\mathscr{E}$-invariant (sub-) Markov kernel is equivalent to an invariant (sub-) Markov kernel, i.e. a (sub-) Markov kernel M such that

$$M(B|g(x)) = M(B|x)$$

when $B \in \mathscr{B}$, $x \in \mathscr{X}$ and $g \in G$.

Assume in addition that $\mathscr{B}$ distinguishes points of $\mathscr{Y}$. Let s be a measurable function which is $\mathscr{E}$-invariant, i.e. $s \circ g = s$ a.e. P_θ for all $\theta \in \Theta$ and all $g \in G$. Then s is equivalent to a measurable function $\tilde{s}$ which is invariant, i.e. $\tilde{s} \circ g = \tilde{s}$ for all $g \in G$.

Let us apply the theorem and its corollary to some of the situations discussed earlier in this section.

Example 6.5.13 (Comparison of absolutely continuous translation experiments. Continuation of example 6.5.3.). Throughout this example we will assume that $\mathscr{X}$ is a second countable locally compact topological group equipped with its class of Borel subsets. We shall also assume that *the topological group $\mathscr{X}$ is amenable*, i.e. that there is a right (or left) shift invariant mean on $\mathscr{A}$.

Say that a finite measure is absolutely continuous if it is absolutely continuous with respect to (right or left) Haar measure.

The above theorems imply that if P is absolutely continuous then the deficiency of the experiment $\mathscr{E}_P$ with respect to any other translation experiment $\mathscr{E}_Q$ is given by

$$\delta(\mathscr{E}_P, \mathscr{E}_Q) = \min_M \|M * P - Q\|$$

where M runs through the set of all probability measures on $\mathscr{A}$. It is worth noticing that the condition that P is absolutely continuous is equivalent to each of the following four conditions:

(a) $\mathscr{E}_P$ is dominated.
(b) $(P_\theta : \theta \in \Theta) \sim$ Haar measure.
(c) the map $\theta \to P_\theta(A)$ is continuous in θ for each $A \in \mathscr{A}$.
(d) the map $\theta \to P_\theta$ from Θ to $L(P_\theta : \theta \in \Theta)$ is continuous.

Since the implications (d) $\Rightarrow$ (c) $\Rightarrow$ (a) are trivial it suffices to show that (a) $\Rightarrow$ (b) $\Rightarrow$ (d). If (a) holds then $(P_\theta : \theta \in \Theta = \mathscr{X})$ is equivalent to a probability measure π. If $\pi(A\theta_0^{-1}) = 0$ then $P_\theta(A\theta_0^{-1}) \equiv_\theta P(A(\theta\theta_0^{-1})) \equiv_\theta 0$ so that $P_\theta(A) \equiv_\theta 0$ and thus $\pi(A) = 0$. It follows that the experiment $(\pi_\theta : \theta \in \Theta)$ is homogeneous. Letting μ denote right Haar measure, by Fubini's theorem we find that $\int \pi_\theta(A)\mu(\mathrm{d}\theta) = \int \mu(x^{-1}A)\pi(\mathrm{d}x)$. It follows readily that $\mu(A) = 0$ if and only if $\pi(A) = 0$ so that (b) holds. If (b) holds and if $\mathrm{d}P/\mathrm{d}\mu$ is continuous then (d) follows from Scheffe's convergence theorem. In general the implication (b) $\Rightarrow$ (d) may be argued by first approximating $\mathrm{d}P/\mathrm{d}\mu$ in $L_1(\mu)$ with continuous functions having compact supports and then using Scheffe's convergence theorem.

Deficiencies between absolutely continuous translation experiments may conveniently be expressed in terms of minimax risk for estimating θ for invariant loss. In order to clarify this let us note that an invariant non-negative loss function L is of the form $L_\theta(t) = \phi(t\theta^{-1})$ for some non-negative measurable function ϕ on $\mathscr{X}$. Looking for minimax procedures, as argued after corollary 6.5.10 we may restrict our attention to invariant procedures. If the invariant kernel is represented as left convolution multiplication with the probability measure M, then the risk in $\mathscr{E}_P$ becomes

$$r(\theta : M) = \int\int \phi(gx)M(\mathrm{d}g)P(\mathrm{d}x).$$

It follows that as far as minimax risk is concerned, we may restrict our attention to estimators which are left multiplications, i.e. which are of the form $\hat{\theta}(x) \equiv_x gx$ for a fixed element g in $\mathscr{X}$. The greatest lower bound of maximum risks may therefore be expressed as $m(\mathscr{E}_P, \phi) = \inf_g \int \phi(gx)P(\mathrm{d}x)$. This yields the lower bound $2\sup\{m(\mathscr{E}_P, \phi) - m(\mathscr{E}_Q, \phi) : \|\phi\| = 1\}$ for the deficiency $\delta(\mathscr{E}_P, \mathscr{E}_Q)$. Using standard support function arguments (see chapter 8 and Torgersen (1972)) it may be shown that this is also an upper bound so that

$$\delta(\mathscr{E}_P, \mathscr{E}_Q) = 2\sup\{m(\mathscr{E}_P, \phi) - m(\mathscr{E}_Q, \phi) : \|\phi\| = 1\}$$

and

$$\Delta(\mathscr{E}_P,\mathscr{E}_Q) = 2\sup\{|m(\mathscr{E}_P,\phi) - m(\mathscr{E}_Q,\phi)| : \|\phi\| = 1\}$$

when P and Q are absolutely continuous.

The deficiency $\delta(\mathscr{E}_P,\mathscr{E}_Q)$ is at most $\inf_\theta \|_\theta P - Q\|$ which measures the accuracy we may hope for when we restrict ourselves to non-randomized kernels, i.e. to left multiplications. As an illustration of this consider the situation where $\mathscr{X}$ is the additive group of integers. Let P assign masses 3/8, 4/8 and 1/8 to, respectively, -1, 0 and 1, while Q assigns the same masses to, respectively, 1, 0 and -1. Then $\|M * P - Q\| = \|Q * N - P\| = \frac{1}{6}$ when $M(N)$ assigns masses 1/3 and 2/3 to 0 and 1 (-1) respectively. It follows that $\Delta(\mathscr{E}_P,\mathscr{E}_Q) \leqslant \frac{1}{6}$. The inaccuracy of the best approximation obtainable by non-randomized kernels is $\inf_\theta \|_\theta P - Q\| = \frac{1}{2}$ and consequently is at least three times as large as Δ. In spite of such examples it may be shown that if $P_1, P_2, \ldots$ and P are absolutely continuous then $\lim_{n\to\infty} \Delta(\mathscr{E}_{P_n},\mathscr{E}_P) = 0$ if and only if $\lim_{n\to\infty} \inf_\theta \|_\theta P_n - P\| = 0$.

If we do not assume that P is absolutely continuous then, as our next example will show, the convolution formula for the deficiency $\delta(\mathscr{E}_P,\mathscr{E}_Q)$ is not necessarily valid. The assumption, which is in force throughout this example, that $\mathscr{X}$ permits a mean which is right shift invariant on $\mathscr{A}$, implies however, that $\delta(\mathscr{E}_P,\mathscr{E}_Q) = \min_T \|PT - Q\|$ where T runs through the set of invariant transitions from $L(\mathscr{E}_P)$ to $L(\mathscr{E}_Q)$.

Finally let us consider the condition of amenability of $M(\mathscr{X},\mathscr{A})$ in terms of comparison of experiments. The simplest way of stating this condition is in terms of a totally non-informative translation experiment. If we allow ourselves to work with additive probability set functions which might not be σ-additive, then any right shift invariant mean m on $M(\mathscr{X},\mathscr{A})$ defines a totally non-informative experiment. If, however, we want to express the condition in terms of σ-additive probability set functions, then we may do this in terms of sequences $\{P_n\}$ of probability measures which converge strongly to right invariance, i.e. $\|P_{n,\theta} - P\| \to 0$ as $n \to \infty$ for each θ. We refer to Greenleaf (1969) for a proof of this statement. This in turn is equivalent to the condition that $\mathscr{E}_{P_n}|\Theta_0 \to_\Delta \mathscr{M}_i$ for each finite subset Θ_0 of Θ. Thus the condition of amenability is equivalent to the condition that $\mathscr{M}_i$ belongs to the closure of the set of translation experiments for the topology of Δ-convergence on finite subsets of Θ.

The next example shows, even for amenable groups, that in general it does not suffice to restrict our attention to invariant kernels when we are comparing non-coherent experiments.

Example 6.5.14 (Comparison of invariant non-coherent experiments). Let us specify the set-up in the previous example by letting $\mathscr{X} = \mathbb{R}$ be the additive group of real numbers topologized in the usual way. This group, like any other Abelian group, is amenable.

Let τ be any absolutely continuous probability measure on the Borel class $\mathscr{A}$ and let δ_θ, for each $\theta \in \Theta = \mathbb{R}$, be the one-point distribution in θ. Put $P = (1-\lambda)\tau + \lambda\delta$ where $0 \leqslant \lambda \leqslant 1$ and let Q be any probability measure on $\mathscr{A}$ such that $\mathscr{E}_P \geqslant \mathscr{E}_Q$. Then, by theorem 6.5.7, $Q = PT$ for some shift invariant transition T from $L(\mathscr{E}_P)$ to $L(\mathscr{E}_Q)$.The convolution criterion applied to $\mathscr{E}_\tau$ implies that there is a Markov kernel M such that $\tau T = M * T$. Hence $Q = PT \geqslant (1-\lambda)\tau T = (1-\lambda)(M * \tau)$. Conversely assume that $Q \geqslant (1-\lambda)M * \tau$ for some M. Let $\hat{Q}$ be a probability measure such that $Q = (1-\lambda)M * \tau + \lambda\hat{Q}$. Choose a finite subset F of Θ and define a Markov kernel $\tilde{M}$ by putting $\tilde{M}(B|x) = M(B-x)$ or $= \hat{Q}(B-x)$ as $x \notin F$ or $x \in F$. Then $Q_\theta = (1-\lambda)M * \tau_\theta + \lambda\hat{Q}_\theta = (1-\lambda)\tau_\theta\tilde{M} + \lambda\delta_\theta\tilde{M} = P_\theta\tilde{M}$ when $\theta \in F$. It follows then by the randomization criterion that $\mathscr{E}_P|F \geqslant \mathscr{E}_Q|F$. Hence $\mathscr{E}_P \geqslant \mathscr{E}_Q$.

Altogether this shows that $\mathscr{E}_P \geqslant \mathscr{E}_Q$ if and only if $Q \geqslant (1-\lambda)M * \tau$ for some probability measure M.

If P is a convolution factor in Q then, by the Markov kernel criterion, $\mathscr{E}_P \geqslant \mathscr{E}_Q$. However, we may have $\mathscr{E}_P \geqslant \mathscr{E}_Q$ in cases where P (which necessarily is not absolutely continuous) is not a convolution factor of Q. Put e.g. $Q = (1-\lambda)(\rho * \tau) + \lambda\delta$ for some probability distribution $\rho \neq \delta$. Then $Q \geqslant (1-\lambda)(\rho * \tau)$ so that $\mathscr{E}_P \geqslant \mathscr{E}_Q$. If P is a convolution factor of P, say $Q = M * P$, then

$$(1-\lambda)(M * \tau) + \lambda M = (1-\lambda)(\rho * \tau) + \lambda\delta.$$

Assume now that $0 < \lambda < 1$. Comparing the masses in 0 we find $\lambda M(0) = \lambda$ so that $M = \delta$. Then $\tau = \rho * \tau$ contradicting the assumption that $\rho \neq \delta$.

This construction also yields a counterexample to another 'likely' hypothesis. If P, Q and π are probability distributions on the real line and if $\mathscr{E}_P \geqslant \mathscr{E}_Q$ then, by the convolution criterion, $\mathscr{E}_{P*\pi} \geqslant \mathscr{E}_{Q*\pi}$. (If the set of zeros of the characteristic function of π does not have interior points then this implication may, by the same criterion, be reversed.)

Let P and Q be as above with $0 < \lambda < 1$. Thus $P = (1-\lambda)\tau + \lambda\delta$ and $Q = (1-\lambda)(\rho * \tau) + \lambda\delta$ where we now assume that the probability distribution ρ differs from δ. Then, since $Q \geqslant (1-\lambda)\rho * \tau$, the informational inequality $\mathscr{E}_P \geqslant \mathscr{E}_Q$ holds. Let π be any absolutely continuous probability distribution on $\mathbb{R}$ such that the set of zeros of its characteristic function does not possess interior points. We cannot then have $\mathscr{E}_{P*\pi} \geqslant \mathscr{E}_{Q*\pi}$ since by the convolution criterion this leads to the contradiction that P is a convolution factor in Q.

Example 6.5.15 (Invariant performance functions. Hunt–Stein's theorem. Continuation of example 6.5.5). Assume that we are given an experiment $\mathscr{F} = (T, \mathscr{S}; Q_\theta : \theta \in \Theta)$ on the decision space $(T, \mathscr{S})$ and a (generalized) decision rule ρ from $\mathscr{E} = (\mathscr{X}, \mathscr{A}; P_\theta : \theta \in \Theta)$ to $(T, \mathscr{S})$. Assume also that the pair $(\mathscr{E}, \mathscr{F})$ is invariant under a group G acting by maps $g \to g_{\mathscr{X}}$, $g \to g_T$ and $g \to g_\Theta$. We have just studied conditions ensuring the existence of an invariant (generalized) decision rule $\bar{\rho}$ from $\mathscr{E}$ to $\mathscr{F}$ which is 'fairly' independent of $\mathscr{F}$ such that

$$\|P_\theta\bar{\rho} - Q_\theta\| \leqslant \sup_g \|P_{g(\theta)}\rho - Q_{g(\theta)}\|; \qquad \theta \in \Theta.$$

Now consider the particular case where Θ may be invariantly decomposed as $\Theta = \Theta_1 \cup \Theta_2 \cup \cdots \cup \Theta_r$ and where the problem is to decide which of the sets Θ_i contain the true value of the parameter. When we say '*invariantly decomposed*' we simply mean that the sets $\Theta_1, \ldots, \Theta_r$ are disjoint and that $g[\Theta_i] = \Theta_i$; $g \in G; i = 1, \ldots, r$. Let us identify each set Θ_i with its label i. Thus our decision space becomes the finite set $T_r = \{1, \ldots, r\}$. The ideal decision is then $t = i$ when $\theta \in \Theta_i$. The requirement of invariance of the correct decision is met here provided we let each $g \in G$ act on T by the identity map.

Let Q_θ assign mass 1 to $t = i$ when $\theta \in \Theta_i$. Then the experiment $(Q_\theta : \theta \in \Theta)$ is invariant under g, i.e. $Q_\theta = Q_{g(\theta)}$ when $\theta \in \Theta$ and $g \in G$.

For simplicity let us assume that $\mathscr{E}$ is coherent and let ρ be a decision rule from $\mathscr{E}$ to $(T, \mathscr{S})$. Assume that there is a right shift invariant mean m on the right shift invariant space spanned by the functions $g \to \mu g^{-1}\rho_i$ where $\mu \in L(\mathscr{E})$. Then by theorem 6.5.7, there is an $\mathscr{E}$-invariant decision rule $\bar{\rho}$ such that

$$\|P_\theta\bar{\rho} - Q_\theta\| \leqslant \sup_g \|P_{g(\theta)}\rho - Q_{g(\theta)}\|; \qquad \theta \in \Theta.$$

Writing out the statistical distances involved here we see that these inequalities may be written $P_\theta\bar{\rho}(i) \geqslant \inf_g P_{g(\theta)}\rho(i)$ when $\theta \in \Theta_i$. Thus the minimum probability of arriving at the correct decision for any given θ belonging to an invariant subset of Θ is at least as large for $\bar{\rho}$ as it is for ρ. Actually we saw that $\bar{\rho}$ might be chosen so that $E_\theta\bar{\rho}_i = \int [E_{g(\theta)}\rho_i] m(dg)$ for all $\theta \in \Theta$ and all $i = 1, 2, \ldots, r$. This implies that if L is any invariant loss function, i.e. $L_{g(\theta)} = L_\theta$ when $\theta \in \Theta$ and $g \in G$, then $r(\theta : \bar{\rho}, L) = \int r(g(\theta) : \rho, L) m(dg)$. Hence

$$\min_{\theta \in \tilde{\Theta}} r(\theta : \rho, L) \leqslant r(\theta; \bar{\rho}, L) \leqslant \sup_{\theta \in \tilde{\Theta}} r(\theta : \rho, L); \qquad \theta \in \tilde{\Theta}$$

for any G-invariant subset $\tilde{\Theta}$ of Θ.

If we put $r = 2$ then we obtain $P_\theta\bar{\rho}(2) \geqslant \inf_g P_{g(\theta)}\bar{\rho}(2); \theta \in \Theta_2$ while $P_\theta\bar{\rho}(2) = (1 - P_\theta\bar{\rho}(1)) \leqslant (1 - \inf_g P_{g(\theta)}\rho(1)) = \sup_g P_{g(\theta)}\rho(2); \theta \in \Theta_1$.

It follows that if ρ is a test for the hypothesis '$\theta \in \Theta_1$' against the alternative '$\theta \in \Theta_2$' with level of significance α and with power being at least β on Θ_2, then the $\mathscr{E}$-invariant test $\bar{\rho}$ has the same properties. Thus, provided $\mathscr{E}$ is coherent and provided right shift invariant means exist on the appropriate space of functions on G, maximin tests can be chosen $\mathscr{E}$-invariant.

Furthermore, if G is a measurable group possessing a non-null σ-finite measure τ such that $\tau(Ug) = 0$ when $\tau(U) = 0$ and $g \in G$ and if $g(x)$ is jointly measurable in (g, x), then maximin tests may be chosen invariant.

This is essentially the content of the Hunt–Stein theory (Hunt & Stein, 1946) for testing invariant hypotheses as described in e.g. Lehmann, 1986.

The following example shows that if the groups are not amenable then invariant transitions may not suffice.

Example 6.5.16 (A 'counterexample' for non-amenable groups by Stein. Continuation of example 6.5.5). Let us consider the example from Lehmann (1986, example 9 in chapter 9) to which we referred in example 6.5.5. In this example non-trivial minimax tests cannot be chosen invariant. Referring to corollary 6.5.10 we see that this is also a situation where comparison within invariant pairs of experiments cannot be expressed in terms of invariant kernels.

Stein's example may be described as follows.

Let $r \geqslant 2$ be an integer and let G be the group of all $r \times r$ nonsingular matrices. Equip G with the topology and with the measurability structure such that it has a subset of $\mathbb{R}^{(r^2)}$. Let Σ denote the set of positive definite symmetric $r \times r$ matrices and take the cartesian product $]0, \infty[\times \Sigma$ as the parameter set Θ. Consider two independent random vectors $X^{(1)}$ and $X^{(2)}$ in $\mathbb{R}^r$ such that $X^{(1)}$ is normal $(0, \sigma)$ while $X^{(2)}$ is normal $(0, \Delta\sigma)$ where $(\Delta, \sigma) \in \Theta$. Let $\mathscr{E} = (\mathscr{X}, \mathscr{A}; P_\theta : \theta \in \Theta)$ be the experiment which is realized by the pair $(X^{(1)}, X^{(2)})$. Thus $\mathscr{X} = \mathbb{R}^r \times \mathbb{R}^r$, $\mathscr{A} =$ the class of Borel subsets of $\mathscr{X}$ and $P_\theta = N(0, \sigma) \times N(0, \Delta\sigma)$ when $\theta = (\Delta, \sigma) \in \Theta$.

If $(X^{(1)}, X^{(2)})$ is distributed according to $\theta = (\Delta, \sigma)$ and $g \in G$ then $(gX^{(1)}, gX^{(2)})$ is distributed according to $\bar{\theta} = (\Delta, g\sigma g') \in \Theta$.

Thus this $\mathscr{E}$ becomes G-invariant if G acts on $\mathscr{X} = \mathbb{R}^r \times \mathbb{R}^r$ and on Θ by

$$g(x^{(1)}, x^{(2)}) = (gx^{(1)}, gx^{(2)})$$

and

$$g(\Delta, \sigma) = (\Delta, g\sigma g')$$

when $(x^{(1)}, x^{(2)}) \in \mathscr{X}$, $(\Delta, \sigma) \in \Theta$ and $g \in G$.

Let $x^{(1)}, x^{(2)}, u^{(1)}$ and $u^{(2)} \in \mathbb{R}^r$. Then there exists a $g \in G$ such that $gx^{(1)} = u^{(1)}$ and $gx^{(2)} = u^{(2)}$ provided $x^{(1)}$ and $x^{(2)}$ are linearly independent. It follows that any invariant test δ is equivalent to a constant. Hence, by theorem 6.5.11, any $\mathscr{E}$-invariant test is also equivalent to a constant.

Note next that if $F = [X_1^{(2)}]^2/[X_1^{(1)}]^2$ then F/Δ has an F-distribution.

Consider the problem of testing the hypothesis H: '$\Delta \leqslant \Delta_0$' against the alternative '$\Delta \geqslant \Delta_1$' where $\Delta_0 < \Delta_1$. Then any non-trivial test with rejection region '$F \geqslant$ constant' has a power function which is strictly increasing in Δ and which does not depend on σ.

Any partitioning of Θ based on Δ alone is clearly invariant. Thus we might expect there to be an invariant test δ with minimum power on the region '$\Delta \geqslant \Delta_1$' which is strictly greater than its level of significance. The fact that any invariant test has a constant power function tells us, however, that this is impossible.

Looking over the results proved and checking the assumed regularity conditions, we are forced to conclude that there is no mean on G which is right (or left) shift invariant on the class of Borel subsets.

We shall now consider two situations where we do not need any assumptions about existence of invariant means. First consider the informational inequality $\mathscr{E} \geqslant \mathscr{F}$ when $\mathscr{E}$ is boundedly complete.

Example 6.5.17 (Estimation of shift for invariant loss. Continuation of example 6.5.13). Consider an absolutely continuous translation experiment $\mathscr{E}_P$ on a finite dimensional inner product space $\mathscr{X}$. We shall again consider the problem of estimating the shift θ for the loss function $(\theta, t) \to \phi(t - \theta)$ where ϕ is non-negative and measurable. By invariance, as explained in example 6.5.13, as far as minimax risk is concerned, we can restrict our attention to estimators $\hat{\theta}$ of the form $\hat{\theta}(x) \equiv_x x + g$ for a fixed vector g in $\mathscr{X}$. The risk of this estimator is $\int \phi(x + g)P(\mathrm{d}x) = \int_0^\infty [1 - P(A_c - g)]\,\mathrm{d}c$ where $A_c = \{x : \phi(x) \leqslant c\}$.

Now there are various known conditions which ensure that $P(A_c - g) \leqslant P(A_c)$. By Anderson's inequality (Anderson, 1955) this is the case if the density of P with respect to Lebesgue measure may be specified quasi-concave and symmetric with respect to reflection about the origin, while ϕ is quasi-convex and symmetric in the same sense. In particular this applies to the situation where P is a (nonsingular) normal distribution and ϕ is a seminorm. In the case of quadratic loss these estimators are then inadmissible when the dimension of $\mathscr{X}$ is greater than two (Stein, 1956).

Theorem 6.5.18 (Majorization by boundedly complete experiments). *Let* $\mathscr{E} = (\mathscr{X}, \mathscr{A}; P_\theta : \theta \in \Theta)$ *and* $\mathscr{F} = (\mathscr{Y}, \mathscr{B}; Q_\theta : \theta \in \Theta)$ *be two experiments.*

Assume that the pair $(\mathscr{E}, \mathscr{F})$ *of experiments is invariant under G and that* $\mathscr{E}$ *is boundedly complete and coherent.*

Assume that the actions of G are given by maps $g \to g_{\mathscr{X}}$, $g \to g_{\mathscr{Y}}$ *and* $g \to g_\Theta$ *where the first two maps are homomorphisms from G to the group of Borel isomorphisms on the indicated spaces.*

Then $\mathscr{E} \geqslant \mathscr{F}$ *if and only if* $Q_\theta \equiv P_\theta T$ *for some invariant transition T from* $L(\mathscr{E})$ *to* $L(\mathscr{F})$.

Remark 1. It is not difficult to show that $\mathscr{E} \geqslant \mathscr{F}$ whenever $\mathscr{E} \geqslant_2 \mathscr{F}$ and $\mathscr{E}$ is coherent as well as boundedly complete.

Remark 2. If conditions (i), (ii) and (iii) of theorem 6.5.11 are satisfied then T may be represented by a uniformly $\mathscr{E}$-invariant kernel.

Proof of the theorem. The Markov kernel criterion implies that $Q_\theta \equiv_\theta P_\theta T$ for some transition T from $L(\mathscr{E})$ to $L(\mathscr{F})$. Then, by invariance, $Q_\theta \equiv_{\theta,g} P_\theta T^g$ so that, for each $B \in \mathscr{B}$, $\int [T^g]^* I_B \,\mathrm{d}P_\theta \equiv_\theta \int T^* I_B \,\mathrm{d}P_\theta$ when $g \in G$. Hence $[T^g]^* I_B \equiv_{g,B} T^* I_B$ so that $[T^g]^* \equiv_g T^*$, i.e. $T^g \equiv_g T$. □

The other situation where we do not need any assumption on invariant means is in the problem of the exchangeability of the principles of invariance and of sufficiency. Before giving one of the results in the paper by Hall, Wijsmann and Ghosh (1965) let us make a few remarks on the terminology. Some background information may be found in example 6.5.6, and in proposition 6.5.1 and the development leading up to this proposition. Suppose we are given an experiment $\mathscr{E} = (\mathscr{X}, \mathscr{A}; P_\theta : \theta \in \Theta)$. Recall that a sub σ-algebra $\mathscr{S}$ of $\mathscr{A}$ is sufficient if and only if for each event A, $P_\theta(A|\mathscr{S})$ may be specified independently of θ. If $\mathscr{S}_1$ and $\mathscr{S}_2$ are two sub σ-algebras then we shall say that *$\mathscr{S}_1$ is contained in $\mathscr{S}_2$ up to $\mathscr{E}$-null sets* if to each $S_1 \in \mathscr{S}_1$ there corresponds an $S_2 \in \mathscr{S}_2$ such that $P_\theta(S_1 \Delta S_2) \equiv_\theta 0$. The notation for this is $\mathscr{S}_1 \leqslant \mathscr{S}_2$. If $\mathscr{S}_1 \leqslant \mathscr{S}_2$ and $\mathscr{S}_2 \leqslant \mathscr{S}_1$ then we shall say that $\mathscr{S}_1$ and $\mathscr{S}_2$ are equivalent (up to $\mathscr{E}$-null sets) and write this $\mathscr{S}_1 \sim \mathscr{S}_2$. The relation '$\leqslant$' is clearly a partial ordering and '$\sim$' is the associated equivalence relation. The 'closure' $\bar{\mathscr{S}}$ of $\mathscr{S}$ is the σ-algebra of events A such that $P_\theta(A \Delta S) \equiv_\theta 0$ for some $S \in \mathscr{S}$. Clearly $\bar{\bar{\mathscr{S}}} = \bar{\mathscr{S}}$. Finally a σ-algebra $\mathscr{S}$ is called minimal sufficient if it is sufficient and if $\mathscr{S} \leqslant \mathscr{S}_1$ whenever $\mathscr{S}_1$ is sufficient. Clearly $\mathscr{S}$ is (minimal) sufficient if and only if $\bar{\mathscr{S}}$ is (minimal) sufficient. There is at most one minimal sufficient closed sub σ-algebra $\mathscr{S}$ and it may be shown that any coherent experiment possesses minimal sufficient sub σ-algebras.

Let the group G act on the sample space $(\mathscr{X}, \mathscr{A})$ of the experiment $\mathscr{E} = (\mathscr{X}, \mathscr{A}; P_\theta : \theta \in \Theta)$ by the homomorphism $g \to g_{\mathscr{X}}$ from G to the group of Borel isomorphisms of $(\mathscr{X}, \mathscr{A})$ onto itself. Assume that G acts on Θ by the map $g \to g_\Theta$ from G to the set of maps of Θ into itself.

Finally assume that $\mathscr{E}$ is invariant under these actions of G i.e. that $P_\theta g^{-1} \equiv_\theta P_{g(\theta)}$; $g \in G$.

For any sub σ-algebra $\mathscr{S}$ of $\mathscr{A}$ let the σ-algebra of invariant sets in $\mathscr{S}$ be denoted by $\mathscr{S}_I$, i.e.

$$\mathscr{S}_I = \{S : S \in \mathscr{S} \text{ and } g[S] \underset{g}{\equiv} S\}.$$

Let t be any measurable function (statistic) from $(\mathscr{X}, \mathscr{A})$ to the measurable space $(\mathscr{Y}, \mathscr{B})$. Then t is $\mathscr{S}_I$-measurable if and only if t is $\mathscr{S}$-measurable and $t(g(x)) \sim t(x) \bmod(\mathscr{B})$ whenever $x \in \mathscr{X}$ and $g \in G$.

A function Φ on $\mathscr{X}$ is called a *maximal invariant* if $\Phi(x_1) = \Phi(x_2)$ if and only if $x_2 = g(x_1)$ for some $g \in G$. Thus Φ is a maximal invariant if and only if its contours are precisely the orbits of G. The prominent maximal invariant is thus the map which to each point $x \in \mathscr{X}$ assigns the orbit passing through it.

Algebras of invariant events and invariant statistics may be characterized in terms of a given maximal invariant Φ as follows.

Let $\mathscr{S}$ be a sub σ-algebra of $\mathscr{A}$ and let t be a measurable function from $(\mathscr{X}, \mathscr{A})$ to $(\mathscr{Y}, \mathscr{B})$. Assume that $\mathscr{B}$ distinguishes points of $\mathscr{Y}$. Then $\mathscr{S}_I = \{S : S \in \mathscr{S}$

and $S = \Phi^{-1}[C]$ for some set $C\}$ and t is $\mathscr{S}_I$-measurable if and only if t is $\mathscr{S}$-measurable and $t = \gamma \circ \Phi$ for some function γ.

Assume now that $\mathscr{S}$ is a sufficient sub σ-algebra of $\mathscr{A}$. For each $A \in \mathscr{A}$, let $M(A|\cdot)$ be a version of $P_\theta(A|\mathscr{S})$. Put $M^g(A|\cdot) = M(g[A]|g)$. Then

$$\int_{g^{-1}[S]} I_A \, \mathrm{d}P_\theta = \int (I_S \circ g)(I_{g[A]} \circ g) \, \mathrm{d}P_\theta = \int I_S I_{g[A]} \, \mathrm{d}P_{g(\theta)} = \int_S M(g[A]|\cdot) \, \mathrm{d}P_{g(\theta)}$$

$$= \int_{g^{-1}[S]} M^g(A|\cdot) \, \mathrm{d}P_\theta$$

when $A \in \mathscr{A}$ and $S \in \mathscr{S}$. Hence $M^g(A|\cdot) = P_\theta(A|g^{-1}(\mathscr{S}))$. Thus $g^{-1}(\mathscr{S})$ is also sufficient. In particular this implies the statement of the following proposition.

Proposition 6.5.19. *Let the experiment $\mathscr{E} = (\mathscr{X}, \mathscr{A}; P_\theta : \theta \in \Theta)$ be invariant under the actions of the group G as described above. Let $\mathscr{S}$ be a sub σ-algebra of $\mathscr{A}$. Then:*

(i) *if $g \in G$ then $g^{-1}(\mathscr{S})$ is sufficient, minimal sufficient and closed according to whether $\mathscr{S}$ is, respectively, sufficient, minimal sufficient and closed.*
(ii) *if $\mathscr{S}$ is minimal sufficient then $g^{-1}(\mathscr{S}) \sim \mathscr{S}$ for all $g \in G$. If, in addition, $\mathscr{S}$ is closed, then $\mathscr{S}$ respects G, i.e. $g^{-1}(\mathscr{S}) = \mathscr{S}$ for all $g \in G$.*

If $\mathscr{E}$ is coherent then, as we shall see in the next chapter, $\mathscr{A}$ always contains minimal sufficient σ-algebras. The closures of these σ-algebras are all the same and this closure is the minimal sufficient closed sub σ-algebra of $\mathscr{A}$. In general, by the proposition, any σ-algebra $\mathscr{S}$ which is minimal sufficient and closed respects the group G.

Considering a general closed and sufficient sub σ-algebra $\mathscr{S}$ which respects G we obtain the following result, due to Hall, Wijsman and Ghosh (1965).

Theorem 6.5.20 (Exchangeability of the principles of sufficiency and of invariance). *Let the experiment $\mathscr{E} = (\mathscr{X}, \mathscr{A}; P_\theta : \theta \in \Theta)$ be invariant under the group G as described above. Let $\mathscr{S}$ be a sufficient and closed sub σ-algebra of $\mathscr{A}$ which respects G. Then the invariance reduction $\mathscr{S}_I$ of $\mathscr{S}$ is sufficient for the invariance reduction $\mathscr{A}_I$ of $\mathscr{A}$ provided both of the following conditions are satisfied:*

(i) *G is a measurable group possessing a non-null σ-finite measure τ such that $\tau(Ug) = 0$ whenever $\tau(U) = 0$ and $g \in G$.*
(ii) *$g(x)$ is jointly measurable in $(g, x) \in G \times \mathscr{X}$.*

Remark. We refer to proposition 6.5.1 for the significance of the condition that $\mathscr{S}$ respects G.

Proof of the theorem. Let us use the same notation as in the proof of proposition 6.5.19. We saw in that proof that $M^g(A|\cdot) = P_\theta(A|g^{-1}(\mathscr{S}))$ when

$M(A|\cdot) = P_\theta(A|\mathscr{S})$. Hence, since $\mathscr{S}$ respects G, $M(A|\cdot) = M(A|g)$ a.e. P_θ for each θ when $A \in \mathscr{A}_I$. It follows that $M(A|\cdot)$ is $\mathscr{E}$-invariant for each $A \in \mathscr{A}_I$. Hence, by corollary 6.5.12, for each $A \in \mathscr{A}_I$ there is an invariant test function $\tilde{M}(A|\cdot)$ such that $M(A|\cdot) = \tilde{M}(A|\cdot)$ a.e. P_θ: $\theta \in \Theta$. Since $\mathscr{S}$ is closed the function $\tilde{M}(A|\cdot)$ is necessarily $\mathscr{S}$-measurable. It follows that, for each $A \in \mathscr{A}_I$, $M(A|\cdot)$ may be specified $\mathscr{S}_I$-measurable and then for each θ, $M(A|\cdot)$ is a version of $P_\theta(A|\mathscr{S}_I)$. □

6.6 References

Anderson, T. W. 1955. The integral of a symmetric unimodal function over a symmetric convex set and some probability inequalities. *Proc. Amer. Math. Soc.* **6**, pp. 170–6.

Blackwell, D. 1951. Comparison of experiments. *Proc. Second Berkeley Sympos. Math. Statist. Probab.* pp. 93–102.

— 1953. Equivalent comparisons of experiments. *Ann. Math. Statist.* **24**, pp. 265–72.

Bohnenblust, H. F., Shapley, L. S. & Sherman, S. 1949. Reconnaissance in game theory. *Unpublished Rand Research Memorandum*, **208**, pp. 1–18.

Boll, C. 1955. Comparison of experiments in the infinite case. Ph. D. thesis, Stanford University.

Bondar, J. V. & Milnes, P. 1981. Amenability: A survey for statistical applications of Hunt–Stein and related conditions on groups. *Z. Wahrscheinlichkeitstheorie verw. Geb.* **57**, pp. 103–28.

Day, M. 1961. Fixed point theorems for compact convex sets. *Illinois J. Math.* **5**, pp. 585–89.

Greenleaf, F. 1969. Invariant means on topological groups. *Van Nostrand Mathematical Studies.* Van Nostrand, Princeton.

Grünbaum, B. 1967. *Convex polytopes.* Interscience, London.

Hall, W. J., Wijsman, R. A. & Ghosh, J. K. 1965. The relationship between sufficiency and invariance with applications in sequential analysis. *Ann. Math. Statist.* **36**, 575–614.

Halmos, P. R. 1958. *Naive set theory. D.* Van Nostrand, Princeton.

Hardy, G. H., Littlewood, J. E. & Polya, G. 1973. *Inequalities.* Cambridge Univ. Press.

Helgeland, J. 1982. Additional observations and statistical information in the case of 1-parameter exponential distributions. *Z. Wahrscheinlichkeitstheorie verw. Geb.* **59**, pp. 77–100.

Hodges, J. L. Jr. & LeCam, L. 1960. The Poisson approximation to the Poisson Binomial distribution. *Ann. Math. Statist.* **31**, pp. 737–40.

Hunt, G. & Stein, C. 1946. *Most stringent tests of statistical hypotheses.* Unpublished.

Khintchin, A. Ya. 1933. *Asymptotiche Gesetze der Wahrscheinlichkeitsrechnung*, Springer–Verlag, Berlin.

LeCam, L. 1964. Sufficiency and approximate sufficiency. *Ann. Math. Statist.* **35**, pp. 1419–55.

— 1974. On the information contained in additional observations. *Ann. Statist.* **2**, pp. 630–49.

— 1986. *Asymptotic methods in statistical decision theory.* Springer–Verlag, Berlin.

Lehmann, E. L. 1986. *Testing statistical hypotheses.* Wiley, New York.

Lindqvist, B. 1977. How fast does a Markov chain forget the initial state? A decision theoretical approach. *Scand. J. Statist.* **4**, pp. 145–52.

— 1978a. On the loss of information incurred by lumping states of a Markov chain. *Scand. J. Statist.* **5**, pp. 92–8.

— 1978b. A decision theoretical characterization of weak ergodicity. *Z. Wahrscheinlichkeitstheorie verw. Geb.* **44**, 155–8.

— 1984a. On the memory of an absorbing Markov chain. *Revue Roumaine de mathématiques pures et appliquées.* **29**, pp. 171–82.

— 1984b. On the memory of a countable Markov chain. A decision theoretical approach. *Stat. research report*, University of Trondheim, NTH.

Loève, M. 1977. *Probability theory I, II.* Springer–Verlag, New York.

Mammen, E. 1984. Local optimal Gaussian approximation of an exponential family. *Sonderforschungsbereich* **123**, p. 299, Heidelberg.
— 1986. The statistical information contained in additional observations. *Ann. Statist.* **14**, pp. 665–78.
Morse, N. & Sacksteder, R. 1966. Statistical isomorphism. *Ann. Math. Statist.* **37**, pp. 203–14.
Prohorov, Yu. V. 1953. Asymptotic behaviour of the binomial distribution. *Russian mathematical surveys*, **8**, pp. 135–42.
Rockafellar, R. 1970. *Convex analysis.* Princeton University Press.
Scheffé, H. 1947. A useful convergence theorem for probability distributions. *Ann. Math. Statist.* **18**, pp. 434–8.
Sheu, S. S. 1984. The Poisson approximation to the Binomial distribution. *The American Statistician*, **38**, pp. 206–7.
Stein, C. 1956. Inadmissibility of the usual estimator for the mean of a multivariate normal distribution. *Proc. 3rd Berkeley Symp. on Math. Statist. and Prob.* **1**, pp. 197–206, University of California Press, Berkeley.
Torgersen, E. 1970. Comparison of experiments when the parameter space is finite. *Z. Wahrscheinlichkeitstheorie verw. Geb.* **16**, pp. 219–49.
— 1972. Comparison of translation experiments. *Ann. Math. Statist.* **43**, pp. 1383–99.
Wagon, S. 1985. The Banach–Tarski paradox. *Encyclopedia of mathematics and its applications.* Cambridge University Press.

7

Equivalence, Representations and Functionals of Experiments

7.1 Introduction

As the notion of equivalence for experiments suggests, a prime object of this theory is to study properties and functionals which respect this equivalence. In order to simplify the language we introduced (following LeCam) the notion of the *type* of an experiment as the totality (not a set) of all experiments which are equivalent to the given experiment. We argued earlier that the collection of types may be considered as a well defined set. One of the purposes of this chapter is to provide various specific descriptions of this set.

In general a rule $\mathscr{E} \to \kappa(\mathscr{E})$ is called *a characterization of experiments* (with parameter set Θ) if:

(i) there is a set such that $\kappa(\mathscr{E})$ belongs to this set for each experiment $\mathscr{E}$ with parameter set Θ;

and

(ii) experiments $\mathscr{E}$ and $\mathscr{F}$ which have the same parameter set Θ are equivalent if and only if $\kappa(\mathscr{E}) = \kappa(\mathscr{F})$.

Characterizations whose values are experiments equivalent to the original experiments are called *representations.* Thus a representation of experiments with parameter set Θ is a rule $\mathscr{E} \to \tilde{\mathscr{E}}$ which to each experiment $\mathscr{E}$ assigns another experiment $\tilde{\mathscr{E}}$ with the same parameter set Θ such that:

(i) there is a set such that $\tilde{\mathscr{E}}$ belongs to this set for any experiment $\mathscr{E}$ with parameter set Θ.
(ii) $\mathscr{E} \sim \tilde{\mathscr{E}}$ for all experiments $\mathscr{E}$ with parameter set Θ.
(iii) if $\mathscr{E}$ and $\mathscr{F}$ are equivalent experiments with parameter set Θ then $\tilde{\mathscr{E}} = \tilde{\mathscr{F}}$.

If follows that if $\mathscr{E} \to \tilde{\mathscr{E}}$ is a representation then experiments $\mathscr{E}$ and $\mathscr{F}$ are equivalent if and only if $\tilde{\mathscr{E}} = \tilde{\mathscr{F}}$. In particular $\tilde{\tilde{\mathscr{E}}} = \tilde{\mathscr{E}}$.

In this chapter we will discuss equivalence of experiments in general and also consider some particular representations and characterizations of experiments.

This also leads naturally to the study of internal properties such as minimal sufficiency and completeness.

As has been shown by LeCam, two important structures which may be used to characterize the type of an experiment $\mathscr{E} = (P_\theta : \theta \in \Theta)$ are:

(i) the closure of the linear space in $L(\mathscr{E})$ which is generated by the measure $P_\theta : \theta \in \Theta$. This space may be called *the Banach space generated by* $\mathscr{E}$ and it will be denoted as $B(\mathscr{E})$. Thus a measure μ belongs to $B(\mathscr{E})$ if and only if it is the limit, for the total variation norm, of a sequence $\sum_\theta a_\theta^{(n)} P_\theta$ where $a^{(1)}, a^{(2)}, \dots$ are finite measures on Θ having finite supports.

(ii) the closure of the vector lattice in $L(\mathscr{E})$ which is generated by the measures $P_\theta : \theta \in \Theta$. This space will be denoted as $V(\mathscr{E})$ or as $V(P_\theta : \theta \in \Theta)$. Thus $V(\mathscr{E})$ is an L-space with the set $\{P_\theta : \theta \in \Theta\}$ as a generator. We shall call $V(\mathscr{E})$ *the vector lattice of* $\mathscr{E}$.

A typical measure in $V(\mathscr{E})$ is a limit, for the total variation norm, of measures of the form $\bigvee_{i=1}^k \mu_i - \bigvee_{i=1}^s \nu_i$ where the measures $\mu_1, \dots, \mu_k, \nu_1, \dots, \nu_s$ are all finite linear combinations of the measures P_θ.

We may also write

$$B(\mathscr{E}) = \bigcup_{\Theta_0} \{B(P_\theta : \theta \in \Theta_0)\}$$

and

$$V(\mathscr{E}) = \bigcup_{\Theta_0} \{V(P_\theta : \theta \in \Theta_0)\}$$

where Θ_0 runs through the class of all countable subsets of Θ.

Furthermore, if Θ is countable and if λ is a probability distribution on Θ such that $\lambda_\theta > 0$ for all θ and if $P_\lambda = \sum_\theta \lambda_\theta P_\theta$, then $V(\mathscr{E})$ consists of all finite measures μ which are absolutely continuous with respect to P_λ and have densities with respect to P_λ which are measurable functions of $f = (\mathrm{d}P_\theta/\mathrm{d}P_\lambda : \theta \in \Theta)$. This may be checked by observing that the class W of measures satisfying this condition constitutes a closed vector lattice containing the distributions $P_\theta : \theta \in \Theta$. Thus $V(\mathscr{E}) \subseteqq W$. On the other hand if λ and f are as above then any measure whose density with respect to P_λ is of the form $\psi_1(f) - \psi_2(f)$ where $\psi_1, \psi_2 \in \Psi$, is in $V(\mathscr{E})$. Hence, since these measures are dense in $L_1(P_\lambda f^{-1})$, we find that $W \subseteqq V(\mathscr{E})$.

The inclusions

$$B(\mathscr{E}) \subseteqq V(\mathscr{E}) \subseteqq L(\mathscr{E})$$

hold trivially for any experiment.

Consider the particular case where $\mathcal{E}$ is totally non-informative. Then $B(\mathcal{E}) = V(\mathcal{E})$ and the L-space $V(\mathcal{E})$ is isomorphic to $\mathbb{R}$ equipped with the Euclidean distance norm. If, in addition, the sample space of $\mathcal{E}$ contains a single point, then $B(\mathcal{E}) = V(\mathcal{E}) = L(\mathcal{E}) \cong \mathbb{R}$. Thus equality may hold in each of the two inclusions. In particular it follows that the equation '$V(\mathcal{E}) = L(\mathcal{E})$' does not respect equivalence. On the other hand, as we shall see, the requirement '$B(\mathcal{E}) = V(\mathcal{E})$' does respect equivalence. Actually this condition signifies that $\mathcal{E}$ permits a 'boundedly complete and sufficient' reduction or, equivalently, that $\mathcal{E}$ cannot be expressed as a 'proper mixture' of non-equivalent experiments. The requirement of equality in the last set inequality, i.e. the requirement that $V(\mathcal{E}) = L(\mathcal{E})$, signifies that $\mathcal{E}$ is 'irreducible' by sufficiency.

If $\mathcal{E}$ is coherent then we may delete the quotation marks here. As in LeCam's 1964 paper, the general case may either be treated directly by vector lattice methods or it may be reduced to the coherent case, by isomorphisms.

If Θ is finite then any experiment is equivalent to a particular experiment with the set of prior probability distributions on Θ as sample space (Blackwell, 1951). These experiments are called standard experiments and the first section of this chapter will demonstrate the importance of these experiments. It turns out that a standard experiment is determined by one single and natural probability measure called the standard probability measure.

Deficiencies, products and mixtures have simple descriptions in terms of these measures. Furthermore, the topologies of the deficiency distances Δ_k turn out to all be the same as the (compact) weak topology for standard probability measures.

Among the many topics treated in the first section, here we shall only mention that the last part provides characterizations of various functionals (affine, monotone, multiplicative, ...) which may be defined on the set of types of experiments when the parameter set Θ is finite.

The next section covers properties of the basic concepts of sufficiency and completeness within the framework of LeCam. In particular it will be shown that the algebra in $M(\mathcal{E})$ generated by the minimal non-negative Radon–Nikodym derivatives $dP_{\theta_2}/dP_{\theta_1}$ provides the maximal reduction of $\mathcal{E} = (P_\theta : \theta \in \Theta)$. In other words this algebra represents the smallest reduction of $\mathcal{E}$ which is obtainable by the principle of sufficiency. An experiment which is not reduced by this algebra is, following LeCam, said to be in minimal form. These are precisely the experiments which satisfy the condition $V(\mathcal{E}) = L(\mathcal{E})$ mentioned above.

We refer the reader to LeCam's works concerning the important generalization of the notion of sufficiency to the notion of insufficiency (LeCam 1974, 1986).

The third section is devoted to various generalizations to general parameter sets of the results described in the first section. The central concept here is the concept of the conical measure of an experiment. Conical measures were introduced by Choquet (1969) and their usefulness for experiments was observed by LeCam. It turns out that the Δ-topology is no longer compact, although it is complete, in the general case. However, there is a very useful topology, the weak experiment topology, which is always compact. We do not pursue the topological problems far here, but again refer the reader to LeCam's works.

As likelihood functions which differ by a known constant are usually considered equivalent we could take a projective space as the natural sample space for our 'canonical' experiments. Thus we arrive at a concept of likelihood experiments and this is the main topic of the fourth section.

Complements in chapter 10 which are of interest in connection with the topics treated in this chapter are nos. 2, 4–8, 11, 13–17, 19–28, 33–35, 38, 41–43, 47 and 49.

7.2 Standard experiments. Functionals of experiments

If the parameter set of an experiment is finite then the likelihoods may be normalized in order to ensure that this random function of θ adds to 1 in θ and thus represents a point probability function on Θ. The random distribution obtained in this way is the posterior distribution of θ for the uniform prior distribution. This sufficient statistic maps $\mathscr{E}$ into another experiment $\hat{\mathscr{E}}$ which is equivalent to $\mathscr{E}$ and has the set of probability distributions on Θ as its sample space. Following Blackwell (1951) this experiment is called *the standard experiment of $\mathscr{E}$*. As the posterior distribution is determined up to sets of $\sum_{\theta \in \Theta} P_\theta$-measure zero, the standard experiment of $\mathscr{E}$ does not depend on how this posterior distribution is specified. The standard experiment is a very convenient tool as we now shall see.

Let us write out explicitly the standard experiment of an experiment $\mathscr{E} = (P_\theta : \theta \in \Theta)$ with finite parameter set Θ. Here, as elsewhere, we will find it convenient to write $\sum_\theta$ instead of $\sum_{\theta \in \Theta}$. Put $P_{\Theta} = \sum_\theta P_\theta$ and $f_\theta = \mathrm{d}P_\theta/\mathrm{d}P_{\Theta}$ when $\theta \in \Theta$. We can always choose our specifications of these densities such that $f_\theta \geqslant 0; \theta \in \Theta$ and $\sum_\theta f_\theta = 1$. Then $f = (f_\theta : \theta \in \Theta)$ is a version of the point probability function of the posterior distribution of θ for the uniform prior. Let K_{Θ} denote the set of probability distributions on Θ identified as the fundamental probability simplex in $\mathbb{R}^{\Theta}$. Equip K_{Θ} with its class of Borel subsets and consider f as a map from the sample space of $\mathscr{E}$ to K_{Θ}. Then f is measurable and the experiment $(P_\theta f^{-1} : \theta \in \Theta)$ it induces on K_{Θ} is just the standard experiment of $\mathscr{E}$. In this section we shall use the notation $\hat{\mathscr{E}}$ for the standard experiment of an experiment $\mathscr{E}$. Thus with the notation introduced above

$$\hat{\mathscr{E}} = (P_\theta f^{-1} : \theta \in \Theta).$$

The measure $\sum_\theta P_\theta f^{-1} = P_\Theta f^{-1}$ is called *the standard measure of* $\mathscr{E}$. This measure, as well as $\hat{\mathscr{E}}$, does not depend on how f is specified.

The total mass of the standard measure equals the number of points in Θ. Thus a standard measure is not a probability measure when $\#\Theta \geqslant 2$. Normalizing the standard measure we obtain the *standard probability measure (Blackwell measure)* $[\#\Theta]^{-1} P_\Theta f^{-1}$ of $\mathscr{E}$.

Characteristic properties of standard experiments, standard measures and standard probability measures are collected in the following proposition.

Proposition 7.2.1 (Characterizing properties of standard experiments). *Assume that the parameter set Θ is finite and that $S_\theta : \theta \in \Theta$ are probability measures on K_Θ. Let S and Π be non-negative measures on K_Θ. Then:*

(i) *$(S_\theta : \theta \in \Theta)$ is a standard experiment if and only if the projection $x \to x_\theta$ of K_Θ into the θth coordinate space of $\mathbb{R}^\Theta$, for each θ, is a version of $\mathrm{d}S_\theta / \mathrm{d} \sum_\theta S_\theta$.*

If this condition is satisfied then $(S_\theta : \theta \in \Theta)$ is its own standard experiment.

(ii) *S is a standard measure if and only if $\int x_\theta S(\mathrm{d}x) \equiv_\theta 1$.*

If this requirement is satisfied then S is the standard measure of the standard experiment $(S_\theta : \theta \in \Theta)$ where $S_\theta(B) \equiv_\theta \int_B x_\theta S(\mathrm{d}x)$ for all Borel subsets B of K_Θ.

(iii) *Π is a standard probability measure if and only if $\int x_\theta \Pi(\mathrm{d}x) \equiv_\theta [\#\Theta]^{-1}$.*

If this requirement is satisfied then Π is the standard probability measure of the standard experiment $(S_\theta : \theta \in \Theta)$ where $S_\theta(B) \equiv_\theta [\#\Theta] \int_B x_\theta \Pi(\mathrm{d}x)$ for all Borel subsets B of K_Θ.

Proof. Assume that $(S_\theta : \theta \in \Theta)$, S and Π are, respectively, the standard experiment, the standard measure and the standard probability measure of an experiment $\mathscr{E} = (P_\theta : \theta \in \Theta)$. Put $f_\theta = \mathrm{d}P_\theta / \mathrm{d} \sum_\theta P_\theta$ and put $f = (f_\theta : \theta \in \Theta)$. Then $S_\theta \equiv_\theta P_\theta f^{-1}$, $S = \sum_\theta P_\theta f^{-1}$ and $\Pi = [\#\Theta]^{-1} S$. It follows that $S_\theta(B) = \int I_B(f) \,\mathrm{d}P_\theta = \int I_B(f) f_\theta \,\mathrm{d} \sum_\theta P_\theta = \int_B x_\theta \,\mathrm{d} \sum_\theta S_\theta$. This proves the 'only if' parts of the first statements in (i), (ii) and (iii). The 'if' parts follow readily by checking that the last statement follows directly from the definition. □

Any experiment having a finite parameter set has a representation which is a standard experiment.

Proposition *7.2.2. Let $\mathscr{E}$ be an experiment with finite parameter set Θ. Then $\mathscr{E} \sim \hat{\mathscr{E}} = \hat{\hat{\mathscr{E}}}$.*

Proof. If ψ is sublinear on $\mathbb{R}^\Theta$ then the maximal utility function $\mathscr{E} \to \int \psi(d\mathscr{E})$ has the same value in $\mathscr{E}$ as in $\hat{\mathscr{E}}$. Finally $\hat{\mathscr{E}} = \hat{\hat{\mathscr{E}}}$ by the last statement in part (i) of the previous proposition. □

An experiment $\mathscr{E}$ is equivalent to at most one standard experiment, by the following theorem.

Theorem 7.2.3. *Standard experiments which are equivalent for testing problems are equal.*

Proof. Let $\mathscr{E} = (S_\theta : \theta \in \Theta)$ and $\mathscr{F} = (T_\theta : \theta \in \Theta)$ be standard experiments such that $\mathscr{E} \sim_2 \mathscr{F}$. Put $S = \sum_\theta S_\theta$ and $T = \sum_\theta T_\theta$. Then, by theorem 6.3.4, $\int \langle a, x\rangle^+ S(dx) = \int \langle a, x\rangle^+ T(dx)$ for any vector $a \in \mathbb{R}^\Theta$. Equating the right partial derivatives with respect to a_θ at a given point $a \in \mathbb{R}^\Theta$ we find

$$S_\theta(\{x : \langle a, x\rangle \geqslant 0\}) \underset{a}{\equiv} T_\theta(\{x : \langle a, x\rangle \geqslant 0\}); \qquad \theta \in \Theta.$$

Let e be the vector in $\mathbb{R}^\Theta$ such that $e_\theta \equiv 1$. If $a \subset \mathbb{R}^\Theta$ and b is a constant then

$$S_\theta(\{x : \langle a, x\rangle \geqslant b\}) = S_\theta(\{x : \langle a - be, x\rangle \geqslant 0\}) = T_\theta(\{x : \langle a, x\rangle \geqslant b\}).$$

It follows that any linear functional on $\mathbb{R}^\Theta$ has the same distribution under S_θ as under T_θ. Hence S_θ and T_θ have the same characteristic function so that $S_\theta = T_\theta$. It follows that $\mathscr{E} = \mathscr{F}$. □

When Θ is finite it follows that the standard experiment provides a representation of experiments and that standard (probability) measures are characterizations. Remember that for any experiment $\mathscr{E} = (P_\theta : \theta \in \Theta)$ we defined the spaces $B(\mathscr{E})$ and $V(\mathscr{E})$ as:

$B(\mathscr{E})$ = the closed linear subspace of $L(\mathscr{E})$ spanned by $(P_\theta : \theta \in \Theta)$

and

$V(\mathscr{E})$ = the closed sub-vector lattice of $L(\mathscr{E})$ generated by $(P_\theta : \theta \in \Theta)$.

Note that $V(\mathscr{E})$ is a sub L-space of $L(\mathscr{E})$ and that isomorphisms between Banach spaces (in particular between L-spaces) are isometries by definition.

Theorem 7.2.4 (Criteria for equivalence). *The following conditions are equivalent for experiments $\mathscr{E}$ and $\mathscr{F}$ having the same, not necessarily finite, parameter set Θ.*

(i) $\mathscr{E} \sim \mathscr{F}$.

(ii_1) $\mathscr{E} \sim_k \mathscr{F}$ *for some* $k \geqslant 2$.

(ii_2) $\mathscr{E} \sim_2 \mathscr{F}$.

(iii) *The Banach spaces $B(\mathscr{E})$ and $B(\mathscr{F})$ are isomorphic by an isomorphism from $B(\mathscr{E})$ to $B(\mathscr{F})$ which, for each θ, maps P_θ into Q_θ.*

(iv) *The L-spaces $V(\mathscr{E})$ and $V(\mathscr{F})$ are isomorphic by an isomorphism from $V(\mathscr{E})$ to $V(\mathscr{F})$ which, for each θ, maps P_θ into Q_θ.*

If Θ is finite then these conditions are equivalent to each of the following conditions:

(v) $\hat{\mathscr{E}} = \hat{\mathscr{F}}$.
(vi) *$\mathscr{E}$ and $\mathscr{F}$ have the same standard measure.*
(vii) *$\mathscr{E}$ and $\mathscr{F}$ have the same standard probability measure.*

Remark 1. The equivalence of conditions (i), (ii_1) and (ii_2) amounts to the statement that the equivalences '$\sim_k$', $k = 2, 3, \ldots, \infty$ for experiments are all the same. In other words these equivalences all reduce to the equality relation for types.

Remark 2. The words 'isomorphic' and 'isomorphism' in (iii) may be replaced by the words 'isometric' and 'isometry'. As noted by LeCam, this follows from a general result of Mazur and Ulam (1932) which states that an isometry between normed linear spaces which maps the null vectors into each other is necessarily linear. Condition (iv) may also be found in LeCam (1964).

It should be noted that in (iv) it does not suffice to require that the vector lattices $V(\mathscr{E})$ and $V(\mathscr{F})$ are merely isomorphic, unless one of them has a boundedly complete and sufficient reduction.

Criterion (v), together with the previous results on standard experiments, establishes that standard experiments are indeed a representation when Θ is finite. Similarly (vi) and (vii) establish standard probability measures and standard measures as characterizations.

Proof. It suffices to consider the case of a finite parameter set Θ. The implications (i) $\Rightarrow$ (ii_1) $\Rightarrow$ (ii_2) were established in section 6.2. The implications (ii_2) $\Rightarrow$ (v) $\Leftrightarrow$ (vi) $\Leftrightarrow$ (vii) $\Rightarrow$ (i) have been noted earlier in this section. Thus conditions (i), (ii_1), (ii_2), (v), (vi) and (vii) are all equivalent. Furthermore, the implication (iv) $\Rightarrow$ (iii) is trivial while (iii) is just a rephrasing of the sublinear function criterion for equivalence for testing problems. Therefore it remains to show that (iv) holds when $\mathscr{E} \sim \mathscr{F}$. Assume then that $\mathscr{E}$ and $\mathscr{F}$ are equivalent. Then $\hat{\mathscr{E}} = \hat{\mathscr{F}}$ so that $V(\hat{\mathscr{E}}) = V(\hat{\mathscr{F}})$. It suffices therefore to show that $V(\mathscr{E}) \equiv V(\hat{\mathscr{E}})$ by a correspondence such that, for all θ, P_θ corresponds to S_θ where $(S_\theta : \theta \in \Theta)$ is the standard experiment of $\mathscr{E}$. Put $f_\theta = \mathrm{d}P_\theta/\mathrm{d}P_\Theta$; $\theta \in \Theta$ and $f = (f_\theta : \theta \in \Theta)$ so that $S_\theta = P_\theta f^{-1}$ when $\theta \in \Theta$. Let μ be a measure in $V(\mathscr{E})$. Then $\mathrm{d}\mu/\mathrm{d}P_\Theta$ is of the form $\phi(f)$ for some Borel function ϕ on K_Θ. Map μ into $\nu \in V(\hat{\mathscr{E}})$ where $\mathrm{d}\nu/\mathrm{d}S_\Theta = \phi$. Then $\mu \leftrightarrow \nu$ is the required correspondence. □

Corollary 7.2.5 (A 'norm transform'). *Let Θ be any non-empty set. Then the transform which to each distribution a on Θ with finite support assigns the norm*

$\|\sum_\theta a_\theta P_\theta\|$ *for the experiment* $\mathscr{E} = (P_\theta : \theta \in \Theta)$ *determines the type of* $\mathscr{E}$ *and, consequently, is a characterization.*

It is known from the theory of sufficiency that pairwise sufficiency implies sufficiency in the dominated case. Considering this as a statement on pairwise equivalence we might wonder whether, or when, pairwise equivalence implies equivalence. In general however, we may have experiments which are pairwise equivalent but not equivalent. Even worse, we may have experiments $\mathscr{E}$ and $\mathscr{F}$ such that the restricted experiments $\mathscr{E}|\Theta_0$ and $\mathscr{F}|\Theta_0$ are equivalent for any proper subset Θ_0 of Θ. This can of course only occur when Θ is finite. Here is an example for a general finite Θ.

Example 7.2.6 (Equivalence for all proper subsets of Θ does not imply equivalence). Assume that the set Θ contains m points where $2 \leqslant m < \infty$. Take the class $\mathscr{X}$ of all subsets of Θ as the sample space of both experiments $\mathscr{E}$ and $\mathscr{F}$. For each $x \in \mathscr{X}$ and each $\theta \in \Theta$ put

$$P_\theta(x) = \begin{cases} 2^{2-m} & \text{if } x \text{ is even and } \theta \in x \\ 0 & \text{otherwise;} \end{cases}$$

and

$$Q_\theta(x) = \begin{cases} 2^{2-m} & \text{if } x \text{ is odd and } \theta \in x \\ 0 & \text{otherwise.} \end{cases}$$

If m is even the $\|_\wedge P_\theta\| = {}_\wedge P_\theta(\Theta) = 2^{2-m} \neq 0$ while ${}_\wedge Q_\theta = 0$. If m is odd then, on the other hand, ${}_\wedge P_\theta = 0$ while ${}_\wedge Q_\theta \neq 0$. It follows that the L-spaces $V(\mathscr{E})$ and $V(\mathscr{F})$ cannot be isomorphic so that $\mathscr{E} \nsim \mathscr{F}$.

Choose a point $\theta_0 \in \Theta$ and define a map ϕ from $\mathscr{X}$ to $\mathscr{X}$ by putting

$$\phi(x) = \begin{cases} x \cup \{\theta_0\} & \text{if } \theta_0 \notin x \\ x - \{\theta_0\} & \text{if } \theta_0 \in x. \end{cases}$$

Then ϕ is 1–1 and, consequently, permutes $\mathscr{X}$. If $\theta \neq \theta_0$ then $P_\theta \phi^{-1} = Q_\theta$. Thus the restrictions of $\mathscr{E}$ and $\mathscr{F}$ to $\Theta - \{\theta_0\}$ are equivalent. It follows that the restrictions of $\mathscr{E}$ and $\mathscr{F}$ to any proper subset Θ_0 of Θ are equivalent.

Later on in this section we will see that this example may be used to characterize those experiments $\mathscr{E}$ which satisfy the cancellation law for products.

We shall now see that ordered experiments which are pairwise equivalent are equivalent. In particular it follows that the experiments $\mathscr{E}$ and $\mathscr{F}$ considered in the last example cannot be ordered.

Theorem 7.2.7. *Let $\mathscr{E}$ and $\mathscr{F}$ be experiments such that either $\mathscr{E} \geqslant \mathscr{F}$ or $\mathscr{F} \geqslant \mathscr{E}$. Then $\mathscr{E}$ and $\mathscr{F}$ are equivalent if and only if they are pairwise equivalent.*

Proof. Suppose $\mathscr{E} \geqslant \mathscr{F}$ and that $\mathscr{E}_{\{\theta_1,\theta_2\}} \sim \mathscr{F}_{\{\theta_1,\theta_2\}}$ for each two-point subset $\{\theta_1, \theta_2\}$ of Θ. We may assume without loss of generality that Θ is finite and then we may as well assume that $\mathscr{E} = (S_\theta : \theta \in \Theta)$ and $\mathscr{F} = (T_\theta : \theta \in \Theta)$ are standard experiments. Then, by the randomization criterion, $T_\theta \equiv_\theta S_\theta M$ for some randomization M from K_Θ to K_Θ. Put $U_\theta = S_\theta \times M$; $\theta \in \Theta$ and let $\mathscr{H}$ denote the experiment $(U_\theta : \theta \in \Theta)$. Then the first projection is sufficient in $\mathscr{H}$ so that $\mathscr{H} \sim \mathscr{E}$. Hence the second projection is pairwise sufficient in $\mathscr{H}$. Thus $\mathscr{H} \sim \mathscr{F}$ by proposition 6.2.28 so that $\mathscr{E} \sim \mathscr{F}$. □

The standard measures of homogeneous experiments have all their mass distributed in the relative interior ri K_Θ of the probability simplex K_Θ in $\mathbb{R}^\Theta$. Let us agree to say that experiments $\mathscr{E}$ and $\mathscr{F}$ having the same finite parameter set Θ are *homogeneously equivalent* if their standard measures agree on the interior of K_Θ. Thus homogeneous experiments (with the same finite parameter set) are equivalent if and only if they are homogeneously equivalent. Non-homogeneous experiments may easily be homogeneously equivalent without being equivalent. However, full equivalence may always be expressed in terms of homogeneous equivalence.

Proposition 7.2.8 (Equivalence and homogeneous equivalence). *Experiments $\mathscr{E}$ and $\mathscr{F}$, having the same parameter set Θ, are equivalent if and only if the restrictions $\mathscr{E}|\Theta_0$ and $\mathscr{F}|\Theta_0$ are homogeneously equivalent for each finite non-empty subset Θ_0 of Θ.*

Proof. We may assume without loss of generality that Θ is finite. If $\#\Theta = 1$ then all experiments are equivalent and the statement is trivial. Assume that the statement is verified when $\#\Theta < m$ and consider the case '$\#\Theta = m$'. Let $\mathscr{E} = (P_\theta : \theta \in \Theta)$ and $\mathscr{F} = (Q_\theta : \theta \in \Theta)$ be experiments such that $\mathscr{E}|\Theta_0$ and $\mathscr{F}|\Theta_0$ are homogeneously equivalent for all non-empty subsets Θ_0 of Θ. Let U_r denote the union of relative interiors of faces of $K = K_\Theta$ which have dimensions $\geqslant r$. Thus $K = U_0 \supseteqq U_1 \supseteqq \cdots \supseteqq U_{m-1} = \text{ri } K$. By assumption $S|U_{m-1} = T|U_{m-1}$.

Assume $S|U_{r+1} = T|U_{r+1}$ where $0 \leqslant r \leqslant m-2$. Let $B \subseteqq \text{ri } A$ where B is a Borel set and A is an r-dimensional face of K. Then $A = \langle e^\theta : \theta \in F\rangle$ for an $(r+1)$-point set F. Put $s(x) = \sum_F x_\theta$, $\Pi(x) = (x_\theta/s(x) : \theta \in F)$ and $\hat{B} = \{x : s(x) > 0 \text{ and } \sum_F (x_\theta/s(x))e^\theta \in B\}$. Note that $s(x) = 1$ and $\Pi(x) = (x_\theta : \theta \in F)$ when $x \in A$. Furthermore, $\hat{B} = \Pi^{-1}(\Pi[B])$ and $x_\theta > 0$ when $\theta \in F$ and $x \in B$. Hence $B = \hat{B} - U_{r+1}$. The standard measures of $\mathscr{E}|F$ and $\mathscr{F}|F$ are, respectively, $(sS)\Pi^{-1}$ and $(sT)\Pi^{-1}$. Furthermore the mapping Π restricted to A is a homeomorphism from A onto the probability simplex K_F in $\mathbb{R}^F$. Thus $\prod[B] \subseteqq \text{ri } K_F$. Hence, by the primary induction hypothesis, $(sS)(\prod^{-1}[\prod[B]]) = (sT)[\prod^{-1}[B]]$. Thus $\int_{\hat{B}} s\, dS = \int_{\hat{B}} s\, dT$. Also $\int_{\hat{B}} I_{U_{r+1}} s\, dS = \int_{\hat{B}} I_{U_{r+1}} s\, dT$ since S and T agree on U_{r+1}. Hence $S(B) = \int_B s\, dS = \int_{\hat{B}} s\, dS - \int_{\hat{B}} I_{U_{r+1}} s\, dS =$ the same expression in $T = T(B)$. Thus S and T agree on U_r. Induction on r yields $S = T$. Finally the proof is completed by induction on m. □

We have seen that provided Θ is finite, experiments may be identified with certain measures. A natural problem which now poses itself is the following.

How do the concepts (e.g. mixtures, products, deficiencies) which we have studied so far, look in terms of these probability measures?

Closely related is the converse problem of what known concepts for probability measures, (e.g. convex combinations, products, convolutions, functionals, distances) tell us about experiments.

Let us first compare distances. Recall the following notions for comparison of metrics on a set $\mathscr{X}$.

A *metric* d_1 is called *at most as coarse as the metric* d_2 (d_2 is at least as fine as d_1) if the uniformity defined by d_1 is contained in the uniformity defined by d_2. Thus d_1 is at most as coarse as d_2 if to each $\varepsilon > 0$ there corresponds a $\delta > 0$ such that $d_1(x, y) < \varepsilon$ whenever $d_2(x, y) < \delta$. If d_1 is at most as coarse as d_2 and d_2 is at most as coarse as d_1 then we shall say that *d_1 and d_2 are equivalent.* Thus metrics are equivalent if they define the same uniformities.

If we only consider the topologies then we say that *the metric d_1* is topologically at most as coarse as d_2 (d_2 is topologically at *least as fine as* d_1) if the topology defined by d_1 is contained in the topology defined by d_2. Thus d_1 is topologically at least as coarse as d_2 if and only if to each $x \in S$ and each $\varepsilon > 0$ there corresponds a $\delta > 0$ such that $d_1(x, y) < \varepsilon$ whenever y is such that $d_2(x, y) < \delta$. If d_1 is topologically at most as coarse as d_2 and d_2 is topologically at most as coarse as d_1 then we shall say that *d_1 and d_2 are topologically equivalent.* Thus d_1 and d_2 are topologically equivalent if and only if they define the same topologies.

Let $\mathscr{P}$ denote the set of all standard probability measures on K_{Θ}. Note that the Δ-topology coincides with the coarsest topology on $\mathscr{P}$ which makes $\int \psi \, dS$ continuous in $S \in \mathscr{P}$ for each sublinear function ψ on $\mathbb{R}^{\Theta}$. This may be seen by observing that the topology of pointwise convergence on functions $\psi \in \Psi$ coincides (by the inequality $\psi(x) - \psi(y) \leqslant \bigvee_\theta |x_\theta - y_\theta| \sum_\theta \psi(e^\theta) \bigvee \psi(-e^\theta)$) with the topology of uniform convergence on bounded sets.

It follows that the set Ψ^B of sublinear functions ψ on $\mathbb{R}^{\Theta}$ which satisfies the inequality $\sum_\theta \psi(e^\theta) \bigvee \psi(-e^\theta) \leqslant 1$ is compact for the topology of uniform convergence on bounded sets. This in turn implies that the topology of functionwise convergence to standard (probability) measures on Ψ^B coincides with the topology of uniform convergence on Ψ^B, i.e. with the Δ-topology. Furthermore, the set of standard probability measures is clearly compact for the usual weak convergence of probability measures. Hence this topology is the same as the Δ-topology.

It follows that any Hausdorff topology (uniformity) which is at most as coarse as the topology (uniformity) of Δ is in fact equal to the topology (uniformity) of Δ. In particular any metric which is topologically at most as coarse as Δ is in fact equivalent to Δ. However, we can easily find metrics on $\mathscr{P}$ which are strictly finer than Δ. A useful one is the total variation distance. The ultimate, but not very useful example, is of course any metric which metricizes the discrete topology.

Now consider any standard probability measure Q on K_Θ. Then there is a sequence $(Q_n, n = 1, 2, \ldots)$ of probability measures with finite supports such that Q_n converges weakly to Q as $n \to \infty$. Put $a_{n,\theta} = \int x_\theta Q_n(\mathrm{d}x)$ and let $Q_{n,\theta}$ be the probability measure whose density with respect to Q_n is $x \to x_\theta / a_{n,\theta}$. Then $\int \psi(\mathrm{d}Q_{n,\theta} : \theta \in \Theta) = \int \psi(x_\theta / a_{n,\theta}; \theta \in \Theta) Q_n(\mathrm{d}x) \to \int \psi(x) Q(\mathrm{d}x)$ for all $\psi \in \Psi$. Thus $(Q_{n,\theta} : \theta \in \Theta) \to_\Delta \mathscr{E}$ when $\mathscr{E}$ has standard probability measure Q. Altogether we have proved the following theorem.

Theorem 7.2.9 (The Δ-topology when Θ is finite). *Assume that Θ is finite. Then the Δ-topology for types of experiments is the same as the topology of weak convergence of standard (probability) measures. It follows that the set of types of experiments is a compact (and hence complete) metric space for the metric Δ. Any metric which is topologically at most as coarse as Δ is equivalent to Δ.*

In particular it follows that the metrics $\Delta_2, \Delta_3, \ldots$ and Δ are all equivalent when Θ is finite. Furthermore, the set of types of experiments with finite sample spaces constitutes a dense subset.

The following example gives an indication on what can go wrong in the infinite case.

Example 7.2.10 (Counterexamples on convergence when Θ is infinite). Take $\mathscr{X} = \{0, 1\}$ as the sample space and let $\Theta_0 = \{1, 2, \ldots\}$ be an infinite subset of Θ. For each integer n, put $P_{n,n} = \delta_0$ and $P_{\theta,n} = \delta_1$ if $\theta \neq n$. Then $\Delta(\mathscr{E}_m, \mathscr{E}_n) = 1$ when $m \neq n$. Thus Δ is not separable and therefore not compact when Θ is infinite. Furthermore, if Θ is infinite and $\delta(\mathscr{E}, \mathscr{M}_a) < 1$ then there is an estimator $\hat{\theta}$ in $\mathscr{E}$ such that $P_\theta\ (\hat{\theta} = \theta) > \frac{1}{2}$ for all θ. Thus the experiment permits a non-trivial infinite decomposition $\{[\hat{\theta} = \theta] : \theta \in \Theta\}$ of its sample space. It follows that $\mathscr{M}_a$ is not approximable by experiments with finite sample spaces in this case.

In spite of these examples, we shall see later in the general case that Δ-convergence at least, is complete.

Metrics described by joint distributions are particularly convenient to work with. Suppose e.g. that X and Y are random vectors having a joint distribution such that the distribution of X is the standard probability measure $\bar{S}$ while the distribution of Y is the standard probability measure $\bar{T}$. Put $\|x\| = \sup_\theta |x_\theta|$ when $x \in \mathbb{R}^\Theta$ and let $\psi \in \Psi$ satisfy $\sum_\theta \psi(e^\theta) \vee \psi(-e^\theta) \leqslant 1$. Then $|\psi(x) - \psi(y)| \leqslant \|x - y\|$; $x, y \in \mathbb{R}^\Theta$. Hence $|\int \psi\, \mathrm{d}\bar{S} - \int \psi\, \mathrm{d}\bar{T}| = |E[\psi(X) - \psi(Y)]| \leqslant E\|X - Y\| \leqslant \varepsilon + \Pr(\|X - Y\| \geqslant \varepsilon)$.

Two distances which offer themselves here are the Prohorov distance and the dual Lipschitz distance (Dudley distance). They may be defined for probability measures P and Q on the class of Borel sets of a metric space $(\mathscr{X}, d)$ as follows.

The *Prohorov distance* $\Pi(P, Q)$ between the probability distributions P and Q is the greatest lower bound of all numbers $\varepsilon > 0$ such that $P(A) \leqslant Q(A^\varepsilon) + \varepsilon$ for any Borel set A. Here $A^\varepsilon = \{x : d(x, A) \leqslant \varepsilon\}$. The *dual Lipschitz distance* $D(P, Q)$ is the least upper bound of all numbers $\int f \, dP - \int f \, dQ$ where $|f(x) - f(y)| \leqslant d(x, y)$ for all $x, y \in \mathscr{X}$. We refer to Huber (1981) and Dudley (1989) for information on these distances. A particularly important and interesting paper is Strassen (1965).

Here we will need the fact that these metrics are equivalent and that they metrizice the topology of weak convergence when the metric space $(\mathscr{X}, d)$ is complete and separable and that they then have the following joint distribution characterizations.

(i) $\Pi(P, Q) \leqslant \varepsilon$ if and only if there are random variables X and Y on some probability space such that

$$\mathscr{L}(X) = P, \qquad \mathscr{L}(Y) = Q \quad \text{and} \quad \Pr(d(X, Y) \geqslant \varepsilon) \leqslant \varepsilon.$$

(ii) If the metric d is bounded by 1 then $D(P, Q) \leqslant \varepsilon$ if and only if there are random variables X and Y on some probability space such that

$$\mathscr{L}(X) = P, \qquad \mathscr{L}(Y) = Q \quad \text{and} \quad E[d(X, Y)] \leqslant \varepsilon.$$

A general joint distribution result due to Strassen, which yields (i) as a particular case, is proved in section 9.6.

These characterisations yield inequalities for standard probability measures.

Theorem 7.2.11 (Inequalities for metrics for standard probability measures). *Let Θ be a finite set containing m points.*

Assume that $\mathscr{E}$ and $\mathscr{F}$ are experiments with standard probability measures $\bar{S}$ and $\bar{T}$ respectively. Let D and $\prod$ denote, respectively, dual Lipschitz distance and Prohorov distance for the supremum distance on $\mathbb{R}^\Theta$. (The supremum distance on $\mathbb{R}^\Theta$ is the metric $(x, y) \to \bigvee_\theta |x_\theta - y_\theta|$.) Then

$$\Delta(\mathscr{E}, \mathscr{F}) \leqslant mD(\bar{S}, \bar{T}) \leqslant 2m \prod(\bar{S}, \bar{T}).$$

Furthermore, $\Delta(\mathscr{E}, \mathscr{F}) \leqslant m\|\bar{S} - \bar{T}\|$ where $\|\bar{S} - \bar{T}\|$ is the total variation of the difference measure $\bar{S} - \bar{T}$.

Remark. The role of the constant m in these inequalities indicates that a comparison of metrics for experiments having the same parameter set Θ may be much more involved when Θ is infinite than when it is finite.

Proof. The last statement follows directly by applying the randomization criterion, with the kernel being the identity function, to the standard experiments. □

Multiplying experiments amounts to convoluting standard measures for pointwise multiplication in the projective likelihood space.

Theorem 7.2.12 (Standard measures of mixtures and products). *Let $\mathscr{E}_1, \mathscr{E}_2, \ldots, \mathscr{E}_n$ be experiments with the same finite parameter set Θ. Let $\pi_1, \pi_2, \ldots, \pi_n$ be non-negative numbers such that $\pi_1 + \cdots + \pi_n = 1$. Let S_i; $i = 1, \ldots, n$ be the standard measure of $\mathscr{E}_i$.*

Then $\sum_{i=1}^n \pi_i S_i$ is the standard measure of the π-mixture $\sum_{i=1}^n \pi_i \mathscr{E}_i$ of $\mathscr{E}_1, \ldots, \mathscr{E}_n$.

The standard measure T of $\prod_{i=1}^n \mathscr{E}_i$ is given by

$$\int \psi(x) T(\mathrm{d}x) = \int \psi(x_\theta^1 x_\theta^2 \ldots x_\theta^n : \theta \in \Theta) \left(\prod_{i=1}^n S_i \right) (\mathrm{d}(x^1, x^2, \ldots, x^n))$$

when $\psi \in \Psi_\Theta$.

It follows that the convexity structure for standard measures corresponds precisely to the structure of experiments for mixtures. Now a compact convex set in $\mathbb{R}^\Theta$ and, more generally, in a locally convex linear topological space, is the closed convex hull of the set of its extreme points. Thus the set of types of experiments with finite parameter sets is the 'closed convex hull' of extremal experiments, i.e. experiments with extreme standard measures. Actually a theorem due to Choquet, (see Torgersen, 1970) shows how any experiment with a finite parameter set may be obtained as a possibly infinite mixture of extremal experiments. It is therefore of considerable interest to determine precisely the nature of these experiments. This characterization in fact possesses a very recognizable statistical content. It turns out that an experiment is extremal if and only if it permits a boundedly complete and sufficient statistic. Except for 'technicalities' (they may not be simple) this generalizes to experiments with arbitrary parameter sets. If Θ is finite then an experiment is extremal if and only if its standard measure is supported by the vertices of a sub-simplex of K_Θ. This is a consequence of the following proposition.

Proposition 7.2.13 (Extreme distributions with given expectations). *Let $t \in \mathbb{R}^\Theta$. Consider the set $\mathscr{P}_t$ of probability measures S on the Borel class in $\mathbb{R}^\Theta$ such that $\int x_\theta S(\mathrm{d}x) = t_\theta$; $\theta \in \Theta$. This set is convex and S is extreme within this set if and only if it is supported by the vertices of a simplex. If S is extreme, then it is uniquely determined by its smallest support.*

Proof. Suppose S is concentrated on the vertices of a simplex $\langle a^1, \ldots, a^k \rangle$. Put $S(\{a^i\}) = \alpha_i$; $i = 1, \ldots, k$. Thus $\sum_i \alpha_i = 1$ and $t = \int x S(\mathrm{d}x) = \sum \alpha_i a^i$. Suppose $S = (1 - \theta)Q + \theta P$ where $Q, P \in \mathscr{P}_t$ and $0 < \theta < 1$. Then, since $S \gg Q$, Q is supported by $\{a^1, \ldots, a^k\}$. Put $Q(\{a^i\}) = \alpha_{i'}'$; $i = 1, \ldots, k$. Then

$$t = \int xQ(\mathrm{d}x) = \sum \alpha_i' a^i.$$

Hence, since $0 = t - t = \sum_i (\alpha_i - \alpha_i')a_i$ and $\sum_i (\alpha_i - \alpha_i') = 0$,

$$\alpha_i = \alpha_i'; \qquad i = 1, \ldots, k.$$

Thus $Q = S = P$. It follows that S is extreme.

Suppose that S is extreme. Let $p^1, \ldots, p^k$ be points of increase for S. (A point $p \in \mathbb{R}^\Theta$ is called a point of increase for a probability measure S if $S(V) > 0$ for any measurable neighbourhood of p.) We may then construct a measurable partition $\{V_1, \ldots, V_k\}$ of $\mathbb{R}^\Theta$ such that V_i is a neighbourhood of p^i; $i = 1, \ldots, k$. Put $\lambda_i = S(V_i)$. Then $\lambda_1, \ldots, \lambda_k > 0$ and $\sum \lambda_i = 1$. Put $f_i = \lambda_i^{-1} I_{V_i}$ so that $\int f_i \, \mathrm{d}S = 1$. $S \in \mathscr{P}_t$ implies

$$t = \int xS(\mathrm{d}x) = \sum_i \lambda_i \int xf_i(x)S(\mathrm{d}x) = \sum_i \lambda_i v^i$$

where $v^i = \int xf_i(x)S(\mathrm{d}x)$. Let $\mu_1, \ldots, \mu_k \geqslant 0$ be constants such that $\sum \mu_i = 1$ and $\sum_i \mu_i v^i = t$. Then

$$\int x(\textstyle\sum \mu_i f_i(x))S(\mathrm{d}x) = \sum \mu_i v^i = t$$

and

$$\int (\textstyle\sum \mu_i f_i(x))S(\mathrm{d}x) = 1.$$

It follows that $Q = (\sum \mu_i f_i)S \in \mathscr{P}_t$. Let $a > 1$ be a constant such that $\sum \mu_i f_i \leqslant a$. Put $\theta = (a-1)/a$ and $P = (1/\theta)[S - (1-\theta)Q]$. Then $S = (1-\theta)Q + \theta P$, $0 < \theta < 1$ and $P, Q \in \mathscr{P}_t$. Hence, since S is extreme, $P = Q = S$. Thus $\sum \mu_i f_i = 1$ a.e. S. Then, since $S(V_i) > 0$, there is an $x \in V_i$ such that

$$1 = \sum \mu_i f_i(x) = \mu_i \lambda_i^{-1}.$$

Hence

$$\mu_i = \lambda_i; \qquad i = 1, \ldots, k.$$

We have shown that

$$\mu_1, \ldots, \mu_k \geqslant 0, \qquad \sum \mu_i = 1, \qquad \sum \mu_i v^i = t \Leftrightarrow \mu_i = \lambda_i; \qquad i = 1, \ldots, k.$$

This implies that $v^1, \ldots, v^k$ are geometrically independent. Consequently $k \leqslant m + 1$ where $m = \#\Theta$. It follows that S is supported by a k-point set $\{q^1, \ldots, q^k\}$ where $k \leqslant m + 1$. By the same reasoning, $q^1, \ldots, q^k$ are geometrically independent. Let $\lambda_1, \ldots, \lambda_k$ be the weights assigned to $q^1, \ldots, q^k$ by S. Then we have

$$e = \int xS(\mathrm{d}x) = \sum_i \lambda_i q^i.$$

It follows that $\lambda_1, \ldots, \lambda_k$ and, consequently, S is determined by $q^1, \ldots, q^k$. □

Corollary 7.2.14 (The extreme standard measures). *Any standard measure whose support is the set of vertices of a simplex is extreme. To any simplex which contains* $[\#\Theta]^{-1}(1,\dots,1)$ *and is contained in* K_Θ *there corresponds an extreme standard measure which is supported by the set of vertices of the simplex. This correspondence is one to one and onto between the set of simplices which contains* $[\#\Theta]^{-1}(1,\dots,1)$ *and is contained in* K_Θ *on the one hand, and the set of extreme standard measures on the other.*

The set of standard measures is a metrizable compact convex subset of $C(K_\Theta)^*$. Under these circumstances the set of extreme points is the intersection of a countable family of open sets. A simple proof may be found in Phelps (1966). Thus, as we could have inferred directly in this case, the set $\mathscr{V}$ of extreme standard measures is a Borel measurable subset of the set of all standard measures. Furthermore, by a theorem due to Choquet (see Phelps, 1966) each standard measure can be represented by a probability distribution $\prod$ on $\mathscr{V}$ in the sense that $\alpha(S) = \int \alpha(V)\prod(\mathrm{d}V)$ for each continuous affine function α on the set of standard measures. This may be re-written as $S_\theta(B) \equiv_\theta \int V_\theta(B)\prod(\mathrm{d}V)$ for each Borel set B. Another way of expressing this is by the equivalence $\mathscr{E} \sim \int \mathscr{E}_V \prod(\mathrm{d}V)$ where $\mathscr{E}$ has standard measure S and the mixture $\int \mathscr{E}_V \prod(\mathrm{d}V)$ is defined in the obvious way.

It was mentioned above that the experiments which are extreme for mixtures are precisely those which permit a sufficient and (boundedly) complete reduction. If Θ is finite then, by theorem 7.2.12, this amounts to the assertion that the latter characterization is equivalent to extremeness of standard measures.

Corollary 7.2.15 (Extremeness and completeness). *Let* $\mathscr{E} = ((\mathscr{X},\mathscr{A}),(P_\theta:\theta\in\Theta))$ *be an experiment with finite parameter set* Θ *and standard measures* S. *Then*:

(i) *S is extreme provided $\mathscr{E}$ is boundedly complete.*
(ii) *$\mathscr{E}$ is complete provided S is extreme and $\mathscr{A}$ is minimal sufficient.*

Remark. It follows that $\mathscr{E}$ is extreme and $\mathscr{A}$ is minimal sufficient if and only if $\mathscr{A}$ is boundedly complete. It also follows that $\mathscr{E}$ is complete if and only if it is boundedly complete. If Θ is infinite then, in general, bounded completeness does not imply completeness.

Proof of the corollary. Let S be the standard measure of $\mathscr{E}$ and suppose $\mathscr{E}$ is boundedly complete. Let U and V be standard measures such that $S = \frac{1}{2}U + \frac{1}{2}V$. Put $u = \mathrm{d}U/\mathrm{d}S$. Then $e = \int xU(\mathrm{d}x) = \int xu(x)S(\mathrm{d}x)$ so that $\int u\circ f\,\mathrm{d}P_\theta = 1$ where $f = (\mathrm{d}P_\theta/\mathrm{d}(\sum_\theta P_\theta):\theta\in\Theta)$. By assumption, $u\circ f = 1$ a.e. $\sum_\theta P_\theta$ so that $u = 1$ a.e. S. Hence $U = V = S$. It follows that S is extreme.

Suppose that S is extreme and $\mathscr{A}$ is minimal sufficient. Let h be a $\sum_\theta P_\theta$-integrable function satisfying $\int h\,\mathrm{d}P_\theta \equiv_\theta 0$. Since $\mathscr{A}$ is minimal sufficient we may assume that h is of the form $g \circ f$. Then $\int g(x)S(dx) = 0$. By geometrical independence of the vertices in the simplex corresponding to S, $g = 0$ a.e. S. Hence $h = 0$ a.e. $\sum_\theta P_\theta$. □

The randomization criterion for ε-deficiency shows, in the particular case where $\varepsilon = 0$, how we may decrease information by passing from one experiment to another experiment by a transition. There is an interesting dual way of expressing this which tells us how we may increase information by passing from one experiment to another experiment by certain density preserving transitions. In the case of dominated experiments with Euclidean sample spaces this may be expressed as follows.

Let the experiments under consideration be $\mathscr{E} = (\mathscr{X}, \mathscr{A}; P_\theta : \theta \in \Theta)$ and $\mathscr{F} = (\mathscr{Y}, \mathscr{B}; Q_\theta : \theta \in \Theta)$. Assume that the sample spaces $(\mathscr{X}, \mathscr{A})$ and $(\mathscr{Y}, \mathscr{B})$ are both Euclidean.

Let λ be a prior probability distribution on some σ-algebra of subsets of Θ which makes $P_\theta(A)$ and $Q_\theta(B)$ measurable in θ when $A \in \mathscr{A}$ and $B \in \mathscr{B}$. Put $P_\lambda = \int P_\theta \lambda(d\theta)$ and $Q_\lambda = \int Q_\theta \lambda(d\theta)$ and let us assume that P_λ dominates $\mathscr{E}$ and that Q_λ dominates $\mathscr{F}$.

Suppose now that $\mathscr{E} \geqslant \mathscr{F}$. The randomization criterion then yields a Markov kernel M from $(\mathscr{X}, \mathscr{A})$ to $(\mathscr{Y}, \mathscr{B})$ such that $P_\theta M \equiv_\theta Q_\theta$. Then $P_\lambda M = Q_\lambda$. Construct the probability measure $P_\lambda \times M$ on $(\mathscr{X}, \mathscr{A}) \times (\mathscr{Y}, \mathscr{B})$ and let D be a regular conditional distribution of the first coordinate given the second coordinate for this distribution. Then $P_\lambda \times M = D \times Q_\lambda$ and $P_\lambda = DQ_\lambda$, i.e. $P_\lambda(A) \equiv_A \int D(A|y)Q_\lambda(dy)$. The interesting features of the transition D are that it maps Q_λ into P_λ and that it carries the densities $(dP_\theta/dP_\lambda : \theta \in \Theta)$ in $\mathscr{E}$ into the densities $(dQ_\theta/dQ_\lambda : \theta \in \Theta)$ in $\mathscr{F}$. The checking of this runs as follows

$$\begin{aligned}
&\int_B \{[dP_\theta/dP_\lambda]_x D(dx|y)\} Q_\lambda(dy) \\
&\qquad = \int [I_B \otimes (dP_\theta/dP_\lambda)] \, d(D \times Q_\lambda) \\
&\qquad = \int [I_B \otimes (dP_\theta/dP_\lambda)] \, d(P_\lambda \times M) \\
&\qquad = \int [dP_\theta/dP_\lambda]_x \left[\int I_B(y) M(dy|x) \right] P_\lambda(dx) \\
&\qquad = \int M(B|x) P_\theta(dx) = (P_\theta M)(B) = Q_\theta(B)
\end{aligned}$$

when $B \in \mathscr{B}$ and $\theta \in \Theta$.

Conversely, if D has these properties then

$$Q_\theta(B) = \int_B \{[\mathrm{d}P_\theta/\mathrm{d}P_\lambda]_x D(\mathrm{d}x|y)\} Q_\lambda(\mathrm{d}y).$$

The computations above now show that $Q_\theta \equiv_\theta P_\theta M$ so that $\mathscr{E} \geqslant \mathscr{F}$. □

Similar computations show that the deficiency ordering of experiments may naturally be expressed in Bayesian terms.

Theorem 7.2.16 (Bayesian criterion for being more informative). *Let $\mathscr{E} = (\mathscr{X}, \mathscr{A}; P_\theta : \theta \in \Theta)$ and $\mathscr{F} = (\mathscr{Y}, \mathscr{B}; Q_\theta : \theta \in \Theta)$ be two experiments and let λ be a prior probability measure on some σ-algebra $\mathscr{U}$ of subsets of Θ.*

Assume that $\mathscr{E}$ is coherent and that $(\mathscr{Y}, \mathscr{B})$ is Euclidean. Assume also that the following regularity conditions are satisfied:

(i) *$P_\theta(A)$ and $Q_\theta(B)$ are measurable in θ when $A \in \mathscr{A}$ and $B \in \mathscr{B}$.*
(ii) *Θ possesses a topology which makes $P_\theta(A)$ and $Q_\theta(B)$ continuous in θ when $A \in \mathscr{A}$ and $B \in \mathscr{B}$.*
(iii) *any open non-empty measurable subset of Θ has positive λ-measure (i.e. Θ does not contain any proper closed subset of λ-measure 1).*

Say that a triple (U, V, W) of random variables on some probability space is a Markov triple *if U and W are conditionally independent given V.*

Then $\mathscr{E} \geqslant \mathscr{F}$ if and only if there is a Markov triple (T, X, Y) such that $\mathscr{L}(T) = \lambda$, $\mathscr{L}(X|T) = P_T$ and $\mathscr{L}(Y|T) = Q_T$.

Remark. The condition that (U, V, W) is a Markov triple may also be phrased: for each bounded (non-negative) measurable function g on the range space of W, $E(g(W)|U, V)$ may be specified independently of U. If the range space of W is Euclidean this amounts to requiring that the conditional distribution of W given (U, V) may be specified independently of U. Furthermore, since (W, V, U) is a Markov triple whenever (U, V, W) is, here we may interchange the roles of U and W.

Thus provided $(\Theta, \mathscr{U})$ is Euclidean, we may phrase the condition that (T, X, Y) is a Markov triple as:

> 'the posterior distribution of T given (X, Y) may be specified independently of Y'.

Proof. The regularity conditions imply that $\mathscr{E} \geqslant \mathscr{F}$ if and only if $Q_\theta \equiv_\theta P_\theta M$ for some Markov kernel M. If $Q_\theta \equiv_\theta P_\theta M$ for the Markov kernel M then we may define T, X and Y such that $\mathscr{L}(T) = \lambda$, $\mathscr{L}(X|T) = P_T$ and $\mathscr{L}(Y|T, X) = M(\cdot|X)$. Then (T, X, Y) is a Markov triple and it remains to show that $\mathscr{L}(Y|T) = Q_T$. If g is a bounded measurable function on Θ and $B \in \mathscr{B}$ then we obtain

$$\begin{aligned}
Eg(T)Q_T(B) &= Eg(T)\left[\int M(B|x)P_T(\mathrm{d}x)\right]\\
&= Eg(T)E[M(B|X)|T]\\
&= Eg(T)M(B|X)\\
&= Eg(T)E(I_B(Y)|T,X)\\
&= Eg(T)I_B(Y).
\end{aligned}$$

Thus $Q_T(B)$ is a version of $Pr(Y \in B|T)$.

Conversely assume that such a Markov triple (T, X, Y) exists. Let $M(\cdot|X)$ be a regular conditional distribution of Y given (T, X) which does not depend on T. Then M is a Markov kernel. Put $\tilde{Q}_\theta(B) = (P_\theta M)(B)$ when $\theta \in \Theta$ and $B \in \mathscr{B}$. Let g be a bounded measurable function on T. Then the above computations hold for any $B \in \mathscr{B}$ provided $Q_T(B)$ is replaced by $\tilde{Q}_T(B)$. Thus

$$\begin{aligned}
Eg(T)\tilde{Q}_T(B) &= Eg(T)I_B(Y)\\
&= Eg(T)E[I_B(Y)|T]\\
&= Eg(T)Q_T(B); \qquad B \in \mathscr{B}.
\end{aligned}$$

Hence $\tilde{Q}_\theta(B) = Q_\theta(B)$ for almost all θ when $B \in \mathscr{B}$. Put $N_B = \{\theta : \tilde{Q}(B) \neq Q_\theta(B)\}$. Then N_B is an open measurable set with λ-measure zero. Thus $N_B = \emptyset$ for each $B \in \mathscr{B}$ so that

$$Q_\theta \equiv \tilde{Q}_\theta \equiv P_\theta M$$

yielding $\mathscr{E} \geqslant \mathscr{F}$. □

Assume now that $\mathscr{E}$ and $\mathscr{F}$ are standard experiments and that λ is the uniform distribution. Let S and T be the standard probability measures of $\mathscr{E}$ and $\mathscr{F}$ respectively. Then $S = P_\lambda$, $T = Q_\lambda$ and $[\mathrm{d}P_\theta/\mathrm{d}P_\lambda]_x = [\mathrm{d}Q_\theta/\mathrm{d}Q_\lambda]_x = mx_\theta$ where $m = \#\Theta$. The requirements on D are then:

(i) $\int x_\theta D(\mathrm{d}x|y) \equiv_\theta y_\theta$ for T-almost all y.
(ii) $S = DT$.

A transition from $\mathbb{R}^\Theta$ to $\mathbb{R}^\Theta$ such that $\int x_\theta D(\mathrm{d}x|y) \equiv_\theta y_\theta$ for all $y \in \mathscr{Y}$ is called a *dilation* (of $\mathbb{R}^\Theta$). Intuitively it dilates a point y by replacing it by a probability distribution with expectation y. The only non-randomized dilation is thus the identity map.

Dilations enter naturally when we construct the concave envelope $\hat{f}$ of a real valued function f on $\mathbb{R}^\Theta$ where Θ is finite. Assume f is majorized by some affine function. Then this function may either be written as $\hat{f} = \inf\{l : l \geqslant f, l \text{ is affine}\}$ or alternatively as

$$\hat{f}(y) = \min\left\{\sum_{i=1}^{m+1} p_i f(y_i) : p_1, \ldots, p_{m+1} \geqslant 0, \sum_{i=1}^{m+1} p_i = 1, \sum p_i y_i = y\right\}.$$

The assumption $\sum_{i=1}^{m+1} p_i y_i = y$ within the bracket $\{\ \}$ tells that the expectation of the distribution which assigns mass p_i to y_i; $i = 1, \ldots, m+1$, has expectation y.

Assume now that the probability distribution S on $\mathbb{R}^\Theta$ is obtained from another probability distribution T on $\mathbb{R}^\Theta$ by a dilation D and that $\int |y_\theta| T(\mathrm{d}y) < \infty$ for all θ. Let ϕ be a convex function on $\mathbb{R}^\Theta$. Then, by Jensen's inequality,

$$\int \phi(x) S(\mathrm{d}x) = \int \left[\int \phi(x) D(\mathrm{d}x|y)\right] T(\mathrm{d}y)$$

$$\geqq \int \phi\left(\int x D(\mathrm{d}x|y)\right) T(\mathrm{d}y) = \int \phi(y) T(\mathrm{d}y).$$

In particular $\int x_\theta S(\mathrm{d}x) \equiv_\theta \int y_\theta T(\mathrm{d}y)$. Thus dilations preserve expectations.

Theorem 7.2.17 (The dilation criterion). *Let Θ be a finite set and let S and T be probability distributions on $\mathbb{R}^\Theta$ having the same finite first order moments, i.e. $\int x_\theta S(\mathrm{d}x) = \int y_\theta T(\mathrm{d}y) \in \mathbb{R}$ for each θ. For each θ, let S_θ be the distribution on $\mathbb{R}^\Theta$ which has density $x \to x_\theta$ with respect to S. Similarly T_θ, for each θ, denotes the distribution which has density $x \to x_\theta$ with respect to T. Then the following conditions are equivalent:*

(i) *there is a Markov kernel M from $\mathbb{R}^\Theta$ to $\mathbb{R}^\Theta$ such that $SM = T$ and $S_\theta M \equiv_\theta T_\theta$.*
(ii) *there is a dilation D from $\mathbb{R}^\Theta$ to $\mathbb{R}^\Theta$ such that $T = DS$.*
(iii) *$\int \phi \, \mathrm{d}S \geqslant \int \phi \, \mathrm{d}T$ for all convex functions ϕ on $\mathbb{R}^\Theta$.*

These conditions imply the following condition:

(iv) *$\int |a_0 + \sum_\theta a_\theta x_\theta| S(\mathrm{d}x) \geqslant \int |a_0 + \sum_\theta a_\theta x_\theta| T(\mathrm{d}x)$ for all constants $a_0 \in \mathbb{R}$ and all vectors $a = (a_\theta : \theta \in \Theta) \in \mathbb{R}^\Theta$; and this condition in turn implies:*
(v) *T is supported by the closed convex hull generated by the minimal closed support of S.*

If $t \in \mathbb{R}^\Theta$ and if S is extreme within the convex set of probability distributions P such that $\int x_\theta P(\mathrm{d}x) \equiv_\theta t_\theta$, then all the conditions (i)–(v) *are equivalent.*

Proof. If (i) holds and if D is a regular conditional distribution of X given Y when X has distribution S, and Y given X has distribution $M(\cdot|X)$, then after regularizing D on an exceptional null set, we obtain a dilation satisfying (ii). Conversely, if D satisfies (ii) and the Markov kernel M satisfies $S \times M = D \times T$, then M satisfies (i). Thus conditions (i) and (ii) are equivalent and, by

Jensen's inequality, they imply (iii). Assume therefore that (iii) is satisfied. If S and T are concentrated on $]0, \infty[^\Theta$ then by the sublinear functional criterion, this tells us that the experiment $(S_\theta/\|S_\theta\|; \theta \in \Theta)$ is at least as informative as the experiment $(T_\theta/\|T_\theta\|; \theta \in \Theta)$. Thus by the randomization criterion we obtain a Markov kernel M satisfying (i). The general case may be argued similarly by observing that the arguments leading to the Markov kernel criterion do not require that the underlying distributions are non-negative.

We obtain (iv) from (iii) by specialization. Let U and V be the minimal closed supports of S and T respectively. If v is not contained in the closure of $\langle U \rangle$ then by strong hyperplane separation there is a vector $a \in \mathbb{R}^\Theta - \{0\}$ and a constant $b \in \mathbb{R}$ such that $\langle a, v \rangle > b = \sup\{\langle a, x \rangle : x \in U\}$. Then $\langle a, x \rangle - b \leqslant 0$ for all $x \in U$ so that

$$\begin{aligned} 0 &= \int [\langle a, x \rangle - b]^+ S(\mathrm{d}x) \\ &= \tfrac{1}{2} \int [\langle a, x \rangle - b] S(\mathrm{d}x) + \tfrac{1}{2} \int |\langle a, x \rangle - b| S(\mathrm{d}x) \\ &= \tfrac{1}{2} \int [\langle a, x \rangle - b] T(\mathrm{d}x) + \tfrac{1}{2} \int |\langle a, x \rangle - b| S(\mathrm{d}x) \end{aligned}$$

since $\int x_\theta S(\mathrm{d}x) \equiv_\theta \int x_\theta T(\mathrm{d}x)$. If, in addition (iv) holds, then

$$0 = \int [\langle a, x \rangle - b]^+ S(\mathrm{d}x) \geqq \int [\langle a, x \rangle - b]^+ T(\mathrm{d}x).$$

It follows that the closed set $C = \{x : \langle a, x \rangle \leqslant b\}$ is a support of T. Hence $v \in V \subseteqq C$ yielding the contradiction $\langle a, v \rangle \leqslant b$. Thus (iv) $\Rightarrow$ (v).

Finally assume that condition (v) is satisfied and that S is extreme within $\{P : \int x P(\mathrm{d}x) = t\}$. Then S is supported by the vertices of a simplex $\langle v^1, v^2, \ldots, v^r \rangle$ where $r \leqslant \#\Theta + 1$. If $y \in V$ then there are non-negative numbers $m(y|1), m(y|2), \ldots, m(y|r)$ such that $y = \sum_j m(y|j) v^j$ and $\sum_j m(y|j) = 1$. Let λ_j be the S-mass of v^j and for each Borel set $B \subseteqq \mathbb{R}^\Theta$ put $\hat{m}(B|j) = \lambda_j^{-1} \int_B m(y|j) T(\mathrm{d}y)$. Then $\hat{m}(V|j) \equiv_j \hat{m}(\mathbb{R}^\Theta|j)$ and

$$\begin{aligned} t &= \int y T(\mathrm{d}y) = \int \sum_j m(y|j) v^j T(\mathrm{d}y) \\ &= \sum_j \left[\int \lambda_j^{-1} m(y|j) v^j T(\mathrm{d}y) \right] \lambda_j = \sum_j \hat{m}(V|j) v^j \lambda_j. \end{aligned}$$

Furthermore, $\sum_j \hat{m}(V|j)\lambda_j = \sum_j \int_V m(y|j) T(\mathrm{d}y) = \int_V 1 T(\mathrm{d}y) = 1$. Thus we have

$$\sum_j \hat{m}(V|j) v^j \lambda_j = t = \sum v^j \lambda_j$$

and

$$\sum \hat{m}(V|j)\lambda_j = 1 = \sum_j \lambda_j.$$

Hence, by geometrical independence $\hat{m}(V|j) \equiv_j 1$. Identifying 'j' with v^j we see that $\hat{m}$ defines a Markov kernel from $\{v^1, \ldots, v^r\}$ to $\mathbb{R}^\Theta$. Applying this kernel to the measures $S_\theta : \theta \in \Theta$ and to S we find for any Borel set B

$$(\hat{m}S_\theta)(B) = \sum_j \hat{m}(B|j)S_\theta(v^j) = \sum_j \hat{m}(B|j)v^j_\theta\lambda_j$$

$$= \sum_j \int_B \hat{m}(y|j)T(\mathrm{d}y)v^j_\theta = \int_B y_\theta T(\mathrm{d}y) = T_\theta(B)$$

and

$$(\hat{m}S)(B) = \sum_j \hat{m}(B|j)S(v^j) = \sum_j \hat{m}(B|j)\lambda_j$$

$$= \sum_j \int_B \hat{m}(y|j)T(\mathrm{d}y) = \int_B 1T(\mathrm{d}y) = T(B).$$

Hence condition (i) is satisfied. □

Corollary 7.2.18 (The dilation criterion for 'being more informative'). *Assume that the parameter set Θ is finite. Let the experiments $\mathscr{E}$ and $\mathscr{F}$ have standard (probability) measures S and T. Then the following conditions are equivalent:*

(i) $\mathscr{E} \geqslant \mathscr{F}$.
(ii) $S = DT$ *for a dilation* D.
(iii) $\int \phi\, \mathrm{d}S \geqslant \int \phi\, \mathrm{d}T$ *for all convex functions* ϕ *on* K_Θ.

These conditions imply the following condition:

(iv) $\mathscr{E} \geqslant_{(2)} \mathscr{F}$;

and this condition in turn implies:

(v) *T is supported by the closed convex hull generated by the minimal closed support of S.*

If $\mathscr{E}$ permits a complete and sufficient sub σ-algebra then all the conditions (i)–(v) *are equivalent.*

Remark 1. Sherman (1951) and Stein (1951) proved that conditions (ii) and (iii) are equivalent when the sample spaces are finite, i.e. when S and T are finitely supported. This was generalized by Strassen (1965) without the finiteness condition and within a Banach space setting.

Remark 2. Although convex functions on K_Θ may not be continuous at the boundary they are at least upper semicontinuous there, and hence measurable, throughout K_Θ.

Proof. Although the proof follows from the last theorem, we could observe directly that (iii) is essentially the sublinear functional criterion for comparison, since any continuous convex function on K_Θ may be extended to a (not necessarily real valued) sublinear function on $\mathbb{R}^\Theta$. Furthermore, if $(S_\theta : \theta \in \Theta)$ and $(T_\theta : \theta \in \Theta)$ are the standard experiments of $\mathscr{E}$ and $\mathscr{F}$ and if M is a Markov kernel satisfying the condition $S_\theta M \equiv_\theta T_\theta$ then, since $S = \sum S_\theta$ and $T = \sum T_\theta$, we also have $SM = T$. We saw earlier that S is extreme if and only if $\mathscr{E}$ permits a sufficient and complete sub σ-algebra. This, together with the last statement of the theorem, proves the last statement of the corollary.

□

Any experiment having a finite parameter set may be approximated from below by experiments having finite sample spaces. Indeed if $\mathscr{E} = (\mathscr{X}, \mathscr{A}; P_\theta : \theta \in \Theta)$ and if $\mathscr{A}_1 \subseteqq \mathscr{A}_2 \subseteqq \cdots$ are sub σ-algebras of $\mathscr{A}$ such that the σ-algebra generated by $\bigcup_n \mathscr{A}_n$ is sufficient, then, by the sublinear function criterion for comparison, $\mathscr{E}|\mathscr{A}_n \uparrow \mathscr{E}$. Choosing all $\mathscr{A}_n$ finite, as we always can, we obtain 'finite' approximations from below.

For standard measures this implies the not so surprising result that the set of standard measures having finite supports is dense. What is more interesting is that by working with standard experiments we may easily visualize finite sample space approximations which are either from below or from above.

Assume that we have 'triangulated' the fundamental probability simplex in $\mathbb{R}^\Theta$ by partitioning it as a union of relatively open simplices $U_1, \ldots, U_r$. Letting D be the unique dilation which assigns any point to the vertices of the simplex containing it, we obtain a finite sample space approximation DS from above for any standard measure S. On the other hand we may decrease the information in S by replacing its restriction to U_i, for each i, by the one-point distribution assigning mass $S(U_i)$ to the centre of gravity $\xi_i = \int_{U_i} xS(\mathrm{d}x)/S(U_i)$ of U_i. If ψ is sublinear then, by Jensen's inequality, $\int \psi\, \mathrm{d}S \geqslant \sum_i S(U_i)\psi(\xi_i)$ so that this is an approximation from below.

Using the supremum distance $x, y \to \sup_\theta |x_\theta - y_\theta|$ on the fundamental probability simplex, and letting τ denote the maximal diameter of sets U_i, we find for a sublinear function ψ and for a standard measure S and its approximation $\hat{S}$, that in both cases

$$\left|\int \psi\, \mathrm{d}S - \int \psi\, \mathrm{d}\hat{S}\right| \leqslant \sum \left|\int_{U_i} \psi\, \mathrm{d}S - \int_{U_i} \psi\, \mathrm{d}\hat{S}\right|$$
$$\leqslant \tau \sum_\theta \psi(e^\theta) \vee \psi(-e^\theta) \# \Theta$$

so that

$$\Delta(S, \tilde{S}) \leqslant \tau \# \Theta.$$

Decreasing the mesh τ we may ensure that these approximations are as close as we please–and this uniformly in S.

This may be the right time to show by an example that $\mathscr{E}$ need not be at least as informative as $\mathscr{F}$ is, even if $\mathscr{E}$ is more informative than $\mathscr{F}$ for testing problems. As we shall see later this cannot occur when $\#\Theta = 2$. Thus we need at least three points in the parameter set. It follows also from theorem 6.3.17 that we shall need at least three points in the sample space of $\mathscr{F}$. Thus the following example is quite satisfying.

Example 7.2.19. Let $\mathscr{E}$ and $\mathscr{F}$ be given by the Markov matrices:

$\theta \backslash x$	1	2	3	4
1	0	$\frac{1}{6}$	$\frac{2}{6}$	$\frac{3}{6}$
2	$\frac{2}{6}$	0	$\frac{1}{6}$	$\frac{3}{6}$
3	$\frac{1}{6}$	$\frac{2}{6}$	0	$\frac{3}{6}$

and

$\theta \backslash x$	1	2	3
1	$\frac{1}{6}$	$\frac{2}{6}$	$\frac{3}{6}$
2	$\frac{2}{6}$	$\frac{3}{6}$	$\frac{1}{6}$
3	$\frac{3}{6}$	$\frac{1}{6}$	$\frac{2}{6}$

respectively.

The columns of the second matrix are $u + v$, $u + w$ and $v + w$ where u, v, w and $u + v + w$ are the columns of the first matrix. By proposition 6.3.16, $\mathscr{E} \geqslant_2 \mathscr{F}$. Suppose there is a Markov kernel M from $\mathscr{E}$ to $\mathscr{F}$ such that $Q_\theta = PM_\theta$; $\theta \in \Theta$. Then

$$u + v = M(\{1\}|1)u + M(\{1\}|2)v + M(\{1\}|3)w + M(\{1\}|4)(u + v + w)$$

and

$$u + w = M(\{2\}|1)u + M(\{2\}|2)v + M(\{2\}|3)w + M(\{2\}|4)(u + v + w).$$

Hence, by linear independence, $M(\{1\}|1) = M(\{2\}|1) = 1$. This yields the contradiction $1 \geqslant M(\{1, 2\}|1) = 1 + 1$. Hence $\delta_3(\mathscr{E}, \mathscr{F}) = \delta(\mathscr{E}, \mathscr{F}) > 0$.

We may utilize the facts we have established so far for standard experiments, to describe certain important types of functionals on experiments with a finite parameter set. Note first that whether Θ is finite or not, mixtures as well as products respect equivalence and thus are well defined for types of experiments. Furthermore, by theorem 7.2.12 or directly, products behave, up to equivalence, distributively with respect to mixtures.

Consider a real valued functional Ω on the set of types of experiments with a parameter set Θ which is not necessarily finite. We shall say that:

(i) Ω is *non-negative* if $\Omega(\mathscr{E}) \geqslant 0$ for all experiments $\mathscr{E}$.
(ii) Ω is *monotonically increasing* if $\Omega(\mathscr{E}) \geqslant \Omega(\mathscr{F})$ when $\mathscr{E} \geqslant \mathscr{F}$.
(iii) Ω is *monotonically decreasing* if $\Omega(\mathscr{E}) \leqslant \Omega(\mathscr{F})$ when $\mathscr{E} \geqslant \mathscr{F}$.
(iv) Ω is *monotone* if it is either monotonically increasing or monotonically decreasing.
(v) Ω is *convex* if $\Omega((1 - p)\mathscr{E} + p\mathscr{F}) \leqslant (1 - p)\Omega(\mathscr{E}) + p\Omega(\mathscr{F})$ for all $p \in [0, 1]$ and all pairs $(\mathscr{E}, \mathscr{F})$ of experiments.
(vi) Ω is *concave* if $-\Omega$ is convex.
(vii) Ω is *affine* if it is both concave and convex.

(viii) Ω is *τ-continuous* if it is continuous with respect to a given topology τ for the set of types.

(ix) Ω is *multiplicative* if $\Omega(\mathscr{E} \times \mathscr{F}) = \Omega(\mathscr{E}) \cdot \Omega(\mathscr{F})$ for all pairs $(\mathscr{E}, \mathscr{F})$ of experiments.

(x) Ω is *additive* if $\Omega(\mathscr{E} \times \mathscr{F}) = \Omega(\mathscr{E}) + \Omega(\mathscr{F})$ for all pairs $(\mathscr{E}, \mathscr{F})$ of experiments.

We may also consider maps from the set of types into other spaces than the real numbers. The terms introduced above for a real functional may then be used for general maps, provided the corresponding terms have a 'natural' interpretation in the range space of this map. Thus these terms are all well defined for maps from the set of types to any ordered topological vector space.

A map from the set of types of experiments with parameter set Θ into a set of functionals may also be called a *transform.*

Note that a map F from the set of types of experiments with a finite parameter set Θ into a topological space, by compactness, is *convergence determining* if and only if it is continuous and 1–1.

Let us consider a few functionals.

First of all the deficiency $\delta_k(\mathscr{E}, \mathscr{F})$ is monotonically decreasing in $\mathscr{E}$ and monotonically increasing in $\mathscr{F}$. The deficiency $\delta_k(\mathscr{E}, \mathscr{F})$ as well as the deficiency distance $\Delta_k(\mathscr{E}, \mathscr{F})$ are convex in each argument. This follows since $\delta_k(\mathscr{E}, \mathscr{F})$, for fixed $\mathscr{E}$ (for fixed $\mathscr{F}$), is the supremum of affine functionals in $\mathscr{F}(\mathscr{E})$. By theorem 6.3.24, the same holds for the weighted deficiencies and for the lower weighted deficiencies. A study of $\delta(\mathscr{E}, \mathscr{M}_a)$ and $\delta(\mathscr{M}_i, \mathscr{E})$ as measures of information may be found in Torgersen, 1981. We have encountered the functional $\Omega_\lambda \to 1 - \frac{1}{2}\delta(\mathscr{E}, \mathscr{M}_a | \lambda)$ before and have observed that this quantity is the maximal Bayes probability of guessing correctly the true distribution for the prior λ. If we put $\mathscr{E} = (P_\theta : \theta \in \Theta)$ then $\Omega_\lambda(\mathscr{E}) = \|\bigvee_\theta \lambda_\theta P_\theta\| = \int (\bigvee_\theta \lambda_\theta x_\theta) S(\mathrm{d}x)$ where S is the standard measure of $\mathscr{E}$. It was shown by Morse & Sacksteder (1966) that the transform $\lambda \to \Omega_\lambda(\mathscr{E})$ from K_Θ to $[0, \infty[$ is a characterization, i.e. it determines the type of $\mathscr{E}$. This follows by first extending the set of permissible distributions λ to the set of all non-negative distributions on Θ and then by showing that the right hand partial derivative of $\Omega_\lambda(\mathscr{E})$ with respect to λ_{θ_0} is $\int_{A_{\theta_0}} x_{\theta_0} S(\mathrm{d}x)$ where $A_{\theta_0} = \{x : \bigvee_\theta \lambda_\theta x_\theta = \lambda_{\theta_0} x_{\theta_0}\}$. Thus the quantity $P_{\theta_0}(\mathrm{d}P_\theta/\mathrm{d}P_{\theta_0} \leqslant \lambda_{\theta_0}/\lambda_\theta : \theta \in \Theta) = \int_{A_{\theta_0}} x_{\theta_0} S(\mathrm{d}x)$ is determined when $\lambda_\theta > 0$ for all θ. This in turn implies that $\mathscr{L}_{P_{\theta_0}}(\mathrm{d}P_\theta/\mathrm{d}P_{\theta_0} : \theta \in \Theta)$ is determined for each $\theta_0 \in \Theta$. Let $(S_\theta : \theta \in \Theta)$ be the standard experiment of $\mathscr{E}$ so that $S = \sum_\theta S_\theta$ is the standard measure of $\mathscr{E}$. If $\Theta = \{1, \dots, m\}$, $B_\theta = \{x : x_1 = \cdots = x_{\theta-1} = 0,\ x_\theta > 0\}$, and ψ is bounded on compacts and positive homogeneous, then $\int \psi(\mathrm{d}\mathscr{E}) = \int \psi(x) S(\mathrm{d}x) = \sum_\theta \int_{B_\theta} \psi(x | x_\theta) S_\theta(\mathrm{d}x)$ where the θth term for $\theta = \theta_0$ is determined by $\mathscr{L}_{P_{\theta_0}}(\mathrm{d}P_\theta/\mathrm{d}P_{\theta_0} : \theta \in \Theta)$. Thus, the transform $\lambda \to \Omega_\lambda(\mathscr{E})$ determines the type of $\mathscr{E}$.

This proves the following proposition.

Proposition 7.2.20. *Let Θ be any set. The transform which to each probability distribution on Θ with finite support assigns for an experiment $\mathscr{E} = (P_\theta : \theta \in \Theta)$ the maximal probability $\|\bigvee_\theta \lambda_\theta P_\theta\| = 1 - \frac{1}{2}\delta(\mathscr{E}, \mathscr{M}_a|\lambda)$ of guessing correctly the true value of θ for the prior distribution λ, determines the type of $\mathscr{E}$ and, consequently, is a characterization.*

Before discussing functionals in general let us make a short excursion to a related problem. Let us consider a complete separable metric space $(\mathscr{X}, d)$. Let $\mathscr{P}$ denote the set of probability measures on the class of Borel subsets of $\mathscr{X}$. Topologize $\mathscr{P}$ by its weak topology, i.e. the smallest topology which makes $P(f)$ continuous in P when $f \in C(\mathscr{X})$. If $f \in C(\mathscr{X})$ then the functional $P \to P(f)$ is affine as well as continuous. The interesting fact is now that there are no other functionals on $\mathscr{P}$ which are both continuous and affine and, in fact, if Ω is such a functional then $\Omega(P) \equiv_P P(f)$ where $f(x) \equiv_x \Omega(\delta_x)$. The function f is continuous as well as bounded. Using the arguments from Huber, 1981, this may be seen as follows.

Firstly if x_n is any sequence in $\mathscr{X}$ then $(1 - (1/n))\delta_x + (1/n)\delta_{x_n} \to \delta_x$. It follows that $(1/n)\Omega(\delta_{x_n}) \to 0$. Hence f is bounded and it is clearly continuous. Furthermore $\Omega(P) = P(f)$ when P has finite support. If $P \in \mathscr{P}$ then there are distributions P_n; $n = 1, 2, \ldots$ with finite supports such that $P_n \to P$. Hence $\Omega(P) = \lim_n \Omega(P_n) = \lim_n P_n(f) = P(f)$.

We could try to mimic this procedure for functionals of experiments. However there is the difficulty that one-point distributions are not standard probability measures. Fortunately there are particular experiments which we may utilize instead. We describe these in the following example.

Example 7.2.21. As sample space, let us take the set Θ together with a point Υ(e.g. the set Θ itself) which does not belong to Θ. If $\xi \in [0,1]^\Theta$ then we define the experiment $\mathscr{G}_\xi = (P_\theta : \theta \in \Theta)$ by putting $P_\theta(\theta) = 1 - \xi_\theta = 1 - P_\theta(\Upsilon)$. The standard measure of $\mathscr{G}_\xi$ then assigns mass $1 - \xi_\theta$ to each vertex e^θ of K_Θ and the remaining mass $\sum_\theta \xi_\theta$ is assigned to $\xi/\sum_\theta \xi_\theta$. Note that $\mathscr{G}_\xi \sim \mathscr{M}_a$ if and only if there is at most one θ such that $\xi_\theta > 0$ while $\mathscr{G}_\xi \sim \mathscr{M}_i$ if and only if $\xi = (1, 1, \ldots, 1)$. If h is a positively homogeneous function on $[0, \infty[^\Theta$ then $\int h(d\mathscr{G}_\xi) = \sum_\theta (1 - \xi_\theta)h(e^\theta) + h(\xi)$.

If $\xi, \eta \in [0,1]^\Theta$ then $\xi\eta \in [0,1]^\Theta$ and $\mathscr{G}_{\xi\eta} \sim \mathscr{G}_\xi \times \mathscr{G}_\eta$. It follows that $\mathscr{G}_\xi \geqslant \mathscr{G}_\eta$ when $\xi \leqslant \eta$. On the other hand if $\mathscr{G}_\xi \geqslant_2 \mathscr{G}_\eta$ then, by the sublinear function criterion, $(\xi_{\theta_0} - \beta\xi_{\theta_1})^+ - \xi_{\theta_0} \geqslant (\eta_{\theta_0} - \beta\eta_{\theta_1})^+ - \eta_{\theta_0}$ when $\theta_0, \theta_1 \in \Theta$ and $\beta > 0$. $\beta \to \infty$ yields $\xi_{\theta_0} \leqslant \eta_{\theta_0}$ provided $\xi_{\theta_1} > 0$. Thus $\mathscr{G}_\xi \geqslant_2 \mathscr{G}_\eta$ and $\mathscr{G}_\xi \nsim \mathscr{M}_a \Rightarrow \xi \leqslant \eta$. Hence $\mathscr{G}_\xi \geqslant \mathscr{G}_\eta \Leftrightarrow \mathscr{G}_\xi \geqslant_2 \mathscr{G}_\eta \Leftrightarrow \#\{\theta : \xi_\theta > 0\} = 1$ or $\xi \leqslant \eta$.

It may be shown that the class of experiments $\mathscr{G}_\xi$ is contained in the slightly larger class of experiments $\mathscr{E}$ such that $\mathscr{F} \geqslant \mathscr{E}$ if and only if $\mathscr{F} \sim \mathscr{E} \times \mathscr{H}$ for some experiment $\mathscr{H}$. Now consider any experiment $\mathscr{E}$ having standard probability measure $\bar{S}$. Let Q_ξ denote the standard measure of $\mathscr{G}_\xi$. Then $T =$

$\int Q_\xi \bar{S}(\mathrm{d}\xi)$ is the standard measure of some experiment which we may write as $\int \mathscr{G}_\xi \bar{S}(\mathrm{d}\xi)$. If h is non-negatively homogeneous measurable and bounded on compacts then $\int h\,\mathrm{d}T = \int [\sum_\theta h(e^\theta)(1-\xi^\theta) + h(\xi)]\bar{S}(\mathrm{d}\xi) = \sum_\theta h(e^\theta)(1-(1/m)) + \int h(\xi)\bar{S}(\mathrm{d}\xi) = (1-(1/m))\int h(\mathrm{d}\mathscr{M}_a) + (1/m)\int h(\mathrm{d}\mathscr{E})$ where $m = \#\Theta$. Thus

$$\int \mathscr{G}_\xi \bar{S}(\mathrm{d}\xi) \sim \left(1-\frac{1}{m}\right)\mathscr{M}_a + \frac{1}{m}\mathscr{E}.$$

Let Ω be any affine and continuous functional on the set of experiments with parameter set Θ. Define the experiments $\mathscr{G}_\xi$ as in the example and let $\mathscr{E}$ be an experiment with standard measure S and standard probability measure $\bar{S} = S/m$ where $m = \Theta$.

The decomposition given in the example yields

$$\Omega\left(\int \mathscr{G}_\xi \bar{S}(d\xi)\right) = \left(1-\frac{1}{m}\right)\Omega(\mathscr{M}_a) + \frac{1}{m}\Omega(\mathscr{E}).$$

Approximating $\mathscr{E}$ with experiments with finite sample spaces we find, by continuity, that the left hand side may be written $\int \Omega(\mathscr{G}_\xi)\bar{S}(\mathrm{d}\xi)$. Thus

$$\Omega(\mathscr{E}) = \int\left[\Omega(\mathscr{G}_\xi) - \left(1-\frac{1}{m}\right)\Omega(\mathscr{M}_a)\right]S(\mathrm{d}\xi).$$

It follows that a functional Ω on the set of types of experiments is affine and continuous if and only if it is of the form

$$\Omega(\mathscr{E}) = \int h(\mathrm{d}\mathscr{E})$$

for some continuous and positively homogeneous function h on $[0,\infty[^\Theta$. On the other hand, it is clear that any positively homogeneous measurable function h on $[0,\infty[^\Theta$ which is bounded from below by a linear function defines an affine functional Ω by

$$\Omega(\mathscr{E}) = \int h(\mathrm{d}\mathscr{E}).$$

A functional Ω which may be expressed in this form is here called *representable*.

If $h(x) \equiv_x \sum_\theta c_\theta x_\theta$ where $\sum_\theta c_\theta = 0$ (i.e. h is a contrast) then $\int h(\mathrm{d}\mathscr{E}) = 0$. Thus the representing function h is not unique. However, there is no more arbitrariness than that, since h is unique up to an additive contrast. To see this consider a function h such that $\int h(\mathrm{d}\mathscr{E}) = 0$ for all experiments $\mathscr{E}$. Then $\sum_\theta h(e^\theta)(1-\xi_\theta) + h(\xi) \equiv_\xi \int h(\mathscr{G}_\xi) \equiv_\xi 0$. Thus $h(\xi) = \sum_\theta h(e^\theta)(\xi_\theta - 1)$ when $\xi \in [0,1]^\Theta$. $\xi = 0$ yields $0 = h(0) = -\sum_\theta h(e^\theta)$ so that h is a contrast.

Next let us consider those functions h which define non-negative functionals.

Assume that h is a real valued measurable positively homogeneous function (thus $h(tx) = th(x)$ when $t \geqslant 0$) on $[0, \infty[^\Theta$ such that $\int h(d\mathscr{E}) \geqslant 0$ for all experiments $\mathscr{E}$. Then $h(\xi) + \sum_\theta (1 - \xi_\theta) h(e^\theta) = \int h(d\mathscr{G}_\xi) \geqslant 0$ when $\xi \in [0,1]^\Theta$. In particular it follows that h is bounded from below on K_Θ. Let $\underline{h}$ denote the largest continuous convex function on K_Θ which is majorized by h. In particular consider the value of $\underline{h}$ at $e/m = (1/m, \ldots, 1/m)$. Then since $\{(x,y): y \geqslant \underline{h}(x)\}$ is the convex hull of $\{(x,y): y \geqslant h(x)\}$, there are vectors $x^1, \ldots, x^m$ in K_Θ and weights $\lambda_1, \ldots, \lambda_m$ such that $e/m = \sum_{i=1}^m \lambda_i x^i$ and $\underline{h}(e/m) = \sum_{i=1}^m \lambda_i h(x^i)$.

Let S_0 be the standard probability measure which assigns mass λ_i to x^i, $i = 1, \ldots, m$. If S is the standard probability measure of the experiment $\mathscr{E}$ then, since $\mathscr{E} \geqslant \mathscr{M}_i$, $\int \underline{h}(x) S(dx) \geqslant \underline{h}(e/m) = \int h(x) S_0(dx) \geqslant 0$. Thus $\underline{h}$ also defines a non-negative affine functional and $h \geqslant \underline{h}$. Extend $\underline{h}$ to a sublinear functional on $\mathbb{R}^\Theta$ and let $c \in \mathbb{R}^\Theta$ define a supporting hyperplane of $\{(x,y): y \geqslant \underline{h}(x)\}$ at $(e/m, \underline{h}(e/m))$, i.e.

$$\underline{h}(x) \geqslant (c, x) \qquad \text{for all } x \geqslant 0$$

while $0 \leqslant \underline{h}(e) = (c, e)$. We may now decrease c, which does not affect the first inequality, and thereby obtain a vector $a \in \mathbb{R}^\Theta$ such that

$$\underline{h}(x) \geqslant (a, x) \qquad \text{for all } x \geqslant 0$$

while $0 = (a, e)$. Altogether we have found a contrast, $x \to (a, x)$ on $\mathbb{R}^\Theta$ such that $h(x) \geqslant (a, x)$ for all $x \geqslant 0$.

This proves the non-trivial part of the following proposition.

Proposition 7.2.22 (Non-negative representable functionals). *If h is a measurable non-negatively homogeneous function on $[0, \infty[^\Theta$ which is bounded on bounded sets then the affine functional $\mathscr{E} \to \int h(d\mathscr{E})$ is non-negative if and only if h is minorized by a contrast on $\mathbb{R}^\Theta$.*

This yields the following characterization of monotone representable functionals.

Corollary 7.2.23 (Monotone representable functionals). *If h is a measurable non-negatively homogeneous function on $[0, \infty[^\Theta$ which is bounded on bounded sets then the affine functional $\mathscr{E} \to \int h(d\mathscr{E})$ is monotonically increasing (decreasing) if and only if h is sublinear (superlinear) on $\mathbb{R}^\Theta$.*

Remark. h need not be continuous. If we consider the restriction of h to K_Θ then the proposition may be phrased: 'h defines a monotonically increasing (decreasing) functional if and only if $h|K_\Theta$ is convex (concave)'.

Proof. Note first that if h is minorized by a contrast on $\mathbb{R}^\Theta$ and if $h(e^\theta) \equiv_\theta 0$ then this contrast is the zero contrast and thus $h \geqslant 0$.

The 'if' part of the corollary was proved in section 6.3.

Assume that h defines a monotonically increasing functional. Let $v, w \in K_\Theta$ and let $t \in [0, 1[$. Put $u = (1-t)v + tw$. Define a dilation D from K_Θ to K_Θ by

$$D(x|x) = 1 \qquad \text{if } x \neq u$$

$$D(v|u) = 1 - t$$

$$D(w|u) = t.$$

Put $\tilde{h}(x) = \int h(y)D(\mathrm{d}y|x) - h(x)$ when $x \in K_\Theta$. Then $\tilde{h}(e^\theta) = h(e^\theta) - h(e^\theta) = 0$. If S is a standard probability measure we obtain

$$\int \tilde{h}(x)S(\mathrm{d}x) = \int h(y)(DS)(\mathrm{d}y) - \int h(y)S(\mathrm{d}y) \geqslant 0$$

since the experiment defined by DS is at least as informative as the experiment defined by S. Hence, by the previous result, $\tilde{h} \geqslant 0$. In particular $0 \leqslant \tilde{h}(u) = (1-t)h(v) + th(w) - h((1-t)v + tw)$. Hence h is convex. □

Now let us consider representable and multiplicative functionals.

Let $\Omega(\mathscr{E}) = \int h(\mathrm{d}\mathscr{E})$ where h is a positively homogeneous and measurable function on $[0, \infty[^\Theta$. Since $\mathscr{E} \times \mathscr{M}_i \sim \mathscr{E}$ and $\mathscr{E} \times \mathscr{M}_a \sim \mathscr{M}_a$, the assumption of multiplicativity implies that $\Omega(\mathscr{E})\Omega(\mathscr{M}_i) = \Omega(\mathscr{E})$ and $\Omega(\mathscr{E})\Omega(\mathscr{M}_a) = \Omega(\mathscr{M}_a)$ for any experiment $\mathscr{E}$.

Define Ω_0 and Ω_1 by $\Omega_0(\mathscr{E}) \equiv_{\mathscr{E}} 0$ and $\Omega_1(\mathscr{E}) \equiv_{\mathscr{E}} 1$. Then Ω_0 and Ω_1 are both multiplicative. Ω_0 is representable by h if and only if h is a contrast while Ω_1 is representable by h if and only if $h(x) \equiv_x \sum_\theta x_\theta/m + c(x)$ where $c(x)$ is a contrast in x.

Next assume that Ω is multiplicative and that $\Omega \neq \Omega_0$ and $\Omega \neq \Omega_1$. Then there are experiments $\mathscr{E}$ and $\mathscr{F}$ such that $\Omega(\mathscr{E}) \neq 0$ while $\Omega(\mathscr{F}) \neq 1$. Then the identities above imply that $\Omega(\mathscr{M}_i) = 1$ and $\Omega(\mathscr{M}_a) = 0$. If Ω is represented by h then $\Omega(\mathscr{G}_\xi) = \int h(\mathrm{d}\mathscr{G}_\xi) = h(\xi) + \sum_\theta (1 - \xi_\theta)h(e^\theta)$. Adding a suitable contrast we may arrange things so that $h(e^\theta) = \kappa$ does not depend on θ. Then $\Omega(\mathscr{G}_\xi) = h(\xi) + \sum_\theta (1 - \xi_\theta)\kappa$. Putting $\xi = 0$ we find $0 = \Omega(\mathscr{M}_a) = \Omega(\mathscr{G}_0) = h(0) + m\kappa = m\kappa$ so that $\kappa = 0$. Hence $\Omega(\mathscr{G}_\xi) = h(\xi)$ when $\xi \in [0, 1]^\Theta$.

Let $x, y \in [0, \infty]^\Theta$. Then there is a constant $t > 0$ such that $\xi = tx$ and $\eta = ty$ both belong to $[0, 1]^\Theta$. Then $h(xy) = h(\xi\eta/t^2) = h(\xi\eta)/t^2 = \Omega(\mathscr{G}_{\xi\eta})/t^2 = \Omega(\mathscr{G}_\xi \times \mathscr{G}_\eta)/t^2 = [\Omega(\mathscr{G}_\xi)/t]\cdot[\Omega(\mathscr{G}_\eta)/t)] = [h(\xi)/t]\cdot[h(\eta)/t] = h(\xi|t)\cdot h(\eta|t) = h(x)\cdot h(y)$. For each θ define the function h_θ on $[0, \infty[$ by

$$h_\theta(z) = h\left(1, \ldots, \overset{(\theta)}{z}, \ldots, 1\right).$$

Then, since $x = \prod_\theta (1, \ldots, \overset{(\theta)}{x_\theta}, \ldots, 1)$, we find that $h(x) = \prod_\theta h_\theta(x_\theta)$. Furthermore, $h_\theta(z_1 z_2) = h(1, \ldots, z_1 z_2, \ldots, 1) = h(1, \ldots, z_1, \ldots, 1)h(1, \ldots, z_2, \ldots, 1) = h_\theta(z_1)\cdot h_\theta(z_2)$. In particular $h_\theta(z) = (h_\theta(z^{1/2}))^2 \geqslant 0$ for all θ and $z \geqslant 0$. Thus $h \geqslant 0$. If $z_0 > 0$ and if $h_\theta(z_0) = 0$ then $h_\theta(z) = h_\theta((z/z_0)z_0) = h_\theta(z/z_0)h_0(z_0) = 0$

for all $z \geqslant 0$. Hence $h = 0$, contradicting the assumption that $\Omega \neq 0$. It follows that $h_\theta(z) = (h_\theta(z^{1/2}))^2 > 0$ when $z > 0$. On the other hand $h_\theta(0) = h_\theta(0 \cdot 0) = h_\theta(0)^2$ so that $h_\theta(0) = 1$ or $h_\theta(0) = 0$. If $h_\theta(0) = 1$ then $h_\theta(z) = h_\theta(z) \cdot h_\theta(0) = h_\theta(z \cdot 0) = h_\theta(0) - 1$ for all $z \geqslant 0$. Put $\phi_\theta(x) = \log h_\theta(e^x)$ when $x \in \mathbb{R}$. Then $\phi_\theta(x + y) = \phi_\theta(x) + \phi_\theta(y)$ when $x, y \in \mathbb{R}$. Together with the measurability of ϕ_θ this implies that $\phi_\theta(x) \equiv_x t_\theta x$ for some real constant t_θ. (Put $H = \{x : \phi_\theta(x) = x\phi_\theta(1)\}$. Then H is a measurable subgroup of $(\mathbb{R}, +)$ which contains the rational numbers and ϕ_θ is continuous since it is measurable and midconvex (see e.g. Roberts & Varberg, 1973). Thus H is closed so that $H = \mathbb{R}$. It follows that $h_\theta(z) = e^{\phi_\theta(\log z)} = z^{t_\theta}$ when $z > 0$. If $t_\theta < 0$ then $h(1, \ldots, \overset{(\theta)}{z}, \ldots, 1) = h_\theta(z) \to \infty$ as $z \to 0$. This contradicts the integrability of h (finiteness of Ω) with respect to any standard measure. Thus $t_\theta \geqslant 0$. If $a > 0$ then $h(a, a, \ldots, a) = \prod_\theta h_\theta(a) = a^{\sum_\theta t_\theta}$ and $h(a, a, \ldots, a) = ah(1, 1, \ldots, 1) = 1$. Hence $\sum_\theta t_\theta = 1$. What happens to h_θ at $z = 0$? If $t_\theta > 0$ then we cannot have $h_\theta(0) = 1$, since this implies that $h_\theta(z) \equiv_z 1$. Thus $h_\theta(0) = 0$ in this case so that $h_\theta(z) = z^{t_\theta}$ for all $z \geqslant 0$ when $t_\theta > 0$. If $t_\theta = 0$ then continuity dictates that we should have $h_\theta(0) = 1$. However, there is no way of showing that $h_\theta(z)$ is continuous at $z = 0$ when $t_\theta = 0$. In fact we are free (just check it) to interpret 0^0 as 1, as is customary, but also as 0. The first choice makes h continuous, but does not, however make $\int \prod_\theta z^{t_\theta} S(dz)$ continuous as a function of t throughout K_Θ when the experiment $\mathscr{E}$ with standard measure S is non-homogeneous. If on the other hand, we interpret all expressions 0^0 as 0 then h is not continuous when $t = (t_\theta : \theta \in \Theta) \in bdK_\Theta$. However now $\int \prod z_\theta^{t_\theta} S(dz)$ becomes continuous in t for each standard measure S.

Altogether this proves the following proposition.

Proposition 7.2.24 (Multiplicative representable functionals). *Let h be a non-negatively homogeneous real valued measurable function on $[0, \infty[^\Theta$ which is bounded on bounded sets, and put $\Omega(\mathscr{E}) = \int h(d\mathscr{E})$. Then Ω is multiplicative and constant if and only if either $\Omega = 0$ or $\Omega = 1$. The first condition is satisfied if and only if h is a contrast, while the second condition is satisfied if and only if h differs from $x \to (\sum_\theta x_\theta)/\#\Theta$ by a contrast.*

The functional Ω is multiplicative and non-constant if and only if there is a prior distribution t on Θ such that h differs by a contrast from a function $z \to \prod_\theta z_\theta^{t_\theta}$ where the factor $z_\theta^{t_\theta}$, for each θ, should be interpreted either as 0 or as 1 when $z_\theta = t_\theta = 0$.

Any such function h is concave (i.e. superadditive since h is homogeneous) on $[0, \infty[^\Theta$ and thus defines a monotonically decreasing functional.

Remark. The assumption which assured that $t_\theta \geqslant 0$ for all θ was the assumption that Θ was real valued. If we permit the values $+\infty$ for Ω then $\mathscr{E} \to \int h(d\mathscr{E})$ is multiplicative (and affine) for any function h on $\mathbb{R}^\Theta$ of the form $h(x) \equiv_x \prod_\theta x_\theta^{t_\theta}$ where the real constants $t_\theta : \theta \in \Theta$ are only subject to the condition $\sum_\theta t_\theta = 1$.

If we agree to put $0^0 = 1$ for all factors 0^0 appearing in $h(x) \equiv_x \prod_\theta x_\theta^{t_\theta}$ then the map $t \to \int h(d\mathscr{E}) = \int \prod (dP_\theta)^{t_\theta}$ from K_Θ to $[0, 1]$ is the Hellinger transform $H(\cdot|\mathscr{E})$ of $\mathscr{E}$ as defined in chapter 1. As $H(\cdot|\mathscr{E})$ for fixed $\mathscr{E}$, is an integral of log convex functions, it is log convex. If on the other hand, we interpret each factor 0^0 as 0 then we get another convex function $\tilde{H}(\cdot|\mathscr{E})$ on K_Θ.

We summarize some of the properties of these functions in the following proposition.

Proposition 7.2.25 (The Hellinger transform of a given experiment). *Assume that Θ is finite. For each prior distribution t on Θ let $H(t|\mathscr{E})$ denote the Hellinger transform of $\mathscr{E}$, i.e. $H(t|\mathscr{E}) = \int h(d\mathscr{E}, t)$ where $h(x, t) = \prod_\theta x_\theta^{t_\theta}$ when $x \in [0, \infty[^\Theta$. Also put $\tilde{H}(t|\mathscr{E}) = \int \tilde{h}(d\mathscr{E}, t)$ where $\tilde{h}(x, t) = \prod_\theta x_\theta^{t_\theta}$ if $t_\theta > 0$ whenever $x_\theta = 0$ and where $\tilde{h}(x, t) = 0$ otherwise (i.e. if $x_\theta = t_\theta = 0$ for some θ).*

Then:

(i) $H(\cdot|\mathscr{E}) \geqslant \tilde{H}(\cdot|\mathscr{E})$ *for any experiment* $\mathscr{E}$.
(ii) $H(\cdot|\mathscr{E}) = \tilde{H}(\cdot|\mathscr{E})$ *if and only if* $\mathscr{E}$ *is homogeneous.*
(iii) $H(t|\mathscr{E}) = \tilde{H}(t|\mathscr{E})$ *for all* $t \in K_\Theta$ *such that* $t_\theta > 0$ *for all* θ.

Furthermore, both functions $H(\cdot|\mathscr{E})$ and $\tilde{H}(\cdot|\mathscr{E})$ are log convex. $\tilde{H}(\cdot|\mathscr{E})$ is continuous while $H(\cdot|\mathscr{E})$ is continuous if and only if $\mathscr{E}$ is homogeneous.

Proof. If $\mathscr{E} = (P_\theta : \theta \in \Theta)$ has Hellinger transform H and standard measure S then $H(t) = \int_{\text{ri}\,K} \prod_\theta x_\theta^{t_\theta} S(dx)$ when $t \in \text{ri}\,K$ where K is the probability simplex in $\mathbb{R}^\Theta$. Thus $H|\text{ri}\,K$ is determined by $S|\text{ri}\,K$. Writing the integrand as $x_{\theta_0} \exp \sum_{\theta \neq \theta_0} t_\theta \log(x_\theta/x_{\theta_0})$ we see by the uniqueness theorem for Laplace transforms that the converse also holds, i.e. $S|\text{ri}\,K$ is determined by $H|\text{ri}\,K$. In particular it follows that experiments possessing the same Hellinger transform are homogeneously equivalent and consequently, by proposition 7.2.8, are equivalent. Thus experiments are equivalent if and only if they possess the same Hellinger transform. □

If $\mathscr{E} \geqq \mathscr{F}$ then by proposition 7.2.24, $H(t|\mathscr{E}) \leqslant H(t|\mathscr{F})$ for all $t \in K_\Theta$. One might then wonder if the Hellinger transform characterizes the ordering '$\geqslant$' as well. However, this is far from being true. The problem is closely linked to a problem raised by Blackwell concerning replications of experiments. Blackwell showed in his 1951 paper that experiment multiplication is a monotone operation. In particular the nth powers of ordered experiments are themselves ordered for any n. Blackwell then asked if the converse holds. The answer to this question turned out to be negative. After a counter-example was found it soon became clear that it is not at all unusual for non-comparable experiments to yield comparable experiments after they have been replicated the same number of times (see e.g. Hansen & Torgersen, 1974). Thus we may have experiments $\mathscr{E}$ and $\mathscr{F}$ such that $\mathscr{E}^2 \geqslant \mathscr{F}^2$ while $\mathscr{E} \not\geqslant \mathscr{F}$. In that case $H(\cdot|\mathscr{E}) = (H(\cdot|\mathscr{E}^2))^{1/2} \leqslant (H(\cdot|\mathscr{F}^2))^{1/2} = H(\cdot|\mathscr{F})$ so that the Hellinger transforms are ordered although the experiments themselves are not comparable.

Let us summarize some of the established properties of three particular transforms which we have encountered at various places in this section.

Proposition 7.2.26 (Hellinger, Morse–Sacksteder and norm-transforms). *Consider experiments with the same, not necessarily finite, parameter set Θ. Let Λ and $L_f(\Theta)$ denote, respectively, the set of probability distributions on Θ with finite support and the set of measures on Θ with finite support. Consider the following transforms:*

> *the Hellinger transform which to each experiment $\mathscr{E} = (P_\theta : \theta \in \Theta)$ associates the function $t \to H(t|\mathscr{E}) = \int \prod_\theta \mathrm{d}P_\theta^{t_\theta}$ on Λ.*
>
> *the Morse–Sacksteder transform which to each experiment $\mathscr{E} = (P_\theta : \theta \in \Theta)$ associates the function $\lambda \to \|\bigvee_\theta \lambda_\theta P_\theta\|$ on Λ.*
>
> *the norm-transform which to each experiment $\mathscr{E} = (P_\theta : \theta \in \Theta)$ associates the function $a \to \|\sum_\theta a_\theta P_\theta\|$ on $L_f(\Theta)$.*

Then each of these transforms determine the experiment up to equivalence.

Furthermore, all transforms are affine, and induce the topology of Δ-convergence for restrictions to finite sub-parameter sets. The Hellinger transform is multiplicative and monotonically decreasing while the other two transforms are monotonically increasing.

Say that an experiment $\mathscr{E}$ is *regular* if its Hellinger transform never vanishes. Thus $\mathscr{E}$ is regular if and only if $H(t|\mathscr{E}) > 0$ for all prior probability distributions t on Θ with finite support. By the following proposition, the regular experiments are precisely those experiments which satisfy the cancellation law for products.

Proposition 7.2.27 (Regular experiments and the cancellation law). *The following conditions are equivalent for an experiment $\mathscr{E} = (P_\theta : \theta \in \Theta)$:*

(i) *$\mathscr{E}$ is regular.*
(ii) *$\bigwedge\{P_\theta : \theta \in F\} \neq 0$ for all finite subsets F of Θ.*
(iii) *$\mathscr{F} \sim \mathscr{G}$ whenever $\mathscr{F} \times \mathscr{E} \sim \mathscr{G} \times \mathscr{E}$.*

Proof. If the prior distribution t assigns positive mass to each point θ in F and if $\sum_F t_\theta = 1$ then $H(t|\mathscr{E}) = 0$ if and only if $\bigwedge\{P_\theta : \theta \in F\} = 0$. Thus (i) and (ii) are equivalent. If these conditions are satisfied and if $\mathscr{F} \times \mathscr{E} \sim \mathscr{G} \times \mathscr{E}$ then

$$\begin{aligned} H(t|\mathscr{F}) &= [H(t|\mathscr{F}) \cdot H(t|\mathscr{E})]/[H(t|\mathscr{E})] \\ &= [H(t|\mathscr{F} \times \mathscr{E})]/[H(t|\mathscr{E})] \\ &= [H(t|\mathscr{G} \times \mathscr{E})]/[H(t|\mathscr{E})] \\ &= H(t|\mathscr{G}) \quad \text{for all } t. \text{ Thus } \mathscr{F} \sim \mathscr{G}. \end{aligned}$$

Finally assume that condition (ii) is not satisfied. Then $\bigwedge\{P_\theta : \theta \in F\} = 0$ for some finite subset F of Θ. By example 6.2.6 there are experiments $\mathscr{F} = (Q_\theta : \theta \in \Theta)$ and $\mathscr{G} = (R_\theta : \theta \in \Theta)$ such that $\mathscr{F}|\Theta_0 \sim \mathscr{G}|\Theta_0$ for all proper subsets Θ_0 of F while $\mathscr{F}|F \nsim \mathscr{G}|\mathscr{F}$. We may arrange the measures for $\theta \notin F$ so that

$$Q_{\theta_1} \wedge Q_{\theta_2} = Q_{\theta_1} \wedge \sum_F Q_\theta = 0$$

and

$$R_{\theta_1} \wedge R_{\theta_2} = R_{\theta_1} \wedge \sum_F R_\theta = 0$$

when $\theta_1, \theta_2 \in F^c$.

It follows that $H(t|\mathscr{F}) = H(t|\mathscr{G}) = 0$ if the minimal support of t contains two points outside F or if it contains one point outside F and one point inside F. If the minimal support of t is the set F then $H(t|\mathscr{E}) = 0$. Finally $H(t|\mathscr{F}) = H(t|\mathscr{G})$ whenever the minimal support of t is properly contained in F. Thus we find that $H(t|\mathscr{F}) \cdot H(t|\mathscr{E}) = H(t|\mathscr{G}) \cdot H(t|\mathscr{E})$ in any case. Hence $\mathscr{F} \times \mathscr{E} \sim \mathscr{G} \times \mathscr{E}$ although $\mathscr{F} \nsim \mathscr{G}$. □

We may also consider functionals which are additive for experiment multiplication. If Ω has this property then

$$\Omega(\mathscr{E}) + \Omega(\mathscr{M}_a) = \Omega(\mathscr{E} \times \mathscr{M}_a) = \Omega(\mathscr{M}_a)$$

and

$$\Omega(\mathscr{E}) + \Omega(\mathscr{M}_i) = \Omega(\mathscr{E} \times \mathscr{M}_i) = \Omega(\mathscr{E}).$$

It follows that the only real valued additive functional is the 0-functional. However, if we permit infinite values then there are several interesting functionals which are both additive and affine as well as monotone. The most prominent examples are perhaps entropy (Kullback–Leibler information) and Fisher information. Entropy is defined for dichotomies (P_1, P_2) and is the number $-E_{P_1} \log(\mathrm{d}P_2/\mathrm{d}P_1) = -\mathrm{d}/\mathrm{d}t[\int \mathrm{d}P_1^{1-t}\,\mathrm{d}P_2^t]_{t=0}$ where the derivative is the right side derivative at 0. Fisher information is defined for experiments whose parameter set Θ is a subset of $\mathbb{R}^k$. If $\Theta \subseteqq \mathbb{R}^k$ and if $\theta \in \Theta^0$ then, under regularity conditions, $8(1 - \int (\mathrm{d}P_\theta\,\mathrm{d}P_{\theta+h})^{1/2}) = h' I_\theta h + o(\|h\|^2)$ where I_θ is the Fisher information matrix at θ.

We shall not discuss these information concepts further here, but just note that their basic properties of additivity, affinity and monotonicity follow from, respectively, the multiplicativity, affinity and monotonicity of the Hellinger transform.

Another interesting and closely related area is the theory of majorization and Schur convexity. Schur convex functions (Schur increasing functions might have been a better term) are functions on $\mathbb{R}^n$ which are monotonically increasing for the ordering called majorization. This ordering is precisely the ordering of being 'at least as informative' for particular kinds of dichotomies, or rather pseudo-dichotomies, since we do not require that our measures are probability measures. Thus Schur convex functions may be considered as monotonically increasing functionals in essentially the same sense as considered here. We shall return to this topic in chapter 9.

The procedure we have used so far to obtain representations of experiments with finite parameter sets may easily be generalized to the case of countable parameter sets.

Standard experiments are essentially distributions of posterior distributions for the uniform prior. Of course we could consider other prior distributions and then the distributions of the posterior distribution. Let us briefly consider the case of a countable parameter set Θ. Choose positive weights $\pi_\theta : \theta \in \Theta$ such that $\sum_\theta \pi_\theta < \infty$. Let $\mathscr{E} = (\mathscr{X}, \mathscr{A}; P_\theta : \theta \in \Theta)$ be the given experiment. Put $P_\pi = \sum_\theta \pi_\theta P_\theta$, $f_\theta = \mathrm{d}P_\theta/\mathrm{d}P_\pi$; $\theta \in \Theta$ and $f = (f_\theta : \theta \in \Theta)$. Also put $K_\pi = \{x : x \in \mathbb{R}^\Theta; x \geqslant 0 \text{ and } \sum_\theta \pi_\theta x_\theta \leqslant 1\}$ and $H_\pi = \{x : x \in \mathbb{R}^\Theta \text{ and } \sum_\theta \pi_\theta x_\theta = 1\}$.

Equip $\mathbb{R}^\Theta$ with the (metrizable) topology of pointwise convergence. Then K_π is compact and H_π is a G_δ subset of K_π. It follows that the topological space H_π is Polish, i.e. its topology may be described by a complete separable metric. The reader is referred to e.g. theorem 8.7 in volume I of Choquet, 1969 for a proof of this fact.

The map $f = (f_\theta : \theta \in \Theta)$ from $(\mathscr{X}, \mathscr{A})$ to $\mathbb{R}^\Theta$ maps $\mathscr{E}$ into an experiment $\tilde{\mathscr{E}}^\pi$ with sample space H_π. The ψ-function criterion for equivalence, or sufficiency, shows that the experiment $\tilde{\mathscr{E}}^\pi$ is equivalent to $\mathscr{E}$. Generalizing the concept of a standard experiment we may call $\tilde{\mathscr{E}}^\pi$ *the π-standard experiment of $\mathscr{E}$*. A π-standard experiment $\tilde{\mathscr{E}}^\pi = (S_\theta : \theta \in \Theta)$ is determined by the measure $S_\pi = \sum_\theta \pi_\theta S_\theta$ since the projection of H_π into the θth coordinate space of $\mathbb{R}^\Theta$, for each Θ, is a version of $\mathrm{d}S_\theta/\mathrm{d}S_\pi$. Generalizing the concept of a standard measure we may call S_π *the π-standard measure of $\mathscr{E}$*.

Just as for standard experiments we find that:

(i) $\tilde{\mathscr{E}}^\pi = \mathscr{E}$ if $\mathscr{E}$ is a π-standard experiment

and that

(ii) $\mathscr{E} \sim \mathscr{F}$ if and only if $\tilde{\mathscr{E}}^\pi = \tilde{\mathscr{F}}^\pi$.

The Δ-topology for experiments induces a topology for the normed π-standard measures which is much stronger than the usual topology of weak convergence of probability measures on the space H_π. The first topology is not compact, not even complete (i.e. Cauchy sequences may diverge) when Θ is infinite. The latter topology is metrizable (for countable parameter sets) and describes another useful topology for experiments. This topology is the topology of Δ-convergence for restrictions to finite sub-parameter sets. We shall substantiate these claims in connection with our discussion of conical measures.

Using the π-standard measures we can prove LeCam's consistency theorem in the case of a countable parameter set.

Proposition 7.2.28 (The consistency requirements for the restrictions to finite sub-parameter sets. The case of a countable Θ.). *Assume that parameter set Θ is the set $\{1,2,\ldots\}$ of positive integers.* Put $\Theta_m = \{1,2,\ldots,m\}$; $m = 1,2,\ldots$

Let $\mathscr{E}^{(m)}$; $m = 1,2,\ldots$ be an experiment with parameter set Θ_m. Then there is an experiment $\mathscr{E}$ with parameter set Θ such that $\mathscr{E}|\Theta_m \sim \mathscr{E}^{(m)}$; $m = 1,2,\ldots$ if and only if $\mathscr{E}^{(m+1)}|\Theta_m \sim \mathscr{E}^{(m)}$; $m = 1,2,\ldots$

Proof. Write $\mathscr{E}^{(m)} = (P_1^{(m)},\ldots,P_m^{(m)})$ and $\mathscr{F}^{(m)} = (Q_1^{(m)},Q_2^{(m)},\ldots)$ where $Q_i^{(m)} = P_{i\wedge m}^{(m)}$; $i = 1,2,\ldots$ Let $S^{(m)}$ be the π-standard measure of $\mathscr{F}^{(m)}$. By compactness there is a sub-sequence $S^{(m')}$ such that $S^{(m')}$ converges weakly to a measure S on H_π. Then $1 = \int x_i S^{(m')}(\mathrm{d}x) \to \int x_i S(\mathrm{d}x)$; $i = 1, 2, \ldots$ so that $\sum_i \pi_i = \int \sum \pi_i x_i S(\mathrm{d}x)$. Hence, since $\sum \pi_i x_i \leqslant 1$ on K_π, $S(H_\pi) = 1$ so that S is the π-standard measure of an experiment $\mathscr{E}$. If ψ is sublinear on $\mathbb{R}^i$ then

$$\int \psi(\mathrm{d}\mathscr{E}|\Theta_i) = \int \psi(x_1,\ldots,x_i)S(\mathrm{d}x)$$

$$= \lim_{m'} \int \psi(x_1,\ldots,x_i)S^{(m')}(\mathrm{d}x)$$

$$= \lim_{m'} \int \psi(\mathrm{d}\mathscr{E}^{(m')}|\Theta_i).$$

Assume now that $\mathscr{E}^{m+1}|\Theta_m \sim \mathscr{E}_m$; $m = 1,2,\ldots$ Then $\mathscr{E}^n|\Theta_m \sim \mathscr{E}_m$ when $n \geqslant m$. Thus $\int \psi(\mathrm{d}\mathscr{E}^{(m')}|\Theta_i) = \int \psi(\mathrm{d}\mathscr{E}^{(i)})$ when $m' \geqslant i$. Hence $\int \psi(\mathrm{d}\mathscr{E}|\Theta_i) = \int \psi(\mathrm{d}\mathscr{E}^{(i)})$ for all sublinear functions ψ on $\mathbb{R}^i$. It follows that $\mathscr{E}|\Theta_i \sim \mathscr{E}^{(i)}$; $i = 1,2,\ldots$ Conversely, if such an experiment $\mathscr{E}$ exists then $\mathscr{E}^{(m+1)}|\Theta_m \sim (\mathscr{E}|\Theta_{m+1})|\Theta_m \sim \mathscr{E}|\Theta_m \sim \mathscr{E}^{(m)}$; $m = 1,2,\ldots$ □

7.3 Δ-sufficient algebras of events. Sufficiency and completeness

In section 6.2 we saw that the experiment defined by a sub σ-algebra is equivalent to the original one if and only if this sub σ-algebra is pairwise sufficient. Thus it is pairwise sufficiency (i.e. Δ-sufficiency) and not sufficiency which is the fundamental concept from the point of view of deficiencies. This is also reflected by the fact that a sub σ-algebra which contains a pairwise sufficient σ-algebra is itself pairwise sufficient, while the corresponding statement for sufficient σ-algebras is not true, as shown by Burkholder. However minimal pairwise sufficient σ-algebras may not exist, just as minimal sufficient σ-algebras may not exist. LeCam shows in his 1964 paper that if we enlarge our scope and consider all weakly closed subalgebras of the M-space $M(\mathscr{E})$ which contains the unit then there is a particular subalgebra which is the smallest among those which are Δ-sufficient. This algebra may be identified with the M-space $V(\mathscr{E})^*$ and it is actually the one which is generated by the idempotents which support subsets of $V(\mathscr{E})$.

If we restrict our attention to majorized experiments $\mathscr{E}$ then LeCam's minimal Δ-sufficient algebra is representable as the M-space of the restriction of $\mathscr{E}$ to a (necessarily unique) $L(\mathscr{E})$-closed sub σ-algebra. It follows that this σ-algebra is then the smallest $L(\mathscr{E})$-closed and pairwise sufficient σ-algebra. Furthermore, if $\mathscr{E}$ is coherent then this σ-algebra is minimal sufficient. In this case any weakly closed subalgebra of $M(\mathscr{E})$ which contains the unit is the M-space of the restriction to a unique weakly closed sub σ-algebra. The study of Δ-sufficient and weakly closed algebras of $M(\mathscr{E})$ which contain the unit thus reduces to the study of $L(\mathscr{E})$-closed and pairwise sufficient sub σ-algebras. Now as we saw in chapter 5 any abstract L-space is isomorphic to the L-space of a coherent (totally informative and $\sum$-dominated) experiment. It follows that the general case may be reduced to the case of coherent experiments and, in particular, that the existence of a smallest Δ-sufficient algebra in $M(\mathscr{E})$ follows from the fact that coherent experiments always possess minimal sufficient σ-algebras as we shall soon see. This last fact was proved by Hasegawa & Perlman (1974) but it may also be deduced very directly from LeCam's 1964 paper.

A most thorough treatment of pairwise sufficiency may be found in Siebert (1979). The exposition below owes much to this paper. The papers from some decades ago, by Halmos & Savage (1949), Bahadur (1954, 1955) and Burkholder (1961), to mention a few, are still basic references and the reader ought to be aquainted with at least some of them. However, in order to make the treatment here fairly self-contained, we have included a proof in chapter 1 of the factorization criterion for sufficiency in dominated experiments. A useful exposition on sufficiency may also be found in Heyer (1982).

We shall need the following concepts and notation for an experiment $\mathscr{E} = (\mathscr{X}, \mathscr{A}; P_\theta : \theta \in \Theta)$.

First of all a sub σ-algebra $\mathscr{B}$ of $\mathscr{A}$ is called *Δ-sufficient* (for $\mathscr{E}$) if $\mathscr{E} \sim \mathscr{E}|\mathscr{B}$, i.e. $\Delta(\mathscr{E}, (\mathscr{E}|\mathscr{B})) = 0$. We saw in proposition 6.2.28 that $\mathscr{B}$ is Δ-sufficient if and only if $\mathscr{B}$ is sufficient with respect to each pair $(P_{\theta_1}, P_{\theta_2})$ where $\theta_1, \theta_2 \in \Theta$. Thus our concept of Δ-sufficiency is precisely the notion of pairwise sufficiency as used by Halmos & Savage, Bahadur, Siebert and others.

Let μ be a non-negative measure on $\mathscr{A}$. Then $\mathscr{N}_\mu$ denotes the class of μ-null sets A in $\mathscr{A}$, i.e.

$$\mathscr{N}_\mu = \{A : A \in \mathscr{A} \text{ and } \mu(A) = 0\}.$$

If $\mathscr{C}$ is any class of subsets of $\mathscr{X}$ then $\sigma(\mathscr{C})$ denotes the σ-algebra generated by $\mathscr{C}$. In particular $\sigma(\mathscr{N}_\mu) = \{A : A \in \mathscr{N}_\mu \text{ or } A^c \in \mathscr{N}_\mu\}$ when $0 \leqslant \mu$ is a measure on $\mathscr{A}$. If $\mathscr{B}_i$; $i \in I$ is a family of sub σ-algebras of $\mathscr{A}$ then $\sigma(\bigcup_i \mathscr{B}_i)$ is the smallest

σ-algebra containing $\mathscr{B}_i$; $i \in I$. This σ-algebra will also be denoted as $\bigvee_i \mathscr{B}_i$. More generally, if $\mathscr{H}$ is a class of functions from a given set $\mathscr{X}$ to various measurable spaces, then $\sigma(\mathscr{H})$ denotes the smallest σ-algebra of subsets of $\mathscr{X}$ which make all the functions in $\mathscr{H}$ measurable. If $\mathscr{B}$ is a sub σ-algebra and if $0 \leqslant \mu$ is a measure on $\mathscr{A}$ then the *μ-closure* of $\mathscr{B}$ is the σ-algebra $\mathscr{B} \vee \sigma(\mathscr{N}_\mu)$. Thus $\mathscr{B} \vee \sigma(\mathscr{N}_\mu) = \{A : A \in \mathscr{A} \text{ and } \mu(A \,\Delta\, B) = 0 \text{ for some } B \in \mathscr{B}\} = \sigma\{B \cup N : B \in \mathscr{B},\ N \in \mathscr{A},\ \mu(N) = 0\}$. We shall say that $\mathscr{B}$ is *μ-closed* if $\mathscr{B} \vee \sigma(\mathscr{N}_\mu) = \mathscr{B}$, i.e. if $\mathscr{B} \supseteqq \mathscr{N}_\mu$. The class of $\mathscr{E}$-null sets will be denoted by $\mathscr{N}$. Thus $\mathscr{N} = \{A : A \in \mathscr{A} \text{ and } P_\theta(A) \equiv_\theta 0\}$. The *closure (or $\mathscr{E}$-closure) of a sub σ-algebra $\mathscr{B}$* of $\mathscr{A}$ is the σ-algebra $\mathscr{B} \vee \sigma(\mathscr{N}) = \{A : A \in \mathscr{A} \text{ and } P_\theta(A \,\Delta\, B) \equiv_\theta 0$ for some $B \in \mathscr{B}\}$. This σ-algebra will also be denoted as $\bar{\mathscr{B}}$. If $\mathscr{B} \vee \sigma(\mathscr{N}) = \mathscr{B}$, i.e. if $\mathscr{B} \supseteqq \sigma(\mathscr{N})$, then we shall say that $\mathscr{B}$ is *closed* ($\mathscr{E}$-closed). Thus $\mathscr{B}$ is closed if and only if $\mathscr{B} \supseteqq \mathscr{N}$. The $L(\mathscr{E})$-topology on $\mathscr{A}$ is the smallest topology making $\mu(A)$ continuous in A for each $\mu \in L(\mathscr{E})$. This topology, which need not be Hausdorff, is thus obtained from the restriction of the $w(M, L)$-(or w-) topology to the collection of $\mathscr{E}$-equivalence classes of measurable sets. By a slight abuse of terminology, this topology may be described as the $w(M, L)$-topology (or the w-topology) on $\mathscr{A}$. If $\mathscr{B}$ is a sub σ-algebra of $\mathscr{A}$ then its closure for this topology is denoted as $\bar{\mathscr{B}}^w$ and this σ-algebra is called *the weak closure of $\mathscr{B}$*. This σ-algebra is the intersection of the μ-closures $\mathscr{B} \vee \mathscr{N}_\mu$ for all measures μ in L_+. It suffices then to consider measures μ of the form $\mu = P_{\theta_1} + P_{\theta_2}$ where $\theta_1, \theta_2 \in \Theta$. This may be seen by observing that $\bar{\mathscr{B}}^w$ is the class of all sets A in $\mathscr{A}$ such that $P_\theta(A \,\Delta\, B_\theta) \equiv_\theta 0$ for some consistent family $(B_\theta : \theta \in \Theta)$ in $\mathscr{E}|\mathscr{B}$. It follows that $\bar{\mathscr{B}} = \bar{\mathscr{B}}^w$ whenever $\mathscr{E}|\mathscr{B}$ is coherent. We have always

$$\mathscr{B} \subseteqq \bar{\mathscr{B}} \subseteqq \bar{\mathscr{B}}^w = \bigwedge_{\theta_1, \theta_2} [\mathscr{B} \vee \sigma(\mathscr{N}_{\theta_1} \wedge \mathscr{N}_{\theta_2})] \subseteqq \bigwedge_\theta (\mathscr{B} \vee \sigma(\mathscr{N}_\theta)).$$

The sub σ-algebra $\mathscr{B}$ of $\mathscr{A}$ determines the restricted experiment $\mathscr{E}|\mathscr{B} = (\mathscr{X}, \mathscr{B}; P_\theta|\mathscr{B} : \theta \in \Theta)$. The M-space $M(\mathscr{E}|\mathscr{B})$ of $\mathscr{E}|\mathscr{B}$ may be identified with the weakly closed subalgebra $\tilde{M}(\mathscr{E}|\mathscr{B})$ of $M(\mathscr{E})$ consisting of those functionals v which may be represented as consistent families of $\mathscr{B}$-measurable functions. The map which to each $v \in M(\mathscr{E}|\mathscr{B})$ associates the linear functional $\mu \to v(\mu|\mathscr{B})$ is an isomorphism from the M-space $M(\mathscr{E}|\mathscr{B})$ onto the M-space $\tilde{M}(\mathscr{E}|\mathscr{B})$.

It is easily checked that $\tilde{M}(\mathscr{E}|\mathscr{B}) \subseteqq \tilde{M}(\mathscr{E}|\mathscr{C})$ if and only if $\bar{\mathscr{B}}^w \subseteqq \bar{\mathscr{C}}^w$. In particular $\tilde{M}(\mathscr{E}|\mathscr{B}) = \tilde{M}(\mathscr{E}|\bar{\mathscr{B}}^w)$.

To decide whether or not a particular σ-algebra is Δ-sufficient we may compare it with particular σ-algebras described in terms of likelihood ratios. Following Siebert (1979) we may proceed as follows.

Let Λ denote the set of prior distributions on Θ with countable supports. If $\lambda \in \Lambda$ and $\theta \in \Theta$ then $f_{\theta,\lambda}$ denotes a version of $\mathrm{d}P_\theta/\mathrm{d}P_\lambda$ where $P_\lambda = \sum_\theta \lambda(\theta) P_\theta$. We use the standard notation '$\gg$' to denote the partial order of domination of σ-finite measures on the same σ-algebra of sets.

If $\lambda, \lambda_0 \in \Lambda$ and $\lambda \gg \lambda_0$ then $\mathscr{C}_\lambda^{\lambda_0}$ denotes the P_{λ_0} closure of the σ-algebra induced by $(f_{\theta,\lambda} : \lambda_\theta > 0)$. Then $\mathscr{C}_\lambda^{\lambda_0}$ does not depend on how the Radon–Nikodym derivatives $f_{\theta,\lambda}$ are specified. If λ_0 is the one-point distribution in θ then we may write $\mathscr{C}_\lambda^\theta$ instead of $\mathscr{C}_\lambda^{\lambda_0}$. Since $\sigma(f_{\theta,\lambda} : \lambda_\theta > 0)$ is sufficient for $(P_\theta : \lambda_\theta > 0)$, it follows that $\mathscr{C}_\lambda^{\lambda_0} = \bigcap \{\mathscr{C}_\lambda^\theta : \lambda_0(\theta) > 0\}$ when $\lambda_0 \gg \lambda$. Furthermore, $\mathscr{C}_{\lambda_1}^{\lambda_0} \subseteqq \mathscr{C}_{\lambda_2}^{\lambda_0}$ when $\lambda_1 \ll \lambda_2$ while $\mathscr{C}_\lambda^{\lambda_0} \supseteqq \mathscr{C}_\lambda^{\lambda_1}$ when $\lambda_1 \gg \lambda_0 \gg \lambda$. Any σ-algebra $\mathscr{C}_\lambda^{\lambda_0}$ is weakly closed and the family $\{\mathscr{C}_\lambda^{\lambda_0} : \lambda \gg \lambda_0\}$ is 'σ-directed' in the sense that if $\lambda_1, \lambda_2, \ldots \in \Lambda$ then $\bigcup_i \mathscr{C}_{\lambda_i}^{\lambda_0} \subseteqq \mathscr{C}_\lambda^{\lambda_0}$ for some λ (say $\lambda = \sum 2^{-n}\lambda_n$).

Put $\mathscr{D}^{\lambda_0} = \bigcup \{\mathscr{C}_\lambda^{\lambda_0} : \lambda \gg \lambda_0\}$. We may write $\mathscr{D}^\theta$ instead of $\mathscr{D}^{\lambda_0}$ when $\lambda_0(\theta) = 1$. It may then be checked that for each $\lambda \in \Lambda$, $\mathscr{D}^\lambda$ is a weakly closed σ-algebra. The factorization criterion shows that each σ-algebra $\mathscr{D}^\lambda$ is Δ-sufficient. Furthermore, $\mathscr{D}^\lambda = \bigcap \{\mathscr{D}^\theta : \lambda(\theta) > 0\}$. If $\lambda_0 \ll \lambda_1$ then $\mathscr{D}^{\lambda_0} = \bigcup \{\mathscr{C}_\lambda^{\lambda_0} : \lambda \gg \lambda_0\} = \bigcup \{\mathscr{C}_\lambda^{\lambda_0} : \lambda \gg \lambda_1\} \supseteqq \bigcup \{\mathscr{C}_\lambda^{\lambda_1} : \lambda \gg \lambda_1\} = \mathscr{D}^{\lambda_1}$.

Let us use the notation $\mathscr{D}$ for the intersection of all Δ-sufficient and weakly closed sub σ-algebras of $\mathscr{A}$. Then, following Siebert we obtain the following proposition.

Proposition 7.3.1 (Closure criteria for Δ-sufficiency). *The sub σ-algebra $\mathscr{B}$ (of $\mathscr{A}$) is Δ-sufficient if and only if $\mathscr{B} \vee \sigma(\mathscr{N}_{P_\lambda}) \supseteqq \mathscr{D}^\lambda$ for all $\lambda \in \Lambda$, and this in turn holds if and only if $\mathscr{B} \vee \sigma(\mathscr{N}_{P_\lambda}) \supseteqq \mathscr{D}^\lambda$ for all prior distributions λ on Θ whose minimal supports contain at most two points.*

The intersection $\mathscr{D}$ of all Δ-sufficient and weakly closed sub σ-algebras is

$$\mathscr{D} = \bigcap \{\mathscr{D}^\lambda : \lambda \in \Lambda\} = \bigcap \{\mathscr{D}^\theta : \theta \in \Theta\}.$$

Proof. If $\mathscr{B}$ is Δ-sufficient then all densities $f_{\theta,\lambda}$ may be specified $\mathscr{B}$-measurable. Hence $\mathscr{C}_\lambda^{\lambda_0} \subseteqq \mathscr{B} \vee \sigma(\mathscr{N}_{P_{\lambda_0}})$ whenever $\lambda_0 \ll \lambda$. Thus $\mathscr{D}^\lambda \subseteqq \mathscr{B} \vee \sigma(\mathscr{N}_{P_\lambda})$ for all $\lambda \in \Lambda$. Conversely, the validity of this conclusion for the prior distribution which assigns masses $\frac{1}{2}$ to θ_1 and θ_2 implies that $\mathscr{B}$ is sufficient for $(P_{\theta_1}, P_{\theta_2})$. The last statement follows readily from this. □

The intersection $\mathscr{D}$ of the σ-algebras $\mathscr{D}^\lambda$ is the only candidate for a minimal, or a smallest, Δ-sufficient and weakly closed σ-algebra.

Corollary 7.3.2 (Minimal Δ-sufficiency = minimum Δ-sufficiency). *The following conditions are equivalent for a Δ-sufficient and weakly closed sub σ-algebra $\mathscr{B}$.*

(i) *$\mathscr{B}$ is the smallest sub σ-algebra which is Δ-sufficient and weakly closed.*
(ii) *$\mathscr{B}$ does not contain any other sub σ-algebra which is Δ-sufficient and weakly closed.*
(iii) *$\mathscr{B} = \mathscr{D}$.*

Remark. In general we have to distinguish between being smallest and being minimal. Thus an element e of a partially ordered set $(E, \leqslant)$ is a *smallest* element (a minimal element) if and only if $e \leqslant f$ for any $f \in E$ (such that $f \leqslant e$).

However, by the corollary these notions coincide for the inclusion ordering when E is the set of weakly closed and Δ-sufficient σ-algebras of events. By Burkholder (1961) this is also the case for the set of closed sufficient σ-algebras of events.

Proof. It suffices to show that (ii) $\Rightarrow$ (iii). Assume (ii) holds. If $\lambda \gg \lambda_0$ then $f_{\theta,\lambda}$ may be chosen $\mathscr{B}$-measurable and it is automatically $\mathscr{D}^{\lambda_0}$ measurable. Thus $\mathscr{B} \cap \mathscr{D}^{\lambda_0}$ is pairwise sufficient and weakly closed so that, since $\mathscr{B} \cap \mathscr{D}^{\lambda_0} \subseteqq \mathscr{B}$, $\mathscr{B} = \mathscr{B} \cap \mathscr{D}^{\lambda_0} \subseteqq \mathscr{D}^{\lambda_0}$. Hence $\mathscr{B} \subseteqq \bigcap_\lambda \mathscr{D}^\lambda$ while, by pairwise sufficiency, $\mathscr{B} \supseteqq \bigcap_\lambda \mathscr{D}^\lambda$. □

Before proceeding, still following Siebert, we will now show that a smallest Δ-sufficient and weakly closed σ-algebra exists whenever $\mathscr{E}$ is majorized. We shall also see that this σ-algebra is the smallest closed sufficient σ-algebra when $\mathscr{E}$ is coherent.

Notice by the way, that the properties of being majorized and of being dominated become quite similar if we permit the following generalization of the concept of σ-finiteness of the dominating measure. Let μ be any non-negative measure on the measurable space $(\mathscr{X}, \mathscr{A})$ and let ν be a non-negative finite measure. Say that μ is σ-finite with respect to ν if there is a set N in $\mathscr{A}$ such that $\nu(N) = 0$ while the restriction of μ to N^c is σ-finite. Then, by the Radon–Nikodym theorem, there is a real valued measurable function f such that $\nu(A) \equiv_A \int_A f \, d\mu$ if and only if $\mathscr{N}_\mu \subseteqq \mathscr{N}_\nu$ and μ is σ-finite with respect to ν. In particular an experiment $\mathscr{E} = (\mathscr{X}, \mathscr{A}; P_\theta : \theta \in \Theta)$ is majorized by the non-negative measure μ if and only if $\mathscr{N}_\mu \subseteqq \bigcap_\theta \mathscr{N}_{P_\theta}$ and μ is σ-finite with respect to each of the measures P_θ.

Assume now that $\mathscr{E}$ is majorized by the non-negative measure μ and that $P_\theta(A) \equiv_A \int_A f_\theta \, d\mu$. Consider the following classes of functions:

(i) the functions $f_{\theta_1}/f_{\theta_2}$; $\theta_1, \theta_2 \in \Theta$;
(ii) the functions $f_{\theta_1}/(f_{\theta_1} + f_{\theta_2})$; $\theta_1, \theta_2 \in \Theta$;
(iii) the functions $f_\theta / \sum \{f_\theta : \theta \in F\}$ where $\theta \in F$ and F is finite;
(iv) the functions $f_\theta / \sum_\theta \lambda_\theta f_\theta$ where $\lambda \in \Lambda$ and $\lambda_\theta > 0$.

Extend the domain of definition of these functions by letting them assign the value 0 to any point $x \in \mathscr{X}$ where the denominator of the defining formula vanishes. Then these four classes of functions induce the same σ-algebra $\mathscr{S}$. Let $\mathscr{B}_0$ be the weak closure of $\mathscr{S}$. Then $\mathscr{B}_0$ is Δ-sufficient by its definition. Let $\lambda_0 \in \Lambda$. Then $\mathscr{S} \subseteqq \mathscr{C}_\lambda^{\lambda_0}$ whenever $\lambda \gg \lambda_0$. Hence, since $\mathscr{C}_\lambda^{\lambda_0}$ is weakly closed, $\mathscr{B}_0 \subseteqq \mathscr{C}_\lambda^{\lambda_0}$ so that $\mathscr{B}_0 \subseteqq \mathscr{D}^{\lambda_0}$. It follows that $\mathscr{B}_0 \subseteqq \bigcap \{\mathscr{B} : \mathscr{B}$ is pairwise sufficient and weakly closed$\}$ so that $\mathscr{B}_0$ is the smallest weakly closed and pairwise sufficient σ-algebra in this case.

Now assume in addition that $\mathscr{E}$ is coherent and let $\mathscr{B}$ be any weakly closed and pairwise sufficient σ-algebra. Then $\mathscr{B}$ is clearly also closed. Let $A \in \mathscr{A}$. Then $P_\theta(A|\mathscr{B})$; $\theta \in \Theta$ may be specified consistently. It follows that there is a measurable function g such that $P_\lambda(A|\mathscr{B}) = g$ a.e. P_λ for all $\lambda \in \Lambda$. In particular $P_\theta(A|\mathscr{B}) = g$ a.e. P_θ; $\theta \in \Theta$. The function g is clearly $(\mathscr{B} \vee \mathscr{N}_{P_\lambda})$-measurable for each $\lambda \in \Lambda$. Thus g is $\bar{\mathscr{B}}^w = \mathscr{B}$-measurable. It follows that $\mathscr{B}$ is sufficient and closed. Conversely assume that $\mathscr{B}$ is sufficient and closed. Let $B_\theta : \theta \in \Theta$ be a consistent family of sets in $\mathscr{E}|\mathscr{B}$ and let A be a set in $\mathscr{A}$ such that $P_\theta(A \Delta B_\theta) \equiv_\theta 0$. Let g be a common version of $P_\theta(A|\mathscr{B})$; $\theta \in \Theta$ and put $B = [g = 1]$. Then $P_\theta(B \Delta B_\theta) = 0$ so that $P_\theta(A \Delta B) \equiv 0$. Hence $A \in \bar{\bar{\mathscr{B}}} = \mathscr{B}$. Thus weak and strong closures of sufficient σ-algebras coincide when $\mathscr{E}$ is coherent. Altogether this proves the following theorem.

Theorem 7.3.3 (Smallest Δ-sufficient σ-algebra in majorized experiments). *If $\mathscr{E}$ is majorized then the weak closure of the σ-algebra $\mathscr{S}$ described above is the smallest Δ-sufficient and weakly closed sub σ-algebra and, consequently, coincides with $\mathscr{D}$.*

If, in addition, $\mathscr{E}$ is coherent, then this σ-algebra is minimal sufficient.

Corollary 7.3.4 (Sufficiency in the discrete case). *Let $\mathscr{E} = (\mathscr{X}, \mathscr{A}; P_\theta : \theta \in \Theta)$ be a discrete experiment. Thus $\mathscr{A}$ is the class of all subsets of $\mathscr{X}$ and P_θ, for each θ, has countable support in $\mathscr{X}$.*

Say that points x and y in $\mathscr{X}$ are equivalent (notation $x \sim y$) if $P_\theta(x) \equiv_\theta kP_\theta(y)$ for some positive constant k not depending on θ.

Then the σ-algebra $\mathscr{B}$ of events B which are saturated for this equivalence relation is minimal sufficient in $\mathscr{E}$.

Remark. It follows from the proof that if $\mathscr{C}$ is the σ-algebra of saturated events C for an equivalence relation at least as fine than the one given here then $\mathscr{C}$ is sufficient.

Proof. For each $x \in \mathscr{X}$, let $\bar{x}$ denote the equivalence class containing x. Note that the maximal $\mathscr{E}$ null set $N = \{x : P_\theta(x) \equiv_\theta 0\}$ is an equivalence class. If $N \neq \varnothing$ then let τ be a probability distribution on N having countable support. Consider an event A in $\mathscr{A}$ and a point $x \in \mathscr{X}$. Put $\Gamma(A|x) = P_\theta(A|\bar{x})$ if $P_\theta(x) > 0$. Put $\Gamma(A|x) = \tau(A)$ if $x \in N$. Then $\Gamma(A|x)$ is well defined and does not depend on θ. In fact if $P_{\theta_1}(x) > 0$ and $P_{\theta_2}(x) > 0$ and if $z \in \bar{x}$ then $P_{\theta_1}(z) = aP_{\theta_2}(z)$ where a does not depend on z. Thus $P_{\theta_1}(A|\bar{x}) = P_{\theta_1}(A \cap \bar{x})/P_{\theta_1}(\bar{x}) = aP_{\theta_2}(A \cap \bar{x})/aP_{\theta_2}(\bar{x}) = P_{\theta_2}(A|\bar{x})$. Clearly $\Gamma(A|\cdot)$ is $\mathscr{B}$-measurable and for each $x \in \mathscr{X}$, $\Gamma(\cdot|x)$ is a probability measure on $\bar{x}$ having countable support. Since $P_\theta(A \cap \bar{x}) \equiv_\theta \int_{\bar{x}} \Gamma(A|x) P_\theta(dx)$ for all points $x \in \mathscr{X}$, it follows that $\Gamma(A|\cdot)$ is a common version of $P_\theta(A|\mathscr{B})$; $\theta \in \Theta$. Thus $\mathscr{B}$ is sufficient.

For each θ, identify P_θ with the point function (density) $f_\theta : x \to P_\theta(x)$. Let $\mathscr{S}$ be the σ-algebra induced by the likelihood ratios for the densities $f_\theta : \theta \in \Theta$. Then, since $\mathscr{E}$ is coherent, $\bar{\mathscr{S}}^w$ is minimal sufficient. Thus in order to establish minimal sufficiency, it suffices to show that $\mathscr{B} \subseteqq \bar{\mathscr{S}}^w$. For each non-empty finite subset F of Θ, put $P_F = \sum\{P_\theta : \theta \in F\}$.

Let B be any saturated event. Choose a finite subset F of Θ and let

$V = \{v_1, v_2, \ldots\}$ denote the countable set of equivalence classes v such that $P_F(v) > 0$. Then $P_F(B) = \sum\{P_\theta(v_i) : v_i \subseteqq B\} = P_F(B_0)$ where $B_0 = \bigcup\{v_i : v_i \subseteqq B\}$. Since the equivalence classes $v_1, v_2, \ldots$ are distinct, it follows that $v_1, v_2, \ldots$ are also equivalence classes for $\mathscr{E}|\Theta_0$ for a sufficiently large countable set Θ_0. Put $\phi(x) = (P_{\theta_2}(x)/(P_{\theta_1}(x) + P_{\theta_2}(x)) : \theta_1,\ \theta_2 \in \Theta_0)$ where fractions with denominator 0 should be interpreted as 0.

If $x \sim y$ then $\phi(x) = \phi(y)$ so that we may define $\Phi(v)$ for an equivalence class v by $\Phi(v) = \phi(x)$ where $x \in v$. It follows that the families $\Phi(v_1)$, $\Phi(v_2), \ldots$ are all distinct. Thus $B_0 = \{x : P_F(x) > 0$ and $\phi(x) = \Phi(v_i)$ where $v_i \subseteqq B\}$ so that $B_0 \in \mathscr{S}$. Hence, since $P_F(B \,\Delta\, B_0) = 0$, $B \in \mathscr{S} \vee \mathscr{N}_F$ where $\mathscr{N}_F$ is the σ-algebra consisting of the P_F null sets and their complements. It follows that $B \in \bigcap_F \mathscr{S} \vee \mathscr{N}_F = \bar{\mathscr{S}}^w$. □

Example 7.3.5 (Sufficiency reductions of sampling models). Let $\prod$ be an enumerable population of individuals. The unknown parameter θ is a function on $\prod$ which to each individual i associates a characteristic $\theta(i)$ of interest. The parameter set Θ is thus a set of functions on $\prod$.

Let α be a sequence sampling plan, i.e. a probability distribution on the enumerable set of finite sequences in $\prod$.

As described in example 1.2.1, α determines a sequence sampling plan experiment $\mathscr{E} = (\mathscr{X}, \mathscr{A}; P_{\theta,\alpha}; \theta \in \Theta)$ where:

$\mathscr{X}$ = the set of sequences $(i_1, f_1), \ldots, (i_n, f_n)$ where $i_1, \ldots, i_n \in \prod$ and $f_\nu = \theta(i_\nu)$; $\nu = 1, \ldots, n$ for some $\theta \in \Theta$.
$\mathscr{A}$ = the class of all subsets of $\mathscr{X}$.
$P_{\theta,\alpha}((i_1, f_1), \ldots, (i_n, f_n)) = \alpha(i_1, \ldots, i_n)$ or $= 0$ as $f_\nu = \theta(i_\nu)$; $\nu = 1, \ldots, n$ or not.

It follows that the maximal $\mathscr{E}$ null set N is the set of sequences $((i_1, f_1), \ldots, (i_n, f_n))$ in $\mathscr{X}$ such that $\alpha(i_1, \ldots, i_n) = 0$.

For each $x = ((i_1, f_1), \ldots, (i_n, f_n)) \in \mathscr{X}$, let $U(x)$ denote the set $\{i_1, \ldots, i_n\}$ and $G(x)$ denote the function on $U(x)$ which maps $i_1, \ldots, i_n$ into $f_1, \ldots, f_n$ respectively. Then $(U(x), G(x))$ belongs to the set $\mathscr{Y}$ of pairs (u, g) where u is a finite subset of $\prod$ and $g = \theta|u$ for some θ. Equip $\mathscr{Y}$ with the σ-algebra $\mathscr{B}$ of all its subsets.

Then as explained in example 1.2.1, the statistic (U, G) maps $\mathscr{E}_\alpha$ into the set sampling plan experiment $\bar{\mathscr{E}}_{\bar{\alpha}} = (\bar{P}_{\theta,\alpha} : \theta \in \Theta)$ where the set sampling plan $\bar{\alpha}$ is defined by $\bar{\alpha}(u) = \sum\{\alpha(i_1, \ldots, i_n); \{i_1, \ldots, i_n\} = u\}$. The probability distribution $\bar{P}_{\theta,\bar{\alpha}}$ is given by $\bar{P}_{\theta,\bar{\alpha}}(u, g) = \bar{\alpha}(u)$ or $= 0$ as $g = \theta|u$ or not.

Assume that $(U(x), G(x)) = (U(x'), G(x'))$ where $x = ((i_1, f_1), \ldots, (i_m, f_m))$ and $x' = ((j_1, h_1), \ldots, (j_n, h_n))$. Then $\{i_1, \ldots, i_m\} = \{j_1, \ldots, j_n\}$ and $h_\nu = f_\mu$ when $j_\nu = i_\mu$.

If $\alpha(i_1, \ldots, i_m) > 0$ and $\alpha(j_1, \ldots, j_n) > 0$ then this implies that $P_{\theta,\alpha}(x) \equiv_\theta kP_{\theta,\alpha}(x')$ where $k = \alpha(i_1, \ldots, i_m)/\alpha(j_1, \ldots, j_n)$. Thus the statistic (U, G) is sufficient so that $\mathscr{E}_\alpha \sim \bar{\mathscr{E}}_{\bar{\alpha}}$.

This could have been seen directly by evaluating conditional probabilities.

The σ-algebra induced by the statistic (U, G) need not be minimal sufficient. Suppose however that Θ is sufficiently large to ensure that to any triple (i, u, θ) where $i \in \Pi$, $\theta \in \Theta$ and u is a finite subset of $\prod$ not containing i, there corresponds a $\tilde{\theta} \in \Theta$ such that $\tilde{\theta}|u = \theta|u$ while $\tilde{\theta}(i) \neq \theta(i)$.

Now assume that $x = ((i_1, f_1), \ldots, (i_m, f_m))$ and $x' = ((j_1, h_1), \ldots, (j_n, h_n))$ are sample points such that $P_{\theta,\alpha}(x) \equiv_\theta kP_{\theta,\alpha}(x')$ for some positive constant k not depending on θ.

It may then be checked that if $x \notin N$ then also $x' \notin N$ and then $(U(x), G(x)) = (U(x'), G(x'))$. Thus the σ-algebra induced by U, G is minimal sufficient in this case.

Example 7.3.6 (Totally informative experiments). Let $\mathscr{E} = (\mathscr{X}, \mathscr{A}; P_\theta : \theta \in \Theta)$ be totally informative. Then $\mathscr{E}$ is majorized if and only if to each θ there corresponds a set $A_\theta \in \mathscr{A}$ such that $P_\theta(A_\theta) = 1$ while $P_{\theta'}(A_\theta) = 0$ when $\theta' \neq \theta$. If so then $\mathscr{E}$ is majorized by $\sum_\theta P_\theta$. In this case there is consequently a smallest Δ-sufficient and weakly closed sub σ-algebra. However let us quite generally, i.e. without assuming that $\mathscr{E}$ is majorized, describe the σ-algebras $\mathscr{C}_\lambda^\theta$ in this case.

Let $\{A_\theta : \lambda_\theta > 0\}$ be a (countable) partitioning of $\mathscr{X}$ such that $P_\theta(A_\theta) = 1$ when $\lambda_\theta > 0$. Then $\mathscr{C}_\lambda^\theta$, where $\lambda(\theta) > 0$, is the P_θ-closure of the σ-algebra generated by this partition. Thus $\mathscr{C}_\lambda^\theta = \sigma(\mathscr{N}_{P_\theta}) = \mathscr{D}^\theta$ when $\lambda(\theta) > 0$. It follows that the intersection $\mathscr{D}$ of all Δ-sufficient and weakly closed σ-algebras is the class of events which, for all θ, are independent of themselves. Hence a smallest weakly closed and Δ-sufficient σ-algebra exists if and only if θ is identifiable in $(\mathscr{E}|\mathscr{D})$, i.e. if and only if the restriction map $\theta \to P_\theta|\mathscr{D}$ is 1–1. This condition is also equivalent to the condition that the set of $\{0, 1\}$-valued functions $\theta \to P_\theta(A)$, where $A \in \mathscr{A}$, distinguishes the points of Θ.

If this condition is satisfied then $\mathscr{D}$ is the smallest Δ-sufficient weakly closed σ-algebra.

Here is a situation where this condition is not satisfied. (It is an adaptation of an example from Siebert (1979).)

Define $\mathscr{E} = (\mathscr{X}, \mathscr{A}; P_\theta : \theta \in \Theta)$ as follows:

$\mathscr{X} =]-\infty, \infty[$,
$\mathscr{A} =$ the Borel class on $\mathscr{X}$,
$\Theta = [0, \infty[\bigcup \{\alpha, \beta\}$ where $\alpha \neq \beta$ and $\alpha, \beta \notin [0, \infty[$,
$P_\theta(-\theta) = 1 - P_\theta(\theta) \in]0, 1[$ when $\theta \in [0, \infty[$,
P_α is any probability distribution on $[0, \infty[$ which assigns mass 0 to each point $x \in [0, \infty[$ and

$$P_\beta(A) = P_\alpha(-A)$$

when $A \in \mathscr{A}$.

Here $\mathscr{D}$ is the class of all symmetric Borel sets D such that $P_\alpha(D) \in \{0, 1\}$. Thus $P_\alpha|\mathscr{D} = P_\beta|\mathscr{D}$ in this case.

The general problem on sufficiency and Δ-sufficiency may be reduced to the case of totally informative experiments by the following device.

Proposition 7.3.7 (Reduction to totally informative experiments). *Let $\mathscr{E} = (\mathscr{X}, \mathscr{A}; P_\theta : \theta \in \Theta)$ be the given experiment and let $\pi_t : t \in T$ be a maximal family of disjoint probability measures in $V(\mathscr{E})$. For each measure μ in $L(\mathscr{E})$, let $[\mu]_t$ denote the π_t-absolutely continuous part of μ. Then $\{t : [\mu]_t \neq 0\}$ is always countable and $\mu = \sum_t [\mu]_t$ for any $\mu \in L(\mathscr{E})$. Put $\Theta_t = \{\theta : [P_\theta]_t \neq 0\}$ and $P_{\theta,t} = [P_\theta]_t / \|[P_\theta]_t\|$ when $\theta \in \Theta_t$.*

For each $t \in T$, let $\mathscr{E}_t$ denote the dominated experiment $(P_{\theta,t} : \theta \in \Theta_t)$.

Then a sub σ-algebra $\mathscr{B}$ of $\mathscr{A}$ is Δ-sufficient if and only if it is Δ-sufficient for the totally informative experiment $(\pi_t : t \in T)$ as well as for each experiment $\mathscr{E}_t$; $t \in T$.

We leave the proof of this result to the reader. As a particular case consider the situation where a non-sufficient sub σ-algebra $\mathscr{C}$ contains a sufficient σ-algebra $\mathscr{B}$. Then $\mathscr{B}$ is sufficient for the totally informative experiment $(\pi_t : t \in T)$. Furthermore, since the experiments $(P_{\theta,t} : \theta \in \Theta_t) : t \in T$ are all dominated, $\mathscr{C}$ is sufficient for these experiments. Hence $\mathscr{C}$ is not sufficient for the totally informative experiment $(\pi_t : t \in T)$ although it contains the σ-algebra $\mathscr{B}$ which is sufficient for this experiment.

Here is a list of criteria for Δ-sufficiency.

Theorem 7.3.8 (Criteria for Δ-sufficiency). *Let $\mathscr{E} = (\mathscr{X}, \mathscr{A}; P_\theta : \theta \in \Theta)$ be an experiment and let $\mathscr{B}$ be a sub σ-algebra of $\mathscr{A}$. Then the following conditions are equivalent.*

(i) $\mathscr{E} \sim \mathscr{E}|\mathscr{B}$.

(ii) *$\mathscr{B}$ is Δ-sufficient (and hence sufficient for each dominated family $(P_\theta : \theta \in \Theta_0)$ where $\Theta_0 \subseteqq \Theta$).*

(iii) *$\|a_1 P_{\theta_1} + a_2 P_{\theta_2}\| = \|a_1(P_{\theta_1}|\mathscr{B}) + a_2(P_{\theta_2}|\mathscr{B})\|$ whenever $a_1, a_2 \in \mathbb{R}$ and θ_1, $\theta_2 \in \Theta$.*

(iv) *There is a non-negative and linear projection Π from $M(\mathscr{E})$ onto the imbedding $\tilde{M}(\mathscr{E}|\mathscr{B})$ of $M(\mathscr{E}|\mathscr{B})$ such that $P_\theta(\Pi(v)) \equiv_\theta P_\theta(v)$ when $v \in M(\mathscr{E})$.*

(v) *The map $\mu \to \mu|\mathscr{B}$ from $V(\mathscr{E})$ into $L(\mathscr{E}|\mathscr{B})$ is a vector lattice isomorphism from $V(\mathscr{E})$ onto $V(\mathscr{E}|\mathscr{B})$.*

(vi) *$\tilde{M}(\mathscr{E}|\mathscr{B}) \supseteqq H(\mathscr{E})$ where $H(\mathscr{E})$ is the smallest $w(M, L)$-closed subalgebra of $M(\mathscr{E})$ which contains each idempotent supporting a subset of $V(\mathscr{E})$.*

(vii) *If θ_1, $\theta_2 \in \Theta$ and $c > 0$ then the smallest idempotent in $M(\mathscr{E})$ which supports $[P_{\theta_1} - cP_{\theta_2}]^+$ is in $\tilde{M}(\mathscr{E}|\mathscr{B})$.*

Furthermore:

(α) *if the restriction map $\mu \to \mu|\mathscr{B}$ from $V(\mathscr{E})$ to $L(\mathscr{E}|\mathscr{B})$ is a vector lattice isomorphism then it is also an isometry;*

and

(β) *if Π satisfies the conditions in* (vii) *then it is determined by these conditions and $\Pi(vw) = v\Pi(w)$ and $\mu(\Pi(w)) = \mu(w)$ whenever $v \in \tilde{M}(\mathscr{E}|\mathscr{B})$, $w \in M(\mathscr{E})$ and $\mu \in V(\mathscr{E})$.*

Remark. If μ is a measure on a measurable space $(\mathscr{X}, \mathscr{A})$ and A is a set in $\mathscr{A}$, then A is called *a support of* μ if $\mu(B) = 0$ whenever $B \subseteqq A^c$. More generally if W is a set of measures on $\mathscr{A}$ then a set A in $\mathscr{A}$ is called *a support of* W if it is a support of every measure in W. If A is a support of a measure μ (a set W of measures) then we shall also say that A *supports* μ (*supports* W).

If $\mathscr{E} = (\mathscr{X}, \mathscr{A}; P_\theta : \theta \in \Theta)$ is an experiment, then the class of idempotents in $M(\mathscr{E})$ constitutes a natural extension of the class $\mathscr{A}$ of events. If μ is a measure in $L(\mathscr{E})$ and u is an idempotent in $M(\mathscr{E})$ then we shall say that u is a *support of* μ if $|\mu|(I - u) = 0$. More generally, an idempotent u in $M(\mathscr{E})$ is called a *support* of a subset W of $L(\mathscr{E})$ if it is a support of every $\mu \in W$. If the idempotent u is a support of the measure μ in $L(\mathscr{E})$ (the subset W of $L(\mathscr{E})$) then we shall also say that u *supports* μ (*supports* W). It follows that an event $A \in \mathscr{A}$ is a support of $W \subseteqq L(\mathscr{E})$ if and only if the idempotent I_A is a support of W.

A support of a subset W of $L(\mathscr{E})$ may thus be either an event A in $\mathscr{A}$ or an idempotent in $M(\mathscr{E})$.

It is easily seen that the set of measures which is supported by a given idempotent (set) is a band. Actually the correspondence which to each idempotent u in $M(\mathscr{E})$ assigns the band L_u of measures in $L(\mathscr{E})$ which are supported by u is a 1–1 increasing map from the set $\mathscr{U}$ of idempotents onto the set of bands in $L(\mathscr{E})$. It follows that if $u_t : t \in T$ is a family of idempotents and $\underline{u} = \bigwedge_t u_t$ while $\bar{u} = \bigvee_t u_t$ then $L_{\bar{u}}$ is the band $\bigvee_t L_{u_t}$ generated by $L_{u_t} : t \in T$ while $L_{\underline{u}} = \bigwedge_t L_{u_t}$. If L_0 is a band in L then $L_0 = L_u$ where u is the smallest idempotent supporting L_0. This smallest idempotent exists since $M(\mathscr{E})$ is order complete.

Note that not all bands in $L(\mathscr{E})$ are generated by subsets of $V(\mathscr{E})$. Thus if $\mathscr{E}$ is totally informative then a band in $L(\mathscr{E})$ is $V(\mathscr{E}) = B(\mathscr{E})$-generated if and only if it is of the form $L(\mathscr{E}|\Theta_0)$ for some subset Θ_0 of Θ.

In fact a band $L_0 \subseteqq L$ is $V(\mathscr{E})$-generated if and only if its smallest idempotent support equals the smallest idempotent which supports the vector lattice $L_0 \cap V(\mathscr{E})$.

If L_0 is a principal band in $L(\mathscr{E})$ and if the set A in $\mathscr{A}$ is, up to $\mathscr{E}$-equivalence, a smallest support of L_0 within the class of supports of L_0 which belong to

$\mathscr{A}$, then $\dot{I}_A$ is the smallest idempotent supporting L_0. (As usual $\dot{\delta}$ denotes the functional in $M(\mathscr{E})$ defined by the bounded measurable function δ.)

In order to see this let us consider any idempotent u supporting L_0. Represent u as $(I_{A_\theta} : \theta \in \Theta)$ and let Θ_0 be any countable subset of Θ such that $L_0 \subseteqq L(P_\theta : \theta \in \Theta_0)$. Let $A_{\Theta_0} \in \mathscr{A}$ be such that $I_{A_\theta} = I_{A_{\Theta_0}}$ a.e. P_θ when $\theta \in \Theta_0$. Then A_{Θ_0} is a support of L_0. Hence, by minimality, $I_A \leqslant I_{A_{\Theta_0}}$ a.e. $\mathscr{E}$. It follows that $\dot{I}_A \leqslant u$.

Proof of the theorem. The equivalence of (i) and (ii) is essentially a re-statement of the first part of proposition 6.2.28. For a given pair (θ_1, θ_2) and by the sublinear function criterion condition (iii) says that $(P_{\theta_1}, P_{\theta_2}) \sim_2 (P_{\theta_1}|\mathscr{B}, P_{\theta_2}|\mathscr{B})$ and by proposition 6.2.28, this is equivalent to the condition that $\mathscr{B}$ is sufficient for $(P_{\theta_1}, P_{\theta_2})$. Thus conditions (i), (ii) and (iii) are equivalent. Assume that these conditions are satisfied and let $v \in M(\mathscr{E})$. Represent v as the consistent family $(v_\theta : \theta \in \Theta)$ of measurable functions. Put $\tilde{v}_\theta = E_\theta(v_\theta|\mathscr{B})$. Then, by Δ-sufficiency, the family $(\tilde{v}_\theta : \theta \in \Theta)$ is also consistent and thus represents a functional $\Pi(v)$ in $M(\mathscr{E}|\mathscr{B})$. It is easily seen that $\Pi(v)$ does not depend on the choices of the functions v_θ nor on the specifications of their conditional expectations. Thus $v \to \Pi(v)$ is a well defined map from $M(\mathscr{E})$ to $\tilde{M}(\mathscr{E}|\mathscr{B})$ and it is easily checked that Π satisfies the conditions in (iv). Assume next that Π has these properties. Let $w \in M_+(\mathscr{E})$ and let $B \in \mathscr{B}$. Let v be idempotent in $\tilde{M}(\mathscr{E}|\mathscr{B})$. Then $\Pi(vw)(I - v) \leqslant \|w\| v(I - v) = 0$ so that $\Pi(vw) \leqslant v\Pi(vw) \leqslant v\,\Pi(w)$.

Replacing v with $I - v$ we find that $\Pi(vw) \geqslant v\,\Pi(w)$ so that $\Pi(vw) = v\,\Pi(w)$. Thus, since the idempotents are fundamental in $\tilde{M}(\mathscr{E}|\mathscr{B})$, this extends to any $v \in \tilde{M}(\mathscr{E}|\mathscr{B})$ and any $w \in M(\mathscr{E})$. If $\mu \in V_+(P_\theta : \lambda_\theta > 0)$ where $\lambda \in \Lambda$ then $g = \mathrm{d}\mu/\mathrm{d}P_\lambda$ may be specified $\mathscr{B}$-measurable. Put $\mu_n = \mu \wedge nP_\lambda$, $n = 1, 2, \ldots$ Then $\mu(\Pi(w)) = \lim_n \mu_n(\Pi(w)) = \lim_n P_\lambda((\dot{g} \wedge \dot{n})\,\Pi(w)) = \lim_n P_\lambda(\Pi((\dot{g} \wedge \dot{n})w)) = \lim_n P_\lambda((\dot{g} \wedge \dot{n})w) = \lim_n \mu_n(w) = \mu(w)$. Assume that $\tilde{\Pi}$ is another projection which has the properties described in (iv). If $w \in M(\mathscr{E})$ and $v = \Pi(w) - \tilde{\Pi}(w)$ then $P_\theta(v^2) = P_\theta(v\Pi(w) - v\tilde{\Pi}(w)) = P_\theta(\Pi(vw)) - P_\theta(\tilde{\Pi}(vw)) = P_\theta(vw) - P_\theta(vw) = 0$ for all θ. Thus $v = 0$.

This proves that (i), (ii) and (iii) $\Rightarrow$ (iv) as well as statement (β). Assume that condition (iv) is satisfied by the projection Π. For each $A \in \mathscr{A}$ and each $\kappa \in L|\mathscr{B} = L(\mathscr{E}|\mathscr{B})$ put $(\kappa T)(A) = \kappa(\Pi(\dot{I}_A))$. Then T is a transition from $\mathscr{E}|\mathscr{B}$ to $\mathscr{E}$ such that $(P_\theta|\mathscr{B})T \equiv_\theta P_\theta$. Thus $\mathscr{E}|\mathscr{B} \sim \mathscr{E}$. It follows that conditions (i)–(iv) are equivalent. Assume that these conditions are satisfied and let $\mu \in V(\mathscr{E})$. Then $\mu \in V(P_\theta : \lambda_\theta > 0)$ for some $\lambda \in \Lambda$. The factorization criterion implies that $f_{\theta,\lambda} = \mathrm{d}P_\theta/\mathrm{d}P_\lambda$ may be specified $\mathscr{B}$-measurable for each θ. Now, as noted in the introduction to this chapter, $g = \mathrm{d}\mu/\mathrm{d}P_\lambda$ may be specified as a measurable function of $(f_{\theta,\lambda} : \lambda_\theta > 0)$. Then $g = \mathrm{d}(\mu|\mathscr{B})/\mathrm{d}(P_\lambda|\mathscr{B})$ and $f_{\theta,\lambda} = \mathrm{d}(P_\theta|\mathscr{B})/\mathrm{d}(P_\lambda|\mathscr{B})$. Thus $\mu|\mathscr{B} = g(P_\lambda|\mathscr{B}) \in V(\mathscr{E}|\mathscr{B})$ and $\mu^+|\mathscr{B} = g^+(P_\lambda|\mathscr{B})$. Thus this map is a vector lattice isomorphism of $V(\mathscr{E})$ onto $V(\mathscr{E}|\mathscr{B})$. If the restriction map $\mu \to (\mu|\mathscr{B})$ is a vector lattice isomorphism from $V(\mathscr{E})$ to $L(\mathscr{E}|\mathscr{B})$ then $\|(\mu|\mathscr{B})\| = \| |\mu|\mathscr{B}\| = \| |\mu| \| = \|\mu\|$. Thus by the sublinear function criterion, condition (v) implies that statements (i)–(v) are all equivalent. We have also proved statement (α).

Assume that conditions (i)–(v) are satisfied and let u be the smallest idempotent in $M(\mathscr{E})$ which supports the sub-vector lattice V_0 of $V(\mathscr{E})$. Clearly $0 \leqslant \Pi(u) \leqslant I$. Put $\gamma(x) = 0$ or 1 as $x \neq 0$ or $x = 1$. Then $v = \gamma(\Pi(u))$ is idempotent in $\tilde{M}(\mathscr{E}|\mathscr{B})$. Furthermore $\mu(\Pi(u)) = \mu(u) = \|\mu\| = \mu(I)$ when $0 \leqslant \mu \in V_0$. It follows that $\mu(v) = \|\mu\|$ so that v is also a support of V_0. Hence, by minimality, $v \geqslant u$. Thus, since $\Pi(u) \geqslant v$, $\Pi(u) \geqslant u$. Hence, since $P_\theta(\Pi(u) - u) \equiv_\theta 0$ and $\Pi(u) - u \geqslant 0$, $u = \Pi(u) \in \tilde{M}(\mathscr{E}|\mathscr{B})$. Thus (vi) holds. Then since the implication (vi) $\Rightarrow$ (vii) is trivial, it suffices to show that (viii) $\Rightarrow$ (i)–(v).

Let $\theta_0, \theta_1 \in \Theta$ and let $\lambda \in\,]0, 1[$. Consider the problem of testing 'θ_0' against 'θ_1' for (0–1)-loss and for the prior distribution which assigns mass $(1 - \lambda)$ to θ_0 and λ to θ_1. Then the set $A = [\lambda f_{\theta_1} > (1 - \lambda) f_{\theta_0}]$ is the rejection region of a Bayes solution. This set is also a support of the measure $Q = [\lambda P_{\theta_1} - (1 - \lambda) P_{\theta_0}]^+$. By assumption Q possesses a support $B \in \mathscr{B}$ such that $I_B \leqslant I_A$ a.e. $P_{\theta_1} + P_{\theta_2}$. Then $I_B = I_A$ a.e. $P_{\theta_1} + P_{\theta_2}$ on A. Hence $I_B = I_A$ a.e. $P_{\theta_1} + P_{\theta_2}$ so that I_B is also a Bayes solution. It follows that the dichotomies $(P_{\theta_1}|\mathscr{B}, P_{\theta_2}|\mathscr{B})$ and $(P_{\theta_1}, P_{\theta_2})$ are equivalent for testing problems. Hence, by proposition 6.2.28, these dichotomies are equivalent. □

In part (vi) of the theorem we introduced the notation $H(\mathscr{E})$ for the smallest $w(M, L)$-closed subalgebra of $M(\mathscr{E})$ which for each subset V_0 of $V(\mathscr{E})$ contains the smallest idempotent which supports V_0. This vector lattice is thus defined without reference to the underlying σ-algebra $\mathscr{A}$. It may be defined for abstract experiments as well as for experiments in 'measure form'. It is clear from the theorem that if this vector lattice is of the form $\tilde{M}(\mathscr{E}|\mathscr{B}_0)$ for some weakly closed sub σ-algebra $\mathscr{B}_0$, then $\mathscr{B}_0$ is the smallest pairwise sufficient and weakly closed sub σ-algebra. The converse follows from the following result.

Proposition 7.3.9 (Characterization of $H(\mathscr{E})$). *The following conditions are equivalent for a functional $h \in M(\mathscr{E})$:*

(i) *$h \in H(\mathscr{E})$.*
(ii) *$h \in \tilde{M}(\mathscr{E}|\mathscr{B})$ whenever $\mathscr{B}$ is a Δ-sufficient sub σ-algebra.*
(iii) *$hV(\mathscr{E}) \subseteqq V(\mathscr{E})$, i.e. $h\mu \in V(\mathscr{E})$ when $\mu \in V(\mathscr{E})$.*
(iv) *$hP_\theta \in V(\mathscr{E})$; $\theta \in \Theta$.*

If $h \geqslant 0$ then conditions (i)–(iv) *are equivalent to:*

(v) *there are families* $\nu_t : t \in T$ *and* $\mu_t : t \in T$ *in* $V_+(\mathscr{E})$ *such that* h *is the smallest common non-negative Radon–Nikodym derivative in* $M(\mathscr{E})$ *of* $\mathrm{d}\nu_t/\mathrm{d}\mu_t$; $t \in T$.

If h *is idempotent then conditions* (i)–(iv) *are equivalent to*:

(vi) *there is a subset (and hence a sub-vector lattice) of* $V(\mathscr{E})$ *such that* h *is the smallest support of this subset.*

Remark. If $\mu \in L$ where L is an L-space and if $v \in M = L^*$ then $v\mu$ denotes the unique element $\nu \in L$ such that $\nu(\delta) = \mu(v\delta)$; $\delta \in M(\mathscr{E})$. ($\nu$ as defined belongs clearly to $L^{**} = M^*$. The fact that $\nu \in L$ follows since $|\nu| \leqslant \|v\|\,|\mu|$ and since L, after imbedding, is a band in L^{**}.)

Proof. Note that $(h\mu)^+ = h^+\mu$ when $\mu \in L_+$ and $h \in M$. It follows that the set of functionals h satisfying (i)–(iv) are all vector lattices. Furthermore, these vector lattices contain the constants and are closed for the topology of uniform convergence on order intervals of L. It follows that they are also closed for the L-topology on M. Now if H is any $w(M, L)$-closed sub-vector lattice of $M(\mathscr{E})$ which contains the constants, then the idempotents in H are fundamental in H, i.e. the linear space spanned by them is dense in H. (It suffices to check this when $\mathscr{E}$ is coherent and then $H = M(\mathscr{E}|\mathscr{B})$ where $\mathscr{B} = \{B : B \in \mathscr{A}$ and $\dot{I}_B \in H\}$.) It follows that we may restrict our attention to the case where h is idempotent when proving the equivalence of conditions (i)–(iv) and (vi). Identities such as $h(\bigvee_{i=1}^k \sum_{\theta \in F} a_{\theta i} P_\theta) = \bigvee_{i=1}^k \sum_{\theta \in F} a_{\theta i}(hP_\theta)$ which hold for finite subsets F of Θ and constants $a_{\theta i}$ show that (iii) and (iv) are equivalent. Next assume that these conditions are satisfied and put $Q_\theta = hP_\theta$; $\theta \in \Theta$. Then $Q_\theta \in V(\mathscr{E})$ for each θ and h is the minimal support of $(Q_\theta : \theta \in \Theta)$. If u is another idempotent supporting this family then $P_\theta(u \wedge h) = Q_\theta(u \wedge h) = \|Q_\theta\| = Q_\theta(h) = P_\theta(h)$; $\theta \in \Theta$. Thus $P_\theta(h - u \wedge h) \equiv_\theta 0$ so that $h = u \wedge h \leqslant u$ since $h - u \wedge h$ is non-negative. Hence (iii) and (iv) $\Rightarrow$ (vi) and (vi) $\Rightarrow$ (i) by the definition of $H(\mathscr{E})$ when h is idempotent. Furthermore (i) $\Rightarrow$ (ii) by theorem 7.3.8.

Assume that (ii) holds for $h \in M(\mathscr{E})$ and let $\theta_0 \in \Theta$. Let λ_0 be the one-point distribution in θ_0. Then, since $h \in M(\mathscr{E}|\mathscr{D}^{\lambda_0})$, $hP_{\theta_0} = h_{\theta_0}P_{\theta_0}$ where h_{θ_0} is bounded and $\mathscr{D}^{\lambda_0}$-measurable. Hence h_{θ_0} is $\mathscr{C}_\lambda^{\lambda_0}$-measurable for some $\lambda \gg \lambda_0$. Now $\mathrm{d}P_{\theta_0}/\mathrm{d}P_\lambda$, since $\lambda(\theta_0) > 0$, is also $\mathscr{C}_\lambda^{\lambda_0}$-measurable. It then follows from the characterization of the vector lattice $V(\mathscr{E})$ given in the introduction to this chapter, that $hP_{\theta_0} = h_{\theta_0}P_{\theta_0} = h_{\theta_0}(\mathrm{d}P_{\theta_0}/\mathrm{d}P_\lambda)\,\mathrm{d}P_\lambda \in V(\mathscr{E})$. This completes the proof of the full circle (iv) $\Rightarrow$ (iii) $\Rightarrow$ (vi) $\Rightarrow$ (i) $\Rightarrow$ (ii) $\Rightarrow$ (iv). Thus these conditions are all equivalent.

Next assume that these conditions are satisfied for $h \geqslant 0$. Then h is the smallest (in fact the unique) non-negative common Radon–Nikodym derivative $\mathrm{d}Q_\theta/\mathrm{d}P_\theta$; $\theta \in \Theta$ where $Q_\theta = hP_\theta \in V(\mathscr{E})$; $\theta \in \Theta$. Finally assume that $h \geqslant 0$ is the smallest non-negative and common Radon–Nikodym derivative $\mathrm{d}\nu_t/\mathrm{d}\mu_t$; $t \in T$ for the families $(\mu_t : t \in T)$ and $(\nu_t : t \in T)$ in $V_+(\mathscr{E})$. We may assume without loss of generality that $\nu_t \ll \mu_t$ for all t. Let the σ-algebra $\mathscr{B}$ be pairwise sufficient and let Π be the corresponding conditional expectation operator on $M(\mathscr{E})$. If $w \in M(\mathscr{E})$ and $t \in T$ then

$$\mu_t(w\Pi(h)) = \mu_t(\Pi(w\Pi(h))) = \mu_t(\Pi(w)\Pi(h)) = \mu_t(\Pi(w)h)$$

$$= \nu_t(\Pi(w)) = \nu_t(w)$$

so that $d\nu_t/d\mu_t = \Pi(h)$; $t \in T$. Hence $\Pi(h)$ is also a non-negative and common Radon–Nikodym derivative for these families. By minimality $\Pi(h) \geqslant h$, but $P_\theta(\Pi(h) - h) \equiv_\theta P_\theta(\Pi(h) - \Pi(h)) \equiv_\theta 0$ so that $h = \Pi(h) \in M(\mathscr{E}|\mathscr{B})$. Thus (ii) holds. □

Corollary 7.3.10 (Characterizations of $H(\mathscr{E}) \cap \dot{M}(\mathscr{A})$). *The following conditions are equivalent for a bounded measurable function δ on $(\mathscr{X}, \mathscr{A})$.*

(i) $\dot{\delta} \in H(\mathscr{E})$.
(ii) *δ is measurable with respect to all Δ-sufficient and weakly closed sub σ-algebras.*
(iii) $\delta V(\mathscr{E}) \subseteqq V(\mathscr{E})$.
(iv) *$\delta P_\theta \in V(\mathscr{E})$ for each θ.*

If $\delta \geqslant 0$ then conditions (i)–(iv) *are equivalent to*:

(v) *there are families $(\mu_t : t \in T)$ and $(\nu_t : t \in T)$ in $V_+(\mathscr{E})$ such that $\dot{\delta}$ is the smallest common non-negative Radon–Nikodym derivative in $M(\mathscr{E})$ of $d\nu_t/d\mu_t$; $t \in T$.*

If $\delta = I_B$ is the indicator function of a set B then conditions (i)–(iv) *are equivalent to*:

(vi) *$\dot{I}_B$ is the smallest idempotent in $M(\mathscr{E})$ supporting V_0 where V_0 is some subset of $V(\mathscr{E})$.*

Remark. The equivalence of statements (i), (iii), (iv), (v) and (vi) with (ii) yields criteria for the existence of a smallest Δ-sufficient and weakly closed sub σ-algebra. Thus, by (vi), such a σ-algebra exists if and only if the σ-algebra $\mathscr{B}$ of events B such that $\dot{I}_B$ is a minimal support of a subset of $V(\mathscr{E})$, is Δ-sufficient. Of course this is the same as saying that the σ-algebra of events B which defines idempotents in $H(\mathscr{E})$ is Δ-sufficient.

Here are a few more criteria for minimal Δ-sufficiency.

Corollary 7.3.11. *The following conditions are equivalent for a Δ-sufficient σ-algebra $\mathscr{B}$:*

(i) *$\mathscr{B}$ is contained in any weakly closed Δ-sufficient σ-algebra.*
(ii) $V(\mathscr{E}|\mathscr{B}) = L(\mathscr{E}|\mathscr{B})$.
(iii) $\tilde{M}(\mathscr{E}|\mathscr{B}) = H(\mathscr{E})$.

Proof. Suppose (i) is satisfied and let $\kappa \in L(\mathscr{E}|\mathscr{B}) - V(\mathscr{E}|\mathscr{B})$ where '$-$' denotes set difference. Then by strong separation in the locally convex linear topological space $L(\mathscr{E}|\mathscr{B})$, there is a linear functional h in $M(\mathscr{E}|\mathscr{B})$ such that $\kappa(h) = 1$ while $v(h) = 0$ when $v \in V(\mathscr{E}|\mathscr{B})$. Let $\tilde{h} \in \tilde{M}(\mathscr{E}|\mathscr{B})$ be the corresponding functional in $M(\mathscr{E})$, i.e. $\mu(\tilde{h}) = (\mu|\mathscr{B})(h)$ when $\mu \in L(\mathscr{E})$. If $\mu \in V(\mathscr{E})$ then, by pairwise sufficiency, $\mu|\mathscr{B} \in V(\mathscr{E}|\mathscr{B})$ so that $\mu(\tilde{h}) = 0$. By proposition 7.3.9, condition (i) implies that $\tilde{h} \in H(\mathscr{E})$ so that, by the same proposition, $\tilde{h}\mu \in V(\mathscr{E})$ when $\mu \in V(\mathscr{E})$. In particular $\mu(\tilde{h}^2) = (\tilde{h}\mu)(\tilde{h}) = 0$ when $\mu \in V(\mathscr{E})$. Hence $P_\theta \tilde{h}^2 \equiv_\theta 0$ so that $\tilde{h} = 0$, yielding $h = 0$ which contradicts the assumption that $\kappa(h) = 1$. It follows that $L(\mathscr{E}|\mathscr{B}) = V(\mathscr{E}|\mathscr{B})$ so that (ii) holds.

Assume next that (ii) is satisfied and let $B \in \mathscr{B}$ and $\theta \in \Theta$. Then $I_B(P_\theta|\mathscr{B}) \in L(\mathscr{E}|\mathscr{B}) = V(\mathscr{E}|\mathscr{B})$ so that $I_B(P_\theta|\mathscr{B}) = \mu|\mathscr{B}$ where $\mu \in V(\mathscr{E})$. Hence, if $A \in \mathscr{A}$ and if Π is the conditional expectation operator associated with $\mathscr{B}$, then

$$\begin{aligned}(I_B P_\theta)(I_A) &= P_\theta(I_A I_B) = P_\theta(\Pi(I_A I_B)) = P_\theta(\Pi(I_A) I_B) \\ &= (I_B P_\theta)(\Pi(I_A)) = \mu(\Pi(I_A)) = \mu(A).\end{aligned}$$

The last equality follows from the fact that Π is a conditional expectation operator for all of $V(\mathscr{E})$. Thus $I_B P_\theta = \mu \in V(\mathscr{E})$ for all θ so that, by corollary 7.3.10, $\dot{I}_B \in H(\mathscr{E})$. Hence $\tilde{M}(\mathscr{E}|\mathscr{B}) \subseteqq H(\mathscr{E})$ while, by pairwise sufficiency, $\tilde{M}(\mathscr{E}|\mathscr{B}) \supseteqq H(\mathscr{E})$. Thus (ii) $\Rightarrow$ (iii) and finally the implication (iii) $\Rightarrow$ (i) follows from the fact that $H(\mathscr{E}) \subseteqq \tilde{M}(\mathscr{E}|\mathscr{C})$ whenever $\mathscr{C}$ is Δ-sufficient. □

We may now deduce some of LeCam's results on Δ-sufficiency of weakly closed algebras in $M(\mathscr{E})$. Let $\mathscr{H}$ denote the class of all weakly closed subalgebras of $M(\mathscr{E})$ which contains the unit I.

If $\mu \in L(\mathscr{E})$ and $H \in \mathscr{H}$ then $\mu^H = \mu|H$ defines a functional in the L-space H^*. In particular $P_\theta^H : \theta \in \Theta$ defines an abstract experiment $\mathscr{E}^H$ with L-space $L(\mathscr{E}^H) \subseteqq H^*$. It may be checked that $L(\mathscr{E}^H) = L^H$ where L^H denotes the set $\{\mu^H : \mu \in L\}$. If $H = \tilde{M}(\mathscr{E}|\mathscr{B})$ for some σ-algebra $\mathscr{B}$ then this corresponds to the statement that $L(\mathscr{E}|\mathscr{B}) = \{\mu|\mathscr{B} : \mu \in L(\mathscr{E})\}$. It suffices to check the statement in this particular form since any H is of this form when $\mathscr{E}$ is coherent. Any h in H defines an element $\tilde{h}$ in $M(\mathscr{E}^H)$ by evaluation, i.e. $\tilde{h}(\kappa) = \kappa(h)$ when $\kappa \in L^H$. The map $h \to \tilde{h}$ is an isomorphism between the M-spaces $M(\mathscr{E}^H)$ and H. If $H = \tilde{M}(\mathscr{E}|\mathscr{B})$ then this corresponds to the statement that $M(\mathscr{E}|\mathscr{B})$ and its imbedding $\tilde{M}(\mathscr{E}|\mathscr{B})$ in $M(\mathscr{E}|\mathscr{B})$ are isomorphic, as explained earlier in this section.

The map $\mu \to \mu|H$ provides a transition from $\mathscr{E}$ to $\mathscr{E}^H$ which maps P_θ, for each θ, into P_θ^H. Thus $\mathscr{E} \geqslant \mathscr{E}^H$ for all $h \in \mathscr{H}$.

The following theorem gives some of LeCam's results on sufficiency.

Theorem 7.3.12 (Sufficient subalgebras of $M(\mathscr{E})$). *Let $\mathscr{E}$ be an experiment and let $\mathscr{H}$ denote the class of all $L(\mathscr{E})$-closed subalgebras of $M(\mathscr{E})$ which contains the unit I. Then the following statements are equivalent for an algebra $H \in \mathscr{H}$.*

(i) $\mathscr{E}^H \sim \mathscr{E}$ *(i.e. $\mathscr{E}$ may be obtained from $\mathscr{E}^H$ by a transition).*
(ii) $(P_{\theta_1}^H, P_{\theta_2}^H) \sim (P_{\theta_1}, P_{\theta_2})$ *when* $\theta_1, \theta_2 \in \Theta$.
(iii) $\|a_1 P_{\theta_1}^H + a_2 P_{\theta_2}^H\| = \|a_1 P_{\theta_1} + a_2 P_{\theta_2}\|$ *when* $a_1, a_2 \in \mathbb{R}$ *and* $\theta_1, \theta_2 \in \Theta$.
(iv) *There is a non-negative and linear projection* Π *from* $M(\mathscr{E})$ *onto* H *such that* $P_\theta(\Pi(v)) \equiv_\theta P_\theta(v)$ *when* $v \in M(\mathscr{E})$.
(v) *The map* $\mu \to \mu^H$ *from* $V(\mathscr{E})$ *into* $L(\mathscr{E}^H) = L^H$ *is a vector lattice isomorphism from* $V(\mathscr{E})$ *onto* $V(\mathscr{E}^H)$.
(vi) $H \supseteqq H(\mathscr{E})$.
(vii) *If* $\theta_1, \theta_2 \in \Theta$ *and* $c > 0$ *then the smallest idempotent supporting* $[P_{\theta_1} - cP_{\theta_2}]^+$ *is in* H.

Furthermore:

(α) *if the restriction map* $\mu \to \mu^H$ *from* $V(\mathscr{E})$ *to* L^H *is a vector lattice isomorphism then it is also an isometry;*

and

(β) *if* Π *satisfies the conditions in* (iv) *then it is determined by these conditions and* $\Pi(hv) = h\Pi(v)$ *and* $\mu(\Pi(v)) = \mu(v)$ *when* $h \in H$, $v \in M(\mathscr{E})$ *and* $\mu \in V(\mathscr{E})$.

Remark. As shown in LeCam's 1964 paper (iv) may be weakened to:

(iv′) there is a non-negative and linear projection Π from $M(\mathscr{E})$ onto a subset H_1 of H containing the unit I such that $P_\theta(\Pi(v)) \equiv P_\theta(v)$ when $v \in M(\mathscr{E})$.

This may be seen as follows. Firstly H_1 is clearly a linear space. If $\{h_n\}$ is a net in H_1 which converges to $h \in H$ with respect to the topology of uniform convergence on order bounded intervals in L, then $|P_\theta(\Pi(h_n)) - P_\theta(\Pi(h))| \leqslant P_\theta(|\Pi(h_n) - \Pi(h)|) \leqslant P_\theta(\Pi(|h_n - h|)) = P_\theta(|h_n - h|) \to 0$ so that $h_n = \Pi(h_n)$ converges weakly to $\Pi(h)$ as well as to h. Hence $h = \Pi(h) \in H_1$ so that H_1 is closed for the topology of uniform convergence on order bounded subsets of L and, consequently, is also weakly closed. If $h \in H_1$ then $\Pi(h^+) \geqslant \Pi(h) = h$ and $\Pi(h^+) \geqslant 0$ so that $\Pi(h^+) \geqslant h^+$ while $P_\theta(\Pi(h^+) - h^+) = P_\theta(\Pi(h^+)) - P_\theta(h^+) = P_\theta(h^+) - P_\theta(h^+) \equiv_\theta 0$. Thus H_1 is a vector lattice containing the unit I and therefore is an algebra as well. It follows that $H_1 \in \mathscr{H}$ and that H_1 satisfies (iv). Thus H_1 satisfies (vi) so that $H \supseteqq H_1 \supseteqq V(\mathscr{E})$. Hence H also satisfies (iv).

Proof. We may assume without loss of generality that $\mathscr{E}$ is coherent and then this reduces to the statement of theorem 7.3.8. □

Corollary 7.3.11 yields the following characterizations of $H(\mathscr{E})$.

Proposition 7.3.13 (The minimal Δ-sufficient subalgebra). *Let* $\mathscr{H}$ *be as in theorem* 7.3.12 *and let* $H \in \mathscr{H}$ *be Δ-sufficient, i.e.* $\mathscr{E}^H \sim \mathscr{E}$. *Then the following conditions are equivalent:*

(i) *H is the smallest Δ-sufficient algebra in $\mathscr{H}$.*
(ii) $V(\mathscr{E}^H) = L(\mathscr{E}^H)$.
(iii) $H = H(\mathscr{E})$.

Referring to the coherent case we find that $H(\mathscr{E})$ is 'essentially' the dual of $V(\mathscr{E})$ while $V(\mathscr{E})$ is 'essentially' the restriction of L to $V(\mathscr{E})$. The precise statements are as follows.

Proposition 7.3.14 (Relationships between $V(\mathscr{E})$ and $H(\mathscr{E})$). *The L-space $V(\mathscr{E})$ and the M-space $H(\mathscr{E})$ satisfy the following relations:*

(i) *$L(\mathscr{E}|H(\mathscr{E})) = (L|H(\mathscr{E})) \equiv V(\mathscr{E})$ where '$\equiv$' signifies the isomorphism defined by the restriction map $\mu \to \mu|H(\mathscr{E})$ from $V(\mathscr{E})$ onto $L|H(\mathscr{E})$.*
(ii) *$V(\mathscr{E})^* \equiv H(\mathscr{E})$ where '$\equiv$' signifies the isomorphism defined by the restriction map $h \to h|V(\mathscr{E})$ from $H(\mathscr{E})$ onto $V(\mathscr{E})^*$.*
(iii) *$M(\mathscr{E}) = V(\mathscr{E})^\circ \oplus H(\mathscr{E})$ where '$\oplus$' denotes direct sum of vector spaces (not order direct sum of ordered vector spaces) and where $V(\mathscr{E})^\circ$ denotes the set of functionals which vanish on $V(\mathscr{E})$.*

The $H(\mathscr{E})$ component of any vector $v \in M_+(\mathscr{E})$ is non-negative, i.e. it belongs to $H_+(\mathscr{E})$.

Proof. $L^H = L(\mathscr{E}|H) \equiv V(\mathscr{E}|H)$ where $H = H(\mathscr{E})$ by corollary 7.3.11 and $V(\mathscr{E}|H) \equiv V(\mathscr{E})$ by part (v) of theorem 7.3.12. Thus (i) holds so that $V(\mathscr{E})^* \equiv [L^H]^*$. Now (ii) follows since $[L^H]^* \equiv H$ for any $H \in \mathscr{H}$. Let Π be the conditional expectation operator associated with the Δ-sufficient algebra $H(\mathscr{E})$. The decomposition in (iii) is then obtained by decomposing each $v \in M(\mathscr{E})$ as $v = (v - \Pi(v)) + \Pi(v)$. □

Following LeCam we shall say that an experiment $\mathscr{E}$ *is in minimal form* if $V(\mathscr{E}) = L(\mathscr{E})$. Thus the reduction to a Δ-sufficient algebra $H \in \mathscr{H}$ is maximal if and only if $\mathscr{E}^H$ is in minimal form, i.e. if and only if $H = H(\mathscr{E})$. If $\mathscr{E} = (\mathscr{X}, \mathscr{A}; P_\theta : \theta \in \Theta)$ then this is equivalent to the statement that $\mathscr{A}$ does not contain any other pairwise sufficient and weakly closed sub σ-algebras. If so then $\mathscr{A}$ is minimal sufficient, i.e. $\mathscr{A}$ does not contain any other closed sufficient σ-algebra. If $\mathscr{E}$ is coherent then $\mathscr{E}$ is in minimal form if and only if $\mathscr{A}$ is minimal sufficient.

Considering $(P_\theta : \theta \in \Theta)$ as an abstract experiment in the L-space $V(\mathscr{E})$ it is of course automatically in its minimal form. This experiment is called *the minimal form of $\mathscr{E}$.*

The last of the two isomorphism criteria for equivalence which were given in theorem 7.2.4 states that two experiments $\mathscr{E}$ and $\mathscr{F}$ are equivalent if and only if their minimal forms are 'naturally' isomorphic. The first isomorphism criterion expresses the same thing for the Banach spaces $B(\mathscr{E})$ and $B(\mathscr{F})$, but these spaces are not in general vector lattices.

We have so far concentrated our efforts on the equality on the right in:

$$B(\mathscr{E}) \subseteqq V(\mathscr{E}) \subseteqq L(\mathscr{E}).$$

We have seen that '=' on the right amounts to the condition that $\mathscr{A}$ is 'minimal Δ-sufficient'. Let us now turn to the left hand set inequality and discuss the conditions for equality here.

In the next section we will see that $B(\mathscr{E}) = V(\mathscr{E})$ if and only if $\mathscr{E}$ is not equivalent to a proper mixture of non-equivalent experiments. Furthermore, as will be explained in the next section, any experiment is obtainable as a 'weak' limit of finite mixtures of extremal experiments. Anticipating these results we shall say that an experiment $\mathscr{E}$ is *extremal* if $B(\mathscr{E}) = V(\mathscr{F})$.

As noted in the introduction to this chapter, the condition of extremality for experiments respects equivalence and roughly amounts to the condition that a complete and sufficient reduction is possible. Let us agree to say that an experiment $\mathscr{E}$ is *M-complete* if any $v \in M(\mathscr{E})$ is uniquely determined by its expectation function on Θ. Thus $\mathscr{E} = (P_\theta : \theta \in \Theta)$ is M-complete if and only if $v = 0$ is the only functional v in $M(\mathscr{E})$ which satisfies the conditions $P_\theta(v) \equiv_\theta 0$. Clearly any M-complete experiment is boundedly complete. If $\Theta = [0, 1]$, $P_\theta = \delta_\theta$ when $\theta > 0$ while P_0 is the rectangular distribution on $[0, 1]$, then $\mathscr{E}$ is boundedly complete but not M-complete.

The condition of M-completeness as well as the notion of bounded completeness of course does not respect equivalence.

Here we will content ourselves with proving the following criteria for extremality.

Theorem 7.3.15 (Criteria for extremality). *The following conditions are equivalent for an experiment $\mathscr{E} = (\mathscr{X}, \mathscr{A}; P_\theta : \theta \in \Theta)$:*

(i) *$\mathscr{E}$ is extremal, i.e. $B(\mathscr{E}) = V(\mathscr{E})$.*
(ii) *$V(\mathscr{E})^*$ does not possess any non-zero functional which vanishes on $B(\mathscr{E})$.*
(iii) *$V(\mathscr{E})$ possesses no L-norm different from the original one which assigns norm 1 to each measure $P_\theta : \theta \in \Theta$.*
(iv) *$\mathscr{E}$ is $H(\mathscr{E})$-complete, i.e. 0 is the only functional in $H(\mathscr{E})$ which is an unbiased estimator of zero.*
(v) *the restricted experiment $\mathscr{E}^H$, where $H = H(\mathscr{E})$, is M-complete.*
(vi) *$\mathscr{E} \sim \mathscr{F}$ where $\mathscr{F}$ is M-complete.*

If $\mathscr{E}$ possesses a smallest Δ-sufficient and weakly closed sub σ-algebra $\mathscr{B}$ then these conditions are equivalent to the condition that $\mathscr{E}|\mathscr{B}$ is M-complete.

In particular, if $\mathscr{E}$ is coherent then these conditions are equivalent to the condition that the smallest closed sufficient σ-algebra is boundedly complete.

Remark. The main idea of this proof (which is due to LeCam) is to realize that $B(\mathscr{E}) = V(\mathscr{E})$ if and only if each functional w in $M(\mathscr{E})$ which vanishes on $B(\mathscr{E})$, vanishes on all of $V(\mathscr{E})$.

Proof. In view of the other results of this section it suffices to establish the equivalence of conditions (i), (ii) and (iii). If it simplifies things we may also assume without loss of generality that $\mathscr{E}$ is coherent. Furthermore the implication (ii) $\Rightarrow$ (i) follows as noted in the remark. Thus (i) and (ii) are equivalent.

Consider an L-norm γ on $V(\mathscr{E})$. Then γ defines a non-negative linear functional δ on $V(\mathscr{E})$ by

$$\delta(\mu) = \gamma(\mu^+) - \gamma(\mu^-) \qquad \text{when} \quad \mu \in V(\mathscr{E}).$$

The completeness (Cauchy sequences converge) of $V(\mathscr{E})$ implies then that δ is continuous, i.e. that $\delta \in V(\mathscr{E})^*$.

The identity $\gamma(P_\theta) \equiv_\theta 1$ may then be written $P_\theta(\delta) \equiv_\theta 1 \equiv_\theta P_\theta(I)$ where I is the unit in $V(\mathscr{E})^*$. If (i) and (ii) hold then $\delta = I$ so that $\gamma(\mu) = \|\mu\|$ when $\mu \in V(\mathscr{E})$. Hence (i) $\Rightarrow$ (iii). Finally assume that (iii) holds and that $P_\theta(\delta) \equiv 0$ where $\delta \in V(\mathscr{E})^*$. Let c be a constant such that $|\delta| \leqslant cI$. Then $\frac{1}{2}I \leqslant \phi \leqslant \frac{3}{2}I$ where $\phi = (\delta + 2cI)/2c$. Put $\gamma(\mu) = \int \phi \, \mathrm{d}|\mu|$. Then γ defines an L-norm on $V(\mathscr{E})$ and $\gamma(P_\theta) \equiv_\theta 1$. Hence, by (iii), $\gamma(\mu) = \|\mu\|$; $\mu \in V(\mathscr{E})$ so that $\phi = I$ and $\delta = 0$. Thus (iii) $\Rightarrow$ (ii). □

A first justification of the term extremal is obtained by relating it to L-norms.

Corollary 7.3.16 (Extremal experiments define extreme norms). *The experiment $\mathscr{E} = (P_\theta : \theta \in \Theta)$ is extremal if and only if the norm on $V(\mathscr{E})$ is extreme within the convex set of L-norms on $V(\mathscr{E})$ which assigns norm 1 to each measure $P_\theta : \theta \in \Theta$.*

Proof. If $\mathscr{E}$ is extremal then the convex set in question contains just one point and is thus trivially extreme. Assume that the norm on $V(\mathscr{E})$ is extreme. Let ρ be any L-norm on $V(\mathscr{E})$ such that $\rho(P_\theta) \equiv_\theta 1$. Then $1/N\| \ \| \leqslant \rho \leqslant N\| \ \|$ for some constant $N \geqslant 1$. Decompose $\| \ \|$ as $\| \ \| = (1/(N+1))\rho + (N/(N+1))\tilde{\rho}$. Then $\tilde{\rho}$ is also an L-norm on $V(\mathscr{E})$ such that $\tilde{\rho}(P_\theta) \equiv_\theta 1$. Hence, since $\| \ \|$ is extreme, $\rho = \tilde{\rho} = \| \ \|$. Thus $\mathscr{E}$ is extremal by part (iii) of theorem 7.3.15. □

A nice property of extremal experiments is the following.

Theorem 7.3.17 (Majorization by extremal experiments). *If $\mathscr{E}$ is extremal and if $\mathscr{E} \geqslant_2 \mathscr{F}$ then $\mathscr{E} \geqslant \mathscr{F}$.*

Remark. One might easily have $\mathscr{E} \ngeqslant \mathscr{F}$ although $\mathscr{E} \geqslant_2 \mathscr{F}$ when $\mathscr{F}$ is extremal. Example 7.2.19 is an example of this situation.

Proof. We may assume without loss of generality that the experiments are coherent. Let B be any event in $\mathscr{F} = (Q_\theta : \theta \in \Theta)$. Then by coherence, there is a test function δ_B in $\mathscr{E} = (P_\theta : \theta \in \Theta)$ such that $Q_\theta(B) \equiv_\theta P_\theta(\delta_B)$. Completeness yields $\delta_{B_1+B_2+\cdots} = \delta_{B_1} + \delta_{B_2} + \cdots$ a.e. P_θ when $B_1, B_2, \ldots$ are disjoint events in $\mathscr{F}$. Also $\delta_\varnothing = 0$ a.e. P_θ and $\delta_\mathscr{Y} = 1$ a.e. Q_θ where $\mathscr{Y}$ is the sure event in $\mathscr{F}$. Put

$\mu T(B) = \int \delta_B \, d\mu$ when $\mu \in L(\mathscr{E})$. Then T is a transition from $L(\mathscr{E})$ to $L(\mathscr{F})$ which maps P_θ into Q_θ for each Θ. □

The condition of M-completeness amounts to the condition of overall equality in the inequalities $B(\mathscr{E}) \subseteqq V(\mathscr{E}) \subseteqq L(\mathscr{E})$.

Theorem 7.3.18 (Characterization of M-completeness). *An experiment $\mathscr{E}$ is M-complete if and only if $B(\mathscr{E}) = V(\mathscr{E}) = L(\mathscr{E})$.*

If $H \in \mathscr{H}$ and H is Δ-sufficient (i.e. $H \supseteqq H(\mathscr{E})$) then H is minimal Δ-sufficient (i.e. $H = H(\mathscr{E})$) whenever $\mathscr{E}^H$ is M-complete.

Remark. The Behrens–Fisher problem, which involves samples from two normal populations with possibly different and unknown variances, is the most famous counterexample to the converse statement.

Proof. If $\mathscr{E}$ is M-complete then $B(\mathscr{E}) = L(\mathscr{E})$ by hyperplane separation in $L(\mathscr{E})$. Conversely this implies that $\mathscr{E}$ is M-complete. Let $H \in \mathscr{H}$ and assume $\mathscr{E}^H \sim \mathscr{E}$. Let Π be the conditional expectation operator with respect to the sufficient algebra $H(\mathscr{E})$. Let $h \in H$. Then $h - \Pi(h) \in H$ and $P_\theta(h - \Pi(h)) \equiv_\theta 0$. If $\mathscr{E}^H$ is M-complete then this implies that $h = \Pi(h) \in H(\mathscr{E})$ so that $H = H(\mathscr{E})$. □

Characterizations of the extremal experiments are provided by Siebert (1979). Completeness in the dominated case is discussed in Torgersen (1970, 1977).

Problem 7.3.19 (Fragility of completeness). Consider a random variable X in $[0, 1]$ having a truncated gamma distribution given by the density (with respect to Lebesgue measure)

$$f_\theta(x) = \text{constant } x^{p-1} e^{-\theta x}; \qquad 0 < x < 1.$$

Here p, *the shape parameter*, is assumed known, while the scale parameter $\theta > 0$ is unknown. Let $\mathscr{E}$ be the experiment realized by X. Then $\mathscr{E}^n$ is extremal if and only if p is irrational or p is rational $= r/s$ where the integers r and s are relatively prime and the number n of observations is $\leqslant s$ (see Torgersen & Unni, 1984).

7.4 Conical measures and topologies

We defined the standard probability measure of an experiment as the distribution of the posterior distribution for a prior distribution which assigned positive mass to each point θ in Θ. This is clearly impossible when Θ is uncountable–although we may restrict our attention to some particular prior and then to some class of experiments which behaves itself for this prior. In the general case a way out, as noticed by LeCam, is to work directly with the functionals defined by the types of experiments.

In this section we will use the notation $\mathscr{H}$ for the vector lattice of non-negatively homogeneous real valued measurable functions on the likelihood space $Z = [0, \infty[^\Theta$. The projection of Z onto its θth coordinate space will be denoted by p_θ. If $h \in \mathscr{H}$ then the quantity $\int |h|(\mathrm{d}\mathscr{E})$ is well defined for each experiment $\mathscr{E} = (P_\theta : \theta \in \Theta)$. The set of functions h such that $\int |h|(\mathrm{d}\mathscr{E}) < \infty$ constitutes a sub-vector lattice of $\mathscr{H}$. This vector lattice will be denoted as $\mathscr{H}(\mathscr{E})$. The functional $h \to \int h(\mathrm{d}\mathscr{E})$ on $\mathscr{H}(\mathscr{E})$ determines the type of $\mathscr{E}$. Actually we saw in section 7.2 that even the restriction of this functional to Ψ_2 determined the type of $\mathscr{E}$. We shall now see that there is a unique vector lattice homomorphism $h \to h(\mathscr{E})$ from $\mathscr{H}(\mathscr{E})$ onto $V(\mathscr{E})$ which for each θ maps the projection p_θ onto P_θ and which in addition has the property that the total mass of $h(\mathscr{E})$ is $\int h(\mathrm{d}\mathscr{E})$ for each $h \in \mathscr{H}(\mathscr{E})$. Thus we may write

$$\int h(\mathrm{d}\mathscr{E}) = \int \mathrm{d}h(\mathscr{E})$$

when $h \in \mathscr{H}(\mathscr{E})$.

However, we will first make some simple observations on homomorphisms between vector lattices. The main point is that there are vector lattices of real valued functions which are large enough to produce, by 'natural' homomorphisms, any vector lattice we might want to consider. Consider e.g. the vector lattice of real valued functions on $\mathbb{R}^\Theta$ which is generated by the projections on the coordinate spaces. This vector lattice may be expressed as $\Psi_\infty(\Theta) - \Psi_\infty(\Theta)$. Here, by the notation introduced just before theorem 6.3.4, $\Psi_\infty(\Theta)$ is the cone of maxima of finite families of linear functions each depending on a finite set of coordinates only.

If V is any vector lattice generated by a family $(v_\theta : \theta \in \Theta)$ then there is a unique vector lattice homomorphism from $\Psi_\infty(\Theta) - \Psi_\infty(\Theta)$ to V which, for each θ, maps the θth projection in $\mathbb{R}^\Theta$ into v_θ. This, and similar results, may be argued as follows.

Let $L_f(\Theta)$ denote the L-normed space of finite measures on Θ with finite supports. In other words $L_f(\Theta)$ is just the set of real valued functions a on Θ (i.e. vectors in $\mathbb{R}^\Theta$) such that the set of points θ where $a_\theta \neq 0$ is finite. If $v_\theta : \theta \in \Theta$ is a family in a vector lattice V then the sub-vector lattice generated by $(v_\theta : \theta \in \Theta)$ contains all vectors of the form $\bigvee_{i=1}^k \sum_\theta a_\theta^{(i)} v_\theta$ where $a^{(1)}, a^{(2)}, \ldots, a^{(k)} \in L_f(\Theta)$. We shall use the notation $\psi(v_\theta : \theta \in \Theta | a^{(1)}, \ldots, a^{(k)})$ for this vector. If it is clear from the context what family $(v_\theta : \theta \in \Theta)$ we are considering then we may suppress the v_θ's in this notion. If the v_θ's are probability measures then we recognize that these 'ψ-functions' are measures whose total masses are the maximal Bayes utilities for suitably chosen prior distributions and loss functions. If the v_θ's are the projections on the coordinate spaces in $\mathbb{R}^\Theta$ then we may actually delete the quotation marks here.

It is easily checked that the following identities hold for measures $a^{(1)}, \ldots, a^{(r)}, b^{(1)}, \ldots, b^{(s)}$ in $L_f(\Theta)$:

(i) $t\psi(\cdot|a^{(1)},\ldots,a^{(r)}) = \psi(\cdot|ta^{(1)},\ldots,ta^{(r)})$ when $t \geqslant 0$.
(ii) $\psi(\cdot|a^{(1)},\ldots,a^{(r)}) + \psi(\cdot|b^{(1)},\ldots,b^{(s)}) =$
$\psi(\cdot|a^{(1)} + b^{(1)}, a^{(1)} + b^{(2)},\ldots,a^{(r)} + b^{(s)})$.
(iii) $\psi(\cdot|a^{(1)},\ldots,a^{(r)}) \vee \psi(\cdot|b^{(1)},\ldots,b^{(s)}) =$
$\psi(\cdot|a^{(1)},\ldots,a^{(r)},b^{(1)},\ldots,b^{(s)})$.
(iv) $[\psi(\cdot|a^{(1)},\ldots,a^{(r)}) - \psi(b^{(1)},\ldots,b^{(s)})]^{+} =$
$\psi(\cdot|a^{(1)},\ldots,a^{(r)},b^{(1)},\ldots,b^{(s)}) - \psi(\cdot|b^{(1)},\ldots,b^{(s)})$.

It follows that the set of differences of vectors of the form $\psi(\cdot|a^{(1)},\ldots,a^{(r)})$ for measures $a^{(1)},\ldots,a^{(r)}$ in $L_f(\Theta)$ constitutes a vector lattice. Hence this vector lattice is precisely the vector lattice generated by $v_\theta : \theta \in \Theta$.

The conditions for the existence of homomorphisms between vector lattices, which are required to map certain vectors into certain vectors, may now be expressed as follows.

Proposition 7.4.1 (Existence of homomorphisms). *Let V and W be vector lattices which are generated by, respectively, the family $(v_\theta : \theta \in \Theta)$ and the family $(w_\theta : \theta \in \Theta)$. Then there is a vector lattice homomorphism from V onto W which for each θ maps v_θ into w_θ, if and only if $\psi(w_\theta : \theta \in \Theta|a^{(1)},\ldots,a^{(r)}) \geqslant 0$ whenever $a^{(1)},\ldots,a^{(r)} \in L_f(\Theta)$ and $\psi(v_\theta : \theta \in \Theta|a^{(1)},\ldots,a^{(r)}) \geqslant 0$.*

Proof. A vector lattice isomorphism possessing the desired property necessarily maps $\psi(v_\theta : \theta \in \Theta|a^{(1)},\ldots,a^{(r)})$ into $\psi(w_\theta : \theta \in \Theta|a^{(1)},\ldots,a^{(r)})$. Thus the condition is necessary. Assume that the condition is satisfied. Let $a^{(1)},\ldots,a^{(r)},b^{(1)},\ldots,b^{(s)} \in L_f(\Theta)$ and assume $\psi(v_\theta : \theta \in \Theta|a^{(1)},\ldots,a^{(r)}) \geqslant \psi(v_\theta : \theta \in \Theta|b^{(1)},\ldots,b^{(s)})$. Then

$$\psi(v_\theta : \theta \in \Theta|a^{(1)} - b^{(j)},\ldots,a^{(r)} - b^{(j)})$$
$$= \psi(v_\theta : \theta \in \Theta|a^{(1)},\ldots,a^{(s)}) - \psi(v_\theta : \theta \in \Theta|b^{(j)}) \geqslant 0 \qquad j = 1,2,\ldots,s.$$

Hence, by the condition of the proposition,

$$\psi(w_\theta : \theta \in \Theta|a^{(1)},\ldots,a^{(r)}) - \psi(w_\theta : \theta \in \Theta|b^{(j)})$$
$$= \psi(w_\theta : \theta \in \Theta|a^{(1)} - b^{(j)},\ldots,a^{(r)} - b^{(j)}) \geqslant 0 \qquad j = 1,\ldots,s.$$

Thus $\psi(w_\theta : \theta \in \Theta|a^{(1)},\ldots,a^{(r)}) \geqslant \psi(w_\theta : \theta \in \Theta|b^{(1)},\ldots,b^{(s)})$. Using the combination rules for the ψ expressions given above we find more generally that

$$\psi(w_\theta : \theta \in \Theta|a^{(1)},\ldots,a^{(r)}) - \psi(w_\theta : \theta \in \Theta|b^{(1)},\ldots,b^{(s)})$$
$$\geqslant \psi(w_\theta : \theta \in \Theta|c^{(1)},\ldots,c^{(k)}) - \psi(w_\theta : \theta \in \Theta|d^{(1)},\ldots,d^{(t)})$$

if this inequality holds with the w_θ's replaced by the v_θ's. It follows that there is a well defined map from V to W which maps each difference of 'psi vectors' in V onto the corresponding difference of 'psi vectors' in W. This map is clearly linear and the formula $(x - y)^+ = x \vee y - y$ implies that this is a lattice homomorphism as well. □

The requirements of non-negativity may sometimes be reduced to the case of linear combinations.

Proposition 7.4.2. *Let $v_\theta : \theta \in \Theta$ be a family of vectors in the vector lattice V and let $a^{(1)}, \ldots, a^{(r)}$ be measures in $L_f(\Theta)$ such that $\sum_\theta a_\theta v_\theta \geqslant 0$ for some a in the convex hull $\langle a^{(1)}, \ldots, a^{(r)} \rangle$ of $a^{(1)}, \ldots, a^{(r)}$. Then $\psi(v_\theta : \theta \in \Theta | a^{(1)}, \ldots, a^{(r)}) \geqslant 0$.*

Proof. We have $a = \sum_{i=1}^r c_i a^{(i)}$ for non-negative numbers $c_1, \ldots, c_r$ such that $c_1 + \cdots + c_r = 1$. Hence $\psi(v_\theta : \theta \in \Theta | a^{(1)}, \ldots, a^{(r)}) = \bigvee_i \sum_\theta a_\theta^{(i)} v_\theta \geqslant \sum_i c_i \sum_\theta a_\theta^i v_\theta = \sum_\theta a_\theta v_\theta \geqslant 0$. □

The last two propositions may be applied in the particular case where one of the two vector lattices under consideration is a vector lattice of real valued functions.

Proposition 7.4.3 (Existence of homomorphisms from vector lattices of functions). *Let $(v_\theta : \theta \in \Theta)$ be a family of real valued functions on a set $\mathscr{X}$ and let V be the vector lattice generated by this family.*

Let W be a vector lattice generated by a family $w_\theta : \theta \in \Theta$.

Assume that to each finite family $a^{(1)}, \ldots, a^{(r)}$ of measures in $L_f(\Theta)$ there corresponds a non-empty closed convex cone $H(a^{(1)}, \ldots, a^{(r)})$ in $L_f(\Theta)$ such that:

(i) *if $a \in H(a^{(1)}, \ldots, a^{(r)})$ then $\sum_\theta a_\theta w_\theta \geqslant 0$ and $a_\theta = 0$ whenever $a_\theta^{(1)} = \cdots = a_\theta^{(r)} = 0$.*

(ii) *if γ is a real valued function on Θ such that $\sum \gamma(\theta) a_\theta \geqslant 0$ when $a \in H(a^{(1)}, \ldots, a^{(r)})$ and if $\psi(\gamma(\theta) : \theta \in \Theta | a^{(1)}, \ldots, a^{(r)}) < 0$, then there is a positive number t and a point x in $\mathscr{X}$ such that $v_\theta(x) = \gamma(\theta)/t$ when $\sum_i |a_\theta^{(i)}| > 0$.*

Then there is a necessarily unique vector lattice homomorphism from V to W which for each $\theta \in \Theta$ maps v_θ onto w_θ.

The assumption that H is non-empty, for a closed cone H, is equivalent to the condition that $0 \in H$.

Proof. Let $a^{(1)}, \ldots, a^{(r)}$ be measures in $L_f(\Theta)$ such that $\psi(v_\theta : \theta \in \Theta | a^{(1)}, \ldots, a^{(r)}) \geqslant 0$. Put $F = \{\theta : \sum_i |a_\theta^{(i)}| > 0\}$, $H = H(a^{(1)}, \ldots, a^{(r)})$ and $C = \langle a^{(1)}, \ldots, a^{(r)} \rangle$. By proposition 7.4.2 it suffices to show that $H \cap C \neq \emptyset$. This will be proved by showing that the assumption '$H \cap C = \emptyset$' contradicts the assumptions of the proposition. Assume then that $H \cap C = \emptyset$. Then, since $0 \in H$, $F \neq \emptyset$. Using strong hyperplane separation in $\mathbb{R}^F$ we obtain a vector $(\gamma(\theta) : \theta \in F)$ in $\mathbb{R}^F$ and a number κ such that $\sum_F \gamma(\theta) a_\theta \geqslant \kappa > \sum_F \gamma(\theta) a_\theta^{(i)}$ when $a \in H$ and $i = 1, \ldots, r$.

The homogeneity of H implies that we may put $\kappa = 0$. Extend γ to a real valued function on all of Θ. Then $\sum_\theta \gamma(\theta)a_\theta \geqslant 0 > \psi(\gamma(\theta):\theta \in \Theta | a^{(1)},\ldots,a^{(r)})$ when $a \in H$. It follows that there is a number $t > 0$ and an $x \in \mathscr{X}$ such that $v_\theta(x)/t = \gamma(\theta)$ when $\theta \in F$. We may assume without loss of generality that $t = 1$. This yields the contradiction

$$0 \leqslant \psi(v_\theta:\theta \in \Theta | a^{(1)},\ldots,a^{(r)})_x = \psi(\gamma(\theta):\theta \in \Theta | a^{(1)},\ldots,a^{(r)}) < 0. \qquad \square$$

Corollary 7.4.4 *($\mathscr{X} = \prod_\theta [\alpha_\theta, \beta_\theta]$ or $= [0,1]^{\Theta - \{\theta_0\}}$). Let $(v_\theta:\theta \in \Theta)$ be a family of real valued functions on a set $\mathscr{X}$ and let V be the vector lattice generated by this family.*

Let W be a vector lattice generated by a family $(w_\theta:\theta \in \Theta)$.

Then each of the following two conditions ensures the existence of a necessarily unique vector lattice homomorphism from V onto W mapping v_θ onto w_θ for each θ in Θ.

(i) $0 \in \prod_\theta [\alpha_\theta, \beta_\theta] \subseteqq \mathscr{X} \subseteqq \mathbb{R}^\Theta$ *where $w_\theta \geqslant 0$ when $\alpha_\theta = 0$, while $w_\theta \leqslant 0$ when $\beta_\theta = 0$. The function v_θ, for each θ, is the projection of $\mathscr{X}$ into the θth coordinate space.*

(ii) $\mathscr{X} = [0,1]^{\Theta - \{\theta_0\}}$ *where $\theta_0 \in \Theta$ satisfies $0 \leqslant w_\theta \leqslant w_{\theta_0}$; $\theta \in \Theta$. The function v_θ, for each $\theta \neq \theta_0$, is the projection of $\mathscr{X}$ onto the θth coordinate space, while v_{θ_0} is the constant function 1 on $\mathscr{X}$.*

Remark. (i) implies that we can always put $\mathscr{X} = \mathbb{R}^\Theta$. If $w_\theta \geqslant 0$ for all θ, as in the case where $(w_\theta:\theta \in \Theta)$ is an experiment, then $\mathscr{X} = [0,\infty[^\Theta$ will do.

Proof. Let $a^{(1)},\ldots,a^{(r)} \in L_f(\Theta)$ and put $F = \{\theta:\sum_i |a_\theta^{(i)}| > 0\}$.

Proof of (i). Define $H(a^{(1)},\ldots,a^{(r)}) = H$ as the set $\{a : a \in L_f(\Theta), a_\theta \geqslant 0$ when $\theta \in F$ and $\alpha_\theta = 0 < \beta_\theta$, $a_\theta \leqslant 0$ when $\theta \in F$ and $\alpha_\theta < 0 = \beta_\theta$, $a_\theta = 0$ otherwise$\}$. Then $\sum a_\theta w_\theta \geqslant 0$ when $a \in H$. Assume $\sum_\theta \gamma(\theta)a_\theta \geqslant 0$ when $a \in H$. Let $\theta_0 \in F$ and assume $\alpha_{\theta_0} = 0 < \beta_{\theta_0}$. We may then put $a_{\theta_0} = 1$ and $a_\theta = 0$ if $\theta \neq \theta_0$. Hence $\gamma(\theta_0) \geqslant 0$. Similarly $\gamma(\theta_0) \leqslant 0$ when $\theta_0 \in F$ and $\alpha_{\theta_0} < 0 = \beta_{\theta_0}$. If $\theta_0 \in F$ and $\alpha_{\theta_0} = \beta_{\theta_0} = 0$ then there is no restriction on a_{θ_0} while $a_\theta = 0$ when $\theta \neq \theta_0$. Hence $\gamma(\theta_0) = 0$ in this case. It follows that there is a $t > 0$ such that $\alpha_\theta \leqslant \gamma(\theta)/t \leqslant \beta(\theta)$ when $\theta \in F$.

Proof of (ii). Define $H = H(a^{(1)},\ldots,a^{(r)})$ as the set $\{a : a_\theta \geqslant 0$ when $\theta \in F$, $a_\theta = 0$ when $\theta \notin F\}$ when $\theta_0 \notin F$. It is easily checked that this set has the desired properties. Finally assume that $\theta_0 \in F$. We may then put $H = H(a^{(1)},\ldots,a^{(r)})$ as the set $\{a : a_\theta = 0$ when $\theta \notin F$, $a_{\theta_0} \geqslant \sum_{\theta \neq \theta_0} a_\theta^-\}$. If $a \in H$ then $\sum a_\theta w_\theta = a_{\theta_0} w_{\theta_0} + \sum_{\theta \neq \theta_0} a_\theta w_\theta \geqslant (\sum_{\theta \neq \theta_0} a_\theta^-) w_{\theta_0} + \sum_{\theta \neq \theta_0} a_\theta w_\theta \geqslant \sum_{\theta \neq \theta_0} a_\theta^+ w_\theta \geqslant 0$. Let γ be a real valued function on Θ such that $\gamma | F \neq 0$ and $\sum \gamma(\theta)a_\theta \geqslant 0$ when $a \in H$. Then $\gamma(\theta) \geqslant 0$; $\theta \in F$. Putting $a_{\theta_0} = 1$, $a_{\theta_1} = -1$ and $a_\theta = 0$ when

$\theta \notin \{\theta_0, \theta_1\}$ where $\theta_1 \in F - \{\theta_0\}$ we find $\gamma(\theta_0) \geqslant \gamma(\theta_1)$. Thus $\gamma(\theta_0) \geqslant \gamma(\theta) \geqslant 0$ when $\theta \in F$. Since $\gamma|F \neq 0$, it follows that $\gamma(\theta_0) > 0$. Hence $\gamma(\theta)/t = v_\theta(x); \theta \in F$ when $t = \gamma(\theta_0)$ and $x_\theta = \gamma(\theta)/\gamma(\theta_0)$ when $\theta \in F - \{\theta_0\}$ while $x_\theta = 0$ otherwise. □

By theorem 5.7.11 any abstract L-space $L \neq 0$ is isomorphic to the L-space of an experiment. We shall now see how part (ii) of the last corollary yields a simple proof of this important result. Now it was argued in the proof of corollary 5.7.14, without using the representation theory for abstract L- and M-spaces, that the L-space of any experiment is isomorphic to the L-space of a Σ-dominated and totally informative experiment. Thus we obtain another proof of the fact that any abstract L-space $L \neq 0$ is isomorphic to the L-space of a Σ-dominated and totally informative experiment. This proof is independent of the sophisticated machinery introduced in chapter 5.

First consider the case of an L-space possessing a single generator.

Corollary 7.4.5 (Representations of principal bands in L-spaces). *Any non-null L-space which possesses a single generator is isomorphic to the L-space of all finite measures which are absolutely continuous with respect to some probability measure (on some measurable space). If L is separable then the probability measure may be chosen on* $[0, 1]$.

Remark. The proof utilizes the fact that, by the Radon–Nikodym theorem, the L-space of finite measures which are absolutely continuous with respect to a σ-finite measure $\mu \geqslant 0$ is isomorphic to $L_1(\mu)$, i.e. the L-space of μ-equivalence classes of μ-integrable functions.

Proof. Let the L-space L be generated by the element π. Since $L \neq 0$, we may assume that $\pi \geqslant 0$ and that $\|\pi\| = 1$. Let Θ be a dense subset of the order interval $\{\mu : 0 \leqslant \mu \leqslant \pi\}$ which contains the maximal element π. Put $\mathscr{X} = [0, 1]^{\Theta - \{\theta_0\}}$ where $\theta_0 = \pi$. Then by Tychonoff's theorem, $\mathscr{X}$ is compact for the topology of pointwise convergence on $\Theta - \{\theta_0\}$. Equip $\mathscr{X}$ with the σ-algebra $\mathscr{A}$ of Baire sets, i.e. the smallest σ-algebra on $\mathscr{X}$ making the continuous real valued functions measurable.

It follows from part (ii) of corollary 7.4.4 that there is a vector lattice homomorphism H, from the vector lattice V of continuous real valued functions on $\mathscr{X}$ generated by the projections $v_\theta : \theta \in \Theta - \{\theta_0\}$ and the constants, which maps v_θ into θ when $\theta \neq \theta_0$ and maps 1 into π.

Consider V as a sub-vector lattice of the vector lattice $C(\mathscr{X})$ of continuous real valued functions on $\mathscr{X}$. Clearly V distinguishes points and consequently, by the vector lattice version of the Stone–Weierstrass approximation theorem, is dense in $C(\mathscr{X})$ for the topology of uniform convergence. Furthermore $-\pi \leqslant H(v) \leqslant \pi$ when $-1 \leqslant v \leqslant 1$ and $v \in V$. Thus H is a bounded linear map from the space V (with supremum norm) to L. It follows that H has a unique extension which is a bounded linear map from the Banach space $C(\mathscr{X})$ to L. With a slight misuse of the notation we will denote this extension by H. It is now easily checked that H extended in this way becomes a vector lattice homomorphism from $C(\mathscr{X})$ to L. Define a continuous linear functional S on $C(\mathscr{X})$ by putting $S(v) = \|H(v^+)\| - \|H(v^-)\|$ when $v \in C(\mathscr{X})$. Thus $S = I(H)$ where I is the unit on L^*. Clearly S is non-negative and $S(1) = \|\pi\| = 1$. It follows that there is a (unique) probability measure P on $\mathscr{A}$ such that $S(v) = \int v \, dP$; $v \in C(\mathscr{X})$. It remains to show that $(\mathscr{X}, \mathscr{A}, P)$ has the desired property.

Let $v \in C(\mathscr{X})$. Then using the fact that L is an L-space, $\|H(v)\| = \| \, |H(v)| \, \| = \|H(v)^+ + H(v)^- \| = \|H(v)^+ \| + \|H(v)^- \| = \|H(v^+)\| + \|H(v^-)\| = \int v^+ \, dP + \int v^- \, dP = \int |v| \, dP = \|vP\|$ where as usual vP denotes the measure with density v with respect to P. Thus H is a linear isometry from the linear space $C(\mathscr{X})$, equipped with the $L_1(P)$-norm (rather than the supremum norm), into the L-space L. Since $C(\mathscr{X})$ is dense in $L_1(P)$, it follows that H has a unique extension which we shall also denote by H, which is a linear isometry from $L_1(P)$ to L. Furthermore this extension is also a vector lattice homomorphism. Let $0 \leqslant \lambda \in L$. Then $0 \leqslant \lambda_n \leqslant \pi$ where $\lambda_n = (\lambda/n) \wedge \pi$ and $n = 1, 2, \ldots$ The assumption that Θ is dense in $\{\mu : 0 \leqslant \mu \leqslant \pi\}$ implies that for each n, there is a $\theta_n = H(v_n)$ in Θ such that $\|\theta_n - \lambda_n\| \leqslant n^{-2}$. Thus, since π generates L, $\lambda = \lim_n n\lambda_n = \lim_n n\theta_n = \lim_n H(nv_n)$. Hence, since H is an isometry, $\int |nv_n - v| \, dP \to 0$ for some $v \in L_1(P)$. It follows that $\lambda = H(v)$. Hence the range of H contains L_+ and, consequently, contains all of L. The last statement follows from the fact that $[0, 1]^{\Theta - \{\theta_0\}}$ and $[0, 1]$ are Borel isomorphic when Θ is countable. □

This yields another, independent, proof of corollary 5.7.14, which we repeat as the following corollary.

Corollary 7.4.6 (Representation of an L-space as L($\mathscr{E}$)). *To any L-space $L \neq 0$ there corresponds a totally informative and $\sum$-dominated experiment $\mathscr{E} = (\mathscr{X}, \mathscr{A}; P_\theta : \theta \in \Theta)$ such that L is isomorphic to $L(\mathscr{E})$.*

$\mathscr{E}$ may be chosen so that $\mathscr{X}$ possesses an $\mathscr{A}$-measurable partition $(\mathscr{X}_\theta : \theta \in \Theta)$ such that $P_\theta(\mathscr{X}_\theta) \equiv_\theta 1$ and $\mathscr{A} = \{A : A \subseteqq \mathscr{X} \text{ and } A \cap \mathscr{X}_\theta \in \mathscr{A}; \theta \in \Theta\}$.

Proof. Let $\pi_\theta : \theta \in \Theta$ be a maximal family of disjoint non-negative and normalized vectors in L. Thus $\pi_\theta \geqslant 0$, $\|\pi_\theta\| = 1$ for all θ and $\pi_{\theta_1} \wedge \pi_{\theta_2} = 0$ when $\theta_1 \neq \theta_2$. The band generated by the family $(\pi_\theta : \theta \in \Theta)$ is, by maximality, the whole L-space L. Any $\lambda \in L$ has a unique decomposition as $\lambda = \sum_\theta \lambda_\theta$ where λ_θ belongs to the band B_θ generated by π_θ. The subset of Θ consisting of those θ's such that $\lambda_\theta \neq 0$, is necessarily countable since $\sum_\theta \|\lambda_\theta\| < \infty$. It follows from the previous theorem that to each θ there corresponds a probability space $(\mathscr{X}_\theta, \mathscr{A}_\theta, Q_\theta)$ such that B_θ is isomorphic to the band generated by Q_θ by an isomorphism H_θ which maps π_θ into Q_θ. We may choose the sets

$\{\mathscr{X}_\theta : \theta \in \Theta\}$ disjoint. Define the items mentioned in the theorem by putting

$$\mathscr{X} = \bigcup \{\mathscr{X}_\theta : \theta \in \Theta\},$$
$$\mathscr{A} = \{A : A \subseteqq \mathscr{X} \text{ and } A \cap \mathscr{X}_\theta \in \mathscr{A}_\theta \text{ when } \theta \in \Theta\},$$

and

$$P_\theta(A) = Q_\theta(A \cap \mathscr{X}_\theta) \quad \text{when} \quad A \in \mathscr{A} \quad \text{and} \quad \theta \in \Theta.$$

It follows that the partition $(\mathscr{X}_\theta : \theta \in \Theta)$ is $\mathscr{A}$-measurable, that $P_\theta(\mathscr{X}_\theta) \equiv_\theta 1$ and that $A \in \mathscr{A}$ if and only if $A \subseteqq \mathscr{X}$ and $A \cap \mathscr{X}_\theta \in \mathscr{A}$ for all θ. For each $\lambda \in L$, define a measure $H(\lambda)$ by putting $H(\lambda)(A) = \sum_\theta H_\theta(\lambda_\theta)(A \cap \mathscr{X}_\theta)$ when $A \in \mathscr{A}$. The proof is now completed by checking that H defined in this way is a well defined isomorphism of L onto the band generated by $(P_\theta : \theta \in \Theta)$ which, for each θ, maps π_θ onto P_θ. □

If we are concerned with properties of experiments which respect equivalence then, by the corollary and by theorem 7.2.4, we may without loss of generality restrict our attention to Σ-dominated experiments.

Theorem 7.4.7 (Representation by $\sum$-dominated experiments). *Any experiment is equivalent to a $\sum$-dominated experiment.*

As usual let us denote by $\Psi_k(\Theta)$ (or by just Ψ_k) the set of functions on $\mathbb{R}^\Theta$ which are of the form $x \to \psi(x_\theta : \theta \in \Theta | a^{(1)}, \ldots, a^{(k)})$ for measures $a^{(1)}, \ldots, a^{(k)}$ in $L_f(\Theta)$. Let $p_\theta = \psi(\cdot | e^\theta)$, for each θ, be the projection onto the θth coordinate space. Thus $p_\theta \in \Psi_1; \theta \in \Theta$. The class $\bigcup_{k=1}^\infty \Psi_k$ is clearly a cone as well as a lattice for the pointwise operations.

Finally let Φ, or $\Phi(\Theta)$, denote the vector lattice of differences of functions from $\bigcup_{k=1}^\infty \Psi_k$. Applying corollary 7.4.6 with $\mathscr{X} = \mathbb{R}^\Theta$ we see that there is a well defined unique homomorphism from Φ onto a vector lattice generated by a family $(w_\theta : \theta \in \Theta)$ which maps the θth projection onto w_θ. We may thus without ambiguity write $\phi(w_\theta : \theta \in \Theta)$ for the image of ϕ by this homomorphism. If $\phi(x) \equiv_x \bigvee_i \sum_{\theta \in F} a_\theta^{(i)} x_\theta - \bigvee_j \sum_{\theta \in F} b_\theta^{(j)} x_\theta$ for constants $a_\theta^{(i)}$ and $b_\theta^{(j)}$ then, necessarily, $\phi(w_\theta : \theta \in \Theta) = \bigvee_i \sum_{\theta \in F} a_\theta^{(i)} w_\theta - \bigvee_j \sum_{\theta \in F} b_\theta^{(j)} w_\theta$.

In particular we can apply this to an experiment $\mathscr{E} = (\mathscr{X}, \mathscr{A}; P_\theta : \theta \in \Theta)$. It is then convenient to write $\phi(\mathscr{E})$ instead of $\phi(P_\theta : \theta \in \Theta)$. The quantity $\int \phi(d\mathscr{E})$ which was defined previously is just the total mass of the measure $\phi(\mathscr{E})$. Thus

$$\int d\phi(\mathscr{E}) = \int \phi(d\mathscr{E})$$

when $\phi \in \Phi$.

If the vectors $w_\theta : \theta \in \Theta$ are all $\geqslant 0$ then part (i) of corollary 7.4.4 applies, with $\mathscr{X}$ being the likelihood space $Z = [0, \infty[^\Theta$. This proves the first assertion of the next theorem.

Theorem 7.4.8 (Representation of $V(\mathscr{E})$). *Let p_θ, for each θ, denote the projection of the likelihood space $Z = [0, \infty[^\Theta$ onto its θth coordinate space $[0, \infty[$. Denote by Φ the vector lattice of real valued functions generated by these projections.*

Let $\mathscr{E} = (P_\theta : \theta \in \Theta)$ be an experiment with parameter set Θ. Then there is a unique vector lattice homomorphism $\phi \to \phi(\mathscr{E})$ from Φ to $V(\mathscr{E})$, which for each θ maps p_θ onto P_θ. This map has a unique extension as a vector lattice homomorphism $h \to h(\mathscr{E})$ from the vector lattice $\mathscr{H}(\mathscr{E})$ (of the non-negatively homogeneous measurable functions on Z such that $\int |h|(d\mathscr{E}) < \infty$) into $V(\mathscr{E})$, which satisfies the condition

$$\int h(\mathrm{d}\mathscr{E}) = \int \mathrm{d}h(\mathscr{E}); \qquad h \in \mathscr{H}(\mathscr{E}).$$

Remark. It is easily checked that the last homomorphism mentioned is onto $V(\mathscr{E})$. The actual construction of $h(\mathscr{E})$ is described in the proof below.

Proof. The existence of an extension to $\mathscr{H}(\mathscr{E})$ may be seen as follows. Let $h \in \mathscr{H}(\mathscr{E})$. Then $h(z)$ depends on z via $(z_\theta : \theta \in \Theta_0)$ where Θ_0 is countable. Let $\sigma \geqslant 0$ be any σ-finite measure dominating $(P_\theta : \theta \in \Theta_0)$ and put $f_\theta = \mathrm{d}P_\theta/\mathrm{d}\sigma$ when $\theta \in \Theta$. Then the measure $h(\mathscr{E}) = h(f_\theta : \theta \in \Theta)\sigma$ does not depend on the choice of the densities $f_\theta : \theta \in \Theta$, nor on the measure σ nor on the choice of the set Θ_0. It can be checked that the map $h \to h(\mathscr{E})$ is a vector lattice homomorphism and that $\int \mathrm{d}h(\mathscr{E}) = \int h(\mathrm{d}\mathscr{E})$ for all $h \in \mathscr{H}(\mathscr{E})$. Finally assume that the map $h \to \mu_h$ is any map from $\mathscr{H}(\mathscr{E})$ to $L(\mathscr{E})$ having the desired properties. Let $\mathscr{H}_0$ denote the set of functions h in $\mathscr{H}(\mathscr{E})$ such that $\mu_h = h(\mathscr{E})$. Then $\Phi \subseteqq \mathscr{H}_0 \subseteqq \mathscr{H}(\mathscr{E})$. Let $h \in \mathscr{H}(\mathscr{E})$ and suppose $h(z)$ depends on z via $(z_\theta : \theta \in \Theta_0)$ where Θ_0 is countable. Let c be a prior distribution on Θ which is supported by Θ_0 and which assigns positive mass c_θ to each point θ in Θ_0. Then there is a sequence $(\phi_1, \phi_2, \ldots)$ in Φ such that $\phi_n(z)$, for $n = 1, 2, \ldots$, depends on z via $(z_\theta : \theta \in \Theta_0)$, and which converges to h in the sense that $\int |\phi_n(z) - h(z)| S(\mathrm{d}z) \to 0$ where S is the measure induced from $\sum_\theta c_\theta P_\theta$ by the likelihood map $x \to ([\mathrm{d}P_\theta/\mathrm{d}(\sum_\theta c_\theta P_\theta)]_x : \theta \in \Theta)$. Then $\|\mu_h - \phi_n(\mathscr{E})\| = \|\mu_h - \mu_{\phi_n}\| = \|\mu_{(h-\phi_n)}\| = \|\mu_{|h-\phi_n|}\| = \int \mathrm{d}\mu_{|h-\phi_n|} = \int |h - \phi_n|(\mathrm{d}\mathscr{E}) = \int |h - \phi_n| \, \mathrm{d}S \to 0$. Since this holds with μ_h replaced by $h(\mathscr{E})$, it follows that $\mu_h = h(\mathscr{E})$. □

Lebesgue decompositions of measures in $V(\mathscr{E})$ are easily described in terms of this homomorphism. Thus if $h_1 \in \mathscr{H}(\mathscr{E})$ and $0 \leqslant h_2 \in \mathscr{H}(\mathscr{E})$ then the $h_2(\mathscr{E})$-absolutely continuous part of $h_1(\mathscr{E})$ is $h_{1a}(\mathscr{E})$ where $h_{1a} = h_1 I_{]0,\infty[}(|h_2|)$.

Now consider two equivalent experiments $\mathscr{E} = (P_\theta : \theta \in \Theta)$ and $\mathscr{F} = (Q_\theta : \theta \in \Theta)$. Then, for any non-negative h in $\mathscr{H}$,

$$\|h(\mathscr{E})\| = \int h(\mathrm{d}\mathscr{E}) = \int h(\mathrm{d}\mathscr{F}) = \|h(\mathscr{F})\|.$$

It follows that $\mathscr{H}(\mathscr{E}) = \mathscr{H}(\mathscr{F})$ and that the correspondence $h(\mathscr{E}) \to h(\mathscr{F})$, $h \in \mathscr{H}(\mathscr{E})$ is a well defined isomorphism from the L-space $V(\mathscr{E})$ onto the L-space $V(\mathscr{F})$.

We know already from the previous sections that experiments $\mathscr{E} = (P_\theta : \theta \in \Theta)$ and $\mathscr{F} = (Q_\theta : \theta \in \Theta)$ are equivalent if and only if their minimal forms are isomorphic by an isomorphism which, for each θ, carries P_θ onto Q_θ. The point to be made here is that this isomorphism makes $h(\mathscr{E})$ correspond to $h(\mathscr{F})$ for each $h \in \mathscr{H}(\mathscr{E}) = \mathscr{H}(\mathscr{F})$. If $\mathscr{E} = (P_\theta : \theta \in \Theta)$ is majorized by a not necessarily σ-finite measure μ then $h(\mathscr{E}) = h(\mathrm{d}P_\theta/\mathrm{d}\mu : \theta \in \Theta)\,\mathrm{d}\mu$.

The functional $\Gamma_{\mathscr{E}} : \phi \to \int \phi(\mathrm{d}\mathscr{E})$ on the vector lattice Φ generated by the projections on Z is called *the conical measure* of $\mathscr{E}$. The functions in Φ are precisely those functions on Z which are restrictions to Z of functions belonging to the vector lattice generated by the projections on $\mathbb{R}^\Theta$. Since the integral $\int h(\mathrm{d}\mathscr{E})$, for a non-negatively homogeneous function h on $\mathbb{R}^\Theta$, is determined by the restriction of h to Z, it follows that we can consider the conical measures as being defined for functions ϕ belonging to the vector lattice generated by the projections on $\mathbb{R}^\Theta$. This view has the advantage that it leads to more explicit expressions for deficiencies. It may then be checked that most of the expressions given for deficiencies in section 6.3 can be easily expressed in terms of conical measures. Thus, e.g. $\mathscr{E} \geqslant \mathscr{F}$ if and only if $\Gamma_{\mathscr{E}}(\psi) \geqslant \Gamma_{\mathscr{F}}(\psi) : \psi \in \Psi$.

The reader may consult Choquet's *Lectures on Analysis* (1969) for a thorough discussion of conical measures.

Another advantage of working in $\mathbb{R}^\Theta$ rather than in Z is that we need not change the domain when we encounter measures permitting negative masses. However, in this section, unless otherwise stated, we will assume that the functions ϕ in Φ are functions on Z.

The conical measure $\Gamma_{\mathscr{E}}$ of $\mathscr{E}$ is a non-negative linear functional on Φ having the additional property that $\Gamma_{\mathscr{E}}(p_\theta) \equiv_\theta 1$. Here p_θ, for each θ, is the projection of $Z = [0, \infty[^\Theta$ onto its θth coordinate space $[0, \infty]$. If Γ is a linear functional on Φ then the function from Θ to $\mathbb{R}$ which to each θ assigns the number $\Gamma(p_\theta)$ is called *the resultant* of Γ. Thus the conical measure of an experiment is non-negative and has the constant function 1 as its resultant.

Besides these obvious requirements no other requirements are needed in order to ensure that a linear functional on Φ is the conical measure of some experiment.

Theorem 7.4.9 (Conical measures of experiments). *A linear functional on Φ is the conical measure of an experiment if and only if it is non-negative and has the constant function* 1 *(on Θ) as its resultant.*

Proof. The 'only if' was noted above. Assume that Γ is a non-negative linear functional on Φ and that Γ has resultant 1. We shall exhibit an experiment $\mathscr{E}$ such that $\Gamma = \Gamma_{\mathscr{E}}$. Define a pseudo-norm $\| \ \|_\Gamma$ on Φ by putting $\|\phi\|_\Gamma = \Gamma(|\phi|)$

when $\phi \in \Phi$. Write $\phi_1 \sim \phi_2$ if $\|\phi_1 - \phi_2\|_\Gamma = 0$. Thus '$\sim$' is the equivalence relation on Φ induced by the pseudo-norm $\| \ \|_\Gamma$. Let ϕ_1, ϕ_2, ϕ_1' and ϕ_2' be functions in Φ and assume $\phi_i \sim \phi_i'$; $i = 1, 2$. Let a_1 and a_2 be constants. Then $a_1\phi_1 + a_2\phi_2 \sim a_1\phi_1' + a_2\phi_2'$, $\phi_1 \vee \phi_2 \sim \phi_1' \vee \phi_2'$ and $\phi_1 \wedge \phi_2 \sim \phi_1' \wedge \phi_2'$. Let $\tilde{\phi}$ denote the equivalence class associated with ϕ and let $\tilde{\Phi}$ denote the collection of equivalence classes in Φ.

Let a_1 and a_2 be constants and let ϕ_1 and ϕ_2 be functions in the vector lattice Φ. Then we may define the following in $\tilde{\Phi}$ unambiguously.

$$a_1\tilde{\phi}_1 + a_2\tilde{\phi}_2 = \widetilde{(a_1\phi_1 + a_2\phi_2)},$$

$$\tilde{\phi}_1 \vee \tilde{\phi}_2 = \widetilde{(\phi_1 \vee \phi_2)}, \qquad \tilde{\phi}_1 \wedge \tilde{\phi}_2 = \widetilde{(\phi_1 \wedge \phi_2)},$$

$$\tilde{\phi}_1 \leqslant \tilde{\phi}_2 \quad \text{if} \quad \tilde{\phi}_1 \vee \tilde{\phi}_2 = \tilde{\phi}_2 \quad \text{and} \quad \|\tilde{\phi}\|_\Gamma = \|\phi\|_\Gamma.$$

It can then be checked that $\tilde{\Phi}$ with these definitions of linear structure, order structure and norm becomes an L-normed space with norm $\| \ \|_\Gamma$. Furthermore $\tilde{\phi}_1 \vee \tilde{\phi}_2$ and $\tilde{\phi}_1 \wedge \tilde{\phi}_2$ as defined above are actually $\max(\tilde{\phi}_1, \tilde{\phi}_2)$ and $\min(\tilde{\phi}_1, \tilde{\phi}_2)$ as associated with this order structure. We also have $\tilde{p}_\theta \geqslant 0$ and $\|\tilde{p}_\theta\|_\Gamma \equiv 1$ when $\theta \in \Theta$. If we now complete $\tilde{\Phi}$ in the usual way then we obtain an *L-space* L containing $\tilde{\Phi}$ as a dense sub-vector lattice. By theorem 7.4.7 (or by the last section in chapter 5) this L-space is isomorphic to the L-space $L(\mathscr{E})$ of an experiment $\mathscr{E} = (P_\theta : \theta \in \Theta)$ by an isomorphism which, for each θ, maps $\tilde{p}_\theta$ onto P_θ.

If $a^{(1)}, \ldots, a^{(r)}$ are finite measures with finite supports on Θ then

$$\begin{aligned}\Gamma_{\mathscr{E}}([\psi(p_\theta : \theta \in \Theta | a^{(1)}, \ldots, a^{(r)})]^\pm) &= \|\psi(P_\theta : \theta \in \Theta | a^{(1)}, \ldots, a^{(r)})^\pm\| \\ &= \|\psi(\tilde{p}_\theta : \theta \in \Theta | a^{(1)}, \ldots, a^{(r)})^\pm\|_\Gamma \\ &= \|\psi(p_\theta : \theta \in \Theta | a^{(1)}, \ldots, a^{(r)})^\pm\|_\Gamma \\ &= \Gamma(\psi(p_\theta : \theta \in \Theta | a^{(1)}, \ldots, a^{(r)})^\pm).\end{aligned}$$

Hence $\Gamma_{\mathscr{E}}(\phi) = \Gamma(\phi)$ for all functions ϕ on Φ. It follows that $\Gamma = \Gamma_{\mathscr{E}}$ is the conical measure of the experiment $\mathscr{E}$. □

An important consequence of the last result is LeCam's consistency theorem for statistical experiments. The problem may be phrased as follows. Let $\mathscr{U}$ be a family of non-empty subsets of Θ and assume that $\mathscr{E}^U$, for each U in $\mathscr{U}$, is an experiment with parameter set U. Assume that $\mathscr{U}$ is directed for inclusion. The problem is: what conditions are needed in order to ensure that there is an experiment $\mathscr{E}$ such that the restriction $\mathscr{E}|U$, for each set U in $\mathscr{U}$, is equivalent to $\mathscr{E}^U$?

The problem is very similar to the problem on the existence of random processes with given finite dimensional marginal distributions. In this situation Kolmogorov's consistency criterion tells us that the consistency requirements which are obviously necessary are also sufficient. LeCam's consistency theorem does the same thing for experiments.

Theorem 7.4.10 (The consistency requirements for restrictions). *Let $\mathcal{U}$ be a class of non-empty subsets of Θ which is directed for inclusion. Assume that to each set U in $\mathcal{U}$ there corresponds an experiment $\mathscr{E}^U$ with parameter set U. Then there exists an experiment $\mathscr{E}$ with parameter set Θ such that $\mathscr{E}|U \sim \mathscr{E}^U$ for all U in $\mathcal{U}$ if and only if $\mathscr{E}^{U_1}|U_1 \cap U_2 \sim \mathscr{E}^{U_2}|U_1 \cap U_2$ when $U_1, U_2 \in \mathcal{U}$ and $U_1 \cap U_2 = \emptyset$. If these conditions are satisfied and if $\bigcup \{U : U \in \mathcal{U}\} = \Theta$ then $\mathscr{E}$ is determined up to equivalence.*

Remark. A class $\mathcal{U}$ of sets is called *directed for inclusion* if to any two sets U_1 and U_2 in $\mathcal{U}$ there corresponds a $U_3 \in \mathcal{U}$ such that $U_3 \supseteqq U_1 \cup U_2$.

Proof.

1. Assume $\mathscr{E}|U \sim \mathscr{E}^U$; $U \in \mathcal{U}$. Then $\mathscr{E}^{U_1}|U_1 \cap U_2 \sim (\mathscr{E}|U_1)|(U_1 \cap U_2) = \mathscr{E}|U_1 \cap U_2 \sim \mathscr{E}^{U_2}|U_1 \cap U_2$ when $U_1, U_2 \in \mathcal{U}$ and $U_1 \cap U_2 \neq \emptyset$.
2. Assume $\mathscr{E}^{U_1}|U_1 \cap U_2 \sim \mathscr{E}^{U_2}|U_1 \cap U_2$ when $U_1, U_2 \in \mathcal{U}$ and $U_1 \cap U_2 \neq \emptyset$. We may assume without loss of generality that $\bigcup \{U : U \in \mathcal{U}\} = \Theta$.

Let $\psi \in \Psi$. Then $\psi(z)$ depends on $z = (z_\theta : \theta \in \Theta)$ via $(z_\theta : \theta \in F)$ for some finite subset F of Θ. The directedness of $\mathcal{U}$ implies that $F \subseteqq U$ for some $U \in \mathcal{U}$. Thus $\psi(z)$ depends on z via $(z_\theta : \theta \in U)$. If we put $\Gamma(\psi) = \Gamma_{\mathscr{E}^U}(\psi)$ then by the consistency assumption, this number does not depend on how U was chosen. It is now easily checked that Γ is in fact a conical measure with resultant 1. Hence, by theorem 7.4.9, $\Gamma = \Gamma_{\mathscr{E}}$ for some experiment $\mathscr{E}$. The same argument shows that the conical measure of any experiment $\mathscr{E}$ possessing the required properties is unique. Thus $\mathscr{E}$ is determined up to equivalence. □

Another consequence of the simple characterization of conical measures of experiments is the anticipated characterization of extremal experiments. Note first that the conical measure of a $\pi = (\pi_1, \pi_2, \ldots)$ mixture $\sum \pi_i \mathscr{E}_i$ of experiments $\mathscr{E}_1, \mathscr{E}_2, \ldots$ is $\sum \pi_i \Gamma_i$ where Γ_i, for each i, is the conical measure of $\mathscr{E}_i$. This follows from the fact that the functional $\mathscr{E} \to \int d\phi(\mathscr{E})$ is affine in $\mathscr{E}$ whenever $\phi \in \phi$. It follows that the set of conical measures of experiments constitutes a convex subset of the linear space of linear functionals on Φ. Combining this with the characterizations of extremal experiments given in section 7.3, we obtain external criteria for extremality.

Theorem 7.4.11 (The 'extreme' experiments). *The following conditions are equivalent for an experiment $\mathscr{E} = (P_\theta : \theta \in \Theta)$.*

(i) *$\mathscr{E}$ is extremal.*

(ii) *$\mathscr{E}$ is not equivalent to any mixture $\frac{1}{2}\mathscr{F} + \frac{1}{2}\mathscr{G}$ of non-equivalent experiments $\mathscr{E}$ and $\mathscr{F}$.*

(iii) *If $\mathscr{E}$ is equivalent to a π-mixture $\sum_i \pi_i \mathscr{E}_i$ of experiments $\mathscr{E}_1, \mathscr{E}_2, \ldots$ then $\mathscr{E}_i \sim \mathscr{E}$ whenever $\pi_i > 0$.*

(iv) *$\Gamma_{\mathscr{E}}$ is an extreme point within the convex set of conical measures of experiments.*

Proof. Note first that the equivalence of conditions (ii) and (iv) follows from theorem 7.4.9. The equivalence of (ii) and (iii) for mixtures with finite support follows directly from the equivalence of the corresponding characterizations of extreme points of convex sets. It follows that if (ii) holds and $\mathscr{E} \sim \sum_{i=1}^{\infty} \pi_i \mathscr{E}_i = \pi_1 \mathscr{E}_1 + (1 - \pi_1) \sum_{i=2}^{\infty} (\pi_i / 1 - \pi_1) \mathscr{E}_i$ then $\mathscr{E} \sim \mathscr{E}_1$ when $\pi_1 \in]0, 1[$ and clearly also when $\pi_1 = 1$. Thus (ii)$\Leftrightarrow$(iii). It remains to show that (i) is equivalent to conditions (ii), (iii) and (iv). Assume that $\mathscr{E}$ is extremal and that $\mathscr{E} \sim \frac{1}{2}\mathscr{F} + \frac{1}{2}\mathscr{G}$. Then, by the characterization of isomorphisms between the 'V-spaces' of equivalent experiments we conclude that $\int h(\mathrm{d}\mathscr{E}) = \int h(\mathrm{d}(\frac{1}{2}\mathscr{F} + \frac{1}{2}\mathscr{G})) = \frac{1}{2}\int h(\mathrm{d}\mathscr{F}) + \frac{1}{2}\int h(\mathrm{d}\mathscr{G})$ when $0 \leqslant h \in \mathscr{H}(\mathscr{E})$. Hence $\|h(\mathscr{E})\| = \int \mathrm{d}|h(\mathscr{E})| = \int \mathrm{d}|h|(\mathscr{E}) = \int |h|(\mathrm{d}\mathscr{E}) = \frac{1}{2}\|h(\mathscr{F})\| + \frac{1}{2}\|h(\mathscr{G})\|$. In particular $\|h_1(\mathscr{E}) - h_2(\mathscr{E})\| = \int |h_1 - h_2|(\mathrm{d}\mathscr{E}) = \frac{1}{2}\|h_1(\mathscr{F}) - h_2(\mathscr{F})\| + \frac{1}{2}\|h_1(\mathscr{G}) - h_2(\mathscr{G})\|$ when $h_1, h_2 \in \mathscr{H}(\mathscr{E})$. Thus without ambiguity we can define a functional ρ on $V(\mathscr{E})$ by

$$\rho(h(\mathscr{E})) = \tfrac{1}{4}\|h(\mathscr{F})\| + \tfrac{3}{4}\|h(\mathscr{G})\|; \qquad h \in \mathscr{H}(\mathscr{E}).$$

It is then easily checked that ρ is an L-norm for $V(\mathscr{E})$ such that $\rho(P_\theta) \equiv_\theta 1$. By part (iii) of theorem 7.3.15 there is only one such norm. Thus $\rho(h(\mathscr{E})) = \|h(\mathscr{E})\| = \frac{1}{2}\|h(\mathscr{F})\| + \frac{1}{2}\|h(\mathscr{G})\|$; $h \in \mathscr{H}(\mathscr{E})$. Combining the two expressions for $\rho(h(\mathscr{E}))$ we find that $\|h(\mathscr{E})\| = \|h(\mathscr{F})\| = \|h(\mathscr{G})\|$; $h \in \mathscr{H}(\mathscr{E})$. Hence $\int h(\mathrm{d}\mathscr{E}) = \int h(\mathrm{d}\mathscr{F}) = \int h(\mathrm{d}\mathscr{G})$ when $0 \leqslant h \in \mathscr{H}(\mathscr{E})$ so that $\mathscr{E} \sim \mathscr{F} \sim \mathscr{G}$.

Finally assume that the equivalent conditions (ii), (iii) and (iv) are satisfied. Suppose $\| \ \| = \frac{1}{2}\rho_1 + \frac{1}{2}\rho_2$ where ρ_1 and ρ_2 are two L-norms for $V(\mathscr{E})$ such that $\rho_1(P_\theta) \equiv_\theta \rho_2(P_\theta) \equiv_\theta 1$. Put $\Gamma_i(\phi) = \rho_i(\phi^+(\mathscr{E})) - \rho_i(\phi^-(\mathscr{E}))$; $i = 1, 2$ when $\phi \in \Phi$. Then Γ_1 and Γ_2 are conical measures on Φ, both having resultant 1. If $\phi \in \Phi$ then $\Gamma_{\mathscr{E}}(\phi) = \|\phi^+(\mathscr{E})\| - \|\phi^-(\mathscr{E})\| = \frac{1}{2}\Gamma_1(\phi) + \frac{1}{2}\Gamma_2(\phi)$. By theorem 7.4.9 there are experiments $\mathscr{E}_1$ and $\mathscr{E}_2$ such that $\Gamma_{\mathscr{E}_i} = \Gamma_i$; $i = 1, 2$. Then $\Gamma_{\mathscr{E}} = \frac{1}{2}\Gamma_{\mathscr{E}_1} + \frac{1}{2}\Gamma_{\mathscr{E}_2}$ is the conical measure of $\frac{1}{2}\mathscr{E}_1 + \frac{1}{2}\mathscr{E}_2$ so that $\mathscr{E} \sim \frac{1}{2}\mathscr{E}_1 + \frac{1}{2}\mathscr{E}_2$. Hence $\mathscr{E}_1 \sim \mathscr{E}_2 \sim \mathscr{E}$ so that $\Gamma_1 = \Gamma_2 = \Gamma_{\mathscr{E}}$. It follows that $\rho_1(\phi(\mathscr{E})) = \rho_1(|\phi|(\mathscr{E})) = \Gamma_1(|\phi|) = \Gamma(|\phi|) = \rho_2(\phi(\mathscr{E}))$; $\phi \in \Phi$. Hence $\rho_1 = \rho_2$. Thus $\mathscr{E}$ is extremal by corollary 7.3.16. □

We have seen that the set of conical measures of experiments may be characterized as those non-negative linear functions on Φ which have resultant 1. If we equip the set Φ' of linear functionals on Φ with its Φ-topology then by theorem 7.4.9, the set of conical measures of experiments becomes a closed subset of Φ'. Furthermore $|\Gamma_{\mathscr{E}}(\phi)| = |\int \phi(\mathrm{d}\mathscr{E})| \leqslant \int |\phi|(\mathrm{d}\mathscr{E}) \leqslant \#(F) \sup\{|\phi(z)|: z \in Z, \sum_F z_\theta = 1\}$ if ϕ depends on z via $(z_\theta : \theta \in F)$. It follows then, from

Tychonoff's theorem, that the set of conical measures of experiments is a compact subset of Φ'.

Theorem 7.4.12 (The weak experiment topology). *The set of types of experiment with parameter set Θ is a compact Hausdorff space for the weakest topology which makes $\Gamma_{\mathscr{E}}(\phi)$ a continuous function of the type of $\mathscr{E}$ for each function $\phi \in \Phi$.*

This topology coincides with the weakest topology which, for each finite subset F of Θ, makes the restriction map $\mathscr{E} \to \mathscr{E}|F$ continuous as a map to the set of types of experiments with parameter set F, when the latter set is topologized by Δ.

Furthermore, the class of types of experiments $\mathscr{E}$ such that the set $\mathscr{X}$, as well as the set $\{P_\theta : \theta \in \Theta\}$, is finite, constitutes a dense subset.

Proof. The first two statements were argued above and the last statement is an immediate consequence of the corresponding statements for standard experiments. □

The topology described in the previous theorem is called *the weak topology* (for types of experiments with parameter set Θ). If $(\mathscr{E}_n)$ is a net of experiments such that the type of $\mathscr{E}_n$ converges weakly to $\mathscr{E}$ then we may write this $\mathscr{E}_n \to_w \mathscr{E}$. As long as we limit ourselves to experiments belonging to a well defined set of experiments then the notion of weak limits of experiments is derived from a well defined topology. However this topology is not Hausdorff unless each type possesses at most one representative in this set of experiments. We shall, however, here and elsewhere, usually express ourselves as if the weak topology, as well as the Δ_k-topologies, was a topology on a 'space' consisting of all experiments.

As the notation indicates, the weak topology is weaker than the Δ-topology since $\mathscr{E}_n \to_w \mathscr{E}$ whenever $\mathscr{E}_n \to_\Delta \mathscr{E}$.

If Θ is countable then the weak topology is metrizable. A possible metric for types of experiments is then $(\mathscr{E}, \mathscr{F}) \to \sum_{m=1}^{\infty} 2^{-m}\Delta(\mathscr{E}|\Theta_m, \mathscr{F}|\Theta_m)$ where the sets $\Theta_1, \Theta_2, \ldots$ are all finite and $\bigcup_{m=1}^{\infty} \Theta_m = \Theta$.

Here is an example showing that the set of types of experiments is not separable, and therefore not compact, when Θ is uncountable.

Example 7.4.13. Let us take $\mathscr{X} = \{0, 1\}$ as the sample space of all our experiments. Choose a point $\theta_0 \in \Theta$ and let $\mathscr{U}$ denote the class of finite subsets of Θ which contain θ_0. For each $\theta \in \Theta$ and each $F \in \mathscr{U}$ define a probability distribution P_θ^F by putting

$$P_\theta^F = I_F(\theta)\delta_0 + (I - I_F(\theta))\delta_1.$$

Put $\mathscr{E}^F = (P_\theta^F : \theta \in \Theta)$ when $F \in \mathscr{U}$.

Let $\mathscr{U}_0$ be a subclass of $\mathscr{U}$, $F_0 = \{\theta_0, \theta_1\}$ where $\theta_1 \notin \bigcup \{F : F \in \mathscr{U}_0\}$ and

$F_0 \subseteqq G \in \mathscr{U}$. Then

$$1 \geqslant \Delta(\mathscr{E}^F|G, \mathscr{E}^{F_0}|G) = \Delta_2(\mathscr{E}^F|G, \mathscr{E}^{F_0}|G) \geqslant \Delta_2(\mathscr{E}^F|F_0, \mathscr{E}^{F_0}|F_0) = 1.$$

It follows that $\{\mathscr{E}^F : F \in \mathscr{U}_0\}$ is not dense in $\{\mathscr{E}^F : F \in \mathscr{U}\}$ when $\bigcup\{F : F \subset \mathscr{U}_0\} \neq \Theta$. If Θ is uncountable then the last set inequality holds whenever $\mathscr{U}_0$ is countable. Thus $\{\mathscr{E}^F : F \in \mathscr{U}\}$ is not separable for the weak topology when Θ is uncountable.

By the Krein–Milman theorem, (see e.g. Kelley & Namioka, 1961) a compact convex set in a locally convex linear topological Hausdorff space is the closed convex hull of its set of extreme points. This applies to the set of conical measures of experiments.

Theorem 7.4.14. *The set of types of finite mixtures of extremal experiments is weakly dense within the set of all types of experiments with parameter set Θ.*

Remark. By Torgersen (1977) separable experiments may be expressed as possibly infinite mixtures of boundedly complete experiments.

Proof. Consider the set of conical measures of experiments as a convex subset of the set Φ' of all linear functionals on Φ equipped with the smallest topology making $\Gamma(\phi)$ continuous in $\Gamma \in \Phi'$ for each $\phi \in \Phi$. Then the set of conical measures of experiments becomes a compact convex subset of Φ' and the Krein–Milman theorem applies. □

A sequence (generalized or not) of experiments converges weakly if and only if the minimum Bayes risk converges for each finite decision problem and each prior distribution with finite support. It follows from compactness that it suffices to consider testing problems. This and some other convergence criteria are collected in the following theorem.

Theorem 7.4.15 (Criteria for weak convergence). *The following conditions are equivalent for a generalized sequence (net) ($\mathscr{E}_n = (P_{\theta,n} : \theta \in \Theta)$) of experiments:*

(i) *$\mathscr{E}_n$ converges weakly.*
(ii) *$(\Gamma_{\mathscr{E}_n}(\phi))$ converges for all $\phi \in \Phi$.*
(iii) *$(\Gamma_{\mathscr{E}_n}(\psi))$ converges for all $\psi \in \Psi_2$.*
(iv) *$(\|\sum_\theta a_\theta P_{\theta,n}\|)$ converges for all finite measures a on Θ with finite support.*
(v) *$(\|\bigvee_\theta \lambda_\theta P_{\theta,n}\|)$ converges for all prior probability distributions λ on Θ with finite support.*
(vi) *$(H(t|\mathscr{E}_n) = \int \prod_\theta \mathrm{d}P_{\theta,n}^{t_\theta})$ converges for all prior probability distributions t on Θ with finite support.*

If $\mathscr{E}_n \to_w \mathscr{E} = (P_\theta : \theta \in \Theta)$ then $\mathscr{E}$ is determined by each of the following functionals:

$\Gamma_{\mathscr{E}}(\phi) \quad = \lim_n \Gamma_{\mathscr{E}_n}(\phi); \phi \in \Phi.$

$\Gamma_{\mathscr{E}}(\psi) \quad = \lim_n \Gamma_{\mathscr{E}_n}(\psi); \psi \in \Psi_2.$

$\|\sum_\theta a_\theta P_\theta\| = \lim_n \|\sum_\theta a_\theta P_{\theta,n}\|$ *for each finite measure* a *on* Θ *with finite support.*

$\|\bigvee_\theta \lambda_\theta P_\theta\| = \lim_n \|\bigvee_\theta \lambda_\theta P_{\theta,n}\|$ *for each prior probability measure* λ *on* Θ *with finite support.*

$H(t|\mathscr{E}) \quad = \lim_n H(t|\mathscr{E}_n)$ *for each prior probability distribution* t *on* Θ *with finite support.*

Proof. We may assume without loss of generality that the parameter set Θ is finite and then these statements are all proved in the section on standard experiments. □

Along with a compact Hausdorff space goes the unique uniformity possessing this topology. If for each finite non-empty subset F of Θ we let Δ^F denote the pseudo-distance $(\mathscr{E}, \mathscr{F}) \to \Delta(\mathscr{E}|F, \mathscr{F}|F)$, then this uniformity is generated by the pseudo-distances Δ^F.

Although the Δ-topology is not compact when Θ is infinite it is at least complete.

Theorem 7.4.16 (Completeness of Δ_k). *The set of types of experiments with parameter set* Θ *is complete for the metrics* $\Delta_2, \Delta_3, \ldots, \Delta_\infty = \Delta$.

Proof. Let $2 \leqslant k \leqslant \infty$ and suppose $\Delta_k(\mathscr{E}_n, \mathscr{E}_m) \to 0$ as $n, m \to \infty$. Let $(\mathscr{E}_{n'})$ be a sub-net of $(\mathscr{E}_n)$ which converges weakly to an experiment $\mathscr{E}$ and let $\varepsilon > 0$. Then there is an N_ε such that $\Delta_k(\mathscr{E}_{n'}, \mathscr{E}_{m'}) \leqslant \varepsilon$ whenever $n', m' \geqslant N_\varepsilon$. Letting $m' \to \infty$, we see by the sublinear function expressions for Δ_k that $\Delta_k(\mathscr{E}_{n'}, \mathscr{E}) \leqslant \varepsilon$ when $n' \geqslant N_\varepsilon$. Thus $\Delta_k(\mathscr{E}_{n'}, \mathscr{E}) \to 0$ so that $\Delta_k(\mathscr{E}_n, \mathscr{E}) \to 0$. □

In general it is much easier to establish weak convergence than strong convergence, i.e. Δ-convergence. It is therefore important to have criteria which make it possible to conclude Δ-convergence from weak convergence.

Although the assumptions may appear rather strong, the following convergence criterion due to Lindae (1972) is useful.

Theorem 7.4.17 (Convergence to compact experiments). *Let* $(\mathscr{E}_n = (\mathscr{X}_n, \mathscr{A}_n; P_{\theta,n} : \theta \in \Theta))$ *be a net of experiments which converges weakly to an experiment* $\mathscr{E} = (P_\theta : \theta \in \Theta)$ *such that* $(P_\theta : \theta \in \Theta)$ *is a conditionally compact subset of* $L(\mathscr{E})$. *Then* $\mathscr{E}_n \to_\Delta \mathscr{E}$ *if and only if* $\|P_{\theta_1,n} - P_{\theta_2,n}\| \to \|P_{\theta_1} - P_{\theta_2}\|$ *uniformly in* $\theta_1, \theta_2 \in \Theta$.

Remark. The last condition is automatically satisfied if $\Delta_2(\mathscr{E}_n, \mathscr{E}) \to 0$. Thus $\Delta(\mathscr{E}_n, \mathscr{E}) \to 0$ if and only if $\Delta_k(\mathscr{E}_n, \mathscr{E}) \to 0$ for some $k \geqslant 2$.

Proof. The 'only if' follows from, say, theorem 7.2.9. Assume that the condition is satisfied and choose an $\varepsilon > 0$ and an n_ε such that $|\,\|P_{\theta_1,n} - P_{\theta_2,n}\| - \|P_{\theta_1} - P_{\theta_2}\|\,| \leqslant \varepsilon$ for all $\theta_1, \theta_2 \in \Theta$ when $n \geqslant n_\varepsilon$. The compactness assumption ensures total boundedness and thus the existence of a finite subset Θ_0 of Θ such that to each $\theta \in \Theta$ there corresponds a $\theta' \in \Theta_0$ such that $\|P_\theta - P_{\theta'}\| \leqslant \varepsilon$. Weak convergence implies that n_ε may be chosen so that $\Delta(\mathscr{E}_n|\Theta_0, \mathscr{E}|\Theta_0) \leqslant \varepsilon$ when $n \geqslant n_\varepsilon$. Let $\psi \in \Psi$ satisfy the boundedness condition $\sum_\theta [\psi(-e^\theta) \vee \psi(e^\theta)] \leqslant 1$. If $n \geqslant n_\varepsilon$ then we obtain

$$\begin{aligned}
&\left|\int \psi(\mathrm{d}P_{\theta,n} : \theta \in \Theta) - \int \psi(\mathrm{d}P_\theta : \theta \in \Theta)\right| \\
&\quad \leqslant \left|\int \psi(\mathrm{d}P_{\theta,n} : \theta \in \Theta) - \int \psi(\mathrm{d}P_{\theta',n} : \theta \in \Theta)\right| \\
&\qquad + \left|\int \psi(\mathrm{d}P_{\theta',n} : \theta \in \Theta) - \int \psi(\mathrm{d}P_{\theta'} : \theta \in \Theta)\right| \\
&\qquad + \left|\int \psi(\mathrm{d}P_{\theta'} : \theta \in \Theta) - \int \psi(\mathrm{d}P_\theta : \theta \in \Theta)\right| \\
&\quad \leqslant \bigvee_\theta \|P_{\theta,n} - P_{\theta',n}\| + \varepsilon + \bigvee_\theta \|P_\theta - P_{\theta'}\| \\
&\quad \leqslant \bigvee_\theta \big|\|P_{\theta,n} - P_{\theta',n}\| - \|P_\theta - P_{\theta'}\|\big| + \varepsilon + 2\bigvee_\theta \|P_\theta - P_{\theta'}\| \\
&\quad \leqslant \varepsilon + \varepsilon + 2\varepsilon = 4\varepsilon.
\end{aligned}$$

Hence $\Delta(\mathscr{E}_n, \mathscr{E}) \leqslant 4\varepsilon$ when $n \geqslant n_\varepsilon$. □

The prime examples of conditionally compact experiments are those experiments $(P_\theta : \theta \in \Theta)$ such that the set $\{P_\theta : \theta \in \Theta\}$ is finite. By total boundedness any conditionally compact experiment, up to an approximation with any desired degree of accuracy, is of this form. Furthermore strong limits of conditionally compact experiments are themselves conditionally compact. It follows readily that the collection of compact experiments is precisely the closure of the collection of experiments $(P_\theta : \theta \in \Theta)$ such that both the sample space and the set $\{P_\theta : \theta \in \Theta\}$ of possible distributions are finite.

Proposition 7.4.18 (Conditionally compact experiments). *If $\mathscr{E} = (\mathscr{X}, \mathscr{A}; P_\theta : \theta \in \Theta)$ is conditionally compact and if $\varepsilon > 0$ then there is an experiment $\hat{\mathscr{E}} = (\mathscr{X}, \mathscr{A}; \hat{P}_\theta : \theta \in \Theta)$ such that $\hat{P}_\theta$, for each θ, is contained in the closed vector lattice generated by $\{P_\theta : \theta \in \Theta\}$ and has the following further properties:*

(i) $\|\hat{P}_\theta - P_\theta\| < \varepsilon; \theta \in \Theta$;
(ii) *the set $\{\hat{P}_\theta : \theta \in \Theta\}$ is finite;*
(iii) *a finite subalgebra $\hat{\mathscr{A}}$ of $\mathscr{A}$ is sufficient for $\hat{\mathscr{E}}$.*

In particular it follows that for any experiment $\mathscr{F}$ having parameter set Θ and for any integer $k \geqslant \log_2 \# \hat{\mathscr{A}}$

$$\delta(\mathscr{F}, \mathscr{E}) \leqslant \delta_k(\mathscr{F}, \mathscr{E}) + 2\varepsilon.$$

Proof. By total boundedness there is a function g on Θ with finite range $g[\Theta] = \{\theta_1, \ldots, \theta_r\}$ and having the further property that $\|P_\theta - P_{g(\theta)}\| < \varepsilon/2$; $\theta \in \Theta$.

Put $f_i = \mathrm{d}P_{\theta_i}/\mathrm{d}\mu$ where $\mu = P_{\theta_1} + \cdots + P_{\theta_r}$. Then there are non-negative measurable simple functions $\hat{f}_1, \ldots, \hat{f}_r$ on $\mathscr{X}$ such that $\int |f_i - \hat{f}_i| \,\mathrm{d}\mu < \varepsilon/2$; $i = 1, \ldots, r$. The function $\hat{f}_i$, for each i, may be chosen as measurable functions of $f = (f_1, \ldots, f_r)$ such that $\int \hat{f}_i \,\mathrm{d}\mu = 1$. Putting $\hat{P}_\theta = \hat{f}_{g(\theta)}\mu$; $\theta \in \Theta$ and letting $\hat{\mathscr{A}}$ be the algebra induced by $\hat{f}_1, \ldots, \hat{f}_r$ we obtain an experiment having the desired properties. The last inequality is deduced as follows:

$$\begin{aligned} \delta(\mathscr{F}, \mathscr{E}) &\leqslant \delta(\mathscr{F}, \hat{\mathscr{E}}) + \delta(\hat{\mathscr{E}}, \mathscr{E}) \\ &= \delta_k(\mathscr{F}, \hat{\mathscr{E}}) + \delta(\hat{\mathscr{E}}, \mathscr{E}) \\ &\leqslant \delta_k(\mathscr{F}, \mathscr{E}) + 2\Delta(\hat{\mathscr{E}}, \mathscr{E}) \\ &\leqslant \delta_k(\mathscr{F}, \mathscr{E}) + 2\varepsilon. \end{aligned}$$

□

The functionals used to describe the various convergence criteria in theorem 7.4.15 are all given directly by restrictions to finite parameter sets. Thus it is not clear that e.g. minimax risks for weakly converging experiments converge. In fact we shall see in a moment that minimax risk need not converge – even for bounded loss functions. If the experiments are Δ-converging then the randomization criterion implies directly that minimax risk converges provided the loss function is bounded. Even in the case of Δ-convergence, however, convergence need not take place for unbounded loss functions.

The problem is analogous to the problem of convergence of expectations of unbounded functions of variables which converge in distribution. If $P_1, P_2, \ldots$ and P are probability measures on the same measurable space $(\mathscr{Y}, \mathscr{B})$ and if v is measurable and bounded from below on $\mathscr{Y}$, then $\liminf_n \int v \,\mathrm{d}P_n \geqslant \int v \,\mathrm{d}P$ whenever $\int (v \wedge N) \,\mathrm{d}P_n \to \int (v \wedge N) \,\mathrm{d}P$; $N = 1, 2, \ldots$ As shown by Hajek (1970) in the case of normal translation experiments, and later by LeCam in its full generality, a similar liminf statement holds for minimax risks in weakly convergent experiments.

Before giving the general statement, let us consider a few facts on minimax risk. Let $(\mathscr{E}, (T, \mathscr{S}), L)$ be a decision problem where:

$\mathscr{E} = (P_\theta : \theta \in \Theta)$ is an experiment,
$(T, \mathscr{S})$ is a decision space,
L is a loss function on $\Theta \times T$.

Here as elsewhere it is assumed that $L_\theta(t)$ is measurable and bounded from below in t for each θ. If ρ is a decision rule then its risk function is the function $\theta \to P_\theta \rho L_\theta$ on Θ. The maximin risk is then the quantity $\sup_\theta P_\theta \rho L_\theta$. The minimax risk is the quantity $\inf_\rho \sup_\theta \mathrm{P}_\theta \rho L_\theta$ where ρ varies through the set of all generalized decision rules from $\mathscr{E}$ to $(T, \mathscr{S})$. Here we shall use the notation $m(\mathscr{E}, L)$ or just $m(\mathscr{E})$ to denote this minimax risk. The first comment which should be made is that neither the various suprema of risks nor the infimum of these suprema need be obtainable. Of course we encountered this slight inconsistency of terminology in chapter 3.

Note next that if $\mathscr{E}$ is coherent, if T is a compact Hausdorff space equipped with its class $\mathscr{S}$ of Borel subsets and if L is continuous from below in t, then we may restrict our attention to proper decision rules, i.e. Markov kernels. This follows from the fact that to any bounded additive set function μ on $\mathscr{S}$ there corresponds a unique regular Borel measure yielding the same integrals for continuous functions as μ.

If Θ is finite then by theorem 4.5.17 we may restrict our attention to uniformly finitely supported decision functions. The decision problem which was specified in example 4.5.19 provides an example where $\sup_\theta P_\theta \rho L_\theta \geqslant \frac{1}{2}$ for any decision rules of the Markov kernel variety, while on the other hand, $m(\mathscr{E}, L) = 0$.

Let λ vary within the set of prior distributions on Θ with finite support. The lower semicontinuity in ρ of the Bayes risks $\int P_\theta \rho L_\theta \lambda(\mathrm{d}\theta)$ together with the compactness of the set of possible transitions show that

$$\sup_\lambda \inf_\rho \int (P_\theta \rho L_\theta)\lambda(\mathrm{d}\theta) = \inf_\rho \sup_\lambda \int (P_\theta \rho L_\theta)\lambda(\mathrm{d}\theta) = m(\mathscr{E}, L).$$

It follows that

$$m(\mathscr{E}, L) = \sup_F m(\mathscr{E}|F, L|F)$$

where F runs through the class of all finite non-empty subsets of Θ.

Furthermore, we may express $m(\mathscr{E}, L)$ in terms of bounded loss functions by

$$m(\mathscr{E}, L) = \sup_N m(\mathscr{E}, L \wedge N).$$

This last fact also follows from the general minimax theory of chapter 3, since the pay-off function $D(N, \rho) = \int P_\theta \rho(L_\theta \wedge N)\lambda(\mathrm{d}\theta)$ is concave convex in (N, ρ). (We are not claiming that this expression is concave in N, only that the set $\{N\}$ of strategies for player I is concave. This follows from the trivial inequality

$$(1 - \kappa)D(N_1, \rho) + \kappa D(N_2, \rho) \leqslant D(N_1 \vee N_2, \rho).)$$

Thus

$$\sup_N m(\mathscr{E}, L \wedge N) = \sup_N \inf_\rho \sup_\lambda \int P_\theta \rho(L_\theta \wedge N)\lambda(\mathrm{d}\theta)$$

$$= \sup_N \sup_\lambda \inf_\rho \int \cdots$$

$$= \sup_{\lambda} \sup_{N} \inf_{\rho} \int \cdots$$

$$= \sup_{\lambda} \inf_{\rho} \sup_{N} \int \cdots$$

$$= \sup_{\lambda} \inf_{\rho} \int P_\theta \rho L_\theta \lambda(d\theta)$$

$$= m(\mathscr{E}, L).$$

We are now ready for the convergence theorem for minimax risk.

Theorem 7.4.19 (Asymptotic behaviour of minimax risk). *Let L be a loss function on the decision space $(T; \mathscr{S})$.*

Consider a net $(\mathscr{E}_n)$ of experiments which converges weakly to an experiment $\mathscr{E}$. Let $(T, \mathscr{S})$ be a decision space equipped with a loss function L. Then

$$\liminf_n m(\mathscr{E}_n, L) \geqslant m(\mathscr{E}, L)$$

where equality holds if $\mathscr{E}_n \to_\Delta \mathscr{E}$ and $\sup_\theta \|L_\theta\| < \infty$ for all θ. Equality also holds if $\mathscr{E}_n \geqslant \mathscr{E}$; $n = 1, 2, \ldots$

Proof. If Θ is finite and L is bounded then we have Δ-convergence and the statement follows from the randomization criterion. This yields

$$\liminf M(\mathscr{E}_n, L) \geqslant \liminf_n m(\mathscr{E}_n | F, L \wedge N) | F)$$

$$= m(\mathscr{E} | F, (L \wedge N) | F) \uparrow m(\mathscr{E}, L)$$

as $N \uparrow \infty$ and $F \uparrow \Theta$ through the class of finite non-empty subsets of Θ. □

Example 7.4.20 (Non convergence of minimax risk for convergent experiments). Let $\mathscr{E} = (\mathscr{X}, \mathscr{A}; P_\theta : \theta \in \Theta)$ be totally informative and let Q be a probability distribution on $\mathscr{A}$. For each finite non-empty subset F of Θ let $\mathscr{E}^F$ be an experiment of the form $\mathscr{E}^F = (P_{\theta,F} : \theta \in \Theta)$ where $P_{\theta,F} = P_\theta$ when $\theta \in F$ while $P_{\theta,F} = Q$ when $\theta \notin F$. Then $\mathscr{E}^F \to_w \mathscr{E}$ as $F \uparrow \Theta$. Put $T = \Theta$ and let $L_\theta(t) = 0$ or $= 1$ as $\theta = t$ or $\theta \neq t$. Let $\mathscr{S}$ be any σ-algebra making the one-point sets measurable. Then $m(\mathscr{E}^F, L) = 1$ whenever Θ is infinite, while $m(\mathscr{E}, L) = 0$.

A somewhat less pathological example is obtained by putting $\mathscr{E}_\sigma = (N(\theta, \sigma) : \theta \in \mathbb{R})$ when $\sigma > 0$ and $\mathscr{E}_0 = (\delta_\theta : \theta \in \mathbb{R})$. Here the sample space is the real line equipped with its Borel class. Then $\mathscr{E}_\sigma \to_w \mathscr{E}_0$ as $\sigma \downarrow 0$.

Consider the problem of estimating θ with loss function L defined by putting $L_\theta(t) = e^{|t-\theta|^3} - 1$ when $t, \theta \in \mathbb{R}$. The convexity of L_θ together with the invariance of $\mathscr{E}$ for $(\mathbb{R}, +)$ implies that we may restrict our attention to estimators of the form $x \to x + c$ where c is a constant. It follows easily that $m(\mathscr{E}_\sigma, L) = \infty$ when $\sigma > 0$ while, again, $m(\mathscr{E}, L) = 0$.

The reader is referred to the 1983 paper by P. W. Millar for a treatment of asymptotic minimax theory.

Besides the topologies mentioned so far, there are numerous other more or less 'natural' topologies for experiments. Some topologies are directly derived from limit concepts for probability measures. Thus we may say that the net $(\mathscr{E}_n = (\mathscr{X}, \mathscr{A}; P_{\theta,n} : \theta \in \Theta))$ of experiments having the same sample space $(\mathscr{X}, \mathscr{A})$ converges to the experiment $\mathscr{E} = (\mathscr{X}, \mathscr{A}; P_\theta : \theta \in \Theta)$ provided $P_{\theta,n} \to P_\theta$ for all θ for some notion of convergence of probability measures on $(\mathscr{X}, \mathscr{A})$.

If the topology for probability measures is the topology of the total variation norm then we obtain a topology for convergence of experiments which is much stronger than the weak topology. If, in addition, we insist on uniform convergence in θ, then the corresponding topology for experiments is even stronger than the Δ-topology. On the other hand, if $\mathscr{X}$ is a metric space equipped with its Borel (=Baire) sets $\mathscr{A}$ and if $P_{\theta,n}(f) \to P_\theta(f)$ for all θ and for all bounded continuous functions f, then this does not in any way guarantee that $\mathscr{E}_n \to_w \mathscr{E}$.

Example 7.4.21. Take the interval $\mathscr{X} = [0,1]$ as the sample space and the two-point set $\Theta = \{0,1\}$ as the parameter set of our experiments. Put $\mathscr{E}_n = (P_{\theta,n} : \theta \in \Theta)$, $n = 1,2,\ldots$ and $\mathscr{E} = (P_\theta : \theta \in \Theta)$ where $P_{0,n}$ is the one-point distribution in $1/n$, $P_{1,n} = \delta_0$; $n = 1,2,\ldots$ and $P_0 = P_1 = \delta_0$. Then $P_{\theta,n} \to P_\theta$ weakly for all $\theta \in \Theta$. In this case, however, $\mathscr{E}_n \sim \mathscr{M}_a$; $n = 1,2,\ldots$ while $\mathscr{E} \sim \mathscr{M}_i$ so that $\Delta(\mathscr{E}_n, \mathscr{E}) \equiv_n 1$.

Thus we may easily lose information by passing to the weak limit. It is then a very significant fact, as shown by LeCam, that we can never gain information in this way.

Theorem 7.4.22 (Weak distribution convergence never increases information). *Let $(\mathscr{E}_n = (P_{\theta,n} : \theta \in \Theta))$ and $\mathscr{E} = (P_\theta : \theta \in \Theta)$ be, respectively, a net of experiments with sample space $(\mathscr{X}, \mathscr{A})$ and an experiment with the same sample space.*

Let there also be given a sub-vector lattice M_0 of the vector lattice $M(\mathscr{X}, \mathscr{A})$ of bounded measurable functions. Assume that M_0 contains the constant functions and that the σ-algebra $\mathscr{A}$ is the smallest σ-algebra which makes all functions in M_0 measurable.

Suppose $P_{\theta,n}(v) \to P_\theta(v)$ whenever $\theta \in \Theta$ and $v \in M_0$. Then $\mathscr{E} \leqslant \mathscr{F}$ for any experiment $\mathscr{F}$ which is a point of accumulation for the net $(\mathscr{E}_n)$ for the weak topology for types of experiments.

Remark. If we strengthen the assumption on convergence to uniform weak convergence, i.e. that $\sup_\theta |P_{\theta,n}(v) - P_\theta(v)| \to 0$ when $v \in M_0$, and if in addition we require that $\mathscr{E}$ is conditionally compact, then by Müller (1979) we may conclude that $\delta(\mathscr{E}_n, \mathscr{E}) \to 0$.

This may be argued as follows. Note first that by proposition 7.4.18 it suffices to show that $\delta_k(\mathscr{E}_n, \mathscr{E}) \to 0$ for integers k. Letting $\Theta_n = (\theta_1^{(n)}, \ldots, \theta_k^{(n)})$; $n = 1, 2, \ldots$ be a sequence of k-point subsets of Θ we find easily that $\delta_k(\mathscr{E}_n|\Theta_n, \mathscr{E}|\Theta_n) \to 0$.

Indeed if $\|P_{\theta_i^{(n)}} - \pi_i\| \to 0$; $i = 1, \ldots, k$ then, by the theorem

$$\delta(\mathscr{E}_n|\Theta_n, \mathscr{E}|\Theta_n) \leqslant \delta(\mathscr{E}_n|\Theta_n, (\pi_1, \ldots, \pi_k)) + \delta((\pi_1, \ldots, \pi_k), \mathscr{E}|\Theta_n) \to 0.$$

In general this follows from conditional compactness. Thus $\delta(\mathscr{E}_n|\Theta', \mathscr{E}|\Theta') \to 0$ uniformly for finite sets of bounded cardinality. (In a more general setting this is proved in LeCam, 1974.) Choose a finite subalgebra $\hat{\mathscr{A}}$ of $\mathscr{A}$ containing 2^N sets and let the superscript $\hat{}$ denote restriction to $\hat{\mathscr{A}}$. Then of course the smaller quantity $\delta(\mathscr{E}_n|\Theta', \hat{\mathscr{E}}|\Theta')$ also tends to 0 in the same uniform sense. Hence, by theorem 6.2.14 (a consequence of Helly's intersection theorem, theorem 2.3.11) the deficiency $\delta_k(\mathscr{E}_n, \hat{\mathscr{E}}) \to 0$ for each integer k. On the other hand conditional compactness implies that $\Delta(\hat{\mathscr{E}}, \mathscr{E}) \downarrow 0$ as $\hat{\mathscr{A}} \uparrow \mathscr{A}$. Hence, since $\delta_k(\mathscr{E}_n, \mathscr{E}) \leqslant \delta_k(\mathscr{E}_n, \hat{\mathscr{E}}) + \Delta(\hat{\mathscr{E}}, \mathscr{E})$, we find that $\delta_k(\mathscr{E}_n, \mathscr{E}) \to 0$ for each integer k.

Another argument for this may be based on the inequality given in complement 32 in chapter 10.

Proof of the theorem. Let $a^1, a^2, \ldots, a^k$ be non-negative distributions on Θ having supports contained in the finite subset F of Θ. Put $\psi(z) = \bigvee_{i=1}^k \sum_\theta (a_\theta^i z_\theta)$; $z \in Z = [0, \infty[^\Theta$. Then $\int \psi(\mathrm{d}\mathscr{E}) = \|\bigvee_{i=1}^k \mu_i\|$ where $\mu_i = \sum_\theta a_\theta^i P_\theta$; $i = 1, \ldots, k$. Hence $\int \psi(\mathrm{d}\mathscr{E}) = \sup \sum_{i=1}^k \mu_i(\delta_i)$ where the sup is over all k-tuples $(\delta_1, \ldots, \delta_k)$ of non-negative measurable functions such that $\delta_1 + \cdots + \delta_k = 1$. (This is a particular case of formula (4) of theorem 5.4.1.) The assumptions on M_0 imply that this vector lattice is dense in $M(\mathscr{X}, \mathscr{A})$ for the $L_\infty(\sum_i \mu_i)$-topology on $L_1(\sum_i \mu_i)$. It follows that we may assume that the functions $\delta_1, \ldots, \delta_k$ are taken from M_0. Then $\int \psi(\mathrm{d}\mathscr{E}) = \sup \lim_n \sum_i \mu_{i,n}(\delta_i)$ where

$$\mu_{i,n} = \sum_\theta a_\theta^i P_{\theta,n}.$$

Thus

$$\int \psi(\mathrm{d}\mathscr{E}) \leqslant \liminf_n \sup \sum_i \mu_{i,n}(\delta_i) = \liminf_n \int \psi(\mathrm{d}\mathscr{E}_n).$$

It follows that $\int \psi(\mathrm{d}\mathscr{E}) \leqslant \liminf_n \int \psi(\mathrm{d}\mathscr{E}_n)$ for any $\psi \in \Psi$. Hence $\int \psi(\mathrm{d}\mathscr{E}) \leqslant \int \psi(\mathrm{d}\mathscr{F})$ for each $\psi \in \Psi$ and each 'point' $\mathscr{F}$ of accumulation of $(\mathscr{E}_n)$ for the weak topology. □

In terms of convergence of distributions of statistics, the theorem may be phrased as follows.

Corollary 7.4.23. *Let $(\mathscr{E}_n = (P_{\theta,n} : \theta \in \Theta))$ be a net of experiments which converges weakly to an experiment $\mathscr{E} = (P_\theta : \theta \in \Theta)$. Let (Y_n) be a net of statistics* *such that Y_n, for each n, is a transition from $L(\mathscr{E}_n)$ to the L-space* ba$(\mathscr{Y}, \mathscr{B})$ *of bounded additive set functions on the σ-algebra $\mathscr{B}$ of subsets of the set $\mathscr{Y}$. Also let Y be a transition from $\mathscr{E}$ to $(\mathscr{Y}, \mathscr{B})$.*

Assume that we are given a sub-vector lattice M_0 of the vector lattice $M(\mathscr{Y}, \mathscr{B})$ of the bounded measurable functions. We shall assume that M_0 contains the constants and that it induces $\mathscr{B}$.

Finally assume that (Y_n) converges to Y in 'distribution' in the sense that

$$P_{\theta,n} Y_n v \to P_\theta Y v$$

when $\theta \in \Theta$ and $v \in M_0$. Then the experiment $(P_\theta Y : \theta \in \Theta)$ induced from $\mathscr{E}$ by Y is at most as informative as any weak limit experiment of $(P_{\theta,n} Y_n : \theta \in \Theta)$.

Remark 1. If $\mathscr{Y}$ is a topological space and $\mathscr{B}$ is the class of Baire sets, i.e. the σ-algebra induced by the real valued continuous functions, then we may take M_0 as the vector lattice of bounded continuous functions. The convergence $P_{\theta,n} Y_n v \to P_\theta Y v$; $v \in M_0$ is then convergence in distribution, as it is usually defined in this case.

Remark 2. Using abstract experiments we may, as done by LeCam, replace the vector lattice M_0 by any M-normed space Γ containing a unit. The compactness of the set of transitions from a given L-space to Γ^* as well as the compactness of Γ-convergence of probability 'distributions' in Γ^* then lead to simpler statements than we are giving here.

Proof of the corollary. The corollary follows from the theorem by using the compactness of the weak experiment topology. □

Example 7.4.24 (Distinguished statistics. The Hajek–LeCam convolution criterion). With the set-up and the assumptions of corollary 7.4.23 and following LeCam we will say that the net (Y_n) of statistics is *distinguished* if $\mathscr{E} \sim (P_\theta Y : \theta \in \Theta)$. If (Y_n) is distinguished then it is also asymptotically sufficient in the sense that $\delta((P_{\theta,n} Y_n : \theta \in F), (P_{\theta,n} : \theta \in F)) \to 0$ for any finite non-empty subset F of Θ. The justification of the term distinguished is as follows. Suppose (Y_n) is distinguished and that (Z_n) is another net of statistics such that Z_n, for each n, is a transition from $\mathscr{E}_n$ to ba$(\mathscr{Y}, \mathscr{B})$. Assume also that Z is a statistic from $\mathscr{E}$ to ba$(\mathscr{Y}, \mathscr{B})$ and that (Z_n) converges to Z in distribution. Then Y is at least as informative as Z, i.e. $(P_\theta Y : \theta \in \Theta) \geqslant (P_\theta Z : \theta \in \Theta)$. Thus the experiment $(P_\theta Z : \theta \in \Theta)$ may be obtained from $(P_\theta Y : \theta \in \Theta)$ by a transition M, i.e. $P_\theta Z \equiv_\theta P_\theta Y M$. If the experiments $(P_\theta Z : \theta \in \Theta)$ and $(P_\theta Y : \theta \in \Theta)$ are invariant with respect to some group in the sense explained in section 6.5 and if the appropriate conditions are satisfied, then M may be chosen invariant. In the particular case where $(P_\theta Z : \theta \in \Theta)$ and $(P_\theta Y : \theta \in \Theta)$ are absolutely continuous translation experiments on $\mathbb{R}^k$ then, by example 6.5.13, this implies that $P_0 Z = [P_0 Y] * F$ for some distribution F on $\mathbb{R}^k$. This expresses in a sense that the concentration of the estimators Y_n around θ is asymptotically, for each θ, at least as great as the concentration of Z_n around θ.

Except for the various criteria which are needed in order to ensure that important sequences of statistics are distinguished, this is the main content of the Hajek–LeCam convolution criterion.

This is about as far as we will go in the general study of convergence of experiments. The reader is referred to LeCam's works and to Strasser (1985) for further information on this topic.

7.5 Likelihood experiments

The theory of sufficiency tells us that likelihood functions are sufficient, provided regularity conditions are satisfied. Thus we could try to represent experiments by experiments having the set of possible likelihood functions as sample space. Now usually we do not distinguish proportional likelihood functions. Identifying likelihood functions which are equal up to a multiplicative constant we arrive at a projective space, i.e. a space of rays. The precise definitions are as follows.

We shall call the set $Z = [0, \infty[^{\Theta}$ of finite non-negative functions on Θ *the likelihood set* (for the parameter set Θ). Introduce an equivalence relation in the likelihood set by declaring that two likelihood functions z' and z'' in Z are *equivalent* (*positively proportional*) if $z'_\theta \equiv_\theta kz''_\theta$ for some positive constant k. Let $\mathscr{B}$ denote the σ-algebra of Borel subsets of $[0, \infty[$ and let $\mathscr{B}^{\Theta}$ denote the product σ-algebra. Thus $\mathscr{B}^{\Theta}$ is the smallest σ-algebra on Z which makes all the projections into the coordinate spaces measurable. Let $\mathscr{C}$ denote the σ-algebra of sets in $\mathscr{B}^{\Theta}$ which are saturated for the described equivalence relation. Thus a subset C of Z belongs to $\mathscr{C}$ if and only if $C \in \mathscr{B}^{\Theta}$ and $z'' \in C$ whenever $z' \in C$ and z' is positively proportional to z''. A convenient generator for $\mathscr{C}$ is the algebra of finite dimensional homogeneous measurable sets. In this way we obtain a measurable space $(Z, \mathscr{C})$ which in this chapter we shall call the *likelihood space* (with parameter set Θ). We used this term earlier without reference to a particular measurability structure besides the product σ-algebra $\mathscr{B}^{\Theta}$. Note that a function γ from $(Z, \mathscr{C})$ to some measurable space $(U, \mathscr{D})$ is measurable if and only if γ is measurable as a map from $(Z, \mathscr{B}^{\Theta})$ to $(U, \mathscr{D})$ and $\gamma(z') \sim \gamma(z'') \operatorname{mod}(\mathscr{D})$ whenever z' and z'' are positively proportional. (Here we write $u_1 \sim u_2 \operatorname{mod}(\mathscr{D})$ if $I_D(u_1) = I_D(u_2)$ for all $D \in \mathscr{D}$.)

Convenient generators for the σ-algebra $\mathscr{C}$ are the subclasses $\mathscr{C}'$, $\mathscr{C}''$ and $\mathscr{C}'''$, where $\mathscr{C}'$ is the class of all subsets C of Z which are of the form

$$C = \{z : (z_{\theta_1}/z_{\theta_2}) I_{]0,\infty[}(z_{\theta_2}) \in B\} = \{z : z_{\theta_1}/z_{\theta_2} \in B \text{ or } z_{\theta_2} = 0 \in B\}$$

where $\theta_1, \theta_2 \in \Theta$ and B is a Borel subset of $[0, \infty[$. $\mathscr{C}''$ is defined as $\mathscr{C}'$ except that z_{θ_2} in the expression for the set C is replaced by $z_{\theta_1} + z_{\theta_2}$. $\mathscr{C}'''$ is the algebra of subsets C of Z which are of the form

$$C = \{z : ((z_\theta/z_F)I_{]0,\infty[}(z_F) : \theta \in F) \in B_F\}$$

in which $\theta \in F \subseteqq \Theta$, the set F is finite, $z_F = \sum_F z_\theta$ and B_F is a Borel subset of $\mathbb{R}^F$. This last class is the generator consisting of the finite dimensional sets mentioned above.

Consider any experiment $\mathscr{E} = (\mathscr{X}, \mathscr{A}; P_\theta : \theta \in \Theta)$ and a finite measure λ on $\mathscr{A}$. Choose a non-negative finite measure μ which dominates λ. Also, for each $\theta \in \Theta$, choose a non-negative and finite version of $dP_\theta/d\mu$. (As usual $dP_\theta/d\mu$ denotes a Radon–Nikodym derivative of the μ-absolutely continuous part of P_θ with respect to μ.) Then the map which to each $x \in \mathscr{X}$ associates the likelihood function

$$\theta \to [dP_\theta/d\mu]_x$$

is measurable as a map from $(\mathscr{X}, \mathscr{A})$ to $(Z, \mathscr{C})$. The essential thing to note now is that the measure on $\mathscr{C}$ induced from λ by this map does not depend on the choice of the dominating measure μ, nor on how the Radon–Nikodym derivatives were specified. This follows from the factorization $dP_\theta/d\nu = dP_\theta/d\mu \cdot d\mu/d\nu$ a.e. μ which holds whenever the non-negative σ-finite measure ν dominates μ.

Let $\tilde{\lambda}$ be the measure on $\mathscr{C}$ induced from λ by this map, i.e. $\tilde{\lambda}(C) = \lambda(\{(dP_{.}/d\mu) \in C\})$ when $C \in \mathscr{C}$. Since $\tilde{\lambda}$ does not depend on μ as long as $\mu \gg \lambda$, it follows that the map $\lambda \to \tilde{\lambda}$ is a transition. The probability measure P_{θ_0}, for each $\theta_0 \in \Theta$, is mapped into the probability measure $\tilde{P}_{\theta_0}$ which assigns mass $P_{\theta_0}(\{(dP/dP_{\theta_0}) \in C\})$ to the set $C \in \mathscr{C}$. Let us denote the experiment $(\tilde{P}_\theta : \theta \in \Theta)$ by $\tilde{\mathscr{E}}$. Thus the transition $\lambda \to \tilde{\lambda}$ transports $\mathscr{E}$ into another experiment $\tilde{\mathscr{E}}$ with sample space $(Z, \mathscr{C})$. We shall call $\tilde{\mathscr{E}}$ *the likelihood experiment of* $\mathscr{E}$. In particular consider an experiment $\mathscr{E} = (\mathscr{X}, \mathscr{A}; P_\theta : \theta \in \Theta)$ which is majorized by a non-negative measure μ. Let f_θ, for each θ, be a version of $dP_\theta/d\mu$ and let f denote the random function $x \to (f_\theta(x) : \theta \in \Theta)$ considered as a measurable map from $(\mathscr{X}, \mathscr{A})$ to $(Z, \mathscr{C})$. Then $\tilde{\mathscr{E}} = (P_\theta f^{-1} : \theta \in \Theta)$.

Let us collect a few of the basic properties of the transition $\lambda \to \tilde{\lambda}$ in a proposition.

Proposition 7.5.1. *Let $\mathscr{E} = (\mathscr{X}, \mathscr{A}; P_\theta : \theta \in \Theta)$ be an experiment with likelihood experiment $\tilde{\mathscr{E}} = (Z, \mathscr{C}; \tilde{P}_\theta : \theta \in \Theta)$. Then the map $\lambda \to \tilde{\lambda}$ is an L-space isomorphism between the L-spaces $V(\mathscr{E})$ and $V(\tilde{\mathscr{E}})$.*

Furthermore $\widetilde{h(\mathscr{E})} = h(\tilde{\mathscr{E}})$ for any $h \in \mathscr{H}(\mathscr{E})$ and $d[h_1(\tilde{\mathscr{E}})]/d[h_2(\tilde{\mathscr{E}})] = (h_1/h_2)I_{]0,\infty[}(h_2)$ when $h_1, h_2 \in \mathscr{H}(\mathscr{E})$ and $h_2 \geqslant 0$.

Remark. We defined $\mathscr{H}(\mathscr{E})$ in the previous section as the vector lattice of non-negatively homogeneous functions h on Z such that $\int |h|(d\mathscr{E}) < \infty$.

Proof. Let κ be a prior probability distribution on Θ with countable support Θ_0 and assume $\kappa_{\theta_0} > 0$. Let p_θ as usual denote the projection of Z into its θth coordinate space. Put $\pi = \sum_\theta \kappa_\theta P_\theta$, $f_\theta = \mathrm{d}P_\theta/\mathrm{d}\pi$, $f_\kappa = \sum \kappa_\theta f_\theta$, $z_\kappa = \sum \kappa_\theta z_\theta$ and $p_\kappa = \sum \kappa_\theta p_\theta$. Then $\tilde{P}_{\theta_0} = \mathscr{L}(f|P_{\theta_0})$ and $\tilde{\pi} = \sum \kappa_\theta \tilde{P}_\theta$. In particular $\tilde{P}_{\theta_0}(\{z : z_{\theta_0} > 0\}) = 1$. Using $f_\kappa = 1$ a.e. π we find

$$\begin{aligned}\tilde{P}_{\theta_0}(C) = P_{\theta_0}(f \in C) &= \int I_C(f)\,\mathrm{d}P_{\theta_0}\\ &= \int I_C(f) f_{\theta_0} f_\kappa^{-1}\,\mathrm{d}\pi\\ &= \int_C z_{\theta_0} z_\kappa^{-1} \tilde{\pi}(\mathrm{d}z)\end{aligned}$$

when $C \in \mathscr{C}$ so that

$$\mathrm{d}\tilde{P}_{\theta_0}/\mathrm{d}\tilde{\pi} = p_{\theta_0} p_\kappa^{-1} I_{]0,\infty[}(p_\kappa).$$

Now assume that $h \in \mathscr{H}(\mathscr{E})$ and that $h(z)$ depends on z via $(z_\theta : \theta \in \Theta_0)$ and that $\kappa_\theta > 0$ whenever $\theta \in \Theta_0$. Then $h(\mathscr{E}) = h(f)\pi$ while $h(\tilde{\mathscr{E}}) = h(p_\theta p_\kappa^{-1} : \theta \in \Theta)\tilde{\pi}$. On the other hand

$$\begin{aligned}\widetilde{h(\mathscr{E})}(C) &= \int I_C(f) h(f)\,\mathrm{d}\pi\\ &= \int I_C(f) h(f_\theta f_\kappa^{-1} I_{]0,\infty[}(f_\kappa) : \theta \in \Theta)\,\mathrm{d}\pi\\ &= \int_C h(\mathrm{d}\tilde{P}_\theta/\mathrm{d}\tilde{\pi} : \theta \in \Theta)\,\mathrm{d}\tilde{\pi}\\ &= h(\tilde{\mathscr{E}})(C)\end{aligned}$$

when $C \in \mathscr{C}$ so that $\widetilde{h(\mathscr{E})} = h(\tilde{\mathscr{E}})$.

It follows that the restriction of the map $\lambda \to \tilde{\lambda}$ to $V(\mathscr{E})$ maps $h(\mathscr{E})$ in $V(\mathscr{E})$ onto $h(\tilde{\mathscr{E}})$ in $V(\tilde{\mathscr{E}})$. Thus this map is an isomorphism between the L-spaces $V(\mathscr{E})$ and $V(\tilde{\mathscr{E}})$.

The last expression in the proposition, for the Radon–Nikodym derivative, follows from the representation

$$h_i(\mathscr{E}) = h_i((p_\theta/p_\kappa) I_{]0,\infty[}(p_\kappa) : \theta \in \Theta)\tilde{\pi}$$

which is valid for $i = 1, 2$ whenever $h_1, h_2 \in \mathscr{H}(\mathscr{E})$ and $(h_1(z), h_2(z))$ depends on z via $(z_\theta : \kappa_\theta > 0)$. □

Example 7.5.2. The totally non-informative likelihood experiment is obtained by letting P_θ, for each θ, be the restriction to $\mathscr{C}$ of the one-point distribution in $e \in [0, \infty[^\Theta$ where $e_\theta \equiv_\theta 1$.

The totally informative likelihood experiment is obtained by letting P_θ, for each θ, be the restriction to $\mathscr{C}$ of the one-point distribution in e^θ where $e^\theta_{\theta'} = 1$ or $=0$ as $\theta' = \theta$ or $\theta' \neq \theta$. If Θ is uncountable then this experiment is not coherent and therefore not Σ-dominated.

We shall see shortly that likelihood experiments are always majorized.

Let us also give the construction of the likelihood experiment of an exponential experiment $\mathscr{E} = (\mathscr{X}, \mathscr{A}; P_\theta : \theta \in \Theta)$. Assume P_θ, for each θ, has density $hc_\theta \exp[\sum_{i=1}^r Q_i(\theta) T_i]$ with respect to a measure $\mu \geqslant 0$. Here $h \geqslant 0$ is a measurable function and the c_θ's are positive constants. The measure $\tilde{P}_\theta$ is then the measure on $\mathscr{C}$ induced from P_θ by the map which to each $x \in \mathscr{X}$ associates the function (point) $\theta \to c_\theta \exp[\sum_{i=1}^r Q_i(\theta) T_i(x)]$ in Z.

An experiment which is the likelihood experiment of some experiment is of course called a *likelihood experiment.*

Theorem 7.5.3 (Characterizations of likelihood experiments). *The following conditions are equivalent for an experiment $\mathscr{E} = (Z, \mathscr{C}; P_\theta : \theta \in \Theta)$ having the likelihood space $(Z, \mathscr{C})$ as its sample space:*

(i) *$\mathscr{E}$ is a likelihood experiment.*
(ii) $\mathscr{E} = \tilde{\mathscr{E}}$.
(iii) $dP_{\theta_1}/dP_{\theta_2} = [p_{\theta_1}/p_{\theta_2}] I_{]0,\infty[}(p_{\theta_2})$ *a.e. P_{θ_2} for all $\theta_1, \theta_2 \in \Theta$.*
(iv) $dP_{\theta_1}/d(P_{\theta_1} + P_{\theta_2}) = [p_{\theta_1}/(p_{\theta_1} + p_{\theta_2})] I_{]0,\infty[}(p_{\theta_1} + p_{\theta_2})$ *a.e. $P_{\theta_1} + P_{\theta_2}$ for all $\theta_1, \theta_2 \in \Theta$.*

Proof. If $\mathscr{E}$ is a likelihood experiment then (iii) and (iv) hold by the last statement of the previous proposition. The equivalence of conditions (iii) and (iv) follows by checking that the functions on the right, under the given assumption, behave as the Radon–Nikodym derivatives they are claimed to be. If $\mathscr{E} = (Z, \mathscr{C}; P_\theta : \theta \in \Theta)$ satisfies (iii) and $\tilde{\mathscr{E}} = (Z, \mathscr{C}; \tilde{P}_\theta : \theta \in \Theta)$ is its likelihood experiment, then $\tilde{P}_{\theta_0}(C) = P_{\theta_0}(((p_\theta/p_{\theta_0}) I_{]0,\infty[}(p_{\theta_0}) : \theta \in \Theta) \in C) = P_{\theta_0}(C)$ when $\theta_0 \in \Theta$ so that (ii), and hence (i), holds. □

The concept of a likelihood experiment may be viewed as a generalization to general parameter sets of the concept of a standard experiment. Indeed the behaviour of likelihood experiments is quite similar to the behaviour of standard experiments.

Theorem 7.5.4 (Representation by likelihood experiments). *Any likelihood experiment is majorized and is in minimal form. If $\tilde{\mathscr{E}} = (\tilde{P}_\theta : \theta \in \Theta)$, for any experiment $\mathscr{E} = (P_\theta : \theta \in \Theta)$, denotes the likelihood experiment of $\mathscr{E}$, then:*

(i) $\mathscr{E} \sim \tilde{\mathscr{E}}$.
(ii) $\mathscr{E} \sim \mathscr{F} \Leftrightarrow \tilde{\mathscr{E}} = \tilde{\mathscr{F}}$ *(in particular $\tilde{\tilde{\mathscr{E}}} = \tilde{\mathscr{E}}$).*

(iii) *assume that the experiment* $\mathscr{E} = (P_\theta : \theta \in \Theta)$ *is equivalent to the* $\pi = (\pi_1, \pi_2, \ldots)$*-mixture of the experiments* $\mathscr{E}_i = (P_{\theta,i} : \theta \in \Theta); i = 1, 2, \ldots$ *Thus we assume that* $\mathscr{E} \sim \sum_i \pi_i \mathscr{E}_i$. *Then* $\tilde{\mathscr{E}} \sim \sum_i \pi_i \tilde{\mathscr{E}}_i$ *and* $\tilde{P}_\theta \equiv_\theta \sum_i \pi_i \tilde{P}_{\theta,i}$.

(iv) *assume that the experiment* $\mathscr{E} = (P_\theta : \theta \in \Theta)$ *is equivalent to the product of the experiments* $\mathscr{E}_i = (P_{\theta,i} : \theta \in \Theta)$; $i = 1, \ldots, r$. *Thus we assume that* $\mathscr{E} \sim \prod_{i=1}^r \mathscr{E}_i$. *Then* $\tilde{\mathscr{E}} \sim \prod_{i=1}^r \tilde{\mathscr{E}}_i$ *and* $\tilde{P}_\theta$, *for each* θ, *is induced from* $\prod_{i=1}^r \tilde{P}_{\theta,i}$ *by the (pointwise multiplication) map* $(z^1, z^2, \ldots, z^r) \to z^1 z^2 \ldots z^r$ *from* $(Z, \mathscr{C}) \times \cdots \times (Z, \mathscr{C})$ *to* $(Z, \mathscr{C})$.

(v) *if* Θ_0 *is a non-empty subset of* Θ *and* p_{Θ_0} *denotes the projection of* $Z = [0, \infty[^\Theta$ *onto* $Z_{\Theta_0} = [0, \infty[^{\Theta_0}$ *then* $(\widetilde{\mathscr{E}|\Theta_0}) = (\tilde{P}_\theta p_{\Theta_0}^{-1} : \theta \in \Theta_0)$.

If Θ *is finite and* $\mathscr{E}$ *has standard experiment* $(S_\theta : \theta \in \Theta)$ *and likelihood experiment* $(\tilde{P}_\theta : \theta \in \Theta)$, *then* $\tilde{P}_\theta(C) \equiv_\theta S_\theta(C \cap K_\Theta)$ *for all homogeneous Borel sets* C, *while* $S_\theta(B) \equiv_\theta \tilde{P}_\theta(\{z : z/\sum_\theta z_\theta \in B\})$ *for all Borel sets* B.

Remark. It follows from (v) that LeCam's consistency theorem for experiments follows from the analogue of Kolmogorov's consistency theorem for 'projective spaces' $(Z, \mathscr{C})$.

Proof. The transition $\lambda \to \tilde{\lambda}$, for each θ, maps P_θ into $\tilde{P}_\theta$. Thus by the isomorphism criteria for equivalence, the experiments $\mathscr{E}$ and $\tilde{\mathscr{E}}$ are equivalent. Hence $\mathscr{E} \sim \mathscr{F}$ if and only if $\tilde{\mathscr{E}} \sim \tilde{\mathscr{F}}$. Thus, in order to prove (ii), it suffices to show that $\mathscr{E} = \mathscr{F}$ when $\mathscr{E} = (P_\theta : \theta \in \Theta)$ and $\mathscr{F} = (Q_\theta : \theta \in \Theta)$ are equivalent likelihood experiments. Let g be a bounded measurable function. Put $h = p_\theta g$. Then $h \in \mathscr{H}(\mathscr{E})$ so that, by equivalence, $\int g \, dP_\theta = \int h(d\mathscr{E}) = \int h(d\mathscr{F}) = \int g \, dQ_\theta$. Hence $P_\theta \equiv_\theta Q_\theta$, i.e. $\mathscr{E} = \mathscr{F}$.

Since any experiment is equivalent to a majorized experiment and majorization is preserved by the map $\lambda \to \tilde{\lambda}$, it follows that likelihood experiments are majorized. Furthermore, the proof of (ii) shows that if $\mathscr{E} = (P_\theta : \theta \in \Theta)$ is a likelihood experiment and if g is a bounded measurable function on $(Z, \mathscr{C})$, then gP_θ, for each θ, is of the form $h(\mathscr{E})$ for $h = p_\theta g$. Thus, since $h(\mathscr{E}) \in V(\mathscr{E})$ for all $h \in \mathscr{H}(\mathscr{E})$, $L(\mathscr{E}) = V(\mathscr{E})$ so that $\mathscr{E}$ is in minimal form.

Assume that $\mathscr{E}, \mathscr{E}_1, \mathscr{E}_2, \ldots$ are as in (iii). We know from section 5.3 that mixtures respect equivalence, so that $\tilde{\mathscr{E}} \sim \sum \pi_i \tilde{\mathscr{E}}_i$. Let g and $h = p_\theta g$ be as above. Then

$$\begin{aligned}\int g \, d\tilde{P}_\theta &= \int h(d\tilde{\mathscr{E}}) = \int h(d\mathscr{E}) \\ &= \sum_i \pi_i \int h(d\mathscr{E}_i) = \sum_i \pi_i \int h(d\tilde{\mathscr{E}}_i) \\ &= \sum_i \pi_i \int g \, d\tilde{P}_{\theta,i} = \int g \, d \sum_i \pi_i \tilde{P}_{\theta,i}.\end{aligned}$$

Hence $\tilde{P}_\theta \equiv_\theta \sum \pi_i \tilde{P}_{\theta,i}$. Let $\mathscr{E}$ and $\mathscr{E}_1, \mathscr{E}_2, \ldots, \mathscr{E}_r$ be as in (iv). Since products also respect equivalence, we may assume without loss of generality that these experiments are likelihood experiments. (iv) now follows from the definition of likelihood experiments and for the product rule for Radon–Nikodym derivatives.

To prove (v) we may again assume without loss of generality that $\mathscr{E}$ is a likelihood experiment. Put $(\mathscr{E}|\widetilde{\Theta}_0) = (Q_\theta : \theta \in \Theta_0)$ and let $\theta \in \Theta_0$. Then

$$\begin{aligned} Q_\theta(D) &= P_{\theta_0}(((p_\theta/p_{\Theta_0})I_{]0,\infty[}(p_{\Theta_0}) : \theta \in \Theta_0) \in D) \\ &= P_\theta((p_\theta : \theta \in \Theta_0) \in D) = P_\theta p_{\Theta_0}^{-1}(D) \end{aligned}$$

whenever D is a homogeneous measurable subset of $Z_{\theta_0} = [0, \infty[^{\Theta_0}$.

The last statements on standard experiments and likelihood experiments are easily proved by checking the definitions. □

7.6 References

Bahadur, R. R. 1954. Sufficiency and statistical decision functions. *Ann. Math. Statist.* **25**, pp. 423–62.

— 1955. A characterization of sufficiency. *Ann. Math. Statist.* **26**, pp. 286–93.

Blackwell, D. 1951. Comparison of experiments. *Proc. Second Berkeley Symp. Math. Statist. and Prob.* pp. 93–102.

— 1953. Equivalent comparisons of experiments. *Ann. Math. Statist.* **24**, pp. 265–72.

Burkholder, B. L. 1961. Sufficiency in the undominated case. *Ann. Math. Statist.* **32**, pp. 1191–1200.

Choquet, G. 1969. *Lectures on Analysis*, Vol. I–III. W. A. Benjamin, Reading, Mass.

Dudley, R. M. 1989. *Real analysis and probability.* Wadsworth & Brooks/Cole, Pacific Grove, California.

Hajek, J. 1970. A characterization of limiting distributions of regular estimates. *Z. Wahrscheinlichkeitstheorie und verw. Geb.* **14**, pp. 323–30.

Halmos, P. & Savage, L. J. 1949. Application of the Radon–Nikodym theorem to the theory of sufficient statistics. *Ann. Math. Statist.* **20**, pp. 225–41.

Hansen, O. H. & Torgersen, E. 1974. Comparison of linear normal experiments. *Ann. Statist.* **2**, pp. 367–73.

Hasegawa, M. & Perlman, M. D. 1974. On the existence of a minimal sufficient subfield. *Ann. Statist.* **2**, pp. 1049–55.

Heyer, H. 1982. *Theory of statistical experiments.* Springer–Verlag, New York.

Huber, P. J. 1981. *Robust statistics.* Wiley, New York.

Kelley, J. L. & Namioka, I. 1961. *Linear topological spaces.* Van Nostrand, Princeton.

LeCam, L. 1964. Sufficiency and approximate sufficiency. *Ann. Math. Statist.* **35**, pp. 1419–55.

— 1974. *Notes on asymptotic methods in statistical decision theory.* Centre de recherches mathématiques, Univ. de Montreal.

— 1986. *Asymptotic methods in statistics.* Springer–Verlag, Berlin.

Lindae, D. 1972. Distributions of likelihood ratios and convergence of experiments. *Ph. D. thesis.* Univ. of Calif. Berkeley.

Mazur, S. & Ulam, S. 1932. Sur les transformations isométriques d'espace vectoriels normés. *C.R. Acad. Sci. Paris.* **194**, pp. 946–8.

Millar, P. W. 1983. The minimax principle in asymptotic statistical theory. *Lecture notes in statistics.* **976**, pp. 76–265. Springer–Verlag, Berlin.

Morse, N. & Sacksteder, R. 1966. Statistical isomorphism. *Ann. Math. Statist.* **37**, pp. 203–214.

Müller, D. W. 1979. The increase in risk due to inaccurate models. *Preprint*. Institut für angewandte mathematik. Univ. Heidelberg.
Phelps, R. R. 1966. *Lectures on Choquet's theorem.* Van Nostrand, Princeton.
Roberts, A. & Varberg, D. 1973. *Convex functions.* Academic Press, New York.
Sherman, S. 1951. On a theorem of Hardy, Littlewood, Polya and Blackwell. *Proc. Nat. Acad. Sci. U.S.A.* **37**, pp. 826–31.
Siebert, E. 1979. Statistical experiments and their conical measures. *Z. Wahrscheinlichkeitstheorie verw. Geb.* **45**, pp. 247–58.
Stein, S. 1951. Notes on the comparison of experiments (mimeographed). University of Chicago.
Strassen, V. 1965. The existence of probability measures with given marginals. *Ann. Math. Statist.* **36**, pp. 423–39.
Strasser, H. 1985. *Mathematical theory of statistics.* Walter de Gruyter, Berlin.
Torgersen, E. 1970. Comparison of experiments when the parameter space is finite. *Z. Wahrscheinlichkeitstheorie verw. Geb.* **16**, pp. 219–49.
— 1977. Mixtures and products of dominated experiments. *Ann. Statist.* **5**, pp. 44–64.
— 1981. Measures on information based on comparison with total information and with total ignorance. *Ann. Statist.* **9**, pp. 638–57.
Torgersen, E. & Unni, H. 1984. Censored exponential models. *Sankhyā* **46** Series A, Pt. 1, pp. 1–23.

8

Comparison of Linear Models

8.1 Introduction

In this chapter we will consider comparison of some of the most familiar linear models. Here 'linear' means essentially that the expectations of the observations depend linearly on the unknown parameter. The main emphasis will be on linear models with covariances which are either completely known or are known up to a common positive factor. The observations which realize our models may not be normally distributed. However, if we assume they are normally distributed, the conclusions may be drastically strengthened. In this case the Hellinger transform is a very useful tool.

As we do not wish to stretch the generality too far, we shall restrict our attention to models where the dimensions of the parameter spaces and the sample spaces are finite. We shall find it convenient to use a coordinate-free approach and we shall assume that the underlying spaces are finite dimensional inner product spaces, i.e. finite dimensional Hilbert spaces.

One advantage of using this framework rather than the $\mathbb{R}^n$-approach is that the various related linear spaces need not be re-parametrized in order to be in the appropriate form. This framework also appears to make it easier to envisage generalizations to infinite dimensional spaces.

Before proceeding we should state some basic facts on random vectors. Firstly a finite dimensional inner product space, as any metric space, has a natural topology. The Borel class is of course the σ-algebra generated by the open sets (closed sets, compact sets) and this is the smallest σ-algebra making the linear functionals measurable. Representing the vectors by their coordinates with respect to a fixed orthonormal basis we obtain an isomorphism between H and some space $\mathbb{R}^m$ equipped with the usual scalar product.

The exposition in this chapter is based on the assumption that the reader knows the fundamentals of linear algebra, e.g. as in Halmos (1958). Here and there we have employed some elementary results on generalized inverses.

These results follow fairly directly from the definitions, and the definitions are given here. A reader wanting more knowledge on generalized inverses may consult e.g. Campbell & Meyer (1979) or Graybill (1983). An interesting introduction to the coordinate-free approach in statistics is Eaton (1983).

Some standard notation from linear algebra which we shall use can be found at the end of this section. In this chapter, *all inner product spaces which appear as sample spaces of random vectors or as parameter sets are assumed finite dimensional.* The inner product of vectors u and v will be denoted as (u, v) and the length $(u, u)^{1/2}$ of the vector u will be denoted as $\|u\|$.

Some of the basic features of random vectors which will be used may be described as follows.

Let X denote a random vector in the finite dimensional inner product space H. Then there is a vector $\xi \in H$ such that $E(h, X) = (h, \xi)$ for all $h \in H$ if and only if $E\|X\| < \infty$. This vector ξ is called the expectation vector (or just the expectation) of X and is denoted by EX. If, in addition, $E\|X\|^2 < \infty$ then $\xi = EX$ may also be characterized as the unique vector minimizing $E\|X - \xi\|^2$. In that case there is a unique non-negative definite operator V on H such that $\operatorname{Cov}((h_1, X), (h_2, X)) = (h_1, Vh_2)$ when $h_1, h_2 \in H$. This operator is called the covariance operator of X and is denoted as $\operatorname{Cov} X$. It may then be checked that $E\|X - EX\|^2 = \operatorname{trace} \operatorname{Cov} X$.

The range space and the null space of an operator A will be denoted by $R(A)$ and $N(A)$ respectively.

Thus $\operatorname{Var}(h, X) = 0$ if and only if $h \in N(\operatorname{Cov} X)$ while $EX + R(\operatorname{Cov} X)$ is the smallest affine subset of H which supports the distribution of X.

If $(X_1, \dots, X_n)$ is the vector of coordinates of X with respect to a fixed basis of linearly independent vectors then $(EX_1, \dots, EX_n)$ is the vector of coordinates of EX with respect to the same basis. If the basis is orthonormal then the matrix representation of V is just the covariance matrix of $(X_1, \dots, X_n)$. The operational rules for expectations and covariances of vectors of real valued variables may be carried over to the general case by this observation or, just as simply, by direct proofs. We omit the details and here shall only point out the two formulas $ETX = TEX$ and $\operatorname{Cov} TX = T(\operatorname{Cov} X)T^*$, where T is a linear map from H to some other finite dimensional inner product space.

Two random vectors X and Y belonging to possibly different finite dimensional inner product spaces are called *uncorrelated* if $\operatorname{Cov}(l_1(X), l_2(Y)) = 0$ whenever l_1 and l_2 are linear functionals on, respectively, the range space of X and the range space of Y. This implies that $El_1(X)^2 < \infty$ and $El_2(Y)^2 < \infty$ for all linear functionals l_1 and l_2 and hence that $E\|X\|^2 < \infty$ and $E\|Y\|^2 < \infty$.

A random vector X such that $E\|X\|^2 < \infty$ is called a *second order random vector.*

If X is a second order random vector in H, if $EX = 0$ and if Φ is a linear map from H onto some finite dimensional inner product space $\tilde{H}$, then the random vector X may be decomposed as $X = S(\Phi(X)) + \Psi(X)$ a.s. where S is a linear map from $\tilde{H}$ to H and where $\Psi(X)$ is uncorrelated with $\Phi(X)$. In fact this requirement amounts to the condition that $\Phi V = \Phi V \Phi^* S^*$ where $V = \operatorname{Cov} X$. Such a map S exists since $R(\Phi V) = R(\Phi V \Phi^*)$ and S is unique on the support $\Phi[R(V)]$ of $\Phi(X)$ since $S\Phi V\Phi^* = V\Phi^*$.

The probability distribution of an H-valued random vector X is determined by its *characteristic function* which to each $t \in H$ assigns the complex number $E\exp(i(t, X))$. Here $\exp z$ denotes e^z. We shall say that X is *normally distributed* if (a, X) is normally distributed for all $a \in H$. X is normally distributed if and only if its characteristic function is $t \to \exp[i(t, EX) - \frac{1}{2}\operatorname{Var}(t, X)]$. Thus a normal distribution is uniquely given by its expectation vector and its covariance operator. The condition of normality imposes no restrictions on the vector $\xi = EX$ and the non-negative definite operator $V = \operatorname{Cov} X$. The normal distribution on H with expectation vector ξ and covariance operator V will be denoted as $N(\xi, V)$.

We may now describe the features of the models which we shall discuss in this chapter. Firstly our observations should constitute a random vector X in a finite dimensional inner product space H. Thus we may assume without loss of generality that this space H is also our sample space.

The basic linearity assumption tells us that the expectation vector EX depends linearly on an unknown vector parameter β belonging to a finite dimensional inner product space K. This leads to the 'expectation model'

$$EX = A^*\beta$$

where A is a known linear map from H to K.

Here A^* denotes the adjoint of the operator A. Although several of the results we shall develop are valid when β is restricted to a subset of K, e.g. a subset spanning K, here we shall assume that any vector in K is an *a priori* possible value of β.

It is a matter of personal preference whether we use a 'starless' symbol for the map from H to K or if we do so for its adjoint which maps K into H. The style chosen here, which is in the tradition of Scheffé (1959), permits us to write $E(h, X) = (Ah, \beta)$ without using a star, which gives us fewer stars throughout this chapter.

Considering X as a measurement we might think of the residual $X - EX$ as the 'error'. In the case of variance components, however, the parameters of interest are those specifying the distribution of this term. We shall assume throughout that the covariance operator of $X - EX$, i.e. $\operatorname{Cov} X$, exists.

The simplest linear expectation model is completely characterized by the known operator A from H to K and the known covariance operator $V = \operatorname{Cov} X$ on H. This model will be denoted as $\mathscr{L}(A, V)$. We shall say that a random vector X *realizes* $\mathscr{L}(A, V)$ if $EX = A^*\beta$ and $\operatorname{Cov} X = V$ when $\beta \in K$. $\mathscr{L}(A, V)$ need not be considered as an experiment. However we can also consider $\mathscr{L}(A, V)$ as an experiment with parameter set $K \times \mathscr{P}$ where $\mathscr{P}$ is the set of all distributions on H having expectation 0 and covariance operator V. Instead of $K \times \mathscr{P}$ we could also use a smaller set such that any vector in K is the first coordinate of at least one pair in this set. In particular we may require that X is normally distributed. This leads to the linear normal experiment $\mathscr{N}(A, V) = (N(A^*\beta, V) : \beta \in K)$.

We may also consider models where the covariance operator V depends on some unknown parameter γ belonging to a certain set Γ. Thus we may consider a model $\mathscr{L}(A, V_\gamma : \gamma \in \Gamma)$ which is realized by any random vector X such that

$$EX = A^*\beta$$

and

$$\operatorname{Cov} X = V_\gamma.$$

Here the pair (β, γ) of unknown parameters β and γ is, and may be, any point in $K \times \Gamma$.

In this notation $\mathscr{L}$ indicates linearity in β.

Moreover, if the distribution of X is assumed normal then we obtain *the linear normal experiment*

$$\mathscr{N}(A, V_\gamma : \gamma \in \Gamma) = (N(A^*\beta, V_\gamma) : \beta \in K, \gamma \in \Gamma).$$

Here we will mainly consider the case where the covariance operator is known except for possibly an unknown multiplicative scalar σ^2.

This leads to the model $\mathscr{L}(A, \sigma^2 V; \sigma^2 > 0)$ which is realized by X if $EX = A^*\beta$ and $\operatorname{Cov} X = \sigma^2 V$ when $(\beta, \sigma^2) \in K \times]0, \infty[$ prevails. If, in addition, it is known that X is normally distributed, then we arrive at the linear normal model $\mathscr{N}(A, \sigma^2 V; \sigma^2 > 0) = (N(A^*\beta, \sigma^2 V); \beta \in K, \sigma^2 > 0)$.

Just assuming the structure $\mathscr{L}(A, V_\gamma : \gamma \in \Gamma)$ we may consider the problem of estimating functions of β. If the loss is quadratic loss and if only affine real valued functions of β are considered, then the risk may be completely evaluated without knowing the distribution of $X - EX$ except that its covariance operator is V_γ where $\gamma \in \Gamma$. We may then compare linear models by comparing these risk functions. To be more precise, consider two linear models $\mathscr{L}_1 = \mathscr{L}(A, V_\gamma : \gamma \in \Gamma)$ and $\mathscr{L}_2 = \mathscr{L}(B, W_\gamma : \gamma \in \Gamma)$ which are realized by random vectors X and Y respectively. Let H_A and H_B denote, respectively, the sample space of $\mathscr{L}_1$ and the sample space of $\mathscr{L}_2$.

We shall then say that *$\mathscr{L}_1$ is at least as good as $\mathscr{L}_2$ for unbiased linear estimation with quadratic loss* if to each linear parametric function $\beta \to (k, \beta)$ on K and to each unbiased linear estimator (h_B, Y) of (k, β) there corresponds an unbiased linear estimator (h_A, X) of (k, β), such that

$$\operatorname{Var}(h_A, X) \leqslant \operatorname{Var}(h_B, Y);\ \gamma \in \Gamma.$$

If $\mathscr{L}_1$ majorizes $\mathscr{L}_2$ in this sense then we shall denote this by $\mathscr{L}_1 \geqslant_q \mathscr{L}_2$ where q indicates quadratic loss. '$\geqslant_q$' is then a partial ordering of linear models.

If $\mathscr{L}_1 \geqslant_q \mathscr{L}_2$ and $\mathscr{L}_2 \geqslant_q \mathscr{L}_1$ then we shall say that $\mathscr{L}_1$ and $\mathscr{L}_2$ are equivalent for unbiased estimation and write this $\mathscr{L}_1 \sim_q \mathscr{L}_2$. '$\sim_q$' is then clearly an equivalence for models $\mathscr{L}(A, V_\gamma : \gamma \in \Gamma)$ with the same Γ.

Two early papers covering problems related to this ordering when the covariances are known are Elfving (1952) and Ehrenfeld (1955).

Assume now, with this notation, that $\mathscr{L}_1 \geqslant_q \mathscr{L}_2$. If $k \in R(B)$ then $k = Bh_B$ for some $h_B \in H_B$. Then $E(h_B, Y) = (h_B, B^*\beta) = (Bh_B, \beta) = (k, \beta)$ so that (h_B, Y) is an unbiased estimator of (k, β). Hence there is an $h_A \in H_A$ so that $E(h_A, X) \equiv_\beta (k, \beta)$ while $(h_A, V_\gamma h_A) \leqslant (h_B, W_\gamma h_B)$ for all $\gamma \in \Gamma$. The first conclusion implies that $Ah_A = k$, proving that $R(B) \subseteqq R(A)$. If, in addition, $k \in B[N(W_\gamma)]$ for some γ, then we may choose h_B in the set $N(W_\gamma)$ and then $(h_A, V_\gamma h_A) \leqslant 0$, i.e. $h_A \in N(V_\gamma)$. Hence $k = Ah_A \in A[N(V_\gamma)]$.

Altogether this shows that the ordering '$\geqslant_q$' implies the following relationships between ranges and null spaces.

Proposition 8.1.1. *If $\mathscr{L}(A, V_\gamma; \gamma \in \Gamma) \geqslant_q \mathscr{L}(B, W_\gamma; \gamma \in \Gamma)$ then $R(A) \supseteqq R(B)$ and $A[N(V_\gamma)] \supseteqq B[N(W_\gamma)]$ for all $\gamma \in \Gamma$.*

It follows that if we associate the range space $R(A)$ and its family of subsets $(A[N(V_\gamma)]; \gamma \in \Gamma)$ with the linear model $\mathscr{L}(A, V_\gamma; \gamma \in \Gamma)$ then this association respects the equivalence relation $\sim_q$.

As noted by Stepniak (1983) we may omit the restriction to unbiased estimation and also extend the scope to affine estimators.

Proposition 8.1.2 (Extension to possibly biased estimators). *Let X and Y realize, respectively, the linear models $\mathscr{L}_1 = \mathscr{L}(A, V_\gamma : \gamma \in \Gamma)$ and $\mathscr{L}_2 = \mathscr{L}(B, W_\gamma : \gamma \in \Gamma)$. Then $\mathscr{L}_1 \geqslant_q \mathscr{L}_2$ if and only if to each affine function $\phi(\beta)$ of β and to each affine estimator $\hat{\phi}(Y)$ of $\phi(\beta)$ there corresponds an affine estimator $\tilde{\phi}(X)$ such that $E[\tilde{\phi}(X) - \phi(\beta)]^2 \leqslant E[\hat{\phi}(Y) - \phi(\beta)]^2$ for all γ.*

Proof. Let H_A and H_B denote the sample spaces of $\mathscr{L}_1$ and $\mathscr{L}_2$ respectively. If $\phi(\beta) = c + (k, \beta)$, $\hat{\phi}(Y) = d_1 + (h_B, Y)$ and $\tilde{\phi}(X) = d_2 + (h_A, X)$ then

$$E[\hat{\phi}(Y) - \phi(\beta)]^2 = (h_B, W_\gamma h_B) + [d_1 - c + (Bh_B - (k, \beta))]^2$$

and

$$E[\tilde{\phi}(X) - \phi(\beta)]^2 = (h_A, V_\gamma h_A) + [d_2 - c + (Ah_A - (k, \beta))]^2.$$

The 'only if' follows directly from this while the 'if' follows from this and the fact that non-null linear functionals are unbounded functions. □

The following is an immediate but noteworthy consequence of the definition of the ordering $\geqslant_q$.

Proposition 8.1.3. *$\mathscr{L}(A, \sigma^2 V; \sigma^2 > 0) \geqslant_q \mathscr{L}(B, \sigma^2 W; \sigma^2 > 0)$ if and only if*

$$\mathscr{L}(A, V) \underset{q}{\geqslant} \mathscr{L}(B, W).$$

Thus knowledge of σ^2 does not influence comparability for the ordering '$\geqslant_q$' for linear models $\mathscr{L}(A, \sigma^2 V; \sigma^2 > 0)$ and $\mathscr{L}(B, \sigma^2 W; \sigma^2 > 0)$.

However it is not only in this simple situation that this ordering may be reduced to the situation where the covariances are known. Actually, as shown by Stepniak & Torgersen (1981), this ordering may always be expressed in terms of the same ordering for known covariances.

Theorem 8.1.4 (Reduction to the case of known covariances).

$$\mathscr{L}(A, V_\gamma; \gamma \in \Gamma) \underset{q}{\geqslant} \mathscr{L}(B, W_\gamma; \gamma \in \Gamma)$$

if and only if

$$\mathscr{L}(A, V_\lambda) \underset{q}{\geqslant} \mathscr{L}(B, W_\lambda)$$

for each probability distribution λ on Γ with finite support. Here $V_\lambda = \sum_\gamma \lambda(\gamma) V_\gamma$ and $W_\lambda = \sum_\gamma \lambda(\gamma) W_\gamma$.

Remark. The theorem simplifies a bit when Γ is a convex subset of some linear space and the map $\gamma \sim\to V_\gamma$ is affine. Then it follows that $\mathscr{L}(A, V_\gamma : \gamma \in \Gamma) \geqslant_q \mathscr{L}(B, W_\gamma; \gamma \in \Gamma)$ if and only if $\mathscr{L}(A, V_\gamma) \geqslant_q \mathscr{L}(B, W_\gamma)$ for each $\gamma \in \Gamma$.

Proof. The 'only if' is trivial so suppose that $\mathscr{L}(A, V_\lambda) \geqslant_q \mathscr{L}(B, W_\lambda)$ for all $\lambda \in \Lambda$ where Λ is the space of probability distributions on Γ with finite support. Let X realize $\mathscr{L}_1$ and let Y realize $\mathscr{L}_2$. Let H_A and H_B be the sample spaces of, respectively, $\mathscr{L}_1$ and $\mathscr{L}_2$, where $\mathscr{L}_1 = \mathscr{L}(A, V_\gamma : \gamma \in \Gamma)$ and $\mathscr{L}_2 = \mathscr{L}(B, W_\gamma : \gamma \in \Gamma)$.

Assume $E(h_B, Y) \equiv (k, \beta)$ where $h_B \in H_B$ and $k \in K$. Then $(k, \beta) \equiv_\beta (h_B, B^*\beta) \equiv_\beta (Bh_B, \beta)$ so that $Bh_B = k$. Put $f(\gamma) = (h_B, W_\gamma h_B)$ and by a slight misuse of notation, let us denote $\int f(\gamma)\lambda(\mathrm{d}\gamma) = (h_B, W_\lambda h_B)$ by $f(\lambda)$. By assumption, for each $\lambda \in \Lambda$ there is a vector h_A in H_A such that $E(h_A, X) \equiv (k, \beta)$ while $\operatorname{Var}(h_A, X) \leqslant \operatorname{Var}(h_B, Y)$ when $\operatorname{Cov} X = V_\lambda$ and $\operatorname{Cov} Y = W_\lambda$.

Thus $Ah_A = k$ while $(h_A, V_\lambda h_A) \leqslant f(\lambda)$. Here we may assume that h_A belongs to the support $R(V_\lambda) + R(A^*)$ of $\mathscr{L}(A, V_\lambda)$.

Let H_0 denote the affine subset $A^{-1}[\{k\}]$ of H_A.

Let F be a finite subset of Γ and let $\tilde{F}$ be the set of probability distributions supported by F. By assumption $\sup_{\lambda \in \tilde{F}} \inf_{h \in H_0} [(h, V_\lambda h) - f(\lambda)] \leqslant 0$. The bracket is continuous and concave–convex in $(\lambda, h) \in \tilde{F} \times H_0$ and $\tilde{F}$ is compact.

Minimax theory (e.g. theorem 3.2.12) yields

$$\inf_{h \in h_0} \Phi(h) \leqslant 0$$

where

$$\Phi(h) = \sup_{\lambda \in \tilde{F}} [(h, V_\lambda h) - f(\lambda)] = \sup_{\gamma \in \Gamma} [(h, V_\gamma h) - f(\gamma)].$$

It follows that to each integer $n \geqslant 1$ there corresponds an $h_n \in H_0$ such that $\Phi(h_n) \leqslant 1/n$. Let ρ be the orthogonal projection of H onto $R(\bar{V}_F)$ where $\bar{V}_F = \sum_{\gamma \in F} V_\gamma/\#F$. Then $h - \rho(h) \in N(\bar{V}_F) = \bigcap_{\gamma \in F} N(V_\gamma)$. Hence

$$\Phi(h) = \sup_{\gamma \in F} \{(\rho(h), V_\gamma \rho(h)) - f(\gamma)\}$$

so that

$$(\rho(h_n), V_\gamma \rho(h_n)) \leqslant f(\gamma) + 1/n$$

when $\gamma \in F$. In particular $(\rho(h_n), \bar{V}_F \rho(h_n)) \leqslant \bar{f}_F + 1/n$ where $\bar{f}_F = \sum_{\gamma \in F} f(\gamma)/\#F$.

Since $\bar{V}_F$ is positive definite on its range, it follows that $\rho(h_n); n = 1, 2, \ldots$ is bounded. Let h' be the limit of a convergent sub-sequence. Then $(h', V_\gamma h') \leqslant f(\gamma); \gamma \in F$. Furthermore, since $\rho[H_0]$ is closed, $h' \in \rho[H_0]$. It follows that $h' = \rho(h_F)$ for some $h_F \in H_0$ and then $(h_F, V_\gamma h_F) = (h', V_\gamma h') \leqslant f(\gamma)$ when $\gamma \in F$. Now $N(\bar{V}_F) = \bigcap_{\gamma \in F} N(V_\gamma)$ decreases as F increases. It follows from the finite dimensionality of H that there is a finite subset F_0 of Γ such that $\bigcap_{\gamma \in F_0} N(V_\gamma) = \bigcap_\gamma N(V_\gamma)$ and then $N(\bar{V}_F) = N(\bar{V}_{F_0})$ and $R(\bar{V}_F) = R(\bar{V}_{F_0})$ when $F \supseteq F_0$. In the remaining part of this proof, let us restrict F to $\mathscr{F}$ where $\mathscr{F}$ is the class of those finite subsets of Γ which contain F_0.

Let σ be the orthogonal projection of H onto $R(\bar{V}_{F_0})$. We have just seen that for each $F \in \mathscr{F}$ there is an h_F in H_0 such that $(\sigma(h_F), V_\gamma \sigma(h_F)) = (h_F, V_\gamma h_F) \leqslant f(\gamma)$ when $\gamma \in F$. Then, since $F \supseteq F_0$

$$(\sigma(h_F), \bar{V}_{F_0} \sigma(h_F)) \leqslant \bar{f}_{F_0}.$$

It follows that the family $(\sigma(h_F) : F \in \mathscr{F})$ is bounded. Hence, by conditional compactness, there is an h'' such that to each $\varepsilon > 0$ and to each $F_1 \in \mathscr{F}$ there corresponds an $F \in \mathscr{F}$ such that $F \supseteq F_1$, and $\|h'' - \sigma(h_F)\| < \varepsilon$. It follows that $(h'', V_\gamma h'') \leqslant f(\gamma)$ for all γ in Γ. Furthermore, since $\sigma[H_0]$ is closed, $h'' = \sigma(h)$ for some h in H_0. Then $h - \sigma(h) \in \bigcap_\gamma N(V_\gamma)$ so that

$$(h, V_\gamma h) - (h'', V_\gamma h'') \leqslant f(\gamma); \qquad \gamma \in \Gamma. \qquad \square$$

If X realizes $\mathscr{L}(A, V)$ and if T is a linear map from the sample space H_A of $\mathscr{L}(A, V)$ to the finite dimensional inner product space H_B, then $T(X)$ realizes $\mathscr{L}(AT^*, TVT^*)$. Hence $\mathscr{L}(A, V) \geqslant_q \mathscr{L}(AT^*, TVT^*)$.

The examples $\mathscr{L}(A, V) = \mathscr{L}(4, 1)$ and $\mathscr{L}(B, W) = \mathscr{L}(1, 1)$ on the real line reveal that $\mathscr{L}(A, V)$ may be at least as good as $\mathscr{L}(B, W)$ for unbiased estimation with quadratic loss, although $\mathscr{L}(B, W)$ is not obtainable from $\mathscr{L}(A, V)$ by a linear map.

Note however, that by the definition of this ordering, $\mathscr{L}(A, V) \geqslant_q \mathscr{L}(B, W)$ whenever there is a linear map T such that $TA^* = B^*$ while $TVT^* \leqslant W$. Actually if X realizes $\mathscr{L}(A, V)$ and if Y realizes $\mathscr{L}(B, W)$ and if T satisfies these conditions, then $E(T^*h_B, X) \equiv E(h_B, Y)$ while $\operatorname{Var}(T^*h_B, X) = (T^*h_B, VT^*h_B) = (h_B, TVT^*h_B) \leqslant (h_B, Wh_B) = \operatorname{Var}(h_B, Y)$. We will see in section 8.4 that we have now arrived at a complete criterion, since such a T exists whenever $\mathscr{L}(A, V) \geqslant_q \mathscr{L}(B, W)$.

As mentioned earlier, we will mainly be concerned with two cases, i.e. two collections of experiments. These are the collection of linear normal models of the form $\mathscr{N}(A, V)$ and the collection of linear normal models of the form $\mathscr{N}(A, \sigma^2 V; \sigma^2 > 0)$. As the first is obtained from the last by specifying σ^2, we shall refer to the first situation as *the case of known* σ^2. The second situation may be referred to as *the case of unknown* σ^2.

Whether σ is known or not, the collection of linear normal experiments is closed under some of the standard procedures for deriving new experiments from old ones. Thus the collection is closed for products, and linear transformations of the parameter β and of statistics realizing the experiments, lead to experiments of the same form.

Let us take a closer look at the dominated linear normal experiments when σ is known. It will be shown that any such experiment is equivalent to an experiment $\mathscr{N}(A, V)$ where V is non-singular. We can therefore consider a random vector variable X realizing the experiment $\mathscr{N}(A, V) = (N(A^*\beta, V); \beta \in K)$ where V is non-singular. Putting $P_\beta = N(A^*\beta, V)$ we find that a version of $\log(\mathrm{d}P_\beta/\mathrm{d}P_0)$ is Λ_β, where

$$\Lambda_\beta(x) \underset{x}{\equiv} (\beta, AV^{-1}x) - \tfrac{1}{2}(\beta, AV^{-1}A^*\beta).$$

It follows that if $\mathscr{L}(X) = P_0$ then the process $\beta \to \Lambda_\beta(X)$ is Gaussian with expectations and covariances given by

$$E\Lambda_\beta(X) = -\tfrac{1}{2}\operatorname{Var}\Lambda_\beta(X) = \tfrac{1}{2}(\beta, AV^{-1}A^*\beta)$$

while

$$\operatorname{Cov}(\Lambda_{\beta_1}(X), \Lambda_{\beta_2}(X)) = (\beta_1, AV^{-1}A^*\beta_2)$$

when β, β_1 and $\beta_2 \in K$.

In general, homogeneous experiments such that the log likelihood process is Gaussian, are called *Gaussian experiments*. Thus dominated experiments $\mathscr{N}(A, V)$ are Gaussian.

General properties of Gaussian processes imply general properties of Gaussian experiments. Thus products of Gaussian experiments are Gaussian and Gaussian experiments are infinitely divisible. Convolution factors of normal distributions are also normal (Cramer, 1937). This readily implies that factor experiments of Gaussian experiments are Gaussian. Using this we shall see in section 8.2 that factor experiments of dominated linear normal experiments with known σ are also linear normal. This is of interest since some of the criteria for comparison which we shall derive are in terms of factor experiments.

We saw above that the likelihood function of $\mathcal{N}(A, V)$, provided V is non-singular, is the sum of a term which is linear in β and a quadratic form in β. An asymptotic version of this expansion constitutes the crucial condition for ensuring local asymptotic normality. In particular, this condition is satisfied for replicated quadratically differentiable experiments as the number of replications increases. The reader may consult the books by LeCam (1974b, 1986), by Janssen, Milbrodt and Strasser (1984), and by Strasser (1985) and also the paper by Millar (1983) for information on asymptotic normality and on Gaussian experiments.

Several of the formulas related to linear models simplify considerably if the covariance operators are non-singular and thus may be transformed to identity operators. In the general case the formulas may be quite messy. It is therefore fortunate that there is a method of smoothing available which increases the covariance operator without altering the expectation structure. At a certain cost of loss of information, this smoothing procedure can completely remove singularity. This procedure has been employed by several authors (see the papers by Drygas (1983, 1985) and the references there). The first theorem in Stepniak, Wang & Wu (1984) shows that this smoothing preserves the ordering $\geqslant_q$. This result has been applied to linear sufficiency (Stepniak, 1985). The fact that this smoothing is essentially a particular case of smoothing of translation experiments by convolution, appears not to have been recognized.

Roughly the argument is as follows. Consider experiments $\mathscr{E} = (P_\theta : \theta \in \Theta)$ and $\mathscr{F} = (Q_\theta : \theta \in \Theta)$ which are both invariant under the actions of a group G as explained in section 6.5. If $\mathscr{E}$ is ε-deficient with respect to $\mathscr{F}$ for an invariant function ε on Θ then section 6.5 provides conditions ensuring that $\|P_\theta T - Q_\theta\| \leqslant \varepsilon_\theta$ for an invariant transition T from $L(\mathscr{E})$ to $L(\mathscr{F})$. Invariance implies that $\mu g^{-1} T = (\mu T) g^{-1}$ when $\mu \in L(\mathscr{E})$. Thus if π is a probability measure on G having finite support then integration with respect to π yields

$$\left[\int (\mu g^{-1}) \pi(\mathrm{d}g)\right] T = \int \left[(\mu T) g^{-1}\right] \pi(\mathrm{d}g).$$

If T is induced by an invariant kernel and if obvious measurability conditions are satisfied then this extends to any probability distribution π on the appropriate class of measurable subsets of G. Assuming this works for the probability distributions $\pi_\omega : \omega \in \Omega$ on G we find for

$$P_{\theta,\omega} = \int [P_\theta g^{-1}]\pi_\omega(dg) = \int P_{g(\theta)}\pi_\omega(dg)$$

and

$$Q_{\theta,\omega} = \int [Q_\theta g^{-1}]\pi_\omega(dg) = \int Q_{g(\theta)}\pi_\omega(dg)$$

that

$$\begin{aligned}\|P_{\theta,\omega}T - Q_{\theta,\omega}\| &= \left\| \int [(P_\theta T)g^{-1}]\pi_\omega(dg) - \int (Q_\theta g^{-1})\pi_\omega(dg) \right\| \\ &\leqslant \sup_g \|(P_\theta T)g^{-1} - Q_\theta g^{-1}\| \\ &= \|P_\theta T - Q_\theta\| \leqslant \varepsilon_\theta.\end{aligned}$$

We shall apply this idea to linear regression models having parameter set $\Theta = K \times \Gamma$ where K is the finite dimensional inner product space considered above and Γ may be any set. If $\theta \in \Theta$ is represented as $\theta = (\beta, \gamma)$ where $\beta \in K$ and $\gamma \in \Gamma$ then in some situations γ may be considered as a nuisance parameter. In other situations, e.g. in variance component models, it is γ which is the parameter of interest. With this parameter set we have the following general smoothing result.

Theorem 8.1.5 (Smoothing of regression models). *Consider regression experiments $\mathscr{E} = (P_\theta : \theta \in \Theta)$ and $\mathscr{F} = (Q_\theta : \theta \in \Theta)$ realized by, respectively, vector variables X and Y having the following properties:*

(i) *X assumes its values in a finite dimensional inner product space H_A. There is a (known) linear operator A from H_A to K such that $\mathscr{L}(X - A^*\beta|\beta, \gamma)$ depends on γ only.*
(ii) *Y assumes its values in a finite dimensional inner product space H_B. There is a linear operator B from H_B to K such that $\mathscr{L}(Y - B^*\beta|\beta, \gamma)$ depends on γ only.*

Say that an experiment $\mathscr{G} = (R_\theta : \theta \in \Theta)$ having sample space H_A satisfies the dominatedness condition (LD) *if*:

(LD): *the distributions $R_{(0,\gamma)} : \gamma \in \Gamma$ are all dominated by the Lebesgue measure on some subspace of H_A not depending on γ.*

Say also that a Markov kernel M from H_A to H_B is invariant if $M(U|x + A^*\beta) = M(U - B^*\beta|x)$ whenever $x \in H_A$, $\beta \in K$ and U is a Borel subset of H_B. Then:

(a) *If $\mathscr{E}$ satisfies* (LD) *and if $\delta(\mathscr{E}, \mathscr{F}) < 2$ then there is an invariant Markov kernel M from H_A to H_B such that for any $\theta = (\beta, \gamma)$ and for any K-valued* *random vectors Z_X and Z_Y which have the same distributions and which are independent of, respectively, X and Y:*

$$\|\mathscr{L}(X + A^*Z_X)M - \mathscr{L}(Y + B^*Z_Y)\| \leqslant \delta(\mathscr{E}, \mathscr{F}).$$

(b) *Consider the 'smoothed' experiments $\tilde{\mathscr{E}} = (\tilde{P}_\theta : \theta \in \Theta)$ and $\tilde{\mathscr{F}} = (\tilde{Q}_\theta : \theta \in \Theta)$ realized by, respectively, $\tilde{X} = X + Z_X$ and $\tilde{Y} = Y + Z_Y$ where Z_X and Z_Y have the same distribution depending on γ only. Assume that Z_X and Z_Y, for each θ, are independent of, respectively, X and Y. Assume also that the characteristic functions of the distributions $\mathscr{L}(Z_X|\gamma) = \mathscr{L}(Z_Y|\gamma)$; $\gamma \in \Gamma$ have no zeros.*

Then $\mathscr{E} \geqslant \mathscr{F}$ provided $\tilde{\mathscr{E}} \geqslant \tilde{\mathscr{F}}$ and $\mathscr{E}$ satisfies (LD). *If $\mathscr{E}$ and $\tilde{\mathscr{E}}$ both satisfy* (LD) *then $\delta(\mathscr{E}, \mathscr{F}) \geqslant \delta(\tilde{\mathscr{E}}, \tilde{\mathscr{F}})$ and $\mathscr{E} \geqslant \mathscr{F}$ if and only if $\tilde{\mathscr{E}} \geqslant \tilde{\mathscr{F}}$.*

Remark. By example 6.5.14 the dominatedness condition (LD) is not superfluous here.

Proof. Note first that $R(B) \subseteqq R(A)$ whenever $\delta(\mathscr{E}, \mathscr{F}) < 2$. Indeed if $R(B) \subseteqq R(A)$ then $B^*\beta_0 \neq 0$ for a vector β_0 in K such that $A^*\beta_0 = 0$. Restricting β to $[\beta_0] = \{t\beta_0 : -\infty < t < \infty\}$ and keeping γ fixed we obtain a totally non-informative experiment from $\mathscr{E}$ while the restriction of $\mathscr{F}$ majorizes some translation experiment on the real line. Thus $\delta(\mathscr{E}, \mathscr{F}) = 2$ in that case. By the same argument $\delta(\tilde{\mathscr{E}}, \tilde{\mathscr{F}}) = 2$ where $\tilde{\mathscr{E}}$ is defined in (b).

Thus we may assume without loss of generality that $R(B) \subseteqq R(A)$; i.e. that $B^* = TA^*$ for some linear operator T.

Let the group $G = (K, +)$ act on Θ, H_A and H_B by the following actions of $g \in K$:

$$g(\beta, \gamma) = (g + \beta, \gamma), g(x) = (A^*g + x)$$

and

$$g(y) = B^*g + y$$

when $\beta \in K$, $\gamma \in \Gamma$, $x \in H_A$ and $y \in H_B$. Then the experiments $\mathscr{E}$, $\tilde{\mathscr{E}}$, $\mathscr{F}$ and $\tilde{\mathscr{F}}$ are all invariant under G. Furthermore, any linear operator T from H_A to H_B such that $B^* = TA^*$, provides an example of an invariant Markov kernel from H_A to H_B.

Assume that $\mathscr{E}$ satisfies (LD) with $\tilde{H}_A$ being the appropriate subspace of H_A. Then $\tilde{H}_A + R(A^*) = \bigcup \{\tilde{H}_A + A^*\beta : \beta \in K\}$ is a common support of $P_\theta : \theta \in \Theta$.

Put $\tilde{K} = \{\beta : A^*\beta \in \tilde{H}_A\}$ and let the group $\tilde{G} = (\tilde{K}, +)$ act on $\tilde{\Theta} = \tilde{K} \times \Gamma$, $\tilde{H}_A$ and H_B by the actions defined for $G = (K, +)$.

Restricting β to $\tilde{K}$ we find that the restrictions of $\mathscr{E}$ and $\mathscr{F}$ are invariant under $\tilde{G}$. Furthermore, this restriction of $\mathscr{E}$ is dominated by Lebesgue measure on $\tilde{H}_A$. Then it follows from the theory in section 6.5 that there is a Markov kernel $\tilde{M}$ from $\tilde{H}_A$ to H_B which is invariant under $\tilde{G}$ and such that

$\|P_{(\beta,\gamma)}\tilde{M} - Q_{(\beta,\gamma)}\| \leqslant \delta(\mathscr{E},\mathscr{F})$ when $\beta \in \tilde{K}$ and $\gamma \in \Gamma$. Invariance implies that $\tilde{M}(U + B^*\beta|x + A^*\beta) = \tilde{M}(U|x)$ whenever U is a Borel subset of H_B, $\beta \in \tilde{K}$ and $x \in \tilde{H}_A$. Putting $M(U|x + A^*\beta) = \tilde{M}(U - B^*\beta|x)$ for any Borel subset U of H_B and for vectors $x \in \tilde{H}_A$ and $\beta \in K$ we obtain a well defined invariant Markov kernel from $\tilde{H}_A + R(A^*)$ to H_B. Using any linear map T such that $TA^* = B^*$ we may finally extend M to an invariant kernel from all of H_A to H_B. The invariance of M implies that $(\mu * \delta_{A^*\beta})M = (\mu M) * \delta_{B^*\beta}$ for any finite measure μ on H_A. Here δ_ξ as usual denotes the one-point distribution in ξ. Integrating both sides with respect to a finite measure D on K we obtain $[\mu * D(A^*)^{-1}]M = (\mu M) * D(B^*)^{-1}$.

This integration effectively uses the fact that M is a Markov kernel and not merely an invariant transition.

Using $(0,\gamma) \in \tilde{\Theta} = (\tilde{K} \times \Gamma)$ when $\gamma \in \Gamma$ and letting $*$ denote convolution we find that:

$$\begin{aligned}
&\|\mathscr{L}(X + A^*Z_X|\beta,\gamma)M - \mathscr{L}(Y + B^*Z_Y|\beta,\gamma)\| \\
&\quad = \|\mathscr{L}(X + A^*(Z_X + \beta)|0,\gamma)M - \mathscr{L}(Y + B^*(X_Y + \beta)|0,\gamma)\| \\
&\quad = \|\mathscr{L}(X|0,\gamma)M * \mathscr{L}(B^*(Z_Y + \beta)|\gamma) - \mathscr{L}(Y|0,\gamma) * \mathscr{L}(B^*(Z_Y + \beta)|\gamma)\| \\
&\quad \leqslant \|\mathscr{L}(X|0,\gamma)M - \mathscr{L}(Y|0,\gamma)\| \leqslant \delta(\mathscr{E},\mathscr{F}).
\end{aligned}$$

Finally consider the situation described in (b) and assume that $\tilde{\mathscr{E}}$ satisfies (LD) and that $\tilde{\mathscr{E}} \geqslant \tilde{\mathscr{F}}$. By the arguments above

$$\begin{aligned}
\mathscr{L}(X|\beta,\gamma)M * \mathscr{L}(B^*Z_Y|\gamma) &\equiv [\mathscr{L}(X|\beta,\gamma) * \mathscr{L}(A^*Z_X|\gamma)]M \\
&= \mathscr{L}(X + A^*Z_X|\beta,\gamma)M \\
&\equiv \mathscr{L}(Y + B^*Z_Y|\beta,\gamma) \\
&= \mathscr{L}(Y|\beta,\gamma) * \mathscr{L}(B^*Z_Y|\gamma).
\end{aligned}$$

The assumptions on characteristic functions imply that the cancellation law may be applied. Thus $\mathscr{L}(X|\beta,\gamma)M \equiv \mathscr{L}(Y|\beta,\gamma)$ so that $\mathscr{E} \geqslant \mathscr{F}$. □

Applying this to linear normal models we obtain deficiency decreasing and order preserving smoothing operations.

Corollary 8.1.6 (Smoothing of linear normal models). *Consider linear normal models $\mathscr{E} = \mathscr{N}(A, V_\gamma : \gamma \in \Gamma)$ and $\mathscr{F} = \mathscr{N}(B, W_\gamma : \gamma \in \Gamma)$. Then:*

(a) *if $\delta(\mathscr{E},\mathscr{F}) < 2$ and if $R(V_\gamma) : \gamma \in \Gamma$ does not depend on Γ then there is an invariant Markov kernel M from H_A to H_B such that*

$$\|N(A^*\beta, V_\gamma + A^*GA)M - N(B^*\beta, W_\gamma + B^*GB)\| \leqslant \delta(\mathscr{E},\mathscr{F})$$

whenever $\beta \in K$, $\gamma \in \Gamma$ and G is a non-negative definite operator on K.

(b) *assume that $R(W_\gamma) : \gamma \in \Gamma$ does not depend on γ and that $G_\gamma : \gamma \in \Gamma$ is a family of non-negative definite operators on K such that the ranges*

$R(G_\gamma)$; $\gamma \in \Gamma$ do not depend on γ. Then $\mathcal{N}(A, V_\gamma : \gamma \in \Gamma) \geqslant \mathcal{N}(B, W_\gamma : \gamma \in \Gamma)$ whenever $\mathcal{N}(A, V_\gamma + A^*G_\gamma A : \gamma \in \Gamma) \geqslant \mathcal{N}(B, W_\gamma + B^*G_\gamma B : \gamma \in \Gamma)$.

If, in addition, the ranges $R(V_\gamma)$; $\gamma \in \Gamma$ do not depend on γ then

$$\delta(\mathcal{N}(A, V_\gamma; \gamma \in \Gamma), \mathcal{N}(B, W_\gamma : \gamma \in \Gamma))$$
$$\geqslant \delta(\mathcal{N}(A, V_\gamma + A^*G_\gamma A : \gamma \in \Gamma), \mathcal{N}(B, W_\gamma + B^*G_\gamma B : \gamma \in \Gamma)).$$

In particular the two informational inequalities above are equivalent in this situation.

Remark. The assumptions on ranges of covariance operators are trivially satisfied for the linear normal models $\mathcal{N}(A, V)$ and $\mathcal{N}(A, \sigma^2 V; \sigma^2 > 0)$.

Proof. This follows from the theorem above and the convolution theorem for normal distributions. □

Some interesting internal properties of such smoothing devices are established in sections 8.3 and 8.4.

We will begin the next section by giving explicit expressions for Hellinger transforms of families of normal distributions. Then we will derive as many conclusions on the ordering of these experiments as we can by relying almost solely on these transforms. Although the deficiency ordering is in general stronger than the ordering in terms of Hellinger transforms, we know that the Hellinger transform characterizes the type of an experiment. As the Hellinger transforms of these experiments are determined by the affinities, it follows, regardless of whether σ is known or not, that linear normal experiments are equivalent if and only if corresponding affinities are equal.

If, in addition, the space K of possible β's is one of the spaces $\mathbb{R}, \mathbb{R}^2, \ldots$ then the experiments are quadratically differentiable and the Fisher information matrices exist. These matrices determine the affinities so that the types of dominated linear normal experiments, whether σ is known or unknown, are determined by the Fisher information matrices. As any K is isomorphic to some space $\mathbb{R}^p$, equipped with the usual inner product, this applies to the case of a general inner product space, provided we allow ourselves to work with 'Fisher information operators'.

Using these observations on equivalence we find canonical forms of linear normal experiments which later we will see are minimal, i.e. irreducible by sufficiency. If σ is unknown then equivalence may be expressed in terms of invertible linear operators.

If σ is known then it turns out that not only equivalence but also the deficiency ordering is directly expressible in terms of affinities. This ordering coincides then with a variety of other orderings, e.g. pairwise ordering, ordering by testing problems, ordering by factorization and ordering by Fisher information operators. In section 8.4 we will see that this ordering also coincides with the ordering $\geqslant_q$ described here.

Several of the characterizations of the deficiency ordering which are valid when σ is known are not valid when σ is unknown. However the factorization criterion is valid in both cases (Torgersen, 1984). This criterion expresses for linear normal experiments $\mathscr{N}_1$ and $\mathscr{N}_2$ that $\mathscr{N}_1$ is at least as informative as $\mathscr{N}_2$ if and only if $\mathscr{N}_1$ is equivalent to a product $\mathscr{N}_2 \times \mathscr{E}$ for some experiment $\mathscr{E}$. If so then $\mathscr{E}$ may be chosen as a linear normal experiment. Actually the problem may be reduced to the case of non-singular covariance matrices. In that case the experiments are exponential with a common canonical parameter varying freely within some open set. As shown by Janssen (1988) the factorization criterion extends to this situation (see theorem 9.5.6). A general discussion of ordering by factorization may be found in complement 26 in chapter 10.

If σ is unknown then the deficiency ordering of linear normal experiments does not coincide with the corresponding ordering of affinities or of Hellinger transforms. Anticipating later results we show that comparison by factorization amounts precisely to comparison by kernels which are representable by linear maps.

After having discussed conclusions which may be drawn almost solely from the Hellinger transforms, in section 8.3 we consider some well known internal properties of linear normal models. The emphasis will be on the problem of estimating β and σ^2, and on sufficiency reductions. The purpose is mainly to provide tools for the remaining part of this chapter. Thus we do not discuss highly relevant topics such as e.g. testing theory, multiple decision theory, and admissibility. Here the reader may consult Eaton (1983) as well as Rao (1965) on linear statistical inference and also his 1976 paper on admissibility.

The main results in section 8.3 are two versions of the Gauss–Markov theorem which describes how we may find the best linear unbiased estimator (BLUES) of identifiable linear parametric functions.

The parameter β is not identifiable unless the operator A is onto. Nevertheless it is convenient to consider estimators of β even when β is not identifiable.

If X realizes $\mathscr{L}(A, V)$ then a linear estimator $\hat{\beta}(X)$ of β is called a *BLUE generator* for β (in $\mathscr{L}(A, V)$) if $(h, A^*\hat{\beta}(X))$ is a best linear unbiased estimator for (h, EX) for any vector h belonging to the sample space of $\mathscr{L}(A, V)$. Here 'best' should be interpreted as having the smallest variance within the class of linear unbiased estimators.

Among the BLUE generators there is an essentially unique one which takes its values within the range $R(A)$ of A. This BLUE generator is 'projection sufficient' as well as 'second order complete'. If X is normally distributed and if σ is known then it is a complete and sufficient statistic.

The reader can consult Drygas (1983, 1985) for further information on linear sufficiency.

If σ is unknown then this BLUE generator together with the UMVU estimator of σ^2 constitutes a complete and sufficient statistic. In both cases the minimal forms turn out to be identical to the canonical forms encountered in section 8.2.

The results obtained with the help of Hellinger transforms in section 8.2 and the estimation theory of section 8.3 are combined in section 8.4 using a well known concept of conjugacy from convex analysis. Using this we find that the ordering $\geqslant_q$ for linear models of the form $\mathscr{L}(A, V)$ coincides with the usual (deficiency) ordering of linear normal experiments $\mathscr{N}(A, V)$. This enables us to complete the list of equivalent criteria for the ordering of linear normal experiments with known σ. In contrast to the situation where σ is unknown, the ordering of linear normal models with known σ is strictly weaker than the ordering obtained by considering only those randomizations which are representable as linear maps.

In section 8.5 we turn to deficiencies when σ is known. An explicit expression for deficiencies for comparable experiments given by LeCam (1975) and later Swensen (1980), showed how the deficiencies may be found in the general case (provided σ is known). Actually these authors restrict themselves to the situation where the covariance operators are identity operators. However the general case can be reduced to this case and can further be reduced to pairs of experiments being jointly in a canonical form.

In order to be able to return to the general non-canonical case we shall need to consider certain operations related to minimum and maximum for pairs of non-negative definite operators which do not necessarily commute.

In general the deficiencies may be expressed as statistical distances between centred normal distributions.

The main result shows how these statistical distances may be expressed in terms of the min/max operations just mentioned. This result implies that the general case can be reduced to the situation where the experiments are comparable. The optimal randomizations 'giving' these deficiencies may be chosen as conditional normal distributions.

Using the main result we find simple bounds for the deficiency distance in terms of statistical distances between centred normal distributions. Another result is a simple characterization of the deficiency in terms of linear operators.

Finally in section 8.6 we turn to the problem of finding manageable expressions for deficiencies when σ is unknown. In this case the results are not complete and the reader is challenged to fill the gap.

A complete criterion for the ordering of these experiments was obtained in Hansen & Torgersen (1974) using an idea due to Boll (1955). Here we provide another form of this criterion by showing that it expresses that $\mathscr{N}_1$ is at least as informative as $\mathscr{N}_2$ if and only if $\mathscr{N}_2$ may be obtained from $\mathscr{N}_1$ by a linear map. As mentioned above, this is not so when σ is known.

If σ is unknown, then again in contrast to the situation when σ is known, we may easily have non-comparable linear normal experiments $\mathcal{N}_1$ and $\mathcal{N}_2$ such that $\mathcal{N}_1^2 \geqslant \mathcal{N}_2^2$. This implies that the (deficiency) ordering of these experiments is not the ordering by Hellinger transforms. However, it is shown that it is the ordering by factorization.

As in the previous section we may reduce the general problem to a canonical situation with non-singular covariance operators. Proceeding as in section 8.5 the problem may be further reduced to the problem of finding the deficiency of one translation experiment with respect to another. If σ is unknown then the underlying group is not commutative. It is nevertheless amenable and thus, following Swensen (1980), we can apply the theory of translation experiments. In this way we find e.g. the deficiency whenever the experiments are comparable. Another of Swensen's results describes a situation where knowledge of σ does not alter the deficiency and thus may be found as in the previous section.

Here we list some interesting topics which have not been discussed in this chapter.

1. *Extensions to more complex covariance structures e.g. the covariance structures encountered in the analysis of variance components.* The reader may consult La Motte (1982, 1985) and Rao & Kleffe (1988) and the references there for some recent developments on estimation of variance components. It would be interesting to know the implications of these results for the comparability of variance component models.
2. *Relationships to admissibility problems.* Stein's (1956) discovery of the inadmissibility of some apparently sensible estimators has been the starting point for an impressive line of research. The paper by Cohen (1966) and Rao's paper (1976) belong to this field, and some of the conditions for admissibility which may be found in these papers appear to be of the same nature as some of the conditions appearing in sections 8.2, 8.3 and 8.5.
3. *Complete class results for linear models for the orderings considered here.* Suppose we are interested in estimating certain (linear) combinations of the unknown parameters and that our experiment has to be chosen within a given class κ of experiments. Avoiding details we may say that a subclass κ_0 of κ is essentially complete if it majorizes κ for the particular ordering we are considering.

   Completeness results in this direction are obtained in Ehrenfeld (1955, 1966). He considers there models $\mathscr{L}(A, I)$ for the ordering $\geqslant_q$. Now as we shall see in section 8.4, this ordering coincides with the ordering of 'being at least as informative as' for the corresponding linear normal models with known σ. Thus Ehrenfeld's results may also be phrased as complete class theorems for linear normal models for the ordering $\geqslant_q$. In addition to these papers the reader may also consult Elfving's papers (1952, 1959) and Fellmann's thesis (1974).

4. *Insufficiencies and deficiencies of linear statistics.* The obvious way of measuring the loss of information which we suffer by restricting ourselves to a particular statistic (statistics) is to consider the deficiency of the experiment defined by this statistic with respect to the experiment realized by all available observations.

   In his 1974 paper, LeCam (1974a) introduced the measure of insufficiency for this loss of information. Roughly speaking the insufficiency is obtained by restriction to those randomizations which behave as conditional expectations, i.e. as projections. The insufficiency appears mathematically more tractable than the deficiency although its scope is considerably more limited. The insufficiency occurs quite naturally in asymptotic theory since e.g. asymptotic normality there is often established by showing that insufficiencies are negligible. As the insufficiency majorizes the deficiency such a result is at least as strong as showing that the deficiency is negligible. On the other hand we may sometimes show that the deficiency is small by first establishing weak convergence and then using Lindae's theorem given in chapter 7. Are we then permitted to conclude that the insufficiency is negligible? In general the deficiency may be very much smaller than the insufficiency. This is the case even for linear statistics in linear normal models with known covariances. Thus further arguments are needed. It would therefore be interesting to know when these measures of loss of information are close and when they are far apart. In addition to the 1974 paper mentioned above, the reader may also consult LeCam (1986) which provides a detailed study of the situation when the covariances are known and when both experiments are obtained by replicating the same experiment.

Here is some notation and some conventions which are used in this chapter. Those which are not standard are usually more thoroughly explained on their first occurrence.

Notation and conventions related to inner product spaces:

$\dim H$ = the dimension of the space H.
(u, v) = the inner product of the vectors u and v.
$\|u\|$ = $(u, u)^{1/2}$
$u \perp v$ = denotes that u and v are orthogonal, i.e. that $(u, v) = 0$.
$U^{\perp}$ = $\{v : v|u \text{ when } u \in U\}$
I = the identity operator on some space. Although identity operators on different spaces may be considered in the same paragraph, they are usually all denoted by the same symbol I.
Operator = linear map.
(Linear) transformation = linear map from a linear space to itself.
A^* = the adjoint of the operator A.

Non-negative definite transformations are always self adjoint.
$R(A)$ = the range of A.
$N(A)$ = the nullspace of A.
$\det A$ = the determinant of the linear transformation A.
$A^{1/2}$ = the non-negative definite square root of the non-negative definite transformation A.
$[A]$ = the matrix of the operator A with respect to specified bases.
$\operatorname{diag}(\Delta_1, \ldots, \Delta_p)$ = the $p \times p$ diagonal matrix whose (i, i)th element is Δ_i.
If A and B are self-adjoint operators on the same Hilbert space, then '$A \geqslant B$' is short for '$A - B$ is non-negative definite'.

A^- = a generalized (linear) inverse of the linear map A, i.e. A^- is a linear map such that $AA^-A = A$.
A^+ = the Moore–Penrose (M–P) generalized inverse of A, i.e. A^+ is the unique generalized inverse such that $R(A^+) \subseteqq R(A^*)$ and $N(A^+) \supseteqq N(A^*)$.
$H_1 \oplus H_2$ = the direct sum of the inner product spaces H_1 and H_2.

If A_1 is a linear map from H_1 to K and A_2 is a linear map from H_2 to K then the linear map $A_1 \oplus A_2$ maps $(h_1, h_2) \in H_1 \oplus H_2$ into $A_1h_1 + A_2h_2$. If H_1 and H_2 are subspaces of H then we may also write $H = H_1 \oplus H_2$ when $H_1 \cap H_2 = 0$. If so then H is (isomorphic to) the direct sum of H_1 and H_2.

$A^{-1}[M] = \{x : Ax \in M\}$ is defined whether A is invertible or not.
Maximum and minimum operations $\vee$, $\underline{\vee}$, $\wedge$ and $\overline{\wedge}$ are defined in section 5.

Other notation and conventions are:
$\mathscr{E}$, $\mathscr{F}$, $\mathscr{G}$ and $\mathscr{M}$ usually denote experiments.
$\delta(\mathscr{E}, \mathscr{F})$ = deficiency of $\mathscr{E}$ with respect to $\mathscr{F}$.
$\Delta(\mathscr{E}, \mathscr{F}) = \delta(\mathscr{E}, \mathscr{F}) \vee \delta(\mathscr{E}, \mathscr{F})$ = the deficiency distance between $\mathscr{E}$ and $\mathscr{F}$.
$\mathscr{E} \geqslant \mathscr{F}$ expresses that $\delta(\mathscr{E}, \mathscr{F}) = 0$.
$\mathscr{E} \sim \mathscr{F}$ expresses that $\Delta(\mathscr{E}, \mathscr{F}) = 0$.
'$\geqslant_q$' and '$\sim_q$' are defined in this section.
EX and $\operatorname{Cov} X$ for vector valued variables are defined in this section.
$\int h(\mathrm{d}P_1, \ldots, \mathrm{d}P_m) = \int h(\mathrm{d}P_1/\mathrm{d}\mu, \ldots, \mathrm{d}P_m/\mathrm{d}\mu)\,\mathrm{d}\mu$ where h is non-negatively homogeneous and measurable on $\mathbb{R}^m$ and where the probability measures $P_1, \ldots, P_m$ are majorized by the non-negative measure μ.

This notation reflects the fact that neither the existence nor the value of this integral depends on μ.
$\|P - Q\| = \int |\mathrm{d}P - \mathrm{d}Q|$ = the statistical distance between the probability measures P and Q, i.e. the total variation of $P - Q$.
$\mathscr{L}(X), \mathscr{L}(Y), \ldots$ = the probability distributions (laws) of $X, Y, \ldots$

$\mathscr{L}(A, V)$ = the linear model defined by the operators A and V.
$N(\xi, V)$ = the normal distribution with expectation (vector) ξ and (co)variance (operator) V.
$\mathscr{N}(A, V)$ – the linear normal experiment $(N(A^*\beta, V) : \beta \in K)$ defined by the operators A and V.
$\exp z$ = e^z

$\# F$ = the number of elements in F.

8.2 Comparison based on the Hellinger transform

We begin by describing the Hellinger transform of families of normal distributions.

Theorem 8.2.1 (The Hellinger transform for normal families). *Consider a finite family of non-singular normal distributions $N(\xi_i, V_i)$; $i = 1, \ldots, r$ on a finite dimensional inner product space H. Thus $\xi_1, \ldots, \xi_r$ are vectors in H while $V_1, \ldots, V_r$ are positive definite operators on H.*

If $t_1, \ldots, t_r$ are non-negative numbers such that $t_1 + \cdots + t_r = 1$ and if V_t is the 't-harmonic mean' $[\sum_i t_i V_i^{-1}]^{-1}$ of $V_1, \ldots, V_r$ then

$$-2\log \int \prod_{i=1}^{r} \mathrm{d}N(\xi_i, V_i)^{t_i} = \sum_i t_i \log(\det V_i) - \log(\det V_t) + \sum_i t_i(\xi_i, V_i^{-1}\xi_i)$$
$$- \left(\left(\sum_i t_i V_i^{-1}\xi_i\right), V_t\left(\sum_i t_i V_i^{-1}\xi_i\right)\right).$$

Proof. Choose the Haar measure on H such that the density of $N(\xi, V)$ may be expressed as $x \to (\det V)^{-1/2} \exp(-\frac{1}{2}(x - \xi, V^{-1}(x - \xi)))$. Put $\xi = \sum t_i V_t V_i^{-1}\xi_i$ and $G = \int \prod_i \mathrm{d}N(\xi_i, V_i)^{t_i}$. Then

$$G = \int \prod_i \{(\det V_i)^{-1/2} \exp(-\tfrac{1}{2}((x - \xi_i), V_i^{-1}(x - \xi_i)))\}^{t_i}\mu(\mathrm{d}x)$$

$$= \prod_i (\det V_i)^{-(1/2)t_i} \int [\exp(-\tfrac{1}{2}Q(x))]\mu(\mathrm{d}x)$$

where

$$Q(x) = \sum_i t_i((x - \xi_i), V_i^{-1}(x - \xi_i))$$
$$= \sum_i t_i(x, V_i^{-1}x) - 2\sum_i t_i(\xi_i, V_i^{-1}x) + \sum_i t_i(\xi_i, V_i^{-1}\xi_i)$$
$$= (x, V_t^{-1}x) - 2(\xi, V_t^{-1}x) + \sum_i t_i(\xi_i, V_i^{-1}\xi_i)$$
$$= (x - \xi, V_t^{-1}(x - \xi)) + \sum t_i(\xi_i, V_i^{-1}\xi_i) - (\xi, V_t^{-1}\xi).$$

Hence

$$G = \prod_i [\det V_i]^{-(1/2)t_i} \left[\exp\left(-\frac{1}{2}\left(\sum_i t_i(\xi_i, V_i^{-1}\xi_i) - (\xi, V_t^{-1}\xi) \right)\right) \right] G_1$$

where $G_1 = \int [\exp(-\frac{1}{2}((x - \xi), V_t^{-1}(x - \xi)))]\mu(\mathrm{d}x) = (\det V_t)^{1/2}$. □

Many interesting conclusions may be drawn from this expression. Here is one which is a bit outside the main theme of this chapter.

Corollary 8.2.2 (*log det *V is concave in V). log(det V) *is a (strictly) concave function of V as V varies within the convex set of non-negative definite (positive definite) operators on H.*

Proof. If $\xi_1 = \cdots = \xi_r$ and $V_1, \ldots, V_r$ are non-singular then $0 \leqslant -2\log \int \prod_i \mathrm{d}N(\xi_i, V_i)^{t_i} = \sum_i t_i \log(\det V_i) - \log(\det V_t)$. Thus $\log\det V_t^{-1} \geqslant \sum t_i \log\det V_i^{-1}$ proving the desired concavity inequality for positive definite operators $W_i = V_i^{-1}; i = 1, \ldots, r$. Since $\log\det V = -\infty$ when V is singular, it follows that $\log\det V$ is concave in V. The statement on strict concavity follows from the fact that $(N(\xi, V_i); i = 1, \ldots, r)$ is totally non-informative if and only if $V_1 = V_2 = \cdots = V_r$. □

Assuming that the covariance operators are all the same, we may remove the condition of non-singularity as follows.

Corollary 8.2.3 (The case of equal covariances). *Let V be a non-negative definite operator on the finite dimensional Hilbert space H. Consider also vectors $\xi_1, \ldots, \xi_r$ in H and positive weights $t_1, \ldots, t_r$ such that $t_1 + \cdots + t_r = 1$. Then*

$$\int \prod_i \mathrm{d}N(\xi_i, V)^{t_i} > 0$$

if and only if

$$\xi_i - \xi_j \in R(V); \qquad i = 1, \ldots, r; \qquad j = 1, \ldots, r.$$

In that case

$$\begin{aligned} -2\log \int \prod_i \mathrm{d}N(\xi_i, V)^{t_i} &= \sum t_i(\xi_i, V^+\xi_i) - (\xi_t, V^+\xi_t) \\ &= \sum t_i((\xi_i - \xi_t), V^+(\xi_i - \xi_t)) \\ &= \frac{1}{2}\sum_{i,j} t_i t_j((\xi_i - \xi_j), V^+(\xi_i - \xi_j)) \end{aligned}$$

where $\xi_t = \sum_i t_i \xi_i$.

In particular the affinity γ between $N(\xi_1, V)$ and $N(\xi_2, V)$ is positive if and only if $\xi_1 - \xi_2 \in R(V)$ and then $\gamma = \exp(-\frac{1}{8}((\xi_1 - \xi_2), V^+(\xi_1 - \xi_2)))$.

Proof. Put $G = -2\log\int\prod_i dN(\xi_i, V)^{t_i}$ and let X be distributed as $N(\xi_i, V)$ when 'i prevails'. Replacing X with $X - \xi_t$ we see that we may assume without loss of generality that $\xi_t = 0$. Then $\xi_i - \xi_j \in R(V)$ for all pairs (i, j) if and only if $\xi_1, \ldots, \xi_r \in R(V)$. If, say, $\xi_1 \notin R(V)$ then since $\xi_t = 0$, there is an $i \geqslant 2$ such that $\xi_i - \xi_1 \notin R(V)$. Then the support $\xi_1 + R(V)$ of $N(\xi_1, V)$ is disjoint from the support $\xi_i + R(V)$ of $N(\xi_i, V)$. Thus $G = \infty$ in that case. If on the other hand $\xi_1, \ldots, \xi_r \in R(V)$ then we may apply the proposition to $\tilde{H} = R(V)$ and the operator $\tilde{V}$ obtained from V by restricting the domain as well as the range space (not the range) to $R(V)$. Then $\tilde{V}$ is non-singular and $\tilde{V}^{-1}$ is obtained from V^+ as $\tilde{V}$ is obtained from V. □

This applies directly to the models $\mathcal{N}(A, V)$.

Corollary 8.2.4 (The Hellinger transform for $\mathcal{N}(A, V)$). *Let V be a non-negative definite operator on the finite dimensional Hilbert space H and let A be a linear map from H to the finite dimensional inner product space K.*

Consider also vectors $\beta_1, \ldots, \beta_r$ in K and positive weights $t_1, \ldots, t_r$ such that $t_1 + \cdots + t_r = 1$. Then $\int\prod_i dN(A^*\beta_i, V)^{t_i} > 0$ if and only if $(\beta_i - \beta_j) \perp A[N(V)]$; $i = 1, \ldots, r$; $j = 1, \ldots, r$. Putting $\beta_t = \sum_i t_i\beta_i$ we find

$$-2\log\int\prod_i dN(A^*\beta_i, V)^{t_i} = \sum_i t_i(\beta_i, AV^+A^*\beta_i) - (\beta_t, AV^+A^*\beta_t)$$

$$= \sum_i t_i((\beta_i - \beta_t), AV^+A^*(\beta_i - \beta_t))$$

$$= \frac{1}{2}\sum_{i,j} t_i t_j((\beta_i - \beta_j), AV^+A^*(\beta_i - \beta_j)).$$

In particular the affinity γ between $N(A^*\beta_1, V)$ and $N(A^*\beta_2, V)$ is positive if and only if $(\beta_1 - \beta_2) \perp A[N(V)]$ and then $\gamma = \exp(-\frac{1}{8}((\beta_1 - \beta_2), AV^+A^*(\beta_1 - \beta_2)))$.

Corollary 8.2.5 (Affinity ordering = Hellinger ordering when σ is known). *The Hellinger transform of $\mathcal{N}(A, V)$ is $\leqslant$ the Hellinger transform of $\mathcal{N}(B, W)$ if and only if the affinities of $\mathcal{N}(A, V)$ are always $\leqslant$ the corresponding affinities in $\mathcal{N}(B, W)$ and this in turn holds if and only if $A[N(V)] \supseteqq B[N(W)]$ and $PAV^+A^*P \geqslant PBW^+B^*P$ where P is the orthogonal projection (in K) on $[A[N(V)]]^\perp$. If so then $R(A) \supseteqq R(B)$.*

Proof. Only the last assertion needs to be proved. This set inequality expresses that linear parametric functions which are identifiable in $\mathcal{N}(B, W)$ are also identifiable in $\mathcal{N}(A, V)$. We shall return to this in the next section. Here we will argue solely in terms of the Hellinger transform. Assume then that $\beta \perp R(A)$. Then $\beta \perp A[N(V)]$ so that $\beta = P\beta$ and $PAV^+A^*P\beta = PAV^+A^*\beta = 0$ since $A^*\beta = 0$. Hence $PBW^+B^*P\beta = 0$ so that $B^*\beta = B^*P\beta \in N(W) = R(W)^\perp$. Thus $\beta \perp B[R(W)]$ and we know that $\beta \perp B[N(W)]$. Hence $\beta \perp R(B)$ so that $R(A)^\perp \subseteqq R(B)^\perp$, i.e. $R(A) \supseteqq R(B)$. □

Corollary 8.2.6 (Equivalence of experiments $\mathcal{N}(A, V)$). *The experiments $\mathcal{N}(A, V)$ and $\mathcal{N}(B, W)$ are equivalent if and only if $A[N(V)] = B[N(W)]$ and $(\beta, AV^+A^*\beta) = (\beta, BW^+B^*\beta)$ when $\beta \perp A[N(V)]$. If so then $R(A) = R(B)$.*

Corollary 8.2.7 (Canonical form of $\mathcal{N}(A, V)$). *Each experiment $\mathcal{E} = \mathcal{N}(A, V)$ is equivalent to a unique experiment $\mathcal{E}' = \mathcal{N}(B, W)$ with K as sample space and such that B is the orthogonal projection on $R(B) = R(A)$ while $R(W) \subseteqq R(B)$.*

The covariance operator W may be written

$$W = [PAV^+A^*P]^+$$

where P is the orthogonal projection on $[A[N(V)]]^\perp = [A^*]^{-1}[R(V)]$.

Remark. We shall later see that the experiments $\mathcal{E}'$ are all in minimal form.

Proof. Put $W = [PAV^+A^*P]^+$ where P is the orthogonal projection on $[A[N(V)]]^\perp$. Clearly W is self adjoint and non-negative definite. Since the operation of taking M–P inverses of non-negative definite operators does not change null spaces nor ranges, we find that $N(W) = N(PAV^+A^*P) = N(V^+A^*P) = N(VA^*P) \supseteqq N(A^*P)$. However, the last set inequality is actually an equality, since $VA^*P\beta = 0 \Rightarrow P\beta \in N(VA^*) \Rightarrow P\beta \perp R((VA^*)^*) = R(AV) = A[R(V)]$. On the other hand $P\beta \perp A[N(V)]$ by definition, and $R(A) = A[R(V) + N(V)] = A[R(V)] + A[N(V)]$. Thus $VA^*P\beta = 0 \Rightarrow P\beta \perp R(A) \Rightarrow P\beta \in N(A^*) \Rightarrow A^*P\beta = 0$.

It follows that $N(W) = N(A^*P)$ so that $R(W) = R(PA) = P[R(A)] = P\{[A[N(V)] + [R(A) \cap [A[N(V)]]^\perp]\} = R(A) \cap [A[N(V)]]^\perp = R(A) \cap R(P)$. Thus $N(W) = N(A^*) + A[N(V)]$ and $B[N(W)] = B[N(A^*)] + A[N(V)] = A[N(V)]$ since $N(A^*) \perp R(A)$. Now $R(I - B) = R(A)^\perp \subseteqq [A[N(V)]]^\perp$ so that $P(I - B) = I - B$ and thus $A^*P(I - B) = A^*(I - B) = 0$. Hence $A^*P = A^*PB$, giving $BW^+B^* = BPAV^+A^*PB = PAV^+A^*P = W^+$. If $\beta \perp N(W) = A(N(V))$ then $P\beta = \beta$, yielding $(\beta, BW^+B^*\beta) = (\beta, AV^+A^*\beta)$. Altogether this shows that $\mathcal{N}(A, V) \sim \mathcal{N}(B, W)$.

If W_0 is any non-negative definite operator on K such that $\mathcal{N}(B, W_0) \sim \mathcal{N}(B, W)$ and if $R(W_0) \subseteqq R(A)$ then $W_0^+(I - B^*) = W_0^+(I - B) = 0$ since $R(I - B) = N(A^*) = R(A)^\perp \subseteqq R(W_0)^\perp = N(W_0) = N(W_0^+)$. It follows that $BW_0^+B^* = W_0^+$. Hence, by equivalence, $(\beta, W_0^+\beta) = (\beta W^+\beta)$ when $\beta \perp B[N(W_0)] = B[N(W)] = A[N(V)]$. On the other hand $N(W_0) \supseteqq N(B)$ since $R(W_0) \subseteqq R(A) = R(B)$. Hence $N(W_0) = N(B) \oplus [N(W_0) \cap R(B)]$ so that $B[N(W_0)] = N(W_0) \cap R(B) \subseteqq N(W_0)$. Hence $(\beta, W_0^+\beta) = (\beta, W^+\beta) = 0$ when $\beta \in B[N(W_0)]$. Thus $(\beta, W_0^+\beta) = (\beta, W^+\beta)$ for all β so that $W_0^+ = W^+$, i.e. $W_0 = W$. □

Experiments $\mathcal{N}(A, V)$, and also linear models $\mathcal{L}(A, V)$, are called *non-singular* if $\mathcal{N}(A, V) \sim \mathcal{N}(B, W)$ for some $\mathcal{N}(B, W)$ with W non-singular. On the other hand, if there is no such experiment $\mathcal{N}(B, W)$ then we shall say that $\mathcal{N}(A, V)$ and $\mathcal{L}(A, V)$ are *singular*.

Clearly the notions of singularity and non-singularity both respect equivalence for models $\mathcal{N}(A, V)$. Here are some characterizations of non-singularity.

Proposition 8.2.8 (Non-singularity of models $\mathcal{N}(A, V)$). *The following conditions are equivalent for an experiment $\mathcal{N}(A, V)$.*

(i) *$\mathcal{N}(A, V)$ is non-singular.*
(ii) *$\mathcal{N}(A, V) \sim \mathcal{N}(A\tilde{V}^{-1/2}, I)$ where $\tilde{V}$ is the restriction of V to its range $R(V)$ and I is the identity operator on $R(V)$.*
(iii) *$N(V) \subseteqq N(A)$.*
(iv) *$A[N(V)] = 0$.*
(v) *$R(A^*) \subseteqq R(V)$.*
(vi) *$\mathcal{N}(A, V)$ is dominated.*
(vii) *$\mathcal{N}(A, V)$ is homogeneous.*
(viii) *The Hellinger transform of $\mathcal{N}(A, V)$ is positive everywhere.*

Proof. If $\mathcal{N}(A, V) \sim \mathcal{N}(B, W)$ where $\det W \neq 0$ then $A[N(V)] = B[N(W)] = B\{0\} = \{0\}$ so that $N(V) \subseteqq N(A)$. On the other hand, if $N(V) \subseteqq N(A)$ and X realizes $\mathcal{N}(A, V)$ then X belongs to the set $R(V)$, with probability 1.

On the set $R(V)$ the variable X is computationally equivalent to $(V^+)^{1/2}X$ and the statistic $(V^+)^{1/2}X$ realizes the experiment $\mathcal{N}(A\tilde{V}^{-1/2}, I)$ in (ii). It follows that (i) $\sim$ (ii) $\sim$ (iii). Furthermore, (iv) and (v) are both re-statements of (iii). These conditions imply (vii), which is equivalent to (viii), and trivially (vii) $\Rightarrow$ (vi). The proof may now be completed by checking that (vi) $\Rightarrow$ (v), i.e. that X is supported by $R(V)$ when $\mathcal{N}(A, V)$ is dominated. □

An experiment $\mathcal{N}(A, V)$, or a linear model $\mathcal{L}(A, V)$, will be called *deterministic* if $\mathcal{N}(A, V) \sim \mathcal{N}(B, 0)$ for some experiment $\mathcal{N}(B, 0)$. $\mathcal{N}(A, V)$ is then both non-singular and deterministic if and only if $A = 0$, i.e. if and only if it is totally non-informative.

An experiment $\mathcal{N}(A, V)$ is deterministic if and only if we do not gain information by replicating it.

Proposition 8.2.9 (Deterministic models $\mathcal{N}(A, V)$). *The following conditions are equivalent for an experiment $\mathcal{N}(A, V)$.*

(i) *$\mathcal{N}(A, V)$ is deterministic.*
(ii) *$A[N(V)] = R(A)$.*
(iii) *$\mathcal{N}(A, V)$ is idempotent, i.e. $\mathcal{N}(A, V) \times \mathcal{N}(A, V) \sim \mathcal{N}(A, V)$.*
(iv) *The Hellinger transform of $\mathcal{N}(A, V)$ is an indicator function.*

Remark. If X realizes $\mathcal{L}(A, V)$ then $A[N(V)]$ consists precisely of the vectors k such that (k, β) may be determined with probability 1 on the basis of X. For this reason we will call $A[N(V)]$ *the determinable part of* K *for* $\mathscr{L}(A, V)$ (*for* $\mathscr{N}(A, V)$).

The proposition states that $\mathscr{L}(A, V)$, or $\mathscr{N}(A, V)$, is deterministic if and only if each identifiable linear parametric function belongs to the determinable part of K.

Proof. If $A[N(V)] = R(A)$ then $\mathscr{N}(A, V) \sim \mathscr{N}(A\Pi, 0)$ where Π is the orthogonal projection on $N(V)$. On the other hand, if $\mathscr{N}(A, V) \sim \mathscr{N}(B, W)$ where $W = 0$ then $A[N(V)] = B[N(W)] = R(B) = R(A)$. Thus (i) $\Leftrightarrow$ (ii). These conditions imply (iii) since any $\mathscr{N}(A, 0)$ is clearly idempotent. Furthermore (iii) is clearly a restatement of (iv). If the Hellinger transform is idempotent then its logarithm vanishes on its domain of finiteness, i.e. $AV^+A^*\beta = 0$ when $\beta \perp A(N(V))$. But $AV^+A^*\beta = 0 \Leftrightarrow VA^*\beta = 0 \Leftrightarrow A^*\beta \perp R(V) \Leftrightarrow \beta \perp A[R(V)]$. Thus, since $A[N(V)] + A[R(V)] = R(A)$, $\beta \perp R(A)$ when $\beta \perp A[N(V)]$ so that $A[N(V)] = R(A)$. $\square$

Any experiment $\mathscr{N}(A, V)$ is equivalent to a product $\mathscr{N}(B, W) \times \mathscr{N}(C, U)$ of a non-singular experiment $\mathscr{N}(B, W)$ and a deterministic experiment $\mathscr{N}(C, U)$. In fact $\mathscr{N}(A, V) \sim \mathscr{N}(PA, V) \times \mathscr{N}((I - P)A, 0)$ where P is the orthogonal projection on $A[N(V)]^\perp$.

Let us take a closer look at the Hellinger transform of experiments $\mathscr{N}(A, V)$. For all $\beta \in K$, put $\rho(\beta) = \rho(\beta|A, V) = -4 \log \int (\mathrm{d}N(A^*\beta, V)\,\mathrm{d}N(0, V))^{1/2}$. Then, by corollary 8.2.4, $\rho(\beta) = \frac{1}{2}(\beta, AV^+A^*\beta)$ or $= \infty$ as $A^*\beta \in R(V)$ or $A^*\beta \notin R(V)$.

If t is a prior distribution on K with finite support and if we put $\beta_t = \int \beta t(\mathrm{d}\beta)$ then $-2 \log H(t|\mathscr{N}(A, V)) = \int \rho(\beta - \beta_t)t(\mathrm{d}\beta) = \frac{1}{2}\int\int \rho(\beta' - \beta'')t(\mathrm{d}\beta')t(\mathrm{d}\beta'')$.

In particular the affinity between $N(A^*\beta_1, V)$ and $N(A^*\beta_2, V)$ is $\exp(-\frac{1}{4}\rho(\beta_1 - \beta_2))$. Note also that $R(A) = [\rho = 0]^\perp$, i.e. $N(A^*) = [\rho = 0]$ while $A[N(V)] = [\rho < \infty]^\perp$.

Letting $\mathscr{N}(B, W)$ and P be as in corollary 8.2.7 we find

$$\rho(\beta|A, V) = \rho(\beta|B, W) = \tfrac{1}{2}(\beta, W^+\beta) \qquad \text{or } = \infty$$

as $\beta \perp B[N(W)] = R(B) \cap N(W) = R(A) \cap N(W)$ or not.

Now a *partial quadratic form* on a finite dimensional inner product space H is a real valued function f on H such that for some subspace H_0 of H and for some self-adjoint operator V on H which leaves H_0 invariant, we have $f(h) = \frac{1}{2}(h, Vh)$ or $= \infty$ as $h \in H_0$ or not.

Since $B[N(W)] \subseteq N(W) = N(W^+)$, it follows that $\rho = \rho(\cdot|A, V)$ is a partial quadratic form.

We shall now see that there are no other non-negative partial quadratic forms on K than those of the form $\rho(\cdot|A, V)$ for a linear normal model $\mathscr{N}(A, V)$. Consider a non-negative partial quadratic form ρ on K. Then, by the definition, there is a non-negative definite operator U on K and a U-invariant subspace K_0 such that $\rho(\beta) = \frac{1}{2}(\beta, U\beta)$ or $= \infty$ as $\beta \in K_0$

or not. Let P denote the orthogonal projection on K_0. Then U may be replaced with $\tilde{U} = UP = PUP$. Thus we may assume that $R(U) \subseteqq K_0$. Then $K_0^\perp \oplus R(U) = K_0^\perp + R(U)$ and $K_0^\perp + R(U) = [K_0 \cap N(U)]^\perp = [\rho = 0]^\perp$. Hence $[\rho = 0]^\perp = R(U) \oplus K_0^\perp$.

Let B be the orthogonal projection on $[\rho = 0]^\perp = R(U) \oplus K_0^\perp$ and put $W = U^+$. Then $N(B) = [\rho = 0] = N(U) \cap K_0 \subseteqq N(U) = N(W)$ while $R(W) = R(U) \subseteqq K_0$. Hence $R(B) = [\rho = 0]^\perp = R(W) \oplus [R(B) \cap N(W)]$ and $N(W) = N(U) = N(B) \oplus [N(W) \cap R(B)]$. It follows that $B[N(W)] = N(U) \cap K_0^\perp$ so that $[B[N(W)]]^\perp = R(U) + K_0 = K_0$. If $\beta \perp B[N(W)]$, i.e. if $\beta \in K_0$, then $\rho(\beta|B, W) = \frac{1}{2}(\beta, BW^+B^*\beta) = \frac{1}{2}(\beta, U\beta) = \rho(\beta)$. On the other hand, if $\beta \not\perp B[N(W)]$ then $\rho(\beta|B, W) = \infty = \rho(U)$. This and corollary 8.2.5 prove the following proposition.

Proposition 8.2.10 (Types of experiments $\mathcal{N}(A, V)$ and non-negative partial quadratic forms correspond). *The partial quadratic form $\rho(\cdot|A, V)$ associated with an experiment $\mathcal{N}(A, V)$ defines this experiment up to equivalence.*

A non-negative real valued function ρ on K is a partial quadratic form if and only if it is of the form $\rho(\cdot|A, V)$ for some experiment $\mathcal{N}(A, V)$.

Weak convergence of the linear models $\mathcal{N}(A, V)$ is easily described in terms of the associated partial quadratic forms. Check first that the convex cone of non-negative partial quadratic forms is compact for pointwise convergence. Furthermore, pointwise convergence of Hellinger transforms for these experiments is equivalent to convergence of affinities and this, in turn, is easily seen to be equivalent to convergence of the associated partial quadratic forms.

Theorem 8.2.11 (Weak convergence of experiments $\mathcal{N}(A, V)$). *The family of linear normal experiments of the form $\mathcal{N}(A, V)$ is compact for weak convergence.*

A net $\{\mathcal{N}(A_\nu, V_\nu)\}$ converges weakly to $\mathcal{N}(A, V)$ if and only if $\rho(\beta|A_\nu, V_\nu) \to \rho(\beta|A, V)$ for each $\beta \in$ K.

If $\mathcal{N}(A_i, V_i)$; $i = 1, \ldots, r$, has sample space H_i then $\prod_i \mathcal{N}(A_i, V_i)$ has sample space $H_1 \times \cdots \times H_r$. Equipping this set with an inner product space structure in the usual way we see that $\prod_i \mathcal{N}(A_i, V_i) = \mathcal{N}(A_1 \oplus \cdots \oplus A_r, V_1 \oplus \cdots \oplus V_r)$ where $(A_1 \oplus \cdots \oplus A_r)(h_1, \ldots, h_r) = A_1(h_1) + \cdots + A_r(h_r)$ and $(V_1 \oplus \cdots \oplus V_r)(h_1, \ldots, h_r) = V_1(h_1) + \cdots + V_r(h_r)$ when $h_i \in V_i$; $i = 1, \ldots, r$.

Thus the collection of experiments $\mathcal{N}(A, V)$ is closed for products. Using the multiplicativity of the Hellinger transform, or by direct evaluation, we see that $\rho(\cdot|A_1 \oplus \cdots \oplus A_r, V_1 \oplus \cdots \oplus V_r) = \sum_{i=1}^r \rho(\cdot|A_i, V_i)$.

Proposition 8.2.12 (Products of experiments $\mathcal{N}(A, V)$). *The collection of experiments of the form $\mathcal{N}(A, V)$ is closed for products. $\mathcal{N}(A, V) \sim \prod_{i=1}^r \mathcal{N}(A_i, V_i)$ if and only if $\rho(\cdot|A, V) = \sum_{i=1}^r \rho(\cdot|A_i, V_i)$.*

Another interesting property of the class of non-negative partial quadratic forms is that it is closed under the 'formation' of proper differences. More precisely, if ρ_1 and ρ_2 are non-negative partial quadratic forms on K and if $\rho_1 \geqslant \rho_2$ then $\rho_1 = \rho_2 + \rho_3$ for a non-negative partial quadratic form ρ_3. As we shall soon see, this result has important consequences. However, first let us demonstrate that this is actually so. Consider non-negative partial quadratic forms ρ_1 and ρ_2 on K and assume that $\rho_1 \geqslant \rho_2$. Then for $i = 1, 2$, there is a subspace K_i of K and a non-negative definite operator U_i on K such that $\rho_i(\beta) = \frac{1}{2}(\beta, U_i\beta)$ or $= \infty$ as $\beta \in K_i$ or $\beta \notin K_i$. As we saw before, we may choose U_i so that $R(U_i) \subseteqq K_i$. Then $K_1 \subseteqq K_2$ and $(\beta, U_1\beta) \geqslant (\beta, U_2\beta)$ when $\beta \in K_1$. Also $N(U_1) \cap K_1 = [\rho_1 = 0] \subseteqq [\rho_2 = 0] = N(U_2) \cap K_2$. Let Q be the orthogonal projection on K_1 and put $U_3 = Q(U_1 - U_2)Q$. Then $\rho_1 = \rho_2 + \rho_3$ where $\rho_3(\beta) = \frac{1}{2}(\beta, U_3\beta)$ or $= \infty$ as $\beta \in K_1$ or not.

Using this result we obtain the following complete characterization of the ordering of experiments $\mathcal{N}(A, V)$.

Theorem 8.2.13 (The deficiency ordering of experiments $\mathcal{N}(A, V)$). *The following conditions are equivalent for linear normal experiments $\mathcal{N}(A, V)$ and $\mathcal{N}(B, W)$.*

(i) $\mathcal{N}(A, V) \geqslant \mathcal{N}(B, W)$.
(ii) $\mathcal{N}(A, V) \sim \mathcal{N}(B, W) \times \mathcal{N}(C, U)$ *for some* $\mathcal{N}(C, U)$.
(iii) $\rho(\cdot|A, V) \geqslant \rho(\cdot|B, W)$.

Furthermore, if $\mathcal{N}(A, V)$ and $\mathcal{N}(B, W)$ are non-singular then these conditions are equivalent to:

(iv) $AV^+A^* \geqslant BW^+B^*$.

Remark 1. The Fisher information matrix $I(\theta_0)$ of an experiment $\mathscr{E} = (P_\theta : \theta \in \Theta)$ evaluated at a point θ_0 may be determined under regularity conditions by the affinities since $-2\log\int(\mathrm{d}P_{\theta_0}\mathrm{d}P_{\theta_0+\eta})^{1/2} = \frac{1}{4}(\eta, I(\theta_0)\eta) + o(\|\eta\|^2)$. Thus, since $-2\log\int(\mathrm{d}P_\beta\mathrm{d}P_{\beta+\eta})^{1/2} = \frac{1}{4}(\eta, AV^+A^*\eta)$ we recognize AV^+A^* as the Fisher information 'operator' of $\mathcal{N}(A, V)$, in the non-singular case.

Remark 2. Later we will obtain other interesting criteria for this ordering in terms of operators as well as in terms of risk for the problem of estimating β for quadratic loss.

Proof of the theorem. If $\mathcal{N}(A, V)$ and $\mathcal{N}(B, W)$ are non-singular then the equivalence of conditions (iii) and (iv) follows directly from the expression for ρ given immediately after the proof of proposition 8.2.9. If (iii) holds then by the above remarks $\rho(\cdot|A, V) = \rho(\cdot|B, W) + \rho(\cdot|C, U)$ for some $\mathcal{N}(C, U)$. Thus, by proposition 8.2.12, $\mathcal{N}(A, V) \sim \mathcal{N}(B, W) \times \mathcal{N}(C, U)$. Hence (iii) $\Rightarrow$ (ii) while the implication (ii) $\Rightarrow$ (i) is trivial and we know by corollary 8.2.5 that (i) $\Rightarrow$ (iii). □

The theorem states that the usual ordering of experiments and the much stronger ordering of ordering by factorization, coincide for the linear normal models $\mathscr{N}(A, V)$. The question then arises: what are the factors of experiments $\mathscr{N}(A, V)$? We have just seen that $\mathscr{N}(B, W)$ is a factor of $\mathscr{N}(A, V)$ if and only if $\mathscr{N}(B, W) \leqslant \mathscr{N}(A, V)$. Are there other factors? In general there are, since $\mathscr{N}(A, V)$ may be totally informative and then any experiment is a factor of $\mathscr{N}(A, V)$. Ruling out the singular experiments $\mathscr{N}(A, V)$ we shall see in a moment that all factors of $\mathscr{N}(A, V)$ are equivalent to experiments of the form $\mathscr{N}(B, W)$.

Before embarking on this problem we shall make some relevant general remarks on Gaussian experiments. By the definition in section 8.1, an experiment $\mathscr{E} = (\mathscr{X}, A; P_\theta : \theta \in \Theta)$ is *Gaussian* if and only if $\mathscr{L}(\log(\mathrm{d}P_{\theta_1}/\mathrm{d}P_{\theta_0}), \ldots, \log(\mathrm{d}P_{\theta_r}/\mathrm{d}P_{\theta_0})|P_{\theta_0})$ is normal for each choice of points $\theta_0, \theta_1, \ldots, \theta_r$ in Θ. This definition implies in particular that Gaussian experiments are homogeneous. Thus singular experiments $\mathscr{N}(A, V)$ cannot be Gaussian. On the other hand it is easily seen that any non-singular experiment $\mathscr{N}(A, V)$ is Gaussian. The indecomposability of the normal distribution readily implies that factors of Gaussian experiments are Gaussian. This carries over to the class of non-singular experiments $\mathscr{N}(A, V)$.

Theorem 8.2.14 (Indecomposability of the class of non-singular experiments $\mathscr{N}(A, V)$). *If $\mathscr{F}$ and $\tilde{\mathscr{F}}$ are dominated experiments such that $\mathscr{F} \times \tilde{\mathscr{F}} \sim \mathscr{N}(A, V)$ then $\mathscr{F} \sim \mathscr{N}(B, W)$ for suitable operators B and W.*

Proof. Put $\mathscr{F} = (Q_\beta : \beta \in K)$ and $\tilde{\mathscr{F}} = (\tilde{Q}_\beta : \beta \in K)$ so that $\mathscr{N}(A, V) \sim (Q_\beta \times \tilde{Q}_\beta : \beta \in K)$. Since $\mathscr{F}$, $\tilde{\mathscr{F}} \leqslant \mathscr{N}(A, V) = (N(A^*\beta, V) : \beta \in K)$, it follows that Q_β and $\tilde{Q}_\beta$ depend on β via $A^*\beta$. Thus we may assume that $R(A) = K$. Reducing $\mathscr{N}(A, V)$ to its canonical form and then changing parameter we see that we may assume that $A = V = I =$ the identity operator on K.

Choose a finite sequence $s = (\beta_1, \ldots, \beta_r)$ in K such that the set $\{\beta_1, \ldots, \beta_r\}$ spans K. Introduce the three distributions F, $\tilde{F}$ and G of log likelihoods where

$$F = \mathscr{L}(\log(\mathrm{d}Q_{\beta_i}/\mathrm{d}Q_0); i = 1, \ldots, r|Q_0)$$

$$\tilde{F} = \mathscr{L}(\log(\mathrm{d}\tilde{Q}_{\beta_i}/\mathrm{d}\tilde{Q}_0); i = 1, \ldots, r|\tilde{Q}_0)$$

and

$$G = \mathscr{L}(\log(\mathrm{d}N(\beta_i, I)/\mathrm{d}N(0, I)); i = 1, \ldots, r|N(0, I)).$$

Then $F * \tilde{F} = G$ since $\mathscr{F} \times \tilde{\mathscr{F}} \sim \mathscr{N}(I, I)$. The distribution G is normal with expectation vector $-\frac{1}{2}(\|\beta_1\|^2, \ldots, \|\beta_r\|^2)$ and with covariance matrix $C = ((\beta_i, \beta_j); i = 1, \ldots, r, j = 1, \ldots, r)$. From the indecomposability property of the class of normal distributions it follows that F and $\tilde{F}$ are both normal distributions. Furthermore, since $G = F * \tilde{F}$ the covariance matrix C is the sum of the covariance matrices of F and of $\tilde{F}$. Thus $M \leqslant C$ where M is the covariance matrix of F.

Let $\hat{M}$ and $\hat{C}$ be the operators on $\mathbb{R}^r$ defined by, respectively, the covariance matrices M and C. Then $\hat{M} \leqslant \hat{C} = \Gamma^*\Gamma$ where Γ is the operator from $\mathbb{R}^r$ to K which maps $(z_1, \ldots, z_r) \in \mathbb{R}^r$ into $z_1\beta_1 + \cdots + z_r\beta_r \in K$. Putting $D = ((\Gamma^+)^*\hat{M}\Gamma^+)^{1/2}$ we find, since $R(\Gamma^*) \supseteq R(\hat{M})$, that $\hat{M} = \Gamma^*(\Gamma^+)^*\hat{M}\Gamma^+\Gamma = \Gamma^*D^2\Gamma$. Putting $e_i = (0, \ldots, \overset{(i)}{1}, \ldots, 0)$ this yields $M_{i,j} = (e_i, \hat{M}e_j) = (\Gamma e_i, D^2\Gamma e_j) = (\beta_i, D^2\beta_j) = (D\beta_i, D\beta_j)$.

Let us compare the experiments $\mathscr{F}$ and $\mathscr{N}(D, I) = (N(D^*\beta, I); \beta \in K)$. Put

$$H = \mathscr{L}(\log(\mathrm{d}N(D^*\beta_i, I)/\mathrm{d}N(0, I)); i = 1, \ldots, r | N(0, I))$$

where I denotes the identity operator on $\mathbb{R}^r$. Then H is a normal distribution with expectation vector $-\frac{1}{2}(\|D\beta_1\|^2, \ldots, \|D\beta_2\|^2)$ and covariance matrix $U = \{(D\beta_i, D\beta_j)\} = \{M_{ij}\} = M$. Thus F and H have the same covariance matrices. The fact that F is both normal and the distribution of log likelihoods, connects the expectations and the variances in a particular way. In fact in general, if X is real valued and normally distributed and if e^X is a probability density, then $1 = Ee^X = \exp(EX + \frac{1}{2}\operatorname{Var} X)$ so that $EX = -\frac{1}{2}\operatorname{Var} X$. Since F and H are both normal with the same covariance matrix, it follows that F and H have the same expectation. Hence $F = H$. This implies that the restriction of $\mathscr{F}$ to the set $\{\beta_1, \ldots, \beta_r\}$ is equivalent to the restriction of $\mathscr{N}(D, I)$ to the same set. Note however that D depends on the sequence $s = (\beta_1, \ldots, \beta_r)$. In fact D is an operator on $\mathbb{R}^r$ and therefore depends trivially on r. Let us indicate the dependence on s by writing D_s and I_s instead of D and I. Direct the set of finite sequences in K by the inclusion ordering of the sets they define. By compactness there is a subnet $\{D_{s'}\}$ such that $\mathscr{N}(D_{s'}, I_{s'}) \to \mathscr{N}(B, W)$ for some pair (B, W) of linear maps B and W. Then $\mathscr{N}(B, W)|\{\beta_1, \ldots, \beta_r\} \sim \mathscr{F}|\{\beta_1, \ldots, \beta_r\}$ for all finite sequences $(\beta_1, \ldots, \beta_r)$ in K. Hence $\mathscr{N}(B, W) \sim \mathscr{F}$. □

Another consequence of the indecomposability property of the family of normal distributions is a kind of maximality of the class of non-singular models $\mathscr{N}(A, V)$. This may be explained as follows.

Assume that a K-valued statistic $Y = B^*\beta +$ error realizes an experiment which is at least as informative as a non-singular model $\mathscr{N}(A, V)$ where $R(A) = K$. Assume also that the regression operator B^*, as well as the distribution of the error, is known. Then Y is automatically normally distributed.

This may be seen by first noting that we may assume that $\mathscr{N}(A, V)$ is in canonical form and then we may assume also that $A = I =$ the identity operator on K. Then $\mathscr{N}(A, V) = \mathscr{N}(I, V)$ is an absolutely continuous translation family on K. Since $R(B) \supseteqq R(A) = K$, the operator B is necessarily invertible. Replacing Y with $Z = [B^*]^{-1}Y$ we see that the statistic Z defines a translation experiment which is at least as informative as the normal translation experiment $\mathscr{N}(I, V)$. The convolution criterion now tells us that

$\mathscr{L}([B^*]^{-1}Y|\beta = 0)$ is a convolution factor of $N(0, V)$ and therefore is a normal distribution.

In his paper on least informative distributions, Lehmann (1983b) looks at translation experiments $\mathscr{E}$ on the line such that the variance of the translated distribution is fixed and finite. If we replicate $\mathscr{E}$ n times and then consider the minimum risk for estimating the translation parameter for quadratic loss and for invariant (equivariant) estimators, then Lehmann shows that this minimum risk is maximized for a normal translation family $\mathscr{E}$. The result argued above is related to this, since it implies that no non-normal coherent translation family can be more informative than a normal one.

Let us turn to situations where the covariances are known except for a common multiplicative factor. Such a model is realized by a vector valued statistic X such that $(X - A^*\beta)/\sigma$ has expectation $0 \in K$ and known covariance operator V where $\beta \in K$ and $\sigma > 0$ are unknown parameters. If, in addition, we assume that X is normally distributed, then X realizes the linear normal experiment $\mathscr{N}(A, \sigma^2 V; \sigma^2 > 0)$. Again we may learn quite a bit from the Hellinger transforms. However, it turns out that the ordering of these experiments has the property that replicates $\mathscr{E}^n$ and $\mathscr{F}^n$ of $\mathscr{E}$ and $\mathscr{F}$ may be comparable although $\mathscr{E}$ and $\mathscr{F}$ themselves are not comparable. As $H(\cdot|\mathscr{E}^n) = H(\cdot|\mathscr{E})^n$ for any experiment $\mathscr{E}$, such an ordering cannot be the ordering of Hellinger transforms, or of Fisher information matrices. Using statistical tools, in section 8.6 we will obtain a simple complete and explicit description of the ordering of these experiments. However let us now see what the Hellinger transform tells us.

Proposition 8.2.15 (Hellinger transform for normal families with proportional covariances). *Let V be a non-negative definite operator on the finite dimensional inner product space H. Consider also vectors $\xi_1, \ldots, \xi_r$ in H as well as positive numbers $\sigma_1^2, \ldots, \sigma_r^2$ and positive weights $t_1, \ldots, t_r$ such that $t_1 + \cdots + t_r = 1$.*

Associate with the t's and the σ's the positive weights $s_i = (t_i/\sigma_i^2)/\sum_j (t_j/\sigma_j^2)$; $i = 1, \ldots, r$ and put $\xi_s = \sum s_i \xi_i$ and $\sigma_s^2 = \sum s_i \sigma_i^2$. Then $\bigwedge_i N(\xi_i, \sigma_i^2 V) \neq 0$ if and only if $\xi_i - \xi_j \in R(V)$; $i = 1, \ldots, r$, $j = 1, \ldots, r$. If so, then

$$-2\log \int \prod_i dN(\xi_i, \sigma_i^2 V)^{t_i} = [\text{rank } V]\left[\left(\log \sum_i t_i \sigma_i^{-2}\right) - \sum_i t_i \log \sigma_i^{-2}\right]$$

$$+ \sigma_s^{-2}\left[\sum_i s_i(\xi_i V^+ \xi_i) - (\xi_s, V^+ \xi_s)\right].$$

Here the last bracket may be written as

$$\sum_i s_i((\xi_i - \xi_s), V^+(\xi_i - \xi_s)) = \frac{1}{2}\sum_{i,j} s_i s_j((\xi_i - \xi_j), V^+(\xi_i - \xi_j)).$$

In particular the affinity γ *between* $N(\xi_1, \sigma_1^2 V)$ *and* $N(\xi_2, \sigma_2^2 V)$ *is positive if and only if* $\xi_1 - \xi_2 \in R(V)$ *and then*

$$\gamma = [2\sigma_1\sigma_2/(\sigma_1^2 + \sigma_2^2)]^{1/2\,\text{rank}\,V} \exp[-\tfrac{1}{4}((\xi_1 - \xi_2), V^+(\xi_1 - \xi_2))/(\sigma_1^2 + \sigma_2^2)].$$

Corollary 8.2.16 (The Hellinger transform of $\mathcal{N}(A, \sigma^2 V; \sigma^2 > 0)$). *Consider the experiment* $\mathcal{N}(A, \sigma^2 V; \sigma^2 > 0)$ *and the prior distribution* t *which assigns positive mass* t_i; $i = 1, \ldots, r$, *to the parameter point* (σ_i^2, β_i). *Assume* $\sum_i t_i = 1$ *and put*

$$s_i = (t_i/\sigma_i^2)\Big/\sum_j (t_j/\sigma_j^2); \qquad i = 1, \ldots, r$$

$$\sigma_s^2 = \sum_i s_i\sigma_i^2 = 1\Big/\sum_j (t_j/\sigma_j^2)$$

and

$$\beta_s = \sum_i s_i\beta_i.$$

Then the value of the Hellinger transform H *of* $\mathcal{N}(A, \sigma^2 V; \sigma^2 > 0)$ *at* t *is positive if and only if*

$$(\beta_i - \beta_j) \perp A[N(V)]; \qquad i = 1, \ldots, r, \qquad j = 1, \ldots, r$$

and then

$$-2\log H = [\text{rank}\,V]\left[\left(\log\sum_i t_i\sigma_i^{-2}\right) - \sum t_i \log\sigma_i^{-2}\right]$$
$$+ \sigma_s^{-2}\left[\sum_i s_i(\beta_i, AV^+A^*\beta_i) - (\beta_s, AV^+A^*\beta_s)\right].$$

Here the last bracket may be written as

$$\sum s_i((\beta_i - \beta_s), AV^+A^*(\beta_i - \beta_s)) = \frac{1}{2}\sum_{i,j} s_is_j((\beta_i - \beta_j), AV^+A^*(\beta_i - \beta_j))$$
$$= -2\log H(s|\mathcal{N}(A, V)).$$

In particular the affinity γ *between* $N(A^*\beta_1, \sigma_1^2 V)$ *and* $N(A^*\beta_2, \sigma_2^2 V)$ *is positive if and only if* $(\beta_1 - \beta_2) \perp A[N(V)]$ *and then*

$$\gamma = [2\sigma_1\sigma_2/(\sigma_1^2 + \sigma_2^2)]^{1/2\,\text{rank}\,V} \exp[-\tfrac{1}{2}\rho(\beta_1 - \beta_2|A, V)/(\sigma_1^2 + \sigma_2^2)].$$

Corollary 8.2.17 (Hellinger ordering = ordering by affinities when σ is unknown). *The following conditions are equivalent for the experiments* $\mathcal{E} = \mathcal{N}(A, V\sigma^2; \sigma^2 > 0)$ *and* $\mathcal{F} = \mathcal{N}(B, W\sigma^2; \sigma^2 > 0)$.

(i) $H(\cdot|\mathcal{E}) \leqslant H(\cdot|\mathcal{F})$.
(ii) *The affinities in* $\mathcal{E}$ *are always* $\leqslant$ *the corresponding affinities in* $\mathcal{F}$.
(iii) $\mathcal{N}(A, V) \geqslant \mathcal{N}(B, W)$ *and* rank $V \geqslant$ rank W.

If the models $\mathscr{L}(A, V)$ and $\mathscr{L}(B, W)$ are non-singular then these conditions are equivalent to the condition:

(iv) $AV^+A^* \geqslant BW^+B^*$ *and* rank $V \geqslant$ rank W.

Remark 1. Condition (iv) states that the Fisher information operator of $\mathscr{E}$ majorizes the Fisher information operator of $\mathscr{F}$.

Remark 2. If $\mathscr{E} \geqslant \mathscr{F}$ then of course conditions (i)–(iii) hold. As we remarked above we will see that the converse does not hold, i.e. that conditions (i)–(iii) may be satisfied although $\mathscr{E}$ and $\mathscr{F}$ are not comparable.

Although we cannot easily characterize the deficiency ordering by Hellinger transforms, we can at least deduce the following complete and simple criterion for equivalence.

Corollary 8.2.18 (Equivalence when σ is unknown).

$$\mathscr{N}(A, \sigma^2 V; \sigma^2 > 0) \sim \mathscr{N}(B, \sigma^2 W; \sigma^2 > 0)$$

if and only if

$$\mathscr{N}(A, V) \sim \mathscr{N}(B, W)$$

and rank $V =$ rank W.

What does equivalence actually say about the involved operators? Consider first reduction to a canonical form.

Theorem 8.2.19 (Canonical form of $\mathscr{N}(A, \sigma^2 V; \sigma^2 > 0)$). *To each experiment $\mathscr{E} = \mathscr{N}(A, \sigma^2 V; \sigma^2 > 0)$ there corresponds a unique non-negative definite operator W on K with range contained in $R(A)$ and a unique integer s such that*

$$\mathscr{E} \sim \mathscr{E}_1 \times \mathscr{E}_2$$

where $\mathscr{E}_1 = \mathscr{N}(B, \sigma^2 W; \sigma^2 > 0)$ where B is the orthogonal projection on $R(A)$ and where $\mathscr{E}_2$ is realized by a statistic Z such that Z/σ^2 is χ^2-distributed with s degrees of freedom.

The operator W is defined as in corollary 8.2.7, *while* $s = \dim V[N(A)] = \dim[R(V) \cap [N(A) + N(V)]]$.

Remark 1. $s =$ rank $V -$ rank A when $\mathscr{N}(A, V)$ is non-singular.

Remark 2. By sufficiency, the experiment $\mathscr{E}_2$ is equivalent to $\mathscr{N}(0, \sigma^2 I_s)$ where I_s is the identity operator on $\mathbb{R}^s$. It follows that the factorization may also be written

$$\begin{aligned}\mathscr{N}(A, \sigma^2 V; \sigma^2 > 0) &\sim \mathscr{N}(B, \sigma^2 W; \sigma^2 > 0) \times \mathscr{N}(0, \sigma^2 I_s; \sigma^2 > 0)\\ &= \mathscr{N}(B \oplus 0, \sigma^2 (W \oplus I_s); \sigma^2 > 0).\end{aligned}$$

However the last experiment is not in minimal form, while $\mathscr{E}_1 \times \mathscr{E}_2$ is in minimal form, as will be shown in the next section.

Proof. Let B denote the orthogonal projection on $R(A)$. If $\mathscr{E} \sim \mathscr{E}_1 \times \mathscr{E}_2$ then we see, by specifying σ, that $\mathscr{N}(A, V) \sim \mathscr{N}(B, W)$. Thus W is necessarily as defined in corollary 8.2.7. Hence $\mathscr{E}_1$ is unique. Specifying $\beta = 0$ we obtain $(\mathscr{E}|\beta = 0) \sim (\mathscr{E}_1|\beta = 0) \times (\mathscr{E}_2|\beta = 0)$. As these experiments are all homogeneous the type of $(\mathscr{E}_2|\beta = 0)$ is determined by this factorization. The uniqueness of s now follows from the fact that $2\sigma^4/s^2$ is the variance of the UMVU estimator of σ^2 when $s > 0$. Alternatively we may observe, by sufficiency, that $\mathscr{E}_2$ is realized by s independent $N(0, \sigma^2)$-distributed statistics so that the affinity between, say σ and $\sigma = 1$, is $[\int dN(0, \sigma^2)^{1/2} dN(0, 1)^{1/2}]^s$. Altogether this proves uniqueness. It remains to prove existence. Let W be defined as in corollary 8.2.7 and put $s = \dim V[N(A)]$. If $h \in V[N(A)]$ then $h = Vh_0$ where $h_0 \in N(A)$. Decompose h_0 as $h_0 = h_1 + h_2$ where $h_1 \in N(V)$ and $h_2 \in R(V)$. Then $h = V(h_1 + h_2) = Vh_2$ while $h_2 = h_0 - h_1 \in R(V) \cap [N(A) + N(V)]$. Thus the V-image of $R(V) \cap [N(A) + N(V)]$ contains the set $V[N(A)]$. On the other hand, any $h \in R(V) \cap [N(A) + N(V)]$ may be decomposed as $h = g_1 + g_2$ where $g_1 \in N(A)$ and $g_2 \in N(V)$. Then $Vh = Vg_1 \in V[N(A)]$. It follows that the V-images of the spaces $R(V) \cap [N(A) + N(V)]$ and $N(A)$ coincide. Thus, since V is 1–1 on its range, $s = \dim[R(V) \cap [N(A) + N(V)]] = \dim[R(V) \cap [R(A^*) \cap R(V)]^\perp]$. Note next that, by sufficiency, $\mathscr{E}_2 \sim \mathscr{N}(0, \sigma^2 I_s : \sigma^2 > 0)$ where I_s is the identity operator on $\mathbb{R}^s$. Here 0 should be interpreted as the 0-map from $\mathbb{R}^s$ to K. It follows that $\mathscr{E}_1 \times \mathscr{E}_2 \sim \mathscr{F}$ where $\mathscr{F} = \mathscr{N}(B \oplus 0, \sigma^2(W \oplus I_s); \sigma^2 > 0)$. Therefore by corollary 8.2.18 it remains to show that $\operatorname{rank} V = \operatorname{rank}(W \oplus I_s)$. The first requirement of corollary 8.2.18 is satisfied, since $\mathscr{N}(A, V) \sim \mathscr{N}(B, W) \sim \mathscr{N}(B, W) \times \mathscr{N}(0, I_s)$. Furthermore, by corollary 8.2.7, $\operatorname{rank}(W \oplus I_s) = s + \operatorname{rank}(W)$. Let P denote the orthogonal projection on $[A[N(V)]]^\perp = [A^*]^{-1}[R(V)]$. In the proof of corollary 8.2.7 we found that $R(W) = R(PA)$. Thus $\operatorname{rank}(W) = \operatorname{rank}(PA) = \operatorname{rank}((PA)^*) = \operatorname{rank}(A^*P) = \dim R(A^*P)$.

If $h \in N(PA)$ then $Ah \in N(P) = A[N(V)]$ so that $Ah = Ah_0$ for some $h_0 \in N(V)$. Thus $h = h - h_0 + h_0 \in N(A) + N(V)$. Conversely, if $h = h_1 + h_0$ where $h_1 \in N(A)$ and $h_0 \in N(V)$, then $PAh = PAh_0 = 0$ since $Ah_0 \in A[N(V)]$. Thus $N(PA) = N(A) + N(V)$, yielding $R(A^*P) = R(A^*) \cap R(V)$ and thus $\operatorname{rank} W = \dim[R(V) \cap R(A^*)]$ while $s = \dim[R(V) \cap [R(V) \cap R(A^*)]^\perp]$. Thus, putting $H_0 = R(V) \cap R(A^*)$, we find that $s + \operatorname{rank} W = \dim(H_0) + \dim(R(V) \cap H_0^\perp) = \dim[H_0 + R(V) \cap H_0^\perp] = \dim R(V)$. □

Here is what equivalence amounts to in terms of the involved operators.

Theorem 8.2.20 (Operator criteria for equivalence). *Consider the linear normal experiment $\mathscr{E} = \mathscr{N}(A, \sigma^2 V; \sigma^2 > 0)$ and $\mathscr{F} = \mathscr{N}(B, \sigma^2 W; \sigma^2 > 0)$ with sample spaces H_A and H_B respectively. Then $\mathscr{E}$ and $\mathscr{F}$ are equivalent if and only if there is a linear map S from H_A to H_B such that:*

(i) *$SA^* = B^*$ and $SVS^* = W$.*

(ii) *S is 1–1 on $R(A^*) + R(V)$.*

The map S may be chosen such that

(iii) $N(S) = N(V) \cap N(A)$.

If S satisfies conditions (i), (ii) *and* (iii) *then* S^+ *fulfils the same conditions with* $\mathscr{E}$ *and* $\mathscr{F}$ *interchanged.*

Remark. If S satisfies (i) and (ii) then $S[R(A^*) + R(V)] = R(SA^*) + R(SV) = R(B^*) + N(VS^*)^\perp = R(B^*) + N(SVS^*)^\perp = R(B^*) + N(W)^\perp = R(B^*) + R(W)$ so that S is actually a linear pairing of the supports $R(A^*) + R(V)$ and $R(B^*) + R(W)$.

Proof. Assume S satisfies (i) and (ii). Put $\tilde{S}h = Sh$ or $=0$ as $h \in R(A^*) + R(V)$ or $h \in N(A) \cap N(V)$. Then $\tilde{S}$ satisfies all three conditions (i), (ii) and (iii). Assume that S satisfies these conditions. Condition (iii) implies that $R(S^*) = N(S)^\perp = R(A^*) + R(V)$ and that $R(S) = R(B^*) + R(W)$. Thus $h_A = S^+ h_B$ is the unique vector in $R(A^*) + R(V)$ such that $Sh_A = h_B$ when $h_B \in R(B^*) + R(W)$. Thus S^+ is 1–1 on $R(B^*) + R(W)$. We find also that $S^+ B^* = S^+ SA^* = A^*$ and that $S^+ W(S^+)^* = S^+ SVS^*(S^*)^+ = V$. Furthermore, $S^+ h_B = 0$ when $h_B \in N(B) \cap N(W) = R(S)^\perp$ so that $N(S^+) \supseteqq N(B) \cap N(W)$. It follows that $N(S^+) = N(B) \cap N(W)$. Thus S maps $\mathscr{E}$ into $\mathscr{F}$ while S^+ maps $\mathscr{F}$ into $\mathscr{E}$ so that $\mathscr{E} \sim \mathscr{F}$ whenever an S satisfying (i) and (ii) exists.

It remains to show that such an S exists when $\mathscr{E} \sim \mathscr{F}$. Assume then that $\mathscr{E} \sim \mathscr{F}$ and that $V = I_A$ while $W = I_B$, where I_A (I_B) is the identity operator on $H_A(H_B)$. By corollary 8.2.6 $AA^* = BB^*$ and $\dim H_A = \operatorname{rank} V = \operatorname{rank} W = \dim H_B = p$, say. Then $(A^*\beta_1, A^*\beta_2) = (B^*\beta_1, B^*\beta_2)$; $\beta_1, \beta_2 \in K$ so that $R(A^*)$ and $R(B^*)$ are linearly isometric by the correspondence $A^*\beta \leftrightarrow B^*\beta$. Since $\dim H_A = \dim H_B$, this isometry may be extended to an isometry S from H_A onto H_B. The inverse of this isometry is the isometry S^* from H_B onto H_A. Furthermore, $SA^*\beta = B^*\beta$ and $SVS^* = SS^* = I_B = W$. This proves the existence of S in this case. Now assume that $\mathscr{E} \sim \mathscr{F}$ and that $\mathscr{L}(A, V)$ is non-singular. Then $\mathscr{L}(B, W)$ is also non-singular. It follows that $\mathscr{E}$ and $\mathscr{F}$ are supported by $R(V)$ and $R(W)$ respectively. Thus we may assume without loss of generality that $R(V) = H_A$ and $R(W) = H_B$, i.e. that V and W are non-singular. Therefore assume that V and W are non-singular and let X and Y realize, respectively, $\mathscr{E}$ and $\mathscr{F}$. Then $V^{-1/2}X$ and $W^{-1/2}Y$ realize, respectively, $\mathscr{N}(\tilde{A}, \sigma^2 I_A; \sigma^2 > 0) \sim \mathscr{E}$ and $\mathscr{N}(\tilde{B}, \sigma^2 I_B; \sigma^2 > 0) \sim \mathscr{F}$ where $\tilde{A} = AV^{-1/2}$ and $\tilde{B} = BW^{-1/2}$. Applying the existence result proved above we obtain a linear isometry $\tilde{S}$ from H_A onto H_B such that $\tilde{S}\tilde{A}^* = \tilde{B}^*$. Then $S = W^{1/2}\tilde{S}V^{-1/2}$ satisfies (i) and (ii).

Finally consider the general case of equivalent experiments $\mathscr{E} = \mathscr{N}(A, \sigma^2 V; \sigma^2 > 0)$ and $\mathscr{F} = \mathscr{N}(B, \sigma^2 W; \sigma^2 > 0)$. Then $\mathscr{N}(A, V) \sim \mathscr{N}(B, W)$ so that $R(A) = R(B)$ and $A[N(V)] = B[N(W)]$. Let P denote the orthogonal projection on $[A[N(V)]]^\perp$. Then by corollary 8.2.6 $\mathscr{N}(PA, V) \sim \mathscr{N}(PB, W)$

so that, since rank V = rank W, $\mathcal{N}(PA, \sigma^2 V; \sigma^2 > 0) \sim \mathcal{N}(PB, \sigma^2 W; \sigma^2 > 0)$. Hence, since $\mathcal{N}(PA, V)$ and $\mathcal{N}(PB, W)$ are non-singular $N(PA) = N(A) + N(V) \supseteqq N(V)$ and $N(PB) = N(B) + N(W) \supseteqq N(W)$, and by what we have proved so far, we conclude that there is a linear map Γ from H_A to H_B which is 1–1 on $R(V)$ and such that $\Gamma A^*P = B^*P$ and $\Gamma V \Gamma^* = W$. If $A^*\beta + h = 0$ where $h \in R(V)$ then $A^*\beta = -h \perp N(V)$ so that $\beta \perp A[N(V)]$ yielding $\beta = P\beta$. Hence $0 = \Gamma(A^*\beta + h) = \Gamma(A^*P\beta + h) = B^*P\beta + \Gamma(h) = B^*\beta + \Gamma(h)$. Conversely if $0 = B^*\beta + \Gamma(h)$ where $h \in R(V)$, then $\Gamma(h) \in R(W)$ yielding $B^*\beta = -\Gamma(h) \perp N(W)$ so that $\beta \perp B[N(W)]$, i.e. $P\beta = \beta$. Hence $0 = B^*P\beta + \Gamma(h) = \Gamma(A^*P\beta) + \Gamma(h) = \Gamma(A^*P\beta + h) = \Gamma(A^*\beta + h)$ yielding $A^*\beta + h = 0$ since Γ is 1–1 on $R(V)$. It follows that if h is restricted to $R(V)$ then the map $A^*\beta + h \to B^*\beta + \Gamma(h)$ is a well defined linear map and that this map is a pairing of $R(A^*) + R(V)$ and $R(B^*) + R(W)$. Extending this map to a linear map S from H_A to H_B we find that $SA^*\beta = B^*\beta$ when $\beta \in K$, so that $SA^* = B^*$. Finally $R(V) \subseteqq N(S - \Gamma)$ yielding $N(V) \supseteqq R(S^* - \Gamma^*)$. Hence $SVS^* = \Gamma V \Gamma^* = W$. □

We are now in a position to completely describe ordering by factorization *within* the collection of experiments $\mathcal{N}(A, \sigma^2 V; \sigma^2 > 0)$.

Theorem 8.2.21 (Ordering by factorization of experiments $\mathcal{N}(A, \sigma^2 V; \sigma^2 > 0)$). *The following conditions are equivalent for experiments $\mathcal{E} = \mathcal{N}(A, \sigma^2 V; \sigma^2 > 0)$ and $\mathcal{F} = \mathcal{N}(B, \sigma^2 W; \sigma^2 > 0)$.*

(i) $\mathcal{E} \sim \mathcal{F} \times \mathcal{N}(C, \sigma^2 U; \sigma^2 > 0)$ *for some operators C and U.*
(ii) $\mathcal{F} = \mathcal{E}T^{-1}$ *for a linear map T.*

Remark. Turning to other aspects of these experiments, as we mentioned above, we will see later that these conditions completely characterize the (deficiency) ordering for experiments $\mathcal{N}(A, \sigma^2 V; \sigma^2 > 0)$.

Proof. Let H_A and H_B denote the sample spaces of $\mathcal{E}$ and $\mathcal{F}$ respectively. Let X be an H_A-valued statistic which realizes $\mathcal{E}$.

Assume $\mathcal{F} = \mathcal{E}T^{-1}$ for a linear map T from H_A to H_B. Decompose H_A as $H_A = R(T^*) + H_0$ where H_0 is the V-orthogonal complement of $R(T^*)$, i.e. $H_0 = \{h: Vh \perp R(T^*)\} = V^{-1}[N(T)]$. This decomposition is not direct unless $R(T^*) \cap H_0 = R(T^*) \cap N(V) = 0$. Adjusting for this we obtain the direct (but not necessarily orthogonal) decomposition $H_A = R(T^*) + H_0 \cap [N(T) + R(V)]$. For each $h \in H_A$, let $\Psi_1(h)$ and $\Psi_2(h)$ be, respectively, the components of h in $R(T^*)$ and in $H_0 \cap [N(T) + R(V)]$, according to this decomposition. Put $\Phi_i = \Psi_i^*$; $i = 1, 2$. Then $R(\Phi_1^*) = R(\Psi_1) = R(T^*)$ and $R(\Phi_2^*) = R(\Psi_2) \subseteqq H_0$. Hence $N(\Phi_1) = N(T)$ and $(h_1, Vh_2) = 0$ when $h_i \in R(\Phi_i^*)$; $i = 1, 2$. Thus $\Phi_1(X)$ and $T(X)$ are computationally equivalent while $\Phi_1(X)$ and $\Phi_2(X)$ are uncorrelated. Since X is normally distributed it follows that $\Phi_1(X)$ and $\Phi_2(X)$ are independent. Furthermore $X = \Phi_1(X) + \Phi_2(X)$ since

$\|(V^+)^{1/2}(\Pi(X) - \xi)\|^2 \geqslant \|(V^+)^{1/2}(X - \Pi(X))\|^2$ with equality if and only if $(V^+)^{1/2}(\Pi(X) - \xi) = 0$. Since $\Pi(X) - \xi \in R(V) = R((V^+)^{1/2})$ the last condition is satisfied if and only if $\xi = \Pi(X)$. □

Corollary 8.3.7 (BLUEs as maximum likelihood estimators). *If X realizes a non-singular linear normal model $\mathcal{N}(A, V)$ and if densities are specified continuous, then the maximum likelihood estimator of EX is also the* BLUE *of EX.*

In particular $\hat{\beta}(X)$ is a BLUE *generator if and only if it is a linear maximum likelihood estimator of β.*

It has been noted by several authors that it is possible to replace the covariance operator by a larger operator without destroying the space $V[N(A)]^\perp$ of BLUEs. We refer to Drygas (1983, 1985) and the references there for further information on this. In fact Stepniak (1985) pointed out that we may do this so that the new model is non-singular. This may be achieved by replacing V in $\mathcal{L}(A, V)$ with $\tilde{V} = V + A^*\Gamma A$ where Γ is positive definite. Whether Γ is positive definite or just non-negative definite, $\tilde{V}[N(A)] = V[N(A)]$ so that the BLUEs, as well as their expectations, are the same in the two models. As we saw in corollary 8.1.6 this smoothing device is deficiency decreasing as well as order preserving. Some interesting internal properties of this device are given in the following proposition. Others are derived in section 8.4.

Proposition 8.3.8 (Preservation of BLUEness by smoothing). *Let the variables X and $\tilde{X}$ realize the linear models $\mathcal{L}(A, V)$ and $\mathcal{L}(A, \tilde{V})$ respectively, where $\tilde{V} = V + A^*\Gamma A$ for some non-negative definite operator Γ on K.*

Then $V[N(A)] = \tilde{V}[N(A)]$ so that $T(X)$ is BLUE *for $ET(X) = ET(\tilde{X})$ if and only if $T(\tilde{X})$ is* BLUE. *In particular $\hat{\beta}(X)$ is a* BLUE *generator in $\mathcal{L}(A, V)$ if and only if $\hat{\beta}(\tilde{X})$ is a* BLUE *generator in $\mathcal{L}(A, \tilde{V})$. If so, then*

$$\operatorname{Var}(k, \hat{\beta}(\tilde{X})) = \operatorname{Var}(k, \hat{\beta}(X)) + (k, \Gamma k)$$

whenever $k \in R(A)$.

Finally, $\mathcal{L}(A, \tilde{V})$ is non-singular (i.e. $R(A^*) \subseteqq R(\tilde{V}^*)$) if and only if Γ is 1–1 on the determinable part $A[N(V)]$ of *$\mathcal{L}(A, V)$.*

Remark. We will soon see that $\mathcal{N}(A, \tilde{V}) \leqslant \mathcal{N}(A, V)$.

Proof. Note first that $\tilde{V}[N(A)] = (V + A^*\Gamma A)[N(A)] = V[N(A)]$. Thus $T(X)$ is BLUE in $\mathcal{L}(A, V)$ if and only if $T(\tilde{X})$ is BLUE in $\mathcal{L}(A, \tilde{V})$. As $EX = E\tilde{X}$ the conditions for unbiasedness are the same in the two models.

Furthermore, $\operatorname{Var}(k, \hat{\beta}(\tilde{X})) = \inf\{(h, \tilde{V}h) : Ah = \mathrm{k}\} = \inf\{(h, Vh) + hA^*\Gamma Ah : Ah = k\} = \inf\{(h, Vh) : Ah = k\} + (k, \Gamma k) = \operatorname{Var}(k, \hat{\beta}(X)) + (k, \Gamma k)$ when $k \in R(A)$. The last statement of the proposition tells us that Γ is 1–1 on $A[N(V)]$ if and only if $N(\tilde{V}) = N(V) \cap N(\Gamma A) \subseteqq N(A)$, which is easily checked. □

Let us next turn to sufficiency reductions of the models $\mathscr{L}(A, V)$ and $\mathscr{N}(A, V)$. Here is the main result.

Theorem 8.3.9 (Projection sufficiency in $\mathscr{L}(A, V)$). *Let X realize $\mathscr{L}(A, V)$ and let $\Pi(X)$ be* BLUE *for EX. Then $\Pi(X)$ and $X - \Pi(X)$ are uncorrelated and the statistic $\Pi(X)$ is linearly complete in the sense that* $\mathrm{Var}(h, \Pi(X)) = 0$ *whenever $E(h, \Pi(X)) \equiv_\beta 0$. On the other hand, the residual $\tilde{X} = X - \Pi(X)$ is second order ancillary, since neither $E\tilde{X} = 0$ nor* $\mathrm{Cov}\,\tilde{X}$ *depend on β.*

Proof. Let $\Pi(X)$ be BLUE for EX. Then $\Pi V \Pi^* = V\Pi^* = \Pi V$ since $\Pi A^* = A^*$ and $R(V\Pi^*) \subseteqq R(A^*)$. Thus $\mathrm{Cov}((h, \Pi(X)), \tilde{h}(I - \Pi)(X)) = (h, \Pi V(I - \Pi^*)\tilde{h}) = 0$. If $E(h, \Pi(X)) \equiv_\beta 0$ then $(h, \Pi(X))$ is BLUE of zero so that $\mathrm{Var}(h, \Pi(X)) = 0$. Finally $E\tilde{X} = EX - E\Pi(X) = EX - EX = 0$ while $\mathrm{Cov}\,\tilde{X} = (I - \Pi)V(I - \Pi)^* = (I - \Pi)V = V(I - \Pi^*)$. □

Any linear normal model $\mathscr{N}(A, V)$ is extremal.

Theorem 8.3.10 (Complete, sufficient statistics in $\mathscr{N}(A, V)$). *Let X realize $\mathscr{N}(A, V)$ and let $\Pi(X)$ be* BLUE *for EX. Then the statistics $\Pi(X)$ and $X - \Pi(X)$ are independent. The statistic $\Pi(X)$ is complete and sufficient and thus minimal sufficient, while the residual $X - \Pi(X)$ is ancillary.*

In particular the experiments $\mathscr{N}(A, V)$ are all extremal.

Remark. It follows that the $R(A)$-valued BLUE generator is complete and sufficient.

Proof. Uncorrelated vector variables which have a jointly normal distribution are independent. Thus $\Pi(X)$ and $X - \Pi(X)$ are independent and clearly $X - \Pi(X)$ is ancillary. It follows that $\Pi(X)$ is sufficient. Put $\Pi(X) = A^*\hat{\beta}(X)$ for the (essentially unique) $R(A)$-valued BLUE generator $\hat{\beta}(X)$. Then $E\hat{\beta}(X) = B^*\beta$ where $B = B^*$ is the orthogonal projection on $R(A)$. Since $\Pi(X)$ and $\hat{\beta}(X)$ are equivalent statistics, it suffices to show that $\hat{\beta}(X)$ is complete. Decompose $\hat{\beta}(X)$ as $\hat{\beta}(X) = \hat{\beta}_1(X) + \hat{\beta}_2(X)$ where $\hat{\beta}_1(X) \perp A[N(V)]$ while $\hat{\beta}_2(X) \in A[N(V)]$. Letting P denote the orthogonal projection on $[A[N(V)]]^\perp$ we have $\hat{\beta}_1 = P\hat{\beta}$ and $\hat{\beta}_2 = (I - P)\hat{\beta}$. If $k \in A[N(V)]$ then $0 = \mathrm{Var}(k, \hat{\beta}(X)) = (k, \mathrm{Cov}\,\hat{\beta}(X)k)$ and $(k, \hat{\beta}(X)) = (k, \hat{\beta}_2(X))$. Thus $\mathrm{Cov}\,\hat{\beta}_2(X) = 0$ and clearly $E\hat{\beta}_2(X) = (I - P)B\beta = (I - P)\beta$ since $I - P$ is the orthogonal projection on $A[N(V)] \subseteqq R(A) = R(B)$. Thus $Pr(\hat{\beta}_2(X) = (I - P)\beta) \equiv_\beta 1$.

On the other hand if $k \in R(A) \cap R(P)$ and $\mathrm{Var}(k, \hat{\beta}_1(X)) = 0$ then $\mathrm{Var}(k, \hat{\beta}(X)) = 0$ so that the BLUE of (k, β) has variance 0. This in turn implies that $k \in A[N(V)] = R(P)^\perp$ and thus, since $k \in R(P)$, that $k = 0$. It follows that $\mathrm{Cov}\,\hat{\beta}_1(X)$ is non-singular on $R(A) \cap R(P)$. Furthermore, $E\hat{\beta}_1(X) = PB\beta$ may be any vector in $R(PB) = P[R(B)] = P[R(A)] = R(PA) = R(P) \cap R(A)$. Choosing an orthonormal basis for $R(PA)$ and replacing $\hat{\beta}_1(X)$ with its coordinates with respect to this basis we obtain, for some n, a non-singular normally distributed n-tuple $Z = (Z_1, \ldots, Z_n)$ such that each vector in $\mathbb{R}^n$ is a possible expectation vector of Z. The covariance matrix of Z is clearly completely specified by the basis. Thus we have an exponential family with all of $\mathbb{R}^n$ as its parameter set. As is shown in most text books in mathematical statistics, such families of distributions are complete. It follows that $\hat{\beta}_1(X)$ is a complete statistic and thus $\hat{\beta}(X)$ and $\Pi(X)$ are also complete. □

It follows that the experiment induced by the $R(A)$-valued BLUE generator in $\mathcal{N}(A, V)$ is the minimal form of $\mathcal{N}(A, V)$. In fact we met this form as a canonical form in the previous section. We state this as a corollary.

Corollary 8.3.11 (The minimal form of $\mathcal{N}(A, V)$). *Let X realize the linear normal experiment $\mathcal{N}(A, V)$ and let $\hat{\beta}(X)$ be the $R(A)$-valued* BLUE *generator based on X. Then $\mathcal{N}(A, V)\hat{\beta}^{-1}$ (=the experiment induced by $\hat{\beta}$) $= \mathcal{N}(B, W)$ where B is the orthogonal projection on $R(A)$ while $W = (PAV^+A^*P)^+$ where P is the orthogonal projection on $[A[N(V)]]^\perp$.*

The experiments $\mathcal{N}(A, V)$ and $\mathcal{N}(B, W)$ are equivalent and $\mathcal{N}(B, W)$ is in minimal form.

Proof. We have just seen that $\hat{\beta}(X)$ is complete and sufficient and therefore it is minimal sufficient. The proof follows now from corollary 8.2.7. □

An important consequence of theorem 8.3.10 is that the BLUEs in $\mathcal{L}(A, V)$ are UMVUs in $\mathcal{N}(A, V)$. Before proving this we shall make some general remarks on UMVU (uniformly minimum variance unbiased) estimators. Let X realize an experiment $\mathcal{E} = (\mathcal{X}, \mathcal{A}; P_\theta : \theta \in \Theta)$ and let g be a function from Θ to some finite dimensional inner product space U. A U-valued statistic $\delta(X)$ is called unbiased if $E_\theta \delta(X) \equiv_\theta g(\theta)$. Moreover, if $E_\theta \|\delta(X)\|^2 < \infty$ for all θ and if $\operatorname{Cov}_\theta \delta(X) \leqslant \operatorname{Cov}_\theta \tilde{\delta}(X)$ for all θ for any other unbiased estimator $\tilde{\delta}(X)$ then we shall say that $\delta(X)$ is a UMVU estimator of $g(\theta)$. A UMVU estimator is unique up to null sets, and if $c_1, \ldots, c_r$ are constants then $\sum_i c_i \delta_i(X)$ is UMVU of $\sum c_i g_i(\theta)$ if $\delta_i(X)$ is UMVU of $g_i(\theta)$ for each i. The basic Lehmann–Scheffé criterion states that $\delta(X)$ is UMVU (of its expectation) if and only if $\delta(X)$ is uncorrelated with all unbiased estimators of the null vector in U. The general problem of vector valued estimators $g(\theta)$ may be reduced to the case of real valued functions since $\delta(X)$ is a UMVU estimator for $g(\theta)$ if and only if $(u, \delta(X))$ is a UMVU estimator of $(u, g(\theta))$ for each vector u in U. By the linearity property of UMVU estimation, it suffices to consider vectors u belonging to a given basis of U. This implies that if $\delta(X)$ is a UMVU estimator of $g(\theta)$ and if T is a linear map from U to some other finite dimensional inner product space, then $T(\delta(X))$ is a UMVU estimator of $T(g(\theta))$.

Say that a statistic $Y = \psi(X)$ based on X is quadratically complete if $P_\theta(\delta(Y) = 0) \equiv_\theta 1$ for all real valued measurable functions δ on the range space of Y such that $E_\theta \delta(Y) = 0$ and $E_\theta \delta(Y)^2 < \infty$ for all θ. Thus Y is quadratically complete if and only if the only everywhere quadratically integrable unbiased estimator of zero which is based on Y is the null estimator. If in addition, Y

is sufficient then an everywhere quadratically integrable U-valued estimator is UMVU if and only if it is determined by Y, up to null sets.

Various generalizations (or rather, implementations of the general idea underlying the Lehmann–Scheffé principle) may be obtained by restricting the estimators to affine, or even to convex, sets of statistics.

The next corollary is the promised result on UMVU estimation in $\mathcal{N}(A, V)$.

Corollary 8.3.12 (UMVU estimators in $\mathcal{N}(A, V)$). *If X realizes $\mathcal{N}(A, V)$ and if $T(X)$ is* BLUE *then it is also a* UMVU *(uniformly minimum variance unbiased) estimator of its expectation.*

Corollary 8.3.13 (Variances of BLUEs). *If $k \in R(A)$ then the variance of the* BLUE *of (k, β) is (k, Wk) where W is given in corollary* 8.3.11.

The models $\mathcal{N}(A, V)$ are not strictly speaking coherent. (Just consider the family of one-point distributions on the real line and the Borel class.) However, it becomes coherent by supplementing the Borel class of subsets of the sample space with all sets which are completion measurable for each value of the unknown parameter.

So far we have assumed that the covariance operators of our basic observations are completely known. Let us now turn to the situation where the covariance operator is known up to an unknown positive multiplicative factor σ^2. It is then easily checked that a linear map $X \to TX$ is BLUE for its expectation if and only if it is BLUE when σ is known, i.e. if and only if $R(VT^*) \subseteqq R(A^*)$. Thus if we are just considering linear estimation of β (unbiased or not) for quadratic loss, then the choice of estimator need not be influenced by information on σ.

However, we are now also faced with the problem of estimating the pair (β, σ^2). Linear or affine functionals of X have expectations which do not involve σ^2. It follows that if we are looking for unbiased estimators of σ^2 based only on the model $\mathcal{L}(A, \sigma^2 V; \sigma^2 > 0)$ then we may have to turn to polynomials of degree at least 2. On the other hand our assumptions do not involve moments of order greater than 2. The general 'second degree' polynomial is $Q(X) = q + (h, X) + (X, DX)$, where q is a real number, $h \in H$ and D is a linear operator on H. The expectation of this statistic is $q + (Ah, \beta) + (\beta, ADA^*\beta) + \sigma^2 \operatorname{trace}(DV)$ which is unbiased if and only if $q = 0$, $h \in N(A)$, $A(D + D^*)A^* = 0$ and $\operatorname{trace}(DV) = 1$. This last condition is a norming condition, and the other conditions together express that $Q(EX) \equiv_\beta 0$.

A useful fact which is relevant in this context is that if X is vector valued, $EX = 0$ and $\operatorname{Cov} X = V$, then $E(X, V^+X) = \operatorname{trace}(V^+V) = \operatorname{rank} V$. Moreover, if X is normally distributed then (X, V^+X) is χ^2-distributed with rank V degrees of freedom.

It follows that if F is any linear map from the sample space H of

$\mathscr{L}(A, \sigma^2 V; \sigma^2 > 0)$ to another finite dimensional inner product space and if $R(A^*) \subseteqq N(F)$ (i.e. $EFX \equiv_\beta 0$) then $(FX, (FVF^*)^+ FX)/\text{rank}\, FV$ is an unbiased estimator of σ^2. Furthermore, if X is normally distributed then $(FX, (FVF^*)^+ FX)/\sigma^2$ is χ^2-distributed with rank (FV) degrees of freedom so that our estimator gets variance $\sigma^4/2\,\text{rank}\,(FV)$. Thus in this case we should try to make rank (FV) as large as possible.

The largest value of rank (FV) when $R(A^*) \subseteqq N(F)$ is $\dim V[N(A)]$ and this is obtained for $F = I - \Pi$ where $\Pi(X)$ is BLUE for EX. In fact, as we shall see in a moment, this estimator is UMVU for σ^2 when X is normally distributed.

The ascertained maximality of rank (FV) follows from the normal case since this rank is not affected by the condition of normality. This and a few other facts are collected in the following theorem.

Theorem 8.3.14 (Complete sufficient statistics in $\mathscr{N}(A, \sigma^2 V; \sigma^2 > 0)$). *If X realizes $\mathscr{N}(A, \sigma^2 V; \sigma^2 > 0)$ and if $\Pi(X)$ is* BLUE *for EX then the statistics* $\Pi(X)$ and $Y = X - \Pi(X)$ *are independent.*

If $S(X) = (Y, (\text{Cov}\, Y)^+ Y)^{1/2}$, then the statistic $(\Pi(X), S(X))$ is complete and sufficient and, consequently, minimal sufficient. In particular the experiment $\mathscr{N}(A, \sigma^2 V; \sigma^2 > 0)$ is extremal.

Furthermore, $S^2(X)/\sigma^2$ is χ^2-distributed with $\dim V[N(A)] = \dim [R(V) \cap [N(V) + N(A)]]$ degrees of freedom.

Remark 1. Another way of expressing the minimal sufficient reduction of $\mathscr{N}(A, \sigma^2 V; \sigma^2 > 0)$ is by using the statistic $(\hat{\beta}(X), \hat{\sigma}^2(X))$ where $\hat{\beta}(X)$ is the $R(A)$-valued BLUE generator of β while $\hat{\sigma}^2(X)$ is the UMVU estimator σ^2. $\hat{\beta}(X)$ is then also the UMVU estimator of the orthogonal projection of β on $R(A)$.

Remark 2. It follows that if T is linear and if $T(X)$ is BLUE when $\sigma = 1$ then it is UMVU whether σ is known or not.

Proof. Put $U = \sigma^{-2}\,\text{Cov}\, Y$. Independence follows immediately from theorem 8.3.10, and we noted above that $S^2(X)/\sigma^2$ is χ^2-distributed with rank U degrees of freedom. By theorem 8.3.3 $R(V\Pi^*) \subseteqq R(A^*)$ and $\Pi A^* = A^*$. Thus $\Pi V \Pi^* = V\Pi^* = \Pi V$ and therefore also $(I - \Pi)V(I - \Pi)^* = V(I - \Pi)^* = (I - \Pi)V$. It follows that rank $U = \text{rank}[V(I - \Pi^*)]$ but $R(V(I - \Pi^*)) = V[R(I - \Pi^*)] \subseteqq V[N(A)]$ since $R(I - \Pi^*) \subseteqq N(A)$ (i.e. $A = \Pi^* A$). On the other hand, if $h \perp R(V(I - \Pi^*))$ then $h \in N((I - \Pi)V)$ so that $Vh = \Pi V h = V\Pi^* h \in R(A^*)$ since $R(V\Pi^*) \subseteqq R(A^*)$. Thus $Vh \perp N(A)$ which yields $h \perp V[N(A)]$. It follows that $V[N(A)] = R(V(I - \Pi^*)) = R(\text{Cov}(X - \Pi(X)))$ so that rank $U = \dim V[N(A)]$. The other expression for rank U was established in the proof of theorem 8.2.19.

The random vector Y has expectation vector 0 since $E\Pi(X) \equiv_\beta EX$. Then from the factorization criterion it follows that $S^2(X) = (Y, U^+ Y)/\sigma^2$ is sufficient for $Y = X - \Pi(X)$. Hence, since $(\Pi(X), Y)$ is computationally equivalent to X, the pair $(\Pi(X), S(X))$ is sufficient. It remains to show that this statistic is complete, i.e. that $(\hat{\beta}(X), S(X))$ is complete where $\hat{\beta}(X)$ is the $R(A)$-valued BLUE generator. This may be seen by appropriately modifying the arguments used for proving completeness in the proof of theorem 8.3.10. Again put $\hat{\beta}_1(X) = P\hat{\beta}(X)$, $\hat{\beta}_2(X) = (I - P)\hat{\beta}(X)$ and let P and B denote the orthogonal projections on, respectively, $[A^*]^{-1}[R(V)] = [A[N(V)]]^{\perp}$ and $R(A)$. Also, choose an orthonormal basis in $R(PA)$ and consider the coordinates $Z = (Z_1, \ldots, Z_n)$ of $\hat{\beta}_1(X)$. Again EZ may be any vector in $\mathbb{R}^n$ while the covariance matrix of Z is of the form $\sigma^2 C$ where the matrix C is positive definite. Choosing the orthonormal basis as a basis of eigenvectors of $\operatorname{Cov} \hat{\beta}_1(X)$ we see that we may assume that C is a diagonal matrix with positive diagonal elements. The joint distribution of $(Z_1, \ldots, Z_n)$ and $S^2(X)$ is given by an exponential family with $\mathbb{R}^n \times]-\infty, 0[$ as its 'actual' parameter set. Hence $(\hat{\beta}_1(X), S(X))$ is complete. Then since $\hat{\beta}_2(X) = E\hat{\beta}_2(X)$ it follows easily that the whole triple $(\hat{\beta}_1(X), S(X), \hat{\beta}_2(X))$ is complete. Thus the equivalent statistic $(\hat{\beta}(X), S(X))$ is also complete. □

Corollary 8.3.15 (Minimal form of $\mathcal{N}(A, \sigma^2 V; \sigma^2 > 0)$). *If $\hat{\beta}(X)$ is the $R(A)$-valued* BLUE *generator of β, if $S(X)$ is as in theorem* 8.3.14 *and if $\mathscr{E}_1$ and $\mathscr{E}_2$ are the experiments obtained by observing $\hat{\beta}(X)$ and $S^2(X)$ respectively, then:*

$$\mathscr{E} \sim \mathscr{E}_1 \times \mathscr{E}_2$$

and this is the factorization described in theorem 8.2.19. *In particular it follows that the product $\mathscr{E}_1 \times \mathscr{E}_2$ is in minimal form.*

Proof. This follows from the theorem above and theorem 8.2.19. □

It is a consequence of theorems 8.3.10 and 8.3.14 that if X is distributed as $\mathcal{N}(A, \sigma^2 V; \sigma^2 \in J)$ where $\beta \in K$ is completely unknown and J has interior points, then regardless of whether $\sigma > 0$ is known or unknown, the BLUEs of linear functions of β are automatically UMVU estimators. If $\phi(\beta, \sigma)$ possesses an unbiased estimator $a(X)$ such that $E\|a(X)\|^2 < \infty$ for all β and all $\sigma > 0$ then $E[a(X)|\hat{\beta}(X), S(X)]$ is UMVU for $\phi(\beta, \sigma)$. In particular $S^2(X)/\dim V[N(A)]$ is UMVU for σ^2 provided of course that $V[N(A)] \neq 0$, i.e. that $N(A) \subsetneqq N(V)$.

If $V[N(A)] = 0$, i.e. if $R(A^*) \supseteqq R(V)$, then $\hat{\beta}(X)$ alone is complete, regardless of whether σ is known or not. Thus if $E\phi(\hat{\beta}(X)) \equiv \sigma^2$ then $P_{\beta,\sigma^2}(\phi(\hat{\beta}(X)) = \sigma^2 \equiv 1$. This, however, is impossible, since the family $(N(A^*\beta, \sigma^2 V), \sigma^2 > 0)$ of normal distributions is homogeneous for each $\beta \in K$.

It follows that $\mathscr{E} = \mathcal{N}(A, \sigma^2 V; \sigma^2 > 0)$ admits unbiased estimators of σ^2 if and only if $V[N(A)] \neq 0$. If so then this experiment is not in minimal form.

On the other hand, $\mathscr{E}$ is always in minimal form when $V[N(A)] = 0$.

This is as far as we shall go in describing the internal properties of the models

$\mathscr{L}(A, V)$, $\mathscr{L}(A, \sigma^2 V; \sigma^2 > 0)$, $\mathcal{N}(A, V)$ and $\mathcal{N}(A, \sigma^2 V; \sigma^2 > 0)$. As the two first models in a sense are subordinated to the last two we could have omitted the 'purely' linear models. However, we have included the models $\mathscr{L}(A, V)$ and $\mathscr{L}(A, \sigma^2 V; \sigma^2 > 0)$ since they make it easier to keep track of those properties which solely depend on the expectation/covariance structure. Although we will not go into the interesting topics of admissibility and minimaxity of estimators we will describe some of the Bayes machinery.

Assume that X realizes $\mathscr{L}(A, V)$ and that β itself is a random variable with expectation vector $b \in K$ and covariance operator Γ. Then we may employ the following set-up

$$E(X|\beta) = A^*\beta, \qquad \operatorname{Cov}(X|\beta) = V$$

$$EX = A^*b, \qquad \operatorname{Cov} X = E\operatorname{Cov}(X|\beta) + \operatorname{Cov}(E(X|\beta)) = V + A^*\Gamma A.$$

The joint distribution of the pair (β, X) has expectation vector (b, A^*b) in $K \times H$ and the covariance operator of (β, X) is given by

$$\begin{aligned}\operatorname{Cov}((k, \beta) + (h, X), (\tilde{k}, \beta) + (\tilde{h}, X)) &= (k, \Gamma\tilde{k}) + (h, A^*\Gamma\tilde{k}) + (k\Gamma A, \tilde{h}) \\ &\quad + (h, (V + A^*\Gamma A)\tilde{h}) \\ &= (k, h)\begin{pmatrix} \Gamma, & \Gamma A \\ A^*\Gamma, & V + A^*\Gamma A \end{pmatrix}\begin{pmatrix} \tilde{k} \\ \tilde{h} \end{pmatrix}.\end{aligned}$$

It follows that we may decompose β as $\beta = E(\beta|X) + \beta - E(\beta|X)$ where $E(\beta|X) - b$ is linear in X and where the residual $\beta - E(\beta|X)$ is uncorrelated with X. The posterior expectation $E(\beta|X)$ of β given X is

$$E(\beta|X) = b + \Gamma A(V + A^*\Gamma A)^+(X - A^*b)$$

and the covariance operator of the residual is

$$\operatorname{Cov}(\beta - E(\beta|X)) = \Gamma - \Gamma A(V + A^*\Gamma A)A^*\Gamma.$$

If X realizes the linear normal model $\mathcal{N}(A, V)$ and if β is normally distributed then (β, X) is jointly normally distributed. In this case $E(\beta|X)$ is of course the conditional expectation (vector) of β given X. The posterior distribution is then normal with expectation $E(\beta|X)$ given above and covariance operator

$$\operatorname{Cov}(\beta|X) = \Gamma A(V + A^*\Gamma A)^+ A^*\Gamma.$$

If, in addition, $\mathcal{N}(A, V)$ is non-singular and if $R(\Gamma) \supseteqq R(A)$, then we can write out the density of β given X by first working out the joint density of (β, X). If we do that appropriately then we may end up with the slightly different looking expressions

$$E(\beta|X) = b + (\Gamma^+ + AV^+A^*)^+AV^+(X - A^*b)$$

and

$$\mathrm{Cov}(\beta - E(\beta|X)) = (\Gamma^+ + AV^+A^*)^+.$$

That the two sets of expressions for the posterior distributions are actually identical may be checked as follows.

If $\xi \in H$ and $X = (V + A^*\Gamma A)\xi$ then

$$\begin{aligned}\Gamma A(V + A^*\Gamma A)^+X = \Gamma A\xi &= (\Gamma^+ + AV^+A^*)^+(\Gamma^+ + AV^+A^*)\Gamma A\xi \\ &= (\Gamma^+ + AV^+A^*)^+(A + AV^+A^*\Gamma A)\xi \\ &= (\Gamma^+ + AV^+A^*)^+AV^+(V + A^*\Gamma A)\xi \\ &= (\Gamma^+ + AV^+A^*)AV^+X.\end{aligned}$$

Thus

$$\Gamma A(V + A^*\Gamma A)^+ = (\Gamma^+ + AV^+A^*)AV^+$$

on $R(V + A^*\Gamma A) = R(V) + R(A^*)$. This yields immediately that the last expression for $E(\beta|X)$ equals the first one, and furthermore that

$$\begin{aligned}&\Gamma A(A^*\Gamma A + V)^+A^*\Gamma \\ &= (\Gamma^+ + AV^+A^*)^+AV^+A^*\Gamma \\ &= (\Gamma^+ + AV^+A^*)^+(AV^+A^* + \Gamma^+)\Gamma - (\Gamma^+ + AV^+A^*)^+\Gamma^+\Gamma \\ &= \Gamma - (\Gamma^+ + AV^+A^*)^+.\end{aligned}$$

Putting $\Gamma = \tau^2 I$ where $\tau^2 \to \infty$ we find that the BLUEs for (k, β) are extended Bayes and hence, since their risk functions are constants, they are also minimax.

Assuming that V is non-singular and that A is onto we see by Anderson's inequality (see example 6.5.17) that the BLUE generator $\hat{\beta}(X)$ is a minimax estimator whenever the loss function is of the form $(t, \beta) \to \phi(t - \beta)$ where the measurable function ϕ is non-negative, symmetric and quasiconvex.

The reader may check that the expressions for minimum Bayes risk which can be derived from the above results, provide monotonically decreasing functionals of linear normal experiments with known covariances.

8.4 A relationship between variances of UMVU estimators and affinities for linear normal experiments with known covariances

We saw in section 8.2 how the Hellinger transform of $\mathcal{N}(A, V)$ may be expressed in terms of the function $\rho(\cdot|A, V)$ which to each β associates (-4) times the logarithm of the affinity between the distributions at β and at 0.

In section 8.3 we found expressions for the smallest variance of unbiased estimators of a given identifiable linear parametric function.

If (k, β) is a particular such function, i.e. if $k \in R(A)$, then the half of this smallest variance may be written

$$\tau(k) = \tfrac{1}{2}\min((h, Vh) : Ah = k).$$

Extending this notion a little, for any linear parametric function we may consider the greatest lower bound of variances of unbiased estimators. This leads to the function τ defined for all $k \in K$ by

$$\tau(k) = \tau(k|A, V) = \tfrac{1}{2}\inf((h, Vh) : Ah = k).$$

Thus $\tau(k) = \frac{1}{2}(k, (A^g)^* V A^g k) = \frac{1}{2}(k, (PAV^+A^*P)^+ k)$ when $k \in R(A)$, while $\tau(k) = \infty$ if $k \notin R(A)$. The generalized inverse A^g and the projection P occurring in these expressions were defined in theorem 8.3.5 and corollary 8.3.11 respectively.

We recognize that according to the definition given in section 8.2, τ is also a partial quadratic form. On the other hand we see from the canonical form of $\mathcal{N}(A, V)$, which is described in corollary 8.2.7, that any non-negative partial quadratic form on K is of the form $\tau(\cdot|A, V)$ for some $\mathcal{N}(A, V)$. Combining this with proposition 8.2.10 we find that the set of functions of the form $\rho(\cdot|A, V)$ and the set of functions of the form $\tau(\cdot|A, V)$ both coincide with the convex cone of non-negative partial quadratic forms on K.

We defined the function τ in terms of the structure $\mathcal{L}(A, V)$ while the 'natural' definition of ρ in section 8.2 was based on affinities and thus on the linear normal model $\mathcal{N}(A, V)$. As BLUE becomes UMVU when $\mathcal{L}(A, V)$ becomes $\mathcal{N}(A, V)$ of course we could also have defined τ in terms of $\mathcal{N}(A, V)$.

It is not quite apparent how ρ should be 'naturally' expressed in the purely linear model $\mathcal{L}(A, V)$. That this is possible at all is due to the fact that the functions ρ and τ are related by a notion of conjugacy from convex analysis.

Theorem 8.4.1 (Conjugacy of ρ and τ). *For a given experiment $\mathcal{N}(A, V)$ consider the functions ρ and τ given by*

$$\begin{aligned}\rho(\beta) &= -4\log(\text{affinity between law for } \beta \text{ and law for } 0)\\ &= \tfrac{1}{2}(\beta, AV^+A^*\beta) \text{ or } = \infty \text{ as } A^*\beta \in R(V) \text{ or } A^*\beta \notin R(V)\end{aligned}$$

and

$$\begin{aligned}\tau(k) &= \tfrac{1}{2}\inf\{\operatorname{Var}\delta : \delta \text{ is an unbiased estimator of } (k, \beta)\}\\ &= \tfrac{1}{2}\inf\{(h, Vh) : Ah = k\}; \qquad k \in K.\end{aligned}$$

Then

$$\rho(\beta) = \sup_k\,[(k, \beta) - \tau(k)]$$

and

$$\tau(k) = \sup_\beta\,[(k, \beta) - \rho(\beta)].$$

Before proving this theorem, following Rockafellar (1970), we will make some general comments on the type of conjugacy encountered here, and in particular on conjugates of partial quadratic forms.

If f is any function from K to $[-\infty, \infty]$ then a conjugate function $\hat{f}$ from K to $[-\infty, \infty]$ may be defined by

$$\hat{f}(k) = \sup_{k_1} ((k_1, k) - f(k_1)).$$

The function $\hat{f}$ is convex, lower semicontinuous and is either everywhere $> -\infty$ or is everywhere $= -\infty$. A function from K to $[-\infty, \infty]$ has all these properties if and only if it is the pointwise supremum of the affine functions which it majorizes. This, in turn, is equivalent to the condition $f = \hat{\hat{f}}$. These facts and explicit expressions for conjugates of partial quadratic forms may be found in Rockafellar (1970). As the context here is different, we include proofs of the results we shall need.

Proposition 8.4.2 (The conjugate of a quadratic form). *If V is a self adjoint non-negative definite operator on a finite dimensional inner product space H, then*

$$\sup_h ((h_0, h) - \tfrac{1}{2}(h, Vh)) = \tfrac{1}{2}(h_0, V^+h_0) \text{ or } = \infty \text{ as } h_0 \in R(V) \text{ or not.}$$

Remark. In particular this implies that inverting positive definite operators is a convex operation. Indeed if $U^{-1} = (1 - \lambda)V + \lambda W$ where $0 < \lambda < 1$ and V and W are positive definite, then

$$(h_0, h) - \tfrac{1}{2}(h, U^{-1}h) = (1 - \lambda)[(h_0, h) - \tfrac{1}{2}(h, Vh)] + \lambda[(h_0, h) - \tfrac{1}{2}(h, Wh)].$$

Maximizing with respect to h we find

$$[(1 - \lambda)V + \lambda W]^{-1} = U \leqslant (1 - \lambda)V^{-1} + \lambda W^{-1}$$

Proof. Suppose $h_0 \notin R(V)$. Then $h_0 = h_1 + h_2$ where $h_1 \in R(V)$ and $h_2 \in N(V) - \{0\}$. It follows that $(h_0, th_2) - \frac{1}{2}(th_2, Vth_2) = t\|h_2\|^2 \to \infty$ as $t \to \infty$. Suppose that $h_0 \in R(V)$. If $h = V^+h_0$ then $\langle h_0, h\rangle - \frac{1}{2}\langle h, Vh\rangle = \frac{1}{2}(h_0, V^+h_0)$ which proves $\geqslant$. Now any h may be represented as $h = Vh_0^+ + h_1$. Then, using $VV^+h_0 = h_0$, we find $(h_0, h) - \frac{1}{2}(h, Vh) = \frac{1}{2}(h_0, V^+h_0) - \frac{1}{2}(h_1, Vh_1) \leqslant \frac{1}{2}(h_0, V^+h_0)$, showing that maximum is obtained for $h = V^+h_0$. □

Proof of theorem 8.4.1. Using proposition 8.4.2 we find

$$\begin{aligned}\hat{\tau}(k) &= \sup_{k_1 \in R(A)} ((k, k_1) - \tau(k_1)) \\ &= \sup_{k_1 \in R(A)} \sup_{Ah = k_1} ((k, Ah) - \tfrac{1}{2}(h, Vh)) \\ &= \sup_h ((k, Ah) - \tfrac{1}{2}(h, Vh))\end{aligned}$$

$$= \sup_h ((A^*k, h) - \tfrac{1}{2}(h, Vh))$$

$$= \tfrac{1}{2}(A^*k, V^+A^*k) \qquad \text{or } = \infty \text{ as } A^*k \in R(V) \text{ or not.}$$

(i) follows now from the fact that $A^*k \in R(V)$ if and only if $k \perp A[N(V)]$. Using this expression for $\hat{\tau}$ we find

$$\sup_{k_1} ((k, k_1) - \hat{\tau}(k_1)) = \sup_{A^*k_1 \in R(V)} ((k, k_1) - \tfrac{1}{2}(A^*k_1, V^+A^*k_1)).$$

Suppose $k \notin R(A)$. Then $k = k_2 + k_3$ where $k_2 \in R(A)$ and $0 \neq k_3 \in N(A^*)$. Put $k_1 = tk_3$. Then $(k, k_1) - \tfrac{1}{2}(A^*k_1, V^+A^*k_1) = t\|k_3\|^2 \to \infty$ as $t \to \infty$. Finally suppose that $k \in R(A)$. Let $\mathscr{L}(A, V)$ be realized by the H-valued random vector X. Then (k, β) possesses a BLUE (h, X) based on X. Unbiasedness implies that $Ah = k$ while $Vh \in R(A^*)$ since (h, X) is BLUE. Substituting $k = Ah$ we find

$$\begin{aligned}\sup_{k_1} ((k, k_1) - \hat{\tau}(k_1)) &= \sup_{A^*k_1 \in R(V)} ((h, A^*k_1) - \tfrac{1}{2}(A^*k_1, V^+A^*k_1)) \\ &= \sup\{(h_1, h) - \tfrac{1}{2}(h_1, V^+h_1) : h_1 \in R(V) \cap R(A^*)\} \\ &= \sup\{(Vh_2, h) - \tfrac{1}{2}(Vh_2, V^+Vh_2) : Vh_2 \in R(A^*)\} \\ &\geqslant (Vh, h) - \tfrac{1}{2}(Vh, V^+Vh) = \tfrac{1}{2}(h, Vh) = \tau(k).\end{aligned}$$

It remains to show that $h_2 = h$ is the optimal choice of h_2. Now any h_2 may be represented as $h_2 = h + h_3$ where $h_3 = h_2 - h$. Then $(h, Vh) - \tfrac{1}{2}(h_2, Vh_2) = \tfrac{1}{2}(h, Vh) + (h, Vh_3) - (h, Vh_3) - \tfrac{1}{2}(h_3, Vh_3) \leqslant \tfrac{1}{2}(h, Vh)$. □

If f_1 and f_2 are convex functions on K then their *infimal convolution* is the map $k \to \inf(\{f_1(k_1) + f_2(k_2) : k_1 + k_2 = k\})$. Now the operation of infimal convolution is dual to the operation of pointwise addition. Furthermore, we have just seen that the convex cone of non-negative partial quadratic forms may be described both as the class of functions ρ and as the class of dual functions $\tau = \hat{\rho}$. It follows therefore that the class of non-negative quadratic forms is closed for infimal convolutions.

An important consequence of the established duality is that the ordering '$\geqslant_q$' of linear models $\mathscr{L}(A, V)$ with known covariance operators coincides with the ordering '$\geqslant$' of 'being at least as informative as' for the corresponding linear normal models. We include this and also some previously established criteria for ordering of these models, in the following theorem.

Theorem 8.4.3 (The deficiency ordering and the ordering $\geqslant_q$). *Consider linear models $\mathscr{L}(A, V)$ and $\mathscr{L}(B, W)$ and the corresponding linear normal models $\mathscr{N}(A, V)$ and $\mathscr{N}(B, W)$. Then the following conditions are all equivalent:*

(i) $\mathscr{L}(A, V) \geqslant_q \mathscr{L}(B, W)$.
(ii) $\mathscr{N}(A, V) \geqslant \mathscr{N}(B, W)$.

(iii) $TVT^* \leqslant W$ *for a linear map* T *from* $\mathcal{N}(A, V)$ *to* $\mathcal{N}(B, W)$ *such that* $TA^* = B^*$.
(iv) $\mathcal{N}(A, V) \sim \mathcal{N}(B, W) \times G$ *for some experiment* $\mathcal{G}$ *with parameter set* K.
(v) $\rho(\cdot|A, V) \geqslant \rho(\cdot|B, W)$.
(vi) $\tau(\cdot|A, V) \leqslant \tau(\cdot|B, W)$.

If these conditions are satisfied then $\mathcal{G}$ *in* (iv) *may be chosen of the form* $\mathcal{G} = \mathcal{N}(C, U)$ *for some operators* C *and* U.

Remark. Condition (v) implies that $\mathcal{N}(A, V) \geqslant \mathcal{N}(B, W)$ if and only if $\mathcal{N}(A, V) \geqslant \mathcal{N}(B, W)$ pairwise. Hence $\mathcal{N}(A, V) \geqslant \mathcal{N}(B, W)$ if and only if $\mathcal{N}(A, V)$ is at least as informative as $\mathcal{N}(B, W)$ for testing problems.

Proof. Most of the implications we need were established in theorems 8.2.13 and 8.4.1. In particular theorem 8.4.1 states that conditions (v) and (vi) are equivalent. By the definition of the ordering $\geqslant_q$ in section 8.1, condition (vi) is equivalent to (i). The equivalence of conditions (ii) and (v) was established in theorem 8.2.13. Furthermore, by the same theorem, these conditions are equivalent to (iv′), where (iv′) is obtained from (iv) by requiring that $\mathcal{G}$ is of the form $\mathcal{N}(C, U)$. As (iv′) implies condition (ii) we conclude that conditions (i), (ii), (iv), (v) and (vi) are all equivalent and that these conditions imply the last statement of the theorem. We noted in section 8.1 that (iii) implies (i). Therefore it remains to show that (i) implies (iii). Assume then that condition (i) is satisfied. Let X realize $\mathcal{L}(A, V)$ and let $\hat{\beta} = G(X)$ be the essentially unique BLUE generator based on X which takes its values in $R(A)$. Put $T = B^*G$. Then $TA^* = B^*GA^* = B^*$ since $GA^*\beta = EG(X)$ is the orthogonal projection of β on $R(A) \supseteqq R(B)$. Let h_B be a vector in the sample space of $\mathcal{L}(B, W)$ and put $h_A = T^*h_B = G^*Bh_B$. Then, since $\mathcal{L}(A, V) \geqslant_q \mathcal{L}(B, W)$ and since h_A represents the 'best' unbiased estimator of (Bh_B, β) in $\mathcal{L}(A, V)$, we get $(h_B, TVT^*h_B) = (h_A, Vh_A) \leqslant (h_B, Wh_B)$. □

The notion of conjugacy of partial quadratic forms extends directly to a notion of conjugacy for linear models $\mathcal{L}(A, V)$ or for linear normal experiments $\mathcal{N}(A, V)$. In fact theorem 8.4.1 implies that the condition '$\rho(\cdot|A, V) = \tau(\cdot|B, W)$' and the condition '$\tau(\cdot|A, V) = \rho(\cdot|B, W)$' are equivalent for experiments $\mathcal{N}(A, V)$ and $\mathcal{N}(B, W)$. If these conditions are satisfied then we shall say that $\mathcal{E}$ *and* $\mathcal{F}$ *are conjugate*. It follows that this notion of conjugacy is symmetric, i.e. $\mathcal{F}$ and $\mathcal{E}$ are conjugate when $\mathcal{E}$ and $\mathcal{F}$ are. Examples of conjugate pairs are

$$\mathcal{N}(A, I) \quad \text{and} \quad \mathcal{N}(I, AA^*)$$

$$\text{and } \mathcal{N}(I, V) \quad \text{and} \quad \mathcal{N}(\Pi, V^+)$$

where Π is the orthogonal projection on $R(V)$.

Since the notion of conjugacy is based upon the 'invariant' notions of affinity and of variances of UMVU estimators, it follows that this notion of conjugacy for experiments respects equivalence. The correspondence defined by conjugacy for types of experiments of the form $\mathcal{N}(A, V)$ is involutory by theorem 8.4.1. In fact if $(\mathcal{N}(A, V), \mathcal{N}(B, W))$ and $(\mathcal{N}(B, W), \mathcal{N}(C, U))$ are both pairs of conjugates then $\mathcal{N}(A, V)$ and $\mathcal{N}(C, U)$ are equivalent. The last theorem implies that conjugacy for experiments is a monotonically decreasing relation since $\mathcal{N}(B_1, W_1) \leqslant \mathcal{N}(B_2, W_2)$ when $\mathcal{N}(A_1, V_1) \geqslant \mathcal{N}(A_2, V_2)$ and $(\mathcal{N}(A_1, V_1), \mathcal{N}(B_1, W_1))$ and $(\mathcal{N}(A_2, V_2), \mathcal{N}(B_2, W_2))$ are both pairs of conjugates.

In particular this implies that the operation of constructing Moore–Penrose inverses of non-negative definite operators is monotonically decreasing.

If Γ is non-negative definite on K then the variance formula in proposition 8.3.8 may now be written

$$\tau(k|A, V + A^*\Gamma A) = \tau(k|A, V) + \tfrac{1}{2}(k, \Gamma k).$$

The following consequence of the theorem was noted by Stepniak (1985).

Corollary 8.4.4 (Smoothing preserves the ordering $\geqslant_q$). *If Γ is a non-negative definite operator on K then $\mathscr{L}(A, V) \geqslant_q \mathscr{L}(B, W)$ if and only if $\mathscr{L}(A, V + A^*\Gamma A) \geqslant_q \mathscr{L}(B, W + B^*\Gamma B)$. In particular $\mathscr{L}(A, V) \sim_q \mathscr{L}(B, W)$ if and only if $\mathscr{L}(A, V + A^*\Gamma A) \sim_q \mathscr{L}(B, W + B^*\Gamma B)$.*

Proof. This follows from theorem 8.4.3 and corollary 8.1.6. □

Another related topic is the preservation of information by linear operators. We have seen how the ordering of the linear normal models $\mathcal{N}(A, V)$ may be expressed in terms of linear operators. When does an operator preserve information? As the BLUEs of EX are minimal sufficient it is to be expected that an operator preserves information if and only if the operator induces a partitioning of the sample space which is 'essentially' finer than that of the BLUE of EX.

In order to make this precise, note first that if X realizes $\mathcal{N}(A, V)$ and if S is linear on the sample space of $\mathcal{N}(A, V)$ then BLUEs may be based on $S(X)$ if and only if $V^{-1}[R(A^*)] = [V[N(A)]]^{\perp} \subseteqq R(S^*) + N(A) \cap N(V)$. In fact a functional (h, X) is BLUE if and only if $h \perp A[N(V)]$. The condition on S states that such a functional is always equivalent to a functional $(\tilde{h}, S(X))$ where $h - S^*\tilde{h} \in N(A) \cap N(V)$. Passing to orthogonal complements we may also write this condition as $V[N(A)] \supseteqq N(S) \cap [R(V) + R(A^*)]$. If $\mathcal{N}(A, V)$ is non-singular, i.e. if $R(A^*) \subseteqq R(V)$, then these equivalent inclusions may be written $R(A^*) \subseteqq V[R(S^*)]$ or as $V[N(A)] \supseteqq N(S) \cap R(V)$. Finally note that if Γ is a non-negative definite operator on K then $V[N(A)] = (V + A^*\Gamma A)[N(A)]$. This establishes the following criteria for sufficiency.

Theorem 8.4.5 (Sufficiency of linear statistics). *Assume that X realizes the linear normal experiment $\mathcal{N}(A, V)$. Then the following conditions are equivalent for a linear map S on the sample space of $\mathcal{N}(A, V)$.*

(i) *$S(X)$ is sufficient for X.*
(ii) $\mathcal{N}(AS^*, SVS^*) \geqslant \mathcal{N}(A, V)$.
(iii) $V^{-1}[R(A^*)] \subseteqq R(S^*) + N(V) \cap N(A)$.
(iv) $V[N(A)] \supseteqq N(S) \cap [R(V) + R(A^*)]$.
(v) *$S(Y)$ is sufficient for Y where Y realizes $\mathcal{N}(A, V + A^*\Gamma A)$ with Γ being a non-negative definite operator on K.*

Furthermore, if $\mathcal{N}(A, V)$ is non-singular then these conditions are equivalent to

(iii′) $R(A^*) \subseteqq V[R(S^*)]$

and also to

(iv′) $V[N(A)] \supseteqq N(S) \cap R(V)$.

Before proceeding to minimal sufficiency, let us express functional reduction in terms of the induced partitions.

Consider two linear maps S and T which are both defined on the same finite dimensional inner product space H. Assume that there is a measurable function g from $R(T)$ to $R(S)$ such that $S = g(T)$ a.e. with respect to Haar measure on K. The point we want to make is that then g may be chosen linear and thus the qualification a.e. may be omitted. To see this let $X \in H$ realize the experiment $\mathcal{N}(I, I)$ where both I's denote the identity operator on H. Then $T(X)$ and $S(X)$ realize, respectively, $\mathcal{N}(T^*, TT^*)$ and $\mathcal{N}(S^*, SS^*)$. The existence of g implies that $\mathcal{N}(S^*, SS^*) \leqslant \mathcal{N}(T^*, TT^*)$ and thus $R(S^*) \subseteqq R(T^*)$, i.e. $N(S) \supseteqq N(T)$, showing that g may be chosen linear.

Suppose now that X realizes a linear normal experiment $\mathcal{N}(A, V)$ and that S and T are linear maps such that $S(X) = g(T(X))$ a.s. for all $\beta \in K$ for some measurable function g. This may be written $S(X + A^*\beta) = g(T(X + A^*\beta))$ a.s. for all β where X is $N(0, V)$-distributed. Assigning an $N(0, I)$-distribution to β on K we obtain, by Fubini, that $S(Z) = g(T(Z))$ a.s. where Z is distributed according to $N(0, V + A^*A)$. By the arguments above, this in turn shows again that g may be chosen linear and if g is linear then

$$S(x) = g(T(x)) \qquad \text{for all } x \text{ in } R(V) + R(A^*).$$

This factorization of S on $R(V) + R(A^*)$ implies that $N(T) \cap [R(V) + R(A^*)] \subseteqq N(S)$ and conversely this set inequality implies the existence of a linear g such that $S = gT$ on $R(V) + R(A^*)$.

If X realizes $\mathcal{N}(A, V)$ then the BLUE (=UMVU) $\Pi(X)$ of EX and the $R(A)$-valued BLUE generator $\hat{\beta}(X)$ are equivalent statistics, by theorem 8.3.3. Furthermore, these statistics are complete and sufficient by theorem 8.3.10. In particular they are minimal sufficient. It follows that a statistic $S(X)$ is sufficient if and only if $g(S(X)) = \Pi(X)$ a.s. for some measurable function g. As we have just seen, this in turn is equivalent to the condition that $N(S) \cap [R(V) + R(A^*)] \subseteqq N(\Pi)$. By the same argument $S(X)$ is minimal sufficient if and only if $N(S)$ and $N(\Pi)$ have the same vectors in common with the support $R(V) + R(A^*)$.

Therefore it is desirable to have an explicit expression for the space $N(\Pi) \cap [R(V) + R(A^*)]$. Sufficiency implies that $V[N(A)] \supseteqq N(\Pi) \cap [R(V) + R(A^*)]$. On the other hand, since $\Pi(X)$ is BLUE, $R(\Pi^*) \subseteqq [V[N(A)]]^{\perp}$ so that $N(\Pi) \supseteqq V[N(A)] \supseteqq N(\Pi) \cap [R(V) + R(A^*)]$. Altogether this shows that $N(\Pi) \cap [R(V) + R(A^*)] = V[N(A)]$. This, together with the previous theorem, yields the following criteria for minimal sufficiency.

Proposition 8.4.6 (Minimal sufficiency of linear statistics). *If X realizes $\mathcal{N}(A, V)$ then the following conditions are equivalent for a linear map S on the sample space of $\mathcal{N}(A, V)$.*

(i) *$S(X)$ is minimal sufficient for X.*
(ii) *$\mathcal{N}(AS^*, SVS^*) \geqslant \mathcal{N}(A, V)$ and $V^{-1}[R(A^*)] \supseteqq R(S^*)$.*
(iii) *$V^{-1}[R(A^*)] = R(S^*) + N(V) \cap N(A)$.*
(iv) *$V[N(A)] = N(S) \cap [R(V) + R(A^*)]$.*
(v) *$S(Y)$ is minimal sufficient for Y if Y realizes $\mathcal{N}(A, V + A^*\Gamma A)$ with Γ being a non-negative definite operator on K.*

If $\mathcal{N}(A, V)$ is non-singular then these conditions are equivalent to

(iii′) *$R(A^*) = V[R(S^*)]$*

and also to

(iv′) *$V[N(A)] = N(S) \cap R(V)$.*

Remark 1. The difference between (ii) here and (ii) in theorem 8.4.5 illustrates the fact that sufficiency of $S(X)$ may be expressed in terms of deficiency, while minimal sufficiency of $S(X)$ is not expressible in terms of the experiment induced by S.

Remark 2. The reader is referred to the papers by Drygas (1983, 1985) and the references there for further information on sufficiency of linear statistics in linear models.

8.5 Deficiencies between linear normal experiments with known covariances

Before we begin our evaluation of deficiencies we shall formulate a few simple and general principles which may facilitate the computations. The first one is as follows.

Deficiencies depend on the experiments only via their types.

Thus if $\mathscr{E}$ is an experiment then we may reduce $\mathscr{E}$ to its minimal form or we may enlarge $\mathscr{E}$ by employing known chance mechanisms determined by the outcome of a realization of $\mathscr{E}$. Considering an experiment $\mathscr{N}(A, I)$ we see from this principle that we may replace the experiment $\mathscr{N}(A, V)$ with the experiment $\mathscr{N}((AA^*)^{1/2}, I)$. If follows that we can assume also that A is a non-negative definite operator on K.

The next principle is equally evident.

Deficiencies are not increased if the original parameter is assumed to depend functionally on another parameter. If the correspondence between the two parameters is 1–1 and onto, and thus is just another naming, then deficiencies are preserved.

It follows that if $\mathscr{E} = (P_\theta : \theta \in \Theta)$ and $\mathscr{F} = (Q_\theta : \theta \in \Theta)$ are the two experiments and if g maps $\tilde{\Theta}$ into Θ then $\delta(\tilde{\mathscr{E}}, \tilde{\mathscr{F}}) \leqslant \delta(\mathscr{E}, \mathscr{F})$ where $\tilde{\mathscr{E}} = (P_{g(\tilde{\theta})} : \tilde{\theta} \in \tilde{\Theta})$ and $\tilde{\mathscr{F}} = (Q_{g(\tilde{\theta})} : \tilde{\theta} \in \tilde{\Theta})$. Here equality holds provided g maps $\tilde{\Theta}$ onto Θ.

Applying this to experiments $\mathscr{N}(A, V)$ and $\mathscr{N}(B, W)$ we see that $\delta(\mathscr{N}(F^*A, V), \mathscr{N}(F^*B, W)) \leqslant \delta(\mathscr{N}(A, V), \mathscr{N}(B, W))$ where the parameter set of the first two experiments is $\tilde{K}$ and where F maps $\tilde{K}$ linearly into K. Here equality holds if $R(A)$ and $R(B)$ are subsets of $R(F)$.

Deficiencies for restrictions to subparameter sets are monotonically increasing functions of the subparameter set. This yields the third principle.

If $\mathscr{E} = (P_\theta : \theta \in \Theta)$ and $\mathscr{F} = (Q_\theta : \theta \in \Theta)$ are experiments and if Θ_0 is a subset of Θ such that P_θ does not depend on θ when $\theta \in \Theta_0$ then $\delta(\mathscr{E}, \mathscr{F}) \geqslant \delta(\mathscr{M}_i, \mathscr{F}_{\Theta_0})$.

This is applicable when there are parametric functions which are identifiable in $\mathscr{F}$ but not in $\mathscr{E}$.

As an application we shall now see that $\delta(\mathscr{N}(A, V), \mathscr{N}(B, W)) = 2$ whenever $R(B) \subsetneqq R(A)$. Let $k \in R(B) \cap [R(A)]^c$. Then $k \neq 0$ and (k, β) is identifiable in $\mathscr{N}(B, W)$ but not in $\mathscr{N}(A, V)$. It follows that there is a $\beta_0 \in K$ such that $A^*\beta_0 = 0$ while $(k, \beta_0) \neq 0$, and then $B^*\beta_0 \neq 0$. Writing $\mathscr{N}(A, V) = (P_\beta : \beta \in K)$ and $\mathscr{N}(B, W) = (Q_\beta : \beta \in K)$ we find $\delta(\mathscr{N}(A, V), \mathscr{N}(B, W)) \geqslant \delta((P_{\xi\beta_0} : \xi \in \mathbb{R}), (Q_{\xi\beta_0} : \xi \in \mathbb{R})) = \delta(\mathscr{M}_i, \mathscr{N}(\xi B^*\beta_0, W); \xi \in \mathbb{R})$ where $\mathscr{M}_i$ as usual denotes a totally non-informative experiment with the appropriate parameter set.

Let Y realize $(N(\xi B^*\beta_0, W); \xi \in \mathbb{R})$ and let $\hat{\xi} = \hat{\xi}(Y)$ be the UMVU estimator of ξ based on Y. Then $\hat{\xi}$ is distributed as $N(\xi, c^2)$ for some constant $c \geqslant 0$. Using the fact that $\lim_{\xi \to \infty} \|N(\xi, c^2) - M\| = 2$ for any probability distribution M on $\mathbb{R}$ we find

$$\delta(\mathscr{M}_i, (N(\xi B^*\beta_0, W); \xi \in \mathbb{R})) = 2 = \delta(\mathscr{N}(A, V), \mathscr{N}(B, W)).$$

The fourth principle follows from corollary 6.2.34.

If $\mathscr{E}$ and $\mathscr{F}$ are experiments with parameter set Θ and if Θ_0 is an infinite subset of Θ such that $\mathscr{E}|\Theta_0$ is dominated while $\mathscr{F}|\Theta_0$ is totally informative, then $\delta(\mathscr{E},\mathscr{F}) = 2$.

As an application we shall show that $\delta(\mathscr{N}(A, V), \mathscr{N}(B, W)) = 2$ whenever $B[N(W)] \subsetneq A[N(V)]$.

Let $B[N(W)] \ni k_0 \notin A[N(V)]$. Decompose k_0 as $k_0 = k_1 + k_2$ where $k_1 \perp A[N(V)]$ and $k_2 \in A[N(V)]$. Then $k_1 \neq 0$. Restricting β to $[k_1]$ we may write $\beta = \gamma k_1$ where $\gamma \in \mathbb{R}$. Then $(k_0, \beta) = \gamma(k_0, k_1) = \gamma \|k_1\|^2$ which is strictly increasing in γ. As (k_0, β) may be estimated with certainty in $\mathscr{N}(B, W)$ we find that the restriction of $\mathscr{N}(B, W)$ to $[k_1]$ is totally informative. On the other hand $A^*\beta = \gamma A^* k_1 \in \mathbb{R}(V)$ since $A^* k_1 \perp N(V)$. Thus the restriction of $\mathscr{N}(A, V)$ to $[k_1]$ is homogeneous. It follows that $\delta(\mathscr{N}(A, V), \mathscr{N}(B, W)) = 2$ whenever $R(B) \subsetneq R(A)$ or $B[N(W)] \subsetneq A[N(V)]$.

The last principle is related to factorization and it is as follows.

If $\mathscr{E}^{(i)} = (P_\theta^{(i)}; \theta \in \Theta)$; $i = 1, 2$ and $\mathscr{F}^{(i)} = (Q_\theta^{(i)} : \theta \in \Theta)$; $i = 1, 2$ are experiments then $\delta(\mathscr{E}^{(1)} \times \mathscr{E}^{(2)}, \mathscr{F}^{(1)} \times \mathscr{F}^{(2)}) \leqslant \delta(\mathscr{E}^{(1)}, \mathscr{F}^{(1)}) + \delta(\mathscr{E}^{(2)}, \mathscr{F}^{(2)})$ so that $\delta(\mathscr{E}^{(1)} \times \mathscr{E}^{(2)}, \mathscr{F}^{(1)} \times \mathscr{F}^{(2)}) \leqslant \delta(\mathscr{E}^{(1)}, \mathscr{F}^{(1)})$ when $\mathscr{E}^{(2)} \geqslant \mathscr{F}^{(2)}$. Here equality holds if $\mathscr{E}^{(2)} \geqslant \mathscr{F}^{(2)}$ and $\Theta = \Theta_1 \times \Theta_2$ where $P_\theta^{(1)}$ and $Q_\theta^{(1)}$ only depend on the first coordinate of θ, while $P_\theta^{(2)}$ and $Q_\theta^{(2)}$ only depend on the second coordinate of θ.

The first inequality was deduced in chapter 6 and the second follows since the provision $\mathscr{E}^{(2)} \geqslant \mathscr{F}^{(2)}$ merely states that the second term $\delta(\mathscr{E}^{(2)}, \mathscr{F}^{(2)}) = 0$. If Θ may be factorized as described then the inequality $\delta(\mathscr{E}^{(1)} \times \mathscr{E}^{(2)}, \mathscr{F}^{(1)} \times \mathscr{F}^{(2)}) \geqslant \delta(\mathscr{E}^{(1)}, \mathscr{F}^{(1)})$ follows by restricting the second coordinate of the parameter to a particular value.

As an application of the last principle, consider experiments $\mathscr{N}(A_i, I)$ and $\mathscr{N}(B_i, I)$; $i = 0, 1, 2$ such that $A_0 A_0^* = A_1 A_1^* + A_2 A_2^*$ and $B_0 B_0^* = B_1 B_1^* + B_2 B_2^*$ and $A_2 A_2^* \geqslant B_2 B_2^*$. Thus we assume that $\mathscr{N}(A_0, I) = \mathscr{N}(A_1, I) \times \mathscr{N}(A_2, I)$, $\mathscr{N}(B_0, I) = \mathscr{N}(B_1, I) \times \mathscr{N}(B_2, I)$ and that $\mathscr{N}(A_2, I) \geqslant \mathscr{N}(B_2, I)$. Hence $\delta(\mathscr{N}(A_0, I), \mathscr{N}(B_0, I)) \leqslant \delta(\mathscr{N}(A_1, I), \mathscr{N}(B_1, I))$. In addition assume that $R(A_1) = R(B_1) \perp R(A_2) = R(B_2)$. It follows that $R(A_0) = R(A_1) + R(A_2) = R(B_1) + R(B_2) = R(B_0)$. Decompose $\beta \in R(A_0)$ as $\beta = \beta_1 + \beta_2$ where $\beta_1 \in R(A_1)$ and $\beta_2 \in R(A_2)$. Then $A_i^*\beta = A_i^*\beta_i$ and $B_i^*\beta = B_i^*\beta_i$; $i = 1, 2$. Restricting $\beta \in R(A_0)$ by requiring that β_2 is a definite vector in K we find that

$$\delta(\mathscr{N}(A_0, I), \mathscr{N}(B_0, I)) = \delta(\mathscr{N}(A_1, I), (B_1, I)).$$

Finally let us show how we may reduce the general problem of evaluating deficiencies for these experiments to the situation where both experiments are non-singular. Specifically we will show that $\delta(\mathscr{N}(A, V), \mathscr{N}(B, W)) =$

$\delta(\mathscr{N}(PA, V), \mathscr{N}(PB, W))$ where P is the orthogonal projection on $[A[N(V)]]^\perp$. The experiment $\mathscr{N}(PA, V)$ is always non-singular and as we have seen before, $\mathscr{N}(PB, W)$ is also non-singular when $A[N(V)] \supseteqq B[N(W)]$. As we have just seen, the last condition is always fulfilled when these deficiencies are <2.

Restricting β to $R(P)$ we see by the second reduction principle that $\delta(\mathscr{N}(A, V), \mathscr{N}(B, W)) \geqslant \delta(\mathscr{N}(PA, V), \mathscr{N}(PB, W))$. Consider any number $\varepsilon \geqslant \delta(\mathscr{N}(PA, V), \mathscr{N}(PB, W))$. Decompose each $\beta \in K$ as $\beta = \beta_1 + \beta_2$ where $\beta_1 \in R(P)$ and $\beta_2 \in R(I - P) = A[N(V)]$. Restricting β by specifying β_2, for each β_2 in $A[N(V)]$, we obtain experiments $\mathscr{N}(A, V)_{\beta_2}$ and $\mathscr{N}(B, W)_{\beta_2}$. The experiments $\mathscr{N}(A, V)_{\beta_2}$ are all equivalent to $\mathscr{N}(A, V)_0$ since $X + A^*\beta_2$ realizes $\mathscr{N}(A, V)_{\beta_2}$ when X realizes $\mathscr{N}(A, V)_0$. Similarly the experiments $\mathscr{N}(B, W)_{\beta_2}$ are all equivalent to $\mathscr{N}(B, W)_0$. As $P\beta = \beta_1$ the requirement on ε may be stated: $\varepsilon \geqslant \delta(\mathscr{N}(A, V)_{\beta_2}, \mathscr{N}(B, W)_{\beta_2})$ for all $\beta_2 \in A[N(V)]$. Thus for each $\beta_2 \in A[N(V)]$ there is a transition M_{β_2} such that

$$\sup_{\beta_1} \|N(A^*(\beta_1 + \beta_2), V)M_{\beta_2} - N(B^*(\beta_1 + \beta_2), W)\| \leqslant \varepsilon.$$

Actually we can just put $M_{\beta_2} = T_{\beta_2} M_0 S_{\beta_2}$ where S_{β_2} is translation by $-A^*\beta_2$ in the sample space of $\mathscr{N}(A, V)$ while T_{β_2} is translation by $B^*\beta_2$ in the sample space of $\mathscr{N}(B, W)$. The family of measures $(N(A^*(\beta_1 + \beta_2), V); \beta_1 \in R(P))$ is homogeneous and is supported by $R(V) + A^*\beta_2$. These supports are completely disjoint for different choices of β_2 in $A[N(V)]$. It follows that we may piece the transitions together and thereby obtain a single transition not depending on β_2. This proves our assertion.

Proposition 8.5.1 (Reduction to non-singular experiments). *$\delta(\mathscr{N}(A, V), \mathscr{N}(B, W)) = \delta(\mathscr{N}(PA, V), \mathscr{N}(PB, W))$ where P is the orthogonal projection on $[A[N(V)]]^\perp$.*

As explained above, the proposition implies that we may restrict our attention to non-singular experiments and then we can also assume that the covariance operators are all identity operators.

As we shall see shortly, this situation may be reduced further to the canonical case described in the following result due to Swensen (1980).

Proposition 8.5.2 (Deficiencies for canonical forms). *Let X_i and Y_i; $i = 1, \ldots, p$, be normally distributed with variance 1 and expectations $EX_i = \beta_i$ and $EY_i = \Delta_i^{1/2}\beta_i$; $i = 1, \ldots, p$.*

Here $\Delta_1, \ldots, \Delta_p$ are non-negative constants while the unknown parameter $\beta = (\beta_1, \ldots, \beta_p)$ is allowed to vary freely in $\mathbb{R}^p$.

Let $\mathscr{E}_i$ and $\mathscr{F}_i$ denote, respectively, the experiment realized by X_i and the experiment realized by Y_i. Put $\mathscr{E} = \prod_i \mathscr{E}_i$ and $\mathscr{F} = \prod_i \mathscr{F}_i$ so that $\mathscr{E}$ is realized by $(X_1, \ldots, X_p)$ provided $X_1, \ldots, X_p$ are independent, while $\mathscr{F}$ is realized by $(Y_1, \ldots, Y_p)$ provided $Y_1, \ldots, Y_p$ are independent. Then

$$\delta(\mathscr{E},\mathscr{F}) = \delta\left(\prod_{\Delta_i>1} \mathscr{E}_i, \prod_{\Delta_i>1} \mathscr{F}_i\right)$$

$$= \left\| \prod_{\Delta_i>1} N(0,\Delta_i^{-1}) - \prod_{\Delta_i>1} N(0,1) \right\|$$

$$= \left\| \prod_{\Delta_i>1} N(0,1) - \prod_{\Delta_i>1} N(0,\Delta_i) \right\|.$$

Proof. Factorize $\mathscr{E}$ and $\mathscr{F}$ as $\mathscr{E} = \mathscr{E}' \times \mathscr{E}''$ and $\mathscr{F} = \mathscr{F}' \times \mathscr{F}''$ where $\mathscr{E}' = \prod_{\Delta_i>1} \mathscr{E}_i$, $\mathscr{F}' = \prod_{\Delta_i>1} \mathscr{F}_i$, $E'' = \prod_{\Delta_i \leqslant 1} \mathscr{E}_i$ and $\mathscr{F}'' = \prod_{\Delta_i \leqslant 1} \mathscr{F}_i$. Now $\mathscr{E}'' \geqslant \mathscr{F}''$ since $\mathscr{E}_i \geqslant \mathscr{F}_i$ when $\Delta_i \leqslant 1$. From the last reduction principle it follows that $\delta(\mathscr{E},\mathscr{F}) = \delta(\mathscr{E}',\mathscr{F}')$. We may thus assume that $\Delta_i > 1$ for all $i = 1,\ldots,p$. Multiplying Y_i by $\Delta_i^{-1/2}$ we obtain $\mathscr{E} \sim (N(\beta,I); \beta \in \mathbb{R}^p)$ and $\mathscr{F} \sim (N(\beta,\Delta^{-1}); \beta \in \mathbb{R}^p)$ where $\Delta = \operatorname{diag}(\Delta_1,\ldots,\Delta_p)$.

It follows that $\mathscr{E}$ and $\mathscr{F}$ are the translation experiments on $\mathbb{R}^p$ defined by the distributions $N(0,I)$ and $N(0,\Delta^{-1})$ respectively.

By example 6.5.13, the problem is now reduced to that of finding a probability distribution M on $\mathbb{R}^p$ which minimizes the total variation of $\|N(0,I) * M - N(0,\Delta^{-1})\|$. Consider the problem of testing that a vector variable X is distributed as $N(\beta,I)$ for some $\beta \in \mathbb{R}^p$, against the alternative that the distribution of X is $N(0,\Delta^{-1})$. The Neyman–Pearson lemma tells us that the most powerful level α test for testing $N(0,I)$ against $N(0,\Delta^{-1})$ rejects when $\sum_i (\Delta_i - 1)X_i^2 \leqslant c$, where c is adjusted so that the probability of this event is α when X is $N(0,I)$-distributed. The probability of rejecting when X is distributed as $N(\beta,I)$ is then $\pi(\beta) = Pr(\sum_i (\Delta_i - 1)(U_i + \beta_i)^2 \leqslant c)$ where U is $N(0,I)$-distributed. Now $|U_i + \xi|$ is stochastically increasing in $|\xi_i|$. It follows that $\sum_i (\Delta_i - 1)(U_i + \beta_i)^2$ is stochastically smallest when $\beta_1 = \cdots = \beta_p = 0$. Thus $\pi(\beta) \leqslant \pi(0) = \alpha$ so that the one-point distribution in 0 is least favourable for all levels α.

From the theory of translation experiments as it is described in examples 6.5.3 and 6.5.13, it follows that this one-point distribution is the optimal choice of M. □

Remembering that $\|P_1 \times R - P_2 \times R\| = \|P_1 - P_2\|$ for probability measures P_1, P_2 and R such that the last expression is meaningful we see by the proof above that $\inf_M \|N(0,I) * M - N(0,\Delta^{-1})\| = \|\prod_{\Delta_i>1} N(0,1) - \prod_{\Delta_i>1} N(0,\Delta_i^{-1})\|$ is obtained for $M = \prod_{\Delta_i>1} N(0,0) \times \prod_{\Delta_i \leqslant 1} N(0,\Delta_i^{-1} - 1)$.

This implies that the corresponding Markov kernel associates with $X = (X_1,\ldots,X_p)$ a random p-tuple $Z = (Z_1,\ldots,Z_p)$ which, for given X, is distributed as $\prod_{\Delta_i>1} N(X_i,0) \times \prod_{\Delta_i \leqslant 1} N(X_i,\Delta_i^{-1} - 1)$. However we transformed $N(\Delta_i^{1/2}\beta_i, 1)$ into $N(\beta_i,\Delta_i^{-1})$ by division by $\Delta_i^{1/2}$. Reversing this we see that the optimal approximation of $\mathscr{F}$ by a randomization from $\mathscr{E}$ is obtained by letting the conditional distribution of Z given X be

$$\prod_{\Delta_i > 1} N(\Delta_i^{1/2} X_i, 0) \times \prod_{\Delta_i \leqslant 1} N(\Delta_i^{1/2} X_i, 1 - \Delta_i).$$

Then Z is distributed as $\prod_{\Delta_i > 1} N(\Delta_i^{1/2}\beta_i, \Delta_i) \times \prod_{\Delta_i \leqslant 1} N(\Delta_i^{1/2}\beta_i, 1)$. The total variation of the difference between this distribution and the distribution of a statistic Y realizing $\mathscr{F}$ is $\|\prod_{\Delta_i > 1} N(\Delta_i^{1/2}\beta_i, \Delta_i) - \prod_{\Delta_i > 1} N(\Delta_i^{1/2}\beta_i, 1)\|$. This total variation does not depend on β and therefore is the same as the last expression for the deficiency in proposition 8.4.2. Thus we have verified the optimality of the described kernel.

Extending this and using some notation which will be explained in the remark below, we obtain the complete picture when *A^* is an isometry.*

Proposition 8.5.3 (The deficiency when A^* is an isometry). *If $AA^* = I$ then $\delta(\mathscr{N}(A, I), \mathscr{N}(B, I)) = \|N(0, BB^*) - N(0, BB^* \wedge I)\| = \|N(0, B^*B \vee I) - N(0, I)\|$ and this, in turn, is equal to the deficiency in the previous proposition if $\Delta_1, \ldots, \Delta_p$ are the eigenvalues (counting multiplicities) of BB^*.*

If X realizes $\mathscr{N}(A, I)$ then the optimal approximation of $\mathscr{N}(B, I)$ based on X is obtained by observing a random vector Z such that the conditional distribution of Z given X is $N(B^*AX, (I - B^*B) \vee 0)$. This approximation, i.e. the experiment realized by Z, is $\mathscr{N}(B, B^*B \vee I)$.

Remark. The linear space of self adjoint operators on a finite dimensional inner product space H becomes an ordered linear space for the ordering '$\geqslant$' which states that $A \geqslant B$ if $A - B$ is non-negative definite. However, this ordered linear space is not a vector lattice if $\dim H > 1$. Nevertheless, various interesting subspaces are vector lattices for this ordering.

If $D_1, D_2, \ldots, D_r$ are commuting self adjoint operators then there is a smallest (largest) operator $\bigvee_i D_i$ ($\bigwedge_i D_i$) within the space of self adjoint operators which commutes with $D_1, \ldots, D_r$ and which majorizes (minorizes) these operators.

Now any commutative class of self adjoint operators is contained in a maximal commutative class and maximal commutative classes are necessarily linear. Furthermore, for any commutative class $\mathscr{D}$ (maximal or not) there is an orthonormal basis for the underlying vector space such that each vector in the basis is an eigenvector for any operator in $\mathscr{D}$. If $(\rho_1, \ldots, \rho_m)$ is such a basis and if $\lambda_{D,i}$ is the eigenvalue associated with $D \in \mathscr{D}$ and ρ_i (i.e. if $D\rho_i = \lambda_{D,i}\rho_i$) then the map $D \to (\lambda_{D,1}, \ldots, \lambda_{D,m})$ from $\mathscr{D}$ to $\mathbb{R}^m$ is 1–1 and order preserving. If $\mathscr{D}$ is maximal then this map is also linear and thus it is also a vector lattice isomorphism where the lattice operations in $\mathscr{D}$ are as described above.

It follows that if $D_1, \ldots, D_r$ commute then the operators $\bigvee_{i=1}^r D_i$ and $\bigwedge_{i=1}^r D_i$ defined above may be interpreted within the framework of a vector lattice.

Now several of the expressions we will use for operators involve both algebraic operations and the operations $\bigvee$ and $\bigwedge$. In order not to overload the notation with parentheses, we will assume that the algebraic operations should be carried out first.

In particular this explains the notation $BB^* \wedge I$, $B^*B \vee I$ and $(I - B^*B) \vee 0$ used in the proposition.

Note by the way that I in $BB^* \wedge I$ is the identity operator on K while I in the expressions $B^*B \vee I$ and $I - B^*B$ denotes the identity operator on the sample space of $\mathcal{N}(B, I)$.

Unfortunately we are now in a situation where the standard notation A^+ and A^- for the self adjoint operators $A \vee 0$ and $(-A) \vee 0$ conflict with the quite well established use of this notation as notation for generalized inverses. Thus here we use the notations A^+ and A^- to denote, respectively, the Moore–Penrose generalized inverse of A and an unspecified generalized inverse of A.

However no difficulties are likely to arise since we shall avoid the use of A^+ and A^- in the vector lattice sense when A is an operator, and instead use the complete notation $A \vee 0$ and $(-A) \vee 0$ for these operators.

Proof of the proposition. Let X realize $\mathcal{N}(A, V)$ and let Y realize $\mathcal{N}(B, W)$. Then $X \in H_A$ and $Y \in H_B$ where H_A and H_B are the sample spaces of $\mathcal{N}(A, V)$ and $\mathcal{N}(B, W)$ respectively.

Choose an orthonormal basis $(\tau_1, \ldots, \tau_p)$ in K such that each τ_i is an eigenvector of BB^*. Let Δ_i be the eigenvalue associated with τ_i, i.e. $BB^*\tau_i = \Delta_i\tau_i$. We may assume that the enumeration of the τ's is such that $\Delta_1 \geqslant \cdots \geqslant \Delta_p \geqslant 0$.

Let q denote the number of positive Δ's, i.e. $q = \operatorname{rank} B$. Then $\Delta_i > 0$ if and only if $i \leqslant q$.

The vectors $A^*\tau_i$; $i = 1, \ldots, p$ are orthonormal in H_A. It follows that there is an orthonormal basis $(\rho_1, \ldots, \rho_m)$ in H_A such that $\rho_i = A^*\tau_i$; $i = 1, \ldots, p$. Similarly there is a basis $(\tilde{\rho}_1, \ldots, \tilde{\rho}_n)$ in H_B such that $\tilde{\rho}_i = B^*\tau_i/\Delta_i^{1/2}$; $i = 1, \ldots, q$.

Let $(\beta_1, \ldots, \beta_p)$, $(X_1, \ldots, X_m)$ and $(Y_1, \ldots, Y_n)$ be the coordinate vectors of, respectively, β, X and Y with respect to the chosen bases. Thus $\beta = \sum_i \beta_i\tau_i$, $X = \sum X_i\rho_i$ and $Y = \sum Y_i\tilde{\rho}_i$.Then $X_1, \ldots, X_m$ are independent and normally distributed with variance 1 and $EX_i = \beta_i$ or $= 0$ as $i \leqslant p$ or $i > p$. The variables $Y_1, \ldots, Y_n$ are also independent and normally distributed with variance 1 and $EY_i = \Delta_i^{1/2}\beta_i$ or $= 0$ as $i \leqslant q$ or $i > q$. Adding some independent $N(0, 1)$-variables if needed, we may ensure that $n \geqslant p$ and then we may write $EY_i = \Delta_i^{1/2}\beta_i$ or $= 0$ as $i \leqslant p$ or $i > p$.

It follows that $\mathcal{N}(A, V) \sim \mathcal{E}$ and that $\mathcal{N}(B, W) \sim \mathcal{F}$ where $\mathcal{E}$ and $\mathcal{F}$ are defined as in the previous proposition. Hence $\delta(\mathcal{N}(A, V), \mathcal{N}(B, W)) = \delta(\mathcal{E}, \mathcal{F})$ may be expressed as in that proposition.

Let Δ denote the diagonal matrix with diagonal elements $\Delta_1, \ldots, \Delta_p$ and let I_p denote the $p \times p$ unit matrix. Then Δ and I_p are the matrix representations of, respectively, BB^* and $AA^* = I$. It follows that the matrix representation of $BB^* \wedge I$ is $\Delta \wedge I_p$. Hence

$$\|N(0,B^*) - N(0,BB^* \wedge I)\| = \|N(0,\Delta) - N(0,\Delta \wedge I)\|$$
$$= \left\| \prod_i N(0,\Delta_i) - \prod_i N(0,\Delta_i \wedge 1) \right\|$$
$$= \left\| \prod_{\Delta_i > 1} N(0,\Delta_i) - \prod_{\Delta_i > 1} N(0,1) \right\| = \delta(\mathscr{E},\mathscr{F}).$$

Assume that the conditional distribution of Z given X is $N(B^*AX, (I - B^*B) \vee 0)$. Then $EZ = B^*AA^*\beta = B^*\beta = EY$, $\mathrm{Cov}(Z|X) = (I - B^*B) \vee 0 = E(\mathrm{Cov}\, Z|X)$ and $\mathrm{Cov}\, Z = E(\mathrm{Cov}\, Z|X) + \mathrm{Cov}(EZ|X) = [(I - B^*B) \vee 0] + B^*B = I \vee B^*B$. It follows that $\|\mathscr{L}(Z) - \mathscr{L}(Y)\| = \|N(0,I) - N(0,I \vee B^*B)\|$. Now the matrix representations of I and B^*B are, respectively, the $n \times n$ unit matrix and the $n \times n$ diagonal matrix $\mathrm{diag}(\Delta_1,\ldots,\Delta_p,0,\ldots,0)$. Thus the matrix representation of $I \vee B^*B$ is $\mathrm{diag}(\Delta_1 \vee 1,\ldots,\Delta_p \vee 1, 1,\ldots,1)$ so that $\|\mathscr{L}(Y) - \mathscr{L}(Z)\| = \|\prod_{\Delta_i > 1} N(0,1) \times \prod_{\Delta_i \leqslant 1} N(0,1) - \prod_{\Delta_i > 1} N(0,\Delta_i) \times \prod_{\Delta_i \leqslant 1} N(0,1)\| = \|\prod_{\Delta_i > 1} N(0,1) - \prod_{\Delta_i > 1} N(0,\Delta_i)\| = \delta(\mathscr{E},\mathscr{F})$. □

In order to cover the general case we will need notions of max and min for operators which do not necessarily commute.

Proposition 8.5.4 (Max and min for pairs of operators). *Let A and B be non-negative definite operators on a finite dimensional inner product space H. Assume that $R(B) \subseteqq R(A)$ and let Π denote the orthogonal projection on $R(A)$. Then:*

(i) *there is a smallest self adjoint operator S with range contained in $R(A)$ which majorizes A and B and has the additional property that SA^+B is self adjoint.*

Denoting this operator S as $S = A \veebar B$ we have

$$A \veebar B = A^{1/2}(I \vee (A^+)^{1/2}B(A^+)^{1/2})A^{1/2} = A^{1/2}(\Pi \vee (A^+)^{1/2}B(A^+)^{1/2})A^{1/2}$$

(ii) *there is a largest self adjoint operator T which minorizes A and B and has the additional property that TA^+B is self adjoint.*

Denoting this operator T as $T = A \barwedge B$ we have

$$A \barwedge B = A^{1/2}(I \wedge (A^+)^{1/2}B(A^+)^{1/2})A^{1/2} = A^{1/2}(\Pi \wedge (A^+)^{1/2}B(A^+)^{1/2})A^{1/2}.$$

(iii) $A \veebar B + A \barwedge B = A + B$

$A \veebar B = A \Leftrightarrow A \barwedge B = B \Leftrightarrow A \geqslant B$

$A \veebar B = B \Leftrightarrow A \barwedge B = A \Leftrightarrow A \leqslant B.$

(iv) *if $R(B) = R(A)$ then $A \veebar B = B \veebar A$ and $A \barwedge B = B \barwedge A$.*

(v) *if A and B commute then* $A \veebar B = A \vee B$ *and* $A \barwedge B = A \wedge B$.
(vi) $\|N(0,B) - N(0, A \barwedge B)\| = \|N(0, A \veebar B) - N(0,A)\|$.

Remark. Here it is important to observe that A^+ for a self adjoint operator A *does not* denote the operator $A \vee 0$ but the Moore–Penrose generalized inverse of A.

Proof. Understanding of the arguments below may be facilitated by keeping in mind the spectral form of the lattice operations for commuting self adjoint operators.

Put $U_0 = I \vee (A^+)^{1/2}B(A^+)^{1/2}$, $V_0 = I \wedge (A^+)^{1/2}B(A^+)^{1/2}$, $U_1 = \Pi \vee (A^+)^{1/2}B(A^+)^{1/2}$ and $V_1 = \Pi \wedge (A^+)^{1/2}B(A^+)^{1/2}$. Then $U_1 \leqslant U_0 \leqslant U_1 + (I - \Pi)$ and $V_1 \leqslant V_0 \leqslant V_1 + I - \Pi$. Putting $S_0 = A^{1/2}U_0A^{1/2}$ and $T_0 = A^{1/2}V_0A^{1/2}$ we see that these operators are not changed if U_0 is replaced by U_1 and V_0 is replaced by V_1. The latter relations may be inverted, giving $U_1 = (A^+)^{1/2}S_0(A^+)^{1/2}$ and $V_1 = (A^+)^{1/2}T_0(A^+)^{1/2}$. Now U_1 commutes with $(A^+)^{1/2}BA^{1/2}$ so that $(A^+)^{1/2}B(A^+)^{1/2}U_1 = U_1(A^+)^{1/2}B(A^+)^{1/2}$ yielding $(A^+)^{1/2}BA^+S_0(A^+)^{1/2} = (A^+)^{1/2}S_0A^+B(A^+)^{1/2}$, i.e. $BA^+S_0 = S_0A^+B$ so that S_0A^+B is self adjoint. Similarly T_0A^+B is self adjoint and we find directly that $S_0 \geqslant A^{1/2}(A^+)^{1/2}B(A^+)^{1/2}A^{1/2} \geqslant \Pi B \Pi = B$, $S_0 \geqslant A^{1/2}IA^{1/2} = A$, $T_0 \leqslant B$ and $T_0 \leqslant A$.

Consider any self adjoint operator S which majorizes A and B, whose range is contained in $R(A) = R(\Pi)$ and is such that SA^+B is self adjoint. Put $U = (A^+)^{1/2}S(A^+)^{1/2}$. Then $S = A^{1/2}UA^{1/2}$, $U \geqslant (A^+)^{1/2}A(A^+)^{1/2} = \Pi$ and $U \geqslant (A^+)^{1/2}B(A^+)^{1/2}$. The self adjointness of SA^+B may be expressed by the identity $SA^+B = BA^+S$. Substituting $S = A^{1/2}UA^{1/2}$ this yields $A^{1/2}UA^{1/2}A^+B = BA^+A^{1/2}UA^{1/2}$, i.e. $A^{1/2}U(A^+)^{1/2}B = B(A^+)^{1/2}UA^{1/2}$. Pre- and post-multiplication by $(A^+)^{1/2}$ yields $U(A^+)^{1/2}B(A^+)^{1/2} = (A^+)^{1/2}B(A^+)^{1/2}U$ so that U commutes with $(A^+)^{1/2}B(A^+)^{1/2}$. Hence, since $U\Pi = \Pi U = U$, $U \geqslant \Pi \vee (A^+)^{1/2}B(A^+)^{1/2} = U_1$ yielding $S = A^{1/2}UA^{1/2} \geqslant A^{1/2}U_1A^{1/2} = S_0$ proving the asserted minimality of S_0. Similarly $T \leqslant T_0$ whenever $T \leqslant A, B$ and TA^+B is self adjoint. This completes the proof of (i) and (ii).

Furthermore $S_0 + T_0 = A^{1/2}U_0A^{1/2} + A^{1/2}V_0A^{1/2} = A^{1/2}(U_0 + V_0)A^{1/2} = A^{1/2}(I + (A^+)^{1/2}B(A^+)^{1/2}A^{1/2} = A + B$. Thus $S_0 = A \Leftrightarrow T_0 = B$ and $S_0 = B \Leftrightarrow T_0 = A$. If $A \geqslant B$ then $AA^+B = B$ is self adjoint. Thus $A \geqslant S_0 \geqslant A$ yielding $S_0 = A$. If $B \geqslant A$ then, since BA^+B is self adjoint, $B = S_0$. This completes the proof of (iii).

Assume that $R(A) = R(B)$. If S is any self adjoint operator with range contained in $R(A)$ then the operator SA^+B is self adjoint if and only if $SA^+B = BA^+S$. Pre- and post-multiplying with B^+ we see that SA^+B is self adjoint if and only if $B^+SA^+ = A^+SB^+$ and this condition is symmetric in A, B. Thus $S_0 = A \veebar B = B \veebar A$ and $T_0 = A \barwedge B = B \barwedge A$. If A and B commute then $(A \vee B)A^+B$ is self adjoint so that $S_0 \leqslant A \vee B$. On the other hand S_0 commutes with A and B so that $A \vee B \leqslant S_0$, yielding $S_0 = A \vee B$. Hence, by (iii), $T_0 = A + B - S_0 = A + B - (A \vee B) = A \wedge B$.

It remains to prove (vi). Transforming by the operator $(A^+)^{1/2}$ we see that (vi) may be written $\|N(0,(A^+)^{1/2}B(A^+)^{1/2}) - N(0,\Pi \wedge (A^+)^{1/2}B(A^+)^{1/2}\| = \|N(0,\Pi \vee (A^+)^{1/2}B(A^+)^{1/2} - N(0,\Pi)\|$. Deleting the part of H which is outside $R(A)$ we may assume without loss of generality that A is non-singular and that $\Pi = I$. Then (vi) may be written $\|N(0,\Delta) - N(0,I \wedge \Delta)\| = \|N(0,I \vee \Delta) - N(0,I)\|$ where $\Delta = A^{-1/2}BA^{-1/2}$. Choosing an orthonormal basis of eigenvectors for Δ and transforming by the coordinate transformation we see that we may assume that Δ is some $n \times n$ non-negative definite diagonal matrix while I is the $n \times n$ unit matrix. Then $\|N(0,\Delta) - N(0,I \wedge \Delta)\| = \|\prod_{\Delta_i > 1} N(0,\Delta_i) - \prod_{\Delta_i > 1} N(0,1)\| = \|N(0,I \vee \Delta) - N(0,I)\|$. □

Here are some other identities for total variation differences between centred normal distributions.

Proposition 8.5.5 (Statistical distances between centred normal distributions). *If U and V are non-negative definite operators on the same finite dimensional inner product space then the dichotomies $(N(0,U), N(0,V))$ and $(N(0,V^+), N(0,U^+)$ are equivalent. In particular if U and V are non-singular then*

$$\begin{aligned}\|N(0,U) - N(0,V)\| &= \|N(0,U^{-1}) - N(0,V^{-1})\| \\ &= E|1 - (\det U/\det V)^{1/2}\exp(-\tfrac{1}{2}(X,(V^{-1} - U^{-1})X))| \\ &= E|1 - (\det V/\det U)^{1/2}\exp(-\tfrac{1}{2}(Y,(V - U)Y))|\end{aligned}$$

where X is $N(0,U)$-distributed and Y is $N(0,U^{-1})$-distributed.

Proof. We may assume without loss of generality that U and V are non-singular. The claimed equivalence follows from the fact that the non-singular transformation $U^{-1/2}(U^{-1/2}VU^{-1/2})^{-1/2}U^{-1/2}$ maps the dichotomy $(N(0,U), N(0,V))$ into the dichotomy $(N(0,V^{-1}), N(0,U^{-1}))$. The last equality follows from the identity $\|P - Q\| = \int|1 - dQ/dP|\,dP$ which is valid whenever $Q \ll P$. □

The uniformity defined by the metric $U, V \to \|N(0,U) - N(0,V)\|$ may be described as follows.

Proposition 8.5.6 (Uniformity of statistical distance for centred normal distributions). *Let $U_n; n = 1,2,\ldots$ and $V_n; n = 1,2,\ldots$ be non-negative definite self adjoint operators on a common finite dimensional inner product space. Then $\|N(0,U_n) - N(0,V_n)\| \to 0$ if and only if to each $\varepsilon > 0$ there corresponds an n_0 such that $R(U_n) = R(V_n)$ and $(1-\varepsilon)U_n \leqslant V_n \leqslant (1+\varepsilon)U_n$ when $n \geqslant n_0$. This is in turn equivalent to the condition that $R(U_n) = R(V_n)$ when n is sufficiently large* *and that* $(U_n^+)^{1/2}V_n(U_n^+)^{1/2} - \Pi_n \to 0$ *where* Π_n *is the orthogonal projection on* $R(U_n)$.

In particular $N(0, U_n) \to N(0, U)$ *if and only if* $U_n \to U$ *and* $R(U_n) = R(U)$ *when n is sufficiently large.*

Proof. This follows readily from the facts that if $R(U) = R(V)$ then $\|N(0, U) - N(0, V)\| = \|N(0, (U^+)^{1/2}U(U^+)^{1/2}) - N(0, (U^+)^{1/2}V(U^+)^{1/2})\|$ and that $(U^+)^{1/2}U(U^+)^{1/2} = U^+U$ is the orthogonal projection on $R(U)$. □

It would have been desirable to find inequalities relating $\|N(0, U) - N(0, V)\|$ to other familiar 'non-probabilistic' quantities defined for pairs (U, V) of non-negative definite operators. Thus if γ denotes the affinity between non-singular distributions $N(0, U)$ and $N(0, V)$ then by the formula for the affinity given in section 8.2 we find that $-2\log\gamma = [\log\det(\frac{1}{2}U + \frac{1}{2}V)] - \frac{1}{2}[\log\det U] - \frac{1}{2}[\log\det V]$. It follows that the closeness of $N(0, U)$ to $N(0, V)$ is linked to the concavity of the functional log det in the region of interest.

We are now in a position to give a complete description of the deficiency between linear normal experiments $\mathcal{N}(A, I)$ and $\mathcal{N}(B, I)$.

Theorem 8.5.7 (Deficiencies when the covariance operators are identity operators). $\delta(\mathcal{N}(A, I), \mathcal{N}(B, I)) < 2$ *if and only if* $R(B) \subseteqq R(A)$ *and if so then*

$$\begin{aligned}\delta(\mathcal{N}(A, I), \mathcal{N}(B, I)) &= \|N(0, BB^*) - N(0, AA^* \barwedge BB^*)\| \\ &= \|N(0, AA^* \veebar BB^*) - N(0, AA^*)\| \\ &= \|N(0, B^*(AA^*)^+B) - N(0, (B^*(AA^*)^+B) \vee I)\| \\ &= \|N(0, (B^*(AA^*)^+B) \vee I) - N(0, I)\| \\ &= \|N(0, B^*(AA^*)^+B) - N(0, (B^*(AA^*)^+B) \wedge I)\|.\end{aligned}$$

If X realizes $\mathcal{N}(A, I)$ *then Z realizes a best approximation to* $\mathcal{N}(B, I)$ *if the conditional distribution of Z given X is* $N(B^*(AA^*)^+AX, [I - B^*(AA^*)^+B] \vee 0)$. *If* $R(B) \subseteqq R(A)$ *then this approximation is the experiment* $\mathcal{N}(B, (B^*(AA^*)^+B) \vee I)$.

Remark. The chance mechanism producing Z is determined by any BLUE generator $\hat{\beta}(X)$ for β based on X. In fact if $\hat{\beta}(X)$ is a BLUE generator for β then $B^*(AA^*)^+AX = B^*\hat{\beta}(X)$ so that the conditional distribution of Z given X is $N(B^*\hat{\beta}(X), (I - \operatorname{Cov} B^*\hat{\beta}(X)) \vee 0)$. The experiment realized by X may then be expressed as $\mathcal{N}(B, I \vee \operatorname{Cov} B^*\hat{\beta}(X))$.

Proof. Put $\delta = \delta(\mathcal{N}(A, I), \mathcal{N}(B, I))$. At the beginning of this section we argued that $\delta = 2$ when $R(B) \nsubseteqq R(A)$. Thus $R(B) \subseteqq R(A)$ when $\delta < 2$. Assume that $R(B) \subseteqq R(A)$. We shall show that the four total variations $\|\cdot\|$ occurring in the statement of the theorem are all equal to δ. As e.g. $R(AA^* \veebar BB^*) = R(AA^*)$, this will then imply that $\delta < 2$.

Using proposition 8.5.4 we find that

$$\|N(0,BB^*) - N(0,AA^* \barwedge BB^*)\| = \|N(0,AA^* \veebar BB^*) - N(0,AA^*)\|$$

and also that

$$\|N(0,(B^*(AA^*)^+B) \vee I) - N(0,I)\|$$
$$= \|N(0,B^*(AA^*)^+B) - N(0,I \wedge B^*(AA^*)^+B)\|.$$

Furthermore, the transformation $((AA^*)^+)^{1/2}$ maps the dichotomy $(N(0,AA^* \veebar BB^*), N(0,AA^*))$ into the equivalent dichotomy $(N(0,\Pi \vee (((AA^*)^+)^{1/2}BB^*((AA^*)^+)^{1/2})), N(0,\Pi))$ where Π is the orthogonal projection on $R(A)$. It follows that

$$\|N(0,BB^*) - N(0,AA^* \barwedge BB^*)\|$$
$$= \|N(0,((AA^*)^+)^{1/2}BB^*((AA^*)^+)^{1/2})$$
$$- N(0,\Pi \wedge ((AA^*)^+)^{1/2}BB^*((AA^*)^+)^{1/2})\|.$$

Therefore it suffices to show

(i)
$$\delta = \|N(0,((AA^*)^+)^{1/2}BB^*((AA^*)^+)^{1/2}) - N(0,\Pi \wedge ((AA^*)^+)^{1/2}BB^*((AA^*)^+)^{1/2})\|$$
$$= \|N(0,I \vee B^*(AA^*)^+B) - N(0,I)\|.$$

Replacing the parameter β by $\tilde{\beta} = ((AA^*)^{1/2})\beta$ and putting $\tilde{A} = ((AA^*)^+)^{1/2}A$ and $\tilde{B} = ((AA^*)^+)^{1/2}B$ we find $A^*\beta = \tilde{A}^*\tilde{\beta}$, $B^*\beta = \tilde{B}^*\tilde{\beta}$, $R(A) = R(\tilde{A})$, $\tilde{A}\tilde{A}^* = \Pi$ and $\tilde{B}\tilde{B}^* = ((AA^*)^+)^{1/2}BB^*((AA^*)^+)^{1/2}$. It follows that neither the deficiency $\delta(\mathscr{N}(A,I),\mathscr{N}(B,I))$ nor the two statistical distances occurring in (i) are changed if A and B are replaced by, respectively, $\tilde{A}$ *and* $\tilde{B}$. Altogether this implies that without loss of generality we may prove (i) under the assumption that $AA^* = \Pi$. Then (i) may be written

(ii) $\delta = \|N(0,BB^*) - N(0,\Pi \wedge BB^*)\| = \|N(0,I \vee B^*B) - N(0,I)\|.$

Restricting the parameter to $R(A)$ we see also that we may assume that Π is the identity operator on K. (The I occurring here is the identity operator on the sample space of $\mathscr{N}(B,I)$.) Then (ii) follows directly from proposition 8.5.3.

Assume that X and Y realize $\mathscr{N}(A,I)$ and $\mathscr{N}(B,I)$ respectively and that the conditional distribution of Z given X is $N(B^*(AA^*)^+AX,(I - B^*(AA^*)^+B) \vee 0)$. Then

$$EZ = B^*(AA^*)^+AA^*\beta$$
$$= B^*\Pi\beta = B^*\beta = EY$$

and $\operatorname{Cov} Z = B^*(AA^*)^+ AA^*(AA^*)^+ B + (I - B^*(AA^*)^+ B) \vee 0 = I \vee B^*(AA^*)^+ B$ so that $\|\mathscr{L}(Z) - \mathscr{L}(Y)\| \equiv_\beta \delta$. □

As a particular case we obtain deficiencies for comparable experiments.

Corollary 8.5.8 (Deficiencies for comparable experiments). *$\mathscr{N}(A, I) \leqslant \mathscr{N}(B, I)$ if and only if $AA^* \leqslant BB^*$ and then $\delta(\mathscr{N}(A, I), \mathscr{N}(B, I)) = \|N(0, AA^*) - N(0, BB^*)\| = \|N(0, (AA^*)^+) - N(0, (BB^*)^+)\|$.*

Remark. These deficiencies were given by LeCam (1975).

Proof. We know from section 1 that $\mathscr{N}(A, I) \leqslant \mathscr{N}(B, I)$ if and only if $AA^* \leqslant BB^*$. If, in addition $R(B) \subseteqq R(A)$, then the expressions for the deficiency follow directly from the theorem and proposition 8.5.5. If $R(B) \nsubseteqq R(A)$ then all three quantities $= 2$. □

Corollary 8.5.9 (Deficiency distance and statistical distance). *$\frac{1}{2}\|N(0, AA^*) - N(0, BB^*)\| \leqslant \Delta(\mathscr{N}(A, I), \mathscr{N}(B, I)) \leqslant \|N(0, AA^*) - N(0, BB^*)\|$.*

Proof. If $R(A) \neq R(B)$ then the inequalities reduce to $1 \leqslant 2 \leqslant 2$. Thus we may assume that $R(A) = R(B)$. By using the operator $((AA^*)^+)^{1/2}$ and then restricting the range, we may assume also that $AA^* = I$. Let $\Delta_1, \ldots, \Delta_r$ be the eigenvalues (counting multiplicities) of BB^*. Referring to proposition 8.5.2 and the proof of proposition 8.5.3, and separating coordinates according to whether corresponding eigenvalues are > 1 or $\leqq 1$, we find (using obvious notations) that

$$\begin{aligned}
&\|N(0, AA^*) - N(0, BB^*)\| \\
&\quad = \|G_1 \times G_2 - H_1 \times H_2\| \leqslant \|G_1 - H_1\| + \|G_2 - H_2\| \\
&\quad \leqslant 2[\|G_1 - H_1\| \vee \|G_2 - H_2\|] \\
&\quad = 2[\delta(\mathscr{N}(A, I), \mathscr{N}(B, I)) \vee \delta(\mathscr{N}(B, I), \mathscr{N}(A, I))].
\end{aligned}$$ □

At the beginning of this section we saw how general deficiencies may be expressed in terms of deficiencies for non-singular experiments. A non-singular experiment $\mathscr{N}(A, V)$ is equivalent to $\mathscr{N}(\tilde{A}, I)$ where I is the identity operator on $R(V)$ and $\tilde{A}$ is $A(V^+)^{1/2}$ restricted to $R(V)$. It follows that general expressions for deficiencies are easily obtained from the results above. We will not write them out here in detail.

The general randomization criterion implies that $\delta(\mathscr{N}(A, V), \mathscr{N}(B, W)) \leqslant \varepsilon$ if and only if $\mathscr{N}(A, V) \geqslant \mathscr{G}$ for some experiment $\mathscr{G}$ such that the statistical distances between corresponding distributions in $\mathscr{G}$ and in $\mathscr{N}(B, W)$ are always $\leqslant \varepsilon$. It is a particular and interesting feature of the linear normal experiments that the approximating experiment $\mathscr{G}$ may always be chosen as a linear normal experiment with known covariance. In fact if $\varepsilon < 2$ and if U is as in the following theorem, then we may put $\mathscr{G} = \mathscr{N}(B, U)$.

Theorem 8.5.10 (Deficiencies in terms of linear operators). $\delta(\mathcal{N}(A, V), \mathcal{N}(B, W)) < 2$ *if and only if* $R(B) \subseteqq R(A)$ *and* $B[N(W)] \subseteqq A[N(V)]$. *If so then this deficiency is the smallest constant* δ *such that for some linear map* T *from* H_A *to* H_B *and for some non-negative definite operator* U *on* H_B *we have*:

(i) $TA^* = B^*$.
(ii) $TVT^* \leqslant U$.
(iii) $\|N(0, U) - N(0, W)\| \leqslant \delta$.

Remark 1. Assume that $\delta < 2$ and let X and Y realize $\mathcal{N}(A, V)$ and $\mathcal{N}(B, W)$ respectively. Let $k \in R(B)$ and consider any unbiased linear estimator $\psi(Y)$ of (k, β). Then $\phi(X) = \psi(TX)$ is an unbiased estimator of (k, β) in $\mathcal{N}(A, V)$ and $\operatorname{Var}\phi(X) \leqslant (1 + \varepsilon_\delta)\operatorname{Var}\psi(Y)$ where $\varepsilon_\delta \to 0$ as $\delta \to 0$.

In fact we may put $\varepsilon_\delta = \sup\{\|\Pi - M\| : \|N(0, M) - N(0, \Pi)\| \leqslant \delta\}$ where Π is the orthogonal projection on $R(W) = R(U)$.

Remark 2. It is clear from the case '$\delta = 0$' that in general we cannot ensure equality in (ii).

Remark 3. This theorem implies readily that the smoothing device considered in proposition 8.3.8 is deficiency decreasing.

Proof. At the beginning of this section we saw that $R(B) \subseteqq R(A)$ and that $B[N(W)] \subseteqq A[N(V)]$ when $\delta < 2$. We saw also that $\delta = \delta(\mathcal{N}(PA, V), \mathcal{N}(PB, W))$ where P is the orthogonal projection on $[A[N(V)]]^\perp$. If $B[N(W)] \subseteqq A[N(V)]$ then $\mathcal{N}(PA, V)$ and $\mathcal{N}(PB, W)$ are both non-singular and thus $\delta < 2$ by theorem 8.5.7. From here on assume that $R(B) \subseteqq R(A)$ and that $B[N(W)] \subseteqq A[N(V)]$. Note that if $\mathcal{N}(\tilde{A}, \tilde{V}) \sim \mathcal{N}(A, V)$ then we may argue as if $A = \tilde{A}$ and $V = \tilde{V}$. Thus we may assume also that $\mathcal{N}(A, V)$ is in minimal form with A being an orthogonal projection in K and with $R(V) \subseteqq R(A)$. Assume first that $A[N(V)] = 0$. Then $\mathcal{N}(A, V)$ and $\mathcal{N}(B, W)$ are both non-singular and the conclusion follows readily from theorem 8.5.7 and the operator criterion for the ordering of these experiments given in theorem 8.4.3.

Consider the general case when $\delta < 2$. Then we may apply what we have argued so far to $\mathcal{N}(PA, V)$ and $\mathcal{N}(PB, W)$ and thereby obtain T and $\tilde{W}$ such that $TA^*P = B^*P$, $TVT^* \leqslant \tilde{W}$ and $\|N(0, \tilde{W}) - N(0, W)\| \leqslant \delta$. These requirements on T involve only its restriction to $R(V)$. Now $R(A^*(I - P)) = R(I - P) = A[N(V)]$ which is orthogonal to $R(V)$ by the particular form of $\mathcal{N}(A, V)$. Thus we only need to find a linear map $\tilde{T}$ such that $\tilde{T}A^*(I - P) = B^*(I - P)$, since we then may piece T and $\tilde{T}$ together by using T on $R(V)$ and $\tilde{T}$ on $N(V)$. Now $R((I - P)B) \subseteqq R(I - P) = A[N(V)] = R((I - P)A)$ so that $(I - P)B = (I - P)A\tilde{T}^*$ for some linear map $\tilde{T}$ and then $\tilde{T}A^*(I - P) = B^*(I - P)$. □

In the remaining part of this section we will restrict ourselves to some comments on deficiencies which are of a more qualitative nature. Firstly we should note how the deficiencies are expressed in terms of the Fisher information operators and thus are related to the affinities and to variances of the best linear unbiased estimators.

Another interesting feature of these deficiencies is that they are not altered if the experiments involved are replicated the same number of times. As we shall see, this is in strong contrast to the situation where the covariance operator is unknown.

If $\delta(\mathcal{N}(A, V)^n, \mathcal{N}(B, W)^n) = 2$ for some n then it is 2 for all n, and we have seen that this is the case if and only if $R(B) \subsetneqq R(A)$ or $B[N(W)] \subsetneqq A[N(V)]$. On the other hand if $R(B) \subseteqq R(A)$ and $B[N(W)] \subseteqq A[N(V)]$ then $\mathcal{N}(A, V)^n \geqslant \mathcal{N}(B, W)$ if and only if n is at least as great as any eigenvalue of $((PAV^+A^*P)^+)^{1/2}(PBW^+B^*P)((PAV^+A^*P)^+)^{1/2}$, i.e. if and only if $nPAV^+A^*P \geqslant PBW^+B^*P$.

Referring to section 4 of chapter 9 we may ask about local comparison of these experiments. As the ordering 'being at least as informative as' equals the ordering of the Fisher information operators it is clear that the local ordering and the global ordering coincide. However we have not tackled the important problem of comparison when the parameter is restricted to various subsets of K.

Consider local comparison within shrinking neighbourhoods in the parameter set and for an increasing number of replications. If the neighbourhoods are around $\beta_0 \in K$ then we may re-parametrize by putting $\tilde{\beta} = n^{1/2}(\beta - \beta_0)$ so that $\beta = \beta_0 + \tilde{\beta}/n^{1/2}$. In this way the experiment $\mathcal{N}(A, V)$ is changed to the experiment $\mathcal{N}(A/n^{1/2}, V)$. Replicating this n times we obtain $\mathcal{N}(A/n^{1/2}, V)^n \sim \mathcal{N}(A, V)$ so that the effects of shrinking and of replicating cancel each other in this case. This agrees with LeCam's central limit theorem which we mentioned in example 1.4.8, since his result implies that the experiments $\mathcal{N}(A, V)$ are the local limits of replicates of locally re-parametrized quadratically differentiable experiments.

Suppose now that an experiment $\mathscr{E}$ has to be chosen within a given set κ of experiments. Assume that there is an ideal experiment $\mathscr{H}$ such that any experiment $\mathscr{E}$ in κ is at most as informative as $\mathscr{H}$. We may then consider the deficiency $\delta(\mathscr{E}, \mathscr{H})$ where $\mathscr{E} \in \kappa$ as a measure of the lack of information in $\mathscr{E}$. This quantity is clearly monotonically decreasing in $\mathscr{E}$ and it is small if and only if $\mathscr{E}$ contains almost as much information as $\mathscr{H}$. However if this quantity is large then we should be aware that $\mathscr{E}$ may still be close to the ideal experiment for decision problems of particular interest. Thus a full interpretation of $\delta(\mathscr{E}, \mathscr{H})$ for a particular $\mathscr{E}$ should tell us something about the decision problems which are 'responsible' for this deficiency. The extreme case where $\mathscr{H}$ majorizes all experiments, i.e. when $\mathscr{H}$ is totally informative, is discussed in Torgersen (1981).

Instead of working with an experiment majorizing κ we might of course also make a comparison with an undesirable experiment $\mathscr{G}$ and use $\delta(\mathscr{G}, \mathscr{E})$ as a monotonically increasing measure of the amount of information in $\mathscr{E}$. Lindqvist (1977, 1984) used this idea to investigate the loss of memory in a Markov process. The problem of measuring the potential value of additional observations is also of this nature. This problem is discussed in Torgersen (1972), LeCam (1974a, 1975), Swensen (1980), Helgeland (1982) and Mammen (1986). Explicit expressions for deficiencies for the problem of measuring the amount of information in additional observations in regression models may be found in Swensen's paper.

8.6 Deficiencies between linear normal experiments when the covariances are known except for a common multiplicative scalar

As in the previous section, it is convenient to begin by reducing the comparison problems to the non-singular case by delimiting the cases of maximal deficiency.

A slight modification of the arguments used in the proof of proposition 8.5.1 establishes the following analogue.

Proposition 8.6.1 (Reduction to non-singular experiments). *$\delta(\mathscr{N}(A,\sigma^2 V;\sigma^2>0), \mathscr{N}(B,\sigma^2 W;\sigma^2>0)) = \delta(\mathscr{N}(PA,\sigma^2 V;\sigma^2>0), \mathscr{N}(PB,\sigma^2 W;\sigma^2>0))$ where P is the orthogonal projection on $[A[N(V)]]^{\perp}$.*

When delimiting the case of maximal deficiency we must take into account the problem of estimating σ^2. Actually it is simpler to characterize the opposite situation of a non-maximal deficiency.

Theorem 8.6.2 (Non-maximal deficiency). *Consider linear normal experiments $\mathscr{E} = \mathscr{N}(A,\sigma^2 V;\sigma^2>0)$ and $\mathscr{F} = \mathscr{N}(B,\sigma^2 W;\sigma^2>0)$. Let $\mathscr{E}_a$ and $\mathscr{F}_a$ denote, respectively, the experiments obtained from $\mathscr{E}$ and $\mathscr{F}$ by restricting β to $[A[N(V)]]^{\perp} = [A^*]^{-1}[R(V)]$. Then the following conditions are equivalent.*

(i) $\delta(\mathscr{E},\mathscr{F}) < 2$.
(ii) *$\delta(\mathscr{N}(A,V),\mathscr{N}(B,W))<2$ and if $N(A)\subseteqq N(V)$ then $N(W)=B^{-1}[A[N(V)]]$.*
(iii) *$R(B)\subseteqq R(A)$, $B[N(W)]\subseteqq A[N(V)]$ and if $N(A)\subseteqq N(V)$ then $N(W) = B^{-1}[A[N(V)]]$.*
(iv) *any linear function of β which is identifiable (with certainty) in $\mathscr{F}$ is also identifiable (with certainty) in $\mathscr{E}$, and if K_0 is a subspace of K such that σ^2 possesses an unbiased estimator in the experiment obtained from $\mathscr{F}$ by restricting β to K_0, then σ^2 also possesses an unbiased estimator in the same restriction of $\mathscr{E}$.*
(v) *$\mathscr{F}_a$ is non-singular and each linear function of (β,σ^2) which possesses an unbiased estimator in $\mathscr{F}_a$ also possesses an unbiased estimator in $\mathscr{E}_a$.*

Proof. In the light of theorem 8.5.10, condition (ii) is just a rephrasing of condition (iii).

Let P denote the orthogonal projection on $[A^*]^{-1}[R(V)] = [A[N(V)]]^{\perp}$. Say that a subspace K_0 of K is *σ^2-friendly in $\mathscr{E}$ ($\mathscr{F}$)* if the experiment obtained from $\mathscr{E}$ ($\mathscr{F}$) by restricting β to K_0 admits unbiased estimators of σ^2. Say that K_0 is *σ^2-unfriendly in $\mathscr{E}$ ($\mathscr{F}$)* if it is not σ^2-friendly in $\mathscr{E}$ ($\mathscr{F}$). Just after corollary 8.3.15 we argued that K itself is σ^2-unfriendly if and only if $V[N(A)] = 0$, i.e. if and only if $R(V) \subseteqq R(A^*)$. It follows that K_0 is σ^2-unfriendly if and only if $R(V) \subseteqq A^*[K_0]$ (see the next paragraph for details). This in turn is equivalent to the condition that $R(V) \subseteqq R(A^*)$ and $K_0 + N(A^*) \supseteqq R(PA)$ (see next paragraph), and we know that $R(PA) = R(A) \cap R(P) = R(A) \cap [A^*]^{-1}[R(V)]$. It follows that $R(PA)$ is σ^2-unfriendly whenever there is a σ^2-unfriendly K_0 and in that case $R(PA)$ is the smallest σ^2-unfriendly space which is contained in $R(A)$. The details of this argument are as follows.

Assume first that $R(V) \subseteqq A^*[K_0]$. Then $R(V) \subseteqq R(A^*)$. If $\beta \in R(PA)$ then $A^*\beta \in R(V)$ so that $A^*\beta = A^*k_0$ where $k_0 \in K_0$. Hence $\beta \in K_0 + N(A^*)$ so that $R(PA) \subseteqq K_0 + N(A^*)$. Assume that $R(V) \subseteqq R(A^*)$ and that $K_0 + N(A^*) \supseteqq R(PA)$ and let $h \in R(V)$. Then $h = A^*\beta$ for some $\beta \in R(A)$. Hence $\beta \in [A^*]^{-1}[R(V)]$ so that $\beta \in R(PA)$. Thus $\beta = k_0 + \beta_0$ where $k_0 \in K_0$ and $\beta_0 \in N(A^*)$. Hence $h = A^*\beta = A^*k_0 \in A^*[K_0]$. If K_0 is σ^2-unfriendly then $R(V) \subseteqq A^*[K_0] \subseteqq R(A^*)$ and then $A^*[R(PA)] = A^*[[A^*]^{-1}[R(V)]] \supseteqq R(V)$ so that $R(PA)$ is also σ^2-unfriendly. If K_1 is σ^2-unfriendly and if $K_1 \subseteqq R(A)$ then $K_1 + N(A^*) \supseteqq R(PA)$, so that by projecting on $R(A)$ we find that $K_1 \supseteqq R(PA)$. Altogether this shows that $R(PA)$ is σ^2-unfriendly whenever there are σ^2-unfriendly spaces and that $R(PA)$ is then the smallest σ^2-unfriendly subspace of the range $R(A)$ of A.

Assume that condition (iii) is satisfied. The two first statements of (iii) translate easily into the first claim of (iv). Let K_0 be a subspace of K so that σ^2 possesses an unbiased estimator in the restriction of $\mathscr{F}$ to $K_0 \times]0, \infty[$. Then, since K_0 is σ^2-friendly in $\mathscr{F}$, $R(W) \nsubseteqq B^*[K_0]$. Assume now that K_0 is σ^2-unfriendly in $\mathscr{E}$. Then $R(V) \subseteqq A^*[K_0]$ so that $R(V) \subseteqq R(A^*)$ (i.e. $N(V) \supseteqq N(A)$) and thus, by (iii), $N(W) = B^{-1}[A[N(V)]]$). Note now that $N(PB) = B^{-1}[A[N(V)]]$. Thus $R(W) = R(B^*P)$ so that $R(P)$ is σ^2-unfriendly in $\mathscr{F}$. Furthermore $K_0 + N(A^*) \supseteqq R(PA)$. Using the fact that $R(PB) = P[R(B)] \subseteqq P[R(A)] = R(PA) \subseteqq R(A)$ we find that $A^*PB\underset{\sim}{h} = 0 \Leftrightarrow PB\underset{\sim}{h} \perp R(A) \Leftrightarrow PB\underset{\sim}{h} = 0$. It follows that $N(A^*PB) = N(PB)$ so that $R(B^*PA) = R(B^*P)$. Hence also $R(W) = B^*[R(PA)]$. This implies that $R(W) \subseteqq B^*[K_0 + N(A^*)] \subseteqq B^*[K_0] + B^*[N(A^*)] = B^*[K_0]$, since $N(A^*) \subseteqq N(B^*)$. Thus we have arrived at a contradiction and this shows that K_0 is σ^2-friendly in $\mathscr{E}$. Hence (iii) $\Rightarrow$ (iv).

Therefore assume that (iv) holds and let $(k, \beta) + c\sigma^2$ possess an unbiased estimator in $\mathscr{F}_a$. Here $k \in K$ is a known vector and $c > 0$ is a known constant.

Specifying σ we see that (k, β) is identifiable in $\mathscr{F}_a$ and thus it is also identifiable in $\mathscr{E}_a$. It follows that (k, β) possesses an unbiased estimator in $\mathscr{E}_a$ as well as in $\mathscr{F}_a$. Hence $c\sigma^2 = (k, \beta) + c\sigma^2 - (k, \beta)$ also possesses an unbiased estimator in $\mathscr{F}_a$. It follows, by (iv), that $c\sigma^2$ possesses an unbiased estimator in $\mathscr{E}_a$. Hence the linear parametric function $(k, \beta) + c\sigma^2$ possesses an unbiased estimator in $\mathscr{E}_a$.

If $k \in B[N(W)]$ then (k, β) is identifiable with certainty in $\mathscr{F}$ and hence, by assumption, also in $\mathscr{E}$, so that $k \in A[N(V)]$. It follows that $PB[N(W)] \subseteqq PA[N(V)] = 0$ so that $\mathscr{F}_a$ is non-singular.

Now assume that condition (v) is satisfied. Then, by non-singularity, $PB[N(W)] = 0$ so that $B[N(W)] \subseteqq A[N(V)]$. If $k \in R(B)$ then (k, β) possesses an unbiased estimator in $\mathscr{F}$ and hence also in $\mathscr{F}_a$ and in $\mathscr{E}_a$. Thus $R(B) \subseteqq R(PA) \subseteqq R(A)$.

First consider the situation when $N(V) \supseteqq N(A)$, i.e. when $R(A^*) \supseteqq R(V)$. Then $R(A)$ is σ^2-unfriendly in $\mathscr{E}$ so that $R(PA)$, as well as the space K, is σ^2-unfriendly in $\mathscr{E}$. Hence, by condition (v), K is also σ^2-unfriendly in $\mathscr{F}_a$. It follows that $N(W) \supseteqq N(PB) = B^{-1}[A[N(V)]] \supseteqq B^{-1}[B[N(W)]] \supseteqq N(W)$ so that $N(W) = N(PB)$. Hence (v) $\Rightarrow$ (iii) and we have seen that (ii) $\Leftrightarrow$ (iii) $\Rightarrow$ (iv) $\Rightarrow$ (v). It follows that these four conditions are equivalent. Therefore it remains to prove that (i) and (ii) are equivalent. Now by proposition 8.6.1, condition (i) is equivalent to

(i′) $\delta(\mathscr{E}_a, \mathscr{F}_a) < 2$

and from the established equivalence of (ii) and (v) it follows that (ii) remains unchanged if $\mathscr{E}$ and $\mathscr{F}$ are replaced by $\mathscr{E}_a$ and $\mathscr{F}_a$ respectively. Furthermore, conditions (i) and (ii) both imply that $\delta(\mathscr{N}(A, V), \mathscr{N}(B, W)) < 2$ and hence, by theorem 8.5.10, that $\mathscr{E}_a$ and $\mathscr{F}_a$ are both non-singular.

It follows that we may assume without loss of generality that $\mathscr{E}$ and $\mathscr{F}$ are non-singular and then, using $N(A) \supseteqq N(V)$ and $N(B) \supseteqq N(W)$, we see that we may assume also that the operators V and W are non-singular.

Now it remains to show that $\delta(\mathscr{E}, \mathscr{F}) < 2$ if and only if $N(B) = 0$ when $N(A) = 0$ for experiments $\mathscr{E} = \mathscr{N}(A, \sigma^2 V; \sigma^2 > 0)$ and $\mathscr{F} = \mathscr{N}(B, \sigma^2 W; \sigma^2 > 0)$ such that V and W are non-singular and $R(B) \subseteqq R(A)$.

Let the vector valued statistics X and Y realize $\mathscr{E}$ and $\mathscr{F}$ respectively. Let $\hat{\beta}(X)$ be the $R(A)$-valued BLUE generator in $\mathscr{E}$ and let $\tilde{\beta}(Y)$ be the $R(B)$-valued BLUE generator in $\mathscr{F}$. Also put $S(X) = [(X - A^*\hat{\beta}(X), \Gamma^+(X - A^*\hat{\beta}(X)))]^{1/2}$ where $\Gamma = \mathrm{Cov}(X - A^*\hat{\beta}(X))$. Then, by theorem 8.3.14, $\hat{\beta}(X)$ and $S(X)$ are independent and $(\hat{\beta}(X), S(X))$ is complete and sufficient.

The assumption of non-singularity ensures that the family of distributions of $(\hat{\beta}(X), S(X))$ is homogeneous. Define the statistic $\tilde{S}(Y)$, analogous to $S(X)$, by $S(Y) = [(Y - B^*\tilde{\beta}(Y), \tilde{\Gamma}^+(Y - B^*\tilde{\beta}(Y)))]^{1/2}$ where $\tilde{\Gamma} = \mathrm{Cov}[Y - B^*\tilde{\beta}(Y)]$.

Now assume that $\delta(\mathscr{E}, \mathscr{F}) < 2$ and suppose $N(A) = 0$. Then $S(X) = 0$. Let $\tilde{\mathscr{F}}$ be the experiment induced from $\mathscr{F}$ by the non-negative function $\tilde{S}^2(Y)$. Then $\delta(\mathscr{E}, \tilde{\mathscr{F}}) \leqslant \delta(\mathscr{E}, \mathscr{F}) + \delta(\mathscr{F}, \tilde{\mathscr{F}}) \leqslant \delta(\mathscr{E}, \mathscr{F})$ so that $\delta(\mathscr{E}, \tilde{\mathscr{F}}) < 2$. Let $b \in R(A)$ act on $(\hat{\beta}, \tilde{S}, \beta)$ by mapping this triple in $R \times [0, \infty[\times R$ into $(\hat{\beta} + b, \tilde{S}, \beta + b)$. Then the pair $(\mathscr{E}, \tilde{\mathscr{F}})$ is invariant under these actions. A kernel from $\mathscr{E}$ to $\tilde{\mathscr{F}}$ is invariant if and only if it does not depend on $\hat{\beta}$, i.e. on X. Hence $\delta(\mathscr{E}, \tilde{\mathscr{F}}) = \delta(\mathscr{M}_i, \tilde{\mathscr{F}})$ where $\mathscr{M}_i$ is a totally non-informative experiment. It follows that $\delta(\mathscr{M}_i, \tilde{\mathscr{F}}) < 2$. If $N(B) \neq 0$ then $\tilde{S}^2(Y)/\sigma^2$ is χ^2-distributed with $\dim N(B)$ degrees of freedom. Passing to logarithms we see that there is an absolutely continuous translation family $\mathscr{G} = (G * \delta_\theta : \theta \in \mathbb{R})$ on the real line such that $\delta(\mathscr{M}_i, \mathscr{G}) = 2 - \varepsilon$ where $\varepsilon > 0$. Hence $\|G * \delta_\theta - H\| \leqslant 2 - \varepsilon$ for all $\theta \in \mathbb{R}$ for some absolutely continuous probability distribution H. If $\hat{G}$ and $\hat{H}$ are the characteristic functions of G and H respectively, then this implies that $|\hat{G}(t)e^{i\theta t} - \hat{H}(t)| \leqslant 2 - \varepsilon$ for all real numbers θ and t. Assuming $t \neq 0$ and putting $\theta = \pi/t$ we find $|\hat{G}(t) + \hat{H}(t)| \leqslant 2 - \varepsilon$ which gives us a contradiction, since $|\hat{G}(t) + \hat{H}(t)| \to |1 + 1| = 2$ when $t \to 0$. It follows that $N(B) = 0$ when $\delta(\mathscr{E}, \mathscr{F}) < 2$ and $N(A) = 0$.

Finally assume that either $N(A) \neq 0$ or that $N(A) = 0$ and $N(B) = 0$. Let Q_A and Q_B denote, respectively, the orthogonal projections on $R(A)$ and $R(B)$. Then $EQ_B\hat{\beta}(X) = Q_B Q_A \beta = Q_B \beta = E\tilde{\beta}(Y)$. Let M and $\tilde{M}$ denote the covariance operators of, respectively, $\hat{\beta}(X)$ and $\tilde{\beta}(Y)$ when $\sigma = 1$. Then $\operatorname{Cov} Q_B\hat{\beta}(X) = \sigma^2 Q_B M Q_B$ while $\operatorname{Cov} \tilde{\beta}(Y) = \sigma^2 \tilde{M}$. Non-singularity implies that $R(M) = R(A)$ and that $R(\tilde{M}) = R(B)$. Thus $R(Q_B M Q_B) = R(Q_B M) = Q_B[R(M)] = Q_B[R(A)] = R(B) = R(\tilde{M})$. If $N(A) = 0$ then, by the last assumption, $N(B) = 0$ so that $\hat{\beta}(X)$ is sufficient for X while $\tilde{\beta}(Y)$ is sufficient for Y. But

$$\begin{aligned} \|\mathscr{L}(Q_B\hat{\beta}(X)) - \mathscr{L}(\tilde{\beta}(Y))\| &= \|N(Q_B\beta, \sigma^2 Q_B M Q_B) - N(Q_B\beta, \sigma^2\tilde{M})\| \\ &= \|N(0, Q_B M Q_B) - N(0, \tilde{M})\| < 2. \end{aligned}$$

Hence $\delta(\mathscr{E}, \mathscr{F}) < 2$ in this case.

If $N(A) \neq 0$ then we define a map T such that the pair $(\hat{\beta}, S^2)$ is mapped into the pair $(Q_B\hat{\beta}, S^2)$ or into $Q_B\hat{\beta}$ as $N(B) \neq 0$ or $N(B) = 0$. If $N(B) = 0$ then by the same calculations as above we find that $\delta(\mathscr{E}, \mathscr{F}) < 2$. If $N(B) \neq 0$ then

$$\begin{aligned} &\|\mathscr{L}(T(\hat{\beta}(X), S^2(X))) - \mathscr{L}(\tilde{\beta}(Y), \tilde{S}^2(Y))\| \\ &\quad = \|[N(0, Q_B M Q_B) \times D] - [N(0, \tilde{M}) \times \tilde{D}]\| \end{aligned}$$

where D and $\tilde{D}$ are the χ^2-distributions with, respectively, $\dim[N(A)]$ degrees of freedom and $\dim[N(B)]$ degrees of freedom. In particular it follows that the statistical distance $\|\mathscr{L}(T(\hat{\beta}(X), S^2(X))) - \mathscr{L}(\tilde{\beta}(Y), \tilde{S}^2(Y))\|$ does not depend on (β, σ) and that this constant is less than 2. Hence $\delta(\mathscr{E}, \mathscr{F}) < 2$. □

The χ^2-distributions and, more generally, the gamma distributions, play an important role in this section and we shall now describe some relevant features of these distributions.

We shall use the standard notation $\Gamma(\alpha)$ for the integral $\int_0^\infty x^{\alpha-1}e^{-x}\,dx$ when $\alpha > 0$.

If $\alpha > 0$ and $\sigma > 0$ then the distribution with density $x \to \Gamma(\alpha)^{-1}\sigma^{-\alpha}x^{\alpha-1}e^{-x/\sigma}$ on $]0, \infty[$ will be denoted as $\Gamma_{\alpha,\sigma}$. If $\alpha = 0$ then $\Gamma_{\alpha,\sigma} = \Gamma_{0,\sigma}$ will denote the one-point distribution in 0. Thus $\Gamma_{n/2,2}$ denotes the χ^2-distribution with n degrees of freedom.

In general the distribution $\Gamma_{\alpha,\sigma}$ is called *the gamma distribution with shape parameter α and scale parameter σ*. If X is distributed as $\Gamma_{\alpha,\sigma}$ then $EX = \alpha\sigma$, $\operatorname{Var} X = \alpha\sigma^2$ and X/σ is distributed as $\Gamma_{\alpha,1}$.

The family $(\Gamma_{\alpha,\sigma}; \alpha \geqslant 0)$ for each $\sigma > 0$, is a convolution semigroup since $\Gamma_{\alpha_1,\sigma} * \Gamma_{\alpha_2,\sigma} = \Gamma_{\alpha_1+\alpha_2,\sigma}$.

Let $\mathscr{G}_\alpha$ denote the scale experiment $(\Gamma_{\alpha,\sigma}; \sigma > 0)$. Then $(\mathscr{G}_\alpha; \alpha \geqslant 0)$ is a semigroup of experiments in the following sense.

Proposition 8.6.3 (The gamma convolution semigroup). *If $\mathscr{G}_\alpha; \alpha \geqslant 0$, is defined as above then*

$$\mathscr{G}_{\alpha_1} \times \mathscr{G}_{\alpha_2} \sim \mathscr{G}_{\alpha_1+\alpha_2}; \qquad \alpha_1, \alpha_2 \geqslant 0.$$

In particular it follows that $\mathscr{G}_{\alpha_1} \leqslant \mathscr{G}_{\alpha_2}$ if and only if $\alpha_1 \leqslant \alpha_2$ and this in turn is equivalent to the condition that $\mathscr{G}_{\alpha_1}$ is a factor of $\mathscr{G}_{\alpha_2}$.

Proof. For each $\alpha > 0$ let X_α realize $\mathscr{G}_\alpha$. If X_{α_1} and X_{α_2} are independent then $X_{\alpha_1} + X_{\alpha_2}$ is sufficient for $(X_{\alpha_1}, X_{\alpha_2})$. On the other hand $X_{\alpha_1} + X_{\alpha_2}$ has the same distribution as $X_{\alpha_1+\alpha_2}$ and thus realizes $\mathscr{G}_{\alpha_1+\alpha_2}$. This proves the first statement. The second statement follows readily from this and the observation that the variance of the UMVU estimator of σ in $\mathscr{G}_\alpha$ is σ^2/α. □

Deficiencies within this semigroup are readily found.

Proposition 8.6.4 (Deficiencies within the gamma convolution semigroup). *If $\mathscr{G}_\alpha; \alpha \geqslant 0$, is defined as above, then*

$$\delta(\mathscr{G}_{\alpha_1}, \mathscr{G}_{\alpha_2}) = \begin{cases} 0 & \text{if } \alpha_1 \geqslant \alpha_2; \\ \|\Gamma_{\alpha_1,\alpha_2/\alpha_1} - \Gamma_{\alpha_2,1}\| & \text{if } 0 < \alpha_1 \leqslant \alpha_2; \\ 2 & \text{if } 0 = \alpha_1 < \alpha_2. \end{cases}$$

Remark. It follows that if X_{α_1} and X_{α_2} realize $\mathscr{G}_{\alpha_1}$ and $\mathscr{G}_{\alpha_2}$ respectively and if $0 < \alpha_1 \leqslant \alpha_2$, then the optimal approximation to $\mathscr{G}_{\alpha_2}$ on the basis of X_{α_1} is $\tilde{X}_{\alpha_2} = cX_{\alpha_1}$ where c is chosen so that $E\tilde{X}_{\alpha_2} = EX_{\alpha_2}$, i.e. $c = \alpha_2/\alpha_1$.

Proof. Again let X_α realize $\mathscr{G}_\alpha$ for each $\alpha \geqslant 0$. If $\alpha_1 \geqslant \alpha_2$ then $\delta(\mathscr{G}_{\alpha_1}, \mathscr{G}_{\alpha_2}) = 0$ by the previous proposition. If $0 = \alpha_1 < \alpha_2$ then by using characteristic functions as in the proof of theorem 8.6.2 we see that $\delta(\mathcal{G}_{\alpha_1}, \mathcal{G}_{\alpha_2}) = 2$. Therefore it remains to find the deficiency when $0 < \alpha_1 < \alpha_2$. Assume then that $0 < \alpha_1 \leqslant \alpha_2$.

Put $\sigma_0 = \alpha_2/\alpha_1$ and consider the Neyman–Pearson tests for testing the hypothesis that X is distributed as X_{α_1} with $\sigma = \sigma_0$, against the alternative that X is distributed as X_{α_2} with $\sigma = 1$. These tests tell us to reject when $X^{\alpha_2 - 1}\exp(-X) > \text{constant} \cdot X^{\alpha_1 - 1}\exp(-X/\sigma_0)$, i.e. when $(1 - \sigma_0^{-1})X - (\alpha_2 - \alpha_1)\log X < \text{constant}$. As the left hand side is convex in X the rejection regions have the form $c_1 < X < c_2$ where

$$(1 - \sigma_0^{-1})c_1 - (\alpha_2 - \alpha_1)\log c_1 = (1 - \sigma_0^{-1})c_2 - (\alpha_2 - \alpha_1)\log c_2,$$

i.e.

$$(c_2 - c_1)/(\log c_2 - \log c_1) = (\alpha_2 - \alpha_1)(1 - \sigma_0^{-1})^{-1}.$$

The probability $\pi(\sigma)$ of rejection when X is distributed as X_{α_1} for a given $\sigma > 0$ may then be written

$$\pi(\sigma) = \text{constant} \int_{c_1/\sigma}^{c_2/\sigma} x^{\alpha_1 - 1} e^{-x}\, dx$$

so that

$$\pi'(\sigma) = \text{constant}[c_1^{\alpha_1}(\exp(-c_1/\sigma)) - c_2^{\alpha_1}\exp(-c_2/\sigma)]\sigma^{-\alpha_1 - 1}$$

which is greater than, equal to, or less than 0 as σ is less than, equal to or greater than $\alpha_1^{-1}(c_2 - c_1)/(\log c_2 - \log c_1)$. Thus π has a unique maximum for $\sigma = \alpha_1^{-1}(c_2 - c_1)/(\log c_2 - \log c_1) = \sigma_0$. It follows that the one-point distribution in $\sigma_0 = \alpha_2/\alpha_1$ is least favourable for testing $\{\Gamma_{\alpha_1,\sigma}; \sigma > 0\}$ against Γ_{α_2}. Note that the constant σ_0 is determined just so that $E\sigma_0 X_{\alpha_1} = EX_{\alpha_2}$.

Using the method described in example 6.5.3 we find now that $\delta(\mathcal{G}_{\alpha_1}, \mathcal{G}_{\alpha_2}) = \|\mathcal{L}(\sigma_0 X_{\alpha_1}) - \mathcal{L}(X_{\alpha_2})\|$. □

In the following when we use these results we will often find it convenient to re-parametrize $\mathcal{G}_\alpha$ by replacing the parameter σ with σ^2. In this way we obtain, for each $\alpha \geqslant 0$, an experiment $\tilde{\mathcal{G}}_\alpha$ realized by S_α^2 where the statistic S_α^2 is distributed so that S_α^2/σ^2 has distribution $\Gamma_{\alpha,1}$. The experiments $\tilde{\mathcal{G}}_\alpha : \alpha > 0$, in contrast to the experiments $\mathcal{G}_\alpha : \alpha > 0$, are not translation experiments on the multiplicative group of positive numbers.

After a re-parametrization, comparison within pairs of linear normal models may be expressed for pairs having a canonical form.

Following Hansen & Torgersen (1974) and Swensen (1980) we collect the spadework in the following theorem.

Theorem 8.6.5 (Deficiencies for canonical forms). *Let $X_1, \ldots, X_p$ and $S \geqslant 0$ be independent statistics such that X_i is $N(\beta_i, \sigma^2)$-distributed; $i = 1, \ldots, p$, while S^2/σ^2 is gamma distributed with known shape parameter $\alpha_1 \geqslant 0$ and with some known scale parameter.*

Let $Y_1, \ldots, Y_r$ and $T \geqslant 0$ be another set of independent variables where Y_i is $N(\lambda_i \beta_i, \sigma^2)$-distributed; $i = 1, \ldots, r$ while T^2/σ^2 is gamma distributed with known shape parameter $\alpha_2 \geqslant 0$ and with the same scale parameter as S^2/σ^2.

Here $\lambda_1, \ldots, \lambda_r$ are known constants while the unknown parameter is $(\beta_1, \ldots, \beta_p, \sigma^2)$ which a priori may be any point in $\mathbb{R}^p \times]0, \infty[$.

Let $\mathscr{E}$ denote the experiment realized by $(X_1, \ldots, X_p, S)$ and let $\mathscr{F}$ be the experiment realized by $(Y_1, \ldots, Y_r, T)$ and put $v = \#\{i : 1 \leqslant i \leqslant r; |\lambda_i| < 1\}$.

Then $\mathscr{E} \geqslant \mathscr{F}$ if and only if $\alpha_1 \geqslant \alpha_2 + \frac{1}{2}v$ and $|\lambda_1| \leqslant 1, \ldots, |\lambda_r| \leqslant 1$.

In all cases where $\alpha_1 \geqslant \alpha_2 + \frac{1}{2}v$ the deficiency $\delta(\mathscr{E}, \mathscr{F})$ equals the statistical distance between $\mathscr{L}((\lambda_i X_i; 1 \leqslant i \leqslant r, |\lambda_i| > 1))$ and $\mathscr{L}((Y_i; 1 \leqslant i \leqslant r, |\lambda_i| > 1))$ and this is also the deficiency when σ is known.

If either $v = 0 \leqslant \alpha_1 \leqslant \alpha_2$ or if $\alpha_1 = 0$ then $\delta(\mathscr{E}, \mathscr{F})$ is the statistical distance between $\mathscr{L}(\lambda_1 X_1, \ldots, \lambda_r X_r, (\alpha_2/\alpha_1)^{1/2} S)$ and $\mathscr{L}(Y_1, \ldots, Y_r, T)$. In this case the map

$$(x_1, \ldots, x_p, s) \to (\lambda_1 x_1, \ldots, \lambda_r x_r, (\alpha_2/\alpha_1)^{1/2} s)$$

from the sample space of $\mathscr{E}$ to the sample space of $\mathscr{F}$, maps $\mathscr{E}$ into an experiment $\tilde{\mathscr{F}} \leqslant \mathscr{E}$ such that the statistical distance between distributions in $\tilde{\mathscr{F}}$ and in $\mathscr{E}$ which are obtained from the same vector parameter $(\beta_1, \ldots, \beta_p, \sigma)$ equals the deficiency $\delta(\mathscr{E}, \mathscr{F})$.

Here $(\alpha_2/\alpha_1)^{1/2} S$ and $(\alpha_2/\alpha_1)^{1/2} s$ may be interpreted as 0 when $\alpha_1 = 0$.

In general the deficiency $\delta(\mathscr{E}, \mathscr{F})$ is a permutation symmetric function of $(|\lambda_1|, \ldots, |\lambda_r|)$ and is thus not altered if $X_i : i > r$ are deleted from $\mathscr{E}$.

Remark 1. If $0 < \alpha_1 < \alpha_2 + \frac{1}{2}v$ and $v > 0$ then no explicit expression for the deficiency $\delta(\mathscr{E}, \mathscr{F})$ is known.

Remark 2. We have assumed that the gamma distributions of S^2/σ^2 and T^2/σ^2 have the same known (positive) scale parameter. The actual value of this scale parameter does not influence the deficiency.

Remark 3. The UMVU estimator of any one of the parameters $\beta_1, \ldots, \beta_p$ and σ^2, if it exists, is at least as good in $\mathscr{E}$ as in $\mathscr{F}$ if and only if $|\lambda_1| \leqslant 1, \ldots, |\lambda_r| \leqslant 1$ and $\alpha_1 \geqslant \alpha_2$. It follows from the theorem that this is not enough to ensure that $\mathscr{E} \geqslant \mathscr{F}$.

On the other hand, the UMVU estimators of any one of the parameters $\beta_1, \ldots, \beta_p$ and σ^2 in $\mathscr{E}$ if they exist, are not better than the corresponding estimators in $\mathscr{F}$ if and only if $v = 0$ and $\alpha_1 \leqslant \alpha_2$. By the first statement of the theorem, this condition is always fulfilled when $\mathscr{E} \leqslant \mathscr{F}$. In this case the deficiency may be found from the third statement of the theorem.

An investigation of the relationship between comparability of experiments on the one hand and the existence of unbiased estimators on the other may be found in Lehmann (1983a).

Remark 4. If the dichotomy (D_1, D_2) is equivalent to the dichotomy $(\mathscr{L}(\lambda_i X_i; 1 \leqslant i \leqslant r, |\lambda_i| > 1), \mathscr{L}(Y_i; 1 \leqslant i \leqslant r, |\lambda_i| > 1))$ then we may write $\delta(\mathscr{E}, \mathscr{F}) = \|D_1 - D_2\|$ when $\alpha_1 \geqslant \alpha_2 + \frac{1}{2}\nu$, while we may write $\delta(\mathscr{E}, \mathscr{F}) = \|D_1 \times \Gamma_{\alpha_1, \alpha_2/\alpha_1} - D_2 \times \Gamma_{\alpha_2, 1}\|$ when $\nu = 0$ and $\alpha_1 \leqslant \alpha_2$.

It follows that if we let $\prod'$ indicate products over the index set $\{i : 1 \leqslant i \leqslant r, |\lambda_i| > 1\}$ then we may put $(D_1, D_2) = (\prod' N(0,1), \prod' N(0, \lambda_i^{-2}))$ or $= (\prod' N(0, \lambda_i^2), \prod' N(0,1))$.

Proof. In order to facilitate the reading of the proof we begin by indicating its structure. This structure is due to the fact that the theorem is essentially three theorems organized as a single theorem. These theorems are:

(i) $\mathscr{E} \geqslant \mathscr{F}$ if and only if ...
(ii) in all cases where $\alpha_1 \geqslant \alpha_2 + \frac{1}{2}\nu$ the deficiency $\delta(\mathscr{E}, \mathscr{F})$ equals ...
(iii) if either $\nu = 0 \leqslant \alpha_1 \leqslant \alpha_2$ or if $\alpha_1 = 0$ then $\delta(\mathscr{E}, \mathscr{F})$ is ...

Although the proofs of these statements partially overlap, the main structure of this proof is:

the first part where (i) is proved;
the second part where (ii) is proved;
the third part where (iii) is proved.

First part of the proof. Let $\delta_{p,r} = \delta_{p,r}(\lambda_1, \ldots, \lambda_r; \alpha_1, \alpha_2)$ denote the deficiency $\delta(\mathscr{E}, \mathscr{F})$ and put $X = (X_1, \ldots, X_p)$, $X' = (X_i : 1 \leqslant i \leqslant r, |\lambda_i| > 1)$, $Y = (Y_1, \ldots, Y_r)$ and $Y' = (Y_i : 1 \leqslant i \leqslant r, |\lambda_i| > 1)$.

Let $\xi \in \mathbb{R}^{p-r}$ act on $(X_1, \ldots, X_p, S)$ by mapping it into $(X_1, \ldots, X_r, X_{r+1} + \xi_1, \ldots, X_p + \xi_{p-r}, S)$. Let ξ act on $(Y_1, \ldots, Y_r, T)$ by leaving it alone. It is then easily checked that the pair $(\mathscr{E}, \mathscr{F})$ of experiments is invariant under the action of the group $(\mathbb{R}^{p-r}, +)$. Furthermore a kernel from $\mathscr{E}$ to $\mathscr{F}$ is invariant if and only if it depends on $(X_1, \ldots, X_p, S)$ via $(X_1, \ldots, X_r, S)$. Hence $\delta_{p,r} = \delta_{r,r}$ and clearly this deficiency is symmetric in $(|\lambda_1|, \ldots, |\lambda_r|)$. This proves the last statement of the theorem.

It follows that we may assume that $r = p$ and that $\lambda_1 \geqslant \cdots \geqslant \lambda_p \geqslant 0$. If $\lambda_1, \ldots, \lambda_p \leqslant 1$ then this implies that $p - \nu = \{i : \lambda_i = 1\}$ and that $\lambda_i = 1$ if and only if $i \leqslant p - \nu$.

The first claim of the theorem states that $\mathscr{E} \geqslant \mathscr{F}$ if and only if $1 \geqslant \lambda_1 \geqslant \cdots \geqslant \lambda_p$ and $\alpha_1 \geqslant \alpha_2 + \frac{1}{2}\nu$. Now a comparison of the variances of the UMVU estimator of β_i; $i = 1, \ldots, p$ shows that $\lambda_1, \ldots, \lambda_p \leqslant 1$ when $\mathscr{E} \geqslant \mathscr{F}$. Thus we may assume that $1 \geqslant \lambda_1 \geqslant \cdots \geqslant \lambda_p$ when we prove the first statement. We shall also assume in this part of the proof that the common scale factor of $\mathscr{L}(S^2/\sigma^2)$ and $\mathscr{L}(T^2/\sigma^2)$ is 2 so that S^2/σ^2 has distribution $\Gamma_{\alpha_1, 2}$ while T/σ^2 has distribution $\Gamma_{\alpha_2, 2}$.

Assume now that $\alpha_1 \geqslant \alpha_2 + \frac{1}{2}v$ and put $\alpha = \alpha_1 - \frac{1}{2}v$. Then $\alpha \geqslant \alpha_2 \geqslant 0$. Construct independent variables $\tilde{X}_1, \dots, \tilde{X}_p$ on Z such that $\tilde{X}_i$ is distributed as $N(0,0)$ or as $N(0,\sigma^2)$ as $i \leqslant p - v$ or $i > p - v$, while Z/σ^2 is distributed as $\Gamma_{\alpha,2}$. We shall also require that $(\tilde{X}_1, \dots, \tilde{X}_p, Z)$ is independent of (X, S). Consider the experiment realized by $(X_1, \dots, X_p, \tilde{X}_1, \dots, \tilde{X}_p, Z)$. As $\sum \tilde{X}_i^2 + Z$ is sufficient for $(\tilde{X}_1, \dots, \tilde{X}_p, Z)$ and since $(\sum \tilde{X}_i^2 + Z)/\sigma^2$ is distributed as $\Gamma_{v/2+\alpha,2} = \Gamma_{\alpha_1,2}$ we conclude that the experiment realized by $(X_1, \dots, X_p, \tilde{X}_1, \dots, \tilde{X}_p, Z)$ is equivalent to $\mathscr{E}$. Put $\tilde{Y}_i = \lambda_i X_i + (1 - \lambda_i^2)^{1/2} \tilde{X}_i$; $i = 1, \dots, p$. Then $\tilde{Y}_1, \dots, \tilde{Y}_p$ are independent and $\tilde{Y}_i$ is $N(\lambda_i \beta_i, \sigma^2)$-distributed. Furthermore, since $\mathscr{G}_{\alpha_1} \geqslant \mathscr{G}_{\alpha_2}$, we can construct a variable $\tilde{T}$ on the basis of Z having the same distribution as T. In that way we can obtain a realization of $\mathscr{F}$ on the basis of $(X_1, \dots, X_p, \tilde{X}_1, \dots, \tilde{X}_p, Z)$. It follows that $\mathscr{E} \geqslant \mathscr{F}$.

Therefore let us assume that $\mathscr{E} \geqslant \mathscr{F}$ and remember that we are assuming that $1 \geqslant \lambda_1 \geqslant \cdots \geqslant \lambda_p$.

One might now hope that a comparison of the UMVU estimators of σ^2 would complete the proof of the first statement of the theorem. Unfortunately this comparison is not straightforward although, as we now shall see, it does yield the desired conclusion.

Consider the group G of transformations on $\mathbb{R}^p \times [0, \infty[$ of the form $\Phi_{b,c}: (x, s) \to (b + cx, cs)$ where $(b, c) \in \Theta = \mathbb{R}^p \times]0, \infty[$.

Carrying the group operations back to Θ we obtain a group G with the composition rule

$$(b_1, c_1)(b_2, c_2) = (b_1 + c_1 b_2, c_1 c_2).$$

The unit is $(0, 1)$ and the inverse $(b, c)^{-1}$ of (b, c) is $(-b/c, 1/c)$. The subgroup $G_0 = \mathbb{R}^p \times \{1\}$ of G is isomorphic to $\mathbb{R}^p$. It is easily checked that G_0 is normal (i.e. invariant) so that the factor group G/G_0 is well defined.

This factor group is isomorphic to $(\mathbb{R}, +)$ so that G_0 and G/G_0 are commutative. Referring to section 6.5 we see that the group G is amenable.

Note that G considered as a topological group admits a Haar measure and that this measure is σ-finite.

Let G act on the sample space $\mathbb{R}^p \times [0, \infty[$ of $\mathscr{E}$ by left multiplication, i.e. $(b, c)(x, s) = \Phi_{b,c}(x, s) = (b + cx, cs)$, where $(b, c) \in G = \Theta = \mathbb{R}^p \times]0, \infty[$.

If P is the distribution of (X, S) when $\beta = 0$ and $\sigma = 1$, then $(\beta, \sigma)(X, S)$ has the same distribution as (X, S) has when (β, σ) prevails. If $\alpha_1 > 0$ then this implies that $\mathscr{E}$ may be considered as a left translation experiment on G.

If $\lambda_i > 0$ for all $i \leqslant p$ then $\mathscr{F}$ is equivalent to the experiment realized by $(Y_1/\lambda_1, \dots, Y_p/\lambda_p, T)$ and the latter experiment is also a left translation experiment on G when $\alpha_2 > 0$. This will be useful later on. However here we will keep $\mathscr{F}$ as it is and define the action of $(b, c) \in G$ on $(y, t) \in \mathbb{R}^p \times [0, \infty[$ by

$$(b, c)(y, t) = (\lambda_1 b_1 + c y_1, \dots, \lambda_p b_p + c y_p, ct).$$

Again we may check that the distribution of $(\beta,\sigma)(Y,T)$ for $\beta = 0$ and $\sigma = 1$ is the same as the distribution of (Y,T) under (β,σ).

It follows that both $\mathscr{E}$ and $\mathscr{F}$ are invariant.

Finally note that the map $(x_1,\ldots,x_p,s) \to (\lambda_1 x_1,\ldots,\lambda_p x_p,s)$ from the sample space of $\mathscr{E}$ to the sample space of $\mathscr{F}$ is invariant under actions by G. Thus there are invariant kernels.

The theory of invariance, e.g. corollary 6.5.10 and theorem 6.5.11, implies now that $\mathscr{F}$ is obtainable from $\mathscr{E}$ by an invariant Markov kernel M. (Remember we are assuming that $\mathscr{E} \geqslant \mathscr{F}$.) It follows that there is a joint distribution for (X,S,Y,T) such that the distribution of this quadruple under (β,σ) is the same as the distribution of $((\beta,\sigma)(X,S),(\beta,\sigma)(Y,T))$ when (X,S,Y,T) is distributed with $\beta = 0$ and $\sigma = 1$. Thus we may write $X_i = \beta_i + \sigma\hat{X}_i$; $i = 1,\ldots,p$, $S = \sigma\hat{S}$, $Y_i = \lambda_i\beta_i + \sigma\hat{Y}_i$; $i = 1,\ldots,p$ and $T = \sigma\hat{T}$ where $(\hat{X},\hat{S},\hat{Y},\hat{T})$ is distributed as (X,S,Y,T) with $\beta = 0$ and $\sigma = 1$.

Suppose $\alpha_1 > 0$. Then $S > 0$, $\hat{S} > 0$ and

$$\begin{aligned}&((Y_1 - \lambda_1 X_1)/S,\ldots,(Y_p - \lambda_p X_p)/S,\, T/S)\\ &\quad = ((\hat{Y}_1 - \lambda_1\hat{X}_1)/\hat{S},\ldots,(\hat{Y}_p - \lambda_p\hat{X}_p)/\hat{S},\, \hat{T}/\hat{S})\end{aligned}$$

is ancillary while (X,S) is a complete sufficient statistic. Thus by Basu's lemma, or directly, we see that $((Y_1 - \lambda_1 X_1)/S,\ldots,(Y_p - \lambda_p X_p)/S,\, T/S)$ and (X,S) are independent. It follows that $(Y_1 - \lambda_1 X_1,\ldots,Y_p - \lambda_p X_p, T)$ is independent of X. We shall now see that the random variables $Y_1 - \lambda_1 X_1,\ldots,Y_p - \lambda_p X_p, T$ are independent so that all $2p+1$ variables $Y_1 - \lambda_1 X_1,\ldots,Y_p - \lambda_p X_p$, $X_1,\ldots,X_p, T$ are independent. To this end, following Boll (1955), let us turn to characteristic functions.

Let $\rho_1,\ldots,\rho_p$ and κ be real numbers and consider the expectation

$$\begin{aligned}&E\exp\left[\sum_j i\rho_j(Y_j - \lambda_j X_j) + i\kappa T\right]\\ &\quad = \left\{E\exp\left[\sum_j i\rho_j(Y_j - \lambda_j X_j) + i\kappa T\right]\right\}\left\{E\exp\sum_j i\rho_j\lambda_j X_j\right\}\\ &\qquad \times \left\{E\exp\sum_j i\rho_j\lambda_j X_j\right\}^{-1}\\ &\quad = \left\{E\exp\left[\sum_j i\rho_j Y_j + i\kappa T\right]\right\}\left\{E\exp\sum_j i\rho_j\lambda_j X_j\right\}^{-1}.\end{aligned}$$

The last expression factorizes, since Y and T are independent, as

$$\left[\left(E\exp\sum_j i\rho_j Y_j\right)\Big/\left(E\exp\sum_j \rho_j\lambda_j X_j\right)\right]E\exp(i\kappa T).$$

Putting $\kappa = 0$ we find

$$E\exp\sum_j i\rho_j(Y_j - \lambda_j X_j) = \left(E\exp\sum_j i\rho_j Y_j\right)\Big/\left(E\exp\sum_j i\rho_j\lambda_j X_j\right)$$

$$= \prod_j (E\exp(i\rho_j Y_j))/E\exp(i\rho_j\lambda_j X_j)$$

so that

$$E\exp\left[\sum_j i\rho_j(Y_j - \lambda_j X_j) + i\kappa T\right] = \left[E\exp\sum_j i\rho_j(Y_j - \lambda_j X_j)\right][E\exp(i\kappa T)]$$

proving that $Y_1 - \lambda_1 X_1, \ldots, Y_p - \lambda_p X_p$ and T are independent.
Furthermore

$$E\exp[i\rho_j(Y_j - \lambda_j X_j)] = [\exp(i\rho_j\lambda_j\beta_j - \tfrac{1}{2}\rho_j^2\sigma^2)]/[\exp(i\rho_j\lambda_j\beta_j - \tfrac{1}{2}\rho_j^2\lambda_j^2\sigma^2)]$$
$$= \exp(-\tfrac{1}{2}\rho_j^2(1 - \lambda_j^2)\sigma^2)$$

showing that $Y_j - \lambda_j X_j$ is $N(0,(1 - \lambda_j^2)\sigma^2)$-distributed.

If $i > p - v$ then $\lambda_i < 1$ so that $U = [\sum_{i=p-v+1}^{p}(Y_i - \lambda_i X_i)^2/(1 - \lambda_i^2)]^{1/2}$ is well defined. The statistic U is independent of T and U^2/σ^2 has a χ^2-distribution with v degrees of freedom. Putting $\eta = \alpha_2 + \frac{1}{2}v$ we see that the desired inequality may be written $\alpha_1 \geqslant \eta$. As this inequality is trivial when $\eta = 0$ we shall assume that $\eta > 0$. Now $(U^2 + T^2)/\sigma^2$ is distributed as $\Gamma_{\eta,2}$. Hence $(U^2 + T^2)/2\eta$ is an unbiased estimator of σ^2 with variance σ^4/η. By sufficiency it follows that there is also an unbiased estimator of σ^2 based on (X, S). This is only possible if $ES^2 > 0$, i.e. if $\alpha_1 > 0$, and then the UMVU estimator $S^2/2\alpha_1$ of σ^2 has variance σ^4/α_1. Comparing the variances of these estimators of σ^2 we find that $\alpha_1 \geqslant \eta$. This 'almost' completes the proof of the first statement of the theorem. We only venture to claim that we have 'almost' proved this statement, since we have argued as if $\alpha_1 > 0$. Instead of going back and seeing how the arguments may be modified in order to be valid when $\alpha_1 = 0$, we give the whole argument here.

If $\alpha_1 = 0$ then $\mathscr{E}$ admits no unbiased estimator of σ^2. Hence, since $\mathscr{E} \geqslant \mathscr{F}$, $\mathscr{F}$ does not admit any unbiased estimator of σ^2 either, so $\alpha_2 = 0$. The only invariant kernel from $\mathscr{E}$ to $\mathscr{F}$ is then the map which maps $(x_1, \ldots, x_p, 0)$ into $(\lambda_1 x_1, \ldots, \lambda_p x_p, 0)$. Thus

$$\delta(\mathscr{E}, \mathscr{F}) = \left\| \prod_{i=1}^{p} N(\lambda_i\beta, \lambda_i^2\sigma^2) - \prod_{i=1}^{p} N(\lambda_i\beta_i, \sigma^2) \right\|$$

$$= \left\| \prod_{i=1}^{p} N(0, \lambda_i^2) - \prod_{i=1}^{p} N(0,1) \right\| \geqslant \left\| N(0, \lambda_i^2) - N(0,1) \right\|; \quad i = 1, \ldots, p.$$

If $\mathscr{E} \geqslant \mathscr{F}$, i.e. if $\delta(\mathscr{E}, \mathscr{F}) = 0$, then this implies that $\lambda_1 = \cdots = \lambda_p = 1$ so that actually $\mathscr{E} = \mathscr{F}$. This completes the proof of the 'if and only if' statement of the theorem.

Second part of the proof. Assume that $\alpha_1 \geqslant \alpha_2 + \frac{1}{2}\nu$. Put $X'' = (X_i : |\lambda_i| \leqslant 1)$ and $Y'' = (Y_i : |\lambda_i| \leqslant 1)$. Factorize $\mathscr{E}$ as $\mathscr{E} = \mathscr{E}' \times \mathscr{E}''$ and $\mathscr{F}$ as $\mathscr{F} = \mathscr{F}' \times \mathscr{F}''$ where $\mathscr{E}'$ and $\mathscr{F}'$ are realized by X' and Y' respectively, while $\mathscr{E}''$ and $\mathscr{F}''$ are realized by (X'', S) and (Y'', T) respectively. Then $\delta(\mathscr{E}, \mathscr{F}) \leqslant \delta(\mathscr{E}', \mathscr{F}') + \delta(\mathscr{E}'', \mathscr{F}'')$. From the statements proved so far it follows that $\mathscr{E}'' \geqslant \mathscr{F}''$. Hence $\delta(\mathscr{E}, \mathscr{F}) \leqslant \delta(\mathscr{E}', \mathscr{F}')$. Now the statistical distance between $\mathscr{L}(\lambda_i X_i : |\lambda_i| > 1)$ and $\mathscr{L}(Y_i : |\lambda_i| > 1)$ does not depend on (β, σ). It follows that this distance majorizes $\delta(\mathscr{E}', \mathscr{F}')$ and thus, since $\delta(\mathscr{E}, \mathscr{F}) \leqslant \delta(\mathscr{E}', \mathscr{F}')$, it also majorizes $\delta(\mathscr{E}, \mathscr{F})$. On the other hand $\delta(\mathscr{E}, \mathscr{F}) \geqslant \delta(\mathscr{E}_1, \mathscr{F}_1)$ where $\mathscr{E}_1$ and $\mathscr{F}_1$ are obtained from $\mathscr{E}$ and $\mathscr{F}$ respectively, by putting $\sigma = 1$. From theorem 8.5.7 it follows that the deficiency $\delta(\mathscr{E}_1, \mathscr{F}_1)$ is precisely equal to this statistical distance. This completes the proof of the second statement of the theorem.

Third part of the proof. In the remaining part of the proof we will choose 1 as the common scale factor of S^2/σ^2 and T^2/σ^2. Thus S^2/σ^2 and T^2/σ^2 are distributed as $\Gamma_{\alpha_1, 1}$ and $\Gamma_{\alpha_2, 1}$ respectively.

Assume that $0 < \alpha_1 \leqslant \alpha_2$ and that $\nu = 0$. Then $\lambda_1, \ldots, \lambda_p \geqslant 1$.

Putting $\tilde{Y}_i = Y_i/\lambda_i$ we see that $(\tilde{Y}_1, \ldots, \tilde{Y}_p, T)$ realizes an experiment $\tilde{\mathscr{F}}$ which is equivalent to $\mathscr{F}$, and that $\tilde{Y} = (\tilde{Y}_1, \ldots, \tilde{Y}_p)$ is distributed as $N(\beta, \sigma^2 \Lambda^{-2})$.

In the first part of this proof we introduced the notation P for the joint distribution of (X, S) when $\beta = 0$ and $\sigma = 1$. Let Q denote the joint distribution of $(\tilde{Y}, T)$ when $\beta = 0$ and $\sigma = 1$. Then we may write $P = N(0, I) \times \Gamma_{\alpha_1, 1}$ and $Q = N(0, \Lambda^{-2}) \times \Gamma_{\alpha_2, 1}$.

Referring to the group G defined in the first part of the proof we find that the distribution of (X, S) and $(\tilde{Y}, T)$ for a specified parameter (β, σ) is the distribution of, respectively, $(\beta, \sigma)(X, S) = (\beta + \sigma X, \sigma S)$ and $(\beta, \sigma)(\tilde{Y}, T) = (\beta + \sigma \tilde{Y}, \sigma T)$, when $\mathscr{L}(X, S) = P$ and $\mathscr{L}(\tilde{Y}, T) = Q$. Thus $\mathscr{E}$ and $\tilde{\mathscr{F}}$ are both left 'translation' experiments on G.

Furthermore, the measure on $G = \{(b, c) : b \in \mathbb{R}^p; c > 0\}$ whose density with respect to Lebesgue measure is $(b, c) \to 1/c^{p+1}$ is a left Haar measure. Thus Haar measure and Lebesgue measure are mutually absolutely continuous on G. It follows that P is absolutely continuous with respect to Haar measure on G.

The method described in example 6.5.3 suggests that now we should try to test the set of right translates of P against the single alternative Q.

Actually so far we have only utilized the fact that the constants $\lambda_1, \ldots, \lambda_p$ and α_1 are all positive. Now, however, we will also need the fact that $\lambda_1, \ldots, \lambda_p \geqslant 1$ and that $\alpha_2/\alpha_1 \geqslant 1$.

Consider the right $(0, \kappa)$ 'translate' $P * \delta_{(0, \kappa)}$ of P where $\kappa = (\alpha_2/\alpha_1)^{1/2}$. Thus $P * \delta_{(0, \kappa)}$ is the distribution of $(X, \kappa S) = (X, S)(0, \kappa)$ when (X, S) is distributed according to P. The most powerful level ε test for testing $P * \delta_{(0,\kappa)}$ against the alternative Q rejects when

$$\exp\left(-\frac{1}{2}\sum_i \lambda_i^2 X_i^2\right) S^{2\alpha_2-1}\exp(-S^2)$$
$$\geqslant \text{constant}\cdot(\exp - \tfrac{1}{2}\sum X_i^2)S^{2\alpha_1-1}\exp(-S^2/\kappa^2)$$

i.e. when

$$(1-\kappa^{-2})S^2 - (\alpha_2 - \alpha_1)\log S^2 \leqslant c - \frac{1}{2}\sum_i (\lambda_i^2 - 1)X_i^2$$

where the constant c is adjusted to the level ε of significance. Let us denote the probability of rejecting under a general right translate $P * \delta_{(\beta,\sigma)}$ of P, by $\pi(\beta,\sigma)$. The identity $(X,S)(\beta,\sigma) = (X + S\beta, \sigma S)$ implies immediately that $\pi(\beta,\sigma) = Pr((1-\kappa^{-2})(\sigma S)^2 - (\alpha_2-\alpha_1)\log(\sigma S)^2 \leqslant c - \frac{1}{2}\sum(\lambda_i^2 - 1)(X_i + S\beta_i)^2)$. Since the conditional distribution of $\sum(\lambda_i^2-1)(X_i+S\beta_i)^2$ given S is stochastically increasing in $|\beta| = (|\beta_1|,\dots,|\beta_p|)$, we see that $\pi(\beta,\sigma) \leqslant \pi(0,\sigma)$. Putting $\beta = 0$ and conditioning on X we see from the proof of proposition 8.6.4 that $\pi(0,\sigma) \leqslant \pi(0,\kappa)$. It follows that any dichotomy $(P * \delta_{(\beta,\sigma)}, Q)$ is at least as informative as $(P * \delta_{(0,\kappa)}, Q)$.

This implies that

$$\delta(\mathscr{E},\mathscr{F}) = \|P * \delta_{(0,\kappa)} - Q\|$$
$$= \|N(0,I)\times\Gamma_{\alpha_1,\kappa} - N(0,\Lambda^{-2})\times\Gamma_{\alpha_2,1}\|$$

where $\Lambda = \operatorname{diag}(\lambda_1,\dots,\lambda_p)$.

If $\alpha_1 = 0$ then $Pr(S=0) = 1$. Now any invariant Markov kernel from $\mathscr{E}$ to $\mathscr{F}$ maps $(x_1,\dots,x_p,s)$ into $(\lambda_1 x_1,\dots,\lambda_p x_p, s)$ when $s = 0$. It follows that the invariant map $(x_1,\dots,x_p,s)\to(\lambda_1x_1,\dots,\lambda_px_p,s)$ in this case represents the essentially unique invariant Markov kernel (randomization) from $\mathscr{E}$ to $\mathscr{F}$. It follows that the formula for $\delta(\mathscr{E},\mathscr{F})$ which we derived above, with the convention stated in the theorem, is also valid when $\alpha_1 = 0$.

If (X,S) is distributed according to $\delta_{(\beta,\sigma)} * P$ then the distribution of $(\lambda_1X_1, \dots, \lambda_pX_p, (\alpha_2/\alpha_1)^{1/2}S)$ is $N(\Lambda\beta, \sigma^2\Lambda^2)\times\Gamma_{\alpha_1,(\alpha_2/\alpha_1)\sigma^2}$ while $\mathscr{F} = (N(\Lambda\beta,\sigma^2 I)\times\Gamma_{\alpha_2,\sigma^2} : \beta\in\mathbb{R}^p, \sigma^2 > 0)$. The desired expression for the deficiency follows now by

$$\|N(\Lambda\beta,\sigma^2\Lambda^2)\times\Gamma_{\alpha_1,(\alpha_2/\alpha_1)\sigma^2} - N(\Lambda\beta,\sigma^2 I)\times\Gamma_{\alpha_2,\sigma^2}\|$$
$$= \|N(0,\sigma^2\Lambda^2)\times\Gamma_{\alpha_1,(\alpha_2/\alpha_1)\sigma^2} - N(0,\sigma^2 I)\times\Gamma_{\alpha_2,\sigma^2}\|$$
$$= \|N(0,\Lambda^2)\times\Gamma_{\alpha_1,(\alpha_2/\alpha_1)} - N(0,I)\times\Gamma_{\alpha_2,1}\|$$
$$= \|N(0,I)\times\Gamma_{\alpha_1,\alpha_2/\alpha_1} - N(0,\Lambda^{-2})\times\Gamma_{\alpha_2,1}\|$$

where each '=' is implied by an obvious equivalence of dichotomies. □

Combining the first result described in the theorem with the comparability results in section 8.1, we obtain the following 'picture'.

Theorem 8.6.6 (Ordering of linear normal experiments when σ is unknown). *The following conditions are equivalent for experiments $\mathscr{E} = \mathscr{N}(A, \sigma^2 V; \sigma^2 > 0)$ and $\mathscr{F} = \mathscr{N}(B, \sigma^2 W; \sigma^2 > 0)$.*

(i) $\mathscr{E} \geqslant \mathscr{F}$.

(ii) $\mathscr{N}(A, V) \geqslant \mathscr{N}(B, W)$ *and* rank V-rank $W \geqslant \text{rank}[P(AV^+A^* - BW^+B^*)P]$ *where P is the orthogonal projection on $[A[N(V)]]^\perp = [A^*]^{-1}[R(V)]$.*

(iii) *$\mathscr{F}$ is obtainable from $\mathscr{E}$ by a linear map, i.e. $TA^* = B^*$ and $TVT^* = W$ for some linear map T.*

(iv) *$\mathscr{E} \sim \mathscr{F} \times \mathscr{G}$ for some experiment $\mathscr{G}$.*

If these conditions are satisfied then $\mathscr{G}$ in (iv) *may be chosen as a linear normal experiment of the form $\mathscr{G} = \mathscr{N}(C, \sigma^2 U; \sigma^2 > 0)$.*

Remark 1. Consider replicated experiments $\mathscr{E}^n$ and $\mathscr{F}^n; n = 1, 2, \ldots$ where $\mathscr{E} = \mathscr{N}(A, \sigma^2 V; \sigma^2 > 0)$ and $\mathscr{F} = \mathscr{N}(B, \sigma^2 W; \sigma^2 > 0)$. Assuming that $\mathscr{N}(A, V) \geqslant \mathscr{N}(B, W)$ we see that if $\mathscr{E}$ and $\mathscr{F}$ are not equivalent, then $\mathscr{E}^n \geqslant \mathscr{F}^n$ if and only if

$$\text{rank } V > \text{rank } W$$

and

$$n \geqslant \text{rank}[P(AV^+A^* - BW^+B^*)P]/(\text{rank } V - \text{rank } W).$$

Remark 2. By Janssen's convolution theorem for exponential models (theorem 9.5.6) the necessity of the rank inequality in (ii), in the non-singular case, may be derived directly without relying on theorem 8.6.5. In this case the equivalence of conditions (i) and (iv) is a particular case of theorem 9.5.6.

Proof. As all the conditions imply that $\mathscr{N}(A, V) \geqslant \mathscr{N}(B, W)$ we may assume that $\mathscr{N}(A, V) \geqslant \mathscr{N}(B, W)$. Let (iv′) denote the condition that $\mathscr{E} \sim \mathscr{F} \times \mathscr{N}(C, \sigma^2 U; \sigma^2 > 0)$ for some operators C and U. Then the equivalence of conditions (i)–(iv) will follow if we establish the equivalence of (i)–(iv′). Now theorem 8.2.21 states that conditions (iii) and (iv′) are equivalent. Furthermore, by theorem 8.2.22, condition (iii) is also equivalent to condition (ii). As condition (iii) clearly implies condition (i), it suffices to show that condition (i) implies that rank V − rank $W \geqslant \text{rank}[P(AV^+A^* - BW^+B^*)P]$. Therefore let us assume that $\mathscr{E} \geqslant \mathscr{F}$.

Restricting β to $R(P) = [A[N(V)]]^\perp$ we find that $\mathscr{N}(PA, \sigma^2 V; \sigma^2 > 0) \geqslant \mathscr{N}(PB, \sigma^2 W; \sigma^2 > 0)$. Since the last two experiments are non-singular it follows that we may assume that $\mathscr{E}$ and $\mathscr{F}$ are non-singular. Then we may assume also that V and W are identity operators on H_A and H_B

respectively. As $R(B) \subseteqq R(A)$ we may also assume that $R(A) = K$ so that AA^* is invertible. Putting $m = \dim H_A$ and $n = \dim H_B$ we find that the desired inequality may be written $m \geqslant n + \operatorname{rank}(AA^* - BB^*)$. Since A is assumed onto, the operator AA^* is positive definite. Putting $\tilde{A} = (AA^*)^{-1/2}A$ and $\tilde{B} = (AA^*)^{-1/2}B$ we find $\tilde{A}\tilde{A}^* = I$, $\tilde{B}\tilde{B}^* = (AA^*)^{-1/2}BB^*(AA^*)^{-1/2}$ and $I - \tilde{B}\tilde{B}^* = (AA^*)^{-1/2}(AA^* - BB^*)(AA^*)^{-1/2}$. The desired inequality may thus be written $m \geqslant n + \operatorname{rank}(I - \tilde{B}\tilde{B}^*)$.

On the other hand, a change of parameter from β to $(AA^*)^{1/2}\beta$ shows that $\mathcal{N}(\tilde{A}, \sigma^2 I; \sigma^2 > 0) \geqslant \mathcal{N}(\tilde{B}, \sigma^2 I; \sigma^2 > 0)$. Thus we may assume that AA^* is the identity operator, i.e. that A^* is an isometry from K to H_A. We proceed now by using coordinates as in the proof of theorem 8.2.22 and also using the notation there. Let X realize $\mathcal{N}(A, \sigma^2 I; \sigma^2 > 0)$ and let Y realize $\mathcal{N}(B, \sigma^2 I; \sigma^2 > 0)$. Put $X_i = (X, \rho_i)$, $Y_j = (Y, \tilde{\rho}_j)$ and $\beta_i = (\beta, \tau_i)$. Thus X_i, Y_j and β_i are the ρ_i coordinate, the $\tilde{\rho}_j$ coordinate and the τ_i coordinate of, respectively, X, Y and β. Then $EX_i = (EX, \rho_i) = (A^*\beta, \rho_i) = \sum_\nu \beta_\nu (A^*\Gamma_\nu, \rho_i) = \sum_\nu \beta_\nu (\rho_\nu, \rho_i) = \beta_i$ or $= 0$ as $i \leqslant p$ or $i > p$. Furthermore, $\operatorname{Cov}(X_i, X_j) = (\rho_i, I\rho_j)\sigma^2 = 0$ or $= \sigma^2$ as $i \neq j$ or $i = j$.

Hence $X_1, \ldots, X_m$ are independently and normally distributed with the same variance σ^2. Similarly $Y_1, \ldots, Y_n$ are also independently and normally distributed with the same variance σ^2 and $EY_i = (EY, \tilde{\rho}_i) = (B^*\beta, \tilde{\rho}_i) = \sum_\nu \beta_\nu (B^*\tau_\nu, \tilde{\rho}_i) = \sum_{\nu \leqslant p} \beta_\nu (\Delta_\nu^{1/2}\tilde{\rho}_\nu, \tilde{\rho}_i) = \Delta_i^{1/2}\beta_i$ or $= 0$ as $i \leqslant p$ or $i > p$. Note that we have not assumed that $n \geqslant p$.

It follows that we are in precisely the situation described in theorem 8.6.5, with $r = \operatorname{rank} B$, $\lambda_i = \Delta_i^{1/2}$; $i = 1, \ldots, r$, $\alpha_1 = \frac{1}{2}(m - p)$ and $\alpha_2 = \frac{1}{2}(n - r)$. Hence, by theorem 8.6.5, $\Delta_1, \ldots, \Delta_r \leqslant 1$ and $m - p \geqslant n - r + \#\{i : 1 \leqslant i \leqslant r, \Delta_i < 1\}$ so that $m - n \geqslant \#\{i : 1 \leqslant i \leqslant r\} + p - r = \#\{i : 1 \leqslant i \leqslant p; \Delta_i < 1\}$. Now the matrix of $\tilde{B}\tilde{B}^*$ with respect to the chosen coordinate systems is $\operatorname{diag}(\Delta_1, \ldots, \Delta_p)$. Thus $\operatorname{rank}(I - \tilde{B}\tilde{B}^*) = \#\{i : \Delta_i < 1\}$ so that $m \geqslant n + \operatorname{rank}(I - \tilde{B}\tilde{B}^*) = n + \operatorname{rank}(AA^* - BB^*)$. □

Condition (iii) of this theorem, as well as condition (iii) of theorem 8.4.3, shows how comparability of linear normal models $\mathscr{E}$ and $\mathscr{F}$ may be expressed in terms of linear operators. In both cases it is a requirement that the regression operator A^* of $\mathscr{E}$ is mapped into the regression operator of $\mathscr{F}$. The difference between the situation when σ is known and the situation when σ is unknown can be expressed as follows. If σ is known then we require only that the image of the covariance operator in $\mathscr{E}$ is majorized by the covariance operator in $\mathscr{F}$, while when σ is unknown we require that the covariance operator in $\mathscr{F}$ is precisely the image of the covariance operator in $\mathscr{E}$.

These facts become more clear if we express them in terms of statistics X and Y realizing $\mathscr{E}$ and $\mathscr{F}$ respectively. Assume that $\mathscr{E} = \mathcal{N}(A, \sigma^2 V; \sigma^2 > 0)$ and that $\mathscr{F} = \mathcal{N}(B, \sigma^2 W; \sigma^2 > 0)$ when σ is unknown. Then X and Y realize $\mathcal{N}(A, V)$ and $\mathcal{N}(B, W)$ respectively, when it is known that $\sigma = 1$.

Let T be a linear map from the sample space of $\mathscr{E}$ to the sample space of $\mathscr{F}$. Then the requirement of condition (iii) may be phrased

$$ET(X) = EY \qquad \text{and} \qquad \operatorname{Cov} T(X) = \operatorname{Cov} Y$$

while condition (iii) of theorem 8.4.3 may be phrased

$$ET(X) = EY \qquad \text{and} \qquad \operatorname{Cov} T(X) \leqslant \operatorname{Cov} Y.$$

The requirement $ET(X) = EY$, i.e. $T(EX) = EY$, is needed in both cases and, as it stands, is a condition of unbiasedness. However this requirement may also be regarded as a condition of invariance, as we shall explain now.

Define a group G as follows. If σ is known then we let G be the group $(K, +)$. If σ is unknown then G is the set $K \times]0, \infty[$ equipped with the group operation defined by putting $(b, c)(\beta, \sigma) = (b + c\beta, c\sigma)$ when $b, \beta \in K$ and $c, \sigma \in]0, \infty[$.

Identifying K with $\mathbb{R}^p$, we see that if σ is unknown then G is just the group G considered in the proof of theorem 8.6.5.

If σ is unknown then $(b, c) \in G$ acts on x belonging to the sample space of $\mathscr{E}$ by mapping it into $cx + A^*b$. If σ is known then a vector b in $G = K$ acts on x belonging to the sample space of $\mathscr{E}$ by mapping it into $x + A^*b$. Finally let the action of G on itself be left multiplication. Then the pair $(\mathscr{E}, \mathscr{F})$ of experiments is invariant under G.

A linear map T from the sample H_A of $\mathscr{E}$ to the sample space H_B of $\mathscr{F}$ is invariant if and only if $TA^* = B^*$. Thus both these theorems state that comparability may be decided by invariant linear maps.

Of course it is to be expected that it suffices to consider invariant kernels. The point we are making is that we may assume that these invariant kernels are induced by invariant linear maps.

Putting $\xi = EX = A^*\beta$ we can ask for the invariant estimators based on X of (ξ, σ) when σ is unknown, and for the invariant estimators based on X for ξ when σ is known. In the first case $(\hat{\xi}(X), \sigma(X))$ is invariant if and only if $\hat{\xi}(A^*b + cX) = A^*b + c\hat{\xi}(X)$ and $\hat{\sigma}(A^*b + cX) = c\hat{\sigma}(X)$. The last condition is fulfilled if and only if $\hat{\sigma}$ is a non-negatively homogeneous function of the orthogonal projection of X on $N(A)$. Thus UMVU estimators of σ, if they exist, are clearly invariant.

The condition on the estimator $\hat{\xi}(X)$ of $\xi = EX$ amounts to the requirement that $\hat{\xi}(X) =$ the UMVU estimator of EX plus a term which is also a non-negatively homogeneous function of the orthogonal projection of X on $N(A)$. If $\hat{\xi}$ is linear then this is just the condition of unbiasedness.

Finally if σ is known then $\hat{\xi}(X)$ is invariant if and only if $\hat{\xi}(X) =$ the UMVU estimator of $X +$ a function of the orthogonal projection of X on $N(A)$. If $\hat{\xi}$ is linear then this is again just the requirement that $\hat{\xi}$ is unbiased.

By corollary 8.1.6, the smoothing procedure described just before proposition 8.3.8 is order preserving for models $\mathcal{N}(A, \sigma^2 V; \sigma^2 > 0)$. This is easily confirmed by our last theorem.

Sufficiency, or local sufficiency, of linear operators is much easier to handle when σ is unknown than when σ is known. In fact there is essentially only one sufficient linear map, and this is of course the identity map.

Proposition 8.6.7 (Linear sufficiency when σ is unknown). *If X realizes $\mathscr{E} = \mathcal{N}(A, \sigma^2 V; \sigma^2 > 0)$ and if S is a linear map defined on the sample space of $\mathscr{E}$ then $S(X)$ is sufficient for X if and only if S is 1–1 on the support $R(V) + R(A^*)$ of $\mathscr{E}$.*

Moreover, if $\mathscr{E}$ is non-singular and $K = \mathbb{R}^p$ then this also amounts to the condition that $S(X)$ is locally sufficient.

Proof. Suppose S is linear and that $S(X)$ is sufficient. The statistic $S(X)$ realizes the experiment $\mathcal{N}(AS^*, \sigma^2 SVS^*; \sigma^2 > 0)$. By sufficiency, this experiment is equivalent to $\mathcal{N}(A, \sigma^2 V; \sigma^2 > 0)$. By theorem 8.6.6 there is a linear map T such that $TSA^* = A^*$ and $TSVS^*T^* = V$. If $h = A^*\beta + Vg$ and if $Sh = 0$ then $0 = Sh = SA^*\beta + SVg$ so that $0 = TSh = TSA^*\beta + TSVg = A^*\beta + TSVg$. Hence $A^*\beta \in R(TSV) = R(TSVS^*T^*) = R(V)$. It follows that $h = A^*\beta + Vg \in R(V)$.

The map TS is 1–1 on $R(V)$ since $R(TSV) = TS[R(V)] = R(V)$. Therefore, since $h \in R(V)$ and $TSh = 0$, it follows that $h = 0$.

Assume that $\mathcal{N}(A, V)$ is non-singular, i.e. that $R(A^*) \subseteqq R(V)$, and that $S(X)$ is locally sufficient. Then $S(X)$ provides the same Fisher information operator as X, i.e. $AV^+A^* = AS^*(SVS^*)^+SA^*$ and $\dim V[N(A)] = \dim SVS^*[N(AS^*)]$. The first condition amounts to the requirement that $S(X)$ is sufficient when σ is known, and by theorem 8.4.5 this is equivalent to the condition that $V[N(A)] \supseteqq N(S) \cap R(V)$.

Now $SVS^*[N(AS^*)] = \{SVS^*\tilde{h} : AS^*\tilde{h} = 0\} \subseteqq \{SVh : h \in N(A)\} = SV[N(A)]$ so that $SVS^*[N(AS^*)] = SV[N(A)]$ and $\dim SV[N(A)] = \dim V[N(A)]$. Hence S is 1–1 on $V[N(A)]$, i.e. $N(S) \cap V[N(A)] = 0$ so that $N(S) \cap R(V) \subseteqq V[N(A)] \cap N(S) = 0$, proving that S is 1–1 on $R(V)$. □

Proceeding according to the scheme which is outlined in theorem 8.6.5 and following Swensen, we will now provide conditions for deficiencies not to be altered by information on σ.

Theorem 8.6.8 (A situation where $\delta(\mathscr{E}, \mathscr{F}) = \delta((\mathscr{E}|\sigma = 1), (\mathscr{F}|\sigma = 1))$). *Let both the linear normal experiments $\mathscr{E} = \mathcal{N}(A, \sigma^2 V; \sigma^2 > 0)$ and $\mathscr{F} = \mathcal{N}(B, \sigma^2 W; \sigma^2 > 0)$ be non-singular. Assume that*

$$R(B) \subseteqq R(A)$$

and that

$$\operatorname{rank}(V) - \operatorname{rank}(A) \geqslant \operatorname{rank}(W) - \operatorname{rank}(B)$$
$$+ \operatorname{rank}[AV^+A^* - (AV^+A^* \barwedge BW^+B^*)].$$

Then $\delta(\mathscr{E}, \mathscr{F}) = \delta(\mathscr{N}(A, V), \mathscr{N}(B, W))$.

It follows that the deficiency in this case is not influenced by information on σ and thus may be found by theorem 8.5.7.

Proof. Let H_A and H_B denote, respectively, the sample space of $\mathscr{E}$ and the sample space of $\mathscr{F}$.

We may assume without loss of generality that V and W are both identity operators such that $\mathscr{E} = \mathscr{N}(A, \sigma^2 I; \sigma^2 > 0)$ and $\mathscr{F} = \mathscr{N}(B, \sigma^2 I; \sigma^2 > 0)$. Arguing as before we may assume also that A is onto and that A^* is an isometry. Proceeding as in the proof of theorem 8.2.22 and also as in the proof of theorem 8.6.6, we choose orthonormal bases $(\tau_1, \ldots, \tau_p), (\rho_1, \ldots, \rho_m)$ and $(\tilde{\rho}_1, \ldots, \tilde{\rho}_n)$ in K, H_A and H_B respectively according to the prescription of the proof of theorem 8.2.22. Let X realize $\mathscr{E}$ and let Y realize $\mathscr{F}$ and let $(X_1, \ldots, X_m)$, $(Y_1, \ldots, Y_n)$ and $(\beta_1, \ldots, \beta_p)$ be the coordinate vectors of X, Y and β respectively. Then $X_1, \ldots, X_m$ and $Y_1, \ldots, Y_n$ are independent and normally distributed variables which all have variance σ^2. Furthermore, $EX_i = \beta_i$ or $=0$ as $i \leqslant p$ or $i > p$ while $EY_j = \Delta_j^{1/2}\beta_j$ or $=0$ as $j \leqslant r = \operatorname{rank} B$ or $j > r$. Of course, it is essential for the validity of these expressions for EY_j that $\Delta_1 \geqslant \cdots \geqslant \Delta_r > \Delta_{r+1} = \cdots = \Delta_p = 0$. Theorem 8.6.5 now applies, with the same r and with $\alpha_1 = \frac{1}{2}(m-p)$, $\alpha_2 = \frac{1}{2}(n-r)$, $\lambda_j = \Delta_j^{1/2}$; $j = 1, \ldots, r$ and $v = \#\{i : i \leqslant r, \Delta_j < 1\} = \operatorname{rank}(I - I \wedge BB^*) = \operatorname{rank}(AA^* - AA^* \barwedge BB^*)$. □

If the experiment $\mathscr{F} = \mathscr{N}(B, \sigma^2 I; \sigma^2 > 0)$ majorizes $\mathscr{E} = \mathscr{N}(A, \sigma^2 I; \sigma^2 > 0)$ in the sense that UMVU estimators of linear functions of (β, σ^2) are at least as good in $\mathscr{F}$ as in $\mathscr{E}$, then the deficiency $\delta(\mathscr{E}, \mathscr{F})$ may be found from the third result described in theorem 8.6.5. This may also be phrased as follows.

Theorem 8.6.9 ($\delta(\mathscr{E}, \mathscr{F})$ when $\mathscr{E}$ is locally at most as informative as $\mathscr{F}$). *Consider experiments $\mathscr{E} = \mathscr{N}(A, \sigma^2 I; \sigma^2 > 0)$ and $\mathscr{F} = \mathscr{N}(B, \sigma^2 I; \sigma^2 > 0)$ with sample spaces H_A and H_B respectively. Let n_A and n_B denote* $\dim H_A$ *and* $\dim H_B$ *respectively. Also put $\alpha_1 = \frac{1}{2}(n_A -$ rank $A)$ and $\alpha_2 = \frac{1}{2}(n_B -$ rank $B)$.*

Assume that $AA^* \leqslant BB^*$ and that $n_A \leqslant n_B$.

Then $\delta(\mathscr{E}, \mathscr{F}) < 2$ only if rank A = rank B *and then $R(A) = R(B)$ and*

$$\delta(\mathscr{E}, \mathscr{F}) = \|N(0, (AA^*)^+) \times \Gamma_{\alpha_1, \alpha_2/\alpha_1} - N(0, (BB^*)^+) \times \Gamma_{\alpha_2, 1}\|$$
$$= \|N(0, BB^*) \times \Gamma_{\alpha_1, \alpha_2/\alpha_1} - N(0, AA^*) \times \Gamma_{\alpha_2, 1}\|.$$

Here $\Gamma_{\alpha_1, \alpha_2/\alpha_1}$ should be interpreted as the one-point distribution in 0 *when $\alpha_1 = 0$.*

Remark 1. The two assumptions of this theorem state that $\mathscr{E}$ is locally at most as informative as $\mathscr{F}$, by example 9.4.7.

Remark 2. A map which yields the optimal approximation may be described as follows.

Assume that X realizes $\mathscr{E}$ and that Y realizes $\mathscr{F}$. Let $\hat{\beta}(X)$ be the $R(A)$-valued BLUE generator based on X and let $S^2(X)/(n_A - \operatorname{rank} A)$ be the UMVU estimator of σ^2 based on X. Also let $\tilde{\beta}(Y)$ be the $R(B)$-valued BLUE generator based on Y and let $T^2(Y)/[n_B\text{-rank}(B)]$ be the UMVU estimator of σ^2 based on Y.

Then, by sufficiency $(\hat{\beta}(X), S^2(X))$ realizes an experiment $\tilde{\mathscr{E}}$ equivalent to $\mathscr{E}$ and $(\tilde{\beta}(Y), T^2(Y))$ realizes an experiment $\tilde{\mathscr{F}}$ equivalent to $\mathscr{F}$. The optimal approximation is now provided by the map from the sample space of $\tilde{\mathscr{E}}$ to the sample space of $\tilde{\mathscr{F}}$ which maps $(\hat{\beta}, S^2)$ into $(\hat{\beta}, (\alpha_2/\alpha_1)S^2)$.

This follows directly from the first expression for the deficiency.

Proof. The two statistical distances which appear in the theorem are equal since, by proposition 8.5.5, the dichotomies $(N(0,(AA^*)^+), N(0,(BB^*)^+))$ and $(N(0,BB^*), N(0,AA^*))$ are equivalent.

It follows directly from part (iii) of theorem 8.6.2 that $\delta(\mathscr{E},\mathscr{F}) < 2$ if and only if $R(B) \subseteqq R(A)$ and if $N(B) = 0$ when $N(A) = 0$. As $R(A) \subseteqq R(B)$ the conditions '$R(B) \subseteqq R(A)$', '$R(B) = R(A)$' and 'rank A = rank B' are equivalent. Assume that $R(B) = R(A)$ and put $r = \operatorname{rank} A = \operatorname{rank} B$. As in the previous proof, we may reduce the problem to a situation where A^* is an isometry.

Let X realize $\mathscr{E}$ and let Y realize $\mathscr{F}$. Put $m = n_A$, $n = n_B$, $p = r$ and choose bases $(\tau_1,\ldots,\tau_p)$, $(\rho_1,\ldots,\rho_m)$ and $(\tilde{\rho}_1,\ldots,\tilde{\rho}_n)$ as in the proof of theorem 8.6.8. Let X_i, Y_j and β_i be, respectively, the ρ_i coordinate of X, the $\tilde{\rho}_j$ coordinate of Y and the τ_i coordinate of β. Again $X_1,\ldots,X_m$ and $Y_1,\ldots,Y_n$ are independent and normally distributed with the same variance σ^2. Furthermore $EX_i = \beta_i$ or $= 0$ as $i \leqslant p$ or $i > p$, while $EY_i = \Delta_i^{1/2}\beta_i$ or $= 0$ as $i \leqslant p$ or $i > p$. Note that the eigenvalues $\Delta_1,\ldots,\Delta_p$ are all $\geqslant 1$ since $BB^* \geqslant I$.

The theorem now follows directly from the third statement of theorem 8.6.5. □

In order to see which cases are covered by these results and which are still open, let us briefly survey what we have learned about the deficiency of $\mathscr{E} = \mathscr{N}(A,\sigma^2 V;\sigma^2 > 0)$ with respect to $\mathscr{F} = \mathscr{N}(B,\sigma^2 W;\sigma^2 > 0)$.

Let H_A and H_B denote, respectively, the sample space of $\mathscr{E}$ and the sample space of $\mathscr{F}$. Put $n_A = \dim H_A$ and $n_B = \dim H_B$.

Firstly, in theorem 8.6.2 we were able to single out the situations where $\delta(\mathscr{E},\mathscr{F}) < 2$ and then by proposition 8.6.1 the problem may be reduced to the case of non-singular experiments. If $\mathscr{E}$ and $\mathscr{F}$ are non-singular then we may assume also that V and W are identity operators.

Therefore let us restrict our attention to experiments $\mathscr{E} = \mathscr{N}(A,\sigma^2 I;\sigma^2 > 0)$ and $\mathscr{F} = \mathscr{N}(B,\sigma^2 I;\sigma^2 > 0)$. From theorem 8.6.2 it follows that $\delta(\mathscr{E},\mathscr{F}) < 2$ if and only if $R(B) \subseteqq R(A)$ and that B is 1–1 whenever A is 1–1. Assume now that these conditions are satisfied. Put $\nu = \operatorname{rank}(AA^* - AA^* \bar{\wedge} BB^*)$ where $AA^* \bar{\wedge} BB^*$ is defined as in section 8.5.

Now theorem 8.6.6 tells us that $\delta(\mathscr{E}, \mathscr{F}) = 0$, i.e. $\mathscr{E} \geqslant \mathscr{F}$, if and only if $AA^* \geqslant BB^*$ and $n_A - n_B \geqslant \nu$.

Theorem 8.6.8 generalizes this by giving the deficiency whenever $n_A - \operatorname{rank} A \geqslant (n_B - \operatorname{rank} B) + \nu$.

Finally theorem 8.6.9 gives the deficiency whenever $\mathscr{E}$ is locally at most as informative as $\mathscr{F}$, i.e. when $\nu = 0$ and $n_A \leqslant n_B$. If $\nu = 0$ then, since $R(B) \subseteqq R(A)$, $AA^* \leqslant BB^*$, so that actually $R(B) = R(A)$. Thus the inequalities '$n_A \leqslant n_B$' and '$n_A - \operatorname{rank} A \leqslant n_B - \operatorname{rank} B$' are trivially equivalent when $\nu = 0$.

The remaining problem is then to find explicit expressions for the deficiency $\delta(\mathscr{E}, \mathscr{F})$ when the following conditions are all satisfied

$R(B) \subseteqq R(A)$,
B is 1–1 whenever A is 1–1,
$\nu > 0$

and

$$n_A - \operatorname{rank} A < (n_B - \operatorname{rank} B) + \nu.$$

This problem is still open.

8.7 References

Boll, C. 1955. Comparison of experiments in the infinite case. *Ph.D. dissertation*, Stanford Univ.

Campbell, S. L. & Meyer, C. D. 1979. *Generalized inverses of linear transformations.* Pitman, London.

Cohen, A. 1966. All admissible linear estimates of the mean vector. *Ann. Math. Statist.* **37**, pp. 458–63.

Cramer, H. 1937. *Random variables and probability distributions.* Cambridge University Press.

Drygas, H. 1983. Sufficiency and completeness in the general Gauss–Markov model. *Sankhyā.* **45**, pp. 88–98.

— 1985. On the unified theory of least squares. *Probability and Math. Statistics.* **5**, pp. 177–86.

Eaton, M. L. 1983. *Multivariate statistics. A vector space approach.* Wiley, New York.

Ehrenfeld, S. 1955. Complete class theorems in experimental design. *Third Berkeley Symp. Math. Statist. Prob.* **1**, pp. 69–75.

— 1966. On a minimal essentially complete class of experiments. *Ann. Math. Statist.* **32**, pp. 435–40.

Elfving, G. 1952. Optimum allocation in linear regression theory. *Ann. Math. Statist.* **26**, pp. 241–55.

— 1959. Design of linear experiments. Probability and Statistics. *The Harald Cramer Volume*, pp. 58–74. Wiley, New York.

Fellmann, J. 1974. On the allocation of linear observations. *Commentationes Physico-mathematicae*, **44**, pp. 27–78.

Graybill, F. A. 1983. *Matrices with applications in statistics.* Wadsworth, Belmont, California.

Halmos, P. R. 1958. *Finite dimensional vectorspaces.* Van Nostrand, New York.

Hansen, O. H. & Torgersen, E. 1974. Comparison of linear normal experiments. *Ann. Statist.* **2**, pp. 367–73.

Helgeland, J. 1982. Additional observations and statistical information in the case of 1-parameter exponential distributions. *Z. Wahscheinlichkeitstheorie verw. Geb.* **59**, pp. 77–100.
Janssen, A 1988. A convolution theorem for the comparison of exponential families. *Res. report* no **211**, Univ. of Siegen.
Janssen, A., Milbrodt, H. & Strasser, H. 1984. Infinitely divisible experiments. *Lecture notes in statistics.* Springer–Verlag, Berlin.
La Motte, L. R. 1982. Admissibility in linear estimation. *Ann. Statist.* **10**, pp. 245–55.
— 1985. Admissibility, Unbiasedness and Nonnegativity in the balanced, Random, One-Way ANOVA model. *Lecture notes in statistics,* **35**, pp. 184–9, Springer–Verlag, Berlin.
LeCam, L. 1974a. On the information contained in additional observations. *Ann. Statist.* **2**, pp. 630–49.
— 1974b. *Notes on asymptotic methods in statistical decision theory.* Université de Montréal.
— 1975. Distances between experiments. In J. N. Srivastava (Ed.) *A survey of statistical design and linear models,* pp. 383–95, North–Holland, Amsterdam.
— 1986. *Asymptotic methods in statistics.* Springer–Verlag, Berlin.
Lehmann, E. L. 1983a. Comparison of experiments for some multivariate normal situations. *Studies in econometrics, time series and multivariate statistics,* pp. 491–503, Academic Press, New York.
— 1983b. Least informative distributions. *Recent advances in statistics,* pp. 593–9, Academic Press, New York.
Lehmann, E. L. & Scheffé, H. 1950. Completeness, similar regions and unbiased estimation. Part I. *Sankhyā,* **10**, pp. 305–40.
Lindqvist, B. 1977. How fast does a Markov chain forget the initial state? A decision theoretical approach. *Scand. J. Statist.* **4**, pp. 145–52.
— 1984. On the memory of an absorbing Markov chain. *Revue Roumaine de Mathematiques pures et appliquées,* **29**, No. 2, pp. 171–82.
Mammen, E. 1986. The statistical information contained in additional observations. *Ann. Statist.* **14**, pp. 665–78.
Millar, P. W. 1983. The minimax principle in asymptotic statistical theory. *Lecture notes in statistics,* **976**, pp. 74–265, Springer–Verlag, Berlin.
Rao, C. R. 1965. *Linear statistical inference and its applications.* Wiley, New York.
— 1976. Estimation of parameters in linear models. *Ann. Statist.* **4**, pp. 1023–37.
Rao, C. R. & Kleffe, J. 1988. *Estimation of variance components and applications.* North–Holland, Amsterdam.
Rockafellar, R. T. 1970. *Convex analysis.* Princeton University Press.
Scheffé, H. 1959. *The analysis of variance.* Wiley, New York.
Stein, C. 1956. Inadmissibility of the usual estimator for the mean of a multivariate normal distribution. *Proc. 3rd Berkeley Symp. on Math. Statist and Prob.* **1**, pp. 197–206, University of California Press, Berkeley.
Stepniak, C. 1983. Optimal allocation of units in experimental designs with hierarchical and cross classification. *Ann. Inst. Statist. Math.* **35**, pp. 461–73.
— 1985. *Reduction problems in comparison of linear models.* Unpublished.
Stepniak, C. & Torgersen, E. 1981. Comparison of linear models with partially known covariances with respect to unbiased estimation. *Scand. J. Statist.* **8**, pp. 183–4.
Stepniak, C., Wang, S. G. & Wu, C. F. J. 1984. Comparison of linear experiments with known covariances. *Ann. Statist.* **12**, pp. 358–65.
Strasser, H. 1985. *Mathematical theory of statistics.* Walter de Gruyter, Berlin.
Swensen, A. R. 1980. Deficiencies between linear normal experiments. *Ann. Statist.* **8**, pp. 1142–55.
Torgersen, E. 1972. Comparison of translation experiments. *Ann. Math. Stat.* **43**, pp. 1383–99.
— 1981. Measures of information based on comparison with total information and with total ignorance. *Ann. Statist.* **9**, pp. 638–57.
— 1984. Orderings of linear models. *J. of Statist. Planning and Inference* **9**, pp. 1–17.

9

Majorization and Approximate Majorization

9.1 Introduction

We saw in chapter 6 that a variety of natural and apparently different criteria for comparison actually were equivalent. Thus we found equivalent criteria in terms of overall comparison of risk functions, in terms of Bayes risk for a fixed prior distribution, in terms of sublinear functionals, in terms of performance functions of decision rules and in terms of randomizations.

The idea that the underlying distributions which constitute a statistical experiment should be proper probability distributions was of course essential for the interpretation of these results. In particular the interpretation of risk as expected loss depended on this assumption. However if we look over the proofs we see that the arguments do not rely on this. Thus the formal expression of the risk of a decision procedure, as the corresponding bilinear functional evaluated for the underlying distribution and the loss function, remains well defined within a much more general set-up.

We shall now see that such a generalization, or extension, to 'non-proper' experiments actually yields interesting results on 'proper' experiments. Another benefit is a unified approach to several results concerning systems of inequalities which are frequently encountered in mathematical statistics. The theory we shall formulate here, partly based on Torgersen (1969, 1985), may be considered as a generalization of the theory of majorization as it is described in Marshall & Olkin (1979) and in later generalizations of Dahl (1983) and Karlin & Rinott (1983).

At this point let us briefly indicate some directions of applications.

Example 9.1.1 (Comparison for different losses). Assume that all losses are bounded by 1 and let a, b and ε be given functions of the unknown parameter θ.

When are we entitled to claim for all decision problems (of a certain type) and given experiments $\mathscr{E}$ and $\mathscr{F}$ that to each risk function s obtainable in $\mathscr{F}$ there corresponds a risk function r obtainable in $\mathscr{E}$ such that $a(\theta)r(\theta) \leqslant b(\theta)s(\theta) + \varepsilon(\theta)$ for all θ?

If $a(\theta) \equiv_\theta 1$ and $b(\theta) \equiv_\theta 1$ then this is just a description of the situation where $\mathscr{E}$ is ε-deficient with respect to $\mathscr{F}$ according to LeCam's definition of deficiency.

Theoretically at least, this problem for general a, b and ε, is easily described in terms of general families of measures. In the next section we will see that one possible answer may be obtained by a straightforward generalization of LeCam's randomization (Markov kernel) criterion.

Example 9.1.2 (Local (non-asymptotic) comparison of experiments). Consider experiments having the same subset Θ of $\mathbb{R}^m$ as their parameter set. Let θ^0 be an interior point of Θ. If $\mathscr{E} = (P_\theta : \theta \in \Theta)$ is an experiment then, assuming differentiability, we may consider the measure P_{θ^0} along with the partial derivatives $[\partial P_\theta/\partial\theta_i]_{\theta=\theta^0}$; $i = 1, \ldots, m$. This family of measures is not an experiment since the total masses of the measures $[\partial P_\theta/\partial\theta_i]_{\theta=\theta^0}$ are all zero.

If θ is close to θ^0 then the measure P_θ may be approximated by $P_{\theta^0} + \sum_{i=1}^m (\theta - \theta^0)_i[\partial P_\theta/\partial\theta_i]_{\theta=\theta^0}$. Thus it is not surprising that if θ is close to θ^0 and if quantities of smaller order than the distance from θ to θ^0 are considered negligible, then the local behaviour of $\mathscr{E}$ around θ^0 may be completely described in terms of this family of measures. However it is interesting that, as we shall see later, local comparison of experiments within small neighbourhoods of θ^0 may be expressed quite naturally and (at least in theory) simply, in terms of such families of measures (Torgersen, 1972a, b).

Example 9.1.3 (Comparison of measures. Dilations). Systems of inequalities of the following type have received attention in various connections.

Let μ and ν be measures on a measurable space $(\mathscr{X}, \mathscr{A})$. Consider also a convex set H of integrable functions which contains the constants and which is closed under the formation of maxima of finite subsets. What conditions on (H, μ, ν) ensure that the inequality $\int h\,\mathrm{d}\mu \geqslant \int h\,\mathrm{d}\nu$ holds for every functon h in H?

The ordering of experiments is closely related to the case where μ and ν are conical measures of experiments.

We may also consider systems of inequalities of the form $\int h\,\mathrm{d}\mu \geqslant \int h\,\mathrm{d}\nu - \varepsilon_h$ for some given function $h \to \varepsilon_h$ on H. Thus we are led to compare the families $(h\mu : h \in H)$ and $(h\nu : h \in H)$ of measures.

In section 9.4 we will see how the general theory of comparison of families of measures provides a method of attack for such problems.

A related discussion of such problems may be found in Karlin & Rinott (1983).

Example 9.1.4 (Distributions with given marginals). The fundamental papers by LeCam (1964) and Strassen (1965) both show that there is a close relationship between the theory of comparison of experiments on the one hand and various existence problems for joint distributions with given marginals on the other. This relationship is also apparent from earlier works by e.g. Blackwell (1953) and Boll (1955).

Thus the dilation criterion for ordering of experiments, as shown by Strassen, is related to the problem of whether or not a sequence of distributions could be the sequence of distributions of a martingale. In his paper Strassen considered the problem of deciding whether one distribution P is the convolution factor of another distribution Q. Boll had already shown that this is tantamount to the problem of deciding whether the translation experiment defined by P is at least as informative as the translation experiment defined by Q. Using results from LeCam's paper, the statistical significance of Strassen's criterion in terms of minimax risk was clarified by Torgersen (1972c), who also generalized these results to the case of ε-deficiency.

There are other results which bring out the connection between these fields but here we shall mention only the various criteria for stochastic orderings of distributions on a general partially ordered set. The basic criterion in terms of probabilities of 'monotone' events follows directly from Strassen's paper. By Torgersen (1982) this directly provides the criteria for comparison of experiments associated with sampling plans P and Q as described in example 1.3.1. Furthermore, by the results described in the previous example, ε-deficiency implies that one sampling distribution is up to an amount ε stochastically larger than the other sampling distribution.

In section 9.6 we will show that a general result due to Strassen which yields this and other interesting results, may be deduced from the general principles for comparison of families of measures.

Example 9.1.5 (Majorization). The theory of comparison of measure families may be considered as a natural generalization of most of the various theories of majorization. In particular a functional of experiments which is monotonically increasing (decreasing) for the ordering 'being more informative' may be considered as Schur convex (concave). Thus Fisher information and statistical distance may be considered Schur convex while the Hellinger transform is Schur concave.

In order to make this clearer let us comment on the best known particular case which is the case of majorization of vectors in $\mathbb{R}^n$. Two important sources of information for this case are Hardy, Littlewood & Polya (1934) and Marshall & Olkin (1979). Using the notation from the latter we will write $(x_{[1]}, x_{[2]}, \ldots, x_{[n]})$ for the vector in $\mathbb{R}^n$ obtained from the vector $(x_1, \ldots, x_n)$ by arranging the x's in decreasing order.

Let $p = (p_1, \ldots, p_n)$ and $q = (q_1, \ldots, q_n)$ be two vectors in $\mathbb{R}^n$ such that $p_1 + \cdots + p_n = q_1 + \cdots + q_n$. Also consider p and q as $1 \times n$ row matrices. Then the following conditions are known to be equivalent.

(i) $p_{[1]} + \cdots + p_{[j]} \geqslant q_{[1]} + \cdots + q_{[j]}; j = 1, \ldots, n$.
(ii) $\sum_j \phi(p_j) \geqslant \sum \phi(q_j)$ when the function ϕ is convex on $\mathbb{R}$.
(iii) $\sum_j (p_j - c)^+ \geqslant \sum_j (q_j - c)^+; c \in \mathbb{R}$.
(iv) $pM = q$ for a doubly stochastic $n \times n$ matrix M.

Assume now that the vectors p and q are probability vectors. Define probability distributions P_0, Q_0, P_1 and Q_1 on $\{1, \ldots, n\}$ by putting $P_0(j) = 1/n$, $P_1(j) = p_j$, $Q_0(j) = 1/n$ and $Q_1(j) = q_j; j = 1, \ldots, n$. Put $\Theta = \{0, 1\}$ and consider the experiments (dichotomies) $\mathscr{E} = (P_0, P_1)$ and $\mathscr{F} = (Q_0, Q_1)$.

Using the Neyman–Pearson lemma on the problem of testing '$\theta = 0$' against '$\theta = 1$' we see that (i) tells us that $\mathscr{E}$ yields at least as large a power as $\mathscr{F}$ for any level of significance. By Blackwell (1951, 1953) this is equivalent to the condition that $\mathscr{E}$ is at least as informative as $\mathscr{F}$. Multiplying (ii) and (iii) by (-1) we see that condition (ii) is the overall minimum Bayes risk criterion for comparability, and that (iii) is the minimum Bayes risk criterion for testing problems. Finally condition (iv) states that $P_\theta M = Q_\theta; \theta = 0, 1$ for a Markov kernel M from the set $\{1, 2, \ldots, n\}$ into itself. By Blackwell again, these conditions are all equivalent. In particular note that the condition that the Markov matrix M is doubly stochastic implies that the corresponding Markov kernel preserves the uniform distribution, i.e. that $P_0 M = Q_0$.

Replacing the distribution $P_0 = Q_0$ with more general distributions we arrive at more general notions of majorization as e.g. the notion of π-majorization considered in Marshall and Olkin's book.

Now majorization is a partial ordering on all of $\mathbb{R}^n$ and not just on the probability simplex. The validity of the arguments (e.g. the Neyman–Pearson lemma) used above does not, however, depend on the condition that the vectors p and q are probability vectors. We shall see in section 6 that most of the mathematics remain valid, after some straightforward modifications, for general families of measures. Thus, as in Dahl (1983), we may consider ε-deficiency as well as multivariate majorization.

Multivariate majorization was also considered by Karlin & Rinott (1983).

As is amply demonstrated by Marshall and Olkin in their book on inequalities, the concept of majorization in $\mathbb{R}^n$ is a very useful tool for establishing inequalities in statistics and elsewhere.

The point we are making is that this concept fits naturally into the extension of the framework of comparison of experiments which we shall consider here. However majorization in $\mathbb{R}^n$ is a distinguished particular case, with several features which appear difficult to deduce from this general theory of comparison of families of measures.

In sections 9.4–9.7 we will take a closer look at the situations described in the examples of this section. However it is necessary to first become acquainted with some basic principles for comparison of families of measures. These principles are therefore the topic of the next two sections.

Several results on comparison of dichotomies, as well as results on general comparison for testing problems which fits into this framework, are derived in sections 9.3 and 9.7.

Complements in chapter 10 of interest in connection with the topics covered in this chapter are numbers 2, 4–8, 11, 12, 22–24 and 46–48.

9.2 Equivalent criteria for comparison of families of measures

In this section we will be concerned with general principles for comparing families of measures. It will be assumed that all the measures belonging to a given family are real valued and that they are defined on a common measurable space.

We have found it convenient to use the term *measure family* to denote a family of measures satisfying these two requirements. Thus the measures in a measure family are all bounded, but we do not assume that they are non-negative. A measure family having two members, i.e. an ordered pair of finite measures on the same measurable space, will sometimes be called a *measure pair*.

A measure family $\mathscr{E}$ with measurable space $(\mathscr{X}, \mathscr{A})$ may be denoted as $\mathscr{E} = (\mathscr{X}, \mathscr{A}; \mu_\theta : \theta \in \Theta)$ or just as $\mathscr{E} = (\mu_\theta : \theta \in \Theta)$ where $(\mu_\theta : \theta \in \Theta)$ is the family of finite measures which constitutes $\mathscr{E}$. The set Θ is called the *parameter set of* $\mathscr{E}$. This set may be any set but we shall assume, unless otherwise stated, that if two measure families $\mathscr{E}$ and $\mathscr{F}$ are compared, then their parameter sets are the same.

Thus an experiment is a measure family such that all the measures in the family are probability measures.

Extending the notion of the L-space of an experiment we define *the L-space of the measure family* $\mathscr{E} = (\mu_\theta : \theta \in \Theta)$ as the L-space of finite measures which are dominated by finite measures of the form $\sum \{c_\theta |\mu_\theta| : \theta \in \Theta_0\}$ where Θ_0 is countable and the numbers $c_\theta : \theta \in \Theta_0$ are non-negative. It follows that the L-space of $\mathscr{E} = (\mu_\theta : \theta \in \Theta)$ is the band of finite measures generated by $\mathscr{E}$. The L-space of the measure family $\mathscr{E}$ will be denoted as $L(\mathscr{E})$.

Consider now in addition to the parameter set Θ a set T and a family $L = (L_\theta : \theta \in \Theta)$ of real valued functions on T. If the measure families under consideration are experiments then in the following we may think of T as a set of decisions and L as a loss function.

Let us agree to use the notation $\|g\|$ for the supremum norm $\sup_t |g(t)|$ of a real valued function g on T. Risking the slight possibility of confusion in interpretation we shall use the notation $\|\mu\|$ to denote the total variation of a measure μ. Finally we shall use the term *deficiency function* for any non-negative function on the parameter set.

We shall now list four principles for comparing measure families $\mathscr{E} =$

$(\mathscr{X}, \mathscr{A}; \mu_\theta : \theta \in \Theta)$ and $\mathscr{F} = (\mathscr{Y}, \mathscr{B}; \nu_\theta : \theta \in \Theta)$. The comparison will be expressed in terms of a fixed deficiency function ε and a fixed set T.

We precede the statement of each of the listed principles by a subtitle indicating the interpretation of the principles when the measure families are experiments.

It should be noted that the only aspect of the set T which matters here is its cardinality which we shall denote by k.

In the statement of the third of the principles below and also in several other places, we use the convenient notation $\int h(\mathrm{d}\mu_\theta : \theta \in \Theta)$ to denote $\int h(f_\theta : \theta \in \Theta)\,\mathrm{d}\mu$ whenever the finite measure μ_θ, for each θ, has density f_θ with respect to the non-negative measure μ and h is a non-negative homogeneous measurable function on $\mathbb{R}^\theta$. It is then easily checked that neither the existence nor the value of this integral depends on the choice of the majorizing measure μ nor on the specifications of the densities.

The first four principles mentioned in the introduction may be formulated as follows.

(i) *Pointwise comparison of risks.* To each family $L_\theta : \theta \in \Theta$ of real valued functions on T and each Markov kernel σ from $\mathscr{F}$ to T there corresponds a transition ρ from $L(\mathscr{E})$ to T such that

$$\mu_\theta \rho L_\theta \leqslant \nu_\theta \sigma L_\theta + \varepsilon_\theta \|L_\theta\|; \qquad \theta \in \Theta.$$

(ii) *Comparison of Bayes risks.* To each finite subset Θ_0 of Θ and to each family $L_\theta : \theta \in \Theta$ of real valued functions on T and each Markov kernel σ from $\mathscr{F}$ to T there corresponds a Markov kernel ρ from $\mathscr{E}$ to T such that

$$\sum_{\Theta_0} \mu_\theta \rho L_\theta \leqslant \sum_{\Theta_0} \nu_\theta \sigma L_\theta + \sum_{\Theta_0} \varepsilon_\theta \|L_\theta\|.$$

(iii) *Comparison of maximum Bayes utilities.* The sublinear function criterion

$$\int \psi(\mathrm{d}\mu_\theta : \theta \in \Theta_0) \geqslant \int \psi(\mathrm{d}\nu_\theta : \theta \in \Theta_0) - \sum_{\Theta_0} \varepsilon_\theta[\psi(-e^\theta) \vee \psi(e^\theta)]$$

for each finite subset Θ_0 of Θ and for each function ψ on $\mathbb{R}^{\Theta_0}$ which is a maximum of $k = \# T$ linear functionals.

Here $e^\theta = (0, \ldots, \overset{\theta}{1}, \ldots, 0)$ denotes the θth unit vector in $\mathbb{R}^{\theta_0}$.

(iv) *Comparison of performance functions.* To each Markov kernel σ from $\mathscr{F}$ to T there corresponds a transition ρ from $L(\mathscr{E})$ to T such that

$$\|\mu_\theta \rho - \nu_\theta \sigma\| \leqslant \varepsilon_\theta; \qquad \theta \in \Theta.$$

Proceeding as in the proof of theorem 6.3.1 we obtain the following basic theorem.

Theorem 9.2.1 (Equivalent rules for comparison). *Criteria* (i)–(iv) *are all equivalent.*

Restricting our attention in (iii) to functions ψ such that ψ or $-\psi$ are projections on the coordinate spaces of $\mathbb{R}^{\Theta}$, we see that the equivalent conditions (i)–(iv) imply

$$|\mu_\theta(\mathscr{X}) - \nu_\theta(\mathscr{Y})| \leqslant \varepsilon_\theta; \qquad \theta \in \Theta \tag{9.2.1}$$

which is conditions (i)–(iv) when T possesses just one element.

Assuming that the deficiency function ε satisfies (9.2.1), which it always does when $\mathscr{E}$ and $\mathscr{F}$ are experiments, we find that the 'deficiency term' $\sum \varepsilon_\theta[\psi(-e^\theta) \vee \psi(e^\theta)]$ in (iii) may be replaced by the linear (in ψ) term

$$\sum_\theta \varepsilon_\theta[\psi(-e^\theta) + \psi(e^\theta)]/2 + \sum_\theta (\nu_\theta(\mathscr{Y}) - \mu_\theta(\mathscr{X}))[\psi(e^\theta) - \psi(-e^\theta)]/2.$$

We may then even assume that $\psi(-e^\theta) \equiv_\theta \psi(e^\theta)$ and thus get the simple expression $\sum_\theta \varepsilon_\theta \psi(e^\theta)$ for the deficiency term. We may also pass from a general sublinear function ψ on $\mathbb{R}^\theta$ to a sublinear function which is monotonically decreasing (increasing), by replacing $\psi(x)$ with $\psi(x) - \sum_\theta x_\theta \psi(e^\theta)$ (with $\psi(x) + \sum_\theta x_\theta \psi(-e^\theta)$).

Extending the notion of deficiency we shall say that *the measure family $\mathscr{E} = (\mu_\theta : \theta \in \Theta)$ is ε-deficient with respect to the measure family $\mathscr{F} = (\nu_\theta : \theta \in \Theta)$ for k-point sets* if the equivalent conditions (i)–(iv) are satisfied. If $\mathscr{E}$ and $\mathscr{F}$ are experiments then deficiency 'for k-point sets' is the same as deficiency 'for k-decision problems'.

If $\mathscr{E}$ is ε-deficient with respect to $\mathscr{F}$ for k-point sets for $k = 1, 2, \ldots$ then we shall say that *$\mathscr{E}$ is ε-deficient with respect to $\mathscr{F}$ for ∞-point sets* or just say that *$\mathscr{E}$ is ε-deficient with respect to $\mathscr{F}$*. Thus the notion of ε-deficiency for k-point sets is defined for all positive integers k and for $k = \infty$.

It is easily checked that $\mathscr{E}$ is ε'-deficient with respect to $\mathscr{F}$ for k'-point sets problems whenever $\mathscr{E}$ is ε-deficient with respect to $\mathscr{F}$ for k-point sets for $\varepsilon \leqslant \varepsilon'$ and $k \geqslant k'$.

The possibility of expressing deficiencies for k-point sets in terms of deficiencies for ∞-point sets, which we established in theorem 6.3.17, extends, by essentially the same proof, as follows.

Theorem 9.2.2 (Reduction to deficiencies for ∞-point sets). *$\mathscr{E} = (\mu_\theta : \theta \in \Theta)$ is ε-deficient with respect to $\mathscr{F} = (\mathscr{Y}, \mathscr{B}; \nu_\theta : \theta \in \Theta)$ for k-point sets if and only if $\mathscr{E}$ is ε-deficient with respect to $\mathscr{F}|\mathscr{B}_0$ for each subalgebra $\mathscr{B}_0$ of $\mathscr{B}$ containing at most 2^k sets.*

A substantial reduction is available when Θ is a two-point set. In this case the fact that convex polygons may be decomposed as vector sums of triangles and line segments tells us that it suffices to consider three-point sets T and thus functions ψ which are maxima of at most three linear functionals. Actually additional assumptions will often guarantee that it suffices to consider two-point sets T and thus functions ψ which are maxima of at most two linear functionals. As we shall see in the next section, this is the case whenever $\mathscr{E}$ and $\mathscr{F}$ are experiments (i.e. dichotomies) or if μ_θ, $\nu_\theta \geqslant 0$ and $\varepsilon_\theta = 0$ for one of the two points θ in Θ. The latter case covers the usual case of majorization as well as the cases of weak majorization in Marshall & Olkin (1979).

We shall now see, without using any new ideas, that LeCam's randomization criterion is also valid for general measure families.

Remember that if $\mathscr{E} = (\mathscr{X}, \mathscr{A}; \mu_\theta : \theta \in \Theta)$ and $\mathscr{F} = (\mathscr{Y}, \mathscr{B}; \nu_\theta : \theta \in \Theta)$ then any Markov kernel M from $(\mathscr{X}, \mathscr{A})$ to $(\mathscr{Y}, \mathscr{B})$ induces a map $\mu \to \mu M$ from $L(\mathscr{E})$ into the L-space of finite measures on $\mathscr{B}$. This map is a transition, i.e. it is linear, non-negative and preserves total mass. We have noted before that the set of all transitions, by Tychonoff, is compact for the pointwise topology on $L(\mathscr{E}) \times \mathscr{B}$.

Assume that $\mathscr{E} = (\mathscr{X}, \mathscr{A}; \mu_\theta : \theta \in \Theta)$ is ε-deficient with respect to $\mathscr{F} = (\mathscr{Y}, \mathscr{B}; \nu_\theta : \theta \in \Theta)$. Let N denote the set of all triples $n = (\pi, \eta, F)$ where $\pi = (B_1, \ldots, B_k)$ is a measurable partitioning of $\mathscr{Y}$, $\eta = (y_1, \ldots, y_k)$ where $y_i \in B_i$, $i = 1, \ldots, k$ and F is a finite subset of Θ. Direct the set N by defining $n_1 = (\pi_1, \eta_1, F_1) \geqslant n_2 = (\pi_2, \eta_2, F_2)$ whenever $F_1 \supseteq F_2$ and each set in π_2 is a union of sets in π_1. Let $n = (\pi, \eta, F) \in N$ be described as above and consider $\{y_1, \ldots, y_k\}$ as a choice for the set T appearing in the statements of conditions (i)–(iv). Let σ_n denote the Markov kernel (function) from $\mathscr{F}$ to $\{y_1, \ldots, y_k\}$ which maps y into y_i when $y \in B_i$. Then, by (iv), there is a Markov kernel ρ_n from $\mathscr{E}$ to the same space such that $\|\mu_\theta \rho_n - \nu_\theta \sigma_n\| \leqslant \varepsilon_\theta$ when $\theta \in F$. Put $M_n(B|x) = \sum I_B(y_i)\rho_n(y_i|x)$ when $B \in \mathscr{B}$ and $x \in \mathscr{X}$. Then M_n is a Markov kernel from $(\mathscr{X}, \mathscr{A})$ to $(\mathscr{Y}, \mathscr{B})$ and thus defines a transition. By compactness there is a transition M which is a point of accumulation for the net of transitions defined by the net $\{M_n\}$ of Markov kernels. It follows then that $\mu_\theta M g - \nu_\theta g \leqslant \varepsilon_\theta \|g\|$ for all $\theta \in \Theta$ and for all $g \in M(\mathscr{Y}, \mathscr{B})$. Hence $\|\mu_\theta M - \nu_\theta\| \leqslant \varepsilon_\theta$; $\theta \in \Theta$. This proves the promised extension of the randomization criterion.

Theorem 9.2.3 (The randomization criterion for comparison of measure families). *The measure family $\mathscr{E} = (\mu_\theta : \theta \in \Theta)$ is ε-deficient with respect to the measure family $\mathscr{F} = (\nu_\theta : \theta \in \Theta)$ if and only if there is a transition M from the L-space of $\mathscr{E}$ to the L-space of bounded additive set functions on $\mathscr{B}$ such that*

$$\|\mu_\theta M - \nu_\theta\| \leqslant \varepsilon_\theta; \qquad \theta \in \Theta.$$

Remark 1. Still using the same arguments as in LeCam (1964), (see section 6.4), we may always modify M so that it maps $L(\mathscr{E})$ into $L(\mathscr{F})$.

Remark 2. It follows that one possible answer to the problem raised in example 9.1.1 is:

'If and only if there is a transition M such that $\|a_\theta P_\theta M - b_\theta Q_\theta\| \leqslant \varepsilon_\theta$ for all $\theta \in \Theta$'.

If $a_\theta \equiv_\theta 1$ then this implies that $|b_\theta - 1| \leqslant \varepsilon_\theta$; $\theta \in \Theta$. If $b_\theta = 1 - \varepsilon_\theta$ then the 'θth' inequality states that $P_\theta M \geqslant (1 - \varepsilon_\theta)Q_\theta$. On the other hand, if $b_\theta = 1 + \varepsilon_\theta$ then the θth inequality states that $P_\theta M \leqslant (1 + \varepsilon_\theta)Q_\theta$.

If we want to ensure that the transition M appearing in the randomization criterion is representable as a Markov kernel then we can use the same regularity conditions as we used for experiments. Say that a measure family $\mathscr{E} = (\mu_\theta : \theta \in \Theta)$ is *coherent* if all bounded linear functionals on $L(\mathscr{E})$ are representable as bounded measurable functions. If $\mathscr{E} = (\mu_\theta : \theta \in \Theta)$ is coherent and if $(\mathscr{Y}, \mathscr{B})$ is Euclidean then any transition from $L(\mathscr{E})$ to the L-space of finite measures on $(\mathscr{Y}, \mathscr{B})$ is representable as a Markov kernel.

If the families $\mathscr{E}$ and $\mathscr{F}$, as well as the deficiency function ε, are invariant under the actions of an amenable group, then the kernel M may, under regularity conditions, be chosen invariant.

Several of the results on comparison of measure families are more or less straightforward generalizations of known results from the theory of comparison of experiments. Thus we may define *the deficiency* $\delta_k(\mathscr{E}, \mathscr{F})$ *of* $\mathscr{E}$ *with respect to* $\mathscr{F}$ *for k-point sets* (*k-decisions when* $\mathscr{E}$ *and* $\mathscr{F}$ *are experiments*) as the smallest (it exists) constant ε such that $\mathscr{E}$ is ε-deficient with respect to $\mathscr{F}$ for k-point sets. Symmetrizing the deficiency we define *the deficiency distance* $\Delta_k(\mathscr{E}, \mathscr{F})$ *between the measure families* $\mathscr{E} = (\mu_\theta : \theta \in \Theta)$ *and* $\mathscr{F} = (\nu_\theta : \theta \in \Theta)$ *for k-point sets* as the largest of the numbers $\delta_k(\mathscr{E}, \mathscr{F})$ and $\delta_k(\mathscr{F}, \mathscr{E})$.

If $\delta_k(\mathscr{E}, \mathscr{F}) = 0$ then we shall say that $\mathscr{E}$ *majorizes* $\mathscr{F}$ *for k-point sets* and write this $\mathscr{E} \geqslant_k \mathscr{F}$. Then $\geqslant_k$ is an ordering for measure families and this ordering extends the notion of one experiment being at least as informative as another for k-decision problems. Similarly we shall say that $\mathscr{E}$ and $\mathscr{F}$ are equivalent for k-point sets if $\mathscr{E} \geqslant_k \mathscr{F}$ and $\mathscr{F} \geqslant_k \mathscr{E}$, i.e. if $\Delta_k(\mathscr{E}, \mathscr{F}) = 0$.

The qualification 'for k-point sets' as well as the subscript k may be omitted when $k = \infty$.

Although most of the known results from the theory of comparison of experiments generalize easily, there are a few surprises. Thus equivalence for two-point sets (i.e. for testing problems in the case of experiments) no longer implies full equivalence. Equivalence for three-point sets does however imply full equivalence and this in turn is equivalent to the condition that the vector lattices generated by the measure families are isometrically isomorphic by a correspondence (and hence *the* correspondence) making the θth measures correspond to each other for each $\theta \in \Theta$. An exposition of these as well as some other results on measure families is given in the next section.

9.3 Miscellaneous. Standard measures, comparison for two-point sets, sufficiency

In this section we will extend some of the known results for comparison of experiments to the case of general measure families. However, the particular case of experiments was not always discussed earlier in this work. This applies essentially to those results which are related to comparison problems for testing problems. Most of this material is described in example 9.3.13 and in corollaries 9.3.9, 9.3.12, 9.3.20, 9.3.22, 9.3.24, 9.3.26, 9.3.27, 9.3.29, 9.3.30, 9.3.32.

After having looked over these corollaries, the reader might proceed to section 9.4 and then return to this section when some of the results here are needed.

Let us begin by generalizing the concepts of a standard measure and of a standard experiment.

Consider a measure family $\mathscr{E} = (\mathscr{X}, \mathscr{A}; \mu_\theta : \theta \in \Theta)$ with finite parameter set Θ. Put $\mu = \sum_\theta |\mu_\theta|$ and let f_θ, for each θ, denote a version of $\mathrm{d}\mu_\theta/\mathrm{d}\mu$. Put $f = (f_\theta : \theta \in \Theta)$. Thus f is a map from $\mathscr{X}$ to the finite dimensional vector space $\mathbb{R}^\theta$. Let K be the set of functions z in $\mathbb{R}^\theta$ such that $\sum_\theta |z_\theta| = 1$. Thus K may be identified with the set of finite measures on Θ possessing total variation 1.

Clearly $f(x) \in K$ for μ-almost all x. We define *the standard family* $\hat{\mathscr{E}}$ *of* $\mathscr{E}$ as the measure family induced on K by f. Thus we may write $\hat{\mathscr{E}} = (\mu_\theta f^{-1} : \theta \in \Theta)$. The measure $S_\mathscr{E} = \mu f^{-1}$ will be called *the standard measure of* $\mathscr{E}$. It is then easily checked that neither the standard family $\hat{\mathscr{E}}$ of $\mathscr{E}$, nor the standard measure $S_\mathscr{E}$ of $\mathscr{E}$, depends on how the densities f_θ; $\theta \in \Theta$ are specified.

It is also easily checked that a measure family $\mathscr{E} = (S_\theta : \theta \in \Theta)$ on K is a standard family of some measure family if and only if $x \to x_\theta$, for each θ, is a version of $\mathrm{d}S_\theta/\mathrm{d}\sum_\theta |S_\theta|$. In fact if this condition is satisfied then $\mathscr{E}$ is its own standard family. Thus $\hat{\hat{\mathscr{E}}} = \mathscr{E}$ for any measure family $\mathscr{E} = (\mu_\theta : \theta \in \Theta)$ with finite parameter set Θ. Furthermore the standard measure $S_\mathscr{E}$ of $\mathscr{E}$ determines the standard family $(S_\theta : \theta \in \Theta)$ of $\mathscr{E}$ since $S = \sum_\theta |S_\theta|$. Clearly any non-negative and finite measure S on K is the standard measure of some measure family. In fact S is the standard measure of $(S_\theta : \theta \in \Theta)$ where $x \to x_\theta$, for each θ, is a version of $\mathrm{d}S_\theta/\mathrm{d}S$.

Standard measures as well as standard families determine measure families up to Δ-equivalence.

Theorem 9.3.1 (Equivalent standard families are equal). *Let $\mathscr{E} = (\mu_\theta : \theta \in \Theta)$ and $\mathscr{F} = (\nu_\theta : \theta \in \Theta)$ be two measure families having the same finite parameter set Θ. Then the following conditions are all equivalent.*

(i) $\Delta(\mathscr{E}, \mathscr{F}) = 0$.

(ii) *$\mathscr{E}$ and $\mathscr{F}$ have the same standard family.*

(iii) *$\mathscr{E}$ and $\mathscr{F}$ have the same standard measure.*

(iv) $\delta_3(\mathscr{E}, \mathscr{F}) = \delta_2(\mathscr{F}, \mathscr{E}) = 0$.
(v) $\Delta_3(\mathscr{E}, \mathscr{F}) = 0$.

Furthermore, if there are points $\theta_0, \theta_1 \in \Theta$ such that $0 < \mu_{\theta_0} \gg \mu_\theta$ when $\theta \neq \theta_1$, then these conditions are all equivalent to:

(vi) $\Delta_2(\mathscr{E}, \mathscr{F}) = 0$.

Proof. Conditions (ii) and (iii) are equivalent since standard families and standard measures determine each other. Let $\hat{\mathscr{E}} = (S_\theta : \theta \in \Theta)$ and $\hat{\mathscr{F}} = (T_\theta : \theta \in \Theta)$ denote the standard families of $\mathscr{E}$ and $\mathscr{F}$ respectively. Thus $S = \sum_\theta |S_\theta|$ and $T = \sum_\theta |T_\theta|$ are, respectively, the standard measures of $\mathscr{E}$ and $\mathscr{F}$. If ψ is a maximum of a finite set of linear functionals on $\mathbb{R}^\theta$ then $\int \psi(\mathrm{d}\mathscr{E}) = \int \psi(\mathrm{d}\hat{\mathscr{E}})$ and $\int \psi(\mathrm{d}\mathscr{F}) = \int \psi(\mathrm{d}\hat{\mathscr{F}})$. Thus (iii) $\Rightarrow$ (i) and clearly (i) $\Rightarrow$ (v) $\Rightarrow$ (iv).

Therefore, in order to establish the equivalence of conditions (i)–(v), it suffices to show that (iv) $\Rightarrow$ (iii). Assume then that (iv) holds and let l and ψ be two linear functionals on $\mathbb{R}^\Theta$. Put $\phi_b = l^+ \vee b\psi$; $b \geqslant 0$. Then $[\phi_b - \phi_0]/b \to I_{]-\infty,0]}(l)\psi^+$ as $b \downarrow 0$. Using the fact that ϕ_b is a maximum of 3 or 2 linear functionals as $b > 0$ or $b = 0$, we find that $\int \phi_b \, \mathrm{d}S \geqslant \int \phi_b \, \mathrm{d}T$ while $\int \phi_0 \, \mathrm{d}S = \int \phi_0 \, \mathrm{d}T$. Hence $\int [(\phi_b - \phi_0)/b] \, \mathrm{d}(S - T) \geqslant 0$. Letting $b \downarrow 0$ we find that $\int_{l \leqslant 0} \psi^+ \mathrm{d}(S - T) \geqslant 0$. Substituting $x \to \pm x_{\theta_0}$ for ψ this yields $S_{\theta_0}^{\pm}(l \leqslant 0) \geqslant T_{\theta_0}^{\pm}(l \leqslant 0)$. If $l(x) \equiv_x (a, x)$ or $l(x) \equiv_x -(a, x)$ where $a \in \mathbb{R}^\Theta$ then this shows that

$$S_{\theta_0}^+\left(\left\{x : \sum_{\theta \neq \theta_0} a_\theta (x_\theta / x_{\theta_0}) \leqslant -a_{\theta_0}\right\}\right) \geqslant \text{same expression for } \mathscr{F}$$

and

$$S_{\theta_0}^+\left(\left\{x : \sum_{\theta \neq \theta_0} a_\theta (x_\theta^- / x_{\theta_0}) \geqslant -a_{\theta_0}\right\}\right) \geqslant \text{same expression for } \mathscr{F}.$$

If $S_{\theta_0}^+(\{x : \sum_{\theta \neq \theta_0} a_\theta (x_\theta / x_{\theta_0}) = -a_{\theta_0}\}) = 0$ then by adding both sides of the last two inequalities we find that $\|S_{\theta_0}^+\| \geqslant \|T_{\theta_0}^+\|$. Now we know already, by Δ_2-equivalence, that $\|S_{\theta_0}^+\| = \|T_{\theta_0}^+\|$. Thus the last two inequalities, for any choice of $a \in \mathbb{R}^\theta$, are actually equalities. Put $h(x) = (x_\theta / x_{\theta_0} : \theta \neq \theta_0)$. We have then shown that $S_{\theta_0}^+(h \in U) = T_{\theta_0}^+(h \in U)$ for any halfplane U in $\mathbb{R}^{\Theta - \{\theta_0\}}$. Thus this equality extends to any Borel set U in $\mathbb{R}^{\Theta - \{\theta_0\}}$. (We have used the fact that the class of halfplanes is a determining class for finite measures on $\mathbb{R}^\Theta$. This is usually argued by first showing that the Fourier transform of a given finite measure is determined by the masses this measure assigns to the halfplanes, and then using the uniqueness theorem for Fourier transforms.) If $x_{\theta_0} > 0$ then x may be recovered from $h(x)$ since $x_\theta = [1 + \sum_{\theta \neq \theta_0} |x_\theta / x_{\theta_0}|]^{-1} (x_\theta / x_{\theta_0})$. Thus $S_{\theta_0}^+ = T_{\theta_0}^+$. Similarly $S_{\theta_0}^- = T_{\theta_0}^-$ so that $S_{\theta_0} = T_{\theta_0}$. Since θ_0 was an arbitrary point in Θ, it follows that (ii), and hence (iii), holds.

Finally assume that $\Delta_2(\mathscr{E}, \mathscr{F}) = 0$ and that $0 \leqslant \mu_{\theta_0} \gg \mu_\theta$ when $\theta \neq \theta_1$. Then

$$\int \left(\sum a_\theta x_\theta\right)^+ S(\mathrm{d}x) = \int \left(\sum a_\theta x_\theta\right)^+ T(\mathrm{d}x); \qquad a \in \mathbb{R}^\Theta. \tag{9.3.1}$$

Taking, respectively, the right hand and the left hand partial derivative with respect to a_θ, we get

$$\int_{\sum a_\theta x_\theta > 0} x_\theta S(\mathrm{d}x) + \int_{\sum a_\theta x_\theta = 0} x_\theta^+ S(\mathrm{d}x) = \text{same expression in } T \tag{9.3.2}$$

$$\int_{\sum a_\theta x_\theta > 0} x_\theta S(\mathrm{d}x) - \int_{\sum a_\theta x_\theta = 0} x_\theta^- S(\mathrm{d}x) = \text{same expression in } T. \tag{9.3.3}$$

By Δ_2-equivalence and since $S_{\theta_0} \geqslant 0$

$$\int x_{\theta_0}^- T(\mathrm{d}x) = \int x_{\theta_0}^- S(\mathrm{d}x) = 0.$$

Hence $T_{\theta_0} \geqslant 0$. Subtracting (9.3.3) from (9.3.2) yields

$$\int_{\sum a_\theta x_\theta = 0} |x_\theta| \, S(\mathrm{d}x) = \int_{\sum a_\theta x_\theta = 0} |x_\theta| \, T(\mathrm{d}x).$$

In particular

$$\int_{x_{\theta_0} = 0} |x_\theta| \, T(\mathrm{d}x) = \int_{x_{\theta_0} = 0} |x_\theta| \, S(\mathrm{d}x) = 0 \quad \text{when } \theta \neq \theta_1.$$

It follows that $T_{\theta_0} \gg T_\theta$ when $\theta \neq \theta_1$. By (9.3.2)

$$S_{\theta_0}\left(\sum_{\theta \neq \theta_0} a_\theta (x_\theta / x_{\theta_0}) > a_{\theta_0}\right) = \text{same expression in } T$$

for all $a \in \mathbb{R}^\Theta$. It follows, as above, that $x \to x_\theta / x_{\theta_0}$; $\theta \neq \theta_0$ has the same distribution under S_{θ_0} as under T_{θ_0} so that $S_{\theta_0} = T_{\theta_0}$. Hence S and T are equal on $\{x : x_{\theta_0} > 0\}$, and we have seen that

$$x_\theta = 0 \text{ for } (S + T)\text{-almost all } x \text{ on } \{x : x_{\theta_0} = 0\} \text{ when } \theta \neq \theta_1.$$

It follows that the restrictions of S and T to $\{x : x_{\theta_0} = 0\}$ are concentrated on the two-point set $\{v, w\}$ where $v_\theta = w_\theta = 0$ when $\theta \neq \theta_1$ and $v_{\theta_1} = -w_{\theta_1} = 1$. Now

$$S(\{v\}) + S(\{w\}) = S(x_{\theta_0} = 0) = \|S\| - S(x_{\theta_0} > 0).$$

Δ_2-equivalence implies that $\|S\| = \|T\|$. Hence $S(\{v\}) + S(\{w\}) = \|T\| - T(\{x : x_{\theta_0} > 0\}) = T(\{v\}) + T(\{w\})$. Δ_1-equivalence implies that $\int x_{\theta_1} S(\mathrm{d}x) = \int x_{\theta_1} T(\mathrm{d}x)$ so that

$$S(\{v\}) - S(\{w\}) = \int_{x_{\theta_0}=0} x_{\theta_1} S(dx) = \int x_{\theta_1} S(dx) - \int_{x_{\theta_0}>0} x_{\theta_1} S(dx)$$
$$- \int x_{\theta_1} T(dx) - \int_{x_{\theta_0}>0} x_{\theta_1} T(dx) = T(\{v\}) - T(\{w\}).$$

It follows that $S(\{v\}) = T(\{v\})$ and $S(\{w\}) = T(\{w\})$ so that $S = T$. □

The last part of this proof might have been simplified somewhat by appealing to the criteria for Δ_2-equivalence given later as theorem 9.3.16.

Example 9.3.2 (Δ_2-equivalence does not imply equivalence). Let μ_0, μ_1, μ_2, ν_0, ν_1, ν_2 be given by the matrix

	1	2	3	4
μ_0	0	0	0	1
μ_1	0	$\frac{1}{4}$	$-\frac{1}{4}$	0
μ_2	$\frac{1}{4}$	0	$-\frac{1}{4}$	0
ν_0	0	0	0	1
ν_1	0	$-\frac{1}{4}$	$\frac{1}{4}$	0
ν_2	$-\frac{1}{4}$	0	$\frac{1}{4}$	0

Then

$$\|a_0\mu_0 + a_1\mu_1 + a_2\mu_2\| = |a_0| + \|a_1\mu_1 + a_2\mu_2\| = |a_0| + \|a_1\nu_1 + a_2\nu_2\|$$
$$= \|a_0\nu_0 + a_1\nu_1 + a_2\nu_2\|$$

and $\mu_i(\mathscr{X}) = \nu_i(\mathscr{X})$; $i = 0, 1, 2$. It follows that $\Delta_2((\mu_0, \mu_1, \mu_2), (\nu_0, \nu_1, \nu_2)) = 0$. On the other hand $\|\mu_1 \vee \mu_2 \vee 0\| = \frac{1}{2} = 2\|\nu_1 \vee \nu_2 \vee 0\|$ so that

$$\Delta_3((\mu_0, \mu_1, \mu_2), (\nu_0, \nu_1, \nu_2)) \geqslant \Delta_3((\mu_1, \mu_2), (\nu_1, \nu_2)) > 0.$$

This shows not only that Δ_2-equivalence does not imply full (i.e. Δ-) equivalence, but also that the assumption that $\mu_{\theta_0} \gg \mu_\theta$ when $\theta \neq \theta_1$ cannot be deleted in the last statement of the theorem.

Removing the condition that the set Θ is finite, we obtain the following criteria for equivalence.

Corollary 9.3.3 (Equivalence criteria). *The following conditions are equivalent for measure families $\mathscr{E} = (\mu_\theta : \theta \in \Theta)$ and $\mathscr{F} = (\nu_\theta : \theta \in \Theta)$.*

(i) $\Delta(\mathscr{E}, \mathscr{F}) = 0$.
(ii) $\delta_3(\mathscr{E}, \mathscr{F}) = \delta_2(\mathscr{F}, \mathscr{E}) = 0$.
(iii) $\Delta_3(\mathscr{E}, \mathscr{F}) = 0$.

(iv) $\int h(\mathrm{d}\mu_\theta : \theta \in \Theta) = \int h(\mathrm{d}\nu_\theta : \theta \in \Theta)$ *for each non-negatively homogeneous measurable function h such that one of these integrals exists.*

Furthermore, *if there are points* θ_0, θ_1 *in* Θ *such that* $0 \leqslant \mu_{\theta_0} \gg \mu_\theta$ *when* $\theta \neq \theta_1$ *then these conditions are equivalent to*:

(v) $\Delta_2(\mathscr{E}, \mathscr{F}) = 0$.

Proof. The theorem implies that conditions (i), (ii) and (iii) are equivalent and that these conditions are equivalent to (v) under the stated condition. If Θ is finite and if $\mathscr{E}$ and $\mathscr{F}$ possess standard measures S and T respectively, then the integrals in (iv) may be written $\int h\,\mathrm{d}S$ and $\int h\,\mathrm{d}T$ respectively. Thus (iv) states that $S = T$, i.e. condition (iii) in the theorem. This shows also that condition (iv) implies conditions (i)–(iii) in any case. It remains to show that condition (iv) holds provided it is satisfied for the restrictions $\mathscr{E}|\Theta_0$ and $\mathscr{F}|\Theta_0$ for any non-empty finite subset Θ_0. Now measurability implies that there is a countable subset Θ_0 of Θ such that $h(x)$ depends on x via $x|\Theta_0$. It follows that we may assume that $\Theta = \{1, 2, \ldots\}$ and that (iv) holds for Θ replaced by $\Theta_m = \{1, \ldots, m\}$ for any positive integer m. Put $\mu = \sum_{i=1}^{\infty} c_i|\mu_i|$ and $\nu = \sum_{i=1}^{\infty} c_i|\nu_i|$ where $\sum_i c_i\|\mu_i\| = \sum_i c_i\|\nu_i\| < \infty$. Put $f_i = \mathrm{d}\mu_i/\mathrm{d}\mu$ and $g_i = \mathrm{d}\nu_i/\mathrm{d}\nu$. Also put $f = (f_1, f_2, \ldots)$ and $g = (g_1, g_2, \ldots)$. Then our assumptions imply that the 'standard measures' μf^{-1} and νg^{-1} are equal and this in turn implies (iv). □

Although Δ_2-equivalence for general measure families does not imply full equivalence, this is nevertheless so for statistical experiments.

Corollary 9.3.4 (Equivalence of non-negative measure families). *If* $\mathscr{E} = (\mu_\theta : \theta \in \Theta)$ *and* $\mathscr{F} = (\nu_\theta : \theta \in \Theta)$ *are two measure families and if the measures* $\mu_\theta : \theta \in \Theta$ *are all non-negative then* $\Delta(\mathscr{E}, \mathscr{F}) = 0$ *if and only if* $\Delta_2(\mathscr{E}, \mathscr{F}) = 0$.

Proof. Assume that $\Delta_2(\mathscr{E}, \mathscr{F}) = 0$ and that Θ is finite. Put $\mu = \sum_\theta \mu_\theta$, $\nu = \sum_\theta \nu_\theta$, $\tilde{\mathscr{E}} = (\mu, \mu_\theta : \theta \in \Theta)$ *and* $\tilde{\mathscr{F}} = (\nu, \nu_\theta : \theta \in \Theta)$. Then $0 \leqslant \mu \gg \mu_\theta : \theta \in \Theta$ and $\Delta_2(\tilde{\mathscr{E}}, \tilde{\mathscr{F}}) = 0$. Hence $0 = \Delta(\tilde{\mathscr{E}}, \tilde{\mathscr{F}}) \geqslant \Delta(\mathscr{E}, \mathscr{F})$. □

Equivalence of measure families may be expressed in terms of experiments.

Corollary 9.3.5 (Reduction to non-negative measure families). *The measure families* $\mathscr{E} = (\mu_\theta : \theta \in \Theta)$ *and* $\mathscr{F} = (\nu_\theta : \theta \in \Theta)$ *are equivalent if and only if the measure families* $\check{\mathscr{E}} = (\mu_\theta^+ : \theta \in \Theta, \mu_\theta^- : \theta \in \Theta)$ *and* $\check{\mathscr{F}} = (\nu_\theta^+ : \theta \in \Theta, \nu_\theta^- : \theta \in \Theta)$ *are equivalent. Thus* $\mathscr{E}$ *and* $\mathscr{F}$ *are equivalent if and only if* $\check{\mathscr{E}}$ *and* $\check{\mathscr{F}}$ *are equivalent for* 2*-point sets.*

Proof. This follows readily from criterion (iv) in corollary 9.3.3. □

It follows in particular that problems on sufficiency may be reduced to the case of experiments.

Theorem 9.3.6 (Sufficiency reduction of measure families). *Let* $(\mathscr{E} = \mathscr{X}, \mathscr{A}; \mu_\theta : \theta \in \Theta)$ *be a measure family. Then the following conditions are equivalent for a sub σ-algebra* $\mathscr{B}$ *of* $\mathscr{A}$.

(i) $\Delta(\mathscr{E}, \mathscr{E}|\mathscr{B}) = 0$.
(ii) $\Delta_2(\mathscr{E}, \mathscr{E}|\mathscr{B}) = 0$.
(iii) $\Delta_2((\mu_{\theta_1}, \mu_{\theta_2}), (\mu_{\theta_1}|\mathscr{B}, \mu_{\theta_2}|\mathscr{B})) = 0; \theta_1, \theta_2 \in \Theta$.
(iv) $\Delta_2(\tilde{\mathscr{E}}, \tilde{\mathscr{E}}|\mathscr{B}) = 0$ *where* $\tilde{\mathscr{E}} = (\mu_\theta^+ : \theta \in \Theta; \mu_\theta^- : \theta \in \Theta)$.
(v) (a) *The Hahn set of each measure* μ_θ *may be specified* $\mathscr{B}$*-measurable.*
(b) *To each set A in* $\mathscr{A}$ *there corresponds a* $\mathscr{B}$*-measurable function Z such that* $\int_B Z \, d\mu \equiv_\theta \mu_\theta(AB)$ *for all sets B in* $\mathscr{B}$.

If $\mathscr{E}$ *is dominated and if* π *is a dominating probability measure of the form* $\sum_\theta c_\theta |\mu_\theta|$ *where* $c_\theta \geq 0$ *for all* θ, *then these conditions are equivalent to:*

(vi) $d\mu_\theta/d\pi$ *may be specified* $\mathscr{B}$*-measurable for each* θ.

Proof. The implications (i) ⇒ (ii) ⇒ (iii) are clear. Assume that condition (iii) holds. Then it follows from criterion (iv) of theorem 9.3.1 that

$$\Delta((\mu_{\theta_1}, \mu_{\theta_2}), (\mu_{\theta_1}|\mathscr{B}, \mu_{\theta_2}|\mathscr{B})) = 0 \quad \text{when } \theta_1, \theta_2 \in \Theta.$$

By Δ_1-equivalence the total variation of a measure μ_θ is equal to the total variation of its restriction $(\mu_\theta|\mathscr{B})$ of μ_θ to $\mathscr{B}$. Thus the Hahn set of μ_θ may be specified $\mathscr{B}$-measurable. In particular $(\mu_\theta|\mathscr{B})^\pm = (\mu_\theta^\pm|\mathscr{B})$ when $\theta \in \Theta$. By corollary 9.3.5 this implies that

$$\Delta((\mu_{\theta_1}^+, \mu_{\theta_1}^-, \mu_{\theta_2}^+, \mu_{\theta_2}^-), (\mu_{\theta_1}^+|\mathscr{B}, \mu_{\theta_1}^-|\mathscr{B}, \mu_{\theta_2}^+|\mathscr{B}, \mu_{\theta_2}^-|\mathscr{B})) = 0$$

when $\theta_1, \theta_2 \in \Theta$. It follows that the ordered measure families $\tilde{\mathscr{E}}$ and $\tilde{\mathscr{E}}|\mathscr{B}$ are pairwise equivalent where $\tilde{\mathscr{E}} = (\mu_\theta^+ : \theta \in \Theta; \mu_\theta^- : \theta \in \Theta)$. Hence, since ordered experiments which are pairwise equivalent are equivalent, (iv) holds. If condition (iv) is satisfied then, for any θ,

$$\|\mu_\theta\| = \|\mu_\theta^+ - \mu_\theta^-\| = \|(\mu_\theta^+|\mathscr{B}) - (\mu_\theta^-|\mathscr{B})\| = \|(\mu_\theta|\mathscr{B})\|$$

so that the Hahn set of μ_θ may again be specified $\mathscr{B}$-measurable. Hence, by corollary 9.3.5, condition (i) holds. Thus conditions (i)–(iv) are equivalent. The equivalence of conditions (i)–(v) and the last statement of the theorem follow now by appealing to the corresponding statements for experiments. □

Corollary 9.3.7 (Equivalence of comparable measure families). *If* $\mathscr{E} \geq \mathscr{F}$ *are measure families with the same parameter set* Θ *then* $\mathscr{E} \sim \mathscr{F}$ *if and only if* $\mathscr{E}|\{\theta_1, \theta_2\} \sim \mathscr{F}|\{\theta_1, \theta_2\}$ *for each two-point set* $\{\theta_1, \theta_2\}$ *of points from* Θ.

Furthermore, if Θ *itself is a two-point set and* $\mathscr{E} \geq \mathscr{F}$ *then* $\mathscr{E} \sim \mathscr{F}$ *if and only if* $\mathscr{E} \sim_2 \mathscr{F}$.

Proof. We may assume without loss of generality that $\mathscr{E} = (\mathscr{X}, \mathscr{A}; \mu_\theta : \theta \in \Theta)$ and $\mathscr{F} = (\mathscr{Y}, \mathscr{B}; \nu_\theta : \theta \in \Theta)$ where Θ is finite and $(\mathscr{Y}, \mathscr{B})$ is Euclidean. If $\mathscr{E} \geqslant \mathscr{F}$ then $\nu_\theta \equiv_\theta \mu_\theta M$ for a Markov kernel M. Put $\hat{\mathscr{E}} = (\mathscr{X} \times \mathscr{Y}, \mathscr{A} \times \mathscr{B}; \mu_\theta \times M : \theta \in \Theta)$. Then $\mathscr{E} \sim \hat{\mathscr{E}}|(\mathscr{A} \times \{\emptyset, \mathscr{Y}\})$ and $\mathscr{F} \sim \hat{\mathscr{E}}|(\{\emptyset, \mathscr{X}\} \times \mathscr{B})$. Thus $\hat{\mathscr{E}} \geqslant \mathscr{E}$. Let N denote the Markov kernel from $\mathscr{X}$ to $\mathscr{X} \times \mathscr{Y}$ defined by $N(A \times B|x) \equiv_x I_A(x) \cdot M(B|x)$; $A \in \mathscr{A}$, $B \in \mathscr{B}$. Then $(\mu_\theta \times M) \equiv_\theta \mu_\theta N$ so that $\mathscr{E} \geqslant \hat{\mathscr{E}}$. Hence $\hat{\mathscr{E}} \sim \mathscr{E}$. It follows that $\mathscr{A} \times \{\emptyset, \mathscr{Y}\}$ is 'sufficient' for $\hat{\mathscr{E}}$, i.e. $\hat{\mathscr{E}} \sim \hat{\mathscr{E}}|(\mathscr{A} \times \{\emptyset, \mathscr{Y}\})$. If $\mathscr{E}|\{\theta_1, \theta_2\} \sim \mathscr{F}|\{\theta_1, \theta_2\}$; $\theta_1, \theta_2 \in \Theta$ then $\hat{\mathscr{E}}|\{\theta_1, \theta_2\} \sim \mathscr{F}|\{\theta_1, \theta_2\}$; $\theta_1, \theta_2 \in \Theta$ so that, by the theorem, $\{\emptyset, \mathscr{X}\} \times \mathscr{B}$ is sufficient for $\hat{\mathscr{E}}$ and thus $\mathscr{E} \sim \hat{\mathscr{E}} \sim \mathscr{F}$. This proves the first statement of the corollary. The last statement follows by the same construction from the theorem. □

The last corollary implies that if Θ contains two points then comparable measure families are equivalent if and only if they are equivalent for 2-point sets. It is also known that deficiency for dichotomies may be expressed in terms of testing problems only. We shall now prove a result which is applicable not only to experiments but also to some of those pairs of measures which arise in local comparison as well as in the theory of majorization.

Theorem 9.3.8 (Comparison of ordered pairs of measures). *Let $\mathscr{E} = (\mu_1, \mu_2)$ and $\mathscr{F} = (\nu_1, \nu_2)$ be two measure pairs such that the measures μ_1 and ν_1 are both non-negative.*

Let ε_1 and ε_2 be non-negative numbers and assume also that the measure ν_2 is non-negative when $\varepsilon_1 > 0$.

Then $\mathscr{E}$ is ε-deficient with respect to $\mathscr{F}$ if and only if $\mathscr{E}$ is ε-deficient with respect to $\mathscr{F}$ for two-point sets.

Proof. Assume that $\mathscr{E}$ is ε-deficient with respect to $\mathscr{F}$ for two-point sets. First consider the situation where $\nu_2 \geqslant 0$. Let ψ be a sublinear function on $\mathbb{R}^2$ which is a maximum of a finite set H of linear functionals. Then there is a unique minimal subset H_0 of H such that $\psi(x) = \max\{h(x) : h \in H_0\}$ for all $x \geqslant 0$. Let $h_1, \ldots, h_k$ be an enumeration of H_0 such that $h_1(0,1) \leqslant h_2(0,1) \leqslant \cdots \leqslant h_k(0,1)$. Then $h_1(0,1) < h_2(0,1) < \cdots < h_k(0,1)$ and $h_1(1,0) > h_2(1,0) > \cdots > h_k(1,0)$.

Let I_ν be the interval consisting of those numbers $z \geqslant 0$ such that $\psi(1,z) = h_\nu(1,z)$. Then, by convexity and since $h_1(0,1) < \cdots < h_k(0,1)$, $I_\nu = [\zeta_{\nu-1}, \zeta_\nu] \cap [0, \infty[$; $\nu = 1, \ldots, k$ where $\zeta_0 = 0 < \zeta_1 < \zeta_2 < \cdots < \zeta_{k-1} < \zeta_k = \infty$. If $z \in I_\nu$ then $h_1(1,z) \leqslant \cdots \leqslant h_\nu(1,z) \geqslant h_{\nu+1}(1,z) \geqslant \cdots \geqslant h_k(1,z)$. (If $z \in I_\nu$ and if $h_{\nu_1}(1,z) > h_{\nu_2}(1,z)$ where $\nu_1 < \nu_2 < \nu$ then, since $h_{\nu_1}(1,0) > h_{\nu_2}(1,0)$, $h_{\nu_1}(1,\zeta) > h_{\nu_2}(1,\zeta)$ for all $\zeta \in [0,z]$. This contradicts the fact that h_{ν_2} majorizes h_{ν_1} on $I_{\nu_2} \subseteqq [0,z]$. Thus $h_1(1,z) \leqslant \cdots \leqslant h_\nu(1,z)$ when $z \in I_\nu$. Similarly $h_\nu(1,z) \geqslant \cdots \geqslant h_k(1,z)$ when $z \in I_\nu$.) It follows that $\psi(x) = \tilde{\psi}(x)$; $x \geqslant 0$ where $\tilde{\psi} = h_1 + (h_2 - h_1)^+ + \cdots + (h_k - h_{k-1})^+$. If $z \leqslant 0$ then $h_1(1,z) \geqslant h_2(1,z) \geqslant \cdots$ yielding $\tilde{\psi}(1,z) = h_1(1,z) \leqslant \psi(1,z)$. Thus $\tilde{\psi}(x) \leqslant \psi(x)$ whenever $x = (x_1, x_2)$

where $x_1 \geqslant 0$. Furthermore

$$\begin{aligned}\tilde{\psi}(-1,0) &= -h_1(1,0) + [h_1(1,0) - h_2(1,0)] + \cdots + [h_{k-1}(1,0) - h_k(1,0)] \\ &= -h_k(1,0) \\ &= h_k(-1,0) \\ &\leqslant \psi(-1,0).\end{aligned}$$

By assumption and since $\tilde{\psi}$ is a sum of maxima of pairs of linear functions

$$\begin{aligned}\int \tilde{\psi}(\mathrm{d}\mu_1, \mathrm{d}\mu_2) &\geqslant \int \tilde{\psi}(\mathrm{d}\nu_1, \mathrm{d}\nu_2) - \sum_{i=1}^{2} \varepsilon_i[\tilde{\psi}(-e^i) + \tilde{\psi}(e^i)]/2 \\ &\quad - \sum_{i=1}^{2} [\nu_i(\mathscr{Y}) - \mu_i(\mathscr{X})][\tilde{\psi}(e^i) - \tilde{\psi}(-e^i)]/2 \\ &= \int \tilde{\psi}(\mathrm{d}\nu_1, \mathrm{d}\nu_2) - \frac{1}{2}\sum_{i=1}^{2} [\varepsilon_i + \mu_i(\mathscr{X}) - \nu_i(\mathscr{Y})]\tilde{\psi}(-e^i) \\ &\quad - \frac{1}{2}\sum_{i=1}^{2} [\varepsilon_i + \nu_i(\mathscr{Y}) - \mu_i(\mathscr{X})]\tilde{\psi}(e^i)\end{aligned}$$

where $e^1 = (1,0)$ and $e^2 = (0,1)$.

To complete the proof we need to show that $\tilde{\psi}$ may be replaced with ψ in this inequality. Now $\int \psi(\mathrm{d}\mu_1, \mathrm{d}\mu_2) \geqslant \int \tilde{\psi}(\mathrm{d}\mu_1, \mathrm{d}\mu_2)$ since $\psi \geqslant \tilde{\psi}$ on $[0, \infty[\times]-\infty, \infty[$. This argument also shows that $\psi(\pm e^2) \geqslant \tilde{\psi}(\pm e^2)$ and that $\psi(e^1) \geqslant \tilde{\psi}(e^1)$. Furthermore $\int \psi(\mathrm{d}\nu_1, \mathrm{d}\nu_2) = \int \tilde{\psi}(\mathrm{d}\nu_1, \mathrm{d}\nu_2)$ since $\psi = \tilde{\psi}$ on $[0, \infty[\times [0, \infty[$. These inequalities, together with the inequality $\psi(-e^1) \geqslant \tilde{\psi}(-e^1)$, show that it is indeed possible to substitute ψ for $\tilde{\psi}$ in the 'deficiency inequality'.

Next let us replace the assumption that $\nu_2 \geqslant 0$ with the assumption that $\varepsilon_1 = 0$. Consider a finite set G of linear functionals on $\mathbb{R}^2$ and put $\psi = \max\{g : g \in G\}$. Let G_1 be the unique minimal subset of G such that $\psi(x) = \max\{g(x) : g \in G_1\}$ for all $x = (x_1, x_2)$ such that $x_1 \geqslant 0$. Let $g_1, \ldots, g_r$ be an enumeration of G_1 such that $g_1(0,1) \leqslant g_2(0,1) \leqslant \cdots \leqslant g_r(0,1)$. Then $g_1(0,1) < g_2(0,1) < \cdots < g_r(0,1)$. Note however that in this case we are not allowed to conclude that $g_1(1,0) \geqslant \cdots \geqslant g_r(1,0)$. Put $\hat{\psi} = g_1 + (g_2 - g_1)^+ + \cdots + (g_r - g_{r-1})^+$. Then $\hat{\psi}(x) = \psi(x)$ when $x = (x_1, x_2)$ where $x_1 \geqslant 0$. Furthermore, since $\mathscr{E}$ is $(0, \varepsilon_2)$-deficient with respect to $\mathscr{F}$ for two-point sets

$$\begin{aligned}\int \hat{\psi}(\mathrm{d}\mu_1, \mathrm{d}\mu_2) &\geqslant \int \hat{\psi}(\mathrm{d}\nu_1, \mathrm{d}\nu_2) - \tfrac{1}{2}[\varepsilon_2 + \mu_2(\mathscr{X}) - \nu_2(\mathscr{Y})]\hat{\psi}(-e^2) \\ &\quad - \tfrac{1}{2}[\varepsilon_2 + \nu_2(\mathscr{Y}) - \mu_2(\mathscr{X})]\hat{\psi}(e^2).\end{aligned}$$

Thus it remains to show that $\hat{\psi}$ may be replaced with ψ here. This follows since $\int \psi(\mathrm{d}\mu_1, \mathrm{d}\mu_2) = \int \hat{\psi}(\mathrm{d}\mu_1, \mathrm{d}\mu_2)$, $\int \psi(\mathrm{d}\nu_1, \mathrm{d}\nu_2) = \int \hat{\psi}(\mathrm{d}\nu_1, \mathrm{d}\nu_2)$, $\psi(-e^2) = \hat{\psi}(-e^2)$ and $\psi(e^2) = \hat{\psi}(e^2)$. □

Recall that a dichotomy is an experiment $\mathscr{E}$ whose parameter set is a two-point set. Thus the theorem applies to dichotomies.

Corollary 9.3.9 (ε-deficiency of dichotomies). *If $\mathscr{E}$ and $\mathscr{F}$ are dichotomies possessing the same parameter set then $\mathscr{E}$ is ε-deficient with respect to $\mathscr{F}$ if and only if $\mathscr{E}$ is ε-deficient with respect to $\mathscr{F}$ for testing problems.*

As any polygon is a sum of triangles, in the proof of the theorem we can restrict our attention to functions ψ which are maxima of three linear functionals. Thus the number k in the first part of the proof and the number r in the last part of the proof could have been assumed to be either 1, 2 or 3.

Example 9.3.10 (Δ_2-equivalent measure pairs need not be equivalent). If μ_1, μ_2, ν_1, ν_2 are given as in example 9.3.2 then $0 = \Delta_2((\mu_1, \mu_2), (\nu_1, \nu_2)) < \Delta_3((\mu_1, \mu_2), (\nu_1, \nu_2))$. Thus the conclusion of the last theorem may not hold if we omit the assumption that the measures μ_1 and ν_1 are non-negative.

Deficiencies for measure pairs with respect to two-point sets may be expressed conveniently in terms of integrals of concave functions.

Proposition 9.3.11 (Deficiency inequalities for measure pairs). *$\mathscr{E} = (\mathscr{X}, \mathscr{A}, \mu_1, \mu_2)$ is $(\varepsilon_1, \varepsilon_2)$-deficient with respect to $\mathscr{F} = (\mathscr{Y}, \mathscr{B}, \nu_1, \nu_2)$ for two-point sets if and only if $|\mu_i(\mathscr{X}) - \nu_i(\mathscr{Y})| \leqslant \varepsilon_i$; $i = 1, 2$ and*

$$\int (t\, d\mu_1) \wedge d\mu_2 \leqslant \int (t\, d\nu_1) \wedge d\nu_2 + \tfrac{1}{2}\varepsilon_1 |t| + \tfrac{1}{2}\varepsilon_2 + [\mu_1(\mathscr{X}) - \nu_1(\mathscr{Y})]\tfrac{1}{2}t + [\mu_2(\mathscr{X}) - \nu_2(\mathscr{Y})]\tfrac{1}{2}$$

for each real number t.

If the measures μ_1, μ_2, ν_1 and ν_2 are all non-negative and if $\mu_i(\mathscr{X}) = \nu_i(\mathscr{Y})$; $i = 1, 2$, then $\mathscr{E}$ is ε-deficient with respect to $\mathscr{F}$ if and only if

$$\int g(d\mu_2/d\mu_1)\, d\mu_1 \leqslant \int g(d\nu_2/d\nu_1)\, d\nu_1 + \lim_{z\to\infty} [g(z) - g(0) - zg'(\infty)]\tfrac{1}{2}\varepsilon_1 + [g'(0) - g'(\infty)]\tfrac{1}{2}\varepsilon_2$$

for each concave function g on $[0, \infty[$.

If the measures μ_1 and ν_1 are non-negative, $\varepsilon_1 = 0$, $\mu_i(\mathscr{X}) = \nu_i(\mathscr{Y})$; $i = 1, 2$, $\mu_1 \gg \mu_2^-$ and $\nu_1 \gg \nu_2^-$ then $\mathscr{E}$ is ε-deficient with respect to $\mathscr{F}$ if and only if

$$\int g(d\mu_2/d\mu_1)\, d\mu_1 \leqslant \int g(d\nu_2/d\nu_1)\, d\nu_1 + [g'(-\infty) - g'(\infty)]\tfrac{1}{2}\varepsilon_2$$

for each concave function g on the real line.

In the last two statements $g'(\infty)$ should be interpreted as $\lim_{z\to\infty} g'(z)$ *while $g'(-\infty)$ in the last statement should be interpreted as* $\lim_{z\to-\infty} g'(z)$.

Proof. $\mathscr{E}$ is ε-deficient with respect to $\mathscr{F}$ for two-point sets if and only if this is so for one-point sets and

$$\int \psi(\mathrm{d}\mu_1, \mathrm{d}\mu_2) \geqslant \int \psi(\mathrm{d}\nu_1, \mathrm{d}\nu_2) - \sum_{i=1}^{2} \varepsilon_i[\psi(-e^i) + \psi(e^i)]/2$$
$$- \sum_{i=1}^{2} [\nu_i(\mathscr{Y}) - \mu_i(\mathscr{X})][\psi(e^i) - \psi(-e^i)]/2$$

for each maximum ψ of two linear functionals. As equality holds trivially for each linear function ψ we may restrict our attention to functions ψ of the form $\psi(x_1, x_2) \equiv_x (tx_1) \vee x_2$. Subtracting the inequality for this choice of ψ from the equality for $\psi(x) \equiv_x tx_1 + x_2$ we arrive at the first criterion of this proposition.

From now on in this proof assume that $\mathscr{E}$ is ε-deficient with respect to $\mathscr{F}$ and that the measures μ_1 and ν_1 are non-negative and that $\mu_i(\mathscr{X}) = \nu_i(\mathscr{Y})$; $i = 1, 2$. If $\mu_1 \gg \mu_2^-$ and $\nu_1 \gg \nu_2^-$ then

$$\int t\,\mathrm{d}\mu_1 \wedge \mathrm{d}\mu_2 = \int (t \wedge \mathrm{d}\mu_2/\mathrm{d}\mu_1)\,\mathrm{d}\mu_1$$

and

$$\int t\,\mathrm{d}\nu_1 \wedge \mathrm{d}\nu_2 = \int (t \wedge \mathrm{d}\nu_2/\mathrm{d}\nu_1)\,\mathrm{d}\nu_1.$$

Put $g_t(z) = t \wedge z$; $z \geqslant 0$ for all $t \geqslant 0$ and assume that the measures μ_2 and ν_2 are non-negative. Then $\lim_{z\to\infty} [g_t(z) - g_t(0) - zg_t'(\infty)] = t$ while $g_t'(0) - g_t'(\infty) = 1$ when $t > 0$. Thus the second inequality with $g = g_t$ reduces to the first inequality. It follows, by linearity, that the second inequality of the proposition holds for any sum of a linear function and a linear combination of functions g_t with positive coefficients. A concave function on $[0, \infty[$ is of this form if and only if there are points $\xi_0 = 0 < \xi_1 < \cdots < \xi_{r-1} < \xi_r = \infty$ such that g is linear on each interval $[\xi_{i-1}, \xi_i]$; $i = 1, \ldots, r$. (We actually proved this in the proof of theorem 9.3.8.) The second statement follows now by approximating with such functions.

Finally assume that $\mu_1 \gg \mu_2^-$, $\nu_1 \gg \nu_2^-$ and that $\varepsilon_1 = 0$. Put $h_t(z) = z \wedge t$ for all real numbers z and t. Then the last inequality of the proposition with $g = h_t$ reduces to the first inequality. It follows again by linearity that the last inequality of the proposition holds for any sum $f + \sum_{i=1}^{m} c_i h_{t_i}$ where f is linear and the constants c_i are positive. (Note, by letting $t \to \infty$, that $\mu_{2a}(\mathscr{X}) = \nu_{2a}(\mathscr{Y})$ where $\mu_{2a}(\nu_{2a})$ is the μ_1-absolutely (ν_1-absolutely) continuous part of $\mu_2(\nu_2)$.) Thus the last inequality holds for any minimum of a finite number of linear functions on $\mathbb{R}$. $\square$

Writing $\Theta = \{1,2\}$, deficiencies for dichotomies may be described simply in terms of minimum Bayes risk for the problem of testing $\theta = 1$ against $\theta = 2$ for 0–1 loss.

Corollary 9.3.12 (Deficiency inequalities for dichotomies). *If $\mathscr{E} = (P_1, P_2)$ and $\mathscr{F} = (Q_1, Q_2)$ are two dichotomies and $\varepsilon_1, \varepsilon_2 \geqslant 0$ then $\mathscr{E}$ is ε-deficient with respect to $\mathscr{F}$ if and only if*

$$\int (1-\lambda)\,\mathrm{d}P_1 \wedge \lambda\,\mathrm{d}P_2 \leqslant \int (1-\lambda)\,\mathrm{d}Q_1 \wedge \lambda\,\mathrm{d}Q_2 + [(1-\lambda)\varepsilon_1 + \lambda\varepsilon_2]/2$$

for each number $\lambda \in [0,1]$.

This in turn is equivalent to the condition that

$$\int g(\mathrm{d}P_2/\mathrm{d}P_1)\,\mathrm{d}P_1 \leqslant \int g(\mathrm{d}Q_2/\mathrm{d}Q_1)\,\mathrm{d}Q_1 + \lim_{z\to\infty}[g(z) - g(0) - zg'(\infty)]\tfrac{1}{2}\varepsilon_1 + [g'(0) - g'(\infty)]\tfrac{1}{2}\varepsilon_2$$

for each concave function g on $[0,\infty[$.

Remark. The quantity $\int (1-\lambda)\,\mathrm{d}P_1 \wedge \lambda\,\mathrm{d}P_2$ is the minimum Bayes risk in the experiment $\mathscr{E}$ for the prior $(1-\lambda, \lambda)$ in the problem of testing '$\theta = 1$' against '$\theta = 2$' for the 0–1 loss function.

Example 9.3.13 (Testing affinities). Let $\mathscr{E} = (P_0, P_1)$ be a dichotomy and consider the problem of estimating the subscript θ on P for squared error loss, and on the basis of a realization of $\mathscr{E}$. The Bayes estimator for the uniform prior is then simply $\mathrm{d}P_1/\mathrm{d}(P_0 + P_1)$. Thus the minimum Bayes risk for the uniform prior is

$$\frac{1}{2}\int \frac{\mathrm{d}P_0\,\mathrm{d}P_1}{\mathrm{d}P_0 + \mathrm{d}P_1} = \frac{1}{2}\int [\mathrm{d}P_1/\mathrm{d}P_0]/[1 + \mathrm{d}P_1/\mathrm{d}P_0]\,\mathrm{d}P_0.$$

In LeCam (1977), this quantity is called the *testing affinity of* $\mathscr{E}$ and is denoted as $\alpha(\mathscr{E})$. LeCam showed in this paper that $\delta(\mathscr{E},\mathscr{F}) \geqslant 2[\alpha(\mathscr{E}) - \alpha(\mathscr{F})]$. This inequality may be established here by applying corollary 9.3.12 to $g: x \to x/(1+x)$.

Let us consider comparison with respect to two-point sets. If the measure families are experiments then comparison for testing problems may be expressed in terms of the set of available power functions. As we now shall see this generalizes directly to measure families.

For simplicity we will assume that the parameter set Θ is finite. If $\mathscr{E} = (\mu_\theta : \theta \in \Theta)$ then we will write $\tau(\mathscr{E})$ or $\tau(\mu_\theta : \theta \in \Theta)$ for the set of functions in $\mathbb{R}^\Theta$ which are of the form $\theta \to \mu_\theta(\delta)$ for a test function δ. Thus if $\mathscr{E}$ is an experiment then $\tau(\mathscr{E})$ is the set of available power functions of tests. An interesting numerical measure of information in an experiment $\mathscr{E}$ is the $m = \#\Theta$ dimensional volume of the set $\tau(\mathscr{E})$. In example 9.7.6 we will provide closed expressions for the evaluation of the volume of $\tau(\mathscr{E})$ for any measure family $\mathscr{E}$ with finite parameter set. Comparison for two-point sets may be expressed in terms of these sets as follows.

Proposition 9.3.14 (Range criterion for deficiency for two-point sets). *If Θ is finite then $\mathscr{E} = (\mathscr{X}, \mathscr{A}; \mu_\theta : \theta \in \Theta)$ is ε-deficient with respect to $\mathscr{F} = (\mathscr{Y}, \mathscr{B}; \nu_\theta : \theta \in \Theta)$ for two-point sets if and only if*

$$|\mu_\theta(\mathscr{X}) - \nu_\theta(\mathscr{Y})| \leqslant \varepsilon_\theta; \qquad \theta \in \Theta$$

and

$$\tau(\mathscr{E}) + B_{(1/2)\varepsilon} + \tfrac{1}{2}(\nu_.(\mathscr{Y}) - \mu_.(\mathscr{X})) \supseteq \tau(\mathscr{F})$$

where $B_{(1/2)\varepsilon} = \prod_\theta [-\frac{1}{2}\varepsilon_\theta, \frac{1}{2}\varepsilon_\theta]$ and $\mu_.(\mathscr{X}) - \nu_.(\mathscr{Y})$ denotes the vector $(\mu_\theta(\mathscr{X}) - \nu_\theta(\mathscr{Y}) : \theta \in \Theta)$.

In particular it follows that if $\mu_\theta(\mathscr{X}) \equiv_\theta \nu_\theta(\mathscr{Y})$ then $\Delta_2(\mathscr{E}, \mathscr{F})/2$ is the Hausdorff distance between the sets $\tau(\mathscr{E})$ and $\tau(\mathscr{F})$ for the sup norm *distance on $\mathbb{R}^\Theta$.*

Proof. $\mathscr{E}$ is ε-deficient with respect to $\mathscr{F}$ for two-point sets if and only if $|\mu_\theta(\mathscr{X}) - \nu_\theta(\mathscr{Y})| \leqslant \varepsilon_\theta$; $\theta \in \Theta$ and $\int |\sum a_\theta \, d\mu_\theta| \geqslant \int |\sum a_\theta \, d\nu_\theta| - \sum_\theta |a_\theta| \varepsilon_\theta$; $a \in \mathbb{R}^\Theta$. The proposition follows now since the functions $a \to \|\sum_\theta a_\theta \mu_\theta\|$ and $a \to \|\sum_\theta a_\theta \nu_\theta\|$ are the support functions of, respectively, the sets $2\tau(\mathscr{E}) - \mu_.(\mathscr{X})$ and $2\tau(\mathscr{F}) - \nu_.(\mathscr{Y})$. □

The proposition applies in particular to the δ_2-ordering of measure families.

Corollary 9.3.15 (Range criterion for ordering by two-point sets). *If Θ is finite then $\mathscr{E} = (\mathscr{X}, \mathscr{A}; \mu_\theta : \theta \in \Theta) \geqslant_2 \mathscr{F} = (\mathscr{Y}, \mathscr{B}; \nu_\theta : \theta \in \Theta)$ if and only if*

$$\mu_\theta(\mathscr{X}) \underset{\theta}{\equiv} \nu_\theta(\mathscr{Y})$$

and

$$\tau(\mathscr{E}) \supseteqq \tau(\mathscr{F}).$$

Theorem 9.3.16 (Equivalence for two-point sets). *Let $\mathscr{E} = (\mathscr{X}, \mathscr{A}; \mu_\theta : \theta \in \Theta)$ and $\mathscr{F} = (\mathscr{Y}, \mathscr{B}; \nu_\theta : \theta \in \Theta)$ be two measure families with the same finite parameter set Θ. Let S denote the standard measure of $\mathscr{E}$ and let T denote the standard measure of $\mathscr{F}$. Then the following conditions are equivalent.*

(i) $\Delta_2(\mathscr{E}, \mathscr{F}) = 0$.

(ii) $\tau(\mathscr{E}) = \tau(\mathscr{F})$.

(iii) *The normed linear spaces generated by $(\mu_\theta : \theta \in \Theta)$ and by $(\nu_\theta : \theta \in \Theta)$ are isometric by an isometry which preserves total mass and which maps μ_θ into ν_θ for each θ and also maps 0 into 0.*

(iv) *$\mu_\theta(\mathscr{X}) \equiv_\theta \nu_\theta(\mathscr{Y})$ and $\int \phi(d\mu_\theta : \theta \in \Theta) = \int \phi(d\nu_\theta : \theta \in \Theta)$ for each even non-negative homogeneous function ϕ on $\mathbb{R}^\Theta$ which is bounded on bounded sets.*

(v) $\mu_\theta(\mathscr{X}) \equiv_\theta \nu_\theta(\mathscr{Y})$ *and* $\mathscr{E} \oplus (-\mathscr{E}) \sim \mathscr{F} \oplus (-\mathscr{F})$.

(vi) $\int x_\theta S(\mathrm{d}x) \equiv_\theta \int x_\theta T(\mathrm{d}x)$ *and* $T + \bar{T} = S + \bar{S}$ *where* $\bar{T}(A) = T(-A)$ *and* $\bar{S}(A) = S(-A)$ *for each Borel set* A. *(Here* $-A$ *denotes* $\{-x : x \in A\}$ *and not* A^c.*)*

(vii) $T = \bar{\rho}\bar{S} + (1 - \rho)S$ *where* ρ *is a test function such that* $\int \rho(x)x_\theta S(\mathrm{d}x) \equiv 0$ *and* $\bar{\rho}(x) \equiv_x \rho(-x)$ *and where* $\bar{S}$ *is defined in* (vi). *Here* ρ *may always be modified such that* $\bar{\rho} = 1 - \rho$.

Proof. Conditions (i) and (ii) are equivalent by corollary 9.3.15. If these conditions are satisfied then $\int \psi(\mathrm{d}\mu_\theta : \theta \in \Theta) = \int \psi(\mathrm{d}\nu_\theta : \theta \in \Theta)$ when $\psi = l$ or $\psi = |l|$ where l is a linear functional on $\mathbb{R}^\Theta$. The last case provides the isometry while the first case shows that this isometry preserves total mass. Thus (iii) holds. If (iii) holds then, by Mazur & Ulam (1932), the required isometry may be chosen linear. This in turn shows that $\int \psi(\mathrm{d}\mu_\theta : \theta \in \Theta) = \int \psi(\mathrm{d}\nu_\theta : \theta \in \Theta)$ when ψ is a maximum of at most two linear functionals. Thus conditions (i), (ii) and (iii) are all equivalent.

If these conditions are satisfied and if $a \in \mathbb{R}^\Theta$ then $\int x_\theta S(\mathrm{d}x) \equiv_\theta \int x_\theta T(\mathrm{d}x)$ and

$$\int [(\textstyle\sum a_\theta x_\theta + h x_{\theta_0})^+ - (\sum a_\theta x_\theta)^+]/h T(\mathrm{d}x) = \text{the same expression in } S.$$

$h \downarrow 0$ yields

$$T\left(\left\{x : \sum_\theta a_\theta x_\theta > 0\right\}\right) + T_{\theta_0}^+\left(\left\{x : \sum_\theta a_\theta x_\theta = 0\right\}\right) = \text{the same expression in } S.$$

Let $a_\theta : \theta \neq \theta_0$ be fixed and let N denote the countable set of numbers t such that the set $\{x : \sum_{\theta \neq \theta_0} a_\theta x_\theta = t x_{\theta_0}\}$ either has positive $|S_{\theta_0}|$ mass or has positive $|T_{\theta_0}|$ mass. If $t \notin N$ then $T_{\theta_0}(\{x : \sum_{\theta \neq \theta_0} a_\theta x_\theta > t x_{\theta_0}\}) = S_{\theta_0}(\{x : \sum_{\theta \neq \theta_0} a_\theta x_\theta > t x_{\theta_0}\})$. Here the left hand side may be written

$$T_{\theta_0}^+\left(\left\{x : \sum_{\theta \neq \theta_0} a_\theta x_\theta / x_{\theta_0} > t\right\}\right) - T_{\theta_0}^-\left(\left\{x : \sum_{\theta \neq \theta_0} a_\theta x_\theta / x_{\theta_0} < t\right\}\right)$$

$$= T_{\theta_0}^+\left(\left\{x : \sum_{\theta \neq \theta_0} a_\theta x_\theta / x_{\theta_0} > t\right\}\right) - \left[\|T_{\theta_0}^-\| - T_{\theta_0}^-\left(\left\{x : \sum_{\theta \neq \theta_0} a_\theta x_\theta > t\right\}\right)\right]$$

$$= |T_{\theta_0}|\left(\left\{x : \sum_{\theta \neq \theta_0} a_\theta x_\theta / x_{\theta_0} > t\right\}\right) + \|T_{\theta_0}^-\|.$$

Since $\|T_{\theta_0}^-\| = \|S_{\theta_0}^-\|$, it follows that $x \to \sum_{\theta \neq \theta_0} a_\theta x_\theta / x_{\theta_0}$ induces the same distribution (on the line) under $|T_{\theta_0}|$ as under $|S_{\theta_0}|$. Thus, since this holds for all numbers $a_\theta : \theta \in \Theta_0$, the map $x \to x/x_{\theta_0}$ induces the same distribution (on $\mathbb{R}^\Theta$) under $|T_{\theta_0}|$ as under $|S_{\theta_0}|$. If ϕ is a measurable non-negative function on $\mathbb{R}^\Theta$ such that $\phi(x) \equiv_x \phi(-x)$ then, since $\sum_\theta |x_\theta| = 1$ for $(S + T)$-almost all x, this yields

$$\int \phi(x)|T_{\theta_0}|(\mathrm{d}x) = \int \phi\left([x/x_{\theta_0}]/\sum_\theta |x_\theta/x_{\theta_0}|\right)|T_{\theta_0}|(\mathrm{d}x)$$
$$= \int \phi(x)|S_{\theta_0}|(\mathrm{d}x).$$

Hence, since $T = \sum_\theta |T_\theta|$ and $S = \sum_\theta |S_\theta| : \int \phi \, \mathrm{d}T = \int \phi \, \mathrm{d}S$, it follows that condition (iv) is satisfied.

Conversely (iv) implies that $\int \psi(\mathrm{d}\mu_\theta : \theta \in \Theta) = \int \psi(\mathrm{d}\nu_\theta : \theta \in \Theta)$ whenever ψ is a maximum of at most two linear functionals. Hence conditions (i)–(iv) are equivalent. If (iv) holds then $\int \phi \, \mathrm{d}(T + \bar{T}) = \int [\phi(x) + \phi(-x)] T(\mathrm{d}x) = \int \phi \, \mathrm{d}(S + \bar{S})$ for each measurable function $\phi \geqslant 0$. Thus (iv) implies (vi) and the reverse implication follows by noting that $\int \phi \, \mathrm{d}T = \int \phi \, \mathrm{d}\bar{T}$ when ϕ is bounded measurable and $\phi(x) \equiv_x \phi(-x)$. Assume that (vi) holds and put $\rho = \mathrm{d}\bar{T}/\mathrm{d}(T + \bar{T}) = \mathrm{d}\bar{T}/\mathrm{d}(S + \bar{S})$. If ϕ is bounded measurable then

$$\int \phi(\rho + \bar{\rho})\mathrm{d}(S + \bar{S}) = \int \phi(\rho + \bar{\rho})\mathrm{d}(T + \bar{T})$$
$$= \int \phi\rho \, \mathrm{d}(T + \bar{T}) + \int \bar{\phi}\rho \, \mathrm{d}(T + \bar{T})$$
$$= \int (\phi + \bar{\phi})\rho \, \mathrm{d}(T + \bar{T})$$
$$= \int (\phi + \bar{\phi}) \, \mathrm{d}\bar{T}$$
$$= \int \phi \, \mathrm{d}\bar{T} + \int \phi \, \mathrm{d}T$$
$$= \int \phi \, \mathrm{d}(T + \bar{T})$$
$$= \int \phi \, \mathrm{d}(S + \bar{S})$$

so that $\rho + \bar{\rho} = 1$ a.e. $S + \bar{S}$. Hence $T = (1 - \rho)(T + \bar{T}) = (1 - \rho)(S + \bar{S}) = \bar{\rho}\bar{S} + (1 - \rho)S$. Furthermore,

$$\int x_\theta S(\mathrm{d}x) = \int x_\theta T(\mathrm{d}x)$$
$$= \int (1 - \rho(x))x_\theta(S + \bar{S})(\mathrm{d}x)$$

$$= -\int \rho(x)x_\theta S(\mathrm{d}x) - \int \rho(x)x_\theta \bar{S}(\mathrm{d}x)$$

$$= -\int \rho(x)x_\theta S(\mathrm{d}x) + \int \rho(-x)x_\theta S(\mathrm{d}x)$$

$$= \int (1 - 2\rho(x))x_\theta S(\mathrm{d}x)$$

$$= \int x_\theta S(\mathrm{d}x) - 2\int \rho(x)x_\theta S(\mathrm{d}x)$$

so that $\int \rho(x)x_\theta S(\mathrm{d}x) \equiv_\theta 0$. Thus (vii) holds where ρ satisfies the condition $\rho + \bar{\rho} = 1$. Conversely, if (vii) holds for a test function ρ then

$$\int x_\theta T(\mathrm{d}x) = \int x_\theta \bar{\rho}(x)\bar{S}(\mathrm{d}x) + \int x_\theta(1 - \rho(x))S(\mathrm{d}x)$$

$$= -\int x_\theta \rho(x)S(\mathrm{d}x) + \int x_\theta S(\mathrm{d}x) - \int x_\theta \rho(x)S(\mathrm{d}x)$$

$$= \int x_\theta S(\mathrm{d}x)$$

and

$$T + \bar{T} = \bar{\rho}\bar{S} + (1 - \rho)S + \rho S + (1 - \bar{\rho})\bar{S} = S + \bar{S}.$$

Thus (vi) holds. The proof is now completed by noting that (v) is just a reformulation of (vi). □

If we restrict our attention to non-randomized test functions δ then we see that $\tau(\mathscr{E})$ is the closed convex hull of the range of the vector valued measure $(\mu_\theta : \theta \in \Theta)$. Actually we do not need to say 'closed' here since the range of a measure family $(\mu_\theta : \theta \in \Theta)$ is always closed and, consequently, so is its convex hull. We may therefore call $\tau(\mathscr{E})$ *the convex range of* $\mathscr{E}$. If each measure μ_θ is non-atomic and Θ is finite then the range of this vector valued measure is convex and thus coincides with the convex range.

We saw in the previous theorem that the set $\tau(\mathscr{E})$ characterizes $\mathscr{E}$ up to Δ_2-equivalence in general, and up to Δ-equivalence if it is assumed that $\mathscr{E}$ is an experiment. Some combinatorial properties of these sets are collected in the following proposition.

Proposition 9.3.17 (Combinatorial properties of the convex ranges). *Let* $\mathscr{E} = (\mathscr{X}, \mathscr{A}; \mu_\theta : \theta \in \Theta)$ *and* $\mathscr{F} = (\mathscr{Y}, \mathscr{B}; \nu_\theta : \theta \in \Theta)$ *be two measure families having the same finite parameter set* Θ. *Then*:

(i) *$\tau(t\mathscr{E}) = t\tau(\mathscr{E})$ when $t \in \mathbb{R}$ and $t\mathscr{E} = (t\mu_\theta : \theta \in \Theta)$.*

(ii) *if Ω is a finite set and if T is a linear map from $\mathbb{R}^\Theta$ to $\mathbb{R}^\Omega$ then*

$$\tau(T[\mathscr{E}]) = T[\tau(\mathscr{E})]$$

where $T[\mathscr{E}] = (\kappa_\omega : \omega \in \Omega) = T(\mu_\theta : \theta \in \Theta)$.

(iii) *if the direct sum $\mathscr{G} = \mathscr{E} \oplus \mathscr{F}$ is defined as $(\zeta, \mathscr{C}; \rho_\theta : \theta \in \Theta)$ where $\zeta = (\mathscr{X} \times \{1\}) \cup (\mathscr{Y} \times \{2\})$, $\mathscr{C} = \{(A \times \{1\}) \cup (B \times \{2\}); A \in \mathscr{A}, B \in \mathscr{B}\}$ and $\rho_\theta((A \times \{1\}) \cup (B \times \{2\})) \equiv_\theta \mu_\theta(A) + \nu_\theta(B)$, then $\tau(\mathscr{G}) = \tau(\mathscr{E}) + \tau(\mathscr{F})$.*

(iv) *if ϕ is a test function on $(\mathscr{X}, \mathscr{A})$ then*

$$\tau(\mathscr{E}) - (\mu_\theta(\phi) : \theta \in \Theta) = \tau((1 - \phi)\mu_\theta : \theta \in \Theta) + \tau((-\phi)\mu_\theta : \theta \in \Theta).$$

Proof. The proofs of statements (i)–(iii) follow readily from the definition. If δ, δ_1 and δ_2 are test functions on $(\mathscr{X}, \mathscr{A})$ then

$$\mu_\theta(\delta) - \mu_\theta(\phi) = ((1 - \phi)\mu_\theta)(\delta) + ((-\phi)\mu_\theta)(1 - \delta)$$

while

$$((1 - \phi)\mu_\theta)(\delta_1) + ((-\phi)\mu_\theta)(\delta_2) = \mu_\theta((1 - \phi)\delta_1 + \phi(1 - \delta_2)) - \mu_\theta(\phi).$$

This proves statement (iv). □

If δ is a test function then $1 - \delta$ is also a test function. It follows that the convex range $\tau(\mathscr{E})$ of $\mathscr{E} = (\mu_\theta : \theta \in \Theta)$ is symmetric about the point $(\mu_\theta(\mathscr{X})/2; \theta \in \Theta)$. Thus $\tau(\mathscr{E})$ is a compact convex set containing the origin and also a (necessarily unique) point of symmetry. If $\#\Theta \leqslant 2$ then this is all we can say, since as we shall see in a moment, any set with these properties is of the form $\tau(\mathscr{E})$. If $\#\Theta \geqslant 3$ however then the sets $\tau(\mathscr{E})$ are quite exceptional within the class of sets having the properties just described. In general the basic properties of the sets $\tau(\mathscr{E})$ are that they contain the origin and are limits, for the Hausdorff distance, of finite sums of line segments. Following Grünbaum (1967) we shall call a subset of a linear space a *zonotope* if it is a finite sum of line segments. Some properties of the class of zonotopes are collected in the following proposition.

Proposition 9.3.18 (Translates of convex ranges). *Assume that Θ is finite and equip $\mathbb{R}^\Theta$ with its Euclidean distance or any other distance describing the same uniformity.*

Then the closure of the class of zonotopes within the class of closed sets consists precisely of the class of translates of convex ranges $\tau(\mathscr{E})$ of measure families $\mathscr{E} = (\mu_\theta : \theta \in \Theta)$.

If $\#\Theta \leqslant 2$ then this closure coincides with the class of compact convex sets possessing a point of symmetry.

Furthermore, the class of zonotopes is closed as a subclass of the class of polytopes.

Remark. A set in $\mathbb{R}^\Theta$ is a polytope if it is the convex hull of a finite set of points. Some general properties of polytopes are given in chapter 2.

Proof. If $\mathscr{X} = \{1,\ldots,N\}$ and if $\mu_\theta : \theta \in \Theta$ are finite measures on $\mathscr{X}$ then $\tau(\mu_\theta : \theta \in \Theta) = \sum_{j=1}^N \langle 0, \mu_.(j)\rangle$. Putting $N = 2$ we may obtain any line segment containing 0 as a set of the form $\tau(\mathscr{E})$. It follows that any finite sum of such line segments is also of the form $\tau(\mathscr{E})$. Thus any zonotope is the translate of a set $\tau(\mathscr{E})$. On the other hand, consider any closed set C which is the limit for the Hausdorff distance of sets of the form $\tau(\mathscr{E})$. Thus $C = \lim_n \tau(\mathscr{E}_n)$ where the $\mathscr{E}_n$, for each n, have standard measure S_n. Then C is convex and $\int(\sum a_\theta x_\theta)^+ S_n(\mathrm{d}x) \to \sup\{(y,a) : y \in C\}$ for each $a \in \mathbb{R}^\Theta$. It follows that the sequence $S_1, S_2, \ldots$ of measures is bounded. Hence, since these measures are all supported by a common compact set, there is a measure S and a subsequence $S_{n'}$ such that $\int f\,\mathrm{d}S_{n'} \to \int f\,\mathrm{d}S$ for each continuous function f on $\mathbb{R}^\Theta$. Then $\int(\sum_\theta a_\theta x_\theta)^+ S(\mathrm{d}x) = \sup\{(y,a) : y \in C\}$ so that $\tau(\mathscr{E}) = C$ when $\mathscr{E}$ has standard measure S. Thus the class of sets $\tau(\mathscr{E})$ is closed and, consequently, the class of translates of such sets is also closed. Consider a set $\tau(\mathscr{E})$ where $\mathscr{E}$ has standard measure S. Then there are measure families $\mathscr{E}_n$; $n = 1,2,\ldots$ with finitely supported standard measures S_n; $n = 1,2,\ldots$ such that $\int f\,\mathrm{d}S_n \to \int f\,\mathrm{d}S$ when f is continuous on $\mathbb{R}^\Theta$. Then $\tau(\mathscr{E}_n) \to \tau(\mathscr{E})$ and $\tau(\mathscr{E}_n)$, for each n, is a zonotope. This completes the proof of the first statement.

Suppose next that $\tau(\mathscr{E})$ is a polytope with vertices $\xi^1,\ldots,\xi^N$. Then $\xi^j_\theta \equiv_{\theta,j} \mu_\theta(A_j)$ for measurable sets $A_1,\ldots,A_N$. It follows that $\tau(\mathscr{E}) = \tau(\mathscr{E}|\mathscr{B})$ where $\mathscr{B}$ is the set algebra generated by $\{A_1,\ldots,A_N\}$. Any set $\tau(\mathscr{E})$ where $\mathscr{E}$ is defined on a finite measurable space is clearly a zonotope.

The proof will now be completed by showing that any compact convex set C containing 0 and having a point of symmetry is of the form $\tau(\mathscr{E})$ when $\#\Theta = 2$.

Put $b = \inf\{x_1 : (x_1,x_2) \in C\}$ and choose $(x_1^0, x_2^0) \in C$ such that $x_1^0 = b$ and $x_2^0 = \inf\{x_2 : (x_1^0, x_2) \in C\}$. Then $D = C - (x_1^0, x_2^0)$ has the same properties. If $D = \tau(\mathscr{E})$ then $C = \tau(\mathscr{E}) + x^0 = \tau(\mathscr{E}) - (-x^0)$ and by part (iv) of proposition 9.3.17, this set is of the desired form since $0 \in C$ and therefore $-x^0 \in \tau(\mathscr{E})$. It follows that we may assume that $x_1 \geqslant 0$ when $(x_1,x_2) \in C$ and that $x_2 \geqslant 0$ when $(0,x_2) \in C$.

Put $I = [0,a]$ where $a = \sup\{x_1 : (x_1,x_2) \in C$ for some $x_2\}$. Then the point of symmetry of C is $(a/2, \kappa/2)$ for some number κ. We may assume without loss of generality that $a > 0$. Also put $\beta(\alpha) = \sup\{y : (\alpha, y) \in C\}$ when $\alpha \in I$. Then β is concave and continuous on I. It follows that β is absolutely continuous. By abuse of the notation, let β also denote the measure on I which assigns mass $\beta(\alpha)$ to each interval $[0,\alpha]$. Also let λ denote the Lebesgue measure on I. The proof will now be completed by showing that $\tau(\lambda,\beta) = C$.

For each $\alpha \in I$, put $\tilde{\beta}(\alpha) = \sup\{\int \delta(x)\beta(\mathrm{d}x) : \int \delta(x)\lambda(\mathrm{d}x) = \alpha\}$ where δ runs through the class of test functions on $I = [0, a]$. Then $\tilde{\beta}(\alpha) \geqslant \beta(\alpha)$ since $\int I_{[0,\alpha]}\,\mathrm{d}\lambda = \alpha$. The function $\tilde{\beta}$ is also concave and continuous.

If $\int \delta(x)\lambda(\mathrm{d}x) = \alpha$ and $\delta(0) - 1$ then

$$\beta(\alpha) - \int \delta(x)\beta(\mathrm{d}x) = \int (I_{[0,\alpha]}(x) - \delta(x))\beta(\mathrm{d}x)$$

$$= \int (I_{[0,\alpha]}(x) - \delta(x))(\beta'(x) - \beta'(\alpha))\,\mathrm{d}x \geqslant 0$$

since the last integrand is non-negative. It follows that $\tilde{\beta} \leqslant \beta$ so that $\tilde{\beta} = \beta$. Hence

$$\beta(\alpha) = \sup\left\{\int \delta(x)\beta(\mathrm{d}x) : \int \delta(x)\lambda(\mathrm{d}x) = \alpha\right\}.$$

By symmetry

$$\kappa - \beta(a - \alpha) = \kappa - \sup\left\{\int \delta\,\mathrm{d}\beta : \int \delta\,\mathrm{d}\lambda = a - \alpha\right\}.$$

On the other hand

$$\begin{aligned}\kappa - \beta(a - \alpha) &= \kappa - \sup\{y : (a - \alpha, y) \in C\}\\ &= \kappa - \sup\{\kappa - (\kappa - y) : (\alpha, \kappa - y) \in C\}\\ &= \kappa - \sup\{\kappa - y : (\alpha, y) \in C\}\\ &= \inf\{y : (\alpha, y) \in C\}.\end{aligned}$$

$\alpha = 0$ yields $\kappa - \beta(a) = \inf\{y : (0, y) \in C\} = 0$ so that $\kappa = \beta(a) = \beta(I)$. The set $\tau(\lambda, \beta)$ is symmetric about $(\lambda(I), \beta(I))/2 = (a/2, \kappa/2)$. Thus $\tau(\lambda, \beta)$ and C have the same point $(a/2, \kappa/2)$ as a point of symmetry. Hence $\tau(\lambda, \beta) = C = \{(\alpha, y) : \kappa - \beta(a - \alpha) \leqslant y \leqslant \beta(\alpha)\}$. □

Consider now a set C in $\mathbb{R}^\Theta$ which is a limit of a sequence of zonotopes which all contain the origin. By the proposition this set is of the form $C = \tau(\mathscr{E}) - v$ for a vector v and $v \in \tau(\mathscr{E})$ since $0 \in C$. Hence, by part (iv) of proposition 9.3.18 $C = \tau(\mathscr{F})$ for some measure family $\mathscr{F}$. Conversely any set $\tau(\mathscr{E})$, by this proposition, is clearly the limit of a sequence of zonotopes containing 0. This yields the promised characterization of the class of convex ranges of measure families.

Corollary 9.3.19 (The 'range' of a measure family). *The class of convex ranges $\tau(\mathscr{E})$ of measure families $\mathscr{E}$ having the same finite parameter set Θ is the closure of the class of zonotopes which contain* 0.

If $\#\Theta \leqslant 2$ then this class is the class of compact convex sets which contain the origin and have points of symmetry.

A polytope is of the form $\tau(\mathscr{E})$ if and only if it is a zonotope which contains the origin.

Corollary 9.3.20 ('The range of an experiment'). *The class of convex ranges $\tau(\mathscr{E})$ of experiments $\mathscr{E}$ having the same finite parameter set Θ is the closure of the class of zonotopes which are contained in $[0,1]^\Theta$ and which contain the points $(0,\dots,0)$ and $(1,\dots,1)$.*

If $\#\Theta = 2$ then this class is the class of compact convex subsets of $[0,1]^\Theta$ which contain 0 and are symmetric about $(\frac{1}{2},\frac{1}{2})$.

We saw in the proof of proposition 9.3.18 that the structure of measure families with two-point parameter sets is essentially a structure for the compact convex sets in the plane which contain the origin and possess points of symmetry. The fact that this class is closed under the formation of intersections of sets possessing the same point of symmetry yields the following proposition.

Proposition 9.3.21 (Lattice of measure pairs). *The collection of measure pairs (μ_1,μ_2) such that $\mu_1 \geqslant 0$ constitutes an order complete lattice for the ordering $\geqslant$ which states that $(\mu_1,\mu_2) \geqslant (\nu_1,\nu_2)$ when $\nu_i = \mu_i M$; $i = 1,2$ for some transition M.*

Corollary 9.3.22 (The lattice of dichotomies). *The collection of dichotomies with parameter set $\Theta = \{1,2\}$ constitutes an order complete lattice for the ordering 'being at least as informative as'.*

Greatest lower bounds of sets of dichotomies may be described conveniently in terms of greatest lower bounds of powers of most powerful tests. Similarly, smallest upper bounds of sets of dichotomies may be described in terms of greatest lower bounds of minimum Bayes risks.

We shall now consider these representations for general measure pairs and we shall begin by considering 'generalized' power. If we restrict our attention to measure pairs (μ_1,μ_2) such that $\mu_1 \geqslant 0$ and $\mu_1 \gg \mu_2^-$, then we have the following result which was established in the proof of proposition 9.3.18.

Proposition 9.3.23 (Representation by 'maximum generalized power'). *Let C be a compact convex subset of $[0,\infty[\, \times\,]-\infty,\infty[$. Assume that C possesses a point of symmetry, that $(0,0) \in C$ and that $x_2 \geqslant 0$ when $(0,x_2) \in C$.*

Then $C = \tau(\lambda,\beta)$ where λ is the Lebesgue measure on the interval $I = [0,\sup\{x_1;(x_1,x_2) \in C\}]$ and β is the measure on I given by the distribution function $\alpha \to \sup\{x_2 : (\alpha,x_2) \in C\}$.

Remark. If C is any compact convex subset of the plane which contains the origin and has a point of symmetry then, as we saw in the proof of proposition

9.3.17, there is a vector v such that the translate $C - v$ satisfies the requirements of this proposition.

Corollary 9.3.24 (Range representation of dichotomies). *A subset C of the plane $\mathbb{R}^2$ is the range $\tau(P_1, P_2)$ of a dichotomy (P_1, P_2) if and only if C is a closed convex subset of $[0,1]^2$ which contains the origin $(0,0)$ and is symmetric about $(\frac{1}{2}, \frac{1}{2})$.*

If C satisfies these conditions then $C = \tau(\mathscr{E})$ for the dichotomy $\mathscr{E} = (\lambda, \beta)$ where λ is the rectangular distribution on $[0,1]$ while β is the probability measure on $[0,1]$ given by the distribution function $\alpha \to \sup\{x_2 : (\alpha, x_2) \in C\}$.

If C is a compact subset of $\mathbb{R}^2$ and if α is a real number then we shall use the notation $\beta(\alpha|C)$ for the quantity $\sup\{x_2 : (\alpha, x_2) \in C\}$. We are then using the convention that the sup in question should be interpreted as $-\infty$ when α does not belong to the projection of C on the first coordinate axis. Let us call $\beta(\cdot|C)$ *the upper envelope function* of C.

Similarly we may define *the lower envelope function* $\alpha \to \inf\{y : (\alpha, y) \in C\}$ where we put $\inf \emptyset = \infty$. The set C may be recovered from these functions since

$$C = \{(x_1, x_2) : \inf\{y : (x_1, y) \in C\} \leqslant x_2 \leqslant \sup\{y : (x_1, y) \in C\}\}.$$

If, however, C is symmetric about $s = (s_1, s_2)$, then one of these functions determines the other, and thus the set C, since $\inf\{y : (\alpha, y) \in C\} = \inf\{y : (2s_1 - \alpha, 2s_2 - y) \in C\} = 2s_2 - \beta(2s_1 - \alpha|C)$; $\alpha \in \mathbb{R}$.

If $\mathscr{E} = (\mathscr{X}, \mathscr{A}; \mu_1, \mu_2)$ is a measure pair then we may write $\beta(\alpha|\mathscr{E})$ or $\beta(\alpha|\mu_1, \mu_2)$ instead of $\beta(\alpha|\tau(\mathscr{E}))$.

Deficiency for two-point sets for Δ_1-equivalent measure pairs may be expressed by these functions as follows.

Proposition 9.3.25 (Deficiency inequalities for envelope functions). *Let $\mathscr{E} = (\mathscr{X}, \mathscr{A}, \mu_1, \mu_2)$ and $\mathscr{F} = (\mathscr{Y}, \mathscr{B}, \nu_1, \nu_2)$ be two measure pairs such that $\mu_i(\mathscr{X}) = \nu_i(\mathscr{Y})$; $i = 1, 2$. Then $\mathscr{E}$ is $(\varepsilon_1, \varepsilon_2)$-deficient with respect to $\mathscr{F}$ for two-point sets if and only if*

$$\sup\{\beta(\alpha'|\mu_1, \mu_2) : |\alpha' - \alpha| \leqslant \tfrac{1}{2}\varepsilon_1\} + \tfrac{1}{2}\varepsilon_2 \geqslant \beta(\alpha|\nu_1, \nu_2); \qquad \alpha \in \mathbb{R}.$$

Remark. It follows from theorem 9.3.8 that the qualification 'for two-point sets' is superfluous when the measures μ_1 and ν_1 are non-negative and if either ν_2 is non-negative or if $\varepsilon_1 = 0$.

Proof. Put $f(\alpha) = \sup\{\beta(\alpha'|\mu_1, \mu_2) : |\alpha' - \alpha| \leqslant \frac{1}{2}\varepsilon_1)\} + \frac{1}{2}\varepsilon_2$; $\alpha \in \mathbb{R}$. $\mathscr{E}$ is ε-deficient with respect to $\mathscr{F}$ if and only if $\tau(\mathscr{F}) \subseteqq \tau(\mathscr{E}) + [-\frac{1}{2}\varepsilon_1, \frac{1}{2}\varepsilon_1] \times [-\frac{1}{2}\varepsilon_2, \frac{1}{2}\varepsilon_2]$ where both sides of this set inequality have the same point $(\mu_1(\mathscr{X})/2, \mu_2(\mathscr{X})/2)$ as a point of symmetry. The upper envelope function of the left hand side is $\beta(\cdot|\nu_1, \nu_2)$. Therefore it suffices to show that f is the upper envelope of the right hand side. (If C and D are compact convex sets in the plane which possesses the same point of symmetry then $C \supseteqq D$ if and only if the upper envelope function of C majorizes the upper envelope function of D.) Put $B = [-\frac{1}{2}\varepsilon_1, \frac{1}{2}\varepsilon_1] \times [-\frac{1}{2}\varepsilon_2, \frac{1}{2}\varepsilon_2]$. If $(\alpha, y) \in \tau(\mathscr{E}) + B$ then

$$\alpha = \alpha' + \rho_1$$

and

$$y = z + \rho_2$$

where $(\alpha', z) \in \tau(\mathscr{E})$ and $\rho = (\rho_1, \rho_2) \in B$. Thus $y \leqslant \beta(\alpha'|\mu_1, \mu_2) + \frac{1}{2}\varepsilon_2$ and $|\alpha' - \alpha| \leqslant \frac{1}{2}\varepsilon_1$. Then f majorizes the upper envelope function of $\tau(\mathscr{E}) + B$. On the other hand if $f(\alpha) > -\infty$ then $f(\alpha) = \beta(\alpha'_0|\mu_1, \mu_2) + \frac{1}{2}\varepsilon_2$ where $|\alpha'_0 - \alpha| \leqslant \frac{1}{2}\varepsilon_1$. Decomposing $(\alpha, f(\alpha)) = (\alpha, \beta(\alpha^0|\mu_1, \mu_2) + \frac{1}{2}\varepsilon_2)$ as $(\alpha'_0, \beta(\alpha'_0|\mu_1, \mu_2)) + (\alpha - \alpha'_0, \frac{1}{2}\varepsilon_2))$ we see that this point belongs to $\tau(\mathscr{E}) + B$. Thus f is majorized by the upper envelope function of $\tau(\mathscr{E}) + B$. It follows that f is the upper envelope function of $\tau(\mathscr{E}) + B$. □

Corollary 9.3.26 (Probabilities of errors criterion for deficiencies). *Let $\mathscr{E} = (\mu_1, \mu_2)$ and $\mathscr{F} = (\nu_1, \nu_2)$ be non-negative measure pairs such that $\|\mu_i\| = \|\nu_i\|$; $i = 1,2$. Then $\mathscr{E}$ is $(\varepsilon_1, \varepsilon_2)$-deficient with respect to $\mathscr{F}$ if and only if*

$$\beta(\alpha + \tfrac{1}{2}\varepsilon_1|\mathscr{E}) + \tfrac{1}{2}\varepsilon_2 \geqslant \beta(\alpha|\mathscr{F})$$

when $0 \leqslant \alpha \leqslant \|\mu_1\|$.

In particular this criterion is valid for dichotomies $\mathscr{E} = (\mu_1, \mu_2)$ and $\mathscr{F} = (\nu_1, \nu_2)$.

Corollary 9.3.27 (The Levy diagonal distance). *If $\mathscr{E} = (\mu_1, \mu_2)$ and $\mathscr{F} = (\nu_1, \nu_2)$ are non-negative measure pairs such that $\|\mu_i\| = \|\nu_i\|$; $i = 1, 2$ then*

$$\Delta(\mathscr{E}, \mathscr{F}) = \Lambda(\beta(\cdot|\mathscr{E}), \beta(\cdot|\mathscr{F}))/2^{1/2}$$

where $\Lambda(F, G)$ for distribution functions F and G, denotes the Levy diagonal distance between the distribution functions F and G on $[0, \|\mu_1\|]$.

In particular this holds for dichotomies $\mathscr{E} = (\mu_1, \mu_2)$ and $\mathscr{F} = (\nu_1, \nu_2)$.

Proof. This follows from the previous corollary with $\varepsilon_1 = \varepsilon_2$. □

Corollary 9.3.28 (The ordering of measure pairs for two-point sets). *If $\mathscr{E} = (\mathscr{X}, \mathscr{A}, \mu_1, \mu_2)$ and $\mathscr{F} = (\mathscr{Y}, \mathscr{B}, \nu_1, \nu_2)$ are measure pairs then $\mathscr{E} \geqslant_2 \mathscr{F}$ if and only if*

$$\mu_i(\mathscr{X}) = \nu_i(\mathscr{Y}); \qquad i = 1, 2$$

and

$$\beta(\cdot|\mathscr{E}) \geqslant \beta(\cdot|\mathscr{F}).$$

If in addition the measures μ_1 and ν_1 are non-negative then $\geqslant_2$ may be replaced here by $\geqslant$.

Applying this to dichotomies we obtain the following criterion from Blackwell (1953) for ordering of dichotomies.

Corollary 9.3.29 (Probabilities of errors criterion for ordering of dichotomies). *The dichotomy $\mathscr{E} = (P_1, P_2)$ is at least as informative as the dichotomy $\mathscr{F} = (Q_1, Q_2)$ if and only if $\beta(\cdot|\mathscr{E}) \geqslant \beta(\cdot|\mathscr{F})$.*

Corollary 9.3.30 (The Kolmogorov distance). *Let $\mathscr{E} = (\mu_1, \mu_2)$ and $\mathscr{F} = (\nu_1, \nu_2)$ be measure pairs such that $\mu_1 \geqslant 0$ and $\nu_1 \geqslant 0$. Then $2\sup\{|\beta(\alpha|\mathscr{E}) - \beta(\alpha|\mathscr{F})|;\ 0 \leqslant \alpha \leqslant \|\mu_1\|\} = \inf\{\varepsilon : \mathscr{E}$ and $\mathscr{F}$ are mutually $(0,\varepsilon)$-deficient with respect to each other$\}$.*

Infima and suprema of families of measure pairs may be represented as follows.

Theorem 9.3.31 (Lattice operations on envelope functions and support functions). *Let $\mathscr{E}_i = (\mu_{i1}, \mu_{i2})$; $i \in I$ be a family of measure pairs such that the measures μ_{i1}; $i \in I$ are all non-negative and such that the total mass of the measure μ_{i1} as well as the total mass of the measure μ_{i2} does not depend on $i \in I$. Then $\underline{\mathscr{E}} = \bigwedge_i \mathscr{E}_i$ for the ordering $\geqslant$ exists and*

$$\beta(\cdot|\underline{\mathscr{E}}) = \inf_i \beta(\cdot|\mathscr{E}_i).$$

Furthermore a smallest upper bound $\bar{\mathscr{E}} = \bigvee_i \mathscr{E}_i$ exists if and only if $\sup_i \|\mu_{i2}\| < \infty$. If this condition is satisfied and if $\bar{\mathscr{E}} = (\bar{\mu}_1, \bar{\mu}_2)$ then

$$\int (t\,d\bar{\mu}_1 \wedge d\bar{\mu}_2) = \inf_i \int (t\,d\mu_{i1} \wedge d\mu_{i2}).$$

Proof. Put $H = \bigcap_i \tau(\mathscr{E}_i)$ and let C denote the closed convex hull of $\bigcup_{i=1}^{\infty} \tau(\mathscr{E}_i)$. Let s denote the common point of symmetry for the sets $\tau(\mathscr{E}_i)$; $i \in I$. Then s is a point of symmetry for the sets H and C as well. Furthermore the origin belongs to these sets. It follows that $\tau(\underline{\mathscr{E}}) = H$ and that $\tau(\bar{\mathscr{E}}) = C$ provided $\sup_i \|\mu_{i2}\| < \infty$. If $\alpha \in [0, 2s_1]$ then the point $(\alpha, \beta(\alpha|\underline{\mathscr{E}}))$ belongs to $\tau(\mathscr{E}_i)$ for all $i \in I$. Thus $\beta(\alpha|\underline{\mathscr{E}}) \leqslant \inf_i \beta(\alpha|\mathscr{E}_i)$. On the other hand the points $(\alpha, \beta(\alpha|\underline{\mathscr{E}}))$ and $(\alpha, \beta(\alpha|\mathscr{E}_i))$ both belong to $\tau(\mathscr{E}_i)$. Thus the line segment between these points is contained in $\tau(\mathscr{E}_i)$. The point $(\alpha, \inf_i \beta(\alpha|\mathscr{E}_i))$ belongs to this segment and hence also to $\tau(\mathscr{E}_i)$ for every $i \in I$. Thus the last point also belongs to H so that $\inf_i \beta(\alpha|\mathscr{E}_i) \leqslant \beta(\alpha|\underline{\mathscr{E}})$. Assume that $\sup_i \|\mu_{i2}\| < \infty$. Let $t \in \mathbb{R}$. The support functions of the sets C and $\tau(\mathscr{E}_i)$ are, respectively, the maps $(a,b) \to \int (a\,d\bar{\mu}_1 + b\,d\bar{\mu}_2)^+$ and $(a,b) \to \int (a\,d\mu_{i1} + b\,d\mu_{i2})^+$. Thus $\int (a\,d\bar{\mu}_1 + b\,d\bar{\mu}_2)^+ \equiv_{a,b} \sup_i \int (a\,d\mu_{i1} + b\,d\mu_{i2})^+$. It follows that $\int t\,d\bar{\mu}_1 \vee d\bar{\mu}_2 = \sup_i \int t\,d\mu_{i1} \vee d\mu_{i2}$. On the other hand $\int t\,d\bar{\mu}_1 + d\bar{\mu}_2 = \int t\,d\mu_{i1} + d\mu_{i2} = 2ts_1 + 2s_2$. Thus

$$\int (t\,\mathrm{d}\bar{\mu}_1 \wedge \mathrm{d}\bar{\mu}_2) = \int [(t\,\mathrm{d}\bar{\mu}_1 + \mathrm{d}\bar{\mu}_2) - (t\,\mathrm{d}\bar{\mu}_1 \vee \mathrm{d}\bar{\mu}_2)]$$

$$= 2ts_1 + 2s_2 - \sup_i \int (t\,\mathrm{d}\mu_{i1} \vee \mathrm{d}\mu_{i2})$$

$$= \inf_i \int [(t\,\mathrm{d}\mu_{i1} + \mathrm{d}\mu_{i2}) - (t\,\mathrm{d}\mu_{i1} \vee t\,\mathrm{d}\mu_{i2})]$$

$$= \inf_i \int t\,\mathrm{d}\mu_{i1} \wedge \mathrm{d}\mu_{i2}. \qquad \square$$

Corollary 9.3.32 (Power and Bayes risk expressions for lattice operations on dichotomies). *Let $\mathscr{E}_i = (P_{i1}, P_{i2})$; $i \in I$ be a family of dichotomies. Then*

$$\beta\left(\cdot \,\middle|\, \bigwedge_i \mathscr{E}_i\right) = \bigwedge_i \beta(\cdot \,|\, \mathscr{E}_i)$$

while

$$b\left(\cdot \,\middle|\, \bigvee_i \mathscr{E}_i\right) = \bigwedge_i b(\cdot \,|\, \mathscr{E}_i)$$

where we use the notation $b(\lambda|\mathscr{E})$ for the quantity $\int (1-\lambda)\mathrm{d}P \wedge \lambda Q$ for a dichotomy $\mathscr{E} = (P, Q)$ and a number $\lambda \in [0, 1]$.

9.4 Local (fixed sample size) comparison of statistical experiments

Let us turn to local comparison within small neighbourhoods of a given point θ^0 belonging to the interior of Θ. We shall assume throughout this section that the parameter set Θ is a subset of $\mathbb{R}^m$ for some positive integer m. When reading this section the reader could return to section 1.4 and verify the unproven statements in examples 1.4.7 and in the first part of example 1.4.8.

An experiment $\mathscr{E} = (\mathscr{X}, \mathscr{A}; P_\theta : \theta \in \Theta)$ will be called *differentiable* (*in the first mean*) *at* θ^0 if the map $\theta \to P_\theta$ from Θ to the Banach space of finite measures on $\mathscr{A}$ is Frechet differentiable.

According to the definition of Frechet differentiability $\mathscr{E}$ is differentiable at θ^0 if and only if there is a linear map T from $\mathbb{R}^m$ to the space of bounded measures such that $\|P_\theta - P_{\theta^0} - T(\theta - \theta^0)\|/\|\theta - \theta^0\| \to 0$ as $\theta \to 0$. Letting $\dot{P}_{\theta^0,i}$ denote the image of the ith unit vector $(0, \ldots \overset{(i)}{1}, \ldots, 0)$ by this map, we may represent T as the map which maps $h \in \mathbb{R}^m$ into $\sum h_i \dot{P}_{\theta^0,i}$. Thus $\mathscr{E}$ is differentiable in θ^0 if and only if there are finite measures $\dot{P}_{\theta^0,1}, \ldots, \dot{P}_{\theta^0,m}$ such that $\|P_\theta - P_{\theta^0} - \sum(\theta_i - \theta_i^0)\dot{P}_{\theta^0,i}\|/\|\theta - \theta^0\| \to 0$ as $\theta \to \theta^0$. If $m = 1$ then we may write $\dot{P}_{\theta^0}$ instead of $\dot{P}_{\theta^0,i}$. Then the condition of differentiability at θ^0 reduces to the condition that $(P_\theta - P_{\theta^0})/(\theta - \theta^0)$ converges in total variation to some measure $\dot{P}_{\theta^0}$ as $\theta \to \theta^0$. We shall of course call $\dot{P}_{\theta^0}$ the *derivative* of $\mathscr{E}$ (or the map $\theta \to P_\theta$) at θ^0 and we may write $[\mathrm{d}P_\theta/\mathrm{d}\theta]_{\theta=\theta^0}$ instead of $\dot{P}_{\theta^0}$. Similarly $\dot{P}_{\theta^0,i}$ will be called the partial derivative of $\mathscr{E}$ (or of the map $\theta \to P_\theta$) at $\theta = \theta^0$ and this measure may be denoted as $[\partial P_\theta/\partial\theta_i]_{\theta=\theta^0}$.

The notion of differentiability is weaker than the notion of differentiability in quadratic mean. With the above notation the experiment $\mathscr{E} = (P_\theta : \theta \in \Theta)$ is called *differentiable in quadratic mean* if the map $\theta \to (\mathrm{d}P_\theta/\mathrm{d}P_{\theta^0})^{1/2}$ from Θ to $L_2(P_{\theta^0})$ is Frechet differentiable at θ^0 and if $\|P_{\theta,s}\|/\|\theta - \theta^0\|^2 \to 0$ as $\theta \to \theta^0$ where $P_{\theta,s}$ denotes the total mass of the P_{θ^0} singular part of P_θ. The notion of differentiability in quadratic mean leads to basic results concerning the asymptotic behaviour of replicated experiments (see LeCam (1974) and Millar (1983)). If we merely assume differentiability in the first mean, then these results need not hold.

We shall not be concerned with asymptotic theory in this sense here and so the chosen notion of differentiability appears to be appropriate, although there are many other possibilities for weaker as well as for stronger notions.

Let us return to differentiability (in the first mean) as defined above.

Adapting freely from Hajek & Sidak (1967) we obtain the following useful sufficient set of conditions for differentiability in the one-dimensional case.

Theorem 9.4.1 (Differentiability of densities). *Let $\mathscr{E} = (P_\theta : \theta \in \Theta)$ be an experiment where Θ is a subset of the real line. Then the following conditions together ensure that $\mathscr{E}$ is differentiable at an interior point θ^0 of Θ.*

(i) *There exist a positive number c and a positive σ-finite measure μ such that P_θ is defined and dominated by μ when $|\theta - \theta^0| \leqslant c$.*

(ii) *There are non-negative densities*

$$f_\theta = \mathrm{d}P_\theta/\mathrm{d}\mu : |\theta - \theta^0| \leqslant c$$

such that the maps $\theta \to f_\theta(x)$ from $[\theta^0 - c, \theta^0 + c]$ to $[0, +\infty]$ are absolutely continuous for μ-almost all x.

(iii) *The derivative $\dot{f}_\theta(x) = \lim_{h\to 0} [f_{\theta+h}(x) - f_\theta(x)]/h$ exists for μ-almost all x when $|\theta - \theta^0| < c$.*

(iv) $\limsup_{\theta\to\theta^0} \int |\dot{f}_\theta| \,\mathrm{d}\mu \leqslant \int |\dot{f}_{\theta^0}| \,\mathrm{d}\mu < \infty$.

Remark. If we replace condition (iv) with

(iv′) $\limsup_{\theta\to\theta^0} \int (\dot{f}_\theta/f_\theta)^2 \,\mathrm{d}P_\theta \leqslant \int (\dot{f}_{\theta^0}/f_{\theta^0})^2 \,\mathrm{d}P_{\theta^0} < \infty$

then (i), (ii), (iii) and (iv′) together yield a sufficient condition for differentiability in quadratic mean (see Hajek & Sidak, 1967).

Thus continuity of the Fisher information $\int (\dot{f}_\theta/f_\theta)^2 \,\mathrm{d}P_\theta$ together with conditions (i), (ii) and (iii) ensures that $\mathscr{E}$ is differentiable in quadratic mean on $]\theta^0 - c, \theta^0 + c[$.

Proof. It follows from (ii) and (iii) that there is a μ-null set N such that if $x \notin N$ then $\dot{f}_{\theta^0}(x)$ exists and $f_\theta(x)$ is absolutely continuous in $\theta \in [\theta^0 - c, \theta^0 + c]$. This last property ensures that the map $(x, \theta) \to f_\theta(x)$ is jointly measurable on $N^c \times [\theta^0 - c, \theta^0 + c]$. Put $\hat{f}_\theta(x) = \limsup_{n\to\infty} n(f_{\theta_n}(x) - f_\theta(x))$ where $\theta_n = \theta + 1/n$. Then the map $(x, \theta) \to \hat{f}_\theta(x)$ is also jointly measurable on $N^c \times]\theta^0 - c, \theta^0 + c[$. By (ii) $\hat{f}_\theta(x) = \dot{f}_\theta(x)$ for almost all (Lebesgue) θ in $[\theta^0 - c, \theta^0 + c]$, for all $x \in N^c$. For any $\theta \in [\theta^0 - c, \theta^0 + c]$ we have

$$\begin{aligned}\int \left| \frac{f_\theta(x) - f_{\theta^0}(x)}{\theta - \theta^0} \right| \mu(\mathrm{d}x) &= \int \frac{1}{|\theta - \theta^0|} \left| \int_{\langle \theta^0, \theta \rangle} \hat{f}_t(x)\, \mathrm{d}t \right| \mu(\mathrm{d}x) \\ &\leqslant \int \frac{1}{|\theta - \theta^0|} \left[\int_{\langle \theta^0, \theta \rangle} |\hat{f}_t(x)|\, \mathrm{d}t \right] \mu(\mathrm{d}x) \\ &= \frac{1}{|\theta - \theta^0|} \int_{\langle \theta^0, \theta \rangle} \phi(t)\, \mathrm{d}t\end{aligned}$$

(by Fubini) where $\phi(t) = \int |\hat{f}_t(x)| \mu(\mathrm{d}x)$. Here $\langle a, b \rangle = [a, b]$ or $= [b, a]$ as $a \leqslant b$ or $a \geqslant b$.

By (iv) $\phi(\theta) \to \phi(\theta^0)$ as $\theta \to \theta^0$. Hence also $\int_{\langle \theta^0, \theta \rangle} \phi(t)\, \mathrm{d}t / |\theta - \theta^0| \to \phi(\theta^0)$ as $\theta \to \theta^0$.

It follows that

$$\begin{aligned}\limsup_{\theta \to \theta^0} \int [|f_\theta(x) - f_{\theta^0}(x)| / |\theta - \theta^0|] \mu(\mathrm{d}x) &\leqslant \int |\hat{f}_{\theta^0}(x)| \mu(\mathrm{d}x) \\ &= \int |\dot{f}_{\theta^0}(x)| \mu(\mathrm{d}x) < \infty.\end{aligned}$$

Scheffé's convergence theorem (Scheffé, 1947) implies now that $\lim_{\theta\to\theta^0} \|(P_\theta - P_{\theta^0})/(\theta - \theta^0) - \dot{P}_{\theta^0}\| = \lim_{\theta\to\theta^0} \int |(f_\theta(x) - f_{\theta^0}(x))/(\theta - \theta^0) - \dot{f}_{\theta^0}(x)| \mu(\mathrm{d}x) = 0.$ □

Let P be a probability measure on the line and consider the translation experiment $\mathscr{E}_P = (P_\theta : \theta \in \mathbb{R})$ determined by P. Thus $P_\theta(A) = P(A - \theta)$ for each Borel set A. Note first that $\mathscr{E}_P$ is differentiable at some particular value θ^0 of θ if and only if it is differentiable at $\theta = \theta^0$ for every θ^0. If $\mathscr{E}_P$ is differentiable at 0 (and thus everywhere) and if $\dot{P}$ denotes the derivative $\lim_{\theta\to 0} (P_\theta - P)/\theta$ then $\dot{P}_\theta$, for each θ, is just the measure induced from $\dot{P}$ by the translation: $x \to x + \theta$. We may therefore restrict our attention to local approximations at 0.

If $\mathscr{E}_P$ is differentiable then P_θ depends continuously on θ. In particular it follows that $\mathscr{E}_P$ is dominated. As we saw in example 6.5.13, this in turn implies that P is absolutely continuous. Put $f_\theta(x) = f(x - \theta)$. Then $\dot{f}_\theta(x) = \lim_{h\to 0}(f_{\theta+h}(x) - f_\theta(x))/h = -f'(x - \theta)$ exists for almost all x. Hence $\mathscr{E}_P$ is differentiable at any θ. Let h be a density of $\dot{P}$ and put $H(x) = \int_{-\infty}^{x} h(y)\,dy = \dot{P}(]-\infty, x])$. Then

$$H(x) = \lim_{\theta\to 0}[P_\theta(]-\infty, x]) - P(]-\infty, x])]/\theta$$

$$= \lim_{\theta\to 0}[P(]-\infty, x - \theta]) - P(]-\infty, x])]/\theta.$$

Hence $f = -H$ is a density of P. This density is clearly absolutely continuous on bounded intervals and $\int |f'(x)|\,dx < \infty$.

Conversely the existence of a density f of P with these properties implies, by the theorem, that $\mathscr{E}_P$ is differentiable.

We have established the equivalence of the following differentiability conditions.

Proposition 9.4.2 (Differentiability of translation experiments). *Let P be a probability distribution on the real line and let $\mathscr{E}_P = (P_\theta : \theta \in \Theta)$ denote the translation experiment determined by P. Then the following conditions are equivalent:*

(i) *$\mathscr{E}_P$ is differentiable at some point θ^0.*
(ii) *$\mathscr{E}_P$ is differentiable at all points θ^0.*
(iii) *P has a (necessarily unique) density f which is absolutely continuous on bounded intervals and which satisfies the condition $\int |f'(x)|\,dx < \infty$.*

Remark. If 'differentiable' is replaced with 'differentiable in quadratic mean' and if the integral $\int |f'(x)|\,dx$ in (iii) is replaced with the (Fisher information) integral $\int (f'/f)^2\,dP$ then we obtain the basic density criterion for differentiability in quadratic mean of translation experiments.

If we let $e^i = (0,\ldots,\overset{(i)}{1},\ldots,0)$ denote the ith unit vector in $\mathbb{R}^m$ then $\mathscr{E}$ is differentiable at θ^0 if and only if the limits (partial derivatives) $[\partial P_\theta/\partial\theta_i]_{\theta^0} = \lim_{t\to 0}(P_{\theta^0+te^i} - P_{\theta^0})/t$ exist; $i = 1,\ldots,m$ and $\|P_\theta - P_{\theta^0} - \sum_{i=1}^m [\partial P_\theta/\partial\theta_i]_{\theta^0}(\theta_i - \theta_i^0)\|/\|\theta - \theta^0\| \to 0$ as $\theta \to \theta^0$.

If $\mathscr{E} = (P_\theta : \theta \in \Theta)$ is differentiable at θ^0 then the family consisting of the $(m + 1)$ measures $(P_{\theta^0}, [\partial P_\theta/\partial\theta_i]_{\theta=\theta^0}; i = 1,\ldots,m)$ will be called the *first order characterization of $\mathscr{E}$ at θ^0*. Put $\mu_h = P_{\theta^0} + \sum_{i=1}^m h_i[\partial P_\theta/\partial\theta_i]_{\theta=\theta^0}$ when $h \in \mathbb{R}^m$. Then the first order characterization of $\mathscr{E}$ at θ^0 and the family $(\mu_h : h \in \mathbb{R}^m)$ determine each other. If we insist that *the first order (local) approximation* $h \to \mu_h$ of P_{θ^0+h}; $h \in \mathbb{R}^m$ should be an experiment then we can replace μ_h by $|\mu_h|/\|\mu_h\|$ here (see theorem 9.4.3).

We shall find it convenient to write $\dot{P}_{\theta^0,i}$ for the partial derivative of P_θ with respect to θ_i at θ^0, i.e. $\dot{P}_{\theta^0,i} = [\partial P_\theta/\partial\theta_i]_{\theta^0}$. The first order characterization of $\mathscr{E}$

at θ^0 will be denoted as $\dot{\mathscr{E}}_{\theta^0}$. Thus $\dot{\mathscr{E}}_{\theta^0} = (P_{\theta^0}, \dot{P}_{\theta^0,1}, \ldots, \dot{P}_{\theta^0,m})$ if $\mathscr{E} = (P_\theta : \theta \in \Theta)$ is differentiable at $\theta = \theta^0$.

Again let $e_i = (0, \ldots, \overset{(i)}{1}, \ldots, 0)$ denote the ith unit vector of $\mathbb{R}^m$. If $\mathscr{E} = (P_\theta : \theta \in \Theta)$ is differentiable at θ^0 then the experiments $(P_{\theta^0 + te^i} : |t| < \varepsilon)$ are all well defined and differentiable in $t = 0$ provided ε is sufficiently small. Furthermore P_θ is approximable by $\mu_\theta = P_{\theta^0} + \sum_{i=1}^m (P_{\theta^0 + (\theta_i - \theta_i^0)e^i} - P_{\theta^0})$ in the sense that $\|P_\theta - \mu_\theta\| / \|\theta - \theta^0\| \to 0$ as $\theta \to \theta^0$.

Conversely, these two conditions guarantee that $\mathscr{E}$ is differentiable at $\theta = \theta^0$. The usual arguments from calculus imply that $\mathscr{E}$ is differentiable at θ^0 provided the partial derivatives exist and are continuous within some neighbourhood of θ^0.

What kind of a measure family is the first order approximation of a differentiable experiment? Note first that if $\dot{\mathscr{E}}_{\theta^0} = (P_{\theta^0}, \dot{P}_{\theta^0,1}, \ldots, \dot{P}_{\theta^0,m})$ is the first order approximation of $\mathscr{E}$ at θ^0 then $P_{\theta^0} \gg \dot{\mathscr{E}}_{\theta^0}$ and each measure $\dot{P}_{\theta^0,i}$; $i = 1, \ldots, m$, has total mass zero. All general properties of first order approximations may be deduced from these two properties.

Theorem 9.4.3 (Characterization of first order approximations). *A measure family $\mathscr{D} = ((\mathscr{X}, \mathscr{A}), \pi, \sigma_1, \ldots, \sigma_m)$ is the first order approximation in θ^0 of some differentiable experiment $\mathscr{E}$, if and only if $\sigma_1(\mathscr{X}) = \cdots = \sigma_m(\mathscr{X}) = 0$ and π is a probability measure dominating $\sigma_1, \sigma_2, \ldots, \sigma_m$. If so, then $((\mathscr{X}, \mathscr{A}), \pi, \sigma_1, \ldots, \sigma_m)$ is the first order approximation in θ^0 of the experiment $\mathscr{E} = ((\mathscr{X}, \mathscr{A}), P_\theta : \theta \in \Theta)$ where $\theta^0 \in \Theta$ and*

$$P_\theta = \left| \pi + \sum_i (\theta_i - \theta_i^0)\sigma_i \right| \Big/ \left\| \pi + \sum_i (\theta_i - \theta_i^0)\sigma_i \right\|; \qquad \theta \in \Theta.$$

Furthermore, these conditions imply that

$$\left(\left\| \pi + \sum_i (\theta_i - \theta_i^0)\sigma_i \right\| - 1 \right) \Big/ \|\theta - \theta^0\| \to 0 \quad as \quad \theta \to \theta^0.$$

Remark. P_θ is well defined since $\|\pi + \sum_i (\theta_i - \theta_i^0)\sigma_i\| \geqslant 1$.

Proof. We may assume without loss of generality that $\theta^0 = 0$. The conditions are obviously necessary so suppose $\sigma_1(\mathscr{X}) = \cdots = \sigma_m(\mathscr{X}) = 0$ and that π is a probability measure dominating $\sigma_1, \ldots, \sigma_m$. Put $s_i = d\sigma_i / d\pi$. Then

$$\begin{aligned} \|\pi + \textstyle\sum \theta_i \sigma_i\| - 1 &= \int |1 + \textstyle\sum \theta_i s_i| \, d\pi - 1 \\ &= \int (1 + \textstyle\sum \theta_i s_i)^+ \, d\pi + \int (1 + \textstyle\sum \theta_i s_i)^- \, d\pi - \int (1 + \textstyle\sum \theta_i s_i)^- \\ &\quad + \int (1 + \textstyle\sum \theta_i s_i)^- \, d\pi - 1 \end{aligned}$$

$$= \int (1 + \sum \theta_i s_i)\,d\pi + 2\int (1 + \sum \theta_i s_i)^-\,d\pi - 1$$

$$= 2\int (1 + \sum \theta_i s_i)^-\,d\pi \leqslant 2\int_{\sum \theta_i s_i < -1} (1 + \sum |\theta_i|\,|s_i|)\,d\pi$$

$$\leqslant 2\int_{\sum|\theta_i|\max_i|s_i|>1} \left[1 + (\sum|\theta_i|)\max_i |s_i|\right] d\pi$$

$$\leqslant 4\sum|\theta_i| \int_{\max_i|s_i|>(\sum|\theta_i|)^{-1}} \max_i |s_i|\,d\pi.$$

Hence

$$(\|\pi + \sum \theta_i\sigma_i\| - 1)/\sum|\theta_i| \to 0 \quad \text{as} \quad \sum|\theta_i| \to 0$$

and this is equivalent to the last statement of the theorem. Clearly $P_0 = \pi$. It remains to show that $\|P_\theta - \pi - \sum \theta_i\sigma_i\|/\|\theta\| \to 0$ as $\|\theta\| \to 0$. We get successively

$$\|\theta\|^{-1}\|P_\theta - \pi - \sum\theta_i\sigma_i\|$$

$$= \|\theta\|^{-1}\left\| \frac{|\pi + \sum\theta_i\sigma_i|}{\|\pi + \sum\theta_i\sigma_i\|} - \pi - \sum\theta_i\sigma_i \right\|$$

$$\leqslant \|\theta\|^{-1}\left\| \frac{|\pi+\sum\theta_i\sigma_i|}{\|\pi+\sum\theta_i\sigma_i\|} - |\pi+\sum\theta_i\sigma_i| \right\| + \|\theta\|^{-1}\left\| \left|\pi + \sum_i \theta_i\sigma_i\right| - \pi - \sum\theta_i\sigma_i \right\|$$

$$= \|\theta\|^{-1}\|\pi + \sum\theta_i\sigma_i\|\left(1 - \frac{1}{\|\pi + \sum\theta_i\sigma_i\|}\right) + 2\|\theta\|^{-1}\|(\pi + \sum\theta_i\sigma_i)^-\|$$

$$= \|\theta\|^{-1}[\|\pi + \sum\theta_i\sigma_i\| - 1] + \|\theta\|^{-1}[\|\pi + \sum\theta_i\sigma_i\| - 1]$$

$$= 2\|\theta\|^{-1}[\|\pi + \sum\theta_i\sigma_i\| - 1] \to 0 \quad \text{as} \quad \|\theta\| \to 0. \qquad \square$$

Assume now that $\mathscr{E} = (P_\theta : \theta \in \Theta)$ is differentiable at θ^0 and that $\mathscr{E}$ is at least as informative as $\mathscr{F} = (Q_\theta : \theta \in \Theta)$ for testing problems. Then the convergence criteria mentioned just before the theorem, together with the norm criterion for comparison by testing problems and the Cauchy criterion for norm convergence, imply that $\mathscr{F}$ is also differentiable at θ^0. In other words $\mathscr{F}$ is differentiable in θ^0 provided $\mathscr{F} \leqslant_2 \mathscr{E}$ where $\mathscr{E}$ is differentiable in θ^0.

Furthermore, if $\mathscr{F} \leqslant \mathscr{E}$ then $\dot{\mathscr{F}}_{\theta^0}$ is obtained from $\dot{\mathscr{E}}_{\theta^0}$ by any transition mapping $\mathscr{E}$ into $\mathscr{F}$. To be more precise, if $\mathscr{E} = (P_\theta : \theta \in \Theta)$, $\mathscr{F} = (Q_\theta : \theta \in \Theta)$ and $Q_\theta \equiv_\theta P_\theta M$ for a transition M then $\dot{Q}_{\theta^0,i} = \dot{P}_{\theta^0,i}M$; $i = 1,\ldots,m$.

Likewise finite products of experiments which are differentiable at θ^0 are themselves differentiable at θ^0. More precisely, if $\mathscr{E} = (P_\theta : \theta \in \Theta)$ and $\mathscr{F} = (Q_\theta : \theta \in \Theta)$ are differentiable at θ^0 and if $\mathscr{G} = \mathscr{E} \times \mathscr{F}$, then the first order approximation $\mathscr{G}_{\theta^0}$ of $\mathscr{G}$ is the measure family

$$(P_{\theta^0} \times Q_{\theta^0}, P_{\theta^0} \times \dot{Q}_{\theta^0,i} + \dot{P}_{\theta^0,i} \times Q_{\theta^0}; i = 1,\ldots,m).$$

The local properties of the differentiable experiment $\mathscr{E} = (\mathscr{X}, \mathscr{A}; P_\theta : \theta \in \Theta)$ at θ^0, are determined by the distribution $F(\cdot|\theta^0, \mathscr{E})$ of the random vector $(\mathrm{d}\dot{P}_{\theta^0,i}/\mathrm{d}P_{\theta^0}; i = 1, \ldots, m)$ under P_{θ^0} as will be made clear later on.

Clearly $\int xF(\mathrm{d}x|\theta^0, \mathscr{E}) = 0$ and any probability distribution on $\mathbb{R}^m$ having expectation zero is of the form $F(\cdot|\theta^0, \mathscr{E})$ for some differentiable experiment $\mathscr{E}$.

We shall only point out here that if $\mathscr{E}$ and $\mathscr{F}$ are both differentiable at θ^0 and if $\mathscr{G} = \mathscr{E} \times \mathscr{F}$ then $F(\cdot|\theta^0, \mathscr{G}) = F(\cdot|\theta^0, \mathscr{E}) * F(\cdot|\theta^0, \mathscr{F})$ where $*$ denotes convolution.

Later on we will see how $F(\cdot|\theta^0, \mathscr{E})$ and $F(\cdot|\theta^0, \mathscr{F})$ must be related in order to ensure that $\mathscr{E}$ is locally at least as informative as $\mathscr{F}$ at θ^0.

Consider now two experiments $\mathscr{E} = (P_\theta : \theta \in \Theta)$ and $\mathscr{F} = (Q_\theta : \theta \in \Theta)$ which are both differentiable at $\theta = \theta^0$. Equip $\mathbb{R}^m$ with the L-norm $\| \ \| : x \to \sum_i |x_i|$. For each $\varepsilon > 0$, let the restrictions of $\mathscr{E}$ and $\mathscr{F}$ to the ε-ball $N(\theta^0, \varepsilon) = \{\theta : \|\theta - \theta^0\| \leqslant \varepsilon\}$ be denoted by $\mathscr{E}_\varepsilon$ and $\mathscr{F}_\varepsilon$ respectively.

The problem of local comparison of $\mathscr{E}$ and $\mathscr{F}$ within small neighbourhoods of θ^0 may now be discussed in terms of the behaviour of the deficiency of $\delta(\mathscr{E}_\varepsilon, \mathscr{F}_\varepsilon)$ for small values of ε. If $\mathscr{M}_i$ denotes the totally non informative experiment then continuity implies that the deficiencies $\delta(\mathscr{M}_i, \mathscr{F}_\varepsilon)$ and $\delta(\mathscr{M}_i, \mathscr{E}_\varepsilon)$ both $\to 0$ as $\varepsilon \to 0$. Thus $\delta_k(\mathscr{E}_\varepsilon, \mathscr{F}_\varepsilon) \to 0$ as $\varepsilon \to 0$ for each $k = 1, 2, \ldots, \infty$.

Let us next determine the rate of the convergence of $\delta_k(\mathscr{E}_\varepsilon, \mathscr{F}_\varepsilon)$ as $\varepsilon \to 0$. If $\mathscr{E} = (P_\theta : \theta \in \Theta)$ is differentiable in θ^0 then we may expand P_θ as $P_\theta = P_{\theta^0} + \sum(\theta - \theta^0)_i \dot{P}_{\theta^0,i} + \tau(\mathscr{E}, \theta^0, \theta)$ where the measure $\tau(\mathscr{E}, \theta^0, \theta)$ is defined by this expansion. The differentiability assumption implies then that $\|\tau(\mathscr{E}, \theta^0, \theta)\|/\|\theta - \theta^0\| \to 0$ as $\theta \to \theta^0$.

In the following we will find it convenient to utilize the symbol o in the usual way, i.e. o denotes any real valued function on $]0, \infty[$ such that $o(t)/t \to 0$ as $t \to 0$.

If $\mathscr{E} = (P_\theta : \theta \in \Theta)$ and $\mathscr{F} = (Q_\theta : \theta \in \Theta)$ are both differentiable in θ^0 then the smallest (it exists) constant η such that $\dot{\mathscr{E}}_{\theta^0}$ is $(0, \eta, \ldots, \eta)$ deficient with respect to $\dot{\mathscr{F}}_{\theta^0}$ for k-point sets will be called *the local deficiency at θ^0 of $\mathscr{E}$ with respect to $\mathscr{F}$ for k-decision problems.* The local deficiency at θ^0 of $\mathscr{E}$ with respect to $\mathscr{F}$ for k-decision problems will be denoted as $\dot{\delta}_{k,\theta^0}(\mathscr{E}, \mathscr{F})$. Here, as elsewhere, we may omit the qualification 'for k-decision problems' and the subscript k if $k = \infty$.

The local deficiency determines the rate of convergence of $\delta_k(\mathscr{E}_\varepsilon, \mathscr{F}_\varepsilon)/\varepsilon$ as $\varepsilon \to 0$.

Theorem 9.4.4 (Local behaviour of deficiencies within small neighbourhoods). *With the notation introduced above for differentiable experiments $\mathscr{E} = (P_\theta : \theta \in \Theta)$ and $\mathscr{F} = (Q_\theta : \theta \in \Theta)$ we have*

$$\delta_k(\mathscr{E}_\varepsilon, \mathscr{F}_\varepsilon) \leqslant \varepsilon \dot{\delta}_{k,\theta^0}(\mathscr{E}, \mathscr{F}) + o(\varepsilon)$$

where $o(\varepsilon) = \sup\{\|\tau(\mathscr{E}, \theta^0, \theta)\| + \|\tau(\mathscr{F}, \theta^0, \theta)\| : \|\theta - \theta^0\| \leqslant \varepsilon\}$. *Furthermore*

$$\delta_k(\mathscr{E}_\varepsilon, \mathscr{F}_\varepsilon)/\varepsilon \to \dot{\delta}_{k,\theta^0}(\mathscr{E}, \mathscr{F}) \text{ as } \varepsilon \to 0.$$

Remark 1. By the randomization criterion the local deficiency may be expressed as

$$\dot{\delta}_{\theta^0}(\mathscr{E}, \mathscr{F}) = \min\left\{\max_i \|\dot{P}_{\theta^0,i}M - \dot{Q}_{\theta^0,i}\| : P_{\theta^0}M = Q_{\theta^0}\right\}$$

where M varies within the set of transitions (randomizations) from $L(\mathscr{E})$ to the L-space $\Gamma(\mathscr{E})^*$ of bounded additive set functions on the sample space of $\mathscr{F}$.

We may limit our attention to transitions from $L_1(P_{\theta^0})$ to $L_1(Q_{\theta^0})$ (see remark 1 after theorem 9.2.3).

Remark 2. The proof implies that the statements of the theorem remain true, if, for each $\varepsilon > 0$, $\mathscr{E}_\varepsilon$ and $\mathscr{F}_\varepsilon$ are replaced by the restrictions of the experiments $\mathscr{E}$ and $\mathscr{F}$ to the subset of $N(\theta^0, \varepsilon)$ consisting of the $2m$ points (vertices) $\theta^0 \pm (0, \ldots, \overset{(i)}{\varepsilon}, \ldots, 0)$; $i = 1, \ldots, m$.

Proof. First consider the case $k = \infty$. If $P_{\theta^0}M = Q_{\theta^0}$ for a transition M and if $\|\theta - \theta^0\| \leqslant \varepsilon$ then

$$\begin{aligned}
&\|P_\theta M - Q_\theta\| \\
&= \|(P_\theta - P_{\theta^0})M - (Q_\theta - Q_{\theta^0})\| \\
&= \left\|\sum_i (\theta - \theta^0)_i(\dot{P}_{\theta^0,i}M - \dot{Q}_{\theta^0,i}) + \tau(\mathscr{E}, \theta^0, \theta)M - \tau(\mathscr{F}, \theta^0, \theta)\right\| \\
&\leqslant \sum_i |(\theta - \theta^0)_i| \max_i \|\dot{P}_{\theta^0,i}M - \dot{Q}_{\theta^0,i}\| + \|\tau(\mathscr{E}, \theta^0, \theta)\| + \|\tau(\mathscr{F}, \theta^0, \theta)\| \\
&\leqslant \varepsilon \max_i \|\dot{P}_{\theta^0,i}M - \dot{Q}_{\theta^0,i}\| + o(\varepsilon)
\end{aligned}$$

so that $\delta(\mathscr{E}_\varepsilon, \mathscr{F}_\varepsilon) \leqslant \varepsilon\dot{\delta}_{\theta^0}(\mathscr{E}, \mathscr{F}) + o(\varepsilon)$. It follows that $\limsup \delta(\mathscr{E}_\varepsilon, \mathscr{F}_\varepsilon)/\varepsilon \leqslant \dot{\delta}_{\theta^0}(\mathscr{E}, \mathscr{F})$ as $\varepsilon \to 0$.

On the other hand, consider any number $c > \liminf_{\varepsilon \to 0} \delta(\mathscr{E}_\varepsilon, \mathscr{F}_\varepsilon)/\varepsilon$. Then $\delta(\mathscr{E}_\varepsilon, \mathscr{F}_\varepsilon) < c\varepsilon$ for all ε belonging to a sequence $\varepsilon_1, \varepsilon_2, \ldots$ which decreases to zero. Assume that ε belongs to this sub-sequence. The randomization criterion (theorem 9.2.3) then yields a transition M_ε from $L(\mathscr{E})$ to $M(\mathscr{F})$ such that $\|P_\theta M_\varepsilon - Q_\theta\| < c\varepsilon$ when $\theta \in N(\theta^0, \varepsilon)$. By compactness, we may assume that M_ε converges pointwise on $L(\mathscr{E}) \times M(\mathscr{F})$ to a transition M from $L(\mathscr{E})$ to $M(\mathscr{F})$. Expanding around θ^0 we find that

$$\left\|P_{\theta^0}M_\varepsilon - Q_{\theta^0} + \sum_i (\theta_i - \theta_i^0)(\dot{P}_{\theta^0,i}M_\varepsilon - \dot{Q}_{\theta^0,i}) + \tau(\mathscr{E}, \theta^0, \theta)M_\varepsilon - \tau(\mathscr{F}, \theta^0, \theta)\right\| < c\varepsilon$$

when $\|\theta - \theta^0\| \leqslant \varepsilon$. In particular $\|P_{\theta^0}M_\varepsilon - Q_{\theta^0}\| < c\varepsilon$. $\varepsilon \to 0$ then yields $P_{\theta^0}M = Q_{\theta^0}$.

Putting $\theta' = \theta^0 + (0, \ldots, \overset{(i)}{\varepsilon}, \ldots, 0)$ and $\theta'' = \theta^0 - (0, \ldots, \overset{(i)}{\varepsilon}, \ldots, 0)$ we find that

$$\|\varepsilon(\dot{P}_{\theta^0,i}M_\varepsilon - \dot{Q}_{\theta^0,i}) + P_{\theta^0}M_\varepsilon - Q_{\theta^0} + \tau(\mathscr{E}, \theta^0, \theta')M_\varepsilon - \tau(\mathscr{F}, \theta^0, \theta')\| < c\varepsilon$$

and

$$\|\varepsilon(\dot{P}_{\theta^0,i}M_\varepsilon - \dot{Q}_{\theta^0,i}) - (P_{\theta^0}M_\varepsilon - Q_{\theta^0}) - (\tau(\mathscr{E}, \theta^0, \theta'')M_\varepsilon - \tau(\mathscr{F}, \theta^0, \theta''))\| < c\varepsilon.$$

Hence $2\varepsilon\|\dot{P}_{\theta^0,i}M_\varepsilon - Q_{\theta^0,i}\| + o(\varepsilon) < 2c\varepsilon$. Dividing through by ε and letting $\varepsilon \to 0$ we find that $\|\dot{P}_{\theta^0,i}M - \dot{Q}_{\theta^0,i}\| \leqslant c$. Thus $c \geqslant \dot{\delta}_{\theta^0}(\mathscr{E}, \mathscr{F})$. It follows that $\liminf \delta(\mathscr{E}_\varepsilon, \mathscr{F}_\varepsilon)/\varepsilon \geqslant \dot{\delta}_{\theta^0}(\mathscr{E}, \mathscr{F})$ as $\varepsilon \downarrow 0$ so that $\delta(\mathscr{E}_\varepsilon, \mathscr{F}_\varepsilon)/\varepsilon \to \dot{\delta}_{\theta^0}(\mathscr{E}, \mathscr{F})$ as $\varepsilon \to 0$. This completes the proof when $k = \infty$.

The case of a finite k may be reduced to the case $k = \infty$ as follows. Assume that $k < \infty$ and let $(\mathscr{Y}, \mathscr{B})$ be the sample space of $\mathscr{F}$. Let $\mathscr{B}_0$ be a subalgebra of $\mathscr{B}$ containing at most 2^k sets. The '$k = \infty$' part of the theorem implies that $\delta(\mathscr{E}_\varepsilon, \mathscr{F}_\varepsilon | \mathscr{B}_0) \leqslant \varepsilon\dot{\delta}_{\theta^0}(\mathscr{E}, \mathscr{F} | \mathscr{B}_0) + o(\varepsilon)$ where $o(\varepsilon)$ is defined as in the theorem and, consequently, does not depend on $\mathscr{B}_0$. Taking the supremum on both sides for all such algebras $\mathscr{B}_0$ we find by theorem 9.2.2 $\delta_k(\mathscr{E}_\varepsilon, \mathscr{F}_\varepsilon) \leqslant \varepsilon\dot{\delta}_{k,\theta^0}(\mathscr{E}, \mathscr{F}) + o(\varepsilon)$. If follows that $\limsup_{\varepsilon \to 0} \delta_k(\mathscr{E}_\varepsilon, \mathscr{F}_\varepsilon)/\varepsilon \leqslant \dot{\delta}_{k,\theta^0}(\mathscr{E}, \mathscr{F})$.

On the other hand $\liminf_{\varepsilon \to 0} \delta_k(\mathscr{E}_\varepsilon, \mathscr{F}_\varepsilon)/\varepsilon \geqslant \liminf_{\varepsilon \to 0} \delta(\mathscr{E}_\varepsilon, \mathscr{F}_\varepsilon | \mathscr{B}_0)/\varepsilon = \dot{\delta}_{\theta^0}(\mathscr{E}, \mathscr{F} | \mathscr{B}_0)$ for any subalgebra $\mathscr{B}_0$ of $\mathscr{B}$ containing at most 2^k sets.

Taking the supremum for all such algebras $\mathscr{B}_0$ we see that $\liminf_{\varepsilon \to 0} \delta_k(\mathscr{E}_\varepsilon, \mathscr{F}_\varepsilon)/\varepsilon \geqslant \dot{\delta}_{k,\theta^0}(\mathscr{E}, \mathscr{F})$. □

Define the *local deficiency distance at* θ^0 *between* $\mathscr{E}$ *and* $\mathscr{F}$ for k-decision problems as the largest of the numbers $\dot{\delta}_{k,\theta^0}(\mathscr{E}, \mathscr{F})$ and $\dot{\delta}_{k,\theta^0}(\mathscr{F}, \mathscr{E})$. This number will be denoted as $\dot{\Delta}_{k,\theta^0}(\mathscr{E}, \mathscr{F})$. Here the subscript k and the qualification 'for k-decision problems' may be omitted when $k = \infty$. It is easily checked that $\dot{\Delta}_{k,\theta^0}$ is a pseudometric for experiments. The local behaviour of this distance may readily be obtained from the last theorem.

Corollary 9.4.5 (Local behaviour of deficiency distances within small neighbourhoods). *$\Delta_k(\mathscr{E}_\varepsilon, \mathscr{F}_\varepsilon) \leqslant \varepsilon\dot{\Delta}_{k,\theta^0}(\mathscr{E}, \mathscr{F}) + o(\varepsilon)$ where $o(\varepsilon)$ is defined as in the theorem. Furthermore equality holds for some function $o(\varepsilon)$.*

Related to the local deficiencies are the local orderings and the local equivalences of experiments. Thus we shall say that $\mathscr{E}$ *is locally at least as informative as* $\mathscr{F}$ *for k-decision problems at* $\theta = \theta^0$ if $\dot{\delta}_{k,\theta^0}(\mathscr{E}, \mathscr{F}) = 0$. This defines, for $k = 1, 2, \ldots, \infty$, a partial ordering $\geqslant_{k,\theta^0}$ for differentiable experiments. It follows from the transition criterion for comparison that $\mathscr{E}$ is locally at least as informative as $\mathscr{F}$ at θ^0 provided the restriction of $\mathscr{E}$ to any $(m + 1)$-point sub-parameter set containing θ^0 is at least as informative as the same restriction of $\mathscr{F}$.

If $\mathscr{E} \geqslant_{k,\theta^0} \mathscr{F}$ and $\mathscr{F} \geqslant_{k,\theta^0} \mathscr{E}$, i.e. if $\dot{\Delta}_{k,\theta^0}(\mathscr{E},\mathscr{F}) = 0$ then we shall say that $\mathscr{E}$ *and* $\mathscr{F}$ *are locally equivalent for k-decision problems at* $\theta = \theta^0$.

This defines for each $k = 1, 2, \ldots, \infty$ a local equivalence $\sim_{k,\theta^0}$ for differentiable experiments. By corollary 9.3.3 however, the non trivial equivalences $\sim_{k,\theta^0}$; $k = 2, 3, \ldots, \infty$ are all the same. More generally the pseudo-distances $\dot{\Delta}_{k,\theta^0}$; $k = 2, 3, \ldots, \infty$ all define the same notion of convergence.

What is the statistical significance of local deficiencies and related notions? Some insight may be gained from the following characterization in terms of performance functions.

Proposition 9.4.6 (Local comparison of performance functions). *Let* $\mathscr{E} = (P_\theta : \theta \in \Theta)$ *and* $\mathscr{F} = (Q_\theta : \theta \in \Theta)$ *both be differentiable at* $\theta = \theta^0$.

Let $(T, \mathscr{S})$ *be a decision space and consider a decision rule* σ *in* $\mathscr{F}$. *Then there is a decision rule* ρ *in* $\mathscr{E}$ *such that*

$$\limsup_{\theta \to \theta^0} \|P_\theta \rho - Q_\theta \sigma\| / \|\theta - \theta^0\| \leqslant \dot{\delta}_{\theta^0}(\mathscr{E},\mathscr{F}).$$

Furthermore $(T, \mathscr{S})$ *and* σ *may be chosen such that*

$$\limsup_{\theta \to \theta^0} \|P_\theta \rho - Q_\theta \sigma\| / \|\theta - \theta^0\| \geqslant \dot{\delta}_{\theta^0}(\mathscr{E},\mathscr{F})$$

for all decision rules ρ *in* $\mathscr{E}$.

Remark. If $(T, \mathscr{S})$ is a k-decision space then the first inequality may be sharpened by replacing $\dot{\delta}_{\theta^0}(\mathscr{E},\mathscr{F})$ by the usually smaller number $\dot{\delta}_{k,\theta^0}(\mathscr{E},\mathscr{F})$. Furthermore if $(T, \mathscr{S})$ is a k-decision space then σ in $\mathscr{F}$ may be chosen such that

$$\limsup_{\theta \to \theta^0} \|P_\theta \rho - Q_\theta \sigma\| / \|\theta - \theta^0\| \geqslant \dot{\delta}_{k,\theta^0}(\mathscr{E},\mathscr{F})$$

for all decision rules ρ in $\mathscr{E}$.

This may be seen by applying the proposition to the restrictions of $\mathscr{F}$ to algebras of events containing at most 2^k events.

Proof. Note first that for any transition M

$$\limsup_{\theta \to \theta^0} \|P_\theta M - Q_\theta\| / \|\theta - \theta^0\| = \max_i \|\dot{P}_{\theta^0,i} M - \dot{Q}_{\theta^0,i}\| \quad \text{or} \quad = \infty \quad \text{as}$$

$$P_{\theta^0} M = Q_{\theta^0} \qquad \text{or} \qquad P_{\theta^0} M \neq Q_{\theta^0}.$$

The first statement of the proposition follows now by putting $\rho = M\sigma$ where $P_{\theta^0} M = Q_{\theta^0}$ and $\max_i \|\dot{P}_{\theta^0,i} M - \dot{Q}_{\theta^0,i}\| = \dot{\delta}_{\theta^0}(\mathscr{E},\mathscr{F})$. The second statement follows by observing that we may let $(T, \mathscr{S})$ be the sample space of $\mathscr{F}$ and then choose σ as the identity map. □

If ψ is a sublinear function on $\mathbb{R} \times \mathbb{R}^m$ and if $\mathscr{E} = (P_\theta : \theta \in \Theta)$ is differentiable at $\theta = \theta^0$ then $\int \psi(\mathrm{d}P_{\theta^0}, \mathrm{d}\dot{P}_{\theta^0,1}, \ldots, \mathrm{d}\dot{P}_{\theta^0,m}) = \int \psi(1, x_1, \ldots, x_m) F(\mathrm{d}x|\theta^0, \mathscr{E})$. It follows readily that $\mathscr{E} = (P_\theta : \theta \in \Theta)$ is locally at least as informative as $\mathscr{F}$ at $\theta = \theta^0$ if and only if $\int \phi \, \mathrm{d}F(\cdot|\theta^0, \mathscr{E}) \geqslant \int \phi \, \mathrm{d}F(\cdot|\theta^0, \mathscr{F})$ for all convex functions ϕ on $\mathbb{R}^m$. As we shall see in the next section, this is in turn equivalent to the condition that $F(\cdot|\theta^0, \mathscr{E}) = DF(\cdot|\theta^0, \mathscr{F})$ for a dilation D from $\mathbb{R}^m$ to $\mathbb{R}^m$.

The Fisher information matrix $\Gamma(\theta^0, \mathscr{E})$ is the covariance matrix of $F(\cdot|\theta^0, \mathscr{E})$, provided of course that $F(\cdot|\theta^0, \mathscr{E})$ possesses finite second order moments. It follows that if $\mathscr{E}$ is locally at least as informative as $\mathscr{F}$ at θ^0 and if the Fisher information matrix of $\mathscr{E}$ at θ^0 exists, then the Fisher information matrix of $\mathscr{F}$ at θ^0 also exists and then the difference matrix $\Gamma(\theta^0, \mathscr{E}) - \Gamma(\theta^0, \mathscr{F})$ is non-negative definite. This proves the local, and hence the 'global', monotonicity of the Fisher information matrix.

Example 9.4.7 (Local orderings of linear normal models). For each $n_A \times p$ matrix A' let $\mathscr{E}_A$ denote the linear normal experiment $(N(A'\beta, I_A) : \beta \in \mathbb{R}^p)$ where I_A denotes the $n_A \times n_A$ unit matrix. The Fisher information matrix of $\mathscr{E}_A$ is the $p \times p$ matrix AA'. If B is another matrix with p rows and if $\mathscr{E}_A$ is locally at least as informative as $\mathscr{E}_B$ then, by the remarks above, $AA' \geqslant BB'$. The ordering '$\geqslant$' for matrices which is used in this example is the ordering which declares that $M \geqslant N$ if $M - N$ is non-negative definite.

If $AA' \geqslant BB'$ then $\mathscr{E}_A \sim \mathscr{E}_B \times \mathscr{E}_M$ where M is the non-negative definite square root of $AA' - BB'$. It follows that the local orderings, as well as the global orderings of linear normal models with known variances, coincide with the usual ordering of Fisher information matrices.

When we turn to the case of unknown variances then matters are a bit more involved. Let $\mathscr{F}_A$ denote the experiment $(N(A'\beta, \sigma^2 I_A) : \beta \in \mathbb{R}^p, \sigma > 0)$ where A' and I_A are as above. The $(p+1) \times (p+1)$ Fisher information matrix of $\mathscr{F}_A$ with respect to the unknown parameters $\beta_1, \ldots, \beta_p$ and σ is

$$\Gamma(\beta, \sigma, \mathscr{F}_A) = \begin{pmatrix} AA'/\sigma^2, & 0 \\ 0, & 2n_A/\sigma^2 \end{pmatrix}.$$

By theorem 8.6.6 the experiment $\mathscr{F}_A$ is at least as informative as $\mathscr{F}_B$ if and only if $AA' \geqslant BB'$ and $n_A \geqslant n_B + \operatorname{rank}(AA' - BB')$. In fact, by the same theorem, this is equivalent to the condition that $\mathscr{F}_A \sim \mathscr{F}_B \times \mathscr{F}_C$ (i.e. $AA' = BB' + CC'$ and $n_A = n_B + n_C$) for some $n_C \times p$ matrix C'.

The first result was extended by Lehmann (1983) to the case of multivariate regression. In that case $\mathscr{F}_A$, for each $n_A \times p$ matrix A', is realized by observing a random $n_A \times q$ matrix X such that $EX = A'\beta$ for an unknown $p \times q$ matrix β while the rows of X are independent and multinormally distributed with the same unknown non-singular covariance matrix σ^2. (Actually Lehmann (1983) assumes that A is in a 'reduced' form where rank $A = p$.) If we compare the distributions of the minimal sufficient statistics we see again that $\mathscr{F}_A \sim$

$\mathscr{F}_B \times \mathscr{F}_C$ when $AA' = BB' + CC'$ and $n_A = n_B + n_C$. On the other hand if we consider e.g. the first column of β and restrict β and σ^2 such that all the other entries of β are 0 while σ^2 is a diagonal matrix such that all the diagonal elements are known except the first one, then we are back in the univariate case with $q = 1$. It follows that the criteria mentioned above extend directly to the multivariate case.

Returning to the univariate case we see that the Fisher information matrix of $\mathscr{F}_A$ majorizes the Fisher information matrix of $\mathscr{F}_B$ if and only if $AA' \geqslant BB'$ and $n_A \geqslant n_B$. This amounts to the condition that the two restrictions we may obtain from $\mathscr{F}_A$ by assuming that exactly one of the quantities β and σ are known, are at least as informative as the corresponding restrictions of $\mathscr{F}_B$. It follows that the local orderings and the global ordering, as well as the ordering by the Fisher information matrices, are all stronger than the ordering of Fisher information matrices for known σ. Assuming that $AA' \geqslant BB'$ we may distinguish the three cases '$n_A < n_B$', '$n_A \geqslant n_B$' and '$n_A < n_B + \operatorname{rank}(AA' - BB')$' and '$n_A \geqslant n_B + \operatorname{rank}(AA' - BB')$'.

In the first case $\mathscr{F}_A$ and $\mathscr{F}_B$ are not comparable for any of the mentioned orderings. In the last case $\mathscr{F}_B$, as we noted above, is a factor of $\mathscr{F}_A$ so that $\mathscr{F}_A$ majorizes $\mathscr{F}_B$ for the global ordering and thus also for the weaker orderings. In the second case we know that $\mathscr{F}_A$ does not majorize $\mathscr{F}_B$ globally while it does majorize $\mathscr{F}_B$ for the ordering of Fisher information matrices.

Therefore it only remains to consider the problem of local ordering when $n_B + \operatorname{rank}(AA' - BB') > n_A \geqslant n_B$.

As a particular case consider univariate normally distributed variables X and Y such that X is distributed as $N(\xi, \sigma^2)$ while Y is distributed as $N(\xi, \sigma^2/2)$. Assume that ξ and $\sigma > 0$ are both unknown. Replacing X with $2^{1/2}X$ we see that we are in this situation with $n_A = n_B = 1$, $AA' = 2$ and $BB' = 1$. We shall see below that in spite of the fact that the experiments defined by X and Y are not comparable, X is locally more informative than Y at any point (ξ, σ).

If fact we shall now show that *the local ordering of these experiments coincides with* the usual ordering of the Fisher information matrices. Thus let us assume that $AA' \geqslant BB'$ and that $n_A \geqslant n_B$. Using the dilation criterion mentioned above we shall prove our claim by showing that $\dot{\delta}_{\beta^0, \sigma^0}(\mathscr{F}_A, \mathscr{F}_B) = 0$ for all $\beta^0 \in \mathbb{R}^p$ and all $\sigma^0 > 0$.

The consequence that the local orderings do not depend on where localization takes place follows, as we shall see, from the arguments below. However this fact also follows by general considerations on invariance under groups acting transitively and 'smoothly' on the parameter set.

Differentiating the log likelihoods we find that

$$F(\cdot|\sigma, \beta, \mathscr{F}_A) = \mathscr{L}\left((AX)'\Big/\sigma, \left(\sum_1^{n_A} x_i^2 - n_A\right)\Big/\sigma\right)$$

where X is distributed as $N(0, I_A)$. Similarly $F(\cdot|\sigma, \beta, \mathscr{F}_B) = \mathscr{L}((BY)'/\sigma, (\sum_1^{n_B} Y_i^2 - n_B)/\sigma)$ where Y is distributed as $N(0, I_B)$. Thus we must show that $E\phi((AX)'/\sigma, (\sum_1^{n_A} X_i^2 - n_A)/\sigma) \geqslant$ the same expression for $\mathscr{F}_B$ when ϕ is convex on $\mathbb{R}^{p+1}$ and, say, a maximum of a finite number of linear functions. Replacing ϕ with $\phi(\cdot/\sigma)$ we see that we may assume without loss of generality that $\beta = 0$ (in $\mathbb{R}^p$) and $\sigma = 1$. Proceeding as in Hansen & Torgersen (1974) we first consider the case where AA' is the $p \times p$ identity matrix while $BB' = \Lambda$ is a $p \times p$ diagonal matrix. Let λ_i denote the (i, i)th element of Λ. For convenience of formulation we will assume that the diagonal elements of Λ are ordered in decreasing order. If $s = \text{rank}\, B = \text{rank}\, \Lambda$ then $\lambda_i > 0$ or $= 0$ as $i \leqslant s$ or $i > s$.

As the rows of A are orthonormal we may add $n_A - p$ rows to A so that the extended $n_A \times n_A$ matrix $\tilde{A}$ is orthonormal. Putting $\tilde{X} = \tilde{A}X$ we see that $AX = (\tilde{X}_1, \ldots, \tilde{X}_p)'$ and that $\sum X_i^2 = \sum \tilde{X}_i^2$. Thus, since $\tilde{X}$ is also distributed as $N(0, I_A)$, we find that $E\phi((AX)', \sum X_i^2 - n_A) = E\phi(X_1, \ldots, X_p, X_1^2 + \cdots + X_{n_A}^2 - n_A)$.

Likewise the rows of B are orthogonal. The first s rows became orthonormal after having been divided by, respectively, $\lambda_1^{1/2}, \lambda_2^{1/2}, \ldots, \lambda_s^{1/2}$. The $p - s$ remaining rows are all the $1 \times n_B$ zero matrix. Extend the described orthonormal system of row matrices to an $n_B \times n_B$ orthonormal matrix U and put $\tilde{Y} = UY$. Then $\tilde{Y}$ is distributed as Y and $BY = (\lambda_1^{1/2}\tilde{Y}_1, \ldots, \lambda_s^{1/2}\tilde{Y}_s, 0, \ldots, 0)'$ while $\sum Y_i^2 = \sum \tilde{Y}_i^2$. Hence $E\phi((BY)', \sum Y_i^2 - n_B) = E\phi(\lambda_1^{1/2} Y_1, \ldots, \lambda_s^{1/2} Y_s, 0, \ldots, 0, \sum Y_i^2 - n_B)$. Our task is therefore to show that $E\phi(X_1, \ldots, X_p, \sum_1^{n_A} X_i^2 - n_A) \geqslant E\phi(\lambda_1^{1/2} X_1, \ldots, \lambda_p^{1/2} X_p, \sum_1^{n_B} X_i^2 - n_B)$ when $X_1, X_2, \ldots$ are independent $N(0,1)$-variables. Assume first that $n_B \geqslant p$. Then, by Jensen's inequality

$$E\phi\left(X_1, \ldots, X_p, \sum_1^{n_A} X_i^2 - n_A\right) = EE\phi(\ldots|X_1, \ldots, X_{n_B})$$

$$\geqslant E\phi\left(X_1, \ldots, X_p, \sum_1^{n_B} X_i^2 - n_B\right).$$

If, on the other hand, $n_B < p$ then, since $n_B \geqslant s$, the same argument implies that $E\phi(\lambda_1^{1/2} X_1, \ldots, \lambda_p^{1/2} X_p, \sum_1^{n_B} X_i^2 - n_B) \leqslant E\phi(\lambda_1^{1/2} X_1, \ldots, \lambda_p^{1/2} X_p, \sum_1^p X_i^2 - p)$ and that $E\phi(X_1, \ldots, X_p, \sum_1^p X_i^2 - p) \leqslant E\phi(X_1, \ldots, X_p, \sum_1^{n_A} X_i^2 - n_A)$. Putting $t = \max(p, n_B)$ we see that in both cases we will complete the proof if we can show that $E\phi(X_1, \ldots, X_p, \sum_1^t X_i^2 - t) \geqslant E\phi(\lambda_1^{1/2} X_1, \ldots, \lambda_p^{1/2} X_p, \sum_1^t X_i^2 - t)$. Let the variables $\xi_1 = \pm 1, \xi_2 = \pm 1, \ldots, \xi_p = \pm 1$ be independent and independent of $(X_1, \ldots, X_t)$. Assume that $Pr(\xi_i = 1) = (\lambda_i^{1/2} + 1)/2$. (This is feasible since $\lambda_i \in [0, 1]$.) Then $E\xi_i = \lambda_i^{1/2}$. By symmetry and Jensen's inequality we obtain

$$E\phi(X_1, \ldots, X_p, \textstyle\sum_1^t X_i^2) = E\phi(\xi_1 X_1, \ldots, \xi_p X_p, \sum_1^t X_i^2)$$

$$\geqslant \phi(\lambda_1^{1/2} X_1, \ldots, \lambda_p^{1/2} X_p, \textstyle\sum_1^t X_i^2).$$

This establishes the desired inequality when AA' is the $p \times p$ unit matrix and BB' is a diagonal matrix. More generally, if rank $A = p$ then there is a $p \times p$ matrix F such that $FAA'F'$ is the $p \times p$ unit matrix while $FBB'F'$ is a diagonal matrix (with the diagonal elements being in decreasing order). Then $AX = F^{-1}FAX$ and $BY = F^{-1}FBY$ and we are back to the previous case with ϕ replaced by $\tilde{\phi}$ where $\tilde{\phi}(x_1, \ldots, x_p, z) \equiv \phi((F^{-1}x)', z)$.

Finally if rank $A = r < p$ then we may choose a basis $v_{.1}, v_{.2}, \ldots, v_{.r}$ for the column space of A. Any vector in the column space of B belongs to this space since $AA' \geqslant BB'$. If $a_{.j}$ is the jth column in A and $b_{.j}$ is the jth column in B then we may write $a_{.j} = \sum_i s_{ij} v_{.i}$ and $b_{.j} = \sum t_{ij} v_{.i}$. Putting $S = \{s_{ij}\}$ and $T = \{t_{ij}\}$ we obtain $A = VS$ and $B = VT$ where V is the $p \times r$ matrix $(v_{.1}, \ldots, v_{.r})$. Then S and T have, respectively, dimensions $r \times n_A$ and $r \times n_B$. Furthermore rank $S = r$ and $SS' \geqslant TT'$. If $\phi_V(y_1, \ldots, y_r, z) \equiv \phi(Vy, z)$ then $\phi((AX)', \sum X_i^2 - n_A) = \phi_V((SX)', \sum X_i^2 - n_A)$ and $\phi((BY)', \sum Y_i^2 - n_B) = \phi_V((TY)', \sum Y_i^2 - n_B)$. Thus we may apply the previous arguments to $\mathscr{F}_S$ and $\mathscr{F}_T$.

It would be interesting to know if there are general and manageable expressions for the local deficiencies between linear normal models.

The statistical significance of local deficiencies is particularly transparent in the one-dimensional (i.e. $m = 1$) case. In this case, as we saw in section 9.2, the deficiencies $\dot{\delta}_{k,\theta^0}(\mathscr{E}, \mathscr{F})$; $k = 2, 3, \ldots$ are all the same. They may then be expressed in terms of powers of most powerful tests or in terms of slopes of power-functions of locally most powerful tests.

Before stating these results let us remind ourselves of some notation from section 9.3. Let $\mathscr{E} = (P_\theta : \theta \in \Theta)$ be differentiable in θ^0. If $0 \leqslant \alpha \leqslant 1$ then $\beta(\alpha | P_{\theta^0}, \dot{P}_{\theta^0})$ denotes the maximal value of the integral $\dot{P}_{\theta^0}(\delta)$ for all size α tests δ for testing '$\theta = \theta^0$' against '$\theta > \theta^0$'. Thus $\beta(\alpha | P_{\theta^0}, \dot{P}_{\theta^0})$ is the maximal slope at θ^0 within the class of all size α tests for testing '$\theta = \theta^0$' against '$\theta > \theta^0$'. The function $\beta(\cdot | P_{\theta^0}, \dot{P}_{\theta^0})$ restricted to $[0, 1]$ is a concave function on $[0, 1]$, and may be any such function, vanishing at $\alpha = 0$ and $\alpha = 1$. This function characterizes $\mathscr{E}$ up to $\dot{\Delta}_{\theta^0}$-equivalence, i.e. it characterizes $\dot{\mathscr{E}}_{\theta^0}$. We shall therefore use the notation $\kappa(\alpha | \dot{\mathscr{E}}_{\theta^0})$ instead of $\beta(\alpha | P_{\theta^0}, \dot{P}_{\theta^0})$.

By the same notation the maximal power at P_2 which is obtainable for level α tests for testing 'P_1' against 'P_2' is denoted by $\beta(\alpha | P_1, P_2)$. Thus

$$\kappa(\alpha | \dot{\mathscr{E}}_{\theta^0}) = \sup\{\dot{P}_{\theta^0}(\delta) : P_{\theta^0}(\delta) = \alpha\}$$

while

$$\beta(\alpha | P_{\theta^0}, P_{\theta^0 + \varepsilon}) = \sup\{P_{\theta^0 + \varepsilon}(\delta) : P_{\theta^0}(\delta) = \alpha\}.$$

The promised results may now be stated as follows.

Theorem 9.4.8 (Local deficiences in terms of maximal power and in terms of maximal slope). *Assume that the parameter set Θ is a set of real numbers having* *the number θ^0 as an interior point. Let the experiments $\mathscr{E} = (P_\theta : \theta \in \Theta)$ and $\mathscr{F} = (Q_\theta : \theta \in \Theta)$ both be differentiable in θ^0. Then*

$$\dot{\delta}_{\theta^0}(\mathscr{E}, \mathscr{F}) = 2 \lim_{\varepsilon \to 0} \sup_{0 \leqslant \alpha \leqslant 1} [\beta(\alpha | Q_{\theta^0}, Q_{\theta^0+\varepsilon}) - \beta(\alpha | P_{\theta^0}, P_{\theta^0+\varepsilon})]/\varepsilon$$

and

$$\dot{\delta}_{\theta^0}(\mathscr{E}, \mathscr{F}) = \sup_{0 \leqslant \alpha \leqslant 1} [\kappa(\alpha | \dot{\mathscr{F}}_{\theta^0}) - \kappa(\alpha | \dot{\mathscr{E}}_{\theta^0})].$$

Proof. The last expression for $\dot{\delta}_{\theta^0}(\mathscr{E}, \mathscr{F})$ follows directly from the definition and from corollary 9.3.30. Put $A_\varepsilon = 2 \sup_{0 \leqslant \alpha \leqslant 1} [\beta(\alpha | Q_{\theta^0}, Q_{\theta^0+\varepsilon}) - \beta(\alpha | P_{\theta^0}, P_{\theta^0+\varepsilon})]$. Then, by corollary 9.3.30, $(P_{\theta^0}, P_{\theta^0+\varepsilon})$ is $(0, A_\varepsilon)$-deficient with respect to $(Q_{\theta^0}, Q_{\theta^0+\varepsilon})$. It follows that there is a transition M_ε such that $P_{\theta^0} M_\varepsilon = Q_{\theta^0}$ while $\| P_{\theta^0+\varepsilon} M_\varepsilon - Q_{\theta^0+\varepsilon} \| \leqslant A_\varepsilon$. Let M be a transition such that $M_\varepsilon \to M$ for some sequence of numbers ε tending to zero. Then

$$\begin{aligned} \| \dot{P}_{\theta^0} M - \dot{Q}_{\theta^0} \| &\leqslant \liminf_{\varepsilon \to 0} \| \dot{P}_{\theta^0} M_\varepsilon - \dot{Q}_{\theta^0} \| \\ &= \liminf_{\varepsilon \to 0} [\| P_{\theta^0+\varepsilon} M_\varepsilon - Q_{\theta^0} \| / \varepsilon] \\ &\leqslant \liminf_{\varepsilon \to 0} A_\varepsilon / \varepsilon \end{aligned}$$

while $P_{\theta^0} M = Q_{\theta^0}$. Thus $\dot{\delta}_{\theta^0}(\mathscr{E}, \mathscr{F}) \leqslant \liminf_{\varepsilon \to 0} A_\varepsilon/\varepsilon$. On the other hand if $P_{\theta^0} M = Q_{\theta^0}$ then $A_\varepsilon \leqslant \| P_{\theta^0+\varepsilon} M - Q_{\theta^0+\varepsilon} \|$ so that

$$\begin{aligned} \limsup_{\varepsilon \to 0} A_\varepsilon/\varepsilon &\leqslant \limsup_{\varepsilon \to 0} \| [(P_{\theta^0+\varepsilon} - P_{\theta^0})/\varepsilon] M - [(Q_{\theta^0+\varepsilon} - Q_{\theta^0})/\varepsilon] \| \\ &= \| \dot{P}_{\theta^0} M - \dot{Q}_{\theta^0} \|. \end{aligned}$$

Hence $\limsup_{\varepsilon \to 0} A_\varepsilon/\varepsilon \leqslant \dot{\delta}_{\theta^0}(\mathscr{E}, \mathscr{F})$ so that $A_\varepsilon/\varepsilon \to \dot{\delta}_{\theta^0}(\mathscr{E}, \mathscr{F})$ as $\varepsilon \to 0$. □

The function $\kappa = \kappa(\cdot | \dot{\mathscr{E}}_{\theta^0})$ yields a direct representation of $\dot{\mathscr{E}}_{\theta^0}$ since $\dot{\mathscr{E}}_{\theta^0} \sim (\lambda, \kappa)$ where λ is the uniform distribution on $[0, 1]$ and κ denotes the measure on $[0, 1]$ having κ as its distribution function. In particular the Fisher information at $\theta = \theta_0$ may be expressed as $\int_0^1 \kappa'(\alpha)^2 \, d\alpha = \int_0^1 [\kappa(\alpha)/\alpha - \kappa'(\alpha)]^2 \, d\alpha$.

It follows from the theorem that $\mathscr{E} = (P_\theta : \theta \in \Theta)$ is locally at least as informative as $\mathscr{F} = (Q_\theta : \theta \in \Theta)$ at $\theta = \theta^0$ if and only if $\kappa(\cdot | \dot{\mathscr{E}}_{\theta^0}) \geqslant \kappa(\cdot | \dot{\mathscr{F}}_{\theta^0})$. Thus we see again that the function $\kappa(\cdot | \dot{\mathscr{E}}_{\theta^0})$ characterizes $\mathscr{E}$ up to $\dot{\Delta}_{\theta^0}$-equivalence. The local deficiency distance $\dot{\Delta}_{\theta^0}$ then becomes just the supremum norm distance for these functions.

If a uniformly most powerful level α test δ for testing '$\theta = \theta^0$' against '$\theta > \theta^0$' is available in $\mathscr{E} = (P_\theta : \theta \in \Theta)$ then $E_{\theta^0+\varepsilon}\delta = \beta(\alpha | P_{\theta^0}, P_{\theta^0+\varepsilon})$ while $[d(E_\theta \delta)/d\theta]_{\theta = \theta^0} = \kappa(\alpha | \dot{\mathscr{E}}_{\theta^0})$. This works in particular when, say, the distributions $(P_\theta : \theta \geqslant \theta^0)$ have monotone likelihood ratio with respect to some statistic.

Let us consider the particular case of translation experiments on the real line. It is then known (see e.g. Lehmann, 1986) that the translation family determined by a distribution G possessing a continuous density g has monotone likelihood ratio (with respect to the identity function) if and only if $\{x : g(x) > 0\}$ is an interval on which $\log g$ is concave. This implies, as is easily seen, that there is a point ξ on the real line where g achieves its maximum value and that g is monotonically increasing to the left of ξ and monotonically decreasing to the right of ξ.

In general a distribution G (or the measure it induces) is called *unimodal* if there is a point ξ such that G is convex on $]-\infty, \xi[$ and G is concave on $]\xi, \infty[$. It follows that the condition of monotone likelihood ratio in this case implies unimodality. Actually we may say a bit more since, by Ibragimov (1956), the condition of monotone likelihood ratio implies that the convolution of G with any unimodal distribution is again unimodal. A distribution having this property is called *strongly unimodal.* Strongly unimodal distribution functions are clearly unimodal. The converse, however, is far from being true. (The Cauchy distribution provides an example of a unimodal distribution which is not strongly unimodal and, consequently, does not have monotone likelihood ratio.) Strongly unimodal distributions need not be absolutely continuous but if they are absolutely continuous then they determine a differentiable translation family with monotone likelihood ratio.

Note that translates of (strongly) unimodal distributions are again (strongly) unimodal. We may therefore say without ambiguity that a translation experiment $\mathscr{E}_P$ on the real line is *unimodal* or *strongly unimodal* according to whether P is unimodal or strongly unimodal. These experiments are not only interesting in themselves but also because any differentiable experiment $\mathscr{E} = (P_\theta : \theta \in \Theta)$ such that $\dot{P}_{\theta 0} \neq 0$ is locally equivalent to a strongly unimodal translation experiment and this experiment is unique up to equivalence. Before proving this let us collect some basic characterizations of the monotone likelihood ratio property.

Theorem 9.4.9 (Characterizations of strong unimodality). *Let G be a distribution function on the real line possessing a (necessarily unique) continuous density g. Let G_θ and g_θ denote the θ-translate of G and g respectively. Thus $G_\theta(x) \equiv_x G(x - \theta)$ and $g_\theta(x) \equiv_x g(x - \theta)$ when $\theta \in \mathbb{R}$. The following conditions are then equivalent.*

(i) $g(G^{-1}(\cdot))$ *is concave on* $]0, 1[$.
(i′) $g = \kappa(1 - G)$ *where* κ *is concave on* $[0, 1]$.
(ii) $[g > 0]$ *is an interval and* $\log g$ *is concave on this interval.*
(iii) $g_{\theta_2}/g_{\theta_1}$ *is monotonically increasing on* $[g_{\theta_1} > 0] \cup [g_{\theta_2} > 0]$.
(iv) G *is strongly unimodal.*

Furthermore these conditions imply that the translation experiment $\mathscr{E} = (P_\theta : \theta \in \mathbb{R})$ *is differentiable and that the function* κ *appearing in* (i') *is determined by*

$$\kappa(\alpha) = \kappa(\alpha | \dot{\mathscr{E}}_{\theta^0}) = g(G^{-1}(1-\alpha))$$

for any $\alpha \in [0,1]$ *and any* $\theta^0 \in \mathbb{R}$.

Remark. The proof below of the equivalence of (ii) and (iii) is taken largely from Lehmann (1986).

Proof. Assume that (ii) is satisfied and that $\{x : g(x) > 0\} =]t_0, t_1[$. The concavity of $\log g$ implies that $\lim_{x \to t_0} g(x)$ and $\lim_{x \to t_1} g(x)$ exist. By continuity and integrability we find that $0 = \lim_{x \to t_0} g(x) = \lim_{x \to t_1} g(x)$. It follows that there is a point $\xi \in]t_0, t_1[$ such that $g(\xi) = \max_x g(x)$. Furthermore $\log g$, and hence g, is absolutely continuous on any compact sub-interval of $]t_0, t_1[$.

If $\theta_2 > \theta_1$, then $g_{\theta_2}(x)/g_{\theta_1}(x) = 0$, or $= \exp -[\log g(-\theta_1 + x) - \log g(-\theta_2 + x)]$, or $= \infty$ as $x \in [g_{\theta_1} > 0] \cap [g_{\theta_2} = 0]$, or $x \in [g_{\theta_2} > 0] \cap [g_{\theta_1} > 0]$ or $x \in [g_{\theta_1} = 0] \cap [g_{\theta_2} > 0]$. It follows, by concavity, that G_{θ_2} has monotonically increasing likelihood ratio with respect to G_{θ_1} when $\theta_2 > \theta_1$. Hence the test with rejection region $[G^{-1}(1-\alpha), \infty[$ is a level α UMP test for testing $\theta \leqslant 0$ against $\theta > 0$, provided $\alpha > 0$. The power function of this test is the function

$$\theta \to 1 - G(G^{-1}(1-\alpha) - \theta).$$

Differentiating we find that the maximum level α slope at $\theta = 0$ is $g(G^{-1}(1-\alpha))$.

By concavity the set N of points $t \in]t_0, t_1[$ such that $(\log g)'(t)$, and hence $g'(t)$, does not exist, is necessarily countable. Restricting t to N^c we may state that $g'(t) \geqslant 0$ or $\leqslant 0$ according to whether $t \leqslant \xi$ or $t \geqslant \xi$. Then

$$\int_{-\infty}^{\infty} |g'(x)|\,dx = \int_{-\infty}^{\xi} g'(x)\,dx + \int_{\xi}^{\infty} -g'(x)\,dx = g(\xi) + g(\xi) = 2g(\xi) < \infty$$

so that $\mathscr{E}$ is differentiable.

Thus (ii) implies (iii) and that $\kappa(\alpha|\dot{\mathscr{E}}_{\theta^0}) = g(G^{-1}(1-\alpha))$ when $\alpha \in [0,1]$ and $\theta^0 \in \mathbb{R}$. It follows that (ii) implies (i) as well. Assume next that $g_{\theta_2}/g_{\theta_1}$ is monotonically increasing on $[g_{\theta_1} > 0] \cup [g_{\theta_2} > 0]$ when $\theta_2 > \theta_1$. Let $a < b$ be numbers so that $g(a) > 0$ and $g(b) > 0$. Put $x = 0$, $x' = (b-a)/2$, $\theta = -(a+b)/2$ and $\theta' = -a$. Then $x < x'$ and $\theta < \theta'$. Hence

$$\begin{aligned} g(a)/g((a+b)/2) &= g(x - \theta')/g(x - \theta) \\ &\leqslant g(x' - \theta')/g(x' - \theta) \\ &= g((a+b)/2)/g(b). \end{aligned}$$

Hence $g((a+b)/2) > 0$ and

$$\tfrac{1}{2}\log g(a) + \tfrac{1}{2}\log g(b) \leqslant \log g(\tfrac{1}{2}(a + b)).$$

It follows that $[g > 0]$ is an interval and that $\log g$ is concave. Thus (ii) $\Leftrightarrow$ (iii) $\Rightarrow$ (i).

Assume that condition (i′) holds for a given concave function κ. Then $[g>0]$ is again an interval $]t_0, t_1[$. Furthermore $g(G^{-1}(p)) = \kappa(1-p); 0<p<1$ so that condition (i) is satisfied. If $[a,b] \leqslant]t_0, t_1[$ and if X is distributed according to G then

$$\begin{aligned}\int_a^b -\kappa'(1 - G(x))\,dx &= \int_a^b [-\kappa'(1 - G(x))/g(x)]g(x)\,dx \\ &= \int_{G(a)}^{G(b)} [-\kappa'(1 - p)/\kappa(1 - p)]\,dp \\ &= \Big|_{G(a)}^{G(b)} \log \kappa(1 - p) \\ &= \log g(b) - \log g(a).\end{aligned}$$

Thus the measure induced by the distribution function $\log g$ on $]t_0, t_1[$ possesses the monotonically decreasing density $-\kappa'(1 - G)$. It follows that $\log g$ is concave on $]t_0, t_1[$. Hence condition (ii) is satisfied. Finally, assume that (i) is satisfied and put $\kappa(p) = g(G^{-1}(1 - p))$; $p \in]0, 1[$. Extend the concave function κ to $[0, 1]$ by continuity. If $0 < G(x) < 1$ then we obtain $\kappa(1 - G(x)) = g(G^{-1}(G(x))) = g(x)$. Thus $\lim_{x\to-\infty} g(x) = \kappa(1)$ while $\lim_{x\to\infty} g(x) = \kappa(0)$. It follows that $\kappa(0) = \kappa(1) = 0$. The function κ is absolutely continuous on $[0, 1]$ and possesses, by concavity, a monotonically decreasing derivative which is well defined for all points p not belonging to some countable set. If $x_1 < x < x_2$ and $g(x_1), g(x_2) > 0$ then $G(x_1), G(x_2) \in]0, 1[$ so that also $G(x) \in]0, 1[$ and, therefore, $g(x) = \kappa(1 - G(x)) > 0$. Furthermore $g(x) = \kappa(1 - G(x))$ for all x so that condition (i′) is satisfied also.

So far we have shown that conditions (i), (i′), (ii) and (iii) are all equivalent and that they imply the last statement. The proof is now completed by referring to Ibragimov (1956) where it is shown that a non-degenerate distribution G is strongly unimodal if and only if G is absolutely continuous with a density g satisfying (ii). □

We shall now see that the strongly unimodal translation experiments provide a characterization of the 'local' types of differentiable experiments.

Corollary 9.4.10 (Representation by strongly unimodal translation experiments). *Assume that $\Theta \subseteqq \mathbb{R}$ and that θ^0 is an interior point of Θ. Then any experiment $\mathscr{E} = (P_\theta : \theta \in \Theta)$ which is differentiable in θ^0 with $\dot{P}_{\theta^0} \neq 0$ is locally (i.e. Δ_{θ^0}-) equivalent to a strongly unimodal translation experiment $\hat{\mathscr{E}} =$*

$(G_\theta : \theta \in \Theta)$. $\hat{\mathscr{E}}$ is unique up to global (i.e. Δ-) equivalence, i.e. G is determined up to a translation.

The set $\{G_\theta : \theta \in \mathbb{R}\}$ of translates of the cumulative distribution function G together with the infinite translates 0 and 1 constitutes precisely the set of solutions of the differential equation $G' = \kappa(1 - G|\dot{\mathscr{E}}_{\theta^0})$.

Remark. The essential idea of this proof is taken from Hajek & Sidak (1967).

Proof. We may assume without loss of generality that $\Theta = \mathbb{R}$. Let $\mathscr{E} = (P_\theta : \theta \in \Theta)$ be differentiable in $\theta^0 \in \mathbb{R}$ and let us write $\kappa(\alpha)$ instead of $\kappa(\alpha|\dot{\mathscr{E}}_{\theta^0})$. Then κ is concave on $[0,1]$. Furthermore $\kappa(p) > 0$ when $p \in]0,1[$ while $\kappa(0) = \kappa(1) = 0$. Choose a point $p_0 \in]0,1[$ and put $\Psi(p) = \int_{p_0}^{p} dp/\beta(1-p)$ when $0 \leqslant p \leqslant 1$. Then Ψ is continuous and strictly increasing and $-\infty \leqslant \Psi(0) < \Psi(p) < \Psi(1) \leqslant \infty$ when $0 < p < 1$. If $x \in]\Psi(0), \Psi(1)[$ then there is a unique number $G(x) \in]0,1[$ such that $\Psi(G(x)) = x$. Put $G(x) = 0$ when $x \leqslant \Psi(0)$ and put $G(x) = 1$ when $x \geqslant \Psi(1)$. Then G is a continuous cumulative (probability) distribution function on $\mathbb{R}$. G is strictly increasing on $]\Psi(0), \Psi(1)[$ and $G^{-1}(p) = \Psi(p)$ when $p \in]0,1[$. Let μ denote Lebesgue measure on $]\Psi(0), \Psi(1)[$ and let μG^{-1} be the measure induced on $]0,1[$ from μ by the map G. Then

$$
\begin{aligned}
(\mu G^{-1})([a,b[) &= \mu(\{x : a \leqslant G(x) < b\}) \\
&= \mu([\Psi(a), \Psi(b)[) \\
&= \Psi(b) - \Psi(a) \\
&= \int_a^b dp/\kappa(1-p)
\end{aligned}
$$

when $\Psi(0) < a < b < \Psi(1)$. Hence

$$
\begin{aligned}
\int_{-\infty}^{x} \kappa(1 - G(\xi))\, d\xi &= \int_{\Psi(0)}^{x} I_{]0,G(x)]}(G(\xi))\kappa(1 - G(\xi))\, d\xi \\
&= \int_0^{G(x)} \kappa(1-p)\mu G^{-1}(dp) \\
&= \int_0^{G(x)} \kappa(1-p)/\kappa(1-p)\, dp = G(x)
\end{aligned}
$$

when $x \in]\Psi(0), \Psi(1)[$. It follows that $\int_{-\infty}^{x} \kappa(1 - G(\xi))\, d\xi = G(x)$ for all $x \in \mathbb{R}$. Thus G is absolutely continuous with a continuous density $g = \kappa(1 - G)$. Hence G is strongly unimodal by criterion (i′) of the theorem. If $\hat{\mathscr{E}} = (G_\theta : \theta \in \Theta)$ then the last statement of the theorem implies that $\kappa(\cdot|\dot{\hat{\mathscr{E}}}_{\theta^0}) = \kappa = \kappa(\cdot|\dot{\mathscr{E}}_{\theta^0})$ so that $\hat{\mathscr{E}}$ and $\mathscr{E}$ are locally equivalent at θ^0. Assume that at θ^0 $\mathscr{E}$ is also locally equivalent to the translation experiment $\tilde{\mathscr{E}} = (H_\theta : \theta \in \Theta)$ determined by the strongly unimodal distribution function H. Then $\kappa(\cdot|\dot{\mathscr{E}}_{\theta^0}) = \kappa(\cdot|\dot{\tilde{\mathscr{E}}}_{\theta^0}) = \kappa$. Hence, by the theorem, $Z = G$ and $Z = H$ are both solutions of the differential equation $Z' = \kappa(1 - Z)$. Note that the definition of G implies that $G(0) = p_0$. We may also assume without loss of generality that $H(0) - p_0$. (If not then replace H by the suitable translate.)

Now consider any function Z on $\mathbb{R}$ which satisfies the differential equation $Z' = \kappa(1 - Z)$. Then Z is necessarily continuously differentiable on $\mathbb{R}$ and $Z' \geqslant 0$. Put $\pi_0 = \lim_{x\to-\infty} Z(x)$ and $\pi_1 = \lim_{x\to\infty} Z(x)$. The differential equation requires that $0 \leqslant Z \leqslant 1$. The constant functions 0 and 1 are clearly solutions of this differential equation. Let us therefore assume that Z is neither of these two constant functions. Then $0 < Z(x) < 1$ for some x. Hence $\pi_0 < 1$ and $\pi_1 > 0$. If $\pi_1 < 1$ and if $\gamma \leqslant Z(x)$ where $\pi_1 > \gamma > 0$ then $1 > 1 - \gamma \geqslant 1 - Z(x) \geqslant 1 - \pi_1 > 0$ so that $Z'(x) = \kappa(1 - Z(x)) \geqslant \kappa((1 - \gamma) \wedge (1 - \pi_1)) > 0$. This, however, implies the contradiction $1 \geqslant Z(x) \to \infty$ as $x \to \infty$. Thus $\pi_1 = 1$. A similar argument shows that $\pi_0 = 0$. Thus Z is a cumulative distribution function. Note next that the set of solutions of our differential equation is invariant under translations. Let $\tilde{Z}$ be the unique translate of Z such that $\tilde{Z}(0) = p_0$. If $0 < \tilde{Z}(x) < 1$ then $d/dx\Psi(\tilde{Z}(x)) = \tilde{Z}'(x)/\beta(1 - \tilde{Z}(x)) = 1$. It follows that there is a constant c such that $\Psi(\tilde{Z}(x)) = x + c$ when $0 < \tilde{Z}(x) < 1$. Then $c = \Psi(\tilde{Z}(0)) = \Psi(p_0) = 0$. Hence $\Psi(\tilde{Z}(x)) = x$ when $0 < \tilde{Z}(x) < 1$. We may therefore also conclude that $\Psi(0) < x < \Psi(1)$ when $0 < \tilde{Z}(x) < 1$. Thus, by the definition of G, $\tilde{Z}(x) = G(x)$ when $0 < \tilde{Z}(x) < 1$. It follows then that $\tilde{Z} = G$. This proves the last statement of the corollary.

We saw above that if $\mathscr{E}$ is also locally equivalent at θ^0 to the strongly unimodal translation experiment $\tilde{\mathscr{E}} = (H_\theta : \theta \in \Theta)$ determined by the strongly unimodal distribution function H, then $Z = H$ is a solution of the differential equation $Z' = \kappa(1 - Z)$. Hence H is a translate of G. Thus $\Delta(\mathscr{E}, \tilde{\mathscr{E}}) = 0$. □

The local deficiency between strongly unimodal translation experiments may be simply described in terms of the fractile density functions of the translated distributions.

Proposition 9.4.11 (Local deficiencies between strongly unimodal translation experiments). *If G and H are strongly unimodal distribution functions on $\mathbb{R}$ possessing densities g and h respectively, then*

$$\dot{\delta}_{\theta^0}((G_\theta : \theta \in \Theta), (H_\theta : \theta \in \Theta)) = \sup\{h(H^{-1}(p)) - g(G^{-1}(p)); 0 \leqslant p \leqslant 1\}.$$

Proof. This follows directly from theorems 9.4.8 and 9.4.9. □

We have seen that the density–fractile function $p \to G'(G^{-1}(1 - p))$ is a useful characterization of the 'local type' of the strongly unimodal translation experiment $(G_\theta : \theta \in \Theta)$. In fact this function is the slope function $\beta(\cdot|G_{\theta^0}, \dot{G}_{\theta^0})$. If G is no longer strongly unimodal then this is no longer the case. However the density–fractile function, as we now shall see, is still a characteristic of the 'local type' of $\mathscr{E}$ at θ^0.

Proposition 9.4.12 (Local characterization by the 'density–fractile' function). *Let $\mathscr{E} = (G_\theta : \theta \in \mathbb{R})$ be a differentiable translation experiment on the real line. Put $\gamma(p) = G'(G^{-1}(1-p)); 0 < p < 1$. Then $\lim_{p\to 0}\gamma(p) = \lim_{p\to 1}\gamma(p) = 0$ and the continuous extension of γ to $[0,1]$ is absolutely continuous. The pair (λ, γ) where λ is the uniform distribution on $[0,1]$ and γ denotes the measure possessing γ as its distribution function, is (Δ-) equivalent to the first order approximation $\dot{\mathscr{E}}_{\theta^0}$ of $\mathscr{E}$.*

Proof. It is easily checked that $\lim_{p\to 0}\gamma(p) = \lim_{p\to 1}\gamma(p) = 0$. The set of points $p \in [0,1]$ such that either $G^{-1}(1-p) = 0$ or $G''(G^{-1}(1-p))$ fails to exist has Lebesgue measure zero. Furthermore

$$\int_0^1 |G''(G^{-1}(1-p))/G'(G^{-1}(1-p))|\,\mathrm{d}p = \int |G''(x)/G'(x)|\,G(\mathrm{d}x)$$

$$= \int |G''(x)|\,\mathrm{d}x < \infty$$

so that $p \to -G''(G^{-1}(1-p))/\gamma(p)$ is integrable and if $0 < p_1 < p_2 < 1$ then we find

$$\int_{p_1}^{p_2} [-G''(G^{-1}(1-p))/\gamma(p)]\,\mathrm{d}p = \int_{G^{-1}(1-p_2)}^{G^{-1}(1-p_1)} [-G''(x)/G'(x)]\,G(\mathrm{d}x)$$

$$= \int_{G^{-1}(1-p_2)}^{G^{-1}(1-p_1)} (-G''(x))\,\mathrm{d}x$$

$$= \Big|_{G^{-1}(1-p_2)}^{G^{-1}(1-p_1)} - G'(x)$$

$$= \gamma(p_2) - \gamma(p_1)$$

so that γ is absolutely continuous and $\gamma'(p) = -G''(G^{-1}(1-p))/\gamma(p)$ for almost all $p \in [0,1]$. As $\gamma(0) = \gamma(1) = 0$ we find that the Lebesgue measure λ on $[0,1]$, together with the measure defined by γ, is the first order approximation of the 'local type' of some differentiable experiment. Furthermore $\|G + \xi\dot{G}\| = \int |g(x) - \xi g'(x)|\,\mathrm{d}x = \int |1 + \xi(-g'(x)/g(x))|G(\mathrm{d}x) = \int |1 + \xi\gamma'(p)|\,\mathrm{d}p$ for all $\xi \in \mathbb{R}$. Hence, by theorem 9.3.8, we see that (λ, γ) represents the first order approximation of $(G_\theta : \theta \in \mathbb{R})$ and thus characterizes $\mathscr{E}$ up to local equivalence. □

We saw above that $\gamma(0) = \gamma(1) = 0$ and that γ is absolutely continuous. In addition it is easily seen that $1/\gamma$ is integrable on any interval $[\varepsilon, 1-\varepsilon]$ where $\varepsilon > 0$. This, however, is all we can ascertain, since any function γ on $[0,1]$ with these properties may be derived from a differentiable translation experiment

$(G_\theta : \theta \in \mathbb{R})$. In fact there is a unique differentiable translation experiment $(G_\theta : \theta \in \mathbb{R})$ for each number $p_0 \in]0, 1[$ such that $G(0) = p_0$ and such that the open set $[G' > 0]$ is Lebesgue equivalent to an interval.

To see this note first that $G'(G^{-1}(1 - p)) =_p \gamma(p)$ if and only if $G' = \gamma(1 - G)$. Furthermore

$$\int_{G(0)}^{G(x)} \mathrm{d}p/\gamma(1-p) = \int_{G(0)}^{G(x)} \mathrm{d}p/G'(G^{-1}(p))$$

$$= \int_0^x G(\mathrm{d}\xi)/G'(\xi)$$

$$= \int_0^x I_{]0,\infty[}(G'(\xi))\,\mathrm{d}\xi.$$

Thus $\int_{G(0)}^{G(x)} \mathrm{d}p/\gamma(1-p) = x$ if $G' > 0$ a.e. on $\langle 0, x \rangle$. The uniqueness statement follows from this.

Existence follows by defining G such that $\int_{p_0}^{G(x)} \mathrm{d}p/\gamma(1-p) = x$ when $x \in]\int_{p_0}^{0} \mathrm{d}p/\gamma(1-p), \int_{p_0}^{1} \mathrm{d}p/\gamma(1-p)[$ and then proceeding as in the first part of the proof of corollary 9.4.10.

We have proved the following proposition.

Proposition 9.4.13 (Density–fractile functions). *If $(G_\theta : \theta \in \mathbb{R})$ is a differentiable translation experiment then the function $\gamma : p \to g(G^{-1}(1-p))$ on $]0, 1[$ enjoys the following properties:*

(i) *γ may be extended to an absolutely continuous function on $[0, 1]$ such that $\gamma(0) = \gamma(1) = 0$.*

(ii) *$\int_\varepsilon^{1-\varepsilon} \mathrm{d}p/\gamma(p) < \infty$ when $0 < \varepsilon < \frac{1}{2}$.*

Conversely, if γ satisfies conditions (i) *and* (ii) *and if $0 < p_0 < 1$, then γ is of the form $\gamma : p \to g(G^{-1}(1-p))$ for a unique translation experiment $(G_\theta : \theta \in \mathbb{R})$ such that $G(0) = p_0$ and such that $(G' > 0)$ is Lebesgue equivalent to an interval.*

In particular it follows that if γ satisfies conditions (i) *and* (ii) *then γ is of the form $\gamma : p \to g(G^{-1}(1-p))$ for a differentiable translation experiment $(G_\theta : \theta \in \Theta)$ such that $[G' > 0]$ is Lebesgue equivalent to an interval. If so then the distribution G is determined up to a translation.*

Remark. It follows that if we restrict our attention to differentiable translation experiments $G_\theta : \theta \in \mathbb{R}$ such that $[G' > 0]$ is equivalent to an interval, then the function $\gamma : p \to g(G^{-1}(1-p))$ characterizes G up to a translation. If we go beyond distributions having intervals as their minimal support then the most general distribution yielding the same γ as G is obtained from G by translating each of the topological components of the open set $[G' > 0]$ such that they remain separated and ordered as before and, furthermore, such that the mass distributions on these components are translated accordingly.

If strongly unimodal translation experiments are locally equivalent then, as we have seen, the corresponding distributions are equal up to a shift. Hence local equivalence implies global equivalence in this case. As the converse statement is generally true we may conclude that local (i.e. $\dot{\Delta}$-) and global (i.e. Δ-) equivalence coincide for strongly unimodal differentiable translation experiments.

In general, however, translation experiments may be locally equivalent without being 'globally' equivalent. In fact, by the above results, any differentiable translation experiment is everywhere locally equivalent to a strongly unimodal translation experiment.

Extending the fact that Δ and $\dot{\Delta}$ induce the same equivalences for strongly unimodal translation experiments we shall now show that these two distances actually determine the same topologies for these experiments. We shall argue this by first observing that 'ordinary' weak convergence within the set of strongly unimodal distribution functions possessing continuous densities implies uniform convergence of these densities.

Proposition 9.4.14 (Weak topology for strongly unimodal distributions). *Consider strongly unimodal distributions $G, G_1, G_2, \ldots$ on $\mathbb{R}$. We shall assume that G has continuous density g while G_n; $n = 1, 2, \ldots$ has continuous density g_n.*

Let $n \to \infty$. Then the following conditions are equivalent:

(i) $G_n(x) \to G(x)$ *for all* $x \in \mathbb{R}$.
(ii) $G_n(x) \to G(x)$ *uniformly in* $x \in \mathbb{R}$.
(iii) $G_n \to G$ *in total variation.*
(iv) $g_n \to g$*-almost everywhere Lebesgue.*
(v) $g_n(x) \to g(x)$ *uniformly in* $x \in \mathbb{R}$.

Remark. If we replace the condition of strong unimodality with the condition of unimodality then the conclusion of the theorem still holds provided we add the assumption that

$$\limsup_n \sup_x g_n(x) \leqslant \sup_x g(x).$$

Proof. The implication (v) $\Rightarrow$ (iv) is trivial while the implication (iv) $\Rightarrow$ (iii) follows from Scheffé's convergence theorem. Then, since the implications (iii) $\Rightarrow$ (ii) and (ii) $\Rightarrow$ (i) are trivial, it suffices to show that (i) $\Rightarrow$ (v). Let us therefore assume that $G_n(x) \to G(x)$ for all $x \in \mathbb{R}$. Unimodality implies that, for each n, there is a number a_n such that G_n is convex on $]-\infty, a_n[$ while G_n is concave on $]a_n, \infty[$. As a probability cumulative distribution function G necessarily is neither convex nor concave the sequence $a_1, a_2, \ldots$ must be bounded. Furthermore $G_n(x) \to G(x)$ uniformly in x (i.e. (i) $\Leftrightarrow$ (ii)) by a standard argument which we will give here since we will need it again in a later stage of this proof. This argument runs as follows.

If $F_1, F_2, \ldots$ and F are distribution functions on the real line, if F is continuous and if $F_n(x) \to F(x)$ for all x then the family $(F, F_1, F_2, \ldots)$ is uniformly equicontinuous and thus $F_n(x) \to F(x)$ uniformly in x.

We may assume now by compactness that the sequence a_n; $n = 1, 2, \ldots$ converges to a number a. Replacing G_n with $G_n * \delta_{-a_n}$ and G with $G * \delta_{-a}$ we see that we may also assume that $a = a_1 = a_2 = \cdots = 0$. Then $G, G_1, G_2, \ldots$ are all convex on $]-\infty, 0]$ and they are all concave on $[0, \infty[$. Thus $g_n(0) \geqslant g_n(x)$ and $g(0) \geqslant g(x)$; $x \in \mathbb{R}$.

Now g_n is monotonically decreasing on $[0, \infty[$. Hence

$$\liminf_n g_n(x) \geqslant \liminf_n [G_n(x, x + \varepsilon)/\varepsilon]$$

$$= G(x, x + \varepsilon)/\varepsilon \downarrow g(x) \quad \text{as} \quad \varepsilon \downarrow 0.$$

Thus $\liminf_n g_n(x) \geqslant g(x)$ when $x \geqslant 0$. If $x > 0$ then similarly we obtain

$$\limsup_n g_n(x) \leqslant \limsup_n (G_n(x - \varepsilon, x)/\varepsilon)$$

$$= G(x - \varepsilon, \varepsilon)/\varepsilon \downarrow g(x) \quad \text{as} \quad \varepsilon \downarrow 0.$$

Hence $g_n(x) \to g(x)$ when $x > 0$.

Similarly $g_n(x) \to g(x)$ when $x < 0$. *We have so far only used unimodality and not strong unimodality.* If only unimodality is assumed then, as is shown in the next example, we cannot conclude that $g_n(0) \to g(0)$. Using log concavity we find for any ε that

$$\log g_n(\varepsilon) = \log g_n(\tfrac{1}{2} \cdot 0 + \tfrac{1}{2}(2\varepsilon)) \geqslant \tfrac{1}{2} \log g_n(0) + \tfrac{1}{2} \log g_n(2\varepsilon)$$

so that $\log g_n(0) \leqslant 2 \log g_n(\varepsilon) - \log g_n(2\varepsilon)$ provided $0 < \varepsilon$ and $G_n(2\varepsilon) < 1$. Using the fact that g_n is monotonically decreasing on $[0, \infty[$ we find that

$$\limsup_n \log g_n(0) \leqslant 2 \log g(\varepsilon) - \log g(2\varepsilon) \to \log g(0) \quad \text{as} \quad \varepsilon \downarrow 0.$$

Hence $\limsup_n g_n(0) \leqslant g(0)$ so that $g_n(x) \to g(x)$ for all $x \in \mathbb{R}$.

Observe now that the densities $g, g_1, g_2, \ldots$ are all uniformly continuous distribution functions. Furthermore these distribution functions are all monotonically increasing on $]-\infty, 0]$ while they are all monotonically decreasing on $[0, \infty[$. It follows then, by pointwise convergence, that the family $\{g, g_1, g_2, \ldots\}$ is uniformly equicontinuous so that $g_n(x) \to g(x)$ uniformly in x. □

Here is an example showing that unless we add further assumptions the assumption of strong unimodality cannot be weakened to the assumption of unimodality.

Example 9.4.15. Let G be a symmetric strongly unimodal distribution function possessing a continuous density g. Then $g(x) \equiv_x g(-x)$. Let a_n; $n = 1, 2, \ldots$

be a sequence of positive numbers such that $a_n/n \to 0$ while $\liminf_n a_n > 0$. (Thus $a_n = n^{1/2}$, $n = 1, 2, \dots$ is a possibility.) Put

$$C_n = \int_{n|x|>1} g(x)\,\mathrm{d}x + (2/n)g(1/n) + a_n/n$$

and

$$g_n(x) = \begin{cases} g(x)/C_n & \text{when} \quad |x| \geqslant 1/n; \\ [g(1/n) + a_n - na_n|x|]/C_n & \text{when} \quad |x| \leqslant 1/n. \end{cases}$$

Then g_n is an even continuous probability density. g_n is monotonically increasing on $]-\infty, 0]$ while g_n is monotonically decreasing on $[0, \infty[$. It follows that the distribution G_n determined by g_n is unimodal. Furthermore

$$g_n'(x) = \begin{cases} g'(x)/C_n & \text{when} \quad |x| > 1/n; \\ -na_n \operatorname{sgn} x/C_n & \text{when} \quad 0 < |x| < 1/n. \end{cases}$$

Thus $\int |g_n'(x)|\,\mathrm{d}x = C_n^{-1} \int_{n|x|\geqslant 1} |g'(x)|\,\mathrm{d}x + 2a_n/C_n < \infty$ so that the translation experiment $\mathscr{G}_n$ determined by G_n is differentiable. The translation experiment $\mathscr{G}$ determined by G is also differentiable since G is strongly unimodal and has a continuous density.

Letting $n \to \infty$ we find that $\lim_n C_n = 1$ and, consequently, that $\lim_n g_n(x) = g(x)$ when $x \neq 0$. Thus, by Scheffé's convergence theorem, $G_n \to G$ in total variation. The densities $g_1, g_2, \dots$, however, do not converge everywhere pointwise, since $\liminf_n g_n(0) = g(0) + \liminf_n a_n > g(0)$. In particular this sequence does not converge uniformly so that condition (v) is not satisfied in this case.

It follows that we cannot delete the assumption that the distributions $G_1, G_2, \dots$ are strongly unimodal, even when we are assuming that G is strongly unimodal.

The deficiency distance, $\Delta(\mathscr{G}_n, \mathscr{G})$, between $\mathscr{G}_n$ and $\mathscr{G}$ is clearly at most equal to the total variation of $G_n - G$. Thus $\Delta(\mathscr{G}_n, \mathscr{G}) \to 0$. On the other hand

$$\begin{aligned} \liminf_n \|\dot{G}_{n,\theta^0}\| &= \liminf_n \int |g_n'(x)|\,\mathrm{d}x \\ &= \int |g'(x)|\,\mathrm{d}x + 2 \liminf_n a_n \\ &> \int |g'(x)|\,\mathrm{d}x = \|\dot{G}_{\theta^0}\| \end{aligned}$$

so that $\mathscr{G}_n$ does not converge to $\mathscr{G}$ for any distance $\dot{\Delta}_{\theta^0}$ (they do not depend on θ^0 in this case).

Thus we have a situation where unimodal translation experiments $\mathscr{G}_1, \mathscr{G}_2, \dots$ converge globally to a strongly unimodal translation experiment $\mathscr{G}$ although they do not converge locally anywhere.

Let $\dot{\Delta}$ denote either of the equal restrictions of the local deficiency distance $\dot{\Delta}_{\theta^0}$; $\theta^0 \in \mathbb{R}$ to the differentiable translation experiments.

Global and local convergence coincide within the set of strongly unimodal translation experiments.

Theorem 9.4.16 (Equivalence of local and global convergence for strongly unimodal translation experiments). *Consider strongly unimodal distributions* $G, G_1, G_2, \ldots$ *on* $\mathbb{R}$. *We shall assume that* G *has continuous density* g *while* G_n; $n = 1, 2, \ldots$ *has continuous density* g_n.

Let $\mathscr{G}$ *be the translation experiment determined by* G *and let* $\mathscr{G}_n$; $n = 1, 2, \ldots$, *be the translation experiment determined by* G_n. *Then* $\Delta(\mathscr{G}_n, \mathscr{G}) \to 0$ *if and only if* $\dot{\Delta}(\mathscr{G}_n, \mathscr{G}) \to 0$.

Furthermore, these conditions are equivalent to the condition that the equivalent conditions of proposition 9.4.14 *hold for suitable translates of the distributions* $G_1, G_2, \ldots$ *and* G.

Remark. In the proof of this statement we will rely on the following convergence criterion (Torgersen, 1972c) for absolutely continuous translation experiments. Let $P_1, P_2, \ldots$ and P be absolutely continuous distributions on $\mathbb{R}^k$ where $k < \infty$ and let $\mathscr{E}_{P_1}, \mathscr{E}_{P_2}, \ldots$ and $\mathscr{E}_P$ be the corresponding translation experiments. Then $\Delta(\mathscr{E}_{P_n}, \mathscr{E}_P) \to 0$ if and only if there are points $a_1, a_2, \ldots$ in $\mathbb{R}^k$ such that $P_n * \delta_{a_n}$ converges to P in total variation.

Proof. Choose a number $p_0 \in]0, 1[$. We may assume without loss of generality that $G(0) = G_1(0) = G_2(0) = \cdots = p_0$. Put $\kappa_n(p) = g_n(G^{-1}(1 - p))$ and $\kappa(p) = g(G^{-1}(1 - p))$. If $\dot{\Delta}(\mathscr{G}_n, \mathscr{G}) \to 0$ then $\kappa_n(p) \to \kappa(p)$ uniformly in $p \in [0, 1]$. Furthermore $\int_{p_0}^{\tau} dp/\kappa_n(p) = -G_n^{-1}(1 - \tau)$ and $\int_{p_0}^{\tau} dp/\kappa(p) = -G^{-1}(1 - \tau)$. Hence $G_n^{-1}(p) \to G^{-1}(p)$ when $p \in]0, 1[$ so that $G_n(x) \to G(x)$ for all x. Hence, by the proposition, $G_n \to G$ in total variation so that $\Delta(\mathscr{G}_n, \mathscr{G}) \to 0$. Assume that $\Delta(\mathscr{G}_n, \mathscr{G}) \to 0$. Then there are constants $a_1, a_2, \ldots$ such that $G_n * \delta_{a_n} \to G$ in total variation. Then $G_n(x - a_n) - G(x) \to 0$ uniformly in x. Hence $p_0 - G(a_n) = G_n(a_n - a_n) - G(a_n) \to 0$ so that $G(a_n) \to p_0 = G(0)$. It follows that $a_n \to 0$. Hence $G_n(x) \to G(x)$ for all x. Then $G_n^{-1}(1 - p) \to G^{-1}(1 - p)$; $0 < p < 1$ and $g_n(x) \to g(x)$ uniformly in x. Hence $\kappa_n(p) \to \kappa(p)$; $p \in [0, 1]$ and this convergence is necessarily uniform in p. Hence $\dot{\Delta}(\mathscr{G}_n, \mathscr{G}) \to 0$. □

It should be kept in mind that the theory for local comparison which has been outlined here is only concerned with first order approximations. There are important cases of interest where the first order approximations are quite inadequate. Thus any experiment $\mathscr{E} = (P_\theta : \theta \in \Theta)$ such that $[\partial P_\theta / \partial \theta_i]_{\theta^0} = 0$; $i = 1, \ldots, m$ is locally equivalent (for first order approximations) to a totally non-informative experiment. If, however, we want to test '$\theta = \theta^0$' against close alternatives, then such experiments may behave quite differently.

Along with the notion of local information considered here goes a notion of local sufficiency. Thus if $\mathscr{E} = (\mathscr{X}, \mathscr{A}; P_\theta : \theta \in \Theta)$ is differentiable at $\theta = \theta^0$

and if $\mathscr{F}$ is the restriction $\mathscr{E}|\mathscr{B}$ of $\mathscr{E}$ to some sub σ-algebra $\mathscr{B}$ of $\mathscr{A}$ then $\dot{\Delta}_{\theta^0}(\mathscr{E},\mathscr{F}) = 0$ if and only if $\mathrm{d}\dot{P}_{\theta^0,i}/\mathrm{d}P_{\theta^0}$ may be specified $\mathscr{B}$-measurable for all $i = 1,\ldots,m$. Thus these Radon–Nikodym derivatives describe the minimal locally sufficient sub σ-algebra of $\mathscr{A}$.

Local sufficiency of $\mathscr{B}$ may also be described by the requirement that conditional expectations should vary little within small neighbourhoods of θ^0.

9.5 Comparison of measures. Dilations

Orderings of experiments and of measures are often expressed in terms of dilations. The usual notion of a dilation as a kernel which dilates is, however, too narrow and should for our purposes be replaced by some notion of (almost) density preserving kernels. The typical situation is as follows.

Assume that we are given two measure families $\mathscr{E} = (\mathscr{X},\mathscr{A};\mu_\theta:\theta\in\Theta)$ and $\mathscr{F} = (\mathscr{Y},\mathscr{B};\nu_\theta:\theta\in\Theta)$ such that the measurable spaces $(\mathscr{X},\mathscr{A})$ and $(\mathscr{Y},\mathscr{B})$ are both Euclidean.

Assume also that we are given a point $\theta^0\in\Theta$ such that the measures μ_{θ^0} and ν_{θ^0} are non-negative and dominate, respectively, $\mathscr{E}$ and $\mathscr{F}$. Deficiency for deficiency functions vanishing at θ^0 may then be expressed in terms of reversed randomizations.

Theorem 9.5.1 (Almost density preserving kernels). *Assume that the requirements described above are satisfied and that $\varepsilon_{\theta^0} = 0$. For each $\theta\in\Theta$, let f_θ and g_θ be densities of, respectively, μ_θ and ν_θ with respect to μ_{θ^0} and ν_{θ^0} respectively.*

Then $\mathscr{E}$ is ε-deficient with respect to $\mathscr{F}$ if and only if $\mu_{\theta^0} = D\nu_{\theta^0}$ for a Markov kernel D from $(\mathscr{Y},\mathscr{B})$ to $(\mathscr{X},\mathscr{A})$ such that

$$\int\left|\int f_\theta(x)D(\mathrm{d}x|y) - g_\theta(y)\right|\nu_{\theta^0}(\mathrm{d}y) \leqslant \varepsilon_\theta; \qquad \theta\in\Theta.$$

Proof. It follows from theorem 9.2.3 that $\mathscr{E}$ is ε-deficient with respect to $\mathscr{F}$ if and only if there is a Markov kernel M from $(\mathscr{X},\mathscr{A})$ to $(\mathscr{Y},\mathscr{B})$ such that $\|\mu_\theta M - \nu_\theta\| \leqslant \varepsilon_\theta; \theta\in\Theta$. If M has this property then, since $\varepsilon_{\theta^0} = 0$, $\mu_{\theta^0}M = \nu_{\theta^0}$. It follows that there is a joint distribution on $(\mathscr{X}\times\mathscr{Y},\mathscr{A}\times\mathscr{B})$ with marginals μ_{θ^0} and ν_{θ^0} such that the conditional distribution of the second coordinate given $x\in\mathscr{X}$ is $M(\cdot|x)$. Let D be a regular conditional distribution of the first coordinate (in $\mathscr{X}$) given the second coordinate (in $\mathscr{Y}$). Then $\mu_{\theta^0}\times M = D\times\nu_{\theta^0}$. It is easily checked that $y\to\int f_\theta(x)D(\mathrm{d}x|y)$ is a density of $\mu_\theta M$ with respect to ν_{θ^0} so that $\varepsilon_\theta \geqslant \|\mu_\theta M - \nu_\theta\| = \int|\int f_\theta(x)D(\mathrm{d}x|y) - g_\theta(y)|\,\nu_{\theta^0}(\mathrm{d}y)$.

Conversely assume that a randomization D mapping ν_{θ^0} onto μ_{θ^0} and having this property exists. Let M be a regular conditional distribution of the second coordinate (in $\mathscr{Y}$) given the first coordinate (in $\mathscr{X}$) for the distribution $D\times\nu_{\theta^0}$ on $\mathscr{A}\times\mathscr{B}$. Thus $D\times\nu_{\theta^0} = \mu_{\theta^0}\times M$. We find again that

$y \to \int f_\theta(x)D(dx|y)$ is a density of $\mu_\theta M$ with respect to $\nu_{\theta 0}$ and thus $\|\mu_\theta M - \nu_\theta\| = \int |\int f_\theta(x)D(dx|y) - g_\theta(y)| \nu_{\theta 0}(dy) \leqslant \varepsilon_\theta$; $\theta \in \Theta$. □

The assumption that the kernels should map $\nu_{\theta 0}$ into $\mu_{\theta 0}$ (or $\mu_{\theta 0}$ into $\nu_{\theta 0}$) is a generalized form of the condition that a Markov matrix should be doubly Markov (stochastic). We obtain the latter condition if we impose the condition that $\mathscr{X}$ and $\mathscr{Y}$ are the same finite sets and the measures $\mu_{\theta 0}$ and $\nu_{\theta 0}$ are both the uniform distribution on this set.

Let us express these facts in terms of matrices in the slightly more general situation where $\mu_{\theta 0}$ is a non-negative distribution μ on $\{1,\ldots,m\}$ and $\nu_{\theta 0}$ is a non-negative distribution ν on $\{1,\ldots,n\}$.

Let us also assume, for simplicity, that the measure families are finite. Thus we assume that we are given a finite family $\mathscr{E} = (\mu, \mu_1, \ldots, \mu_r)$ of measures on $\{1,\ldots,m\}$ and a finite family $\mathscr{F} = (\nu, \nu_1, \ldots, \nu_r)$ of measures on $\{1,\ldots,n\}$.

Identify $\mathscr{E}$ with (μ, A) where μ is the row matrix $(\mu(1),\ldots,\mu(m))$ and A is the $r \times m$ matrix whose (k,i)th element is the number $\mu_k(i)$; $k = 1,\ldots,r$; $i = 1,\ldots,m$.

Identify $\mathscr{F}$ with (ν, B) where ν is the row matrix $(\nu(1),\ldots,\nu(n))$ and B is the $r \times n$ matrix whose (k,j)th element is $\nu_k(j)$; $k = 1,\ldots,r$; $j = 1,\ldots,n$.

Thus the entries of the row matrices μ and ν are non-negative and we shall assume that they are positive. Otherwise there is no restriction on the (real valued) entries of the matrices μ, A, ν and B.

The theorem yields the following comparison rules for matrices.

Corollary 9.5.2 (Informational inequalities for matrices). Let $\varepsilon_1,\ldots,\varepsilon_r$ *be non-negative numbers and let A and B be matrices with real valued entries and with dimensions $r \times m$ and $r \times n$ respectively.*

Also let μ and ν be row matrices with positive entries and of dimensions m and n respectively. Put $\tilde{A}_{k,i} = A_{k,i}\mu_i$ and $\tilde{B}_{k,j} = B_{k,j}\nu_j$; $k = 1,\ldots,r$; $i = 1,\ldots,m$; $j = 1,\ldots,n$.

Then the following conditions are equivalent.

(i) $\mu M = \nu$ *for an $m \times n$ Markov matrix M such that* $\sum_j |(\tilde{A}M)_{k,j} - \tilde{B}_{k,j}| \leqslant \varepsilon_k$; $k = 1,\ldots,r$.

(ii) $\mu = \nu D$ *for an $n \times m$ Markov matrix D such that* $\sum_j |(AD)_{k,j} - B_{k,j}| \nu_j \leqslant \varepsilon_k$; $k = 1,\ldots,r$.

(iii)
$$\sum_i \psi(\mu_i, A_{1,i}, A_{2,i}, \ldots, A_{r,i})$$
$$\geqslant \sum_j \psi(\nu_j, B_{1,j}, B_{2,j}, \ldots, B_{r,j})$$
$$- \sum_{k=1}^{r} \varepsilon_k[\psi(0, \overset{(1)}{0}, \ldots, \overset{(k)}{-1}, \ldots, \overset{(r)}{0}) \vee \psi(0, \overset{(1)}{0}, \ldots, \overset{(k)}{1}, \ldots, \overset{(r)}{0})]$$

for each sublinear function ψ on $\mathbb{R} \times \mathbb{R}^r$.

Remark. We leave it to the reader to simplify these statements when μ and ν are uniform. If $\varepsilon_1 = \varepsilon_2 = \cdots = \varepsilon_k = 0$ then (iii) may be phrased

(iii′) $\sum_i \phi(A_{1i}/\mu_i, \ldots, A_{ri}/\mu_i)\mu_i \geqslant \sum_j \phi(B_{1j}/\nu_j, \ldots, B_{rj}/\nu_j)\nu_j$ *for each convex function ϕ on $\mathbb{R}^r$.*

Consider a non-negative measure T on $\mathbb{R}^m$ and a Markov kernel D from $\mathbb{R}^m$ to $\mathbb{R}^m$. If $\varepsilon_1, \ldots, \varepsilon_m$ are non-negative numbers and if $\varepsilon = (\varepsilon_1, \ldots, \varepsilon_m)$ then we shall say that *D is a (T, ε) dilation* if $\int |\int x_i D(\mathrm{d}x|y) - y_i| \, T(\mathrm{d}y) \leqslant \varepsilon_i; i = 1, \ldots, m$. By the definition in the previous section a dilation is a Markov kernel D from $\mathbb{R}^m$ to $\mathbb{R}^m$ such that $\int x_i D(\mathrm{d}x|y) \equiv_y y_i; i = 1, \ldots, m$. Clearly a dilation is a $(T, 0)$ dilation and conversely any $(T, 0)$ dilation is T-equivalent to a dilation.

Here is a more general corollary of the theorem which includes some of the standard results on dilations in $\mathbb{R}^m$ as special cases.

Corollary 9.5.3 (ε-dilations in $\mathbb{R}^m$). *Let S and T be non-negative measures on $\mathbb{R}^m$ such that the projections are integrable. Let the measure S_i (the measure T_i) for each $i = 1, \ldots, m$, be the measure having the projection on the ith coordinate space as a density with respect to the measure S (the measure T). Then the following conditions are equivalent for non-negative numbers $\varepsilon_1, \ldots, \varepsilon_k$.*

(i) $SM = T$ *for a Markov kernel M from $\mathbb{R}^m$ to $\mathbb{R}^m$ such that* $\|S_i M - T_i\| \leqslant \varepsilon_i; i = 1, \ldots, m$.
(ii) $S = DT$ *for a (T, ε) dilation D from $\mathbb{R}^m$ to $\mathbb{R}^m$.*
(iii) $\int \psi(1, x) S(\mathrm{d}x) \geqslant \int \psi(1, x) T(\mathrm{d}x)$

$$- \sum_{i=1}^m \varepsilon_i [\psi(\overset{(0)}{0}, \overset{(1)}{0}, \ldots, \overset{(i)}{-1}, \ldots, \overset{(m)}{0}) \vee \psi(\overset{(0)}{0}, \overset{(1)}{0}, \ldots, \overset{(i)}{1}, \ldots, \overset{(m)}{0})]$$

for each sublinear function ψ on $\mathbb{R} \times \mathbb{R}^m$.

Remark. If $\varepsilon_1 = \cdots = \varepsilon_k = 0$ then (ii) may be written:

(ii′) $S = DT$ *for a dilation D.*

In this case (iii) may be written:

(iii′) $\int \phi(x) S(\mathrm{d}x) \geqslant \int \phi(x) T(\mathrm{d}x)$ *for all convex functions ϕ on $\mathbb{R}^m$.*

Proof. Apply the theorem and the sublinear function criterion ((iii) in section 9.2) to the measure families $(S, S_1, \ldots, S_m)$ and $(T, T_1, \ldots, T_m)$. □

If $\mathscr{E} = (\mathscr{X}, \mathscr{A}; \mu_\theta : \theta \in \Theta)$ is a measure family with finite parameter set Θ then in section 9.3 we defined its standard measure as the measure induced from $\mu = \sum_\theta |\mu_\theta|$ by the map $x \to ([\mathrm{d}\mu_\theta/\mathrm{d}\mu]_x; \theta \in \Theta)$ from $\mathscr{X}$ to $\mathbb{R}^\Theta$. Thus the corollary may be phrased as follows.

Corollary 9.5.4 (Ordering in terms of standard measures). *Let S and T denote the standard measures of the measure families* $\mathscr{E} = (\mu_1, \ldots, \mu_m)$ *and* $\mathscr{F} = (\nu_1, \ldots, \nu_m)$ *respectively.*

For each $i = 1, \ldots, m$, *let the measure* S_i (*the measure* T_i) *be the measure on* $\mathbb{R}^m$ *having the projection on the* ith *coordinate space as a density with respect to the measure S (the measure T).*

Then the measure family $(\sum_i |\mu_i|, \mu_1, \ldots, \mu_m)$ *is* $(0, \varepsilon_1, \ldots, \varepsilon_m)$-*deficient with respect to* $(\sum_i |\nu_i|, \nu_1, \ldots, \nu_m)$ *if and only if the (equivalent) conditions* (i), (ii) *and* (iii) *of the previous corollary are satisfied.*

In particular if the measures $\mu_1, \ldots, \mu_m$ *and* $\nu_1, \ldots, \nu_m$ *are all non-negative then* $\mathscr{E}$ *is* 0-*deficient with respect to* $\mathscr{F}$ *if and only if* $S = DT$ *for a dilation D.*

Another application is to differentiable experiments. In particular the result below provides the criterion for 'being locally at least as informative as' which we commented on and utilized in the previous section.

Corollary 9.5.5 (Local deficiencies). *Assume that the parameter set* Θ *is a subset of* $\mathbb{R}^m$ *and that the point* θ^0 *is an interior point of* Θ.

Let the experiments $\mathscr{E} = (P_\theta : \theta \in \Theta)$ *and* $\mathscr{F} = (Q_\theta : \theta \in \Theta)$ *both be differentiable in* θ^0 *and let* $\varepsilon_1, \ldots, \varepsilon_m$ *be non-negative numbers.*

Then each of (the equivalent) conditions (i)–(iii) *of corollary 9.5.3 are equivalent to the condition that* $\dot{\mathscr{E}}_{\theta^0} = (P_{\theta^0}, \dot{P}_{\theta^0,1}, \ldots, \dot{P}_{\theta^0,m})$ *is* $(0, \varepsilon_1, \ldots, \varepsilon_m)$-*deficient with respect to* $\dot{\mathscr{F}}_{\theta^0} = (Q_{\theta^0}, \dot{Q}_{\theta^0,1}, \ldots, \dot{Q}_{\theta^0,m})$.

In particular $\mathscr{E}$ *is locally at least as informative as* $\mathscr{F}$ *at* θ^0 *if and only if* $F(\cdot|\theta^0, \mathscr{E})$ *is a dilation of* $F(\cdot|\theta^0, \mathscr{F})$ *or, equivalently, that* $\int \phi(x) F(\mathrm{d}x|\theta^0, \mathscr{E}) \geqslant \int \phi(x) F(\mathrm{d}x|\theta^0, \mathscr{F})$ *for each convex function* ϕ *on* $\mathbb{R}^m$.

We saw in section 6.5 how comparability of translation experiments may be decided by a convolution criterion. As shown by Janssen (1988), comparison of exponential experiments may also be expressed in terms of convolution. Furthermore, by the same paper, comparison in this case reduces to comparison by factorization. We shall now derive this important result from the dilation criterion, theorem 9.5.1, and by using Laplace transforms as in Janssen's paper.

Theorem 9.5.6 (Convolution criterion for comparison of exponential experiments). *Assume that the parameter set* Θ *is a subset of a finite dimensional inner product space K. Let* $\mathscr{E} = (P_\theta : \theta \in \Theta)$ *and* $\mathscr{F} = (Q_\theta : \theta \in \Theta)$ *be exponential experiments such that for non-negative measures* μ *and* ν *on K and for* $\mathbb{R}^k$-*valued statistics S and T*

$$\mathrm{d}P_\theta/\mathrm{d}\mu = a(\theta) f e^{(\theta, S)}, \qquad \mathrm{d}Q_\theta/\mathrm{d}\nu = b(\theta) g e^{(\theta, T)}.$$

Here f and g are measurable functions not depending on θ.

Consider the following conditions:

(a) $\mathscr{F} \leqslant \mathscr{E}$,

(b) $\mathscr{F} \times \mathscr{G} \sim \mathscr{E}$ *for some experiment* $\mathscr{G}$,

(c) $\mathscr{L}_\theta(T)$ *is a convolution factor of* $\mathscr{L}_\theta(S)$ *for some* $\theta \in \Theta$,

(d) $\mathscr{L}_\theta(T)$ *is a convolution factor of* $\mathscr{L}_\theta(S)$ *for all* $\theta \in \Theta$.

Then conditions (c) *and* (d) *are equivalent and imply* (b) *with* $\mathscr{G}$ *being an exponential experiment in the same form.*

Furthermore, if Θ *has interior points then all four conditions* (a)–(d) *are equivalent.*

Remark 1. In contrast to the convolution criterion for comparison of translation experiments, this criterion applies to a situation where information generally is increased by convoluting a given family of distributions.

Remark 2. Asymptotic factorization was studied in Ehm & Müller (1983). The theorem is close to the results in that paper.

Remark 3. If the set Θ possesses interior points relative to the smallest affine set containing it, then (a) and (b) are still equivalent. Indeed if θ_0 is an interior point in this sense then it is readily checked that all four conditions are equivalent provided in (c) and (d) we replace S and T by their projections on the linear space spanned by $\Theta - \theta_0$.

Proof. We may assume, by sufficiency, that K is the sample space of both $\mathscr{E}$ and $\mathscr{F}$ and that S and T are both the identity map on K. Note also that $\mathrm{d}P_\theta/\mathrm{d}P_{\theta_0} = (a(\theta)/a(\theta_0))\mathrm{e}^{(\theta-\theta_0,S)}$ and that $\mathrm{d}Q_\theta/\mathrm{d}Q_{\theta_0} = (b(\theta)/b(\theta_0))\mathrm{e}^{(\theta-\theta_0,T)}$ when $\theta_0, \theta \in \Theta$. If $P_{\theta_0} = Q_{\theta_0} * \kappa$ for a probability measure κ then it is a matter of checking that $P_\theta \equiv Q_\theta * \kappa_\theta$ where $\mathrm{d}\kappa_\theta/\mathrm{d}\kappa = \mathrm{e}^{(\theta-\theta_0,U)}c(\theta)$ with U being the identity map and $c(\theta)$ an appropriate constant. Furthermore, putting $\mathscr{G} = (\kappa_\theta : \theta \in \Theta)$ and under the same assumption, $T + U$ is sufficient in $\mathscr{F} \times \mathscr{G}$ so that $\mathscr{E} \sim \mathscr{F} \times \mathscr{G}$. This completes the proof of the first statement.

Finally assume that θ_0 is an interior point of Θ and that $\mathscr{F} \leqslant \mathscr{E}$. We may assume without loss of generality that $\theta_0 = 0$. By the 'dilation' criterion, theorem 9.5.1, $P_0 = DQ_0$ for a kernel D such that $\int (a(\theta)/a(0))\mathrm{e}^{(\theta,x)}D(\mathrm{d}x|y) = (b(\theta)/b(0))\mathrm{e}^{(\theta,y)}$ for Q_0-almost all y.

Construct random variables X and Y such that $\mathscr{L}(Y) = Q_0$, $\mathscr{L}(X|Y) = D(\cdot|Y)$. Then $\mathscr{L}(X) = P_0$ and $E\mathrm{e}^{(\theta,X-Y)} = E(E\mathrm{e}^{(\theta,X-Y)}|Y) = (b(\theta)/b(0))(a(0)/a(\theta))$ so that $E\mathrm{e}^{(\theta,X-Y)} \cdot E\mathrm{e}^{(\theta,Y)} = (a(0)/a(\theta)) = E\mathrm{e}^{(\theta,X)}$ for all θ belonging to some open ball containing the origin. It follows, by analytic continuation, that the characteristic function of X is the product of the characteristic functions of $X - Y$ and of Y. Hence $\mathscr{L}(X - Y) * \mathscr{L}(Y) = \mathscr{L}(X)$ so that Q_0 is a convolution factor of P_0. □

The dilation results for standard measures may be viewed as a particular case of dilation criteria for comparison of measures. A fairly general situation may be described as follows.

Theorem 9.5.7 (Orderings of measures). *Let μ and ν be non-negative finite measures on a measurable space $(\mathscr{X}, \mathscr{A})$.*

Consider also a convex set H of $\mu + \nu$ integrable functions containing 0 and having the property that $h_1 \vee h_2 \in H$ when $h_1, h_2 \in H$.

Then $\int h \, \mathrm{d}\mu \geqslant \int h \, \mathrm{d}\nu$ for all $h \in H$ if and only if there is a transition M from $L_1(\mu)$ to $L_1(\nu)$ such that

$$(h\mu)M \geqslant h\nu; \qquad h \in H.$$

If, in addition, H contains the constants and if $(\mathscr{X}, \mathscr{A})$ is Euclidean, then this condition is also equivalent to the condition that $\mu = D\nu$ for a Markov kernel D such that $\int h(x)D(\mathrm{d}x|y) \geqslant h(y)$ for ν-almost all y for all $h \in H$.

Remark 1. If μ is a measure and if $\int h \, \mathrm{d}\mu$ exists then $h\mu$ denotes the measure which to each measurable set A assigns the mass $\int_A h \, \mathrm{d}\mu$.

Remark 2. If H is countable or if H is 'sufficiently separable' then D may be modified so that $\int h(x)D(\mathrm{d}x|y) \geqslant h(y)$ for all $y \in \mathscr{X}$ and for all $h \in H$.

Proof. Consider the measure families $\mathscr{E} = (h\mu : h \in H)$ and $\mathscr{F} = (h\nu : h \in H)$. Let $h_1, \ldots, h_r \in H$ and let ψ be a sublinear function on $\mathbb{R}^r$ which is both monotonically increasing and a maximum of a finite set of linear functionals on $\mathbb{R}^r$. Thus $\psi(x) \equiv_x \bigvee_{t=1}^{n} \sum_{i=1}^{r} a_{ti} x_i$ for non-negative constants a_{ti}. If $a_1, \ldots, a_r$ are non-negative constants then, since $0 \in H$ and H is convex, $(1/N) \sum_{i=1}^{r} a_i h_i \in H$ when $N \geqslant \sum_{i=1}^{r} a_i$. It follows that $(1/N)\psi(h_1, \ldots, h_r) \in H$ when N is sufficiently large. Thus $\int \phi(h_1, \ldots, h_r) \, \mathrm{d}\mu \geqslant \int \psi(h_1, \ldots, h_r) \, \mathrm{d}\nu$. Consider the maximum ψ of an arbitrary finite family of linear functionals on $\mathbb{R}^r$. The map $x \to \psi(x) + \sum_{i=1}^{r} x_i \psi(0, \ldots, \overset{(i)}{-1}, \ldots, 0)$ is then monotonically increasing on $\mathbb{R}^r$. Thus

$$\int \psi(h_1, \ldots, h_r) \, \mathrm{d}\mu \geqslant \int \psi(h_1, \ldots, h_r) \, \mathrm{d}\nu - \sum_{i=1}^{r} \varepsilon(h_i) \psi(0, \ldots, \overset{(i)}{-1}, \ldots, 0)$$

where $\varepsilon(h) = \int h \, \mathrm{d}\mu - \int h \, \mathrm{d}\nu$; $h \in H$. It follows that $\mathscr{E}$ is ε-deficient with respect to $\mathscr{F}$. Hence, by the randomization criterion, there is a transition M from $L_1(\mu)$ to $L_1(\nu)$ such that $\|(h\mu)M - (h\nu)\| \leqslant \varepsilon(h)$; $h \in H$. Now

$$\begin{aligned} \|(h\mu)M - (h\nu)\| &= [(h\mu)M - h\nu](1) + 2\|[(h\mu)M - (h\nu)]^-\| \\ &= \varepsilon(h) + 2\|[(h\mu)M - (h\nu)]^-\|. \end{aligned}$$

Thus $\|[(h\mu)M - (h\nu)]^-\| = 0$ when $h \in H$ so that $(h\mu)M \geqslant h\nu$; $h \in H$. Conversely, these inequalities imply that

$$\int h\,\mathrm{d}\mu = (h\mu)(1)$$
$$= ((h\mu)M)(1)$$
$$\geqslant (h\nu)(1)$$
$$= \int h\,\mathrm{d}\nu$$

when $h \in H$.

Assume that this condition is satisfied, that H contains the constants and that the measurable space $(\mathscr{X}, \mathscr{A})$ is Euclidean. Then $(\pm\mu)M \geqslant \pm\nu$ so that $\mu M = \nu$. Let D be a regular conditional distribution of the first coordinate given the second for the distribution $\mu \times M$. Thus $\mu \times M = \nu \times D$. If $A \in \mathscr{A}$ and $h \in H$ then

$$\int_A \left[\int h(x)D(\mathrm{d}x|y)\right]\nu(\mathrm{d}y) = \int h(x)I_A(y)(D \times \nu)(\mathrm{d}(x, y))$$
$$= \int h(x)I_A(y)(\mu \times M)(\mathrm{d}(x, y))$$
$$= \int M(A|x)h(x)\mu(\mathrm{d}x)$$
$$= ((h\mu)M)(A) \geqslant (h\nu)(A)$$
$$= \int_A h\,\mathrm{d}\nu.$$

Hence $\mu = D\nu$ and $\int h(x)D(\mathrm{d}x|\cdot) \geqslant h$ a.e. ν. This in turn implies, by essentially the same computations, that $(h\mu)M \geqslant h\nu$ where the Markov kernel M satisfies $\mu \times M = D \times \nu$. □

Consider the particular case where H consists of all functions on $\mathbb{R}^m$ which are maxima of a finite set of (monotonically increasing) linear functionals. If $\int h\,\mathrm{d}\mu \geqslant \int h\,\mathrm{d}\nu$ for all $h \in H$ then, by the theorem and remark 2, $\mu = D\nu$ for a Markov kernel D such that $\int xD(\mathrm{d}x|y) \geqslant y$ for all $y \in \mathbb{R}^m$. Conversely the existence of such a D implies, by Jensen's inequality, that $\int h\,\mathrm{d}\mu \geqslant \int h\,\mathrm{d}\nu$, $h \in H$. If, more generally, $S_1, S_2, \ldots$ is a sequence of probability distributions on $\mathbb{R}^m$ such that the sequence $\int h\,\mathrm{d}S_1, \int h\,\mathrm{d}S_2, \ldots$ is monotonically increasing for each $h \in H$ then $S_{n+1} = D_n S_n$ where $\int xD_{n+1}(\mathrm{d}x|y) \geqslant y$; $y \in \mathbb{R}^m$. Let the joint distribution of $(X_1, X_2, \ldots)$ be determined by the requirement that $\mathscr{L}(X_1) = S_1$ and that $D_n(\cdot|X_n)$, for each n, is a conditional distribution of X_{n+1} given X_n. Then the Markov process $X_1, X_2, \ldots$ is a (sub) martingale such that X_n, for each n, has distribution S_n. This and Jensen's inequality prove the following result, due to Strassen (1965).

Corollary 9.5.8 ((Sub) martingales with prescribed marginal distributions). *Assume that the probability distributions $S_1, S_2, \ldots$ on $\mathbb{R}^m$ all possess finite expectations. Then there is a (sub) martingale $X_1, X_2, \ldots$ with, respectively, marginal distributions $S_1, S_2, \ldots$ if and only if $\int h \, dS_1 \leqslant \int h \, dS_2 \leqslant \cdots$ whenever the function h is convex (and monotonically increasing) on $\mathbb{R}^m$.*

Another interesting situation occurs when we assume that the functions in H are monotonically increasing with respect to some partial ordering. In the latter case we obtain the joint distribution characterization of stochastic orderings of distributions on sets with prescribed partial orderings (see e.g. Karlin & Rinott, 1983). As more general situations will be treated in the next section we will not write this out here.

The argument used in the proof of the last part of theorem 9.5.6 required that $\int(-1)\,d\mu \geqslant \int(-1)\,d\nu$. Thus the proof does not work if the functions in H are non-negative. For this situation there is a useful result due to Fischer & Holbrook (1980). The general underlying idea is quite simple and may be expressed as follows.

Proposition 9.5.9 (Existence of minorizing functionals). *Let γ be a real valued convex functional on a convex subset H of a linear space.*

Let $\mathscr{P}$ be a sub convex class of real valued concave functionals on H.

Assume that $\mathscr{P}$ is compact for the weakest topology which makes $P(h)$ lower semicontinuous in $P \in \mathscr{P}$ for each $h \in H$.

Then there is a P in $\mathscr{P}$ such that $\gamma \geqslant P$ on H if and only if $\gamma(h) \geqslant \inf\{P(h) : P \in \mathscr{P}\}$ for all $h \in H$.

Remark. $\mathscr{P}$ is called *sub convex* if to each pair $(P_0, P_1) \in \mathscr{P} \times \mathscr{P}$ and to each $t \in]0, 1[$ there corresponds a $P_t \in \mathscr{P}$ such that $P_t \leqslant (1-t)P_0 + tP_1$. The compactness condition amounts to requiring that to each net (P_α) in $\mathscr{P}$ there corresponds a $P \in \mathscr{P}$ such that $\limsup_\alpha P_\alpha(h) \geqslant P(h)$; $h \in H$.

Proof. The 'only if' is trivial and the 'if' follows by using the minimax theory in chapter 3 on the concave–convex pay-off function $(P, h) \to \gamma(h) - P(h)$. □

Applying this to the situation of corollary 9.5.3 with $\varepsilon_1 = \cdots = \varepsilon_m = 0$, we obtain the following corollary.

Corollary 9.5.10 (Existence of sub Markov kernels). *Let S and T be non-negative measures on $\mathbb{R}^m$ such that the projections are integrable. Let the measure S_i (the measure T_i) for each $i = 1, \ldots, m$ be the measure having the projection on the ith coordinate space as a density with respect to the measure S (the measure T). The following conditions are then equivalent.*

(i) *$SM = T$ for a sub Markov kernel M from $\mathbb{R}^m$ to $\mathbb{R}^m$ such that $S_i M = T_i$; $i = 1, \ldots, m$.*
(ii) *$S \geqslant DT$ for a dilation D from $\mathbb{R}^m$ to $\mathbb{R}^m$.*
(iii) *$\int \phi \, \mathrm{d}S \geqslant \int \phi \, \mathrm{d}T$ for each non-negative convex function ϕ on $\mathbb{R}^m$.*
(iv) *There is a non-negative measure U on $\mathbb{R}^m$ such that $\int \phi \, \mathrm{d}S \geqslant \int \phi \, \mathrm{d}(T + U)$ for all convex functions ϕ on $\mathbb{R}^m$. If so then the measure U may always be chosen as a one-point mass distribution.*

Remark. If S and T are required to possess compact supports then the equivalence of (iii) and (iv) follows from Fischer & Holbrook (1980).

Proof. Suppose (i) is satisfied. Then $\|S\| \geqslant \|SM\| = \|T\|$. If $\|S\| = \|T\|$ then M may be chosen as a proper Markov kernel and then we know from the remark after corollary 9.5.3 that $\int \phi \, \mathrm{d}S \geqslant \int \phi \, \mathrm{d}T$ for all convex functions ϕ on $\mathbb{R}^m$. If $\|S\| > \|T\|$ then we may put $\xi_i = [\|S_i\| - \|T_i\|]/[\|S\| - \|T\|]$; $i = 1, \ldots, m$. Put $\xi = (\xi_1, \ldots, \xi_m)$ and define a Markov kernel $\tilde{M}$ by putting $\tilde{M}(B|x) = M(B|x) + [1 - M(\mathbb{R}^m|x)] I_B(\xi)$. Then $S\tilde{M} = T + U$ where U is the one-point distribution which assigns mass $\|S\| - \|T\|$ to ξ. It is then a matter of checking that $S_i \tilde{M} = (T + U)_i = T_i + U_i$ where, for each measure, κ, κ_i denotes the measure having $x \to x_i$ as density with respect to κ. Thus the remark after corollary 9.5.3 applies again and we get $\int \phi \, \mathrm{d}S \geqslant \int \phi \, \mathrm{d}(T + U) = \int \phi \, \mathrm{d}T + \int \phi \, \mathrm{d}U \geqslant \int \phi \, \mathrm{d}T$ when ϕ is non-negative and convex on $\mathbb{R}^m$. Hence (i) $\Rightarrow$ (iv) and clearly (iv) $\Rightarrow$ (iii). Assume that (ii) is satisfied. If ϕ is non-negative and convex then $\int \phi \, \mathrm{d}S \geqslant \int \phi \, \mathrm{d}(DT) \geqslant \int \phi \, \mathrm{d}T$ where the last '$\geqslant$' follows from the same remark. Thus (ii) $\Rightarrow$ (iii).

Assume next that (iii) is satisfied. Put $\rho(x) = \bigvee_i |x_i|$ when $x = (x_1, \ldots, x_m) \in \mathbb{R}^m$. Then $\|S\| = \int 1 \, \mathrm{d}S \geqslant \int 1 \, \mathrm{d}T = \|T\|$. If $\|S\| = \|T\|$ and if ϕ is convex and bounded from below by the constant b then $\int \phi \, \mathrm{d}S = \int (\phi - b) \, \mathrm{d}S + b\|S\| \geqslant \int (\phi - b) \, \mathrm{d}T + b\|T\| = \int \phi \, \mathrm{d}T$. This readily implies (iv) with $U = 0$.

Put $\Gamma = [\|S\| - \|T\|]^{-1}(S - T)$ when $\|S\| > \|T\|$. Then $\Gamma(\phi) \geqslant 0$ for all non-negative convex functions ϕ on $\mathbb{R}^m$ and $\Gamma(1) = 1$. Let H_N; $N = 1, 2, \ldots$, denote the set of convex functions h such that $h \leqslant \rho$ while $h(x) \geqslant 0$ when $\rho(x) \geqslant N$. Take $\mathscr{P}_N$ as the set of probability distributions on $\{x : \rho(x) \leqslant N\}$. If $h \in H_N$ then $\Gamma(h) \geqslant \inf_x h(x) = \inf\{P(h) : P \in \mathscr{P}_N\}$. It follows, since $\mathscr{P}_N$ is tight, that there is a $P_N \in \mathscr{P}_N$ such that $\Gamma(h) \geqslant P_N(h)$; $h \in H_N$. In particular $\Gamma(\rho) \geqslant P_N(\rho)$. It follows that $(P_1, P_2, \ldots)$ is tight. Let the probability distribution P be a weak limit point for this sequence. If h is convex and if $h \in H_{N_0}$ then $h \in H_N$ when $N \geqslant N_0$. Hence $\Gamma(h) \geqslant \limsup_N P_N(h) \geqslant P(h)$. In particular $\Gamma(h \vee (\rho - N_0)) \geqslant P(h \vee (\rho - N_0))$ when h is convex and $h \leqslant \rho$. $N_0 \to \infty$ yields $\Gamma(h) \geqslant P(h)$. It follows that $\Gamma(\phi) \geqslant P(\phi)$ when ϕ is convex and $\phi \leqslant k\rho$ for some positive constant k. Jensen's inequality tells us that $P(\phi) \geqslant \phi(\int x P(\mathrm{d}x))$ when ϕ is convex. It follows that we may assume that P is a one-point distribution. Altogether this shows that (iv) holds with U being the one-point distribution which assigns mass $\|S\| - \|T\|$ to the point $\int xP(\mathrm{d}x)$. Thus (iii) $\Rightarrow$ (iv).

If (iv) holds then, by the remark after corollary 9.5.3, there is a Markov kernel M such that $SM = T + U$ and $S_i M = T_i + U_i$; $i = 1, \dots, m$. Furthermore $T = (T + U)\hat{M}$ where $\hat{M}$ is the sub Markov kernel $(x, B) \to I_B(x)g(x)$ where g is a version of $\mathrm{d}T/\mathrm{d}(T + U)$ such that $0 \leqslant g \leqslant 1$. Then $SM\hat{M} = T$ and $S_i M\hat{M} = T_i$ for the sub Markov kernel $M\hat{M}$. Hence (iv) $\Rightarrow$ (i). Still assuming (iv) and applying the same remark we find that $S = D(T + U)$ for a dilation D. Thus $S \geqslant DT$ so that (iv) $\Rightarrow$ (ii). The proof is now completed by combining the established implications. □

Returning to matrices again we obtain a further corollary.

Corollary 9.5.11 (Existence of sub Markov matrices). *Let μ and ν be row matrices of dimensions $1 \times m$ and $1 \times n$ respectively. We shall assume that the entries $\mu_1, \dots, \mu_m$ of μ and $\nu_1, \dots, \nu_n$ of ν are all positive.*

Let A and B be matrices with real entries and with dimensions $r \times m$ and $r \times n$ respectively. Then the following statements are equivalent.

(i) $\mu M = \nu$ *for an $m \times n$ sub Markov matrix M such that* $\sum_i A_{k,i}\mu_i M_{i,j} = B_{k,j}\nu_j$; $k = 1, \dots, r$; $j = 1, \dots, n$.

(ii) $\mu \geqslant \nu D$ *for an $n \times m$ Markov matrix D such that $AD = B$.*

(iii) $\sum_i \phi(A_{1,i}/\mu_i, \dots, A_{r,i}/\mu_i)\mu_i \geqslant \sum_j \phi(B_{1,j}/\nu_j, \dots, B_{r,j}/\nu_j)\nu_j$ *for each non-negative convex function ϕ on $\mathbb{R}^r$.*

Proof. Apply the previous corollary to S and T where S assigns mass μ_i to each point $(A_{1,i}, \dots, A_{r,i})$ while T assigns mass ν_j *to each point* $(B_{1,j}, \dots, B_{r,j})$. □

Consider again the situation of theorem 9.5.7, under the additional assumptions that the functions in H are bounded and that the constant functions are in H.

If, in this situation, we weaken the requirement that $\mu(h) \geqslant \nu(h)$ for all $h \in H$, to the requirement that $\mu(h) \geqslant \nu(h)$ for *all non-negative functions* $h \in H$, then there is a non-negative and additive set function κ on $\mathscr{A}$ such that $\mu(h) \geqslant \nu(h) + \kappa(h)$ for all $h \in H$. This may be seen as follows. The assumption that $1 \in H$ implies that $\|\mu\| \geqslant \|\nu\|$. If $\|\mu\| = \|\nu\|$ and c is a lower bound for h then we can conclude that $\frac{1}{2}(h - c) \in H_+$ so that $\frac{1}{2}\mu(h - c) \geqslant \frac{1}{2}\nu(h - c)$ yielding $\mu(h) \geqslant \nu(h)$. Thus we may put $\kappa = 0$ in this case. If $\|\mu\| > \|\nu\|$ then we may apply proposition 9.5.9 to the functional $\Gamma = [\|\mu\| - \|\nu\|]^{-1}(\mu - \nu)$ yielding a finitely additive probability set function P such that $\Gamma(h) \geqslant P(h)$ for all $h \in H$. We may then put $\kappa = [\|\mu\| - \|\nu\|]P$.

If we were now able to show that κ could be chosen countably additive then theorem 9.5.7 would become applicable with ν replaced by $\nu + \kappa$. A particular case where this works is when $\mathscr{A}$ is the class of Borel subsets of a compact Hausdorff space $\mathscr{X}$ and the functions in H are continuous.

Corollary 9.5.12 (Comparison for integrals of non-negative functions). *Let μ and ν be non-negative finite measures on the Borel class $\mathscr{A}$ of a compact metric space $\mathscr{X}$.*

Let H be a convex set of continuous functions on $\mathscr{X}$ such that $h_1 \vee h_2 \in H$ when $h_1 \in H$ and $h_2 \in H$. We shall also assume that H contains the constant functions. Then the following conditions are equivalent.

(i) $\int h \, d\mu \geqslant \int h \, d\nu$ *when* $0 \leqslant h \in H$.
(ii) *There is a non-negative measure κ on $\mathscr{A}$ such that* $\int h \, d\mu \geqslant \int h \, d\nu + \int h \, d\kappa$ *for all* $h \in H$.
(iii) *There is a sub Markov kernel M such that* $(h\mu)M \geqslant h\nu$; $h \in H$.
(iv) $\mu \geqslant D\nu$ *for a Markov kernel D such that* $\int h(x) D(dx|y) \geqslant h(y)$ *for ν-almost all y and for all* $h \in H$.

Remark. If additional assumptions on linear and topological structures are satisfied then the equivalence of (i) and (ii) follows from Fischer & Holbrook (1980).

9.6 Application to a problem on probability distributions with given marginals

Here we will apply the theory of section 9.2 to derive a result in Strassen's 1965 paper on probability distribution with given marginals.

The result is related to the problem of determining the set of possible probabilities $Pr((X, Y) \in S)$ for a specified set S and for prescribed marginal distributions P and Q for X and Y. By convexity this set is clearly an interval and it is also fairly clear that it is closed when we permit joint distributions which are not necessarily countably additive. Thus it suffices, at least under this permission, to find the end points of this interval. As the problem of minimizing $Pr((X, Y) \in S)$ is equivalent to the problem of maximizing $Pr((X, Y) \notin S)$ it should then also suffice to determine the right end point of this interval.

Here, as in Strassen's paper, we will consider the problem of maximizing $Pr((X, Y) \in S)$ when S is closed and the sample spaces of X and Y are both complete separable metric spaces. The right end point of the above interval may then be described as follows.

Let B be any event in the sample space of Y. Then $Pr((X, Y) \in S)$ may be decomposed as

$$Pr((X, Y) \in S \text{ and } Y \notin B) + Pr((X, Y) \in S \text{ and } Y \in B).$$

Hence

$$Pr((X, Y) \in S) \leqslant Pr(Y \notin B) + Pr((X, y) \in S \text{ for some } y \in B)$$

where the number on the right hand side is determined by P and Q.

For each subset B of the sample space $\mathscr{Y}$ of Y let us agree to use the notation $B^{(S)}$ for the set of points x in the sample space $\mathscr{X}$ of X such that $(x, y) \in S$ for some y in B. The upper bound derived above may then be written

$$\inf_B [Q(B^c) + P(B^{(S)})]$$

and it follows from Strassen's paper (under the regularity conditions previously mentioned) that this upper bound is achieved.

Usually it is not required to consider all sets B. If e.g. X and Y take their values in the same partially ordered set $(\mathscr{X}, \leqslant)$ and if $S = \{(x, y): x \geqslant y\}$ then it suffices to consider monotonically increasing sets B (B is called *monotonically increasing (decreasing)* if its indicator function is monotonically increasing (decreasing)). Thus if $\mathscr{X}$ is a subset of the real line with the usual ordering then we obtain the greatest lower bound of all numbers $1 - G(x) + F(x)$; $x \in \mathscr{X}$ where F and G are the distribution functions of X and Y respectively.

In general there is a joint distribution such that $Pr((X, Y) \in S) \geqslant 1 - \varepsilon$ if and only if $Q(B) \leqslant P(B^{(S)}) + \varepsilon$ for each (measurable) subset B of $\mathscr{Y}$.

If ρ is a real valued function on $\mathscr{X} \times \mathscr{Y}$ and if $S = \{(x, y): \rho(x, y) \leqslant \varepsilon\}$ then this amounts to the requirement that $Q(B) \leqslant P(\{x: \rho(x, y) \leqslant \varepsilon \text{ for some } y \in B\}) + \varepsilon$ when B is a (measurable) subset of $\mathscr{Y}$. In particular, as was shown in Strassen (1965), this yields the joint distribution characterization of the Prohorov distance for a given distance ρ on $\mathscr{X}$.

There is a rephrasing of this result which is obtained by relaxing the condition that the distribution of Y should be exactly Q while at the same time strengthening the requirement on $Pr((X, Y) \in S)$ by requiring that $Pr((X, Y) \in S) = 1$. Using this rephrasing we shall now see how this type of result may be deduced from the basic principles for comparing measure families.

The essential idea used in the proof may be grasped by restricting the attention to finite sets $\mathscr{X}$ and $\mathscr{Y}$. The reader might then see that the arguments using measure families may be replaced by simple arguments using well known results on support functions. The point here however is to provide another example of the relationship between the principles for comparison of measure families on the one hand and existence problems for probability distributions with given marginals on the other.

Except for the rephrasing (condition (iii) below), the following result is a particular case of a theorem in Strassen's paper.

Proposition 9.6.1 (Probabilities for a specified set assigned by joint distributions possessing prescribed marginals). *Let $(\mathscr{X}, \mathscr{A}, P)$ and $(\mathscr{Y}, \mathscr{B}, Q)$ be probability* *spaces and let S be a subset of $\mathscr{X} \times \mathscr{Y}$. Assume that $\mathscr{X}$ and $\mathscr{Y}$ are compact metric spaces with Borel classes $\mathscr{A}$ and $\mathscr{B}$ respectively. Assume also that S is a closed subset of $\mathscr{X} \times \mathscr{Y}$. Then the following two conditions are equivalent for a number $\varepsilon \in [0, 1]$.*

(i) *$Q(B) \leqslant P(B^{(S)}) + \varepsilon$ for all Borel sets B.*
(ii) *There is a joint distribution R on $\mathscr{A} \times \mathscr{B}$ with marginals P and Q such that $R(S) \geqslant 1 - \varepsilon$.*

If, furthermore, the projection of S into $\mathscr{X}$ is onto, then these conditions are equivalent to:

(iii) *there is a joint distribution R on $\mathscr{A} \times \mathscr{B}$ with marginals P and μ such that $R(S) = 1$ and $\|\mu - Q\| \leqslant 2\varepsilon$.*

Remark 1. The set $B^{(S)}$ is the projection on $\mathscr{X}$ of the measurable set $(\mathscr{X} \times B) \cap S$ and is consequently analytic and therefore completion measurable.

Remark 2. It is apparent from this proof that several of the assumptions may be weakened. Thus we shall only make use of (i) for closed sets B.

If S is defined by a partial ordering then it suffices to consider monotonically increasing sets B in (i) and then $B^{(S)} = B$. In this case these conditions are equivalent to the condition that $Q(h) \leqslant P(h) + \varepsilon\|h\|$ when the function h is bounded, monotonically increasing and measurable.

Proof. We know already that (ii) $\Rightarrow$ (i). Assume now that (iii) holds. Then by the joint distribution characterization of the statistical distance, there is a joint distribution U on $\mathscr{B} \times \mathscr{B}$ with marginals μ and Q such that $U(\{(y, y) : y \in \mathscr{Y}\}) \geqslant 1 - \varepsilon$. (Let T denote the map $y \to (y, y)$ from $\mathscr{Y}$ to $\mathscr{Y} \times \mathscr{Y}$. Then we may put $U = \mu T^{-1}$ when $\mu = Q$ and put $U = (\mu \wedge Q)T^{-1} + 2\|\mu - Q\|^{-1}[(\mu - Q)^+ \times (\mu - Q)^-]$ when $\mu \neq Q$.) Define a joint distribution V for variables X, Z and Y such that (X, Z) has distribution R while the conditional distribution of Y given (X, Z) is given by a regular conditional distribution of Y given Z in the situation where U is the joint distribution of Y and Z. Then (ii) holds if R is replaced by the joint distribution of (X, Y). Thus in any case (and we did not use compactness) (iii) $\Rightarrow$ (ii) $\Rightarrow$ (i).

Assume now that (i) holds and that the projection of S into $\mathscr{X}$ is onto. For any bounded measurable function h on $\mathscr{Y}$ and $x \in \mathscr{X}$ put $\hat{h}(x) = \sup\{h(y) : y \in S_x\}$. If t is a real number then clearly $[h > t]^{(S)} = [\hat{h} > t]$. Thus, by (i)

$$Q(h) = \int_0^{\|h\|} Q(h > t)\,\mathrm{d}t \leqslant \int_0^{\|h\|} [P(\hat{h} > t) + \varepsilon]\,\mathrm{d}t = P(\hat{h}) + \varepsilon\|h\|$$

when h is non-negative. It follows from the expression for $[\hat{h} > t]$ above and remark 1 that $\hat{h}$ is completion measurable. Using $\hat{c}(x) \equiv c$ for a constant function c we find, since $\widehat{h - c} = \hat{h} - c$, that $Q(h) \leqslant P(\hat{h} + 2\varepsilon\|h\|)$ for any bounded measurable function h. Here we used the assumption that the projection of S into $\mathscr{X}$ is onto. Let h vary through the set $C(\mathscr{Y})$ of continuous functions on $\mathscr{Y}$. Then the linear functional Q is majorized by the sum of the two sublinear functionals $h \to P(\hat{h})$ and $h \to 2\varepsilon\|h\|$. It follows that Q may be decomposed as $Q = \mu + \nu$ where μ and ν are linear functionals (measures) such that $\mu(h) \leqslant P(\hat{h})$; $h \in C(\mathscr{Y})$ while $\nu(h) \leqslant 2\varepsilon\|h\|$. If $h \leqslant 0$ then $\hat{h} \leqslant 0$ and thus $\mu(h) \leqslant 0$. If $h = c$ is a constant then $\mu(c) \leqslant P(\hat{c}) = P(c) = c$. It follows that μ is a probability distribution and that ν is a measure whose total variation is at most 2ε.

Consider the measure families $\mathscr{E} = (\hat{h}P : h \in C(\mathscr{Y}))$ and $\mathscr{F} = (h\mu : h \in C(\mathscr{Y}))$. Put $\eta_h = P(\hat{h}) - \mu(h)$ when $h \in C(\mathscr{Y})$.

Let $h_1, \ldots, h_r \in C(\mathscr{Y})$ and let $h = \psi(h_1, \ldots, h_r)$ where ψ is a sublinear monotonically increasing function on $\mathbb{R}^r$. Put $h = \psi(h_1, \ldots, h_r)$. Then $\int \psi(h_1, \ldots, h_r)\,d\mu = \int h\,d\mu \leqslant \int \hat{h}\,dP \leqslant \int \psi(\hat{h}_1, \ldots, \hat{h}_r)\,dP$. If ψ is any sublinear function on $\mathbb{R}^r$ then $x \to \psi(x) + \sum_i x_i \psi(0, \ldots, \overset{(i)}{-1}, \ldots, 0)$ is monotonically increasing. Then it follows from the sublinear function criterion (iii) of section 9.2 that $\mathscr{E}$ is η-deficient with respect to $\mathscr{F}$. Thus there is a Markov kernel M from $(\mathscr{X}, \mathscr{A})$ to $(\mathscr{Y}, \mathscr{B})$ such that $\|(\hat{h}P)M - h\mu\| \leqslant \eta_h$; $h \in C(\mathscr{Y})$. Thus $P(\hat{h}) - \mu(h) + 2\|[(\hat{h}P)M - h\mu]^-\| \leqslant \eta_h$. $h = 1$ then yields $PM \geqslant \mu$ so that $PM = \mu$. Let D be a regular conditional distribution of 'x' given 'y' for $P \times M$. Then $P \times M = D \times \mu$. If $h \in C(\mathscr{Y})$ and $B \in \mathscr{B}$ then

$$\begin{aligned}
\int_B \left[\int \hat{h}(x) D(dx|y)\right] \mu(dy) &= \int \hat{h}(x) I_B(y) (D \times M)(d(x, y)) \\
&= \int \hat{h}(x) I_B(y) (P \times M)(d(x, y)) \\
&= \int M(B|x) \hat{h}(x)(dx) \\
&= ((\hat{h}P)M)(B) \\
&\geqslant (h\mu)(B) \\
&= \int_B h(y) \mu(dy).
\end{aligned}$$

Thus $\int \hat{h}(x) D(dx|y) \geqslant h(y)$ for μ-almost all y when $h \in C(\mathscr{Y})$. Since $C(\mathscr{Y})$ is separable, we may now modify D so that $\int \hat{h}(x) D(dx|y) \geqslant h(y)$ for all $y \in \mathscr{Y}$ and all $h \in C(\mathscr{Y})$. Since $\hat{h}_n \uparrow \hat{h}$ when $h_n \uparrow h$, this inequality extends to all lower semicontinuous functions h on $\mathscr{Y}$. Let $y_0 \in \mathscr{Y}$ and for $n = 1, 2, \ldots$, let the open ball around y_0 with radius $1/n$ be denoted as B_n. Applying the inequality to the indicator function h_n of B_n we find that $D(B_n^{(S)}|y_0) = \int \hat{h}_n(x)D(\mathrm{d}x|y) \geqslant h_n(y_0) = 1$. Furthermore, since $B_1^{(S)} \cap B_2^{(S)} \cap \cdots = \{x : (x, y_0) \in S\}$, we find that $D(\{x : (x, y_0) \in S\}|y_0) = 1$ for all points $y_0 \in \mathscr{Y}$. Thus $(P \times M)(S) = (D \times \mu)(S) = \int D(\{x : (x, y) \in S\}|y)\mu(\mathrm{d}y) = 1$.

It follows that condition (iii) is satisfied with $R = P \times M$.

Altogether this shows that conditions (i), (ii) and (iii) are equivalent when the projection of S into $\mathscr{X}$ is onto.

It remains to show that (i) $\Rightarrow$ (ii) in the general case. Assume then that (i) holds. Let the metric d which metrizes $\mathscr{Y}$ be bounded by 1. Add a point Ω to $\mathscr{Y}$ and extend d by putting $d(\Omega, \Omega) = 0$ while $d(\Omega, y) = d(y, \Omega) = 1$ for all points $y \in \mathscr{Y}$. Put $\mathscr{Y}_1 = \mathscr{Y} \cup \{\Omega\}$ and $S_1 = S \cup \{(x, \Omega) : x \in \mathscr{X}\}$. Then (i) holds if $\mathscr{Y}$ and S are replaced by $\mathscr{Y}_1$ and S_1 respectively and if Q is replaced by its extension $\tilde{Q}$ to $\mathscr{Y}_1$. Thus, by what we have proved, there is a joint distribution $\tilde{R}$ on $\mathscr{X} \times \mathscr{Y}_1$ with marginals P and $\tilde{Q}$ such that $\tilde{R}(S_1) \geqslant 1 - \varepsilon$. Then, since $\tilde{Q}(\mathscr{Y}) = 1$, $\tilde{R}$ is supported by $\mathscr{X} \times \mathscr{Y}$ and thus (ii) holds with R being the restriction of $\tilde{R}$ to $\mathscr{X} \times \mathscr{Y}$. □

Using the fact that a complete separable metric space is homeomorphic to a G_δ subset of a compact metric space, we obtain the theorem due to Strassen mentioned above. This theorem states that (i) and (ii) are equivalent for probability spaces $(\mathscr{X}, \mathscr{A}, P)$ and $(\mathscr{Y}, \mathscr{B}, Q)$ and a subset S of $\mathscr{X} \times \mathscr{Y}$ such that $\mathscr{X}$ and $\mathscr{Y}$ are complete separable metric spaces and S is closed.

Extensions and related results may be found in Dudley (1968, 1989).

9.7 Majorization

The concepts of majorization defined in Marshall & Olkin (1979) generalize as follows.

Let $\mathscr{E} = (\mathscr{X}, \mathscr{A}; \mu_\theta : \theta \in \Theta)$ and $\mathscr{F} = (\mathscr{Y}, \mathscr{B}; \nu_\theta : \theta \in \Theta)$ be two measure families. Say that $\mathscr{E}$ *weakly supermajorizes* $\mathscr{F}$ if $\mathscr{E}$ is $\nu_{.}(\mathscr{Y}) - \mu_{.}(\mathscr{X})$ deficient with respect to $\mathscr{F}$. Say that $\mathscr{E}$ *weakly sub majorizes* $\mathscr{F}$ if $\mathscr{E}$ is $\mu_{.}(\mathscr{X}) - \nu_{.}(\mathscr{Y})$ deficient with respect to $\mathscr{F}$. Say finally that $\mathscr{E}$ *majorizes* $\mathscr{F}$ if $\mathscr{E}$ is 0-deficient with respect to $\mathscr{F}$.

It is then easily checked that the two kinds of weak majorization, as well as majorization, are partial orderings for measure families.

The basic properties of weak supermajorization are collected in the following proposition.

Proposition 9.7.1 (Weak supermajorization). *The following statements are equivalent for measure families $\mathscr{E} = (\mathscr{X}, \mathscr{A}; \mu_\theta : \theta \in \Theta)$ and $\mathscr{F} = (\mathscr{Y}, \mathscr{B}; \nu_\theta : \theta \in \Theta)$.*

(i) *$\mathscr{E}$ weakly supermajorizes $\mathscr{F}$.*
(ii) *$\mu_\theta M \leqslant \nu_\theta$ for all θ for some transition M from $L(\mathscr{E})$ to $L(\mathscr{F})$.*

(iii) *If $\theta_1, \ldots, \theta_r \in \Theta$ and ψ is a sublinear and monotonically decreasing function on $\mathbb{R}^r$ then*

$$\int \psi(\mathrm{d}\mu_{\theta_1}, \ldots, \mathrm{d}\mu_{\theta_r}) \geqslant \int \psi(\mathrm{d}\nu_{\theta_1}, \ldots, \mathrm{d}\nu_{\theta_r}).$$

Remark 1. If $\mu_\theta M \leqslant \nu_\theta$ for a transition M and if $\mu_\theta(\mathscr{X}) = \nu_\theta(\mathscr{Y})$ then actually $\mu_\theta M = \nu_\theta$. It follows that conditions (i)–(iii) imply that (iii) holds for any sublinear function ψ which is monotonically decreasing in x_i when $\mu_{\theta_i}(\mathscr{X}) \neq \nu_{\theta_i}(\mathscr{Y})$.

Remark 2. If $\theta_0 \in \Theta$ is such that $\mu_{\theta_0}, \nu_{\theta_0} \geqslant 0$, $\|\mu_{\theta_0}\| = \|\nu_{\theta_0}\|$, $\mu_{\theta_0} \gg \mathscr{E}$, $\nu_{\theta_0} \gg \mathscr{F}$ and the sample spaces are Euclidean then, as we saw in section 9.5, we can express weak supermajorization in terms of dilations. In that case (iii) may be replaced by

(iii′) $\int \phi(\mathrm{d}\mu_{\theta_1}/\mathrm{d}\mu_{\theta_0}, \ldots, \mathrm{d}\mu_{\theta_r}/\mathrm{d}\mu_{\theta_0})\,\mathrm{d}\mu_{\theta_0} \geqslant \int \phi(\mathrm{d}\nu_{\theta_1}/\mathrm{d}\nu_{\theta_0}, \ldots, \mathrm{d}\nu_{\theta_r}/\mathrm{d}\nu_{\theta_0})\,\mathrm{d}\nu_{\theta_0}$ *when ϕ is monotonically decreasing and convex on $\mathbb{R}^r$.*

Remark 3. If $\Theta = \{\theta_0, \theta_1\}$, $\mu_{\theta_0} \geqslant 0$, $\nu_{\theta_0} \geqslant 0$ and $\|\mu_{\theta_0}\| = \|\nu_{\theta_0}\|$ then as explained in section 9.3, it suffices to consider comparisons for two-points sets (testing problems) and we find that (iii) is equivalent to:

(iii″) $\mu_{\theta_1}(\mathscr{X}) \leqslant \nu_{\theta_1}(\mathscr{Y})$ *and* $\int [c\,\mathrm{d}\mu_{\theta_0} - \mathrm{d}\mu_{\theta_1}]^+ \geqslant \int [c\mathrm{d}\nu_{\theta_0} - \mathrm{d}\nu_{\theta_1}]^+$; $c \in \mathbb{R}$.

Considering the support function

$$(c_0, c_1) \to \begin{cases} \displaystyle\int [c_0 \mathrm{d}\mu_{\theta_0} + c_1 \mathrm{d}\mu_{\theta_1}]^+ & \text{when } c_1 \leqslant 0; \\ \infty & \text{when } c_1 > 0; \end{cases}$$

we see that (iii) can now be written:

(iii‴) $\mu_{\theta_1}(\mathscr{X}) \leqslant \nu_{\theta_1}(\mathscr{Y})$ *and* $\min\{\mu_{\theta_1}(\delta) : \mu_{\theta_0}(\delta) = \alpha\} \leqslant \min\{\nu_{\theta_1}(\phi) : \nu_{\theta_0}(\phi) = \alpha\}$; $\alpha \geqslant 0$, *where δ (ϕ) runs through the set of test functions in $\mathscr{E}$ ($\mathscr{F}$).*

The minima in (iii‴) may be obtained in the same way as we obtain maximum power by the Neyman–Pearson lemma.

Putting $\alpha = \mu_{\theta_0}(\mathscr{X})$ we see that the condition which appears first in (iii″) and in (iii‴) is superfluous when $\mu_{\theta_0} \gg \mu_{\theta_1}$.

Proof. Assume that $\mathscr{E}$ is $(\nu_{.}(\mathscr{Y}) - \mu_{.}(\mathscr{X}))$-deficient with respect to $\mathscr{F}$. Then there is a transition M such that $\|\mu_\theta M - \nu_\theta\| \leqslant \nu_\theta(\mathscr{Y}) - \mu_\theta(\mathscr{X})$ for all $\theta \in \Theta$. By the identity $\|\lambda\| = \int 1\,\mathrm{d}\lambda + 2\|\lambda^-\|$, this may be written $\nu_\theta \geqslant \mu_\theta M : \theta \in \Theta$. For any monotonically decreasing sublinear function ψ on $\mathbb{R}^r$ this in turn implies that $\int \psi(\mathrm{d}\mu_{\theta_1}, \ldots, \mathrm{d}\mu_{\theta_r}) \geqslant \int \psi(\mathrm{d}(\mu_{\theta_1} M), \ldots, \mathrm{d}(\mu_{\theta_r} M)) \geqslant \int \psi(\mathrm{d}\nu_{\theta_1}, \ldots, \mathrm{d}\nu_{\theta_r})$. Altogether we have shown that (i) $\Rightarrow$ (ii) $\Rightarrow$ (iii). If (iii) holds and ψ is any sublinear function on $\mathbb{R}^r$ then, putting $e^i = (0, \ldots, \overset{(i)}{1}, \ldots, 0)$, we see that $x \to$

$\psi(x) - \int x_i \psi(e^i)$ is monotonically decreasing so that

$$\int \psi(\mathrm{d}\mu_{\theta_1}, \ldots, \mathrm{d}\mu_{\theta_r}) \geqslant \int \psi(\mathrm{d}\nu_{\theta_1}, \ldots, \mathrm{d}\nu_{\theta_r}) - \sum_{i=1}^{r} \psi(e^i)(\nu_{\theta_i}(\mathscr{Y}) - \mu_{\theta_i}(\mathscr{X})).$$

Hence (i) holds. □

We omit the proof of the case of weak sub majorization which is quite similar.

Proposition 9.7.2 (Weak sub majorization). *The following conditions are equivalent for measure families* $\mathscr{E} = (\mathscr{X}, \mathscr{A}; \mu_\theta : \theta \in \Theta)$ and $\mathscr{F} = (\mathscr{Y}, \mathscr{B}; \nu_\theta : \theta \in \Theta)$.

(i) $\mathscr{E}$ *weakly sub majorizes* $\mathscr{F}$.
(ii) $\mu_\theta M \geqslant \nu_\theta$ *for all* θ *for some transition* M *from* $L(\mathscr{E})$ *to* $L(\mathscr{F})$.
(iii) $\int \psi(\mathrm{d}\mu_{\theta_1}, \ldots, \mathrm{d}\mu_{\theta_r}) \geqslant \int \psi(\mathrm{d}\nu_{\theta_1}, \ldots, \mathrm{d}\nu_{\theta_r})$ *when* $\theta_1, \ldots, \theta_r \in \Theta$ *and* ψ *is a sublinear and monotonically increasing function on* $\mathbb{R}^r$.

Remarks. Here we may make remarks similar to those we made after the previous proposition. Thus remarks 1 and 2 apply provided we substitute 'increasing' for 'decreasing'. If $\Theta = \{\theta_0, \theta_1\}$, $\mu_{\theta_0}, \nu_{\theta_0} \geqslant 0$, $\|\mu_{\theta_0}\| = \|\nu_{\theta_0}\|$ then (iii) may be replaced by

$$\mu_{\theta_1}(\mathscr{X}) \geqslant \nu_{\theta_1}(\mathscr{Y}) \quad \text{and} \quad \int [c\,\mathrm{d}\mu_{\theta_0} + \mathrm{d}\mu_{\theta_1}]^+ \geqslant \int [c\,\mathrm{d}\nu_{\theta_0} + \mathrm{d}\nu_{\theta_1}]^+; \quad c \in \mathbb{R}$$

or by

$$\mu_{\theta_1}(\mathscr{X}) \geqslant \nu_{\theta_1}(\mathscr{Y}) \quad \text{and} \quad \max\{\mu_{\theta_1}(\delta) : \mu_{\theta_0}(\delta) = \alpha\} \geqslant \max\{\nu_{\theta_1}(\phi) : \nu_{\theta_0}(\phi) = \alpha\}; \quad \alpha > 0$$

where $\delta(\phi)$ runs through the set of test functions in $\mathscr{E}$ ($\mathscr{F}$). The condition $\mu_{\theta_1}(\mathscr{X}) \geqslant \nu_{\theta_1}(\mathscr{Y})$ is superfluous when $\mu_{\theta_0} \gg \mu_{\theta_1}$.

Combining the criteria for the two kinds of weak majorization, we obtain criteria for majorization.

Theorem 9.7.3 (Majorization criteria for measure families). *The following properties are equivalent for measure families* $\mathscr{E} = (\mu_\theta : \theta \in \Theta)$ *and* $\mathscr{F} = (\nu_\theta : \theta \in \Theta)$.

(i) $\mathscr{E}$ *majorizes* $\mathscr{F}$.
(ii) $\mathscr{E}$ *weakly supermajorizes* $\mathscr{F}$ *and weakly sub majorizes* $\mathscr{F}$.
(iii) $\mu_\theta M \equiv_\theta \nu_\theta$ *for some transition* M *from* $L(\mathscr{E})$ *to* $L(\mathscr{F})$.
(iv) *If* $\theta_1, \ldots, \theta_r \in \Theta$ *and* ψ *is sublinear on* $\mathbb{R}^r$ *then*

$$\int \psi(\mathrm{d}\mu_{\theta_1}, \ldots, \mathrm{d}\mu_{\theta_r}) \geqslant \int \psi(\mathrm{d}\nu_{\theta_1}, \ldots, \mathrm{d}\nu_{\theta_r}).$$

Remark 1. If $\theta_0 \in \Theta$ is such that $\mu_{\theta_0} \geqslant 0$, $\nu_{\theta_0} \geqslant 0$, $\|\mu_{\theta_0}\| = \|\nu_{\theta_0}\|$, $\mu_{\theta_0} \gg \mathscr{E}$ and $\nu_{\theta_0} \gg \mathscr{F}$ then, provided the sample spaces are Euclidean, we may express majorization in terms of dilations. In that case (iv) may be expressed as:

(iv′) $\int \phi(\mathrm{d}\mu_{\theta_1}/\mathrm{d}\mu_{\theta_0}, \ldots, \mathrm{d}\mu_{\theta_r}/\mathrm{d}\mu_{\theta_0})\,\mathrm{d}\mu_{\theta_0} \geqslant \int \phi(\mathrm{d}\nu_{\theta_1}/\mathrm{d}\nu_{\theta_0}, \ldots, \mathrm{d}\nu_{\theta_r}/\mathrm{d}\nu_{\theta_0})\,\mathrm{d}\nu_{\theta_0}$ *when* ϕ *is convex on* $\mathbb{R}^r$.

If in addition μ_{θ_0} and ν_{θ_0} are probability measures, then this is a particular case of corollary 7.2.18.

Remark 2. If $\Theta = \{\theta_0, \theta_1\}$, $\mu_{\theta_0} \geqslant 0$, $\nu_{\theta_0} \geqslant 0$, $\|\mu_{\theta_0}\| = \|\nu_{\theta_0}\|$ and $c_1 \neq 0$ is a fixed number, then (iv) is equivalent to

(iv″) $\mu_{\theta_1}(\mathscr{X}) = \nu_{\theta_1}(\mathscr{Y})$ *and* $\int [c_0\,\mathrm{d}\mu_{\theta_0} + c_1\,\mathrm{d}\mu_{\theta_1}]^+ \geqslant \int [c_0\,\mathrm{d}\nu_{\theta_0} + c_1\,\mathrm{d}\nu_{\theta_1}]^+$ *when* $c_0 \in \mathbb{R}$.

The function $(c_0, c_1) \to \int [c_0\,\mathrm{d}\mu_{\theta_0} + c_1\,\mathrm{d}\mu_{\theta_1}]^+$ is the support function of the closed convex range of $(\mu_{\theta_0}, \mu_{\theta_1})$. Thus conditions (i)–(iv) in this case are also equivalent to the condition that $\mu_{\theta_1}(\mathscr{X}) = \nu_{\theta_1}(\mathscr{Y})$ and that the convex hull of the range of $(\mu_{\theta_0}, \mu_{\theta_1})$ contains the range of $(\nu_{\theta_0}, \nu_{\theta_1})$. Furthermore, the latter condition implies the first when $\mu_{\theta_0} \gg \mu_{\theta_1}$.

If $\mu_{\theta_1}(\mathscr{X}) = \nu_{\theta_1}(\mathscr{Y})$ then $(\mu_{\theta_0}, \mu_{\theta_1})$ majorizes $(\nu_{\theta_0}, \nu_{\theta_1})$ if and only if $(\mu_{\theta_0}, \mu_{\theta_1})$ weakly super (sub) majorizes $(\nu_{\theta_0}, \nu_{\theta_1})$.

If $\Theta = \{\theta_0, \theta_1\}$, $\mathscr{X} = \mathscr{Y} = \{1, \ldots, n\}$ and $\mu_{\theta_0}(i) = \nu_{\theta_0}(i) = 1$; $i = 1, \ldots, n$ then we obtain the equivalence of conditions (i)–(iv) in example 9.1.5.

Example 9.7.4 (Majorization in $\mathbb{R}^n$***).*** Here are some of the basic facts on majorization in $\mathbb{R}^n$. First however, we need a few conventions and definitions.

If $x = (x_1, \ldots, x_n)$ is a point in $\mathbb{R}^n$ then the coordinates of x arranged in monotonically increasing order (respectively: decreasing order) are $x_{(1)} \leqslant \cdots \leqslant x_{(n)}$ (respectively: $x_{[1]} \geqslant \cdots \geqslant x_{[n]}$).

Following Marshall and Olkin we shall say that y *weakly sub majorizes* x if $\sum_{i=1}^k y_{[i]} \geqslant \sum_{i=1}^k x_{[i]}$; $k = 1, \ldots, n$. We shall say that y *weakly supermajorizes* x if $\sum_{i=1}^k y_{(i)} \leqslant \sum_{i=1}^k x_{(i)}$; $k = 1, \ldots, n$.

If y weakly sub majorizes x and if $\sum_i y_i = \sum_i x_i$ then we shall say that y *majorizes* x. Thus y majorizes x if and only if y majorizes x weakly both in the sense of sub majorization and in the sense of supermajorization.

Say also that a function f on $\mathbb{R}^n$ is symmetric if it is symmetric in its arguments, i.e. if $f(x) = f(y)$ whenever the coordinates of y are permutations of the coordinates of x. Furthermore, by the definition in section 2.2 (or section 3.2), a real valued function f on $\mathbb{R}^n$ is quasiconvex if and only if the sets $[f \leqslant a]$; $a \in \mathbb{R}$ are all convex.

Consider the measure pairs $(\mu, x\mu)$ and $(\mu, y\mu)$ of measures on $\{1, \ldots, n\}$ where μ is the uniform probability distribution and where $(x\mu)(i) = x_i$ and $(y\mu)(i) = y_i$; $i = 1, \ldots, n$. Then, by the results of this section:

(i) the following conditions are all equivalent to the condition that y weakly sub majorizes x:

 (a) $(\mu, y\mu)$ weakly sub majorizes $(\mu, x\mu)$.

(b) $My \geqslant x$ for a doubly stochastic matrix M.
(c) $\hat{M}y = x$ for a matrix $\hat{M}$ with non-negative elements such that $\hat{M}_{i,j} \leqslant N_{i,j}$; $i = 1,\ldots,n$; $j = 1,\ldots,n$ for a doubly stochastic matrix N.
(d) $\sum_i \phi(y_i) \geqslant \sum_i \phi(x_i)$ for all monotonically increasing convex functions ϕ on $\mathbb{R}$.
(e) $\sum(y_i - c)^+ \geqslant \sum(x_i - c)^+$; $c \in \mathbb{R}$.
(f) $f(y) \geqslant f(x)$ for all symmetric quasiconvex functions f on $\mathbb{R}^n$ which are monotonically increasing.

(ii) the following conditions are all equivalent to the condition that y weakly supermajorizes x:
(a) the measure pair $(\mu, y\mu)$ weakly sub majorizes the measure pair $(\mu, x\mu)$.
(b) $My \leqslant x$ for a doubly stochastic matrix M.
(c) $\hat{M}y = x$ for a matrix $\hat{M}$ such that $\hat{M}_{i,j} \geqslant N_{i,j}$; $i = 1,\ldots,n$; $j = 1,\ldots,n$ for a doubly stochastic matrix N.
(d) $\sum_i \phi(y_i) \geqslant \sum_i \phi(x_i)$ for all monotonically decreasing convex functions ϕ on $\mathbb{R}$.
(e) $\sum(c - y_i)^+ \geqslant \sum(c - x_i)^+$; $c \in \mathbb{R}$.
(f) $f(y) \geqslant f(x)$ for all symmetric quasiconvex functions f on $\mathbb{R}^n$ which are monotonically decreasing.

(iii) the following conditions are all equivalent to the condition that y majorizes x:
(a) $(\mu, y\mu)$ majorizes $(\mu, x\mu)$.
(b) $My = x$ for a doubly stochastic matrix M.
(c) x is contained in the convex hull of points obtained by permuting the coordinates of y.
(d) $\sum_i y_i = \sum_i x_i$ and y weakly sub (super) majorizes x.
(e) $\sum \phi(y_i) \geqslant \sum \phi(x_i)$ when ϕ is convex on $\mathbb{R}$.
(f) $\sum_i y_i = \sum_i x_i$ and $\sum(y_i - c)^+ \geqslant \sum(x_i - c)^+$; $c \in \mathbb{R}$.
(g) $f(y) \geqslant f(x)$ for all symmetric quasiconvex functions f on $\mathbb{R}^n$.

According to a well established, and unfortunate, terminology, functions which are monotonically increasing for the majorization ordering are called *Schur convex*. Thus symmetric quasiconvex functions are Schur convex.

Using parts (b) of (i) and (ii) we infer that the joint condition of symmetry and quasiconvexity in parts (f) of (i) and (ii) may be replaced by the condition that the function f is Schur convex.

Example 9.7.5 (Lorenz functions and β-functions. The Lorenz ordering of distribution functions on $[0, \infty[$***).*** Let F be any distribution on $[0, \infty[$ having finite positive mean $\mu_F = \int yF(dy) = \int_0^1 F^{-1}(p)\,dp$. The Lorenz function (curve) associated with F is the (graph of the) function L_F on $[0, 1]$ defined by

$$L_F(p) = \int_0^p F^{-1}(p)\,\mathrm{d}p/\mu_F.$$

Lorenz functions, or functionals of these functions, are widely used in econometrics to measure spread (or variation) of population characteristics such as income or wealth.

Let $\mathscr{D}_F$ denote the dichotomy (F_0, F_1) where $F_0 = F$ and F_1 has density $x \to x/\mu_F$ with respect to F_0.

It is easily inferred that a function L is a Lorenz function if and only if it is of the form $L(\alpha) \equiv 1 - \beta(1 - \alpha)$ for a $N\text{–}P$ function β such that $\beta(0) = 0$. If L is in this form with β being the $N\text{–}P$ function of the dichotomy (P_0, P_1), then L is the inverse function of the $N\text{–}P$ function of the reversed dichotomy (P_1, P_0).

It follows that a function is a Lorenz function if and only if it is a continuous convex function from $[0, 1]$ onto $[0, 1]$ having the origin as fixed point.

The point of this example is that the theory of the Lorenz order in econometrics, except for terminology, is simply the theory of comparison of dichotomies.

Consider the case of a population of n individuals enumerated $1, \ldots, n$ having, respectively, yearly incomes $x_i : i = 1, \ldots, n$. Selecting randomly one individual such that each individual has the same chance of being selected, we arrive at the empirical distribution function F based on the sample $x_1, x_2, \ldots, x_n$.

Letting $x_{(l)} \leqslant \cdots \leqslant x_{(n)}$ be the x's arranged in increasing order we find $\mu_F = (1/n)(x_1 + \cdots + x_n) = \bar{x}$ and $F^{-1}(t) = x_{(i)}; t \in](i-1)/n, i/n]; i = 1, \ldots, n$.

It follows that the Lorenz function L_F of F is linear on each interval $[(i-1)/n, i/n]; i = 1, \ldots, n$ and that $L_F(i/n) = [x_{(1)} + \cdots + x_{(i)}]/\bar{x}$.

Now if L_G is the Lorenz function arising from a possibly different assignment $i \to y_i$ of incomes, then $L_F \geqslant L_G$ if and only if $y = (y_1, \ldots, y_n)$ majorizes $x = (x_1, \ldots, x_n)$ as described in the previous example.

In general if F and G are distributions on $[0, \infty[$ such that both have finite positive expectations then, following Arnold (1987), we shall say that *G Lorenz majorizes F* if $L_G \leqslant L_F$. By corollary 9.3.26 this amounts to the condition that $\mathscr{D}_F$ is at least as informative as $\mathscr{D}_G$ where $\mathscr{D}_G$ is defined in terms of G as $\mathscr{D}_F$ was defined above in terms of F.

Consider now non null observation vectors $p = (p_1, \ldots, p_d)$ and $q = (q_1, \ldots, q_d)$ having non negative coordinates. Letting F_p and F_q be the empirical distribution functions based on, respectively, p and q we find that F_p Lorenz majorizes F_q if and only if $p/\sum p_i$ majorizes $q/\sum q_i$.

Proceeding to approximate majorization let us agree to say that a distribution function F having Lorenz function L_F *ε-Lorenz majorizes the distribution function G* having Lorenz function L_G if $L_F \leqslant L_G + \frac{1}{2}\varepsilon$.

By theorem 9.5.1, F ε-Lorenz majorizes G if and only if $\|G_1 - F_1 M\| \leqslant \varepsilon$ for a Markov kernel M such that $G = FM$. Here G_1 is defined in terms of G as F_1 was defined above in terms of F. In other words, F ε-Lorenz majorizes G if and only if there are random variables X and Y having distributions F and G respectively such that

$$E|E((X/EX)|Y) - (Y/EY)| \leqslant \varepsilon.$$

If X and Y are non-negative random variables having finite positive expectations then, by corollary 9.3.12, the distribution of X ε-Lorenz majorizes the distribution of Y if and only if

$$E\varphi(X/EX) \geqslant E\varphi(Y/EY) - \tfrac{1}{2}\varepsilon[\varphi'(\infty) - \varphi'(0)]$$

when φ is convex on $[0, \infty[$ and $\varphi'(\infty) = \lim_{x\to\infty} \varphi'(x)$. By the same corollary it suffices to consider convex functions φ of the form $\varphi(x) \equiv |x - c|$ with $c > 0$ and then the criterion reduces to the requirement that $E(|X - cEX|/EX) \geqslant E(|Y - cEY|/EY) - \varepsilon$ when $c > 0$.

The reader is referred to Arnold (1987) as well as to Marshall & Olkin (1979) for further information on Lorenz functions and on the Lorenz ordering.

Example 9.7.6 (The Gini index. Continuation of example 9.7.5). An interesting numerical measure of spread which has received much attention in econometrics is the Gini index.

For a Lorenz curve (function) the *Gini index* is twice the area of the region enclosed between the curve and the diagonal of the unit square. Thus the Gini index of the Lorenz function L is the number $\gamma = 1 - 2\int_0^1 L(\alpha)\,d\alpha$.

Let us write γ_F for the Gini index of the Lorenz function L_F of a distribution F. It then follows directly from the definitions that $\gamma_F \geqslant \gamma_G - \varepsilon$ when F ε-Lorenz majorizes G.

If we express the Lorenz function L as $L(\alpha) \equiv 1 - \beta(1 - \alpha)$ then the Gini index of L may be expressed in terms of the N–P function β as $\gamma = 2\int_0^1 \beta(\alpha)\,d\alpha - 1$. From this expression we see that G is the area of the convex range of any dichotomy having β as its N–P-function. Getting rid of the unpleasant condition that $L(1) = 1$, i.e. that $\beta(0) = 0$, we may define the *Gini index* of a dichotomy $\mathscr{D} = (P_0, P_1)$ as the area $\gamma = 2\int_0^1 \beta(\alpha|\mathscr{D})\,d\alpha - 1$ of the convex range of $\mathscr{D}$. The Gini index of a dichotomy $\mathscr{D} = (P_0, P_1)$ may be expressed directly in terms of the measures P_0 and P_1 by the formula

$$\gamma = \|(P_0 \times P_1) - (P_1 \times P_0)\|/2 = 1 - \|(P_0 \times P_1) \wedge (P_1 \times P_0)\|.$$

Indeed the quantities on the right do not change when $\mathscr{D} = (P_0, P_1)$ is replaced by another equivalent dichotomy. If $\mathscr{D}$ has N–P function β then it is equivalent to $(R[0,1], \beta)$ and thus $\|(P_0 \times P_1) \wedge (P_1 \times P_0)\| = \int\int[\beta'(\alpha_2) \wedge \beta'(\alpha_1)]\,d\alpha_1\,d\alpha_2 = 2\int\int_{\alpha_1 > \alpha_2} \beta'(\alpha_2)\,d\alpha_1\,d\alpha_2 = 2\int_0^1 (1 - \beta(\alpha_1))\,d\alpha = 1 - \gamma$.

It follows from these formulas that the Gini index of the Lorenz function of the distribution of a non-negative random variable X may be expressed as $E|X_1 - X_2|/(2EX) = 1 - [E(X_1 \wedge X_2)/EX]$ where X_1 and X_2 are independent and identically distributed as X.

These ideas may be generalized to measure families with finite parameter sets. In section 9.3 we considered the convex range $\tau(\mathscr{E})$ of a measure family $\mathscr{E} = (\mu_\theta : \theta \in \Theta)$ when $m = \#\Theta < \infty$. The m-dimensional volume $\gamma(\mathscr{E})$ of $\tau(\mathscr{E})$ may be considered as a multivariate Gini index.

This index provides a monotonically information increasing functional of experiments. Furthermore, by the Brunn–Minkowski theorem, its nth root is concave for mixing. (See e.g. Grünbaum (1967) for references on the Brunn–Minkowski theorem.)

For a given measure family $\mathscr{E} = (\mu_1, \ldots, \mu_n)$ the multivariate Gini index may be expressed as

$$\begin{aligned}\gamma(\mathscr{E}) &= \text{Volume } \tau(\mathscr{E}) \\ &= \frac{1}{n!}\left\| \sum_\pi (\operatorname{sgn} \pi)(\mu_{\pi(1)} \times \cdots \times \mu_{\pi(n)} \right\|\end{aligned}$$

where $\|\cdot\|$ denotes total variation and where π runs through the set of all permutations of $\{1, \ldots, n\}$.

The structure of this formula may become more apparent if we write it as

$$\gamma(\mathscr{E}) = \frac{1}{n!}\left\| \det \begin{pmatrix} \mu_1, \ldots, \mu_n \\ \cdots\cdots\cdots \\ \mu_1, \ldots, \mu_n \end{pmatrix} \right\|.$$

If $\mu_1, \mu_2, \ldots, \mu_n$ are dominated by the non-negative σ-finite measure σ and if $f_i = \mathrm{d}\mu_i/\mathrm{d}\sigma$; $i = 1, \ldots, n$ then $\gamma(\mathscr{E})$ may be computed as the integral

$$\int \frac{1}{n!}\left| \det \begin{pmatrix} f_1(x_1), \ldots, f_n(x_1) \\ \cdots\cdots\cdots\cdots\cdots \\ f_1(x_n), \ldots, f_n(x_n) \end{pmatrix} \right| \sigma^n(\mathrm{d}(x_1, \ldots, x_n)).$$

The formula

$$\text{Volume } \tau(\mathscr{E}) = \frac{1}{n!}\left\| \sum_\pi (\operatorname{sgn} \pi)\mu_{\pi(1)} \times \cdots \times \mu_{\pi(n)} \right\|$$

may be established by verifying that:

(i) if $T = (t_{i,k}; i, k = 1, \ldots, n)$ is a non-singular $n \times n$ matrix and if we replace $\mu_1, \ldots, \mu_n$ with, respectively, $v_1 = \sum_k t_{1k}\mu_k, \ldots, v_n = \sum_k t_{nk}\mu_k$, then both sides of the equality are multiplied by $|\det T|$.

(ii) if $\mathscr{X} = \{1, \ldots, r\}$ and $\mu_i(j) = a_{ij}$; $i = 1, \ldots, n$, $j = 1, \ldots, r$ then the identity reduces to

$$\sum_{1 \le j_1 < \cdots < j_n \le r} (|\det(a_{\cdot,j_1}, \ldots, a_{\cdot,j_n})|) = \text{Volume}(\langle 0, a_{\cdot,1} \rangle + \cdots + \langle 0, a_{\cdot,r} \rangle) \tag{9.7.1}$$

where $\langle\ \rangle$ denotes convex hull.

If the vectors $a_{\cdot,1}, \ldots, a_{\cdot,r}$ are linearly dependent then both sides of (9.7.1) are zero.

If $r = n$ and $a_{\cdot,1}, \ldots, a_{\cdot,r}$ are linearly independent then (9.7.1) may, by (i), be reduced to the statement that the volume of a cube is the product of the lengths of its sides.

The validity of (9.7.1) follows now by induction on r. (Using (i) we may assume that $a_{i,r} = 0$ or $= s \geqslant 0$ as $i < n$ or $i = n$.)

(iii) both sides of the desired equality are continuous for weak convergence of standard measures. Since the set of finitely supported standard measures is dense, it suffices to consider the finite case.

Other possibilities for defining multivariate Gini indices are discussed in Arnold (1987).

9.8 References

Arnold, B. C. 1987. Majorization and the Lorenz Order: A brief introduction. *Lecture notes in Statistics No. 43*, Springer–Verlag, Berlin.

Blackwell, D. 1951. Comparison of experiments. *Proc. Second Berkeley Sympos. Math. Statist. Probab.* pp. 93–102.

— 1953. Equivalent comparisons of experiments. *Ann. Math. Statist.* **24**, pp. 265–72.

Boll, C. 1955. *Comparison of experiments in the infinite case.* Ph. D. thesis. Stanford University.

Dahl, G. 1983. *Pseudo experiments and majorization.* Thesis. Univ. of Oslo. Statistical research report, 1984.

Dudley, R. M. 1968. Distances of probability measures and random variables. *Ann. Math. Statist.* **39**, pp. 1563–72.

Dudley, R. M. 1989. *Real analysis and probability.* Wadsworth & Brooks/Cole, California.

Ehm, W. & Müller, D. W. 1983. Factorizing the information contained in an experiment conditionality on the observed value of a statistic. *Z. Wahrscheinlichkeitstheorie verw. Geb.* **65**, pp. 121–34.

Fischer, P. & Holbrook, J. A. R. 1980. Balayage defined by the non negative convex functions. *Proc. Amer. Math. Soc.* **79**, pp. 445.

Grünbaum, B. 1967. *Convex polytopes.* Interscience, London.

Hajek, J. & Sidak, Z. 1967. *Theory of rank tests.* Academic Press and Academia Publishing house of the Czechoslovak Academy of Sciences.

Hansen, O. H. & Torgersen, E. N. 1974. Comparison of linear normal experiments. *Ann. Statist.* **2**, pp. 367–73.

Hardy, G. H., Littlewood, J. E. & Polya, G. 1934. *Inequalities.* Cambridge University Press.

Ibragimov, J. A. 1956. On the composition of unimodal distributions (Russian). *Teoriya veroyatnostey* **1**, pp. 283–8.

Janssen, A. 1988. A convolution theorem for the comparison of exponential families. *Res. report* no. **211**, Univ. of Siegen.

Karlin, S. & Rinott, Y. 1983. Comparison of measures, multivariate majorization, and applications to statistics. In: Karlin, Amemiya, Goodman. *Studies in econometric time series and multivariate statistics.* Academic Press, New York.

LeCam, L. 1964. Sufficiency and approximate sufficiency. *Ann. Math. Statist.* **35**, pp. 1419–55.
— 1974. *Notes on asymptotic methods in statistical decision theory.* Centre de Recherches Math. Univ. de Montréal.
— 1977. *On the asymptotic normality of estimates.* Proceedings of the symposium to honour J. Neyman. Panstw. Wyslawn. Nauk. Warsaw, pp. 203–17.
Lehmann, E. L. 1983. Comparison of experiments for some multivariate normal situations. In: Karlin, Amemiya, Goodman. *Studies in econometric time series and multivariate statistics.* Academic Press, New York.
— 1986. *Testing statistical hypotheses.* Wiley, New York.
Marshall, A. W. & Olkin, I. 1979. *Inequalities: Theory of majorization and its applications.* Academic Press, New York.
Mazur, S. & Ulam, S. 1932. Sur les transformations isometriques d'espace vectoriels normés. *C.R. Acad. Sci. Paris* **194**, pp. 946–8.
Millar, P. W. 1983. The minimax principle in asymptotic statistical theory. *Lecture notes in Mathematics No. 976*, pp. 76–265. Springer–Verlag, Berlin.
Scheffé, H. 1947. A useful convergence theorem for probability distributions. *Ann. Math. Statist.* **18**, pp. 434–8.
Strassen, V. 1965. The existence of probability measures with given marginals. *Ann. Math. Statist.* **36**, pp. 423–39.
Torgersen, E. 1969. *On ε-comparison of experiments.* Mimeographed notes. Univ. of California, Berkeley.
— 1972a. Local comparison of experiments when the parameter set is one dimensional. *Statist. res. report no.* **4**, Univ. of Oslo.
— 1972b. Local comparison of experiments. *Statist. res. report no.* **5**, Univ. of Oslo.
— 1972c. Comparison of translation experiments. *Ann. Math. Statist.* **43**, pp. 1383–99.
— 1982. Comparison of some statistical experiments associated with sampling plans. *Probability and Mathematical Statistics* **3**, pp. 1–17.
— 1985. Majorization and approximate majorization for families of measures. *Lecture notes in Statistics No. 35*, pp. 247–310, Springer–Verlag, Berlin.

10

Complements: Further Examples, Problems and Comments

10.1 Introduction

Several complements to the main themes are presented here in a mixed example/problem form. Problems with known solutions are usually presented as statements in need of verification. These are often sketched or may be found in cited references. Complements 38–49, supplemented with parts of complement 37, provide a study of monotone likelihood ratio. A more thorough treatment of this material is Torgersen (1989) which also provides applications to the problem on how information is affected by selection.

The table gives a brief description of the content of each complement, and the relevant chapters in each case. The symbol ↑ following the complement number indicates that this complement is partially based on the previous one.

Complement Number	Relevant Chapter(s)	Description
1	1	Power of tests need not increase strictly with the numbers of observations. A characterization of double dichotomies.
2	1, 7, 9	An 'algebra' for concave distribution functions on $[0, 1[$.
3↑	7	Counter example on the algebraic structure.
4	1, 6, 7, 9	Duality between power and minimum Bayes risk.
5↑	1, 6, 7, 9	Maximin power.
6↑	1, 6, 7, 9	Minimum mixed Bayes risk.
7↑	1, 6, 7, 9	Maximin power and minimum mixed Bayes risk.
8↑	1, 6, 7, 9	Maximin power and minimum mixed Bayes risk in replicated experiments.
9	6	Deficiencies for dichotomies (see also complement 11).
10↑	1, 6	The Neyman–Pearson lemma for statistical distance neighbourhoods.
11	1, 6, 7, 9	Tools for dichotomies.
12↑	1, 6, 9	Singular parts, domination.
13↑	6, 7	Contiguity.
14	7	Hellinger transforms of dichotomies.

Complement Number	Relevant Chapter(s)	Description
15	6, 7	Majorization and minorization by double dichotomies.
16	6, 7	Relative nearness of parameter values. The binomial case.
17↑	6, 7	Relative nearness of parameter values. The Poisson case.
18	1, 6	Deficiency distances to total information and to non information.
19	7	Decomposition by conditioning.
20	7	Dichotomies as totally ordered mixtures of double dichotomies.
21	7	Uniqueness of decompositions into mixtures of totally ordered families of extremal experiments.
22	2, 6, 7, 9	Measurable decompositions of sublinear functionals.
23↑	6, 7, 9	Existence of randomizations with prescribed properties.
24↑	7, 9	Existence of dilations.
25↑	7	Minorized decompositions.
26	6, 7, 8	Comparison by factorization.
27	1, 7	A non-separable dominated experiment such that all pairs of parameter values define equivalent dichotomies.
28	1, 7	Representations of separable experiments.
29	6	Two distances for translation experiments.
30↑	6	Translation experiments on the integers.
31	6	Semigroups of translation experiments.
32	6	Inequalities for convex extensions.
33	6, 7	Equivalence of ordered experiments.
34	6, 7	Monotone nets of experiments.
35	7	Equicontinuity and deficiency convergence.
36	1, 7	Hellinger extensions of experiments.
37	6	Comparison for given losses.
38	1,7	Monotone likelihood ratio.
39↑	4	Monotone loss functions.
40	1	A fine point on the Neyman–Pearson lemma.
41↑	6, 7	Monotone likelihood and power diagrams I.
42↑	7	Monotone likelihood and power diagrams II.
43	7	Continuation of complement 41. Monotone likelihood and pairwise equivalence.
44	6	Comparison of a monotone likelihood experiment with respect to another experiment for monotone decision problems.
45↑	6	Lehmann's criterion.
46	1, 6, 9	Composition of β-functions and monotone likelihood. Pairwise order completeness.
47↑	1, 7, 9	Stationary monotone likelihood and strong unimodality.
48	6, 9	Continuation of complement 45. Comparison involving differentiable monotone likelihood.
49	7	Monotone likelihood in terms of betweenness for the total variation distance.

10.2 Complements

Complement 1. Power of tests need not increase strictly with the number of observations. A characterization of double dichotomies

If $\mathscr{D}$ is a dichotomy such that $\beta(\alpha_0|\mathscr{D}^2) = \beta(\alpha_0|\mathscr{D}) \in]0,1[$ then $\mathscr{D}$ is equivalent to the double dichotomy $\begin{pmatrix} 1-\alpha_0 & \alpha_0 \\ 1-\beta_0 & \beta_0 \end{pmatrix}$ where $\beta_0 = \beta(\alpha_0|\mathscr{D})$.

On the other hand if $\mathscr{D}$ is equivalent to the double dichotomy $\begin{pmatrix} 1-p & p \\ 1-q & q \end{pmatrix}$ then $\beta(p^{n-1}|\mathscr{D}^{n-1}) = \beta(p^{n-1}|\mathscr{D}^n) = q^{n-1}$ for all $n = 2, 3\ldots$

The first statement may be proved by showing (using the concavity of β and complement 2(a)) that under the given circumstances the supremum in the expression $\beta(\alpha_0|\mathscr{D}^2) = \sup\{\int \beta(\phi(\xi))\beta(\mathrm{d}\xi) : \int \phi(\xi)\,\mathrm{d}\xi = \alpha_0\}$ is obtained for any function ϕ of the form $\phi(\xi) \equiv_\xi (1-t)I_{[0,\alpha_0]}(\xi) + t\alpha_0$.

Complement 2. An 'algebra' for concave distribution functions on $[0, 1[$

Any concave distribution function on the interval $[0, 1[$ is representable as $\beta(\cdot|\mathscr{D})$ for a dichotomy $\mathscr{D}$ (see chapters 1 and 9) and this dichotomy is unique up to equivalence. In fact we may specify $\mathscr{D}$ as $\mathscr{D} = (\lambda, \beta)$ where λ is the uniform distribution on $[0, 1]$ and where β also denotes the distribution having distribution function β. Conversely if $\mathscr{D}$ is a dichotomy then $\beta(\cdot|\mathscr{D})$ is a concave distribution function on $[0, 1[$.

It follows that the set of concave distribution functions on $[0, 1[$ inherits all the structures of types of dichotomies. Thus the ordering being 'at least as informative' corresponds to the ordering 'being stochastically at most as large as' and this is just the pointwise ordering of distribution functions.

We have also seen, in chapter 9, that the deficiency distance for dichotomies is twice the Levy diagonal distance for the corresponding β functions.

The associative and commutative compositions of multiplication and of mixing lead to commutative and associative compositions in the set of concave distribution functions on $[0, 1[$. Let us use the notation $\otimes$ and $\oplus$ for the compositions derived from multiplication and mixing respectively. It is then easily checked that $\otimes$ is distributive with respect to $\oplus$ and that the one-point distribution in 0 acts as unit for $\otimes$.

(a) Explicitly the product $\beta_1 \otimes \beta_2$ of two concave distribution functions $\beta_1 = \beta(\cdot|\mathscr{D}_1)$ and $\beta_2 = \beta(\cdot|\mathscr{D}_2)$ on $[0, 1[$ is given by

$$\beta_1 \otimes \beta_2(\alpha) = \beta(\alpha|\mathscr{D}_1 \times \mathscr{D}_2) = \sup \int_0^1 \beta_1(\phi(x))\beta_2(\mathrm{d}x)$$

where the supremum is taken over all (monotonically decreasing) test functions ϕ on $[0, 1]$ such that $\int_0^1 \phi(x)\,\mathrm{d}x = \alpha$.

(b) With notation as in (a) we find that

$$\beta_1 \otimes \beta_2(\alpha) \geqslant \beta_1(\beta_2(\alpha))$$

where the right hand side is a *bona fide* β-function (see complement 46 for the interpretation of $\beta_1 \beta_2$).

(c) The mixture $p_1\beta_1 \oplus p_2\beta_2 \oplus \ldots$ of concave distribution functions $\beta_1 = \beta(\alpha|\mathscr{D}_1)$, $\beta_2 = \beta_2(\alpha|\mathscr{D}_2), \ldots$ on $[0, 1[$ for the mixing distribution $(p_1, p_2, \ldots)$ on the positive integers is given by $(p_1\beta_1 \oplus p_2\beta_2 \oplus \ldots)(\alpha) = \beta(\alpha|p_1\mathscr{D}_1 + p_2\mathscr{D}_2 + \cdots) = \sup\{p_1\beta_1(\alpha_1) + p_2\beta_2(\alpha_2) + \cdots\}$ where the supremum is taken over all sequences $\alpha_1, \alpha_2, \ldots$ of non-negative numbers such that $\alpha_1 + \alpha_2 + \cdots = \alpha$.

(d) With notation as in (b) we find that $p_1\beta_1 \oplus p_2\beta_2 \oplus \cdots \geqslant p_1\beta_1 + p_2\beta_2 + \cdots$. Thus the experiment mixture majorizes the pointwise mixture and in general (see below) this majorization is strict.

Remarks. The fact that mixtures of experiments majorize the corresponding pointwise mixtures is due to the general fact that information is usually lost by deletions of ancillary statistics. Thus if the dichotomies $\mathscr{D}_i = (\lambda, \beta_i)$; $i = 1, 2, \ldots$ are realized by $X_1, X_2, \ldots$ respectively, and if I is a positive integer valued random variable which is independent of $(X_1, X_2, \ldots)$ then $\sum Pr(I = i)\mathscr{D}_i$ is realized by the pair (I, X_I). Deleting the (ancillary) index I we are left with the single observation X_I and this observation realizes the dichotomy $(\lambda, \sum_i Pr(I = i)\beta_i)$.

The extreme points for the experiment mixture $\oplus$ are the β-functions derived from the double dichotomies, i.e. the dichotomies which are representable as 2×2 Markov matrices. These β-functions are precisely those whose graphs, together with the diagonal of the unit square, bound a triangle. If the double dichotomy is represented as the Markov matrix $\begin{pmatrix} 1-p & p \\ 1-q & q \end{pmatrix}$ where $p \leqslant q$ then the vertices of this triangle are the points $(0, 0)$, (p, q) and $(1, 1)$.

Any dichotomy is representable as an, essentially unique, totally ordered mixture of double dichotomies (see complement 20).

On the other hand the extreme points for pointwise mixtures of distribution functions are those given by the particular double dichotomies of the form $\begin{pmatrix} 1-p, & p \\ 0 & 1 \end{pmatrix}$. In fact if $0 < \alpha_0 < \beta(\alpha_0) < 1$ then $\beta = (1-\theta)\beta_0 + \theta\beta_1$ where $\beta_0(\alpha) \equiv_\alpha [\beta(0) + 1/\alpha_0(\beta(\alpha_0) - \beta(0))\alpha] \wedge \beta(\alpha)$. Here β_1 is a concave probability distribution function on $[0, 1[$ provided θ is sufficiently close to 1.

Usually the most convenient way of studying algebraic combinations for the operations $\otimes$ and $\oplus$ is to use Hellinger transforms. The Hellinger transform of a dichotomy $\mathscr{D}$ of the form $\mathscr{D} = (\lambda, \beta)$ or, equivalently, of a concave distribution function β on $[0, 1[$, assigns the value $\int_0^1 [\beta'(\alpha)]^t \, d\alpha$ to any non-degenerate distribution $(1-t, t)$ on the two-point parameter set $\{0, 1\}$.

Complement 3↑. Counter example on the algebraic structure

The algebraic structure for experiments defined by mixtures and products may suggest false conclusions concerning the order structure. Thus we know that the informational inequality $\mathscr{E}^2 \geqslant \mathscr{F}^2$ may be satisfied for non-comparable experiments $\mathscr{E}$ and $\mathscr{F}$. Another tempting conjecture may be based on the following computations on Hellinger transforms

$$\begin{aligned} H(\cdot\,|\tfrac{1}{2}\mathscr{E}^2 + \tfrac{1}{2}\mathscr{E}^2) &= \tfrac{1}{2}H(\cdot\,|\mathscr{E})^2 + \tfrac{1}{2}H(\cdot\,|\mathscr{F})^2 \\ &= \tfrac{1}{2}[H(\cdot\,|\mathscr{E}) - H(\cdot\,|\mathscr{F})]^2 + H(\cdot\,|\mathscr{E})H(\cdot\,|\mathscr{F}) \\ &= \tfrac{1}{2}[H(\cdot\,|\mathscr{E}) - H(\cdot\,|\mathscr{F})]^2 + H(\cdot\,|\mathscr{E} \times \mathscr{F}) \geqslant H(\cdot\,|\mathscr{E} \times \mathscr{F}). \end{aligned}$$

In particular affinities and Hellinger distances behave as if $\frac{1}{2}\mathscr{E}^2 + \frac{1}{2}\mathscr{F}^2$ was at most as informative as $\mathscr{E} \times \mathscr{F}$. In spite of this we cannot conclude in general that $\frac{1}{2}\mathscr{E}^2 + \frac{1}{2}\mathscr{F}^2$ is at most as informative as $\mathscr{E} \times \mathscr{F}$.

Counter examples may be constructed by considering statistical distances, or rather the associated 'affinities', for dichotomies $\mathscr{E} = \begin{pmatrix} 1-\xi_1 & \xi_1 & 0 \\ 0 & \xi_2 & 1-\xi_2 \end{pmatrix}$ and $\mathscr{F} = \begin{pmatrix} 1-\eta_1 & \eta_1 & 0 \\ 0 & \eta_2 & 1-\eta_2 \end{pmatrix}$. (Choose e.g. $\xi_1 = \eta_1 = \frac{1}{2}, \xi_2 = 1$ and $\eta_2 = \frac{1}{4}$.)

Additional problem. Let p and q be homogeneous polynomials of degree r in (x, y). Assuming that p and q are both mixtures of products $x^a y^{r-a}$ and that $p(x, y) \geqslant q(x, y)$ when $x, y \geqslant 0$ we may ask: what additional assumptions must p and q satisfy in order that $p(\mathscr{E}, \mathscr{F}) \leqslant q(\mathscr{E}, \mathscr{F})$ for all pairs $(\mathscr{E}, \mathscr{F})$ of experiments?

Complement 4. Duality between power and minimum Bayes risk

As usual we define the β-function $\beta(\cdot\,|\mathscr{D}) = \beta$ and the b-function $b(\cdot\,|\mathscr{D}) = b$ of a dichotomy $\mathscr{D} = (P_0, P_1)$ by

$$\beta(\alpha) = \max\{P_1(\delta) : 0 \leqslant \delta \leqslant 1, P_0(\delta) \leqslant \alpha\}; \qquad 0 \leqslant \alpha \leqslant 1$$

and

$$b(\lambda) = \min\{(1 - \lambda)P_0(\delta) + \lambda(1 - P_1(\delta)) : 0 \leqslant \delta \leqslant 1\} = \|(1 - \lambda)P_0 \wedge \lambda P_1\|; \quad 0 \leqslant \lambda \leqslant 1.$$

Then:

(a) $(1 - \lambda)\alpha + \lambda(1 - \beta(\alpha)) \geqslant b(\lambda)$; $\alpha, \lambda \in [0, 1]$.

(b) for any given α in $[0, 1]$ the number $\beta = \beta(\alpha)$ is the largest number such that $(1 - \lambda)\alpha + \lambda(1 - \beta) \geqslant b(\lambda)$; $0 \leqslant \lambda \leqslant 1$, i.e.

$$\beta(\alpha) = \inf_{\lambda > 0} \frac{1}{\lambda}[(1 - \lambda)\alpha + \lambda - b(\lambda)]$$

where inf may be replaced with min when $\alpha > 0$.

(c) for given λ in $[0, 1]$ the number $b = b(\lambda)$ is the largest number such that $(1-\lambda)\alpha + \lambda(1-\beta(\alpha)) \geqslant b$; $\alpha \in [0, 1]$, i.e.

$$b(\lambda) = \min_{\alpha} [(1-\lambda)\alpha + \lambda(1-\beta(\alpha))].$$

(d) for each given α (given λ) in $[0, 1]$ there is a non-empty closed interval of numbers λ (numbers α) such that

$$(1-\lambda)\alpha + \lambda(1-\beta(\alpha)) = b(\lambda).$$

If α and λ are related in this way then any most powerful level α test is a $(1-\lambda, \lambda)$ Bayes procedure for the testing (estimation) problem with loss $= 0$ or $= 1$ as the claim is right or not.

Complement 5↑. Maximin power

This is the first of five complements on power and on mixed Bayes risk in composite testing problems.

Consider two non-empty sets Π_0 and Π_1 of probability measures on a common measurable space $(\mathscr{X}, \mathscr{A})$. We shall be concerned with the problem of testing the null hypothesis '$P \in \Pi_0$' against the alternative '$P \in \Pi_1$'. Tests below will be identified with the $[0, 1]$-valued measurable functions providing the probabilities of rejection; i.e. with the test functions.

We shall admit as tests anything which from the point of view of any finite selection of distributions in $\Pi_0 \cup \Pi_1$ looks like a test function. To be more specific, in addition to the usual test functions (i.e. measurable functions between 0 and 1) we will also consider limits of such functions for the $L(\Pi_0 \cup \Pi_1)$-topology on $L(\Pi_0 \cup \Pi_1)^*$. This leads to the notion of a generalized test function δ as a non-negative linear functional on $L(\Pi_0 \cup \Pi_1)$ such that $\mu(\delta) \leqslant \|\mu\|$ when $\mu \in L(\Pi_0 \cup \Pi_1)$.

If $\Pi = \Pi_0 \cup \Pi_1$ is coherent (in particular when $\Pi = \Pi_0 \cup \Pi_1$ is dominated or discrete) then all generalized test functions are equivalent to usual test functions.

If $\alpha \in [0, 1]$ then a generalized test function (and thus also a usual test function) is called a level α test if $P_0(\delta) \leqslant \alpha$ when $P_0 \in \Pi_0$. The set of generalized level α tests may be much larger than the closure of the set of usual level α tests.

Our goal is to find tests δ making the minimum power (probability of rejection) $\inf_{P_1 \in \Pi_1} P_1(\delta)$ as large as possible subject to the condition that δ is a level α test.

This minimum power can never be larger than the number $\beta(\alpha|\Pi_0, \Pi_1) = \sup_{\delta \in D_\alpha} \inf_{P_1 \in \Pi_1} P_1(\delta)$ where D_α is the set of level α generalized test functions δ. A generalized test function δ (and thus also a usual test function) is called a maximin level α test if it has level α and minimum power $\inf_{P_1 \in \Pi_1} P_1(\delta) = \beta(\alpha|\Pi_0, \Pi_1)$.

By weak compactness generalized maximin level α tests exist, i.e. the sup in the definition of $\beta(\alpha|\Pi_0, \Pi_1)$ is obtained and is thus a maximum.

It is readily checked that if δ is a maximin level α test and if $\beta(\alpha|\Pi_0,\Pi_1) < 1$ then δ has size α, i.e. $\sup_{P_0 \in \Pi_0} P_0(\delta) = \alpha$.

Maximin power may be expressed in terms of usual test functions by

$$\beta(\alpha|\Pi_0,\Pi_1) = \inf\{\beta(\alpha|\tilde{\Pi}_0,\tilde{\Pi}_1) : \tilde{\Pi}_0 \subseteqq \Pi_0, \tilde{\Pi}_1 \subseteqq \Pi_1 \text{ and } \tilde{\Pi}_0 \cup \tilde{\Pi}_1 \text{ is coherent}\}$$

$$= \inf\{\beta(\alpha|\tilde{\Pi}_0,\tilde{\Pi}_1) : \tilde{\Pi}_0 \subseteqq \Pi_0, \tilde{\Pi}_1 \subseteqq \Pi_1 \text{ and } \tilde{\Pi}_0 \cup \tilde{\Pi}_1 \text{ is finite}\}.$$

(a) $\beta(\alpha|\Pi_0,\Pi_1)$ is not altered if Π_i; $i = 0, 1$, is increased by adding probability measures belonging to the closure of the convex hull $\langle\Pi_i\rangle$ of Π_i for the L^*-topology on $L = L(\Pi_0 \cup \Pi_1)$.

(b) Furthermore $\beta(\alpha|\Pi_0,\Pi_1)$ may be expressed in terms of maximin powers for completely specified alternatives by

$$\beta(\alpha|\Pi_0,\Pi_1) = \inf_{P_1 \in \langle\Pi_1\rangle} \beta(\alpha|\Pi_0,P_1)$$

(c) $\beta(\alpha|\Pi_0,\Pi_1)$ is monotonically decreasing in Π_0, Π_1 so that in particular $\beta_0(\alpha|\Pi_0,\Pi_1) \leqslant \inf\{\beta(\alpha|P_0,P_1) : P_0 \in \Pi_0, P_1 \in \Pi_1\}$. We shall see in a moment that we have equality if Π_0 and Π_1 are convex or if (Π_0,Π_1) admits a least favourable pair as defined below.

(d) If $\mathscr{E} = (P_\theta : \theta \in \Theta)$ is an experiment then $\beta(\alpha|\{P_\theta : \theta \in \Theta_0\}, \{P_\theta : \theta \in \Theta_1\})$ depends on $\mathscr{E}$ by way of its type.

(e) A generalized level α test (and thus a usual level α test) δ is a maximin level α test provided it is a most powerful level α test for testing 'P_0' against 'P_1' where $P_0 \in \Pi_0$ and $P_1 \in \Pi_1$ and provided minimum power in Π_1 is obtained at P_1. In that case $\beta(\alpha|\Pi_0,\Pi_1) = \beta(\alpha|P_0,P_1)$ so that $\beta(\alpha|P_0,P_1) \leqslant \beta(\alpha|Q_0,Q_1)$ whenever $Q_0 \in \Pi_0$ and $Q_1 \in \Pi_1$. We may then say that (P_0,P_1) is a least favourable pair at level α.

(f) The function $\alpha \to \beta(\alpha|\Pi_0,\Pi_1)$ is the β-function of a dichotomy, i.e. $\beta(\cdot|\Pi_0,\Pi_1)$ is a continuous concave distribution function on $[0,1]$.

Remark. If in the definition of $\beta(\cdot|\Pi_0,\Pi_1)$ we restrict our attention to ordinary test functions, then $\beta(\cdot|\Pi_0,\Pi_1)$ is still concave on $[0,1]$. In particular $\beta(\cdot|\Pi_0,\Pi_1)$ is still continuous on $]0,1]$. Although we expect that this function need not be continuous at $\alpha = 0$, and thus not equal to the β-function of a dichotomy, we have not come across an example of this phenomenon. An example, if it exists, must necessarily be derived from a pair (Π_0,Π_1) such that $\Pi_0 \cup \Pi_1$ is not coherent.

Complement 6↑. Minimum mixed Bayes risk

When we consider risk here (and below) it is for the 0–1 loss function which prescribes loss $= 1$ whenever you classify wrongly the true distribution with respect to the classification (Π_0,Π_1) and which prescribes loss $= 0$ when you classify correctly. Decision rules are described by the test functions they define, i.e. by the probabilities of deciding '$P \in \Pi_1$'. Thus if the generalized test function δ is used then the risk is $P_0(\delta)$ when $P_0 \in \Pi_0$ while it is $1 - P_1(\delta)$ when $P_1 \in \Pi_1$. This yields the two 'maximum' risks $\sup_{P_0 \in \Pi_0} P_0(\delta)$ and $1 - \inf_{P_1 \in \Pi_1} P_1(\delta)$. If nature chooses Π_0 and Π_1 according to the probabilities $1 - \lambda$ and λ then we get the (maximum) *mixed Bayes risk*

$$(1-\lambda) \sup_{P_0 \in \Pi_0} P_0(\delta) + \lambda\left(1 - \inf_{P_1 \in \Pi_1} P_1(\delta)\right)$$

$$= \sup_{P_0 \in \Pi_0} \sup_{P_1 \in \Pi_1} [(1-\lambda)P_0(\delta) + \lambda(1 - P_1(\delta))].$$

The *minimum* $(1-\lambda, \lambda)$ *mixed Bayes risk* is thus obtained by minimizing this expression with respect to δ. This quantity will be denoted as $b(\lambda|\Pi_0, \Pi_1)$. Thus

$$b(\lambda|\Pi_0, \Pi_1) = \inf_{\delta} \sup_{P_0 \in \Pi_0} \sup_{P_1 \in \Pi_1} [(1-\lambda)P_0(\delta) + \lambda(1 - P_1(\delta))].$$

A generalized test function δ is called a $(1-\lambda, \lambda)$ *mixed Bayes test* if $b(\lambda|\Pi_0, \Pi_1) = \sup_{P_0 \in \Pi_0} \sup_{P_1 \in \Pi_1} [(1-\lambda)P_0(\delta) + \lambda(1 - P_1(\delta)]$.

By weak compactness the inf is obtained, i.e. minimum $(1-\lambda, \lambda)$ mixed Bayes decision rules exist. If $\Pi_0 \cup \Pi_1$ is coherent then these rules may be chosen as ordinary decision rules. Otherwise we may have to resort to generalized Bayes decision rules.

Minimum mixed Bayes risk may be reduced to the coherent case by

$$b(\lambda|\Pi_0, \Pi_1) = \sup\{b(\lambda|\tilde{\Pi}_0, \tilde{\Pi}_1) : \tilde{\Pi}_0 \subseteq \Pi_0, \tilde{\Pi}_1 \subseteq \Pi_1, \tilde{\Pi}_0 \cup \tilde{\Pi}_1 \text{ is coherent}\}$$

$$= \sup\{b(\lambda|\tilde{\Pi}_0, \tilde{\Pi}_1) : \tilde{\Pi}_0 \subseteq \Pi_0, \tilde{\Pi}_1 \subseteq \Pi_1, \tilde{\Pi}_0 \cup \tilde{\Pi}_1 \text{ is finite}\}.$$

(a) $b(\lambda|\Pi_0, \Pi_1)$ is not altered if Π_i; $i = 0, 1$ is increased by adding probability measures belonging to the closure of the convex hull $\langle \Pi_i \rangle$ of Π_i for the L^*-topology on $L = L(\Pi_0 \cup \Pi_1)$.

(b) If $\mathscr{E} = (P_\theta : \theta \in \Theta)$ is an experiment then $b(\lambda|\{P_\theta : \theta \in \Theta_0\}, \{P_\theta : \theta \in \Theta_1\})$ depends only on the type of $\mathscr{E}$.

(c) $b(\cdot|\Pi_0, \Pi_1)$ is the b-function of a dichotomy, i.e. $b(\cdot|\Pi_0, \Pi_1)$ is nonnegative and concave on $[0, 1]$ and it is majorized by $\lambda \to (1-\lambda) \wedge \lambda$.

(d) $b(\lambda|\Pi_0, \Pi_1)$ is monotonically increasing in (Π_0, Π_1). In particular $b(\lambda|\Pi_0, \Pi_1) \geqslant \sup_{P_0 \in \Pi_0} \sup_{P_1 \in \Pi_1} b(\lambda|P_0, P_1)$ where equality holds when Π_0 and Π_1 are convex.

Complement 7↑. Maximin power and minimum mixed Bayes risk

We have seen that dichotomies may be described in terms of their β-functions as well as in terms of their b-functions. By complement 4 there is a duality between these functions. This duality is preserved for the (Π_0, Π_1) testing (classification) problem considered in the previous complements. This is a consequence of results described in complement 4 and the following relationship between the minimum mixed Bayes risk and maximin power

$$b(\lambda|\Pi_0,\Pi_1) = \inf_{\alpha} [(1-\lambda)\alpha + \lambda(1-\beta(\alpha|\Pi_0,\Pi_1))].$$

It follows that $\beta(\cdot|\Pi_0,\Pi_1)$ is the β-function of a given dichotomy if and only if $b(\cdot|\Pi_0,\Pi_1)$ is the b-function of the same dichotomy.

(a) Any level α maximin test is a $(1-\lambda,\lambda)$ mixed Bayes test when $(1-\lambda)\alpha + \lambda(1-\beta(\alpha(\Pi_0,\Pi_1))) = b(\lambda|\Pi_0,\Pi_1)$. (There is, as we have seen, a non-empty closed interval of such λ's.)
(b) If $\lambda > 0$ then any $(1-\lambda,\lambda)$ mixed Bayes test is a maximin level α test where $\alpha = \sup_{P_0 \in \Pi_0} P_0(\delta)$.
(c) By complement 6(d) we find that $\beta(\alpha|\Pi_0,\Pi_1) = \inf\{\beta(\alpha|P_0,P_1): P_0 \in \Pi_0, P_1 \in \Pi_1\}$ when the sets Π_0 and Π_1 are both convex.
(d) The following conditions are equivalent for a pair $(P_0,P_1) \in \Pi_0 \times \Pi_1$:
 (i) $b(\cdot|P_0,P_1) \geqslant b(\cdot|Q_0,Q_1)$; $Q_0 \in \Pi_0, Q_1 \in \Pi_1$;
 (ii) $\beta(\cdot|P_0,P_1) \leqslant \beta(\cdot|Q_0,Q_1)$; $Q_0 \in \Pi_0, Q_1 \in \Pi_1$.
 If, furthermore, Π_0 and Π_1 are convex, then these conditions are equivalent to each of:
 (iii) $b(\cdot|\Pi_0,\Pi_1) = b(\cdot|P_0,P_1)$;
 (iv) $\beta(\cdot|\Pi_0,\Pi_1) = \beta(\cdot|P_0,P_1)$.
 A pair (P_0,P_1) satisfying the equivalent conditions (i) and (ii) is called *least favourable*.
(e) Least favourable pairs may sometimes be found directly from the Neyman–Pearson lemma. Thus (P_0,P_1) is a least favourable pair if $P_0 \in \Pi_0$, $P_1 \in \Pi_1$ and if the level α most powerful test δ_α for testing 'P_0' against 'P_1' for *each* α may be chosen such that

$$\max_{Q_0 \in \Pi_0} Q_0(\delta_\alpha) \leqslant P_0(\delta_\alpha)$$

 and

$$\min_{Q_1 \in \Pi_1} Q_1(\delta_\alpha) \geqslant P_1(\delta_\alpha).$$

 The test δ_α, for each α, is then a maximin level α test.

This condition is satisfied whenever (P_0,P_1) admits a specification g of the Radon–Nikodym derivative $\mathrm{d}P_1/\mathrm{d}(P_0+P_1)$ so that g is stochastically largest (smallest) under Π_0 for P_0 (under Π_1 for P_1). Of course the density g may not be an ordinary function but an element of $M = L(\Pi_0 \cup \Pi_1)^*$. That this is so follows immediately from the Neyman–Pearson lemma.

Assume in this situation that (Q_0,Q_1) is also a least favourable pair for (Π_0,Π_1). If h is a version of $\mathrm{d}Q_1/\mathrm{d}(Q_0+Q_1)$ then $[h > 1-\lambda]$ is a $(1-\lambda,\lambda)$ Bayes test for separating Q_0 and Q_1. But $(1-\lambda)Q_0(g > 1-\lambda) +$

$\lambda(1 - Q_1(g > 1 - \lambda)) \leqslant (1 - \lambda)P_0(g > 1 - \lambda) + \lambda(1 - P_1(g > 1 - \lambda)) = b(\lambda|P_0, P_1) = b(\lambda|Q_0, Q_1)$ and thus '$\leqslant$' may be replaced with '$=$'. It follows that the events $[g > 1 - \lambda]$ and $[h > 1 - \lambda]$ are $(Q_0 + Q_1)$-equivalent provided $(Q_0 + Q_1)(h = 1 - \lambda) = 0$. Hence $g = h$ a.e. $Q_0 + Q_1$ and $\mathscr{L}(g|Q_i) = \mathscr{L}(g|P_i)$; $i = 0, 1$. It follows (as noted by Rieder (1977)) that g has the same properties with respect to (Q_0, Q_1) as with respect to (P_0, P_1). In particular $Q_i = P_i$; $i = 0, 1$ whenever Q_0 and Q_1 both belong to the vector lattice generated by (P_0, P_1).

(f) If (P_0, P_1) is a least favourable pair and if Π_0 and Π_1 are convex and dominated by $P_0 + P_1$ then any density $g = \mathrm{d}P_1/\mathrm{d}(P_0 + P_1)$ has the properties described in (e). Indeed then

$$b(\lambda|\Pi_0, \Pi_1) = \inf_{\delta} \sup_{Q_0 \in \Pi_0} \sup_{Q_1 \in \Pi_1} [(1 - \lambda)Q_0(\delta) + \lambda(1 - Q_1(\delta))]$$

$$= \sup_{Q_0 \in \Pi_0} \sup_{Q_1 \in \Pi_1} b(\lambda|Q_0, Q_1) = b(\lambda|P_0, P_1).$$

Let δ_λ, for each λ, be a minimum $(1 - \lambda, \lambda)$ mixed Bayes decision rule and let $\tilde{\delta}_\lambda$ be the test defined by the rejection region $[g > 1 - \lambda]$. Then $\tilde{\delta}_\lambda$ is a $(1 - \lambda, \lambda)$ Bayes test for the pair (P_0, P_1). The above expression for $b(\lambda|\Pi_0, \Pi_1)$ implies that δ_λ is also a $(1 - \lambda, \lambda)$ Bayes test for (P_0, P_1) so that $\delta_\lambda = \tilde{\delta}_\lambda$ a.e. when $(P_0 + P_1)(g = 1 - \lambda) = 0$. By the same equality, provided $0 \leqslant \lambda \leqslant 1$, we find that $P_0(\delta_\lambda) \geqslant Q_0(\delta_\lambda)$ and that $P_1(\delta_\lambda) \leqslant Q_1(\delta_\lambda)$ when $Q_0 \in \Pi_0$ and $Q_1 \in \Pi_1$. It follows that g is stochastically largest (smallest) under $\Pi_0(\Pi_1)$ for $P_0(P_1)$.

Remarks. Least favourable pairs do not exist in general. It was shown by Strassen (1965) and Huber & Strassen (1973–74) that the crucial condition is that the sets Π_0 and Π_1 both belong to the class of sets Π of the form

$$\Pi = \{P : P(A) \leqslant v(A); A \in \mathscr{A}\}$$

where v is a monotonically increasing set function (2-alternating capacity) such that $v(\emptyset) = 0$, $v(\mathscr{X}) = 1$ and $v(A \cap B) + v(A \cup B) \leqslant v(A) + v(B)$ when $A, B \in \mathscr{A}$. In addition there are some regularity conditions of an apparently more technical nature.

Besides the papers mentioned above, the reader is referred to Rieder (1977, 1981), Huber (1981) and Bednarski (1982) for further information on capacities and maximin tests.

We shall see later how Huber's robust tests for statistical distance contaminated hypotheses may be derived from the general theory of deficiencies between experiments.

Complement 8↑. Maximin power and minimum mixed Bayes risk in replicated experiments

Assume that r independent observations $X_1, \ldots, X_r$ are available and that it is known that the distribution of X_i, for each i, is either in the set $\Pi_{0,i}$ or in the set $\Pi_{1,i}$. Furthermore it is assumed that the classification based on the first index does not depend on the second index i. Thus we assume that either X_i, for each i, is distributed according to a distribution in $\Pi_{0,i}$ or that X_i, for each i, is distributed according to a distribution in $\Pi_{1,i}$. Putting

$$\Pi_0 = \{P_1 \times \cdots \times P_r : P_i \in \Pi_{0,i}; i = 1,\ldots,r\}$$

and

$$\Pi_1 = \{P_1 \times \cdots \times P_r : P_i \in \Pi_{1,i}; i = 1,\ldots,r\}$$

these assumptions amount to the claim that $(X_1,\ldots,X_r)$ is either distributed according to a distribution in Π_0 or is distributed according to a distribution in Π_1.

Assuming that $(P_{0,i}, P_{1,i})$ is a least favourable pair for the pair $(\Pi_{0,i}, \Pi_{1,i})$ for each i, we find by the monotonicity of experiment multiplication that $(P_{0,1} \times \cdots \times P_{0,r}, P_{1,i} \times \cdots \times P_{1,r})$ is least favourable for (Π_0, Π_1).

Assume furthermore that the measures $P_{0,i}$ and $P_{1,i}$ for each $i = 1,\ldots,r$ have, respectively, densities $p_{0,i}$ and $p_{1,i}$ with respect to some majorizing non-negative σ-finite measure μ_i such that

$$Q_{0,i}(p_{1,i} \geqslant \xi p_{0,i}) \leqslant P_{0,i}(p_{1,i} \geqslant \xi p_{0,i})$$

and

$$Q_{1,i}(p_{1,i} \geqslant \xi p_{0,i}) \geqslant P_{1,i}(p_{1,i} > \xi p_{0,i})$$

when $\xi \geqslant 0$, and $Q_{0,i} \in \Pi_{0,i}$ and $Q_{1,i} \in \Pi_{1,i}$.

Using the fact that $p_{0,1} \otimes \cdots \otimes p_{0,r}$ and $p_{1,1} \otimes \cdots \otimes p_{1,r}$ are the densities of, respectively, $P_{0,1} \times \cdots \times P_{0,r}$ and $P_{1,1} \times \cdots \times P_{1,r}$ with respect to $\mu_1 \times \cdots \times \mu_r$ we find that

$$\beta(\alpha|\Pi_0, \Pi_1) = \beta(\alpha|P_{0,1} \times \cdots \times P_{0,r}, P_{1,1} \times \cdots \times P_{1,r})$$

and that

$$b(\lambda|\Pi_0, \Pi_1) = b(\lambda|P_{0,1} \times \cdots \times P_{0,r}, P_{1,1} \times \cdots \times P_{1,r}).$$

It follows, under these circumstances, that maximin power and minimum mixed Bayes risk behave under combination of independent observations just as if each set $\Pi_{0,i}$ consisted of the single measure $P_{0,i}$ and each set $\Pi_{1,i}$ consisted of the single measure $P_{1,i}$. In other words the functions β and b behave as they do under multiplication of dichotomies.

If in particular the distributions $P_{0,i}$ and $P_{1,i}$ do not depend on i then maximin power and minimum mixed Bayes risk behave as in the case of replicated dichotomies.

Complement 9. Deficiencies for dichotomies (see also complement 11)
In the case of dichotomies the randomizations whose existence is guaranteed by the randomization criterion may be assumed to satisfy a rather interesting side condition. This is a consequence of:

(a) *if* $\varepsilon_0 + \varepsilon_1 > 0$ *then the dichotomy* $\mathscr{E} = (P_0, P_1)$ *is* $(\varepsilon_0, \varepsilon_1)$*-deficient with respect to the dichotomy* $\mathscr{F} = (Q_0, Q_1)$ *if and only if the triple* $(P_0, P_1, (\varepsilon_0 P_1 + \varepsilon_1 P_0)/(\varepsilon_0 + \varepsilon_1))$ *is* $(\varepsilon_0, \varepsilon_1, 0)$*-deficient with respect to* $(Q_0, Q_1, (\varepsilon_1 Q_0 + \varepsilon_0 Q_1)/(\varepsilon_0 + \varepsilon_1))$.

Indeed if $\mathscr{E}$ is $(\varepsilon_0, \varepsilon_1)$-deficient with respect to $\mathscr{F}$ then the sublinear function criterion implies that $\|a_0 P_0 + a_1 P_1\| \geqslant \|a_0 Q_0 + a_1 Q_1\| - |a_0 \varepsilon_0 - a_1 \varepsilon_1|$ whenever a_0 and a_1 have opposite sign. If a_0 and a_1 have the same sign, however, then this inequality is trivial. Thus we have improved the deficiency term $|a_0|\varepsilon_0 + |a_1|\varepsilon_1$ by replacing it with the usually smaller quantity $|a_0 \varepsilon_0 - a_1 \varepsilon_1|$.

Putting $\psi(x_0, x_1) = |a_0 x_0 + a_1 x_1|$ the inequality may be written

$$\int \psi(dP_0, dP_1) \geqslant \int \psi(dQ_0, dQ_1) - \frac{1}{2}[\psi(\varepsilon_0, -\varepsilon_1) + \psi(-\varepsilon_0, \varepsilon_1)]. \quad (10.2.1)$$

As (10.2.1) holds trivially for linear functions we find that it holds for all functions on $\mathbb{R}^2$ which are maxima of two linear functionals.

The only parts of ψ which matter in (10.2.1) are the values it assigns to the points in the halfplane $\{(x_0, x_1) : \varepsilon_1 x_0 + \varepsilon_0 x_1 \geqslant 0\}$. As any sublinear function on $\mathbb{R}^2$ may be approximated (uniformly on compacts) by sums of maxima of pairs of linear functionals we conclude that (10.2.1) holds for any sublinear function ψ on $\mathbb{R}^2$.

Put $P_2 = \varepsilon_1 P_0 + \varepsilon_0 P_1/(\varepsilon_0 + \varepsilon_1)$ and $Q_2 = \varepsilon_1 Q_0 + \varepsilon_0 Q_1/(\varepsilon_0 + \varepsilon_1)$ and let γ be sublinear on $\mathbb{R}^3$. Define ψ on $\mathbb{R}^2$ by putting $\psi(x_0, x_1) = \gamma(x_0, x_1, \varepsilon_1 x_0 + \varepsilon_0 x_1/(\varepsilon_0 + \varepsilon_1))$ when $x_0, x_1 \in \mathbb{R}$.

It follows from (10.2.1) that

$$\begin{aligned} \int \gamma(dP_0, dP_1, dP_2) &= \int \psi(dP_0, dP_1) \\ &\geqslant \int \psi(dQ_0, dQ_1) - \frac{1}{2}[\psi(\varepsilon_0, -\varepsilon_1) + \psi(-\varepsilon_0, \varepsilon_1)] \\ &= \int \gamma(dQ_0, dQ_1, dQ_2) - \frac{1}{2}[\psi(\varepsilon_0, -\varepsilon_1) + \psi(-\varepsilon_0, \varepsilon_1)]. \end{aligned}$$

Putting $e_0 = (1, 0, 0)$ and $e_1 = (0, 1, 0)$ we find that

$$\psi(\varepsilon_0, -\varepsilon_1) = \gamma(\varepsilon_0, -\varepsilon_1, 0) \leqslant \varepsilon_0 \gamma(e_0) + \varepsilon_1 \gamma(-e_1)$$

and

$$\psi(-\varepsilon_0, \varepsilon_1) = \gamma(-\varepsilon_0, \varepsilon_1, 0) \leqslant \varepsilon_0 \gamma(-e_0) + \varepsilon_1 \gamma(e_1).$$

Thus

$$\int \gamma(dP_0, dP_1, dP_2) \geqslant \int \gamma(dQ_0, dQ_1, dQ_2) - \frac{1}{2}\varepsilon_0[\gamma(-e_0) + \gamma(e_0)] - \tfrac{1}{2}\varepsilon_1[\gamma(-e_1) + \gamma(e_1)]$$

for all sublinear functionals γ on $\mathbb{R}^2$. Thus (P_0, P_1, P_2) is $(\varepsilon_0, \varepsilon_1, 0)$-deficient with respect to (Q_0, Q_1, Q_2).

(b) *If the dichotomy (P_0, P_1) is $(\varepsilon_0, \varepsilon_1)$-deficient with respect to (Q_0, Q_1) then there is a Markov kernel M such that $\|P_i M - Q_i\| \leqslant \varepsilon_i$; $i = 0, 1$ and which in addition satisfies the condition*:

$$(\varepsilon_1 P_0 + \varepsilon_0 P_1)M = \varepsilon_1 Q_0 + \varepsilon_0 Q_1.$$

(No restrictions on ε_0 and ε_1 are needed.)

(c) *Consider a dichotomy $\mathscr{D} = (P_0, P_1)$ and let $\varepsilon_0, \varepsilon_1 \geqslant 0$. Then there are probability distributions $\hat{P}_0$ and $\hat{P}_1$ in the vector lattice $V(P_0, P_1)$ generated by $\{P_0, P_1\}$ such that*:
  (i) $\|\hat{P}_i - P_i\| \leqslant \varepsilon_i$; $i = 0, 1$;
  (ii) $\varepsilon_0 \hat{P}_1 + \varepsilon_1 \hat{P}_0 = \varepsilon_0 P_1 + \varepsilon_1 P_0$;
  (iii) *the dichotomy $\hat{\mathscr{D}} = (\hat{P}_0, \hat{P}_1)$ is at most as informative as any dichotomy which is $(\varepsilon_0, \varepsilon_1)$-deficient with respect to $\mathscr{D}$. If, furthermore, $\varepsilon_0 + \varepsilon_1 \leqslant \|P_1 - P_0\|$ then* (i), (ii) *and* (iii) *together imply that equality that holds in* (i).

Indeed a dichotomy $\mathscr{D}^*$ represented by the β-function β^* on $[0, 1]$ is $(\varepsilon_0, \varepsilon_1)$-deficient with respect to $\mathscr{D}$ if and only if β^* majorizes $\beta(\cdot - \frac{1}{2}\varepsilon_0 | P_0, P_1) - \frac{1}{2}\varepsilon_1$.

Thus the set of types of dichotomies $\mathscr{D}^*$ which are $(\varepsilon_0, \varepsilon_1)$-deficient with respect to $\mathscr{D}$ has as its smallest element the type represented by the smallest concave function $\hat{\beta}$ on $[0, 1]$ which majorizes $\beta(\cdot - \frac{1}{2}\varepsilon_0 | P_0, P_1) - \frac{1}{2}\varepsilon_1$. By the randomization criterion and (b) above this least informative dichotomy may also be represented as a dichotomy $(\hat{P}_0, \hat{P}_1)$ such that (i)–(iii) hold.

If $\varepsilon_0 + \varepsilon_1 \leqslant \|P_1 - P_0\|$ and if, say $\|\hat{P}_0 - P_0\| < \varepsilon_0$, then we may consider the dichotomy $\mathscr{D}_* = ((1 - \lambda)\hat{P}_0 + \lambda \hat{P}_1, \hat{P}_1)$ where $0 \leqslant \lambda \leqslant 1$. As $\|(1 - \lambda)\hat{P}_0 + \lambda \hat{P}_1 - \hat{P}_0\| = \lambda \|\hat{P}_1 - \hat{P}_0\|$ it is clear that $\mathscr{D}_*$ is $(\lambda \|\hat{P}_1 - \hat{P}_0\|, 0)$-deficient with respect to $\hat{\mathscr{D}}$. Hence $\mathscr{D}_*$ is $(\lambda \|\hat{P}_1 - \hat{P}_0\| + \|\hat{P}_0 - P_0\|, \varepsilon_1)$-deficient with respect to $\mathscr{D}$. If λ is sufficiently small, say when $\lambda < \lambda_0$, then $\lambda \|\hat{P}_1 - \hat{P}_0\| +$

$\|\hat{P}_0 - P_0\| < \varepsilon_0$ so that then $\mathscr{D}_*$ is $(\varepsilon_0, \varepsilon_1)$-deficient with respect to $\mathscr{D}$. By minimality, i.e. (iii), we conclude that $\mathscr{D}_* \geqslant \hat{\mathscr{D}}$. On the other hand it is clear that $\hat{\mathscr{D}} \geqslant \mathscr{D}_*$ so that, in fact, $\hat{\mathscr{D}} \sim \mathscr{D}_*$. Considering the statistical distances we find that $(1 - \lambda)\|\hat{P}_1 - \hat{P}_0\| = \|(1 - \lambda)\hat{P}_0 + \lambda\hat{P}_1 - \hat{P}_1\| = \|\hat{P}_0 - \hat{P}_1\|$ so that $\|\hat{P}_0 - \hat{P}_1\| = 0$, i.e. $\hat{P}_0 = \hat{P}_1 = (1/(\varepsilon_0 + \varepsilon_1))(\varepsilon_0 P_1 + \varepsilon_1 P_0)$. It follows that $\|\hat{P}_0 - P_0\| = (\varepsilon_0/(\varepsilon_0 + \varepsilon_1))\|P_0 - P_1\| < \varepsilon_0$, contradicting the assumption that $\varepsilon_0 + \varepsilon_1 \leqslant \|P_1 - P_0\|$.

We shall see in the next complement that these results on deficiencies for dichotomies are closely related to Huber's constructions of robust tests as given in Huber (1965, 1981).

Complement 10↑. The Neyman–Pearson lemma for statistical distance neighbourhoods

Consider probability measures P_0 and P_1 on the same measurable space $(\mathscr{X}, \mathscr{A})$. If we want to test the hypothesis 'P_0' against 'P_1' and if we want our tests to have probabilities of errors not above prescribed bounds, even if the model is slightly off the mark, then we might want to consider the problem of testing some neighbourhood of P_0 against some neighbourhood of P_1.

In this section we will assume that these neighbourhoods are statistical distance neighbourhoods. It will be convenient to use the notation $S(P, \varepsilon)$ for the set of probability distributions Q whose statistical distances $\|Q - P\|$ from P do not exceed ε.

Consider numbers $\varepsilon_0, \varepsilon_1 \geqslant 0$ and the neighbourhoods $S(P_0, \varepsilon_0)$ and $S(P_1, \varepsilon_1)$ of P_0 and P_1 respectively. The following conditions are then equivalent.

(a_1) A totally non informative dichotomy is $(\varepsilon_0, \varepsilon_1)$-deficient with respect to (P_0, P_1).

(a_2) $S(P_0, \varepsilon_0) \cap S(P_1, \varepsilon_1) \neq \varnothing$.

(a_3) $\|P_0 - P_1\| \leqslant \varepsilon_0 + \varepsilon_1$.

For this reason we will restrict our attention in the sequel to the situation where these conditions are not satisfied, i.e. when P_0 and P_1 are sufficiently far apart, or $\varepsilon_0 + \varepsilon_1$ is sufficiently small, in order that $\|P_0 - P_1\| > \varepsilon_0 + \varepsilon_1$.

Let us keep the notation of the previous complement. The important observation is:

(b) *any dichotomy* $\hat{\mathscr{D}} = (\hat{P}_0, \hat{P}_1)$ *satisfying* (i)–(iii) *of point* (c) *in complement 9 is least favourable for testing* $S(P_0, \varepsilon_0)$ *against* $S(P_1, \varepsilon_1)$ *according to the definition in complement* 7.

Indeed if $\|Q_0 - P_0\| \leqslant \varepsilon_0$ and $\|Q_1 - P_1\| \geqslant \varepsilon_1$ then (Q_0, Q_1) is $(\varepsilon_0, \varepsilon_1)$-deficient with respect to (P_0, P_1) and thus, by minimality, $(\hat{P}_0, \hat{P}_1) \leqslant (Q_0, Q_1)$.

(c) Let $\lambda \to b(\lambda) = \|(1-\lambda)P_0 \wedge \lambda P_1\|$ and $\lambda \to \hat{b}(\lambda) = \|(1-\lambda)\hat{P}_0 \wedge \lambda \hat{P}_1\|$ be the Bayes risk functions of $\hat{\mathscr{D}}$ and of $\mathscr{D}$. Then

$$\hat{b}(\lambda) = (1-\lambda) \wedge \lambda \wedge [b(\lambda) + (1-\lambda)\tfrac{1}{2}\varepsilon_0 + \lambda\cdot\tfrac{1}{2}\varepsilon_1]; \qquad 0 \leqslant \lambda \leqslant 1.$$

A dichotomy given by a 'b-function' b^* is $(\varepsilon_0, \varepsilon_1)$-deficient with respect to (P_0, P_1) if and only if $(1-\lambda)\frac{1}{2}\varepsilon_0 + \lambda\cdot\frac{1}{2}\varepsilon_1 \geqslant b^*(\lambda) - b(\lambda)$ where b is the 'b-function' associated with $\mathscr{D}$, i.e. $b(\lambda) \equiv_\lambda \|(1-\lambda)P_0 \wedge \lambda P_1\|$. Hence, by the minimality of $\hat{b}$:

$$\text{(d)} \quad \hat{b}(\lambda) = \begin{cases} \lambda & \text{when } \lambda \leqslant 1-\mu_1 \\ b(\lambda) + (1-\lambda)\tfrac{1}{2}\varepsilon_0 + \lambda\cdot\tfrac{1}{2}\varepsilon_1 & \text{when } 1-\mu_1 \leqslant \lambda \leqslant 1-\mu_0 \\ 1-\lambda & \text{when } \lambda \geqslant 1-\mu_0. \end{cases}$$

Here $\lambda = 1-\mu_0$ is the smallest solution of the equation $b(\lambda) + (1-\lambda)\frac{1}{2}\varepsilon_0 + \lambda\cdot\frac{1}{2}\varepsilon_1 = 1-\lambda$ in $[0,1]$ while $\lambda = 1-\mu_1$ is the largest solution of the equation $b(\lambda) + (1-\lambda)\frac{1}{2}\varepsilon_0 + \lambda\cdot\frac{1}{2}\varepsilon_1 = \lambda$ in $[0,1]$. Actually $1-\mu_0 > \frac{1}{2} > 1-\mu_1$ so that $\mu_0 < \frac{1}{2} < \mu_1$.

This may be seen as follows. Put $t(\lambda) = b(\lambda) + \lambda(\frac{1}{2}(\varepsilon_1 - \varepsilon_0) + 1)$ and $u(\lambda) = b(\lambda) + \lambda(\frac{1}{2}(\varepsilon_1 - \varepsilon_0) - 1)$. The two equations may then, respectively, be written as $t(\lambda) = 1 - \frac{1}{2}\varepsilon_0$ and $u(\lambda) = -\frac{1}{2}\varepsilon_0$.

Now

$$\begin{aligned} t(\tfrac{1}{2}) &= 1 - \tfrac{1}{4}\|P_1 - P_0\| + \tfrac{1}{4}(\varepsilon_1 - \varepsilon_0) \\ &< 1 - \tfrac{1}{4}(\varepsilon_0 + \varepsilon_1) + \tfrac{1}{4}(\varepsilon_1 - \varepsilon_0) \\ &= 1 - \tfrac{1}{2}\varepsilon_0 \end{aligned}$$

while

$$\begin{aligned} u(\tfrac{1}{2}) &= \tfrac{1}{2}\|P_0 \wedge P_1\| - \tfrac{1}{2} + \tfrac{1}{4}(\varepsilon_1 - \varepsilon_0) \\ &= -\tfrac{1}{4}\|P_1 - P_0\| + \tfrac{1}{4}(\varepsilon_1 - \varepsilon_0) \\ &< -\tfrac{1}{4}(\varepsilon_1 + \varepsilon_0) + \tfrac{1}{4}(\varepsilon_1 - \varepsilon_0) \\ &= -\tfrac{1}{2}\varepsilon_0. \end{aligned}$$

Furthermore $t(1) = \frac{1}{2}(\varepsilon_1 - \varepsilon_0) + 1 \geqslant 1 - \frac{1}{2}\varepsilon_0$ while $u(0) = 0 \geqslant -\frac{1}{2}\varepsilon_0$. The existence of solutions $\lambda = 1-\mu_0$ and $\lambda = 1-\mu_1$ as described follows now by continuity and then the form of $\hat{b}$ is a consequence of the fact that b and $\hat{b}$ are concave.

Concavity also implies that the solution $\lambda = 1-\mu_0$ is the only solution of the first equation when $\varepsilon_1 > 0$, while the solution $\lambda = 1-\mu_1$ is the unique solution of the second equation when $\varepsilon_0 > 0$.

(e) The density $\hat{g} = \mathrm{d}\hat{P}_1/\mathrm{d}(\hat{P}_0 + \hat{P}_1)$ may now be expressed in terms of $g = \mathrm{d}P_1/\mathrm{d}(P_0 + P_1)$ by

$$\hat{g} = \begin{cases} \mu_0 & \text{when} \quad g \leqslant \mu_0 \\ g & \text{when} \quad \mu_0 \leqslant g \leqslant \mu_1 \\ \mu_1 & \text{when} \quad g \geqslant \mu_1. \end{cases}$$

Furthermore $\hat{g}$ is stochastically largest under $S(P_0, \varepsilon_0)$ for $\hat{P}_0$ while it is stochastically smallest under $S(P_1, \varepsilon_1)$ for $\hat{P}_1$.

This may be argued as follows. Firstly the test 'Accept P_1 when $g > 1 - \lambda$' is Bayes for the prior $(1 - \lambda, \lambda)$ and for the (0–1) loss function in the dichotomy $\mathscr{D} = (P_0, P_1)$. If $\lambda \in [1 - \mu_1, 1 - \mu_0]$ then this yields

$$\begin{aligned} & b(\lambda) + (1 - \lambda)\tfrac{1}{2}\varepsilon_0 + \lambda \cdot \tfrac{1}{2}\varepsilon_1 \\ & = \hat{b}(\lambda) \leqslant (1 - \lambda)\hat{P}_0(g > 1 - \lambda) + \lambda \hat{P}_1(g \leqslant 1 - \lambda) \\ & \leqslant (1 - \lambda)[P_0(g > 1 - \lambda) + \tfrac{1}{2}\varepsilon_0] + \lambda[P_1(g \leqslant 1 - \lambda) + \tfrac{1}{2}\varepsilon_1] \\ & = b(\lambda) + (1 - \lambda) \cdot \tfrac{1}{2}\varepsilon_0 + \lambda \cdot \tfrac{1}{2}\varepsilon_1. \end{aligned}$$

Thus equality prevails throughout here. It follows that this test is also Bayes for the same prior and for the same loss function in $\hat{\mathscr{D}} = (\hat{P}_0, \hat{P}_1)$. By uniqueness we are forced to conclude that if $(\hat{P}_0 + \hat{P}_1)(\hat{g} = 1 - \lambda) = 0$ then the events $[g > 1 - \lambda]$ and $[\hat{g} > 1 - \lambda]$ are equivalent for the measure $\hat{P}_0 + \hat{P}_1$. Thus the sets $[g > \xi]$ and $[\hat{g} > \xi]$ are $(\hat{P}_0 + \hat{P}_1)$-equivalent whenever $\mu_0 \leqslant \xi \leqslant \mu_1$.

On the other hand $\hat{b}(1 - \mu_0) = \mu_0$, i.e. $\|\mu_0\hat{P}_0 \wedge (1 - \mu_0)\hat{P}_1\| = \|\mu_0\hat{P}_0\|$ showing that $\mu_0\hat{P}_0 \wedge (1 - \mu_0)\hat{P}_1 = \mu_0\hat{P}_0$, i.e. that $(1 - \mu_0)\hat{P}_1 \geqslant \mu_0\hat{P}_0$ and this in turn amounts to the condition that $(1 - \mu_0)\hat{g} \geqslant \mu_0(1 - \hat{g})$ a.e. $\hat{P}_0 + \hat{P}_1$ or, equivalently, that $\hat{g} \geqslant \mu_0$ a.e. $\hat{P}_0 + \hat{P}_1$. Similarly the equation $\hat{b}(1 - \mu_1) = 1 - \mu_1$ amounts to the requirement that $\hat{g} \leqslant \mu_1$ a.e. $\hat{P}_0 + \hat{P}_1$.

Altogether this shows that $\hat{g}$ may be specified as required. Furthermore if $\lambda \in [1 - \mu_1, 1 - \mu_0]$ then we saw above that $\hat{P}_0(g > 1 - \lambda) = P_0(g > 1 - \lambda) + \tfrac{1}{2}\varepsilon_0$ when $\lambda < 1$ while $\hat{P}_1(g \leqslant 1 - \lambda) = P_1(g \leqslant 1 - \lambda) + \tfrac{1}{2}\varepsilon_1$ when $\lambda > 0$. If (Q_0, Q_1) is a dichotomy such that $\|Q_i - P_i\| \leqslant \varepsilon_i$; $i = 1, 2$ then this implies that $Q_0(g > 1 - \lambda) \leqslant P_0(g > 1 - \lambda) + \tfrac{1}{2}\varepsilon_0 = \hat{P}_0(g > 1 - \lambda)$ when $\lambda > 0$. The claims on stochastic optimality follow now by observing that $[g > \xi]$ and $[\hat{g} > \xi]$ are the same events when $\mu_0 \leqslant \xi < \mu_1$.

(f) If $\hat{g}$ is specified as in (e) and if the constants $c \geqslant 0$ and $\gamma \in [0, 1]$ are chosen such that $\hat{P}_0(\hat{g} > c) + \gamma\hat{P}(\hat{g} = c) = \alpha$ then the test $\delta = I_{\hat{g} > c} + \gamma I_{\hat{g} = c}$ is a maximin level α test for testing $S(P_0, \varepsilon_0)$ against $S(P_1, \varepsilon_1)$. If $\mu_0 < c < \mu_1$ then this test is also a most powerful level $P_0(\delta)$ test for testing 'P_0' against 'P_1'.

(g) The probability measures $\hat{P}_0$ and $\hat{P}_1$ whose existences are guaranteed by complement 9(c), are unique.

Indeed if $0 < \varepsilon_0 + \varepsilon_1 \geqslant \|P_1 - P_0\|$ then necessarily $\hat{P}_0 = \hat{P}_1 = (1/(\varepsilon_0 + \varepsilon_1))(\varepsilon_0 P_1 + \varepsilon_1 P_0)$ while $\hat{P}_i = P_i; i = 1,2$ when $\varepsilon_0 = \varepsilon_1 = 0$. Thus it suffices to consider the case where $0 < \varepsilon_0 + \varepsilon_1 < \|P_1 - P_0\|$ and then the above results apply. If, in addition $\varepsilon_0 > 0$ and $\varepsilon_1 > 0$ then

$$\mathrm{d}\hat{P}_i/\mathrm{d}(\varepsilon_1 P_0 + \varepsilon_0 P_1) = \mathrm{d}\hat{P}_i/\mathrm{d}(\varepsilon_1 \hat{P}_0 + \varepsilon_0 \hat{P}_1)$$
$$= (1-g)^{1-i} g^i / [\varepsilon_1(1-g) + \varepsilon_0 g]; \qquad i = 0, 1.$$

If $\varepsilon_0 = 0$ then $\hat{P}_0 = P_0$ and $\mathrm{d}\hat{P}_1/\mathrm{d}(P_0 + \hat{P}_1) = \hat{g}$ so that $\hat{P}_1$ is determined on the set $[\hat{g} < 1]$. If, in addition, $(P_0 + P_1)(\hat{g} = 1) > 0$ then $\mu_1 = 1$ and $[\hat{g} = 1] = [g = 1]$. As $\mathrm{d}\hat{P}_1/\mathrm{d}(P_0 + P_1)$ is a (measurable) function of g and since $\hat{P}_1(\hat{g} = 1) = 1 - \hat{P}_1(g < 1)$ this shows that $\hat{P}_1$ is also determined in this case, regardless of whether $(P_0 + P_1)(\hat{g} = 1) = 0$ or not. If $\varepsilon_1 = 0$ then $\hat{P}_1 = P_1$ and arguing as above we find that $\hat{P}_0$ is also determined.

If $0 < \varepsilon_0 + \varepsilon_1 < \|P_1 - P_0\|$ then $\hat{P}_0$ and $\hat{P}_1$ may be explicitly expressed as follows. Let $P_{0,s}(P_{1,s})$ denote the $P_1(P_0)$-singular part of $P_0(P_1)$. Then

$$\hat{P}_0 = P_0 - \tfrac{1}{2}\varepsilon_0 \Gamma \quad \text{and} \quad \hat{P}_1 = P_1 + \tfrac{1}{2}\varepsilon_1 \Gamma$$

where $\Gamma = \Gamma_0 - \Gamma_1$ and where

$$\Gamma_0 = 2\frac{[\mu_0 P_0 - (1-\mu_0)P_1]^+}{\varepsilon_1(1-\mu_0) + \varepsilon_0\mu_0} \quad \text{if} \quad \varepsilon_1(1-\mu_0) + \varepsilon_0\mu_0 > 0,$$

while

$$\Gamma_0 = P_{0,s}/\|P_{0,s}\| \quad \text{when} \quad \varepsilon_1(1-\mu_0) + \varepsilon_0\mu_0 = 0, \quad \text{(i.e. when } \varepsilon_1 = \mu_0 = 0),$$

and

$$\Gamma_1 = 2\frac{[(1-\mu_1)P_1 - \mu_1 P_0]^+}{\varepsilon_1(1-\mu_1) + \varepsilon_0\mu_0} \quad \text{if} \quad \varepsilon_1(1-\mu_1) + \varepsilon_0\mu_1 > 0,$$

while

$$\Gamma_1 = P_{1,s}/\|P_{1,s}\| \quad \text{when} \quad \varepsilon_1(1-\mu_1) + \varepsilon_0\mu_1 = 0,$$
$$\text{(i.e. when } \varepsilon_0 = 0 \text{ and } \mu_1 = 1).$$

The measures Γ_0 and Γ_1 are disjoint probability measures living on the sets $[g \leqslant \mu_0]$ and $[g \geqslant \mu_1]$ respectively. This may be verified by comparing the densities of the measures involved. It would be interesting to know to what extent other robust tests and methods may be derived by similar arguments, e.g. by using the extended deficiency concept developed in chapter 9.

Complement 11. Tools for dichotomies

Let $\mathscr{D} = (P, Q)$ be a dichotomy and put $M = \frac{1}{2}(P + Q)$. Four probability measures associated with $\mathscr{D}$ are:

$$K = \mathscr{L}(\mathrm{d}Q/\mathrm{d}P|P) \qquad \text{on } [0, \infty[,$$

$$F = \mathscr{L}(\log(\mathrm{d}Q/\mathrm{d}P)|P) \qquad \text{on } [-\infty, \infty[,$$

$$\tfrac{1}{2}S = \mathscr{L}(\mathrm{d}Q/\mathrm{d}(P + Q)|M) \quad \text{on } [0, 1],$$

and

β on $[0, 1]$ where $\beta[0, \alpha]$, for each $\alpha \in [0, 1]$, is the power of the most powerful level α test for testing 'P' against 'Q'.

In addition to these four measures we have the minimum Bayes risk function

$$b(\lambda) = \|(1 - \lambda)P \wedge \lambda Q\|; \qquad 0 \leqslant \lambda \leqslant 1$$

and the Hellinger transform

$$H(t) = \int \mathrm{d}P^{1-t}\,\mathrm{d}Q^t; \qquad 0 \leqslant t \leqslant 1.$$

Any one of these objects determines the dichotomy up to equivalence and is thus a representation. Below we list various facts concerning these representations. Proofs of some of them may be found in this book. The proofs of the others are left to the reader.

(a) The map $\mathscr{D} \to K$ is 1–1 and onto the set of probability measures on $]0, \infty[$ such that $\int xK(\mathrm{d}x) \leqslant 1$.

This map is also affine for mixtures and it is order preserving if functions K are ordered by dilations.

Furthermore $\mathscr{D} \sim (K, L) \to K$ where $[\mathrm{d}L/\mathrm{d}K]_x \equiv_x x$ and $L(\{\infty\}) = 1 - \int xK(\mathrm{d}x)$.

(b) The map $\mathscr{D} \to F$ is 1–1 and onto the set of probability measures on $[-\infty, \infty[$ satisfying $\int e^x F(\mathrm{d}x) \leqslant 1$.

This map is also affine and it is order preserving for the ordering '$\geqslant$' where $F_1 \geqslant F_2$ if and only if $F_1 = DF_2$ for a Markov kernel D such that $\int e^z D(\mathrm{d}z|x) \equiv_x e^x$.

Furthermore $\mathscr{D} \sim (F, G) \to F$ where $x \to e^x$ is a version of $\mathrm{d}G/\mathrm{d}F$ and where $G(\{\infty\}) = 1 - \int e^x F(\mathrm{d}x)$.

This map has the additional property of converting multiplication of experiments into convolutions of measures F.

(c) The map $\mathscr{D} \to \frac{1}{2}S$ is 1–1 and onto the set of probability measures on $[0, 1]$ having expectation $\frac{1}{2}$.

This map is affine and it is order preserving if measures are ordered by dilations.

Furthermore $\mathscr{D} \sim (S_1, S_2) \to S$ where $[\mathrm{d}S_1/\mathrm{d}S]_x \equiv_x 1 - x$ while $[\mathrm{d}S_2/\mathrm{d}S]_x \equiv_x x$. This measure is the projection of the standard measure of $\mathscr{D}$ on the 'first' axis.

(d) The map $\mathscr{D} \to \beta$ is 1–1 and onto the set of concave continuous probability distribution functions on $[0, 1]$.

This map is order preserving if functions β are ordered pointwise.

Furthermore $\mathscr{D} \sim (\lambda, \beta) \to \beta$ where λ is the uniform distribution on $[0, 1]$.

(e) The map $\mathscr{D} \to b$ is 1–1 and onto the set of non-negative concave functions b on $[0, 1]$ such that $b(\lambda) \leqslant \lambda \wedge (1 - \lambda)$; $\lambda \in [0, 1]$.

This map is order reversing and affine if functions b are ordered pointwise.

(f) Some relationships between these quantities are:

(i) $K = \mathscr{L}(e^X | F) = \mathscr{L}((X/1 - X) | S_1) = \mathscr{L}_\lambda(\beta')$.

(ii) $\|\xi P - Q\| = 1 - \xi + 2\int_0^\xi K[0, x]\,\mathrm{d}x$.

(iii) $F = \mathscr{L}(\log X | K)$.

(iv) $\beta(\alpha) = 1 - \int_\alpha^1 K^{-1}(1 - p)\,\mathrm{d}p = \inf_\lambda (1/\lambda)[(1 - \lambda)\alpha + \lambda - b(\lambda)]$; $0 \leqslant \alpha \leqslant 1$.

(v) $b(\lambda) = (1 - \lambda) - \int_0^{1-\lambda} S[0, x]\,\mathrm{d}x = \inf_\alpha [(1 - \lambda)\alpha + \lambda(1 - \beta(\alpha))]$; $0 \leqslant \lambda \leqslant 1$.

(vi) $H(t) = (1 - t)t \int_0^1 b(\lambda)(1 - \lambda)^{t-2} \lambda^{-t-1}\,\mathrm{d}\lambda$; $0 < t < 1$.

(g) $\beta(\alpha | P, Q) = 1 - \beta^{-1}(1 - \alpha | Q, P)$; $0 \leqslant \alpha \leqslant 1$ where the exponent '-1' indicates the left fractile function.

$$b(\lambda | P, Q) = b(1 - \lambda | Q, P); \qquad 0 \leqslant \lambda \leqslant 1.$$

$$H(t | P, Q) = H(1 - t | Q, P); \qquad 0 \leqslant t \leqslant 1.$$

(h) Let $\mathscr{D} = (P, Q)$ and $\tilde{\mathscr{D}} = (\tilde{P}, \tilde{Q})$ be two dichotomies. Employ the notation K, F, S, β and b for $\mathscr{D}$ and use the analogous notation $\tilde{K}$, $\tilde{F}$, $\tilde{S}$, $\tilde{\beta}$ and $\tilde{b}$ for $\tilde{\mathscr{D}}$.

Let ε_1 and ε_2 be non-negative numbers. Then the following conditions are all equivalent to the condition that $\mathscr{D}$ is $(\varepsilon_1, \varepsilon_2)$-deficient with respect to $\tilde{\mathscr{D}}$:

(i) $\varepsilon_1 \xi + \varepsilon_2 \geqslant 2\int_0^\xi (\tilde{K} - K)[0, x]\,\mathrm{d}x$; $\xi > 0$.

(ii) $(1 - \lambda)\varepsilon_1 + \lambda\varepsilon_2 \geqslant 2\int_0^{1-\lambda} (\tilde{S} - S)[0, x]\,\mathrm{d}x$; $0 \leqslant \lambda \leqslant 1$.

(iii) $\beta(\alpha + \frac{1}{2}\varepsilon_1) + \frac{1}{2}\varepsilon_2 \geqslant \tilde{\beta}(\alpha)$; $0 \leqslant \alpha \leqslant 1$.

(iv) $\frac{1}{2}\varepsilon_1(1 - \lambda) + \frac{1}{2}\varepsilon_2\lambda \geqslant b(\lambda) - \tilde{b}(\lambda)$; $0 \leqslant \lambda \leqslant 1$.

(v) $\int f\,\mathrm{d}K \leqslant \int f\,\mathrm{d}\tilde{K} + \frac{1}{2}\varepsilon_1(f(\infty) - f(0)) + \frac{1}{2}\varepsilon_2 f'(0)$ for each concave function f on $[0, \infty[$ such that $f(x)/x \to 0$ as $x \to \infty$.

(i) From (h) we get several interesting expressions and criteria for deficiencies, deficiency distances and convergence. For each of (i)–(iv) in

(h) the reader may deduce the corresponding expressions for $\delta(\mathscr{D}, \tilde{\mathscr{D}})$ and $\Delta(\mathscr{D}, \tilde{\mathscr{D}})$. In particular, what criteria for ordering of dichotomies do these expressions for the deficiency provide? How do the expressions reduce when $\mathscr{D}$ or $\tilde{\mathscr{D}}$ is totally informative (non informative)?

Express convergence of dichotomies in terms of the distributions K, S, β and for the functions b.

(j) Deficiencies and deficiency distances for dichotomies may be expressed in terms of ordered dichotomies by the relations

$$\delta(\mathscr{D}, \tilde{\mathscr{D}}) = \delta(\mathscr{D} \wedge \tilde{\mathscr{D}}, \tilde{\mathscr{D}}) = \delta(\mathscr{D}, \mathscr{D} \vee \tilde{\mathscr{D}})$$

and

$$\Delta(\mathscr{D}, \tilde{\mathscr{D}}) = \delta(\mathscr{D} \wedge \tilde{\mathscr{D}}, \mathscr{D} \vee \tilde{\mathscr{D}}).$$

(k) Referring to the information ordering '$\geqslant$' for dichotomies we may define liminf and limsup for a net $(\mathscr{D}_n)$ of dichotomies by

$$\liminf_n \mathscr{D}_n = \sup_n \inf_{m \geqslant n} \mathscr{D}_m$$

and

$$\limsup_n \mathscr{D}_n = \inf_n \sup_{m \geqslant n} \mathscr{D}_m.$$

Liminf is readily expressed pointwise for β-functions by

$$\beta\left(\alpha \,\middle|\, \liminf_n \mathscr{D}_n\right) \underset{\alpha}{\equiv} \liminf_n \beta(\alpha | \mathscr{D}_n).$$

Limsup is readily expressed pointwise in terms of minimum Bayes risk by

$$b\left(\lambda \,\middle|\, \limsup_n \mathscr{D}_n\right) \underset{\lambda}{\equiv} \liminf_n b(\lambda | \mathscr{D}_n).$$

(l) The following hold for a net $(\mathscr{D}_n)$ of dichotomies:
  (i) $\liminf_n \mathscr{D}_n \leqslant \limsup_n \mathscr{D}_n$.
  (ii) for any cluster dichotomy $\mathscr{D}$ of $(\mathscr{D}_n)$

$$\liminf_n \mathscr{D}_n \leqslant \mathscr{D} \leqslant \limsup_n \mathscr{D}_n.$$

  (iii) $\text{Liminf}_n \mathscr{D}_n$ and $\limsup_n \mathscr{D}_n$ are cluster dichotomies of $(\mathscr{D}_n)$.
  (iv) $(\mathscr{D}_n)$ converges if and only if the dichotomies $\liminf_n \mathscr{D}_n$ and $\limsup_n \mathscr{D}_n$ are equivalent and if so then $(\mathscr{D}_n)$ converges to any of these dichotomies.

Operations on dichotomies, on β-functions, on b-functions, on laws of likelihoods and on Hellinger transforms correspond to each other as outlined in the following table.

Glossary

Dichotomies $\mathscr{D}=(P,Q)$	β-functions $=N-P$ functions $\beta(\alpha)\sup\{\int\delta\,dP:\int\delta\,dQ=\alpha\}$; $0\leqq\alpha\leqq 1$	b-functions $=$ dual $N-P$ functions $b(\lambda)=\|(1-\lambda)P\wedge\lambda Q\|$; $0\leqq\lambda\leqq 1$	Laws of likelihood $K=\mathscr{L}_P(dQ/dP)$	Hellinger transform $H(t)=\int dP^{1-t}\,dQ^t$; $0\leqq t\leqq 1$
Deficiencies	(h) : (iii)	(h) : (ii)	(h) : (i), (v)	
Deficiency distance	Paul–Levy diagonal distance/$2^{1/2}$	2 sup norm distance		
Ordering	pointwise ordering	pointwise anti-ordering	ordering by dilations	pointwise anti-ordering is weaker
Equivalence	equality	equality	equality	equality
Convergence	pointwise convergence on $]0,1]=$ uniform convergence on intervals $[\alpha,1]$; $0<\alpha\leqq 1$	pointwise convergence $=$ uniform convergence	weak* convergence	pointwise convergence
Informational infimum	pointwise infimum	concave envelope of pointwise supremum		
Informational supremum	concave envelope of pointwise supremum	pointwise infimum		
Mixtures	'mixed supremal' convolutions complement 2	convex combinations	convex combinations	convex combinations
Products	complement 2	complement 37, point (d)	multiplicative convolutions	pointwise products
Formation of monotone likelihood	functional composition complement 46.			

Complement 12↑. Singular parts, domination

If P and Q are probability measures on the same measurable space then P may be decomposed as $P = P_a + P_s$ and Q as $Q = Q_a + Q_s$ where P_a and Q_a are, respectively, the Q-absolutely continuous part of P and the P-absolutely continuous part of Q. Thus P_s and Q_s are, respectively, the Q-singular part of P and the P-singular part of Q.

(a) Using the notation in complement 11 we obtain the following expressions for the total masses of the measures P_a, P_s, Q_a and Q_s.

$$\|P_s\| = 1 - \|P_a\| = K(\{0\}) = F(\{-\infty\}) = S(\{0\})$$
$$= 1 - \inf\{\alpha : \beta(\alpha) = 1\} = 1 + b'(1) = 1 - H(0+)$$

and

$$\|Q_s\| = 1 - \|Q_a\| = 1 - \int xK(\mathrm{d}x) = 1 - \int e^x F(\mathrm{d}x) = S(\{1\})$$
$$= \beta(0) = 1 - b'(0) = 1 - H(1-).$$

(b) The following conditions are all equivalent to the condition '$P \gg Q$'.
  (i) $\int_0^\infty xK(\mathrm{d}x) = 1$.
  (ii) $\int_{-\infty}^\infty e^x F(\mathrm{d}x) = 1$.
  (iii) $S(\{1\}) = 0$.
  (iv) $\beta(0) = 0$.
  (v) $b'(0) = 1$.
  (vi) $H(1-) = 1$.

(c) The following conditions are all equivalent to the condition '$Q \gg P$'.
  (i) $K(\{0\}) = 0$.
  (ii) $F(\{-\infty\}) = 0$.
  (iii) $S(\{0\}) = 0$.
  (iv) $\beta(\alpha) < 1$ when $\alpha < 1$.
  (v) $b'(1) = -1$.
  (vi) $H(0+) = 1$.

(d) If $\mathscr{E} \geqslant \mathscr{F}$ and if $\mathscr{E}$ is dominated or separable or differentiable (in quadratic mean) at $\theta = \theta^0$, then $\mathscr{F}$ has the same property.

Complement 13↑. Contiguity

Consider a sequence $\mathscr{D}_n = (\mathscr{X}_n, \mathscr{A}_n, P_n, Q_n)$; $n = 1, 2, \ldots$ of dichotomies.

Following LeCam (1960) (see also Hajek & Sidak (1967) and Roussas (1972)) we will say that *the sequence* $\{Q_n\}$ *is contiguous with respect to the sequence* $\{P_n\}$ if Q is absolutely continuous with respect to P for each limit dichotomy (P, Q) of $(\mathscr{D}_n)$. If (Q_n) is contiguous with respect to (P_n) then this will be denoted $(P_n)]](Q_n)$.

Let K_n, F_n, S_n, β_n, b_n and H_n be related to (P_n, Q_n) as K, F, S, β, b and H were related to (P, Q) in complement 11.

(a) Contiguity, by complement 12, has simple characterizations in terms of each of the sequences (K_n), (F_n), (S_n), (β_n), (b_n) and (H_n).

(b) If $(P_n)]](Q_n)$ then the total mass of the P_n-absolutely continuous part of Q_n converges to 0. (Q_n) need not however be contiguous with respect to (P_n) although P_n and Q_n are mutually absolutely continuous for each n. (Indeed $(P, Q)^n \to$ any totally informative dichotomy when $P \neq Q$.)

(c) $(P_n)]](Q_n)$ if and only if $Q_n(A_n) \to 0$ when $P_n(A_n) \to 0$.

(d) If $(P_n)]](Q_n)$ then $(P_{n'})]](Q_{n'})$ for any sub-sequence $(\mathscr{D}_{n'})$ of $(\mathscr{D}_n)$.

(e) Let T_n, for each n, be a statistic on $\mathscr{D}_n$ taking its values in a Polish space $(\mathscr{Y}, \mathscr{B})$.

If $(P_n)]](Q_n)$ and $\{\mathscr{L}(T_n|P_n)\}$ is tight then $\{\mathscr{L}(T_n|Q_n)\}$ is also tight.

(f) Let T_n be as in (e) and assume that $(P_n)]](Q_n)$. Assume also that $\mathscr{L}(\log(dQ_n/dP_n), T_n|P_n) \to \mathscr{L}(X, T)$ where X is $[-\infty, \infty[$-valued and T is $\mathscr{Y}$-valued and measurable with respect to the σ-algebra $\mathscr{B}$ on $\mathscr{Y}$. Then $\mathscr{L}(\log(dQ_n/dP_n), T_n|Q_n) \to e^X \mathscr{L}(X, T)$ regardless of how dQ_n/dP_n; $n = 1, 2, \ldots$ are specified.

(g) If $\mathscr{L}(\log(dQ_n/dP_n)|P_n) \to N(\xi, \sigma^2)$ then necessarily $\xi \leqslant -\frac{1}{2}\sigma^2$.

(h) Assume $\mathscr{L}(\log(dQ_n/dP_n)|P_n) \to N(\xi, \sigma^2)$. Then $(P_n)]](Q_n)$ if and only if $\xi = -\frac{1}{2}\sigma^2$. If so then $\mathscr{L}(\log(dQ_n/dP_n)|Q_n) \to N(\frac{1}{2}\sigma^2, \sigma^2)$.

Complement 14. Hellinger transforms of dichotomies

The domain of the Hellinger transform of a dichotomy $\mathscr{D} = (P_0, P_1)$ may be extended by applying the defining formula $H(t) = \int dP_0^{1-t} dP_1^t$ for all real numbers t. If $t \notin [0, 1]$ then of course there is no guarantee that the quantity $H(t)$ is finite. In fact if P_0 does not dominate P_1 then $H(t) = \infty$ when $t > 1$ while $H(t) = \infty$ when $t < 0$ and P_1 does not dominate P_0. The multiplicative property is still valid, i.e.

$$H(t|\mathscr{D}_1 \times \mathscr{D}_2) = H(t|\mathscr{D}_1) \cdot H(t|\mathscr{D}_2); \qquad t \in \mathbb{R}$$

for any two dichotomies $\mathscr{D}_1$ and $\mathscr{D}_2$. These considerations extend without difficulty to any finite parameter set. What is interesting is that the map $x \to x^t$ is concave or convex on $[0, \infty[$ as $t \in [0, 1]$ or not. Thus $H(t|\mathscr{D})$ is informationally decreasing or increasing in $\mathscr{D}$ as $t \in [0, 1]$ or not. In some cases we may get a complete description of ordering of dichotomies this way. In particular this is so for majorization by double dichotomies. Here a *double dichotomy* is an experiment which is equivalent to an experiment having two-point sets both as the parameter set and as the set of sample points.

If $\mathscr{D}$ is the double dichotomy $\begin{pmatrix} 1 - p_0, & p_0 \\ 1 - p_1, & p_1 \end{pmatrix}$ then

$$H(t|\mathscr{D}) = (1 - p_0)\left(\frac{1 - p_1}{1 - p_0}\right)^t + p_0\left(\frac{p_1}{p_0}\right)^t.$$

The dominating term here is $p_0(p_1/p_0)^t$ when t is large and $p_1 > p_0 > 0$, while it is $(1 - p_0)[(1 - p_1)/(1 - p_0)]^t$ when t is small (negative) and $1 > p_1 > p_0$. It follows readily that if $\mathscr{D}_1$ and $\mathscr{D}_2$ are double dichotomies then $\mathscr{D}_1 \geqslant \mathscr{D}_2$ if and only if $H(t|\mathscr{D}_1) \geqslant H(t|\mathscr{D}_2)$ when $|t|$ is sufficiently large. More generally if $\mathscr{D}$ is a double dichotomy and if $H(t|\mathscr{D}) \geqslant H(t|\mathscr{E})$ when $|t|$ is sufficiently large then this implies that $\mathscr{D} \geqslant \mathscr{D}_1$ for any double dichotomy $\mathscr{D}_1 \leqslant \mathscr{E}$. As $\mathscr{E}$ is the supremum of such dichotomies $\mathscr{D}_1$ we find that $\mathscr{D} \geqslant \mathscr{E}$ if and only if $H(t|\mathscr{D}) \geqslant H(t|\mathscr{E})$ when $|t|$ is sufficiently large.

This implies that if n is a positive integer and if $\mathscr{D}$ is a double dichotomy then $\mathscr{D}^n \geqslant \mathscr{E}^n$ if and only if $\mathscr{D} \geqslant \mathscr{E}$.

More generally if $\mathscr{D}$ is a double dichotomy and if $\mathscr{E}$ is any dichotomy and if $\pi_0, \pi_1, \ldots, \pi_n \geqslant 0$; $\pi_0 + \cdots + \pi_n = 1$ and $\pi_0 < 1$ the mixture $\sum_{k=0}^n \pi_k \mathscr{D}^k$ is at least as informative as a mixture $\sum_{k=0}^n \pi_k \mathscr{E}^k$ if and only if $\mathscr{D} \geqslant \mathscr{E}$. Indeed if $\sum \pi_k \mathscr{D}^k \geqslant \sum \pi_k \mathscr{E}^k$ then $\sum_k \pi_k H(t|\mathscr{D})^k \geqslant \sum_k \pi_k H(t|\mathscr{E})^k$ and thus $H(t|\mathscr{D}) \geqslant H(t|\mathscr{E})$ when $t \notin [0, 1]$.

A much simpler may of expressing majorization by double dichotomies is described in the next complement.

Complement 15. Majorization and minorization by double dichotomies

In general majorization by extremal experiments with finite parameter sets, by theorem 7.2.17, is simply expressed as inclusion of closed convex hulls of supports of laws of likelihoods. In the case of dichotomies $\mathscr{D} = (P_0, P_1)$ the smallest closed interval containing any P_1-maximal version of $\mathrm{d}P_1/\mathrm{d}P_0$ is the interval $J(\mathscr{D}) = [\underline{J}(\mathscr{D}), \bar{J}(\mathscr{D})]$ where

$$\underline{J}(\mathscr{D}) = \sup\{\xi : P_0(\xi \leqslant \mathrm{d}P_1/\mathrm{d}P_0) = 1\}$$

and

$$\bar{J}(\mathscr{D}) = \inf\{\xi : P_0(\xi \geqslant \mathrm{d}P_1/\mathrm{d}P_0) = 1\}$$

when $P_0 \gg P_1$ and $= \infty$ otherwise.

In these expressions $\mathrm{d}P_1/\mathrm{d}P_0$ may be any version of the Radon–Nikodym derivative of the P_0-absolutely continuous part of P_1 with respect to P_0.

The smallest closed convex set supporting the standard measure of $\mathscr{D}$ is then the set of pairs $(1 - x, x)$ where $x \in [\underline{J}(\mathscr{D})/(1 + \underline{J}(\mathscr{D})), \bar{J}(\mathscr{D})/(1 + \bar{J}(\mathscr{D}))]$. As $\underline{J}(\mathscr{D}) \leqslant 1 \leqslant \bar{J}(\mathscr{D})$ the number 1 is always in the interval $J(\mathscr{D})$.

Expressed in terms of the β-function $\beta = \beta(\cdot|\mathscr{D}) : \alpha \to \sup\{P_1(\delta) : 0 \leqslant \delta \leqslant 1, P_0(\delta) \leqslant \alpha\}$ we find that

$$\underline{J}(\mathscr{D}) = \lim_{\alpha \to 1} \frac{1 - \beta(\alpha)}{1 - \alpha} = \inf_{\alpha < 1} \frac{1 - \beta(\alpha)}{1 - \alpha} = \beta'(1)$$

and

$$\bar{J}(\mathscr{D}) = \lim_{\alpha \to 0} \frac{\beta(\alpha)}{\alpha} = \sup_{\alpha > 0} \frac{\beta(\alpha)}{\alpha} = \beta'(0) \text{ or } = \infty \text{ as } \beta(0) = 0 \quad \text{or} \quad \beta(0) > 0.$$

Here $\beta'(1)$ denotes the left hand derivative of β at $\alpha = 1$ while $\beta'(0)$ is the right hand derivative of β at $\alpha = 0$.

The bounds $\underline{J}(\mathscr{D})$ and $\bar{J}(\mathscr{D})$ behave multiplicatively under products, i.e.

$$\underline{J}(\mathscr{D}_1 \times \mathscr{D}_2) = \underline{J}(\mathscr{D}) \cdot \underline{J}(\mathscr{D}_2)$$

and

$$\bar{J}(\mathscr{D}_1 \times \mathscr{D}_2) = \bar{J}(\mathscr{D}_1) \cdot \bar{J}(\mathscr{D}_2).$$

Thus, with the obvious notation

$$J(\mathscr{D}_1 \times \mathscr{D}_2) = J(\mathscr{D}_1) \cdot J(\mathscr{D}_2).$$

Also if $\mathscr{D}_1, \mathscr{D}_2, \ldots$ is a finite or infinite sequence of dichotomies and $\pi_1, \pi_2, \ldots > 0$ and $\pi_1 + \pi_2 + \cdots = 1$ then

$$\underline{J}\left(\sum_k \pi_k \mathscr{D}_k\right) = \bigwedge_k \underline{J}(\mathscr{D}_k)$$

and

$$\bar{J}\left(\sum_k \pi_k \mathscr{D}_k\right) = \bigvee_k \bar{J}(\mathscr{D}_k).$$

By theorem 7.2.17, or directly, we find that if $\mathscr{D}$ is a double dichotomy and $\mathscr{E}$ is any dichotomy, then $\mathscr{D} \geqslant \mathscr{E}$ if and only if $J(\mathscr{D}) \supseteqq J(\mathscr{E})$.

This implies in this situation and for a given integer $n \geqslant 1$, that $\mathscr{D}^n \geqslant \mathscr{E}^n$ if and only if $\mathscr{D} \geqslant \mathscr{E}$. More generally if $\pi_0, \ldots, \pi_n \geqslant 0$ and if $\pi_0 < 1$ and $\pi_0 + \cdots + \pi_n = 1$ then we find that $\sum_{i=0}^n \pi_i \mathscr{D}^i \geqslant \sum_{i=0} \pi_i \mathscr{E}^i$ if and only if $\mathscr{D} \geqslant \mathscr{E}$. Does this extend to infinite mixtures?

A dichotomy $\mathscr{E}$ is at least as informative as the double dichotomy $\begin{pmatrix} 1 - \alpha_0, & \alpha_0 \\ 1 - \beta_0, & \beta_0 \end{pmatrix}$, where $\alpha_0 \leqslant \beta_0$, if and only if $\beta(\alpha_0 | \mathscr{E}) \geqslant \beta_0$, i.e. if and only if $\mathscr{E}$ permits a level α_0 test with power β_0 for testing 'P_0' against 'P_1'. Furthermore a dichotomy is the supremum of the double dichotomies which it majorizes.

Complement 16. Relative nearness of parameter values. The binomial case

A given experiment $\mathscr{E} = (P_\theta : \theta \in \Theta)$, as we have seen, usually defines many interesting structures on the parameter set Θ.

In particular $\mathscr{E}$ provides notions of relative nearness which are finer than those defined by the statistical distance or by the Hellinger distance. One interesting possibility is to say that θ_0 *is closer to* θ_1 *than* $\tilde{\theta}_0$ *is to* $\tilde{\theta}_1$ if $(P_{\theta_0}, P_{\theta_1}) \geqslant (P_{\tilde{\theta}_0}, P_{\tilde{\theta}_1})$.

Assume e.g. that Θ is a subset of the real line and that $(P_\theta : \theta \in \Theta)$ has monotone likelihood ratio. Then clearly $(P_{\theta_0}, P_{\theta_1}) \geqslant (P_{\tilde{\theta}_0}, P_{\tilde{\theta}_1})$ whenever the pairs (θ_0, θ_1) and $(\tilde{\theta}_0, \tilde{\theta}_1)$ are similarly ordered and the interval $\langle \theta_0, \theta_1 \rangle$ with end points θ_0 and θ_1 contains $\tilde{\theta}_0$ and $\tilde{\theta}_1$.

However this condition may be much too strong in particular cases. Thus if $(P_\theta : \theta \in \mathbb{R})$ is a translation family on the real line, or more generally if $(P_\theta : \theta \in \mathbb{R}) \sim (P_{\theta+a} : \theta \in \mathbb{R})$ for any number a, then $(P_{\theta_0}, P_{\theta_1}) \geqslant (P_{\tilde{\theta}_0}, P_{\tilde{\theta}_1})$ if and only if $(P_0, P_{\theta_1 - \theta_0}) \geqslant (P_0, P_{\tilde{\theta}_1 - \tilde{\theta}_0})$. If $P = P_0$ is strongly unimodal then this requirement is satisfied whenever (θ_0, θ_1) and $(\tilde{\theta}_0, \tilde{\theta}_1)$ are similarly ordered and $|\theta_1 - \theta_0| \geqslant |\tilde{\theta}_1 - \tilde{\theta}_0|$. If, in addition, P has a point of symmetry, then the similarity condition is superfluous, i.e. θ_0 is closer to θ_1 than $\tilde{\theta}_0$ is to $\tilde{\theta}_1$ if and only if $|\theta_1 - \theta_0| \geqslant |\tilde{\theta}_1 - \tilde{\theta}_0|$.

Considering a double dichotomy $\mathscr{D} = \begin{pmatrix} 1 - p_0, & p_0 \\ 1 - p_1, & p_1 \end{pmatrix}$ such that $p_0 \leqslant p_1$ we find that $\beta(\alpha|\mathscr{D}) = (p_1/p_0)\alpha$ when $0 < \alpha \leqslant p_0$ while $\beta(\alpha|\mathscr{D}) = 1 - (1 - \alpha)[(1 - p_1)/(1 - p_0)]$ when $p_0 \leqslant \alpha \leqslant 1$. Thus the graph of $\beta(\cdot|\mathscr{D})$ is the upper boundary of the triangle $\langle (0,0), (p_0, p_1), (1,1) \rangle$.

If $B_{n,p}$ denotes the binomial distribution corresponding to n trials and success parameter p and $\mathscr{D} \sim \begin{pmatrix} 1 - p_0, & p_0 \\ 1 - p_1, & p_1 \end{pmatrix}$ then, by sufficiency, $\mathscr{D}^n \sim (B_{n,p_0}, B_{n,p_1})$. We find easily that $\beta(\alpha|\mathscr{D}^n) = \alpha(p_1/p_0)^n$ when $0 \leqslant \alpha \leqslant p_0^n$ while $\beta(\alpha|\mathscr{D}^n) = 1 - (1 - \alpha)[(1 - p_1)/(1 - p_0)]^n$ when $1 - (1 - p_0)^n \leqslant \alpha \leqslant 1$.

The graphs of the β-functions show that if $p_0 \leqslant p_1$ and $q_0 \leqslant q_1$ then the double dichotomy $\mathscr{D}_1 = \begin{pmatrix} 1 - p_0, & p_0 \\ 1 - p_1, & p_1 \end{pmatrix}$ is at least as informative as the double dichotomy $\mathscr{D}_2 = \begin{pmatrix} 1 - q_0, & q_0 \\ 1 - q_1, & q_1 \end{pmatrix}$ if and only if the point (q_0, q_1) lies within the triangle with vertices (p_0, p_1), $(0,0)$ and $(1,1)$. Agreeing to interpret fractions $0/0$ as 1 this may be written

$$\frac{1 - p_1}{1 - p_0} \leqslant \frac{1 - q_1}{1 - q_0} \leqslant \frac{q_1}{q_0} \leqslant \frac{p_1}{p_0}.$$

Considering the slopes of $\beta(\cdot|\mathscr{D}_1^n)$ and $\beta(\cdot|\mathscr{D}_2^n)$ for small and large values of α we see that if $n \geqslant 1$ then $\mathscr{D}_1^n \geqslant \mathscr{D}_2^n$ if and only if $\mathscr{D}_1 \geqslant \mathscr{D}_2$.

Arguing as in the previous complement we again find that if $\mathscr{D}$ is a double dichotomy then $\mathscr{D}^n \geqslant \mathscr{E}^n$ if and only if $\mathscr{D} \geqslant \mathscr{E}$.

Complement 17↑. Relative nearness of parameter values. The Poisson case

Let B_λ, for each $\lambda \geqslant 0$, denote the Poisson distribution with expectation λ. Thus $B_\lambda(x) = (\lambda^x/x!)e^{-\lambda}$; $x = 0, 1, 2, \ldots$

If a certain phenomenon occurs according to a Poisson point process with unknown intensity λ then the number of occurrences in an interval of length

s is a sufficient statistic for the process on that interval and, as is well known, is Poisson distributed with parameter $s\lambda$. Increasing the period of observation from s to $t > s$ we obtain more information. It follows that the experiment $(B_{\lambda t} : \lambda \geqslant 0)$ is at least as informative as $(B_{\lambda s} : \lambda \geqslant 0)$ when $t \geqslant s$. This may also be seen by considering independent observations X and Y which are distributed as $B_{\lambda s}$ and $B_{\lambda(t-s)}$ respectively. Then $Z = X + Y$ is $B_{\lambda t}$-distributed and is sufficient for the pair (X, Y).

Another simple observation is that if X and Y are independent observations such that X is B_λ-distributed and Y is B_κ-distributed where κ is known then $X + Y$ is at most as informative as (X, Y) which in turn is just as informative as X. Thus $(B_{\lambda+\kappa} : \lambda \geqslant 0) \leqslant (B_\lambda : \lambda \geqslant 0)$ when $\kappa \geqslant 0$. This of course, by the randomization criterion, is also a consequence of the convolution identity $B_\lambda * B_\kappa \equiv_\lambda B_{\lambda+\kappa}$.

Combining these observations we find that $(B_\lambda : \lambda \geqslant 0) \geqslant (B_{\lambda s+\kappa} : \lambda \geqslant 0)$ whenever $0 \leqslant s \leqslant 1$ and $\kappa \geqslant 0$. In particular $(B_{\lambda_0}, B_{\lambda_1}) \geqslant (B_{\mu_0}, B_{\mu_1})$ where $\mu_0 = s\lambda_0 + \kappa$, $\mu_1 = s\lambda_1 + \kappa$; $0 \leqslant s \leqslant 1$ and $\kappa \geqslant 0$.

Assuming that μ_0 and μ_1 depend on (λ_0, λ_1) in this way and that $\lambda_0 \leqslant \lambda_1$ we find that $0 \leqslant \mu_1 - \mu_0 = s(\lambda_1 - \lambda_0) \leqslant \lambda_1 - \lambda_0$ and that $\lambda_1\mu_0 - \lambda_0\mu_1 = \kappa(\lambda_1 - \lambda_0) \geqslant 0$. Hence $\lambda_1 - \lambda_0 \geqslant \mu_1 - \mu_0$ and, furthermore, $\lambda_1/\lambda_0 \geqslant \mu_1/\mu_0$ provided these fractions are defined.

Conversely if these inequalities hold then $\mu_i = s\lambda_i + \kappa$; $i = 0, 1$ for some numbers $s \in [0, 1]$ and $\kappa \geqslant 0$.

It follows that $(B_{\lambda_0}, B_{\lambda_1}) \geqslant (B_{\mu_0}, B_{\mu_1})$ provided (λ_0, λ_1) and (μ_0, μ_1) are similarly ordered and if (μ_0, μ_1) belongs to the infinite triangle with vertices $(0, 0)$, (λ_0, λ_1) and direction $(1 : 1)$.

Assume that $(B_{\lambda_0}, B_{\lambda_1}) \geqslant (B_{\mu_0}, B_{\mu_1})$. Then $J(B_{\lambda_0}, B_{\lambda_1}) \supseteqq J(B_{\mu_0}, B_{\mu_1})$ where these intervals were defined in complement 15. Now it is easily checked that $J(B_{\lambda_0}, B_{\lambda_1}) = [\exp(-(\lambda_1 - \lambda_0)), \infty]$ or $= [0, \exp(-(\lambda_1 - \lambda_0))]$ as $\lambda_1 > \lambda_0$ or $\lambda_1 < \lambda_0$. It follows that (λ_0, λ_1) and (μ_0, μ_1) are similarly ordered. Assume for definiteness that $\lambda_0 \leqslant \lambda_1$ and that $\mu_0 \leqslant \mu_1$.

Note that $H(t|B_{\lambda_0}, B_{\lambda_1}) = \int (\mathrm{d}B_{\lambda_0}/\mathrm{d}B_{\lambda_1})^t \, \mathrm{d}B_{\lambda_0} = \exp[\lambda_0^{1-t}\lambda_1^t - (1-t)\lambda_0 - t\lambda_1]$. Hence

$$\lambda_0^{1-t}\lambda_1^t - (1-t)\lambda_1 - t\lambda_1 \geqslant \mu_0^{1-t}\mu_1^t - (1-t)\mu_0 - t\mu_1 \quad \text{when} \quad t \notin [0, 1].$$

If $\lambda_0 + \lambda_1 > 0$ and $\mu_0 > 0$ then letting $t \to \pm\infty$ in the inequality $\lambda_0^{1-t}\lambda_1^t - (1-t)\lambda_0 - t\lambda_1 \geqslant \mu_0^{1-t}\mu_1^t - (1-t)\mu_0 - t\mu_1$ we find that $\lambda_1/\lambda_0 \geqslant \mu_1/\mu_0$ and that $\lambda_1 - \lambda_0 \geqslant \mu_1 - \mu_0$. Replacing μ_0 and μ_1 with $\mu_0 + \varepsilon$ and $\mu_1 + \varepsilon$ where $\varepsilon \downarrow 0$, as above, we find that $\lambda_1/\lambda_0 \geqslant \mu_1/\mu_0$ whenever these fractions are defined.

Altogether this shows that for the Poisson family $(B_\lambda : \lambda \geqslant 0)$, μ_0 is nearer to μ_1 than λ_0 is to λ_1 if and only if (μ_0, μ_1) is within the infinite triangle with vertices at $(0, 0)$, (λ_0, λ_1) and direction $(1 : 1)$. This in turn is equivalent to the condition that (λ_0, λ_1) and (μ_0, μ_1) are similarly ordered and $|\lambda_1 - \lambda_0| \geqslant$

$|\mu_1 - \mu_0|$ and that $(\lambda_0 \vee \lambda_1)/\lambda_0 \wedge \lambda_1 \geqslant (\mu_0 \vee \mu_1)/\mu_0 \wedge \mu_1$ when these fractions are both defined.

As $(B_{\lambda_0}, B_{\lambda_1})^n \sim (B_{n\lambda_0}, B_{n\mu_1})$ it is clear that $(B_{\lambda_0}, B_{\lambda_1})^n \geqslant (B_{\mu_0}, B_{\mu_1})^n$ if and only if $(B_{\lambda_0}, B_{\lambda_1}) \geqslant (B_{\mu_0}, B_{\mu_1})$. On the other hand it follows that if $0 < \lambda_0 < \lambda_1$ and $0 < \mu_0 < \mu_1$ then no power of $(B_{\lambda_0}, B_{\lambda_1})$ majorizes (B_{μ_0}, B_{μ_1}) unless $\lambda_1/\lambda_0 \geqslant \mu_1/\mu_0$. If $\lambda_1/\lambda_0 \geqslant \mu_1/\mu_0$ then $(B_{\lambda_0}, B_{\lambda_1})^n \geqslant (B_{\mu_0}, B_{\mu_1})^n$ if and only if $n \geqslant (\mu_1 - \mu_0)/(\lambda_1 - \lambda_0)$.

Another consequence which is not that apparent is that if $0 \leqslant s \leqslant 1$ then $(B_{\lambda_0^s}, B_{\lambda_1^s}) \geqslant (B_{\mu_0^s}, B_{\mu_1^s})$ whenever $(B_{\lambda_0}, B_{\lambda_1}) \geqslant (B_{\mu_0}, B_{\mu_1})$. This follows from the fact that the inequalities: $\lambda_1^s - \lambda_2^s \geqslant \mu_1^s - \mu_0^s \geqslant 0$ hold for all $s \in [0,1]$ provided they hold for $s = 1$ and provided $\lambda_1/\lambda_0 \geqslant \mu_1/\mu_0$ when these fractions are defined.

Indeed if these provisions are satisfied then, since $x \to x^s$ is concave, we find that

$$\mu_1^s - \mu_0^s = (\mu_0 + \mu_1 - \mu_0)^s - \mu_0^s \leqslant (\mu_0 + \lambda_1 - \lambda_0)^s - \mu_0^s \leqslant \lambda_1^s - \lambda_0^s$$

provided $\lambda_0 \leqslant \mu_0$ and also that

$$\mu_1^s - \mu_0^s = \left(\mu_0 \frac{\mu_1}{\mu_0}\right)^s - \mu_0^s \leqslant \left(\mu_0 \frac{\lambda_1}{\lambda_0}\right)^s - \mu_0^s = \mu_0^s[(\lambda_1/\lambda_0)^s - 1]$$

$$\leqslant \lambda_0^s[(\lambda_1/\lambda_0)^s - 1] = \lambda_1^s - \lambda_0^s$$

provided $\lambda_0 \geqslant \mu_0$.

Complement 18. Deficiency distances to total information and to non information

Letting $\mathscr{M}_i$ and $\mathscr{M}_a$ denote, respectively, a totally non informative dichotomy and a totally informative dichotomy, we find for any dichotomy $\mathscr{D}$ that:

(a) $\delta(\mathscr{D}, \mathscr{M}_a) = 2\sup_\lambda b(\lambda|\mathscr{D}) = 2\alpha_0$ where $\alpha = \alpha_0$ is the unique solution of the equation $\beta(\alpha|\mathscr{D}) + \alpha = 1$.

(b) $\delta(\mathscr{M}_i, \mathscr{D}) = 2\sup_\lambda [(1 - \lambda) \wedge \lambda - b(\lambda|\mathscr{D})] = 1 - 2b(\frac{1}{2}|\mathscr{D}) = \sup_\alpha [\beta(\alpha|\mathscr{D}) - \alpha]$.

(c) the intuitive notion that $\mathscr{D}$ is close to or far away from $\mathscr{M}_a$ according to whether $\mathscr{D}$ is far away from or close to $\mathscr{M}_i$ may be substantiated by the inequalities

$$1 - \delta(\mathscr{M}_i, \mathscr{D}) \leqslant \delta(\mathscr{D}, \mathscr{M}_a) \leqslant 2[1 - \delta(\mathscr{M}_i, \mathscr{D})]/[2 - \delta(\mathscr{M}_i, \mathscr{D})].$$

These inequalities are both symmetric in the information measures $\delta(\mathscr{M}_i, \mathscr{D})$ and $\delta(\mathscr{D}, \mathscr{M}_a)$ and they are both sharp for a given value of one of these measures.

(d) if $t \to H(t) = \int dP_0^{1-t} dP_1^t$ is the Hellinger transform of $\mathscr{D} = (P_0, P_1)$ then

$$1 - \delta(\mathscr{D}, \mathscr{M}_i) \leqslant H(t)$$

and

$$\delta(\mathscr{D}, \mathscr{M}_a) \leqslant 2H(t)/[1 + H(t)]$$

when $t \in [0, 1]$.

(e) The $1/n^{1/2}$ rate. Let $\mathscr{D}_n = (P_n, Q_n)$; $n = 1, 2, \ldots$ be dichotomies and for each n, let $\gamma_n = \int dP_n^{1/2}\, dQ_n^{1/2}$ and $D_n = [2(1 - \gamma_n)]^{1/2}$ denote, respectively, the affinity and the Hellinger distance between P_n and Q_n. Then:

(i) $(\mathscr{D}_n^n)$ is bounded away from $\mathscr{M}_i$, i.e. $\inf_n \Delta(\mathscr{D}_n^n, \mathscr{M}_i) > 0$, if and only if $\sup_n \gamma_n^n < 1$ and this in turn is equivalent to the condition that $D_n > c_0/n^{1/2}$, $n = 1, 2, \ldots$ for some positive constant c_0.

(ii) $(\mathscr{D}_n^n)$ is bounded away from $\mathscr{M}_a$, i.e. $\inf_n \Delta(\mathscr{D}_n^n, \mathscr{M}_a) > 0$, if and only if $\inf_n \gamma_n^n > 0$ and this in turn is equivalent to the condition that $D_n < 2^{1/2} \wedge (c_1/n^{1/2})$; $n = 1, 2, \ldots$ for some constant $c_1 < \infty$.

Remarks. The left hand side inequality of (c) follows from the triangular inequality for the deficiency distance applied to the 'triangle' $(\mathscr{M}_i, \mathscr{D}, \mathscr{M}_a)$.

The right hand side inequality in (c) may be written

$$b(\lambda) \leqslant 2b(\tfrac{1}{2})/[1 + 2b(\tfrac{1}{2})]; \qquad 0 \leqslant \lambda \leqslant 1.$$

As $b(\lambda) = \|(1 - \lambda)P_0 \wedge \lambda P_1\| \leqslant \lambda$ this holds trivially when $\lambda \leqslant 2b(\frac{1}{2})/[1 + 2b(\frac{1}{2})]$. On the other hand if $2b(\frac{1}{2})/[1 + 2b(\frac{1}{2})] < \lambda \leqslant \frac{1}{2}$ then, by concavity, $b(\lambda) \leqslant 2(1 - \lambda)b(\frac{1}{2}) \leqslant 2b(\frac{1}{2})/[1 + 2b(\frac{1}{2})]$. By symmetry, since $b(\lambda | P_0, P_1) = b(1 - \lambda | P_1, P_0)$, the inequality extends to $\lambda \in [\frac{1}{2}, 1]$ as well.

Sharpness may be established by considering the dichotomies $\begin{pmatrix} 1-\xi & \xi & 0 \\ 0 & \eta & 1-\eta \end{pmatrix}$ for various values of $\xi, \eta \in [0, 1]$.

Finally the second inequality of (d), in view of (c), is a consequence of the first inequality, and this follows by putting $\lambda = \frac{1}{2}$ in the inequality $b(\lambda) = \|(1 - \lambda)P_0 \wedge \lambda P_1\| \leqslant (1 - \lambda)^{1-t}\lambda^t H(t)$; $0 \leqslant t \leqslant 1$.

In general if the parameter set Θ contains m points then

$$\frac{1}{m(m-1)}\delta(\mathscr{M}_i, \mathscr{E}) \leqslant \delta(\mathscr{M}_i, \mathscr{M}_a) - \delta(\mathscr{E}, \mathscr{M}_a) \leqslant \delta(\mathscr{M}_i, \mathscr{E})$$

and

$$\frac{1}{m(m-1)^2}\delta(\mathscr{E}, \mathscr{M}_a) \leqslant \delta(\mathscr{M}_i, \mathscr{M}_a) - \delta(\mathscr{M}_i, \mathscr{E}) \leqslant \delta(\mathscr{E}, \mathscr{M}_a).$$

Here $\mathscr{E}$ is any experiment with parameter set Θ and $\mathscr{M}_i$ and $\mathscr{M}_a$ are, respectively, a totally non informative experiment with parameter set Θ and a totally informative experiment with parameter set Θ. Thus $\delta(\mathscr{M}_i, \mathscr{M}_a) = 2 - (2/m)$.

The right hand side inequality follows again from the triangular inequality, and proofs of the left hand side inequalities are given in Torgersen (1976). These proofs may also be found in Heyer (1982).

Convergence to $\mathscr{M}_a$ of replicated experiments is discussed in Torgersen (1981). If Θ is infinite then the left hand side inequalities above are of no use and indeed it may then easily happen that $\delta(\mathscr{M}_i, \mathscr{E}^n) \equiv_n 2$ while $\delta(\mathscr{E}^n, \mathscr{M}_a) \equiv_n 1$.

(e) warrants many remarks but here we shall only mention two aspects. Firstly if $\mathscr{D}_n$; $n = 1, 2, \ldots$ satisfies conditions (i) and (ii), i.e. if $\varepsilon_0 < \|P_n^n - Q_n^n\| < 2 - \varepsilon_1$; $n = 1, 2, \ldots$ for positive constants ε_0 and ε_1, then all limit dichotomies of $\mathscr{D}_n^n$ are neither totally informative nor are they totally non informative. It may be shown that the possible limit dichotomies are all infinitely divisible. Conversely if $\mathscr{D} \not\sim \mathscr{M}_i$ and $\mathscr{D} \not\sim \mathscr{M}_a$ and $\mathscr{D}$ is infinitely divisible then $\mathscr{D} \equiv_n (\mathscr{D}^{1/n})^n$ and $\mathscr{D}^{1/n}$; $n = 1, 2, \ldots$ clearly satisfy conditions (i) and (ii). The reader is referred to LeCam (1969, 1986) as well as to Janssen, Milbrodt & Strasser (1984) and Strasser (1985) for further information on infinitely divisible experiments. See also complement 31.

The other remark concerning (e) is that if $\mathscr{E} = (P_\theta : \theta \in \Theta)$ is any experiment and if (θ_n) and $(\tilde{\theta}_n)$ are parameter points then conditions (i) and (ii) provide the precise conditions ensuring that $\mathscr{E}^n | \{\theta_n, \tilde{\theta}_n\}$; $n = 1, 2, \ldots$ is bounded away from $\mathscr{M}_i$ and $\mathscr{M}_a$. The $1/n^{1/2}$ rate may then not be the appropriate rate for a given metric on Θ but, as we have seen, it is so for the metric derived from Hellinger distances in $\mathscr{E}$.

Complement 19. Decomposition by conditioning

Let the homogeneous experiment $\mathscr{E}$ be realized by observing $X = (Y, Z)$ where Y and Z take values in measurable spaces $(\mathscr{Y}, \mathscr{B})$ and $(\mathscr{Z}, \mathscr{C})$ respectively. Suppose further that there are σ-finite measures μ and ν on $\mathscr{B}$ and $\mathscr{C}$ respectively such that X has density $(y, z) \to f_\theta(y, z)$ with respect to $\mu \times \nu$.

Then the marginal distribution of Z has density $g_\theta(z) \equiv_z \int f_\theta(y, z)\mu(dy)$ with respect to ν.

A regular version of the conditional distribution of Y given Z is given by the density $f_\theta(y, Z)/g_\theta(Z)$ with respect to μ.

Assuming for simplicity that $g_\theta(z) > 0$ for all θ and for all z we may describe the conditional experiment $\mathscr{E}^z$ given $Z = z$ by the family $f_\theta(\cdot, z)/g_\theta(z)$ of densities with respect to μ.

If t is a prior distribution on Θ with finite support then we obtain the following decomposition of the Hellinger transform of $\mathscr{E}$ evaluated at t

$$H(t|\mathscr{E}) = \int \prod_\theta g_\theta(z)^{t_\theta} H(t|\mathscr{E}^z)\nu(dz).$$

Suppose now that $\mathscr{E}^z \geqslant \mathscr{E}$ for ν-almost all z. Then this decomposition yields

$$H(t|\mathscr{E}) \leqslant \int \prod_\theta g_\theta(z)^{t_\theta} H(t|\mathscr{E})\nu(dz) = H(t|\mathscr{E}Z^{-1})H(t|\mathscr{E}) \leqslant H(t|\mathscr{E})$$

so that $H(t|\mathscr{E}Z^{-1}) = 1$ when $H(t|\mathscr{E}) > 0$.

Thus, by homogeneity, the assumption that $\mathscr{E}^z \geqslant \mathscr{E}$ for ν-almost all z implies that Z is ancillary and thus that $\mathscr{E}$ is the $P_\theta Z^{-1}$ mixture of $\mathscr{E}^z : z \in \mathscr{Z}$. It follows also that $\mathscr{E}^Z \sim \mathscr{E}$-almost surely.

This implies in particular that if Z is not ancillary then there is a subset C of $\mathscr{Z}$ with positive ν-measure such that $\mathscr{E}^z \not\geqslant \mathscr{E}$ when $z \in C$.

On the other hand $\mathscr{E}^Z \sim \mathscr{E}$-almost surely for all Θ when Z is sufficient. It follows that the informational inequalities $\mathscr{E}^z \geqslant \mathscr{E}$; $z \in \mathscr{Z}$ cannot be simultaneously reversed here.

Complement 20. Dichotomies as totally ordered mixtures of double dichotomies

Any dichotomy is a mixture of a totally ordered family of double dichotomies and this representation is essentially unique. The existence of a totally ordered decomposition may be based on the following facts.

(a) If P and Q are probability measures on the interval $[a, b]$ and if P and Q assign the same masses to each one-point set, then the statistic $x \to P[a, x] - Q[a, x]$ has the same distribution under P as under Q.

(b) If $\mathscr{D} = (P_0, P_1)$ is a dichotomy such that $P_0 \gg P_1$ then $\beta(0|\mathscr{D}) = 0$ and thus $T : \alpha \to \beta(\alpha|\mathscr{D}) - \alpha$ is ancillary for the representation $(\lambda, \beta(\cdot|\mathscr{D}))$. (Here λ is the uniform distribution on $[0, 1]$ and $\beta(\cdot|\mathscr{D})$ also denotes the measure having distribution function $\beta(\cdot|\mathscr{D})$.) In that case the decomposition obtained by conditioning on T provides a totally ordered decomposition of $\mathscr{D}$. This procedure may be generalized to cover the case $\beta(0|\mathscr{D}) > 0$ as well.

The decomposition obtained, see Torgersen (1970), is the essentially unique decomposition of $\mathscr{D}$ as a mixture of a totally ordered set of double dichotomies.

Complement 21. Uniqueness of decompositions into mixtures of totally ordered families of extremal experiments

Let $\mathscr{E}_1 > \mathscr{E}_2 > \cdots$ and $\mathscr{F}_1 > \mathscr{F}_2 > \cdots$ be two strictly decreasing sequences of extremal experiments.

If $p_1, p_2, \ldots, q_1, q_2, \ldots > 0$ and $1 = \sum p_i = \sum q_i$ then the mixtures $\sum p_i \mathscr{E}_i$ and $\sum q_i \mathscr{F}_i$ are not equivalent unless $\mathscr{E}_i \sim \mathscr{F}_i$ and $p_i = q_i$; $i = 1, 2, \ldots$

In fact if $\mathscr{E} \sim \mathscr{F}$ where $\mathscr{E} = \sum p_i \mathscr{E}_i$ and $\mathscr{F} = \sum q_i \mathscr{F}_i$ then $\mathscr{E}_1 \geqslant \mathscr{E} \sim \sum q_i \mathscr{F}_i$ so that, by complement 25, $\mathscr{E}_1 \sim \sum q_i \mathscr{G}_i$ where $\mathscr{G}_i \geqslant \mathscr{F}_i$; $i = 1, 2, \ldots$ Since $\mathscr{E}_1$ is extremal, it follows that $\mathscr{E}_1 \sim \mathscr{G}_1 \sim \mathscr{G}_2 \sim \cdots$ and in particular that $\mathscr{E}_1 \geqslant \mathscr{F}_1$. Similarly $\mathscr{F}_1 \geqslant \mathscr{E}_1$ so that $\mathscr{E}_1 \sim \mathscr{F}_1$. If $p_1 > q_1$ then this yields

$$\frac{p_1 - q_1}{1 - q_1}\mathscr{E}_1 + \frac{p_2}{1 - q_1}\mathscr{E}_2 + \cdots \sim \frac{q_2}{1 - q_1}\mathscr{F}_2 + \cdots$$

Repeating the arguments we find that $\mathscr{E}_1 \sim \mathscr{F}_2$. Hence, since $\mathscr{E}_1 \sim \mathscr{F}_1$, we find also that $\mathscr{F}_1 \sim \mathscr{F}_2$ which contradicts the assumption that $\mathscr{F}_1 > \mathscr{F}_2$. It follows that $p_1 \leqslant q_1$ and, similarly, $q_1 \leqslant p_1$ so that $p_1 = q_1$. The proof of the above assertion now follows readily by induction.

Complement 22. Measurable decompositions of sublinear functionals

It is a most useful fact that the support function of a finite algebraic sum of sets in $\mathbb{R}^n$ is the sum of the support functions of the summands (see chapter 2). Furthermore, compact convex sets are completely determined by their support functions. These results extend to general locally convex linear spaces. In particular weakly compact convex subsets of the algebraic dual V' of a general linear space V are characterized by their support functions on V.

If $C_1, C_2, \ldots, C_r$ are subsets of a linear space V then a vector v belongs to the algebraic sum $C_1 + \cdots + C_r$ if and only if it may be decomposed as $v = v_1 + \cdots + v_r$ where $v_i \in C_i$; $i = 1, \ldots, r$.

Trying to extend this to a general family C_t; $t \in T$ of sets we might consider 'integrals' $\int v_t \pi(dt)$ where $v_t \in C_t$; $t \in T$ and π is a probability measure on T. These integrals may then 'tentatively' constitute the 'set' $\int C_t \pi(dt)$. If, on the other hand, we define a π-mixture of convex sets C_t; $t \in T$ by integrating their support functions with respect to π, then it may not be that obvious that a given vector v in $\int C_t \pi(dt)$ is actually of the form $\int v_t \pi(dt)$ where $v_t \in C_t$ when $t \in T$.

Using the 'measurable' Hahn–Banach theorem, corollary 2.2.14, we obtain the following useful decomposition result due to Strassen (1965).

Theorem. *Let V be a separable Banach space equipped with a norm $\|\cdot\|$.*

Consider as a candidate for a mixing distribution a probability measure π on some given measurable space $(T, \mathscr{S})$. Let $h_t : t \in T$ be a family of continuous sublinear functionals on V such that $h_t(v)$ is measurable in t for each $v \in V$.

Assume that the norm $\|h_t\| = \sup\{|h_t(v)| : \|v\| = 1\}$ is π-integrable in t. Then any linear functional l on V such that $l(v) \leqslant \int h_t(v)\pi(dt)$ when $v \in V$ admits a decomposition

$$l(v) \underset{v}{\equiv} \int l_t(v)\pi(dt)$$

where $l_t(v)$ is measurable in t and linear in v and where $l_t \leqslant h_t$; $t \in T$.

Roughly the argument is as follows. Let $\tilde{C}$ denote the class of all linear functionals which admit a representation of the desired form. Then $\tilde{C}$ is clearly convex and our task is to show that $l \in \tilde{C}$ when $l \leqslant h$ where $h(v) \equiv_v \int h_t(v)\pi(dt)$. If this were not so then we might hope to argue that l and $\tilde{C}$ could be strongly separated by an evaluation. Indeed, as we shall see in a moment, the set $\tilde{C}$ is weakly compact for the V-topology of V^* and, furthermore, the linear functionals on V^* which are continuous for this topology are precisely the evaluations. Assuming that we have established strong separation we may conclude that there is a vector v_0 in V such that the $l(v_0) > \sup\{\tilde{l}(v_0) : \tilde{l} \in \tilde{C}\}$. We know however from the measurable version of the Hahn–Banach theorem, i.e. corollary 2.2.14, that there is a measurable family $\hat{l}_t : t \in T$ of linear functionals such that $\hat{l}_t \leqslant h_t$ for all t and such that $\hat{l}_t(v_0) \equiv_t h_t(v_0)$. Putting $\hat{l} = \int \hat{l}_t(v)\pi(\mathrm{d}t)$ we find that $l \leqslant \int h_t \pi(\mathrm{d}t) = h$ yielding the contradiction

$$l(v_0) = \int \hat{l}_t(v_0)\pi(\mathrm{d}t) = \int h_t(v_0)\pi(\mathrm{d}t) = h(v_0)$$
$$\geqslant l(v_0) > \sup\{\tilde{l}(v_0) : \tilde{l} \in \tilde{C}\}.$$

Thus it remains only to establish that $\tilde{C}$ as defined above is compact for the V-topology (weak topology) on V^*. The required strong separation follows from the general fact that pairs of disjoint closed convex sets in a locally convex linear topological space may be strongly separated by continuous linear functionals provided one of the sets is compact.

By Tychonoff's theorem $\tilde{C}$ is clearly conditionally compact. Thus it remains to show that $\tilde{C}$ is closed for the V-topology on V^*. This may be proved as follows.

Consider a linear functional $\tilde{l}$ in the closure of $\tilde{C}$ for this topology. We must show that $\tilde{l}$ admits a decomposition as required for membership in $\tilde{C}$.

By assumption $\tilde{l}(v) \equiv_v \lim_\nu \tilde{l}_\nu(v)$ for a net $\{\tilde{l}_\nu\}$ in $\tilde{C}$. Each $\tilde{l}_\nu$ admits a decomposition $\tilde{l}_\nu(v) \equiv_v \int l_\nu(v|t)\pi(\mathrm{d}t)$ where $l_\nu(v|t)$ is linear in v and measurable in t and where $l_\nu(v|t) \leqslant h_t(v)$; $v \in V$. By the weak compactness lemma, and Tychonoff's theorem, we may assume that $l_\nu(v|\cdot)$ converges for the $L_1(\pi)^* = L_\infty(\pi)$-topology on $L_1(\pi)$ for each $v \in V$.

Let $V_0 = \{v_1, v_2, \ldots\}$ be a countable dense subset of V and let $t \to f_j(t)$ be a real valued representation of $\lim_\nu l_\nu(v_j|\cdot)$. If $\alpha_1, \alpha_2, \ldots, \alpha_r$ are real numbers then $\sum \alpha_j f_j = \lim_\nu \sum \alpha_j l_\nu(v_j|\cdot) = \lim_\nu l_n(\sum \alpha_j v_j|\cdot) \leqslant h_.(\sum \alpha_j v_j)$ in $L_1(\pi)$. In particular $\sum \alpha_j f_j(t) \leqslant h_t(\sum \alpha_j v_j)$ for π-almost all t.

It follows that there is a set T_0 in $\mathscr{S}$ such that $\pi(T) = 1$ and such that $\sum \alpha_j f_j(t) \leqslant h_t(\sum \alpha_j v_j)$ when $t \in T_0$ and $\alpha_1, \ldots, \alpha_r$ are rational. By continuity this extends to each linear combination $\sum \alpha_j v_j$ such that $\sum \alpha_j f_j(t) \leqslant h_t(\sum \alpha_j v_j)$ when $\alpha_1, \ldots, \alpha_r \in \mathbb{R}$ and $t \in T_0$. If $\sum \alpha_j v_j = \sum \beta_j v_j$ and $t \in T_0$ then this yields

$$\sum \alpha_j f_j(t) - \sum \beta_j f_j(t) = \sum (\alpha_j - \beta_j) f_j(t)$$
$$\leqslant h_t\left(\sum (\alpha_j - \beta_j) v_j\right) = h_t(0) = 0$$

and, by symmetry, also that $\sum \beta_j f_j(t) - \sum \alpha_j f_j(t) \leqslant 0$. Thus $\sum \alpha_j f_j(t) = \sum \beta_j f_j(t)$ when $\sum \alpha_j v_j = \sum \beta_j v_j$ and $t \in T_0$.

Therefore, provided $t \in T_0$, we may define a linear functional $l(\cdot|t)$ on the linear span $[V_0]$ of V_0 by putting $l(\sum \alpha_j v_j | t) = \sum \alpha_j f_j(t)$ when $\alpha_1, \ldots, \alpha_r \in \mathbb{R}$. By continuity and since V_0 is dense in V, $l(\cdot|t)$ may be extended to a linear functional on all of V. Then $l(v|t)$ is measurable on T_0 for all $v \in V$.

By corollary 2.2.14 we may extend $l(\cdot|t)$ to all $t \in T$ such that $l(\cdot|t) \leqslant h_t$; $t \in T$ and such that $l(v|t)$ is linear in v and measurable in t.

If $v \in V$ then $v = \lim_n v_{i_n}$ for a sequence $i_1, i_2 \ldots$ of indices. Then $\tilde{l}(v) = \lim_n \tilde{l}(v_{i_n}) \overset{n}{=} \lim_n \lim_\nu \tilde{l}_\nu(v_{i_n}) = \lim_n \lim_\nu \int l_\nu(v_{i_n}|t)\pi(\mathrm{d}t) = \lim_n \int f_{i_n}(t)\pi(\mathrm{d}t) = \lim_n \int l(v_{i_n}|t)\pi(\mathrm{d}t) = \int l(v|t)\pi(\mathrm{d}t)$.

Thus $\tilde{l}(v) \equiv_v \int l(v|t)\pi(\mathrm{d}t)$ so that $\tilde{l} \in \tilde{C}$. Altogether this shows that $\tilde{C}$ is in fact closed for the weak topology on V^*.

Complement 23↑. Existence of randomizations with prescribed properties

Strassen's decomposition theorem for sublinear functionals may be used to establish the existence of kernels having particular properties, as shown in Strassen (1965).

Consider e.g. a probability measure P on some measurable space along with convex sets $C_t : t \in T$ of probability measures on the same space. We may then enquire whether P may be decomposed as $P = \int M_t \pi(\mathrm{d}t)$ for a given mixing probability measure π on T and for probability measures $M_t : t \in T$ such that $M_t \in C_t$ for π-almost all t.

Here $P = \int M_t \pi(\mathrm{d}t)$ is short for $\int g \,\mathrm{d}P = \int (\int g \,\mathrm{d}M_t)\pi(\mathrm{d}t)$ when g is bounded and measurable or, equivalently, that $P(A) = \int M_t(A)\pi(\mathrm{d}t)$ for each measurable set A.

Putting $\psi_t(g) = \sup\{\int g \,\mathrm{d}M : M \in C_t\}$ we find trivially that if P admits such a decomposition and if g is bounded and measurable then

$$\int g \,\mathrm{d}P \leqslant \int \psi_t(g)\pi(\mathrm{d}t) \tag{10.2.2}$$

provided $\psi_t(g)$ is measurable in t.

What is most interesting here is that under suitable regularity conditions we may turn this around and conclude that for sufficiently many functions g, (10.2.2) ensures such a decomposition of P. Loosely this may be argued as follows. Assume that (10.2.2) holds for all bounded measurable functions belonging to a sufficiently large linear space V of functions g. Noting that $\int g \,\mathrm{d}P$ is linear in g while $\psi_t(g)$, for each t, is sublinear in g, we may conclude that

$$\int g \,\mathrm{d}P = \int l_t(g)\pi(\mathrm{d}t)$$

for linear functionals $l_t \leqslant \psi_t$; $t \in T$. Now by the representation theory for continuous linear functionals, there is some hope that we may conclude that l_t, for each t, is representable by a measure M_t, i.e. that $l_t(g) \equiv_g \int g \,\mathrm{d}M_t$. If

M_t does not belong to the convex set C_t then we might hope to conclude that M_t is strongly separable from C_t by a linear functional which, by a final hope, is the evaluation by one of the functions g under consideration. This, however, is impossible since for all $g \in V$, $l_t(g) \leqslant \sup\{\int g \, \mathrm{d}M : M \in C_t\}$. Admitting the possibility that this reasoning may be wrong for t's belonging to some null set we might venture to hope that $M_t \in C_t$ for π-almost all t.

Following Strassen we may verify this when conditions (i)–(iv) below are satisfied:

(i) P is a probability measure on a measurable space $(\mathscr{X}, \mathscr{A})$ where $\mathscr{X}$ is a Borel subset of a complete separable metric space $\mathscr{Y}$ and $\mathscr{A}$ is the class of Borel sets contained in $\mathscr{X}$. $V =$ the space of bounded uniformly continuous functions on $\mathscr{X}$.

(ii) the mixing distribution π is a probability measure on a measurable space $(T, \mathscr{S})$.

(iii) C_t, for each $t \in T$, is a convex set of probability measures on $\mathscr{A}$. We shall assume that C_t, for each t, is closed for the weak topology for probability measures on $\mathscr{X}$. (We do not assume that C_t is compact for this topology.)

(iv) $\psi_t(g) = \sup\{\int g \, \mathrm{d}M : M \in C_t\}$ is measurable in t when $g \in V$.

Any complete separable metric space may be topologically imbedded as a G_δ subset of a compact metric space. It follows readily that we may assume without loss of generality, that $\mathscr{Y}$ is a compact metric space.

Then $C(\mathscr{Y})$ is separable and $f|\mathscr{X} \in V$ when $f \in C(\mathscr{Y})$. By assumption $\int f \, \mathrm{d}P \leqslant \int \sup\{\int f \, \mathrm{d}M : M \in C_t\}\pi(\mathrm{d}t)$; $f \in C(\mathscr{Y})$. The argument above shows now that

$$P = \int M_t \pi(\mathrm{d}t)$$

where M_t, for each t, belongs to the closure of C_t within the class of probability measures on $\mathscr{Y}$. On the other hand $1 = P(\mathscr{X}) = \int M_t(\mathscr{X})\pi(\mathrm{d}t)$ so that $M_t(\mathscr{X}) = 1$ when $t \in T_0$ where T_0 is a measurable subset of T having π-probability 1. It follows that the restriction of M_t to $\mathscr{A}$ belongs to C_t when $t \in T_0$.

In the next complement, following Strassen, we will use this result to derive anew the dilation criterion for the ordering of standard measures.

The reader is referred to Strassen's 1965 paper for several other interesting applications of this decomposition result and also for related results.

Complement 24↑. Existence of dilations

Consider experiments $\mathscr{E}$ and $\mathscr{F}$ having the same finite parameter set Θ and having standard measures S and T respectively. We argued in chapter 5, without using the randomization criterion, that $\mathscr{E} \geqslant \mathscr{F}$ if and only if $\int \phi \, \mathrm{d}S \leqslant \int \phi \, \mathrm{d}T$ when ϕ is concave on the fundamental probability simplex $K =$

$\{x : x \in \mathbb{R}^\Theta, x \geqslant 0 \sum_\theta x_\theta = 1\}$. If $\hat{f}$ is the smallest concave function majorizing a given function f then this requirement may be written

$$\int f \, \mathrm{d}S \leqslant \int \hat{f} \, \mathrm{d}T$$

for all continuous functions f on K.

Now $\hat{f}(y)$ is sublinear and continuous in f for each $y \in K$. It follows that we may decompose S as $S = \int D_y T(\mathrm{d}y)$ where, for each y, $\int f \, \mathrm{d}D_y \leqslant \hat{f}(y)$ for all continuous f. If $f \leqslant 0$ then $\int f \, \mathrm{d}D_y \leqslant \hat{f}(y) \leqslant 0$ so that the measures D_y are non-negative. If f is affine then $\hat{f} = f$ so that $\int f \, \mathrm{d}D_y \leqslant f(y)$ and, by replacing f with $-f$, we find that $\int f \, \mathrm{d}D_y = f(y)$. It follows that D_y is in fact a probability measure on K having expectation y. Thus $D : y \to D_y$ is a dilation and $S = DT$.

Conversely if $S = DT$ then, by Jensen's inequality, $\int \phi \, \mathrm{d}S \leqslant \int \phi \, \mathrm{d}T$ when ϕ is concave.

This is the proof of the dilation criterion given in Strassen's 1965 paper, but in a less general form.

It follows in particular that if $\mathscr{E} \geqslant \mathscr{F} \sim \int \mathscr{F}_\alpha \pi(\mathrm{d}\alpha)$ then $\mathscr{E}$ may be decomposed as $\mathscr{E} = \int \mathscr{E}_\alpha \pi(\mathrm{d}\alpha)$ where, for each α, $\mathscr{E}_\alpha \geqslant \mathscr{F}_\alpha$. Indeed letting T_α be the standard measure of $\mathscr{F}_\alpha$ we find for a dilation D that $S = DT = D \int T_\alpha \pi(\mathrm{d}\alpha) = \int S_\alpha \pi(\mathrm{d}\alpha)$ where, for each α, $S_\alpha = DT_\alpha$.

Complement 25↑. Minorized decompositions

Consider an experiment $\mathscr{F}$ which is equivalent to a given mixture $\sum_{i=1}^\infty p_i \mathscr{F}_i$ of experiments $\mathscr{F}_1, \mathscr{F}_2, \ldots$

If $\mathscr{E}_i$; $i = 1, 2, \ldots$ are experiments such that $\mathscr{E}_i \geqslant \mathscr{F}_i$; $i = 1, 2, \ldots$ then, by the sublinear function criterion for comparison,

$$\sum p_i \mathscr{E}_i \geqslant \mathscr{F}.$$

It is interesting that any experiment which is at least as informative as $\mathscr{F}$ is in fact of this form. If Θ is finite then, as argued in the previous complement, this follows directly from the dilation criterion for comparison. By weak compactness this generalizes to any Θ.

The informational inequalities '$\leqslant$' cannot be reversed here. Consider e.g. the double dichotomies $\mathscr{E}_1 = \begin{pmatrix} 1 & 0 \\ \frac{1}{3} & \frac{2}{3} \end{pmatrix}$, $\mathscr{E}_2 = \begin{pmatrix} \frac{2}{3} & \frac{1}{3} \\ 0 & 1 \end{pmatrix}$ and $\mathscr{F} = \begin{pmatrix} \frac{5}{6} & \frac{1}{6} \\ \frac{1}{6} & \frac{5}{6} \end{pmatrix}$. Then $\mathscr{F} \leqslant \frac{1}{2}\mathscr{E}_1 + \frac{1}{2}\mathscr{E}_2$ but it is not possible to decompose $\mathscr{F}$ as $\mathscr{F} = \frac{1}{2}\mathscr{F}_1 + \frac{1}{2}\mathscr{F}_2$ where $\mathscr{F}_i \leqslant \mathscr{E}_i$; $i = 1, 2$. In fact putting $\mathscr{E} = \frac{1}{2}\mathscr{E}_1 + \frac{1}{2}\mathscr{E}_2$ we find

$$b(\lambda|\mathscr{E}_1) = \tfrac{1}{3}\lambda; \qquad \lambda \leqslant \tfrac{3}{4}$$

$$b(\lambda|\mathscr{E}_2) = \lambda; \qquad \lambda \leqslant \tfrac{1}{4}$$

and thus

$$b(\lambda|\mathscr{E}) = \tfrac{2}{3}\lambda; \qquad \lambda \leqslant \tfrac{1}{4}.$$

Furthermore $b(\lambda|\mathscr{F}) = \lambda$ or $= \frac{1}{6}$ as $\lambda \leqslant \frac{1}{6}$ or $\lambda \in [\frac{1}{6}, \frac{5}{6}]$. If $\mathscr{F} = \frac{1}{2}\mathscr{F}_1 + \frac{1}{2}\mathscr{F}_2$ where $\mathscr{F}_i \leqslant \mathscr{E}_i$; $i = 1, 2$ and $\lambda \in [\frac{1}{6}, \frac{1}{4}]$ then this yields

$$b(\lambda|\mathscr{F}_2) \geqslant b(\lambda|\mathscr{E}_2) = \lambda$$

so that $b(\lambda|\mathscr{F}_2) = \lambda$ and

$$b(\lambda|\mathscr{F}_1) = 2b(\lambda|\mathscr{F}) - b(\lambda|\mathscr{F}_2) = \tfrac{1}{3} - \lambda.$$

However, since $b(\lambda|\mathscr{F}_1)$ is concave, the last requirement implies the impossibility that $b(\lambda|\mathscr{F}) < 0$ when $\lambda > \frac{1}{3}$.

Complement 26. Comparison by factorization

A product of experiments is clearly at least as informative as any of the factor experiments occurring in the product. Although this situation is exceptional, nevertheless ordering of experiments may be established surprisingly often on the basis of this simple observation.

In general we shall say that an experiment $\mathscr{E}$ *is a factor experiment of an experiment* $\mathscr{F}$ if $\mathscr{E} \times \mathscr{G} \sim \mathscr{F}$ for some experiment $\mathscr{G}$. Thus $\mathscr{E}$ is a factor experiment of $\mathscr{F}$ if and only if the ratio $H_{\mathscr{F}}/H_{\mathscr{E}}$ of Hellinger transforms, by a proper quantification of 0/0, is also a Hellinger transform.

Let us agree to use the notation $\mathscr{E}|\mathscr{F}$ to denote that $\mathscr{E}$ is a factor experiment of $\mathscr{F}$. It is readily checked that this defines a relationship between types and, in fact, is a partial ordering of types of experiments. Furthermore this ordering, as we noted above, is stronger than the ordering 'being at most as informative as'.

It is shown in chapter 8 that these orderings coincide for the standard linear normal models with known or with unknown variances.

By Janssen (1988), theorem 9.5.6, this extends to comparison of exponential experiments with a common canonical parameter varying freely within an open set. As mentioned in remark 1 to that theorem, asymptotic factorization of information is treated in Ehm & Müller (1983).

(a) The general problem of comparison by factorization may be reduced to the case of experiments having finite parameter sets. Indeed the experiment $\mathscr{E} = (P_\theta : \theta \in \Theta)$ is a factor of the experiment $\mathscr{F} = (Q_\theta : \theta \in \Theta)$ if and only if $(P_\theta : \theta \in \Theta_0)$ is a factor of $(Q_\theta : \theta \in \Theta_0)$ for all non-empty finite subsets Θ_0 of Θ. This may be deduced from the compactness of the weak experiment topology.

For each of the orderings 'being at most as informative as' and 'being a factor experiment of' the collection of experiments which are majorized by a given experiment is compact for the weak experiment topology. In other words:

(b) if $\mathscr{F}$ is any given experiment then the collections $\{\mathscr{E} : \mathscr{E} \leqslant \mathscr{F}\}$ and $\{\mathscr{E} : \mathscr{E}|\mathscr{F}\}$ are both compact for the weak experiment topology.

Let us turn to experiments having the same *finite* parameter set Θ. Identifying standard measures with the experiments they define we obtain two partial orderings of standard measures. The first one is the ordering '$\leqslant$' which declares $S \leqslant T$ for standard measures S and T if this inequality holds when S and T are replaced by experiments having standard measures S and T respectively. The second ordering 'being a factor of' declares that S is a factor of T, notation $S|T$, if this is so for experiments having standard measures S and T respectively.

In order to describe and compare these orderings of standard measures we shall find it convenient to use the following notation.

K $= \{x : x \in \mathbb{R}^\Theta, x \geqslant 0, \sum_\theta x_\theta = 1\}$ is the fundamental probability simplex in $\mathbb{R}^\Theta$.
m = the number of points in Θ.
$C(K)$ = the set of real valued continuous functions on K.
$\hat{f}$ = the smallest concave function majorizing a given real valued function f.
$S, T, U, \ldots$ denotes standard measures.
$f_S(x)$ $= \int f((x_\theta y_\theta/\sum_\theta x_\theta y_\theta : \theta \in \Theta))\sum_\theta x_\theta y_\theta S(\mathrm{d}x)$ when $f \in C(K)$ and S is a standard measure.
e $= (1, \ldots, 1)$ in $\mathbb{R}^\Theta$.
$S * T$ is the standard measure of $\mathscr{E} \times \mathscr{F}$ when $\mathscr{E}$ and $\mathscr{F}$ have standard measures S and T respectively.

The results established in chapter 7 on ordering of experiments imply:

(c) $S \leqslant T$ if and only if $T(f) \leqslant S(\hat{f})$; $f \in C(K)$. In fact

$$S(\hat{f}) = \sup_D \int \left[\int f(y) D(\mathrm{d}y|x) \right] S(\mathrm{d}x) = \sup \{T(f) : T \geqslant S\}; f \in C(K)$$

where D runs through the set of dilations from K to K.

The analogous statement holds for the ordering 'being a factor of' by the following statements.

(d) $S|T$ if and only if $T(f) \leqslant m\hat{f}_S(e/m)$; $f \in C(K)$. In fact

$$m\hat{f}_S(e/m) = \sup_U S * U(f) = \sup\{T(f) : S|T\}; \qquad f \in C(K)$$

where U runs through the set of standard measures on K.

(e) Let Ω be a point outside Θ and put $\mathscr{X} = \Theta \cup \{\Omega\}$. If $0 \leqslant a \leqslant 1$ is a function on Θ then for each $\theta \in \Theta$ we may define a probability distribution P_θ on $\mathscr{X}$ by putting

$$P_\theta(\Omega) = 1 - P_\theta(\theta) = a_\theta.$$

Let $\mathscr{S}_a$ denote the experiment $(P_\theta : \theta \in \Theta)$. Thus $\mathscr{S}_0$ and $\mathscr{S}_e$ are, respectively, totally informative and totally non informative. The Hellinger transform of $\mathscr{S}_a$ is given by $H(t|\mathscr{S}_a) = \prod_\theta a_\theta^{t_\theta}$ provided $t \notin \text{ext}\, K$.

(f) If a and b are functions on Θ such that $0 \leqslant a(\theta) \leqslant 1$ and $0 \leqslant b(\theta) \leqslant 1$ for all $\theta \in \Theta$ then

$$a \geqslant b \Rightarrow \mathscr{S}_a | \mathscr{S}_b \Leftrightarrow \mathscr{S}_a \leqslant \mathscr{S}_b$$

where the first $\Rightarrow$ may be replaced with $\Leftrightarrow$ provided $\mathscr{S}_b$ is not totally informative, i.e. provided $b \neq 0$.

(g) Let $\mathscr{E} = (P_\theta : \theta \in \Theta)$ be any experiment with parameter set Θ and let a be any function on Θ such that $0 \leqslant a(\theta) \leqslant 1$ when $\theta \in \Theta$. For each θ, let κ_θ be the $\sum\{P_{\theta'} : \theta' \neq \theta\}$-absolutely continuous part of P_θ. Then

$$a(\theta) \geqslant \|\kappa_\theta\|; \qquad \theta \in \Theta \Leftrightarrow \mathscr{S}_a | \mathscr{E} \Leftrightarrow \mathscr{S}_a \leqslant \mathscr{E}.$$

There are other experiments $\mathscr{F}$ than those which are equivalent to experiments $\mathscr{S}_a$ which permit the implication $\mathscr{F} \leqslant \mathscr{E} \Rightarrow \mathscr{F} | \mathscr{E}$. They are described in detail in Torgersen (1974).

Complement 27. A non-separable dominated experiment such that all pairs of parameter values define equivalent dichotomies

An experiment $\mathscr{E} = (P_\theta : \theta \in \Theta)$ with these properties may be constructed as follows.

Let (μ, ν) be any dichotomy such that $\mu \gg \nu \neq \mu$ and let Θ be any uncountable set. Put $\mu_{\theta',\theta} = \mu$ or $= \nu$ as $\theta' \neq \theta$ or $\theta' = \theta$. Then an experiment $\mathscr{E} = (P_\theta : \theta \in \Theta)$ with the above properties is obtained by putting $P_\theta = \prod_{\theta'} \mu_{\theta',\theta}$ when $\theta \in \Theta$. Indeed $\mathscr{E}$ is dominated by $\prod_{\theta'} \mu$ and $(P_{\theta_0}, P_{\theta_1}) \sim (\mu \times \nu, \nu \times \mu)$ when $\theta_0 \neq \theta_1$.

Complement 28. Representations of separable experiments

Any separable experiment $\mathscr{E}$ is equivalent to an experiment $(P_\theta : \theta \in \Theta)$ where P_θ, for each θ, is an atomless probability measure on $[0, 1]$ and where P_{θ_0}, for a given $\theta_0 \in \Theta$, is the uniform distribution on $[0, 1]$.

Assuming that $\mathscr{E} = (Q_\theta : \theta \in \Theta)$ and that $\#\Theta \geqslant 2$ this may be seen as follows.

Let Θ_0 be a countable subset of Θ such that $\{Q_\theta : \theta \in \Theta_0\}$ is dense in $\{Q_\theta : \theta \in \Theta\}$. Choose a prior distribution μ on Θ_0 such that $\mu(\theta) > 0$ for all $\theta \in \Theta_0$ and put $Q_\mu = \int Q_\theta \mu(\mathrm{d}\theta)$. Also put $f_\theta = \mathrm{d}Q_\theta/\mathrm{d}Q_\mu$ and $f = (f_\theta : \theta \in \Theta_0)$. The densities $f_\theta : \theta \in \Theta$ may be modified so that f belongs to the subset H of $[0, \infty[^{\Theta_0}$ consisting of those non-negative functions x on Θ_0 which satisfy the condition $\int x \, \mathrm{d}\mu = 1$. Then f maps $\mathscr{E}$ onto an equivalent experiment $\mathscr{E}f^{-1}$

on H. As H is Borel isomorphic to the interval $[-1,0[$ we can also assume that the measures $Q_\theta : \theta \in \Theta$ are all on $[-1,0[$. Put $A = \{x : \sum_\theta Q_\theta(x) > 0\}$. Then, by separability, A is an enumerable subset of $[-1,0[$. Let $a_0, a_1, a_2, \ldots$ be an enumeration of its elements. Put $\mathscr{Y} = [-1,0[\cup \{[\nu, \nu+1[. a_\nu \text{ is defined}\}$. Then $\mathscr{Y} = [-1,\infty[$ or $= [-1, n+1[$ as $\# A = \infty$ or $\# A = n+1$. Let λ_ν denote the uniform distribution on $[\nu, \nu+1[$ and put

$$V_\theta(B) = Q_\theta(B \cap A^c) + \sum_\nu Q_\theta(a_\nu)\lambda_\nu(B \cap [\nu, \nu+1[)$$

for each Borel subset B of $\mathscr{Y}$.

It is easily checked that $\|\sum c_\theta Q_\theta\| = \|\sum c_\theta V_\theta\|$ for each $c \in \mathbb{R}^\Theta$ such that $\#\{\theta : c_\theta \neq 0\}$ is finite. By the testing criterion for equivalence the experiment $\mathscr{E}$ is equivalent to $(V_\theta : \theta \in \Theta)$.

The measures $V_\theta : \theta \in \Theta$ are obviously continuous, i.e. atomless. Let $V_{\theta,a}$ and $V_{\theta,s}$ denote, respectively, the V_{θ_0}-absolutely continuous part of V_θ and the V_{θ_0}-singular part of V_θ. Thus

$$V_{\theta_0,a} = V_{\theta_0}, \quad V_{\theta_0,s} = 0, \quad V_{\theta,a} \ll V_{\theta_0}, \quad V_{\theta,a} \wedge V_{\theta_0} = 0 \quad \text{and} \quad V_\theta = V_{\theta,a} + V_{\theta,s}.$$

Let $F : x \to V_{\theta_0}(]-\infty, x])$ be the distribution function of V_{θ_0} and let $x \to \rho(x)$ be a 1–1 bimeasurable map from $\mathbb{R}$ onto a Borel subset of $[0,1]$ having Lebesgue measure zero. Finally, for each θ define a measure P_θ on $[0,1]$ by

$$P_\theta = V_{\theta,a}F^{-1} + V_{\theta,s}\rho^{-1}.$$

It is then readily checked that P_θ, for each θ, is an atomless probability measure on $[0,1]$ and that P_{θ_0} is the uniform distribution on $[0,1]$. Again let c be a function on Θ such that $\{\theta : c_\theta \neq 0\}$ is finite and put $P_c = \sum_\theta c_\theta P_\theta$. Then $P_c = V_{c,a}F^{-1} + V_{c,s}\rho^{-1}$ where $V_{c,a} = \sum_\theta c_\theta V_{\theta,a}$ and $V_{c,s} = \sum_\theta c_\theta V_{\theta,s}$ are, respectively, the V_{θ_0}-absolutely continuous part of $V_c = \sum_\theta c_\theta V_\theta$ and the V_{θ_0}-singular part of V_c. Thus

$$\|P_c\| = \|V_{c,a}F^{-1}\| + \|V_{c,s}\rho^{-1}\| = \|V_{c,a}\| + \|V_{c,s}\| = \|V_c\| = \|Q_c\|.$$

Using the testing criterion for equivalence once more we see that $\mathscr{E} \sim (P_\theta : \theta \in \Theta)$.

Remark. Let us say that a countable subset Θ_0 of Θ is a *separant* for the (necessarily separable) experiment $\mathscr{E} = (P_\theta : \theta \in \Theta)$ if the set $\{P_\theta : \theta \in \Theta_0\}$ is dense in $\{P_\theta : \theta \in \Theta\}$ for statistical distance. Then any experiment $\mathscr{E}$ having separant Θ_0 may be decomposed as

$$\mathscr{E} \sim \int \mathscr{E}_t \pi(\mathrm{d}t)$$

where $\mathscr{E}_t$, for each t, is an extremal experiment having Θ_0 as separant (see Torgersen, 1977).

Complement 29. Two distances for translation experiments

A probability measure P on a second countable locally compact topological group $\mathscr{X}$ defines a (right) translation experiment $\mathscr{E}_P = (P_\theta : \theta \in \Theta)$ where $P_\theta(A) = P(A\theta^{-1})$ for each Borel set A. If the probability measures P and Q are absolutely continuous with respect to Haar measure on $\mathscr{X}$ then, by invariance, the deficiency distance is $\delta(\mathscr{E}_P, \mathscr{E}_Q) = \inf_M \|M * P - Q\|$ where $*$ denotes convolution and M runs through the class of all probability measures on the Borel class $\mathscr{A}$ of subsets of $\mathscr{X}$.

The deficiency distance $\Delta(\mathscr{E}_P, \mathscr{E}_Q) = \delta(\mathscr{E}_P, \mathscr{E}_Q) \vee \delta(\mathscr{E}_Q, \mathscr{E}_P)$ defines, as a function of (P, Q), a pseudo-metric on the set of absolutely continuous probability measures. By convenient misuse of notation here we will write $\Delta(P, Q)$ instead of $\Delta(\mathscr{E}_P, \mathscr{E}_Q)$.

Restricting our attention to non-randomized kernels we arrive at the generally larger distance

$$\tau(P, Q) = \inf_\theta \|\delta_\theta * P - Q\|$$

between the absolutely continuous probability measures P and Q. Then:

(a) $\Delta(P, Q) = 0 \Leftrightarrow \tau(P, Q) = 0 \Leftrightarrow \delta_\theta * P = Q$ for some θ.
(b) Δ and τ are topologically equivalent.
(c) Δ and τ determine the same Cauchy sequences.
(d) it is not known whether or not Δ and τ are equivalent as metrics, i.e. if they determine the same uniformities.

Remark. Using the general facts on translation experiments which were derived in section 4 of chapter 6, the reader may try to establish (a) to (c) for the additive group of integers and then consult Torgersen (1972).

Complement 30↑. Translation experiments on the integers

Let the experiment $\mathscr{E}$ be realized by an integer valued variable X such that $X = \theta + U$ where θ is an unknown integer and U is an unobservable variable (error) having a known distribution given by

$$Pr(U = u) = f(u); \qquad u = \ldots, -1, 0, 1, \ldots$$

If, in particular, $f(u) = 1$ or $= 0$ as $u = 0$ or $u \neq 0$, then the experiment $\mathscr{E}$ obtained in this way is totally informative. Let $\mathscr{M}_a$ denote any totally informative experiment on the integers.

(a) Using invariance we find that

$$\delta(\mathscr{E}^n, \mathscr{M}_a) = 2\left[1 - \sum_{x_2, \ldots, x_n} \max_x P(x)P(x + x_2) \ldots P(x + x_n)\right]$$

$$\equiv_\theta 2P_\theta^n(\hat{\theta}_n \neq \theta)$$

for any translation invariant maximum likelihood estimator $\hat{\theta}_n$ of θ based on n observations $X_1,\ldots,X_n$ of X.

(b) If $f(u) = \text{constant } a^{|u|}$; $u = \ldots, -1,0,1,\ldots$ where $0 < a < 1$ and if $X_1,\ldots,X_n$ are as in (a) then any empirical median is a maximum likelihood estimator for θ based on $X_1,\ldots,X_n$. In particular it follows that

$$\delta(\mathscr{E},\mathscr{M}_a) = \delta(\mathscr{E}^2,\mathscr{M}_a) = 2(1-c) = 4a/(1+a).$$

(c) For any probability distribution f on the integers there are constants $c \geqslant 0$ and $\rho < 1$ such that $\delta(\mathscr{E}^n,\mathscr{M}_a) \leqslant c\rho^n$; $n = 1,2,\ldots$

(d) Assume $f(u) = 1/N$; $u = 1,2,2^2,\ldots,2^{N-1}$ where $N \geqslant 1$ is an integer. Then $\delta(\mathscr{E}^n,\mathscr{M}_a) = (N-1)/N^n$; $n = 1,2,\ldots$ In particular it follows that $\mathscr{E}$ may be chosen so that $P_\theta(\hat{\theta} \neq \theta) \leqslant 10^{-200}$ for all θ for some translation invariant estimator $\hat{\theta}$ based on two observations X_1, X_2 while on the other hand, $P_\theta(\theta^* \neq \theta) \geqslant 1 - 10^{-200}$ for some θ for any estimator θ^* based on one observation.

Further discussion and more examples of deficiencies with respect to total information may be found in Torgersen (1981).

Complement 31. Semigroups of translation experiments

Adapting terminology from the theory of random processes we shall say that a family $(\mathscr{H}_t : t > 0)$ of experiments constitutes a semigroup if $\mathscr{H}_s \times \mathscr{H}_t \sim \mathscr{H}_{s+t}$ when $s, t > 0$.

A semigroup $(\mathscr{H}_t : t > 0)$ of experiments is always monotonically increasing for the information ordering, i.e. $\mathscr{H}_s \leqslant \mathscr{H}_t$ when $0 < s \leqslant t$.

Semigroups of experiments provide interesting and accessible examples of replicated experiments, since $\mathscr{H}_t^n \sim \mathscr{H}_{nt}$; $t > 0, n = 1,2,\ldots$ when $(\mathscr{H}_t : t > 0)$ is a semigroup.

In particular

$$\delta(\mathscr{H}_t^n, \mathscr{H}_t^{n+a}) = \delta(\mathscr{H}_{nt}, \mathscr{H}_{nt+at}); \qquad t > 0, \quad a > 0, \quad n = 1,2,\ldots$$

where the left hand side quantifies the amount of information gained by increasing the number of independent realizations of $\mathscr{H}_t$ from n to $n + a$.

A semigroup $(\mathscr{H}_t : t > 0)$ may always be represented as $\mathscr{H}_t = \mathscr{H}^t : t > 0$ for a necessarily infinitely divisible experiment $\mathscr{H}$. Here we use the notation $\mathscr{E}^t$ for an experiment, if it exists, which has Hellinger transform $H(\cdot|\mathscr{E})^t$. Indeed by the semigroup property and by monotonicity, $\mathscr{H}_t \sim \mathscr{H}_1^t$; $t > 0$. Conversely if $\mathscr{H}$ is infinitely divisible then $(\mathscr{H}^t : t > 0)$ is a well defined semigroup. See the references given at the end of complement 18 for information on infinitely divisible experiments.

Consider random variables U_t and V_t having densities (with respect to Lebesgue measure) given by

$$x \to \text{constant } x^{t-1} I_{]0,1]}(x) \qquad \text{on} \qquad]0,\infty[$$

and

$$x \to \text{constant} \exp(-tx^2/2) \qquad \text{on} \qquad]-\infty, \infty[,$$

respectively.

Put $X_t = \sigma U_t$ and $Y_t = \theta + V_t$ where $\sigma > 0$ and θ are unknown parameters. Then X_t and Y_t realize, respectively, experiments $\mathscr{E}_t$ and $\mathscr{F}_t = (N(\theta, t^{-1/2})$: $-\infty < \theta < \infty)$ and:

(a) assuming that the variables $(X_t : t > 0)$ are independent and also that the variables $(Y_t : t > 0)$ are independent we find that $X_s \vee X_t$ and $(sY_s + tY_t)/(s + t)$ are sufficient for (X_s, X_t) and (Y_s, Y_t) respectively.

(b) the families $\mathscr{E}_t : t > 0$ and $\mathscr{F}_t : t > 0$ are both semigroups.

(c) if $0 < s \leqslant t$ then $\delta(\mathscr{E}_s, \mathscr{E}_t) = 2(1 - (s/t))(s/t)^{s/(t-s)}$ and

$$\delta(\mathscr{F}_s, \mathscr{F}_t) = \|N(0, s/t) - N(0, 1)\|.$$

(d) as $n \to \infty$ we find that

$$\delta(\mathscr{E}_s^n, \mathscr{E}_s^{n+a}) \sim \frac{2}{e}\frac{a}{n}$$

and

$$\delta(\mathscr{F}_s^n, \mathscr{F}_s^{n+a}) \sim \left(\frac{2}{\pi e}\right)^{1/2} \frac{a}{n}.$$

Here '$\sim$' signifies that the ratio between the two sides $\to 1$ as $n \to \infty$.

Remarks. The deficiencies in (c) may be found by using the method of least favourable prior distributions, described in section 6.5. Actually the least favourable distributions in both cases are one-point distributions.

In section 8.6 we encountered the semigroup of scale experiments derived from the densities $x \to \text{constant}\, x^{t-1}e^{-x}$ on $]0, \infty]$. Still more semigroups may be formed from this semigroup and from the two semigroups given above by using group isomorphisms such as $x \to x^{-1}$, $x \to -x$, $x \to \log x$ and $x \to e^x$.

Comments on the interesting fact that the deficiencies in (d) stabilize at the rate $1/n$, rather than $1/n^{1/2}$ which might be expected, were given in chapters 1 and 6. Among the many interesting contributions to the problem of measuring the information in additional observations, here we shall only draw the reader's attention to Mammen (1986) and the references there. Mammen shows that deficiencies $\delta(\mathscr{E}^n, \mathscr{E}^{n+a})$ are of the order $1/n$ provided the set Θ is finite dimensional according to a dimensionality concept which was introduced in this context by LeCam. This dimensionality concept assigns dimension $\leqslant D$ to Θ if every subset of Θ having diameter δ for the Hellinger distance may be covered by 2^D sets which all have diameter $\leqslant \delta/2$ for this distance.

Complement 32. Inequalities for convex extensions

If $\mathscr{V} = (V_\theta : \theta \in \Theta)$ is an experiment and if $\lambda_i : i \in I$ is a family of probability distributions on Θ then, under regularity conditions, $(\int V_\theta \lambda_i(\mathrm{d}\theta); i \in I)$ is a well defined experiment with parameter set I. By the randomization criterion this experiment is at most as informative as the experiment $(\lambda_i : i \in I)$. In fact if the distributions λ_i; $i \in I$ are all discrete then any experiment $(U_i : i \in I)$ which is at most as informative as $(\lambda_i : i \in I)$ may be expressed in this form for measures V_θ in $L(U_i : i \in I)$.

Let Λ denote the set of prior distributions on Θ having finite support. Let us also for any experiment $\mathscr{V} = (V_\theta : \theta \in \Theta)$ define the *convex extension* $\hat{\mathscr{V}}$ of $\mathscr{V}$ by $\hat{\mathscr{V}} = (V_\lambda : \lambda \in \Lambda)$ where $V_\lambda = \int V_\theta \lambda(\mathrm{d}\theta)$; $\lambda \in \Lambda$.

Using convex extensions we shall now provide bounds for deficiencies for k-decision problems which are in terms of experiments having at most k points in their parameter sets.

Consider two experiments $\mathscr{E} = (P_\theta : \theta \in \Theta)$ and $\mathscr{F} = (Q_\theta : \theta \in \Theta)$ along with their convex extensions $\hat{\mathscr{E}} = (P_\lambda : \lambda \in \Lambda)$ and $\hat{\mathscr{F}} = (Q_\lambda : \lambda \in \Lambda)$. Then, for $k = 1, 2, \ldots, \infty$, we have:

$$\delta_k(\hat{\mathscr{E}}, \hat{\mathscr{F}}) = \delta_k(\mathscr{E}, \mathscr{F}) \leqslant (k-1)\sup\{\delta_k(\hat{\mathscr{E}}|\Lambda', \hat{\mathscr{F}}|\Lambda' : \Lambda' \subseteqq \Lambda; \#\Lambda' = k\} \quad (10.2.3)$$

where '$\leqslant$' may be replaced with '$=$' when $k \leqslant 2$.

It is readily checked that this statement necessarily remains true if deficiencies δ_k are replaced by deficiency distances Δ_k throughout.

(10.2.3) may be argued as follows. Firstly the equality on the left is a direct consequence of corollary 6.3.3. If $k = 2 = \#\Lambda'$ then this yields: $\delta_2(\mathscr{E}, \mathscr{F}) = \delta_2(\hat{\mathscr{E}}, \hat{\mathscr{F}}) \geqslant \delta_2(\hat{\mathscr{E}}|\Lambda', \hat{\mathscr{F}}|\Lambda')$. It remains to prove, for $k = 2, 3, \ldots$, that $\delta_k(\mathscr{E}, \mathscr{F}) \leqslant (k-1)\eta$ where $\eta = \sup\{\delta_k(\hat{\mathscr{E}}|\Lambda', \hat{\mathscr{F}}|\Lambda') : \Lambda' \subseteqq \Lambda;\ \#\Lambda' = k\}$. We may then assume without loss of generality that Θ is finite. Thus we must show that $\int \psi(\mathrm{d}\mathscr{F}) - \int \psi(\mathrm{d}\mathscr{E}) \leqslant \frac{1}{2}(k-1)\eta \sum_\theta [\psi(e^\theta) + \psi(-e^\theta)]$ where $\psi(x) \equiv_x \bigvee_{i=1} \langle a_i, x\rangle$ for vectors $a_1, \ldots, a_k \in \mathbb{R}^\Theta$. As adding a linear functional to ψ does not alter either side of our last inequality we may assume that $\psi(-e^\theta) = -\bigwedge_i a_i(\theta) = 0$; $\theta \in \Theta$.

The desired inequality may then be written

$$\int \psi(\mathrm{d}\mathscr{F}) - \int \psi(\mathrm{d}\mathscr{E}) \leqslant \frac{1}{2}(k-1)\eta \sum_\theta \vee a_i(\theta).$$

Now a_i, for each i, may be expressed as $a_i = c_i \lambda_i$ for the constant $c_i = \sum_\theta a_i(\theta)$ and for a prior probability distribution λ_i. Putting $\Lambda' = \{\lambda_1, \ldots, \lambda_k\}$ and using that $\bigwedge_i a_i(\theta) \equiv_\theta 0$ we find that

$$\int \psi(\mathrm{d}\mathscr{F}) - \int \psi(\mathrm{d}\mathscr{E}) = \int \bigvee_i c_i \,\mathrm{d}Q_{\lambda_i} - \int \bigvee_i c_i P_{\lambda_i} \leqslant \frac{1}{2} \sum_i c_i \delta(\hat{\mathscr{E}}|\Lambda', \hat{\mathscr{F}}|\Lambda')$$

$$\leqslant \frac{1}{2}\eta \sum_\theta \sum_i a_i(\theta) \leqslant \frac{1}{2}\eta \sum_\theta (k-1) \bigvee_i a_i(\theta).$$

Complement 33. Equivalence of ordered experiments

Consider a probability space $(\mathscr{X}, \mathscr{A}, P)$ and a sub σ-algebra $\mathscr{B}$ of $\mathscr{A}$. Let $\bar{\mathscr{B}}$ be the P-closure of $\mathscr{B}$ within $\mathscr{A}$, i.e.

$$\bar{\mathscr{B}} = \{B : B \in \mathscr{A} \text{ and } P(B \,\Delta\, B_0) = 0 \quad \text{for some} \quad B_0 \in \mathscr{B}\}.$$

(a) Let ϕ be a strictly convex (concave) function on an interval I of the real line. If X is an I-valued random variable on $(\mathscr{X}, \mathscr{A})$ and if $E|X| < \infty$ then X is $\bar{\mathscr{B}}$-measurable provided $E\phi(X) = E\phi(E(X|\mathscr{B}))$.

(b) A real valued integrable random variable X on $(\mathscr{X}, \mathscr{A})$ is $\bar{\mathscr{B}}$-measurable if and only if X has the same distribution as $E(X|\mathscr{B})$.

(c) Let ϕ be strictly convex (concave) on $[0, 1]$ and assume that $\mathscr{E} = (P_\theta : \theta \in \Theta) \leqslant \mathscr{F} = (Q_\theta : \theta \in \Theta)$.

Then $\mathscr{E} \sim \mathscr{F}$ if and only if

$$\int \phi(\mathrm{d}P_{\theta_2}/\mathrm{d}(P_{\theta_1} + P_{\theta_2}))\,\mathrm{d}(P_{\theta_1} + P_{\theta_2})$$

$$= \int \phi(\mathrm{d}Q_{\theta_2}/\mathrm{d}(Q_{\theta_1} + Q_{\theta_2}))\,\mathrm{d}(Q_{\theta_1} + Q_{\theta_2})$$

when $\theta_1, \theta_2 \in \Theta$.

Remarks. By Jensen's inequality (see chapter 2) $E(\phi(X)|\mathscr{B}) \leqslant \phi(E(X|\mathscr{B}))$ a.c. The equality in (a) states that both sides of this inequality have the same expectation.

If ϕ is the convex function $x \to |x - \frac{1}{2}|$ then the equalities in (c) may be written

$$\|P_{\theta_1} - P_{\theta_2}\| = \|Q_{\theta_1} - Q_{\theta_2}\|; \qquad \theta_1, \theta_2 \in \Theta.$$

The function ϕ is not strictly convex however, and in fact these equalities may all hold for ordered non-equivalent (dichotomies) experiments $\mathscr{E}$ and $\mathscr{F}$.

Complement 34. Monotone nets of experiments

By weak compactness any monotone net of experiments is weakly convergent.

Consider now an experiment $\mathscr{E} = (\mathscr{X}, \mathscr{A}; P_\theta : \theta \in \Theta)$ where we have modified the usual assumptions by only requiring that $\mathscr{A}$ is an algebra of subsets of $\mathscr{X}$ and that the set functions $P_\theta : \theta \in \Theta$ are finitely additive probability set functions. The P_θ's are then all in the L-space of bounded additive set functions on $\mathscr{A}$. (This L-space is the dual of the M-normed space, for the supremum norm, of step-functions $\sum_{i=1}^r c_i I_{A_i}$ where $c_1, \ldots, c_r \in \mathbb{R}$ and $A_1, \ldots, A_r \in \mathscr{A}$.) It follows that $\mathscr{E}$ is a well defined experiment.

Let $\hat{\mathscr{E}} = (\hat{P}_\theta : \theta \in \Theta)$ denote the standard experiment of $\mathscr{E}$ when Θ is finite.

(a) Assume that the algebra $\mathscr{A}$ is the union of a collection κ of subalgebras $\mathscr{B}$ of $\mathscr{A}$. Assume also that κ is directed for inclusion. Then

$$\mathscr{E} = \lim \mathscr{E}|\mathscr{B} = \sup \mathscr{E}|\mathscr{B}$$

where $\mathscr{B}$ runs through κ.

In particular this is so if $\mathscr{B}$ runs through all finite subalgebras of $\mathscr{A}$. This may be verified by considering the statements below.

(b) Assume that Θ if finite and that $\psi = \max\{l_i : 1 \leqslant i \leqslant r\}$ where $l_1, \ldots, l_r$ are linear functionals. Let $K = \{x : x \in \mathbb{R}^\Theta, x \geqslant 0, \sum_\theta x_\theta = 1\}$ denote the probability simplex in $\mathbb{R}^\Theta$. Then

$$\int \psi(\mathrm{d}\mathscr{E}) = \sup_g \sum_{i=1}^r \int g_i \,\mathrm{d}l_i(\hat{P}_\theta : \theta \in \Theta) = \sup_g \sum_{i=1}^r \psi\left(\int g_i \,\mathrm{d}\hat{P}_\theta : \theta \in \Theta\right)$$

$$= \sup_\pi \sum_{i=1}^r l_i(P_\theta(A_i) : \theta \in \Theta) = \sup_\pi \psi(P_\theta(A_i) : \theta \in \Theta)$$

where $g = (g_1, \ldots, g_r)$ runs through the set of r-tuples $(g_1, \ldots, g_r)$ of non-negative continuous functions of K such that $g_1 + \cdots + g_r = 1$ and where $\pi = (A_1, \ldots, A_r)$ runs through the set of measurable r-partitions of $\mathscr{X}$.

Here a partition $(A_1, \ldots, A_r)$ of $\mathscr{A}$, whether $\mathscr{A}$ is a σ-algebra or not, is called measurable provided the sets $A_1, \ldots, A_r$ all belong to $\mathscr{A}$.

(c) If Θ is finite and if ϕ is a convex continuous function on the likelihood space which depends on a finite number of coordinates only, then then

$$\int \phi(\mathrm{d}P_\theta : \theta \in \Theta) = \sup_g \sum_i \phi\left(\int g_i \,\mathrm{d}\hat{P}_\theta : \theta \in \Theta\right)$$

$$= \sup_\pi \sum_i \phi(P_\theta(A_i) : \theta \in \Theta).$$

Here g runs through the set of all finite sequences $g = (g_1, \ldots, g_r)$ of non-negative continuous functions on the likelihood space such that $g_1 + \cdots + g_r = 1$ while $\pi = (A_1, \ldots, A_r)$ runs through the set of all finite measurable partitions of $\mathscr{X}$.

(d) If t is a prior distribution on Θ with finite support then the value of the Hellinger transform of $\mathscr{E}$ at t may be expressed as

$$H(t|\mathscr{E}) = \int \prod_\theta \mathrm{d}P_\theta^{t_\theta} = \inf_g \sum_\theta \left(\int g_i \,\mathrm{d}\hat{P}_\theta\right)^{t_\theta}$$

$$= \inf_\pi \sum_i \prod_\theta P_\theta(A_i)^{t_\theta}$$

where g and π are restricted as in (c).

(e) Assume that the probabilities $P_\theta : \theta \in \Theta$ are σ-additive on $\mathscr{A}$.

In this situation let P_θ, by misuse of notation, also denote the unique extension of P_θ to a probability measure on $\sigma(\mathscr{A})$. Then the experiments $\mathscr{E} = (\mathscr{X}, \mathscr{A}; P_\theta : \theta \in \Theta)$ and $\underline{\mathscr{E}} = (\mathscr{X}, \sigma(\mathscr{A}); P_\theta : \theta \in \Theta)$ are equivalent.

(f) Assume again that the algebra $\mathscr{A}$ is the union of a collection κ of subalgebras $\mathscr{B}$ of $\mathscr{A}$ and that this collection is directed for inclusion.

Then for any experiment $\mathscr{G}$ with parameter set Θ and for any integer $k = 1, 2, \ldots$ the deficiency $\delta_k(\mathscr{G}, \mathscr{E})$ may be expressed as

$$\delta_k(\mathscr{G},\mathscr{E}) = \sup_{\mathscr{B}} \delta_k(\mathscr{G},\mathscr{E}|\mathscr{B}).$$

Letting $k \to \infty$ this yields

$$\delta(\mathscr{G},\mathscr{E}) = \sup_{\mathscr{B}} \delta(\mathscr{G},\mathscr{E}|\mathscr{B}).$$

Complement 35. Equicontinuity and deficiency convergence

Consider a sequence $\mathscr{E}_n = (P_{\theta,n} : \theta \in \Theta)$ of experiments having the same parameter set Θ. Assume that $\mathscr{E}_n|\Theta_0 \to (Q_\theta : \theta \in \Theta_0)$ in the weak experiment topology for a subset Θ_0 of Θ which is dense in Θ for a metric d. Then $(\mathscr{E}_n)$ converges in the weak experiment topology provided the sequence is equicontinuous in the sense that to any $\theta \in \Theta$ and any $\varepsilon > 0$ there correspond numbers $b(\theta, \varepsilon)$ and $N(\theta, \varepsilon)$ such that $\|P_{t,n} - P_{\theta,n}\| < \varepsilon$ when $d(t, \theta) < \varepsilon$ and $n \geqslant N(\theta, \varepsilon)$.

Indeed under this provision there is a unique extension $\mathscr{E} = (P_\theta : \theta \in \Theta)$ of $(Q_\theta : \theta \in \Theta_0)$ such that $\mathscr{E}_n \to \mathscr{E}$ in the weak experiment topology.

The experiment $\mathscr{E}$ may be constructed as follows.

Let $\theta \in \Theta$ and consider a sequence $(t(m))$ in Θ_0 converging to θ for the distance d. Choosing $b(\theta, \varepsilon)$ and $N(\theta, \varepsilon)$ as above we see that if $d(t(m'), \theta) < \varepsilon$ and $d(t(m''), \theta) < \varepsilon$ then $\|P_{t(m'),n} - P_{t(m''),n}\| < 2\varepsilon$ when $n \geqslant N(\theta, \varepsilon)$. By weak convergence we see that these requirements on $t(m')$ and $t(m'')$ imply that $\|Q_{t(m')} - Q_{t(m'')}\| < 2\varepsilon$. It follows that $(Q_{t(m)})$ is a Cauchy sequence for the total variation norm. Thus $P_\theta = \lim_m Q_{t(m)}$ exists and it is readily checked that this limit does not depend on how the sequence $t(m)$ was chosen, provided of course that $d(t(m), \theta) \to 0$. In particular $P_\theta = Q_\theta$ when $\theta \in \Theta_0$ so that $\mathscr{E} = (P_\theta : \theta \in \Theta)$ is an extension of $(Q_\theta : \theta \in \Theta_0)$. Using the testing criterion for weak convergence it is now readily checked that $\mathscr{E}_n \to \mathscr{E}$ and also that $\mathscr{E}$ is the unique extension of $(Q_\theta : \theta \in \Theta_0)$ having this property.

Complement 36. Hellinger extensions of experiments

An alternative to the usual notion of a convex extension is as follows.

Let Λ denote the set of prior probability distributions on Θ having finite supports. If $\mathscr{E} = (P_\theta : \theta \in \Theta)$ is an experiment then the usual convex extension of $\mathscr{E}$ is the experiment $\hat{\mathscr{E}} = (P_\lambda : \lambda \in \Lambda)$ where $P_\lambda = \int P_\theta \lambda(\mathrm{d}\theta)$ when $\lambda \in \Lambda$. If $\mathscr{E}$ is regular in the sense that the Hellinger transform of $\mathscr{E}$ never vanishes, then for each $\lambda \in \Lambda$ we can construct the probability measure $\check{P}_\lambda = \prod_\theta \mathrm{d}P_\theta^{\lambda_\theta}/H(\lambda|\mathscr{E})$ where $H(\lambda|\mathscr{E}) = \int \prod_\theta \mathrm{d}P_\theta^{\lambda_\theta}$. If P_θ has density f_θ with respect to some majorizing measure μ when $\lambda_\theta > 0$, then $\check{P}_\lambda$ has density $\prod_\theta f_\theta^{\lambda_\theta} / \int \prod_\theta f_\theta^{\lambda_\theta} \,\mathrm{d}\mu$ with respect to μ.

Clearly $\check{P}_\lambda = P_\theta$ when λ is the one-point distribution in θ.

The *Hellinger extension of* $\mathscr{E}$ is the experiment $\check{\mathscr{E}} = (\check{P}_\lambda : \lambda \in \Lambda)$. Here it is only defined for regular experiments $\mathscr{E}$. Some properties of this extension are:

(a) if $\mathscr{E}$ is regular then $\check{\mathscr{E}}$ is also regular.
(b) if z is a prior distribution on Λ having finite support then

$$H(z|\check{\mathscr{E}}) = H\left(\int \lambda z(\mathrm{d}\lambda)|\mathscr{E}\right)\Big/\prod_{\lambda} H(\lambda|\mathscr{E})^{z(\lambda)}.$$

(c) if $\mathscr{F} = \mathscr{E}_1 \times \cdots \times \mathscr{E}_m$ where the factor experiments $\mathscr{E}_1, \ldots, \mathscr{E}_m$ are regular then $\check{\mathscr{F}} = \check{\mathscr{E}}_1 \times \cdots \times \check{\mathscr{E}}_m$.
(d) if $\mathscr{E}$ is regular and infinitely divisible then $\check{\mathscr{E}}$ is also infinitely divisible.
(e) $\check{\mathscr{E}}$ is Gaussian (see definition in chapter 7) if and only if $\mathscr{E}$ is Gaussian.
(f) $\check{\mathscr{E}}$ is a Poisson experiment (see LeCam (1986) for the definition) if and only if $\mathscr{E}$ is a Poisson experiment.

Using Hellinger extensions it may be shown that if $(\mu_\theta : \theta \in \Theta)$ is a finite family of finite non-negative measures on a common measurable space then

$$\lim_{n\to\infty} \left\| \bigwedge_\theta \mu_\theta^n \right\|^{1/n} = \sup_n \left\| \bigwedge_\theta \mu_\theta^n \right\|^{1/n} = \inf_\Lambda \int \prod_\theta (\mathrm{d}\mu_\theta)^{\lambda_\theta}$$

(see Torgersen, 1981). This generalizes the results described in chapter 1 concerning the Chernoff exponential rate of convergence of replicated dichotomies, (Chernoff, 1952) by providing the asymptotic exponential rate of convergence of performance characteristics of optimal confidence sets having a fixed cardinality.

Complement 37. Comparison for given losses

Most of the general results described in sections 6.2–6.4 pertained to comparison for all loss functions on a given decision space. Thus we considered comparison for k-decision problems for $k = 1, 2, \ldots$ as well as overall comparison. In section 6.5 we considered comparison for invariant loss functions and in chapter 8 we considered comparison for linear estimation problems.

Indeed the deficiency may be expected to be produced by a 'worst case' given by either a single loss function or by a sequence of loss functions. Furthermore the deficiency of one experiment with respect to another may be considered as the deficiency for any class of loss functions which may be judged responsible for that deficiency.

Although it may not be correct to put all the blame on any single loss function, we might nevertheless hope for substantial, even complete, removal of deficiency by restricting our attention to whatever class of loss functions is judged relevant. A very interesting example can be found in Lehmann (1988).

Here we will proceed to the opposite extreme of overall comparison, by restricting comparison to one single general loss function. Results obtainable in this situation are applicable to comparison for any class of loss functions.

In order to ensure the existence of risk we shall assume, besides measurability, that all losses which may be incurred by a given loss function for a given state θ of nature are bounded from below.

Consider then a decision space $(T, \mathscr{S})$ along with a loss function $L = (L_\theta(t) : \theta \in \Theta, t \in T)$ on $\Theta \times T$. We shall follow the set-up of section 4.5. Thus we shall permit ourselves to work freely with generalized decision rules, i.e. transitions from the L-space of $\mathscr{E}$ to the L-space of bounded additive set functions on $\mathscr{S}$. By the notation of that section the risk at θ incurred by the decision rule ρ is denoted as $r(\theta : \rho)$ and this quantity may be expressed as $r(\theta : \rho) = \sup_N P_\theta \rho(L_\theta \wedge N)$. If L_θ is bounded or ρ is σ-continuous then this simplifies as the integral

$$r(\theta : \rho) = \int L_\theta \, \mathrm{d}(P_\theta \rho) = P_\theta \rho L_\theta.$$

Denote the class of prior distributions with finite supports by Λ. For any given prior distribution λ in Λ the minimum Bayes risk is $\inf_\rho \int r(\theta : \rho)\lambda(\mathrm{d}\theta)$ and by the notation in section 6.3, this quantity is denoted as $b_{\mathscr{E}}(\lambda | L)$.

If s is a function on the parameter set such that $s(\theta) \geqslant r(\theta : \rho)$ for all θ for some decision rule ρ then clearly $\sum_\theta s(\theta)\lambda(\theta) \geqslant b_{\mathscr{E}}(\lambda | L)$ for all $\lambda \in \Lambda$. LeCam's fundamental support function characterization of the risk set, theorem 4.5.18, states that this may be turned around. Thus if $\sum_\theta s(\theta)\lambda(\theta) \geqslant b_{\mathscr{E}}(\lambda | L)$ for all $\lambda \in \Lambda$ then there is a decision procedure ρ such that $s(\theta) \geqslant r(\theta : \rho)$ for all θ.

A simple device which may be used to reduce general problems to problems involving only finite decision spaces is inherent in the following.

Note first that if T_0 is any subset of T and $L | T_0$ denotes the restriction of L to $\Theta \times T_0$ then $b_{\mathscr{E}}(\lambda | L) \geqslant b_{\mathscr{E}}(\lambda | (L | T_0))$. (The measurability structure on T_0 here is given by the σ-algebra $T_0 \cap \mathscr{S}$ of subsets of T_0.)

Let Θ_0 be any finite subset of Θ such that $\lambda(\theta) = 0$ when $\theta \notin \Theta_0$. Let $\tilde{L}$ be the restriction of L to $\Theta_0 \times T$. We may then consider the set of functions $(\tilde{L}_{.}(t) : t \in T)$ as a subset of the finite dimensional space of real valued functions on Θ_0. It follows that there is a countable subset T_0 of T such that the subset $(\tilde{L}_{.}(t) : t \in T_0)$ is dense within the larger set $(\tilde{L}_{.}(t) : t \in T)$. Hence $\inf_t \sum_{\theta \in \Theta_0} L_\theta(t) z_\theta = \inf_{t \in T_0} \sum_{\theta \in \Theta_0} L_\theta(t) z_\theta$ whenever $z_\theta \geqslant 0$ when $\theta \in \Theta_0$.

Referring to section 4.5, and in particular to theorem 4.5.17, we conclude that

$$\begin{aligned} b_{\mathscr{E}}(\lambda | L) &= \min_\rho \sum r(\theta : \rho)\lambda(\theta) \\ &= \int \bigwedge_t \sum_\theta \lambda_\theta L_\theta(t) \, \mathrm{d}P_\theta \\ &= b_{\mathscr{E}}(\lambda | (L | T_0)) \end{aligned}$$

where t in $\bigwedge_t$ may be restricted to T_0 as long as $\lambda(\theta) = 0$ when $\theta \notin \Theta_0$. In particular it follows that $b_{\mathscr{E}}(\lambda | L) = \inf_{T_1} b_{\mathscr{E}}(\lambda | (L | T_1))$ where T_1 runs through the class of all finite subsets of T. Now consider two loss functions L and W on $\Theta \times T$. If $\mathscr{E}$ and $\mathscr{F}$ are experiments then we may attempt to compare the decision problems $(\mathscr{E}, L)$ and $(\mathscr{F}, W)$. We may do this by considering deficiencies as we did in chapter 6. Thus if κ is a non-negative function on Θ then we may say that *the decision problem* $(\mathscr{E}, L)$ *is* κ*-deficient with respect to* $(\mathscr{F}, W)$ if to each risk function s in $(\mathscr{F}, W)$ there corresponds a risk function r in $(\mathscr{E}, L)$ such that $r(\theta) \leqslant s(\theta) + \kappa(\theta)$; $\theta \in \Theta$.

(a) Proceeding as in chapter 6 we may introduce deficiencies, deficiency distances, ordering and equivalences for comparison of decision problems.
(b) The decision problem $(\mathscr{E}, L)$ is κ-deficient with respect to the decision problem $(\mathscr{F}, W)$ if and only if $b_{\mathscr{E}}(\lambda | L) \leqslant b_{\mathscr{F}}(\lambda | W) + \sum_\theta \kappa(\theta)\lambda(\theta)$ whenever $\lambda \in \Lambda$.
(c) If the decision problem $(\mathscr{E}_i, L)$; $i = 1, 2, \ldots$, is κ-deficient with respect to the decision problem $(\mathscr{F}_i, W)$ and if $p_1, p_2, \ldots$ are non-negative numbers (mixing probabilities) such that $\sum_i p_i = 1$ then $(\sum_i p_i \mathscr{E}_i, L)$ is $\sum p_i \kappa_i$-deficient with respect to $(\sum_i p_i \mathscr{F}_i, W)$.
(d) If X and Y are independent variables realizing $\mathscr{E}$ and $\mathscr{F}$ respectively and if the prior distribution λ has finite support, then using the usual Bayes set-up, we find for the product experiment $\mathscr{G} = \mathscr{E} \times \mathscr{F}$ that

$$b_{\mathscr{G}}(\lambda | L) = Eb_{\mathscr{E}}(\lambda(\cdot | Y) | L) = Eb_{\mathscr{F}}(\lambda(\cdot | X) | L)$$

where $\lambda(\cdot | X)$ and $\lambda(\cdot | Y)$ are the appropriate posterior distributions.
(e) If the decision problem $(\mathscr{E}_i, L)$; $i = 1, \ldots, r$, is κ_i-deficient with respect to $(\mathscr{F}_i, L)$ then, by (d) and Fubini, $(\prod_{i=1}^r \mathscr{E}_i, L)$ is $\sum_i \kappa_i$-deficient with respect to $(\prod_{i=1}^r \mathscr{F}_i, L)$.

In particular it follows that if $\mathscr{E}_i$; $i = 1, \ldots, r$, is at least as informative as $\mathscr{F}_i$ for the loss function L then the product experiment $\prod_i \mathscr{E}_i$ is at least as informative as $\prod_i \mathscr{F}_i$ for the loss function L.

As an application of these ideas let us compare a monotone likelihood experiment $\mathscr{E}$ (complement 38) with respect to another experiment $\mathscr{F}$ for monotone decision problems (complement 39). In that case the parameter set Θ is the set of real numbers.

By complement 44, comparison in that case reduces to pairwise comparison. By the 'principle' established in (d) we may conclude that if a monotone likelihood experiment $\mathscr{E}$ is pairwise at least as informative as $\mathscr{F}$, then $\mathscr{E}^n$ is at least as informative as $\mathscr{F}^n$ for monotone decision problems. Thus if we e.g. consider the problem of testing the null hypothesis '$\theta \leqslant \theta_0$' against the alternative '$\theta \geqslant \theta_1$' for given quantities θ_0 and θ_1 and for a given level of significance, then maximin power in $\mathscr{F}^n$ is not greater than maximin power in $\mathscr{E}^n$.

Complement 38. Monotone likelihood ratio

It is assumed throughout this complement that the parameter Θ is a set of real numbers.

Consider an experiment $\mathscr{E} = (\mathscr{X}, \mathscr{A}; P_\theta : \theta \in \Theta)$ along with a real valued statistic Z on the sample space $(\mathscr{X}, \mathscr{A})$ of $\mathscr{E}$. We shall say that *$\mathscr{E}$ has monotonically increasing (decreasing) likelihood ratio in Z* if to each pair $(\theta_1, \theta_2) \in \Theta \times \Theta$ such that $\theta_2 > \theta_1$ there corresponds a monotonically increasing (decreasing) function ϕ_{θ_2,θ_1} on $\mathbb{R}$ such that $\phi_{\theta_2,\theta_1}(Z)$ is a P_{θ_2}-maximal version of $dP_{\theta_2}/dP_{\theta_1}$.

Note that maximality implies that $\phi_{\theta_2,\theta_1}(Z) = \infty$ a.s. P_{θ_2} on any P_{θ_1} null set N.

Clearly $\mathscr{E}$ has monotonically increasing likelihood ratio in Z if and only if $\mathscr{E}$ has monotonically decreasing likelihood ratio in $-Z$. Thus in many problems we may restrict our attention to monotonically increasing likelihood ratios.

Consider now the particular case of a totally informative experiment, i.e. an experiment $\mathscr{E} = (P_\theta : \theta \in \Theta)$ such that P_{θ_1} and P_{θ_2} are mutually singular when $\theta_1 \neq \theta_2$. If the sample space $\mathscr{X}$ permits a measurable partitioning A_θ: $\theta \in \Theta$ such that $P_\theta(A_\theta) \equiv 1$ then $\mathscr{E}$ has monotonically increasing likelihood ratio in Z where $Z(x) = \theta$ when $x \in A_\theta$ and $\theta \in \Theta$. Although this construction is feasible whenever Θ is countable, and thus for any restriction to a countable sub-parameter set, there may not be any statistic Z such that $\mathscr{E}$ has monotonically increasing likelihood ratio in Z. This is e.g. the case when $\Theta = [0, 1]$, $P_0 =$ the uniform distribution on $[0, 1]$ and P_θ is the one-point distribution in θ when $\theta > 0$.

Thus we are confronted with a situation where $\mathscr{E}$ does not have monotonically increasing likelihood ratio in any statistic although all restricted experiments $\mathscr{E}|\Theta_0$, with Θ_0 finite, have monotonically increasing likelihood ratio in some statistic. In order to obtain smooth statements, and indeed a smooth theory, it is important that the monotone likelihood ratio property is bestowed on such experiments. We shall therefore say that an experiment $\mathscr{E}$ has *the monotone likelihood ratio ($M - L$) property*, or equivalently that *$\mathscr{E}$ is an $M - L$-experiment*, if for any finite sub-parameter set Θ_0 there is a real valued statistic Z_0 such that the restricted experiment $\mathscr{E}|\Theta_0$ has monotonically increasing (decreasing) likelihood ratio in Z_0.

Later we will see that it suffices to check this for all three-point subsets Θ_0 of Θ. The advantage of introducing concepts in this way is that it often enables us to reduce general problems on statistical experiments to the same problems for experiments having finite parameter sets. In addition the above definition captures the statistically essential properties of monotone likelihood ratio.

Monotone likelihood may be studied conveniently in terms of a partial ordering $\rfloor$ defined on the set of non-negative function Θ, i.e. the set of possible likelihood functions.

This relation is defined by declaring that $v \rfloor w$ if $v_{\theta_2}/v_{\theta_1} \geqslant w_{\theta_2}/w_{\theta_1}$ whenever

$\theta_2 > \theta_1$ and $v_{\theta_1} > 0$. It is then assumed that we put $w_{\theta_2}/w_{\theta_1} = 0$ when $w_{\theta_1} = w_{\theta_2} = 0$.

(a) If v and w are non-negative functions on the likelihood space then each of the following two conditions are equivalent to the relationship $v \,\underline{]}\, w$.

(i) $\det\begin{pmatrix} v_{\theta_2}, & w_{\theta_2} \\ v_{\theta_1}, & w_{\theta_1} \end{pmatrix} \geqslant 0$ when $\theta_2 > \theta_1$.

(ii) $v_{\theta_2}/w_{\theta_2} \geqslant v_{\theta_1}/w_{\theta_1}$ when $\theta_2 > \theta_1$, $v_{\theta_1} > 0$ and $w_{\theta_2} > 0$.

(b) The relation $\underline{]}$ is not transitive. Indeed $v \,\underline{]}\, 0 \,\underline{]}\, w$ for any likelihood functions v and w. Disregarding the null function however, this relation is a partial order. Furthermore, equivalence for non-null likelihood function amounts to positive proportionality.

(c) If V is a set of non-null functions which is totally ordered for $\underline{]}$ then so is its pointwise closure.

If Θ is finite and if $\mathscr{E}$ has the $(M - L)$ property then the standard measure of $\mathscr{E}$ is supported by a set which is totally ordered for $\underline{]}$.

(d) Let V be a set of non-negative functions on Θ such that to each θ there corresponds at least one function v in V such that $v(\theta) > 0$. Assume also that V is totally ordered for the relation $\underline{]}$. Then zeros and positive values of a function v in V appear as $0, \ldots, 0, +, \ldots, +, 0, \ldots, 0$ where one or both sequences of zeros may be empty. In other words, if $\theta_1 < \theta < \theta_2$ then $v_\theta > 0$ if $v_{\theta_1} > 0$ and $v_{\theta_2} > 0$; i.e. there are no zeros between positive values.

(e) If $v \,\underline{]}\, w$ for non-negative functions v and w on Θ and if both functions respect the sign rule described in (d) then, passing from v to w, the end points of intervals of positivity are either not moved or are pushed towards the right.

(f) Assume that $\Theta = \{1, \ldots, m\}$, that V is totally ordered for $\underline{]}$ and that the null function is not in V.

We may then decompose V as $V = \bigcup_{i=1}^m V_i$ where

$$V_i = \{v : v(\theta) = 0 \text{ when } \theta < i \text{ while } v(i) > 0\}.$$

Furthermore each set V_i may be decomposed as $V_i = \bigcup_{j=i}^m V_{i,j}$ where

$$\begin{aligned} V_{i,j} = \{v : v \in V,\ & v(\theta) = 0 \text{ when } \theta < i, \\ & v(\theta) > 0 \text{ when } i \leqslant \theta \leqslant j, \\ & v(\theta) = 0 \text{ when } \theta > j\}. \end{aligned}$$

Note that if $v, w \in V_{i,j}$ and $w_j/w_i = v_j/v_i$ then v and w are positively proportional.

There exists a measurable real valued function T on V which is strictly increasing in the sense that if $v, w \in V$ and $v \,\underline{[}\, w$ then $T(v) \leqslant T(w)$ with equality if and only if v and w are positively proportional.

(g) Let Θ and V be as in (f). Let T be any real valued function on V which is strictly increasing in the sense described in (f). Then T is measurable and, furthermore, if $\theta_1 < \theta_2$ then there is a monotonically increasing function ϕ_{θ_1,θ_2} such that $v_{\theta_2}/v_{\theta_1} = \phi_{\theta_1,\theta_2}(T(v))$ whenever $v \in V$ and $v_{\theta_1} + v_{\theta_2} > 0$.

(h) Monotone likelihood ratio is a property of type, i.e. $\mathscr{E}$ is an $(M - L)$-experiment if $\mathscr{E} \sim \mathscr{F}$ where $\mathscr{F}$ is an $(M - L)$-experiment. (By complement 43, it suffices to require that $\mathscr{E}$ and $\mathscr{F}$ are pairwise equivalent.)

(i) $\mathscr{E}$ has the $(M - L)$ property provided $\mathscr{E}|\Theta'$ has the $(M - L)$ property for all three- or four-point subsets Θ' of Θ.

(j) Assume that $\Theta = \{1, \ldots, m\}$ and that v and w are non-negative functions on Θ which both respect the sign rule described in (d). Assume also that $(v_i, v_{i+1}) \underline{]} (w_i, w_{i+1})$; $i = 1, \ldots, m - 1$. Then either $v \underline{]} w$ or there is an index $t \in [2, m - 1]$ such that v and w may be expressed as

$$v = (v_1, \ldots, v_{t-1}, 0, \ldots, 0)$$

and

$$w = (0, \ldots 0, w_{t+1} \ldots, w_m).$$

(In the latter case $w \underline{]} v$.)

(k) Assume that Θ is a four-point set, say $\Theta = \{1, 2, 3, 4\}$. Put $\Theta_i = \Theta - \{i\}$; $i = 1, 2, 3, 4$. Let $\mathscr{E} = (P_\theta : \theta \in \Theta)$ be an experiment such that the restricted experiments $\mathscr{E}_i = (P_\theta : \theta \in \Theta_i)$ all have the $(M - L)$ property. Let W_i be a totally ordered support of the standard measure of $\mathscr{E}_i$. Let V be the set of non-negative functions v on Θ such that $v|\Theta_i \in W_i$; $i = 1, 2, 3, 4$. Then V is totally ordered and supports the standard measure of $\mathscr{E}$.

(l) $\mathscr{E} = (P_\theta : \theta \in \Theta)$ has the $(M - L)$ property if and only if $\mathscr{E}|\Theta'$ has the $(M - L)$ property for all three-point subsets Θ' of Θ.

Complement 39↑. Monotone loss functions

Assume that the parameter set Θ is a set of real numbers and that the decision space T is a Borel subset of the real line (being equipped with the relativized measurability structure). Then a loss function $L = (L_\theta(t) : \theta \in \Theta, t \in T)$ is called *monotone* if:

(i) to each $\theta \in \Theta$ there corresponds a decision $\tau(\theta) \in T$ such that $L_\theta(t)$ is monotonically decreasing or increasing in t as $t \leqslant \tau(\theta)$ or $t \geqslant \tau(\theta)$.

(ii) τ is monotonically increasing.

(iii) $\tau[\Theta] = T$, i.e. any decision is optimal for some θ.

Consider e.g. the problem of testing the (null hypotheses) '$\theta \leqslant \theta_0$' against $\theta > \theta_0$ with (0–1) loss, i.e.

$$L_\theta(0) = \begin{cases} 0 & \text{when} \quad \theta \leqslant \theta_0; \\ 1 & \text{when} \quad \theta > \theta_0; \end{cases}$$

while

$$L_\theta(1) = \begin{cases} 1 & \text{when} \quad \theta \leqslant \theta_0; \\ 0 & \text{when} \quad \theta > \theta_0. \end{cases}$$

This loss function is monotone, with $\tau(\theta) = 0$ or $= 1$ as $\theta \leqslant \theta_0$ or $\theta > \theta_0$.

Another example is the problem of estimating θ by a loss function of the form $L_\theta(t) \equiv_{t,\theta} h(t - \theta)$ where h is monotonically decreasing on $]-\infty, 0]$ and monotonically increasing on $[0, \infty[$. In that case $\tau(\theta) \equiv_\theta \theta$.

If conditions (i) and (iii) hold while in (ii) τ is monotonically decreasing, then we may replace Θ with $\tilde{\Theta} = -\Theta$ and L with $\tilde{L}$ defined by $\tilde{L}_{\tilde{\theta}}(t) = L_{-\tilde{\theta}}(t)$ when $\tilde{\theta} \in \tilde{\Theta}$ and $t \in T$. Then $\tilde{L}$ is monotone with τ replaced by $\tilde{\tau}$ given by $\tilde{\tau}(\tilde{\theta}) = \tau(-\tilde{\theta})$; $\tilde{\theta} \in \tilde{\Theta}$.

This applies in particular to the problem of testing a null hypothesis '$\theta \geqslant \theta_0$' against the alternative '$\theta < \theta_0$'. Alternatively we may let 1 and 0 represent, respectively, non-rejection of null hypothesis and rejection of null hypothesis.

The problem of testing '$\theta = \theta_0$' against the alternative '$\theta \neq \theta_0$' with (0–1) loss is not monotone. In this case monotonicity cannot be achieved by renaming parameters or decisions.

If L is monotone and the decision space T is finite, say $T = \{t_1, \ldots, t_k\}$ with $t_1 < \cdots < t_k$, then we may decompose Θ as $\Theta = \Theta_1 \cup \cdots \cup \Theta_k$ where $\Theta_i = \{\theta : \tau(\theta) = t_i\}$. Then $\Theta_1, \ldots, \Theta_k$ are disjoint intervals in Θ and $\theta_1 < \cdots < \theta_k$ whenever $\theta_i \in \Theta_i$; $i = 1, \ldots, k$. If, in addition, Θ is finite, then there are parameter points $\theta_1 < \theta_2 < \cdots < \theta_{k-1}$ such that

$$\Theta_1 = \{\theta : \theta \in \Theta \text{ and } \theta \leqslant \theta_1\}, \Theta_2 = \{\theta : \theta \in \Theta \text{ and } \theta_1 < \theta \leqslant \theta_2\}, \ldots,$$
$$\Theta_{k-1} = \{\theta \in \Theta \text{ and } \theta_{k-2} < \theta \leqslant \theta_{k-1}\}, \Theta_k = \{\theta : \theta \in \Theta \text{ and } \theta > \theta_{k-1}\}.$$

A general monotone decision problem may be approximated by a finite monotone decision problem by the following device. Let (T, L) be monotone as described above and consider any non-empty subset T_0 of T and any non-empty subset Θ_0 of Θ. Then there are subsets T_1 of T and Θ_1 of Θ such that $T_0 \subseteqq T_1$, $\Theta_0 \subseteqq \Theta_1$ and $L|\Theta_1 \times T_1$ is monotone. In fact we may put $T_1 = T_0 \cup \tau[\Theta_0]$ and $\Theta_1 = \Theta_0 \cup \{\theta_t : t \in T_1 - \tau[\Theta_0]\}$ where $\theta = \theta_t$, for each t, is a solution of the equation $\tau(\theta) = t$. Then $\tau[\Theta_1] = T_1$ so that $L|\Theta_1 \times T_1$ is monotone.

Note that the sets Θ_1 and T_1 are both finite when the sets Θ_0 and T_0 are finite.

If the experiment $\mathscr{E} = (P_\theta : \theta \in \Theta)$ has monotonically increasing likelihood ratio in a statistic Z and if the loss function is monotone then, by Karlin & Rubin (1956), we may restrict our attention to decision rules δ which are *monotone* in the sense that $\delta([t, \infty[|Z = z) = 1$ whenever $\delta([t, \infty[|Z = z') > 0$ for some $z' < z$.

If δ is non-randomized this amounts to the condition that $\delta(z)$ is monotonically increasing in z. For randomized procedures the above definition of monotonicity amounts to a stronger requirement than the requirement that $\delta(\cdot\,|z)$ increases monotonically in z for the stochastic ordering of distributions.

Complement 40↑. A fine point on the Neyman–Pearson lemma
Let $\mathscr{D} = (P_0, P_1)$ be a dichotomy and consider possible levels $\alpha_1 < \alpha_2 < \alpha_3$ of significance for testing the null hypothesis '$\theta = 0$' against '$\theta = 1$'. Assume that the most powerful tests δ_1 and δ_3 having levels α_1 and α_3 respectively, are chosen so that $\delta_1 \leqslant \delta_3$. Then the most powerful level α_2 test, δ_2, may be chosen so that $\delta_1 \leqslant \delta_2 \leqslant \delta_3$.

Complement 41↑. Monotone likelihood and power diagrams I
In this complement it is assumed that the parameter set Θ is a set of real numbers. Furthermore we admit as a test function any non-negative functional in the M-space of $\mathscr{E}$ which is majorized by the unit of that M-space. This amounts to accepting as a power function any pointwise limit of power functions of usual test functions.

Let $\mathscr{E} = (P_\theta : \theta \in \Theta)$ be an experiment with parameter set Θ. Consider the set $\Pi_{\mathscr{E}}$ of power functions π of test functions such that whenever $\theta_1 < \theta_2$ and $\pi(\theta_1) > 0$ then $\pi(\theta_2)$ is the power of the most powerful level $\pi(\theta_1)$ test for testing '$\theta = \theta_1$' against '$\theta = \theta_2$'. Then $\Pi_{\mathscr{E}}$ is a set of monotonically increasing functions from Θ to $[0, 1]$ having the following additional properties.

(i*) If $\theta_0 \in \Theta$ and $0 < \alpha < 1$ then there is at most one function π in $\Pi_{\mathscr{E}}$ such that $\pi(\theta_0) = \alpha$.
(ii) If $\theta_1 < \theta_2$ then $\pi(\theta_2)$ is a concave function of $\pi(\theta_1)$ on the set $\{\pi : \pi \in \Pi_{\mathscr{E}}, \pi(\theta_1) > 0\}$.
(iii) $\Pi_{\mathscr{E}}$ is closed for pointwise convergence on Θ.

Any set Π of monotonically increasing functions satisfying the first two conditions (i*) and (ii) is totally ordered for the pointwise ordering.

If $\mathscr{E}$ has the monotone likelihood property then (i*) may be strengthened to:

(i) if $\theta_0 \in \Theta$ and $0 < \alpha < 1$ then there is exactly one function π in $\Pi_{\mathscr{E}}$ such that $\pi(\theta_0) = \alpha$.

The set $\Pi_{\mathscr{E}}$ characterizes an $(M - L)$-experiment up to equivalence. This is the main point of this complement. We shall see in the next complement that properties (i) and (ii) together completely characterize the $(M - L)$ property.

Condition (iii) does not amount to much. Indeed if Π is any set of monotonically increasing functions from Θ to $[0, 1]$ satisfying conditions (i) and (ii) then (iii) amounts to the condition that Π contains all indicator functions which are pointwise limits of functions from Π.

Consider any experiment $\mathscr{E}$ having parameter set Θ. We shall argue below that there is a monotone assignment $\pi \to \delta_\pi$ from $\Pi_{\mathscr{E}}$ to the set of test functions such that $\pi(\theta) \equiv_\theta E_\theta \delta_\pi$ when $\pi \in \Pi_{\mathscr{E}}$. Accepting this consider an $(M - L)$ experiment $\mathscr{F} = (Q_\theta : \theta \in \Theta)$ along with another experiment $\mathscr{E} = (P_\theta : \theta \in \Theta)$ such that $\Pi_{\mathscr{E}} = \Pi_{\mathscr{F}}$. Assume that $\mathscr{F}$ has monotonically increasing likelihood ratio in the statistic Z. If z is a constant then the function $\pi_z : \theta \to Q_\theta(Z > z)$ is in $\Pi_{\mathscr{F}}$ and thus, since $\Pi_{\mathscr{E}} = \Pi_{\mathscr{F}}$, this function is also in $\Pi_{\mathscr{E}}$. Hence there is a monotonically decreasing assignment $z \to \delta_z$ from $\mathbb{R}$ to the set of test functions in $\mathscr{E}$ such that $E_\theta \delta_z \equiv_{\theta,z} \pi_z(\theta) \equiv_{\theta,z} Q_\theta(Z > z)$.

Regularizing this assignment we obtain a Markov kernel M from $\mathscr{E}$ to $\mathscr{F}$ such that δ_z, for each z, may be represented as $\delta_z = M(]z,\infty[|\cdot)$. Then $Q_\theta(Z > z) = \int M(]z,\infty[|\cdot)\,\mathrm{d}P_\theta = (P_\theta M)(]z,\infty[)$ when $z \in \mathbb{R}$ and $\theta \in \Theta$. Hence $Q_\theta \equiv_\theta P_\theta M$. Thus $\mathscr{F} \leqslant \mathscr{E}$. On the other hand $\mathscr{F}$ and $\mathscr{E}$ are pairwise equivalent. Hence, by theorem 7.2.7 $\mathscr{E}$ and $\mathscr{F}$ are equivalent. In particular, by complement 38, $\mathscr{E}$ is also an $(M - L)$-experiment. We have proved:

if $\mathscr{E}$ and $\mathscr{F}$ are experiments such that $\Pi_{\mathscr{E}} \supseteq \Pi_{\mathscr{F}}$ and if $\mathscr{F}$ has

the $(M - L)$ property then $\mathscr{E}$ and $\mathscr{F}$ are equivalent.

In complement 43, we will see that the requirement $\Pi_{\mathscr{E}} \supseteq \Pi_{\mathscr{F}}$ may be replaced by the condition that the experiments $\mathscr{E}$ and $\mathscr{F}$ are pairwise equivalent.

Consider any experiment $\mathscr{E}$ whose parameter set Θ is a set of real numbers. Then to any $\pi \in \Pi_{\mathscr{E}}$ there corresponds a test δ_π such that $E_\theta \delta_\pi \equiv \pi(\theta)$. The fact that the assignment $\pi \to \delta_\pi$ may be chosen monotone may be argued as follows.

(a) If $\pi \in \Pi_{\mathscr{E}}$ and δ is a test such that $E_{\theta_0}\delta \leqslant \pi(\theta_0)$ where $\pi(\theta_0) > 0$, then $E_\theta \delta \leqslant \pi(\theta)$ when $\theta \leqslant \theta_0$.

If $\pi \in \Pi_{\mathscr{E}}$ and δ is a test such that $E_{\theta_1}\delta \geqslant \pi(\theta_1)$ where $\pi(\theta_1) < 1$ then $E_\theta \delta \geqslant \pi(\theta)$ whenever $\theta \leqslant \theta_1$.

Amalgamating these 'principles' we conclude that if $\pi \in \Pi$ and if $0 < \pi(\theta_0) \leqslant \pi(\theta_1) < 1$ and if δ is a test such that $E_{\theta_i}\delta_i = \pi(\theta_i)$; $i = 1, 2$ then $E_\theta \delta = \pi(\theta)$ whenever $\theta_0 \leqslant \theta \leqslant \theta_1$.

(b) Assume that $\mathscr{E}$ is homogeneous and that either $\theta_0 \leqslant \theta \leqslant \theta_1$ for all $\theta \in \Theta$ or that $\mathscr{L}(\mathrm{d}P_b|\mathrm{d}P_a|\theta = a)$ is non-atomic for any pair (a, b) of points in Θ. In the latter case let (θ_0, θ_1) be any pair in Θ such that $\theta_0 < \theta_1$. In both cases a power function π in $\Pi_{\mathscr{E}}$ is the power function of the unique most powerful level $\pi(\theta_0)$ test δ_π for testing '$\theta \leqslant \theta_0$' against '$\theta \leqslant \theta_1$' which is based on $\mathrm{d}P_{\theta_1}/\mathrm{d}P_{\theta_0}$.

(c) Assume from now on that $\Theta = \{1, \ldots, m\}$ and for $1 \leqslant i \leqslant j \leqslant m$, let $\Pi^{i,j}$ be the set of power functions π in $\Pi_{\mathscr{E}}$ such that $\pi(\theta) = 0$ when $\theta < i$, $0 < \pi(\theta) < 1$ when $i \leqslant \theta \leqslant j$, and $\pi(\theta) = 1$ when $\theta > j$. Then there is a unique test δ_π having power function π which is measurable with respect to the minimal sufficient σ-algebra.

If $f_\theta = \mathrm{d}P_\theta/\mathrm{d}\Sigma_\theta P_\theta$ and $A = \bigcup\{[f_\theta > 0] : \theta < i\}$, $B = \bigcup\{[f_\theta > 0] :$

$\theta \geqslant j\}$ and $C = [A \cup B]^c$ then there are constants $k \in [0, \infty]$ and $\gamma \in [0, 1]$ such that

$$\delta_\pi = \begin{cases} 0 & \text{on} \quad A \cup \{[\mathrm{d}P_i/\mathrm{d}P_j > k] \cap C\} \\ \gamma & \text{on} \quad [\mathrm{d}P_i/\mathrm{d}P_j = k] \cap C \\ 1 & \text{on} \quad B \cup \{[\mathrm{d}P_i/\mathrm{d}P_j < k] \cap C\}. \end{cases}$$

The assignment $\pi \to \delta_\pi$ is then monotone on $\Pi^{i,j}$.

(d) Let Π denote the set of non-indicator power functions in $\Pi_{\mathscr{E}}$. Then Π may be decomposed as $\Pi = \bigcup_{i=1}^m \Pi^i$ where Π^i is the set of power functions π in Π such that $\pi(\theta) = 0$ when $\theta < i$ while $0 < \pi(i) < 1$. Each set Π^i may be further decomposed as $\Pi^i = \bigcup_{j=i}^m \Pi^{i,j}$.

These decompositions are all monotonically decreasing in the sense that $\pi_1 \geqslant \pi_2 \geqslant \cdots \geqslant \pi_m$ and $\sigma_i \geqslant \sigma_{i+1} \geqslant \cdots \geqslant \sigma_m$ whenever $\pi_i \in \Pi^i$; $i = 1, \ldots, m$ and $\sigma_j \in \Pi^{i,j}$ when $j \geqslant i$.

(e) Let ϕ_i be the indicator function of the event $[f_\theta = 0 \text{ when } \theta < i]$. Then $1 = \phi_1 \geqslant \phi_2 \geqslant \cdots \geqslant \phi_m \geqslant \phi_{m+1} = 0$. If δ is a test with power function in Π^i then $\phi_i \geqslant \delta \geqslant \phi_{i+1}$.

Let ψ_j be the indicator function of the event $\{f_\theta < 0 \text{ for some } \theta \geqslant j\}$. Then $1 = \psi_1 \geqslant \cdots \geqslant \psi_m$. Finally put $\phi_{i,j} = \phi_i \cdot \phi_j$ so that $\phi_i = \phi_{i,i} \geqslant \phi_{i,i+1} \geqslant \cdots \geqslant \phi_{i,m} \geqslant \phi_{i,m+1}$.

If δ is a test with power function in $\Pi^{i,j}$ then $\phi_{i,j} \geqslant \delta \geqslant \phi_{i,j+1}$.

(f) In (c) we provided a monotone map $\pi \to \delta_\pi$ for each set $\Pi^{i,j}$. Piecing these maps together, using (d) and (e), we obtain a monotone assignment $\pi \to \delta_\pi$ such that $\pi(\theta) = E_\theta \delta_\pi$ when $\pi \in \Pi$ and $\theta \in \Theta$. For a given power function $\pi \in \Pi$ the test (function) δ_π is the unique test having power function π and which is measurable with respect to the minimal sufficient σ-algebra. Furthermore any such test δ_π assumes at most one other value besides 0 and 1.

Complement 42↑. Monotone likelihood and power diagrams II

We saw in complement 41 that types of $(M - L)$-experiments and sets $\Pi_{\mathscr{E}}$ of power functions are the same thing. The question then naturally arises: what sets Π of monotonically increasing functions from Θ to $[0, 1]$ are of the form $\Pi_{\mathscr{E}}$ for some $(M - L)$-experiment $\mathscr{E}$? We have seen that any such set $\Pi = \Pi_{\mathscr{E}}$ satisfies the following conditions.

(i) If $\theta_0 \in \Theta$ and $0 < \alpha < 1$ then there is a unique $\pi \in \Pi_{\mathscr{E}}$ such that $\pi(\theta_0) = \alpha$.
(ii) If $\theta_1 < \theta_2$ then $\pi(\theta_2)$ is a concave function of $\pi(\theta_1)$ as long as $\pi(\theta_1) > 0$.
(iii) Π is closed for pointwise convergence.

We shall now see that if Π is a set of monotonically increasing functions satisfying (i)–(iii) then $\Pi = \Pi_{\mathscr{E}}$ for some $(M - L)$-experiment $\mathscr{E}$. Consider first the case where the only indicator functions in Π are the constant functions 0 and 1. (This amounts to requiring that the constructed experiment $\mathscr{E}$ should be homogeneous.) Choose a point $\theta_0 \in \Theta$ and for each number $\alpha \in]0, 1[$, let $\pi = \pi(\cdot|\theta_0, \alpha)$ be the unique function π in Π such that $\pi(\theta_0) = \alpha$. Put $F_\theta(\alpha) = \pi(\theta|\theta_0, \alpha)$. Then F_θ is a continuous distribution function on $[0, 1]$. In this case our assertion is verified by checking that $\mathscr{F} = (F_\theta : \theta \in \Theta)$ is an $(M - L)$-experiment such that $\Pi_{\mathscr{F}} = \Pi$. Indeed $\mathscr{F}$ has monotonically decreasing likelihood ratio in the identity function $T(\alpha) \equiv \alpha$.

Consider a general finite parameter set Θ, say $\Theta = \{1, \ldots, m\}$. Let Π satisfy conditions (i)–(iii). For each pair $(i, i + 1)$; $i = 1, \ldots, m - 1$ there is then a β-function β_i such that $\pi(i + 1) = \beta_i(\pi(i))$ when $\pi(i) > 0$. Put $\mathscr{F} = (F_1, \ldots, F_m)$ where $F_1, \ldots, F_m$ are defined recursively by letting F_1 be the uniform distribution on $[0, 1]$ and by letting F_{i+1} have distribution function given by $F_{i+1}(x) = \beta_i(F_i(x))$; $x \geqslant -2(i - 1)$ while $F_{i+1}(x) = \frac{1}{2}\beta_i(0)(x + 2i)$; $-2i \leqslant x \leqslant -2(i - 1)$. Then $\mathscr{F}$ has monotonically decreasing likelihood ratio in $T(x) \equiv x$ and $\Pi_{\mathscr{F}} = \Pi$.

Finally consider the case of a general subset Θ of $\mathbb{R}$ and a set Π of functions satisfying conditions (i)–(iii). If Θ_0 is a finite non-empty subset of Θ then there is an experiment $\mathscr{F}(\Theta_0)$ producing $\Pi|\Theta_0$. Extending $\mathscr{F}(\Theta_0)$ to an experiment $\mathscr{F}'(\Theta_0)$ having parameter set Θ we obtain the desired $(M - L)$-experiment as a 'cluster point' for the net $(\mathscr{F}'(\Theta_0))$ for the weak experiment topology.

Complement 43. Continuation of complement 41. Monotone likelihood and pairwise equivalence

Monotone likelihood is a triplewise property in the sense that whether or not an experiment is an $(M - L)$-experiment is determined by the restrictions to subsets of the parameter set having three points. On the other hand, since dichotomies are $(M - L)$-experiments, monotone likelihood is not a pairwise property.

Notwithstanding these facts, the question of whether or not an experiment is equivalent to an $(M - L)$-experiment depends solely on pairwise behaviour. We shall see below that if an experiment $\mathscr{E}$ is pairwise equivalent to an $(M - L)$-experiment $\mathscr{F}$ then $\mathscr{E}$ and $\mathscr{F}$ are actually equivalent.

We saw in complement 39 that this was so provided $\mathscr{E}$ and $\mathscr{F}$ were triplewise equivalent. It follows that we may assume without loss of generality that Θ is a three-point set, say $\Theta = \{1, 2, 3\}$, and thus we may describe the pairwise equivalent experiments $\mathscr{E}$ and $\mathscr{F}$ as $\mathscr{E} = (P_1, P_2, P_3)$ and $\mathscr{F} = (Q_1, Q_2, Q_3)$. It suffices to argue that $\Pi_{\mathscr{F}} \subseteq \Pi_{\mathscr{E}}$ since then, by the $(M - L)$ property of $\mathscr{F}$, $\Pi_{\mathscr{F}} = \Pi_{\mathscr{E}}$ and thus, by complement 41, $\mathscr{E}$ and $\mathscr{F}$ are equivalent.

Say that a power function π in $\Pi_{\mathscr{F}}$ is representable if it is the power function of a test in $\mathscr{E}$. By pairwise equivalence, it suffices to argue that any given power function π in $\Pi_{\mathscr{F}}$ is representable.

(a) If π is one of the (at most four) possible monotonically increasing indicator functions on Θ then, by pairwise equivalence, π is representable.

(b) If $0 < \pi(1)$ and $\pi(3) < 1$ then, by pairwise equivalence and by optimality, $E_\theta \delta \equiv_\theta \pi(\theta)$ for any most powerful level $\pi(1)$ test in $\mathscr{E}$ for testing $\theta = 1$ against $\theta = 3$.

(c) If $\pi(1) = 0$ and $\pi(3) = 1$ then Q_1 and Q_2 are mutually singular. Hence, by pairwise equivalence, P_1 and P_3 are also mutually singular. Put $\mu = \Sigma_\theta P_\theta$, $f_\theta = dP_\theta/d\mu$; $\theta = 1, 2, 3$ and $\delta = (1 - \lambda)I_{f_3 > 0} + \lambda I_{f_1 = 0}$. Then $E_\theta \delta = \pi(\theta)$; $\theta = 1, 3$. Furthermore

$$P_2(f_1 = 0) \geqslant \int_{f_1 = 0} \psi f_2 \, d\mu = \int \psi f_2 \, d\mu = E_2 \psi = \pi(2)$$

for any test function ψ in $\mathscr{E}$ such that $E_\theta \psi = \pi(\theta)$; $\theta = 1, 2$. (Then $\psi = 0$ a.e. μ when $f_1 > 0$.) Similarly

$$P_3(f_3 > 0) = \int_{f_3 > 0} f_2 \, d\mu = \int_{f_3 > 0} \phi f_2 \, d\mu \leqslant E_2 \phi = \pi(2)$$

for any test function ϕ in $\mathscr{E}$ such that $E_\theta \phi = \pi(\theta)$; $\theta = 2, 3$. (Then $\phi = 1$ a.e. μ when $f_3 > 0$.)

(d) If $0 = \pi(1)$ and $\pi(3) < 1$ then, for each $n \geqslant 2$, there is a power function π_n in $\Pi_{\mathscr{F}}$ such that $\pi_n(3) = 1/n$. Then $\pi_n \downarrow \pi^* \in \Pi_{\mathscr{F}}$ and clearly $\pi^* \geqslant \pi$ while $\pi^*(1) = \pi(1) = 0$. If $\pi^*(3) = 1$ then π^* is representable by (c).

If $\pi^*(3) < 1$ then $\pi_n(3) < 1$ for n sufficiently large and then by (b), π_n is representable. By weak compactness π^* is also representable in this case. Thus, in any case, $E_\theta \delta^* \equiv_\theta \pi^*(\theta)$ for a test function δ^*. The inequalities $\pi^*(2) \geqslant \pi(2)$ and $\pi^*(3) \geqslant \pi(3)$ imply, by complement 40, that there is a test function $\tilde{\delta}$ in $\mathscr{E}$ such that $\tilde{\delta} \leqslant \delta^*$ while $E_\theta \tilde{\delta} = \pi(\theta)$; $\theta = 2, 3$. Then $E_\theta \tilde{\delta} \equiv_\theta \pi(\theta)$ so that π is representable.

(e) The remaining case, $0 < \pi(1) \leqslant \pi(3) = 1$, may be treated similarly by choosing π_n; $n = 2, 3 \ldots$ so that $\pi_n(3) = 1 - (1/n)$.

Complement 44. Comparison of a monotone likelihood experiment with respect to another experiment for monotone decision problems

Consider an $(M - L)$-experiment $\mathscr{E}$ along with another experiment $\mathscr{F}$ and a non-negative function ε on Θ.

It is implicit in the definition of deficiency in chapter 6 that the size of the loss function $L = (L_\theta : \theta \in \Theta)$ for specified θ is expressed by the quantity $\|L_\theta\| = \sup_t |L_\theta(t)|$. When considering monotone decision problems we shall find it convenient to replace $\|L_\theta\|$ by $\|L_\theta\|_*$ defined by

$$\|L_\theta\|_* = \frac{1}{2}\left[\sup_{t_1 < t_2} (L_\theta(t_1) - L(t_2))^+ + \sup_{t_1 > t_2} (L_\theta(t_1) - L(t_2))^+\right].$$

As $\|L_\theta\|_* \leqslant 2\|L_\theta\|$ and $\sup_{t_1,t_2}|L_\theta(t_1) - L_\theta(t_2)| \leqslant 2\|L_\theta\|_*$ this replacement may not be considered revolutionary.

The main aim of this complement is to establish the equivalence of the following four conditions for deficiency.

Condition 1. To each monotone loss function and to each risk function s available in $\mathscr{F}$ there corresponds a risk function r available in $\mathscr{E}$ such that $r(\theta) \leqslant s(\theta) + \varepsilon_\theta \|L_\theta\|_*$ for all θ.

Condition 2. $\mathscr{E}_{\{\theta_1,\theta_2\}}$ is $(\varepsilon_{\theta_1,\theta_2})$-deficient with respect to $\mathscr{F}_{\{\theta_1,\theta_2\}}$ for all pairs (θ_1, θ_2) of points in Θ.

Condition 3. For any $\theta_0 \in \Theta$, $\mathscr{E}$ is $\frac{1}{2}\varepsilon$-deficient with respect to $\mathscr{F}$ for the problem of testing the null hypothesis '$\theta \leqslant \theta_0$' against '$\theta > \theta_0$' with (0–1) loss.

Condition 4. To any power function σ of a test in $\mathscr{F}$ there corresponds a power function ρ of a test in $\mathscr{E}$ such that $\rho(\theta) \leqslant \sigma(\theta) + \frac{1}{2}\varepsilon_\theta$ or $\geqslant \sigma(\theta) - \frac{1}{2}\varepsilon_\theta$ as $\theta \leqslant \theta_0$ or $\theta > \theta_0$.

If these conditions are satisfied then the decision procedure in $\mathscr{E}$ producing the risk function r in condition 1 may be chosen independently of the monotone loss function L and thus only depending on the decision procedure in $\mathscr{F}$ producing the risk function s.

The equivalence of conditions 1–4 may be inferred as follows.

(a) The problem of testing '$\theta \leqslant \theta_0$' against '$\theta > \theta_0$' with (0–1) loss is monotone. Thus condition 1 $\Rightarrow$ condition 3 and, trivially, condition 3 $\Leftrightarrow$ condition 4 $\Rightarrow$ condition 2.

(b) Assume that condition 2 holds and let σ be the power function of a test in $\mathscr{F}$. Let $\theta_0 \in \Theta$. The existence of a power function ρ in $\mathscr{E}$ such that $\rho(\theta) \leqslant \sigma(\theta) + \frac{1}{2}\varepsilon_\theta$ or $\geqslant \sigma(\theta) - \frac{1}{2}\varepsilon_\theta$ as $\theta \leqslant \theta_0$ or $\theta > \theta_0$ may be argued as follows. Firstly we may assume without loss of generality that Θ is finite. If Θ contains at most two points then the claim is trivial. Proceeding by induction let us assume that the claim is true whenever Θ contains at most n points. Let $\Theta = \{0, 1, \ldots, n\}$ be an $(n+1)$-point set and let $\theta_0 = j$; $j = 0, 1, \ldots, n$. Consider first the case where $j = 0$. Deleting $\theta = n - 1$ from Θ we conclude that there is a power function ρ^* in $\Pi_{\mathscr{E}}$ such that $\rho^*(0) \leqslant \sigma(0) + \frac{1}{2}\varepsilon_0$ while $\rho^*(i) \geqslant \sigma(i) - \frac{1}{2}\varepsilon_i$; $i = 1, \ldots, n-2, n$. If $\rho^*(n-1) \geqslant \sigma(n-1) - \frac{1}{2}\varepsilon_{n-1}$ then the claim is proved. Therefore assume that $\rho^*(n-1) < \sigma(n-1) - \frac{1}{2}\varepsilon_{n-1}$. By assumption, i.e. condition 2, there is a ρ in $\Pi_{\mathscr{E}}$ such that $\rho(0) \leqslant \sigma(0) + \frac{1}{2}\varepsilon_0$ while $\rho(n-1) \geqslant \sigma(n-1) - \frac{1}{2}\varepsilon_{n-1} > \rho(n-1)$. As noted in complement 41 the class $\Pi_{\mathscr{E}}$ is totally ordered. It follows that $\rho \geqslant \rho^*$ so that ρ provides the required power function.

Now assume that $\theta_0 = j \geqslant 1$. By the induction hypothesis there is a power function ρ^* in $\Pi_{\mathscr{E}}$ such that $\rho^*(i) \leqslant \sigma(i) + \frac{1}{2}\varepsilon_i$ when $i \leqslant j - 1$, while $\rho^*(i) \geqslant \sigma(i) - \frac{1}{2}\varepsilon_i$ when $i \geqslant j+1$. If $\rho^*(j) \leqslant \sigma(j) + \frac{1}{2}\varepsilon_j$ then the claim is proved. Therefore assume that $\rho^*(j) > \sigma(j) + \frac{1}{2}\varepsilon_j$. The induction hypothesis implies that there is a power function ρ in $\Pi_{\mathscr{E}}$ such that $\rho(j) \leqslant \sigma(j) + \frac{1}{2}\varepsilon_j$ while $\rho(i) \geqslant \sigma(i) - \frac{1}{2}\varepsilon_i$ when $i > j$. The inequality $\rho(j) < \rho^*(j)$ implies that $\rho \leqslant \rho^*$ and thus that ρ satisfies the desired inequalities. Altogether we have so far shown that conditions 2, 3 and 4 are equivalent.

(c) Finally, assume that conditions 2–4 are satisfied. Consider a monotone loss function L on a decision space T. We may assume without loss of generality that the sets Θ and T are both finite.

If $T = \{t_1, t_2, \ldots, t_k\}$ with $t_1 < t_2 < \cdots < t_k$ then there are parameter points $\theta_1 < \theta_2 < \cdots < \theta_{k-1}$ such that

$$L_\theta(t_1) \leqslant L_\theta(t_2) \leqslant \cdots \leqslant L_\theta(t_k); \qquad \theta \leqslant \theta_1,$$

$$L_\theta(t_1) \geqslant \cdots \geqslant L_\theta(t_i) \leqslant \cdots \leqslant L_\theta(t_k); \ \theta_{i-1} < \theta \leqslant \theta_i; \qquad i = 2, \ldots, k-1$$

and

$$L_\theta(t_k) \leqslant L_\theta(t_{k-1}) \leqslant \cdots \leqslant L_\theta(t_1); \ \theta > \theta_{k-1}.$$

If δ is a decision rule in $\mathscr{F} = (Q_\theta : \theta \in \Theta)$ then the risk at θ is $s(\theta) = \sum_t L_\theta(t)(Q_\theta\delta)(t)$. Putting $\delta_i = \delta(\{t_{i+1}, \ldots, t_k\}|\cdot)$; $i = 0, 1, 2, \ldots, k-1$ and $\delta_k = 0$ we find

$$\begin{aligned} s(\theta) &= \sum_{i=1}^{k} L_\theta(t_i)[E_\theta\delta_{i-1} - E_\theta\delta_i] \\ &= \sum_{i=0}^{k-1} L_\theta(t_{i+1})E_\theta\delta_i - \sum_{i=1}^{k} L_\theta(t_i)E_\theta\delta_i \\ &= \sum_{i=0}^{k-1} [L_\theta(t_{i+1}) - L_\theta(t_i)]E_\theta\delta_i \end{aligned}$$

where $L_\theta(t_0) \equiv_\theta 0$.

By assumption, for $i = 1, 2, \ldots, k-1$, there is a power function ρ_i in $\Pi_{\mathscr{E}}$ such that

$$\rho_i(\theta) \leqslant E_\theta\delta_i + \tfrac{1}{2}\varepsilon_\theta \quad \text{when} \quad \theta \leqslant \theta_i$$

while

$$\rho_i(\theta) \geqslant E_\theta\delta_i - \tfrac{1}{2}\varepsilon_\theta \quad \text{when} \quad \theta > \theta_i.$$

Remembering that $\Pi_{\mathscr{E}}$ is totally ordered we may construct $\rho_1, \ldots, \rho_{k-1}$ such that $\rho_1 \geqslant \rho_2 \geqslant \cdots \geqslant \rho_{k-1}$. We may achieve this by replacing ρ_i; $i = 1, \ldots, k-1$, with $\tilde{\rho}_i = \rho_i \vee \cdots \vee \rho_{k-1}$. If $\theta \leqslant \theta_i$ and $k-1 \geqslant j \geqslant i$ then $\theta \leqslant \theta_j$ and $\rho_j(\theta) \leqslant E_\theta\delta_j + \frac{1}{2}\varepsilon_\theta$. Hence $\tilde{\rho}_i(\theta) \leqslant E_\theta\delta_i + \frac{1}{2}\varepsilon_\theta$ when $\theta \leqslant \theta_i$. If $\theta > \theta_i$ then $\tilde{\rho}_i(\theta) \geqslant \rho_i(\theta) \geqslant E_\theta\delta_i + \frac{1}{2}\varepsilon_\theta$. Let us therefore assume that $\rho_0 \geqslant \rho_1 \geqslant \rho_2 \geqslant \cdots \geqslant$

$\rho_{k-1} \geqslant \rho_k$ where $\rho_0 = 1$ and $\rho_k = 0$. Then there are test functions $1 = \phi_0 \geqslant \phi_1 \geqslant \cdots \geqslant \phi_{k-1} \geqslant \phi_k = 0$ such that $E_\theta\phi_i \equiv_\theta \rho_i(\theta)$.

Finally put $\psi_i = \phi_{i-1} - \phi_i$; $i = 1,\ldots,k$. Then $\psi_1,\ldots,\psi_k \geqslant 0$ and $\psi_1 + \cdots + \psi_k = 1$. The test functions $\psi_1,\ldots,\psi_k$ define the decision procedure ψ in $\mathscr{E} = (P_\theta : \theta \in \Theta)$ given by $\psi(t_i|\cdot) = \psi_i$; $i = 1,\ldots,k$. The risk function r of ψ may be expressed as $r(\theta) \equiv_\theta \sum_{i=0}^{k-1}[L_\theta(t_{i+1}) - L_\theta(t_i)]E_\theta\phi_i$. Hence $r(\theta) - s(\theta) = \sum_{i=1}^{k-1}[L_\theta(t_{i+1}) - L_\theta(t_i)](E_\theta\phi_i - E_\theta\delta_i)$. (The 0th term may be disregarded since $E_\theta\phi_0 = E_\theta\delta_0 = 1$.)

Assume that $\theta_{j-1} < \theta \leqslant \theta_j$. If $i < j$ then $L_\theta(t_i) \geqslant L_\theta(t_{i+1})$ and, since $\theta > \theta_i$, $E_\theta\phi_i \geqslant E_\theta\delta_i - \frac{1}{2}\varepsilon_\theta$. Hence

$$\begin{aligned}[L_\theta(t_{i+1}) - L_\theta(t_i)](E_\theta\phi_i - E_\theta\delta_i) &= [L_\theta(t_i) - L_\theta(t_{i+1})](E_\theta\delta_i - E_\theta\phi_i) \\ &\leqslant \tfrac{1}{2}\varepsilon_\theta[L_\theta(t_i) - L_\theta(t_{i+1})]\end{aligned}$$

when $i < j$. If $i \geqslant j$ then $L_\theta(t_{i+1}) \geqslant L_\theta(t_i)$ and, since $\theta \leqslant \theta_i$, $E_\theta\phi_i \leqslant E_\theta\delta + \frac{1}{2}\varepsilon_\theta$. Hence

$$[L_\theta(t_{i+1}) - L_\theta(t_i)](E_\theta\phi_i - E_\theta\delta_i) \leqslant \tfrac{1}{2}\varepsilon_\theta[L_\theta(t_{i+1}) - L_\theta(t_i)]$$

when $i \geqslant j$. It follows that

$$\begin{aligned}r(\theta) - s(\theta) &= \sum_{i=1}^{k-1} = \sum_{i=1}^{j-1} + \sum_{i=j}^{k-1} \\ &\leqslant \tfrac{1}{2}\varepsilon_\theta[(L_\theta(t_i) - L_\theta(t_j)) + (L_\theta(t_k) - L_\theta(t_j))] \\ &\leqslant \varepsilon_\theta\|L_\theta\|_*.\end{aligned}$$

By a similar analysis we conclude that $r(\theta) - s(\theta) \leqslant \varepsilon_\theta\|L_\theta\|_*$ when $\theta \leqslant \theta_1$ or $\theta > \theta_{k-1}$.

The particular case where $\varepsilon_\theta \equiv 0$, i.e. the case of being at least as informative for monotone decision problems, is considered in complement 45.

Complement 45↑. Lehmann's criterion

Assume that Θ is a set of real numbers. By Lehmann (1988) an $(M - L)$-experiment $\mathscr{E}$ is pairwise at least as informative as another $(M - L)$-experiment $\mathscr{F}$ if and only if $\mathscr{E}$ is pairwise at least as informative as $\mathscr{F}$ for monotone decision problems. This is also a consequence of complement 44 with $\varepsilon_\theta \equiv 0$.

Without requiring experiment $\mathscr{F}$ to have any particular properties, we conclude from complement 44 that the following conditions are equivalent for an $(M - L)$-experiment $\mathscr{E}$ and an experiment $\mathscr{F}$.

(i) $\mathscr{E}$ is at least as informative as $\mathscr{F}$ for monotone decision problems.
(ii) For any $\theta_0 \in \Theta$, $\mathscr{E}$ is at least as informative as $\mathscr{F}$ for testing '$\theta \leqslant \theta_0$' against '$\theta > \theta_0$' for (0–1) loss.

(iii) To any power function σ of a test in $\mathscr{F}$ and to any $\theta_0 \in \Theta$ there corresponds a power function ρ of a test in $\mathscr{E}$ such that $\rho(\theta) \leqslant \sigma(\theta)$ or $\rho(\theta) \geqslant \sigma(\theta)$ as $\theta \leqslant \theta_0$ or $\theta \geqslant \theta_0$.
(iv) $\mathscr{E}$ is pairwise at least as informative as $\mathscr{F}$.

Lehmann's criterion applies to experiments $\mathscr{E} = (F_\theta : \theta \in \Theta)$ and $\mathscr{F} = (G_\theta : \theta \in \Theta)$ defined by univariate distributions F_θ and G_θ. We shall provide criteria which ensure that $\mathscr{E}$ is pairwise at least as informative as $\mathscr{F}$.

In all the cases considered here the general situation may be reduced to the case where Θ is a two-point set, say $\{0, 1\}$.

The statements below involve fractiles of distributions. These fractiles are the usual minimal ones with the exception that any 0-fractile should be chosen maximal. We shall also find it convenient to let X denote the identity function on $\mathbb{R}$, i.e. $X(x) \equiv_x x$. Then:

(a) if $\mathscr{E}$ is pairwise at least as informative as $\mathscr{F}$ and if $\mathscr{E}$ has monotone likelihood ratio in X then $F_\theta^{-1}G_\theta(x)$ is monotonically increasing in θ for all x.
(b) if $F_\theta^{-1}G_\theta(x)$ is monotonically increasing in θ for all x then $\mathscr{E}$ is pairwise at least as informative as $\mathscr{F}$ provided $\mathscr{F}$ has monotonically increasing likelihood ratio in X and provided all distributions $F_\theta : \theta \in \Theta$ are non-atomic.
(c) *Lehmann's criterion.* Assume that both experiments $\mathscr{E}$ and $\mathscr{F}$ have monotone likelihood ratio in X and also that the distributions $F_\theta : \theta \in \Theta$ are non-atomic.

 Then $\mathscr{E}$ is pairwise at least as informative as $\mathscr{F}$ if and only if $F_\theta^{-1}G_\theta(x)$ is monotonically decreasing in θ for all x.
(d) even when $\mathscr{E}$ and $\mathscr{F}$ both have monotone likelihood ratio the condition that $F_\theta^{-1}G_\theta(x)$ decreases monotonically in θ for all x does not entail that $\mathscr{E}$ is pairwise at least as informative as $\mathscr{F}$.

Choose e.g. a β-function γ such that $\gamma(0) = 0$ along with numbers p_0 and p_1 such that $0 < p_0 < p_1 < 1$. Let $F_\theta; \theta = 0, 1$ be the distribution assigning masses $1 - p_\theta$ and p_θ to 0 and 1 respectively. Let G_0 be the uniform distribution on $[0, 1]$ and let G_1 be the distribution on $[0, 1]$ given by the distribution function $\alpha \to 1 - \gamma(1 - \alpha)$. Then $F_\theta^{-1}G_\theta(x)$ is monotonically decreasing in θ if and only if the power of any level 'p_0' test for testing '$\theta = 0$' against '$\theta = 1$' in $\mathscr{F}$ is at most p_1.

On the other hand $\mathscr{E} \geqslant \mathscr{F}$ if and only if $\gamma(\alpha) \leqslant (p_1/p_0)\alpha$ when $0 \leqslant \alpha \leqslant p_0$ while $\gamma(\alpha) \leqslant [(1 - p_1)/(1 - p_0)](\alpha - p_0) + p_1$ when $\alpha \geqslant p_0$. However the first condition ensures only that the latter conditions are satisfied for α sufficiently close to p_0.

As a particular case let us consider the location and scale family generated by a continuous distribution function F on the real line. Thus our experiment $\mathscr{F} = \mathscr{E}_\sigma = (F_{\theta,\sigma} : \theta \in \mathbb{R})$ is realized by observing $X = \theta + \sigma U$ where U has distribution F and where the positive scale factor σ is known.

The experiment $\mathscr{E}_\sigma$ has monotonically increasing likelihood ratio in X for all $\sigma > 0$ if and only if this is so for $\sigma = 1$ and this, by complement 47, amounts to the condition that F is strongly unimodal, i.e. that F is absolutely continuous with a log concave density.

Considering the effect of increasing σ as increasing the 'noise' we might suspect that information decreases with increasing σ. By the convolution criterion this amounts to the condition that $\mathscr{L}(X|\sigma = 1)$ is a convolution factor of $\mathscr{L}(X|\sigma)$ when $\sigma > 1$. By Loève (1960) this is just one way of stating that F belongs to the class of self decomposable distributions. The prime examples are of course the normal distributions. An interesting example of a non self decomposable strongly unimodal distribution is the uniform distribution on, say, the unit interval. It is a most curious fact that by Lehmann (1988), in this case X provides less information for a given $\sigma > 0$ than for $\sigma = 1$ if and only if σ is a positive integer.

Considering rejection regions of the form $[w, \infty[$ we infer that

$$\beta(\alpha|\mathscr{L}(X|\theta_1,\sigma), \mathscr{L}(X|\theta_2,\sigma)) \geqslant 1 - F\left(F^{-1}\left(1 - \alpha - \frac{1}{\sigma}(\theta_2 - \theta_1)\right)\right)$$

with equality when F is strongly unimodal and $\theta_2 > \theta_1$. Thus if F is strongly unimodal and σ increases, then information always decreases pairwise and therefore information decreases for monotone decision problems.

This may also be seen directly from Lehmann's criterion since $F_{\theta,\sigma_1}^{-1}(F_{\theta,\sigma_2}(x)) = (1 - (\sigma_1/\sigma_2))x + (\sigma_1/\sigma_2)x$ when $0 < F_{\theta,\sigma_2}(x) < 1$.

Complement 46. Composition of β-functions and monotone likelihood. Pairwise order completeness

If $\mathscr{E} = (P_\theta : \theta \in \Theta)$ is an $(M - L)$-experiment then necessarily

$$\beta(\beta(\alpha|P_{\theta_1}, P_{\theta_2})|P_{\theta_2}, P_{\theta_3}) \underset{\alpha}{\equiv} \beta(\alpha|P_{\theta_1}, P_{\theta_3})$$

whenever $\theta_1 \leqslant \theta_2 \leqslant \theta_3$. Here as usual, for any dichotomy, $\beta(\alpha|P,Q)$ denotes the maximal power among level tests for testing 'P' against 'Q'.

Conversely assume that we are given a family $\{\beta(\cdot|\theta_1,\theta_2) : \theta_1 < \theta_2\}$ of β-functions such that $\beta(\beta(\cdot|\theta_1,\theta_2)|\theta_2,\theta_3) \equiv_\alpha \beta(\alpha|\theta_1,\theta_3)$ whenever $\theta_1 < \theta_2 < \theta_3$. Let Π be the set of functions π from Θ to $[0, 1]$ such that

(i) $\beta(\pi(\theta_1)|\theta_1,\theta_2) = \pi(\theta_2)$ when $\theta_1 < \theta_2$ and $\pi(\theta_1) > 0$

and

(ii) $\pi(\theta) > 0$ whenever there is a $\theta' > \theta$ such that $1 > \pi(\theta') > \beta(0|\theta,\theta')$.

If $\theta_0 \in \Theta$ and $0 < \alpha < 1$ then we may construct a function π in Π such that $\pi(\theta_0) = \alpha$, by putting

$$\pi(\theta) = \begin{cases} \beta(\alpha|\theta_0, \theta) & \text{when } \theta > \theta_0 \\ \alpha & \text{when } \theta = \theta_0 \\ \text{the unique } x \text{ such that } \beta(x|\theta, \theta_0) = \alpha & \text{when } \theta < \theta_0 \\ & \text{and } \beta(0|\theta, \theta_0) \leqslant \alpha \\ 0 & \text{otherwise.} \end{cases}$$

It is readily checked that $\pi \in \Pi$ and that there is no other function in Π mapping θ_0 into α. Furthermore the set Π is closed for pointwise convergence. It follows, by complements 42 and 43, that there is an $(M - L)$-experiment $\mathscr{E} = (P_\theta : \theta \in \Theta)$, which is unique up to equivalence, such that $\Pi_{\mathscr{E}} = \Pi$. Hence $\beta(\alpha|\theta_1, \theta_2) \equiv_\alpha \beta(\alpha|P_{\theta_1}, P_{\theta_2})$ whenever $\theta_1 < \theta_2$.

Using this relationship we may often reduce general problems on $(M - L)$-experiments to corresponding problems for dichotomies. Some examples of this are:

(a) an experiment $\mathscr{E} = (P_\theta : \theta \in \Theta)$ is an $(M - L)$-experiment if and only if

$$\beta(\beta(\alpha|P_{\theta_1}, P_{\theta_2})|P_{\theta_2}, P_{\theta_3}) \underset{\alpha}{\equiv} \beta(\alpha|P_{\theta_1}, P_{\theta_3}).$$

(b) the collection of $(M - L)$-experiments is closed (and thus compact) for the weak experiment topology. (Remember that, by complement 43, any experiment which is pairwise equivalent to an $(M - L)$-experiment is itself an $(M - L)$-experiment.)

(c) a net $\{\mathscr{E}_n\}$ of $(M - L)$-experiments converges weakly to an experiment $\mathscr{E}$ if and only if $\mathscr{E}_n$ converges pairwise to $\mathscr{E}$. If so then the limit experiment $\mathscr{E}$ is necessarily an $(M - L)$-experiment.

(d) the pairwise supremum of a family $\mathscr{E}^t : t \in T$ of $(M - L)$-experiments may be constructed as follows.

Express $\mathscr{E}^t$ as $\mathscr{E}^t = (P_\theta^t : \theta \in \Theta)$ and put $\bar{\beta}(\cdot|\theta_1, \theta_2) = \sup_t \beta(\cdot|P_{\theta_1}^t, P_{\theta_2}^t)$ where sup should be interpreted for the informational (and not the pointwise) ordering of β-functions.

The family $\{\bar{\beta}(\cdot|\theta_1, \theta_2) : \theta_1 < \theta_2\}$ does *not* in general satisfy the composition rule. We have however obtained something since

$$\bar{\beta}(\bar{\beta}(\alpha|\theta_1, \theta_2)|\theta_2, \theta_3) \geqslant \bar{\beta}(\alpha|\theta_1, \theta_3)$$

whenever $\alpha \in [0, 1]$ and $\theta_1 \leqslant \theta_2 \leqslant \theta_3$.

Starting with the β-functions $\{\bar{\beta}(\cdot|\theta_1, \theta_2); \theta_1 < \theta_2\}$ with each finite subset F of Θ we may associate a β-function $\bar{\beta}_F$ as follows. Arrange the distinct numbers in F in increasing order as $F = \{a_0, a_1, \ldots, a_m\}$ and let $\bar{\beta}_F$ be defined as the composition $\bar{\beta}_F = \beta_{a_{m-1}, a_m} \beta_{a_{m-2}, a_{m-1}} \cdots \beta_{a_0, a_1}$. By the last inequality $\bar{\beta}_F \leqslant \bar{\beta}_G$ whenever $F \subseteq G$. Finally put $\beta^*(\cdot|\theta_1, \theta_2) =$

sup$\{\beta_F$: the smallest number in F is θ_1 and the largest number in F is $\theta_2\}$ when $\theta_1 < \theta_2$. As $\bar{\beta}_F \leqslant \bar{\beta}_G$ when $F \subseteq G$ it does not matter whether this sup is interpreted pointwise or for the information ordering.

If $\theta_1 < \theta_2 < \theta_3$ then the sets F appearing in the definition of $\beta^*(\cdot|\theta_1, \theta_3)$ may be chosen such that they all contain θ_2. It follows readily that the family $\{\beta^*(\cdot|\theta_1, \theta_2): \theta_1 < \theta_2\}$ satisfies the composition rule. Thus there is an $(M-L)$-experiment $\mathscr{E}^* = (P_\theta^*: \theta \in \Theta)$ such that $\beta^*(\alpha|\theta_1, \theta_2) \equiv_\alpha \beta(\alpha|P_{\theta_1}^*, P_{\theta_2}^*)$.

It is straightforward to check that $\mathscr{E}^* \geqslant \mathscr{E}^t$ pairwise for all $t \in T$ and furthermore that $\mathscr{E}^* \leqslant \mathscr{F}$ pairwise for any other $(M-L)$-experiment $\mathscr{F}$ such that $\mathscr{F} \geqslant \mathscr{E}^t$ pairwise for all t.

(e) replacing sup with inf in (d) we obtain the analogous construction of the pairwise infimum of a family of $(M-L)$-experiments.

(f) the collection of types of $(M-L)$-experiments (having the same parameter set Θ) is order complete for the pairwise ordering.

(g) (open problem). Characterize the smallest (largest) $(M-L)$-experiment which pairwise majorizes (minorizes) the Cauchy translation family. Translation families which are $(M-L)$-experiments are discussed in complement 47.

If we replace the composition rule with the multiplication rule $\beta(\cdot|\theta_1, \theta_3) \equiv \beta(\cdot|\theta_1, \theta_3) \otimes \beta(\cdot|\theta_2, \theta_3)$ then we arrive at another interesting class of experiments. (If β_1 and β_2 are β-functions then $\beta_1 \otimes \beta_2$ is the β-function of the product of a dichotomy having β-function β_1 and a dichotomy having β-function β_2.)

In Jansen, Milbrodt & Strasser (1984), Strasser introduces the notion of independent increments for experiments and then shows that an experiment possesses independent increments if and only if it obeys the above multiplication rule.

Complement 47↑. Stationary monotone likelihood and strong unimodality

As explained in complement 46, a monotone likelihood experiment may, up to equivalence, be identified with the families $(\beta(\cdot|\theta_1, \theta_2): \theta < \theta_2)$ of β-functions satisfying the basic composition rule

$$\beta(\cdot|\theta_1, \theta_2) = \beta((\cdot|\theta_1, \theta_2)|\theta_2, \theta_3).$$

The identification is then that if $\mathscr{E} = (P_\theta: \theta \in \Theta)$ is an $(M-L)$-experiment then $\beta(\alpha|P_{\theta_1}, P_{\theta_2}) \equiv_\alpha \beta(\cdot|\theta_1, \theta_2)$ when $\theta_1 < \theta_2$.

(It goes without saying that θ with or without subscript belongs to the parameter set Θ.)

Let us agree to say that $\mathscr{E}$ has stationary monotone likelihood if $\beta(\cdot|P_{\theta_1}, P_{\theta_2})$, for $\theta_1 < \theta_2$, depends on (θ_1, θ_2) only via $\theta_2 - \theta_1$. If so then this is the case for all pairs (θ_1, θ_2) of parameter points.

What are the stationary $(M - L)$ types? In other words, up to equivalence, what does the most general $(M - L)$-experiment having stationary monotone likelihood look like?

Note first that any totally informative experiment as well as any totally non informative experiment possesses stationary monotone likelihood.

Next consider a strongly unimodal distribution function F on the real line. Thus F is either a one-point distribution or it is absolutely continuous with a log concave density. Let F_θ be the distribution of $X + \theta$ when X has distribution F. In both cases the translation experiment $\mathscr{E} = (F_\theta : \theta \in \mathbb{R})$ is an $(M - L)$-experiment with $(N - P)$ functions given by $\beta(\alpha | F_{\theta_1}, F_{\theta_2}) = 1 - F(F^{-1}(1 - \alpha) - (\theta_2 - \theta_1))$ when $0 < \alpha < 1$ and $\theta_1 < \theta_2$. The case where F is a one-point distribution provides a particular version of a totally informative experiment. In any case we infer that the $(M - L)$-experiment $\mathscr{E}$ has stationary monotone likelihood.

We shall now see that there are no other types of stationary monotone likelihood on the real line.

Let $(\beta(\cdot | \theta_1, \theta_2) : -\infty < \theta_1 < \theta_2 < \infty)$ be a family of β-functions satisfying the basic composition rule

$$\beta(\cdot | \theta_1, \theta_3) = \beta(\beta(\cdot | \theta_1, \theta_2) | \theta_2, \theta_3).$$

Assume that this family is stationary, i.e. that when $\theta_1 < \theta_2$, $\beta(\cdot | \theta_1, \theta_2)$ depends on θ_1, θ_2 only via $\theta_2 - \theta_1$. Put $\gamma_\theta = \beta(\cdot | 0, \theta)$ when $\theta > 0$ and let γ_0 be the identity function on $[0, 1]$. Then:

(a) $\gamma_\theta : \theta \geqslant 0$ is a semigroup of β-functions obeying the composition rule

$$\gamma_{\theta_1}(\gamma_{\theta_2}) = \gamma_{\theta_1 + \theta_2}$$

whenever $\theta_1, \theta_2 \geqslant 0$.

The β-functions $\beta(\cdot | \theta_1, \theta_2)$ are recovered from this semigroup by the identity $\beta(\cdot | \theta_1, \theta_2) = \gamma_{\theta_2 - \theta_1}; \theta_1 < \theta_2$.

(b) if $\gamma_\theta = \gamma_0$ for some $\theta > 0$ then $\gamma_{n\theta} = \gamma_\theta \gamma_\theta \cdots \gamma_\theta = \gamma_0$ for all $n = 1, 2, \ldots$ and then $\gamma_\theta = \gamma_0$ for all θ. In this case we have arrived at total non informativity.

(c) if $h \downarrow 0$ then $\gamma_h(\alpha) \downarrow \gamma_{0+}(\alpha) \geqslant \gamma_0(\alpha) = \alpha$. If $\gamma_{0+}(\alpha) > \alpha$ for some $\alpha > 0$ then for any $h > 0$ and $k = h/n$ we find that $\gamma_h(\alpha) = \gamma_k \gamma_k \cdots \gamma_k(\alpha) \geqslant \gamma_{0+} \gamma_{0+} \cdots \gamma_{0+}(\alpha) \geqslant (\gamma_{0+} \otimes \cdots \otimes \gamma_{0+})(\alpha)$ where $\otimes$ indicates products of experiments, or rather the corresponding operation on β-functions. Letting $n \to \infty$, the last term $\to 1$ whenever $0 < \alpha < 1$. Thus $\gamma_h(\alpha) \equiv_\alpha 1$ when $h > 0$. Hence also $\beta(\alpha | \theta_1, \theta_2) \equiv_\alpha 1$ when $\theta_1 < \theta_2$.

It follows that if $\gamma_{0+}(\alpha) > \alpha$ for some $\alpha > 0$ then the family $(\beta(\cdot | \theta_1, \theta_2) : \theta_1 < \theta_2)$ comes from a totally informative experiment. This experiment may be realized as the translation experiment of a one-point distribution on the line, and thus as a strongly unimodal translation family.

(d) It remains to consider the 'regular' case, i.e. the case where

$$\gamma_h(\alpha) > \alpha \quad \text{when} \quad h > 0 \quad \text{and} \quad 0 < \alpha < 1 \tag{10.2.4}$$

and

$$\gamma_h(\alpha) \downarrow \gamma_{0+}(\alpha) \equiv \alpha \quad \text{when} \quad 0 \leqslant \alpha \leqslant 1. \tag{10.2.5}$$

By Dini's lemma, the convergence in (10.2.5) is uniform in α.

Let $\mathscr{E} = (P_\theta : \theta \in \Theta)$ be any experiment 'producing' the family $(\beta(\cdot|\theta_1, \theta_2) : \theta_1 < \theta_2)$. Then $\beta(\alpha|P_{\theta_1}, P_{\theta_2}) \equiv_\alpha \gamma_{\theta_2 - \theta_1}(\alpha)$ when $\theta_1 \leqslant \theta_2$ and thus $\|P_{\theta_2} - P_{\theta_1}\| = 2\sup_\alpha[\gamma_{\theta_2-\theta_1}(\alpha) - \alpha] \to 0$ as $0 \leqslant \theta_2 - \theta_1 \to 0$. It follows that P_θ is uniformly continuous in θ. By example 6.5.13 this shows that if $\mathscr{E}$ is a translation family then the distributions P_θ; $\theta \in \Theta$ are all absolutely continuous.

Define a function F on the real line by choosing a number $\alpha_0 \in]0, 1[$ and then putting

$$F(x) = \begin{cases} 1 - \gamma_{-x}(\alpha_0) & \text{when } x \leqslant 0 \\ 1 - y & \text{where } \gamma_x(y) = \alpha_0 \text{ when } x \geqslant 0 \text{ and } \gamma_x(0) \leqslant \alpha_0 \\ 1 & \text{when } x \geqslant 0 \text{ and } \gamma_x(0) \geqslant \alpha_0. \end{cases}$$

Then F is a continuous cumulative probability distribution function. Furthermore, by the semigroup property, $1 - F(x - h) = \gamma_h(1 - F(x))$ when $h \geqslant 0$ and $F(x) < 1$. It follows that if $h \geqslant 0$ and $F(a) < 1$ then

$$F(a - h) = 1 - \gamma_h(1 - F(a)) = \int_{-\infty}^{a} \gamma_h'(1 - F)\,\mathrm{d}F.$$

Let F_θ, for any $\theta \in \mathbb{R}$, denote the θ-translate of F, i.e. $F_\theta(x) \equiv_x F(x - \theta)$. The last identity may then be written

$$F_\theta(a) = \int_{-\infty}^{a} \gamma_\theta'(1 - F)\,\mathrm{d}F$$

when $F(a) < 1$ and $\theta \geqslant 0$. It follows that $\mathrm{d}F_\theta/\mathrm{d}F$ may be specified as $[\mathrm{d}F_\theta/\mathrm{d}F]_x = \gamma_\theta'(1 - F(x))$ when $F(x) < 1$ while $[\mathrm{d}F_\theta/\mathrm{d}F]_x = \infty$ when $F(x) = 1$.

Let I be the open interval $I = \{x : 0 < F(x) < 1\}$. Then, assuming $\theta \geqslant 0$, F_θ is F-absolutely continuous on I. Both distributions F and F_θ assign mass 0 to $\{x : F(x) = 0\}$. It follows that the above specification of $\mathrm{d}F_\theta/\mathrm{d}F$ is F_θ-maximal.

This, in turn, implies that $\mathscr{E} = (F_\theta : \theta \in \Theta)$ has monotonically increasing likelihood ratio in $T(x) \equiv_x x$.

Testing '$\theta \leqslant 0$' against '$\theta > 0$' at level α we obtain, for the most powerful test, power $\gamma_\theta(\alpha)$ at the alternative $\theta > 0$. Thus $\mathscr{E}$ produces the family $(\beta(\cdot|\theta_1, \theta_2))$ of β-functions. Hence, as noted above, F is necessarily absolutely continuous. If $x - h$, $x \in I$ and $h > 0$ then $1 - F(x - h) = \gamma_h(1 - F(x)) > 1 - F(x)$ so that F is strictly increasing on I.

Specify the density f of F by putting

$$f(x) \underset{x}{\equiv} \limsup_{\delta \to 0} [F(x + \delta) - F(x)]/\delta.$$

If $x - \theta$, $x \in I$ and $\theta \geqslant 0$ then this yields

$$\begin{aligned} f(x - \theta) &= \limsup_{\delta \to 0} \frac{F(x + \delta - \theta) - F(x)}{F(x + \delta) - F(x)} \frac{F(x + \delta) - F(x)}{\delta} \\ &= \gamma_\theta'(1 - F(x)) f(x) \end{aligned}$$

provided $f(x) < \infty$ or $\gamma_\theta'(1 - F(x)) > 0$.

If $x \in I$ and $f(x) = 0$ or $f(x) = \infty$ then for any $y = x - \theta \leqslant x$ such that $y \in I$ and $0 < f(y) < \infty$ we find that $\gamma_\theta'(1 - F(x)) = 0$ and thus $\gamma_\theta'(1 - F(z)) = 0$ when $z \leqslant x$, so that

$$F(y) = F_\theta(x) = \int_{-\infty}^{x} \gamma_\theta'(1 - F) \, \mathrm{d}F = 0.$$

This, however, contradicts the assumption that $F(x) > 0$.

Thus f is finite and positive on I and $f(x - \theta) = \gamma_\theta'(1 - F(x)) f(x)$ whenever $x \in I$ and $\theta \geqslant 0$. It follows that $\log f$ is concave on I and thus that F is strongly unimodal.

Complement 48. Continuation of complement 45. Comparison involving differentiable monotone likelihood

Consider continuously differentiable experiments $\mathscr{E} = (P_\theta : \theta \in \Theta)$ and $\mathscr{F} = (Q_\theta : \theta \in \Theta)$ having an open interval Θ on the real line as their common parameter set. At any given point $\theta_0 \in \Theta$ we may consider the problem of testing the null hypothesis '$\theta = \theta_0$' against the alternative '$\theta > \theta_0$'. Let $\kappa(\alpha | \theta_0, \mathscr{E})$ (respectively, $\kappa(\alpha | \theta_0, \mathscr{F})$) be the maximal slope at $\theta = \theta_0$ among all size (= obtained significance level) α tests in $\mathscr{E}$ (respectively, in $\mathscr{F}$).

If the experiment $\mathscr{E}$ is pairwise at least as informative as $\mathscr{F}$ then

$$\begin{aligned} \kappa(\alpha | \theta_0, \mathscr{E}) &= \lim_{h \to 0} \frac{1}{h} [\beta(\alpha | P_{\theta_0}, P_{\theta_0 + h}) - \alpha] \\ &\geqslant \lim_{h \to 0} \frac{1}{h} [\beta(\alpha | Q_{\theta_0}, Q_{\theta_0 + h}) - \alpha] \\ &= \kappa(\alpha | \theta_0, \mathscr{F}). \end{aligned}$$

In general this cannot be reversed. If, however, $\mathscr{E}$ possesses the monotone likelihood property, then $\mathscr{E}$ is pairwise at least as informative as $\mathscr{F}$ if and only if $\mathscr{E}$ is (everywhere) locally at least as informative as $\mathscr{F}$, i.e. if and only if $\kappa(\cdot|\theta,\mathscr{E}) \geqslant \kappa(\cdot|\theta,\mathscr{F})$ for all $\theta \in \Theta$.

This may be argued from the existence theory for first order differential equations as follows. Assume that the (M L) experiment $\mathscr{E}$ majorizes $\mathscr{F}$ locally, i.e. that $\kappa(\cdot|\theta,\mathscr{E}) \geqslant \kappa(\cdot|\theta,\mathscr{F})$; $\theta \in \Theta$. Put $\kappa_\varepsilon(\alpha|\theta) = \kappa(\alpha|\theta,\mathscr{E}) + \varepsilon[\alpha \wedge (1-\alpha)]$. Then $\kappa_\varepsilon(\cdot|\theta)$, for each θ, is a possible slope function.

Choose a $\theta_0 \in \Theta$ and a number $\alpha_0 \in]0,1[$. Let π be the unique power function in $\Pi_{\mathscr{E}}$ (defined in complement 41) such that $\pi(\theta_0) = \alpha_0$. Consider also a power function σ of a test in $\mathscr{F}$ such that $\sigma(\theta_0) = \alpha_0$. By optimality $\dot{\pi}(\theta) \equiv_\theta \kappa(\pi(\theta)|\theta,\mathscr{E})$ where as usual the dot denotes differentiation with respect to θ. We shall argue that $\pi(\theta) \leqslant \sigma(\theta)$ or $\geqslant \sigma(\theta)$ as $\theta \leqslant \theta_0$ or $\theta \geqslant \theta_0$.

By the theory of differential equations there is an interval $]\theta_0 - r, \theta_0 + r[$ and a function z_ε from $]\theta_0 - r, \theta_0 + r[$ to $]0,1[$ such that $z_\varepsilon(\theta_0) = \alpha_0$ and $\dot{z}_\varepsilon(\theta) = \kappa_\varepsilon(z_\varepsilon(\theta)|\theta)$ when $\theta \in]\theta_0 - r, \theta_0 + r[$. It may be checked that z_ε can be extended to a unique monotonically increasing function from Θ to $[0,1]$ such that $\dot{z}_\varepsilon(\theta) = \kappa_\varepsilon(z_\varepsilon(\theta)|\theta)$ for all θ.

If $\theta > \theta_0$ and if θ is sufficiently close to θ_0 then, since $z_\varepsilon(\theta_0) = \pi(\theta_0) = \alpha$ while $\dot{z}_\varepsilon(\theta_0) > \dot{\pi}(\theta_0)$, $z_\varepsilon(\theta) > \pi(\theta)$. At any point $\theta > \theta_0$ such that $z_\varepsilon(\theta) = \pi(\theta) \in]0,1[$ we must conclude that

$$\dot{z}_\varepsilon(\theta) = \kappa_\varepsilon(z_\varepsilon(\theta)|\theta) = \kappa_\varepsilon(\pi(\theta)|\theta) > \kappa(\pi(\theta)|\theta,\mathscr{E}) = \dot{\pi}(\theta).$$

At a closest point $\theta > \theta_0$ of equality however, we must have $\dot{z}_\varepsilon(\theta) \leqslant \dot{\pi}(\theta)$.

It follows that $z_\varepsilon(\theta) \geqslant \pi(\theta)$ when $\theta \geqslant \theta_0$. Likewise we find that $z_\varepsilon(\theta) \leqslant \pi(\theta)$ when $\theta \leqslant \theta_0$.

By the same argument $z_\varepsilon(\theta)\downarrow$ in ε when $\theta \geqslant \theta_0$ while $z_\varepsilon(\theta)\uparrow$ when $\theta \leqslant \theta_0$. Putting $z(\theta) = \lim_{\varepsilon\to 0} z_\varepsilon(\theta)$ we find by Dini's lemma that

$$\begin{aligned} z(\theta) = \lim_{\varepsilon\to 0} z_\varepsilon(\theta) &= \lim_{\varepsilon\to 0}\left[\alpha_0 + \int_{\theta_0}^{\theta} \kappa_\varepsilon(z_\varepsilon(t)|t)\,dt\right] \\ &= \lim_{\varepsilon\to 0}\left[\alpha_0 + \int_{\theta_0}^{\theta} \kappa(z_\varepsilon(t)|t,\mathscr{E})\,dt + \varepsilon\int_{\theta_0}^{\theta}[z_\varepsilon(t) \wedge (1 - z_\varepsilon(t))]\,dt\right] \\ &= \alpha_0 + \int_{\theta_0}^{\theta} \kappa(z(t)|t,\mathscr{E})\,dt; \qquad \theta \in \Theta \end{aligned}$$

so that $\dot{z}(\theta) \equiv_\theta \kappa(z(\theta)|\theta,\mathscr{E})$ while $z(\theta_0) = \alpha_0$ and thus, by uniqueness, $\pi(\theta) \equiv_\theta z(\theta)$.

Returning to the power function σ in $\mathscr{F}$ we find that

$$\dot{\sigma}(\theta) \leqslant \kappa(\sigma(\theta)|\theta,\mathscr{F}) \leqslant \kappa(\sigma(\theta)|\theta,\mathscr{E}) \leqslant \kappa_\varepsilon(\sigma(\theta)|\theta).$$

Then $\sigma(\theta) \geqslant z_\varepsilon(\theta)$ or $\leqslant z_\varepsilon(\theta)$ as $\theta \leqslant \theta_0$ or $\geqslant \theta_0$. Letting $\varepsilon \to 0$ we find that $\sigma(\theta) \geqslant \pi(\theta)$ or $\leqslant \pi(\theta)$ as $\theta \leqslant \theta_0$ or $\theta \geqslant \theta_0$. If $\theta = \theta_1 > \theta_2$ then σ may be chosen so that $\sigma(\theta_1) = \beta(\alpha_0|Q_{\theta_0}, Q_{\theta_1})$ and thus

$$\beta(\alpha_0|Q_{\theta_0}, Q_{\theta_1}) \leqslant \pi(\theta_1) = \beta(\alpha_0|P_{\theta_0}, P_{\theta_1}).$$

It follows that $\mathscr{E}$ is pairwise more informative than $\mathscr{F}$.

Complement 49. Monotone likelihood in terms of betweenness for the total variation distance

An interesting notion of betweenness for metric spaces declares v to be between u and w for the metric d if $d(u, v) + d(v, w) = d(u, w)$. We mention in passing that A. Wald contributed to the theory of betweenness (see Menger, 1952).

In view of the triangular inequality, v is between u and w for the metric d if and only if $d(u, v) + d(v, w) \leqslant d(u, w)$. If u, v and w are real numbers and d is the usual distance then this is just the usual notion of betweenness. Indeed if $a_1, \ldots, a_r$ are real numbers then $|a_1 + \cdots + a_r| = |a_1| + \cdots + |a_r|$ if and only if there is no conflict of sign in the sequence $a_1, \ldots, a_r$, i.e. $a_i a_j \geqslant 0; i, j = 1, \ldots, r$. This generalizes immediately to the L_1-distance for measure spaces. Thus if $a_1, \ldots, a_r$ are integrable functions then $\int |a_1 + \cdots + a_r| = \int |a_1| + \cdots + \int |a_r|$ if and only if there is a null set such that there is no sign conflict in the sequence $(a_1, \ldots, a_r)$ outside this null set. Expressed for finite (signed) measures μ_1, $\mu_2, \ldots, \mu_r$ this says that the total variation of $(\mu_1 + \mu_2 + \cdots + \mu_r)$ is the sum of the total variations of $\mu_1, \ldots, \mu_r$ if and only if there is a common Hahn set for all these measures.

Combining this with the order characterization of monotone likelihood provided in complement 38, we shall derive a between characterization of monotone likelihood.

(a) Let $\mathscr{E} = (P_1, P_2, P_3)$ be an $(M - L)$-experiment. Put $f_i = \mathrm{d}P_i/\mathrm{d}\sum_i P_i$; $i = 1, 2, 3$ where we may assume that $f_1, f_2, f_3 \geqslant 0$ and that $f_1 + f_2 + f_3 = 1$. We may also assume that the set of vectors $f = (f_1, f_2, f_3)$ is totally ordered for the ordering $\underline{]}$ defined in complement 38. Let r be a non-negative number and put $t = \inf\{f_3(x)/f_2(x) : f_2(x)/f_1(x) > r\}$. Assuming $r < \text{essentialsup}_{P_1+P_2} \mathrm{d}P_2/\mathrm{d}P_1$ this quantity is finite.

The definition of t implies directly that $f_3 \geqslant tf_2$ when $f_2 > rf_1$. On the other hand if $f_2(x) < rf_1(x)$ and $f_2(y)/f_1(y) > r$ then, since the ordering is total, $f(y) \underline{]} f(x)$ so that $f_3(y)/f_2(y) \geqslant f_3(x)/f_2(x)$. Varying y we find then that $t \geqslant f_3(x)/f_2(x)$ so that $f_3(x) \leqslant tf_2(x)$. In any case $(f_2(x) - rf_1(x))(f_3(x) - tf_2(x)) \geqslant 0$. Thus

$$\|P_2 - rP_1\| + \|P_3 - tP_2\| = \|(P_2 - rP_1) + (P_3 - tP_2)\|. \quad (10.2.6)$$

(b) Let $\mathscr{E} = (P_1, P_2, P_3)$ and $f = (f_1, f_2, f_3)$ be as in (a). Assume that for each positive rational number r less than $\text{essentialsup}_{P_1+P_2}(\mathrm{d}P_2/\mathrm{d}P_1)$ there is a number $t = t_r$ so that (10.2.6) holds. Modifying densities on a null set we may then ensure that $(f_2 - rf_1)(f_3 - t_r f_2) \geqslant 0$ for all such rational numbers r.

Consider points x and y in the sample space of $\mathscr{E}$. Assume that the vectors $f(x)$ and $f(y)$ are not comparable for the ordering $\underline{]}$. We may assume without loss of generality that $(f_2(x), f_3(x)) \underline{]} (f_2(y), f_3(y))$. By point (i) in complement 38 the inequality $(f_1(x), f_2(x)) \underline{]} (f_1(y), f_2(y))$ cannot hold. Thus $f_1(x) > 0$, $f_2(y) > 0$ and $f_2(x)/f_1(x) < f_2(y)/f_1(y)$. Choose a rational number r in $]f_2(x)/f_1(x), f_2(y)/f_1(y)[$. Then $f_2(x) < rf_1(x)$ while $f_2(y) > rf_1(y)$. The assumption on signs tells us that $f_3(x) \leqslant t_r f_2(x)$ and that $f_3(y) \geqslant t_r f_2(y)$. If $f_2(x) = 0$ then this implies that also $f_3(x) = 0$ which is impossible, since then $f(y) \underline{[} f(x)$.

Thus $f_2(x) > 0$ and hence $f_3(x)/f_2(x) \leqslant t_r \leqslant f_3(y)/f_2(y)$. However, we assumed that $(f_2(x), f_3(x)) \underline{]} (f_2(y), f_3(y))$. Thus $f_3(x)/f_2(x) = t_r = f_3(y)/f_2(y)$. Combining this with the inequality $f_2(x)/f_1(x) < f_2(y)/f_1(y)$ we find that $f(x) \underline{[} f(y)$, contradicting our assumptions.

It follows that $f(x)$ and $f(y)$ are comparable for all choices of x and y in the sample space of $\mathscr{E}$. Thus $\mathscr{E}$ has the $(M - L)$ property.

(c) If $\Theta \subseteq \mathbb{R}$ then $\mathscr{E} = (P_\theta : \theta \in \Theta)$ has the $(M - L)$ property if and only if to each triple $\theta_1 < \theta_2 < \theta_3$ of points in Θ and to each number r in the open interval $]0, \operatorname{essentialsup}_{P_{\theta_1} + P_{\theta_2}} dP_{\theta_2}/dP_{\theta_1}[$ there corresponds a number $t \in [0, \infty[$ such that

$$\|P_{\theta_2} - rP_{\theta_1}\| + \|P_{\theta_3} - tP_{\theta_2}\| = \|(P_{\theta_2} - rP_{\theta_1}) + (P_{\theta_3} - tP_{\theta_2})\|.$$

10.3 References

Bednarski, T. 1982. Binary experiments, minimax tests and 2-alternating capacities. *Ann. Statist.* **10**, pp. 226–32.

Chernoff, H. 1952. A measure of asymptotic efficiency for tests of a hypothesis based on the sum of observations. *Ann. Math. Statist.* **23**, pp. 493–507.

Ehm, W. & Müller, O. W. 1983. Factorization information contained in an experiment, conditionally on the observed value of a statistic. *Z. Wahrscheinlichkeitstheorie verw. Geb.* **65**, pp. 121–34.

Hajek, J. & Sidak, Z. 1967. *Theory of rank tests.* Academic Press, New York.

Heyer, H. 1982. *Theory of statistical experiments.* Springer–Verlag, New York.

Huber, P. J. 1965. A robust version of the probability ratio test. *Ann. Math. Statist.* **36**, pp. 1753–8.

— 1981. *Robust Statistics.* Wiley, New York.

Huber, P. J. & Strassen, V. 1973–74. Minimax tests and the Neyman Pearson Lemma for capacities. *Ann. Statist.* **1**, pp. 223–4.

Janssen, A. 1988. A convolution theorem for the comparison of exponential families. *Res. report no.* **211**. University of Siegen.

Janssen, A., Milbrodt, H. & Strasser, H. 1984. Infinitely divisible statistical experiments. *Lecture notes in statistics.* Springer–Verlag, Berlin.

Karlin, S. & Rubin, H. 1956. The theory of decision procedures for distributions with monotone likelihood ratio. *Ann. Math. Statist.* **27**, pp. 272–99.

LeCam, L. 1960. Locally asymptotically normal families of distributions. *Univ. of Calif. Publ. in Stat.* **3**, pp. 32–98.

— 1969. Théorie asymptotique de la décision statistique. *Presses de l'Université de Montréal.*

— 1986. *Asymptotic methods in Statistical Decision Theory.* Springer–Verlag, New York.

Lehmann, E. 1988. Comparing location experiments. *Ann. Statist.* **16**, pp. 521–33.
Loève, M. 1960. *Probability theory.* D. van Nostrand, Princeton.
Mammen, E. 1986. The statistical information contained in additional observations. *Ann. Statist.* 1986, pp. 665–78.
Menger, K. 1952. The formative years of Abraham Wald and his work in geometry. *Ann. Math. Statist.* **23**, pp. 14–20.
Rieder, H. 1977. Least favourable pairs for special capacities. *Ann. Statist.* 1977, pp. 909–21.
— 1981. Robustness of one- and two-sample rank tests against gross errors. *Ann. Statist.* 1981, pp. 245–65.
Roussas, G. G. 1972. *Contiguity of probability measures: some applications in statistics.* Cambridge University Press.
Strassen, V. 1965. The existence of probability measures with given marginals. *Ann. Math. Statist.* **36**, pp. 423–39.
Strasser, H. 1985. *Mathematical theory of statistics.* W. de Gruyter, Berlin.
Torgersen, E. 1970. Comparison of experiments when the parameter space is finite. *Z. Wahrscheinlichkeitstheorie verw. Geb.* **16**, pp. 219–49.
— 1972. Comparison of translation experiments. *Ann. Math. Statist.* **43**, pp. 1383–99.
— 1974. Comparison of experiments by factorization. *Stat. res. report.* Univ. of Oslo.
— 1976. Deviations from total information and from total ignorance as measures of information. *Stat. res. report.* Univ. of Oslo.
— 1977. Mixtures and products of dominated experiments. *Ann. Statist.* **5**, pp. 44–64.
— 1981. Measures of information based on comparison with total information and with total ignorance. *Ann. Statist.* **9**, pp. 638–57.
— 1989. Monotone likelihood, power function diagrams and selection. *Stat. res. report* No. **3**, Univ. of Oslo.

LIST OF SYMBOLS

See section 8.1 for notation and conventions concerning inner product spaces.

AUTHOR INDEX

Additional References

Amari, S., Barndorff-Nielsen, O., Kass, R., Lauritzen, S. & Rao, C. 1987. Differential geometry in statistical inference. *I.M.S. Lecture notes no.* **10**.

Barndorff-Nielsen, O. 1977. *Information and exponential families in statistical theory*. Wiley, New York.

Barndorff-Nielsen, O., Hoffmann-Jørgensen, J. & Pedersen, K. 1975. On the minimal sufficiency of the likelihood function. Dept. of theoretical statistics, Univ. of Aarhus, *Research Reports no.* **11**, 1975.

Bartoszynski, R. & Pleszczynska, E. 1976. Reducibility of statistical structures and decision problems. Inst. of Math. *Polish Academy of Sciences, Preprint no.* **93**.

Bayarri, M. J. & De Groot, M. H. 1986. Information in Selection Models. Probab. and Bayesian Statistics. Ed. R. Viertl. Plenum Press, New York, pp. 39–51.

— 1989. Comparison of Experiments with Weighted Distributions. *Technical report No.* **450**, Dept. of Statistics. Carnegie Mellon University, Pittsburgh.

Birgé, L. 1984. Sur un théoreme de minimax et son application aux tests. *Probab. Math. Statist.* **3**, pp. 259–82.

Bradt, R. N. & Karlin, S. 1956. On the design and comparison of certain dichotomous experiments. *Ann. Math. Statist.* **27**, pp. 390–409.

Brown, L. D. 1986. Fundamentals of Statistical Exponential Families. *I.M.S. Lecture notes no.* **9**.

Cencov, N. N. 1972. Statistical decision rules and optimal inference. Translations of math. monographs. *American Math. Soc.* 1980, *Vol.* **53**.

Csiszar, I. 1972. A class of measures of informativity of observation channels. *Periodica Mathematica Hungarica*, **2** (1–4), pp. 191–213.

De Groot, M. H. 1966. Optimal allocation of observations. *Ann. Inst. Statist. Math.* **18**, pp. 23–8.

De Groot, M. H. & Goel, P. K. 1981. *Information about hyperparameters in hierarchical models*, **76**, pp. 140–47.

— 1979. Comparison of experiments and information measures. *Ann. Statist.* **7**, pp. 1066–77.

Feldman, D. 1972. Some properties of Bayesian orderings of experiments. *Ann. Math. Statist.* **43**, pp. 1428–40.

Goel, P. K. 1987. Comparison of experiments and information in Censored Data. *Proc. of 4th Purdue Symp. on Statistical Decision theory and related topics*, **2**, Academic Press, New York.

Heyer, H. 1986. Bayesian aspects in the theory of comparison of statistical experiments. *Probab. and Bayesian Statistics.* Ed. R. Viertl. Plenum Press, New York, pp. 247–56.

Janssen, A. 1986a. Limits of translation invariant experiments. *Journal of Multivariate Analysis* **20**, pp. 129–42.

— 1986b. Asymptotic properties of Neyman–Pearson tests for infinite Kullback–Leibler information. *Ann. Statist.* **14**, pp. 1068–79.

— 1988. The role of extreme order statistics for exponential families. *Res. Report No.* **203**, Univ. of Siegen.
Janssen, A. & Reiss, R. D. 1986. Comparison of location models of Weibull type samples and extreme value processes. *Res. Report No.* **181**, Univ. of Siegen.
Kudo, H. 1955. On minimax invariant estimates of the transformation parameter. *Natur. Sci. Rep. Ochanomizu Univ.* **6**, pp. 31–73.
— 1970. On an approximation to a sufficient statistic including a concept of asymptotic sufficiency. *Jour. of the Faculty of Science, Univ. of Tokyo Sec I*, **XVII**, pp. 273–90.
LeCam, L, 1979. A reduction theorem for certain sequential experiments II. *Ann. Statist.* **7**, pp. 847–59.
Lehmann, E. L. 1983. Estimation with inadequate information. *J. Amer. Stat. Ass.* **78**, pp. 624–7.
Lindley, D. V. 1956. On a measure of the information provided by an experiment. *Ann. Math. Statist.* **27**, pp. 986–1005.
Luschgy, H. 1987. Comparison of shift experiments on a Banach space. In *Mathematical Statistics and Probability Theory*, Vol. A. Ed. M. L. Puri *et al.* Reidel, Dordrecht, pp. 217–30.
— 1987. Elimination of randomization and Hunt Stein type theorems in invariant statistical decision problems. *Statistics.* **18**, pp. 99–111.
— 1988. Pairwise sufficiency and invariance. *Osaka J. of Math.* **25**, 785–94.
— 1989. Ordering of regression models of Gaussian processes. *Res. Report.* Univ. of Münster.
Luschgy, H. & Mussmann, D. 1985. Equivalent properties and completion of statistical experiments. *Sankhyā.* **47**, *ser. A*, pp. 174–95.
— 1986. Products of majorized statistical experiments. Statistics and Decisions **4**, pp. 321–335.
— A characterization of weakly dominated statistical experiments by compactness of the set of decision rules. *Sankhyā, ser. A*, **49**, pp. 388–94.
Luschgy, H., Mussmann, D. & Yamada, S. 1988. Minimal L space and Halmos–Savage criterion for majorized experiments. *Osaka J. of Math.* **25**, pp. 795–803.
Mussmann, D. 1972. Vergleich von Experimenten im schwach dominierten Fall. *Zeit. verw. Geb.* **24**, pp. 295–308.
Müller, D. W. 1979. Asymptotically multinomial experiments and the extension of a theorem of Wald. *Z. Wahrscheinlichkeitstheorie verw. Geb.* **50**, pp. 179–204.
Österreicher, F. 1972. An information-type measure of difference of probability distributions based on testing statistical hypotheses. *Colloqvia mathematica societatis Janos Bolyai* **9**. European meeting of statisticians, Budapest.
— 1978. On the Construction of least favorable pairs of distributions. *Z. Wahrscheinlichkeitstheorie verw. Geb.* **43**, pp. 49–55.
Pfanzagl, J. & Weber, J. 1973. Nonlinear operators and sufficiency. *Ann. Math. Statist.* pp. 183–7.
Pfanzagl, J. & Wefelmeyer, W. 1982. Contributions to a general asymptotic statistical theory. *Lecture notes in statistics, No 13.* Springer–Verlag, New York.
Rüschendorf, L. 1981. Ordering of distributions and rearrangement of functions. *Ann. Prob.* **9**, pp. 276–83.
— 1985. The Wasserstein Distance and Approximation Theorems. *Z. Wahrscheinlichkeitstheorie verw. Geb.* **70**, pp. 117–29.
Rüchendorf, L. & Rachev, S. T. 1989. A characterization of random variables with minimum L^2-distance. *Research report.* Universität Münster.
Strasser, H. 1985. Scale invariance of statistical experiments. *Probab. Math. Statist.* **15**, pp. 1–20.
Suzuki, T. 1978. Asymptotic sufficiency up to higher orders and its application to statistical tests and estimates. *Osaka J. of Math.* **15**, pp. 575–88.
Suzuki, T. 1980. A remark on asymptotic sufficiency up to higher orders in multi-dimensional parameter case. *Osaka J. of Math.* **17**, pp. 242–52.
— 1981. A characterization of Approximate Sufficiency. *Memoirs of the School of Science and Engineering. Waseda Univ.* **45**, pp. 113–22.
— 1988. On a relation between higher order asymptotic risk sufficiency and higher order asymptotic sufficiency in a local sense. *Osaka J. of Math.* pp. 115–22.
Tong, Y. L. 1982. Inequalities in Statistics and Probability. *I.M.S. Lecture notes no.* **5**.

SUBJECT INDEX